W9-AVM-044

EDITION
11

Introduction to Management Science

Bernard W. Taylor III
Virginia Polytechnic Institute and State University

PEARSON

Boston Columbus Indianapolis New York San Francisco Upper Saddle River
Amsterdam Cape Town Dubai London Madrid Milan Munich Paris Montréal Toronto
Delhi Mexico City São Paulo Sydney Hong Kong Seoul Singapore Taipei Tokyo

Editorial Director: Sally Yagan
Editor in Chief: Donna Battista
Senior Acquisitions Editor: Chuck Synovec
Editorial Project Manager: Mary Kate Murray
Editorial Assistant: Ashlee Bradbury
Director of Marketing: Maggie Moylan
Executive Marketing Manager: Anne Fahlgren
Senior Managing Editor: Judy Leale
Production Project Manager: Jane Bonnell
Senior Operations Supervisor: Arnold Vila
Operations Specialist: Cathleen Petersen
Creative Director: Blair Brown
Senior Art Director/Supervisor: Janet Slowik
Art Director: Steve Frim
Cover Design and Interior Design: Mike Fruhbeis
Cover Image: Fotolia/George Bailey
Associate Media Project Manager: Sarah Peterson
Media Project Manager: John Cassar
Composition/Full-Service Project Management: PreMediaGlobal
Printer/Binder: Edwards Brothers
Cover Printer: Lehigh-Phoenix Color/Hagerstown
Text Font: 10/12 Times Roman

Credits and acknowledgments borrowed from other sources and reproduced, with permission, in this textbook appear on the appropriate page within text.

Library of Congress Cataloging-in-Publication Data

Taylor, Bernard W.
Introduction to management science / Bernard W. Taylor III.—Ed. 11.
p. cm.
ISBN-13: 978-0-13-275191-9
ISBN-10: 0-13-275191-7
1. Management science. I. Title.
T56.T38 2011
658.5—dc23

2011031153

10 9 8 7 6 5 4 3 2 1

ISBN 10: 0-13-275191-7
ISBN 13: 978-0-13-275191-9

To Diane, Kathleen, and Lindsey
To the memory of my grandfather,
Bernard W. Taylor, Sr.

Brief Contents

Contents

4 Linear Programming: Modeling Examples 111

5 Integer Programming 184

6 Transportation, Transshipment, and Assignment Problems 233

7 Network Flow Models 289

Preface

The objective of management science is to solve the decision-making problems that confront and confound managers in both the public and the private sector by developing mathematical models of those problems. These models have traditionally been solved with various mathematical techniques, all of which lend themselves to specific types of problems. Thus, management science as a field of study has always been inherently mathematical in nature, and as a result sometimes complex and rigorous. When I began writing the first edition of this book in 1979, my main goal was to make these mathematical topics seem less complex and thus more palatable to undergraduate business students. To achieve this goal I started out by trying to provide simple, straightforward explanations of often difficult mathematical topics. I tried to use lots of examples that demonstrated in detail the fundamental mathematical steps of the modeling and solution techniques. Although in the past three decades the emphasis in management science has shifted away from strictly mathematical to mostly computer solutions, my objective has not changed. I have provided clear, concise explanations of the techniques used in management science to model problems, and provided many examples of how to solve these models on the computer while still including some of the fundamental mathematics of the techniques.

The stuff of management science can seem abstract, and students sometimes have trouble perceiving the usefulness of quantitative courses in general. I remember that when I was a student, I could not foresee how I would use such mathematical topics (in addition to a lot of the other things I learned in college) in any job after graduation. Part of the problem is that the examples used in books often do not seem realistic. Unfortunately, examples must be made simple to facilitate the learning process. Larger, more complex examples reflecting actual applications would be too complex to help the student learn the modeling technique. The modeling techniques presented in this text are, in fact, used extensively in the business world, and their use is increasing rapidly because of computer and information technology. Therefore, the chances that students will use the modeling techniques that they learn from this text in a future job are very great indeed.

Even if these techniques are not used on the job, the logical approach to problem solving embodied in management science is valuable for all types of jobs in all types of organizations. Management science consists of more than just a collection of mathematical modeling techniques; it embodies a philosophy of approaching a problem in a logical manner, as does any science. Thus, this text not only teaches specific techniques but also provides a very useful method for approaching problems.

My primary objective throughout all revisions of this text is readability. The modeling techniques presented in each chapter are explained with straightforward examples that avoid lengthy written explanations. These examples are organized in a logical step-by-step fashion that the student can subsequently apply to the problems at the end of each chapter. I have tried to avoid complex mathematical notation and formulas wherever possible. These various factors will, I hope, help make the material more interesting and less intimidating to students.

New to This Edition

Management science is the application of mathematical models and computing technology to help decision makers solve problems. Therefore, new text revisions like this one tend to focus on the latest technological advances used by businesses and organizations for solving problems, as well as new features that students and instructors have indicated would be helpful to them in learning about management science. Following is a list of the substantial new changes made for this 11th edition of the text:

- This revision incorporates the latest version of Excel 2010, and includes more than 175 new spreadsheet screenshots.
- More than 50 new exhibit screenshots have been added to show the latest versions of Microsoft Project 2010, QM for Windows, Excel QM, TreePlan, and Crystal Ball.
- This edition includes 55 new end-of-chapter homework problems and 2 new cases, so it now contains more than 800 homework problems and 64 cases.
- All 800-plus Excel homework files on the Instructor's Web site have been replaced with new Excel 2010 files.
- A new feature for this revision is "Chapter Web links" for every chapter. More than 550 Web links are provided to access tutorials, summaries, and notes available on the Internet for the various topics in the chapters. Also included are links to YouTube videos that provide additional learning resources. These Web links also replace the end-of chapter references.

- Over 30% of the "Management Science Application" boxes are new for this edition. All of these new boxes provide current, updated applications of management science techniques by companies and organizations.
- New "Sensitivity Analysis" sections have been added throughout Chapter 4 to complement the presentation of the different types of linear programming models, and in Chapter 6.

Learning Features

This 11th edition of *Introduction to Management Science* includes many features that are designed to help sustain and accelerate a student's learning of the material. Several of the strictly mathematical topics—such as the simplex and transportation solution methods—are included as chapter modules on the Companion Web site, at **www.pearsonhighered.com/taylor**. This frees up text space for additional modeling examples in several of the chapters, allowing more emphasis on computer solutions such as Excel spreadsheets, and added additional homework problems. In the following sections, we will summarize these and other learning features that appear in the text.

Text Organization

An important objective is to have a well-organized text that flows smoothly and follows a logical progression of topics, placing the different management science modeling techniques in their proper perspective. The first 10 chapters group together chapters related to mathematical programming that can be solved using Excel spreadsheets, including linear, integer, nonlinear, and goal programming, as well as network techniques.

Within these mathematical programming chapters, the traditional simplex procedure for solving linear programming problems mathematically is located on the Companion Web site, at **www.pearsonhighered.com/taylor**, that accompanies this text. It can still be covered by the student on the computer as part of linear programming, or it can be excluded, without leaving a "hole" in the presentation of this topic. The integer programming mathematical branch and bound solution method is also on the Companion Web site. In Chapter 6, on the transportation and assignment problems, the strictly mathematical solution approaches, including the northwest corner, VAM, and stepping-stone methods, are also on the accompanying Companion Web site. Because transportation and assignment problems are specific types of network problems, the two chapters that cover network flow models and project networks that can be solved with linear programming, as well as traditional model-specific solution techniques and software, follow Chapter 6 on transportation and assignment problems. In addition, in Chapter 10, on nonlinear programming, the traditional mathematical solution techniques, including the substitution method and the method of Lagrange multipliers, are on the Companion Web site.

Chapters 11 through 14 include topics generally thought of as being probabilistic, including probability and statistics, decision analysis, queuing, and simulation. A module on Markov analysis is on the Companion Web site. Also, a module on game theory is on the Companion Web site. Forecasting in Chapter 15 and inventory management in Chapter 16 are both unique topics related to operations management.

Excel Spreadsheets

This new edition continues to emphasize Excel spreadsheet solutions of problems. Spreadsheet solutions are demonstrated in all the chapters in the text (except for Chapter 2, on linear programming modeling and graphical solution), for virtually every management science modeling technique presented. These spreadsheet solutions are presented in optional subsections, allowing the instructor to decide whether to cover them. The text includes more than 175 new Excel spreadsheet screenshots for Excel 2010. Most of these screenshots include reference callout boxes that describe the solution steps within the spreadsheet. Files that include all the Excel spreadsheet model solutions for the examples in the text are included on the Companion Web site and can be easily downloaded by the student to determine how the spreadsheet was set up and the solution derived, and to use as templates to work homework problems. In addition, Appendix B at the end of the text provides a tutorial on how to set up and edit spreadsheets for problem solution. Following is an example of one of the Excel spreadsheet files (from Chapter 3) that is available on the Companion Web site accompanying the text.

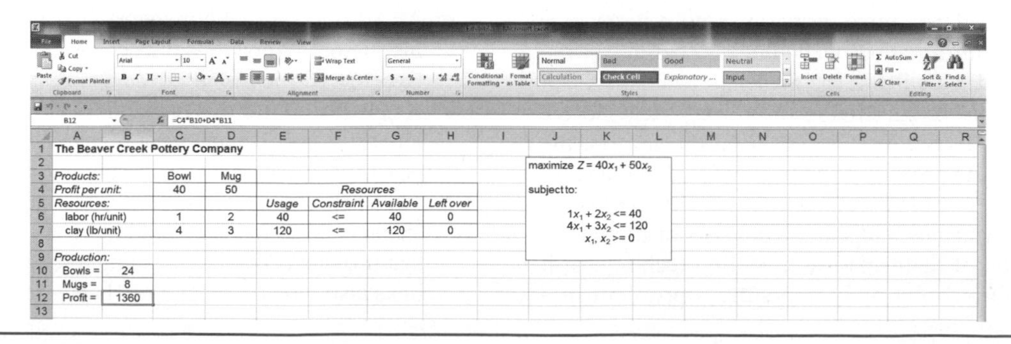

Spreadsheet Add-Ins

Several spreadsheet add-in packages are available with this book, often in trial and premium versions. For complete information on options for downloading each package, please visit **www.pearsonhighered.com/taylor**.

Excel QM

For some management science topics, the Excel formulas that are required for solution are lengthy and complex and thus are very tedious and time-consuming to type into a spreadsheet. In several of these instances in the book, including Chapter 6 on transportation and assignment problems, Chapter 12 on decision analysis, Chapter 13 on queuing, Chapter 15 on forecasting, and Chapter 16 on inventory control, a spreadsheet "add-in" called Excel QM is demonstrated. These add-ins provide a generic spreadsheet setup with easy-to-use dialog boxes and all of the formulas already typed in for specific problem types. Unlike other "black box" software, these add-ins allow users to see the formulas used in each cell. The input, results, and the graphics are easily seen and can be easily changed, making this software ideal for classroom demonstrations and student explorations. Following below is an example of an Excel QM file (from Chapter 13) that is on the Companion Web site that accompanies the text.

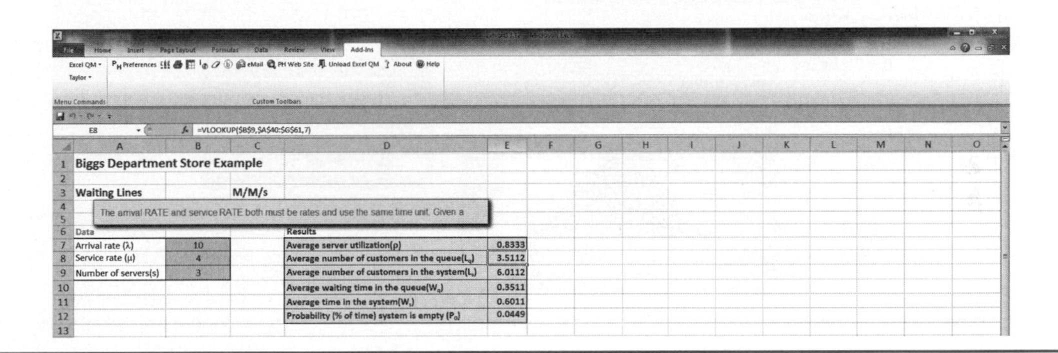

Risk Solver Platform for Education

This program is a tool for risk analysis, simulation, and optimization in Excel. Using the access code, found on the inside front cover of this book, you can download this software for free.

TreePlan

Another spreadsheet add-in program that is demonstrated in the text is TreePlan, a program that will set up a generic spreadsheet for the solution of decision-tree problems in Chapter 12 on decision analysis. This is also available on the Companion Web site. Following is an example of one of the **TreePlan** files (from Chapter 12) that is on the text Companion Web site.

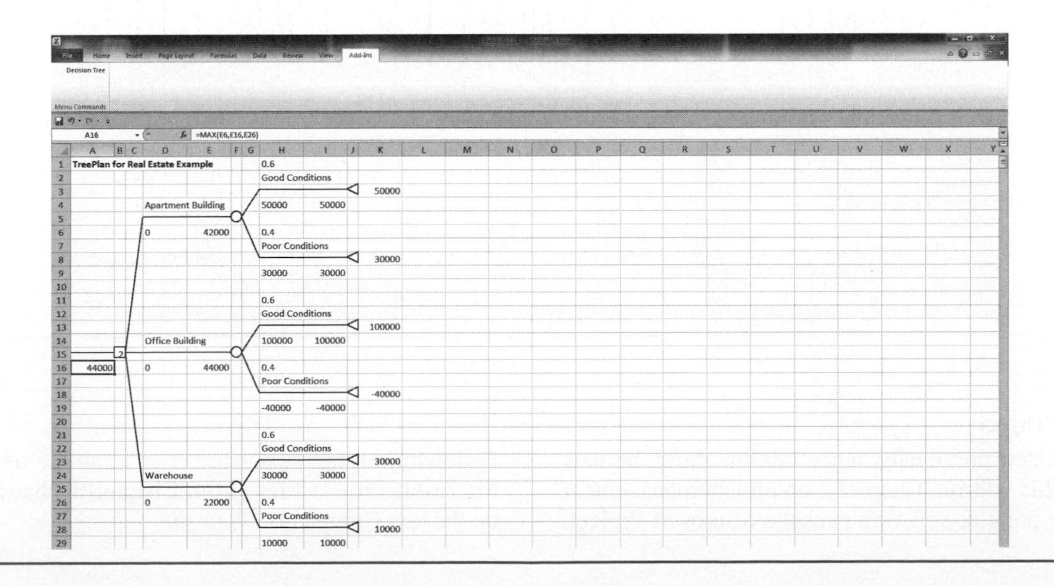

Crystal Ball

Still another spreadsheet add-in program is Crystal Ball by Oracle. Crystal Ball is demonstrated in Chapter 14 on simulation and shows how to perform simulation analysis for certain types of risk analysis and forecasting problems. Following is an example of one of the **Crystal Ball** files (from Chapter 14) that is on the Companion Web site. Using the access code, found on the inside front cover of this book, you can download a 140-day trial version of this software for free.

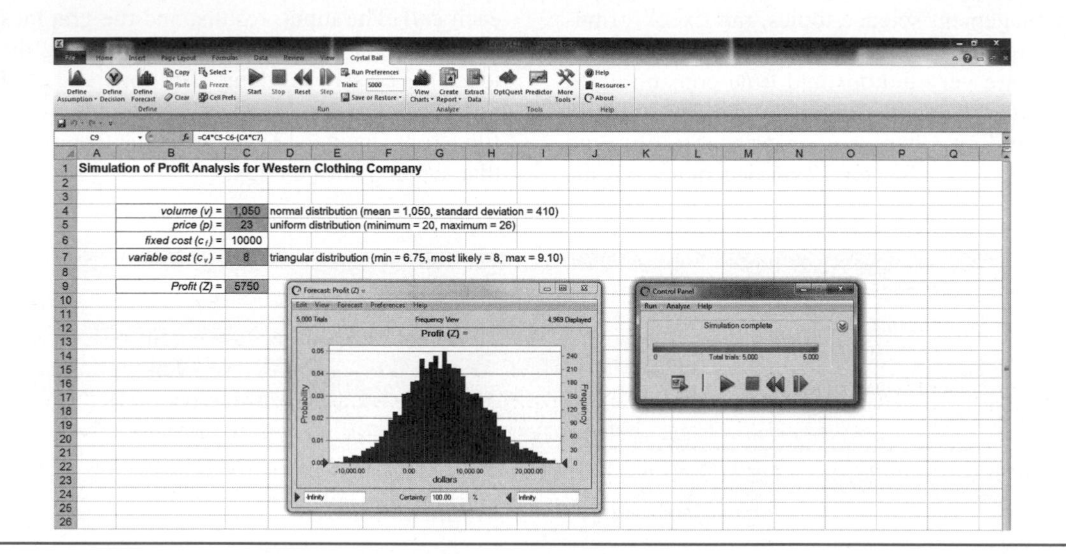

QM for Windows Software Package

QM for Windows is a computer package that is included on the text Companion Web site, and many students and instructors will prefer to use it with this text. This software is very user friendly, requiring virtually no preliminary instruction except for the "help" screens that can be accessed directly from the program. It is demonstrated throughout the text in conjunction with virtually every management science modeling technique, except simulation. The text includes 50 QM for Windows screens used to demonstrate example problems. Thus, for most topics problem solution is demonstrated via both Excel spreadsheets and QM for Windows. Files that include all the QM for Windows solutions for examples in the text are included on the accompanying Companion Web site. Following is an example of one of the QM for Windows files (from Chapter 4) that is on the Companion Web site.

Original Problem w/answers

Product Mix Example Solution

	X1	X2	X3	X4		RHS	Dual
Maximize	90	125	45	65			
Processing time (hrs)	.1	.25	.08	.21	<=	72	233.3333
Shipping capacity (boxes)	3	3	1	1	<=	1,200	22.2222
Budget ($)	36	48	25	35	<=	25,000	0
Blank sweats (dozens)	1	1	0	0	<=	500	0
Blank T's (dozens)	0	0	1	1	<=	500	4.1111
Solution->	175.5556	57.7778	500	0	Optimal Z->	45,522.22	

Microsoft Project

As we indicated previously, when talking about the new features in this edition, Chapter 8 on project management includes the popular software package Microsoft Project. Following on the next page is an example of one of the Microsoft Project files (from Chapter 8) that is available on the text Companion Web site.

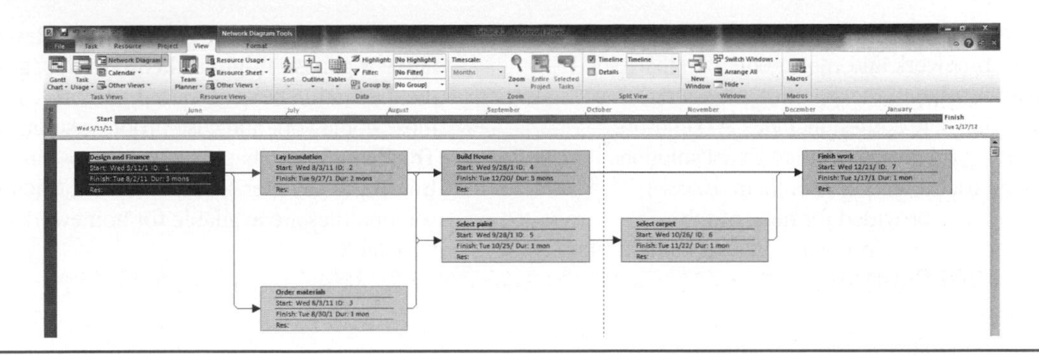

New Problems and Cases

Previous editions of the text always provided a substantial number of homework questions, problems, and cases to offer students practice. This edition includes more than 800 homework problems, 55 of which are new, and 64 end-of-chapter cases, 2 of which are new. In addition, 4 additional spreadsheet modeling cases are provided on the Companion Web site.

"Management Science Application" Boxes

These boxes are located in every chapter in the text. They describe how a company, an organization, or an agency uses the particular management science technique being presented and demonstrated in the chapter to compete in a global environment. There are 52 of these boxes, 16 of which are new, throughout the text. They encompass a broad range of business and public-sector applications, both foreign and domestic.

Marginal Notes

Notes in the margins of this text serve the same basic function as notes that students themselves might write in the margin. They highlight certain topics to make it easier for students to locate them, summarize topics and important points, and provide brief definitions of key terms and concepts.

Examples

The primary means of teaching the various quantitative modeling techniques presented in this text is through examples. Thus, examples are liberally inserted throughout the text, primarily to demonstrate how problems are solved with the different quantitative techniques and to make them easier to understand. These examples are organized in a logical step-by-step solution approach that the student can subsequently apply to the homework problems.

Solved Example Problems

At the end of each chapter, just prior to the homework questions and problems, is a section that provides solved examples to serve as a guide for doing the homework problems. These examples are solved in a detailed, step-by-step fashion.

New Chapter Web Links

A new feature in this edition is a file on the Companion Web site that contains Chapter Web links for every chapter in the text. These Web links access tutorials, summaries, and notes available on the Internet for the various techniques and topics in every chapter in the text. Also included are YouTube videos that provide additional learning resources and tutorials about many of the topics and techniques, links to the development and developers of the techniques in the text, and links to the Web sites for the companies and organizations that are featured in the "Management Science Application" boxes in every chapter. The "Chapter Web links" file includes more than 550 Web links.

Instructor Resources

Instructor's Resource Center

Reached through a link at **www.pearsonhighered.com/taylor**, the Instructor's Resource Center contains the electronic files for the complete Instructor's Solutions Manual, the Test Item File, and Lecture PowerPoint presentations.

- **Register, Redeem, Log In** At **www.pearsonhighered.com/irc**, instructors can access a variety of print, media, and presentation resources that are available with this book in downloadable digital format. Resources are also available for course management platforms such as Blackboard, WebCT, and CourseCompass.
- **Need Help?** Pearson Education's dedicated technical support team is ready to assist instructors with questions about the media supplements that accompany this text. Visit **http://247pearsoned.com** for answers to frequently asked questions and toll-free user support phone numbers. The supplements are available to adopting instructors. Detailed descriptions are provided on the Instructor's Resource Center.

- **Excel Homework Solutions** Almost every end-of-chapter homework and case problem in this text has a corresponding Excel solution file for the instructor. This new edition includes 800 end-of-chapter homework problems, and Excel solutions are provided for all but a few of them. Excel solutions are also provided for most of the 64 end-of-chapter case problems. These solution files can be accessed from the Instructor's Resource Center, at **www.pearsonhighered.com/taylor**, as shown in the illustration below. These Excel files also include those homework and case problem solutions using TreePlan (from Chapter 12) and those using Crystal Ball (from Chapter 14). In addition, Microsoft Project solution files are available for homework problems in Chapter 8.

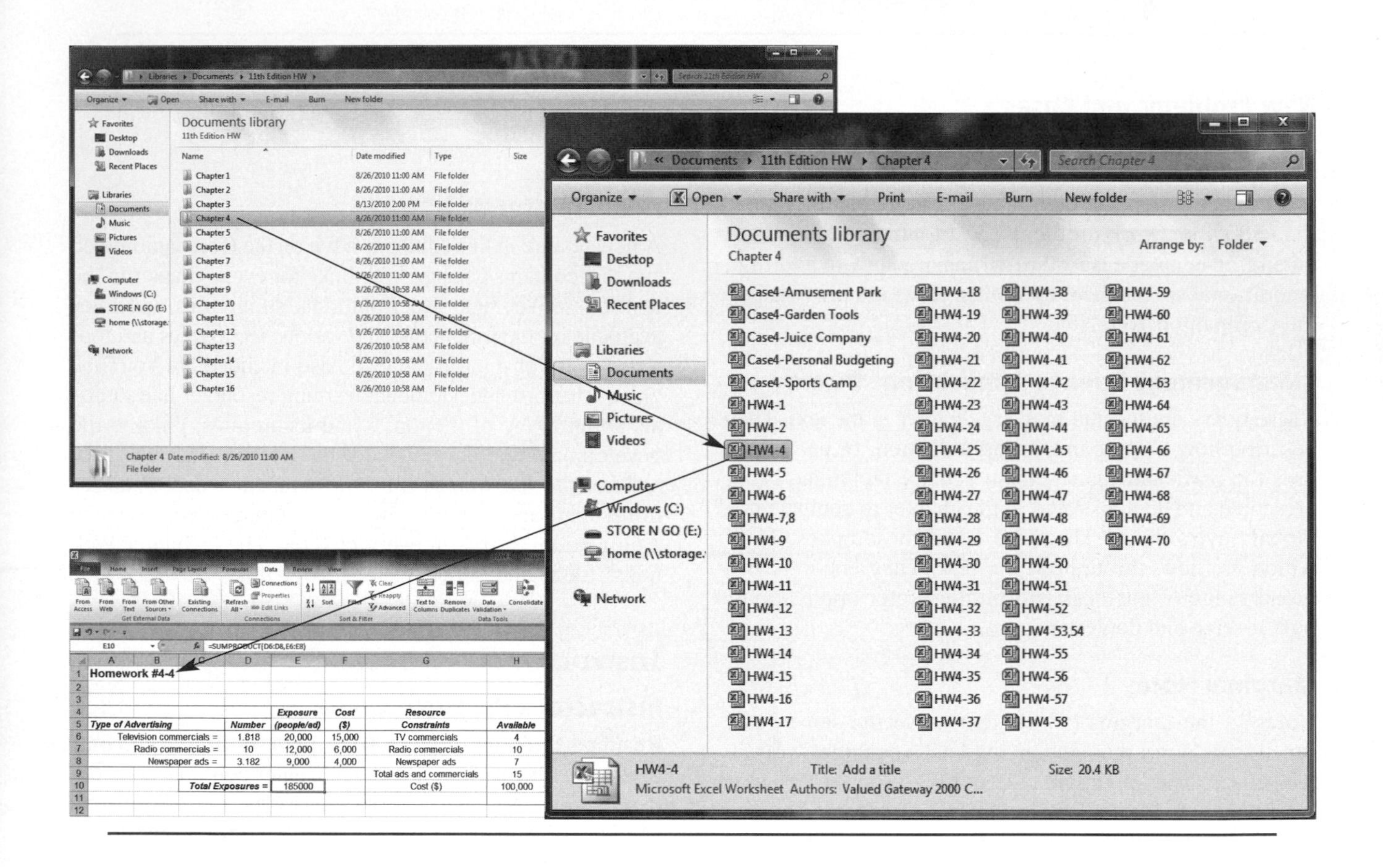

- **Instructor's Solutions Manual** The Instructor's Solutions Manual contains detailed solutions for all end-of-chapter exercises and cases. Each solution has been reviewed for accuracy. The Instructor's Solutions Manual is available for download by visiting **www.pearsonhighered.com/taylor**.
- **Test Item File** The Test Item File contains more than 2,000 questions, including a variety of true/false, multiple-choice, and problem-solving questions for each chapter. Each question is followed by the correct answer, the page references, the main headings, difficulty rating, and key words. It has been reviewed for accuracy. The test item file is available for download by visiting **www.pearsonhighered.com/taylor**.
- **TestGen** Pearson Education's test-generating software is available from **www.pearsonhighered.com/irc**. The software is PC and Mac compatible and preloaded with all of the Test Item File questions. You can manually or randomly view test questions and drag and drop to create a test. You can add or modify test bank questions as needed.
- **Learning Managment Systems** Our TestGens are converted for use in BlackBoard and WebCT. These conversions can be found on the Instructor's Resource Center. Conversions to Moodle, D2L,

or Angel can be requested through your local Pearson Sales Representative.

- **PowerPoint Presentations** PowerPoint presentations are available for every chapter to enhance lectures. They feature figures, tables, Excel, and main points from the text. They are available for download by visiting **www.pearsonhighered.com/taylor**.

Student Resources

Companion Web Site

The Companion Web site for this text (**www.pearsonhighered.com/taylor**) contains the following:

- **Chapter Web Links**—provide access to tutorials, summaries, notes, and YouTube videos.
- **Exhibit Files**—are found throughout the text; these exhibits demonstrate example problems, using Crystal Ball, Excel, Excel QM, Microsoft Project, QM for Windows, and TreePlan.
- **Online Modules**—PDF files of the online modules listed in the table of contents.
- **TreePlan**—link to a free version
- **Excel QM & QM for Windows**—link to a free version
- **Risk Solver Platform**— link to a free trial version
- **Crystal Ball**— link to a free trial version
- **Microsoft Project**— link to a free trial version
- **Subscription Content**—Using the access code, found on the inside front cover of this book, you can download the following software for free:
  - *Risk Solver Platform for Education (RSPE)*—This is a special version of Frontline Systems' Risk Solver Platform software for Microsoft Excel.
  - *Crystal Ball*—free 140-day trial of Crystal Ball software compliments of the Crystal Ball Education Alliance.

For complete information on these items and how to access them, please also visit **www.pearsonhighered.com/taylor.**

CourseSmart

CourseSmart eText books were developed for students looking to save money on required or recommended textbooks. Students simply select their eText by title or author and purchase immediate access to the content for the duration of the course using any major credit card. With a CourseSmart eText, students can search for specific keywords or page numbers, take notes online, print out reading assignments that incorporate lecture notes, and bookmark important passages for later review. For more information or to purchase a CourseSmart eText book, visit **www.coursesmart.com**.

Acknowledgments

As with any other large project, the revision of a textbook is not accomplished without the help of many people. The 11th edition of this book is no exception, and I would like to take this opportunity to thank those who have contributed to its preparation.

I thank the reviewers of this edition: Russell McGee (Texas A&M University), Kefeng Xu (University of Texas at San Antonio), Anthony Narsing (Macon State College), Jaya Singhal (University of Baltimore), Daniel Solow (Case Western Reserve University), Edward Williams (University of Michigan–Dearborn), Zuopeng Zhang (State University of New York at Plattsburgh), and Michael E. Salassi (Louisiana State University).

I also remain indebted to the reviewers of the previous editions: Dr. B. S. Bal, Nagraj Balakrishnan, Edward M. Barrow, Ali Behnezhad, Weldon J. Bowling, Rod Carlson, Petros Christofi, Yar M. Ebadi, Richard Ehrhardt, Warren W. Fisher, James Flynn, Wade Furgeson, Soumen Ghosh, James C. Goodwin, Jr., Richard Gunther, Dewey Hemphill, Ann Hughes, Shivaji Khade, David A. Larson, Sr., Shao-ju Lee, Robert L. Ludke, Peter A. Lyew, Robert D. Lynch, Dinesh Manocha, Mildred Massey, Abdel-Aziz Mohamed, Thomas J. Nolan, Susan W. Palocsay, David W. Pentico, Cindy Randall, Christopher M. Rump, Roger Schoenfeldt, Charles H. Smith, Lisa Sokol, Dothang Truong, John Wang, Barry Wray, Hulya Julie Yazici, and Ding Zhang.

I am also very grateful to Tracy McCoy at Virginia Tech for her valued assistance. I would like to thank my production editor, Jane Bonnell, at Pearson, for her valuable assistance and patience. I very much appreciate the help and hard work of Andrea Stefanowicz and all the folks at PreMediaGlobal, Inc., who produced this edition, and the text's accuracy checker, Annie Puciloski. Finally, I would like to thank my editor, Chuck Synovec, and project manager, Mary Kate Murray, at Pearson, for their continued help and patience.

CHAPTER

1

Management Science

Management science is the application of a scientific approach to solving management problems in order to help managers make better decisions. As implied by this definition, management science encompasses a number of mathematically oriented techniques that have either been developed within the field of management science or been adapted from other disciplines, such as the natural sciences, mathematics, statistics, and engineering. This text provides an introduction to the techniques that make up management science and demonstrates their applications to management problems.

Management science is a scientific approach to solving management problems.

Management science is a recognized and established discipline in business. The applications of management science techniques are widespread, and they have been frequently credited with increasing the efficiency and productivity of business firms. In various surveys of businesses, many indicate that they use management science techniques, and most rate the results to be very good. Management science (also referred to as *operations research, quantitative methods, quantitative analysis*, and *decision sciences*) is part of the fundamental curriculum of most programs in business.

Management science can be used in a variety of organizations to solve many different types of problems.

As you proceed through the various management science models and techniques contained in this text, you should remember several things. First, most of the examples presented in this text are for business organizations because businesses represent the main users of management science. However, management science techniques can be applied to solve problems in different types of organizations, including services, government, military, business and industry, and health care.

Second, in this text all of the modeling techniques and solution methods are mathematically based. In some instances the manual, mathematical solution approach is shown because it helps one understand how the modeling techniques are applied to different problems. However, a computer solution is possible for each of the modeling techniques in this text, and in many cases the computer solution is emphasized. The more detailed mathematical solution procedures for many of the modeling techniques are included as supplemental modules on the companion Web site for this text.

Management science encompasses a logical approach to problem solving.

Finally, as the various management science techniques are presented, keep in mind that management science is more than just a collection of techniques. Management science also involves the philosophy of approaching a problem in a logical manner (i.e., a scientific approach). The logical, consistent, and systematic approach to problem solving can be as useful (and valuable) as the knowledge of the mechanics of the mathematical techniques themselves. This understanding is especially important for those readers who do not always see the immediate benefit of studying mathematically oriented disciplines such as management science.

The Management Science Approach to Problem Solving

*The steps of the **scientific method** are (1) observation, (2) problem definition, (3) model construction, (4) model solution, and (5) implementation.*

As indicated in the previous section, management science encompasses a logical, systematic approach to problem solving, which closely parallels what is known as the **scientific method** for attacking problems. This approach, as shown in Figure 1.1, follows a generally recognized and ordered series of steps: (1) observation, (2) definition of the problem, (3) model construction, (4) model solution, and (5) implementation of solution results. We will analyze each of these steps individually.

Observation

The first step in the management science process is the identification of a problem that exists in the system (organization). The system must be continuously and closely observed so that problems can be identified as soon as they occur or are anticipated. Problems are not always the result of a crisis that must be reacted to but, instead, frequently involve an anticipatory or planning situation. The person who normally identifies a problem is the manager because managers work in places where problems might occur. However, problems can often be identified by a

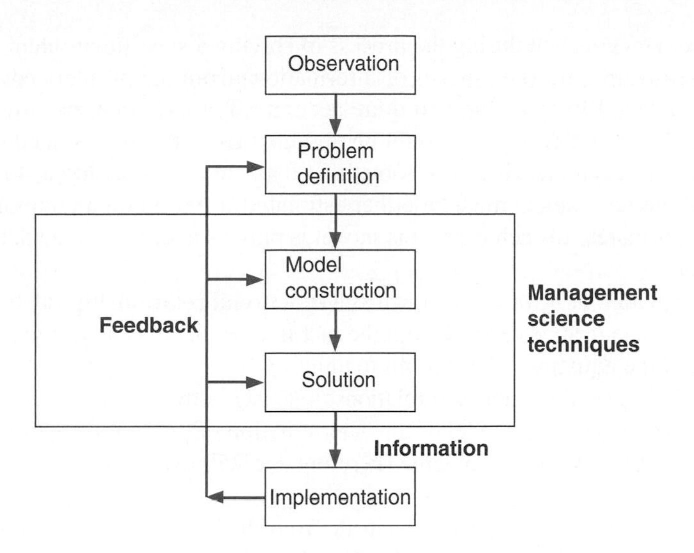

FIGURE 1.1
The management science process

*A **management scientist** is a person skilled in the application of management science techniques.*

management scientist, a person skilled in the techniques of management science and trained to identify problems, who has been hired specifically to solve problems using management science techniques.

Definition of the Problem

Once it has been determined that a problem exists, the problem must be clearly and concisely *defined*. Improperly defining a problem can easily result in no solution or an inappropriate solution. Therefore, the limits of the problem and the degree to which it pervades other units of the organization must be included in the problem definition. Because the existence of a problem implies that the objectives of the firm are not being met in some way, the goals (or objectives) of the organization must also be clearly defined. A stated objective helps to focus attention on what the problem actually is.

Model Construction

*A **model** is an abstract mathematical representation of a problem situation.*

A management science **model** is an abstract representation of an existing problem situation. It can be in the form of a graph or chart, but most frequently a management science model consists of a set of mathematical relationships. These mathematical relationships are made up of numbers and symbols.

As an example, consider a business firm that sells a product. The product costs \$5 to produce and sells for \$20. A model that computes the total profit that will accrue from the items sold is

$$Z = \$20x - 5x$$

*A **variable** is a symbol used to represent an item that can take on any value.*

In this equation, x represents the number of units of the product that are sold, and Z represents the total profit that results from the sale of the product. The *symbols* x and Z are *variables*. The term **variable** is used because no set numeric value has been specified for these items. The number of units sold, x, and the profit, Z, can be any amount (within limits); they can vary. These two variables can be further distinguished. Z is a *dependent variable* because its value is dependent on the number of units sold; x is an *independent variable* because the number of units sold is *not* dependent on anything else (in this equation).

***Parameters** are known, constant values that are often coefficients of variables in equations.*

The numbers \$20 and \$5 in the equation are referred to as **parameters**. Parameters are constant values that are generally coefficients of the variables (symbols) in an equation. Parameters usually remain constant during the process of solving a specific problem. The parameter values are derived from **data** (i.e., pieces of information) from the problem environment. Sometimes the data are readily available and quite accurate. For example, presumably the selling price of $20 and product cost of $5 could be obtained from the firm's accounting department and would be very accurate. However, sometimes data are not as readily available to the manager or firm, and the parameters must be either estimated or based on a combination of the available data and estimates. In such cases, the model is only as accurate as the data used in constructing the model.

__Data__ are pieces of information from the problem environment.

The equation as a whole is known as a **functional relationship** (also called *function and relationship*). The term is derived from the fact that profit, Z, is a *function* of the number of units sold, x, and the equation *relates* profit to units sold.

A model is a __functional relationship__ that includes variables, parameters, and equations.

Because only one functional relationship exists in this example, it is also the *model*. In this case the relationship is a model of the determination of profit for the firm. However, this model does not really replicate a problem. Therefore, we will expand our example to create a problem situation.

Let us assume that the product is made from steel and that the business firm has 100 pounds of steel available. If it takes 4 pounds of steel to make each unit of the product, we can develop an additional mathematical relationship to represent steel usage:

$$4x = 100 \text{ lb. of steel}$$

This equation indicates that for every unit produced, 4 of the available 100 pounds of steel will be used. Now our model consists of two relationships:

$$\begin{aligned} Z &= \$20x - 5x \\ 4x &= 100 \end{aligned}$$

We say that the profit equation in this new model is an *objective function*, and the resource equation is a *constraint*. In other words, the objective of the firm is to achieve as much profit, Z, as possible, but the firm is constrained from achieving an infinite profit by the limited amount of steel available. To signify this distinction between the two relationships in this model, we will add the following notations:

$$\begin{aligned} &\text{maximize } Z = \$20x - 5x \\ &\text{subject to} \\ &\qquad 4x = 100 \end{aligned}$$

This model now represents the manager's problem of determining the number of units to produce. You will recall that we defined the number of units to be produced as x. Thus, when we determine the value of x, it represents a potential (or recommended) *decision* for the manager. Therefore, x is also known as a *decision variable*. The next step in the management science process is to solve the model to determine the value of the decision variable.

Model Solution

A management science technique usually applies to a specific model type.

Once models have been constructed in management science, they are solved using the management science techniques presented in this text. A management science solution technique usually applies to a specific type of model. Thus, the model type and solution method are both part of the management science technique. We are able to say that *a model is solved* because the model represents a problem. When we refer to model solution, we also mean problem solution.

Time Out for Pioneers in Management Science

Throughout this text TIME OUT boxes introduce you to the individuals who developed the various techniques that are described in the chapters. This will provide a historical perspective on the development of the field of management science. In this first instance we will briefly outline the development of management science.

Although a number of the mathematical techniques that make up management science date to the turn of the twentieth century or before, the field of management science itself can trace its beginnings to military operations research (OR) groups formed during World War II in Great Britain circa 1939. These OR groups typically consisted of a team of about a dozen individuals from different fields of science, mathematics, and the military, brought together to find solutions to military-related problems. One of the most famous of these groups—called "Blackett's circus" after its leader, Nobel Laureate P. M. S. Blackett of the University of Manchester and a former naval officer—included three physiologists, two mathematical physicists, one astrophysicist, one general physicist, two mathematicians, an Army officer, and a surveyor. Blackett's group and the other OR teams made significant contributions in improving Britain's early-warning radar system (which was instrumental in their victory in the Battle of Britain), aircraft gunnery, antisubmarine warfare, civilian defense, convoy size determination, and bombing raids over Germany.

The successes achieved by the British OR groups were observed by two Americans working for the U.S. military, Dr. James B. Conant and Dr. Vannevar Bush, who recommended that OR teams be established in the U.S. branches of the military. Subsequently, both the Air Force and Navy created OR groups.

After World War II the contributions of the OR groups were considered so valuable that the Army, Air Force, and Navy set up various agencies to continue research of military problems. Two of the more famous agencies were the Navy's Operations Evaluation Group at MIT and Project RAND, established by the Air Force to study aerial warfare. Many of the individuals who developed operations research and management science techniques did so while working at one of these agencies after World War II or as a result of their work there.

As the war ended and the mathematical models and techniques that were kept secret during the war began to be released, there was a natural inclination to test their applicability to business problems. At the same time, various consulting firms were established to apply these techniques to industrial and business problems, and courses in the use of quantitative techniques for business management began to surface in American universities. In the early 1950s the use of these quantitative techniques to solve management problems became known as management science, and it was popularized by a book of that name by Stafford Beer of Great Britain.

For the example model developed in the previous section,

$$\text{maximize } Z = \$20x - 5x$$
$$\text{subject to}$$
$$4x = 100$$

the solution technique is simple algebra. Solving the constraint equation for x, we have

$$\begin{aligned} 4x &= 100 \\ x &= 100/4 \\ x &= 25 \text{ units} \end{aligned}$$

Substituting the value of 25 for x into the profit function results in the total profit:

$$\begin{aligned} Z &= \$20x - 5x \\ &= 20(25) - 5(25) \\ &= \$375 \end{aligned}$$

A management science solution can be either a recommended decision or information that helps a manager make a decision.

Thus, if the manager decides to produce 25 units of the product and all 25 units sell, the business firm will receive \$375 in profit. Note, however, that the value of the decision variable does not constitute an actual decision; rather, it is *information* that serves as a recommendation or guideline, helping the manager make a decision.

Some management science techniques do not generate an answer or a recommended decision. Instead, they provide *descriptive results:* results that describe the system being modeled.

For example, suppose the business firm in our example desires to know the average number of units sold each month during a year. The monthly *data* (i.e., sales) for the past year are as follows:

Month	Sales	Month	Sales
January	30	July	35
February	40	August	50
March	25	September	60
April	60	October	40
May	30	November	35
June	25	December	50
		Total	480 units

Monthly sales average 40 units (480 ÷ 12). This result is not a decision; it is information that describes what is happening in the system. The results of the management science

Management Science Application

Room Pricing with Management Science at Marriott

Marriott International, Inc., headquartered in Bethesda, Maryland, has more than 140,000 employees working at more than 3,300 hotels in 70 countries. Its hotel franchises include Marriott, JW Marriott, The Ritz-Carlton, Renaissance, Residence Inn, Courtyard, TownePlace Suites, Fairfield Inn, and Springhill Suites. *Fortune* magazine ranks Marriott as the lodging industry's most admired company and one of the best companies to work for.

Marriott uses a revenue management system for individual hotel bookings. This system provides forecasts of customer demand and pricing controls, makes optimal inventory allocations, and interfaces with a reservation system that handles more than 75 million transactions each year. The system makes a demand forecast for each rate category and length of stay for each arrival day up to 90 days in advance, and it provides inventory allocations to the reservation system. This inventory of hotel rooms is then sold to individual customers through channels such as Marriott.com, the company's toll-free reservation number, the hotels directly, and global distribution systems.

One of the most significant revenue streams for Marriott is for group sales, which can contribute more than half of a full-service hotel's revenue. However, group business has challenging characteristics that introduce uncertainty and make modeling it difficult, including longer booking windows (as compared to those for individuals), price negotiation as part of the booking process, demand for blocks of rooms, and lack of demand data. For a group request, a hotel must know if it has sufficient rooms and determine a recommended rate. A key challenge is estimating the value of the business the hotel is turning away if the room inventory is given to a group rather than being held for individual bookings.

© David Zanzinger/Alamy

To address the group booking process, Marriott developed a decision support system, Group Pricing Optimizer (GPO), that provides guidance to Marriott personnel on pricing hotel rooms for group customers. GPO uses various management science modeling techniques and tools, including simulation, forecasting, and optimization techniques, to recommend an optimal price rate. Marriott estimates that GPO provided an improvement in profit of over $120 million derived from $1.3 billion in group business in its first 2 years of use.

Source: Based on S. Hormby, J. Morrison, P. Dave, M. Myers, and T. Tenca, "Marriott International Increases Revenue by Implementing a Group Pricing Optimizer," *Interfaces* 40, no. 1 (January–February 2010): 47–57.

techniques in this text are examples of the two types shown in this section: (1) solutions/decisions and (2) descriptive results.

Implementation

Implementation is the actual use of a model once it has been developed.

The final step in the management science process for problem solving described in Figure 1.1 is implementation. **Implementation** is the actual use of the model once it has been developed or the solution to the problem the model was developed to solve. This is a critical but often overlooked step in the process. It is not always a given that once a model is developed or a solution found, it is automatically used. Frequently the person responsible for putting the model or solution to use is not the same person who developed the model, and thus the user may not fully understand how the model works or exactly what it is supposed to do. Individuals are also sometimes hesitant to change the normal way they do things or to try new things. In this situation the model and solution may get pushed to the side or ignored altogether if they are not carefully explained and their benefit fully demonstrated. If the management science model and solution are not implemented, then the effort and resources used in their development have been wasted.

Model Building: Break-Even Analysis

Break-even analysis is a modeling technique to determine the number of units to sell or produce that will result in zero profit.

In the previous section we gave a brief, general description of how management science models are formulated and solved, using a simple algebraic example. In this section we will continue to explore the process of building and solving management science models, using **break-even analysis**, also called *profit analysis*. Break-even analysis is a good topic to expand our discussion of model building and solution because it is straightforward, relatively familiar to most people, and not overly complex. In addition, it provides a convenient means to demonstrate the different ways management science models can be solved—mathematically (by hand), graphically, and with a computer.

The purpose of break-even analysis is to determine the number of units of a product (i.e., the volume) to sell or produce that will equate total revenue with total cost. The point where total revenue equals total cost is called the *break-even point*, and at this point profit is zero. The break-even point gives a manager a point of reference in determining how many units will be needed to ensure a profit.

Components of Break-Even Analysis

The three components of break-even analysis are volume, cost, and profit. *Volume* is the level of sales or production by a company. It can be expressed as the number of units (i.e., quantity) produced and sold, as the dollar volume of sales, or as a percentage of total capacity available.

Fixed costs are independent of volume and remain constant.

Two type of costs are typically incurred in the production of a product: fixed costs and variable costs. **Fixed costs** are generally independent of the volume of units produced and sold. That is, fixed costs remain constant, regardless of how many units of product are produced within a given range. Fixed costs can include such items as rent on plant and equipment, taxes, staff and management salaries, insurance, advertising, depreciation, heat and light, and plant maintenance. Taken together, these items result in total fixed costs.

Variable costs depend on the number of items produced.

Variable costs are determined on a per-unit basis. Thus, total variable costs depend on the number of units produced. Variable costs include such items as raw materials and resources, direct labor, packaging, material handling, and freight.

Total variable costs are a function of the *volume* and the *variable cost per unit*. This relationship can be expressed mathematically as

$$\text{total variable cost} = vc_v$$

where c_v = variable cost per unit and v = volume (number of units) sold.

Total cost (TC) equals the fixed cost (c_f) plus the variable cost per unit (c_v) multiplied by volume (v).

The **total cost** of an operation is computed by summing total fixed cost and total variable cost, as follows:

$$\text{total cost} = \text{total fixed cost} + \text{total variable cost}$$

or

$$TC = c_f + vc_v$$

where c_f = fixed cost.

As an example, consider Western Clothing Company, which produces denim jeans. The company incurs the following monthly costs to produce denim jeans:

$$\text{fixed cost} = c_f = \$10{,}000$$
$$\text{variable cost} = c_v = \$8 \text{ per pair}$$

If we arbitrarily let the monthly sales volume, v, equal 400 pairs of denim jeans, the total cost is

$$TC = c_f + vc_v = \$10{,}000 + (400)(8) = \$13{,}200$$

Profit is the difference between total revenue (volume multiplied by price) and total cost.

The third component in our break-even model is **profit**. Profit is the difference between *total revenue* and total cost. Total revenue is the volume multiplied by the price per unit,

$$\text{total revenue} = vp$$

where p = price per unit.

For our clothing company example, if denim jeans sell for \$23 per pair and we sell 400 pairs per month, then the total monthly revenue is

$$\text{total revenue} = vp = (400)(23) = \$9{,}200$$

Now that we have developed relationships for total revenue and total cost, profit (Z) can be computed as follows:

$$\begin{aligned}\text{total profit} &= \text{total revenue} - \text{total cost}\\ Z &= vp - (c_f + vc_v)\\ &= vp - c_f - vc_v\end{aligned}$$

Computing the Break-Even Point

For our clothing company example, we have determined total revenue and total cost to be \$9,200 and \$13,200, respectively. With these values, there is no profit but, instead, a loss of \$4,000:

$$\text{total profit} = \text{total revenue} - \text{total cost} = \$9{,}200 - 13{,}200 = -\$4{,}000$$

We can verify this result by using our total profit formula,

$$Z = vp - c_f - vc_v$$

and the values $v = 400$, $p = \$23$, $c_f = \$10{,}000$, and $c_v = \$8$:

$$\begin{aligned}Z &= vp - c_f - vc_v\\ &= \$(400)(23) - 10{,}000 - (400)(8)\\ &= \$9{,}200 - 10{,}000 - 3{,}200\\ &= -\$4{,}000\end{aligned}$$

Obviously, the clothing company does not want to operate with a monthly loss of \$4,000 because doing so might eventually result in bankruptcy. If we assume that price is static because of market conditions and that fixed costs and the variable cost per unit are not subject to change, then the only part of our model that can be varied is *volume*. Using the modeling terms we developed earlier in this chapter, price, fixed costs, and variable costs are parameters, whereas the volume, v, is a decision variable. In break-even analysis we want to compute the value of v that will result in zero profit.

*The **break-even point** is the volume (v) that equates total revenue with total cost where profit is zero.*

At the **break-even point**, where total revenue equals total cost, the profit, Z, equals zero. Thus, if we let profit, Z, equal zero in our total profit equation and solve for v, we can determine the break-even volume:

$$\begin{aligned} Z &= vp - c_f - vc_v \\ 0 &= v(23) - 10{,}000 - v(8) \\ 0 &= 23v - 10{,}000 - 8v \\ 15v &= 10{,}000 \\ v &= 666.7 \text{ pairs of jeans} \end{aligned}$$

In other words, if the company produces and sells 666.7 pairs of jeans, the profit (and loss) will be zero and the company will *break even*. This gives the company a point of reference from which to determine how many pairs of jeans it needs to produce and sell in order to gain a profit (subject to any capacity limitations). For example, a sales volume of 800 pairs of denim jeans will result in the following monthly profit:

$$\begin{aligned} Z &= vp - c_f - vc_v \\ &= \$(800)(23) - 10{,}000 - (800)(8) = \$2{,}000 \end{aligned}$$

In general, the break-even volume can be determined using the following formula:

$$\begin{aligned} Z &= vp - c_f - vc_v \\ 0 &= v(p - c_v) - c_f \\ v(p - c_v) &= c_f \\ v &= \frac{c_f}{p - c_v} \end{aligned}$$

For our example,

$$\begin{aligned} v &= \frac{c_f}{p - c_v} \\ &= \frac{10{,}000}{23 - 8} \\ &= 666.7 \text{ pairs of jeans} \end{aligned}$$

Graphical Solution

It is possible to represent many of the management science models in this text graphically and use these graphical models to solve problems. Graphical models also have the advantage of providing a "picture" of the model that can sometimes help us understand the modeling process better than mathematics alone can. We can easily graph the break-even model for our Western Clothing Company example because the functions for total cost and total revenue are *linear*. That means we can graph each relationship as a straight line on a set of coordinates, as shown in Figure 1.2.

In Figure 1.2, the fixed cost, c_f, has a constant value of \$10,000, regardless of the volume. The total cost line, *TC*, represents the sum of variable cost and fixed cost. The total cost line increases because variable cost increases as the volume increases. The total revenue line also increases as volume increases, but at a faster rate than total cost. The point where these two lines intersect indicates that total revenue equals total cost. The volume, v, that corresponds to this point is the *break-even volume*. The break-even volume in Figure 1.2 is 666.7 pairs of denim jeans.

FIGURE 1.2
Break-even model

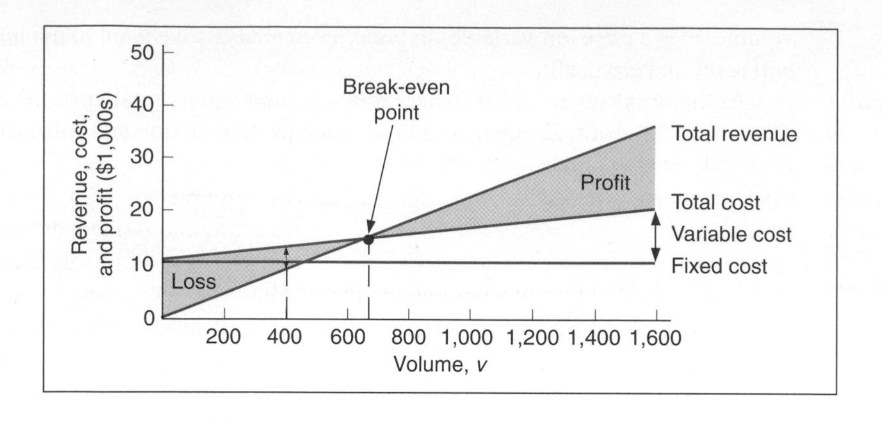

Sensitivity Analysis

We have now developed a general relationship for determining the break-even volume, which was the objective of our modeling process. This relationship enables us to see how the level of profit (and loss) is directly affected by changes in volume. However, when we developed this model, we assumed that our parameters, fixed and variable costs and price, were constant. In reality such parameters are frequently uncertain and can rarely be assumed to be constant, and changes in any of the parameters can affect the model solution. The study of changes on a management science model is called **sensitivity analysis**—that is, seeing how sensitive the model is to changes.

Sensitivity analysis sees how sensitive a management model is to changes.

Sensitivity analysis can be performed on all management science models in one form or another. In fact, sometimes companies develop models for the primary purpose of experimentation to see how the model will react to different changes the company is contemplating or that management might expect to occur in the future. As a demonstration of how sensitivity analysis works, we will look at the effects of some changes on our break-even model.

In general, an increase in price lowers the break-even point, all other things held constant.

The first thing we will analyze is price. As an example, we will increase the price for denim jeans from $23 to $30. As expected, this increases the total revenue, and it therefore reduces the break-even point from 666.7 pairs of jeans to 454.5 pairs of jeans:

$$v = \frac{c_f}{p - c_v}$$

$$= \frac{10{,}000}{30 - 8} = 454.5 \text{ pairs of denim jeans}$$

The effect of the price change on break-even volume is illustrated in Figure 1.3.

FIGURE 1.3
Break-even model with an increase in price

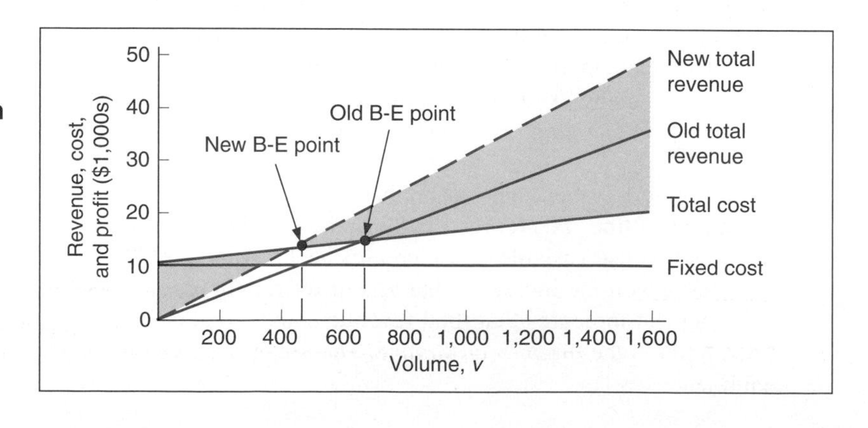

Although a decision to increase price looks inviting from a strictly analytical point of view, it must be remembered that the lower break-even volume and higher profit are *possible* but not guaranteed. A higher price can make it more difficult to sell the product. Thus, a change in price often must be accompanied by corresponding increases in costs, such as those for advertising, packaging, and possibly production (to enhance quality). However, even such direct changes as these may have little effect on product demand because price is often sensitive to numerous factors, such as the type of market, monopolistic elements, and product differentiation.

In general, an increase in variable costs will increase the break-even point, all other things held constant.

When we increased price, we mentioned the possibility of raising the quality of the product to offset a potential loss of sales due to the price increase. For example, suppose the stitching on the denim jeans is changed to make the jeans more attractive and stronger. This change results in an increase in variable costs of \$4 per pair of jeans, thus raising the variable cost per unit, c_v, to \$12 per pair. This change (in conjunction with our previous price change to \$30) results in a new break-even volume:

$$v = \frac{c_f}{p - c_v}$$

$$= \frac{10{,}000}{30 - 12} = 555.5 \text{ pairs of denim jeans}$$

This new break-even volume and the change in the total cost line that occurs as a result of the variable cost change are shown in Figure 1.4.

FIGURE 1.4
Break-even model with an increase in variable cost

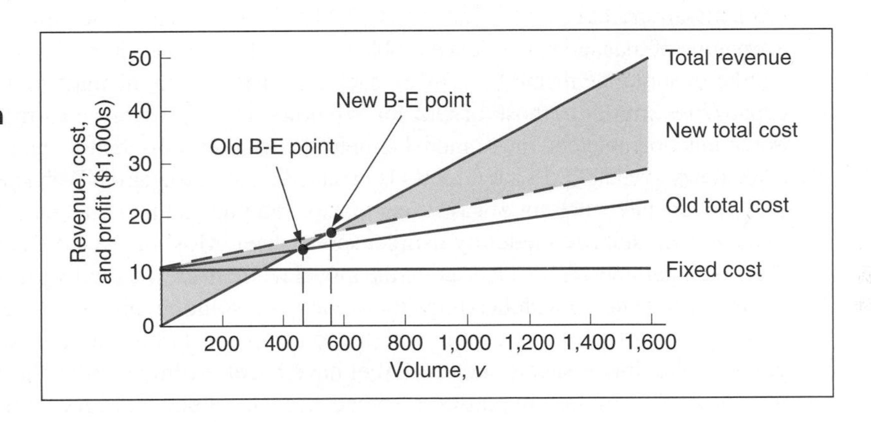

Next let's consider an increase in advertising expenditures to offset the potential loss in sales resulting from a price increase. An increase in advertising expenditures is an addition to fixed costs. For example, if the clothing company increases its monthly advertising budget by \$3,000, then the total fixed cost, c_f, becomes \$13,000. Using this fixed cost, as well as the increased variable cost per unit of \$12 and the increased price of \$30, we compute the break-even volume as follows:

$$v = \frac{c_f}{p - c_v}$$

$$= \frac{13{,}000}{30 - 12}$$

$$= 722.2 \text{ pairs of denim jeans}$$

In general, an increase in fixed costs will increase the break-even point, all other things held constant.

This new break-even volume, representing changes in price, fixed costs, and variable costs, is illustrated in Figure 1.5. Notice that the break-even volume is now higher than the original volume of 666.7 pairs of jeans, as a result of the increased costs necessary to offset the potential loss in sales. This indicates the necessity to analyze the effect of a change in one of the break-even

FIGURE 1.5
Break-even model with a change in fixed cost

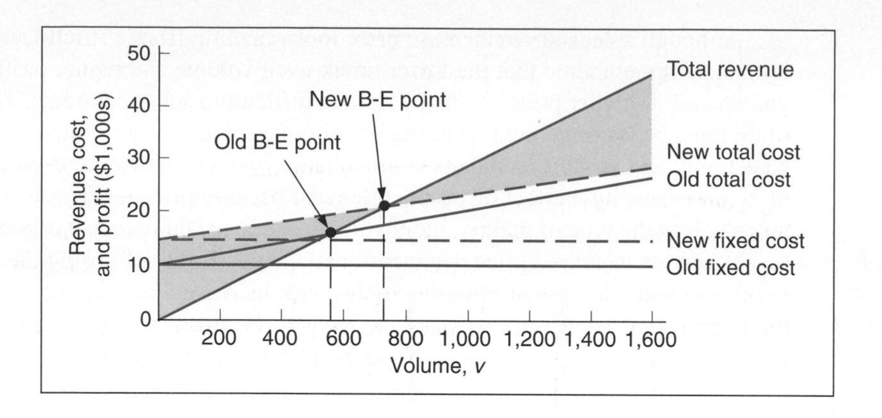

components on the whole break-even model. In other words, generally it is not sufficient to consider a change in one model component without considering the overall effect.

Computer Solution

Throughout the text we will demonstrate how to solve management science models on the computer by using Excel spreadsheets and QM for Windows, a general-purpose quantitative methods software package by Howard Weiss. QM for Windows has program modules to solve almost every type of management science problem you will encounter in this book. There are a number of similar quantitative methods software packages available on the market, with characteristics and capabilities similar to those of QM for Windows. In most cases you simply input problem data (i.e., model parameters) into a model template, click on a solve button, and the solution appears in a Windows format. QM for Windows is included on the companion Web site for this text.

Spreadsheets are not always easy to use, and you cannot conveniently solve every type of management science model by using a spreadsheet. Most of the time you must not only input the model parameters but also set up the model mathematics, including formulas, as well as your own model template with headings to display your solution output. However, spreadsheets provide a powerful reporting tool in which you can present your model and results in any format you choose. Spreadsheets such as Excel have become almost universally available to anyone who owns a computer. In addition, spreadsheets have become very popular as a teaching tool because they tend to guide the student through a modeling procedure, and they can be interesting and fun to use. However, because spreadsheets are somewhat more difficult to set up and apply than is QM for Windows, we will spend more time explaining their use to solve various types of problems in this text.

One of the difficult aspects of using spreadsheets to solve management science problems is setting up a spreadsheet with some of the more complex models and formulas. For the most complex models in the text we will show how to use Excel QM, a supplemental spreadsheet macro that is included on the companion Web site for this text. A *macro* is a template or an overlay that already has the model format with the necessary formulas set up on the spreadsheet so that the user only has to input the model parameters. We will demonstrate Excel QM in six chapters, including this chapter, Chapter 6 ("Transportation, Transshipment, and Assignment Problems"), Chapter 12 ("Decision Analysis"), Chapter 13 ("Queuing Analysis"), Chapter 15 ("Forecasting"), and Chapter 16 ("Inventory Management").

Later in this text we will also demonstrate two spreadsheet add-ins, TreePlan and Crystal Ball, which are also included on the companion Web site for this text. TreePlan is a program for setting up and solving decision trees that we use in Chapter 12 ("Decision Analysis"), whereas Crystal Ball is a simulation package that we use in Chapter 14 ("Simulation"). Also, in Chapter 8 ("Project Management") we will demonstrate Microsoft Project.

In this section, we will demonstrate how to use Excel, Excel QM, and QM for Windows, using our break-even model example for Western Clothing Company.

Excel Spreadsheets

To solve the break-even model using Excel, you must set up a spreadsheet with headings to identify your model parameters and variables and then input the appropriate mathematical formulas into the cells where you want to display your solution. Exhibit 1.1 shows the spreadsheet for the Western Clothing Company example. Setting up the different headings to describe the parameters and the solution is not difficult, but it does require that you know your way around Excel a little. Appendix B provides a brief tutorial titled "Setting Up and Editing a Spreadsheet" for solving management science problems.

EXHIBIT 1.1

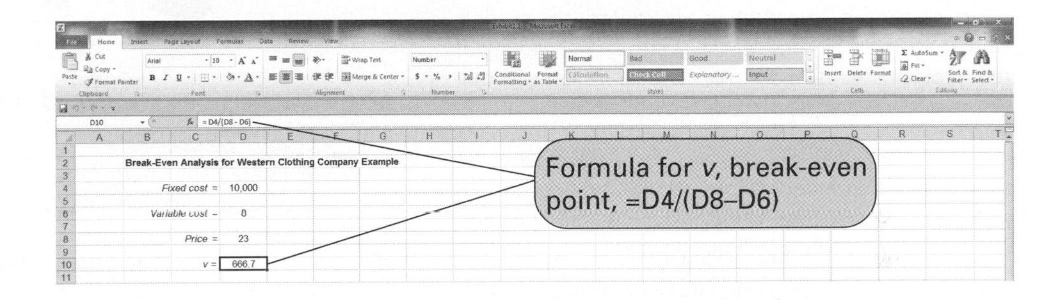

Notice that cell D10 contains the break-even formula, which is displayed on the toolbar near the top of the screen. The fixed cost of \$10,000 is typed in cell D4, the variable cost of \$8 is in cell D6, and the price of \$23 is in cell D8.

As we present more complex models and problems in the chapters to come, the spreadsheets we will develop to solve these problems will become more involved and will enable us to demonstrate different features of Excel and spreadsheet modeling.

The Excel QM Macro for Spreadsheets

Excel QM is included on the companion Web site for this text. You can install Excel QM onto your computer by following a brief series of steps displayed when the program is first accessed.

After Excel is started, Excel QM is normally accessed from the computer's program files, where it is usually loaded. When Excel QM is activated, "Add-Ins" will appear at the top of the spreadsheet (as indicated in Exhibit 1.2). Clicking on "Excel QM" or "Taylor" will pull down a

EXHIBIT 1.2

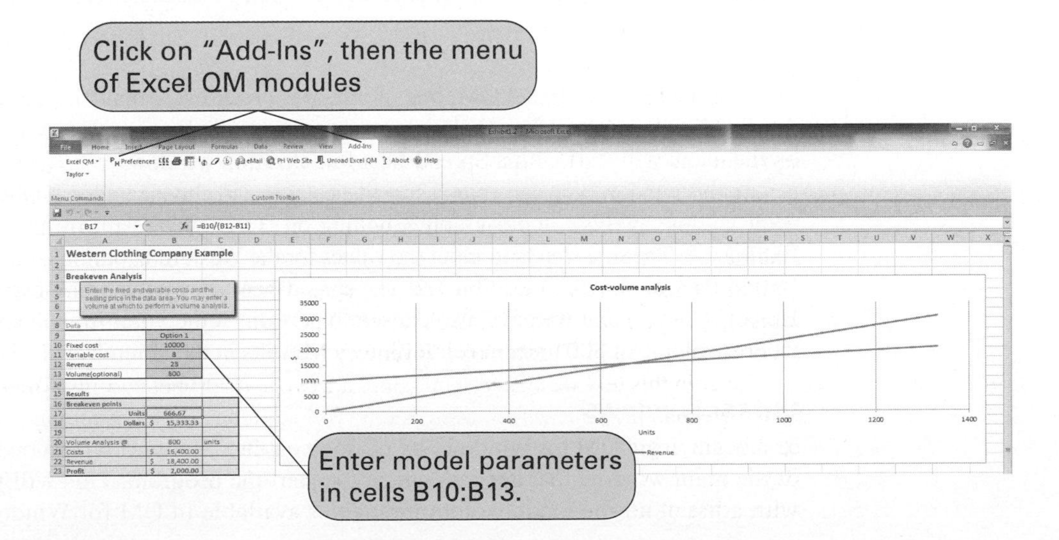

Management Science Application

The Application of Management Science with Spreadsheets

Excel spreadsheets have become an increasingly important management science tool because of their ability to support numerous software add-ins for various management science techniques, their ability to effectively convey complex models to clients, their general availability on virtually every computer, their flexibility and ease of use, and the fact that they are inexpensive. As a result, spreadsheets are used for the application of management science techniques to a wide variety of different problems across many diverse organizations; following are just a few examples of these applications.

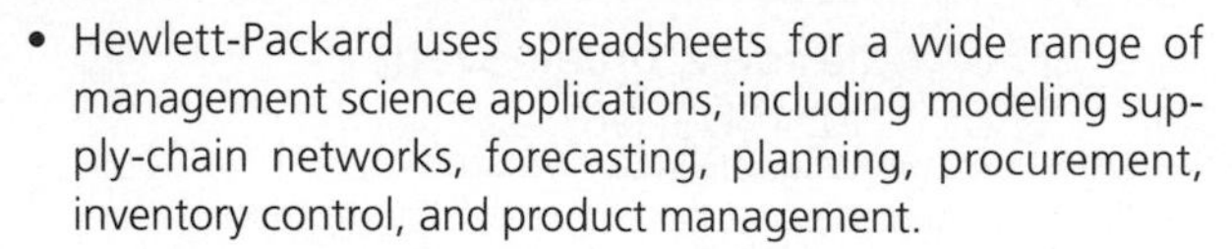

- Hewlett-Packard uses spreadsheets for a wide range of management science applications, including modeling supply-chain networks, forecasting, planning, procurement, inventory control, and product management.
- Procter & Gamble also uses spreadsheets for supply-chain management and specifically inventory control, to which it has attributed over $350 million in inventory reductions.
- Lockheed Martin Space Systems Company uses spreadsheets to apply mathematical programming techniques for project selection.
- The Centers for Disease Control and Prevention (CDC), use Excel spreadsheets to provide people at county health departments in the United States (who have minimal management science skills) with tools using queuing techniques to plan for dispensing medications and vaccines during emergencies, such as epidemics and terrorist attacks.
- A spreadsheet application for the Canadian Army allowed it to reduce annual over-budget expenditures for ammunition for its training programs from over $24 million to $1.3 million in a 2-year period.

Used with permission from Microsoft

- Hypo Real Estate Bank International in Stuttgart, Germany, uses an Excel-based simulation model to assess the potential impact of economic events on the default risk of its portfolio of over €40 billion in real estate loans around the world.
- Business students at the University of Toronto created an Excel spreadsheet model for assigning medical residents in radiology to on-call and emergency rotations at the University of Vermont's College of Medicine.
- The American Red Cross uses Excel spreadsheets to apply data envelopment analysis (DEA) and linear programming techniques for allocating resources and evaluating the performance of its 1,000 chapters.

These are just a few of the many applications of management science techniques worldwide using Excel spreadsheets.

Source: Based on L. LeBlanc and T. Grossman, "Introduction: The Use of Spreadsheet Software in the Application of Management Science and Operations Research," *Interfaces* 38, no. 4 (July–August 2008): 225–27.

menu of the topics in Excel QM, one of which is break-even analysis. Clicking on "Break-Even Analysis" will result in the window for spreadsheet initialization. Every Excel QM macro listed on the menu will start with a Spreadsheet Initialization window.

In this window, you can enter a spreadsheet title and choose under "Options" whether you also want volume analysis and a graph. Clicking on "OK" will result in the spreadsheet shown in Exhibit 1.2. The first step is to input the values for the Western Clothing Company example in cells B10 to B13, as shown in Exhibit 1.2. The spreadsheet shows the break-even volume in cell B17. However, notice that we have also chosen to perform some volume analysis by entering a hypothetical volume of 800 units in cell B13, which results in the volume analysis in cells B20 to B23.

QM for Windows

You begin using QM for Windows by clicking on the "Module" button on the toolbar at the top of the main window that appears when you start the program. This will pull down a window with a list of all the model solution modules available in QM for Windows. Clicking on the

"Break-even Analysis" module will access a new screen for typing in the problem title. Clicking again will access a screen with input cells for the model parameters—that is, fixed cost, variable cost, and price (or revenue). Next, clicking on the "Solve" button at the top of the screen will provide the solution to the Western Clothing Company example, as shown in Exhibit 1.3.

EXHIBIT 1.3

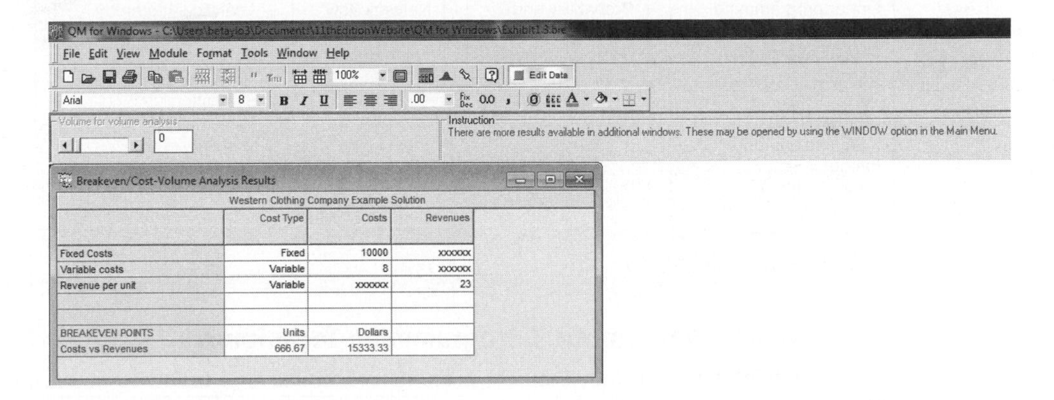

Breakeven/Cost-Volume Analysis Results

Western Clothing Company Example Solution

	Cost Type	Costs	Revenues
Fixed Costs	Fixed	10000	xxxxxx
Variable costs	Variable	8	xxxxxx
Revenue per unit	Variable	xxxxxx	23
BREAKEVEN POINTS	Units	Dollars	
Costs vs Revenues	666.67	15333.33	

You can also get the graphical model and solution for this problem by clicking on "Window" at the top of the solution screen and selecting the menu item for a graph of the problem. The break-even graph for the Western Clothing example is shown in Exhibit 1.4.

EXHIBIT 1.4

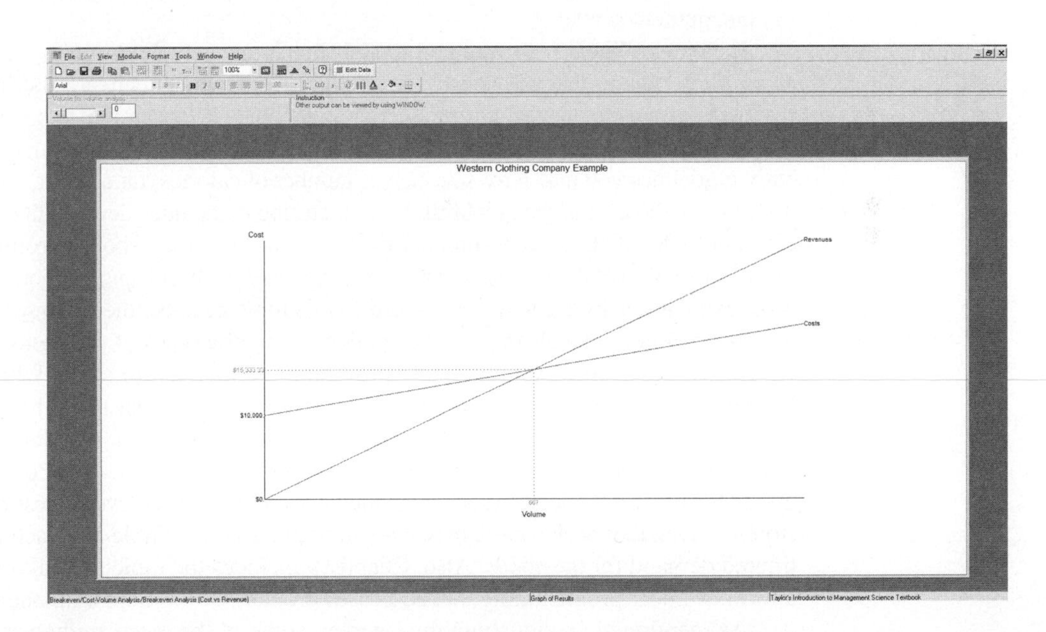

Management Science Modeling Techniques

This text focuses primarily on two of the five steps of the management science process described in Figure 1.1—model construction and solution. These are the two steps that use the management science techniques. In a textbook, it is difficult to show how an unstructured real-world problem is identified and defined because the problem must be written out. However, once a problem statement has been given, we can show how a model is constructed and a solution is derived. The techniques presented in this text can be loosely classified into four categories, as shown in Figure 1.6.

FIGURE 1.6 Classification of management science techniques

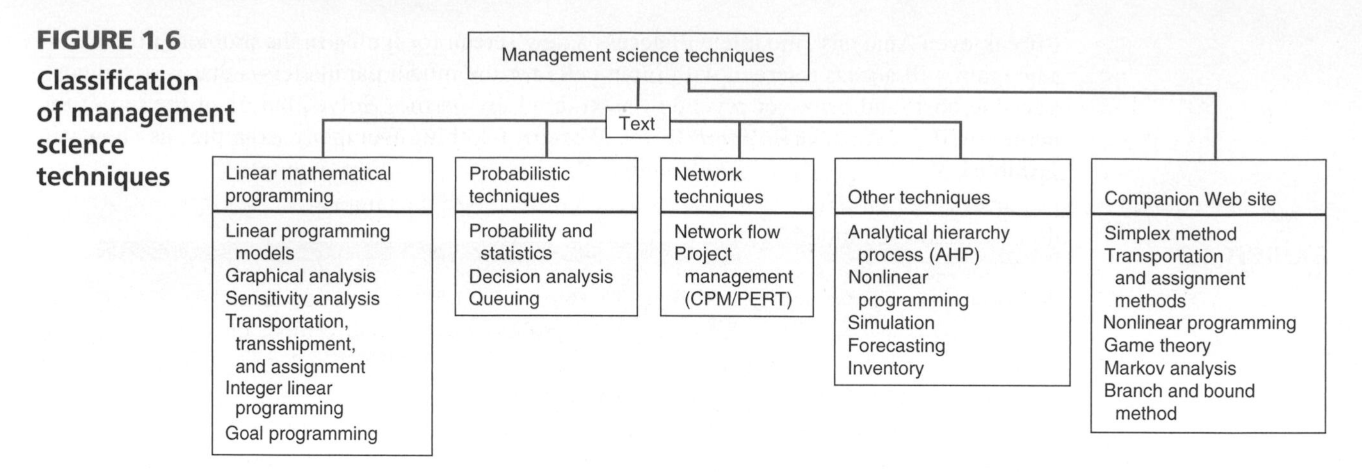

Linear Mathematical Programming Techniques

Chapters 2 through 6 and 9 present techniques that together make up *linear mathematical programming*. (The first example used to demonstrate model construction earlier in this chapter is a very rudimentary linear programming model.) The term *programming* used to identify this technique does not refer to computer programming but rather to a predetermined set of mathematical steps used to solve a problem. This particular class of techniques holds a predominant position in this text because it includes some of the more frequently used and popular techniques in management science.

In general, linear programming models help managers determine solutions (i.e., make decisions) for problems that will achieve some objective in which there are restrictions, such as limited resources or a recipe or perhaps production guidelines. For example, you could actually develop a linear programming model to help determine a breakfast menu for yourself that would meet dietary guidelines you may have set, such as number of calories, fat content, and vitamin level, while minimizing the cost of the breakfast. Manufacturing companies develop linear programming models to help decide how many units of different products they should produce to maximize their profit (or minimize their cost), given scarce resources such as capital, labor, and facilities.

Six chapters in this text are devoted to this topic because there are several variations of linear programming models that can be applied to specific types of problems. Chapter 4 is devoted entirely to describing example linear programming models for several different types of problem scenarios. Chapter 6, for example, focuses on one particular type of linear programming application for transportation, transshipment, and assignment problems. An example of a transportation problem is a manager trying to determine the lowest-cost routes to use to ship goods from several sources (such as plants or warehouses) to several destinations (such as retail stores), given that each source may have limited goods available and each destination may have limited demand for the goods. Also, Chapter 9 includes the topic of goal programming, which is a form of linear programming that addresses problems with more than one objective or goal.

As mentioned previously in this chapter, some of the more mathematical topics in the text are included as supplementary modules on the companion Web site for the text. Among the linear programming topics included on the companion Web site are modules on the simplex method; the transportation and assignment solution methods; and the branch and bound solution method for integer programming models. Also included on the companion Web site are modules on nonlinear programming, game theory, and Markov analysis.

Probabilistic Techniques

Probabilistic techniques are presented in Chapters 11 through 13. These techniques are distinguished from mathematical programming techniques in that the results are probabilistic. Mathematical programming techniques assume that all parameters in the models are known with *certainty*. Therefore, the solution results are assumed to be known with certainty, with no probability that other solutions might exist. A technique that assumes certainty in its solution is referred to as **deterministic**. In contrast, the results from a probabilistic technique *do* contain uncertainty, with some possibility that alternative solutions might exist. In the model solution presented earlier in this chapter, the result of the first example ($x = 25$ units to produce) is deterministic, whereas the result of the second example (estimating an average of 40 units sold each month) is probabilistic.

A deterministic technique assumes certainty in the solution.

An example of a probabilistic technique is decision analysis, the subject of Chapter 12. In decision analysis, it is shown how to select among several different decision alternatives, given uncertain (i.e., probabilistic) future conditions. For example, a developer may want to decide whether to build a shopping mall, build an office complex, build condominiums, or not build anything at all, given future economic conditions that might be good, fair, or poor, each with a probability of occurrence. Chapter 13, on queuing analysis, presents probabilistic techniques for analyzing waiting lines that might occur, for example, at the grocery store, at a bank, or at a movie. The results of waiting line analysis are statistical averages showing, among other things, the average number of customers in line waiting to be served or the average time a customer might have to wait for service.

Network Techniques

Networks, the topic of Chapters 7 and 8, consist of models that are represented as diagrams rather than as strictly mathematical relationships. As such, these models offer a pictorial representation of the system under analysis. These models represent either probabilistic or deterministic systems.

For example, in shortest-route problems, one of the topics in Chapter 7 ("Network Flow Models"), a network diagram can be drawn to help a manager determine the shortest route among a number of different routes from a source to a destination. For example, you could use this technique to determine the shortest or quickest car route from St. Louis to Daytona Beach for a spring break vacation. In Chapter 8 ("Project Management"), a network is drawn that shows the relationships of all the tasks and activities for a project, such as building a house or developing a new computer system. This type of network can help a manager plan the best way to accomplish each of the tasks in the project so that it will take the shortest amount of time possible. You could use this type of technique to plan for a concert or an intramural volleyball tournament on your campus.

Other Techniques

Some topics in the text are not easily categorized; they may overlap several categories, or they may be unique. The analytical hierarchy process (AHP) in Chapter 9 is such a topic that is not easily classified. It is a mathematical technique for helping the decision maker choose between several alternative decisions, given more than one objective; however, it is not a form of linear programming, as is goal programming, the shared topic in Chapter 9, on multicriteria decision making. The structure of the mathematical models for nonlinear programming problems in Chapter 10 is similar to the linear programming problems in Chapters 2 through 6; however, the mathematical equations and functions in nonlinear programming can be nonlinear instead of linear, thus requiring the use of calculus to solve them. Simulation, the subject of Chapter 14, is probably the single most unique topic in the text. It has the capability to solve probabilistic and deterministic problems and is often the technique of last resort when no other management science technique will work. In simulation, a mathematical model is constructed (typically using a computer) that replicates a real-world system under analysis, and then that simulation model is used to solve problems in the "simulated" real-world system. For example, with simulation you could build a model to simulate the traffic patterns of vehicles at a busy intersection to determine how to set the traffic light signals.

Forecasting, the subject of Chapter 15, and inventory management, in Chapter 16, are topics traditionally considered to be part of the field of operations management. However, because they are both important business functions that also rely heavily on quantitative models for their analysis, they are typically considered important topics in the study of management science as well. Both topics also include probabilistic as well as deterministic aspects. In Chapter 15, we will look at several different quantitative models that help managers predict what the future demand for products and services will look like. In general, historical sales and demand data are used to build a mathematical function or formula that can be used to estimate product demand in the future. In Chapter 16, we will look at several different quantitative models that help organizations determine how much inventory to keep on hand in order to minimize inventory costs, which can be significant.

Business Usage of Management Science Techniques

Not all management science techniques are equally useful or equally used by business firms and other organizations. Some techniques are used quite frequently by business practitioners and managers; others are used less often. The most frequently used techniques are linear and integer programming, simulation, network analysis (including critical path method/project evaluation and review technique [CPM/PERT]), inventory control, decision analysis, and queuing theory, as well as probability and statistics. An attempt has been made in this text to provide a comprehensive treatment of all the topics generally considered within the field of management science, regardless of how frequently they are used. Although some topics may have limited direct applicability, their study can reveal informative and unique means of approaching a problem and can often enhance one's understanding of the decision-making process.

The variety and breadth of management science applications and of the potential for applying management science, not only in business and industry but also in government, health care, and service organizations, are extensive. Areas of application include project planning, capital budgeting, production planning, inventory analysis, scheduling, marketing planning, quality control, plant location, maintenance policy, personnel management, and product demand forecasting, among others. In this text the applicability of management science to a variety of problem areas is demonstrated via individual chapter examples and the problems that accompany each chapter.

A small portion of the thousands of applications of management science that occur each year are recorded in various academic and professional journals. Frequently, these journal articles are as complex as the applications themselves and are very difficult to read. However, one particular journal, *Interfaces*, is devoted specifically to the application of management science and is written not just for college professors but for businesspeople, practitioners, and students as well. *Interfaces* is published by INFORMS (Institute for Operations Research and Management Sciences), an international professional organization whose members include college professors, businesspeople, scientists, students, and a variety of professional people interested in the practice and application of management science and operations research.

Interfaces regularly publishes articles that report on the application of management science to a wide variety of problems. The chapters that follow present examples of applications of management science from *Interfaces* and other professional journals. These applications are from a variety of U.S. and overseas companies and organizations in business, industry, services, and government. These examples, as presented here, do not detail the actual models and the model components. Instead, they briefly indicate the type of problem the company or organization faced, the objective of the solution approach developed to solve the problem, and the benefits derived from the model or technique (i.e., what was accomplished). The interested reader who desires more detailed information about these and other management science applications is

Management Science Application

Management Science in Health Care

More than 16% of the U.S. GDP is spent on health care each year (i.e., more than $2 trillion in 2006), making it the single largest industry in the United States. However, it is estimated that as much as 30% of health care costs result from waste through inefficient processes. Management science is really good at making inefficient processes more efficient. Thus, it is not surprising that one of the most frequent recent areas of application of management science techniques is in health care. Following are three brief examples of its many successful applications.

Each year, approximately 2 million patients contract health-care-associated infections in hospitals in the United States. Each year, more than 100,000 of these patients die, and more than $30 billion is spent on treating these infections. This problem is complicated by the emergence of pathogens (i.e., germs) that are resistant to antibiotics. Researchers at John H. Stroger, Jr. Hospital of Cook County in Chicago used management science, specifically simulation (Chapter 14), to address this problem. The simulation modeled the process of pathogens, patients, and visitors entering an intensive care unit; interacting with health care workers and each other; infecting; becoming infected; being cured; being discharged; and being assigned costs. Besides identifying critical issues and interrelationships in hospital and health care procedures, the results specifically indicated that both isolation wards for infected patients and hand hygiene are critical policies in reducing infections.

East Carolina University (ECU) Student Health Service is a health care clinic that serves the 23,000 students at this public university located in Greenville, South Carolina. The Student Health Service facility encompasses 16,000 square feet of space and includes 25 exam rooms, 3 observation beds, a pharmacy, a lab, and radiology resources. On any given day, it is staffed by 5 to 10 doctors and physician assistants to diagnose and treat campus patients. Almost all patients schedule appointments in advance. In a recent year, slightly more than 35,000 appointments were scheduled, but approximately 3,800 were no-shows. The problem of no-shows at health care clinics is not unique, but it is a significant problem, with estimated costs at the ECU clinic of over $400,000 per year resulting from reduced patient access.

Researchers at ECU used a combination of several management science techniques, including forecasting (Chapter 15), decision analysis (Chapter 12), and simulation (Chapter 14), to develop a solution approach employing an overbooking policy (similar to what airlines do for flights). In the first semester, the clinic implemented the policy, appointment times were overbooked by 7.3%, with few patients overscheduled and an estimated cost savings of about $95,000.

Photo courtesy of Memorial Sloan-Kettering Cancer Center

In the United States, prostate cancer is the second most prevalent cancer killer of men, with more than 220,000 new cases each year and a 12% mortality rate. At Memorial Sloan-Kettering Cancer Center in New York, researchers formulated sophisticated models using management science techniques—and specifically linear integer programming (Chapter 5)—for the treatment of prostate cancer using brachytherapy, the placement of radioactive "seeds" inside a tumor. The procedure results in significantly safer and more reliable treatment outcomes, has the potential for saving hundreds of millions of dollars through the elimination of various related procedures, and improves post-treatment quality of life by reducing complications up to 45% to 60%.

Sources: Based on M. Carter, B. Golden, and E. Wasil, "Introduction: Applications of Management Science and Operations Research Models and Methods to Problems in Health Care," *Interfaces* 39, no. 3 (May–June 2009): 183–85; R. Hagtvedt, P. Griffin, P. Keskinocak, and R. Roberts, "A Simulation Model to Compare Strategies for the Reduction of Health-Care-Associated Infections," *Interfaces* 39, no. 3 (May–June 2009): 256–70; J. Kros, S. Dellana, and D. West, "Overbooking Increases Patient Access at East Carolina University's Student Health Services Clinic," *Interfaces* no. 3 (May–June 2009): 271–87; and E. Lee and M. Zaider, "Operations Research Advances Cancer Therapeutics," *Interfaces* 38, no. 1 (January–February 2008): 5–25.

encouraged to go to the library and peruse *Interfaces* and the many other journals that contain articles on the application of management science.

Management Science Models in Decision Support Systems

Historically, management science models have been applied to the solution of specific types of problems; for example, a waiting line model is used to analyze a specific waiting line system at a store or bank. However, the evolution of computer and information technology has enabled the development of expansive computer systems that combine several management science models and solution techniques in order to address more complex, interrelated organizational problems. A **decision support system (DSS)** is a computer-based system that helps decision makers address complex problems that cut across different parts of an organization and operations.

A DSS is normally *interactive*, combining various databases and different management science models and solution techniques with a user interface that enables the decision maker to ask questions and receive answers. In its simplest form any computer-based software program that helps a decision maker make a decision can be referred to as a DSS. For example, an Excel spreadsheet like the one shown for break-even analysis in Exhibit 1.1 or the QM for Windows model shown in Exhibit 1.3 can realistically be called a DSS. Alternatively, enterprise-wide DSSs can encompass many different types of models and large data warehouses, and they can serve many decision makers in an organization. They can provide decision makers with interrelated information and analyses about almost anything in a company.

Figure 1.7 illustrates the basic structure of a DSS with a database component, a modeling component, and a user interface with the decision maker. As noted earlier, a DSS can be small and singular, with one analytical model linked to a database, or it can be very large and complex, linking many models and large databases. A DSS can be primarily a data-oriented system, or it can be a model-oriented system. A new type of DSS, called an *online analytical processing* system, or *OLAP*, focuses on the use of analytical techniques such as management science models and statistics for decision making. A desktop DSS for a single user can be a spreadsheet program such as Excel to develop specific solutions to individual problems. Exhibit 1.1 includes all the components of a DSS—cost, volume, and price data, a break-even model, and the opportunity for the user to manipulate the data and see the results (i.e., a user interface). Expert Choice is another example of a desktop DSS that uses the analytical hierarchy process (AHP) described in Chapter 9 to structure complex problems by establishing decision criteria, developing priorities, and ranking decision alternatives.

FIGURE 1.7
A decision support system

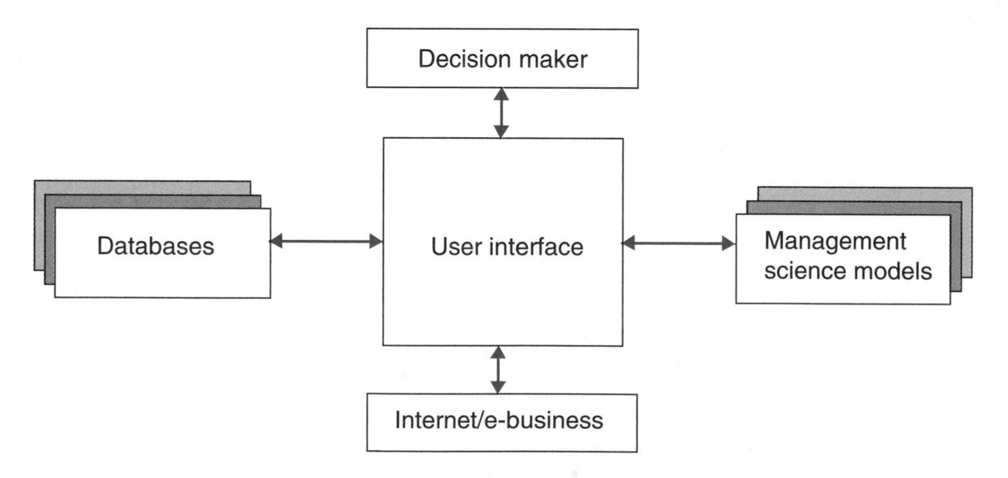

On the other end of the DSS spectrum, an *enterprise resource planning (ERP)* system is software that can connect the components and functions of an entire company. It can transform data, such as individual daily sales, directly into information that supports immediate decisions in other parts of the company, such as ordering, manufacturing, inventory, and distribution. A large-scale DSS such as an ERP system in a company might include a forecasting model (Chapter 15) to analyze sales data and help determine future product demand; an inventory model (Chapter 16) to determine how much inventory to keep on hand; a linear programming model (Chapters 2–5) to determine how much material to order and product to produce, and when to produce it; a transportation model (Chapter 6) to determine the most cost-effective method of distributing a product to customers; and a network flow model (Chapter 7) to determine the best delivery routes. All these different management science models and the data necessary to support them can be linked in a single enterprisewide DSS that can provide many decisions to many different decision makers.

In addition to helping managers answer specific questions and make decisions, a DSS may be most useful in answering What if? questions and performing sensitivity analysis. In other words, a DSS provides a computer-based laboratory to perform experiments. By linking various management science models together with different databases, a user can change a parameter in one model related to one company function and see what the effect will be in a model related to a different operation in the company. For example, by changing the data in a forecasting model, a manager could see the impact of a hypothetical change in product demand on the production schedule, as determined by a linear programming model.

Advances in information and computer technology have provided the opportunity to apply management science models to a broad array of complex organizational problems by linking different models to databases in a DSS. These advances have also made the application of management science models more readily available to individual users in the form of desktop DSSs that can help managers make better decisions relative to their day-to-day operations. In the future it will undoubtedly become even easier to apply management science to the solution of problems with the development of newer software, and management science will become even more important and pervasive as an aid to decision makers as managers are linked within companies with sophisticated computer systems and to other companies via the Internet.

Many companies now interface with new types of DSS over the Internet. In e-business applications, companies can link to other business units around the world through computer systems called *intranets*, with other companies through systems called *extranets*, and over the Internet. For example, electronic data interchange (EDI) and point-of-sale data (through bar codes) can provide companies with instantaneous records of business transactions and sales at retail stores that are immediately entered into a company's DSS to update inventory and production scheduling, using management science models. Internet transportation exchanges enable companies to arrange cost-effective transportation of their products at Web sites that match shipping loads with available trucks at the lowest cost and fastest delivery speed, using sophisticated management science models.

Summary

Management science is an art.

In the chapters that follow, the model construction and solutions that constitute each management science technique are presented in detail and illustrated with examples. In fact, the primary method of presenting the techniques is through examples. Thus, the text offers you a broad spectrum of knowledge of the mechanics of management science techniques and the types of problems to which these techniques are applied. However, the ultimate test of a management scientist or a manager who uses management science techniques is the ability to transfer textbook knowledge to the business world. In such instances there is an *art* to the application of management science, but it is an art predicated on practical experience and sound textbook knowledge. Providing the first of these necessities is beyond the scope of textbooks; providing the second is the objective of this text.

Problems

1. The Willow Furniture Company produces tables. The fixed monthly cost of production is $8,000, and the variable cost per table is $65. The tables sell for $180 apiece.
   a. For a monthly volume of 300 tables, determine the total cost, total revenue, and profit.
   b. Determine the monthly break-even volume for the Willow Furniture Company.
2. The Retread Tire Company recaps tires. The fixed annual cost of the recapping operation is $60,000. The variable cost of recapping a tire is $9. The company charges $25 to recap a tire.
   a. For an annual volume of 12,000 tires, determine the total cost, total revenue, and profit.
   b. Determine the annual break-even volume for the Retread Tire Company operation.
3. The Rolling Creek Textile Mill produces denim. The fixed monthly cost is $21,000, and the variable cost per yard of denim is $0.45. The mill sells a yard of denim for $1.30.
   a. For a monthly volume of 18,000 yards of denim, determine the total cost, total revenue, and profit.
   b. Determine the annual break-even volume for the Rolling Creek Textile Mill.
4. Evergreen Fertilizer Company produces fertilizer. The company's fixed monthly cost is $25,000, and its variable cost per pound of fertilizer is $0.15. Evergreen sells the fertilizer for $0.40 per pound. Determine the monthly break-even volume for the company.
5. Graphically illustrate the break-even volume for the Retread Tire Company determined in Problem 2.
6. Graphically illustrate the break-even volume for the Evergreen Fertilizer Company determined in Problem 4.
7. Andy Mendoza makes handcrafted dolls, which he sells at craft fairs. He is considering mass-producing the dolls to sell in stores. He estimates that the initial investment for plant and equipment will be $25,000, whereas labor, material, packaging, and shipping will be about $10 per doll. If the dolls are sold for $30 each, what sales volume is necessary for Andy to break even?
8. If the maximum operating capacity of the Retread Tire Company, as described in Problem 2, is 8,000 tires annually, determine the break-even volume as a percentage of that capacity.
9. If the maximum operating capacity of the Rolling Creek Textile Mill described in Problem 3 is 25,000 yards of denim per month, determine the break-even volume as a percentage of capacity.
10. If the maximum operating capacity of Evergreen Fertilizer Company described in Problem 4 is 120,000 pounds of fertilizer per month, determine the break-even volume as a percentage of capacity.
11. If the Retread Tire Company in Problem 2 changes its pricing for recapping a tire from $25 to $31, what effect will the change have on the break-even volume?
12. If Evergreen Fertilizer Company in Problem 4 changes the price of its fertilizer from $0.40 per pound to $0.60 per pound, what effect will the change have on the break-even volume?
13. If Evergreen Fertilizer Company changes its production process to add a weed killer to the fertilizer in order to increase sales, the variable cost per pound will increase from $0.15 to $0.22. What effect will this change have on the break-even volume computed in Problem 12?
14. If Evergreen Fertilizer Company increases its advertising expenditures by $14,000 per year, what effect will the increase have on the break-even volume computed in Problem 13?
15. Pastureland Dairy makes cheese, which it sells at local supermarkets. The fixed monthly cost of production is $4,000, and the variable cost per pound of cheese is $0.21. The cheese sells for $0.75 per pound; however, the dairy is considering raising the price to $0.95 per pound. The

dairy currently produces and sells 9,000 pounds of cheese per month, but if it raises its price per pound, sales will decrease to 5,700 pounds per month. Should the dairy raise the price?

16. For the doll-manufacturing enterprise described in Problem 7, Andy Mendoza has determined that $10,000 worth of advertising will increase sales volume by 400 dolls. Should he spend the extra amount for advertising?

17. Andy Mendoza in Problem 7 is concerned that the demand for his dolls will not exceed the break-even point. He believes he can reduce his initial investment by purchasing used sewing machines and fewer machines. This will reduce his initial investment from $25,000 to $17,000. However, it will also require his employees to work more slowly and perform more operations by hand, thus increasing variable cost from $10 to $14 per doll. Will these changes reduce his break-even point?

18. The General Store at State University is an auxiliary bookstore located near the dormitories that sells academic supplies, toiletries, sweatshirts and T-shirts, magazines, packaged food items, and canned soft drinks and fruit drinks. The manager of the store has noticed that several pizza delivery services near campus make frequent deliveries. The manager is therefore considering selling pizza at the store. She could buy premade frozen pizzas and heat them in an oven. The cost of the oven and freezer would be $27,000. The frozen pizzas cost $3.75 each to buy from a distributor and to prepare (including labor and a box). To be competitive with the local delivery services, the manager believes she should sell the pizzas for $8.95 apiece. The manager needs to write up a proposal for the university's director of auxiliary services.
    a. Determine how many pizzas would have to be sold to break even.
    b. If The General Store sells 20 pizzas per day, how many days would it take to break even?
    c. The manager of the store anticipates that once the local pizza delivery services start losing business, they will react by cutting prices. If after a month (30 days) the manager has to lower the price of a pizza to $7.95 to keep demand at 20 pizzas per day, as she expects, what will the new break-even point be, and how long will it take the store to break even?

19. Kim Davis has decided to purchase a cellular phone, but she is unsure about which rate plan to select. The "regular" plan charges a fixed fee of $55 per month for 1,000 minutes of airtime plus $0.33 per minute for any time over 1,000 minutes. The "executive" plan charges a fixed fee of $100 per month for 1,200 minutes of airtime plus $0.25 per minute over 1,200 minutes.
    a. If Kim expects to use the phone for 21 hours per month, which plan should she select?
    b. At what level of use would Kim be indifferent between the two plans?

20. Annie Russell, a student at Tech, plans to open a hot dog stand inside Tech's football stadium during home games. There are seven home games scheduled for the upcoming season. She must pay the Tech athletic department a vendor's fee of $3,000 for the season. Her stand and other equipment will cost her $4,500 for the season. She estimates that each hot dog she sells will cost her $0.35. She has talked to friends at other universities who sell hot dogs at games. Based on their information and the athletic department's forecast that each game will sell out, she anticipates that she will sell approximately 2,000 hot dogs during each game.
    a. What price should she charge for a hot dog in order to break even?
    b. What factors might occur during the season that would alter the volume sold and thus the break-even price Annie might charge?
    c. What price would you suggest that Annie charge for a hot dog to provide her with a reasonable profit while remaining competitive with other food vendors?

21. Hannah Byers and Kathleen Taylor are considering the possibility of teaching swimming to kids during the summer. A local swim club opens its pool at noon each day, so it is available to rent during the morning. The cost of renting the pool during the 10-week period for which Hannah and Kathleen would need it is $1,700. The pool would also charge Hannah and Kathleen an admission, towel service, and life guarding fee of $7 per pupil, and Hannah and Kathleen estimate

an additional $5 cost per student to hire several assistants. Hannah and Kathleen plan to charge $75 per student for the 10-week swimming class.

a. How many pupils do Hannah and Kathleen need to enroll in their class to break even?
b. If Hannah and Kathleen want to make a profit of $5,000 for the summer, how many pupils do they need to enroll?
c. Hannah and Kathleen estimate that they might not be able to enroll more than 60 pupils. If they enroll this many pupils, how much would they need to charge per pupil in order to realize their profit goal of $5,000?

22. The College of Business at Tech is planning to begin an online MBA program. The initial start-up cost for computing equipment, facilities, course development, and staff recruitment and development is $350,000. The college plans to charge tuition of $18,000 per student per year. However, the university administration will charge the college $12,000 per student for the first 100 students enrolled each year for administrative costs and its share of the tuition payments.
    a. How many students does the college need to enroll in the first year to break even?
    b. If the college can enroll 75 students the first year, how much profit will it make?
    c. The college believes it can increase tuition to $24,000, but doing so would reduce enrollment to 35. Should the college consider doing this?

23. The Star Youth Soccer Club helps to support its 20 boys' and girls' teams financially, primarily through the payment of coaches. The club puts on a tournament each fall to help pay its expenses. The cost of putting on the tournament is $8,000, mainly for development, printing, and mailing of the tournament brochures. The tournament entry fee is $400 per team. For every team that enters, it costs the club about $75 to pay referees for the three-game minimum each team is guaranteed. If the club needs to clear $60,000 from the tournament, how many teams should it invite?

24. A group of developers is opening a health club near a new housing development. The health club—which will have exercise and workout equipment, basketball courts, swimming pools, an indoor walking/running track, and tennis courts—is one of the amenities the developers are building to attract new homebuyers. However, they want the health club to at least break even the first year or two. The annual fixed cost for the building, equipment, utilities, staff, and so on is $875,000, and annual variable costs are $200 per member for things like water, towels, laundry, soap, shampoo, and other member services. The membership fee is $225 per month. How many members will the club need to break even? If the club doubles its break-even membership after a year, what will its profit be?

25. The Tech Student Government Association (SGA) has several campus projects it undertakes each year and its primary source of funding to support these projects is a T-shirt sale in the fall for what is known as the "orange effect" football game (with orange being one of Tech's colors). The club's publicized (media) objective is for everyone in the stadium to wear orange. The club's financial goal is to make a profit of $150,000, but in order to have a significant number of fans buy the shirts and wear them to the game, it doesn't want to price the T-shirts much more than $6. The stadium seats 62,000 fans, and the SGA would like to sell approximately 45,000 orange T-shirts to achieve the desired orange effect, which it's relatively confident it can do. It will cost $100,000 to purchase, silk-screen print, and ship this many T-shirts. The SGA sells the shirts through three sources: online, the two Tech bookstores, and a local independent bookstore. While the bookstores don't expect to share in the profits from the sale of the shirts, they do expect for their direct costs to be covered, including labor, space, and other costs. The two Tech bookstores charge the SGA $0.35 per shirt, and the local independent store charges $0.50 per shirt. The cost per sale online (including handling, packaging, and shipping) is $2.30 per shirt. The SGA estimates that it will sell 50% of the shirts at the two Tech bookstores, 35% at the local bookstore, and 15% online. If the SGA sells the T-shirts for $6 and if it sells all the shirts it orders, will it make enough profit to achieve its financial goal? If not, at what price would the SGA need to sell the T-shirts, or how many would the SGA have to sell to achieve its financial goal?

26. The owners of Backstreets Italian Restaurant are considering starting a delivery service for pizza and their other Italian dishes in the small college town where they are located. They can purchase a used delivery van and have it painted with their name and logo for $21,500. They can hire part-time drivers who will work in the evenings from 5 P.M. to 10 P.M. for $8 per hour. The drivers are mostly college students who study at the restaurant when they are not making deliveries. During the day, there are so few deliveries that the regular employees can handle them. The owners estimate that the van will last 5 years (365 days per year) before it has to be replaced and that each delivery will cost about $1.35 in gas and other maintenance costs (including tires, oil, scheduled service, etc.). They also estimate that on average each delivery order will cost $15 for direct labor and ingredients to prepare and package, and will generate $34 in revenue.
    a. How many delivery orders must Backstreets make each month in order for the service to break even?
    b. The owners believe that if they have approximately the break-even number of deliveries during the week, they will at least double that number on Fridays, Saturdays, and Sundays. If that's the case, how much profit will they make, at a minimum, from their delivery service each month (4 weeks per month)?

27. Kathleen Taylor is a high school student who has been investigating the possibility of mowing lawns for a summer job. She has a couple of friends she thinks she could hire on an hourly basis per job. The equipment, including two new lawnmowers and weed-eaters, would cost her $500, and she estimates her cost per lawn, based on the time required to pay her friends to mow an average residential lawn (and not including her own labor) and gas for driving to the jobs and mowing, would be about $14.
    a. If she charges customers $30 per lawn, how many lawns would she need to mow to break even?
    b. Kathleen has 8 weeks available to mow lawns before school starts again, and she estimates that she can get enough customers to mow at least three lawns per day, 6 days per week. How much money can she expect to make over the summer?
    c. Kathleen believes she can get more business if she lowers her price per lawn. If she lowers her price to $25 per lawn and increases her number of jobs to four per day (which is about all she can handle anyway), should she make this decision?

28. The Weemow Lawn Service mows its customers' lawns and provides lawn maintenance starting in the spring, through the summer, and into the early fall. During the winter, the service doesn't operate, and Weemow's owners, Jeff and Julie Weems, find part-time jobs. They are considering the possibility of doing snow removal during the winter. A snowblower and a shovel would cost them $700. Since Jeff would do all the work, with occasional help from Julie, their cost per job would be about $3.
    a. If they charge $35 to clear a normal-size home driveway, how many jobs would they need to break even?
    b. Based on past winters, Jeff and Julie believe they can expect about six major snowfalls in the winter and would be able to work all day for the two days immediately following the snows, when people want their driveways cleared. If they are able to do about 10 snow removal jobs per day (and they believe they will have that much demand because of their existing customer base), how much money can they expect to make?
    c. Another possibility for Weemow is to remove snow from business parking lots. Weemow would need a small tractor with a snow plow, which costs $1,800, and would have to hire someone on an hourly basis to help, which with gas would cost about $28 per job. Jeff and Julie estimate that they could do four of these large jobs per day. If they charged $150 per job, would this be a better alternative than clearing individuals' driveways?
    d. If Weemow wanted to do both (b) and (c), Jeff and Julie would need to hire one more person to do the driveways, while Jeff worked with the other person on the parking lots. This would add $15 in cost per driveway job. Should they do this?

29. In the example used to demonstrate model construction in this chapter (p. 4), a firm sells a product, x, for \$20 that costs \$5 to make, it has 100 pounds of steel to make the product, and it takes 4 pounds of steel to make each unit. The model that was constructed is

$$\begin{aligned} &\text{maximize } Z = 20x - 5x \\ &\text{subject to} \\ &\quad 4x = 100 \end{aligned}$$

Now suppose that there is a second product, y, that has a profit of \$10 and requires 2 pounds of steel to make, such that the model becomes

$$\begin{aligned} &\text{maximize } Z = 15x + 10y \\ &\text{subject to} \\ &\quad 4x + 2y = 100 \end{aligned}$$

Can you determine a solution to this new model that will achieve the objective? Explain your answer.

30. Consider a model in which two products, x and y, are produced. There are 100 pounds of material and 80 hours of labor available. It requires 2 pounds of material and 1 hour of labor to produce a unit of x, and 4 pounds of material and 5 hours of labor to produce a unit of y. The profit for x is \$30 per unit, and the profit for y is \$50 per unit. If we want to know how many units of x and y to produce to maximize profit, the model is

$$\begin{aligned} &\text{maximize } Z = 30x + 50y \\ &\text{subject to} \\ &\quad 2x + 4y = 100 \\ &\quad x + 5y = 80 \end{aligned}$$

Determine the solution to this problem and explain your answer.

31. Maria Eagle is a Native American artisan. She works part time making bowls and mugs by hand from special pottery clay and then sells her items to the Beaver Creek Pottery Company, a Native American crafts guild. She has 60 hours available each month to make bowls and mugs, and it takes her 12 hours to make a bowl and 15 hours to make a mug. She uses 9 pounds of special clay to make a bowl, and she needs 5 pounds to make a mug; Maria has 30 pounds of clay available each month. She makes a profit of \$300 for each bowl she delivers, and she makes \$250 for each mug. Determine all the possible combinations of bowls and mugs Maria can make each month, given her limited resources, and select the most profitable combination of bowls and mugs Maria should make each month. Develop an Excel spreadsheet and a graph to help solve this problem.

32. The Easy Drive Car Rental Agency needs 500 new cars in its Nashville operation and 300 new cars in Jacksonville, and it currently has 400 new cars in both Atlanta and Birmingham. It costs \$30 to move a car from Atlanta to Nashville, \$70 to move a car from Atlanta to Jacksonville, \$40 to move a car from Birmingham to Nashville, and \$60 to move a car from Birmingham to Jacksonville. The agency wants to determine how many cars should be transported from the agencies in Atlanta and Birmingham to the agencies in Nashville and Jacksonville in order to meet demand while minimizing the transport costs. Develop a mathematical model for this problem and use logic to determine a solution.

33. Ed Norris has developed a Web site for his used textbook business at State University. To sell advertising he needs to forecast the number of site visits he expects in the future. For the past 6 months he has had the following number of site visits:

Month	1	2	3	4	5	6
Site Visits	6,300	10,200	14,700	18,500	25,100	30,500

Determine a forecast for Ed to use for month 7 and explain the logic used to develop your forecast.

34. When Tracy McCoy wakes up on Saturday morning, she remembers that she promised the PTA she would make some cakes and/or homemade bread for its bake sale that afternoon. However, she does not have time to go to the store and get ingredients, and she has only a short time to bake things in her oven. Because cakes and breads require different baking temperatures, she cannot bake them simultaneously, and she has only 3 hours available to bake. A cake requires 3 cups of flour, and a loaf of bread requires 8 cups; Tracy has 20 cups of flour. A cake requires 45 minutes to bake, and a loaf of bread requires 30 minutes. The PTA will sell a cake for $10 and a loaf of bread for $6. Tracy wants to decide how many cakes and loaves of bread she should make. Identify all the possible solutions to this problem (i.e., combinations of cakes and loaves of bread Tracy has the time and flour to bake) and select the best one.

35. The local Food King grocery store has eight possible checkout stations with registers. On Saturday mornings customer traffic is relatively steady from 8 A.M. to noon. The store manager would like to determine how many checkout stations to staff during this time period. The manager knows from information provided by the store's national office that each minute past 3 minutes a customer must wait in line costs the store on average $50 in ill will and lost sales. Alternatively, each additional checkout station the store operates on Saturday morning costs the store $60 in salary and benefits. The following table shows the waiting time for the different staff levels.

Registers Staffed	1	2	3	4	5	6	7	8
Waiting Time (min.)	20.0	14.0	9.0	4.0	1.7	1.0	0.5	0.1

How many registers should the store staff and why?

36. A furniture manufacturer in Roanoke, Virginia, must deliver a tractor trailer load of furniture to a retail store in Washington, DC. There are a number of different routes the truck can take from Roanoke to DC, as shown in the following road network, with the distance for each segment shown in miles.

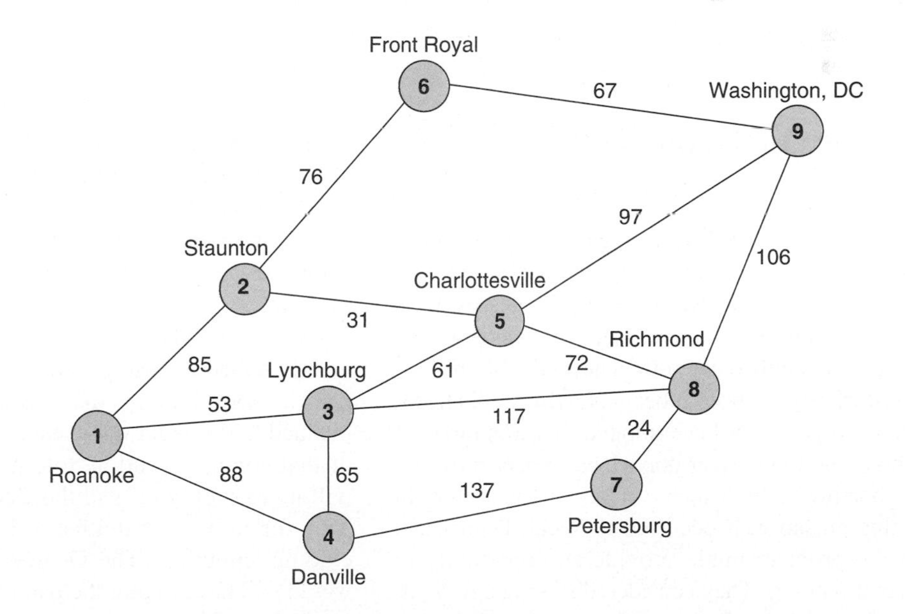

Determine the shortest route the truck can take from Roanoke to Washington, DC.

Case Problem

The Clean Clothes Corner Laundry

When Molly Lai purchased the Clean Clothes Corner Laundry, she thought that because it was in a good location near several high-income neighborhoods, she would automatically generate good business if she improved the laundry's physical appearance. Thus, she initially invested a lot of her cash reserves in remodeling the exterior and interior of the laundry. However, she just about broke even in the year following her acquisition of the laundry, which she didn't feel was a sufficient return, given how hard she had worked. Molly didn't realize that the dry-cleaning business is very competitive and that success is based more on price and quality service, including quickness of service, than on the laundry's appearance.

In order to improve her service, Molly is considering purchasing new dry-cleaning equipment, including a pressing machine that could substantially increase the speed at which she can dry-clean clothes and improve their appearance. The new machinery costs $16,200 installed and can clean 40 clothes items per hour (or 320 items per day). Molly estimates her variable costs to be $0.25 per item dry-cleaned, which will not change if she purchases the new equipment. Her current fixed costs are $1,700 per month. She charges customers $1.10 per clothing item.

A. What is Molly's current monthly volume?
B. If Molly purchases the new equipment, how many additional items will she have to dry-clean each month to break even?
C. Molly estimates that with the new equipment she can increase her volume to 4,300 items per month. What monthly profit would she realize with that level of business during the next 3 years? After 3 years?
D. Molly believes that if she doesn't buy the new equipment but lowers her price to $0.99 per item, she will increase her business volume. If she lowers her price, what will her new break-even volume be? If her price reduction results in a monthly volume of 3,800 items, what will her monthly profit be?
E. Molly estimates that if she purchases the new equipment and lowers her price to $0.99 per item, her volume will increase to about 4,700 units per month. Based on the local market, that is the largest volume she can realistically expect. What should Molly do?

Case Problem

The Ocobee River Rafting Company

Vicki Smith, Penny Miller, and Darryl Davis are students at State University. In the summer they often go rafting with other students down the Ocobee River in the nearby Blue Ridge Mountain foothills. The river has a number of minor rapids but is not generally dangerous. The students' rafts basically consist of large rubber tubes, sometimes joined together with ski rope. They have noticed that a number of students who come to the river don't have rubber rafts and often ask to borrow theirs, which can be very annoying. In discussing this nuisance, it occurred to Vicki, Penny, and Darryl that the problem might provide an opportunity to make some extra money. They considered starting a new enterprise, the Ocobee River Rafting Company, to sell rubber rafts at the river. They determined that their initial investment would be about $3,000 to rent a small parcel of land next to the river on which to make and sell the rafts; to purchase a tent to operate out of; and to buy some small equipment such as air pumps and a rope cutter. They estimated that the labor and material cost per raft will be about $12, including the purchase and shipping costs for the rubber tubes and rope. They plan to sell the rafts for $20 apiece, which they think is about the maximum price students will pay for a preassembled raft.

Soon after they determined these cost estimates, the newly formed company learned about another rafting company in North Carolina that was doing essentially what they planned to do. Vicki got in touch with one of the operators of that company, and he told her the company would be willing to supply rafts to the Ocobee River Rafting Company for an initial fixed fee of $9,000 plus $8 per raft, including shipping. (The Ocobee River Rafting Company would still have to rent the parcel of riverside land and tent for $1,000.) The rafts would already be inflated and assembled. This alternative appealed to Vicki, Penny, and Darryl because it would reduce the amount of time they would have to work pumping up the tubes and putting the rafts together, and it would increase time for their schoolwork.

Although the students prefer the alternative of purchasing the rafts from the North Carolina company, they are concerned about the large initial cost and worried about whether they will lose money. Of course, Vicki, Penny, and Darryl realize that their profit, if any, will be determined by how many rafts they sell. As such, they believe that they first need to determine how many rafts they must sell with each alternative in order to make a profit and which alternative would be best given different levels of demand. Furthermore, Penny has conducted a brief sample survey of people at the river and estimates that demand for rafts for the summer will be around 1,000 rafts.

Perform an analysis for the Ocobee River Rafting Company to determine which alternative would be best for different levels of demand. Indicate which alternative should be selected if demand is approximately 1,000 rafts and how much profit the company would make.

Case Problem

Constructing a Downtown Parking Lot in Draper

The town of Draper, with a population of 20,000, sits adjacent to State University, which has an enrollment of 27,000 students. Downtown Draper merchants have long complained about the lack of parking available to their customers. This is one primary reason for the steady migration of downtown businesses to a mall several miles outside town. The local chamber of commerce has finally convinced the town council to consider the construction of a new multilevel indoor parking facility downtown. Kelly Mattingly, the town's public works director, has developed plans for a facility that would cost $4.5 million to construct. To pay for the project, the town would sell municipal bonds with a duration of 30 years at 8% interest. Kelly also estimates that five employees would be required to operate the lot on a daily basis, at a total annual cost of $140,000. It is estimated that each car that enters the lot would park for an average of 2.5 hours and pay an average fee of $3.20. Further, it is estimated that each car that parks in the lot would (on average) cost the town $0.60 in annual maintenance for cleaning and repairs to the facility. Most of the downtown businesses (which include a number of restaurants) are open 7 days per week.

A. Using break-even analysis, determine the number of cars that would have to park in the lot on an annual basis to pay off the project in the 30-year time frame.
B. From the results in (A), determine the approximate number of cars that would have to park in the lot on a daily basis. Does this seem to be a reasonable number to achieve, given the size of the town and college population?

CHAPTER

2

Linear Programming: Model Formulation and Graphical Solution

Objectives of a business frequently are to maximize profit or minimize cost.

Many major decisions faced by a manager of a business focus on the best way to achieve the objectives of the firm, subject to the restrictions placed on the manager by the operating environment. These restrictions can take the form of limited resources, such as time, labor, energy, material, or money; or they can be in the form of restrictive guidelines, such as a recipe for making cereal or engineering specifications. One of the most frequent objectives of business firms is to gain the most profit possible or, in other words, to *maximize* profit. The objective of individual organizational units within a firm (such as a production or packaging department) is often to *minimize* cost. When a manager attempts to solve a general type of problem by seeking an objective that is subject to restrictions, the management science technique called **linear programming** is frequently used.

***Linear programming** is a model that consists of linear relationships representing a firm's decision(s), given an objective and resource constraints.*

There are three steps in applying the linear programming technique. First, the problem must be identified as being solvable by linear programming. Second, the unstructured problem must be formulated as a mathematical model. Third, the model must be solved by using established mathematical techniques. The linear programming technique derives its name from the fact that the functional relationships in the mathematical model are *linear*, and the solution technique consists of predetermined mathematical steps—that is, a *program*. In this chapter we will concern ourselves with the formulation of the mathematical model that represents the problem and then with solving this model by using a graph.

Model Formulation

A linear programming model consists of certain common components and characteristics. The model components include decision variables, an objective function, and model constraints, which consist of decision variables and parameters. **Decision variables** are mathematical symbols that represent levels of activity by the firm. For example, an electrical manufacturing firm desires to produce x_1 radios, x_2 toasters, and x_3 clocks, where x_1, x_2, and x_3 are symbols representing unknown variable quantities of each item. The final values of x_1, x_2, and x_3, as determined by the firm, constitute a *decision* (e.g., the equation $x_1 = 100$ radios is a decision by the firm to produce 100 radios).

***Decision variables** are mathematical symbols that represent levels of activity.*

The **objective function** is a linear mathematical relationship that describes the objective of the firm in terms of the decision variables. The objective function always consists of either *maximizing* or *minimizing* some value (e.g., maximize the profit or minimize the cost of producing radios).

*The **objective function** is a linear relationship that reflects the objective of an operation.*

The **model constraints** are also linear relationships of the decision variables; they represent the restrictions placed on the firm by the operating environment. The restrictions can be in the form of limited resources or restrictive guidelines. For example, only 40 hours of labor may be available to produce radios during production. The actual numeric values in the objective function and the constraints, such as the 40 hours of available labor, are **parameters**.

*A **model constraint** is a linear relationship that represents a restriction on decision making.*

__Parameters__ are numerical values that are included in the objective functions and constraints.

The next section presents an example of how a linear programming model is formulated. Although this example is simplified, it is realistic and represents the type of problem to which linear programming can be applied. In the example, the model components are distinctly identified and described. By carefully studying this example, you can become familiar with the process of formulating linear programming models.

A Maximization Model Example

Beaver Creek Pottery Company is a small crafts operation run by a Native American tribal council. The company employs skilled artisans to produce clay bowls and mugs with authentic Native American designs and colors. The two primary resources used by the company are special pottery clay and skilled labor. Given these limited resources, the company desires to know how many bowls and mugs to produce each day in order to maximize profit. This is generally referred to as a *product mix* problem type. This scenario is illustrated in Figure 2.1.

Time Out for George B. Dantzig

Linear programming, as it is known today, was conceived in 1947 by George B. Dantzig while he was the head of the Air Force Statistical Control's Combat Analysis Branch at the Pentagon. The military referred to its plans for training, supplying, and deploying combat units as "programs." When Dantzig analyzed Air Force planning problems, he realized that they could be formulated as a system of linear inequalities—hence his original name for the technique, "programming in a linear structure," which was later shortened to "linear programming."

FIGURE 2.1
Beaver Creek Pottery Company

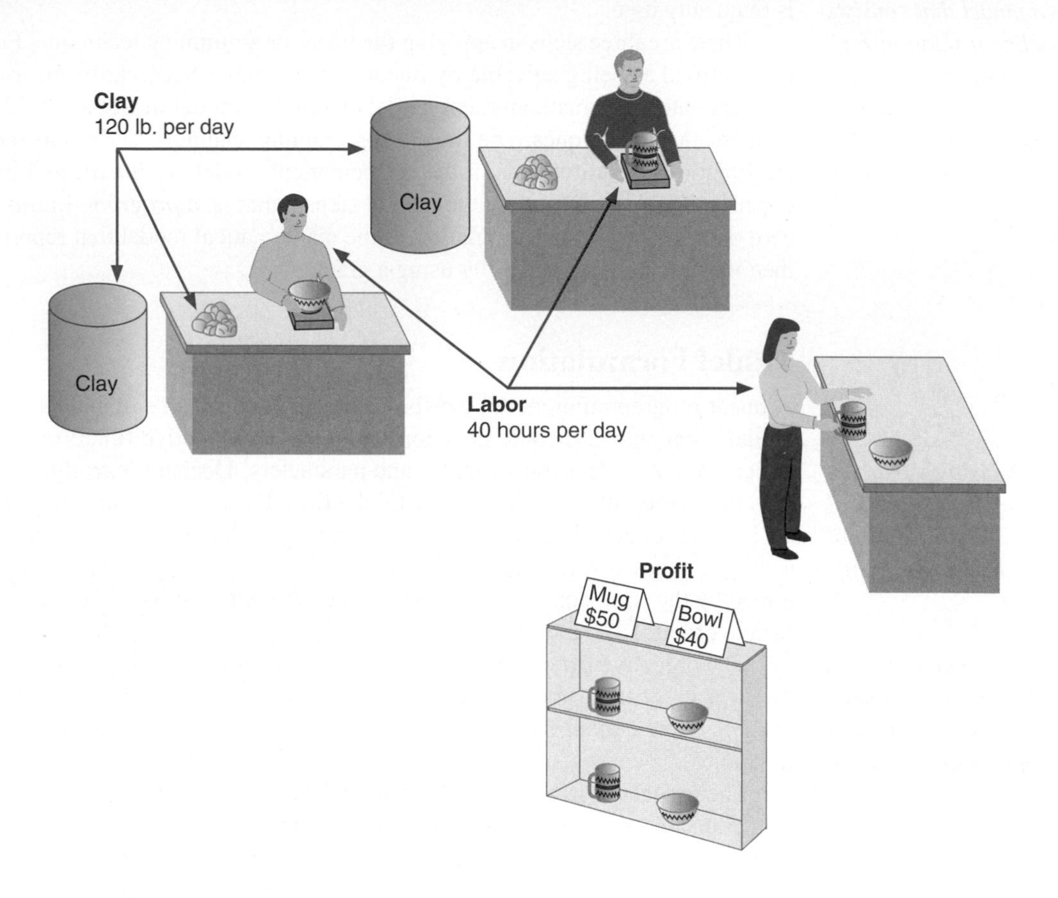

The two products have the following resource requirements for production and profit per item produced (i.e., the model parameters):

	Resource Requirements		
Product	Labor (hr./unit)	Clay (lb./unit)	Profit ($/unit)
Bowl	1	4	40
Mug	2	3	50

There are 40 hours of labor and 120 pounds of clay available each day for production. We will formulate this problem as a linear programming model by defining each component of the

A linear programming model consists of decision variables, an objective function, and constraints.

model separately and then combining the components into a single model. The steps in this formulation process are summarized as follows:

Summary of LP Model Formulation Steps

Step 1: Define the decision variables
How many bowls and mugs to produce

Step 2: Define the objective function
Maximize profit

Step 3: Define the constraints
The resources (clay and labor) available

Decision Variables

The decision confronting management in this problem is how many bowls and mugs to produce. The two decision variables represent the number of bowls and mugs to be produced on a daily basis. The quantities to be produced can be represented symbolically as

$$x_1 = \text{number of bowls to produce}$$
$$x_2 = \text{number of mugs to produce}$$

The Objective Function

The objective of the company is to maximize total profit. The company's profit is the sum of the individual profits gained from each bowl and mug. Profit derived from bowls is determined by multiplying the unit profit of each bowl, \$40, by the number of bowls produced, x_1. Likewise, profit derived from mugs is derived from the unit profit of a mug, \$50, multiplied by the number of mugs produced, x_2. Thus, total profit, which we will define symbolically as Z, can be expressed mathematically as $\$40x_1 + \$50x_2$. By placing the term *maximize* in front of the profit function, we express the objective of the firm—to maximize total profit:

$$\text{maximize } Z = \$40x_1 + 50x_2$$

where

$$Z = \text{total profit per day}$$
$$\$40x_1 = \text{profit from bowls}$$
$$\$50x_2 = \text{profit from mugs}$$

Model Constraints

In this problem, two resources are used for production—labor and clay—both of which are limited. Production of bowls and mugs requires both labor and clay. For each bowl produced, 1 hour of labor is required. Therefore, the labor used for the production of bowls is $1x_1$ hours. Similarly, each mug requires 2 hours of labor; thus, the labor used to produce mugs every day is $2x_2$ hours. The total labor used by the company is the sum of the individual amounts of labor used for each product:

$$1x_1 + 2x_2$$

However, the amount of labor represented by $1x_1 + 2x_2$ is limited to 40 hours per day; thus, the complete labor constraint is

$$1x_1 + 2x_2 \leq 40 \text{ hr.}$$

The "less than or equal to" ($\leq$) inequality is employed instead of an equality ($=$) because the 40 hours of labor is a maximum limitation that *can be used*, not an amount that *must be used*. This constraint allows the company some flexibility; the company is not restricted to using exactly 40 hours but can use whatever amount is necessary to maximize profit, up to and including 40 hours. This means that it is possible to have idle, or excess, capacity (i.e., some of the 40 hours may not be used).

The constraint for clay is formulated in the same way as the labor constraint. Because each bowl requires 4 pounds of clay, the amount of clay used daily for the production of bowls is $4x_1$ pounds; and because each mug requires 3 pounds of clay, the amount of clay used daily for mugs is $3x_2$. Given that the amount of clay available for production each day is 120 pounds, the material constraint can be formulated as

$$4x_1 + 3x_2 \leq 120 \text{ lb.}$$

A final restriction is that the number of bowls and mugs produced must be either zero or a positive value because it is impossible to produce negative items. These restrictions are referred to as **nonnegativity constraints** and are expressed mathematically as

__Nonnegativity constraints__ restrict the decision variables to zero or positive values.

$$x_1 \geq 0, x_2 \geq 0$$

The complete linear programming model for this problem can now be summarized as follows:

$$\begin{aligned} \text{maximize } Z &= \$40x_1 + 50x_2 \\ \text{subject to} & \\ 1x_1 + 2x_2 &\leq 40 \\ 4x_1 + 3x_2 &\leq 120 \\ x_1, x_2 &\geq 0 \end{aligned}$$

The solution of this model will result in numeric values for x_1 and x_2 that will maximize total profit, Z. As *one possible* solution, consider $x_1 = 5$ bowls and $x_2 = 10$ mugs. First, we will substitute this hypothetical solution into each of the constraints in order to make sure that the solution does not require more resources than the constraints show are available:

$$\begin{aligned} 1(5) + 2(10) &\leq 40 \\ 25 &\leq 40 \end{aligned}$$

and

$$\begin{aligned} 4(5) + 3(10) &\leq 120 \\ 50 &\leq 120 \end{aligned}$$

Because neither of the constraints is violated by this hypothetical solution, we say the solution is **feasible** (i.e., possible). Substituting these solution values in the objective function gives $Z = 40(5) + 50(10) = \$700$. However, for the time being, we do not have any way of knowing whether \$700 is the *maximum* profit.

A __feasible solution__ does not violate any of the constraints.

Now consider a solution of $x_1 = 10$ bowls and $x_2 = 20$ mugs. This solution results in a profit of

$$\begin{aligned} Z &= \$40(10) + 50(20) \\ &= 400 + 1{,}000 \\ &= \$1{,}400 \end{aligned}$$

An __infeasible problem__ violates at least one of the constraints.

Although this is certainly a better solution in terms of profit, it is **infeasible** (i.e., not possible) because it violates the resource constraint for labor:

$$\begin{aligned} 1(10) + 2(20) &\leq 40 \\ 50 &\not\leq 40 \end{aligned}$$

Management Science Application

Allocating Seat Capacity on Indian Railways Using Linear Programming

Indian Railways, with more than 1,600 trains, serves more than 7 million passengers each day. Its reservation system books passengers in three coach classes: reserved air-conditioned, reserved non-air conditioned, and unreserved non-air-conditioned. A train can make multiple stops from its origin to its destination, and thus passengers can book many combinations of tickets from a train's origin to its destination or to and/or from intermediate stations. Because passengers can depart the train or board en route, multiple passengers can occupy a single seat during a train's journey from origin to destination. This also means that a seat might be vacant for some segments of the trip and thus will not earn any revenue. If there are an abnormally high number of reservations for intermediate trips, then there might be a high number of partially vacant seats, which might deny subsequent passengers the ability to book complete origin-to-destination trips. This results in suboptimal utilization of a train's capacity. However, in many cases, passenger demand is not high at the train's origin station, and the highest passenger demand occurs en route, at intermediate stations. As a result, the railway has traditionally allocated various seat quotas to intermediate stations and limited the seats allocated for end-to-end trips, in order to maximize capacity utilization and the number of confirmed seat reservations and reduce the number of passengers wait-listed at intermediate stations.

© Neil McAllister/Alamy

In this application, a linear programming model was formulated, with an objective of minimizing the total seats required to fill all possible seat demand (in a specific coach class) between any two major stations, subject to constraints for station-to-station quotas based on historical seat demand. In a test case of 17 trains in the Western Railway zone (based in Mumbai) of Indian Railways, revenue was increased between 2.6% and 29.3%, and the number of passengers carried increased from 8.4% to 29%.

Source: Based on R. Gopalakrishnan and N. Rangaraj, "Capacity Management on Long-Distance Passenger Trains of Indian Railways," *Interfaces* 40, no. 4 (July–August 2010): 291–302.

The solution to this problem must maximize profit without violating the constraints. The solution that achieves this objective is $x_1 = 24$ bowls and $x_2 = 8$ mugs, with a corresponding profit of \$1,360. The determination of this solution is shown using the graphical solution approach in the following section.

Graphical Solutions of Linear Programming Models

Graphical solutions are limited to linear programming problems with only two decision variables.

Following the formulation of a mathematical model, the next stage in the application of linear programming to a decision-making problem is to find the solution of the model. A common solution approach is to solve algebraically the set of mathematical relationships that form the model either manually or using a computer program, thus determining the values for the decision variables. However, because the relationships are *linear*, some models and solutions can be illustrated *graphically*.

The graphical method is realistically limited to models with only two decision variables, which can be represented on a graph of two dimensions. Models with three decision variables can be graphed in three dimensions, but the process is quite cumbersome, and models of four or more decision variables cannot be graphed at all.

The graphical method provides a picture of how a solution is obtained for a linear programming problem.

Although the graphical method is limited as a solution approach, it is very useful at this point in our presentation of linear programming in that it gives a picture of how a solution is derived. Graphs can provide a clearer understanding of how the computer and mathematical solution approaches presented in subsequent chapters work and, thus, a better understanding of the solutions.

Graphical Solution of a Maximization Model

The product mix model will be used to demonstrate the graphical interpretation of a linear programming problem. Recall that the problem describes Beaver Creek Pottery Company's attempt to decide how many bowls and mugs to produce daily, given limited amounts of labor and clay. The complete linear programming model was formulated as

$$\text{maximize } Z = \$40x_1 + 50x_2$$

subject to

$$x_1 + 2x_2 \leq 40 \text{ hr. of labor}$$
$$4x_1 + 3x_2 \leq 120 \text{ lb. of clay}$$
$$x_1, x_2 \geq 0$$

where

$x_1 =$ number of bowls produced
$x_2 =$ number of mugs produced

Figure 2.2 is a set of coordinates for the decision variables x_1 and x_2, on which the graph of our model will be drawn. Note that only the positive quadrant is drawn (i.e., the quadrant where x_1 and x_2 will always be positive) because of the nonnegativity constraints, $x_1 \geq 0$ and $x_2 \geq 0$.

FIGURE 2.2
Coordinates for graphical analysis

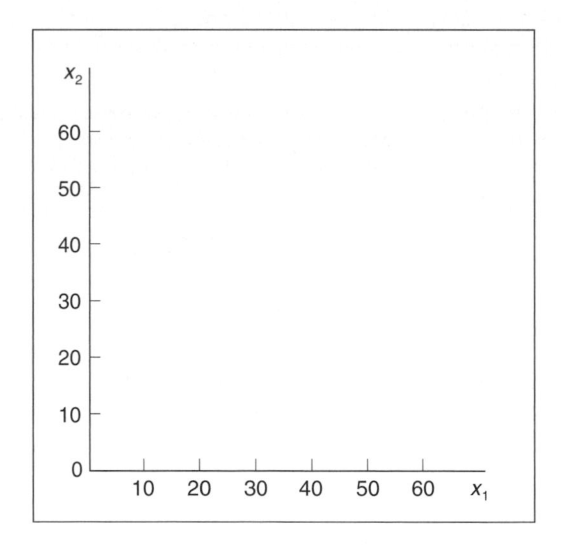

Constraint lines are plotted as equations.

The first step in drawing the graph of the model is to plot the constraints on the graph. This is done by treating both constraints as equations (or straight lines) and plotting each line on the graph. Let's consider the labor constraint line first:

$$x_1 + 2x_2 = 40$$

A simple procedure for plotting this line is to determine two points that are on the line and then draw a straight line through the points. One point can be found by letting $x_1 = 0$ and solving for x_2:

$$(0) + 2x_2 = 40$$
$$x_2 = 20$$

Thus, one point is at the coordinates $x_1 = 0$ and $x_2 = 20$. A second point can be found by letting $x_2 = 0$ and solving for x_1:

$$x_1 + 2(0) = 40$$
$$x_1 = 40$$

Now we have a second point, $x_1 = 40$, $x_2 = 0$. The line on the graph representing this equation is drawn by connecting these two points, as shown in Figure 2.3. However, this is only the graph of the constraint *line* and does not reflect the entire constraint, which also includes the values that are less than or equal to (≤) this line. The *area* representing the entire constraint is shown in Figure 2.4.

FIGURE 2.3
Graph of the labor constraint line

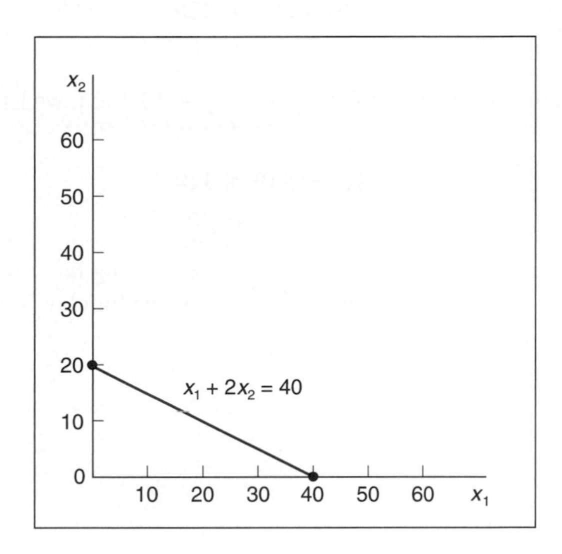

FIGURE 2.4
The labor constraint area

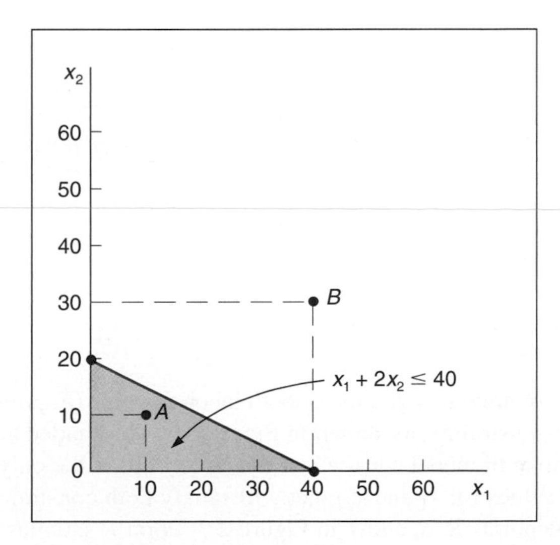

To test the correctness of the constraint area, we check any two points—one inside the constraint area and one outside. For example, check point *A* in Figure 2.4, which is at the intersection of $x_1 = 10$ and $x_2 = 10$. Substituting these values into the following labor constraint,

$$10 + 2(10) \leq 40$$
$$30 \leq 40 \text{ hr.}$$

shows that point A is indeed within the constraint area, as these values for x_1 and x_2 yield a quantity that does not exceed the limit of 40 hours. Next, we check point B at $x_1 = 40$ and $x_2 = 30$:

$$40 + 2(30) \leq 40$$
$$100 \nleq 40 \text{ hr.}$$

Point B is obviously outside the constraint area because the values for x_1 and x_2 yield a quantity (100) that exceeds the limit of 40 hours.

We draw the line for the clay constraint the same way as the one for the labor constraint— by finding two points on the constraint line and connecting them with a straight line. First, let $x_1 = 0$ and solve for x_2:

$$4(0) + 3x_2 = 120$$
$$x_2 = 40$$

Performing this operation results in a point, $x_1 = 0$, $x_2 = 40$. Next, we let $x_2 = 0$ and then solve for x_1:

$$4x_1 + 3(0) = 120$$
$$x_1 = 30$$

This operation yields a second point, $x_1 = 30$, $x_2 = 0$. Plotting these points on the graph and connecting them with a line gives the constraint line and area for clay, as shown in Figure 2.5.

FIGURE 2.5
The constraint area for clay

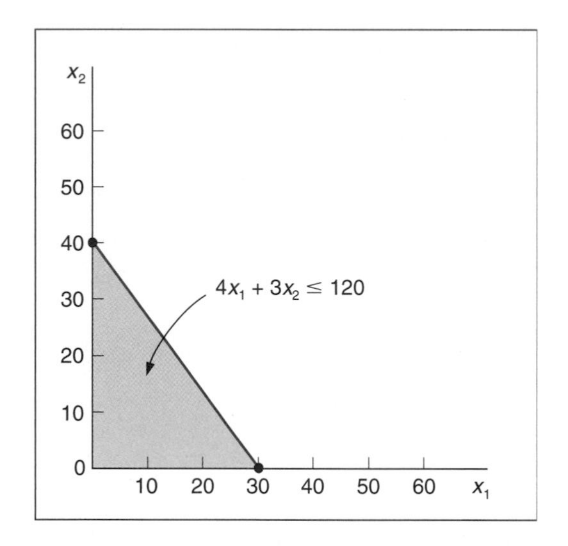

Combining the two individual graphs for both labor and clay (Figures 2.4 and 2.5) produces a graph of the model constraints, as shown in Figure 2.6. The shaded area in Figure 2.6 is the area that is common to both model constraints. Therefore, this is the only area on the graph that contains points (i.e., values for x_1 and x_2) that will satisfy both constraints simultaneously. For example, consider the points R, S, and T in Figure 2.7. Point R satisfies both constraints; thus, we say it is a *feasible* solution point. Point S satisfies the clay constraint ($4x_1 + 3x_2 \leq 120$) but exceeds the labor constraint; thus, it is infeasible. Point T satisfies neither constraint; thus, it is also infeasible.

The feasible solution area is an area on the graph that is bounded by the constraint equations.

The shaded area in Figure 2.7 is referred to as the *feasible solution area* because all the points in this area satisfy both constraints. Some point within this feasible solution area will result in *maximum profit* for Beaver Creek Pottery Company. The next step in the graphical solution approach is to locate this point.

FIGURE 2.6
Graph of both model constraints

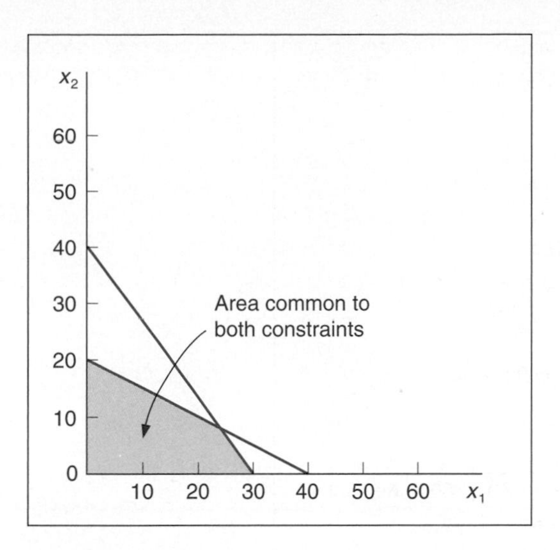

FIGURE 2.7
The feasible solution area constraints

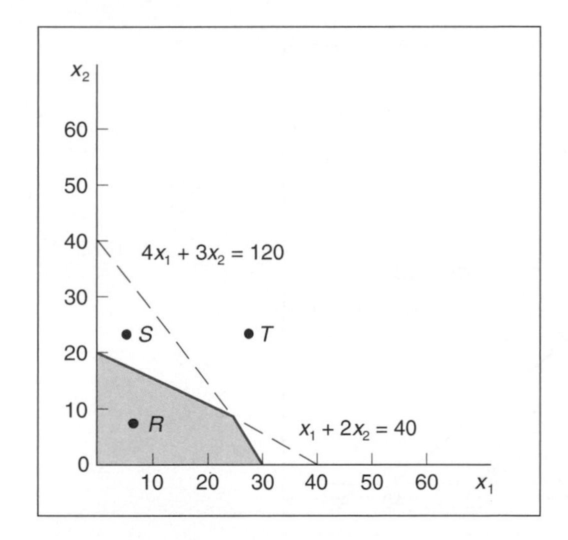

The Optimal Solution Point

The second step in the graphical solution method is to locate the point in the feasible solution area that will result in the greatest total profit. To begin the solution analysis, we first plot the objective function line for an *arbitrarily* selected level of profit. For example, if we say profit, Z, is \$800, the objective function is

$$\$800 = 40x_1 + 50x_2$$

Plotting this line just as we plotted the constraint lines results in the graph shown in Figure 2.8. Every point on this line is in the feasible solution area and will result in a profit of \$800 (i.e., every combination of x_1 and x_2 on this line will give a Z value of \$800). However, let us see whether an even greater profit will still provide a feasible solution. For example, consider profits of \$1,200 and \$1,600, as shown in Figure 2.9.

A portion of the objective function line for a profit of \$1,200 is outside the feasible solution area, but part of the line remains within the feasible area. Therefore, this profit line indicates that there are feasible solution points that give a profit greater than \$800. Now let us increase profit again, to \$1,600. This profit line, also shown in Figure 2.9, is completely outside the feasible solution area. The fact that no points on this line are feasible indicates that a profit of \$1,600 is not possible.

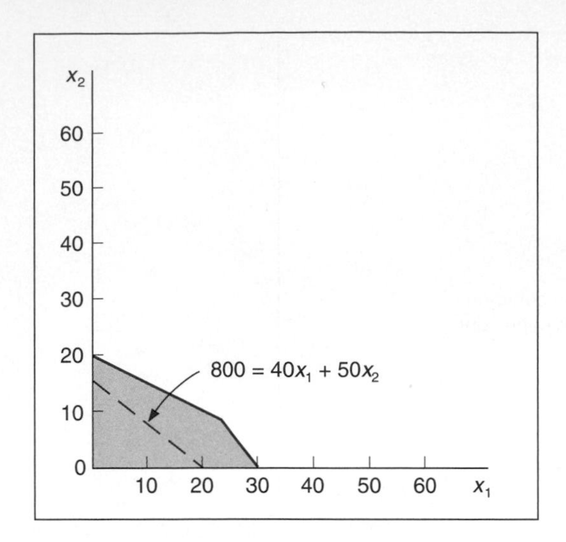

FIGURE 2.8
Objective function line for Z = \$800

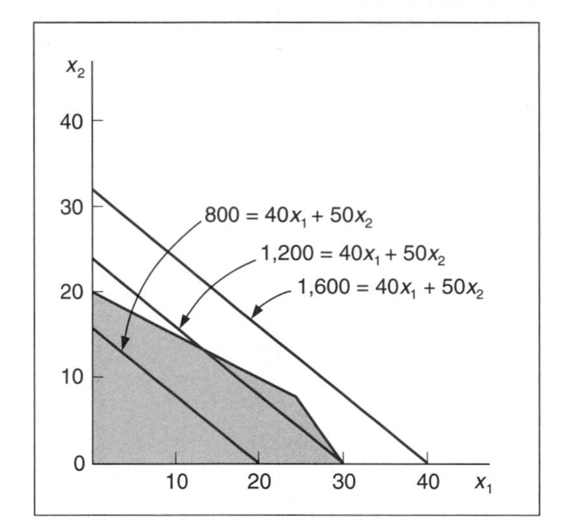

FIGURE 2.9
Alternative objective function lines for profits, Z, of \$800, \$1,200, and \$1,600

Because a profit of \$1,600 is too great for the constraint limitations, as shown in Figure 2.9, the question of the maximum profit value remains. We can see from Figure 2.9 that profit increases as the objective function line moves away from the origin (i.e., the point $x_1 = 0, \ x_2 = 0$). Given this characteristic, the maximum profit will be attained at the point where the objective function line is farthest from the origin *and* is still touching a point in the feasible solution area. This point is shown as point B in Figure 2.10.

The optimal solution is the best feasible solution.

To find point B, we place a straightedge parallel to the objective function line $\$800 = 40x_1 + 50x_2$ in Figure 2.10 and move it outward from the origin as far as we can without losing contact with the feasible solution area. Point B is referred to as the **optimal** (i.e., best) **solution**.

The Solution Values

The third step in the graphical solution approach is to solve for the values of x_1 and x_2 once the optimal solution point has been found. It is possible to determine the x_1 and x_2 coordinates of point B in Figure 2.10 directly from the graph, as shown in Figure 2.11. The graphical coordinates corresponding to point B in Figure 2.11 are $x_1 = 24$ and $x_2 = 8$. This is the optimal solution for the decision

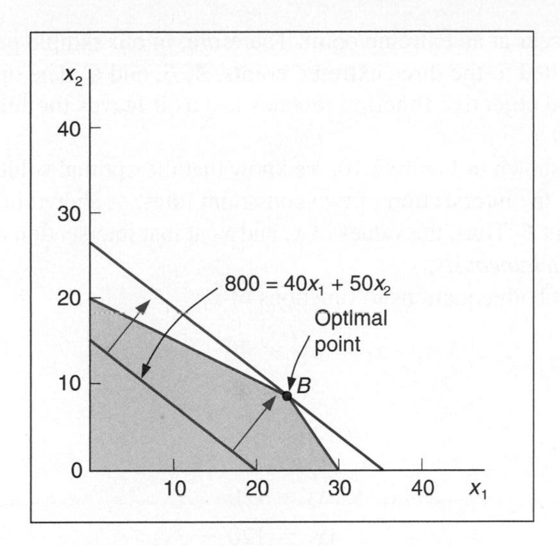

FIGURE 2.10
Identification of optimal solution point

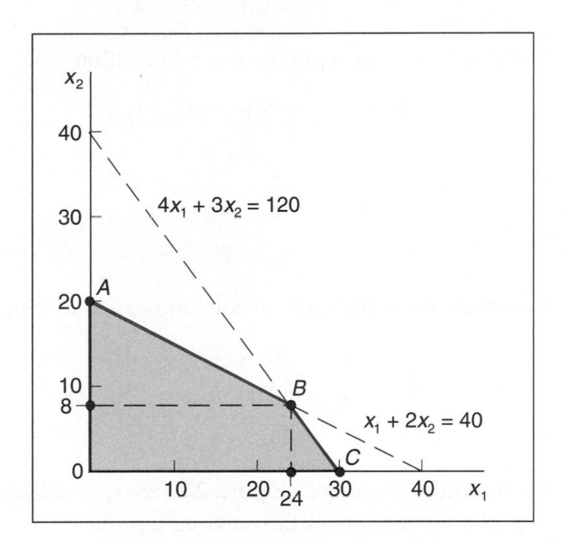

FIGURE 2.11
Optimal solution coordinates

variables in the problem. However, unless an absolutely accurate graph is drawn, it is frequently difficult to determine the correct solution directly from the graph. A more exact approach is to determine the solution values mathematically once the optimal point on the graph has been determined. The mathematical approach for determining the solution is described in the following pages. First, however, we will consider a few characteristics of the solution.

The optimal solution point is the last point the objective function touches as it leaves the feasible solution area.

In Figure 2.10, as the objective function was increased, the last point it touched in the feasible solution area was on the boundary of the feasible solution area. The solution point is always on this boundary because the boundary contains the points farthest from the origin (i.e., the points corresponding to the greatest profit). This characteristic of linear programming problems reduces the number of possible solution points considerably, from all points in the solution area to just those points on the boundary. However, the number of possible solution points is reduced even more by another characteristic of linear programming problems.

***Extreme points** are corner points on the boundary of the feasible solution area.*

The solution point will be on the boundary of the feasible solution area and at one of the *corners* of the boundary where two constraint lines intersect. (The graphical axes, you will recall, are also constraints because $x_1 \geq 0$ and $x_2 \geq 0$.) These corners (points *A*, *B*, and *C* in Figure 2.11) are protrusions, or *extremes*, in the feasible solution area; they are called **extreme points**. It has been proven mathematically that the optimal solution in a linear programming model will always occur at an extreme point. Therefore, in our sample problem the possible solution points are limited to the three extreme points, *A*, *B*, and *C*. The optimal extreme point is the extreme point the objective function touches last as it leaves the feasible solution area, as shown in Figure 2.10.

Constraint equations are solved simultaneously at the optimal extreme point to determine the variable solution values.

From the graph shown in Figure 2.10, we know that the optimal solution point is *B*. Because point *B* is formed by the intersection of two constraint lines, as shown in Figure 2.11, these two lines are *equal* at point *B*. Thus, the values of x_1 and x_2 at that intersection can be found by solving the two equations *simultaneously*.

First, we convert both equations to functions of x_1:

$$x_1 + 2x_2 = 40$$
$$x_1 = 40 - 2x_2$$

and

$$4x_1 + 3x_2 = 120$$
$$4x_1 = 120 - 3x_2$$
$$x_1 = 30 - (3x_2/4)$$

Now, we let x_1 in the first equation equal x_1 in the second equation,

$$40 - 2x_2 = 30 - (3x_2/4)$$

and solve for x_2:

$$5x_2/4 = 10$$
$$x_2 = 8$$

Substituting $x_2 = 8$ into either one of the original equations gives a value for x_1:

$$x_1 = 40 - 2x_2$$
$$x_1 = 40 - 2(8)$$
$$= 24$$

Thus, the optimal solution at point *B* in Figure 2.11 is $x_1 = 24$ and $x_2 = 8$. Substituting these values into the objective function gives the maximum profit,

$$Z = \$40x_1 + 50x_2$$
$$Z = \$40(24) + 50(8)$$
$$= \$1{,}360$$

In terms of the original problem, the solution indicates that if the pottery company produces 24 bowls and 8 mugs, it will receive $1,360, the maximum daily profit possible (given the resource constraints).

Given that the optimal solution will be at one of the extreme corner points, *A*, *B*, or *C*, we can also find the solution by testing each of the three points to see which results in the greatest profit, rather than by graphing the objective function and seeing which point it last touches as it moves out of the feasible solution area. Figure 2.12 shows the solution values for all three points, *A*, *B*, and *C*, and the amount of profit, *Z*, at each point.

As indicated in the discussion of Figure 2.10, point *B* is the optimal solution point because it is the last point the objective function touches before it leaves the solution area. In other words, the objective function determines which extreme point is optimal. This is because the objective function designates the profit that will accrue from each combination of x_1 and x_2 values at the extreme points. If the objective function had had different coefficients (i.e., different x_1 and x_2 profit values), one of the extreme points other than *B* might have been optimal.

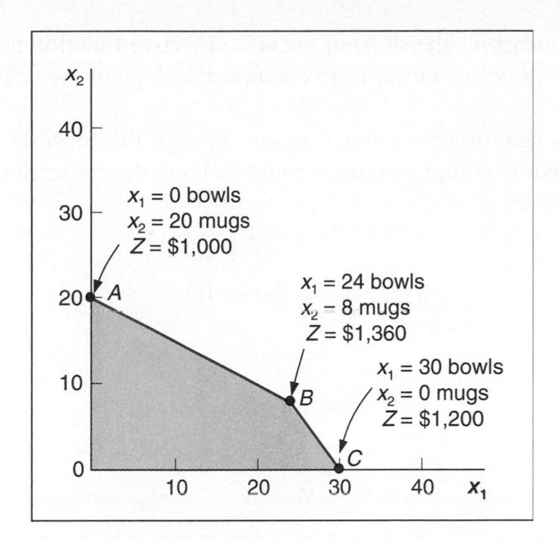

FIGURE 2.12
Solutions at all corner points

Let's assume for a moment that the profit for a bowl is \$70 instead of \$40, and the profit for a mug is \$20 instead of \$50. These values result in a new objective function, $Z = \$70x_1 + 20x_2$. If the model constraints for labor or clay are not changed, the feasible solution area remains the same, as shown in Figure 2.13. However, the location of the objective function in Figure 2.13 is different from that of the original objective function in Figure 2.10. The reason for this change is that the new profit coefficients give the linear objective function a new *slope*.

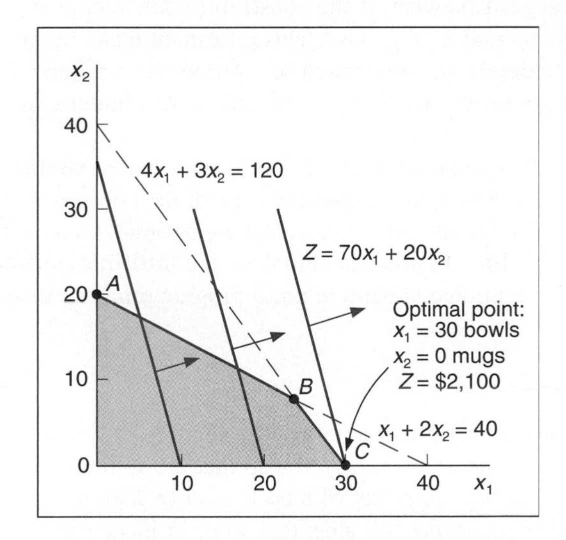

FIGURE 2.13
The optimal solution with $Z = 70x_1 + 20x_2$

The slope is computed as the "rise" over the "run."

The **slope** can be determined by transforming the objective function into the general equation for a straight line, $y = a + bx$, where y is the dependent variable, a is the y intercept, b is the slope, and x is the independent variable. For our sample objective function, x_2 is the dependent variable corresponding to y (i.e., it is on the vertical axis), and x_1 is the independent variable. Thus, the objective function can be transformed into the general equation of a line as follows:

$$Z = 70x_1 + 20x_2$$

$$20x_2 = Z - 70x_1$$

$$\underset{\substack{\uparrow \\ y}}{x_2} = \underset{\substack{\uparrow \\ a}}{\frac{Z}{20}} - \underset{\substack{\uparrow \\ b}}{\frac{7}{2}}x_1$$

This transformation identifies the slope of the new objective function as $-7/2$ (the minus sign indicates that the line slopes downward). In contrast, the slope of the original objective function was $-4/5$.

If we move this new objective function out through the feasible solution area, the last extreme point it touches is point C. Simultaneously solving the constraint lines at point C results in the following solution:

$$x_1 = 30$$
$$4x_1 + 3x_2 = 120$$

and

$$x_2 = 40 - (4x_1/3)$$
$$x_2 = 40 - 4(30)/3$$
$$x_2 = 0$$

Thus, the optimal solution at point C in Figure 2.13 is $x_1 = 30$ bowls, $x_2 = 0$ mugs, and $Z = \$2,100$ profit. Altering the objective function coefficients results in a new solution.

This brief example of the effects of altering the objective function highlights two useful points. First, the optimal extreme point is determined by the objective function, and an extreme point on one axis of the graph is as likely to be the optimal solution as is an extreme point on a different axis. Second, the solution is sensitive to the values of the coefficients in the objective function. If the objective function coefficients are changed, as in our example, the solution may change. Likewise, if the constraint coefficients are changed, the solution space and solution points may change also. This information can be of consequence to the decision maker trying to determine how much of a product to produce. **Sensitivity analysis**—the use of linear programming to evaluate the effects of changes in model parameters—is discussed in Chapter 3.

Sensitivity analysis is used to analyze changes in model parameters.

It should be noted that some problems do not have a single extreme point solution. For example, when the objective function line parallels one of the constraint lines, an entire line segment is bounded by two adjacent corner points that are optimal; there is no single extreme point on the objective function line. In this situation there are **multiple optimal solutions**. This and other irregular types of solution outcomes in linear programming are discussed at the end of this chapter.

***Multiple optimal solutions** can occur when the objective function is parallel to a constraint line.*

Slack Variables

Once the optimal solution was found at point B in Figure 2.12, simultaneous equations were solved to determine the values of x_1 and x_2. Recall that the solution occurs at an extreme point where constraint equation lines intersect with each other or with the axis. Thus, the model constraints are considered as *equations* (=) rather than $\leq$ or $\geq$ inequalities.

A slack variable is added to a $\leq$ constraint to convert it to an equation (=).

A slack variable represents unused resources.

There is a standard procedure for transforming $\leq$ inequality constraints into equations. This transformation is achieved by adding a new variable, called a **slack variable**, to each constraint. For the pottery company example, the model constraints are

$$x_1 + 2x_2 \leq 40 \text{ hr. of labor}$$
$$4x_1 + 3x_2 \leq 120 \text{ lb. of clay}$$

The addition of a unique slack variable, s_1, to the labor constraint and s_2 to the constraint for clay results in the following equations:

$$x_1 + 2x_2 + s_1 = 40 \text{ hr. of labor}$$
$$4x_1 + 3x_2 + s_2 = 120 \text{ lb. of clay}$$

The slack variables in these equations, s_1 and s_2, will take on any value necessary to make the left-hand side of the equation equal to the right-hand side. For example, consider a hypothetical solution of $x_1 = 5$ and $x_2 = 10$. Substituting these values into the foregoing equations yields

$$\begin{aligned} x_1 + 2x_2 + s_1 &= 40 \text{ hr. of labor} \\ 5 + 2(10) + s_1 &= 40 \text{ hr. of labor} \\ s_1 &= 15 \text{ hr. of labor} \end{aligned}$$

and

$$\begin{aligned} 4x_1 + 3x_2 + s_2 &= 120 \text{ lb. of clay} \\ 4(5) + 3(10) + s_2 &= 120 \text{ lb. of clay} \\ s_2 &= 70 \text{ lb. of clay} \end{aligned}$$

In this example, $x_1 = 5$ bowls and $x_2 = 10$ mugs represent a solution that does not make use of the total available amount of labor and clay. In the labor constraint, 5 bowls and 10 mugs require only 25 hours of labor. This leaves 15 hours that are not used. Thus, s_1 represents the amount of *unused* labor, or slack.

In the clay constraint, 5 bowls and 10 mugs require only 50 pounds of clay. This leaves 70 pounds of clay unused. Thus, s_2 represents the amount of *unused* clay. In general, slack variables represent the amount of *unused resources.*

The ultimate instance of unused resources occurs at the origin, where $x_1 = 0$ and $x_2 = 0$. Substituting these values into the equations yields

$$\begin{aligned} x_1 + 2x_2 + s_1 &= 40 \\ 0 + 2(0) + s_1 &= 40 \\ s_1 &= 40 \text{ hr. of labor} \end{aligned}$$

and

$$\begin{aligned} 4x_1 + 3x_2 + s_2 &= 120 \\ 4(0) + 3(0) + s_2 &= 120 \\ s_2 &= 120 \text{ lb. of clay} \end{aligned}$$

Because no production takes place at the origin, all the resources are unused; thus, the slack variables equal the total available amounts of each resource: $s_1 = 40$ hours of labor and $s_2 = 120$ pounds of clay.

What is the effect of these new slack variables on the objective function? The objective function for our example represents the profit gained from the production of bowls and mugs,

$$Z = \$40x_1 + \$50x_2$$

A slack variable contributes nothing to the objective function value.

The coefficient \$40 is the contribution to profit of each bowl; \$50 is the contribution to profit of each mug. What, then, do the slack variables s_1 and s_2 contribute? They contribute *nothing* to profit because they represent unused resources. Profit is made only after the resources are put to use in making bowls and mugs. Using slack variables, we can write the objective function as

$$\text{maximize } Z = \$40x_1 + \$50x_2 + 0s_1 + 0s_2$$

As in the case of decision variables (x_1 and x_2), slack variables can have only nonnegative values because negative resources are not possible. Therefore, for this model formulation, $x_1, x_2, s_1, \text{ and } s_2 \geq 0$.

The complete linear programming model can be written in what is referred to as *standard form* with slack variables as follows:

$$\text{maximize } Z = \$40x_1 + \$50x_2 + 0s_1 + 0s_2$$
subject to
$$x_1 + 2x_2 + s_1 = 40$$
$$4x_1 + 3x_2 + s_2 = 120$$
$$x_1, x_2, s_1, s_2 \geq 0$$

The solution values, including the slack at each solution point, are summarized as follows:

Solution Summary with Slack

Point	Solution Values	Z	Slack
A	x_1 = 0 bowls, x_2 = 20 mugs	\$1,000	s_1 = 0 hr.; s_2 = 60 lb.
B	x_1 = 24 bowls, x_2 = 8 mugs	\$1,360	s_1 = 0 hr.; s_2 = 0 lb.
C	x_1 = 30 bowls, x_2 = 0 mugs	\$1,200	s_1 = 10 hr.; s_2 = 0 lb.

Figure 2.14 shows the graphical solution of this example, with slack variables included at each solution point.

FIGURE 2.14
Solutions at points *A*, *B*, and *C* with slack

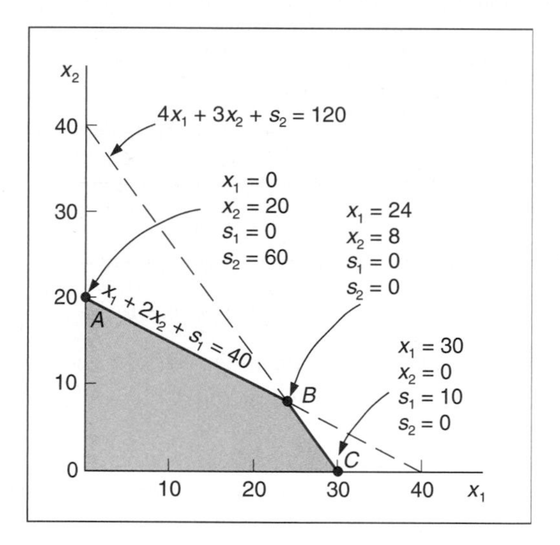

Summary of the Graphical Solution Steps

The steps for solving a graphical linear programming model are summarized here:

1. Plot the model constraints as equations on the graph; then, considering the inequalities of the constraints, indicate the feasible solution area.
2. Plot the objective function; then, move this line out from the origin to locate the optimal solution point.
3. Solve simultaneous equations at the solution point to find the optimal solution values.

Or

2. Solve simultaneous equations at each corner point to find the solution values at each point.
3. Substitute these values into the objective function to find the set of values that results in the maximum *Z* value.

Management Science Application

Improving Customer Service at Amazon.com

Amazon.com started as an Internet book retailer in 1995 and 10 years later was a Fortune 500 company with annual sales of more than $7 billion. Amazon.com is now an Internet retailer that offers new and used products in a variety of categories including music and video, consumer electronics, food, clothing, furniture, and appliances. From 2001 through 2003 Amazon.com achieved the highest score ever recorded in any service industry by the American Customer Satisfaction Index. The company's success is partly a result of its strong customer service operations. Customer service is provided through features on the Web site or via telephone or e-mail. Customer service representatives are available at internally or externally managed contact centers 24 hours a day. Because of its sales growth and the seasonality of its sales (which generally decline in the summer), Amazon must make accurate decisions about the capacity of its contact centers, including the number of customer service representatives to hire and train at its internally managed centers, and the volume of voice calls and e-mails to allocate to external service providers (referred to as ***cosourcers***). The service times for calls and e-mails are categorized depending on the product type, the customer type, and the purchase type. Amazon's objective is to process calls and e-mails at target service levels for different categories; for voice calls the target is that a specific percentage of callers wait no more than a specific time, and for e-mails the target is that a percentage of them receive a response within some time period. Amazon uses a linear programming model to optimize its capacity decisions at its customer service contact centers. The approach provides a minimum-cost capacity plan that provides the number of representatives to hire and train at Amazon's own centers and the volume of contacts to allocate to each cosourcer each week for a given planning horizon. Objective function costs include the number of normal and overtime hours, the number of new representatives hired and transferred, and the cost of contracting with cosourcers. There are several categories of constraints related to the number of voice calls and e-mails and for normal and overtime hours at both the internally managed contact centers and at cosourcers. For a 1-year (52-week) planning horizon the linear programming model consists of approximately 134,000 constraints and almost 16,000 variables.

DYLAN MARTINEZ/Reuters/Landov

Source: Based on M. Keblis and M. Chen, "Improving Customer Service Operations at Amazon.com," *Interfaces* 36, no. 5 (September–October 2006): 433–45.

A Minimization Model Example

As mentioned at the beginning of this chapter, there are two types of linear programming problems: maximization problems (like the Beaver Creek Pottery Company example) and minimization problems. A minimization problem is formulated the same basic way as a maximization problem, except for a few minor differences. The following sample problem will demonstrate the formulation of a minimization model.

A farmer is preparing to plant a crop in the spring and needs to fertilize a field. There are two brands of fertilizer to choose from, Super-gro and Crop-quick. Each brand yields a specific amount of nitrogen and phosphate per bag, as follows:

	Chemical Contribution	
Brand	Nitrogen (lb./bag)	Phosphate (lb./bag)
Super-gro	2	4
Crop-quick	4	3

The farmer's field requires at least 16 pounds of nitrogen and at least 24 pounds of phosphate. Super-gro costs $6 per bag, and Crop-quick costs $3. The farmer wants to know how many bags of each brand to purchase in order to minimize the total cost of fertilizing. This scenario is illustrated in Figure 2.15.

The steps in the linear programming model formulation process are summarized as follows:

> ***Summary of LP Model Formulation Steps***
>
> ***Step 1: Define the decision variables***
>
> How many bags of Super-gro and Crop-quick to buy
>
> ***Step 2: Define the objective function***
>
> Minimize cost
>
> ***Step 3: Define the constraints***
>
> The field requirements for nitrogen and phosphate

FIGURE 2.15
Fertilizing farmer's field

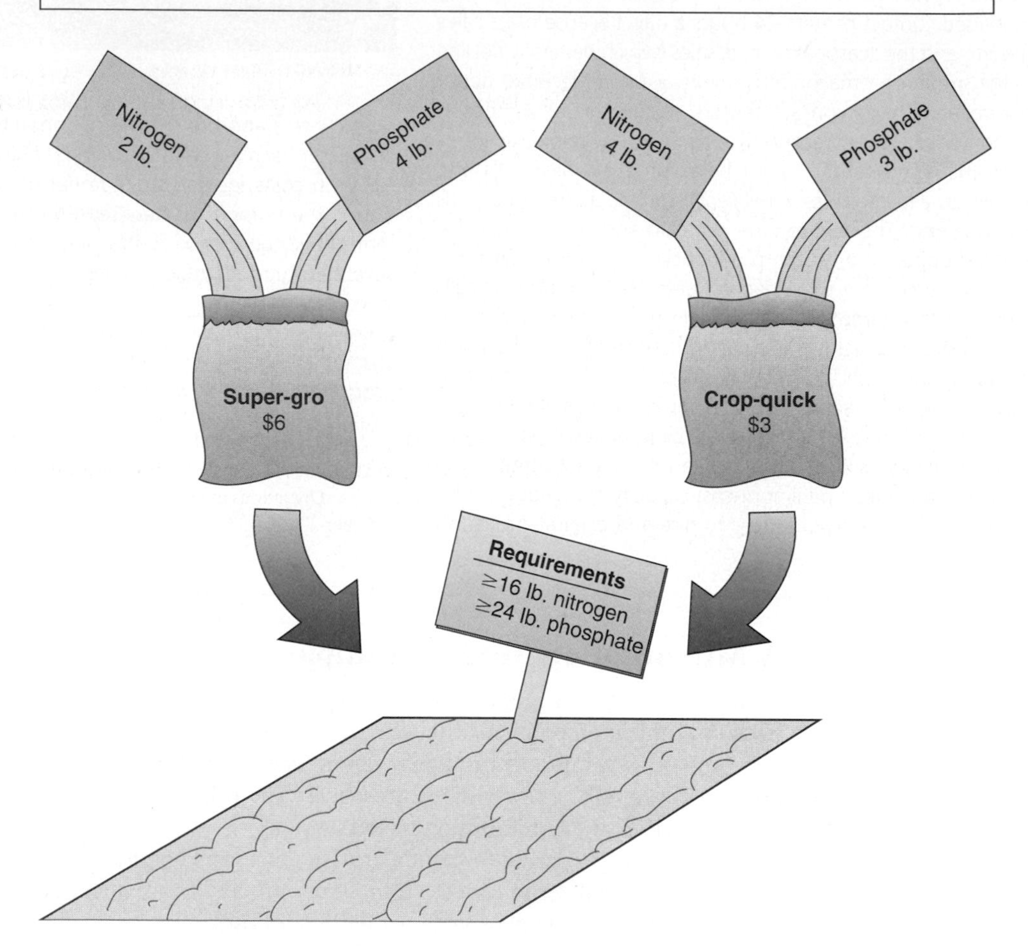

Decision Variables

This problem contains two decision variables, representing the number of bags of each brand of fertilizer to purchase:

$$x_1 = \text{bags of Super-gro}$$
$$x_2 = \text{bags of Crop-quick}$$

The Objective Function

The farmer's objective is to minimize the total cost of fertilizing. The total cost is the sum of the individual costs of each type of fertilizer purchased. The objective function that represents total cost is expressed as

$$\text{minimize } Z = \$6x_1 + 3x_2$$

where

$\$6x_1 =$ cost of bags of Super-gro

$\$3x_2 =$ cost of bags of Crop-quick

Model Constraints

The requirements for nitrogen and phosphate represent the constraints of the model. Each bag of fertilizer contributes a number of pounds of nitrogen and phosphate to the field. The constraint for nitrogen is

$$2x_1 + 4x_2 \geq 16 \text{ lb.}$$

where

$2x_1 =$ the nitrogen contribution (lb.) per bag of Super-gro

$4x_2 =$ the nitrogen contribution (lb.) per bag of Crop-quick

Rather than a $\leq$ (less than or equal to) inequality, as used in the Beaver Creek Pottery Company model, this constraint requires a $\geq$ (greater than or equal to) inequality. This is because the nitrogen content for the field is a minimum requirement specifying that at least 16 pounds of nitrogen be deposited on the farmer's field. If a minimum cost solution results in more than 16 pounds of nitrogen on the field, that is acceptable; however, the amount cannot be less than 16 pounds.

The constraint for phosphate is constructed like the constraint for nitrogen:

$$4x_1 + 3x_2 \geq 24 \text{ lb.}$$

The three types of linear programming constraints are $\leq$, $=$, and $\geq$.

With this example, we have shown two of the three types of linear programming model constraints, $\leq$ and $\geq$. The third type is an exact equality, $=$. This type specifies that a constraint requirement must be exact. For example, if the farmer had said that the phosphate requirement for the field was exactly 24 pounds, the constraint would have been

$$4x_1 + 3x_2 = 24 \text{ lb.}$$

As in our maximization model, there are also nonnegativity constraints in this problem to indicate that negative bags of fertilizer cannot be purchased:

$$x_1, x_2 \geq 0$$

The complete model formulation for this minimization problem is

$$\begin{aligned} &\text{minimize } Z = \$6x_1 + 3x_2 \\ &\text{subject to} \\ &2x_1 + 4x_2 \geq 16 \text{ lb. of nitrogen} \\ &4x_1 + 3x_2 \geq 24 \text{ lb. of phosphate} \\ &x_1, x_2 \geq 0 \end{aligned}$$

Graphical Solution of a Minimization Model

We follow the same basic steps in the graphical solution of a minimization model as in a maximization model. The fertilizer example will be used to demonstrate the graphical solution of a minimization model.

The first step is to graph the equations of the two model constraints, as shown in Figure 2.16. Next, the feasible solution area is chosen, to reflect the ≥ inequalities in the constraints, as shown in Figure 2.17.

FIGURE 2.16
Constraint lines for fertilizer model

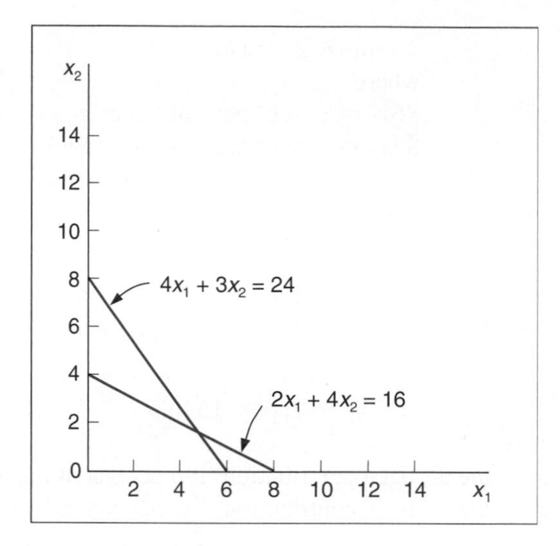

FIGURE 2.17
Feasible solution area

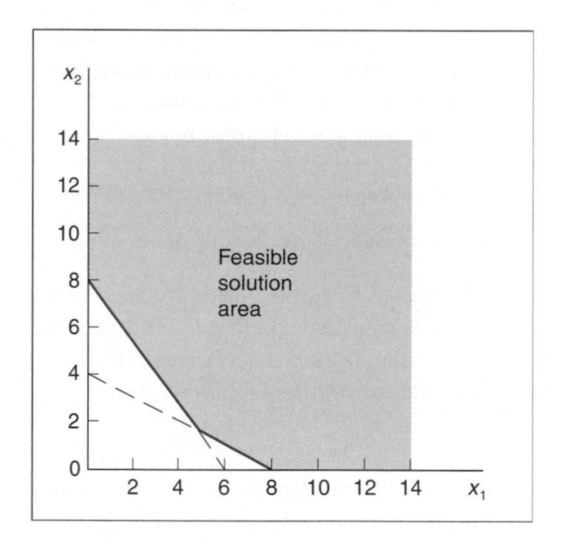

The optimal solution of a minimization problem is at the extreme point closest to the origin.

After the feasible solution area has been determined, the second step in the graphical solution approach is to locate the optimal point. Recall that in a maximization problem, the optimal solution is on the boundary of the feasible solution area that contains the point(s) farthest from the origin. The optimal solution point in a minimization problem is also on the boundary of the feasible solution area; however, the boundary contains the point(s) *closest* to the origin (zero being the lowest cost possible).

As in a maximization problem, the optimal solution is located at one of the extreme points of the boundary. In this case, the corner points represent extremities in the boundary of the feasible solution area that are *closest* to the origin. Figure 2.18 shows the three corner points—*A, B,* and *C*—and the objective function line.

As the objective function edges *toward* the origin, the last point it touches in the feasible solution area is *A*. In other words, point *A* is the closest the objective function can get to the origin without encompassing infeasible points. Thus, it corresponds to the lowest cost that can be attained.

FIGURE 2.18
The optimal solution point

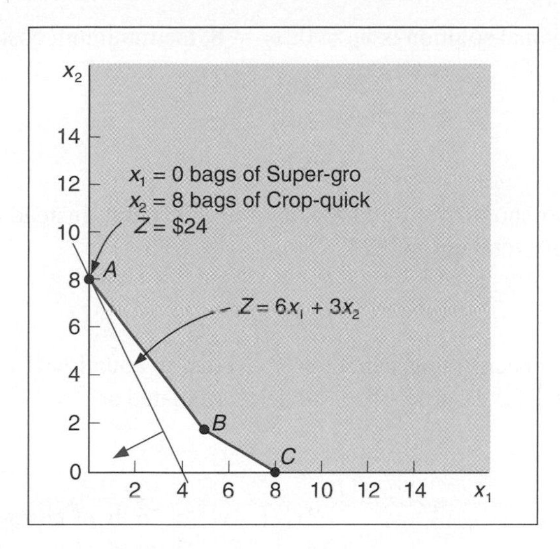

The final step in the graphical solution approach is to solve for the values of x_1 and x_2 at point A. Because point A is on the x_2 axis, $x_1 = 0$; thus,

$$4(0) + 3x_2 = 24$$
$$x_2 = 8$$

Management Science Application

Determining Optimal Fertilizer Mixes at Soquimich (South America)

Soquimich, a Chilean fertilizer manufacturer, is the leading producer and distributor of specialty fertilizers in the world, with revenues of almost US$0.5 billion in more than 80 countries. Soquimich produces four main specialty fertilizers and more than 200 fertilizer blends, depending on the needs of its customers. Farmers want the company to quickly recommend optimal fertilizer blends that will provide the appropriate quantity of ingredients for their particular crop at the lowest possible cost. A farmer will provide a company sales representative with information about previous crop yields and his or her target yields and then company representatives will visit the farm to obtain soil samples, which are analyzed in the company labs. A report is generated, which indicates the soil requirements for nutrients, including nitrogen, phosphorus, potassium, boron, magnesium, sulfur, and zinc. Given these soil requirements, company experts determine an optimal fertilizer blend, using a linear programming model that includes constraints for the nutrient quantities required by the soil (for a particular crop) and an objective function that minimizes production costs. Previously the company determined fertilizer blend recommendations by using a time-consuming manual procedure conducted by experts. The linear programming model enables the company to provide accurate, quick, low-cost (discounted) estimates to its customers, which has helped the company gain new customers and increase its market share.

Alex Havret/Dorling Kindersley

Source: Based on A. M. Angel, L. A. Taladriz, and R. Weber, "Soquimich Uses a System Based on Mixed-Integer Linear Programming and Expert Systems to Improve Customer Service," *Interfaces* 33, no. 4 (July–August 2003): 41–52.

Given that the optimal solution is $x_1 = 0$, $x_2 = 8$, the minimum cost, Z, is

$$\begin{aligned} Z &= \$6x_1 + \$3x_2 \\ Z &= 6(0) + 3(8) \\ &= \$24 \end{aligned}$$

This means the farmer should not purchase any Super-gro but, instead, should purchase eight bags of Crop-quick, at a total cost of $24.

Surplus Variables

Greater-than or equal-to constraints cannot be converted to equations by adding slack variables, as with $\leq$ constraints. Recall our fertilizer model, formulated as

$$\begin{aligned} &\text{minimize } Z = \$6x_1 + \$3x_2 \\ &\text{subject to} \\ &\quad 2x_1 + 4x_2 \geq 16 \text{ lb. of nitrogen} \\ &\quad 4x_1 + 3x_2 \geq 24 \text{ lb. of phosphate} \\ &\quad x_1, x_2 \geq 0 \end{aligned}$$

where

x_1 = bags of Super-gro fertilizer
x_2 = bags of Crop-quick fertilizer
Z = farmer's total cost ($) of purchasing fertilizer

A surplus variable is subtracted from a $\geq$ constraint to convert it to an equation (=).

*A **surplus variable** represents an excess above a constraint requirement level.*

Because this problem has $\geq$ constraints as opposed to the $\leq$ constraints of the Beaver Creek Pottery Company maximization example, the constraints are converted to equations a little differently.

Instead of adding a slack variable as we did with a $\geq$ constraint, we subtract a **surplus variable**. Whereas a slack variable is added and reflects unused resources, a surplus variable is subtracted and reflects the excess above a minimum resource requirement level. Like a slack variable, a surplus variable is represented symbolically by s_1 and must be nonnegative.

For the nitrogen constraint, the subtraction of a surplus variable gives

$$2x_1 + 4x_2 - s_1 = 16$$

The surplus variable s_1 transforms the nitrogen constraint into an equation.

As an example, consider the hypothetical solution

$$\begin{aligned} x_1 &= 0 \\ x_2 &= 10 \end{aligned}$$

Substituting these values into the previous equation yields

$$\begin{aligned} 2(0) + 4(10) - s_1 &= 16 \\ -s_1 &= 16 - 40 \\ s_1 &= 24 \text{ lb. of nitrogen} \end{aligned}$$

In this equation, s_1 can be interpreted as the *extra* amount of nitrogen above the minimum requirement of 16 pounds that would be obtained by purchasing 10 bags of Crop-quick fertilizer.

In a similar manner, the constraint for phosphate is converted to an equation by subtracting a surplus variable, s_2:

$$4x_1 + 3x_2 - s_2 = 24$$

As is the case with slack variables, surplus variables contribute nothing to the overall cost of a model. For example, putting additional nitrogen or phosphate on the field will not affect the farmer's cost; the only thing affecting cost is the number of bags of fertilizer purchased. As such, the standard form of this linear programming model is summarized as

$$\begin{aligned} &\text{minimize } Z = \$6x_1 + 3x_2 + 0s_1 + 0s_2 \\ &\text{subject to} \\ &\quad 2x_1 + 4x_2 - s_1 = 16 \\ &\quad 4x_1 + 3x_2 - s_2 = 24 \\ &\quad x_1, x_2, s_1, s_2 \geq 0 \end{aligned}$$

Figure 2.19 shows the graphical solutions for our example, with surplus variables included at each solution point.

FIGURE 2.19
Graph of the fertilizer example

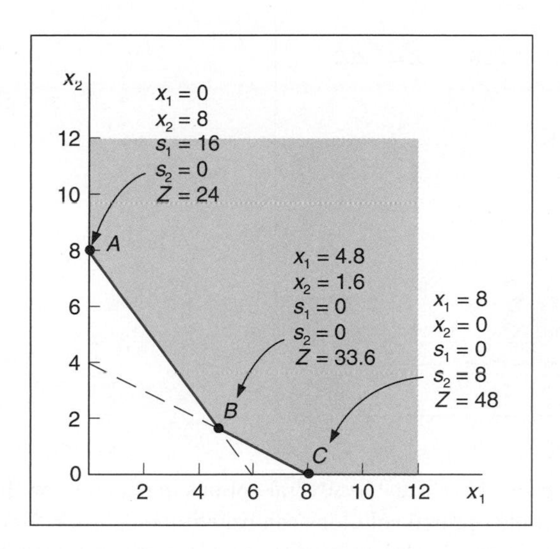

Irregular Types of Linear Programming Problems

For some linear programming models, the general rules do not always apply.

The basic forms of typical maximization and minimization problems have been shown in this chapter. However, there are several special types of atypical linear programming problems. Although these special cases do not occur frequently, they will be described so that you can recognize them when they arise. These special types include problems with more than one optimal solution, infeasible problems, and problems with unbounded solutions.

Multiple Optimal Solutions

Consider the Beaver Creek Pottery Company example, with the objective function changed from $Z = 40x_1 + 50x_2$ to $Z = 40x_1 + 30x_2$:

$$\begin{aligned} &\text{maximize } Z = 40x_1 + 30x_2 \\ &\text{subject to} \\ &\quad x_1 + 2x_2 \leq 40 \text{ hr. of labor} \\ &\quad 4x_1 + 3x_2 \leq 120 \text{ lb. of clay} \\ &\quad x_1, x_2 \geq 0 \end{aligned}$$

where

x_1 = bowls produced
x_2 = mugs produced

Alternate optimal solutions are at the endpoints of the constraint line segment that the objective function parallels.

The graph of this model is shown in Figure 2.20. The slight change in the objective function makes it now *parallel* to the constraint line, $4x_1 + 3x_2 = 120$. Both lines now have the same slope of $-4/3$. Therefore, as the objective function edge moves outward from the origin, it touches the whole line segment BC rather than a single extreme corner point before it leaves the feasible solution area. This means that every point along this line segment is optimal (i.e., each point results in the same profit of $Z = \$1{,}200$). The endpoints of this line segment, B and C, are typically referred to as the **alternate optimal solutions**. It is understood that these points represent the endpoints of a range of optimal solutions.

FIGURE 2.20
Graph of the Beaver Creek Pottery Company example with multiple optimal solutions

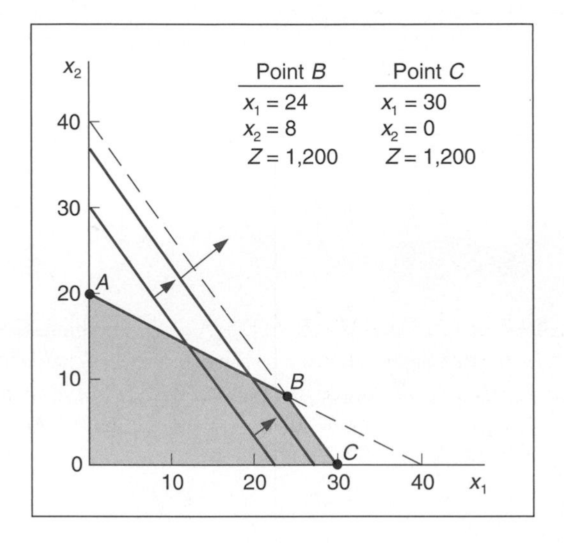

Multiple optimal solutions provide greater flexibility to the decision maker.

The pottery company, therefore, has several options in deciding on the number of bowls and mugs to produce. Multiple optimal solutions can benefit the decision maker because the number of decision options is enlarged. The multiple optimal solutions (along the line segment BC in Figure 2.20) allow the decision maker greater flexibility. For example, in the case of Beaver Creek Pottery Company, it may be easier to sell bowls than mugs; thus, the solution at point C, where only bowls are produced, would be more desirable than the solution at point B, where a mix of bowls and mugs is produced.

*An **infeasible problem** has no feasible solution area; every possible solution point violates one or more constraints.*

An Infeasible Problem

In some cases, a linear programming problem has no feasible solution area; thus, there is no solution to the problem. An example of an infeasible problem is formulated next and depicted graphically in Figure 2.21:

$$\begin{aligned} &\text{maximize } Z = 5x_1 + 3x_2 \\ &\text{subject to} \\ &\quad 4x_1 + 2x_2 \le 8 \\ &\quad x_1 \ge 4 \\ &\quad x_2 \ge 6 \\ &\quad x_1, x_2 \ge 0 \end{aligned}$$

Point A in Figure 2.21 satisfies only the constraint $4x_1 + 2x_2 \le 8$, whereas point C satisfies only the constraints $x_1 \ge 4$ and $x_2 \ge 6$. Point B satisfies none of the constraints. The three constraints do not overlap to form a feasible solution area. Because no point satisfies all three constraints simultaneously, there is no solution to the problem. Infeasible problems do not typically occur, but when they do, they are usually a result of errors in defining the problem or in formulating the linear programming model.

FIGURE 2.21
Graph of an infeasible problem

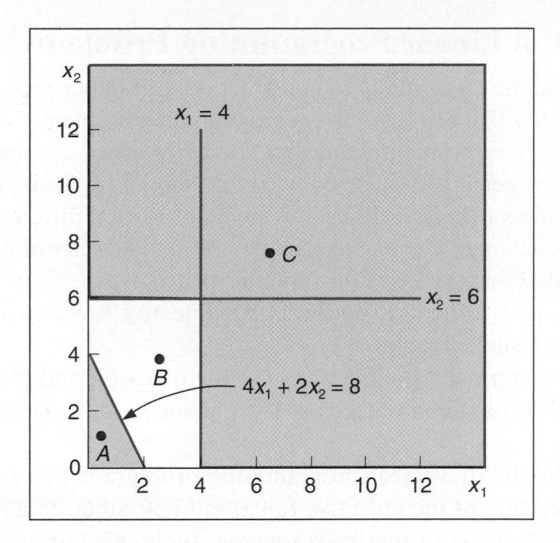

In an unbounded problem the objective function can increase indefinitely without reaching a maximum value.

An Unbounded Problem

In some problems, the feasible solution area formed by the model constraints is not closed. In these cases it is possible for the objective function to increase indefinitely without ever reaching a maximum value because it never reaches the boundary of the feasible solution area.

An example of this type of problem is formulated next and shown graphically in Figure 2.22:

$$\text{maximize } Z = 4x_1 + 2x_2$$
subject to
$$x_1 \geq 4$$
$$x_2 \leq 2$$
$$x_1, x_2 \geq 0$$

FIGURE 2.22
An unbounded problem

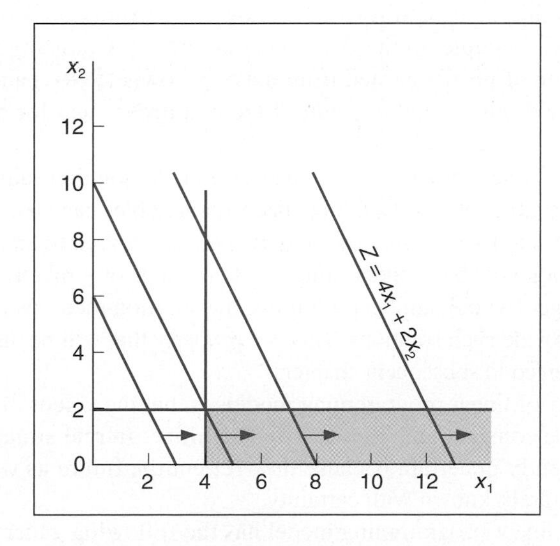

The solution space is not completely closed in.

In Figure 2.22, the objective function is shown to increase without bound; thus, a solution is never reached.

Unlimited profits are not possible in the real world; an unbounded solution, like an infeasible solution, typically reflects an error in defining the problem or in formulating the model.

Characteristics of Linear Programming Problems

The components of a linear programming model are an objective function, decision variables, and constraints.

Now that we have had the opportunity to construct several linear programming models, let's review the characteristics that identify a linear programming problem.

A linear programming problem requires a choice between alternative courses of action (i.e., a decision). The decision is represented in the model by decision variables. A typical choice task for a business firm is deciding how much of several different products to produce, as in the Beaver Creek Pottery Company example presented earlier in this chapter. Identifying the choice task and defining the decision variables is usually the first step in the formulation process because it is quite difficult to construct the objective function and constraints without first identifying the decision variables.

The problem encompasses an objective that the decision maker wants to achieve. The two most frequently encountered objectives for a business are maximizing profit and minimizing cost.

A third characteristic of a linear programming problem is that restrictions exist, making unlimited achievement of the objective function impossible. In a business firm these restrictions often take the form of limited resources, such as labor or material; however, the sample models in this chapter exhibit a variety of problem restrictions. These restrictions, as well as the objective, must be definable by mathematical functional relationships that are linear. Defining these relationships is typically the most difficult part of the formulation process.

Properties of Linear Programming Models

In addition to encompassing only linear relationships, a linear programming model also has several other implicit properties, which have been exhibited consistently throughout the examples in this chapter. The term *linear* not only means that the functions in the models are graphed as a straight line; it also means that the relationships exhibit **proportionality**. In other words, the rate of change, or slope, of the function is constant; therefore, changes of a given size in the value of a decision variable will result in exactly the same relative changes in the functional value.

__Proportionality__ means the slope of a constraint or objective function line is constant.

The terms in the objective function or constraints are __additive__.

Linear programming also requires that the objective function terms and the constraint terms be **additive**. For example, in the Beaver Creek Pottery Company model, the total profit (Z) must equal the sum of profits earned from making bowls (\$$40x_1$) and mugs (\$$50x_2$). Also, the total resources used must equal the sum of the resources used for each activity in a constraint (e.g., labor).

The values of decision variables are continuous or __divisible__.

Another property of linear programming models is that the solution values (of the decision variables) cannot be restricted to integer values; the decision variables can take on any fractional value. Thus, the variables are said to be *continuous* or **divisible**, as opposed to *integer* or *discrete*. For example, although decision variables representing bowls or mugs or airplanes or automobiles should realistically have integer (whole number) solutions, the solution methods for linear programming will not necessarily provide such solutions. This is a property that will be discussed further as solution methods are presented in subsequent chapters.

All model parameters are assumed to be known with __certainty__.

The final property of linear programming models is that the values of all the model parameters are assumed to be constant and known with **certainty**. In real situations, however, model parameters are frequently uncertain because they reflect the future as well as the present, and future conditions are rarely known with certainty.

To summarize, a linear programming model has the following general properties: linearity, proportionality, additivity, divisibility, and certainty. As various linear programming solution methods are presented throughout this book, these properties will become more obvious, and their impact on problem solution will be discussed in greater detail.

Summary

The two example problems in this chapter were formulated as linear programming models in order to demonstrate the modeling process. These problems were similar in that they concerned achieving some objective subject to a set of restrictions or requirements. Linear programming models exhibit certain common characteristics:

- An objective function to be maximized or minimized
- A set of constraints
- Decision variables for measuring the level of activity
- Linearity among all constraint relationships and the objective function

The graphical approach to the solution of linear programming problems is not a very efficient means of solving problems. For one thing, drawing accurate graphs is tedious. Moreover, the graphical approach is limited to models with only two decision variables. However, the analysis of the graphical approach provides valuable insight into linear programming problems and their solutions.

In the graphical approach, once the feasible solution area and the optimal solution point have been determined from the graph, simultaneous equations are solved to determine the values of x_1 and x_2 at the solution point. In Chapter 3 we will show how linear programming solutions can be obtained using computer programs.

Example Problem Solutions

As a prelude to the problems, this section presents example solutions to two linear programming problems.

Problem Statement

Moore's Meatpacking Company produces a hot dog mixture in 1,000-pound batches. The mixture contains two ingredients—chicken and beef. The cost per pound of each of these ingredients is as follows:

Ingredient	Cost/lb.
Chicken	\$3
Beef	\$5

Each batch has the following recipe requirements:

a. At least 500 pounds of chicken
b. At least 200 pounds of beef

The ratio of chicken to beef must be at least 2 to 1. The company wants to know the optimal mixture of ingredients that will minimize cost. Formulate a linear programming model for this problem.

Solution

Step 1: Identify Decision Variables

Recall that the problem should not be "swallowed whole." Identify each part of the model separately, starting with the decision variables:

$$x_1 = \text{lb. of chicken}$$
$$x_2 = \text{lb. of beef}$$

Step 2: Formulate the Objective Function

$$\text{minimize } Z = \$3x_1 + \$5x_2$$

where

$$\begin{aligned} Z &= \text{cost per 1,000-lb batch} \\ \$3x_1 &= \text{cost of chicken} \\ \$5x_2 &= \text{cost of beef} \end{aligned}$$

Step 3: Establish Model Constraints

The constraints of this problem are embodied in the recipe restrictions and (not to be overlooked) the fact that each batch must consist of 1,000 pounds of mixture:

$$\begin{aligned} x_1 + x_2 &= 1{,}000 \text{ lb.} \\ x_1 &\geq 500 \text{ lb. of chicken} \\ x_2 &\geq 200 \text{ lb. of beef} \\ x_1/x_2 &\geq 2/1 \text{ or } x_1 - 2x_2 \geq 0 \end{aligned}$$

and

$$x_1, x_2 \geq 0$$

The Model

$$\begin{aligned} &\text{minimize } Z = \$3x_1 + \$5x_2 \\ &\text{subject to} \\ &\quad x_1 + x_2 = 1{,}000 \\ &\quad x_1 \geq 500 \\ &\quad x_2 \geq 200 \\ &\quad x_1 - 2x_2 \geq 0 \\ &\quad x_1, x_2 \geq 0 \end{aligned}$$

Problem Statement

Solve the following linear programming model graphically:

$$\begin{aligned} &\text{maximize } Z = 4x_1 + 5x_2 \\ &\text{subject to} \\ &\quad x_1 + 2x_2 \leq 10 \\ &\quad 6x_1 + 6x_2 \leq 36 \\ &\quad x_1 \leq 4 \\ &\quad x_1, x_2 \geq 0 \end{aligned}$$

Solution

Step 1: Plot the Constraint Lines as Equations

A simple method for plotting constraint lines is to set one of the constraint variables equal to zero and solve for the other variable to establish a point on one of the axes. The three constraint lines are graphed in the following figure:

The constraint equations

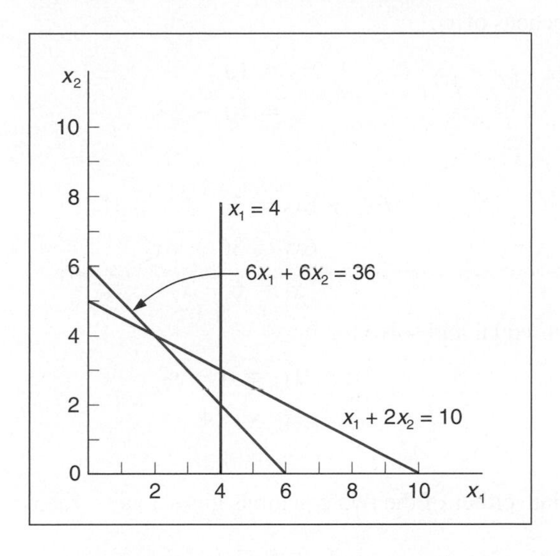

Step 2: Determine the Feasible Solution Area

The feasible solution area is determined by identifying the space that jointly satisfies the ≤ conditions of all three constraints, as shown in the following figure:

The feasible solution space and extreme points

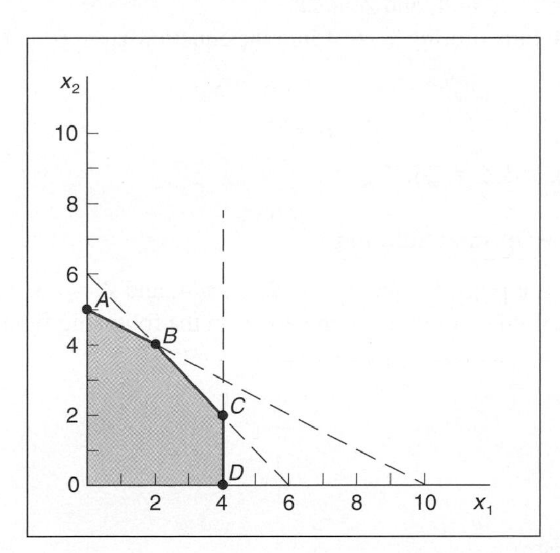

Step 3: Determine the Solution Points

The solution at point A can be determined by noting that the constraint line intersects the x_2 axis at 5; thus, $x_2 = 5$, $x_1 = 0$, and $Z = 25$. The solution at point D on the other axis can be determined similarly; the constraint intersects the axis at $x_1 = 4$, $x_2 = 0$, and $Z = 16$.

The values at points B and C must be found by solving simultaneous equations. Note that point B is formed by the intersection of the lines $x_1 + 2x_2 = 10$ and $6x_1 + 6x_2 = 36$. First, convert both of these equations to functions of x_1:

$$x_1 + 2x_2 = 10$$
$$x_1 = 10 - 2x_2$$

and

$$6x_1 + 6x_2 = 36$$
$$6x_1 = 36 - 6x_2$$
$$x_1 = 6 - x_2$$

Now, set the equations equal and solve for x_2:

$$10 - 2x_2 = 6 - x_2$$
$$-x_2 = -4$$
$$x_2 = 4$$

Substituting $x_2 = 4$ into either of the two equations gives a value for x_1:

$$x_1 = 6 - x_2$$
$$x_1 = 6 - (4)$$
$$x_1 = 2$$

Thus, at point B, $x_1 = 2$, $x_2 = 4$, and $Z = 28$.

At point C, $x_1 = 4$. Substituting $x_1 = 4$ into the equation $x_1 = 6 - x_2$ gives a value for x_2:

$$4 = 6 - x_2$$
$$x_2 = 2$$

Thus, $x_1 = 4$, $x_2 = 2$, and $Z = 26$.

Step 4: Determine the Optimal Solution

The optimal solution is at point B, where $x_1 = 2$, $x_2 = 4$, and $Z = 28$. The optimal solution and solutions at the other extreme points are summarized in the following figure:

Optimal solution point

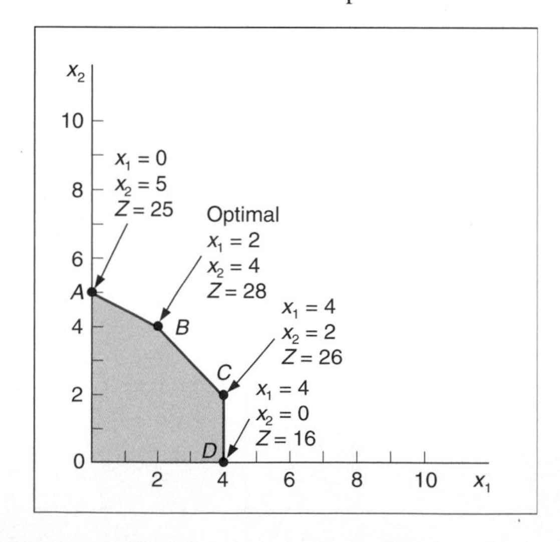

Problems

1. In Problem 34 in Chapter 1, when Tracy McCoy wakes up Saturday morning, she remembers that she promised the PTA she would make some cakes and/or homemade bread for its bake sale that afternoon. However, she does not have time to go to the store to get ingredients, and she has only a short time to bake things in her oven. Because cakes and breads require different baking temperatures, she cannot bake them simultaneously, and she has only 3 hours available to bake. A cake requires 3 cups of flour, and a loaf of bread requires 8 cups; Tracy has 20 cups of flour. A cake requires 45 minutes to bake, and a loaf of bread requires 30 minutes. The PTA will sell a cake for $10 and a loaf of bread for $6. Tracy wants to decide how many cakes and loaves of bread she should make.
   a. Formulate a linear programming model for this problem.
   b. Solve this model by using graphical analysis.

2. A company produces two products that are processed on two assembly lines. Assembly line 1 has 100 available hours, and assembly line 2 has 42 available hours. Each product requires 10 hours of processing time on line 1, while on line 2 product 1 requires 7 hours and product 2 requires 3 hours. The profit for product 1 is $6 per unit, and the profit for product 2 is $4 per unit.
   a. Formulate a linear programming model for this problem.
   b. Solve this model by using graphical analysis.

3. The Munchies Cereal Company makes a cereal from several ingredients. Two of the ingredients, oats and rice, provide vitamins A and B. The company wants to know how many ounces of oats and rice it should include in each box of cereal to meet the minimum requirements of 48 milligrams of vitamin A and 12 milligrams of vitamin B while minimizing cost. An ounce of oats contributes 8 milligrams of vitamin A and 1 milligram of vitamin B, whereas an ounce of rice contributes 6 milligrams of A and 2 milligrams of B. An ounce of oats costs $0.05, and an ounce of rice costs $0.03.
   a. Formulate a linear programming model for this problem.
   b. Solve this model by using graphical analysis.

4. What would be the effect on the optimal solution in Problem 3 if the cost of rice increased from $0.03 per ounce to $0.06 per ounce?

5. The Kalo Fertilizer Company makes a fertilizer using two chemicals that provide nitrogen, phosphate, and potassium. A pound of ingredient 1 contributes 10 ounces of nitrogen and 6 ounces of phosphate, while a pound of ingredient 2 contributes 2 ounces of nitrogen, 6 ounces of phosphate, and 1 ounce of potassium. Ingredient 1 costs $3 per pound, and ingredient 2 costs $5 per pound. The company wants to know how many pounds of each chemical ingredient to put into a bag of fertilizer to meet the minimum requirements of 20 ounces of nitrogen, 36 ounces of phosphate, and 2 ounces of potassium while minimizing cost.
   a. Formulate a linear programming model for this problem.
   b. Solve this model by using graphical analysis.

6. The Pinewood Furniture Company produces chairs and tables from two resources—labor and wood. The company has 80 hours of labor and 36 board-ft. of wood available each day. Demand for chairs is limited to 6 per day. Each chair requires 8 hours of labor and 2 board-ft. of wood, whereas a table requires 10 hours of labor and 6 board-ft. of wood. The profit derived from each chair is $400 and from each table, $100. The company wants to determine the number of chairs and tables to produce each day in order to maximize profit.
   a. Formulate a linear programming model for this problem.
   b. Solve this model by using graphical analysis.

7. In Problem 6, how much labor and wood will be unused if the optimal numbers of chairs and tables are produced?

8. In Problem 6, explain the effect on the optimal solution of changing the profit on a table from $100 to $500.

9. The Crumb and Custard Bakery makes coffee cakes and Danish pastries in large pans. The main ingredients are flour and sugar. There are 25 pounds of flour and 16 pounds of sugar available, and the demand for coffee cakes is 5. Five pounds of flour and 2 pounds of sugar are required to make a pan of coffee cakes, and 5 pounds of flour and 4 pounds of sugar are required to make a pan of Danish. A pan of coffee cakes has a profit of $1, and a pan of Danish has a profit of $5. Determine the number of pans of cakes and Danish to produce each day so that profit will be maximized.
   a. Formulate a linear programming model for this problem.
   b. Solve this model by using graphical analysis.

10. In Problem 9, how much flour and sugar will be left unused if the optimal numbers of cakes and Danish are baked?

11. Solve the following linear programming model graphically:

$$\begin{aligned} \text{maximize } Z &= 3x_1 + 6x_2 \\ \text{subject to} & \\ 3x_1 + 2x_2 &\leq 18 \\ x_1 + x_2 &\geq 5 \\ x_1 &\leq 4 \\ x_1, x_2 &\geq 0 \end{aligned}$$

12. The Elixer Drug Company produces a drug from two ingredients. Each ingredient contains the same three antibiotics, in different proportions. One gram of ingredient 1 contributes 3 units, and 1 gram of ingredient 2 contributes 1 unit of antibiotic 1; the drug requires 6 units. At least 4 units of antibiotic 2 are required, and the ingredients contribute 1 unit each per gram. At least 12 units of antibiotic 3 are required; a gram of ingredient 1 contributes 2 units, and a gram of ingredient 2 contributes 6 units. The cost for a gram of ingredient 1 is $80, and the cost for a gram of ingredient 2 is $50. The company wants to formulate a linear programming model to determine the number of grams of each ingredient that must go into the drug in order to meet the antibiotic requirements at the minimum cost.
   a. Formulate a linear programming model for this problem.
   b. Solve this model by using graphical analysis.

13. A jewelry store makes necklaces and bracelets from gold and platinum. The store has 18 ounces of gold and 20 ounces of platinum. Each necklace requires 3 ounces of gold and 2 ounces of platinum, whereas each bracelet requires 2 ounces of gold and 4 ounces of platinum. The demand for bracelets is no more than four. A necklace earns $300 in profit and a bracelet, $400. The store wants to determine the number of necklaces and bracelets to make in order to maximize profit.
   a. Formulate a linear programming model for this problem.
   b. Solve this model by using graphical analysis.

14. In Problem 13, explain the effect on the optimal solution of increasing the profit on a bracelet from $400 to $600. What will be the effect of changing the platinum requirement for a necklace from 2 ounces to 3 ounces?

15. In Problem 13:
   a. The maximum demand for bracelets is four. If the store produces the optimal number of bracelets and necklaces, will the maximum demand for bracelets be met? If not, by how much will it be missed?
   b. What profit for a necklace would result in no bracelets being produced, and what would be the optimal solution for this profit?

16. A clothier makes coats and slacks. The two resources required are wool cloth and labor. The clothier has 150 square yards of wool and 200 hours of labor available. Each coat requires

3 square yards of wool and 10 hours of labor, whereas each pair of slacks requires 5 square yards of wool and 4 hours of labor. The profit for a coat is $50, and the profit for slacks is $40. The clothier wants to determine the number of coats and pairs of slacks to make so that profit will be maximized.

a. Formulate a linear programming model for this problem.
b. Solve this model by using graphical analysis.

17. In Problem 16, what would be the effect on the optimal solution if the available labor were increased from 200 to 240 hours?

18. The Weemow Lawn Service wants to start doing snow removal in the winter when there are no lawns to maintain. Jeff and Julie Weems, who own the service, are trying to determine how much equipment they need to purchase, based on the various job types they have. They plan to work themselves and hire some local college students on a per-job basis. Based on historical weather data, they estimate that there will be six major snowfalls next winter. Virtually all customers want their snow removed no more than 2 days after the snow stops falling. Working 10 hours per day (into the night), Jeff and Julie can remove the snow from a normal driveway in about 1 hour, and it takes about 4 hours to remove the snow from a business parking lot and sidewalk. The variable cost (mainly for labor and gas) per job is $12 for a driveway and $47 for a parking lot. Using their lawn service customer base as a guideline, they believe they will have demand of no more than 40 homeowners and 25 businesses. They plan to charge $35 for a home driveway and $120 for a business parking lot, which is slightly less than the going rate. They want to know how many jobs of each type will maximize their profit.

a. Formulate a linear programming model for this problem.
b. Solve this model graphically.

19. In Problem 18:

a. If Jeff and Julie pay $3,700 for snow removal equipment, will they make any money?
b. If Jeff and Julie reduce their prices to $30 for a driveway and $100 for a parking lot, they will increase demand to 55 for driveways and 32 for businesses. Will this affect their possible profit?
c. Alternatively, hiring additional people on a per-job basis will increase Jeff and Julie's variable cost to $16 for a driveway and $53 for a parking lot, but it will lower the time it takes to clear a driveway to 40 minutes and a parking lot to 3 hours. Will this affect their profit?
d. If Jeff and Julie combine the two alternatives suggested in (b) and (c), will this affect their profit?

20. Solve the following linear programming model graphically:

$$\begin{aligned} \text{maximize } Z &= 1.5x_1 + x_2 \\ \text{subject to} & \\ x_1 &\leq 4 \\ x_2 &\leq 6 \\ x_1 + x_2 &\leq 5 \\ x_1, x_2 &\geq 0 \end{aligned}$$

21. Transform the model in Problem 20 into standard form and indicate the value of the slack variables at each corner point solution.

22. Solve the following linear programming model graphically:

$$\begin{aligned} \text{maximize } Z &= 5x_1 + 8x_2 \\ \text{subject to} & \\ 3x_1 + 5x_2 &\leq 50 \\ 2x_1 + 4x_2 &\leq 40 \\ x_1 &\leq 8 \\ x_2 &\leq 10 \\ x_1, x_2 &\geq 0 \end{aligned}$$

23. Transform the model in Problem 22 into standard form and indicate the value of the slack variables at each corner point solution.

24. Solve the following linear programming model graphically:

$$\begin{aligned} \text{maximize } Z &= 6.5x_1 + 10x_2 \\ \text{subject to} & \\ 2x_1 + 4x_2 &\leq 40 \\ x_1 + x_2 &\leq 15 \\ x_1 &\geq 8 \\ x_1, x_2 &\geq 0 \end{aligned}$$

25. In Problem 24, if the constraint $x_1 \geq 8$ is changed to $x_1 \leq 8$, what effect does this have on the feasible solution space and the optimal solution?

26. Universal Claims Processors processes insurance claims for large national insurance companies. Most claim processing is done by a large pool of computer operators, some of whom are permanent and some of whom are temporary. A permanent operator can process 16 claims per day, whereas a temporary operator can process 12 per day, and on average the company processes at least 450 claims each day. The company has 40 computer workstations. A permanent operator generates about 0.5 claim with errors each day, whereas a temporary operator averages about 1.4 defective claims per day. The company wants to limit claims with errors to 25 per day. A permanent operator is paid $64 per day, and a temporary operator is paid $42 per day. The company wants to determine the number of permanent and temporary operators to hire in order to minimize costs.
   a. Formulate a linear programming model for this problem.
   b. Solve this model by using graphical analysis.

27. In Problem 26, explain the effect on the optimal solution of changing the daily pay for a permanent claims processor from $64 to $54. Explain the effect of changing the daily pay for a temporary claims processor from $42 to $36.

28. In Problem 26, what would be the effect on the optimal solution if Universal Claims Processors decided not to try to limit the number of defective claims each day?

29. In Problem 26, explain the effect on the optimal solution if the minimum number of claims the firm processes each day increased from 450 to at least 650.

30. Solve the following linear programming model graphically:

$$\begin{aligned} \text{minimize } Z &= 8x_1 + 6x_2 \\ \text{subject to} & \\ 4x_1 + 2x_2 &\geq 20 \\ -6x_1 + 4x_2 &\leq 12 \\ x_1 + x_2 &\geq 6 \\ x_1, x_2 &\geq 0 \end{aligned}$$

31. Solve the following linear programming model graphically:

$$\begin{aligned} \text{minimize } Z &= 3x_1 + 6x_2 \\ \text{subject to} & \\ 3x_1 + 2x_2 &\leq 18 \\ x_1 + x_2 &\geq 5 \\ x_1 &\leq 4 \\ x_2 &\leq 7 \\ x_2/x_1 &\leq 7/8 \\ x_1, x_2 &\geq 0 \end{aligned}$$

32. In Problem 31, what would be the effect on the solution if the constraint $x_2 \leq 7$ were changed to $x_2 \geq 7$?

33. Solve the following linear programming model graphically:

$$\begin{aligned} \text{minimize } Z &= 5x_1 + x_2 \\ \text{subject to} & \\ 3x_1 + 4x_2 &= 24 \\ x_1 &\leq 6 \\ x_1 + 3x_2 &\leq 12 \\ x_1, x_2 &\geq 0 \end{aligned}$$

34. Solve the following linear programming model graphically:

$$\begin{aligned} \text{maximize } Z &= 3x_1 + 2x_2 \\ \text{subject to} & \\ 2x_1 + 4x_2 &\leq 22 \\ -x_1 + 4x_2 &\leq 10 \\ 4x_1 - 2x_2 &\leq 14 \\ x_1 - 3x_2 &\leq 1 \\ x_1, x_2 &\geq 0 \end{aligned}$$

35. Solve the following linear programming model graphically:

$$\begin{aligned} \text{minimize } Z &= 8x_1 + 2x_2 \\ \text{subject to} & \\ 2x_1 - 6x_2 &\leq 12 \\ 5x_1 + 4x_2 &\geq 40 \\ x_1 + 2x_2 &\geq 12 \\ x_2 &\leq 6 \\ x_1, x_2 &\geq 0 \end{aligned}$$

36. Gillian's Restaurant has an ice-cream counter where it sells two main products, ice cream and frozen yogurt, each in a variety of flavors. The restaurant makes one order for ice cream and yogurt each week, and the store has enough freezer space for 115 gallons total of both products. A gallon of frozen yogurt costs \$0.75 and a gallon of ice cream costs \$0.93, and the restaurant budgets \$90 each week for these products. The manager estimates that each week the restaurant sells at least twice as much ice cream as frozen yogurt. Profit per gallon of ice cream is \$4.15, and profit per gallon of yogurt is \$3.60.
    a. Formulate a linear programming model for this problem.
    b. Solve this model by using graphical analysis.

37. In Problem 36, how much additional profit would the restaurant realize each week if it increased its freezer capacity to accommodate 20 extra gallons total of ice cream and yogurt?

38. Copperfield Mining Company owns two mines, each of which produces three grades of ore—high, medium, and low. The company has a contract to supply a smelting company with at least 12 tons of high-grade ore, 8 tons of medium-grade ore, and 24 tons of low-grade ore. Each mine produces a certain amount of each type of ore during each hour that it operates. Mine 1 produces 6 tons of high-grade ore, 2 tons of medium-grade ore, and 4 tons of low-grade ore per hour. Mine 2 produces 2, 2, and 12 tons, respectively, of high-, medium-, and low-grade ore per hour. It costs Copperfield \$200 per hour to mine each ton of ore from mine 1, and it costs \$160 per hour to mine each ton of ore from mine 2. The company wants to determine the number of hours it needs to operate each mine so that its contractual obligations can be met at the lowest cost.
    a. Formulate a linear programming model for this problem.
    b. Solve this model by using graphical analysis.

39. A canning company produces two sizes of cans—regular and large. The cans are produced in 10,000-can lots. The cans are processed through a stamping operation and a coating operation. The company has 30 days available for both stamping and coating. A lot of regular-size cans requires 2 days to stamp and 4 days to coat, whereas a lot of large cans requires 4 days to stamp and 2 days to coat. A lot of regular-size cans earns \$800 profit, and a lot of large-size cans earns \$900 profit. In order to fulfill its obligations under a shipping contract, the company must produce at least nine lots. The company wants to determine the number of lots to produce of each size can (x_1 and x_2) in order to maximize profit.
    a. Formulate a linear programming model for this problem.
    b. Solve this model by using graphical analysis.

40. A manufacturing firm produces two products. Each product must undergo an assembly process and a finishing process. It is then transferred to the warehouse, which has space for only a limited number of items. The firm has 80 hours available for assembly and 112 hours for finishing, and it can store a maximum of 10 units in the warehouse. Each unit of product 1 has a profit of \$30 and requires 4 hours to assemble and 14 hours to finish. Each unit of product 2 has a profit of \$70 and requires 10 hours to assemble and 8 hours to finish. The firm wants to determine the quantity of each product to produce in order to maximize profit.
    a. Formulate a linear programming model for this problem.
    b. Solve this model by using graphical analysis.

41. Assume that the objective function in Problem 40 has been changed from $Z = 30x_1 + 70x_2$ to $Z = 90x_1 + 70x_2$. Determine the slope of each objective function and discuss what effect these slopes have on the optimal solution.

42. The Valley Wine Company produces two kinds of wine—Valley Nectar and Valley Red. The wines are produced from 64 tons of grapes the company has acquired this season. A 1,000-gallon batch of Nectar requires 4 tons of grapes, and a batch of Red requires 8 tons. However, production is limited by the availability of only 50 cubic yards of storage space for aging and 120 hours of processing time. A batch of each type of wine requires 5 cubic yards of storage space. The processing time for a batch of Nectar is 15 hours, and the processing time for a batch of Red is 8 hours. Demand for each type of wine is limited to seven batches. The profit for a batch of Nectar is \$9,000, and the profit for a batch of Red is \$12,000. The company wants to determine the number of 1,000-gallon batches of Nectar (x_1) and Red (x_2) to produce in order to maximize profit.
    a. Formulate a linear programming model for this problem.
    b. Solve this model by using graphical analysis.

43. In Problem 42:
    a. How much processing time will be left unused at the optimal solution?
    b. What would be the effect on the optimal solution of increasing the available storage space from 50 to 60 cubic yards?

44. Kroeger supermarket sells its own brand of canned peas as well as several national brands. The store makes a profit of \$0.28 per can for its own peas and a profit of \$0.19 for any of the national brands. The store has 6 square feet of shelf space available for canned peas, and each can of peas takes up 9 square inches of that space. Point-of-sale records show that each week the store never sells more than half as many cans of its own brand as it does of the national brands. The store wants to know how many cans of its own brand of peas and how many cans of the national brands to stock each week on the allocated shelf space in order to maximize profit.
    a. Formulate a linear programming model for this problem.
    b. Solve this model by using graphical analysis.

45. In Problem 44, if Kroeger discounts the price of its own brand of peas, the store will sell at least 1.5 times as much of the national brands as its own brand, but its profit margin on its own brand will be reduced to \$0.23 per can. What effect will the discount have on the optimal solution?

46. Shirtstop makes T-shirts with logos and sells them in its chain of retail stores. It contracts with two different plants—one in Puerto Rico and one in The Bahamas. The shirts from the plant in Puerto Rico cost \$0.46 apiece, and 9% of them are defective and can't be sold. The shirts from The Bahamas cost only \$0.35 each, but they have an 18% defective rate. Shirtstop needs 3,500 shirts. To retain its relationship with the two plants, it wants to order at least 1,000 shirts from each. It would also like at least 88% of the shirts it receives to be salable.
    a. Formulate a linear programming model for this problem.
    b. Solve this model by using graphical analysis.

47. In Problem 46:
    a. Suppose Shirtstop decided it wanted to minimize the defective shirts while keeping costs below \$2,000. Reformulate the problem with these changes and solve graphically.
    b. How many fewer defective items were achieved with the model in (a) than with the model in Problem 46?

48. Angela and Bob Ray keep a large garden in which they grow cabbage, tomatoes, and onions to make two kinds of relish—chow-chow and tomato. The chow-chow is made primarily of cabbage, whereas the tomato relish has more tomatoes than does the chow-chow. Both relishes include onions, and negligible amounts of bell peppers and spices. A jar of chow-chow contains 8 ounces of cabbage, 3 ounces of tomatoes, and 3 ounces of onions, whereas a jar of tomato relish contains 6 ounces of tomatoes, 6 ounces of cabbage, and 2 ounces of onions. The Rays grow 120 pounds of cabbage, 90 pounds of tomatoes, and 45 pounds of onions each summer. The Rays can produce no more than 24 dozen jars of relish. They make \$2.25 in profit from a jar of chow-chow and \$1.95 in profit from a jar of tomato relish. The Rays want to know how many jars of each kind of relish to produce to generate the most profit.
    a. Formulate a linear programming model for this problem.
    b. Solve this model graphically.

49. In Problem 48, the Rays have checked their sales records for the past 5 years and have found that they sell at least 50% more chow-chow than tomato relish. How will this additional information affect their model and solution?

50. A California grower has a 50-acre farm on which to plant strawberries and tomatoes. The grower has available 300 hours of labor per week and 800 tons of fertilizer, and he has contracted for shipping space for a maximum of 26 acres' worth of strawberries and 37 acres' worth of tomatoes. An acre of strawberries requires 10 hours of labor and 8 tons of fertilizer, whereas an acre of tomatoes requires 3 hours of labor and 20 tons of fertilizer. The profit from an acre of strawberries is \$400, and the profit from an acre of tomatoes is \$300. The farmer wants to know the number of acres of strawberries and tomatoes to plant to maximize profit.
    a. Formulate a linear programming model for this problem.
    b. Solve this model by using graphical analysis.

51. In Problem 50, if the amount of fertilizer required for each acre of strawberries were determined to be 20 tons instead of 8 tons, what would be the effect on the optimal solution?

52. The admissions office at Tech wants to determine how many in-state and how many out-of-state students to accept for next fall's entering freshman class. Tuition for an in-state student is \$7,600 per year, whereas out-of-state tuition is \$22,500 per year. A total of 12,800 in-state and 8,100 out-of-state freshmen have applied for next fall, and Tech does not want to accept more than 3,500 students. However, because Tech is a state institution, the state mandates that it can accept no more than 40% out-of-state students. From past experience the admissions office knows that 12% of in-state students and 24% of out-of-state students will drop out during their first year. Tech wants to maximize total tuition while limiting the total attrition to 600 first-year students.
    a. Formulate a linear programming model for this problem.
    b. Solve this model by using graphical analysis.

53. The Robinsons are planning a wedding and reception for their daughter, Rachel. Some of the most expensive items served at the reception and dinner are wine and beer. The Robinsons are planning on 200 guests at the reception, and they estimate that they need at least four servings (i.e., a glass of wine or bottle of beer) for each guest in order to be sure they won't run out. A bottle of wine contains five glasses. They also estimate that 50% more guests will prefer wine to beer. A bottle of wine costs $8, and a bottle of beer costs $0.75. The Robinsons have budgeted $1,200 for wine and beer. Finally, the Robinsons want to minimize their waste (i.e., unused wine and beer). The caterer has advised them that typically 5% of the wine and 10% of the beer will be left over. How many bottles of wine and beer should the Robinsons order?
    a. Formulate a linear programming model for this problem.
    b. Solve this model graphically.

54. Suppose that in Problem 53, it turns out that twice as many guests prefer wine as beer. Will the Robinsons have enough wine with the amount they ordered in the Problem 53 solution? How much waste will there be with the solution in Problem 53?

55. Xara Stores in the United States imports the designer-inspired clothes it sells from suppliers in China and Brazil. Xara estimates that it will have 45 orders in a year, and it must arrange to transport orders (in less-than-full containers) by container ship with shippers in Hong Kong and Buenos Aires. The shippers Xara uses have a travel time of 32 days from Buenos Aires and 14 days from Hong Kong, and Xara wants its orders to have an average travel time of no more than 21 days. About 10% of the annual orders from the shipper in Hong Kong are damaged, and the shipper in Buenos Aires damages about 4% of all orders annually. Xara wants to receive no more than 6 damaged orders each year. Xara does not want to be dependent on suppliers from just one country, so it wants to receive at least 25% of its orders from each country. It costs $3,700 per order from China and $5,100 per order to ship from Brazil. Xara wants to know how many orders it should ship from each port in order to minimize shipping costs.
    a. Formulate a linear programming model for this problem.
    b. Solve this model by using graphical analysis.

56. In Problem 55, the Chinese shipper would like to gain more shipping orders from Xara because it's a prestigious company and would enhance the shipper's reputation. It has therefore made the following proposals to Xara:
    a. Would Xara give the Chinese shipper more orders if it reduced its shipping costs to $2,500 per shipment?
    b. Would Xara give the Chinese shipper more orders if it reduced its damaged orders to 5%.
    c. Would Xara give the shipper more of its orders if it reduced its travel time to 28 days?

57. Janet Lopez is establishing an investment portfolio that will include stock and bond funds. She has $720,000 to invest, and she does not want the portfolio to include more than 65% stocks. The average annual return for the stock fund she plans to invest in is 18%, whereas the average annual return for the bond fund is 6%. She further estimates that the most she could lose in the next year in the stock fund is 22%, whereas the most she could lose in the bond fund is 5%. To reduce her risk, she wants to limit her potential maximum losses to $100,000.
    a. Formulate a linear programming model for this problem.
    b. Solve this model by using graphical analysis.

58. Professor Smith teaches two sections of business statistics, which combined will result in 120 final exams to be graded. Professor Smith has two graduate assistants, Brad and Sarah, who will grade the final exams. There is a 3-day period between the time the exam is administered and when final grades must be posted. During this period Brad has 12 hours available and Sarah has 10 hours available to grade the exams. It takes Brad an average of 7.2 minutes to grade an exam, and it takes Sarah 12 minutes to grade an exam; however, Brad's exams will have errors that will

require Professor Smith to ultimately regrade 10% of the exams, while only 6% of Sarah's exams will require regrading. Professor Smith wants to know how many exams to assign to each graduate assistant to grade in order to minimize the number of exams to regrade.

a. Formulate a linear programming model for this problem.
b. Solve this model by using graphical analysis.

59. In Problem 58, if Professor Smith could hire Brad or Sarah to work 1 additional hour, which should she choose? What would be the effect of hiring the selected graduate assistant for 1 additional hour?

60. Starbright Coffee Shop at the Galleria Mall serves two coffee blends it brews on a daily basis, Pomona and Coastal. Each is a blend of three high-quality coffees from Colombia, Kenya, and Indonesia. The coffee shop has 6 pounds of each of these coffees available each day. Each pound of coffee will produce sixteen 16-ounce cups of coffee. The shop has enough brewing capacity to brew 30 gallons of these two coffee blends each day. Pomona is a blend of 20% Colombian, 35% Kenyan, and 45% Indonesian, while Coastal is a blend of 60% Colombian, 10% Kenyan, and 30% Indonesian. The shop sells 1.5 times more Pomona than Coastal each day. Pomona sells for $2.05 per cup, and Coastal sells for $1.85 per cup. The manager wants to know how many cups of each blend to sell each day in order to maximize sales.

a. Formulate a linear programming model for this problem.
b. Solve this model by using graphical analysis.

61. In Problem 60:

a. If Starbright Coffee Shop could get 1 more pound of coffee, which one should it be? What would be the effect on sales of getting 1 more pound of this coffee? Would it benefit the shop to increase its brewing capacity from 30 gallons to 40 gallons?
b. If the shop spent $20 per day on advertising that would increase the relative demand for Pomona to twice that of Coastal, should it be done?

62. Solve the following linear programming model graphically and explain the solution result:

$$\begin{aligned} \text{minimize } Z &= \$3{,}000x_1 + 1{,}000x_2 \\ \text{subject to} & \\ 60x_1 + 20x_2 &\geq 1{,}200 \\ 10x_1 + 10x_2 &\geq 400 \\ 40x_1 + 160x_2 &\geq 2{,}400 \\ x_1, x_2 &\geq 0 \end{aligned}$$

63. Solve the following linear programming model graphically and explain the solution result:

$$\begin{aligned} \text{maximize } Z &= 60x_1 + 90x_2 \\ \text{subject to} & \\ 60x_1 + 30x_2 &\leq 1{,}500 \\ 100x_1 + 100x_2 &\geq 6{,}000 \\ x_2 &\geq 30 \\ x_1, x_2 &\geq 0 \end{aligned}$$

64. Solve the following linear programming model graphically and explain the solution result:

$$\begin{aligned} \text{maximize } Z &= 110x_1 + 75x_2 \\ \text{subject to} & \\ 2x_1 + x_2 &\geq 40 \\ -6x_1 + 8x_2 &\leq 120 \\ 70x_1 + 105x_2 &\geq 2{,}100 \\ x_1, x_2 &\geq 0 \end{aligned}$$

Case Problem

METROPOLITAN POLICE PATROL

The Metropolitan Police Department was recently criticized in the local media for not responding to police calls in the downtown area rapidly enough. In several recent cases, alarms had sounded for break-ins, but by the time the police car arrived, the perpetrators had left, and in one instance a store owner had been shot. Sergeant Joe Davis was assigned by the chief as head of a task force to find a way to determine the optimal patrol area (dimensions) for their cars that would minimize the average time it took to respond to a call in the downtown area.

Sergeant Davis solicited help from Angela Maris, an analyst in the operations area for the police department. Together they began to work through the problem.

Joe noted to Angela that normal patrol sectors are laid out in rectangles, with each rectangle including a number of city blocks. For illustrative purposes he defined the dimensions of the sector as x in the horizontal direction and as y in the vertical direction. He explained to Angela that cars traveled in straight lines either horizontally or vertically and turned at right angles. Travel in a horizontal direction must be accompanied by travel in a vertical direction, and the total distance traveled is the sum of the horizontal and vertical segments. He further noted that past research on police patrolling in urban areas had shown that the average distance traveled by a patrol car responding to a call in either direction was one-third of the dimensions of the sector, or $x/3$ and $y/3$. He also explained that the travel time it took to respond to a call (assuming that a car left immediately upon receiving the call) is simply the average distance traveled divided by the average travel speed.

Angela told Joe that now that she understood how average travel time to a call was determined, she could see that it was closely related to the size of the patrol area. She asked Joe if there were any restrictions on the size of the area sectors that cars patrolled. He responded that for their city, the department believed that the perimeter of a patrol sector should not be less than 5 miles or exceed 12 miles. He noted several policy issues and staffing constraints that required these specifications. Angela wanted to know if any additional restrictions existed, and Joe indicated that the distance in the vertical direction must be at least 50% more than the horizontal distance for the sector. He explained that laying out sectors in that manner meant that the patrol areas would have a greater tendency to overlap different residential, income, and retail areas than if they ran the other way. He said that these areas were layered from north to south in the city, so if a sector area was laid out east to west, all of it would tend to be in one demographic layer.

Angela indicated that she had almost enough information to develop a model, except that she also needed to know the average travel speed the patrol cars could travel. Joe told her that cars moving vertically traveled an average of 15 miles per hour, whereas cars traveled horizontally an average of 20 miles per hour. He said that the difference was due to different traffic flows.

Develop a linear programming model for this problem and solve it by using the graphical method.

Case Problem

"THE POSSIBILITY" RESTAURANT

Angela Fox and Zooey Caulfield were food and nutrition majors at State University, as well as close friends and roommates. Upon graduation Angela and Zooey decided to open a French restaurant in Draperton, the small town where the university was located. There were no other French restaurants in Draperton, and the possibility of doing something new and somewhat risky intrigued the two friends. They purchased an old Victorian home just off Main Street for their new restaurant, which they named "The Possibility."

Angela and Zooey knew in advance that at least initially they could not offer a full, varied menu of dishes. They had no idea what their local customers' tastes in French cuisine would be, so they decided to serve only two full-course meals each night, one with beef and the other with fish. Their chef, Pierre, was confident he could make each dish so exciting and unique that two meals would be sufficient, at least until they could assess which menu items were most popular. Pierre indicated that with each meal he could experiment with different appetizers, soups, salads, vegetable dishes, and desserts until they were able to identify a full selection of menu items.

The next problem for Angela and Zooey was to determine how many meals to prepare for each night so they could shop for ingredients and set up the work schedule. They could not afford too much waste. They estimated that they would sell a maximum of 60 meals each night. Each fish dinner, including all accompaniments, requires 15 minutes to prepare, and each beef dinner takes twice as long. There is a total of 20 hours of kitchen staff labor available each day. Angela and Zooey believe that because of the health consciousness of their potential clientele, they will sell at least three fish dinners for every two beef dinners. However, they also believe that at least 10% of their customers will order beef dinners. The profit from each fish dinner will be approximately $12, and the profit from a beef dinner will be about $16.

Formulate a linear programming model for Angela and Zooey that will help them estimate the number of meals they should prepare each night and solve this model graphically.

If Angela and Zooey increased the menu price on the fish dinner so that the profit for both dinners was the same, what effect would that have on their solution? Suppose Angela and Zooey reconsidered the demand for beef dinners and decided that at least 20% of their customers would purchase beef dinners. What effect would this have on their meal preparation plan?

Case Problem

Annabelle Invests in the Market

Annabelle Sizemore has cashed in some treasury bonds and a life insurance policy that her parents had accumulated over the years for her. She has also saved some money in certificates of deposit and savings bonds during the 10 years since she graduated from college. As a result, she has $120,000 available to invest. Given the recent rise in the stock market, she feels that she should invest all of this amount there. She has researched the market and has decided that she wants to invest in an index fund tied to S&P stocks and in an Internet stock fund. However, she is very concerned about the volatility of Internet stocks. Therefore, she wants to balance her risk to some degree.

She has decided to select an index fund from Shield Securities and an Internet stock fund from Madison Funds, Inc. She has also decided that the proportion of the dollar amount she invests in the index fund relative to the Internet fund should be at least one-third but that she should not invest more than twice the amount in the Internet fund that she invests in the index fund. The price per share of the index fund is $175, whereas the price per share of the Internet fund is $208. The average annual return during the last 3 years for the index fund has been 17%, and for the Internet stock fund it has been 28%. She anticipates that both mutual funds will realize the same average returns for the coming year that they have in the recent past; however, at the end of the year she is likely to reevaluate her investment strategy anyway. Thus, she wants to develop an investment strategy that will maximize her return for the coming year.

Formulate a linear programming model for Annabelle that will indicate how much money she should invest in each fund and solve this model by using the graphical method.

Suppose Annabelle decides to change her risk-balancing formula by eliminating the restriction that the proportion of the amount she invests in the index fund to the amount that she invests in the Internet fund must be at least one-third. What will the effect be on her solution? Suppose instead that she eliminates the restriction that the proportion of money she invests in the Internet fund relative to the stock fund not exceed a ratio of 2 to 1. How will this affect her solution?

If Annabelle can get $1 more to invest, how will that affect her solution? $2 more? $3 more? What can you say about her return on her investment strategy, given these successive changes?

CHAPTER

3

Linear Programming: Computer Solution and Sensitivity Analysis

Chapter 2 demonstrated how a linear programming model is formulated and how a solution can be derived from a graph of the model. Graphing can provide valuable insight into linear programming and linear programming solutions in general. However, the fact that this solution method is limited to problems with only two decision variables restricts its usefulness as a *general* solution technique.

In this chapter, we will show how linear programming problems can be solved using several personal computer software packages. We will also describe how to use a computer solution result to experiment with a linear programming model to see what effect parameter changes have on the optimal solution, referred to as *sensitivity analysis*.

Computer Solution

The ***simplex method*** *is a procedure involving a set of mathematical steps to solve linear programming problems.*

See Web site Module A for a chapter on "The Simplex Solution Method."

When linear programming was first developed in the 1940s, virtually the only way to solve a problem was by using a lengthy manual mathematical solution procedure called the **simplex method**. However, during the next six decades, as computer technology evolved, the computer was used more and more to solve linear programming models. The mathematical steps of the simplex method were simply programmed in prewritten software packages designed for the solution of linear programming problems. The ability to solve linear programming problems quickly and cheaply on the computer, regardless of the size of the problem, popularized linear programming and expanded its use by businesses. There are currently dozens of software packages with linear programming capabilities. Many of these are general-purpose management science or quantitative methods packages with linear programming modules, among many other modules for other techniques. There are also numerous software packages that are devoted exclusively to linear programming and its derivatives. These packages are generally cheap, efficient, and easy to use.

As a result of the easy and low-cost availability of personal computers and linear programming software, the simplex method has become less of a focus in the teaching of linear programming. Thus, at this point in our presentation of linear programming, we focus exclusively on computer solution. However, knowledge of the simplex method is useful in gaining an overall, in-depth understanding of linear programming for those who are interested in this degree of understanding. As noted, computer solution itself is based on the simplex method. Thus, while we present linear programming in the text in a manner that does not require use of the simplex method, we also provide in-depth coverage of this topic on the companion Web site for this text.

In the next few sections, we demonstrate how to solve linear programming problems by using Excel spreadsheets and QM for Windows, a typical general-purpose quantitative methods software package.

Excel Spreadsheets

Excel can be used to solve linear programming problems, although the data input requirements can be more time-consuming and tedious than with a software package like QM for Windows that is specifically designed for the purpose. A spreadsheet requires that column and row headings for the specific model be set up and that constraint and objective function formulas be input in their entirety, as opposed to just the model parameters, as with QM for Windows. However, this is also an advantage of spreadsheets, in that it enables the problem to be set up in an attractive format for reporting and presentation purposes. In addition, once a spreadsheet is set up for one problem, it can often be used as a template for others. Exhibit 3.1 shows an Excel spreadsheet set up for our Beaver Creek Pottery Company example introduced in Chapter 2. Appendix B at the end of this text is the tutorial "Setting Up and Editing a Spreadsheet," using Exhibit 3.4 as an example.

The values for bowls and mugs and for profit are contained in cells B10, B11, and B12, respectively. These cells are currently empty or zero because the model has not yet been solved. The objective function for profit, **=C4*B10+D4*B11**, is embedded in cell B12 shown in bar. This formula is essentially the same as $Z = 40x_1 + 50x_2$, where B10 and B11 represent x_1 and

EXHIBIT 3.1

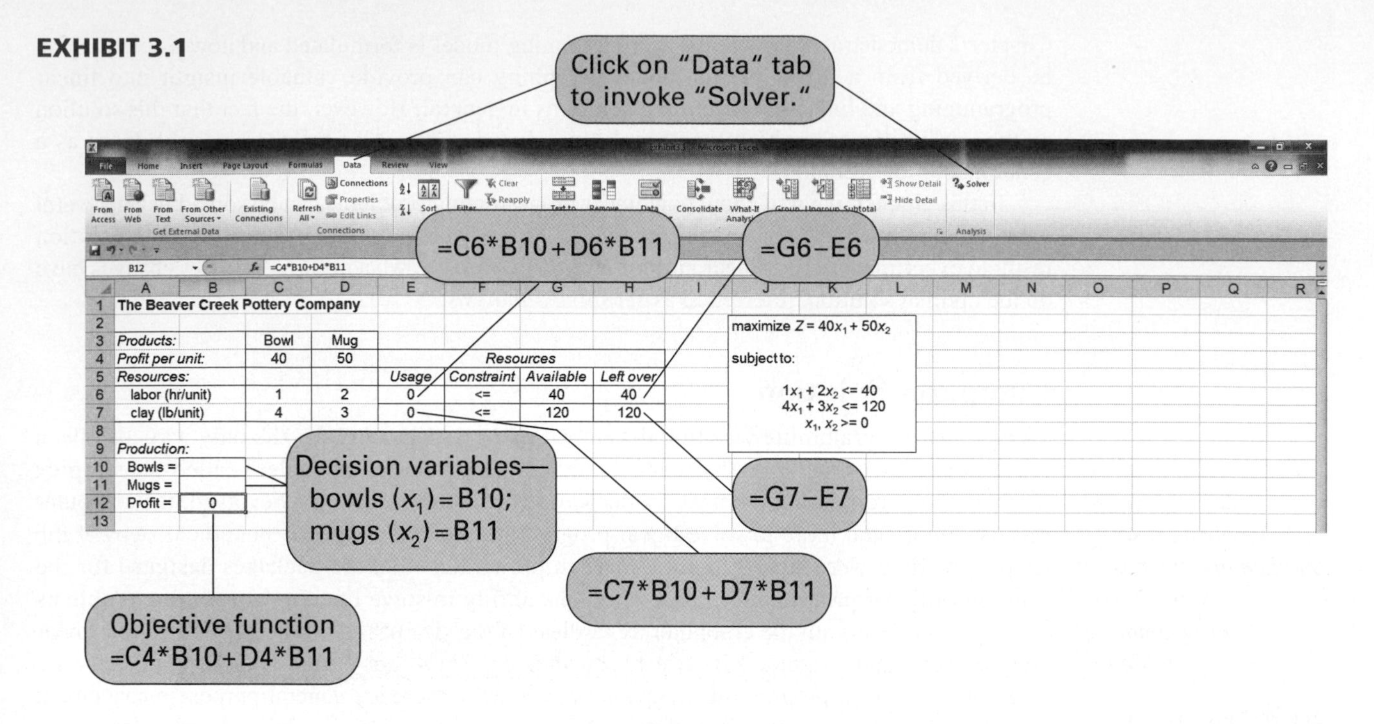

x_2, and B12 equals Z. The objective function coefficients, 40 and 50, are in cells C4 and D4. Similar formulas for the constraints for labor and clay are embedded in cells E6 and E7. For example, in cell E6 we input the formula **=C6*B10+D6*B11**. The < = signs in cells F6 and F7 are for cosmetic purposes only; they have no real effect.

To solve this problem, first click on the "Data" tab on the toolbar at the top of the screen and then click on "Solver" on the far right side of the Data toolbar. The window Solver Parameters will appear, as shown in Exhibit 3.2. Initially all the windows on this screen are blank, and we must input the objective function cell, the cells representing the decision variables, and the cells that make up the model constraints.

When inputting the Solver parameters as shown in Exhibit 3.2, we first "Set Objective:" which is B12 for our example. (Excel automatically inserts the $ sign next to cell addresses; you should not type it in.) Next we indicate that we want to maximize the objective by clicking on "Max." We achieve our objective by changing cells B10 and B11, which represent our model decision variables. The designation "B10:B11" means all the cells between B10 and B11, inclusive. We next input our model constraints by clicking on "Add," which will access the screen shown in Exhibit 3.3.

Exhibit 3.3 shows our labor constraint. Cell E6 contains the constraint formula for labor (**=C6*B10+D6*B11**), whereas cell G6 contains the labor hours available (i.e., 40). We continue to add constraints until the model is complete. Note that we could have input our constraints by adding a single constraint formula, **E6:E7< = G6:G7**, which means that the constraints in cells E6 and E7 are less than or equal to the values in cells G6 and G7, respectively. It is also not necessary to input the nonnegativity constraints for our decision variables, **B10:B11> = 0**. This can be done in the Solver Parameters screen (Exhibit 3.2).

Click on "OK" on the Add Constraint window after all constraints have been added. This will return us to the Solver Parameters screen. There are two more necessary steps before proceeding to solve the problem. On the Solver Parameters screen, check where it says, "Make Unconstrained Variables Non-Negative," and where it says "Select a Solving Method," select "Simplex LP." This will ensure that Solver uses the simplex procedure to solve the model and not some other numeric method (which Excel has available). This is not too important for now, but later it will ensure that we get the right reports for sensitivity analysis, a topic we will take up next.

EXHIBIT 3.2

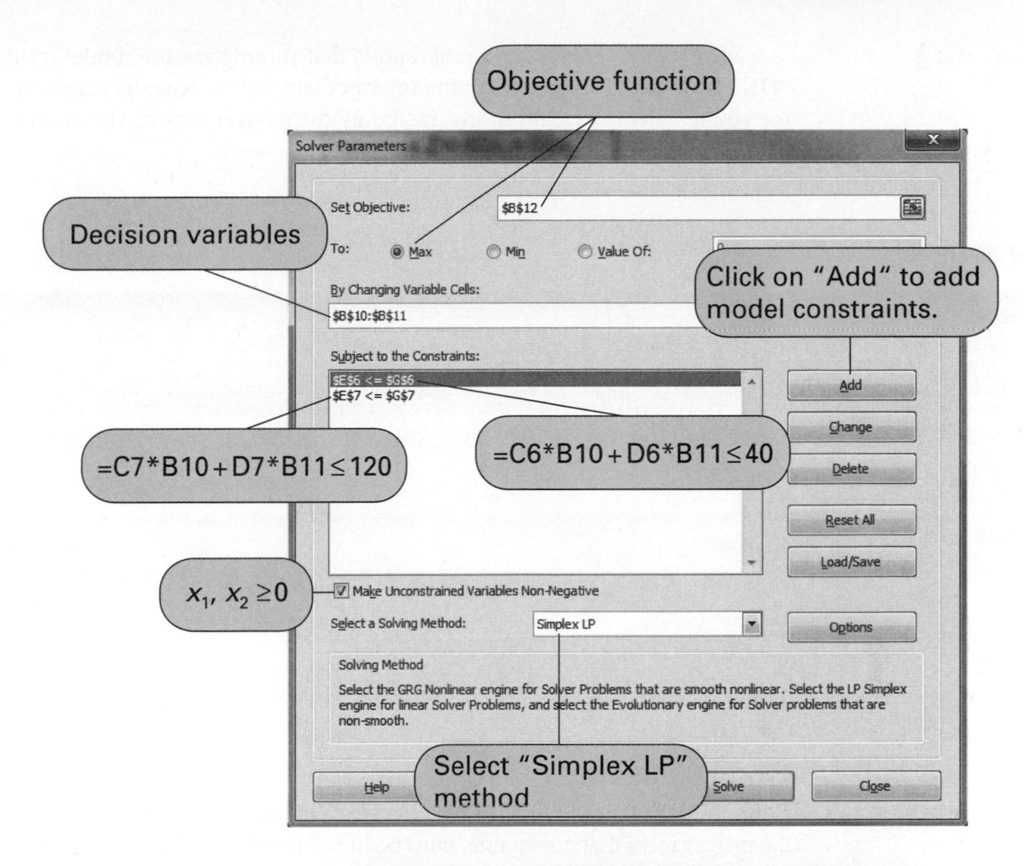

EXHIBIT 3.3

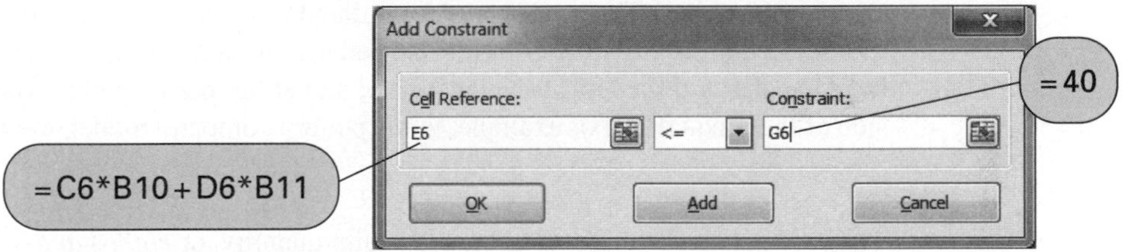

Once the complete model is input, click on "Solve" at the bottom of the Solver Parameters screen (Exhibit 3.2). First, a screen will appear, titled Solver Results, which will provide you with the opportunity to select the reports you want and then when you click on "OK," the solution screen shown in Exhibit 3.4 will appear.

If there had been any extra, or *slack*, left over for labor or clay, it would have appeared in column H on our spreadsheet, under the heading "Left Over." In this case, there are no slack resources left over.

EXHIBIT 3.4

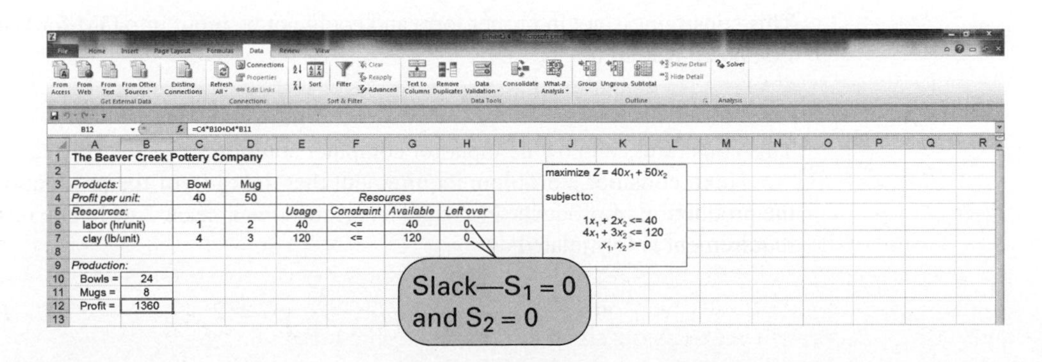

We can also generate several reports that summarize the model results. When you click on "OK" from the Solver screen, the intermediate Solver Results screen provides an opportunity for you to select several reports, including the answer report, shown in Exhibit 3.5. This report provides a summary of the solution results.

EXHIBIT 3.5

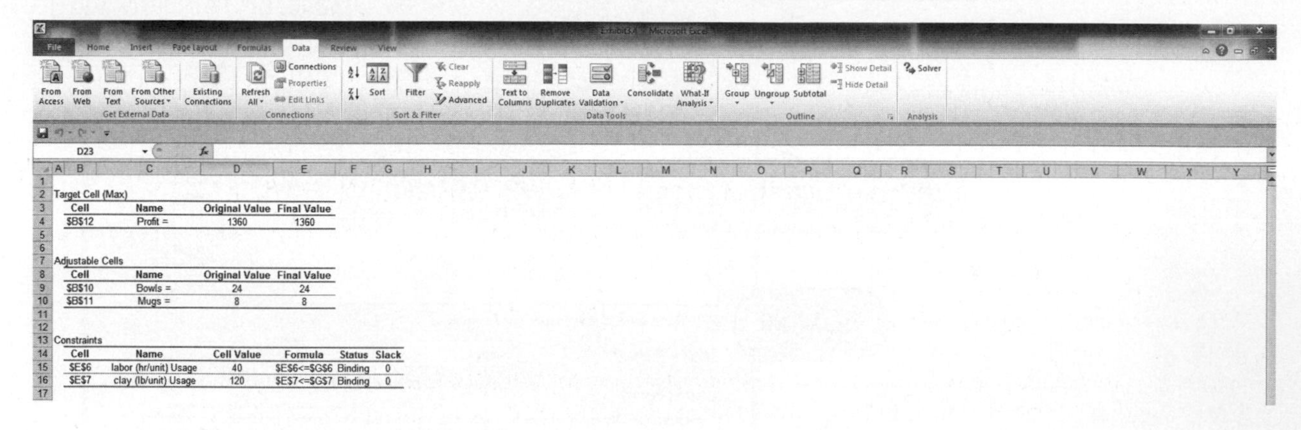

QM for Windows

Before demonstrating how to use QM for Windows, we must first make a few comments about the proper form that constraints must be in before a linear programming model can be solved with QM for Windows. The constraints formulated in the linear programming models presented in Chapter 2 and in this chapter have followed a consistent form. All the variables in the constraint have appeared to the left of the inequality, and all numerical values have been on the right-hand side of the inequality. For example, in the pottery company model, the constraint for labor is

$$x_1 + 2x_2 \leq 40$$

The value, 40, is referred to as the constraint quantity, or *right-hand-side*, value.

The standard form for a linear programming problem requires constraints to be in this form, with variables on the left side and numeric values to the right of the inequality or equality sign. This is a necessary condition to input problems into some computer programs, and specifically QM for Windows, for linear programming solution.

Consider a model requirement that states that the production of product 3 (x_3) must be as much as or more than the production of products 1 (x_1) and 2 (x_2). The model constraint for this requirement is formulated as

$$x_3 \geq x_1 + x_2$$

This constraint is not in proper form and could not be input into QM for Windows as it is. It must first be converted to

$$x_3 - x_1 - x_2 \geq 0$$

This constraint can now be input for computer solution.

Next, consider a problem requirement that the ratio of the production of product 1 (x_1) to the production of products 2 (x_2) and 3 (x_3) must be at least 2 to 1. The model constraint for this requirement is formulated as

$$\frac{x_1}{x_2 + x_3} \geq 2$$

Management Science Application

Optimizing Production Quantities at GE Plastics

© Kristoffer Tripplaar/Alamy

GE Plastics is a $5 billion global business that supplies plastics and raw materials from plants around the world to companies such as automotive, appliance, computer, and medical equipment companies. Among its seven major divisions, the High Performance Polymers (HPP) division is the fastest growing. HPP is a very heat tolerant polymer that is used in the manufacture of microwave cookware, fire helmets, utensils, and aircraft. HPP has a supply chain that consists of two levels of manufacturing plants and distribution channels that is similar to those in all GE Plastics divisions. First-level plants convert feedstocks into resins and then ship them to finishing plants, where they are combined with additives to produce different grades of end products. The products are then shipped to GE Polymerland (a commercial front for GE plastics), which distributes them to customers. Each physical plant has multiple plant lines that operate as independent manufacturing facilities. HPP has 8 resin plant lines that feed 21 finishing plant lines, which produce 24 grades of HPP products. HPP uses a linear programming model to optimize production along this supply chain of manufacturing plant lines and distribution channels. The linear programming model objective function maximizes the total contribution margin, which consists of revenues minus the sum of manufacturing, additive, and distribution costs, subject to demand, manufacturing capacity, and flow constraints. The decision variables are the amount of each resin produced at each resin plant line that will be used at each finishing plant line and the amount of each product at each finishing plant line. The company uses the model to develop a four-year planning horizon by solving 4 single-year models; each single-year LP model has 3,100 variables and 1,100 constraints. The model is solved within an Excel framework, using LINGO, a commercial optimization solver.

Source: Based on R. Tyagi, P. Kalish, K. Akbay, and G. Munshaw, "GE Plastics Optimizes the Two-Echelon Global Fulfillment Network at Its High Performance Polymers Division," *Interfaces* 34, no. 5 (September–October 2004): 359–66.

Fractional relationships between variables in constraints must be eliminated.

Although this constraint does meet the condition that variables be on the left side of the inequality and numeric values on the right, it is not in proper form. The fractional relationship of the variables, $x_1/(x_2 + x_3)$, cannot be input into the most commonly used linear programming computer programs in that form. It must be converted to

$$x_1 \geq 2(x_2 + x_3)$$

and then

$$x_1 - 2x_2 - 2x_3 \geq 0$$

We will demonstrate how to use QM for Windows by solving our Beaver Creek Pottery Company example. The linear programming module in QM for Windows is accessed by clicking on "Module" at the top of the initial window. This will bring down a menu with all the program modules available in QM for Windows. By clicking on "Linear Programming," a window for this program will come up on the screen, and by clicking on "File" and then "New," a screen for inputting problem dimensions will appear. Exhibit 3.6 shows the data entry screen with the model type and the number of constraints and decision variables for our Beaver Creek Pottery Company example.

Exhibit 3.7 shows the data table for our example, with the model parameters, including the objective function and constraint coefficients, and the constraint quantity values. Notice that we also customized the row headings for the constraints by typing in "Labor (hrs)" and "Clay (lbs)."

EXHIBIT 3.6

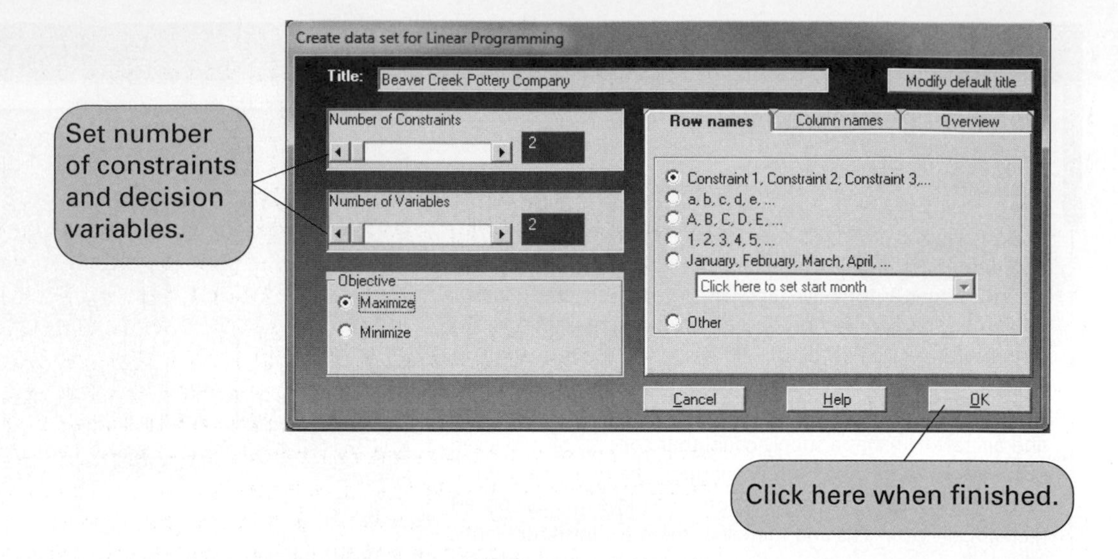

EXHIBIT 3.7

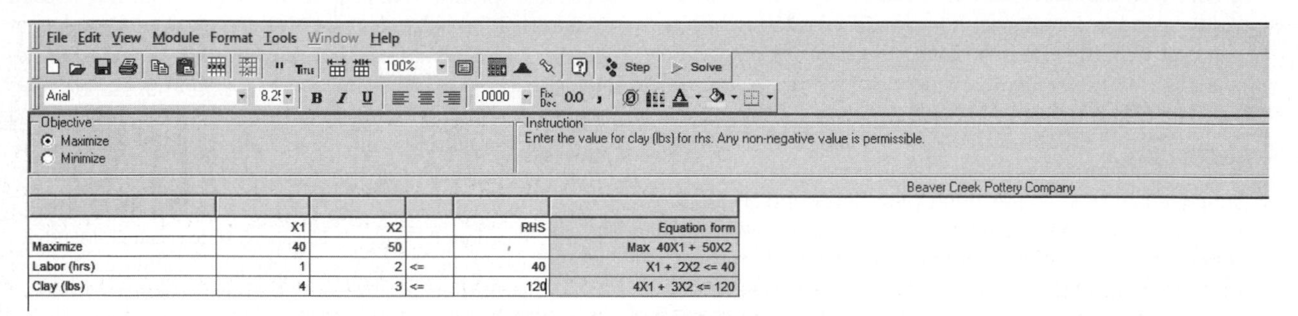

	X1	X2		RHS	Equation form
Maximize	40	50			Max 40X1 + 50X2
Labor (hrs)	1	2	<=	40	X1 + 2X2 <= 40
Clay (lbs)	4	3	<=	120	4X1 + 3X2 <= 120

Once the model parameters have been input, click on "Solve" to get the model solution, as shown in Exhibit 3.8. It is not necessary to put the model into standard form by adding slack variables.

EXHIBIT 3.8

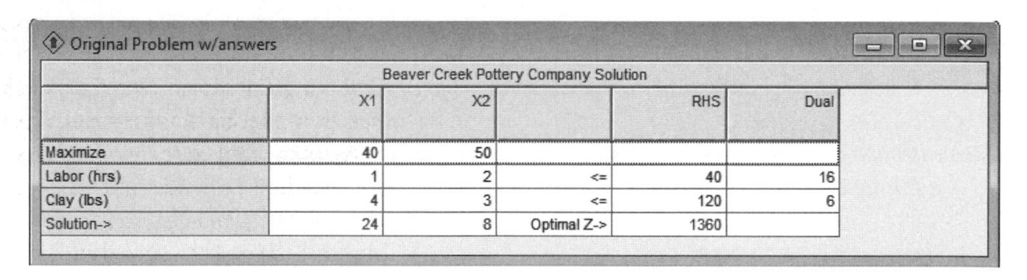

Beaver Creek Pottery Company Solution

	X1	X2		RHS	Dual
Maximize	40	50			
Labor (hrs)	1	2	<=	40	16
Clay (lbs)	4	3	<=	120	6
Solution->	24	8	Optimal Z->	1360	

*The **marginal value** is the dollar amount one would be willing to pay for one additional resource unit.*

Notice the values 16 and 6 under the column labeled "Dual" for the "Labor" and "Clay" rows. These dual values are the **marginal values** of labor and clay in our problem. This is useful information that is provided in addition to the normal model solution values when you solve a linear programming model. We talk about dual values in more detail later in this chapter, but for now it is sufficient to say that the marginal value is the maximum dollar amount the company would be willing to pay for one additional unit of a resource. For example, the dual value of 16 for the labor constraint means that if 1 additional hour of labor could be obtained by the company, it would increase profit by $16. Likewise, if 1 additional pound of clay could be obtained, it would increase profit by $6. Thus, the company would be willing to pay up to $16 for 1 more hour of labor and up to $6 for 1 more pound of clay. The dual value is not the purchase price of one of these resources; it is the maximum amount the company would pay to get more of the resource. These dual values are helpful to the company in making decisions about acquiring additional resources.

Management Science Application

Improving Profitability at Norske Skog with Linear Programming

Luuk van der Lee/Hollandse Hoogte

Norwegian paper (and newsprint) manufacturer Norske Skog, with annual revenues of $3.9 billion and more than 6,400 employees working at 16 paper mills in 12 countries, is the fourth-largest paper manufacturer in the world. Its production lines at paper mills, consisting of pulping plants, paper machines, and a finishing section, can cost more than $500 million. These production lines produce paper 24 hours per day, year round, closing only for routine maintenance; its daily production could create a tabloid publication that would stretch to the moon and back. However, during the past 10-plus years, the demand for paper has declined as electronic media has replaced newsprint publications, jeopardizing the company's financial health and forcing the company to cut costs by closing some of its plants and paper machines. Norske Skog has employed a linear programming model to help it make these difficult and complex decisions. The linear programming model objective function maximized sales revenue minus operating and fixed costs. It included 312 binary variables representing the closure of plants and shutdown of machines; 47,000 non-integer variables representing the amounts of different paper products made on different machines for different customers and raw materials used on different machines; and 2,600 constraints. The model solution called for the closing of one paper mill in Korea and one paper mill in the Czech Republic, as well as the shutdown of one paper machine in Norway. This resulted in a reduction of 450,000 tons of production capacity and total savings of $100 million annually—about 3% of Norske Skog's annual revenue. These decisions recommended by the linear programming model placed the company in a more stable financial situation. In addition, because of the analytical nature of the modeling and decision process, stakeholders and the public trusted the decision more than they had previous, nonquantifiable, decisions.

Source: Based on G. Everett, A. Philpott, K. Vatn, and R. Gjessing, "Norske Skog Improves Global Profitability Using Operations Research," *Interfaces* 40, no. 1 (January–February 2010): 58–70.

QM for Windows provides solution results in several different formats, including a graphical solution by clicking on "Window" and then selecting "Graph." Exhibit 3.9 shows the graphical solution for our Beaver Creek Pottery Company example.

EXHIBIT 3.9

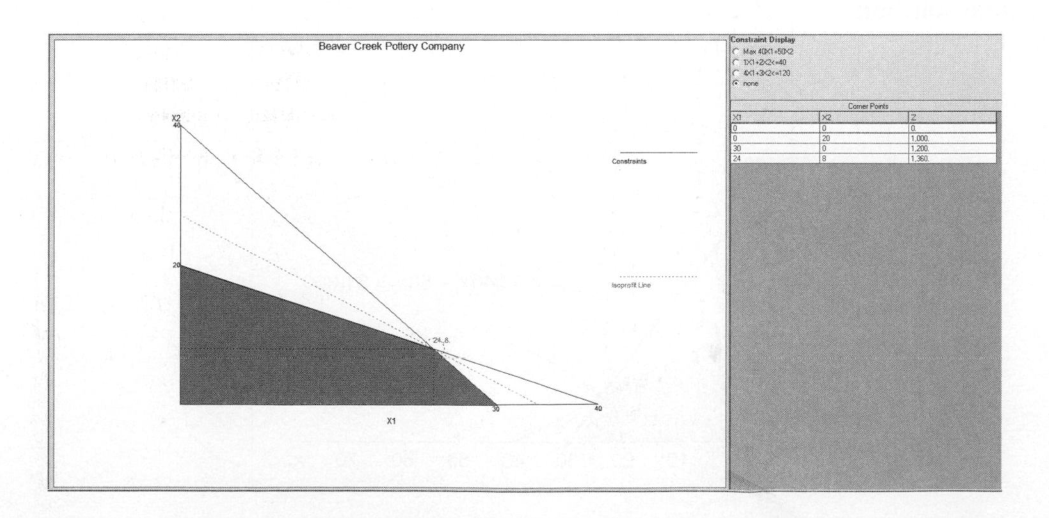

Sensitivity Analysis

When linear programming models were formulated in Chapter 2, it was implicitly assumed that the *parameters* of the model were known with certainty. These parameters include the objective function coefficients, such as profit per bowl; model constraint quantity values, such as available hours of labor; and constraint coefficients, such as pounds of clay per bowl. In the examples presented so far, the models have been formulated as if these parameters are known exactly or with certainty. However, rarely does a manager know all these parameters exactly. In reality, the model parameters are simply estimates (or "best guesses") that are subject to change. For this reason, it is of interest to the manager to see what effect a change in a parameter will have on the solution to the model. Changes may be either reactions to anticipated uncertainties in the parameters or reactions to information. The analysis of parameter changes and their effects on the model solution is known as **sensitivity analysis**.

Sensitivity analysis is the analysis of the effect of parameter changes on the optimal solution.

The most obvious way to ascertain the effect of a change in the parameter of a model is to make the change in the original model, *re-solve* the model, and compare the solution results with the original. However, as we will demonstrate in this chapter, in some cases the effect of changes on the model can be determined without solving the problem again.

Changes in Objective Function Coefficients

The first model parameter change we will analyze is a change in an objective function coefficient. We will use our now-familiar Beaver Creek Pottery Company example to illustrate this change:

$$\begin{aligned} \text{maximize } Z &= \$40x_1 + 50x_2 \\ \text{subject to} & \\ x_1 + 2x_2 &\leq 40 \text{ hr. of labor} \\ 4x_1 + 3x_2 &\leq 120 \text{ lb. of clay} \\ x_1, x_2 &\geq 0 \end{aligned}$$

The graphical solution for this problem is shown in Figure 3.1.

In Figure 3.1, the optimal solution point is shown to be at point $B(x_1 = 24 \text{ and } x_2 = 8)$, which is the last point the objective function, denoted by the dashed line, touches as it leaves the feasible solution area. However, what if we changed the profit of a bowl, x_1, from \$40 to \$100? How would that affect the solution identified in Figure 3.1? This change is shown in Figure 3.2.

FIGURE 3.1
Optimal solution point

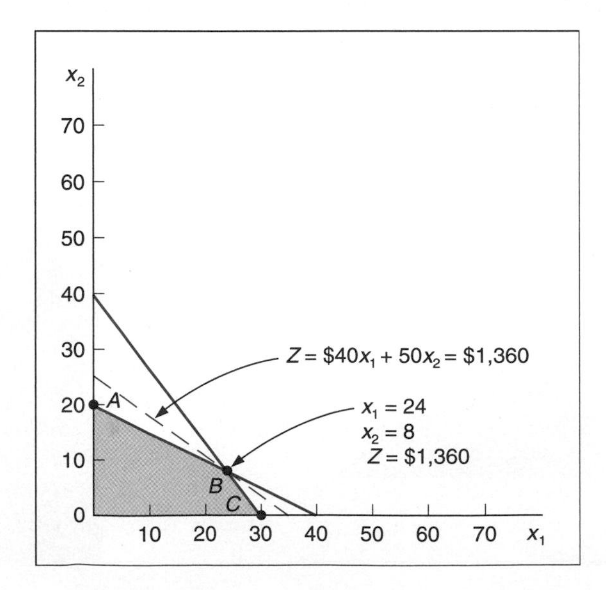

FIGURE 3.2
Changing the objective function x_1 coefficient

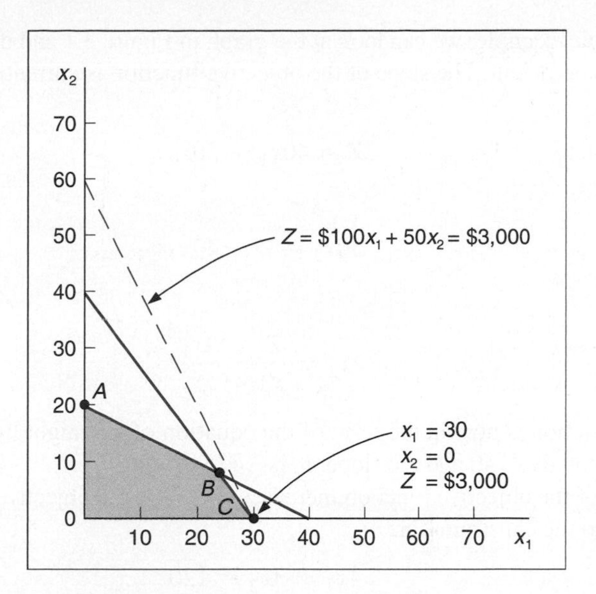

Increasing profit for a bowl (i.e., the x_1 coefficient) from \$40 to \$100 makes the objective function line steeper, so much so that the optimal solution point changes from point *B* to point *C*. Alternatively, if we increased the profit for a mug, the x_2 coefficient, from \$50 to \$100, the objective function line would become flatter, to the extent that point *A* would become optimal, with

$$x_1 = 0, x_2 = 20, \text{and } Z = \$2{,}000$$

This is shown in Figure 3.3.

FIGURE 3.3
Changing the objective function x_2 coefficient

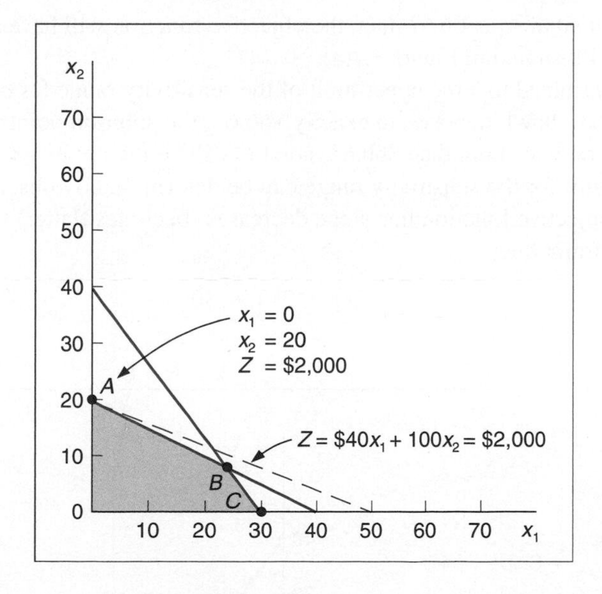

*The **sensitivity range** for an objective coefficient is the range of values over which the current optimal solution point will remain optimal.*

The objective of sensitivity analysis in this case is to determine the range of values for a specific objective function coefficient over which the optimal solution point, x_1 and x_2, will remain optimal. For example, the coefficient of x_1 in the objective function is originally \$40, but at some value greater than \$40, point *C* will become optimal, and at some value less than \$40, point *A* will become optimal. The focus of sensitivity analysis is to determine those two values, referred to as the **sensitivity range** for the x_1 coefficient, which we will designate as c_1.

For our simple example, we can look at the graph in Figure 3.1 and determine the sensitivity range for the x_1 coefficient. The slope of the objective function is currently $-4/5$, determined as follows:

$$Z = 40x_1 + 50x_2$$

or

$$50x_2 = Z - 40x_1$$

and

$$x_2 = \frac{Z}{50} - \frac{4x_1}{5}$$

The objective function is now in the form of the equation of a straight line, $y = a + bx$, where the intercept, a, equals $Z/50$ and the slope, b, is $-4/5$.

If the slope of the objective function increases to $-4/3$, the objective function line becomes exactly parallel to the constraint line,

$$4x_1 + 3x_2 = 120$$

and point C becomes optimal (along with B). The slope of this constraint line is $-4/3$, so we ask ourselves what objective function coefficient for x_1 will make the objective function slope equal $-4/3$? The answer is determined as follows, where c_1 is the objective function coefficient for x_1:

$$\frac{-c_1}{50} = \frac{-4}{3}$$

$$-3c_1 = -200$$

$$c_1 = \$66.67$$

If the coefficient of x_1 is 66.67, then the objective function will have a slope of $-66.67/50$, or $-4/3$. This is illustrated in Figure 3.4(a).

We have determined that the upper limit of the sensitivity range for c_1, the x_1 coefficient, is 66.67. If profit for a bowl increases to exactly \$66.67, the solution points will be both B and C. If the profit for a bowl is more than \$66.67, point C will be the optimal solution point.

The lower limit for the sensitivity range can be determined by observing Figure 3.4(b). In this case, if the objective function line slope decreases (becomes flatter) from $-4/5$ to the same slope as the constraint line,

$$x_1 + 2x_2 = 40$$

FIGURE 3.4
Determining the sensitivity range for c_1

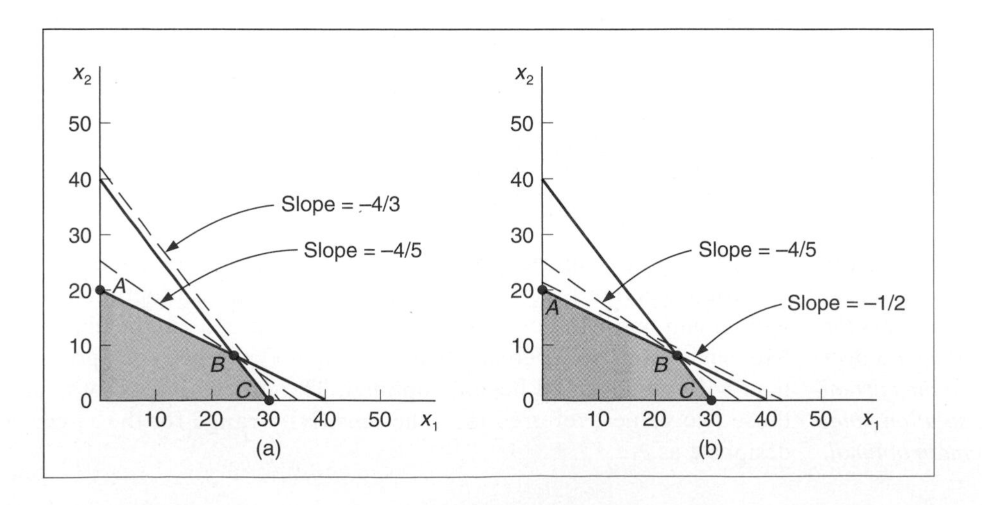

then point A becomes optimal (along with B). The slope of this constraint line is $-1/2$, that is, $x_2 = 20 - (1/2)x_1$. In order to have an objective function slope of $-1/2$, the profit for a bowl would have to decrease to \$25, as follows:

$$\frac{-c_1}{50} = \frac{-1}{2}$$
$$-2c_1 = -50$$
$$c_1 = \$25$$

This is the lower limit of the sensitivity range for the x_1 coefficient.

The complete sensitivity range for the x_1 coefficient can be expressed as

$$25 \leq c_1 \leq 66.67$$

This means that the profit for a bowl can vary anywhere between \$25.00 and \$66.67, and the optimal solution point, $x_1 = 24$ and $x_2 = 8$, will not change. Of course, the total profit, or Z value, will change, depending on whatever value c_1 actually is.

For the manager, this is useful information. Changing the production schedule in terms of how many items are produced can have a number of ramifications in an operation. Packaging, logistical, and marketing requirements for the product might need to be altered. However, with the preceding sensitivity range, the manager knows how much the profit, and hence the selling price and costs, can be altered without resulting in a change in production.

Performing the same type of graphical analysis will provide the sensitivity range for the x_2 objective function coefficient, c_2. This range is $30 \leq c_2 \leq 80$. This means that the profit for a mug can vary between \$30 and \$80, and the optimal solution point, B, will not change. However, for this case and the range for c_1, the sensitivity range generally applies only if one coefficient is varied and the other held constant. Thus, when we say that profit for a mug can vary between \$30 and \$80, this is true only if c_1 remains constant.

Simultaneous changes can be made in the objective function coefficients as long as the changes taken together do not change the optimal solution point. However, determining the effect of these simultaneous changes is overly complex and time-consuming using graphical analysis. In fact, using graphical analysis is a tedious way to perform sensitivity analysis in general, and it is impossible when the linear programming model contains three or more variables, thus requiring a three-dimensional graph. However, Excel and QM for Windows provide sensitivity analysis for linear programming problems as part of their standard solution output. Determining the effect of simultaneous changes in model parameters and performing sensitivity analysis in general are much easier and more practical using the computer. Later in this chapter we will show the sensitivity analysis output for Excel and QM for Windows.

However, before moving on to computer-generated sensitivity analysis, we want to look at one more aspect of the sensitivity ranges for objective function coefficients. Recall that the model for our fertilizer minimization model from Chapter 2 is

$$\begin{aligned}
&\text{minimize } Z = \$6x_1 + 3x_2 \\
&\text{subject to} \\
&2x_1 + 4x_2 \geq 16 \\
&4x_1 + 3x_2 \geq 24 \\
&x_1, x_2 \geq 0
\end{aligned}$$

and the solution shown graphically in Figure 3.5 is $x_1 = 0, x_2 = 8$, and $Z = 24$.

The sensitivity ranges for the objective function coefficients are

$$4 \leq c_1 \leq \infty$$
$$0 \leq c_2 \leq 4.5$$

Notice that the upper bound for the x_1 coefficient range is infinity. The reason for this upper limit can be seen in the graphical solution of the problem shown in Figure 3.5.

FIGURE 3.5
Fertilizer example: sensitivity range for c_1

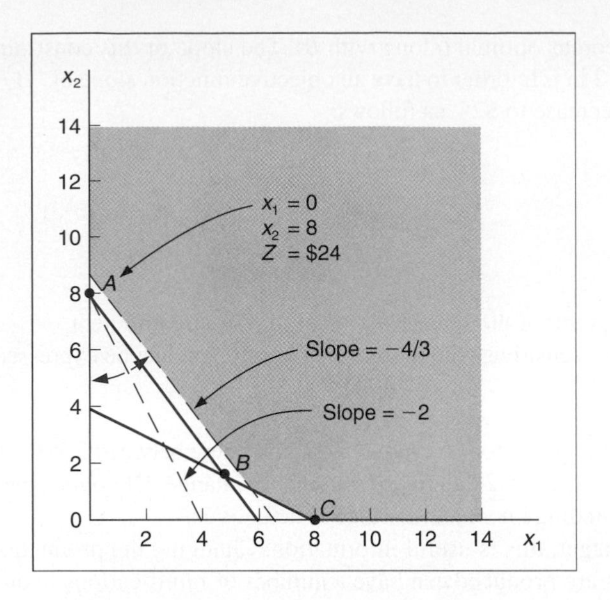

As the objective function coefficient for x_1 decreases from \$6, the objective function slope of -2 decreases, and the objective function line gets flatter. When the coefficient, c_1, equals \$4, then the slope of the objective function is $-4/3$, which is the same as the constraint line, $4x_1 + 3x_2 = 24$. This makes point *B* optimal (as well as *A*). Thus, the lower limit of the sensitivity range for c_1 is \$4. However, notice that as c_1 increases from \$6, the objective function simply becomes steeper and steeper as it rotates toward the x_2 axis of the graph. The objective function will not come into contact with another feasible solution point. Thus, no matter how much we increase cost for Super-gro fertilizer (x_1), point *A* will always be optimal.

Objective Function Coefficient Ranges with the Computer

When we provided the Excel spreadsheet solution for the Beaver Creek Pottery Company example earlier in this chapter, we did not include sensitivity analysis. However, Excel will also generate a sensitivity report that provides the sensitivity ranges for the objective function coefficients. When you click on "Solve" in the Solver Parameters window, you will momentarily go to a Solver Results screen, shown in Exhibit 3.10, which provides you with an opportunity to select different reports before proceeding to the solution. This is how we selected our answer report earlier. The sensitivity report for our Beaver Creek Pottery Company example is shown in Exhibit 3.11.

EXHIBIT 3.10

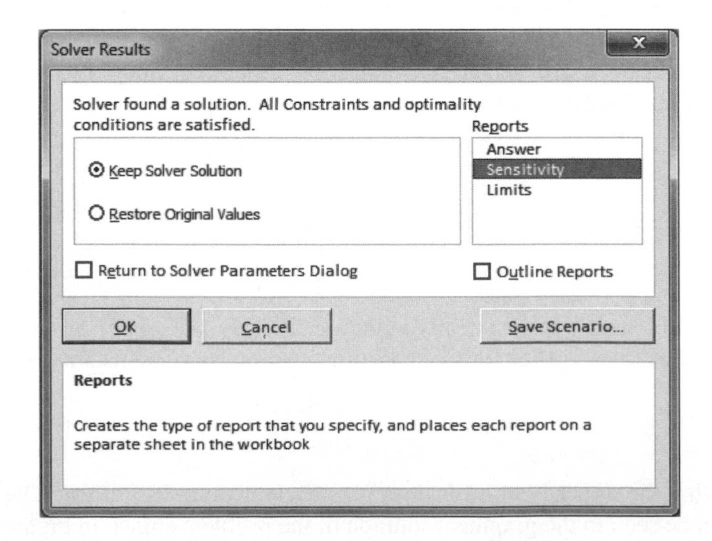

EXHIBIT 3.11

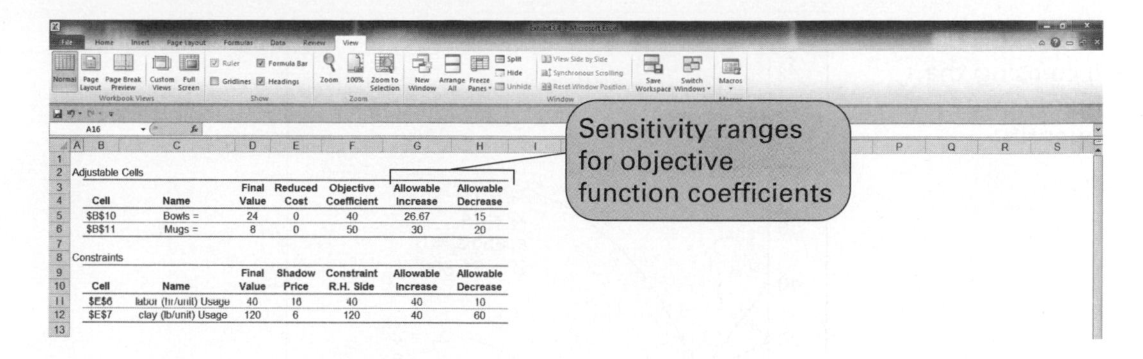

Adjustable Cells

Cell	Name	Final Value	Reduced Cost	Objective Coefficient	Allowable Increase	Allowable Decrease
$B$10	Bowls =	24	0	40	26.67	15
$B$11	Mugs =	8	0	50	30	20

Constraints

Cell	Name	Final Value	Shadow Price	Constraint R.H. Side	Allowable Increase	Allowable Decrease
$E$6	labor (hr/unit) Usage	40	16	40	40	10
$E$7	clay (lb/unit) Usage	120	6	120	40	60

Notice that the sensitivity ranges for the objective function coefficients (40 and 50) are not provided as an upper and lower limit but instead show an allowable increase and an allowable decrease. For example, for the coefficient $40 for bowls (B10), the allowable increase of 26.67 results in an upper limit of 66.67, whereas the allowable decrease of 15 results in a lower limit of 25.

The sensitivity ranges for the objective function coefficients for our example are presented in QM for Windows in Exhibit 3.12. Notice that this output provides the upper and lower limits of the sensitivity ranges for both variables, x_1 and x_2.

EXHIBIT 3.12

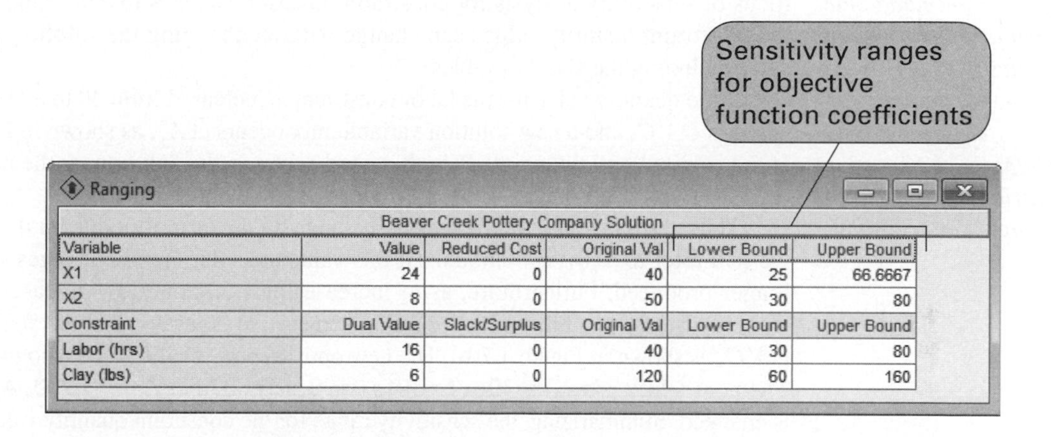

Beaver Creek Pottery Company Solution

Variable	Value	Reduced Cost	Original Val	Lower Bound	Upper Bound
X1	24	0	40	25	66.6667
X2	8	0	50	30	80
Constraint	**Dual Value**	**Slack/Surplus**	**Original Val**	**Lower Bound**	**Upper Bound**
Labor (hrs)	16	0	40	30	80
Clay (lbs)	6	0	120	60	160

Changes in Constraint Quantity Values

The second type of sensitivity analysis we will discuss is the sensitivity ranges for the constraint quantity values—that is, the values to the right of the inequality signs in the constraints. For our Beaver Creek Pottery Company model,

$$\text{maximize } Z = \$40x_1 + 50x_2$$

subject to

$$x_1 + 2x_2 + s_1 = 40 \text{ (labor, hr.)}$$
$$4x_1 + 3x_2 + s_2 = 120 \text{ (clay, lb.)}$$
$$x_1, x_2 \geq 0$$

the constraint quantity values are 40 and 120.

Consider a change in which the manager of the pottery company can increase the labor hours from 40 to 60. The effect of this change in the model is graphically displayed in Figure 3.6.

Increasing the available labor hours from 40 to 60 causes the feasible solution space to change. It was originally *OABC*, and now it is *OA'B'C*. *B'* is the new optimal solution, instead of *B*. However, the important consideration in this type of sensitivity analysis is that the solution *mix* (or variables that do not have zero values), including slack variables, did not change, even

FIGURE 3.6
Increasing the labor constraint quantity

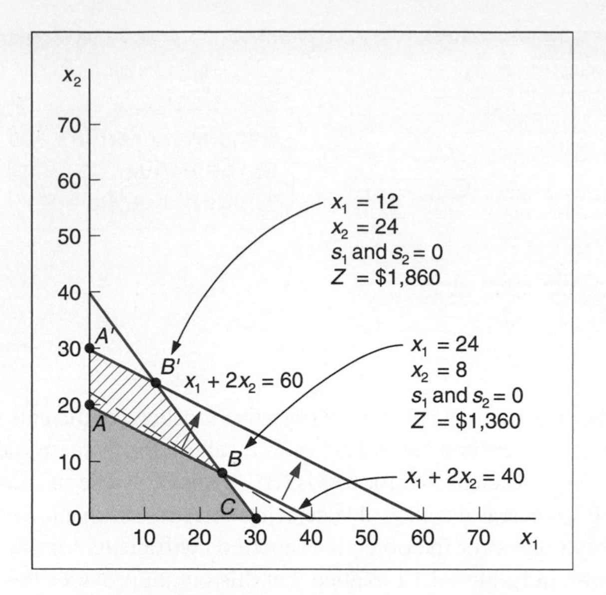

The sensitivity range for a right-hand-side value is the range of values over which the quantity values can change without changing the solution variable mix, including slack variables.

though the values of x_1 and x_2 did change (from $x_1 = 24, x_2 = 8$ to $x_1 = 12, x_2 = 24$). The focus of sensitivity analysis for constraint quantity values is to determine the range over which the constraint quantity values can change without changing the solution variable mix, specifically including the slack variables.

If the quantity value for the labor constraint is increased from 40 to 80 hours, the new solution space is $OA'C$, and a new solution variable mix occurs at A', as shown in Figure 3.7(a). Whereas at the original optimal point, *B*, both x_1 and x_2 are in the solution, at the new optimal point, A', only x_2 is produced (i.e., $x_1 = 0, x_2 = 40, s_1 = 0, s_2 = 0$).

Thus, the upper limit of the sensitivity range for the quantity value for the first constraint, which we will refer to as q_1, is 80 hours. At this value the solution mix changes such that bowls are no longer produced. Furthermore, as q_1 increases past 80 hours, s_1 increases (i.e., slack hours are created). Similarly, if the value for q_1 is decreased to 30 hours, the new feasible solution space is $OA'C$, as shown in Figure 3.7(b). The new optimal point is at *C*, where no mugs (x_2) are produced. The new solution is $x_1 = 30, x_2 = 0, s_1 = 0, s_2 = 0$, and $Z = \$1{,}200$. Again, the variable mix is changed. Summarizing, the sensitivity range for the constraint quantity value for labor hours is

$$30 \leq q_1 \leq 80 \text{ hr.}$$

FIGURE 3.7
Determining the sensitivity range for labor quantity

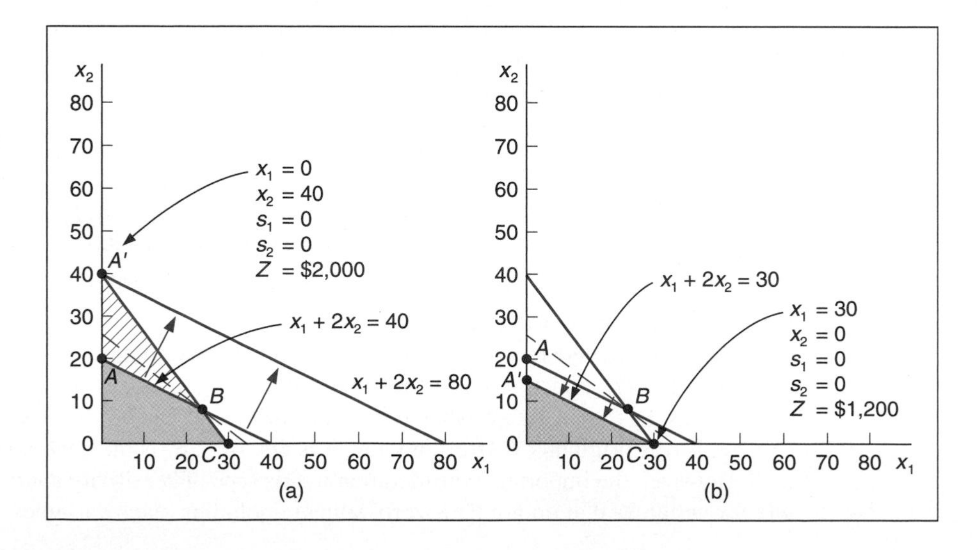

The sensitivity range for clay can be determined graphically in a similar manner. If the quantity value for the clay constraint, $4x_1 + 3x_2 \leq 120$, is increased from 120 to 160, shown in Figure 3.8(a), then a new solution space, OAC', results, with a new optimal point, C'. Alternatively, if the quantity value is decreased from 120 to 60, as shown in Figure 3.8(b), the new solution space is OAC', and the new optimal point is $A(x_1 = 0, x_2 = 20, s_1 = 0, s_2 = 0, Z = \$800)$.

Summarizing, the sensitivity ranges for q_1 and q_2 are

$$30 \leq q_1 \leq 80 \text{ hr.}$$
$$60 \leq q_2 \leq 160 \text{ lb.}$$

FIGURE 3.8
Determining the sensitivity range for clay quantity

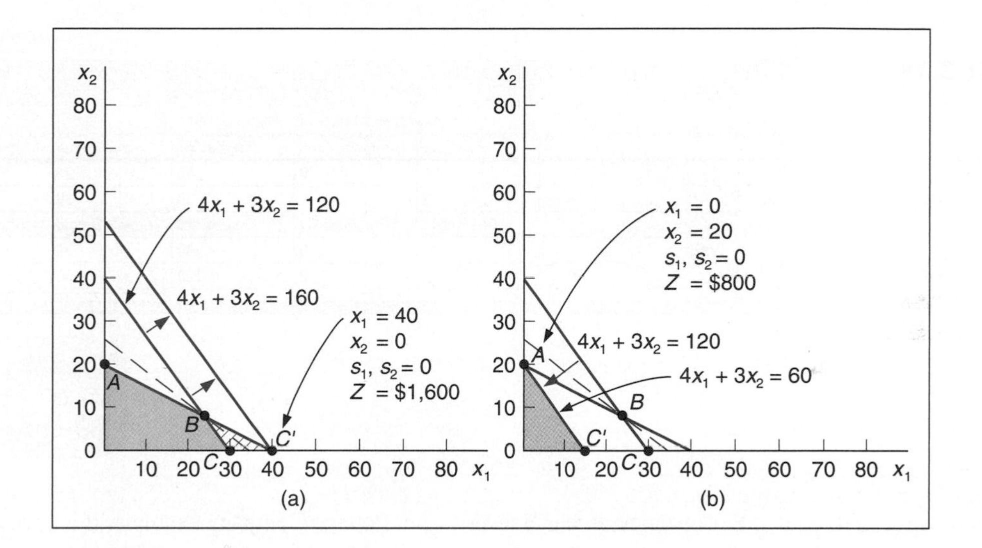

As was the case with the sensitivity ranges for the objective function coefficient, these sensitivity ranges are valid for only one q_i value. All other q_i values are assumed to be held constant. However, simultaneous changes can occur, as long as they do not change the variable mix.

These ranges for constraint quantity values provide useful information for the manager, especially regarding production scheduling and planning. If resources are reduced at the pottery company, then at some point one of the products will no longer be produced, and the support facilities and apparatus for that production will not be needed, or extra hours of resources will be created that are not needed. A similar result, albeit a better one, will occur if resources are increased because profit will be more than with a reduction in resources.

Constraint Quantity Value Ranges with Excel and QM for Windows

Previously we showed how to generate the sensitivity report for Excel, resulting in Exhibit 3.11. This report is shown again in Exhibit 3.13, with the sensitivity ranges for the constraint quantity values indicated. As mentioned previously, the ranges are expressed in terms of an allowable increase and decrease instead of upper and lower limits.

The sensitivity ranges for the constraint quantity values with QM for Windows can be observed in Exhibit 3.14.

Other forms of sensitivity analysis include changing constraint parameter values, adding new constraints, and adding new variables.

Other Forms of Sensitivity Analysis

Excel and QM for Windows provide sensitivity analysis ranges for objective function coefficients and constraint quantity values as part of the standard solution output. However, there are other forms of sensitivity analysis, including changing individual constraint parameters, adding new constraints, and adding new variables.

EXHIBIT 3.13

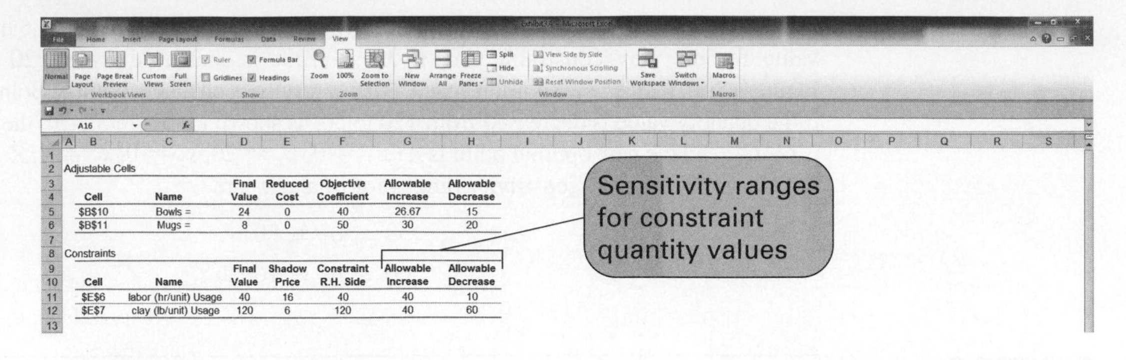

Adjustable Cells

Cell	Name	Final Value	Reduced Cost	Objective Coefficient	Allowable Increase	Allowable Decrease
$B$10	Bowls =	24	0	40	26.67	15
$B$11	Mugs =	8	0	50	30	20

Constraints

Cell	Name	Final Value	Shadow Price	Constraint R.H. Side	Allowable Increase	Allowable Decrease
$E$6	labor (hr/unit) Usage	40	16	40	40	10
$E$7	clay (lb/unit) Usage	120	6	120	40	60

EXHIBIT 3.14

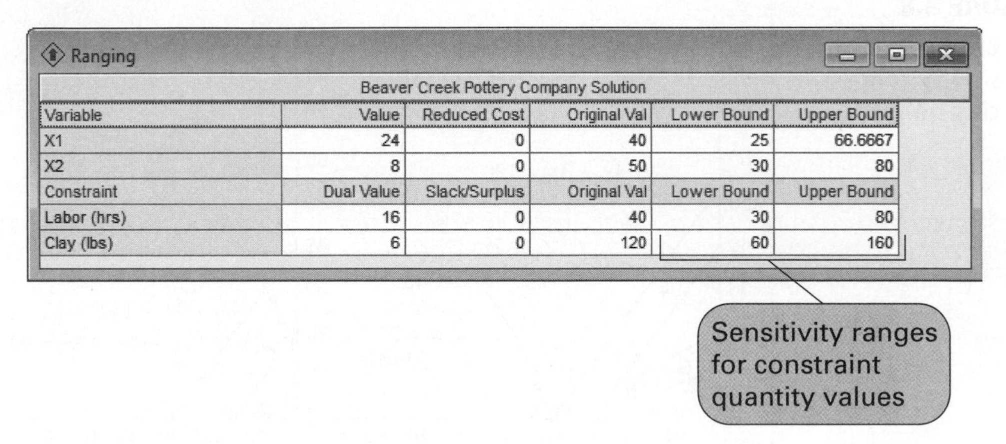

Ranging

Beaver Creek Pottery Company Solution

Variable	Value	Reduced Cost	Original Val	Lower Bound	Upper Bound
X1	24	0	40	25	66.6667
X2	8	0	50	30	80
Constraint	Dual Value	Slack/Surplus	Original Val	Lower Bound	Upper Bound
Labor (hrs)	16	0	40	30	80
Clay (lbs)	6	0	120	60	160

For instance, in our Beaver Creek Pottery Company example, if a new, less-experienced artisan were hired to make pottery, it might take this individual 1.33 hours to produce a bowl instead of 1 hour. Thus, the labor constraint would change from $x_1 + 2x_2 \leq 40$ to $1.33x_1 + 2x_2 \leq 40$. This change is illustrated in Figure 3.9.

Note that a change in the coefficient for x_1 in the labor constraint rotates the constraint line, changing the solution space from *OABC* to *OAC*. It also results in a new optimal solution point, *C*, where $x_1 = 30, x_2 = 0$, and $Z = \$1{,}200$. Then 1.33 hours would be the logical upper limit for this constraint coefficient. However, as we pointed out, this type of sensitivity analysis for

FIGURE 3.9
Changing the x_1 coefficient in the labor constraint

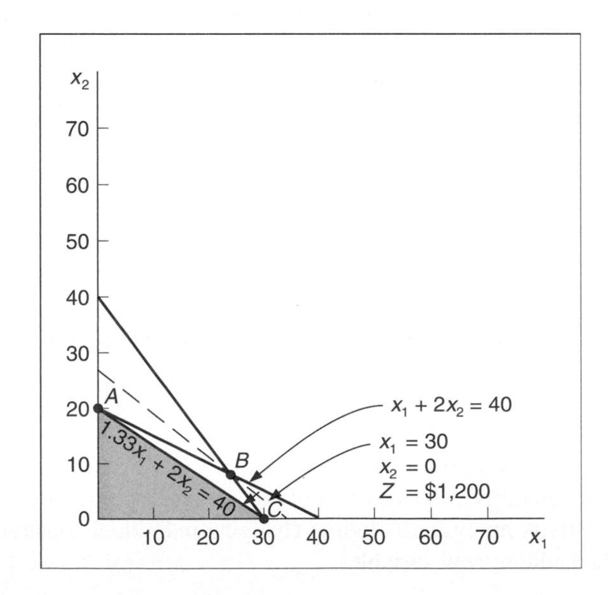

constraint variable coefficients is not typically provided in the standard linear programming computer output. As a result, the most logical way to ascertain the effect of this type of change is simply to run the computer program with different values.

Other types of sensitivity analyses are to add a new constraint to the model or to add a new variable. For example, suppose the Beaver Creek Pottery Company added a third constraint for packaging its pottery, as follows:

$$0.20x_1 + 0.10x_2 \leq 5 \text{ hr.}$$

This would require the model to be solved again with the new constraint. This new constraint does, in fact, change the optimal solution, as shown in the Excel spreadsheet in Exhibit 3.15. This spreadsheet requires a new row (8) added to the original spreadsheet (first shown in Exhibit 3.1) with our new constraint parameter values, and a new constraint, $E8 \leq G8$, added to Solver. In the original model, the solution (shown in Exhibit 3.4) was 24 bowls and 8 mugs, with a profit of $1,360. With the new constraint for packaging added, the solution is now 20 bowls and 10 mugs, with a profit of $1,300.

EXHIBIT 3.15

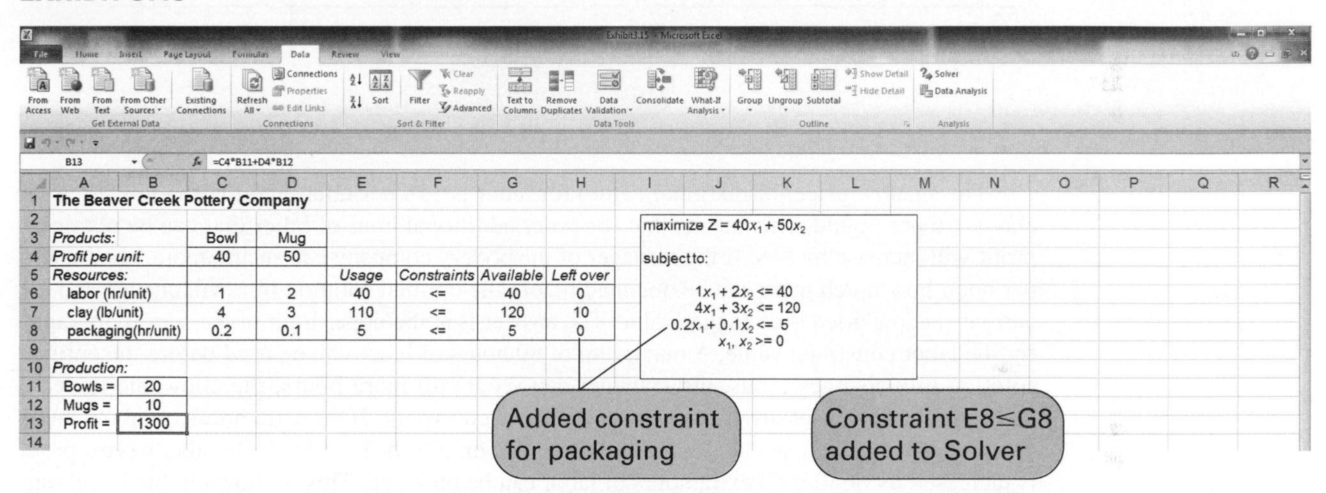

	A	B	C	D	E	F	G	H
1	The Beaver Creek Pottery Company							
2								
3	Products:		Bowl	Mug				
4	Profit per unit:		40	50				
5	Resources:				Usage	Constraints	Available	Left over
6	labor (hr/unit)		1	2	40	<=	40	0
7	clay (lb/unit)		4	3	110	<=	120	10
8	packaging(hr/unit)		0.2	0.1	5	<=	5	0
9								
10	Production:							
11	Bowls =	20						
12	Mugs =	10						
13	Profit =	1300						

If a new variable is added to the model, this would also require that the model be solved again to determine the effect of the change. For example, suppose the pottery company were contemplating producing a third product, cups, requiring 1.2 hours of labor and 2 pounds. of clay. It can secure no additional resources, and the profit for a cup is estimated to be $30. This change is reflected in the following model reformulation:

$$\begin{aligned}
&\text{maximize } Z = \$40x_1 + 50x_2 + 30x_3 \\
&\text{subject to} \\
&x_1 + 2x_2 + 1.2x_3 \leq 40 \ \text{(labor, hr.)} \\
&4x_1 + 3x_2 + 2x_3 \leq 120 \ \text{(clay, lb.)} \\
&x_1, x_2, x_3 \geq 0
\end{aligned}$$

Solving this new formulation with the computer will show that this prospective change will have no effect on the original solution—that is, the model is not sensitive to this change. The estimated profit from cups was not enough to offset the production of bowls and mugs, and the solution remained the same.

Shadow Prices

We briefly discussed dual values (also called *shadow prices*) earlier in this chapter, in our discussion of QM for Windows. You will recall that a dual value was defined as the marginal value of one additional unit of resource. We need to mention shadow prices again at this point in our discussion of sensitivity analysis because decisions are often made regarding resources by considering the marginal value of resources in conjunction with their sensitivity ranges.

Consider again the Excel sensitivity report for the Beaver Creek Pottery Company example shown in Exhibit 3.16.

EXHIBIT 3.16

Adjustable Cells

Cell	Name	Final Value	Reduced Cost	Objective Coefficient	Allowable Increase	Allowable Decrease
\$B\$10	Bowls =	24	0	40	26.67	15
\$B\$11	Mugs =	8	0	50	30	20

Constraints

Cell	Name	Final Value	Shadow Price	Constraint R.H. Side	Allowable Increase	Allowable Decrease
\$E\$6	labor (hr/unit) Usage	40	16	40	40	10
\$E\$7	clay (lb/unit) Usage	120	6	120	40	60

Shadow prices (dual values)

The shadow price (or marginal value) for labor is \$16 per hour, and the shadow price for clay is \$6 per pound. This means that for every additional hour of labor that can be obtained, profit will increase by \$16. If the manager of the pottery company can secure more labor at \$16 per hour, how much more can be obtained before the optimal solution mix will change and the current shadow price is no longer valid? The answer is at the upper limit of the sensitivity range for the labor constraint value. A maximum of 80 hours of labor can be used before the optimal solution mix changes. Thus, the manager can secure 40 more hours, the allowable increase shown in the Excel sensitivity output for the labor constraint. If 40 extra hours of labor can be obtained, what is its total value? The answer is (\$16/hr.)(40 hr.) $=$ \$640. In other words, profit is increased by \$640 if 40 extra hours of labor can be obtained. This is shown in the Excel output in Exhibit 3.17, where increasing the labor constraint from 40 to 80 hours has increased profit from \$1,360 to \$2,000, or by \$640.

EXHIBIT 3.17

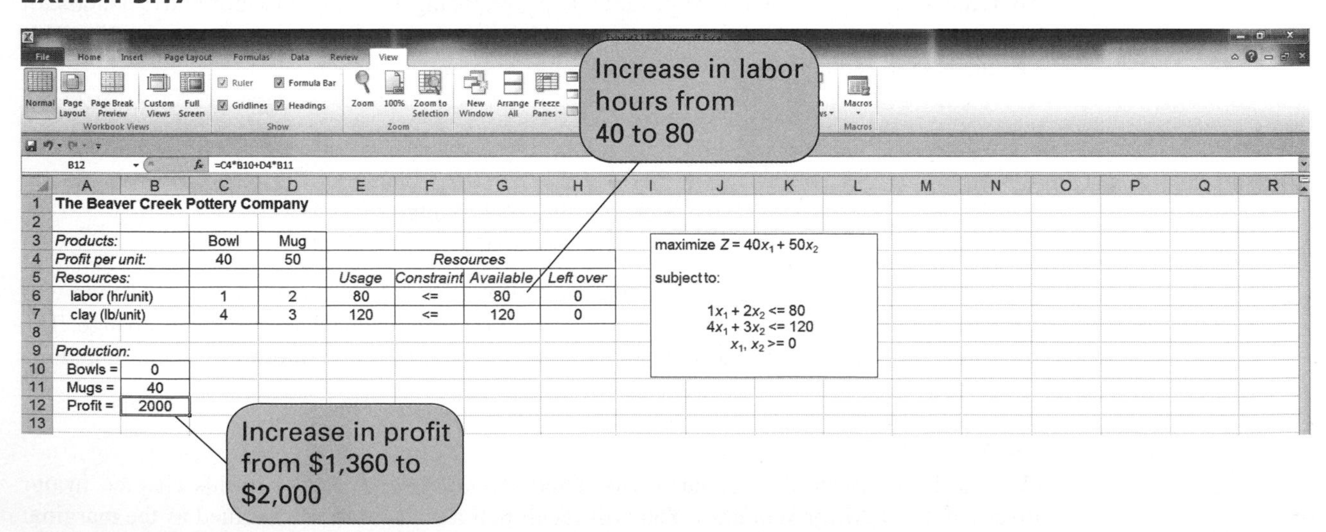

Looking back to Figure 3.7(a), this is the solution point at A′. Increasing the labor hours to more than 80 hours (the upper limit of the sensitivity range) will not result in an additional increase in profit or a new solution; it will result only in slack hours of labor.

Alternatively, what would be the effect on profit if one of the Native American artisans were sick one day during the week, and the available labor hours decreased from 40 to 32? Profit in this case would decrease by $16 per hour, or a total amount of $128. Thus, total profit would fall from $1,360 to $1,232.

Similarly, if the pottery company could obtain only 100 pounds of clay instead of its normal weekly allotment of 120 pounds, what would be the effect on profit? Profit would decrease by $6 per pound for 20 pounds, or a total of $120. This would result in an overall reduction in profit from $1,360 to $1,240.

The sensitivity range for a constraint quantity value is also the range over which the shadow price is valid.

Thus, another piece of information that is provided by the sensitivity ranges for the constraint quantity values is the range over which the shadow price remains valid. When q_i increases past the upper limit of the sensitivity range or decreases below the lower limit, the shadow price will change, specifically because slack (or surplus) will be created. Therefore, the sensitivity range for the constraint quantity value is the range over which the shadow price is valid.

The shadow price of $16 for 1 hour of labor is not necessarily what the manager would *pay* for an hour of labor. This depends on how the objective function is defined. In the Beaver Creek Pottery Company example, we are assuming that all the resources available, 40 hours of labor and 120 pounds of clay, are already paid for. Even if the company does not use all the resources, it still must pay for them. These are *sunk* costs. Therefore, the individual profit values in the objective function for each product are not affected by how much of a resource is actually used; the total profit is independent of the resources used. In this case, the shadow prices are the maximum amounts the manager would pay for additional units of resource. The manager would pay up to $16 for 1 extra hour of labor and up to $6 for an extra pound of clay.

Alternatively, if each hour of labor and each pound of clay were purchased separately, and thus were not sunk costs, profit would be a function of the cost of the resources. In this case, the shadow price would be the additional amount, over and above the original cost of the resource, that would be paid for one more unit of the resource.

Summary

This chapter has focused primarily on the computer solution of linear programming problems. This required us first to show how a linear programming model is put into standard form, with the addition of slack variables or the subtraction of surplus variables. Computer solution also enabled us to consider the topic of sensitivity analysis, the analysis of the effect of model parameter changes on the solution of a linear programming model. In the next chapter, we will provide some examples of more complex linear programming model formulations than the simple ones we have described so far.

Example Problem Solution

This example demonstrates the transformation of a linear programming model into standard form, sensitivity analysis, computer solution, and shadow prices.

Problem Statement

The Xecko Tool Company is considering bidding on a job for two airplane wing parts. Each wing part must be processed through three manufacturing stages—stamping, drilling, and finishing—for which the company has limited available hours. The linear programming model to determine how many of part 1 (x_1) and part 2 (x_2) the company should produce in order to maximize its profit is as follows:

$$\text{maximize } Z = \$650x_1 + 910x_2$$
subject to
$$4x_1 + 7.5x_2 \le 105 \text{ (stamping, hr.)}$$
$$6.2x_1 + 4.9x_2 \le 90 \text{ (drilling, hr.)}$$
$$9.1x_1 + 4.1x_2 \le 110 \text{ (finishing, hr.)}$$
$$x_1, x_2 \ge 0$$

A. Solve the model graphically.
B. Indicate how much slack resource is available at the optimal solution point.
C. Determine the sensitivity ranges for the profit for wing part 1 and the stamping hours available.
D. Solve this model by using Excel.

Solution

A.

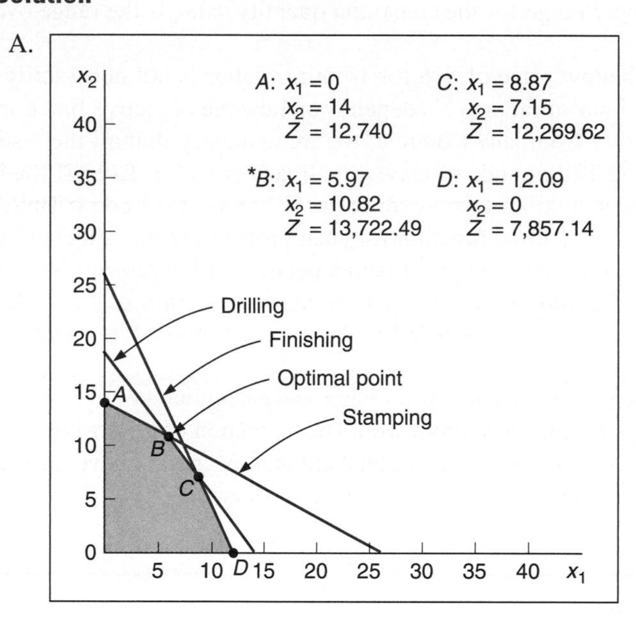

B. The slack at point B, where $x_1 = 5.97$ and $x_2 = 10.82$, is computed as follows:

$$4(5.97) + 7.5(10.82) + s_1 = 105 \text{ (stamping, hr.)}$$
$$s_1 = 0 \text{ hr.}$$
$$6.2(5.97) + 4.9(10.82) + s_2 = 90 \text{ (drilling, hr.)}$$
$$s_2 = 0 \text{ hr.}$$
$$9.1(5.97) + 4.1(10.82) + s_3 = 110 \text{ (finishing, hr.)}$$
$$s_3 = 11.35 \text{ hr.}$$

C. The sensitivity range for the profit for part 1 is determined by observing the graph of the model and computing how much the slope of the objective function must increase to make the optimal point move from B to C. This is the upper limit of the range and is determined by computing the value of c_1 that will make the slope of the objective function equal with the slope of the constraint line for drilling, $6.2x_1 + 4.9x_2 = 90$:

$$\frac{-c_1}{910} = \frac{-6.2}{4.9}$$

$$c_1 = 1{,}151.43$$

The lower limit is determined by computing the value of c_1 that will equate the slope of the objective function with the slope of the constraint line for stamping, $4x_1 + 7.5x_2 = 105$:

$$\frac{-c_1}{910} = \frac{-4}{7.5}$$

$$c_1 = 485.33$$

Summarizing,

$$485.33 \leq c_1 \leq 1{,}151.43$$

The upper limit of the range for stamping hours is determined by first computing the value for q_1 that would move the solution point from B to where the drilling constraint intersects with the x_2 axis, where $x_1 = 0$ and $x_2 = 18.37$:

$$4(0) + 7.5(8.37) = q_1$$

$$q_1 = 137.76$$

The lower limit of the sensitivity range occurs where the optimal point B moves to C, where $x_1 = 8.87$ and $x_2 = 7.15$:

$$4(8.87) + 7.5(7.15) = q_1$$

$$q_1 = 89.10$$

Summarizing, $89.10 \leq q_1 \leq 137.76$.

D. The Excel spreadsheet solution to this example problem is as follows:

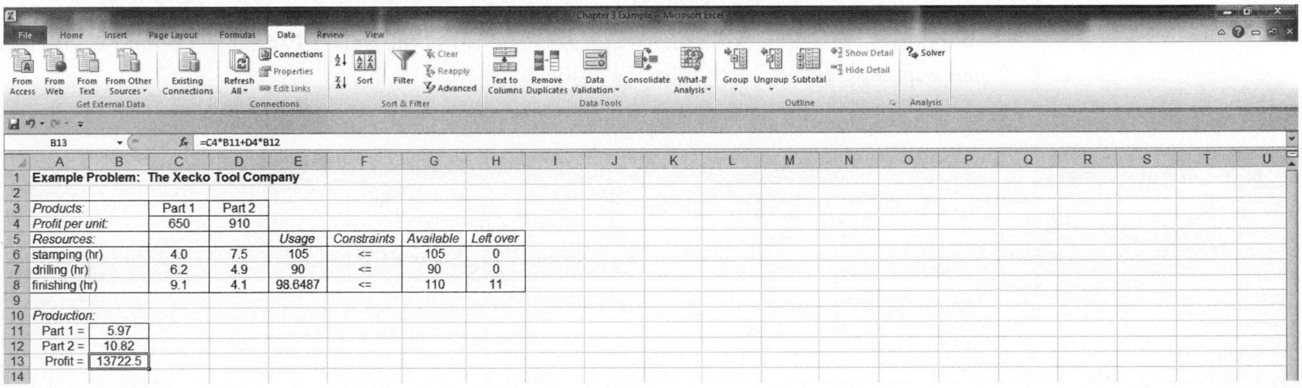

B13 =C4*B11+D4*B12

	A/B	C	D	E	F	G	H
1	Example Problem: The Xecko Tool Company						
2							
3	Products:	Part 1	Part 2				
4	Profit per unit:	650	910				
5	Resources:			Usage	Constraints	Available	Left over
6	stamping (hr)	4.0	7.5	105	<=	105	0
7	drilling (hr)	6.2	4.9	90	<=	90	0
8	finishing (hr)	9.1	4.1	98.6487	<=	110	11
9							
10	Production:						
11	Part 1 = 5.97						
12	Part 2 = 10.82						
13	Profit = 13722.5						
14							

Problems

1. Given the following QM for Windows computer solution of a linear programming model, graph the problem and identify the solution point, including variable values and slack, from the computer output:

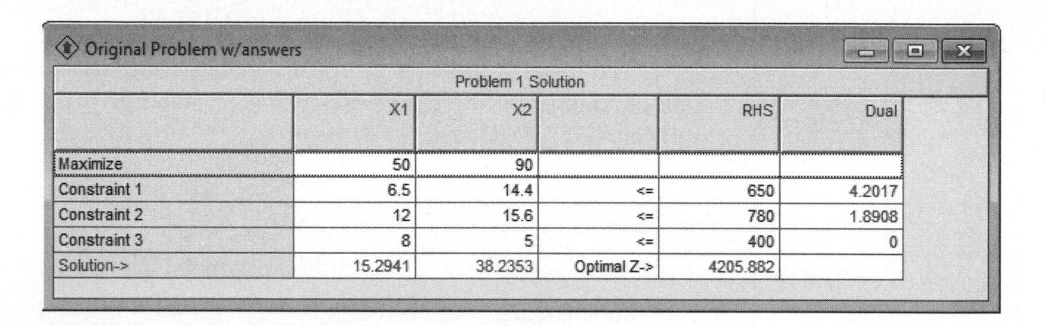

Original Problem w/answers

Problem 1 Solution

	X1	X2		RHS	Dual
Maximize	50	90			
Constraint 1	6.5	14.4	<=	650	4.2017
Constraint 2	12	15.6	<=	780	1.8908
Constraint 3	8	5	<=	400	0
Solution->	15.2941	38.2353	Optimal Z->	4205.882	

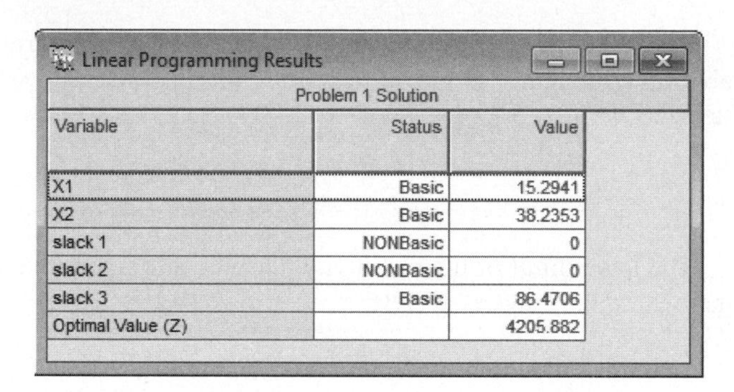

Linear Programming Results

Problem 1 Solution

Variable	Status	Value
X1	Basic	15.2941
X2	Basic	38.2353
slack 1	NONBasic	0
slack 2	NONBasic	0
slack 3	Basic	86.4706
Optimal Value (Z)		4205.882

2. Explain the primary differences between a software package such as QM for Windows and Excel spreadsheets for solving linear programming problems.

3. Given the following Excel spreadsheet for a linear programming model and Solver Parameters window, indicate the formula for cell B13 and fill in the Solver Parameters window with the appropriate information to solve the problem:

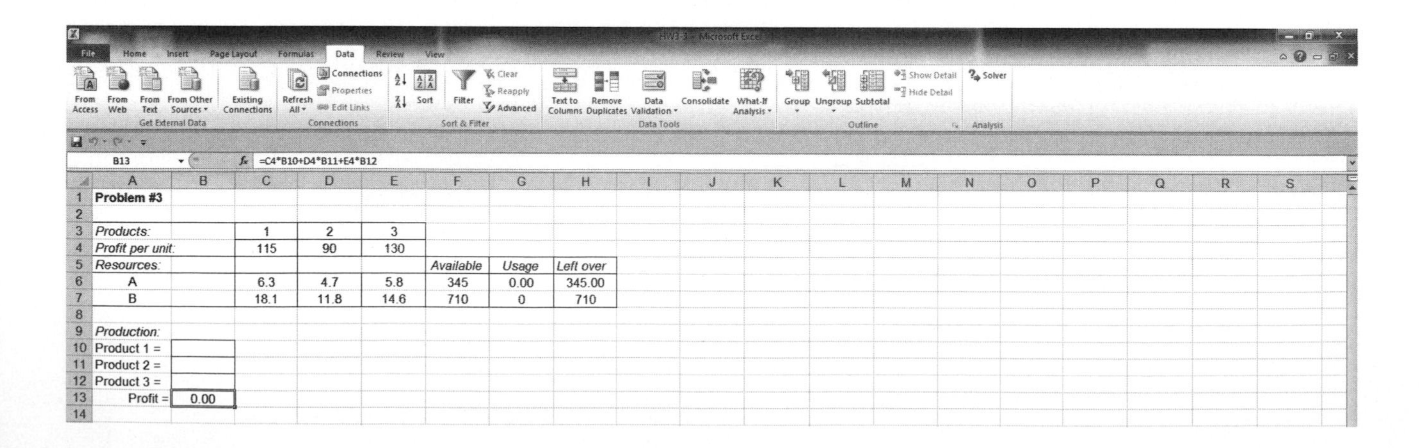

B13 =C4*B10+D4*B11+E4*B12

	A	B	C	D	E	F	G	H
1	Problem #3							
2								
3	Products:		1	2	3			
4	Profit per unit:		115	90	130			
5	Resources:					Available	Usage	Left over
6	A		6.3	4.7	5.8	345	0.00	345.00
7	B		18.1	11.8	14.6	710	0	710
8								
9	Production:							
10	Product 1 =							
11	Product 2 =							
12	Product 3 =							
13	Profit =	0.00						
14								

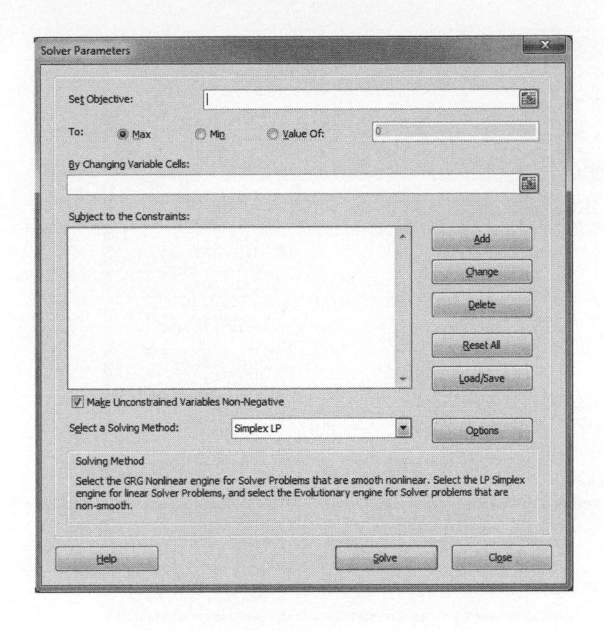

4. Solve the following model from Chapter 2 (Problem 11) using Excel:

$$
\begin{aligned}
\text{maximize } Z &= 3x_1 + 6x_2 \text{ (\$, profit)} \\
\text{subject to} & \\
3x_1 + 2x_2 &\le 18 \text{ (units, resource 1)} \\
x_1 + x_2 &\ge 5 \text{ (units, resource 2)} \\
x_1 &\le 4 \text{ (units, resource 3)} \\
x_1, x_2 &\ge 0
\end{aligned}
$$

5. The following model was solved graphically in Chapter 2 (Problem 22):

$$
\begin{aligned}
\text{maximize } Z &= 5x_1 + 8x_2 \text{ (\$, profit)} \\
\text{subject to} & \\
3x_1 + 5x_2 &\le 50 \text{ (units, resource 1)} \\
2x_1 + 4x_2 &\le 40 \text{ (units, resource 2)} \\
x_1 &\le 8 \text{ (units, resource 3)} \\
x_2 &\le 10 \text{ (units, resource 4)} \\
x_1, x_2 &\ge 0
\end{aligned}
$$

Given the following Excel spreadsheet for this model, indicate the formulas in cells F6, F7, F8, F9, G6, G7, G8, G9, and B14, and fill in the Solver Parameters window with the necessary information to solve the model. Solve the model using Excel.

	A	B	C	D	E	F	G
1	Homework #3-5						
2							
3	Items:		1	2			
4	Profit per item:		5	8			
5	Constraints				Available	Usage	Left over
6	1		3	5	50	0	50
7	2		2	4	40	0	40
8	3		1	0	8	0	8
9	4		0	1	10	0	10
10							
11	Production:						
12	1=						
13	2=						
14	Z=						
15							

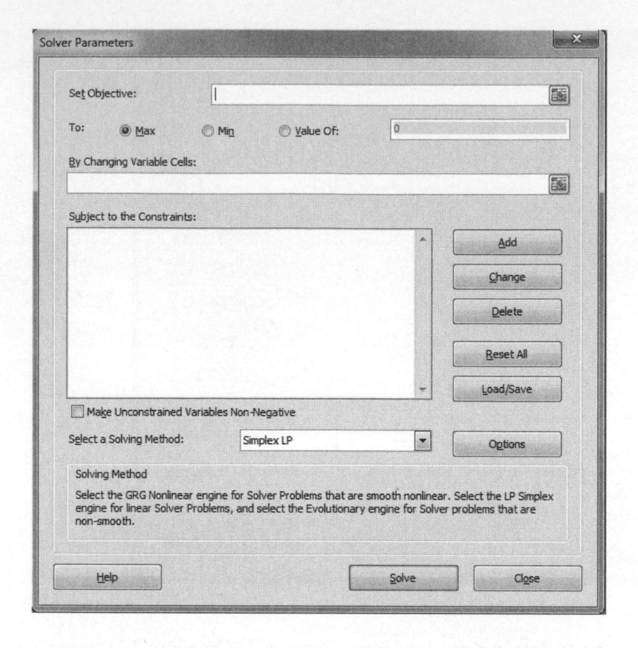

6. Given the following graph of a linear programming model with a single constraint and the objective function maximize $Z = 30x_1 + 50x_2$, determine the optimal solution point:

   Determine the values by which c_1 and c_2 must decrease or increase in order to change the current solution point to the other extreme point.

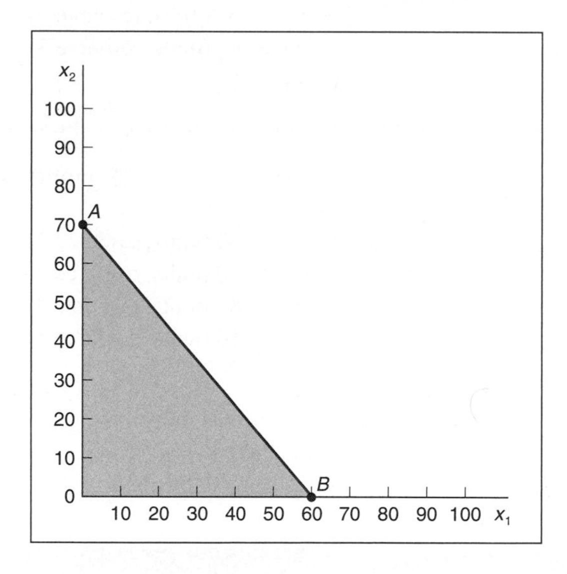

7. Southern Sporting Goods Company makes basketballs and footballs. Each product is produced from two resources—rubber and leather. The resource requirements for each product and the total resources available are as follows:

	Resource Requirements per Unit	
Product	**Rubber (lb.)**	**Leather (ft.^{2})**
Basketball	3	4
Football	2	5
Total resources available	500 lb.	800 ft.^{2}

Each basketball produced results in a profit of $12, and each football earns $16 in profit.

a. Formulate a linear programming model to determine the number of basketballs and footballs to produce in order to maximize profit.
b. Transform this model into standard form.

8. Solve the model formulated in Problem 7 for Southern Sporting Goods Company graphically.
   a. Identify the amount of unused resources (i.e., slack) at each of the graphical extreme points.
   b. What would be the effect on the optimal solution if the profit for a basketball changed from $12 to $13? What would be the effect if the profit for a football changed from $16 to $15?
   c. What would be the effect on the optimal solution if 500 additional pounds of rubber could be obtained? What would be the effect if 500 additional square feet of leather could be obtained?

9. For the linear programming model for Southern Sporting Goods Company, formulated in Problem 7 and solved graphically in Problem 8:
   a. Determine the sensitivity ranges for the objective function coefficients and constraint quantity values, using graphical analysis.
   b. Verify the sensitivity ranges determined in (a) by using the computer.
   c. Using the computer, determine the shadow prices for the resources and explain their meaning.

10. A company produces two products, A and B, which have profits of $9 and $7, respectively. Each unit of product must be processed on two assembly lines, where the required production times are as follows:

	Hours/Unit	
Product	Line 1	Line 2
A	12	4
B	4	8
Total hours	60	40

   a. Formulate a linear programming model to determine the optimal product mix that will maximize profit.
   b. Transform this model into standard form.

11. Solve Problem 10 graphically.
   a. Identify the amount of unused resources (i.e., slack) at each of the graphical extreme points.
   b. What would be the effect on the optimal solution if the production time on line 1 was reduced to 40 hours?
   c. What would be the effect on the optimal solution if the profit for product B was increased from $7 to $15? to $20?

12. For the linear programming model formulated in Problem 10 and solved graphically in Problem 11:
   a. Determine the sensitivity ranges for the objective function coefficients, using graphical analysis.
   b. Verify the sensitivity ranges determined in (a) by using the computer.
   c. Using the computer, determine the shadow prices for additional hours of production time on line 1 and line 2 and indicate whether the company would prefer additional line 1 or line 2 hours.

13. Irwin Textile Mills produces two types of cotton cloth—denim and corduroy. Corduroy is a heavier grade of cotton cloth and, as such, requires 7.5 pounds of raw cotton per yard, whereas denim requires 5 pounds of raw cotton per yard. A yard of corduroy requires 3.2 hours of processing time; a yard of denim requires 3.0 hours. Although the demand for denim is practically unlimited, the maximum demand for corduroy is 510 yards per month. The manufacturer has 6,500 pounds of cotton and 3,000 hours of processing time available

each month. The manufacturer makes a profit of $2.25 per yard of denim and $3.10 per yard of corduroy. The manufacturer wants to know how many yards of each type of cloth to produce to maximize profit.

a. Formulate a linear programming model for this problem.
b. Transform this model into standard form.

14. Solve the model formulated in Problem 13 for Irwin Textile Mills graphically.
    a. How much extra cotton and processing time are left over at the optimal solution? Is the demand for corduroy met?
    b. What is the effect on the optimal solution if the profit per yard of denim is increased from $2.25 to $3.00? What is the effect if the profit per yard of corduroy is increased from $3.10 to $4.00?
    c. What would be the effect on the optimal solution if Irwin Mills could obtain only 6,000 pounds of cotton per month?

15. Solve the linear programming model formulated in Problem 13 for Irwin Mills by using the computer.
    a. If Irwin Mills can obtain additional cotton or processing time, but not both, which should it select? How much? Explain your answer.
    b. Identify the sensitivity ranges for the objective function coefficients and for the constraint quantity values. Then explain the sensitivity range for the demand for corduroy.

16. United Aluminum Company of Cincinnati produces three grades (high, medium, and low) of aluminum at two mills. Each mill has a different production capacity (in tons per day) for each grade, as follows:

	Mill	
Aluminum Grade	**1**	**2**
High	6	2
Medium	2	2
Low	4	10

The company has contracted with a manufacturing firm to supply at least 12 tons of high-grade aluminum, 8 tons of medium-grade aluminum, and 5 tons of low-grade aluminum. It costs United $6,000 per day to operate mill 1 and $7,000 per day to operate mill 2. The company wants to know the number of days to operate each mill in order to meet the contract at the minimum cost.

Formulate a linear programming model for this problem.

17. Solve the linear programming model formulated in Problem 16 for United Aluminum Company graphically.
    a. How much extra (i.e., surplus) high-, medium-, and low-grade aluminum does the company produce at the optimal solution?
    b. What would be the effect on the optimal solution if the cost of operating mill 1 increased from $6,000 to $7,500 per day?
    c. What would be the effect on the optimal solution if the company could supply only 10 tons of high-grade aluminum?

18. Solve the linear programming model formulated in Problem 16 for United Aluminum Company by using the computer.
    a. Identify and explain the shadow prices for each of the aluminum grade contract requirements.
    b. Identify the sensitivity ranges for the objective function coefficients and the constraint quantity values.
    c. Would the solution values change if the contract requirements for high-grade aluminum were increased from 12 tons to 20 tons? If yes, what would the new solution values be?

19. The Bradley family owns 410 acres of farmland in North Carolina on which they grow corn and tobacco. Each acre of corn costs $105 to plant, cultivate, and harvest; each acre of tobacco costs $210. The Bradleys have a budget of $52,500 for next year. The government limits the number of acres of tobacco that can be planted to 100. The profit from each acre of corn is $300; the profit from each acre of tobacco is $520. The Bradleys want to know how many acres of each crop to plant in order to maximize their profit.

    Formulate a linear programming model for this problem.

20. Solve the linear programming model formulated in Problem 19 for the Bradley family farm graphically.
    a. How many acres of farmland will not be cultivated at the optimal solution? Do the Bradleys use the entire 100-acre tobacco allotment?
    b. What would the profit for corn have to be for the Bradleys to plant only corn?
    c. If the Bradleys can obtain an additional 100 acres of land, will the number of acres of corn and tobacco they plan to grow change?
    d. If the Bradleys decide not to cultivate a 50-acre section as part of a crop recovery program, how will it affect their crop plans?

21. Solve the linear programming model formulated in Problem 19 for the Bradley farm by using the computer.
    a. The Bradleys have an opportunity to lease some extra land from a neighbor. The neighbor is offering the land to them for $110 per acre. Should the Bradleys lease the land at that price? What is the maximum price the Bradleys should pay their neighbor for the land, and how much land should they lease at that price?
    b. The Bradleys are considering taking out a loan to increase their budget. For each dollar they borrow, how much additional profit would they make? If they borrowed an additional $1,000, would the number of acres of corn and tobacco they plant change?

22. The manager of a Burger Doodle franchise wants to determine how many sausage biscuits and ham biscuits to prepare each morning for breakfast customers. The two types of biscuits require the following resources:

Biscuit	Labor (hr.)	Sausage (lb.)	Ham (lb.)	Flour (lb.)
Sausage	0.010	0.10	—	0.04
Ham	0.024	—	0.15	0.04

    The franchise has 6 hours of labor available each morning. The manager has a contract with a local grocer for 30 pounds of sausage and 30 pounds of ham each morning. The manager also purchases 16 pounds of flour. The profit for a sausage biscuit is $0.60; the profit for a ham biscuit is $0.50. The manager wants to know the number of each type of biscuit to prepare each morning in order to maximize profit.

    Formulate a linear programming model for this problem.

23. Solve the linear programming model formulated in Problem 22 for the Burger Doodle restaurant graphically.
    a. How much extra sausage and ham is left over at the optimal solution point? Is there any idle labor time?
    b. What would the solution be if the profit for a ham biscuit were increased from $0.50 to $0.60?
    c. What would be the effect on the optimal solution if the manager could obtain 2 more pounds of flour?

24. Solve the linear programming model developed in Problem 22 for the Burger Doodle restaurant by using the computer.
    a. Identify and explain the shadow prices for each of the resource constraints.
    b. Which of the resources constraints profit the most?
    c. Identify the sensitivity ranges for the profit of a sausage biscuit and the amount of sausage available. Explain these sensitivity ranges.

25. Rucklehouse Public Relations has been contracted to do a survey following an election primary in New Hampshire. The firm must assign interviewers to conduct the survey by telephone or in person. One person can conduct 80 telephone interviews or 40 personal interviews in a single day. The following criteria have been established by the firm to ensure a representative survey:
    - At least 3,000 interviews must be conducted.
    - At least 1,000 interviews must be by telephone.
    - At least 800 interviews must be personal.

    An interviewer conducts only one type of interview each day. The cost is $50 per day for a telephone interviewer and $70 per day for a personal interviewer. The firm wants to know the minimum number of interviewers to hire in order to minimize the total cost of the survey.

    Formulate a linear programming model for this problem.

26. Solve the linear programming model formulated in Problem 25 for Rucklehouse Public Relations graphically.
    a. Determine the sensitivity ranges for the daily cost of a telephone interviewer and the number of personal interviews required.
    b. Does the firm conduct any more telephone and personal interviews than are required, and if so, how many more?
    c. What would be the effect on the optimal solution if the firm was required by the client to increase the number of personal interviews conducted from 800 to a total of 1,200?

27. Solve the linear programming model formulated in Problem 25 for Rucklehouse Public Relations by using the computer.
    a. If the firm could reduce the minimum interview requirement for either telephone or personal interviews, which should the firm select? How much would a reduction of one interview in the requirement you selected reduce total cost? Solve the model again, using the computer, with the reduction of this one interview in the constraint requirement to verify your answer.
    b. Identify the sensitivity ranges for the cost of a personal interview and the number of total interviews required.

28. The Bluegrass Distillery produces custom-blended whiskey. A particular blend consists of rye and bourbon whiskey. The company has received an order for a minimum of 400 gallons of the custom blend. The customer specified that the order must contain at least 40% rye and not more than 250 gallons of bourbon. The customer also specified that the blend should be mixed in the ratio of two parts rye to one part bourbon. The distillery can produce 500 gallons per week, regardless of the blend. The production manager wants to complete the order in 1 week. The blend is sold for $5 per gallon.

    The distillery company's cost per gallon is $2 for rye and $1 for bourbon. The company wants to determine the blend mix that will meet customer requirements and maximize profits.

    Formulate a linear programming model for this problem.

29. Solve the linear programming model formulated in Problem 28 for the Bluegrass Distillery graphically.
    a. Indicate the slack and surplus available at the optimal solution point and explain their meanings.
    b. What increase in the objective function coefficients in this model would change the optimal solution point? Explain your answer.

30. Solve the linear programming model formulated in Problen 28 for the Bluegrass Distillery by using the computer.
    a. Identify the sensitivity ranges for the objective function oefficients and explain what the upper and lower limits are.
    b. How much would it be worth to the distillery to obtain additional production capacity?
    c. If the customer decided to change the blend requirement for s custom-made whiskey to a mix of three parts rye to one part bourbon, how would this chane the optimal solution?

31. Prissy Pauline's Barbecue Restaurant in Greensboro makes two priary food items fresh each morning: pulled pork barbecue and beef brisket barbecue. These tw items are sold as part of different menu items, including sandwich and barbecue plates, with a riety of different sides. The maximum amount of barbecue Prissy Pauline's has ever sold in a ay is 250 pounds, and the least amount is 120 pounds. It generally sells more pork barbecue tn beef barbecue—as much as twice as much pork as beef barbecue, and as little as 20% mor ork than beef. The restaurant's meat supplier delivers 150 pounds of pork and 110 pound of beef each day. Regardless of how it is served on the menu, the restaurant makes about $8.65 er pound for pork barbecue and $10.95 per pound for beef barbecue. The restaurant wants t now how many pounds of each type of barbecue to cook each day in order to maximize profit.
    a. Formulate a linear programming model for this problem.
    b. Solve this model graphically.

32. Solve the linear programming model formulated in Problem 31 for Prissy Paulin's Barbecue Restaurant using the computer.
    a. If the restaurant could get additional pork or beef from the meat supplier, whi hould it get? Why?
    b. If the restaurant orders 10 more pounds of beef from its supplier, how much would i crease its profit? Is this something it should do?
    c. The restaurant believes that if it reduces the price of its pork barbecue menu items su hat the profit from pork barbecue is $7.50 per pound, it could sell two and a half times mor rk than beef barbecue. Is this something the restaurant should do?

33. Xara Stores in the United States stocks a particular type of designer denim jeans that is ma factured in China and Brazil and imported to the Xara distribution center in the United States. orders 500 pairs of jeans each month from its two suppliers. The Chinese supplier charges Xara $11 per pair of jeans, and the Brazilian supplier charges $16 per pair (and then Xara marks them up almost 1,000%). Although the jeans from China are less expensive, they also have more defects than those from Brazil. Based on past data, Xara estimates that 7% of the Chinese jeans will be defective compared to only 2% from Brazil, and Xara does not want to import any more than 5% defective items. However, Xara does not want to rely only on a single supplier, so it wants to order at least 20% from each supplier every month.

    Formulate a linear programming model for this problem.

34. Solve the linear programming model formulated in Problem 33 for Xara Stores graphically and by using the computer.
    a. If the Chinese supplier were able to reduce its percentage of defective pairs of jeans from 7% to 5%, what would be the effect on the solution?
    b. If Xara Stores decided to minimize its defective items while budgeting $7,000 for purchasing the jeans, what would be the effect on the solution?

35. Alexis Harrington received an inheritance of $95,000, and she is considering two speculative investments—the purchase of land and the purchase of cattle. Each investment would be for 1 year. Under the present (normal) economic conditions, each dollar invested in land will return the principal plus 20% of the principal; each dollar invested in cattle will return the principal plus 30%. However, both investments are relatively risky. If economic conditions were to deteriorate, there is an 18% chance she would lose everything she invested in land and a 30%

chance she would lose everything she invested in cattle. Alexis does not want to lose more than $20,000 (on average). She wants to know how much to invest in each alternative to maximize the cash value of the investments at the end of 1 year.

Formulate a linear programming model for this problem.

36. Solve the linear programming model formulated in Problem 35 for Alexis Harrington graphically.
    a. How much would the return for cattle have to increase in order for Alexis to invest only in cattle?
    b. Should all of Alexis's inheritance be invested according to the optimal solution?
    c. How much "profit" would the optimal solution earn Alexis over and above her investment?

37. Solve the linear programming model formulated in Problem 35 for Alexis Harrington by using the computer.
    a. If Alexis decided to invest some of her own savings along with the money from her inheritance, what return would she realize for each dollar of her own money that she invested? How much of her own savings could she invest before this return would change?
    b. If the risk of losing the investment in land increased to 30%, how would this change the optimal investment mix?

38. Transform the following linear programming model into standard form and solve by using the computer:

$$\begin{aligned}
\text{maximize } Z &= 140x_1 + 205x_2 + 190x_3 \\
\text{subject to} & \\
10x_1 + 15x_2 + 8x_3 &\leq 610 \\
\frac{x_1}{x_2} &\leq 3 \\
x_1 &\geq 0.4\,(x_1 + x_2 + x_3) \\
x_2 &\geq x_3 \\
x_1, x_2, x_3 &\geq 0
\end{aligned}$$

39. Chemco Corporation produces a chemical mixture for a specific customer in 1,000-pound batches. The mixture contains three ingredients—zinc, mercury, and potassium. The mixture must conform to formula specifications that are supplied by the customer. The company wants to know the amount of each ingredient it needs to put in the mixture that will meet all the requirements of the mix and minimize total cost.

    The customer has supplied the following formula specifications for each batch of mixture:

    - The mixture must contain at least 200 pounds of mercury.
    - The mixture must contain at least 300 pounds of zinc.
    - The mixture must contain at least 100 pounds of potassium.
    - The ratio of potassium to the other two ingredients cannot exceed 1 to 4.

    The cost per pound of mercury is $400; the cost per pound of zinc, $180; and the cost per pound of potassium, $90.
    a. Formulate a linear programming model for this problem.
    b. Solve the model formulated in (a) by using the computer.

40. The following linear programming model formulation is used for the production of four different products, with two different manufacturing processes and two different material requirements:

$$\begin{aligned}
\text{maximize } Z &= \$50x_1 + 58x_2 + 46x_3 + 62x_4 \\
\text{subject to} & \\
4x_1 + 3.5x_2 + 4.6x_3 + 3.9x_4 &\leq 600 \text{ hr. (process 1)} \\
2.1x_1 + 2.6x_2 + 3.5x_3 + 1.9x_4 &\leq 500 \text{ hr. (process 2)} \\
15x_1 + 23x_2 + 18x_3 + 25x_4 &\leq 3{,}600 \text{ lb. (material A)}
\end{aligned}$$

$$8x_1 + 12.6x_2 + 9.7x_3 + 10.5x_4 \leq 1{,}700 \text{ lb. (material B)}$$
$$\frac{x_1 + x_2}{x_1 + x_2 + x_3 + x_4} \geq .60$$
$$x_1, x_2, x_3, x_4 \geq 0$$

a. Solve this problem by using the computer.
b. Identify the sensitivity ranges for the objective function coefficients and the constraint quantity values.
c. Which is the most valuable resource to the firm?
d. One of the four products is not produced in the optimal solution. How much would the profit for this product have to be for it to be produced?

41. Island Publishing Company publishes two types of magazines on a monthly basis: a restaurant and entertainment guide and a real estate guide. The company distributes the magazines free to businesses, hotels, and stores on Hilton Head Island in South Carolina. The company's profits come exclusively from the paid advertising in the magazines. Each of the restaurant and entertainment guides distributed generates $0.50 per magazine in advertising revenue, whereas the real estate guide generates $0.75 per magazine. The real estate magazine is a more sophisticated publication that includes color photos, and accordingly it costs $0.25 per magazine to print, compared with only $0.17 for the restaurant and entertainment guide. The publishing company has a printing budget of $4,000 per month. There is enough rack space to distribute at most 18,000 magazines each month. In order to entice businesses to place advertisements, Island Publishing promises to distribute at least 8,000 copies of each magazine. The company wants to determine the number of copies of each magazine it should print each month in order to maximize advertising revenue.

    Formulate a linear programming model for this problem.

42. Solve the linear programming model formulation in Problem 41 for Island Publishing Company graphically.
    a. Determine the sensitivity range for the advertising revenue generated by the real estate guide.
    b. Does the company spend all of its printing budget? If not, how much slack is left over?
    c. What would be the effect on the optimal solution if the local real estate agents insisted that 12,000 copies of the real estate guide be distributed instead of the current 8,000 copies, or they would withdraw their advertising?

43. Solve the linear programming model formulated in Problem 41 for Island Publishing Company by using the computer.
    a. How much would it be worth to Island Publishing Company to obtain enough additional rack space to distribute 18,500 copies instead of the current 18,000 copies? 20,000 copies?
    b. How much would it be worth to Island Publishing to reduce the requirement to distribute the entertainment guide from 8,000 to 7,000 copies?

44. Mega-Mart, a discount store chain, is to build a new store in Rock Springs. The parcel of land the company has purchased is large enough to accommodate a store with 140,000 square feet of floor space. Based on marketing and demographic surveys of the area and historical data from its other stores, Mega-Mart estimates its annual profit per square foot for each of the store's departments to be as shown in the table on the following page.

    Each department must have at least 15,000 square feet of floor space, and no department can have more than 20% of the total retail floor space. Men's, women's, and children's clothing plus housewares keep all their stock on the retail floor; however, toys, electronics, and auto supplies keep some items (such as bicycles, televisions, and tires) in inventory. Thus, 10% of the total retail floor space devoted to these three departments must be set aside outside the retail area for stocking inventory. Mega-Mart wants to know the floor space that should be devoted to each department in order to maximize profit.

    a. Formulate a linear programming model for this problem.
    b. Solve this model by using the computer.

Department	Profit per ft.^{2}
Men's clothing	$4.25
Women's clothing	5.10
Children's clothing	4.50
Toys	5.20
Housewares	4.10
Electronics	4.90
Auto supplies	3.80

45. a. In Problem 44, Mega-Mart is considering purchasing a parcel of land adjacent to the current site on which it plans to build its store. The cost of the parcel is $190,000, and it would enable Mega-Mart to increase the size of its store to 160,000 square feet. Discuss whether Mega-Mart should purchase the land and increase the planned size of the store.
    b. Suppose that the profit per square foot will decline in all departments by 20% if the store size increases to 160,000 square feet. (If the stock does not turn over as fast, increasing inventory costs will reduce profit.) How might this affect Mega-Mart's decision in (a)?

46. The Food Max grocery store sells three brands of milk in half-gallon cartons—its own brand, a local dairy brand, and a national brand. The profit from its own brand is $0.97 per carton, the profit from the local dairy brand is $0.83 per carton, and the profit from the national brand is $0.69 per carton. The total refrigerated shelf space allotted to half-gallon cartons of milk is 36 square feet per week. A half-gallon carton takes up 16 square inches of shelf space. The store manager knows that each week Food Max always sells more of the national brand than of the local dairy brand and its own brand combined and at least three times as much of the national brand as its own brand. In addition, the local dairy can supply only 10 dozen cartons per week. The store manager wants to know how many half-gallon cartons of each brand to stock each week in order to maximize profit.
    a. Formulate a linear programming model for this problem.
    b. Solve this model by using the computer.

47. a. If Food Max in Problem 46 could increase its shelf space for half-gallon cartons of milk, how much would profit increase per carton?
    b. If Food Max could get the local dairy to increase the amount of milk it could supply each week, would it increase profit?
    c. Food Max is considering discounting its own brand in order to increase sales. If it were to do so, it would decrease the profit margin for its own brand to $0.86 per carton, but it would cut the demand for the national brand relative to its own brand in half. Discuss whether the store should implement the price discount.

48. John Hoke owns Hoke's Spokes, a bicycle shop. Most of John's bicycle sales are customer orders; however, he also stocks bicycles for walk-in customers. He stocks three types of bicycles—road-racing, cross-country, and mountain. A road-racing bike costs $1,200, a cross-country bike costs $1,700, and a mountain bike costs $900. He sells road-racing bikes for $1,800, cross-country bikes for $2,100, and mountain bikes for $1,200. He has $12,000 available this month to purchase bikes. Each bike must be assembled; a road-racing bike requires 8 hours to assemble, a cross-country bike requires 12 hours, and a mountain bike requires 16 hours. He estimates that he and his employees have 120 hours available to assemble bikes. He has enough space in his store to order 20 bikes this month. Based on past sales, John wants to stock at least twice as many mountain bikes as the other two combined because mountain bikes sell better.

    Formulate a linear programming model for this problem.

49. Solve the linear programming model formulated in Problem 48 for Hoke's Spokes by using the computer.
    a. Should John Hoke try to increase his budget for purchasing bikes, increase space to stock bikes, or increase labor hours to assemble bikes? Why?
    b. If John hired an additional worker for 30 hours at $10 per hour, how much additional profit would he make, if any?
    c. If John purchased a cheaper cross-country bike for $1,200 and sold it for $1,900, would this affect the original solution?

50. Metro Food Services Company delivers fresh sandwiches each morning to vending machines throughout the city. The company makes three kinds of sandwiches—ham and cheese, bologna, and chicken salad. A ham and cheese sandwich requires a worker 0.45 minutes to assemble, a bologna sandwich requires 0.41 minutes, and a chicken salad sandwich requires 0.50 minutes to make. The company has 960 available minutes each night for sandwich assembly. Vending machine capacity is available for 2,000 sandwiches each day. The profit for a ham and cheese sandwich is $0.35, the profit for a bologna sandwich is $0.42, and the profit for a chicken salad sandwich is $0.37. The company knows from past sales records that its customers buy as many ham and cheese sandwiches as the other two sandwiches combined, if not more so, but customers need a variety of sandwiches available, so Metro stocks at least 200 of each. Metro management wants to know how many of each sandwich it should stock to maximize profit.

    Formulate a linear programming model for this problem.

51. Solve the linear programming model formulated in Problem 50 for Metro Food Services Company by using the computer.
    a. If Metro Food Services could hire another worker and increase its available assembly time by 480 minutes or increase its vending machine capacity by 100 sandwiches, which should it do? Why? How much additional profit would your decision result in?
    b. What would the effect be on the optimal solution if the requirement that at least 200 sandwiches of each kind be stocked was eliminated? Compare the profit between the optimal solution and this solution. Which solution would you recommend?
    c. What would the effect be on the optimal solution if the profit for a ham and cheese sandwich was increased to $0.40? to $0.45?

52. Mountain Laurel Vineyards produces three kinds of wine—Mountain Blanc, Mountain Red, and Mountain Blush. The company has 17 tons of grapes available to produce wine this season. A cask of Blanc requires 0.21 ton of grapes, a cask of Red requires 0.24 ton, and a cask of Blush requires 0.18 ton. The vineyard has enough storage space in its aging room to store 80 casks of wine.

    The vineyard has 2,500 hours of production capacity, and it requires 12 hours to produce a cask of Blanc, 14.5 hours to produce a cask of Red, and 16 hours to produce a cask of Blush. From past sales the vineyard knows that demand for the Blush will be no more than half of the sales of the other two wines combined. The profit for a cask of Blanc is $7,500, the profit for a cask of Red is $8,200, and the profit for a cask of Blush is $10,500.

    Formulate a linear programming model for this problem.

53. Solve the linear programming model formulated in Problem 52 for Mountain Laurel Vineyards by using the computer.
    a. If the vineyard determined that the profit from Red was $7,600 instead of $8,200, how would that affect the optimal solution?
    b. If the vineyard could secure one additional unit of any of the resources used in the production of wine, which one should it select?
    c. If the vineyard could obtain 0.5 ton more of grapes, 500 more hours of production capacity, or enough storage capacity to store 4 more casks of wine, which should it choose?
    d. All three wines are produced in the optimal solution. How little would the profit for Blanc have to be for it to no longer be produced?

54. Exeter Mines produces iron ore at four different mines; however, the ores extracted at each mine are different in their iron content. Mine 1 produces magnetite ore, which has a 70% iron content; mine 2 produces limonite ore, which has a 60% iron content; mine 3 produces pyrite ore, which has a 50% iron content; and mine 4 produces taconite ore, which has only a 30% iron content. Exeter has three customers that produce steel—Armco, Best, and Corcom. Armco needs 400 tons of pure (100%) iron, Best requires 250 tons of pure iron, and Corcom requires 290 tons. It costs $37 to extract and process 1 ton of magnetite ore at mine 1, $46 to produce 1 ton of limonite ore at mine 2, $50 per ton of pyrite ore at mine 3, and $42 per ton of taconite ore at mine 4. Exeter can extract 350 tons of ore at mine 1; 530 tons at mine 2; 610 tons at mine 3; and 490 tons at mine 4. The company wants to know how much ore to produce at each mine in order to minimize cost and meet its customers' demand for pure (100%) iron.

    Formulate a linear programming model for this problem.

55. Solve the linear programming model formulated in Problem 54 for Exeter Mines by using the computer.
    a. Do any of the mines have slack capacity? If yes, which one(s)?
    b. If Exeter Mines could increase production capacity at any one of its mines, which should it be? Why?
    c. If Exeter decided to increase capacity at the mine identified in (b), how much could it increase capacity before the optimal solution point (i.e., the optimal set of variables) would change?
    d. If Exeter determined that it could increase production capacity at mine 1 from 350 tons to 500 tons, at an increase in production costs to $43 per ton, should it do so?

56. Given the following linear programming model:

$$\begin{aligned}
&\text{minimize } Z = 8.2x_1 + 7.0x_2 + 6.5x_3 + 9.0x_4 \\
&\text{subject to} \\
&6x_1 + 2x_2 + 5x_3 + 7x_4 \geq 820 \\
&\frac{x_1}{x_1 + x_2 + x_3 + x_4} \geq 0.3 \\
&\frac{x_2 + x_3}{x_1 + x_4} \leq 0.2 \\
&x_3 \geq x_1 + x_4 \\
&x_1, x_2, x_3, x_4 \geq 0
\end{aligned}$$

    transform the model into standard form and solve by using the computer.

57. Tracy McCoy has committed to the local PTA to make some items for a bake sale on Saturday. She has decided to make some combination of chocolate cakes, loaves of white bread, custard pies, and sugar cookies. Thursday evening she goes to the store and purchases 20 pounds of flour, 10 pounds of sugar, and 3 dozen eggs, which are the three main ingredients in all the baked goods she is thinking about making. The following table shows how much of each of the main ingredients is required for each baked good:

	Ingredient			
	Flour (cups)	Sugar (cups)	Eggs	Baking Time (min.)
Cake	2.5	2	2	45
Bread	9	0.25	0	35
Pie	1.3	1	5	50
Cookies	2.5	1	2	16

There are 18.5 cups in a 5-pound bag of flour and 12 cups in a 5-pound bag of sugar. Tracy plans to get up and start baking on Friday morning after her kids leave for school and finish before they return after soccer practice (8 hours). She knows that the PTA will sell a chocolate cake for $12, a loaf of bread for $8, a custard pie for $10, and a batch of cookies for $6. Tracy wants to decide how many of each type of baked good she should make in order for the PTA to make the most money possible.

Formulate a linear programming model for this problem.

58. Solve the linear programming model formulated in Problem 57 for Tracy McCoy.
    a. Are any of the ingredients left over?
    b. If Tracy could get more of any ingredient, which should it be? Why?
    c. If Tracy could get 6 more eggs, 20 more cups of flour, or 30 more minutes of oven time, which should she choose? Why?
    d. The solution values for this problem should logically be integers. If the solution values are not integers, discuss how Tracy should decide how many of each item to bake. How do total sales for this integer solution compare with those in the original, non-integer solution?

59. The Wisham family lives on a farm in South Georgia on which it produces a variety of crops and livestock, including pecans. It has 5 acres of pecan trees that yield approximately 1,000 pounds of unshelled pecans per acre each year. The family uses all of its pecan harvest to produce pecan pies, cookies, 1-pound bags of shelled pecans, and 5-pound bags of unshelled pecans, which it sells in town at the local farmers' market. The family sells pies for $5, packages of a dozen cookies for $3, bags of shelled pecans for $7, and bags of unshelled pecans for $16. A shelled pecan is half the weight of an unshelled pecan. It requires 4 ounces of shelled pecans to make a pie, and 6 ounces of shelled pecans to make a dozen cookies. The pies and cookies are baked in the family oven, and there are 120 hours of baking time available. It takes 55 minutes to bake a batch of 4 pies and 15 minutes to bake a batch of 2 dozen cookies. It requires family members 6 minutes to shell the pecans for a pie and package it, 4 minutes to shell the pecans for cookies and to package them, 10 minutes to shell the pecans for a 1 lb. bag of shelled pecans and package them, and 1 minute to package a bag of unshelled pecans; and there are 300 hours available from family members for shelling and packaging. The Wisham family wants to know how many pecan pies, dozens of cookies, and bags of shelled and unshelled pecans to produce in order to maximize its sales revenues.

Formulate a linear programming model for this problem.

60. Solve the linear programming model formulated in Problem 59 for the Wisham family farm using the computer.
    a. Are there any extra (slack) resources available?
    b. If the Wisham family could obtain additional resources, which one would be of most value to them? How much would they be willing to pay for this resource? Why?
    c. If the family could obtain an additional 500 pounds of pecans or 30 hours of oven time, which should they choose?
    d. The family is thinking about buying a bigger oven for $3,000. If they do so, they could make a batch of 5 pies or a batch 3 dozen cookies at one time. Should they buy the oven? Explain your answer.

61. During the winter WeeMow Lawn Service contracts with customers for cutting lawns starting in the spring. WeeMow provides service to residential and commercial customers. The service has identified 50 possible residential customers and 15 potential commercial customers it has contacted. WeeMow services a customer once every 2 weeks during the growing season. A residential lawn on average takes 1.2 hours to cut, and a commercial property requires 5 hours to cut; the service's available work time is 8 hours, 6 days per week. The profit for a residential lawn is $23, and the profit for a commercial property is $61. The service has established a weekly budget of $350 for management, gas, and other materials, plus equipment repair. A residential

lawn averages $12 in management, gas, material, and repair costs, and a commercial property averages $20 in costs. WeeMow wants to know how many residential and commercial jobs it should contract for from among its potential customers in order to maximize profits.

a. Formulate a linear programming model for this problem.
b. Solve this model graphically.

62. Solve the liner programming model for WeeMow Lawn Service in Problem 61 using the computer.
    a. Which resources constrain how many jobs the service can contract for?
    b. If WeeMow could increase its weekly budget by $50 per week, or increase its workday to 9 hours, which would be more profitable?
    c. The solution to this problem should logically be integer, that is, whole jobs; however, the optimal solution is not integer. How would you suggest that WeeMow address this discrepancy? Do you think there is a way to handle this problem with the computer program you are using to solve this problem?

Case Problem

Mossaic Tiles, Ltd.

Gilbert Moss and Angela Pasaic spent several summers during their college years working at archaeological sites in the Southwest. While at those digs, they learned how to make ceramic tiles from local artisans. After college they made use of their college experiences to start a tile manufacturing firm called Mossaic Tiles, Ltd. They opened their plant in New Mexico, where they would have convenient access to a special clay they intend to use to make a clay derivative for their tiles. Their manufacturing operation consists of a few relatively simple but precarious steps, including molding the tiles, baking, and glazing.

Gilbert and Angela plan to produce two basic types of tile for use in home bathrooms, kitchens, sunrooms, and laundry rooms. The two types of tile are a larger, single-colored tile and a smaller, patterned tile. In the manufacturing process, the color or pattern is added before a tile is glazed. Either a single color is sprayed over the top of a baked set of tiles or a stenciled pattern is sprayed on the top of a baked set of tiles.

The tiles are produced in batches of 100. The first step is to pour the clay derivative into specially constructed molds. It takes 18 minutes to mold a batch of 100 larger tiles and 15 minutes to prepare a mold for a batch of 100 smaller tiles. The company has 60 hours available each week for molding. After the tiles are molded, they are baked in a kiln: 0.27 hour for a batch of 100 larger tiles and 0.58 hour for a batch of 100 smaller tiles. The company has 105 hours available each week for baking. After baking, the tiles are either colored or patterned and glazed. This process takes 0.16 hour for a batch of 100 larger tiles and 0.20 hour for a batch of 100 smaller tiles. Forty hours are available each week for the glazing process. Each batch of 100 large tiles requires 32.8 pounds of the clay derivative to produce, whereas each batch of smaller tiles requires 20 pounds. The company has 6,000 pounds of the clay derivative available each week.

Mossaic Tiles earns a profit of $190 for each batch of 100 of the larger tiles and $240 for each batch of 100 smaller patterned tiles. Angela and Gilbert want to know how many batches of each type of tile to produce each week to maximize profit. In addition, they have some questions about resource usage they would like answered.

A. Formulate a linear programming model for Mossaic Tiles, Ltd., and determine the mix of tiles it should manufacture each week.
B. Transform the model into standard form.
C. Solve the linear programming model graphically.
D. Determine the resources left over and not used at the optimal solution point.
E. Determine the sensitivity ranges for the objective function coefficients and constraint quantity values by using the graphical solution of the model.
F. For artistic reasons, Gilbert and Angela prefer to produce the smaller, patterned tiles. They also believe that in the long run, the smaller tiles will be a more successful product. What must the profit be for the smaller tiles in order for the company to produce only the smaller tiles?
G. Solve the linear programming model by using the computer and verify the sensitivity ranges computed in (E).

H. Mossaic believes it may be able to reduce the time required for molding to 16 minutes for a batch of larger tiles and 12 minutes for a batch of smaller tiles. How will this affect the solution?

I. The company that provides Mossaic with clay has indicated that it can deliver an additional 100 pounds each week. Should Mossaic agree to this offer?

J. Mossaic is considering adding capacity to one of its kilns to provide 20 additional glazing hours per week, at a cost of $90,000. Should it make the investment?

K. The kiln for glazing had to be shut down for 3 hours, reducing the available kiln hours from 40 to 37. What effect will this have on the solution?

Case Problem

"The Possibility" Restaurant—Continued

In "The Possibility" Restaurant case problem in Chapter 2, Angela Fox and Zooey Caulfield opened a French restaurant called "The Possibility." Initially, Angela and Zooey could not offer a full, varied menu, so their chef, Pierre, prepared two full-course dinners with beef and fish each evening. In the case problem, Angela and Zooey wanted to develop a linear programming model to help determine the number of beef and fish meals they should prepare each night. Solve Zooey and Angela's linear programming model by using the computer.

A. Angela and Zooey are considering investing in some advertising to increase the maximum number of meals they serve. They estimate that if they spend $30 per day on a newspaper ad, it will increase the maximum number of meals they serve per day from 60 to 70. Should they make the investment?

B. Zooey and Angela are concerned about the reliability of some of their kitchen staff. They estimate that on some evenings they could have a staff reduction of as much as 5 hours. How would this affect their profit level?

C. The final question they would like to explore is raising the price of the fish dinner. Angela believes the price for a fish dinner is a little low and that it could be closer to the price of a beef dinner without affecting customer demand. However, Zooey has noted that Pierre has already made plans based on the number of dinners recommended by the linear programming solution. Angela has suggested a price increase that will increase profit for the fish dinner to $14. Would this be acceptable to Pierre, and how much additional profit would be realized?

Case Problem

Julia's Food Booth

Julia Robertson is a senior at Tech, and she's investigating different ways to finance her final year at school. She is considering leasing a food booth outside the Tech stadium at home football games. Tech sells out every home game, and Julia knows, from attending the games herself, that everyone eats a lot of food. She has to pay $1,000 per game for a booth, and the booths are not very large. Vendors can sell either food or drinks on Tech property, but not both. Only the Tech athletic department concession stands can sell both inside the stadium. She thinks slices of cheese pizza, hot dogs, and barbecue sandwiches are the most popular food items among fans and so these are the items she would sell.

Most food items are sold during the hour before the game starts and during half time; thus it will not be possible for Julia to prepare the food while she is selling it. She must prepare the food ahead of time and then store it in a warming oven. For $600 she can lease a warming oven for the six-game home season. The oven has 16 shelves, and each shelf is 3 feet by 4 feet. She plans to fill the oven with the three food items before the game and then again before half time.

Julia has negotiated with a local pizza delivery company to deliver 14-inch cheese pizzas twice each game—2 hours before the game and right after the opening kickoff. Each pizza will cost her $6 and will include 8 slices. She estimates it will cost her $0.45 for each hot dog and $0.90 for each barbecue sandwich if she makes the barbecue herself the night before. She measured a hot dog and found it takes up about 16 square inches of space, whereas a barbecue sandwich takes up about 25 square inches. She plans to sell a slice of pizza and a hot dog for $1.50 apiece and a barbecue sandwich for $2.25. She has $1,500 in cash available to purchase and prepare the food items for the first home game; for the remaining five games she will purchase her ingredients with money she has made from the previous game.

Julia has talked to some students and vendors who have sold food at previous football games at Tech as well as at other universities. From this she has discovered that she can expect to sell at least as many slices of pizza as hot dogs and barbecue sandwiches combined. She also anticipates that she will probably sell at least twice as many hot dogs as barbecue sandwiches. She believes that she will sell everything she can stock and develop a customer base for the season if she follows these general guidelines for demand.

If Julia clears at least $1,000 in profit for each game after paying all her expenses, she believes it will be worth leasing the booth.

A. Formulate and solve a linear programming model for Julia that will help you advise her if she should lease the booth.
B. If Julia were to borrow some more money from a friend before the first game to purchase more ingredients, could she increase her profit? If so, how much should she borrow and how much additional profit would she make? What factor constrains her from borrowing even more money than this amount (indicated in your answer to the previous question)?
C. When Julia looked at the solution in (A), she realized that it would be physically difficult for her to prepare all the hot dogs and barbecue sandwiches indicated in this solution. She believes she can hire a friend of hers to help her for $100 per game. Based on the results in (A) and (B), is this something you think she could reasonably do and should do?
D. Julia seems to be basing her analysis on the assumption that everything will go as she plans. What are some of the uncertain factors in the model that could go wrong and adversely affect Julia's analysis? Given these uncertainties and the results in (A), (B), and (C), what do you recommend that Julia do?

CHAPTER

4

Linear Programming: Modeling Examples

In Chapters 2 and 3, two basic linear programming models, one for a maximization problem and one for a minimization problem, were used to demonstrate model formulation, graphical solution, computer solution, and sensitivity analysis. Most of these models were very straightforward, consisting of only two decision variables and two constraints. They were necessarily simple models so that the linear programming topics being introduced could be easily understood.

In this chapter, more complex examples of model formulation are presented. These examples have been selected to illustrate some of the more popular application areas of linear programming. They also provide guidelines for model formulation for a variety of problems and computer solutions with Excel and QM for Windows.

You will notice as you go through each example that the model formulation is presented in a systematic format. First, decision variables are identified, then the objective function is formulated, and finally the model constraints are developed. Model formulation can be difficult and complicated, and it is usually beneficial to follow this set of steps in which you identify a specific model component at each step instead of trying to "see" the whole formulation after the first reading.

A Product Mix Example

Quick-Screen is a clothing manufacturing company that specializes in producing commemorative shirts immediately following major sporting events such as the World Series, Super Bowl, and Final Four. The company has been contracted to produce a standard set of shirts for the winning team, either State University or Tech, following a college football bowl game on New Year's Day. The items produced include two sweatshirts, one with silk-screen printing on the front and one with print on both sides, and two T-shirts of the same configuration. The company has to complete all production within 72 hours after the game, at which time a trailer truck will pick up the shirts. The company will work around the clock. The truck has enough capacity to accommodate 1,200 standard-size boxes. A standard-size box holds 12 T-shirts, and a box of 12 sweatshirts is three times the size of a standard box. The company has budgeted $25,000 for the production run. It has 500 dozen blank sweatshirts and T-shirts each in stock, ready for production. This scenario is illustrated in Figure 4.1.

The resource requirements, unit costs, and profit per dozen for each type of shirt are shown in the following table:

	Processing Time (hr.) per Dozen	Cost per Dozen	Profit per Dozen
Sweatshirt—F	0.10	$36	$90
Sweatshirt—B/F	0.25	48	125
T-shirt—F	0.08	25	45
T-shirt—B/F	0.21	35	65

The company wants to know how many dozen (boxes) of each type of shirt to produce in order to maximize profit.

FIGURE 4.1
Quick-Screen Shirts

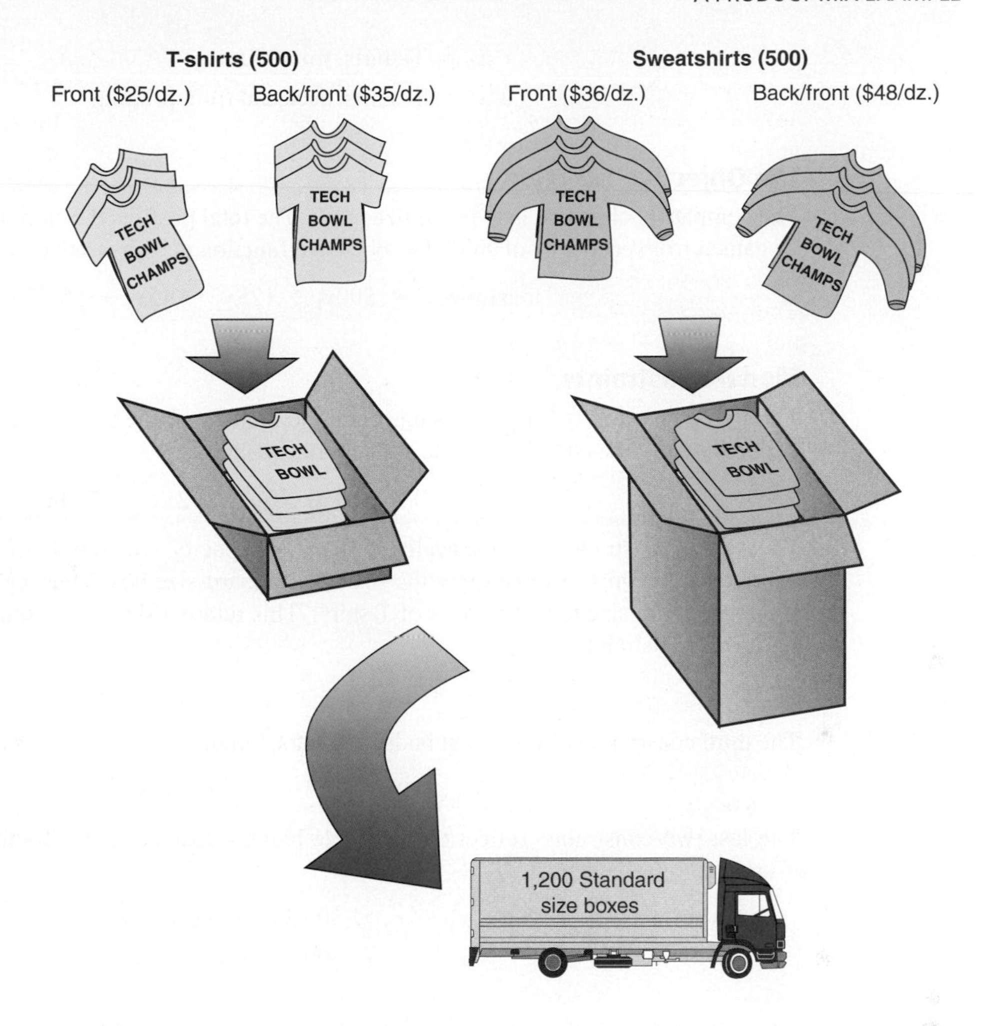

Following is a review of the model formulation steps for this problem:

> ## *Summary of Linear Programming Model Formulation Steps*
>
> ***Step 1: Define the decision variables***
>
> How many (dozens of) T-shirts and sweatshirts of each type to produce
>
> ***Step 2: Define the objective function***
>
> Maximize profit
>
> ***Step 3: Define the constraints***
>
> The resources available, including processing time, blank shirts, budget, and shipping capacity

Decision Variables

This problem contains four decision variables, representing the number of dozens (boxes) of each type of shirt to produce:

$$x_1 = \text{sweatshirts, front printing}$$
$$x_2 = \text{sweatshirts, back and front printing}$$

x_3 = T-shirts, front printing

x_4 = T-shirts, back and front printing

The Objective Function

The company's objective is to maximize profit. The total profit is the sum of the individual profits gained from each type of shirt. The objective function is expressed as

$$\text{maximize } Z = \$90x_1 + 125x_2 + 45x_3 + 65x_4$$

Model Constraints

The first constraint is for processing time. The total available processing time is the 72-hour period between the end of the game and the truck pickup:

$$0.10x_1 + 0.25x_2 + 0.08x_3 + 0.21x_4 \leq 72 \text{ hr}$$

The second constraint is for the available shipping capacity, which is 1,200 standard-size boxes. A box of sweatshirts is three times the size of a standard-size box. Thus, each box of sweatshirts is equivalent in size to three boxes of T-shirts. This relative size differential is expressed in the following constraint:

$$3x_1 + 3x_2 + x_3 + x_4 \leq 1{,}200 \text{ boxes}$$

The third constraint is for the cost budget. The total budget available for production is $25,000:

$$\$36x_1 + 48x_2 + 25x_3 + 35x_4 \leq \$25{,}000$$

The last two constraints reflect the available blank sweatshirts and T-shirts the company has in storage:

$$x_1 + x_2 \leq 500 \text{ dozen sweatshirts}$$
$$x_3 + x_4 \leq 500 \text{ dozen T-shirts}$$

Model Summary

The linear programming model for Quick-Screen is summarized as follows:

This model can be input as shown for computer solution.

$$\begin{aligned} \text{maximize } Z &= 90x_1 + 125x_2 + 45x_3 + 65x_4 \\ \text{subject to} & \\ 0.10x_1 + 0.25x_2 + 0.08x_3 + 0.21x_4 &\leq 72 \\ 3x_1 + 3x_2 + x_3 + x_4 &\leq 1{,}200 \\ 36x_1 + 48x_2 + 25x_3 + 35x_4 &\leq 25{,}000 \\ x_1 + x_2 &\leq 500 \\ x_3 + x_4 &\leq 500 \\ x_1, x_2, x_3, x_4 &\geq 0 \end{aligned}$$

Computer Solution with Excel

The Excel spreadsheet solution for this product mix example is shown in Exhibit 4.1. The decision variables are located in cells **B14:B17**. The profit is computed in cell B18, and the formula for profit, **=B14*D5+B15*E5+B16*F5+B17*G5**, is shown on the formula bar at the top of the spreadsheet. The constraint formulas are embedded in cells H7 through H11, under the column titled "Usage." For example, the constraint formula for processing time in cell H7 is **=D7*B14+E7*B15+F7*B16+G7*B17**. Cells H8 through H11 have similar formulas.

Cells K7 through K11 contain the formulas for the leftover resources, or slack. For example, cell K7 contains the formula **=J7–H7**. These formulas for leftover resources enable us to

EXHIBIT 4.1

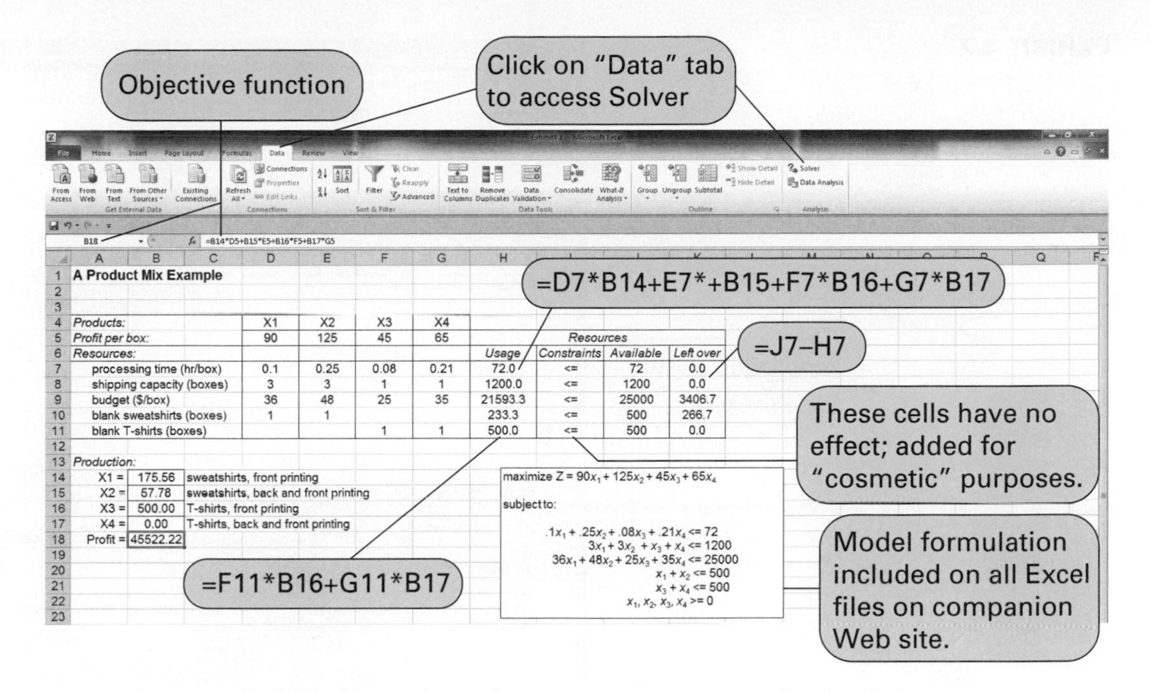

demonstrate a spreadsheet operation that can save you time in developing the spreadsheet model. First, enter the formula for leftover resources, **=J7–H7**, in cell K7, as we have already shown. Next, using the right mouse button, click on "Copy." Then cover cells K8:K11 with the cursor (by holding the left mouse button down). Click the right mouse button again and then click on "Paste." This will automatically insert the correct formulas for leftover resources in cells K8 through K11 so that you do not have to type them all in individually. This copying operation can be used when the variables in the formula are all in the same row or column. The copying operation simply increases the row number for each cell that the formulas are copied into (i.e., J8 and H8, J9 and H9, J10 and H10, and J11 and H11).

Also, note the model formulation in the box in the lower right-hand corner of the spreadsheet in Exhibit 4.1. The model formulations for the remaining linear programming models in this and other chapters are included on the Excel files on the companion website accompanying this text.

The Solver Parameters window for this model is shown in Exhibit 4.2. Notice that we were able to insert all five constraint formulas with one line in the "Subject to the Constraints:" box. We used the constraint **H7:H11 <= J7:J11**, which means that all the constraint usage values computed in cells H7 through H11 are less than the corresponding available resource values computed in cells J7 through J11.

Computer Solution with QM for Windows

The QM for Windows solution for this problem is shown in Exhibit 4.3.

Solution Analysis

The model solution is

$$x_1 = 175.56 \text{ boxes of front-only sweatshirts}$$
$$x_2 = 57.78 \text{ boxes of front and back sweatshirts}$$
$$x_3 = 500 \text{ boxes of front-only T-shirts}$$
$$Z = \$45{,}522.22 \text{ profit}$$

The manager of Quick-Screen might have to round off the solution to send whole boxes—for example, 175 boxes of front-only sweatshirts, 57 of front and back sweatshirts, and 500 of front-only

EXHIBIT 4.2

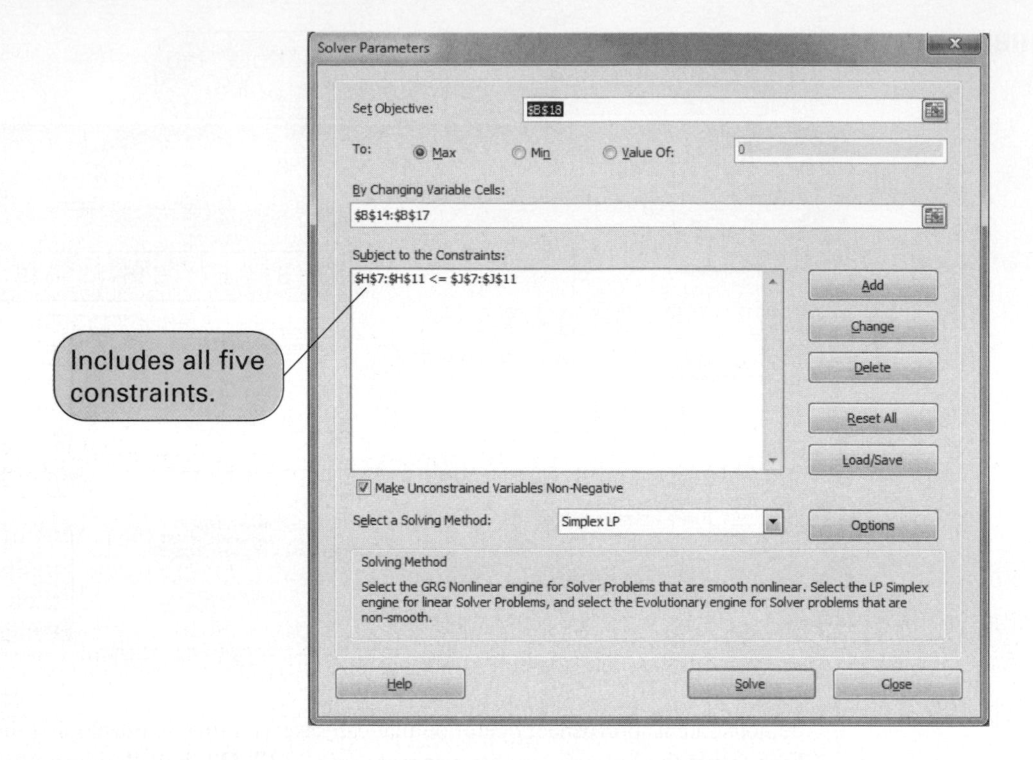

EXHIBIT 4.3

Original Problem w/answers

Product Mix Example Solution

	X1	X2	X3	X4		RHS	Dual
Maximize	90	125	45	65			
Processing time (hrs)	.1	.25	.08	.21	<=	72	233.3333
Shipping capacity (boxes)	3	3	1	1	<=	1,200	22.2222
Budget ($)	36	48	25	35	<=	25,000	0
Blank sweats (dozens)	1	1	0	0	<=	500	0
Blank T's (dozens)	0	0	1	1	<=	500	4.1111
Solution->	175.5556	57.7778	500	0	Optimal Z->	45,522.22	

T-shirts. This would result in a profit of $45,375.00, which is only $147.22 less than the optimal profit value of $45,522.22.

We will discuss how to achieve "integer" solutions in greater detail in chapter 5.

Sensitivity Analysis

After formulating and solving this model, Quick-Screen might decide that it needs to produce and ship at least some of each type of shirt. Management could evaluate this possibility by adding four constraints that establish minimum levels of production for each type of shirt, including front and back T-shirts, x_4, none of which are produced in the current solution. The manager might also like to experiment with the constraints to see the effect on the solution of adding resources. For example, looking at the Ranging window for QM for windows in Exhibit 4.4, the dual value for processing time shows profit would increase by $233.33 per hour (up to 98.33 hours, the upper limit

Time Out for George B. Dantzig

After developing the simplex method for solving linear programming problems, George Dantzig needed a good problem to test it on. The problem he selected was the "diet problem" formulated in 1945 by Nobel economist George Stigler. The problem was to determine an adequate nutritional diet at minimum cost (which was an important military and civilian issue during World War II). Formulated as a linear programming model, the diet problem consisted of 77 unknowns and nine equations. It took nine clerks using hand-operated (mechanical) desk calculators 120 person-days to obtain the optimal simplex solution: a diet consisting primarily of wheat flour, cabbage, and dried navy beans that cost $39.69 per year (in 1939 prices). The solution developed by Stigler using his own numerical method was only 24 cents more than the optimal solution.

EXHIBIT 4.4

Ranging

Product Mix Example Solution

Variable	Value	Reduced Cost	Original Val	Lower Bound	Upper Bound
X1	175.5556	0	90	50	101.9231
X2	57.7778	0	125	113.0769	138.2143
X3	500	0	45	40.8889	Infinity
X4	0	10.3333	65	-Infinity	75.3333
Constraint	Dual Value	Slack/Surplus	Original Val	Lower Bound	Upper Bound
Processing time (hrs)	233.3333	0	72	63.3333	98.3333
Shipping capacity (boxes)	22.2222	0	1,200	884	1,460
Budget ($)	0	3,406.666	25,000	21,593.33	Infinity
Blank sweats (dozens)	0	266.6667	500	233.3333	Infinity
Blank T's (dozens)	4.1111	0	500	0	685.7144

of the sensitivity range for this constraint quality value). Although the 72-hour limit seems pretty strict, it might be possible to reduce individual processing times and achieve the same result.

A Diet Example

Breathtakers, a health and fitness center, operates a morning fitness program for senior citizens. The program includes aerobic exercise, either swimming or step exercise, followed by a healthy breakfast in the dining room. Breathtakers' dietitian wants to develop a breakfast that will be high in calories, calcium, protein, and fiber, which are especially important to senior citizens, but low in fat and cholesterol. She also wants to minimize cost. She has selected the following possible food items, whose individual nutrient contributions and cost from which to develop a standard breakfast menu are shown in the following table:

Breakfast Food	Calories	Fat (g)	Cholesterol (mg)	Iron (mg)	Calcium (mg)	Protein (g)	Fiber (g)	Cost
1. Bran cereal (cup)	90	0	0	6	20	3	5	$0.18
2. Dry cereal (cup)	110	2	0	4	48	4	2	0.22
3. Oatmeal (cup)	100	2	0	2	12	5	3	0.10
4. Oat bran (cup)	90	2	0	3	8	6	4	0.12
5. Egg	75	5	270	1	30	7	0	0.10
6. Bacon (slice)	35	3	8	0	0	2	0	0.09
7. Orange	65	0	0	1	52	1	1	0.40
8. Milk—2% (cup)	100	4	12	0	250	9	0	0.16
9. Orange juice (cup)	120	0	0	0	3	1	0	0.50
10. Wheat toast (slice)	65	1	0	1	26	3	3	0.07

The dietitian wants the breakfast to include at least 420 calories, 5 milligrams of iron, 400 milligrams of calcium, 20 grams of protein, and 12 grams of fiber. Furthermore, she wants to limit fat to no more than 20 grams and cholesterol to 30 milligrams.

Decision Variables

This problem includes 10 decision variables, representing the number of standard units of each food item that can be included in each breakfast:

$$
\begin{aligned}
x_1 &= \text{cups of bran cereal} \\
x_2 &= \text{cups of dry cereal} \\
x_3 &= \text{cups of oatmeal} \\
x_4 &= \text{cups of oat bran} \\
x_5 &= \text{eggs} \\
x_6 &= \text{slices of bacon} \\
x_7 &= \text{oranges} \\
x_8 &= \text{cups of milk} \\
x_9 &= \text{cups of orange juice} \\
x_{10} &= \text{slices of wheat toast}
\end{aligned}
$$

The Objective Function

The dietitian's objective is to minimize the cost of a breakfast. The total cost is the sum of the individual costs of each food item:

$$
\begin{aligned}
\text{minimize } Z = {} & \$0.18x_1 + 0.22x_2 + 0.10x_3 + 0.12x_4 + 0.10x_5 + 0.09x_6 + 0.40x_7 \\
& + 0.16x_8 + 0.50x_9 + 0.07x_{10}
\end{aligned}
$$

Model Constraints

The constraints are the requirements for the nutrition items:

$$
\begin{aligned}
90x_1 + 110x_2 + 100x_3 + 90x_4 + 75x_5 + 35x_6 + 65x_7 + 100x_8 + 120x_9 + 65x_{10} &\ge 420 \text{ calories} \\
2x_2 + 2x_3 + 2x_4 + 5x_5 + 3x_6 + 4x_8 + x_{10} &\le 20 \text{ g of fat} \\
270x_5 + 8x_6 + 12x_8 &\le 30 \text{ mg of cholesterol} \\
6x_1 + 4x_2 + 2x_3 + 3x_4 + x_5 + x_7 + x_{10} &\ge 5 \text{ mg of iron} \\
20x_1 + 48x_2 + 12x_3 + 8x_4 + 30x_5 + 52x_7 + 250x_8 + 3x_9 + 26x_{10} &\ge 400 \text{ mg of calcium} \\
3x_1 + 4x_2 + 5x_3 + 6x_4 + 7x_5 + 2x_6 + x_7 + 9x_8 + x_9 + 3x_{10} &\ge 20 \text{ g of protein} \\
5x_1 + 2x_2 + 3x_3 + 4x_4 + x_7 + 3x_{10} &\ge 12 \text{ g of fiber}
\end{aligned}
$$

Model Summary

The linear programming model for this problem can be summarized as follows:

$$
\begin{aligned}
\text{minimize } Z = {} & 0.18x_1 + 0.22x_2 + 0.10x_3 + 0.12x_4 + 0.10x_5 + 0.09x_6 + 0.40x_7 \\
& + 0.16x_8 + 0.50x_9 + 0.07x_{10}
\end{aligned}
$$

subject to

$$
\begin{aligned}
90x_1 + 110x_2 + 100x_3 + 90x_4 + 75x_5 + 35x_6 + 65x_7 + 100x_8 + 120x_9 + 65x_{10} &\ge 420 \\
2x_2 + 2x_3 + 2x_4 + 5x_5 + 3x_6 + 4x_8 + x_{10} &\le 20 \\
270x_5 + 8x_6 + 12x_8 &\le 30 \\
6x_1 + 4x_2 + 2x_3 + 3x_4 + x_5 + x_7 + x_{10} &\ge 5 \\
20x_1 + 48x_2 + 12x_3 + 8x_4 + 30x_5 + 52x_7 + 250x_8 + 3x_9 + 26x_{10} &\ge 400 \\
3x_1 + 4x_2 + 5x_3 + 6x_4 + 7x_5 + 2x_6 + x_7 + 9x_8 + x_9 + 3x_{10} &\ge 20 \\
5x_1 + 2x_2 + 3x_3 + 4x_4 + x_7 + 3x_{10} &\ge 12 \\
x_i &\ge 0
\end{aligned}
$$

Computer Solution with Excel

The solution to our diet example using an Excel spreadsheet is shown in Exhibit 4.5. The decision variables (i.e., "servings" of each menu item) for our problem are contained in cells **C5:C14** inclusive, and the constraint formulas are in cells F15 through L15. Cells F17 through L17 contain the constraint (right-hand-side) values. For example, cell F15 contains the constraint formula for calories, **=SUMPRODUCT(C5:C14,F5:F14)**. Then, when the Solver Parameters window is accessed, the constraint **F15>=F17** will be added in. This constraint formula in cell F15 could have been constructed by multiplying each of the values in column C by each of the corresponding column F cell values. The equivalent constraint formula in cell F15 to our SUMPRODUCT formula is **=C5*F5+C6*F6+C7*F7+C8*F8+C9*F9+C10*F10+C11*F11+C12*F12+C13*F13+C14*F14**.

EXHIBIT 4.5

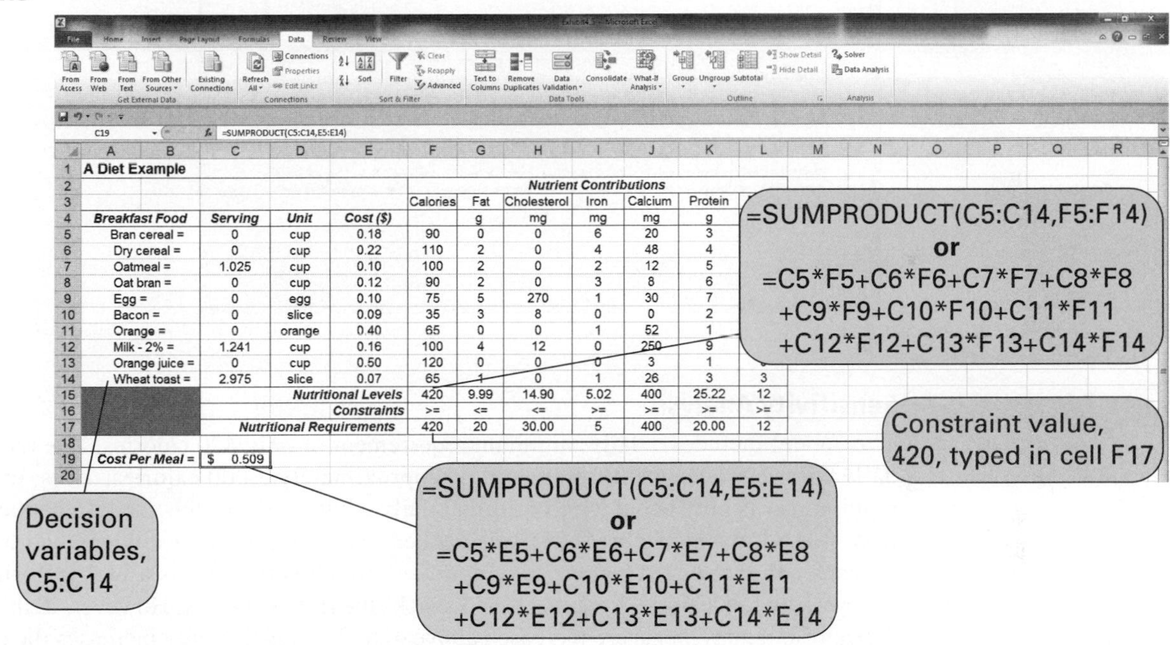

A Diet Example

Breakfast Food	Serving	Unit	Cost ($)	Calories	Fat g	Cholesterol mg	Iron mg	Calcium mg	Protein g	
Bran cereal =	0	cup	0.18	90	0	0	6	20	3	
Dry cereal =	0	cup	0.22	110	2	0	4	48	4	
Oatmeal =	1.025	cup	0.10	100	2	0	2	12	5	
Oat bran =	0	cup	0.12	90	2	0	3	8	6	
Egg =	0	egg	0.10	75	5	270	1	30	7	
Bacon =	0	slice	0.09	35	3	8	0	0	2	
Orange =	0	orange	0.40	65	0	0	1	52	1	
Milk - 2% =	1.241	cup	0.16	100	4	12	0	250	9	
Orange juice =	0	cup	0.50	120	0	0	0	3	1	
Wheat toast =	2.975	slice	0.07	65	1	0	1	26	3	3
			Nutritional Levels	420	9.99	14.90	5.02	400	25.22	12
			Constraints	>=	<=	<=	>=	>=	>=	>=
			Nutritional Requirements	420	20	30.00	5	400	20.00	12
Cost Per Meal =	$ 0.509									

The objective function formula is contained in cell C19, **=SUMPRODUCT(C5:C14,E5:E14)**. In this case, the objective function value is also computed by using the Excel SUMPRODUCT command rather than by individually multiplying each cell value in column C by each cell value in column E and then summing them. This would be a tedious formula to type into cell C19. Alternatively, the SUMPRODUCT() formula multiplies all the values in cells C5 through C14 by all the corresponding values in cells E5 through E14 and then sums them.

The Solver Parameters dialog box for this problem is shown in Exhibit 4.6.

Solution Analysis

The solution is

$$x_3 = 1.025 \text{ cups of oatmeal}$$
$$x_8 = 1.241 \text{ cups of milk}$$
$$x_{10} = 2.975 \text{ slices of wheat toast}$$
$$Z = \$0.509 \text{ cost per meal}$$

The result of this simplified version of a real menu planning model is interesting in that it suggests a very practical breakfast menu. This would be a healthy breakfast for anyone.

EXHIBIT 4.6

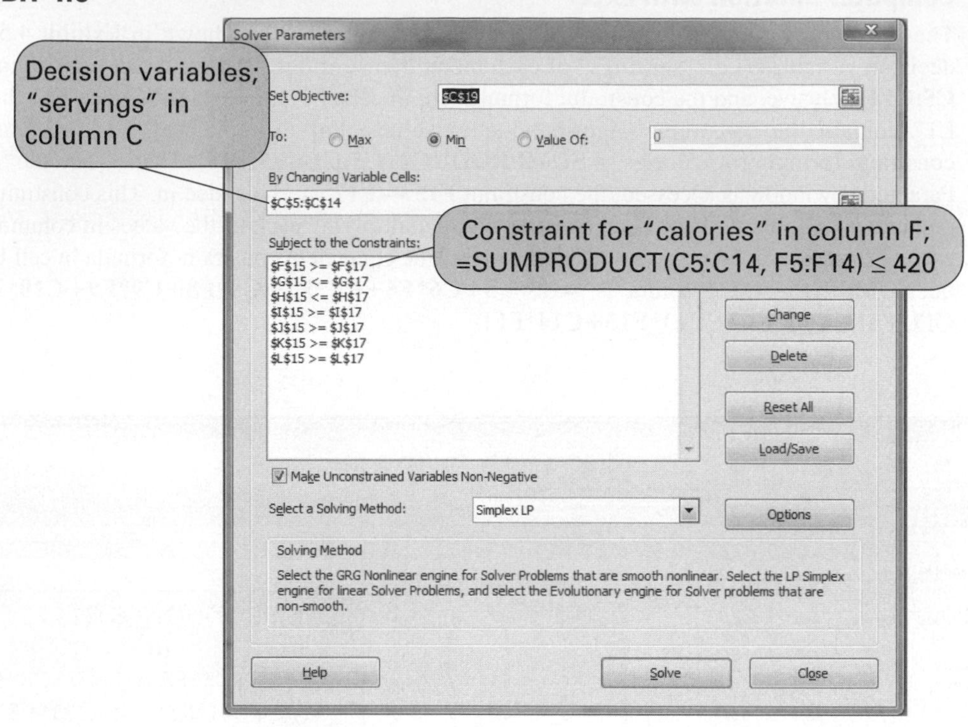

Sensitivity Analysis

This model includes a daily minimum requirement of only 420 calories. The recommended daily calorie requirement for an adult is approximately 2,000 calories. Thus, the breakfast requirement is only about 21% of normal daily adult needs. In this model, the dietitian must have felt a low-calorie breakfast was needed because the senior citizens had high-calorie lunches and dinners. An alternative approach to a healthy diet is a high-calorie breakfast followed by low-calorie, light meals and snacks the rest of the day. However, in this model, as the calorie requirements are increased above 420, the model simply increases the cups of oatmeal. For example, a 700-calorie requirement results in about 5 or 6 cups of oatmeal—not a very appetizing breakfast. This difficulty can be alleviated by establishing upper limits on the servings for each food item and solving the model again with a higher calorie requirement.

An Investment Example

Kathleen Allen, an individual investor, has $70,000 to divide among several investments. The alternative investments are municipal bonds with an 8.5% annual return, certificates of deposit with a 5% return, treasury bills with a 6.5% return, and a growth stock fund with a 13% annual return. The investments are all evaluated after 1 year. However, each investment alternative has a different perceived risk to the investor; thus, it is advisable to diversify. Kathleen wants to know how much to invest in each alternative in order to maximize the return.

The following guidelines have been established for diversifying the investments and lessening the risk perceived by the investor:

1. No more than 20% of the total investment should be in municipal bonds.
2. The amount invested in certificates of deposit should not exceed the amount invested in the other three alternatives.

3. At least 30% of the investment should be in treasury bills and certificates of deposit.
4. To be safe, more should be invested in CDs and treasury bills than in municipal bonds and the growth stock fund, by a ratio of at least 1.2 to 1.

Kathleen wants to invest the entire \$70,000.

Decision Variables

Four decision variables represent the monetary amount invested in each investment alternative:

$$x_1 = \text{amount (\$) invested in municipal bonds}$$
$$x_2 = \text{amount (\$) invested in certificates of deposit}$$
$$x_3 = \text{amount (\$) invested in treasury bills}$$
$$x_4 = \text{amount (\$) invested in growth stock fund}$$

The Objective Function

The objective of the investor is to maximize the total return from the investment in the four alternatives. The total return is the sum of the individual returns from each alternative. Thus, the objective function is expressed as

$$\text{maximize } Z = \$0.085x_1 + 0.05x_2 + 0.065x_3 + 0.130x_4$$

where

$$Z = \text{total return from all investments}$$
$$0.085x_1 = \text{return from the investment in municipal bonds}$$
$$0.05x_2 = \text{return from the investment in certificates of deposit}$$
$$0.065x_3 = \text{return from the investment in treasury bills}$$
$$0.130x_4 = \text{return from the investment in growth stock fund}$$

Model Constraints

In this problem, the constraints are the guidelines established for diversifying the total investment. Each guideline is transformed into a mathematical constraint separately.

The first guideline states that no more than 20% of the total investment should be in municipal bonds. The total investment is \$70,000; 20% of \$70,000 is \$14,000. Thus, this constraint is

$$x_1 \leq \$14{,}000$$

The second guideline indicates that the amount invested in certificates of deposit should not exceed the amount invested in the other three alternatives. Because the investment in certificates of deposit is x_2 and the amount invested in the other alternatives is $x_1 + x_3 + x_4$, the constraint is

$$x_2 \leq x_1 + x_3 + x_4$$

Standard form requires all variables to be to the left of the inequality and numeric values to the right.

This constraint is not in what we referred to in Chapter 3 as **standard form** for a computer solution. In standard form, all the variables would be on the left-hand side of the inequality ($\leq$), and all the numeric values would be on the right side. This type of constraint can be used in Excel just as it is shown here; however, for solution with QM for Windows, all constraints must be in standard form. We will go ahead and convert this constraint and others in this model to standard form, but when we solve this model with Excel, we will explain how the model constraints could be used in their original (nonstandard) form. To convert this constraint to standard form, $x_1 + x_3 + x_4$ must be subtracted from both sides of the $\leq$ sign to put this constraint in proper form:

$$x_2 - x_1 - x_3 - x_4 \leq 0$$

The third guideline specifies that at least 30% of the investment should be in treasury bills and certificates of deposit. Because 30% of \$70,000 is \$21,000 and the amount invested in certificates of deposit and treasury bills is represented by $x_2 + x_3$, the constraint is

$$x_2 + x_3 \geq \$21{,}000$$

The fourth guideline states that the ratio of the amount invested in certificates of deposit and treasury bills to the amount invested in municipal bonds and the growth stock fund should be at least 1.2 to 1:

$$[(x_2 + x_3)/(x_1 + x_4)] \geq 1.2$$

Standard form requires that fractional relationships between variables be eliminated.

This constraint is not in standard linear programming form because of the fractional relationship of the decision variables, $(x_2 + x_3)/(x_1 + x_4)$. It is converted as follows:

$$x_2 + x_3 \geq 1.2\,(x_1 + x_4)$$
$$-1.2x_1 + x_2 + x_3 - 1.2x_4 \geq 0$$

Finally, the investor wants to invest the entire \$70,000 in the four alternatives. Thus, the sum of all the investments in the four alternatives must *equal* \$70,000:

$$x_1 + x_2 + x_3 + x_4 = \$70{,}000$$

Model Summary

The complete linear programming model for this problem can be summarized as

$$\begin{aligned}
\text{maximize } Z &= \$0.085x_1 + 0.05x_2 + 0.065x_3 + 0.130x_4\\
\text{subject to}&\\
x_1 &\leq 14{,}000\\
x_2 - x_1 - x_3 - x_4 &\leq 0\\
x_2 + x_3 &\geq 21{,}000\\
-1.2x_1 + x_2 + x_3 - 1.2x_4 &\geq 0\\
x_1 + x_2 + x_3 + x_4 &= 70{,}000\\
x_1, x_2, x_3, x_4 &\geq 0
\end{aligned}$$

Computer Solution with Excel

The Excel spreadsheet solution for the investment example is shown in Exhibit 4.7, and its Solver Parameters window is shown in Exhibit 4.8. The spreadsheet is set up very similarly to the

EXHIBIT 4.7

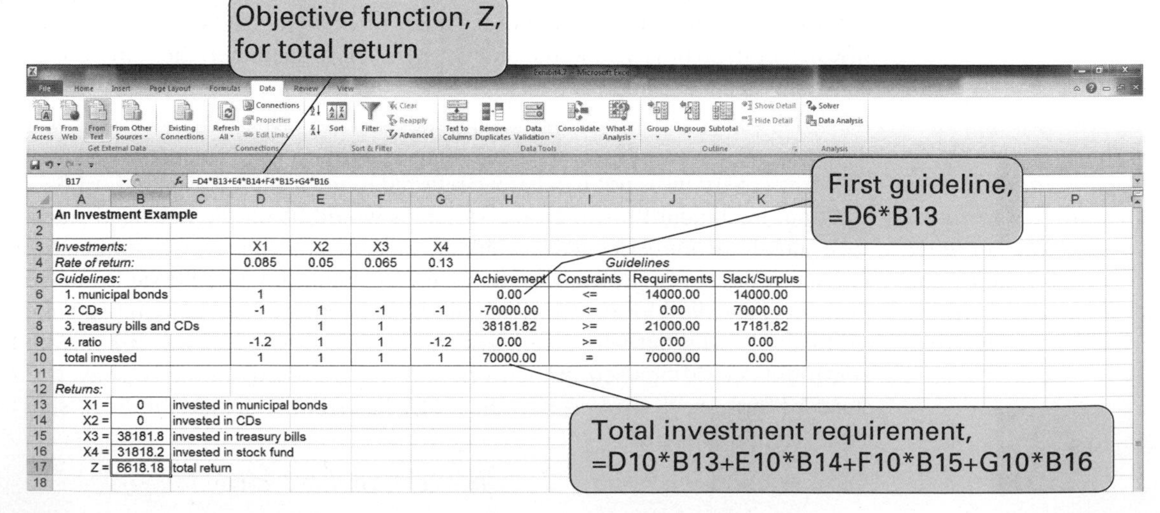

	A	B	C	D	E	F	G	H	I	J	K
1	An Investment Example										
2											
3	Investments:			X1	X2	X3	X4				
4	Rate of return:			0.085	0.05	0.065	0.13		Guidelines		
5	Guidelines:							Achievement	Constraints	Requirements	Slack/Surplus
6	1. municipal bonds			1				0.00	<=	14000.00	14000.00
7	2. CDs			-1	1	-1	-1	-70000.00	<=	0.00	70000.00
8	3. treasury bills and CDs				1	1		38181.82	>=	21000.00	17181.82
9	4. ratio			-1.2	1	1	-1.2	0.00	>=	0.00	0.00
10	total invested			1	1	1	1	70000.00	=	70000.00	0.00
11											
12	Returns:										
13	X1 =	0	invested in municipal bonds								
14	X2 =	0	invested in CDs								
15	X3 =	38181.8	invested in treasury bills								
16	X4 =	31818.2	invested in stock fund								
17	Z =	6618.18	total return								
18											

EXHIBIT 4.8

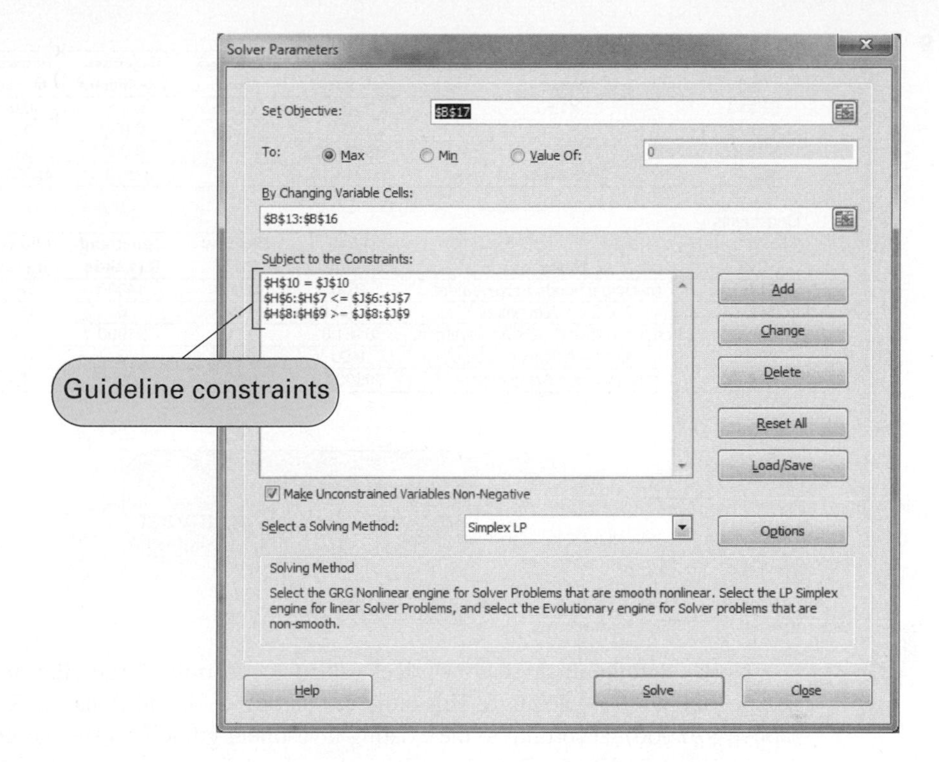

spreadsheet for our product mix example in Exhibit 4.1. The decision variables are located in cells **B13:B16**. The total return (Z) is computed in cell B17, and the formula for Z is shown on the formula bar at the top of the spreadsheet. The constraint formulas for the investment uidelines are embedded in cells H6 through H10. For example, the first guideline formula, in cell H6, is =**D6*B13**, and the second guideline formula, in cell H7, is =**D7*B13+E7*B14+F7*B15+G7*B16**. (Note that it would probably have been easier just to type the guideline formulas directly into cells H6 through H10 rather than create the array of constraint coefficients in **D6:G10**; however, for demonstration purposes, we wanted to show all the parameter values.)

As mentioned earlier, it is not necessary to convert the original model constraints into standard form to solve this model using Excel. For example, the constraint for certificates of deposit, $x_2 \leq x_1 + x_3 + x_4$, could be entered in the spreadsheet in row 7 in Exhibit 4.7 as follows. The value 1, the coefficient for x_2, could be entered in cell E7, **=E7*B14** could be entered in cell J7, and the remainder of the constraint to the right side of the inequality could be entered as **=B13+B15+B16** in cell H7. In the Solver Parameters window, this constraint is entered, with the municipal bonds constraint, as **H6:H7<=J6:J7**. The fourth guideline constraint could be entered in its original form similarly.

Solution Analysis

The solution is

$$x_3 = \$38{,}181.80 \text{ invested in treasury bonds}$$
$$x_4 = \$31{,}818.20 \text{ invested in a growth stock fund}$$
$$Z = \$6{,}818.18$$

Sensitivity Analysis

The sensitivity report for our Excel spreadsheet solution to this problem is shown in Exhibit 4.9.

EXHIBIT 4.9

Adjustable Cells

Cell	Name	Final Value	Reduced Cost	Objective Coefficient	Allowable Increase	Allowable Decrease
$B$13	X1 =	0	-0.045	0.085	0.045	1E+30
$B$14	X2 =	0	-0.015	0.050	0.015	1E+30
$B$15	X3 =	38181.82	0	0.065	0.065	0.015
$B$16	X4 =	31818.18	0	0.13	45354.805	0.045

Constraints

Cell	Name	Final Value	Shadow Price	Constraint R.H. Side	Allowable Increase	Allowable Decrease
$H$6	1. municipal bonds Achievement	0.00	0.000	14000	1E+30	14000
$H$7	2. CDs Achievement	-70000.00	0.000	0	1E+30	70000
$H$8	3. treasury bills and CDs Achievement	38181.82	0.000	21000	17181.82	1E+30
$H$9	4. ratio Achievement	0.00	-0.030	0	70000	37800
$H$10	total invested Achievement	70000.00	0.095	70000	1E+30	31500

Shadow price for the amount available to invest

Notice that the dual (shadow price) value for constraint 5 (i.e., the sum of the investments must equal $70,000) is 0.095. This indicates that for each additional $1 Kathleen Allen invests (above $70,000), according to the existing investment guidelines she has established, she could expect a return of 9.5%. The sensitivity ranges show that there is no upper bound on the amount she could invest and still receive this return.

An interesting variation of the problem is to not specify that the entire amount available (in this case, $70,000) must be invested. This changes the constraints for the first and third guidelines and the constraint that requires that the entire $70,000 be invested.

Recall that the first guideline is "no more than 20% of the total investment should be in municipal bonds." The total investment is no longer exactly $70,000 but the sum of all four investments, $x_1 + x_2 + x_3 + x_4$. The constraint showing that the amount invested in municipal bonds, x_1, as a percentage (or ratio) of this total cannot exceed 20% is written as

$$\frac{x_1}{x_1 + x_2 + x_3 + x_4} \leq 0.20$$

Rewriting this constraint in a form more consistent with a linear programming solution (to eliminate the fractional relationship between variables) results in

$$x_1 \leq 0.2(x_1 + x_2 + x_3 + x_4)$$

and

$$0.8x_1 - 0.2x_2 - 0.2x_3 - 0.2x_4 \leq 0$$

The constraint for the third guideline, which stipulates that at least 30% of the total investment $(x_1 + x_2 + x_3 + x_4)$ should be in treasury bills and CDs $(x_2 + x_3)$, is formulated similarly as

$$\frac{x_2 + x_3}{x_1 + x_2 + x_3 + x_4} \geq 0.30$$

and

$$-0.3x_1 + 0.7x_2 + 0.7x_3 - 0.3x_4 \geq 0$$

Because the entire $70,000 does not have to be invested, the last constraint becomes

$$x_1 + x_2 + x_3 + x_4 \leq 70{,}000$$

The complete linear programming model is summarized as

$$
\begin{aligned}
\text{maximize } Z = \$0.085x_1 + 0.05x_2 + 0.065x_3 + 0.130x_4 \\
\text{subject to} \\
0.8x_1 - 0.2x_2 - 0.2x_3 - 0.2x_4 &\leq 0 \\
x_2 - x_1 - x_3 - x_4 &\leq 0 \\
-0.3x_1 + 0.7x_2 + 0.7x_3 - 0.3x_4 &\geq 0 \\
-1.2x_1 + x_2 + x_3 - 1.2x_4 &\geq 0 \\
x_1 + x_2 + x_3 + x_4 &\leq 70{,}000 \\
x_1, x_2, x_3, x_4 &\geq 0
\end{aligned}
$$

The solution to this altered model is exactly the same as the solution to our original model, wherein the entire \$70,000 must be invested. This is logical because only positive returns are achieved from investing, and thus the investor would leave no money not invested. However, if losses could be realized from some investments, then it might be a good idea to construct the model so that the entire amount would not have to be invested.

Management Science Application

A Linear Programming Model for Optimal Portfolio Selection at GE Asset Management

GE Asset Management is a wholly owned subsidiary of General Electric Company that manages investment portfolios of the assets (worth billions of dollars) of various GE units and other clients, including Genworth Financial and GE Insurance, insurance units of GE that own other insurance businesses. According to modern portfolio theory, the goal of portfolio optimization is to manage risk through diversification and obtain an optimal risk–return trade-off. Thus the objective of portfolio management at GE Asset Management is to maximize return or minimize risk while matching the characteristics of asset portfolios with corresponding liabilities. Assets include cash flows for various insurance and financial products, primarily in the form of corporate and government bonds.

Linear programming is used to determine asset allocations for a portfolio. The portfolio optimization model GE Asset Management formulated uses the variance of economic surplus (the difference between the market values of assets and liabilities), as a measure of total risk. GE initially used its linear programming portfolio model to optimize more than 30 portfolios valued at over \$30 billion. Based on an analysis of these portfolios, GE estimates that the potential benefits of the linear programming model could be approximately \$75 million over a 5-year period.

GE is also using this model to develop strategies and benchmarks for new cash investments, as well as to manage life insurance portfolios, property and equity portfolios, and mortgage insurance portfolios. The linear programming model allows portfolio managers to perform portfolio analytics (e.g., sensitivity analysis) much faster than previous systems, enabling them to respond more quickly to changing market conditions.

Source: Based on K. Chalermkraivuth, S. Bollapragada, M. Clark, J. Deaton, L. Kiaer, J. Murdzek, W. Neeves, B. Scholz, and D. Toledano, "GE Asset Management, Genworth Financial, and GE Insurance Use a Sequential-Linear-Programming Algorithm to Optimize Portfolios," *Interfaces* 35, no. 5 (September–October 2005): 370–80.

A Marketing Example

The Biggs Department Store chain has hired an advertising firm to determine the types and amount of advertising it should invest in for its stores. The three types of advertising available are television and radio commercials and newspaper ads. The retail chain desires to know the number of each type of advertisement it should purchase in order to maximize exposure. It is estimated that each ad or commercial will reach the following potential audience and cost the following amount:

	Exposure (people/ad or commercial)	Cost
Television commercial	20,000	$15,000
Radio commercial	12,000	6,000
Newspaper ad	9,000	4,000

The company must consider the following resource constraints:

1. The budget limit for advertising is $100,000.
2. The television station has time available for 4 commercials.
3. The radio station has time available for 10 commercials.
4. The newspaper has space available for 7 ads.
5. The advertising agency has time and staff available for producing no more than a total of 15 commercials and/or ads.

Decision Variables

This model consists of three decision variables that represent the number of each type of advertising produced:

$$x_1 = \text{number of television commercials}$$
$$x_2 = \text{number of radio commercials}$$
$$x_3 = \text{number of newspaper ads}$$

The Objective Function

The objective of this problem is different from the objectives in the previous examples, in which only profit was to be maximized (or cost minimized). In this problem, profit is not to be maximized; instead, audience exposure is to be maximized. Thus, this objective function demonstrates that although a linear programming model must either maximize or minimize some objective, the objective itself can be in terms of any type of activity or valuation.

For this problem the objective audience exposure is determined by summing the audience exposure gained from each type of advertising:

$$\text{maximize } Z = 20{,}000x_1 + 12{,}000x_2 + 9{,}000x_3$$

where

$$Z = \text{total level of audience exposure}$$
$$20{,}000x_1 = \text{estimated number of people reached by television commercials}$$
$$12{,}000x_2 = \text{estimated number of people reached by radio commercials}$$
$$9{,}000x_3 = \text{estimated number of people reached by newspaper ads}$$

Model Constraints

The first constraint in this model reflects the limited budget of \$100,000 allocated for advertisement:

$$\$15{,}000x_1 + 6{,}000x_2 + 4{,}000x_3 \leq \$100{,}000$$

where

$$\$15{,}000x_1 = \text{amount spent for television advertising}$$
$$6{,}000x_2 = \text{amount spent for radio advertising}$$
$$4{,}000x_3 = \text{amount spent for newspaper advertising}$$

The next three constraints represent the fact that television and radio commercials are limited to 4 and 10, respectively, and newspaper ads are limited to 7:

$$x_1 \leq 4 \text{ television commercials}$$
$$x_2 \leq 10 \text{ radio commercials}$$
$$x_3 \leq 7 \text{ newspaper ads}$$

The final constraint specifies that the total number of commercials and ads cannot exceed 15 because of the limitations of the advertising firm:

$$x_1 + x_2 + x_3 \leq 15 \text{ commercials and ads}$$

Model Summary

The complete linear programming model for this problem is summarized as

$$\begin{aligned} &\text{maximize } Z = 20{,}000x_1 + 12{,}000x_2 + 9{,}000x_3 \\ &\text{subject to} \\ &\$15{,}000x_1 + 6{,}000x_2 + 4{,}000x_3 \leq \$100{,}000 \\ &x_1 \leq 4 \\ &x_2 \leq 10 \\ &x_3 \leq 7 \\ &x_1 + x_2 + x_3 \leq 15 \\ &x_1, x_2, x_3 \geq 0 \end{aligned}$$

Computer Solution with Excel

The solution to our marketing example using Excel is shown in Exhibit 4.10. The model decision variables are contained in cells **D6:D8**. The formula for the objective function in cell E10 is shown on the formula bar at the top of the screen. When Solver is accessed, as shown in Exhibit 4.11, it is necessary to use only one formula to enter the model constraints: **H6:H10 <= J6:J10**.

Solution Analysis

The solution shows

$$x_1 = 1.82 \text{ television commercials}$$
$$x_2 = 10 \text{ radio commercials}$$
$$x_3 = 3.18 \text{ newspaper ads}$$
$$Z = 185{,}000 \text{ audience exposure}$$

This is a case where a non-integer solution can create difficulties. It is not realistic to round 1.82 television commercials to 2 television commercials, with 10 radio commercials and 3 newspaper ads. Some quick arithmetic with the budget constraint shows that such a solution will exceed the

EXHIBIT 4.10

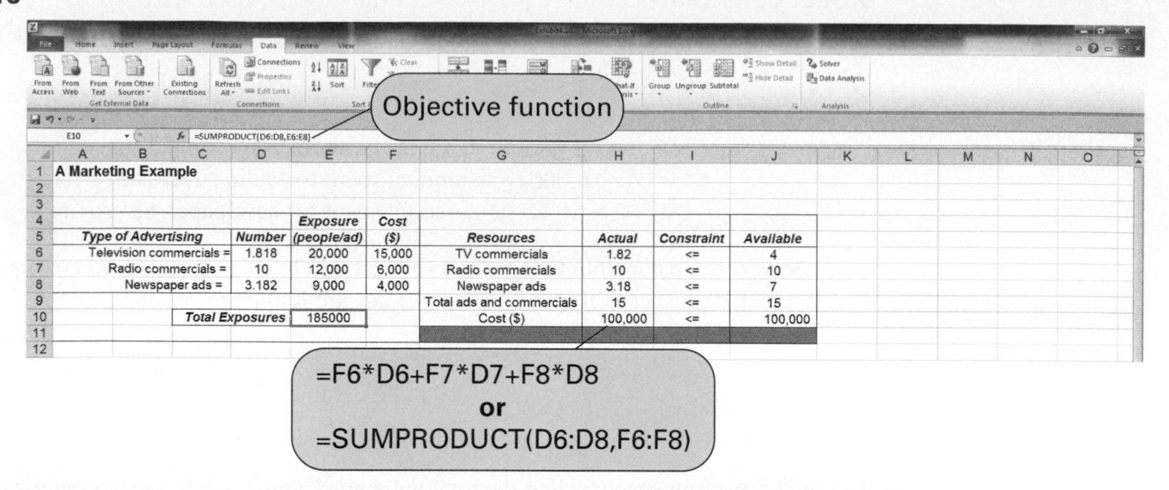

EXHIBIT 4.11

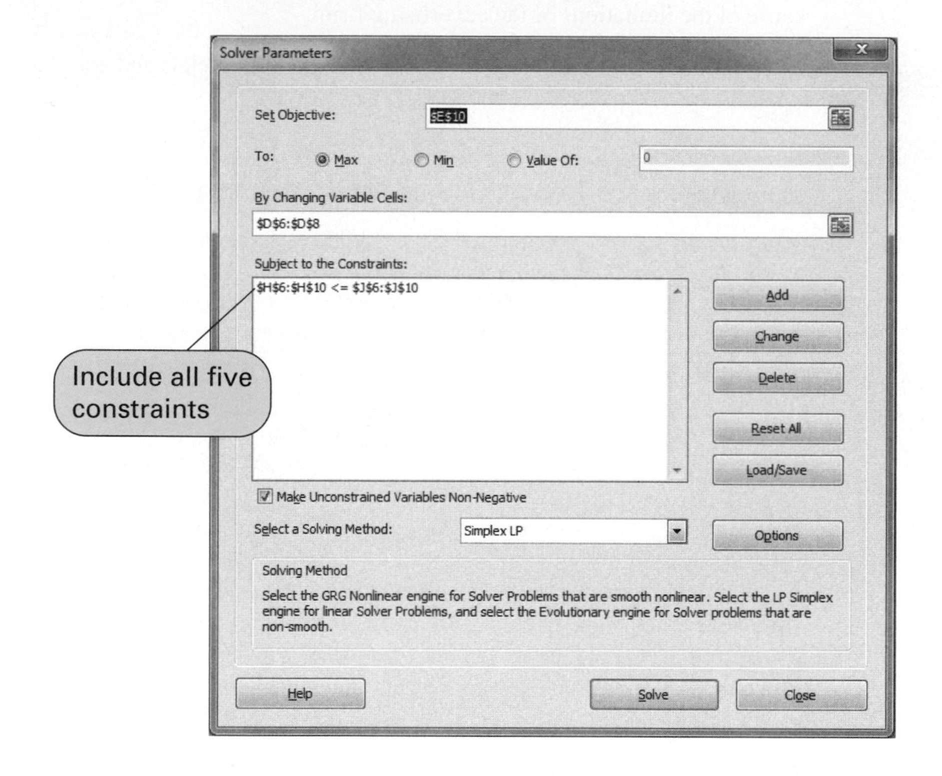

$100,000 budget limitation, although by only $2,000. Thus, the store must either increase its advertising or plan for 1 television commercial, 10 radio commercials, and 3 newspaper ads. The audience exposure for this solution will be 167,000 people, or 18,000 fewer than the optimal number, almost a 10% decrease. There may, in fact, be a better solution than this "rounded-down" solution.

The integer linear programming technique, which restricts solutions to integer values, should be used. Although we will discuss the topic of integer programming in more detail in Chapter 5, for now we can derive an integer solution by using Excel with a simple change when we input our constraints in the Solver Parameters window. We specify that our decision variable cells, **D6:D8**, are integers in the Change Constraint window, as shown in Exhibits 4.12 and 4.13. This will result in the spreadsheet solution in Exhibit 4.14 when the problem is solved, which you will notice is better (i.e., 17,000 more total exposures) than the rounded-down solution.

EXHIBIT 4.12

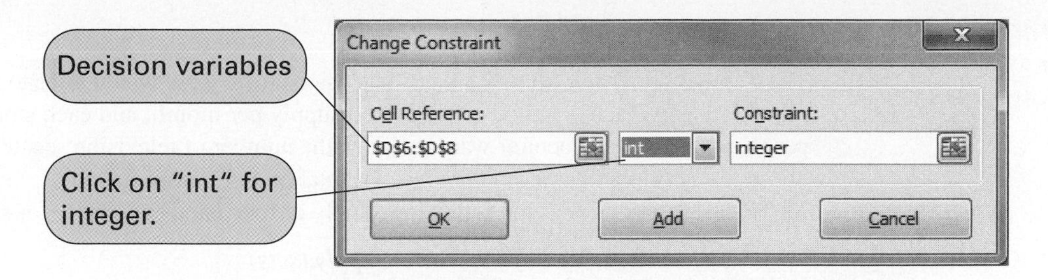

EXHIBIT 4.13

EXHIBIT 4.14

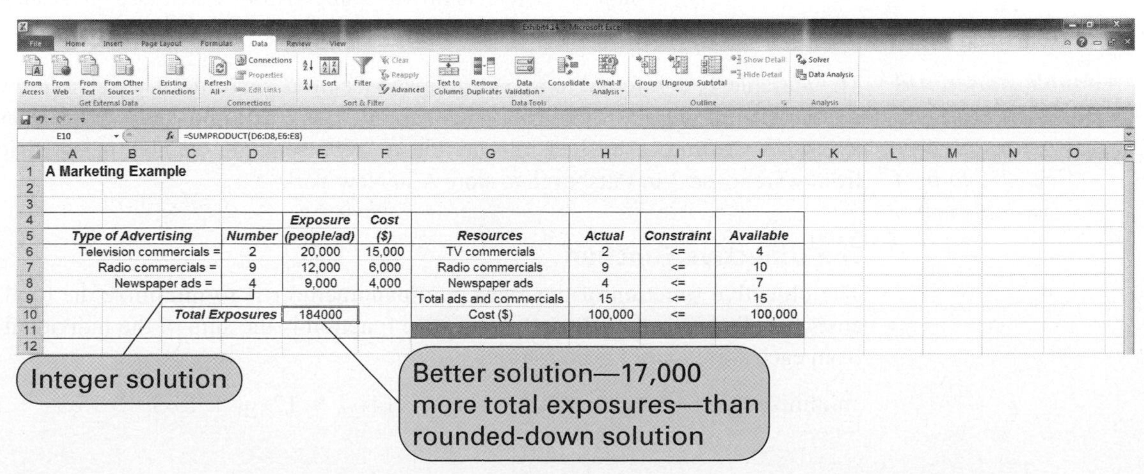

Type of Advertising	Number	Exposure (people/ad)	Cost ($)	Resources	Actual	Constraint	Available
Television commercials =	2	20,000	15,000	TV commercials	2	<=	4
Radio commercials =	9	12,000	6,000	Radio commercials	9	<=	10
Newspaper ads =	4	9,000	4,000	Newspaper ads	4	<=	7
				Total ads and commercials	15	<=	15
Total Exposures		184000		Cost ($)	100,000	<=	100,000

A Transportation Example

The Zephyr Television Company ships televisions from three warehouses to three retail stores on a monthly basis. Each warehouse has a fixed supply per month, and each store has a fixed demand per month. The manufacturer wants to know the number of television sets to ship from each warehouse to each store in order to minimize the total cost of transportation.

Each warehouse has the following supply of televisions available for shipment each month:

Warehouse	*Supply (sets)*
1. Cincinnati	300
2. Atlanta	200
3. Pittsburgh	200
	700

Each retail store has the following monthly demand for television sets:

Store	*Demand (sets)*
A. New York	150
B. Dallas	250
C. Detroit	200
	600

Costs of transporting television sets from the warehouses to the retail stores vary as a result of differences in modes of transportation and distances. The shipping cost per television set for each route is as follows:

From Warehouse	To Store		
	A	B	C
1	\$16	\$18	\$11
2	14	12	13
3	13	15	17

Decision Variables

The model for this problem consists of nine decision variables, representing the number of television sets transported from each of the three warehouses to each of the three stores:

$$x_{ij} = \text{number of television sets shipped from warehouse } i \text{ to store } j$$
$$\text{where } i = 1, 2, 3, \text{ and } j = \text{A, B, C}$$

A double-subscripted variable is simply another form of variable name.

The variable x_{ij} is referred to as a **double-subscripted variable**. The subscript, whether double or single, simply gives a "name" to the variable (i.e., distinguishes it from other decision variables). For example, the decision variable x_{3A} represents the number of television sets shipped from warehouse 3 in Pittsburgh to store A in New York.

The Objective Function

The objective function of the television manufacturer is to minimize the total transportation costs for all shipments. Thus, the objective function is the sum of the individual shipping costs from each warehouse to each store:

$$\text{minimize } Z = \$16x_{1A} + 18x_{1B} + 11x_{1C} + 14x_{2A} + 12x_{2B} + 13x_{2C} + 13x_{3A} + 15x_{3B} + 17x_{3C}$$

Model Constraints

The constraints in this model are the number of television sets available at each warehouse and the number of sets demanded at each store. There are six constraints—one for each warehouse's supply and one for each store's demand. For example, warehouse 1 in Cincinnati is able to supply 300 television sets to any of the three retail stores. Because the number shipped to the three stores is the sum of x_{1A}, x_{1B}, and x_{1C}, the constraint for warehouse 1 is

$$x_{1A} + x_{1B} + x_{1C} \leq 300$$

In a ***"balanced" transportation model****, supply equals demand such that all constraints are equalities; in an* ***"unbalanced" transportation model****, supply does not equal demand, and one set of constraints is* $\leq$.

This constraint is a $\leq$ inequality for two reasons. First, no more than 300 television sets can be shipped because that is the maximum available at the warehouse. Second, fewer than 300 can be shipped because 300 are not needed to meet the total demand of 600. That is, total demand is less than total supply, which equals 700. To meet the total demand at the three stores, all that can be supplied by the three warehouses is not needed. Thus, the other two supply constraints for warehouses 2 and 3 are also inequalities:

$$x_{2A} + x_{2B} + x_{2C} \leq 200$$
$$x_{3A} + x_{3B} + x_{3C} \leq 200$$

The three demand constraints are developed in the same way as the supply constraints, except that the variables summed are the number of television sets supplied from each of the three warehouses. Thus, the number shipped to one store is the sum of the shipments from the three warehouses. They are equalities because all demands can be met:

$$x_{1A} + x_{2A} + x_{3A} = 150$$
$$x_{1B} + x_{2B} + x_{3B} = 250$$
$$x_{1C} + x_{2C} + x_{3C} = 200$$

Model Summary

The complete linear programming model for this problem is summarized as follows:

$$\text{minimize } Z = \$16x_{1A} + 18x_{1B} + 11x_{1C} + 14x_{2A} + 12x_{2B} + 13x_{2C} + 13x_{3A} + 15x_{3B} + 17x_{3C}$$

subject to

$$\begin{aligned} x_{1A} + x_{1B} + x_{1C} &\leq 300 \\ x_{2A} + x_{2B} + x_{2C} &\leq 200 \\ x_{3A} + x_{3B} + x_{3C} &\leq 200 \\ x_{1A} + x_{2A} + x_{3A} &= 150 \\ x_{1B} + x_{2B} + x_{3B} &= 250 \\ x_{1C} + x_{2C} + x_{3C} &= 200 \\ x_{ij} &\geq 0 \end{aligned}$$

Computer Solution with Excel

The computer solution for this model was achieved using Excel and is shown in Exhibit 4.15. Notice that the objective function contained in cell C11 is shown on the formula bar at the top of the screen. The constraints for supply are included in cells F5, F6, and F7, whereas the demand constraints are in cells C8, D8, and E8. Thus, a typical supply constraint for Cincinnati would be **C5+D5+E5≤H5**. The Solver Parameters window is shown in Exhibit 4.16.

EXHIBIT 4.15

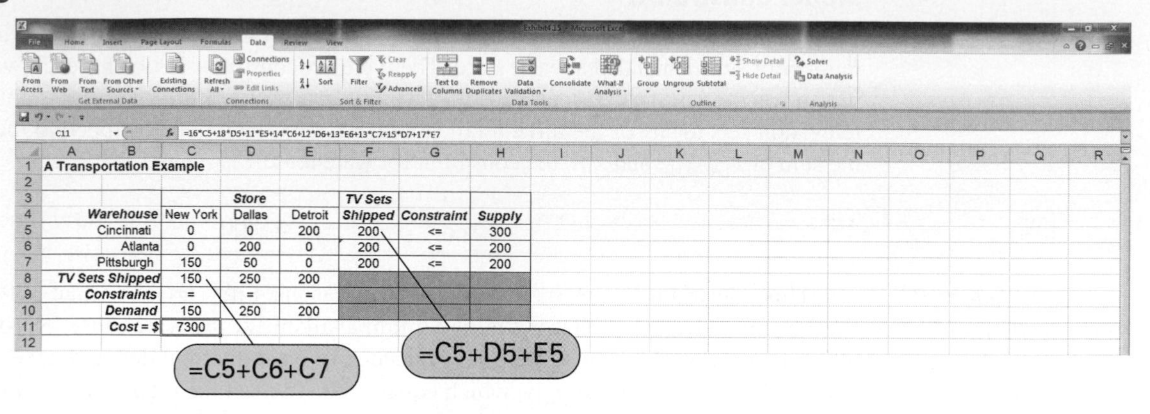

Row	B	C	D	E	F	G	H
1	A Transportation Example						
3			Store		TV Sets		
4	Warehouse	New York	Dallas	Detroit	Shipped	Constraint	Supply
5	Cincinnati	0	0	200	200	<=	300
6	Atlanta	0	200	0	200	<=	200
7	Pittsburgh	150	50	0	200	<=	200
8	TV Sets Shipped	150	250	200			
9	Constraints	=	=	=			
10	Demand	150	250	200			
11	Cost = $	7300					

EXHIBIT 4.16

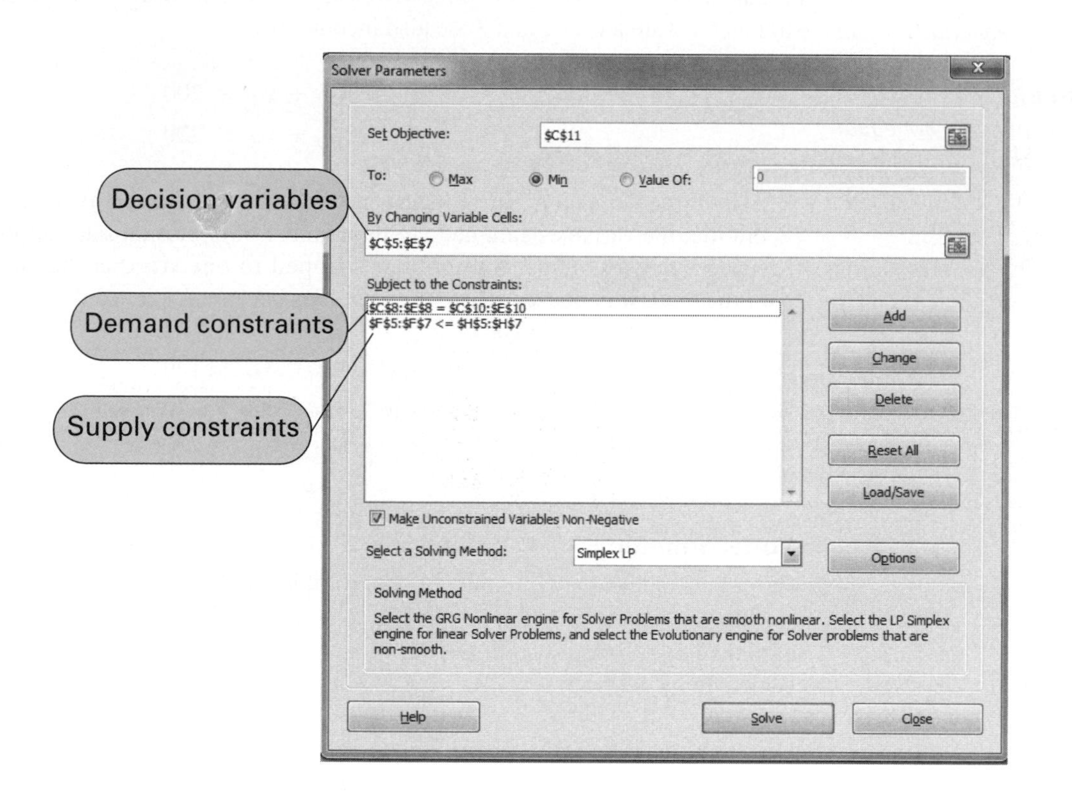

Solution Analysis

The solution is

$$
\begin{aligned}
x_{1C} &= 200 \text{ TVs shipped from Cincinnati to Detroit} \\
x_{2B} &= 200 \text{ TVs shipped from Atlanta to Dallas} \\
x_{3A} &= 150 \text{ TVs shipped from Pittsburgh to New York} \\
x_{3B} &= 50 \text{ TVs shipped from Pittsburgh to Dallas} \\
Z &= \$7{,}300 \text{ shipping cost}
\end{aligned}
$$

Note that the surplus for the first constraint, which is the supply at the Cincinnati warehouse, equals 100 TVs. This means that 100 TVs are left in the Cincinnati warehouse, which has inventory and storage implications for the manager.

This is an example of a type of linear programming model known as a *transportation problem*, which is the topic of Chapter 6. Notice that all the solution values are integer values. This is always the case with transportation problems because of the unique characteristic that all the coefficient parameter values are ones and zeros.

A Blend Example

A petroleum company produces three grades of motor oil—super, premium, and extra—from three components. The company wants to determine the optimal mix of the three components in each grade of motor oil that will maximize profit. The maximum quantities available of each component and their cost per barrel are as follows:

Component	Maximum Barrels Available/Day	Cost/Barrel
1	4,500	$12
2	2,700	10
3	3,500	14

To ensure the appropriate blend, each grade has certain general specifications. Each grade must have a minimum amount of component 1 plus a combination of other components, as follows:

Grade	Component Specifications	Selling Price/Barrel
Super	At least 50% of 1	$23
	Not more than 30% of 2	
Premium	At least 40% of 1	20
	Not more than 25% of 3	
Extra	At least 60% of 1	18
	At least 10% of 2	

The company wants to produce at least 3,000 barrels of each grade of motor oil.

Decision Variables

The decision variables for this problem must specify the quantity of each of the three components used in each grade of motor oil. This requires nine decision variables, as follows: x_{ij} = barrels of component i used in motor oil grade j per day, where $i = 1, 2, 3$ and $j = s$ (super), p (premium), e (extra). For example, the amount of component 1 in super motor oil is x_{1s}. The total amount of each grade of motor oil will be

$$\text{super: } x_{1s} + x_{2s} + x_{3s}$$
$$\text{premium: } x_{1p} + x_{2p} + x_{3p}$$
$$\text{extra: } x_{1e} + x_{2e} + x_{3e}$$

The Objective Function

"Profit" is maximized in the objective function by subtracting cost from revenue.

The company's objective is to maximize profit. This requires that the cost of each barrel be subtracted from the revenue obtained from each barrel. Revenue is determined by multiplying the selling price of each grade of motor oil by the total barrels of each grade produced.

Cost is achieved by multiplying the cost of each component by the total barrels of each component used:

$$\text{maximize } Z = \$23(x_{1s} + x_{2s} + x_{3s}) + 20(x_{1p} + x_{2p} + x_{3p}) + 18(x_{1e} + x_{2e} + x_{3e}) - 12(x_{1s} + x_{1p} + x_{1e}) - 10(x_{2s} + x_{2p} + x_{2e}) - 14(x_{3s} + x_{3p} + x_{3e})$$

Combining terms results in the following objective function:

$$\text{maximize } Z = 11x_{1s} + 13x_{2s} + 9x_{3s} + 8x_{1p} + 10x_{2p} + 6x_{3p} + 6x_{1e} + 8x_{2e} + 4x_{3e}$$

Model Constraints

This problem has several sets of constraints. The first set reflects the limited amount of each component available on a daily basis:

$$x_{1s} + x_{1p} + x_{1e} \leq 4{,}500 \text{ bbl.}$$
$$x_{2s} + x_{2p} + x_{2e} \leq 2{,}700 \text{ bbl.}$$
$$x_{3s} + x_{3p} + x_{3e} \leq 3{,}500 \text{ bbl.}$$

The next group of constraints is for the blend specifications for each grade of motor oil. The first specification is that super contain at least 50% of component 1, which is expressed as

$$\frac{x_{1s}}{x_{1s} + x_{2s} + x_{3s}} \geq 0.50$$

Standard form requires that fractional relationships between variables be eliminated.

This constraint says that the ratio of component 1 in super to the total amount of super produced, $x_{1s} + x_{2s} + x_{3s}$, must be at least 50%. Rewriting this constraint in a form more consistent with linear programming solution results in

$$x_{1s} \geq 0.50(x_{1s} + x_{2s} + x_{3s})$$

and

$$0.50x_{1s} - 0.50x_{2s} - 0.50x_{3s} \geq 0$$

This is the general form a linear programming constraint must be in before you can enter it for computer solution. All variables are on the left-hand side of the inequality, and only numeric values are on the right-hand side.

The constraint for the other blend specification for super grade, not more than 30% of component 2, is developed in the same way:

$$\frac{x_{2s}}{x_{1s} + x_{2s} + x_{3s}} \leq 0.30$$

and

$$0.70x_{2s} - 0.30x_{1s} - 0.30x_{3s} \leq 0$$

The two blend specifications for premium motor oil are

$$0.60x_{1p} - 0.40x_{2p} - 0.40x_{3p} \geq 0$$
$$0.75x_{3p} - 0.25x_{1p} - 0.25x_{2p} \leq 0$$

The two blend specifications for extra motor oil are

$$0.40x_{1e} - 0.60x_{2e} - 0.60x_{3e} \geq 0$$
$$0.90x_{2e} - 0.10x_{1e} - 0.10x_{3e} \geq 0$$

The final set of constraints reflects the requirement that at least 3,000 barrels of each grade be produced:

$$x_{1s} + x_{2s} + x_{3s} \geq 3{,}000 \text{ bbl.}$$
$$x_{1p} + x_{2p} + x_{3p} \geq 3{,}000 \text{ bbl.}$$
$$x_{1e} + x_{2e} + x_{3e} \geq 3{,}000 \text{ bbl.}$$

Model Summary

The complete linear programming model for this problem is summarized as follows:

$$\text{maximize } Z = 11x_{1s} + 13x_{2s} + 9x_{3s} + 8x_{1p} + 10x_{2p} + 6x_{3p} + 6x_{1e} + 8x_{2e} + 4x_{3e}$$

subject to

$$\begin{aligned}
x_{1s} + x_{1p} + x_{1e} &\leq 4{,}500 \\
x_{2s} + x_{2p} + x_{2e} &\leq 2{,}700 \\
x_{3s} + x_{3p} + x_{3e} &\leq 3{,}500 \\
0.50x_{1s} - 0.50x_{2s} - 0.50x_{3s} &\geq 0 \\
0.70x_{2s} - 0.30x_{1s} - 0.30x_{3s} &\leq 0 \\
0.60x_{1p} - 0.40x_{2p} - 0.40x_{3p} &\geq 0 \\
0.75x_{3p} - 0.25x_{1p} - 0.25x_{2p} &\leq 0 \\
0.40x_{1e} - 0.60x_{2e} - 0.60x_{3e} &\geq 0 \\
0.90x_{2e} - 0.10x_{1e} - 0.10x_{3e} &\geq 0 \\
x_{1s} + x_{2s} + x_{3s} &\geq 3{,}000 \\
x_{1p} + x_{2p} + x_{3p} &\geq 3{,}000 \\
x_{1e} + x_{2e} + x_{3e} &\geq 3{,}000 \\
x_{ij} &\geq 0
\end{aligned}$$

Computer Solution with Excel

The Excel spreadsheet solution for this blend example is shown in Exhibit 4.17. Solver is shown in Exhibit 4.18. The decision variables are located in cells **B7:B15** in Exhibit 4.17. The total profit is computed in cell C16, using the formula **=SUMPRODUCT(B7:B15,C7:C15)**, which is also shown on the formula bar at the top of the spreadsheet. The constraint formulas are embedded in cells H6 through H17. Notice that we did not develop an array of constraint coefficients on the spreadsheet for this model; instead, we typed the constraint formulas directly into cells H6:H17, which seemed easier. For example, the constraint formula in cell H6

EXHIBIT 4.17

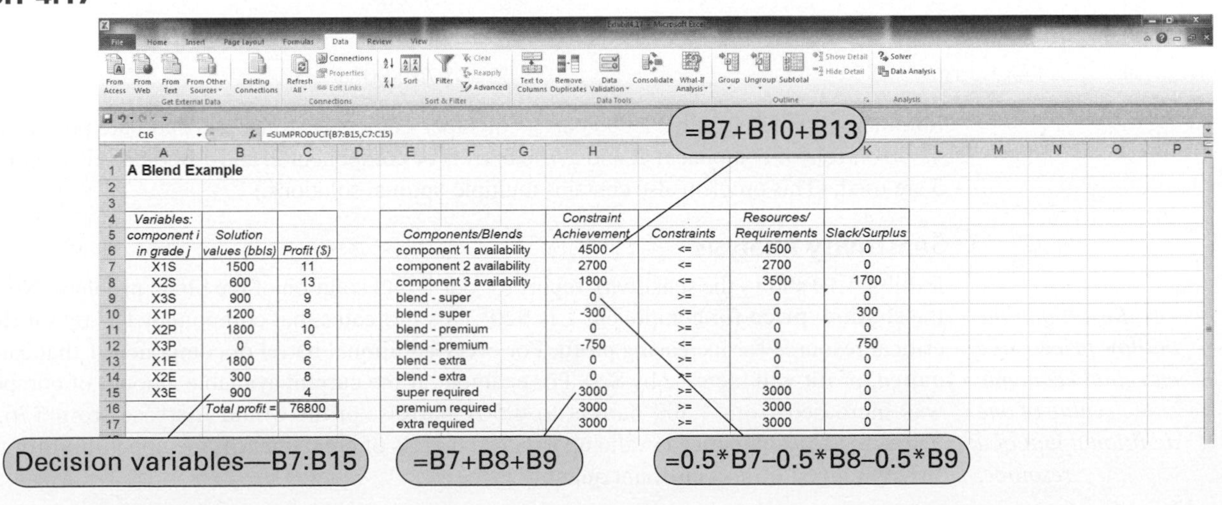

Variables: component i in grade j	Solution values (bbls)	Profit ($)
X1S	1500	11
X2S	600	13
X3S	900	9
X1P	1200	8
X2P	1800	10
X3P	0	6
X1E	1800	6
X2E	300	8
X3E	900	4
	Total profit =	76800

Components/Blends	Constraint Achievement	Constraints	Resources/ Requirements	Slack/Surplus
component 1 availability	4500	<=	4500	0
component 2 availability	2700	<=	2700	0
component 3 availability	1800	<=	3500	1700
blend - super	0	>=	0	0
blend - super	-300	<=	0	300
blend - premium	0	>=	0	0
blend - premium	-750	<=	0	750
blend - extra	0	>=	0	0
blend - extra	0	>=	0	0
super required	3000	>=	3000	0
premium required	3000	>=	3000	0
extra required	3000	>=	3000	0

EXHIBIT 4.18

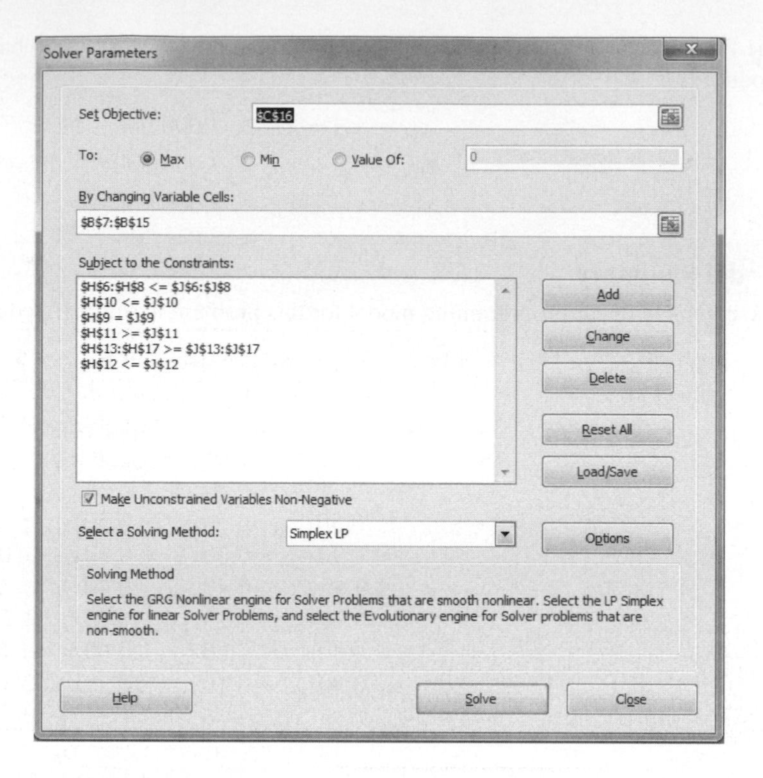

is **=B7+B10+B13**, and the constraint formula in cell H9 is **=.5*B7–.5*B8–.5*B9**. The remaining cells in column H have similar constraint formulas.

Solution Analysis

The solution is

$$
\begin{aligned}
x_{1s} &= 1{,}500 \text{ bbl.}\\
x_{2s} &= 600 \text{ bbl.}\\
x_{3s} &= 900 \text{ bbl.}\\
x_{1p} &= 1{,}200 \text{ bbl.}\\
x_{2p} &= 1{,}800 \text{ bbl.}\\
x_{1e} &= 1{,}800 \text{ bbl.}\\
x_{2e} &= 300 \text{ bbl.}\\
x_{3e} &= 900 \text{ bbl.}\\
Z &= \$76{,}800
\end{aligned}
$$

Summarizing these results, 3,000 barrels of super grade, premium, and extra are produced. Also, 4,500 barrels of component 1, and 2,700 barrels of component 2, and 1,800 barrels of component 3 are used. (This problem also contains multiple optimal solutions.)

Sensitivity Analysis

*Recall that the **shadow price** is the marginal economic value of one additional unit of a resource.*

Exhibit 4.19 shows the sensitivity report for our Excel solution of the blend problem. Notice that the **shadow price** for component 1 is $20. This indicates that component 1 is by far the most critical resource for increasing profit. For every additional barrel of component 1 that can be acquired, profit will increase by $20. For example, if the current available amount of component 1 was increased from 4,500 barrels to 4,501 barrels, profit would increase from $76,800 to $76,820. (This dual price is valid up to 6,200 barrels of component 1, the upper limit for the sensitivity range for this constraint quantity value.)

EXHIBIT 4.19

Adjustable Cells

Cell	Name	Final Value	Reduced Cost	Objective Coefficient	Allowable Increase	Allowable Decrease
$B$7	X1S values (bbls)	1500	0	11	1E+30	3
$B$8	X2S values (bbls)	600	0	13	0	0
$B$9	X3S values (bbls)	900	0	9	0	0
$B$10	X1P values (bbls)	1200	0	8	3	1E+30
$B$11	X2P values (bbls)	1800	0	10	2	0
$B$12	X3P values (bbls)	0	0	6	0	1E+30
$B$13	X1E values (bbls)	1800	0	6	3	1E+30
$B$14	X2E values (bbls)	300	0	8	0	1E+30
$B$15	X3E values (bbls)	900	0	4	22.67	0

Constraints

Cell	Name	Final Value	Shadow Price	Constraint R.H. Side	Allowable Increase	Allowable Decrease
$H$13	blend - extra Achievement	0	-18	0	1.1E-05	8.5E+02
$H$14	blend - extra Achievement	0	0	0	6.0E+02	3.0E+02
$H$15	super required Achievement	3000	0	3000	2.2E-05	1.0E+30
$H$16	premium required Achievement	3000	-1	3000	2.7E-05	8.3E+02
$H$17	extra required Achievement	3000	-7	3000	1.8E-05	3.0E+03
$H$12	blend - premium Achievement	-750	0	0	1.0E+30	7.5E+02
$H$11	blend - premium Achievement	0	-18	0	1.1E-05	6.0E+02
$H$9	blend - super Achievement	0	-18	0	1.1E-05	8.5E+02
$H$10	blend - super Achievement	-300	0	0	1.0E+30	3.0E+02
$H$6	component 1 availability Achievement	4500	20	4500	1.7E+03	1.1E-05
$H$7	component 2 availability Achievement	2700	4	2700	3.0E+02	6.0E+02
$H$8	component 3 availability Achievement	1800	0	3500	1.0E+30	1.7E+03

The shadow price for component 1 is $20.

The upper limit for the sensitivity range for component 1 is 4500 + 1700 = 6200.

In the refinery industry, different grade stocks of oil and gasoline are available based on the makeup and quality of the crude oil that is received. Thus, as crude oil properties change, it is necessary to change blend requirements. Component availability changes as well. The general structure of this model can be used on a daily basis to plan production based on component availability and blend specification changes.

A Multiperiod Scheduling Example

PM Computers assembles its own brand of personal computers from component parts it purchases overseas and domestically. PM sells most of its computers locally to different departments at State University as well as to individuals and businesses in the immediate geographic region.

PM has enough regular production capacity to produce 160 computers per week. It can produce an additional 50 computers with overtime. The cost of assembling, inspecting, and packaging a computer during regular time is $190. Overtime production of a computer costs $260. Furthermore, it costs $10 per computer per week to hold a computer in inventory for future delivery. PM wants to meet all customer orders, with no shortages, to provide quality service. PM's order schedule for the next 6 weeks is as follows:

Week	*Computer Orders*
1	105
2	170
3	230
4	180
5	150
6	250

Management Science Application

Linear Programming Blending Applications in the Petroleum Industry

By the early 1950s, private enterprise had started using George Dantzig's linear programming methods. One of the first wide-scale important applications of linear programming was in the oil industry. As Dantzig later reminisced, "They started out with the simple problem of how to blend the gasoline for the right flash point, the right viscosity and the right octane and try to do it in the cheapest way possible." In 1955 Gifford Symonds wrote *Linear Programming: The Solution of Refinery Problems*, the first formal account of the application of linear programming in the petroleum industry (at Esso), which included such problems as blending aviation gasoline, the selection of crude oils to meet production requirements, and the determination of production rates and inventory levels to meet variable seasonal requirements. Today, refinery blending and production planning problems are still some of the most popular and effective applications of linear programming.

Different grades of gasoline are blended from a variety of available stocks that are intermediate refinery products, such as distilled gasoline, reformate gasoline, and catalytically cracked gasoline, plus additives such as octane enhancers. A typical refinery might have as many as 20 different components that it blends into 4 or more grades of gasoline. The properties or attributes of a blend are determined by a combination of properties that exist in the gasoline stocks (emanating from the original crude oil) and can include such things as vapor pressure, sulfur content, and octane value. A linear programming model would typically determine the volume of each blend subject to constraints for stock availability, demand, and the property specifications (e.g., recipe) for each blend. A typical linear programming model involving 4 blends might include more than 40 variables and more than 70 constraints.

Source: Based on C. E. Bodington and T. E. Baker, "A History of Mathematical Programming in the Petroleum Industry," *Interfaces* 20, no. 4 (July–August 1990): 117–27.

PM Computers wants to determine a schedule that will indicate how much regular and overtime production it will need each week to meet its orders at the minimum cost. The company wants no inventory left over at the end of the 6-week period.

Decision Variables

This model consists of three different sets of decision variables for computers produced during regular time each week, overtime production each week, and extra computers carried over as inventory each week. This results in a total of 17 decision variables. Because this problem contains a large number of decision variables, we will define them with names that will make them a little easier to keep up with:

$$r_j = \text{regular production of computers per week } j\ (j = 1, 2, 3, 4, 5, 6)$$
$$o_j = \text{overtime production of computers per week } j\ (j = 1, 2, 3, 4, 5, 6)$$
$$i_j = \text{extra computers carried over as inventory in week } j\ (j = 1, 2, 3, 4, 5)$$

This will result in 6 decision variables for regular production (r_1, r_2, etc.) and 6 decision variables for overtime production. There are only 5 decision variables for inventory because the problem stipulates that there is to be no inventory left over at the end of the 6-week period.

The Objective Function

The objective is to minimize total production cost, which is the sum of regular and overtime production costs and inventory costs:

$$\text{minimize } Z = \$190(r_1 + r_2 + r_3 + r_4 + r_5 + r_6) + 260(o_1 + o_2 + o_3 + o_4 + o_5 + o_6) + 10(i_1 + i_2 + i_3 + i_4 + i_5)$$

Model Constraints

This model includes three sets of constraints. The first two sets reflect the available capacity for regular and overtime production, and the third set establishes the production schedule per week.

The available regular production capacity of 160 computers for each of the 6 weeks results in six constraints, as follows:

$$r_j \leq 160 \text{ computers in week } j \ (j = 1, 2, 3, 4, 5, 6)$$

The available overtime production capacity of 50 computers per week also results in six constraints:

$$o_j \leq 50 \text{ computers in week } j \ (j = 1, 2, 3, 4, 5, 6)$$

The next set of six constraints shows the regular and overtime production and inventory necessary to meet the order requirement each week:

$$\begin{aligned}
\textit{week } 1&: r_1 + o_1 - i_1 = 105\\
\text{week } 2&: r_2 + o_2 + i_1 - i_2 = 170\\
\text{week } 3&: r_3 + o_3 + i_2 - i_3 = 230\\
\text{week } 4&: r_4 + o_4 + i_3 - i_4 = 180\\
\text{week } 5&: r_5 + o_5 + i_4 - i_5 = 150\\
\text{week } 6&: r_6 + o_6 + i_5 = 250
\end{aligned}$$

For example, 105 computers have been ordered in week 1. Regular production, r_1, and overtime production, o_1, will meet this order, whereas any extra production that might be needed in a future week to meet an order, i_1, is subtracted because it does not go to meet the order. Subsequently, in week 2 the order of 170 computers is met by regular and overtime production, $r_2 + o_2$, plus inventory carried over from week 1, i_1.

Model Summary

The complete model for this problem is summarized as follows:

$$\text{minimize } Z = \$190(r_1 + r_2 + r_3 + r_4 + r_5 + r_6) + \$260(o_1 + o_2 + o_3 + o_4 + o_5 + o_6) + 10(i_1 + i_2 + i_3 + i_4 + i_5)$$

subject to

$$\begin{aligned}
r_j &\leq 160 \ (j = 1, 2, 3, 4, 5, 6)\\
o_j &\leq 50 \ (j = 1, 2, 3, 4, 5, 6)\\
r_1 + o_1 - i_1 &= 105\\
r_2 + o_2 + i_1 - i_2 &= 170\\
r_3 + o_3 + i_2 - i_3 &= 230\\
r_4 + o_4 + i_3 - i_4 &= 180\\
r_5 + o_5 + i_4 - i_5 &= 150\\
r_6 + o_6 + i_5 &= 250\\
r_j, o_j, i_j &\geq 0
\end{aligned}$$

Management Science Application

Employee Scheduling with Management Science

One of the largest areas of application of management science models is employee scheduling, which is often generally referred to as the *capacity-planning problem*. The overall problem, and the subsequent model development, involves a staffing component—that is, how many, when, and where employees are needed—and a scheduling component, which determines when and where each employee is assigned to work. This problem is generally solved in two parts. The staffing problem is solved first using such management science modeling techniques as demand forecasting, queuing analysis, simulation, and/or inventory management. Once a staffing plan has been developed, a schedule is constructed that assigns employees to specific jobs or locations at specific times, using, most often, linear programming. A specific example of this type of problem is the nurse scheduling problem, in which a specific number of nurses in a hospital must be assigned to days and shifts over a specified time period. The solution objective of this problem is generally to minimize the number of nurses in order to avoid waste (and save money) while making sure there is adequate service for patient care. A good solution must also be able to satisfy hospital, legislative, and union policies and maintain nurse and patient morale and retention. Many solutions have been developed for the employee scheduling problem—and the nurse scheduling problem specifically—during the past 40 years (basically since high-speed computers became available). In a recent survey it was estimated that more than 570 systems were currently available from system software development vendors for the nurse scheduling problem alone.

© Richard Green/Alamy

Source: Based on D. Kellogg and S. Walczak, "Nurse Scheduling: From Academia to Implementation or Not?" *Interfaces* 37, no. 4 (July–August 2007): 355–69.

Computer Solution with Excel

The Excel spreadsheet for this multiperiod scheduling example is shown in Exhibit 4.20, and the accompanying Solver Parameters window is shown in Exhibit 4.21. The formula for the objective function is embedded in cell C13 and is shown on the formula bar at the top of the screen. The decision

EXHIBIT 4.20

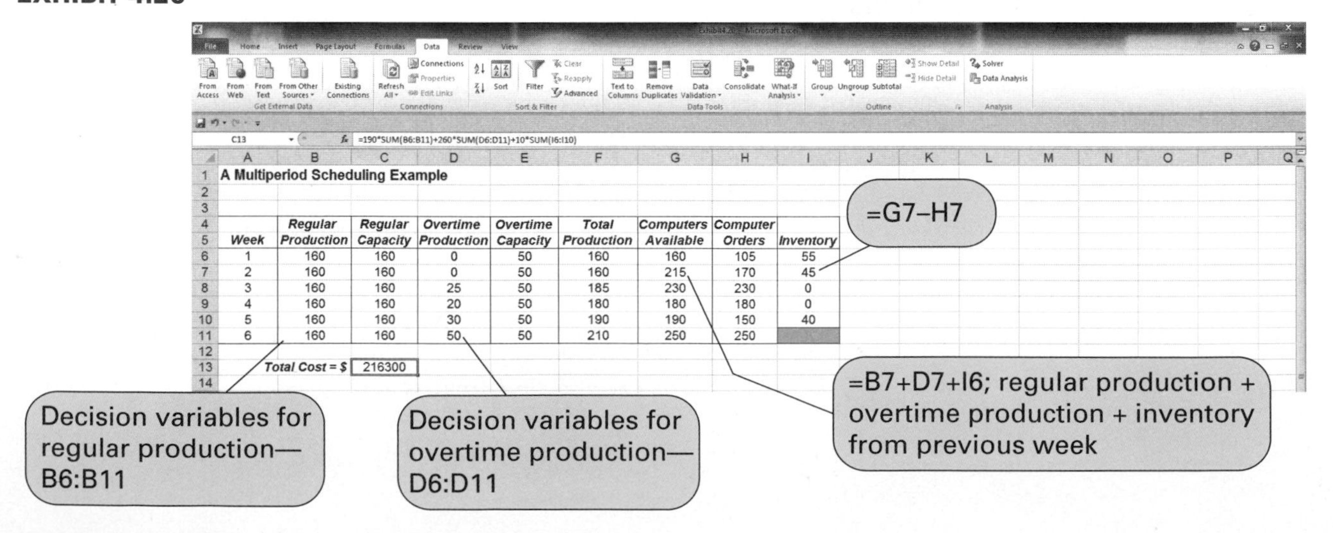

C13 =190*SUM(B6:B11)+260*SUM(D6:D11)+10*SUM(I6:I10)

A Multiperiod Scheduling Example

Week	Regular Production	Regular Capacity	Overtime Production	Overtime Capacity	Total Production	Computers Available	Computer Orders	Inventory
1	160	160	0	50	160	160	105	55
2	160	160	0	50	160	215	170	45
3	160	160	25	50	185	230	230	0
4	160	160	20	50	180	180	180	0
5	160	160	30	50	190	190	150	40
6	160	160	50	50	210	250	250	

Total Cost = $ 216300

EXHIBIT 4.21

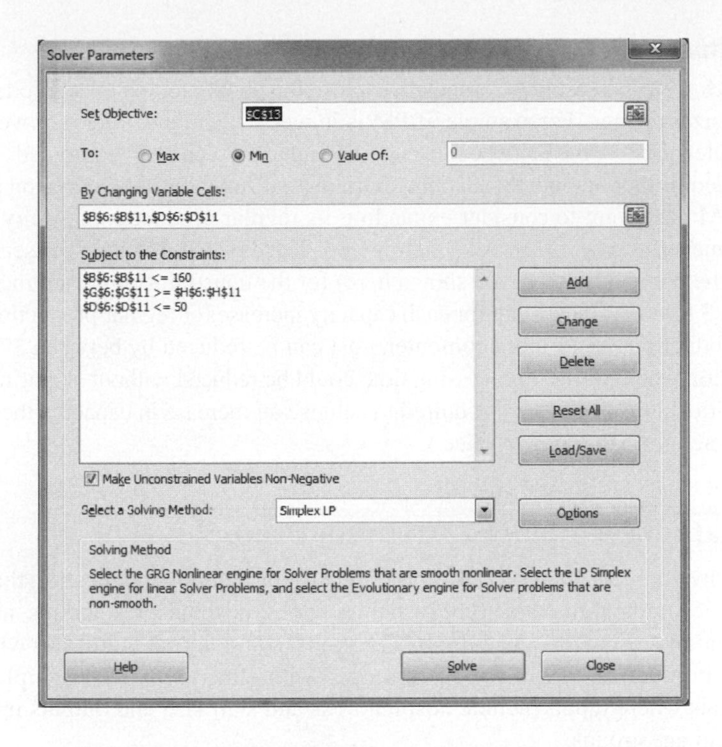

variables for regular production are included in cells **B6:B11**, whereas the variables for overtime production are in cells **D6:D11**. The computers available each week in column G includes the regular production plus overtime production plus the inventory from the previous week. For example, cell G7 (computers available in week 2) has the formula **=B7+D7+I6**. Cell I7 (the inventory left over in week 2) includes the formula **=G7–H7** (i.e., computers available minus computers ordered).

The actual constraint for week 2 would be **G7>=H7**. It is also necessary to enter the constraints for available regular and overtime production capacities into Solver. For example, the constraint for regular capacity would be **B6:B11≤160**.

Solution Analysis

The solution is

r_1 = 160 computers produced in regular time in week 1
r_2 = 160 computers produced in regular time in week 2
r_3 = 160 computers produced in regular time in week 3
r_4 = 160 computers produced in regular time in week 4
r_5 = 160 computers produced in regular time in week 5
r_6 = 160 computers produced in regular time in week 6
o_3 = 25 computers produced with overtime in week 3
o_4 = 20 computers produced with overtime in week 4
o_5 = 30 computers produced with overtime in week 5
o_6 = 50 computers produced with overtime in week 6
i_1 = 55 computers carried over in inventory in week 1
i_2 = 45 computers carried over in inventory in week 2
i_5 = 40 computers carried over in inventory in week 5

PM Computers must operate with regular production at full capacity during the entire 6-week period. It must also resort to overtime production in weeks 3, 4, 5, and 6 in order to meet heavier order volume in weeks 3 through 6.

Sensitivity Analysis

As each week passes, PM Computers can use this model to update its production schedule should order sizes change. For example, if PM is in week 2, and the orders for week 5 change from 150 computers to 200, the model can be altered and a new schedule computed. The model can also be extended further out into the planning horizon to develop future production schedules.

PM may want to consider expanding its regular production capacity because of the heavy overtime requirements in weeks 3 through 6. The dual values (from the sensitivity analysis report from the computer output not shown here) for the constraints representing regular production in weeks 3 through 6 show that for each capacity increase for regular production time that will enable the production of one more computer, cost can be reduced by between \$70 and \$80. (This total reduction would occur if processing time could be reduced without acquiring additional labor, etc. If additional resources were acquired to achieve an increase in capacity, the cost would have to be subtracted from the shadow price.)

A Data Envelopment Analysis Example

Data envelopment analysis (DEA) is a linear programming application that compares a number of service units of the same type—such as banks, hospitals, restaurants, and schools—based on their inputs (resources) and outputs. The model solution result indicates whether a particular unit is less productive, or inefficient, compared with other units. For example, DEA has compared hospitals where inputs include hospital beds and staff size and outputs include patient days for different age groups.

As an example of a DEA linear programming application, consider a town with four elementary schools—Alton, Beeks, Carey, and Delancey. The state has implemented a series of standards of learning (SOL) tests in reading, math, and history that all schools are required to administer to all students in the fifth grade. The average test scores are a measurable output of the school's performance. The school board has identified three key resources, or inputs, that affect a school's SOL scores—the teacher-to-student ratio, supplementary funds per student (i.e., funding generated by the PTA and other private sources over and above the normal budget), and the average educational level of the parents (where 12 = high school graduate, 16 = college graduate, etc.). These inputs and outputs are summarized as follows:

input 1 = teacher-to-student ratio
input 2 = supplementary funds/student
input 3 = average educational level of parents
output 1 = average reading SOL score
output 2 = average math SOL score
output 3 = average history SOL score

The actual input and output values for each school are:

	Inputs			Outputs		
School	1	2	3	1	2	3
Alton	.06	\$260	11.3	86	75	71
Beeks	.05	320	10.5	82	72	67
Carey	.08	340	12.0	81	79	80
Delancey	.06	460	13.1	81	73	69

For example, at Alton, the teacher-to-student ratio is 0.06 (or approximately 1 teacher per 16.67 students), there is \$260 per student supplemental funds, and the average parent educational grade level is 11.3. The average scores on Alton's reading, math, and history tests are 86, 75, and 71, respectively.

The school board wants to identify the school or schools that are less efficient in converting their inputs to outputs relative to the other elementary schools in the town. The DEA linear programming model will compare one particular school with all the others. For a complete analysis, a separate model is necessary for each school. For this example, we will evaluate the Delancey school as compared with the other schools.

Decision Variables

The formulation of a DEA linear programming model is not as readily apparent as the previous example models in this chapter. It is particularly difficult to define the decision variables in a meaningful way.

In a DEA model, the decision variables are defined as a price per unit of each output and each input. These are not the actual prices that inputs or outputs would be valued at. In economic terms, these prices are referred to as *implicit prices*, or *opportunity costs*. These prices are a relative valuation of inputs and outputs among the schools. They do not have much meaning for us, as will become apparent as the model is developed. For now, the decision variables will simply be defined as

$$x_i = \text{a price per unit of each output where } i = 1, 2, 3$$
$$y_i = \text{a price per unit of each input where } i = 1, 2, 3$$

The Objective Function

The objective of our model is to determine whether Delancey is efficient. In a DEA model, when we are attempting to determine whether a unit (i.e., a school) is efficient, it simplifies things to scale the input prices so that the total value of a unit's inputs equals 1. The efficiency of the unit will then equal the value of the unit's outputs. For this example, this means that once we scale Delancey's input prices so that they equal 1 (which we will do by formulating a constraint), the efficiency of Delancey will equal the value of its outputs. The objective, then, is to maximize the value of Delancey's outputs, which also equals efficiency:

$$\text{maximize } Z = 81x_1 + 73x_2 + 69x_3$$

Because the school's inputs will be scaled to 1, the maximum value that Delancey's outputs, as formulated in the objective function, can take on is 1. If the objective function equals 1, the school is efficient; if the objective function is less than 1, the school is inefficient.

Model Constraints

The constraint that will scale the Delancey school's inputs to 1 is formulated as

$$.06y_1 + 460y_2 + 13.1y_3 = 1$$

This sets the value of the inputs to 1 and ultimately forces the value of the outputs to 1 or less. This also has the effect of making the values of the decision variables even less meaningful.

The values of the outputs are forced to be 1 or less by the next set of constraints. In general terms, the efficiency of a school (or any productive unit) can be defined as

$$\text{efficiency} = \frac{\text{value of outputs}}{\text{value of inputs}}$$

It is not possible for a school or any service unit to be more than 100% efficient; thus, the efficiency of the school must be less than or equal to 1 and

$$\frac{\text{value of school's outputs}}{\text{value of school's inputs}} \leq 1$$

Converting this to standard linear form,

$$\text{value of school's outputs} \leq \text{value of school's inputs}$$

Substituting the model's decision variables and parameters for inputs and outputs into this general constraint form results in four constraints, one for each school:

$$86x_1 + 75x_2 + 71x_3 \leq .06y_1 + 260y_2 + 11.3y_3$$
$$82x_1 + 72x_2 + 67x_3 \leq .05y_1 + 320y_2 + 10.5y_3$$
$$81x_1 + 79x_2 + 80x_3 \leq .08y_1 + 340y_2 + 12.0y_3$$
$$81x_1 + 73x_2 + 69x_3 \leq .06y_1 + 460y_2 + 13.1y_3$$

Management Science Application

Evaluating American Red Cross Chapters Using DEA

The American Red Cross (ARC) is one of the world's largest nonprofit social service organizations. It provides disaster-relief services, armed forces emergency communications, and health and safety training to the general population. It is composed of approximately 1,000 chapters and 8 service-area offices that oversee the chapters, and its national headquarters is in Washington, DC. Chapters conduct local fund-raising campaigns to generate revenue, receive donations, receive fees for providing health and safety courses.

ARC chapters receive grants and contracts that support their service infrastructure, including facilities, materials, equipment, and employee recruiting, training and retention. Accordingly, the ARC has pressure from the public to demonstrate that it is responsible and accountable for the funds and resources it receives in delivering services and achieving outcomes. However, prior to 2001, while chapters reported performance data to the national headquarters, they were not individually evaluated or given feedback for improvement, and little information was provided regarding customer satisfaction. The ARC wanted to help its chapters by providing quantifiable performance measurement and assessment of how their inputs (i.e., resources, staff hours, money used to deliver services) are effectively converted into outputs (i.e., measures of the outcomes their service provided). To this end, a data envelopment analysis (DEA) model was developed.

The ARC's DEA model compares chapters according to their efficiency in transforming inputs into outputs across four activity areas: fund-raising, capacity creation and utilization,

the quantity and quality of services delivered, and outcome effectiveness. Each DEA model requires running a linear programming model for each chapter for these four activities each year—a total of 4,000 model runs—and the LP models are run using Excel. In addition to annual savings estimated at $700,000 for data analysis and reporting, the new DEA-based performance measurement and reporting system enabled the ARC to make recommendations to individual chapters for improvements based on comparisons with other chapters, and it also provided visible and transparent results that are understandable to nontechnical parties.

Source: Based on K. Pasupathy and A. Medina-Borja, "Integrating Excel, Access, and Visual Basic to Deploy Performance Measurement and Evaluation at the American Red Cross," *Interfaces* 38, no. 4 (July–August 2008): 324–37.

Model Summary

The complete linear programming model for determining the efficiency of the Delancey school is

$$
\begin{aligned}
&\text{maximize } Z = 81x_1 + 73x_2 + 69x_3\\
&\text{subject to}\\
&.06y_1 + 460y_2 + 13.1y_3 = 1\\
&\quad 86x_1 + 75x_2 + 71x_3 \leq .06y_1 + 260y_2 + 11.3y_3\\
&\quad 82x_1 + 72x_2 + 67x_3 \leq .05y_1 + 320y_2 + 10.5y_3\\
&\quad 81x_1 + 79x_2 + 80x_3 \leq .08y_1 + 340y_2 + 12.0y_3\\
&\quad 81x_1 + 73x_2 + 69x_3 \leq .06y_1 + 460y_2 + 13.1y_3\\
&\qquad\qquad\qquad x_i, y_i \geq 0
\end{aligned}
$$

The objective of this model is to determine whether Delancey is inefficient: If the value of the objective function equals 1, the school is efficient; if it is less than 1, it is inefficient. As mentioned previously, the values of the decision variables, x_i and y_i, have little meaning for us. They are the implicit prices of converting an input into an output, but they have been scaled to 1 to simplify the model. The model solution selects values of x_i and y_i that will maximize the school's efficiency, the maximum of which is 1, but these values have no easily interpretable meaning beyond that.

Computer Solution with Excel

The Excel spreadsheet for this DEA example is shown in Exhibit 4.22. The Solver Parameters window is shown in Exhibit 4.23. The decision variables are located in cells **B12:B14** and **D12:D14**.

The objective function value, Z, which is also the indicator of whether Delancey is inefficient, is shown in cell C16. This value is computed using the formula for the value of the Delancey school's output, **=B8*B12+C8*B13+D8*B14**, which is also shown in cell H8 and is a measure of the school's efficiency.

The output and input values for each school are embedded in cells **H5:H8** and **J5:J8**, respectively. These values are used to develop the model constraints shown in Solver as **H5:H8 ≤ J5:J8**. The scaling constraint is shown as **J8=1**, where the formula for the value of the inputs for the Delancey school, **=E8*D12+F8*D13+G8*D14**, is embedded in cell J8.

EXHIBIT 4.22

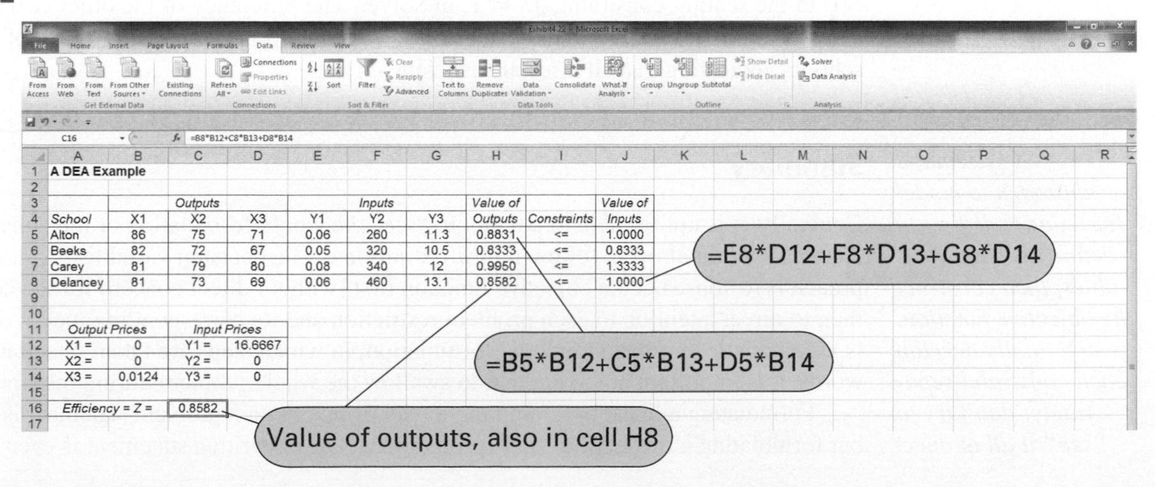

A DEA Example

School	X1	X2	X3	Y1	Y2	Y3	Value of Outputs	Constraints	Value of Inputs
Alton	86	75	71	0.06	260	11.3	0.8831	<=	1.0000
Beeks	82	72	67	0.05	320	10.5	0.8333	<=	0.8333
Carey	81	79	80	0.08	340	12	0.9950	<=	1.3333
Delancey	81	73	69	0.06	460	13.1	0.8582	<=	1.0000

Output Prices		Input Prices	
X1 =	0	Y1 =	16.6667
X2 =	0	Y2 =	0
X3 =	0.0124	Y3 =	0

Efficiency = Z = 0.8582

EXHIBIT 4.23

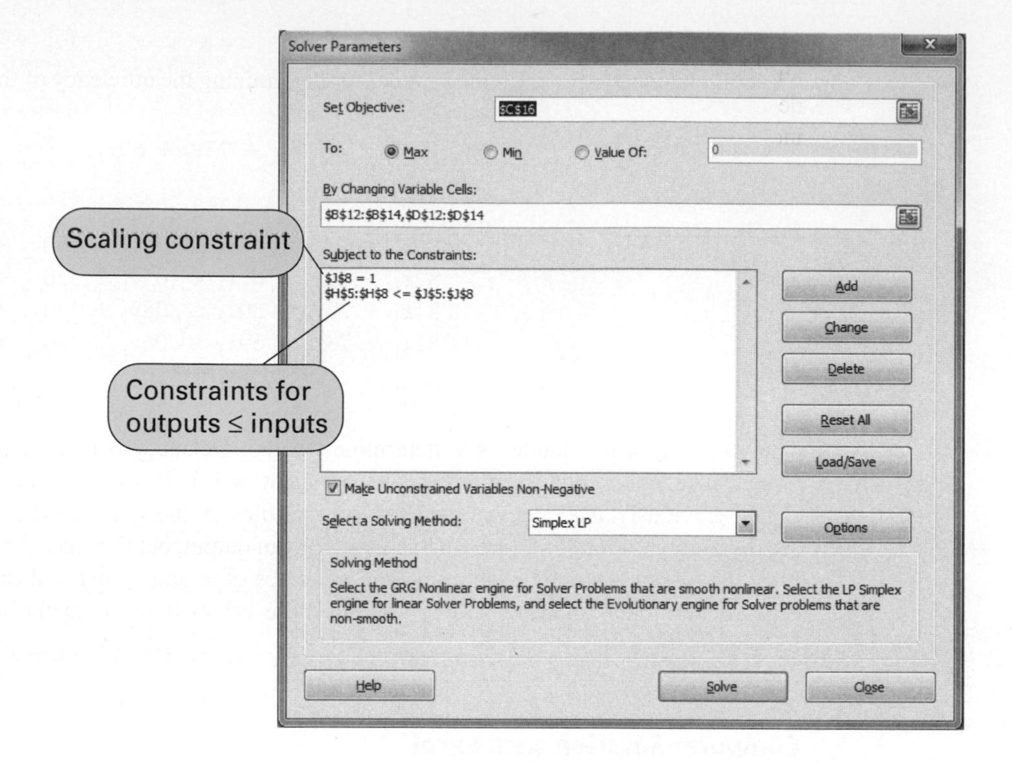

Solution Analysis

The only relevant solution value is the value of the objective function:

$$Z = 0.8582$$

Because this value is less than 1, the Delancey school is inefficient relative to the other schools. It is less efficient at converting its resources into outputs than the other schools. This means that a combination of efficient schools can achieve at least the same level of output as the Delancey school achieved with fewer input resources than required by the Delancey school. In retrospect, we can see that this is a logical result by looking back at the school's inputs and outputs. Although Delancey's input values are among the highest, its output test scores are among the lowest.

It is a simple process to access the efficiency of the other three schools in the town. For example, to ascertain the efficiency of the Alton school, make the value of its output in cell H5 the objective function in Solver, and make the formula for the value of the Alton school's inputs in cell I5 the scaling constraint, **J5 = 1**, in Solver. The efficiency of the other two schools can be assessed similarly. Doing so indicates an objective function value of 1 for each of the three other schools, indicating that all three are efficient.

Summary

A systematic approach to model formulation is first to define decision variables, then construct the objective function, and finally develop each constraint separately; don't try to "see" it all at once.

Generally, it is not feasible to attempt to "see" the whole formulation of the constraints and objective function at once, following the definition of the decision variables. A more prudent approach is to construct the objective function first (without direct concern for the constraints) and then to direct attention to each problem restriction and its corresponding model constraint. This is a systematic approach to model formulation, in which steps are taken one at a time. In other words, it is important not to attempt to swallow the whole problem during the first reading.

Formulating a linear programming model from a written problem statement is often difficult, but formulating a model of a "real" problem that has no written statement is even more difficult.

The steps for model formulation described in this section are generally followed; however, the problem must first be defined (i.e., a problem statement or some similar descriptive apparatus must be developed). Developing such a statement can be a formidable task, requiring the assistance of many individuals and units within an organization.

Developing the parameter values that are presented as givens in the written problem statements of this chapter frequently requires extensive data collection efforts. The objective function and model constraints can be very complex, requiring much time and effort to develop. Simply making sure that all the model constraints have been identified and no important problem restrictions have been omitted is difficult. Finally, the problems that one confronts in actual practice are typically much larger than those presented in this chapter. It is not uncommon for linear programming models of real problems to encompass hundreds of functional relationships and decision variables. Unfortunately, it is not possible in a textbook to re-create a realistic problem environment with no written problem statement and a model of large dimensions. What is possible is to provide the fundamentals of linear programming model formulation and solution—prerequisite to solving linear programming problems in actual practice.

Example Problem Solution

As a prelude to the problems, this section presents an example solution of the formulation and computer solution of a linear programming problem.

Problem Statement

Bark's Pet Food Company produces canned cat food called Meow Chow and canned dog food called Bow Chow. The company produces the pet food from horse meat, ground fish, and a cereal additive. Each week the company has 600 pounds of horse meat, 800 pounds of ground fish, and 1,000 pounds of cereal additive available to produce both kinds of pet food. Meow Chow must be at least half fish, and Bow Chow must be at least half horse meat. The company has 2,250 16-ounce cans available each week. A can of Meow Chow earns $0.80 in profit, and a can of Bow Chow earns $0.96 in profit. The company wants to know how many cans of Meow Chow and Bow Chow to produce each week in order to maximize profit.

A. Formulate a linear programming model for this problem.
B. Solve the model by using the computer.

Solution

A. Model Formulation

Step 1: Define the Decision Variables

This problem encompasses six decision variables, representing the amount of each ingredient i in pet food j:

x_{ij} = ounces of ingredient i in pet food j per week, where $i = h$ (horse meat), f (fish), and c (cereal), and $j = m$ (Meow Chow) and b (Bow Chow)

To determine the number of cans of each pet food produced per week, the total ounces of Meow Chow produced, $x_{hm} + x_{fm} + x_{cm}$, and the total amount of Bow Chow produced, $x_{hb} + x_{fb} + x_{cb}$, would each be divided by 16 ounces.

Step 2: Formulate the Objective Function

The objective function is to maximize the total profit earned each week, which is determined by multiplying the number of cans of each pet food produced by the profit per can. However, because the decision variables are defined in terms of ounces, they must be converted to equivalent cans by dividing by 16 ounces, as follows:

$$\text{maximize } Z = \frac{0.80}{16}(x_{hm} + x_{fm} + x_{cm}) + \frac{0.96}{16}(x_{hb} + x_{fb} + x_{cb})$$

or

$$\text{maximize } Z = \$0.05(x_{hm} + x_{fm} + x_{cm}) + 0.06(x_{hb} + x_{fb} + x_{cb})$$

Step 3: Formulate the Model Constraints

The first set of constraints represents the amount of each ingredient available each week. The problem provides these in terms of pounds of horse meat, fish, and cereal additives. Thus, because the decision variables are expressed as ounces, the ingredient amounts must be converted to ounces by multiplying each pound by 16 ounces. This results in these three constraints:

$$\begin{aligned} x_{hm} + x_{hb} &\leq 9{,}600 \text{ oz. of horse meat} \\ x_{fm} + x_{fb} &\leq 12{,}800 \text{ oz. of fish} \\ x_{cm} + x_{cb} &\leq 16{,}000 \text{ oz. of cereal additive} \end{aligned}$$

Next, there are two recipe requirements specifying that at least half of Meow Chow be fish and at least half of Bow Chow be horse meat. The requirement for Meow Chow is formulated as

$$\frac{x_{fm}}{x_{hm} + x_{fm} + x_{cm}} \geq \frac{1}{2}$$

or

$$-x_{hm} + x_{fm} - x_{cm} \geq 0$$

The constraint for Bow Chow is developed similarly:

$$\frac{x_{hb}}{x_{hb} + x_{fb} + x_{cb}} \geq \frac{1}{2}$$

or

$$x_{hb} - x_{fb} - x_{cb} \geq 0$$

Finally, the problem indicates that the company has 2,250 16-ounce cans available each week. These cans must also be converted to ounces to conform to our decision variables, which results in the following constraint:

$$x_{hm} + x_{fm} + x_{cm} + x_{hb} + x_{fb} + x_{cb} \leq 36{,}000 \text{ oz.}$$

Step 4: The Model Summary

The complete model is summarized as

$$\text{maximize } Z = \$0.05x_{hm} + 0.05x_{fm} + 0.05x_{cm} + 0.06x_{hb} + 0.06x_{fb} + 0.06x_{cb}$$

subject to

$$\begin{aligned} x_{hm} + x_{hb} &\leq 9{,}600 \\ x_{fm} + x_{fb} &\leq 12{,}800 \\ x_{cm} + x_{cb} &\leq 16{,}000 \\ -x_{hm} + x_{fm} - x_{cm} &\geq 0 \\ x_{hb} - x_{fb} - x_{cb} &\geq 0 \\ x_{hm} + x_{fm} + x_{cm} + x_{hb} + x_{fb} + x_{cb} &\leq 36{,}000 \\ x_{ij} &\geq 0 \end{aligned}$$

B. Computer Solution

Original Problem w/answers

Example Problem Solution

	Xhm	Xfm	Xcm	Xhb	Xfb	Xcb		RHS	Dual
Maximize	.05	.05	.05	.06	.06	.06			
Horse meat (oz)	1	0	0	1	0	0	<=	9,600	.02
Fish (oz)	0	1	0	0	1	0	<=	12,800	0
Cereal additive (oz)	0	0	1	0	0	1	<=	16,000	0
Meow Chow recipe	-1	1	-1	0	0	0	>=	0	0
Bow Chow recipe	0	0	0	1	-1	-1	>=	0	-.01
Cans (oz)	1	1	1	1	1	1	<=	36,000	.05
Solution->	0	8,400	8,400	9,600	4,400	5,200	Optimal Z->	1,992	

The computer solution for this model was generated by using the QM for Windows computer software package.

The solution is

$$x_{hm} = 0$$
$$x_{fm} = 8{,}400$$
$$x_{cm} = 8{,}400$$
$$x_{hb} = 9{,}600$$
$$x_{fb} = 4{,}400$$
$$x_{cb} = 5{,}200$$
$$Z = \$1{,}992$$

To determine the number of cans of each pet food, we must sum the ingredient amounts for each pet food and divide by 16 ounces (the size of a can):

$$x_{hm} + x_{fm} + x_{cm} = 0 + 8{,}400 + 8{,}400 = 16{,}800 \text{ oz. of Meow Chow}$$

or

$$16{,}800 \div 16 = 1{,}050 \text{ cans of Meow Chow}$$
$$x_{hb} + x_{fb} + x_{cb} = 9{,}600 + 4{,}400 + 5{,}200 = 19{,}200 \text{ oz. of Bow Chow}$$

or

$$19{,}200 \div 16 = 1{,}200 \text{ cans of Bow Chow}$$

Note that this model has multiple optimal solutions. An alternate optimal solution is $x_{fm} = 10{,}400$, $x_{cm} = 6{,}400$, $x_{hb} = 9{,}600$, and $x_{cb} = 9{,}600$. This converts to the same number of cans of each pet food; however, the ingredient mix per can is different.

Problems

1. In the product mix example in this chapter, Quick-Screen is considering adding some extra operators who would reduce processing times for each of the four clothing items by 10%. This would also increase the cost of each item by 10% and thus reduce unit profits by this same amount (because an increase in selling price would not be possible). Can this type of sensitivity analysis be evaluated using only original solution output, or will the model need to be solved again? Should Quick-Screen undertake this alternative?

   In this problem, the profit per shirt is computed from the selling price less fixed and variable costs. The computer solution output shows the shadow price for T-shirts to be $4.11. If Quick-Screen decided to acquire extra T-shirts, could the company expect to earn an additional $4.11 for each extra T-shirt it acquires above 500, up to the sensitivity range limit of T-shirts?

If Quick-Screen produced equal numbers of each of the four shirts, how would the company reformulate the linear programming model to reflect this condition? What is the new solution to this reformulated model?

2. In the diet example in this chapter, what would be the effect on the optimal solution of increasing the minimum calorie requirement for the breakfast to 500 calories? to 600 calories?
   Increase the breakfast calorie requirement to 700 calories and reformulate the model to establish upper limits on the servings of each food item to what you think would be realistic and appetizing. Determine the solution for this reformulated model.

3. In the investment example in this chapter, how would the solution be affected if the requirement that the entire $70,000 be invested was relaxed such that it is the maximum amount available for investment?
   If the entire amount available for investment does not have to be invested and the amount available is increased by $10,000 (to $80,000), how much will the total optimal return increase? Will the entire $10,000 increase be invested in one alternative?

4. For the marketing example in this chapter, if the budget is increased by $20,000, how much will audience exposures be increased?
   If Biggs Department Store wanted the same total number of people exposed to each of the three types of advertisements, how should the linear programming model be reformulated? What would be the new solution for this reformulated model?

5. For the transportation example in this chapter, suppose that television sets not shipped were to incur storage costs of $9 at Cincinnati, $6 at Atlanta, and $7 at Pittsburgh. How would these storage costs be reflected in the linear programming model for this example problem, and what would the new solution be, if any?
   The Zephyr Television Company is considering leasing a new warehouse in Memphis. The new warehouse would have a supply of 200 television sets, with shipping costs of $18 to New York, $9 to Dallas, and $12 to Detroit. If the total transportation cost for the company (ignoring the cost of leasing the warehouse) is less than with the current warehouses, the company will lease the new warehouse. Should the warehouse be leased?
   If supply could be increased at any one warehouse, which should it be? What restrictions would there be on the amount of the increase?

6. For the blend example in this chapter, if the requirement that "at least 3,000 barrels of each grade of motor oil" was changed to *exactly* 3,000 barrels of each grade, how would this affect the optimal solution?
   If the company could acquire more of one of the three components, which should it be? What would be the effect on the total profit of acquiring more of this component?

7. On their farm, the Friendly family grows apples that they harvest each fall and make into three products—apple butter, applesauce, and apple jelly. They sell these three items at several local grocery stores, at craft fairs in the region, and at their own Friendly Farm Pumpkin Festival for 2 weeks in October. Their three primary resources are cooking time in their kitchen, their own labor time, and the apples. They have a total of 500 cooking hours available, and it requires 3.5 hours to cook a 10-gallon batch of apple butter, 5.2 hours to cook 10 gallons of applesauce, and 2.8 hours to cook 10 gallons of jelly. A 10-gallon batch of apple butter requires 1.2 hours of labor, a batch of sauce takes 0.8 hour, and a batch of jelly requires 1.5 hours. The Friendly family has 240 hours of labor available during the fall. They produce about 6,500 apples each fall. A batch of apple butter requires 40 apples, a 10-gallon batch of applesauce requires 55 apples, and a batch of jelly requires 20 apples. After the products are canned, a batch of apple butter will generate $190 in sales revenue, a batch of applesauce will generate a sales revenue of $170, and a batch of jelly will generate sales revenue of $155. The

Friendlys want to know how many batches of apple butter, applesauce, and apple jelly to produce in order to maximize their revenues.

a. Formulate a linear programming model for this problem.
b. Solve the model by using the computer.

8. a. If the Friendlys in Problem 7 were to use leftover apples to feed livestock, at an estimated cost savings worth $0.08 per apple in revenue, how would this affect the model and solution?
   b. Instead of feeding the leftover apples to the livestock, the Friendlys are thinking about producing apple cider. Cider will require 1.5 hours of cooking, 0.5 hour of labor, and 60 apples per batch, and it will sell for $45 per batch. Should the Friendlys use all their apples and produce cider along with their other three products?

9. A hospital dietitian prepares breakfast menus every morning for the hospital patients. Part of the dietitian's responsibility is to make sure that minimum daily requirements for vitamins A and B are met. At the same time, the cost of the menus must be kept as low as possible. The main breakfast staples providing vitamins A and B are eggs, bacon, and cereal. The vitamin requirements and vitamin contributions for each staple follow:

	Vitamin Contributions			
Vitamin	**mg/Egg**	**mg/Bacon Strip**	**mg/Cereal Cup**	**Minimum Daily Requirements**
A	2	4	1	16
B	3	2	1	12

An egg costs $0.04, a bacon strip costs $0.03, and a cup of cereal costs $0.02. The dietitian wants to know how much of each staple to serve per order to meet the minimum daily vitamin requirements while minimizing total cost.

a. Formulate a linear programming model for this problem.
b. Solve the model by using the computer.

10. Lakeside Boatworks is planning to manufacture three types of molded fiberglass recreational boats—a fishing (bass) boat, a ski boat, and a small speedboat. The estimated selling price and variable cost for each type of boat are summarized in the following table:

Boat	Variable Cost	Selling Price
Bass	$12,500	$23,000
Ski	8,500	18,000
Speed	13,700	26,000

The company has incurred fixed costs of $2,800,000 to set up its manufacturing operation and begin production. Lakeside has also entered into agreements with several boat dealers in the region to provide a minimum of 70 bass boats, 50 ski boats, and 50 speedboats. Alternatively, the company is unsure of what actual demand will be, so it has decided to limit production to no more than 120 of any one boat. The company wants to determine the number of boats that it must sell to break even while minimizing its total variable cost.

a. Formulate a linear programming model for this problem.
b. Solve the model by using the computer.

11. The Pyrotec Company produces three electrical products—clocks, radios, and toasters. These products have the following resource requirements:

	Resource Requirements	
	Cost/Unit	Labor Hours/Unit
Clock	$7	2
Radio	10	3
Toaster	5	2

The manufacturer has a daily production budget of $2,000 and a maximum of 660 hours of labor. Maximum daily customer demand is for 200 clocks, 300 radios, and 150 toasters. Clocks sell for $15, radios for $20, and toasters for $12. The company wants to know the optimal product mix that will maximize profit.

a. Formulate a linear programming model for this problem.
b. Solve the model by using the computer.

12. Betty Malloy, owner of the Eagle Tavern in Pittsburgh, is preparing for Super Bowl Sunday, and she must determine how much beer to stock. Betty stocks three brands of beer—Yodel, Shotz, and Rainwater. The cost per gallon (to the tavern owner) of each brand is as follows:

Brand	Cost/Gallon
Yodel	$1.50
Shotz	0.90
Rainwater	0.50

The tavern has a budget of $2,000 for beer for Super Bowl Sunday. Betty sells Yodel at a rate of $3.00 per gallon, Shotz at $2.50 per gallon, and Rainwater at $1.75 per gallon. Based on past football games, Betty has determined the maximum customer demand to be 400 gallons of Yodel, 500 gallons of Shotz, and 300 gallons of Rainwater. The tavern has the capacity to stock 1,000 gallons of beer; Betty wants to stock up completely. Betty wants to determine the number of gallons of each brand of beer to order so as to maximize profit.

a. Formulate a linear programming model for this problem.
b. Solve the model by using the computer.

13. The Kalo Fertilizer Company produces two brands of lawn fertilizer—Super Two and Green Grow—at plants in Fresno, California, and Dearborn, Michigan. The plant at Fresno has resources available to produce 5,000 pounds of fertilizer daily; the plant at Dearborn has enough resources to produce 6,000 pounds daily. The cost per pound of producing each brand at each plant is as follows:

	Plant	
Product	Fresno	Dearborn
Super Two	$2	$4
Green Grow	2	3

The company has a daily budget of $45,000 for both plants combined. Based on past sales, the company knows the maximum demand (converted to a daily basis) is 6,000 pounds for Super

Two and 7,000 pounds for Green Grow. The selling price is $9 per pound for Super Two and $7 per pound for Green Grow. The company wants to know the number of pounds of each brand of fertilizer to produce at each plant in order to maximize profit.

a. Formulate a linear programming model for this problem.
b. Solve the model by using the computer.

14. Grafton Metalworks Company produces metal alloys from six different ores it mines. The company has an order from a customer to produce an alloy that contains four metals according to the following specifications: at least 21% of metal A, no more than 12% of metal B, no more than 7% of metal C, and between 30% and 65% of metal D. The proportion of the four metals in each of the six ores and the level of impurities in each ore are provided in the following table:

	Metal (%)					
Ore	A	B	C	D	**Impurities (%)**	**Cost/Ton**
1	19	15	12	14	40	$27
2	43	10	25	7	15	25
3	17	0	0	53	30	32
4	20	12	0	18	50	22
5	0	24	10	31	35	20
6	12	18	16	25	29	24

When the metals are processed and refined, the impurities are removed.

The company wants to know the amount of each ore to use per ton of the alloy that will minimize the cost per ton of the alloy.

a. Formulate a linear programming model for this problem.
b. Solve the model by using the computer.

15. The Roadnet Transport Company expanded its shipping capacity by purchasing 90 trailer trucks from a bankrupt competitor. The company subsequently located 30 of the purchased trucks at each of its shipping warehouses in Charlotte, Memphis, and Louisville. The company makes shipments from each of these warehouses to terminals in St. Louis, Atlanta, and New York. Each truck is capable of making one shipment per week. The terminal managers have indicated their capacity for extra shipments. The manager at St. Louis can accommodate 40 additional trucks per week, the manager at Atlanta can accommodate 60 additional trucks, and the manager at New York can accommodate 50 additional trucks. The company makes the following profit per truckload shipment from each warehouse to each terminal. The profits differ as a result of differences in products shipped, shipping costs, and transport rates:

	Terminal		
Warehouse	St. Louis	Atlanta	New York
Charlotte	$1,800	$2,100	$1,600
Memphis	1,000	700	900
Louisville	1,400	800	2,200

The company wants to know how many trucks to assign to each route (i.e., warehouse to terminal) to maximize profit.

a. Formulate a linear programming model for this problem.
b. Solve the model by using the computer.

16. The Hickory Cabinet and Furniture Company produces sofas, tables, and chairs at its plant in Greensboro, North Carolina. The plant uses three main resources to make furniture—wood, upholstery, and labor. The resource requirements for each piece of furniture and the total resources available weekly are as follows:

	Resource Requirements		
	Wood (board ft.)	Upholstery (yd.)	Labor (hr.)
Sofa	7	12	6
Table	5	—	9
Chair	4	7	5
Total available resources	2,250	1,000	240

The furniture is produced on a weekly basis and stored in a warehouse until the end of the week, when it is shipped out. The warehouse has a total capacity of 650 pieces of furniture. Each sofa earns $400 in profit, each table, $275, and each chair, $190. The company wants to know how many pieces of each type of furniture to make per week to maximize profit.

a. Formulate a linear programming model for this problem.
b. Solve the model by using the computer.

17. Lawns Unlimited is a lawn care and maintenance company. One of its services is to seed new lawns as well as bare or damaged areas in established lawns. The company uses three basic grass seed mixes it calls Home 1, Home 2, and Commercial 3. It uses three kinds of grass seed—tall fescue, mustang fescue, and bluegrass. The requirements for each grass mix are as follows:

Mix	Mix Requirements
Home 1	No more than 50% tall fescue
	At least 20% mustang fescue
Home 2	At least 30% bluegrass
	At least 30% mustang fescue
	No more than 20% tall fescue
Commercial 3	At least 50% but no more than 70% tall fescue
	At least 10% bluegrass

The company believes it needs to have at least 1,200 pounds of Home 1 mix, 900 pounds of Home 2 mix, and 2,400 pounds of Commercial 3 seed mix on hand. A pound of tall fescue costs the company $1.70, a pound of mustang fescue costs $2.80, and a pound of bluegrass costs $3.25. The company wants to know how many pounds of each type of grass seed to purchase to minimize cost.

a. Formulate a linear programming model for this problem.
b. Solve this model by using the computer.

18. Alexandra Bergson has subdivided her 2,000-acre farm into three plots and has contracted with three local farm families to operate the plots. She has instructed each sharecropper to plant three crops: corn, peas, and soybeans. The size of each plot has been determined by the capabilities of each local farmer. Plot sizes, crop restrictions, and profit per acre are given in the following tables:

Plot	Acreage
1	500
2	800
3	700

	Maximum Acreage	Profit/Acre
Corn	900	$600
Peas	700	450
Soybeans	1,000	300

Any of the three crops may be planted on any of the plots; however, Alexandra has placed several restrictions on the farming operation. At least 60% of each plot must be under cultivation. Further, to ensure that each sharecropper works according to his or her potential and resources (which determined the acreage allocation), she wants the same proportion of each plot to be under cultivation. Her objective is to determine how much of each crop to plant on each plot to maximize profit.

a. Formulate a linear programming model for this problem.
b. Solve the model by using the computer.

19. As a result of a recently passed bill, a congressman's district has been allocated $4 million for programs and projects. It is up to the congressman to decide how to distribute the money. The congressman has decided to allocate the money to four ongoing programs because of their importance to his district—a job training program, a parks project, a sanitation project, and a mobile library. However, the congressman wants to distribute the money in a manner that will please the most voters, or, in other words, gain him the most votes in the upcoming election. His staff's estimates of the number of votes gained per dollar spent for the various programs are as follows:

Program	Votes/Dollar
Job training	0.02
Parks	0.09
Sanitation	0.06
Mobile library	0.04

In order also to satisfy several local influential citizens who financed his election, he is obligated to observe the following guidelines:

- None of the programs can receive more than 40% of the total allocation.
- The amount allocated to parks cannot exceed the total allocated to both the sanitation project and the mobile library.
- The amount allocated to job training must at least equal the amount spent on the sanitation project.

Any money not spent in the district will be returned to the government; therefore, the congressman wants to spend it all. The congressman wants to know the amount to allocate to each program to maximize his votes.

a. Formulate a linear programming model for this problem.
b. Solve the model by using the computer.

20. Anna Broderick is the dietitian for the State University football team, and she is attempting to determine a nutritious lunch menu for the team. She has set the following nutritional guidelines for each lunch serving:

- Between 1,500 and 2,000 calories
- At least 5 mg of iron
- At least 20 but no more than 60 g of fat
- At least 30 g of protein
- At least 40 g of carbohydrates
- No more than 30 mg of cholesterol

She selects the menu from seven basic food items, as follows, with the nutritional contribution per pound and the cost as given:

	Calories (per lb.)	Iron (mg/lb.)	Protein (g/lb.)	Carbohydrates (g/lb.)	Fat (g/lb.)	Cholesterol (mg/lb.)	$/lb.
Chicken	520	4.4	17	0	30	180	0.80
Fish	500	3.3	85	0	5	90	3.70
Ground beef	860	0.3	82	0	75	350	2.30
Dried beans	600	3.4	10	30	3	0	0.90
Lettuce	50	0.5	6	0	0	0	0.75
Potatoes	460	2.2	10	70	0	0	0.40
Milk (2%)	240	0.2	16	22	10	20	0.83

The dietitian wants to select a menu to meet the nutritional guidelines while minimizing the total cost per serving.

a. Formulate a linear programming model for this problem.
b. Solve the model by using the computer.
c. If a serving of each of the food items (other than milk) was limited to no more than a half pound, what effect would this have on the solution?

21. The Midland Tool Shop has four heavy presses it uses to stamp out prefabricated metal covers and housings for electronic consumer products. All four presses operate differently and are of different sizes. Currently the firm has a contract to produce three products. The contract calls for 400 units of product 1; 570 units of product 2; and 320 units of product 3. The time (in minutes) required for each product to be produced on each machine is as follows:

	Machine			
Product	**1**	**2**	**3**	**4**
1	35	41	34	39
2	40	36	32	43
3	38	37	33	40

Machine 1 is available for 150 hours, machine 2 for 240 hours, machine 3 for 200 hours, and machine 4 for 250 hours. The products also result in different profits, according to the machine they are produced on, because of time, waste, and operating cost. The profit per unit per machine for each product is summarized as follows:

	Machine			
Product	**1**	**2**	**3**	**4**
1	$7.8	$7.8	$8.2	$7.9
2	6.7	8.9	9.2	6.3
3	8.4	8.1	9.0	5.8

The company wants to know how many units of each product to produce on each machine in order to maximize profit.

a. Formulate this problem as a linear programming model.
b. Solve the model by using the computer.

22. The Cabin Creek Coal (CCC) Company operates three mines in Kentucky and West Virginia, and it supplies coal to four utility power plants along the East Coast. The cost of shipping coal from each mine to each plant, the capacity at each of the three mines, and the demand at each plant are shown in the following table:

	Plant				
Mine	**1**	**2**	**3**	**4**	**Mine Capacity (tons)**
1	$7	$9	$10	$12	220
2	9	7	8	12	170
3	11	14	5	7	280
Demand (tons)	110	160	90	180	

The cost of mining and processing coal is $62 per ton at mine 1, $67 per ton at mine 2, and $75 per ton at mine 3. The percentage of ash and sulfur content per ton of coal at each mine is as follows:

Mine	% Ash	% Sulfur
1	9	6
2	5	4
3	4	3

Each plant has different cleaning equipment. Plant 1 requires that the coal it receives have no more than 6% ash and 5% sulfur; plant 2 coal can have no more than 5% ash and sulfur combined; plant 3 can have no more than 5% ash and 7% sulfur; and plant 4 can have no more than 6% ash and sulfur combined. CCC wants to determine the amount of coal to produce at each mine and ship to its customers that will minimize its total cost.

a. Formulate a linear programming model for this problem.
b. Solve this model by using the computer.

23. Ampco is a manufacturing company that has a contract to supply a customer with parts from April through September. However, Ampco does not have enough storage space to store the parts during this period, so it needs to lease extra warehouse space during the 6-month period. Following are Ampco's space requirements:

Month	Required Space (ft.^{2})
April	47,000
May	35,000
June	52,000
July	27,000
August	19,000
September	15,000

The rental agent Ampco is dealing with has provided it with the following cost schedule for warehouse space. This schedule shows that the longer the space is rented, the cheaper it is. For example, if Ampco rents space for all 6 months, it costs \$1.00 per square foot per month, whereas if it rents the same space for only 1 month, it costs \$1.70 per square foot per month:

Rental Period (months)	\$/ft.^{2}/Month
6	\$1.00
5	1.05
4	1.10
3	1.20
2	1.40
1	1.70

Ampco can rent any amount of warehouse space on a monthly basis at any time for any number of (whole) months. Ampco wants to determine the least costly rental agreement that will exactly meet its space needs each month and avoid having any unused space.

a. Formulate a linear programming model for this problem.
b. Solve the model by using the computer.
c. Suppose that Ampco relaxed its restriction that it rent exactly the space it needs every month such that it would rent excess space if it was cheaper. How would this affect the optimal solution?

24. Brooks City has three consolidated high schools, each with a capacity of 1,200 students. The school board has partitioned the city into five busing districts—north, south, east, west, and central—each with different high school student populations. The three schools are located in the central, west, and south districts. Some students must be bused outside their districts, and the school board wants to minimize the total bus distance traveled by these students. The average distances from each district to the three schools and the total student population in each district are as follows:

	Distance (miles)			
District	Central School	West School	South School	Student Population
North	8	11	14	700
South	12	9	—	300
East	9	16	10	900
West	8	—	9	600
Central	—	8	12	500

The school board wants to determine the number of students to bus from each district to each school to minimize the total busing miles traveled.

a. Formulate a linear programming model for this problem.
b. Solve the model by using the computer.

25. a. In Problem 24 the school board decided that because all students in the north and east districts must be bused, then at least 50% of the students who live in the south, west, and central districts must also be bused to another district. Reformulate the linear programming model to reflect this new set of constraints and solve by using the computer.
b. The school board has further decided that the enrollment at all three high schools should be equal. Formulate this additional restriction in the linear programming model and solve by using the computer.

26. The Southfork Feed Company makes a feed mix from four ingredients—oats, corn, soybeans, and a vitamin supplement. The company has 300 pounds of oats, 400 pounds of corn, 200 pounds of soybeans, and 100 pounds of vitamin supplement available for the mix. The company has the following requirements for the mix:
    - At least 30% of the mix must be soybeans.
    - At least 20% of the mix must be the vitamin supplement.
    - The ratio of corn to oats cannot exceed 2 to 1.
    - The amount of oats cannot exceed the amount of soybeans.
    - The mix must be at least 500 pounds.

    A pound of oats costs $0.50; a pound of corn, $1.20; a pound of soybeans, $0.60; and a pound of vitamin supplement, $2.00. The feed company wants to know the number of pounds of each ingredient to put in the mix in order to minimize cost.
    a. Formulate a linear programming model for this problem.
    b. Solve the model by using the computer.

27. The United Charities annual fund-raising drive is scheduled to take place next week. Donations are collected during the day and night, by telephone, and through personal contact. The average donation resulting from each type of contact is as follows:

	Phone	Personal
Day	$2	$4
Night	3	7

The charity group has enough donated gasoline and cars to make at most 300 personal contacts during one day and night combined. The volunteer minutes required to conduct each type of interview are as follows:

	Phone (min.)	Personal (min.)
Day	6	15
Night	5	12

The charity has 20 volunteer hours available each day and 40 volunteer hours available each night. The chairperson of the fund-raising drive wants to know how many different types of contacts to schedule in a 24-hour period (i.e., 1 day and 1 night) to maximize total donations.
a. Formulate a linear programming model for this problem.
b. Solve the model by using the computer.

28. Ronald Thump is interested in expanding his firm. After careful consideration, he has determined three areas in which he might invest additional funds: (1) product research and development, (2) manufacturing operations improvements, and (3) advertising and sales promotion. He has $500,000 available for investment in the firm. He can invest in its advertising and sales promotion program every year, and each dollar invested in this manner is expected to yield a return of the amount invested plus 20% yearly. He can invest in manufacturing operations improvements every 2 years, with an expected return of the investment plus 30% (at the end of each 2-year period). An investment in product research and development would be for a 3-year period, with an expected return of the investment plus 50% (at the end of the 3-year period). To diversify the total initial investment, he wishes to include the requirement that at least $30,000 must be invested in the advertising and sales promotion program, at least $40,000 in

manufacturing operations improvements, and at least $50,000 in product research and development initially (at the beginning of the first year). Ronald wants to know how much should be invested in each of the three alternatives, during each year of a 4-year period, to maximize the total ending cash value of the initial $500,000 investment.

a. Formulate a linear programming model for this problem.
b. Solve the model by using the computer.

29. Iggy Olweski, a professional football player, is retiring, and he is thinking about going into the insurance business. He plans to sell three types of policies—homeowner's insurance, auto insurance, and life insurance. The average amount of profit returned per year by each type of insurance policy is as follows:

Policy	Yearly Profit/Policy
Homeowner's	$35
Auto	20
Life	58

Each homeowner's policy will cost $14 to sell and maintain; each auto policy, $12; and each life insurance policy, $35. Iggy has projected a budget of $35,000 per year. In addition, the sale of a homeowner's policy will require 6 hours of effort; the sale of an auto policy, 3 hours; and the sale of a life insurance policy, 12 hours. Based on the number of working hours he and several employees could contribute, Iggy has estimated that he would have available 20,000 hours per year. Iggy wants to know how many of each type of insurance policy he would have to sell each year in order to maximize profit.

a. Formulate a linear programming model for this problem.
b. Solve the model by using the computer.

30. A publishing house publishes three weekly magazines—*Daily Life, Agriculture Today*, and *Surf's Up*. Publication of one issue of each of the magazines requires the following amounts of production time and paper:

	Production (hr.)	Paper (lb.)
Daily Life	0.01	0.2
Agriculture Today	0.03	0.5
Surf's Up	0.02	0.3

Each week the publisher has available 120 hours of production time and 3,000 pounds of paper. Total circulation for all three magazines must exceed 5,000 issues per week if the company is to keep its advertisers. The selling price per issue is $2.25 for *Daily Life*, $4.00 for *Agriculture Today*, and $1.50 for *Surf's Up*. Based on past sales, the publisher knows that the maximum weekly demand for *Daily Life* is 3,000 issues; for *Agriculture Today*, 2,000 issues; and for *Surf's Up*, 6,000 issues. The production manager wants to know the number of issues of each magazine to produce weekly in order to maximize total sales revenue.

a. Formulate a linear programming model for this problem.
b. Solve the model by using the computer.

31. The manager of a department store in Seattle is attempting to decide on the types and amounts of advertising the store should use. He has invited representatives from the local radio station,

television station, and newspaper to make presentations in which they describe their audiences. The television station representative indicates that a TV commercial, which costs $15,000, would reach 25,000 potential customers. The breakdown of the audience is as follows:

	Male	Female
Senior	5,000	5,000
Young	5,000	10,000

The newspaper representative claims to be able to provide an audience of 10,000 potential customers at a cost of $4,000 per ad. The breakdown of the audience is as follows:

	Male	Female
Senior	4,000	3,000
Young	2,000	1,000

The radio station representative says that the audience for one of the station's commercials, which costs $6,000, is 15,000 customers. The breakdown of the audience is as follows:

	Male	Female
Senior	1,500	1,500
Young	4,500	7,500

The store has the following advertising policy:

- Use at least twice as many radio commercials as newspaper ads.
- Reach at least 100,000 customers.
- Reach at least twice as many young people as senior citizens.
- Make sure that at least 30% of the audience is female.

Available space limits the number of newspaper ads to seven. The store wants to know the optimal number of each type of advertising to purchase to minimize total cost.

a. Formulate a linear programming model for this problem.
b. Solve the model by using the computer.
c. Suppose a second radio station approaches the department store and indicates that its commercials, which cost $7,500, reach 18,000 customers with the following demographic breakdown:

	Male	Female
Senior	2,400	3,600
Young	4,000	8,000

If the store considered this station along with the other media alternatives, how would this affect the solution?

32. The Mill Mountain Coffee Shop blends coffee on the premises for its customers. It sells three basic blends in 1-pound bags, Special, Mountain Dark, and Mill Regular. It uses four different types of coffee to produce the blends—Brazilian, mocha, Colombian, and mild. The shop has the following blend recipe requirements:

Blend	Mix Requirements	Selling Price/Pound
Special	At least 40% Columbian, at least 30% mocha	$6.50
Dark	At least 60% Brazilian, no more than 10% mild	5.25
Regular	No more than 60% mild, at least 30% Brazilian	3.75

The cost of Brazilian coffee is $2.00 per pound, the cost of mocha is $2.75 per pound, the cost of Colombian is $2.90 per pound, and the cost of mild is $1.70 per pound. The shop has 110 pounds of Brazilian coffee, 70 pounds of mocha, 80 pounds of Colombian, and 150 pounds of mild coffee available per week. The shop wants to know the amount of each blend it should prepare each week to maximize profit.

a. Formulate a linear programming model for this problem.
b. Solve this model by using the computer.

33. Toyz is a large discount toy store in Valley Wood Mall. The store typically has slow sales in the summer months that increase dramatically and rise to a peak at Christmas. During the summer and fall, the store must build up its inventory to have enough stock for the Christmas season. To purchase and build up its stock during the months when its revenues are low, the store borrows money.

Following is the store's projected revenue and liabilities schedule for July through December (where revenues are received and bills are paid at the first of each month):

Month	Revenues	Liabilities
July	$20,000	$60,000
August	30,000	60,000
September	40,000	80,000
October	50,000	30,000
November	80,000	30,000
December	100,000	20,000

At the beginning of July, the store can take out a 6-month loan that carries an 11% interest rate and must be paid back at the end of December. The store cannot reduce its interest payment by paying back the loan early. The store can also borrow money monthly at a rate of 5% interest per month.

Money borrowed on a monthly basis must be paid back at the beginning of the next month. The store wants to borrow enough money to meet its cash flow needs while minimizing its cost of borrowing.

a. Formulate a linear programming model for this problem.
b. Solve this model by using the computer.
c. What would the effect be on the optimal solution if Toyz could secure a 9% interest rate for a 6-month loan from another bank?

34. The Skimmer Boat Company manufactures the Water Skimmer bass fishing boat. The company purchases the engines it installs in its boats from the Mar-gine Company, which

specializes in marine engines. Skimmer has the following production schedule for April, May, June, and July:

Month	Production
April	60
May	85
June	100
July	120

Mar-gine usually manufactures and ships engines to Skimmer during the month the engines are due. However, from April through July, Mar-gine has a large order with another boat customer, and it can manufacture only 40 engines in April, 60 in May, 90 in June, and 50 in July. Mar-gine has several alternative ways to meet Skimmer's production schedule. It can produce up to 30 engines in January, February, and March and carry them in inventory at a cost of $50 per engine per month until it ships them to Skimmer. For example, Mar-gine could build an engine in January and ship it to Skimmer in April, incurring $150 in inventory charges. Mar-gine can also manufacture up to 20 engines in the month they are due on an overtime basis, with an additional cost of $400 per engine. Mar-gine wants to determine the least costly production schedule that will meet Skimmer's schedule.

a. Formulate a linear programming model for this problem.
b. Solve this model by using the computer.
c. If Mar-gine were able to increase its production capacity in January, February, and March from 30 to 40 engines, what would the effect be on the optimal solution?

35. The Donnor meat processing firm produces wieners from four ingredients: chicken, beef, pork, and a cereal additive. The firm produces three types of wieners: regular, beef, and all-meat. The company has the following amounts of each ingredient available on a daily basis:

	Pounds/Day	Cost/Pound
Chicken	200	$.20
Beef	300	.30
Pork	150	.50
Cereal additive	400	.05

Each type of wiener has certain ingredient specifications, as follows:

	Specifications	Selling Price/Pound
Regular	Not more than 10% beef and pork combined	$0.90
	Not less than 20% chicken	
Beef	Not less than 75% beef	1.25
All-meat	No cereal additive	1.75
	Not more than 50% beef and pork combined	

The firm wants to know how many pounds of wieners of each type to produce to maximize profits.

a. Formulate a linear programming model for this problem.
b. Solve the model by using the computer.

36. Joe Henderson runs a small metal parts shop. The shop contains three machines—a drill press, a lathe, and a grinder. Joe has three operators, each certified to work on all three machines. However, each operator performs better on some machines than on others. The shop has contracted to do a big job that requires all three machines. The times required by the various operators to perform the required operations on each machine are summarized as follows:

Operator	Drill Press (min.)	Lathe (min.)	Grinder (min.)
1	22	18	35
2	41	30	28
3	25	36	18

Joe Henderson wants to assign one operator to each machine so that the total operating time for all three operators is minimized.

a. Formulate a linear programming model for this problem.
b. Solve the model by using the computer.
c. Joe's brother, Fred, has asked him to hire his wife, Kelly, who is a machine operator. Kelly can perform each of the three required machine operations in 20 minutes. Should Joe hire his sister-in-law?

37. Green Valley Mills produces carpet at plants in St. Louis and Richmond. The plants ship the carpet to two outlets in Chicago and Atlanta. The cost per ton of shipping carpet from each of the two plants to the two warehouses is as follows:

	To	
From	Chicago	Atlanta
St. Louis	$40	$65
Richmond	70	30

The plant at St. Louis can supply 250 tons of carpet per week, and the plant at Richmond can supply 400 tons per week. The Chicago outlet has a demand of 300 tons per week; the outlet at Atlanta demands 350 tons per week. Company managers want to determine the number of tons of carpet to ship from each plant to each outlet in order to minimize the total shipping cost.

a. Formulate a linear programming model for this problem.
b. Solve the model by using the computer.

38. Dr. Maureen Becker, the head administrator at Jefferson County Regional Hospital, must determine a schedule for nurses to make sure there are enough of them on duty throughout the day. During the day, the demand for nurses varies. Maureen has broken the day into twelve 2-hour periods. The slowest time of the day encompasses the three periods from 12:00 A.M. to 6:00 A.M., which, beginning at midnight, require a minimum of 30, 20, and 40 nurses, respectively. The demand for nurses steadily increases during the next four daytime periods. Beginning with the 6:00 A.M.–8:00 A.M. period, a minimum of 50, 60, 80, and 80 nurses are required for these four periods, respectively. After 2:00 P.M. the demand for nurses decreases during the afternoon and evening hours. For the five 2-hour periods beginning at 2:00 P.M. and ending at midnight, 70, 70, 60, 50, and 50 nurses are required, respectively. A nurse reports for duty at the beginning of one of the 2-hour periods and works 8 consecutive hours (which is required in the nurses' contract).

Dr. Becker wants to determine a nursing schedule that will meet the hospital's minimum requirements throughout the day while using the minimum number of nurses.

a. Formulate a linear programming model for this problem.
b. Solve this model by using the computer.

39. A manufacturer of bathroom fixtures produces fiberglass bathtubs in an assembly operation that consists of three processes—molding, smoothing, and painting. The number of units that can undergo each process in an hour is as follows:

Process	Output (units/hr.)
Molding	7
Smoothing	12
Painting	10

(*Note:* The three processes are continuous and sequential; thus, no more units can be smoothed or painted than have been molded.) The labor costs per hour are $8 for molding, $5 for smoothing, and $6.50 for painting. The company's labor budget is $3,000 per week. A total of 120 hours of labor is available for all three processes per week. Each completed bathtub requires 90 pounds of fiberglass, and the company has a total of 10,000 pounds of fiberglass available each week. Each bathtub earns a profit of $175. The manager of the company wants to know how many hours per week to run each process to maximize profit.

a. Formulate a linear programming model for this problem.
b. Solve the model by using the computer.

40. The admissions office at State University wants to develop a planning model for next year's entering freshman class. The university has 4,500 available openings for freshmen. Tuition is $8,600 for an in-state student and $19,200 for an out-of-state student. The university wants to maximize the money it receives from tuition, but by state mandate it can admit no more than 47% out-of-state students. Also, each college in the university must have at least 30% in-state students in its freshman class. In order to be ranked in several national magazines, it wants the freshman class to have an average SAT score of 1150. Following are the average SAT scores for last year's freshman class for in-state and out-of-state students in each college in the university plus the maximum size of the freshman class for each college:

	Average SAT Scores		
College	In-State	Out-of-State	Total Capacity
1. Architecture	1350	1460	470
2. Arts and Sciences	1010	1050	1,300
3. Agriculture	1020	1110	240
4. Business	1090	1180	820
5. Engineering	1360	1420	1,060
6. Human Resources	1000	1400	610

a. Formulate and solve a linear programming model to determine the number of in-state and out-of-state students that should enter each college.
b. If the solution in (a) does not achieve the maximum freshman class size, discuss how you might adjust the model to reach this class size.

41. A manufacturing firm located in Chicago ships its product by railroad to Detroit. Several different routes are available, as shown in the following diagram, referred to as a network:

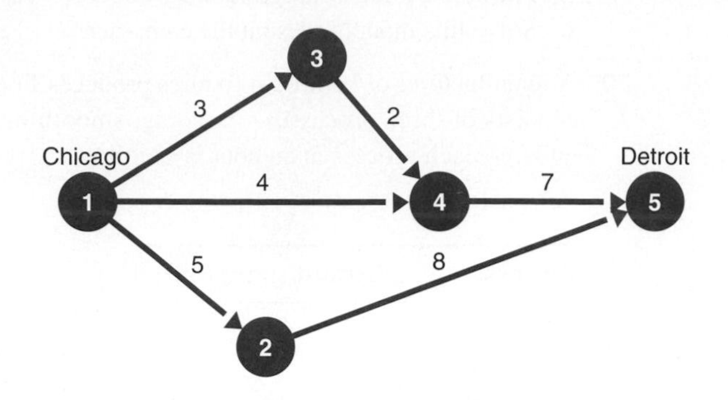

Each circle in the network represents a railroad junction. Each arrow is a railroad branch between two junctions. The number above each arrow is the cost ($1,000s) necessary to ship 1 ton of product from junction to junction. The firm wants to ship 5 tons of its product from Chicago to Detroit at the minimum cost.

a. Formulate a linear programming model for this problem.
b. Solve the model by using the computer.

42. A refinery blends four petroleum components into three grades of gasoline—regular, premium, and diesel. The maximum quantities available of each component and the cost per barrel are as follows:

Component	Maximum Barrels Available/Day	Cost/Barrel
1	5,000	$ 9
2	2,400	7
3	4,000	12
4	1,500	6

To ensure that each gasoline grade retains certain essential characteristics, the refinery has put limits on the percentages of the components in each blend. The limits as well as the selling prices for the various grades are as follows:

Grade	Component Specifications	Selling Price/Barrel
Regular	Not less than 40% of 1	$12
	Not more than 20% of 2	
	Not less than 30% of 3	
Premium	Not less than 40% of 3	18
Diesel	Not more than 50% of 2	10
	Not less than 10% of 1	

The refinery wants to produce at least 3,000 barrels of each grade of gasoline. Management wishes to determine the optimal mix of the four components that will maximize profit.

a. Formulate a linear programming model for this problem.
b. Solve the model by using the computer.

43. Melissa Beadle is a student at Tech, and she wants to decide how many hours each day to allocate to the following activities: class; studying; leisure and fun stuff; personal activities such as eating, bathing, cleaning, laundry, and so on; and sleeping. She has established weights indicating how her different activity hours relate to raising her grade point average. Each hour she spends in class will raise her GPA by 0.3 points, 1 hour of studying will raise her GPA by 0.2 points, leisure and fun activities will raise her GPA by 0.05 points per hour, personal activities will raise her GPA by 0.10 points per hour, and sleep will increase her GPA by 0.15 points per hour. Melissa has only 4 hours of class each day, and she knows she won't study more than 8 hours in a day. She thinks she could spend up to 10 hours a day on leisure and having fun, she'll spend at least 2 hours but not more than 3 hours on personal stuff, and she'll sleep at least 3 hours per day but not more than 10 hours.
    a. Formulate a linear programming model that will allocate Melissa's hours each day so that her GPA (on a 4.0 scale) will be maximized.
    b. Reformulate the model to reflect your own personal preferences relative to your daily activities.

44. Tech, with a student population of 30,000, is located in a small college town in Virginia. DirectCast Cable TV has a small service staff that is sufficient to handle installations and TV hookups for almost the entire year. However, for the month-long period right before and during the beginning of fall semester in August, when all the students return, the cable TV service is overwhelmed and must bring in an additional 14 technicians from other company offices. The DirectCast offices in the following seven cities have available service technicians to loan, as shown:

City	Available Technicians	Distance (mi.)	Cost ($)
Richmond	4	211	3,400
Charlotte	3	173	3,920
Atlanta	5	410	4,760
Greensboro	2	152	3,560
DC	6	263	4,980
Nashville	3	414	4,050
Norfolk	2	302	3,240

This table also shows the distances (in miles) from the cities to the town where Tech is located, and the cost for relocating a technician for 1 month, which includes monthly salary and a bonus. The national DirectCast office incurs this monthly cost. Since most of the technicians will commute to their homes at least once a week for a few days, the company wants to minimize the total distance incurred by its employees, but it would also like to keep its labor costs to $60,000 or less.
    a. Formulate a linear programming model for this problem.
    b. Solve the model by using the computer.
    c. If DirectCast wants to minimize its cost and not the distance traveled, how would this affect the solution?

45. The Cash and Carry Building Supply Company has received the following order for boards in three lengths:

Length	Order (quantity)
7 ft.	700
9 ft.	1,200
10 ft.	300

The company has 25-foot standard-length boards in stock. Therefore, the standard-length boards must be cut into the lengths necessary to meet order requirements. Naturally, the company wishes to minimize the number of standard-length boards used. The company must therefore determine how to cut up the 25-foot boards to meet the order requirements and minimize the number of standard-length boards used.

a. Formulate a linear programming model for this problem.
b. Solve the model by using the computer.
c. When a board is cut in a specific pattern, the amount of board left over is referred to as "trim loss." Reformulate the linear programming model for this problem, assuming that the objective is to minimize trim loss rather than to minimize the total number of boards used, and solve this model. How does this affect the solution?

46. An investment firm has $1 million to invest in stocks, bonds, certificates of deposit, and real estate. The firm wishes to determine the mix of investments that will maximize the cash value at the end of 6 years.

 Opportunities to invest in stocks and bonds will be available at the beginning of each of the next 6 years. Each dollar invested in stocks will return $1.20 (a profit of $0.20) 2 years later; the return can be immediately reinvested in any alternative. Each dollar invested in bonds will return $1.40 3 years later; the return can be reinvested immediately.

 Opportunities to invest in certificates of deposit will be available only once, at the beginning of the second year. Each dollar invested in certificates will return $1.80 four years later. Opportunities to invest in real estate will be available at the beginning of the fifth and sixth years. Each dollar invested will return $1.10 one year later.

 To minimize risk, the firm has decided to diversify its investments. The total amount invested in stocks cannot exceed 30% of total investments, and at least 25% of total investments must be in certificates of deposit.

 The firm's management wishes to determine the optimal mix of investments in the various alternatives that will maximize the amount of cash at the end of the sixth year.

 a. Formulate a linear programming model for this problem.
 b. Solve the model by using the computer.

47. The Jones, Jones, Smith, and Rodman commodities trading firm knows the prices at which it will be able to buy and sell a certain commodity during the next 4 months. The buying price (c_i) and selling price (p_i) for each of the given months (i) are as follows:

	Month i			
	1	2	3	4
c_i	$5	$6	$7	$8
p_i	4	8	6	7

 The firm's warehouse has a maximum capacity of 10,000 bushels. At the beginning of the first month, 2,000 bushels are in the warehouse. The trading firm wants to know the amounts that should be bought and sold each month in order to maximize profit. Assume that no storage costs are incurred and that sales are made at the beginning of the month, followed by purchases.

 a. Formulate a linear programming model for this problem.
 b. Solve the model by using the computer.

48. The production manager of Videotechnics Company is attempting to determine the upcoming 5-month production schedule for video recorders. Past production records indicate that 2,000

recorders can be produced per month. An additional 600 recorders can be produced monthly on an overtime basis. Unit cost is $10 for recorders produced during regular working hours and $15 for those produced on an overtime basis. Contracted sales per month are as follows:

Month	Contracted Sales (units)
1	1,200
2	2,100
3	2,400
4	3,000
5	4,000

Inventory carrying costs are $2 per recorder per month. The manager does not want any inventory carried over past the fifth month. The manager wants to know the monthly production that will minimize total production and inventory costs.

a. Formulate a linear programming model for this problem.
b. Solve the model by using the computer.

49. The manager of the Ewing and Barnes Department Store has four employees available to assign to three departments in the store—lamps, sporting goods, and linens. The manager wants each of these departments to have at least one employee, but not more than two. Therefore, two departments will be assigned one employee, and one department will be assigned two. Each employee has different areas of expertise, which are reflected in the daily sales each employee is expected to generate in each department, as follows:

	Department Sales		
Employee	Lamps	Sporting Goods	Linens
1	$130	$150	$ 90
2	275	300	100
3	180	225	140
4	200	120	160

The manager wishes to know which employee(s) to assign to each department in order to maximize expected sales.

a. Formulate a linear programming model for this problem.
b. Solve the model by using the computer.
c. Suppose that the department manager plans to assign only one employee to each department and to lay off the least productive employee. Formulate a new linear programming model that reflects this new condition and solve by using the computer.

50. Tidewater City Bank has four branches with the following inputs and outputs:

input 1 = teller hours (100s)
input 2 = space (100s ft.^{2})
input 3 = expenses ($1,000s)

output 1 = deposits, withdrawals, and checks processed (1,000s)
output 2 = loan applications
output 3 = new accounts (100s)

The monthly output and input values for each bank are as follows:

	Outputs			**Inputs**		
Bank	1	2	3	1	2	3
A	76	125	12	16	22	12
B	82	105	8	12	19	10
C	69	98	9	17	26	16
D	72	117	14	14	18	14

Using data envelopment analysis (DEA), determine which banks, if any, are inefficient.

51. Carillon Health Systems owns three hospitals, and it wants to determine which, if any, of the hospitals are inefficient. The hospital inputs each month are (1) hospital beds (100s), (2) nonphysician labor (1,000s hrs.), and (3) dollar value of supplies ($1000s). The hospital outputs are patient days (100s) for three age groups—(1) under 15, (2) 15 to 65, and (3) over 65. The output and input values for each hospital are as follows:

	Outputs			**Inputs**		
Hospital	1	2	3	1	2	3
A	9	5	18	7	12	40
B	6	8	9	5	14	32
C	5	10	12	8	16	47

Using DEA, determine which hospitals are inefficient.

52. Managers at the Transcontinent Shipping and Supply Company want to know the maximum tonnage of goods they can transport from city A to city F. The firm can contract for railroad cars on different rail routes linking these cities via several intermediate stations, shown in the following diagram; all railroad cars are of equal capacity:

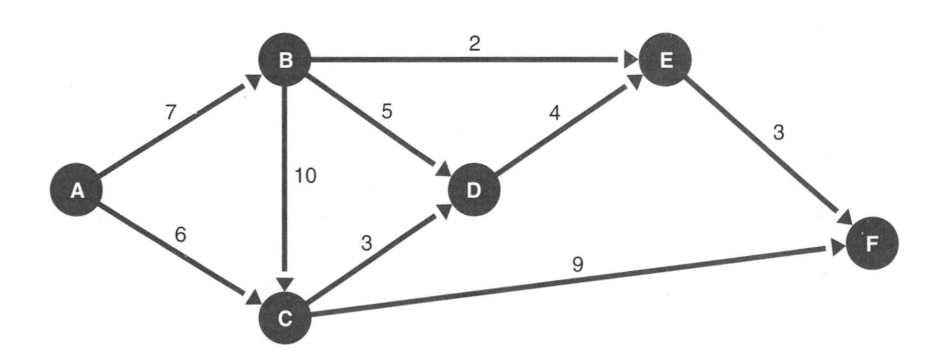

The firm can transport a maximum amount of goods from point to point, based on the maximum number of railroad cars shown on each route segment. Managers want to determine the maximum tonnage that can be shipped from city A to city F.

a. Formulate a linear programming model for this problem.
b. Solve the model by using the computer.

53. The law firm of Smith, Smith, Smith, and Jones is recruiting at law schools for new lawyers for the coming year. The firm has developed the following estimate of the number of hours of casework it will need its new lawyers to handle each month for the following year:

Month	Casework (hr.)	Month	Casework (hr.)
January	650	July	750
February	450	August	900
March	600	September	800
April	500	October	650
May	700	November	700
June	650	December	500

Each new lawyer the firm hires is expected to handle 150 hours per month of casework and to work all year. All casework must be completed by the end of the year. The firm wants to know how many new lawyers it should hire for the year.

a. Formulate a linear programming model for this problem.
b. Solve this model by using the computer.

54. In Problem 53, the optimal solution results in a fractional (i.e., non-integer) number of lawyers being hired. Explain how you would go about logically determining a new solution with a whole (integer) number of lawyers being hired and discuss the difference in results between this new solution and the optimal non-integer solution obtained in Problem 53.

55. The Goliath Tool and Machine Shop produces a single product that consists of three subcomponents that are assembled to form the product. The three components are manufactured in an operation that involves two lathes and three presses. The production time (in minutes per unit) for each machine for the three components is as follows:

	Production Time (min.)		
	Component 1	Component 2	Component 3
Lathe	10	8	6
Press	9	21	15

The shop splits the lathe workload evenly between the two lathes, and it splits the press workload evenly among the three presses. In addition, the firm wishes to produce quantities of components that will balance the daily loading among lathes and presses so that, on the average, no machine is operated more than 1 hour per day longer than any other machine. Each machine operates 8 hours per day.

The firm also wishes to produce a quantity of components that will result in completely assembled products, without any partial assemblies (i.e., in-process inventories). The objective of the firm is to maximize the number of units of assembled product per day.

a. Formulate a linear programming model for this problem.
b. Solve the model by using the computer.
c. The production policies established by the Goliath Tool and Machine Shop are relatively restrictive. If the company were to relax either its machine balancing requirement (that no machine be operated more than an hour longer than any other machine) or its restriction on in-process inventory, which would have the greatest impact on production output? What would be the impact if both requirements were relaxed?

56. A ship has two cargo holds, one fore and one aft. The fore cargo hold has a weight capacity of 70,000 pounds and a volume capacity of 30,000 cubic feet. The aft hold has a weight capacity of 90,000 pounds and a volume capacity of 40,000 cubic feet. The shipowner has contracted to carry loads of packaged beef and grain. The total weight of the available beef is 85,000 pounds; the total weight of the available grain is 100,000 pounds. The volume per mass of the beef is 0.2 cubic foot per pound, and the volume per mass of the grain is 0.4 cubic foot per pound. The profit for shipping beef is $0.35 per pound, and the profit for shipping grain is $0.12 per pound. The shipowner is free to accept all or part of the available cargo; he wants to know how much meat and grain to accept to maximize profit.
    a. Formulate a linear programming model for this problem.
    b. Solve the model by using the computer.

57. Eyewitness News is shown on channel 5 Monday through Friday evenings from 5:00 P.M. to 6:00 P.M. During the hour-long broadcast, 18 minutes are allocated to commercials. The remaining 42 minutes of airtime are allocated to single or multiple time segments for local news and features, national news, sports, and weather. The station has learned through several viewer surveys that viewers do not consistently watch the entire news program; they focus on some segments more closely than others. For example, they tend to pay more attention to the local weather than the national news (because they know they will watch the network news following the local broadcast). As such, the advertising revenues generated for commercials shown during the different broadcast segments are $850 per minute for local news and feature segments, $600 per minute for national news, $750 per minute for sports, and $1,000 per minute for the weather. The production cost for local news is $400 per minute, the cost for national news is $100 per minute, the cost for sports is $175 per minute, and for weather it's $90 per minute. The station budgets $9,000 per show for production costs. The station's policy is that the broadcast time devoted to local news and features must be at least 10 minutes but not more than 25 minutes, whereas national news, sports, and weather must have segments of at least 5 minutes each but not more than 10 minutes. Commercial time must be limited to no more than 6 minutes for each of the four broadcast segment types. The station manager wants to know how many minutes of commercial time and broadcast time to allocate to local news, national news, sports, and weather to maximize advertising revenues.
    a. Formulate a linear programming model for this problem.
    b. Solve by using the computer.

58. The Douglas family raises cattle on their farm in Virginia. They also have a large garden in which they grow ingredients for making two types of relish—chow-chow and tomato. These they sell in 16-ounce jars at local stores and craft fairs in the fall. The profit for a jar of chow-chow is $2.25, and the profit for a jar of tomato relish is $1.95. The main ingredients in each relish are cabbage, tomatoes, and onions. A jar of chow-chow must contain at least 60% cabbage, 5% onions, and 10% tomatoes, and a jar of tomato relish must contain at least 50% tomatoes, 5% onions, and 10% cabbage. Both relishes contain no more than 10% onions. The family has enough time to make no more than 700 jars of relish. In checking sales records for the past 5 years, they know that they will sell at least 30% more chow-chow than tomato relish. They will have 300 pounds of cabbage, 350 pounds of tomatoes, and 30 pounds of onions available. The Douglas family wants to know how many jars of relish to produce to maximize profit.
    a. Formulate a linear programming model for this problem.
    b. Solve by using the computer.

59. The White Horse Apple Products Company purchases apples from local growers and makes applesauce and apple juice. It costs $0.60 to produce a jar of applesauce and $0.85 to produce a bottle of apple juice. The company has a policy that at least 30% but not more than 60% of its output must be applesauce.

    The company wants to meet but not exceed the demand for each product. The marketing manager estimates that the demand for applesauce is a maximum of 5,000 jars, plus an additional

3 jars for each $1 spent on advertising. The maximum demand for apple juice is estimated to be 4,000 bottles, plus an additional 5 bottles for every $1 spent to promote apple juice. The company has $16,000 to spend on producing and advertising applesauce and apple juice. Applesauce sells for $1.45 per jar; apple juice sells for $1.75 per bottle. The company wants to know how many units of each to produce and how much advertising to spend on each to maximize profit.

a. Formulate a linear programming model for this problem.
b. Solve the model by using the computer.

60. Mazy's Department Store has decided to stay open on a 24-hour basis. The store manager has divided the 24-hour day into six 4-hour periods and determined the following minimum personnel requirements for each period:

Time	Personnel Needed
Midnight–4:00 A.M.	90
4:00–8:00 A.M.	215
8:00–Noon	250
Noon–4:00 P.M.	65
4:00–8:00 P.M.	300
8:00–Midnight	125

Personnel must report for work at the beginning of one of these times and work 8 consecutive hours. The store manager wants to know the minimum number of employees to assign for each 4-hour segment to minimize the total number of employees.

a. Formulate a linear programming model for this problem.
b. Solve the model by using the computer.

61. Venture Systems is a consulting firm that develops e-commerce systems and Web sites for its clients. It has six available consultants and eight client projects under contract. The consultants have different technical abilities and experience, and as a result, the company charges different hourly rates for its services. Also, the consultants' skills are more suited for some projects than others, and clients sometimes prefer some consultants over others. The suitability of a consultant for a project is rated according to a 5-point scale, in which 1 is the worst and 5 is the best. The following table shows the rating for each consultant for each project, as well as the hours available for each consultant and the contracted hours and maximum budget for each project.

		Project								
Consultant	Hourly Rate	1	2	3	4	5	6	7	8	Available Hours
A	$155	3	3	5	5	3	3	3	3	450
B	140	3	3	2	5	5	5	3	3	600
C	165	2	1	3	3	2	1	5	3	500
D	300	1	3	1	1	2	2	5	1	300
E	270	3	1	1	2	2	1	3	3	710
F	150	4	5	3	2	3	5	4	3	860
Project Hours		500	240	400	475	350	460	290	200	
Contract Budget ($1,000s)		100	80	120	90	65	85	50	55	

The company wants to know how many hours to assign each consultant to each project in order to best utilize the consultants' skills while meeting the clients' needs.

a. Formulate a linear programming model for this problem.
b. Solve the model by using the computer.
c. If the company's objective is to maximize revenue while ignoring client preferences and consultant compatibility, will this change the solution in (b)?

62. Great Northwoods Outfitters is a retail phone-catalog company that specializes in outdoor clothing and equipment. A phone station at the company will be staffed with either full-time operators or temporary operators 8 hours per day. Full-time operators, because of their experience and training, process more orders and make fewer mistakes than temporary operators. However, temporary operators are cheaper because they receive a lower wage rate and are not paid benefits. A full-time operator can process about 360 orders per week, whereas a temporary operator can process about 270 orders per week. A full-time operator averages 1.1 defective orders per week, and a part-time operator incurs about 2.7 defective orders per week. The company wants to limit defective orders to 200 per week. The cost of staffing a station with full-time operators is $610 per week, and the cost of a station with part-time operators is $450 per week. Using historical data and forecasting techniques, the company has developed estimates of phone orders for an 8-week period, as follows:

Week	Orders	Week	Orders
1	19,500	5	33,400
2	21,000	6	29,800
3	25,600	7	27,000
4	27,200	8	31,000

The company does not want to hire or dismiss full-time employees after the first week (i.e., the company wants a constant group of full-time operators over the 8-week period). The company wants to determine how many full-time operators it needs and how many temporary operators to hire each week to meet weekly demand while minimizing labor costs.

a. Formulate a linear programming model for this problem.
b. Solve this model by using the computer.

63. In Problem 62, Great Northwoods Outfitters is going to alter its staffing policy. Instead of hiring a constant group of full-time operators for the entire 8-week planning period, it has decided to hire and add full-time operators as the 8-week period progresses, although once it hires full-time operators, it will not dismiss them. Reformulate the linear programming model to reflect this altered policy and solve to determine the cost savings (if any).

64. Blue Ridge Power and Light Company generates electrical power at four coal-fired power plants along the eastern seaboard in Virginia, North Carolina, Maryland, and Delaware. The company purchases coal from six producers in southwestern Virginia, West Virginia, and Kentucky. Blue Ridge has fixed contracts for coal delivery from the following three coal producers:

Coal Producer	Tons	Cost/Ton	Million BTUs/Ton
ANCO	190,000	$23	26.2
Boone Creek	305,000	28	27.1
Century	310,000	24	25.6

The power-producing capabilities of the coal produced by these suppliers differs according to the quality of the coal. For example, coal produced by ANCO provides 26.2 million BTUs per ton, while coal produced at Boone Creek provides 27.1 million BTUs per ton. Blue Ridge also purchases coal from three backup auxiliary suppliers, as needed (i.e., it does not have fixed contracts with these producers). In general, the coal from these backup suppliers is more costly and lower grade.

Coal Producer	Available Tons	Cost/Ton	Million BTUs/Ton
DACO	125,000	$31	21.4
Eaton	95,000	29	19.2
Franklin	190,000	34	23.6

The demand for electricity at Blue Ridge's four power plants is as follows (note that it requires approximately 10 million BTUs to generate 1 megawatt hour):

Power Plant	Electricity Demand (million BTUs)
1. Afton	4,600,000
2. Surrey	6,100,000
3. Piedmont	5,700,000
4. Chesapeake	7,300,000

For example, the Afton plant must produce at least 4,600,000 million BTUs next year, which translates to approximately 460,000 megawatt hours.

Coal is primarily transported from the producers to the power plants by rail, and the cost of processing coal at each plant is different. Following are the combined transportation and processing costs for coal from each supplier to each plant.

	Power Plant			
Coal Producer	1. Afton	2. Surrey	3. Piedmont	4. Chesapeake
ANCO	$12.20	$14.25	$11.15	$15.00
Boone Creek	10.75	13.70	11.75	14.45
Century	15.10	16.65	12.90	12.00
DACO	14.30	11.90	16.35	11.65
Eaton	12.65	9.35	10.20	9.55
Franklin	16.45	14.75	13.80	14.90

Formulate and solve a linear programming model to determine how much coal should be purchased and shipped from each supplier to each power plant in order to minimize cost.

65. Valley United Soccer Club has 16 boys' and girls' travel soccer teams. The club has access to three town fields, which its teams practice on in the fall during the season. Field 1 is large enough to accommodate two teams at one time, and field 3 can accommodate three teams,

whereas field 2 has enough room for only one team. The teams practice twice per week, either on Monday and Wednesday from 3 P.M. to 5 P.M. or 5 P.M. to 7 P.M., or on Tuesday and Thursday from 3 P.M. to 5 P.M. or 5 P.M. to 7 P.M. Field 3 is in the worst condition of all the fields, so teams generally prefer the other fields; teams also do not like to practice at field 3 because it can get crowded with three teams. In general, the younger teams like to practice right after school, while the older teams like to practice later in the day. In addition, some teams must practice later because their coaches are available only after work. Some teams also prefer specific fields because they're closer to their players' homes. Each team has been asked by the club field coordinator to select three practice locations and times, in priority order, and they have responded as follows:

	Priority		
Team	1	2	3
U11B	2, 3–5M	1, 3–5M	3, 3–5M
U11G	1, 3–5T	2, 3–5T	3, 3–5T
U12B	2, 3–5T	1, 3–5T	3, 3–5T
U12G	1, 3–5M	1, 3–5T	2, 3–5M
U13B	2, 3–5T	2, 3–5M	1, 3–5M
U13G	1, 3–5M	2, 3–5M	1, 3–5T
U14B	1, 5–7M	1, 5–7T	2, 5–7T
U14G	2, 3–5M	1, 3–5M	2, 3–5T
U15B	1, 5–7T	2, 5–7T	1, 5–7M
U15G	2, 5–7M	1, 5–7M	1, 5–7T
U16B	1, 5–7T	2, 5–7T	3, 5–7T
U16G	2, 5–7T	1, 5–7T	3, 5–7T
U17B	2, 5–7M	1, 5–7T	1, 5–7M
U17G	1, 5–7T	2, 5–7T	1, 5–7M
U18B	2, 5–7M	2, 5–7T	1, 5–7M
U18G	1, 5–7M	1, 5–7T	2, 5–7T

For example, the under-11 boys' age group team has selected field 2 from 3 P.M. to 5 P.M. on Monday and Wednesday as its top priority, field 1 from 3 P.M. to 5 P.M. on Monday and Wednesday as its second priority, and so on.

Formulate and solve a linear programming model to optimally assign the teams to fields and times, according to their priorities. Are any of the teams not assigned to one of their top three selections? If not, how might you modify or use the model to assign these teams to the best possible time and location? How could you make sure that the model does not assign teams to unacceptable locations and times—for example, a team whose coach can be at practice only at 5 P.M.?

66. The city of Salem has four police stations, with the following inputs and outputs:

input 1 = number of police officers
input 2 = number of patrol vehicles
input 3 = space (100s ft.^{2})

output 1 = calls responded to (100s)
output 2 = traffic citations (100s)
output 3 = convictions

The monthly output and input data for each station are

	Outputs			Inputs		
Police Station	**1**	**2**	**3**	**1**	**2**	**3**
A	12.7	3.6	35	34	18	54
B	14.2	4.9	42	29	22	62
C	13.8	5.2	56	38	16	50
D	15.1	4.2	39	35	24	57

Help the city council determine which of the police stations are relatively inefficient.

67. USAir South Airlines operates a hub at the Pittsburgh International Airport. During the summer, the airline schedules 7 flights daily from Pittsburgh to Orlando and 10 flights daily from Orlando to Pittsburgh, according to the following schedule:

Flight	Leave Pittsburgh	Arrive Orlando	Flight	Leave Orlando	Arrive Pittsburgh
1	6 A.M.	9 A.M.	A	6 A.M.	9 A.M.
2	8 A.M.	11 A.M.	B	7 A.M.	10 A.M.
3	9 A.M.	Noon	C	8 A.M.	11 A.M.
4	3 P.M.	6 P.M.	D	10 A.M.	1 P.M.
5	5 P.M.	8 P.M.	E	Noon	3 P.M.
6	7 P.M.	10 P.M.	F	2 P.M.	5 P.M.
7	8 P.M.	11 P.M.	G	3 P.M.	6 P.M.
			H	6 P.M.	9 P.M.
			I	7 P.M.	10 P.M.
			J	9 P.M.	Midnight

The flight crews live in Pittsburgh or Orlando, and each day a new crew must fly one flight from Pittsburgh to Orlando and one flight from Orlando to Pittsburgh. A crew must return to its home city at the end of each day. For example, if a crew originates in Orlando and flies a flight to Pittsburgh, it must then be scheduled for a return flight from Pittsburgh back to Orlando. A crew must have at least 1 hour between flights at the city where it arrives. Some scheduling combinations are not possible; for example, a crew on flight 1 from Pittsburgh cannot return on flights A, B, or C from Orlando. It is also possible for a flight to ferry one additional crew to a city in order to fly a return flight, if there are not enough crews in that city.

The airline wants to schedule its crews in order to minimize the total amount of crew ground time (i.e., the time the crew is on the ground between flights). Excessive ground time for a crew lengthens its workday, is bad for crew morale, and is expensive for the airline. Formulate a linear programming model to determine a flight schedule for the airline and solve by using the computer. How many crews need to be based in each city? How much ground time will each crew experience?

68. The National Cereal Company produces a Light-Snak cereal package with a selection of small pouches of four different cereals—Crunchies, Toasties, Snakmix, and Granolies. Each cereal is produced at a different production facility and then shipped to three packaging facilities, where the four different cereal pouches are combined into a single box. The boxes are then sent to one of three distribution centers, where they are combined to fill customer orders and shipped.

The following diagram shows the weekly flow of the product through the production, packaging, and distribution facilities (referred to as a "supply chain"):

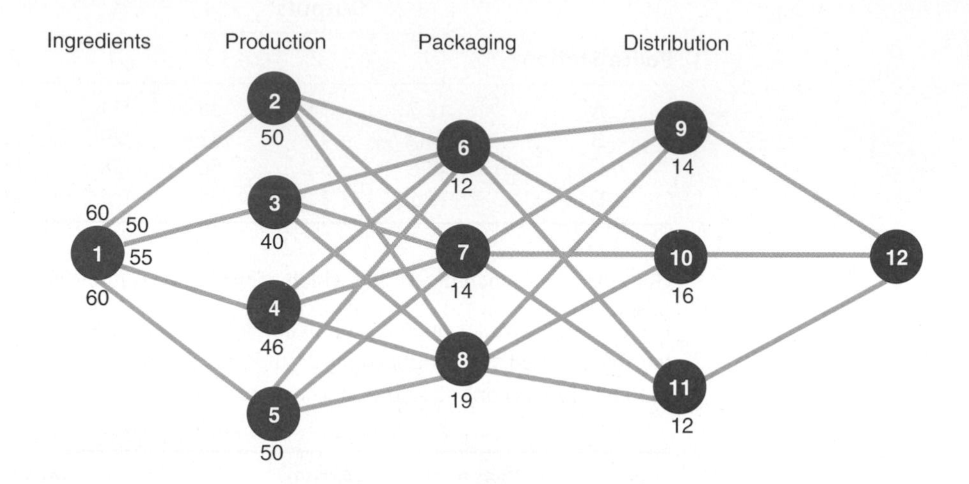

Ingredients capacities (per 1,000 pouches) per week are shown along branches 1–2, 1–3, 1–4, and 1–5. For example, ingredients for 60,000 pouches are available at the production facility, as shown on branch 1–2. The weekly production capacity at each plant (in 1,000s of pouches) is shown at nodes 2, 3, 4, and 5. The packaging facilities at nodes 6, 7, and 8 and the distribution centers at nodes 9, 10, and 11 have capacities for boxes (1,000s) as shown.

The various production, packaging, and distribution costs per unit at each facility are shown in the following table:

Facility	**2**	**3**	**4**	**5**	**6**	**7**	**8**	**9**	**10**	**11**
Unit cost	$.17	.20	.18	.16	.26	.29	.27	.12	.11	.14

Weekly demand for the Light-Snak product is 37,000 boxes.

Formulate and solve a linear programming model that indicates how much product must be produced at each facility to meet weekly demand at the minimum cost.

69. Valley Fruit Products Company has contracted with apple growers in Ohio, Pennsylvania, and New York to purchase apples that the company then ships to its plants in Indiana and Georgia, where they are processed into apple juice. Each bushel of apples produces 2 gallons of apple juice. The juice is canned and bottled at the plants and shipped by rail and truck to warehouses/distribution centers in Virginia, Kentucky, and Louisiana. The shipping costs per bushel from the farms to the plants and the shipping costs per gallon from the plants to the distribution centers are summarized in the following tables:

	Plant		
Farm	**4. Indiana**	**5. Georgia**	**Supply (bushels)**
1. Ohio	.41	.57	24,000
2. Pennsylvania	.37	.48	18,000
3. New York	.51	.60	32,000
Plant Capacity	48,000	35,000	

Plant	Distribution Centers		
	6. Virginia	7. Kentucky	8. Louisiana
4. Indiana	.22	.10	.20
5. Georgia	.15	.16	.18
Demand (gal.)	9,000	12,000	15,000

Formulate and solve a linear programming model to determine the optimal shipments from the farms to the plants and from the plants to the distribution centers in order to minimize total shipping costs.

70. In the event of a disaster situation at Tech from weather, an accident, or terrorism, victims will be transported by emergency vehicles to three area hospitals: Montgomery Regional, Radford Memorial, and Lewis Galt. Montgomery Regional is (on average) 10 minutes away from Tech, Radford Memorial is 20 minutes away, and Lewis Galt is 35 minutes away. Tech wants to analyze a hypothetical disaster situation in which there are 15 victims with different types of injuries. The emergency facilities at Montgomery Regional can accommodate, at most, 8 victims; Radford Memorial can handle 10 victims; and Lewis Galt can admit 7 victims. A priority has been assigned for each victim according to the hospital that would best treat that victim's type of injury, as shown in the following table (where 1 reflects the best treatment).

Hospital	Patient														
	1	2	3	4	5	6	7	8	9	10	11	12	13	14	15
Montgomery Regional	1	1	2	2	2	1	3	3	3	1	3	3	2	1	3
Radford Memorial	2	2	3	3	1	3	3	1	1	1	3	3	2	2	3
Lewis Galt	3	3	1	1	3	2	1	2	2	2	1	1	1	2	1

For example, for victim 1's type of injury, the best hospital is Montgomery Regional, the next best is Radford Memorial, and Lewis Galt is the third best.

a. Formulate and solve a linear programming model that will send the victims to the hospital best suited to administer to their specific injuries while keeping the average transport time to 22 minutes or less.
b. Formulate and solve a linear programming model that will minimize the average transport time for victims while achieving an average hospital priority of at least 1.50 or better.

Case Problem

Summer Sports Camp at State University

Mary Kelly is a scholarship soccer player at State University. During the summer, she works at a youth all-sports camp that several of the university's coaches operate. The sports camp runs for 8 weeks during July and August. Campers come for a 1-week period, during which time they live in the State dormitories and use the State athletic fields and facilities. At the end of a week, a new group of kids comes in. Mary primarily serves as one of the camp soccer instructors. However, she has also been placed in charge of arranging for sheets for the beds the campers will sleep on in the dormitories. Mary has been instructed to develop a plan for purchasing and cleaning sheets each week of camp at the lowest possible cost.

Clean sheets are needed at the beginning of each week, and the campers use the sheets all week. At the end of the week, the campers strip their beds and place the sheets in large bins. Mary must arrange either to purchase new sheets or to clean old sheets. A set of new sheets costs $10. A local laundry has indicated that it will clean a set of sheets for $4. Also, a couple of Mary's friends have asked her to let them clean some of the sheets. They have told her they will charge only $2 for each set of sheets they clean. However, while the laundry will provide cleaned sheets in a week,

Mary's friends can deliver cleaned sheets only in 2 weeks. They are going to summer school and plan to launder the sheets at night at a neighborhood Laundromat.

The accompanying table lists the number of campers who have registered during each of the 8 weeks the camp will operate. Based on discussions with camp administrators from previous summers and on some old camp records and receipts, Mary estimates that each week about 20% of the cleaned sheets that are returned will have to be discarded and replaced. The campers spill food and drinks on the sheets, and sometimes the stains do not come out during cleaning. Also, the campers occasionally tear the sheets, or the sheets get torn at the cleaners. In either case, when the sheets come back from the cleaners and are put on the beds, 20% are taken off and thrown away.

At the beginning of the summer, the camp has no sheets available, so initially sheets must be purchased. Sheets are thrown away at the end of the summer.

Week	Registered Campers
1	115
2	210
3	250
4	230
5	260
6	300
7	250
8	190

Mary's major at State is management science, and she wants to develop a plan for purchasing and cleaning sheets by using linear programming. Help Mary formulate a linear programming model for this problem and solve it by using the computer.

Case Problem

SPRING GARDEN TOOLS

The Spring family has owned and operated a garden tool and implements manufacturing company since 1952. The company sells garden tools to distributors and also directly to hardware stores and home improvement discount chains. The Spring Company's four most popular small garden tools are a trowel, a hoe, a rake, and a shovel. Each of these tools is made from durable steel and has a wooden handle. The Spring family prides itself on its high-quality tools.

The manufacturing process encompasses two stages. The first stage includes two operations—stamping out the metal tool heads and drilling screw holes in them. The completed tool heads then flow to the second stage, which includes an assembly operation where the handles are attached to the tool heads, a finishing step, and packaging. The processing times per tool for each operation are provided in the following table:

	Tool (hr./unit)				Total Hours Available per Month
Operation	*Trowel*	*Hoe*	*Rake*	*Shovel*	
Stamping	0.04	0.17	0.06	0.12	500
Drilling	0.05	0.14	—	0.14	400
Assembly	0.06	0.13	0.05	0.10	600
Finishing	0.05	0.21	0.02	0.10	550
Packaging	0.03	0.15	0.04	0.15	500

The steel the company uses is ordered from an iron and steel works in Japan. The company has 10,000 square feet of sheet steel available each month. The metal required for each tool and the monthly contracted production volume per tool are provided in the following table:

	Sheet Metal (ft.^{2})	Monthly Contracted Sales
Trowel	1.2	1,800
Hoe	1.6	1,400
Rake	2.1	1,600
Shovel	2.4	1,800

The primary reasons the company has survived and prospered are its ability always to meet customer demand on time and its high quality. As a result, the Spring Company will produce on an overtime basis in order to meet its sales requirements, and it also has a long-standing arrangement with a local tool and die company to manufacture its tool heads. The Spring Company feels comfortable subcontracting the first-stage operations because it is easier to detect defects prior to assembly and finishing. For the same reason, the company will not subcontract for the entire tool because defects would be particularly hard to detect after the tool was finished and packaged. However, the company does have 100 hours of overtime available each month for each operation in both stages. The regular production and overtime costs per tool for both stages are provided in the following table:

	Stage 1		Stage 2	
	Regular Cost	Overtime Cost	Regular Cost	Overtime Cost
Trowel	$6.00	$6.20	$3.00	$3.10
Hoe	10.00	10.70	5.00	5.40
Rake	8.00	8.50	4.00	4.30
Shovel	10.00	10.70	5.00	5.40

The cost of subcontracting in stage 1 adds 20% to the regular production cost.

The Spring Company wants to establish a production schedule for regular and overtime production in each stage and for the number of tool heads subcontracted, at the minimum cost. Formulate a linear programming model for this problem and solve the model using the computer. Which resources appear to be most critical in the production process?

Case Problem

Susan Wong's Personal Budgeting Model

After Susan Wong graduated from State University with a degree in management science, she went to work for a computer systems development firm in the Washington, DC, area. As a student at State, Susan paid her normal monthly living expenses for apartment rent, food, and entertainment out of a bank account set up by her parents. Each month they would deposit a specific amount of cash into Susan's account. Her parents also paid her gas, telephone, and bank credit card bills, which were sent directly to them. Susan never had to worry about things like health, car, homeowners', and life insurance; utilities; driver's and car licenses; magazine subscriptions; and so on. Thus, while she was used to spending within a specific monthly budget in college, she was unprepared for the irregular monthly liabilities she encountered once she got a job and was on her own.

In some months Susan's bills would be modest and she would spend accordingly, only to be confronted the next month with a large insurance premium, or a bill for property taxes on her condominium, or a large credit card bill, or a bill for a magazine subscription, and so on the next month. Such unexpected expenditures would result in months when she could not balance her checking account; she would have to pay her bills with her bank credit card and then pay off her accumulated debt in installments while incurring high interest charges. By the end of her first year out of school, she had hoped to have some money saved to begin an investment program, but instead she found herself in debt.

Frustrated by her predicament, Susan decided to get her financial situation in order. First, she sold the condominium that her parents had helped her purchase and moved into a cheaper apartment. This gave her enough cash to clear her outstanding debts, with $3,800 left over to start the new year with. Susan then decided to use some of the management science she had learned in college to help develop a budget. Specifically, she decided to develop a linear programming model to help her decide how much she should put aside each month in short-term investments to meet the demands of irregular monthly liabilities and save some money.

First, Susan went through all her financial records for the year and computed her expected monthly liabilities for the coming year, as shown in the following table:

Month	Bills	Month	Bills
January	$2,750	July	$3,050
February	2,860	August	2,300
March	2,335	September	1,975
April	2,120	October	1,670
May	1,205	November	2,710
June	1,600	December	2,980

Susan's after-taxes-and-benefits salary is $29,400 per year, which she receives in 12 equal monthly paychecks that are deposited directly into her bank account.

Susan has decided that she will invest any money she doesn't use to meet her liabilities each month in either 1-month, 3-month, or 7-month short-term investment vehicles rather than just leaving the money in an interest-bearing checking account. The yield on 1-month investments is 6% per year nominal; on 3-month investments, the yield is 8% per year nominal; and on a 7-month investment, the yield is 12% per year nominal. As part of her investment strategy, any time one of the short-term investments comes due, she uses the principal as part of her budget, but she transfers any interest earned to another long-term investment (which she doesn't consider in her budgeting process).

For example, if she has $100 left over in January that she invests for 3 months, in April, when the investment matures, she uses the $100 she originally invested in her budget, but any interest on the $100 is invested elsewhere. (Thus, the interest is not compounded over the course of the year.)

Susan wants to develop a linear programming model to maximize her investment return during the year so she can take that money and reinvest it at the end of the year in a longer-term investment program. However, she doesn't have to confine herself to short-term investments that will all mature by the end of the year; she can continue to put money toward the end of the year in investments that won't mature until the following year. Her budgeting process will continue to the next year, so she can take out any surplus left over after December and reinvest it in a long-term program if she wants to.

A. Help Susan develop a model for the year that will meet her irregular monthly financial obligations while achieving her investment objectives and solve the model.
B. If Susan decides she doesn't want to include all her original $3,800 in her budget at the beginning of the year, but instead she wants to invest some of it directly in alternative longer-term investments, how much does she need to develop a feasible budget?

This case is based on T. Lewis, "Personal Operations Research: Practicing OR on Ourselves," *Interfaces* 26, no. 5 (September–October 1996): 34–41.

Case Problem

WALSH'S JUICE COMPANY

Walsh's Juice Company produces three products from unprocessed grape juice—bottled juice, frozen juice concentrate, and jelly. It purchases grape juice from three vineyards near the Great Lakes. The grapes are harvested at the vineyards and immediately converted into juice at plants at the vineyard sites and stored there in refrigerated tanks. The juice is then transported to four different plants in Virginia, Michigan, Tennessee, and Indiana, where it is processed into bottled grape juice, frozen juice concentrate, and jelly. Vineyard output typically differs each month in the harvesting season, and the plants have different processing capacities.

In a particular month the vineyard in New York has 1,400 tons of unprocessed grape juice available, whereas the vineyard in Ohio has 1,700 tons and the vineyard in Pennsylvania has 1,100 tons. The processing capacity per month is 1,200 tons of unprocessed juice at the plant in Virginia, 1,100 tons of juice at the plant in Michigan, 1,400 tons at the plant in Tennessee, and 1,400 tons at the plant in Indiana. The cost per ton of transporting unprocessed juice from the vineyards to the plant is as follows:

	Plant			
Vineyard	Virginia	Michigan	Tennessee	Indiana
New York	$850	$720	$910	$750
Pennsylvania	970	790	1,050	880
Ohio	900	830	780	820

The plants are different ages, have different equipment, and have different wage rates; thus, the cost of processing each product at each plant ($/ton) differs, as follows:

	Plant			
Product	Virginia	Michigan	Tennessee	Indiana
Juice	$2,100	$2,350	$2,200	$1,900
Concentrate	4,100	4,300	3,950	3,900
Jelly	2,600	2,300	2,500	2,800

This month the company needs to process a total of 1,200 tons of bottled juice, 900 tons of frozen concentrate, and 700 tons of jelly at the four plants combined. However, the production process for frozen concentrate results in some juice dehydration, and the process for jelly includes a cooking stage that evaporates water content. To process 1 ton of frozen concentrate requires 2 tons of unprocessed juice; 1 ton of jelly requires 1.5 tons of unprocessed juice; and 1 ton of bottled juice requires 1 ton of unprocessed juice.

Walsh's management wants to determine how many tons of grape juice to ship from each of the vineyards to each of the plants and the number of tons of each product to process at each plant. Thus, management needs a model that includes both the logistical aspects of this problem and the production processing aspects. It wants a solution that will minimize total costs, including the cost of transporting grape juice from the vineyards to the plants and the product processing costs. Help Walsh's solve this problem by formulating a linear programming model and solve it by using the computer.

Case Problem

THE KING'S LANDING AMUSEMENT PARK

King's Landing is a large amusement theme park located in Virginia. The park hires high school and college students to work during the summer months of May, June, July, August, and September. The student employees operate virtually all the highly mechanized, computerized rides; perform as entertainers; perform most of the custodial work during park hours; make up the workforce for restaurants, food services, retail shops, and stores; drive trams; and park cars. Park management has assessed the park's monthly needs based on previous summers' attendance at the park and the expected available workforce. Park attendance is relatively low in May, until public schools are out, and then it increases through June, July, and August, and decreases dramatically in September, when schools reopen after Labor Day. The park is open 7 days a week through the summer, until September, when it cuts back to weekends only. Management estimates that it will require 22,000 hours of labor in each of the first 2 weeks of May, 25,000 hours during the third week of May, and 30,000 hours during the last week in May. During the first 2 weeks of June, it will require at least 35,000 hours of labor and 40,000 hours during the last 2 weeks in June. In July 45,000 hours will be required each week, and in August 45,000 hours will be needed each week. In September the park will need only 12,000 hours in the first week, 10,000 hours in each of the second and third weeks, and 8,000 hours the last week of the month.

The park hires new employees each week from the first week in May through August. A new employee mostly trains the first week by observing and helping more experienced employees; however, he or she works approximately 10 hours under the supervision of an experienced employee. An employee is considered experienced after completing 1 week on the job. Experienced employees are considered part-time and are scheduled to work 30 hours per week in order to eliminate overtime and reduce the cost of benefits, and to give more students the opportunity to work. However, no one is ever laid off or will be scheduled for fewer (or more) than 30 hours, even if more employees are available than needed. Management believes this is a necessary condition of employment because many of the student employees move to the area during the summer just to work in the park and live near the beach nearby. If these employees were sporadically laid off and were stuck with lease payments and other expenses, it would be bad public relations and hurt employment efforts in future summers. Although no one is laid off, 15% of all experienced employees quit each week for a variety of reasons, including homesickness, illness, and other personal reasons, plus some are asked to leave because of very poor job performance.

Park management is able to start the first week in May with 700 experienced employees who worked in the park in previous summers and live in the area. These employees are generally able to work a lot of hours on the weekends and then some during the week; however, in May attendance is much heavier on the weekends, so most of the labor hours are needed then. The park expects to have a pool of 1,500 available applicants to hire for the first week in May. After the first week, the pool is diminished by the number of new employees hired the previous week, but each week through June the park gets 200 new job applicants, which decreases to 100 new applicants each week for the rest of the summer. For example, the available applicant pool in the second week in May would be the previous week's pool, which in week 1 is 1,500, minus the number of new employees hired in week 1 plus 200 new applicants. At the end of the last week in August, 75% of all the experienced employees will quit to go back to school, and the park will not hire any new employees in September. The park must operate in September, using experienced employees who live in the area, but the weekly attrition rate for these employees in September drops to 10%.

Formulate and solve a linear programming model to assist the park's management to plan and schedule the number of new employees it hires each week in order to minimize the total number of new employees it must hire during the summer.

CHAPTER

5

Integer Programming

In the linear programming models formulated and solved in the previous chapters, the implicit assumption was that solutions could be fractional or real numbers (i.e., non-integer). However, non-integer solutions are not always practical.

When only integer solutions are practical or logical, it is sometimes assumed that non-integer solution values can be "rounded off" to the nearest feasible integer values. This method would cause little concern if, for example, $x_1 = 8{,}000.4$ nails were rounded off to 8,000 nails because nails cost only a few cents apiece. However, if we are considering the production of jet aircraft and $x_1 = 7.4$ jet airliners, rounding off could affect profit (or cost) by millions of dollars. In this case we need to solve the problem so that an *optimal integer solution* is guaranteed. In this chapter the different forms of integer linear programming models are presented.

Integer Programming Models

The three types of integer programming models are total, 0–1, and mixed.

There are three basic types of integer linear programming models—a total integer model, a 0–1 integer model, and a mixed integer model. In a **total integer model**, all the decision variables are required to have integer solution values. In a **0–1 integer model**, all the decision variables have integer values of zero or one. Finally, in a **mixed integer model**, some of the decision variables (but not all) are required to have integer solutions. The following three examples demonstrate these types of integer programming models.

A Total Integer Model Example

*In a **total integer model**, all decision variables have integer solution values.*

The owner of a machine shop is planning to expand by purchasing some new machines—presses and lathes. The owner has estimated that each press purchased will increase profit by \$100 per day and each lathe will increase profit by \$150 daily. The number of machines the owner can purchase is limited by the cost of the machines and the available floor space in the shop. The machine purchase prices and space requirements are as follows:

Machine	Required Floor Space (ft.²)	Purchase Price
Press	15	\$8,000
Lathe	30	4,000

The owner has a budget of \$40,000 for purchasing machines and 200 square feet of available floor space. The owner wants to know how many of each type of machine to purchase to maximize the daily increase in profit.

The linear programming model for an integer programming problem is formulated in exactly the same way as the linear programming examples in Chapters 2, 3, and 4. The only difference is that in this problem, the decision variables are restricted to integer values because the owner cannot purchase a fraction, or portion, of a machine. The linear programming model follows:

$$\text{maximize } Z = \$100x_1 + 150x_2$$

subject to

$$\$8{,}000x_1 + 4{,}000x_2 \leq \$40{,}000$$
$$15x_1 + 30x_2 \leq 200 \text{ ft.}^2$$
$$x_1, x_2 \geq 0 \text{ and integer}$$

where

x_1 = number of presses
x_2 = number of lathes

The decision variables in this model are restricted to whole machines. The fact that *both* decision variables, x_1 and x_2, can assume any integer value greater than or equal to zero is what gives this model its designation as a total integer model.

A 0–1 Integer Model Example

*In a **0–1 integer model**, the solution values of the decision variables are zero or one.*

A community council must decide which recreation facilities to construct in its community. Four new recreation facilities have been proposed—a swimming pool, a tennis center, an athletic field, and a gymnasium. The council wants to construct facilities that will maximize the expected daily usage by the residents of the community, subject to land and cost limitations. The expected daily usage and cost and land requirements for each facility follow:

Recreation Facility	Expected Usage (people/day)	Cost	Land Requirements (acres)
Swimming pool	300	$35,000	4
Tennis center	90	10,000	2
Athletic field	400	25,000	7
Gymnasium	150	90,000	3

The community has a $120,000 construction budget and 12 acres of land. Because the swimming pool and tennis center must be built on the same part of the land parcel, however, only one of these two facilities can be constructed. The council wants to know which of the recreation facilities to construct to maximize the expected daily usage. The model for this problem is formulated as follows:

$$\text{maximize } Z = 300x_1 + 90x_2 + 400x_3 + 150x_4$$

subject to

$$\$35{,}000x_1 + 10{,}000x_2 + 25{,}000x_3 + 90{,}000x_4 \leq \$120{,}000$$
$$4x_1 + 2x_2 + 7x_3 + 3x_4 \leq 12 \text{ acres}$$
$$x_1 + x_2 \leq 1 \text{ facility}$$
$$x_1, x_2, x_3, x_4 = 0 \text{ or } 1$$

where

x_1 = construction of a swimming pool
x_2 = construction of a tennis center
x_3 = construction of an athletic field
x_4 = construction of a gymnasium

In this model, the decision variables can have a solution value of either *zero* or *one*. If a facility is not selected for construction, the decision variable representing it will have a value of zero. If a facility is selected, its decision variable will have a value of one.

The last constraint, $x_1 + x_2 \leq 1$, reflects the *contingency* that either the swimming pool (x_1) or the tennis center (x_2) can be constructed, but not both. In order for the sum of x_1 and x_2 to be less than or equal to one, either of the variables can have a value of one, or both variables can equal zero. This is also referred to as a *mutually exclusive constraint*.

If the community had specified that either the swimming pool (x_1) or the tennis center (x_2) *must* be built, but not both, then the last constraint would become an equation, $x_1 + x_2 = 1$. This would result in a solution that would include $x_1 = 1$ or $x_2 = 1$, but both would not equal one (nor would both equal zero). In this manner, the model forces a choice between the two facilities. For this reason, it is often called a *multiple-choice constraint*.

A variation of the multiple-choice constraint can be used to formulate a situation in which some specific number of facilities out of the total must be constructed. For example, if the community council had specified that exactly two of the four facilities must be built, this constraint would be formulated as

$$x_1 + x_2 + x_3 + x_4 = 2$$

If, alternatively, the council had specified that no more than two facilities must be constructed, the constraint would be

$$x_1 + x_2 + x_3 + x_4 \leq 2$$

Another type of 0–1 model constraint is a *conditional constraint*. In a conditional constraint, the construction of one facility is conditional upon the construction of another. Suppose, for example, that the pet project of the head of the community council is the swimming pool, and she also believes the tennis center is frivolous. The council head is very influential, so the rest of the council knows that the tennis center has no chance of being selected if the pool is not selected first. However, even if the pool is selected, there is no guarantee that the tennis center will also be selected. Thus, the tennis center (x_2) is conditional upon construction of the swimming pool (x_1). This condition is formulated as

$$x_2 \leq x_1$$

Notice that the tennis center (x_2) cannot equal one (i.e., be selected) unless the pool (x_1) equals one. If the pool (x_1) equals zero (i.e., it is not selected), then the tennis center (x_2) must also equal zero. However, this condition does allow the pool (x_1) to equal one and be selected and the tennis center to equal zero and not be selected.

A variation of this type of conditional constraint is the *corequisite constraint*, wherein if one facility is constructed, the other one will also be constructed and vice versa. For example, suppose the council worked out a political deal among themselves, wherein if the pool is accepted, the tennis center must also be selected and vice versa. This constraint is written as

$$x_2 = x_1$$

This constraint makes x_1 and x_2 equal the same value, either zero or one.

A Mixed Integer Model Example

Nancy Smith has $250,000 to invest in three alternative investments—condominiums, land, and municipal bonds. She wants to invest in the alternatives that will result in the greatest return on investment after 1 year.

*In a **mixed integer model**, some solution values for decision variables are integers and others can be non-integer.*

Each condominium costs $50,000 and will return a profit of $9,000 if sold at the end of 1 year; each acre of land costs $12,000 and will return a profit of $1,500 at the end of 1 year; and each municipal bond costs $8,000 and will result in a return of $1,000 if sold at the end of 1 year. In addition, there are only 4 condominiums, 15 acres of land, and 20 municipal bonds available for purchase.

The linear programming model for this problem is formulated as follows:

$$\begin{aligned} &\text{maximize } Z = \$9,000x_1 + 1,500x_2 + 1,000x_3 \\ &\text{subject to} \\ &\$50,000x_1 + 12,000x_2 + 8,000x_3 \leq \$250,000 \\ &\qquad x_1 \leq 4 \text{ condominiums} \\ &\qquad x_2 \leq 15 \text{ acres} \\ &\qquad x_3 \leq 20 \text{ bonds} \\ &\qquad x_2 \geq 0 \\ &\qquad x_1, x_3 \geq 0 \text{ and integer} \end{aligned}$$

where

$$\begin{aligned} x_1 &= \text{condominiums purchased} \\ x_2 &= \text{acres of land purchased} \\ x_3 &= \text{bonds purchased} \end{aligned}$$

Management Science Application

Allocating Operating Room Time at Toronto's Mount Sinai Hospital

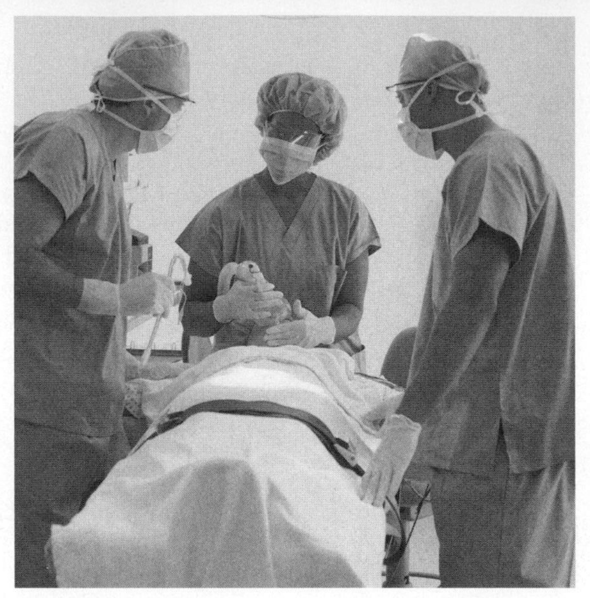

Jupiterimages/Comstock/Thinkstock

Mount Sinai Hospital in Toronto has 14 operating rooms that serve five surgical departments, including surgery, gynecology, ophthalmology, otolaryngology, and oral surgery. The largest of these departments, surgery, consists of five subareas, including orthopedics, general surgery, plastic surgery, vascular surgery, and urology. The hospital has 397.50 operating room hours available per week, and it employs an integer programming model to develop a master surgical schedule that allocates operating room time to its five primary surgical departments. A daily operating room schedule includes the actual cases to be performed, start and end times, attending physicians, and anesthetists, and it is similar to a production schedule in a manufacturing plant. Operating room times are assigned as blocks; whenever possible, only one department is assigned to an operating room on a single day, thus constituting a block. It is possible to split blocks into morning and afternoon sessions and to alternate the same block between departments in different weeks, although the hospital attempts to minimize these types of allocations and keep the schedule as consistent as possible from week to week.

In the integer programming model, the decision variables, x_{ijk}, equal the number of time blocks of operating room *i*, assigned to department *j*, on day *k*. The objective is to minimize the sum of the differences, or shortfalls, between the assigned operating room times and target times, represented as penalty functions, for all departments. A primary constraint is that the sum of all times allocated to a department, *j*, over all days, *k*, be (as nearly as possible) equal to a target number of hours. Additional constraints include daily and weekly bounds on the number of rooms assigned to a department, limits on the amount of under- or overallocation of time to any particular department, and daily room availability. The model solution provides an allocation of time blocks to departments that minimizes the shortage of time to each. Since its implementation, the model has reduced the clerical time for producing a schedule from days (using a manual approach) to 1 or 2 hours, and has resulted in roughly \$20,000 per year in savings derived from the reduced time required by the operating room manager for schedule development. In addition, the model has provided hospital management with greater flexibility, increased ability to explore creative scheduling options, and generally improved quality schedules. However, perhaps the model's greatest benefit has been to provide unbiased, equitable schedules through a consistent, identifiable process, thus reducing conflict between departments and surgeons.

Source: Based on J. T. Blake and J. Donald, "Mount Sinai Hospital Uses Integer Programming to Allocate Operating Room Time," *Interfaces* 32, no. 2 (March–April 2002): 63–73.

Notice that in this model, the solution values for condominiums (x_1) and municipal bonds (x_3) must be integers. It is not possible to invest in a fraction of a condominium or to purchase part of a bond. However, it is possible to purchase less than an acre of land (i.e., a portion of an acre). Thus, two of the decision variables (x_1 and x_3) are restricted to integer values, whereas the other variable (x_2) can take on any real value greater than or equal to zero.

Integer Programming Graphical Solution

It may seem logical that an easy way to solve integer programming problems is to *round off* fractional solution values to integer values. However, that can result in less-than-optimal (i.e., *suboptimal*) results. This outcome can be seen by using graphical analysis. As an example, consider the total integer model for the machine shop formulated in the previous section:

$$\begin{aligned} \text{maximize } Z &= 100x_1 + 150x_2 \\ \text{subject to} & \\ 8{,}000x_1 + 4{,}000x_2 &\leq 40{,}000 \\ 15x_1 + 30x_2 &\leq 200 \\ x_1, x_2 &\geq 0 \text{ and integer} \end{aligned}$$

First, we will use Excel to solve this model as a regular linear programming model without the integer requirements, as shown in Exhibit 5.1.

EXHIBIT 5.1

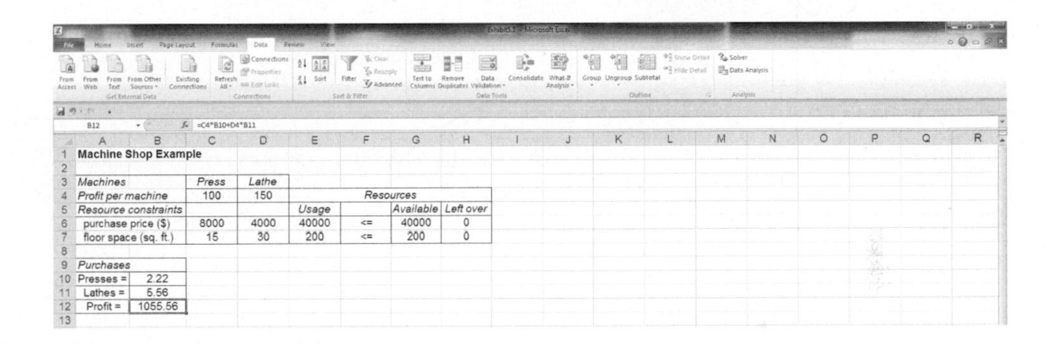

Machine Shop Example						
Machines	Press	Lathe				
Profit per machine	100	150		Resources		
Resource constraints			Usage		Available	Left over
purchase price ($)	8000	4000	40000	<=	40000	0
floor space (sq. ft.)	15	30	200	<=	200	0
Purchases						
Presses =	2.22					
Lathes =	5.56					
Profit =	1055.56					

Rounding non-integer solution values up to the nearest integer value can result in an infeasible solution.

Notice that this model results in a non-integer solution of 2.22 presses and 5.56 lathes, or $x_1 = 2.22$ and $x_2 = 5.56$. Because the solution values must be integers, let us first round off these two values to the *closest integer values*, which are $x_1 = 2$ and $x_2 = 6$. However, if we substitute these values into the second model constraint, we find that this integer solution violates the constraint and thus is infeasible:

$$\begin{aligned} 15x_1 + 30x_2 &\leq 200 \\ 15(2) + 30(6) &\leq 200 \\ 210 &\nleq 200 \end{aligned}$$

A feasible solution is ensured by rounding down non-integer solution values.

In a model in which the constraints are all $\leq$ (and the constraint coefficients are positive), a feasible solution is always ensured by *rounding down*. Thus, a feasible integer solution for this problem is

$$\begin{aligned} x_1 &= 2 \\ x_2 &= 5 \\ Z &= \$950 \end{aligned}$$

However, one of the difficulties of simply rounding down non-integer values is that another integer solution may result in a higher profit (i.e., in this problem, there may be an integer solution that will result in a profit higher than $950). In order to determine whether a better integer solution exists, let us analyze the graph of this model, which is shown in Figure 5.1.

A rounded-down integer solution can result in a less-than-optimal (suboptimal) solution.

In Figure 5.1, the dots indicate integer solution points, and the point $x_1 = 2$, $x_2 = 5$ is the rounded-down solution. Notice that as the objective function edge moves outward through the feasible solution space, there is an *integer* solution point that yields a greater profit than the rounded-down solution. This solution point is $x_1 = 1$, $x_2 = 6$. At this point, $Z = \$1{,}000$, which is $50 more profit per day for the machine shop than is realized by the rounded-down integer solution.

This graphical analysis explicitly demonstrates the error that can result from solving an integer programming problem by simply rounding down. In the machine shop example,

FIGURE 5.1
Feasible solution space with integer solution points

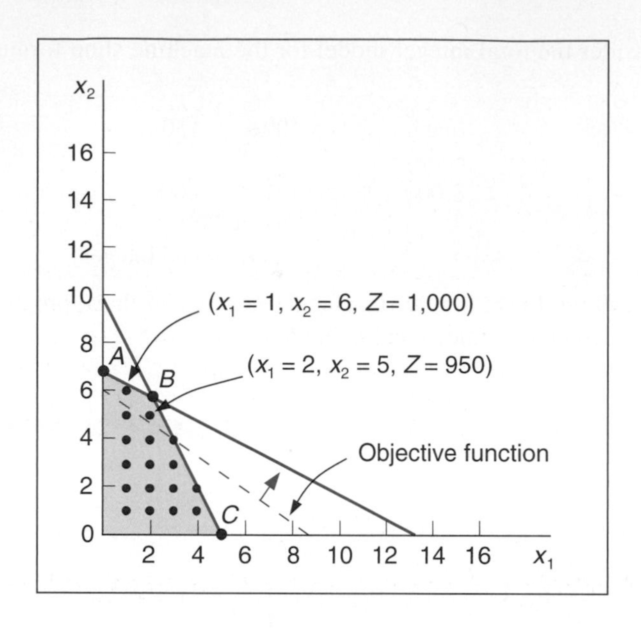

the optimal integer solution is $x_1 = 1, x_2 = 6$, instead of the rounded-down solution, $x_1 = 2$, $x_2 = 5$ (which is often called the *suboptimal* solution or result). Because erroneous results are caused by rounding down regular solutions, a more direct approach for solving integer problems is required.

See Web site Module C for a chapter on "Integer Programming: The Branch and Bound Method."

The traditional approach for solving integer programming problems is the *branch and bound method*. It is a mathematical solution approach that can be applied to a number of different types of problems. The branch and bound method is based on the principle that the total set of feasible solutions (such as the feasible area in Figure 5.1) can be partitioned into smaller subsets of solutions. These smaller subsets can then be evaluated systematically until the best solution is found. The branch and bound method is a tedious and often complex mathematical process. Fortunately, both Excel and QM for Windows have the capability to solve integer programming problems, and thus we will rely on the computer to solve the different types of integer models demonstrated in this chapter. However, for the interested reader, the companion Web site that accompanies this text includes a supplementary module, "Integer Programming: The Branch and Bound Method," that demonstrates the branch and bound method in detail.

Computer Solution of Integer Programming Problems with Excel and QM for Windows

Integer programming problems can be solved using Excel spreadsheets and QM for Windows. In this section we demonstrate both of these computer solution approaches, using the examples for 0–1, total, and mixed integer programming problems developed in the previous sections.

Solution of the 0–1 Model with Excel

Recall our recreational facilities example, formulated on page 186:

$$\begin{aligned}
&\text{maximize } Z = 300x_1 + 90x_2 + 400x_3 + 150x_4\\
&\text{subject to}\\
&\$35{,}000x_1 + 10{,}000x_2 + 25{,}000x_3 + 90{,}000x_4 \le \$120{,}000\\
&\qquad\qquad 4x_1 + 2x_2 + 7x_3 + 3x_4 \le 12 \text{ acres}\\
&\qquad\qquad x_1 + x_2 \le 1 \text{ facility}\\
&\qquad\qquad x_1, x_2, x_3, x_4 = 0 \text{ or } 1
\end{aligned}$$

where

x_1 = construction of a swimming pool
x_2 = construction of a tennis center
x_3 = construction of an athletic field
x_4 = construction of a gymnasium
Z = total expected usage (people) per day

Exhibit 5.2 shows this example model set up in Excel spreadsheet format. The decision variables for the facilities are in cells **C12:C15**, and the objective function (Z) is embedded in cell C16. The objective function is shown on the formula bar at the top of the spreadsheet. The model constraints are contained in cells G7, G8, and G9. For example, cell G7 includes the cost constraint, **=C7*C12+D7*C13+E7*C14+F7*C15**, and the available budget of $120,000 is contained in cell I7. Thus, the cost constraint will be written in Solver as **G7≤I7**.

EXHIBIT 5.2

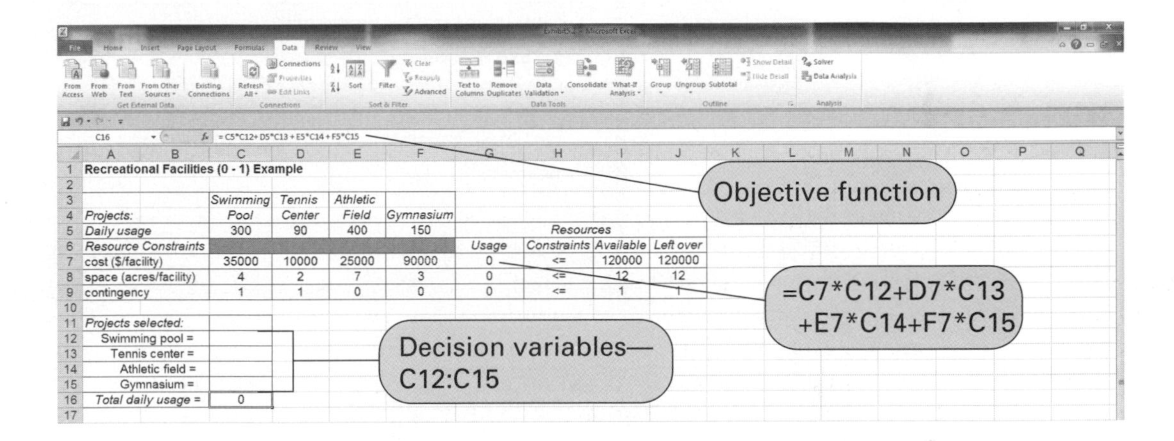

Exhibit 5.3 shows the Solver Parameters screen for our spreadsheet example. Notice that we have established the 0–1 condition of our variables by constraining the decision variable cells to be binary (i.e., 0 or 1), as shown in the "Add Constraint" window in Exhibit 5.4.

Time Out for Ralph E. Gomory

In 1958, Ralph Gomory was the first individual to develop a systematic (algorithmic) approach for solving linear integer programming problems. His "cutting plane method" is an algebraic approach based on the systematic addition of new constraints (or cuts), which are satisfied by an integer solution but not by a continuous variable solution. An alternative to an algebraic approach like Gomory's is an enumerative approach, which is a means of searching through all possible integer solutions (but in an intelligent manner) to limit the extent of the search. One such search method is the branch and bound technique, introduced in 1960 by A. H. Land and A. C. Doig of the London School of Economics and Political Science.

EXHIBIT 5.3

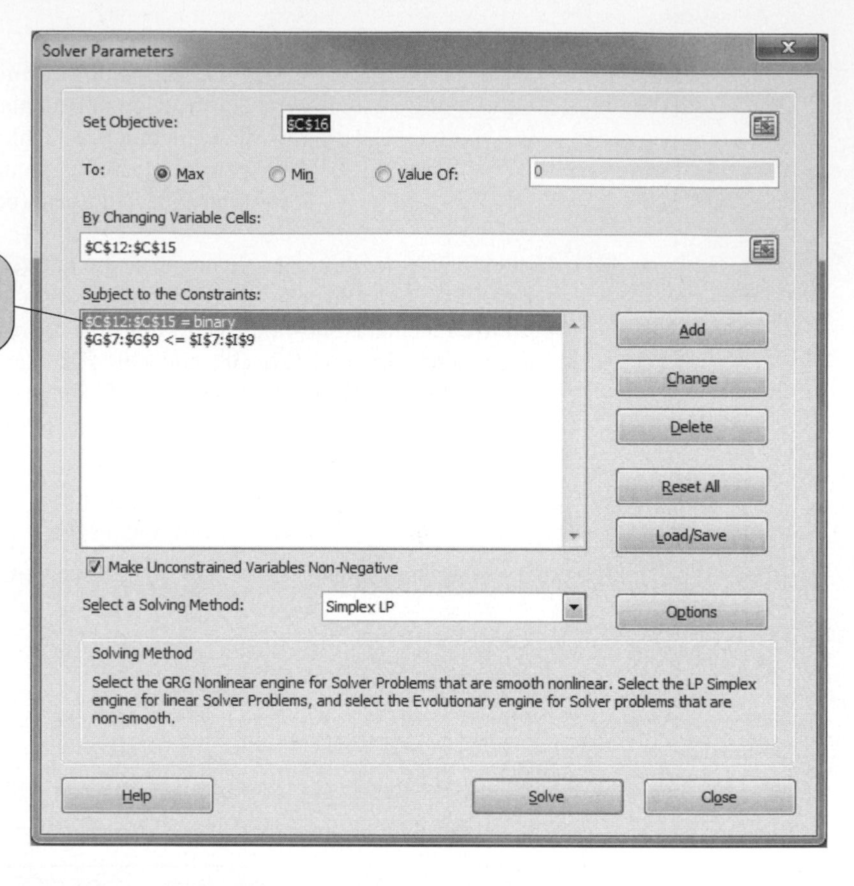

EXHIBIT 5.4

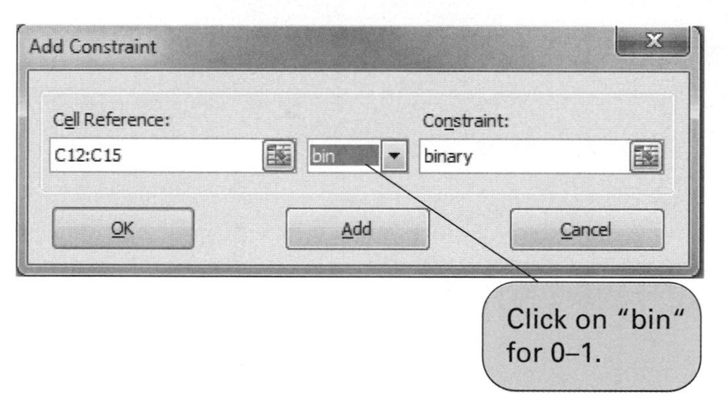

Before solving the problem, click on the "Options" tab on the Solver Parameters screen (see Exhibit 5.3). When the Options window shown in Exhibit 5.5 comes up, make sure "Ignore Integer Constraints" is not activated (with a check). Return to the Solver Parameters window and click on "Solve," which results in the solution shown in Exhibit 5.6.

Solution of the 0–1 Model with QM for Windows

The 0–1 integer programming problem is solved in the "Integer and Mixed Integer" module of QM for Windows. We will demonstrate this module by using our example for selecting recreational facilities that we just solved with Excel.

Exhibit 5.7 shows the data input screen for our recreational facilities example problem. At the bottom of the screen, when you click on "Variable Type" for a variable, a menu will be displayed from which you can indicate if the variable is 0–1, integer, or real. In the case of this

EXHIBIT 5.5

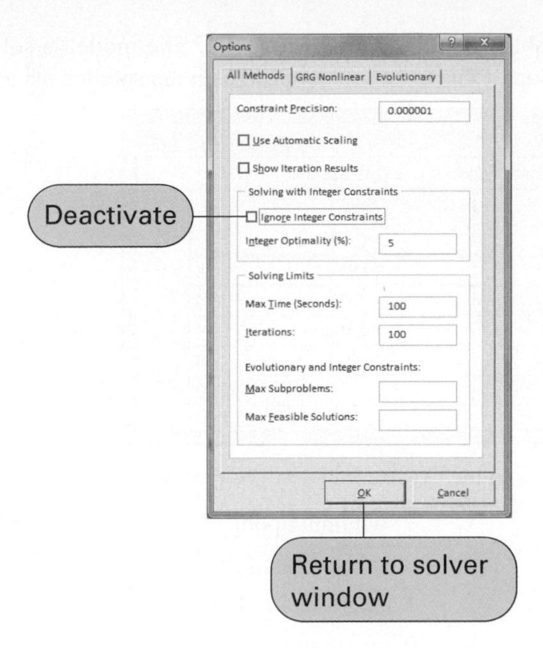

EXHIBIT 5.6

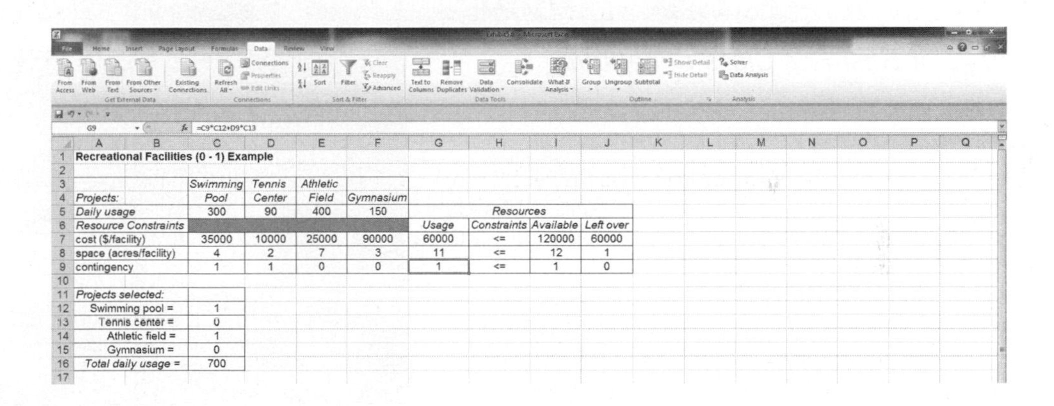

Recreational Facilities (0 - 1) Example

	Swimming Pool	Tennis Center	Athletic Field	Gymnasium				
Projects:								
Daily usage	300	90	400	150		Resources		
Resource Constraints					Usage	Constraints	Available	Left over
cost ($/facility)	35000	10000	25000	90000	60000	<=	120000	60000
space (acres/facility)	4	2	7	3	11	<=	12	1
contingency	1	1	0	0	1	<=	1	0

Projects selected:	
Swimming pool =	1
Tennis center =	0
Athletic field =	1
Gymnasium =	0
Total daily usage =	700

EXHIBIT 5.7

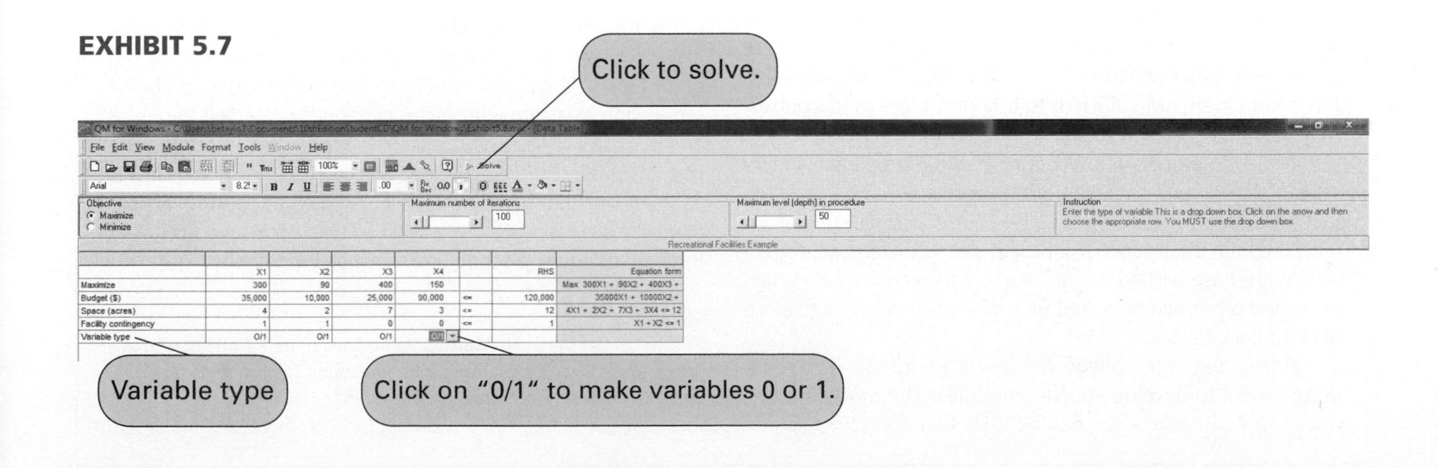

	X1	X2	X3	X4		RHS	Equation form
Maximize	300	90	400	150			Max 300X1 + 90X2 + 400X3 +
Budget ($)	35,000	10,000	25,000	90,000	<=	120,000	35000X1 + 10000X2 +
Space (acres)	4	2	7	3	<=	12	4X1 + 2X2 + 7X3 + 3X4 <= 12
Facility contingency	1	1	0	0	<=	1	X1 + X2 <= 1
Variable type	0/1	0/1	0/1	0/1			

example, all the variables should be designated 0–1. The model is solved by clicking on the "Solve" button at the top of this screen, which results in the solution screen shown in Exhibit 5.8.

EXHIBIT 5.8

Integer & Mixed Integer Programming Res...

Recreational Facilities Example Solution		
Variable	Type	Value
X1	0/1	1
X2	0/1	0
X3	0/1	1
X4	Integer	0
Solution value		700

The solution shown in the QM for Windows output is

$$x_1 = 1 \text{ swimming pool}$$
$$x_2 = 0 \text{ tennis center}$$
$$x_3 = 1 \text{ athletic field}$$
$$x_4 = 0 \text{ gymnasium}$$
$$Z = 700 \text{ people per day expected usage}$$

Management Science Application

College of Business Class Scheduling at Ohio University

The College of Business at Ohio University in Athens, Ohio, encompasses four academic departments—Accounting, Finance, Management Information Systems, and Marketing—with approximately 65 to 75 instructors. Each academic term, the college departments offer 110 to 130 sections of different courses. The college's home building is Copeland Hall, which includes 14 to 16 classrooms. Prior to 1998 the college used a manual process to schedule courses and instructors into classrooms at different times during the day. An associate dean would allocate the classrooms to the different departments, and the department chair would assign courses to instructors and schedule the course sections to the rooms, sometimes taking into account instructor preferences. If a chairperson did not use all of the department's allocated classrooms, they were shared with other departments, and if a department's available classrooms were not sufficient to accommodate all its courses, they were assigned to classrooms outside the building.

While this process was workable, it was not optimal; and instructors' preferences weren't often considered, and they often taught outside the building because course sections couldn't be matched with classrooms of sufficient size. For example, instructors might have wished to only teach certain courses at certain times and on certain days, and they might have wanted to avoid back-to-back classes.

Since 1998 the college has used an integer programming model to develop course schedules. The model assigns instructors to courses in classrooms during specific time slots, based on instructor preference forms developed within the departments. The typical integer programming model for course scheduling in a semester includes more than 2,500 variables and almost 2,000 constraints. The model has improved the use of classroom space, instructors have to teach fewer courses outside Copeland Hall (despite increasing enrollments), instructors are more satisfied with their schedules, and the time to develop schedules has been cut in half.

Rick Fatica/Ohio University

Source: Based on C. H. Martin, "Ohio University's College of Business Uses Integer Programming to Schedule Classes," *Interfaces* 34, no. 6 (November–December 2004): 460–65.

Solution of the Total Integer Model with Excel

We will demonstrate the Excel solution of a total integer programming problem for the machine shop example we solved previously using the branch and bound method. Recall that this model was formulated as

$$\begin{aligned}
\text{maximize } Z &= \$100x_1 + 150x_2 \\
\text{subject to} & \\
8{,}000x_1 + 4{,}000x_2 &\leq \$40{,}000 \\
15x_1 + 30x_2 &\leq 200 \text{ ft.}^2 \\
x_1, x_2 &\geq 0 \text{ and integer}
\end{aligned}$$

where

x_1 = number of presses
x_2 = number of lathes

The machine shop example set up in spreadsheet format is shown in Exhibit 5.9. This is the same basic format as a linear programming model. The essential difference between solving a regular linear programming model and an integer programming model is that we must designate the cells representing the decision variables as being "integer." We accomplish this by adding a constraint within Solver that establishes **B10:B11** as integers, as shown in Exhibit 5.11. The complete Solver, with all constraints, is shown in Exhibit 5.10. Clicking on the "Solve" button will generate the spreadsheet solution, as shown in Exhibit 5.12.

EXHIBIT 5.9

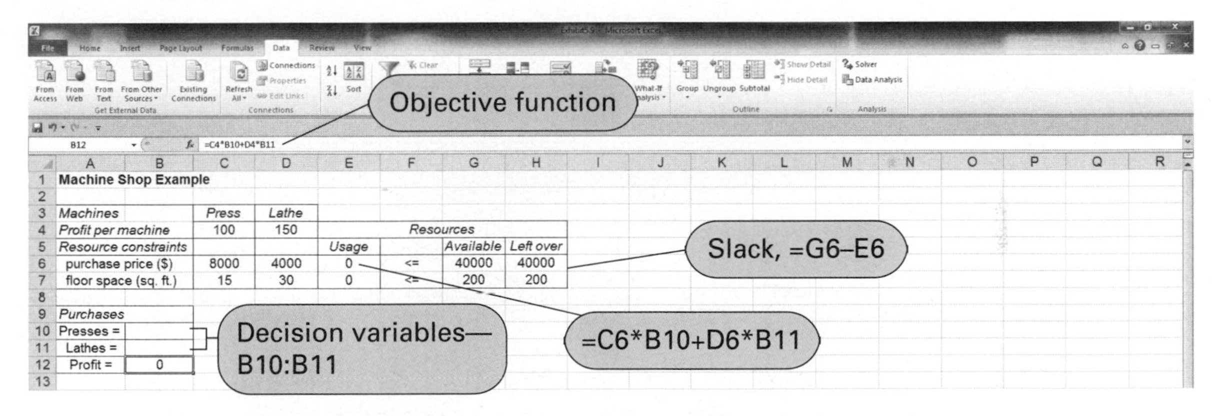

Solution of the Mixed Integer Model with Excel

The solution of a mixed integer programming model with Excel will be demonstrated by using the example for investments formulated earlier in the chapter:

$$\begin{aligned}
\text{maximize } Z &= \$9{,}000x_1 + 1{,}500x_2 + 1{,}000x_3 \\
\text{subject to} & \\
\$50{,}000x_1 + 12{,}000x_2 + 8{,}000x_3 &\leq \$250{,}000 \\
x_1 &\leq 4 \text{ condominiums} \\
x_2 &\leq 15 \text{ acres} \\
x_3 &\leq 20 \text{ bonds} \\
x_2 &\geq 0 \\
x_1, x_3 &\geq 0 \text{ and integer}
\end{aligned}$$

where

x_1 = condominiums purchased
x_2 = acres of land purchased
x_3 = bonds purchased

EXHIBIT 5.10

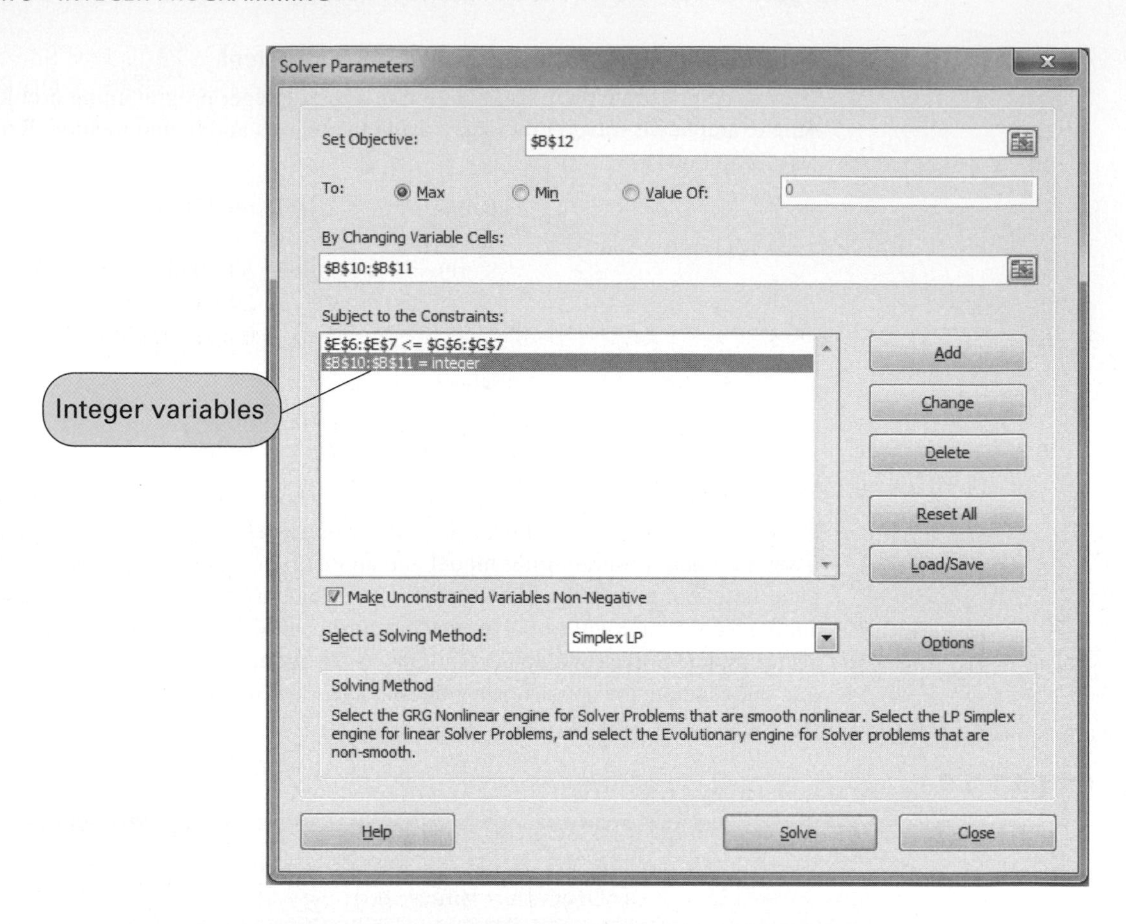

EXHIBIT 5.11

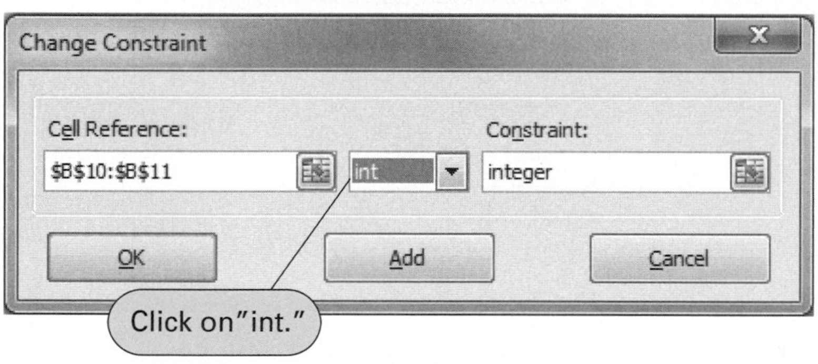

EXHIBIT 5.12

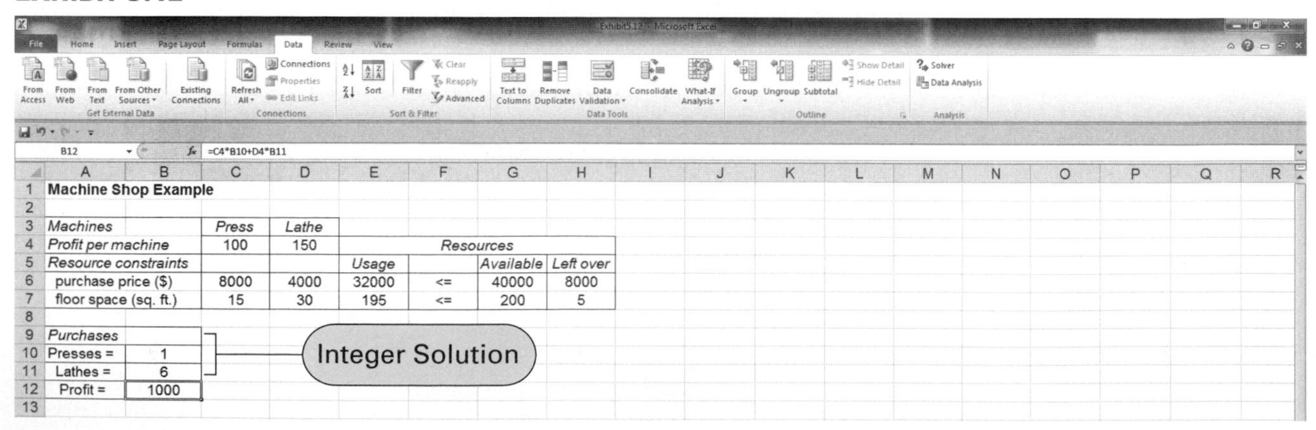

The Excel spreadsheet solution is shown in Exhibit 5.13, and the Solver Parameters window is shown in Exhibit 5.14. The decision variables are contained in cells **B8:B10**. The objective function formula is in cell B11. Notice that we did not include the constraint values for the

EXHIBIT 5.13

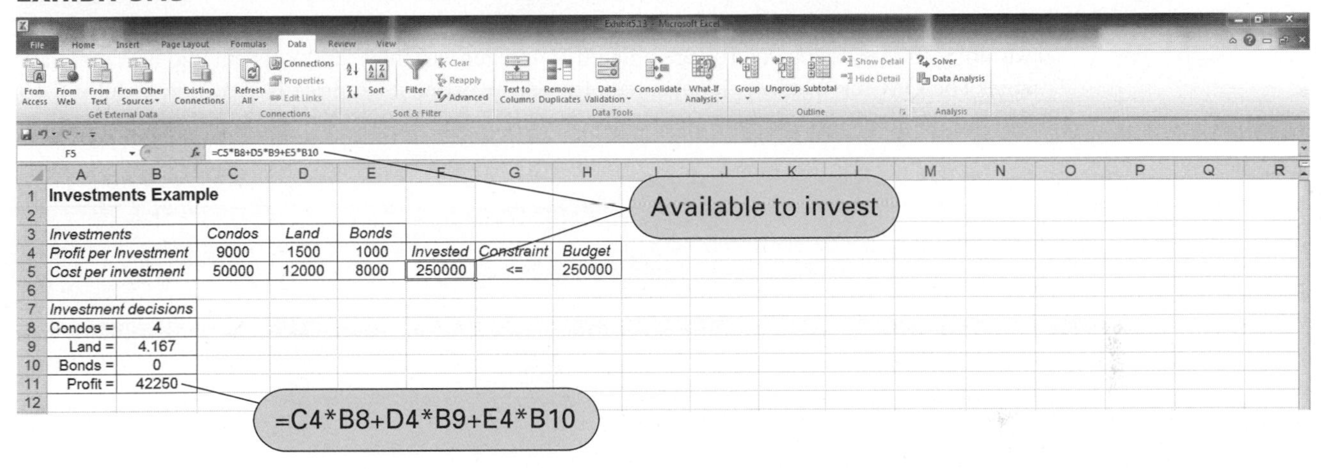

	A	B	C	D	E	F	G	H
1	Investments Example							
2								
3	Investments		Condos	Land	Bonds			
4	Profit per Investment		9000	1500	1000	Invested	Constraint	Budget
5	Cost per investment		50000	12000	8000	250000	<=	250000
6								
7	Investment decisions							
8	Condos =	4						
9	Land =	4.167						
10	Bonds =	0						
11	Profit =	42250						

EXHIBIT 5.14

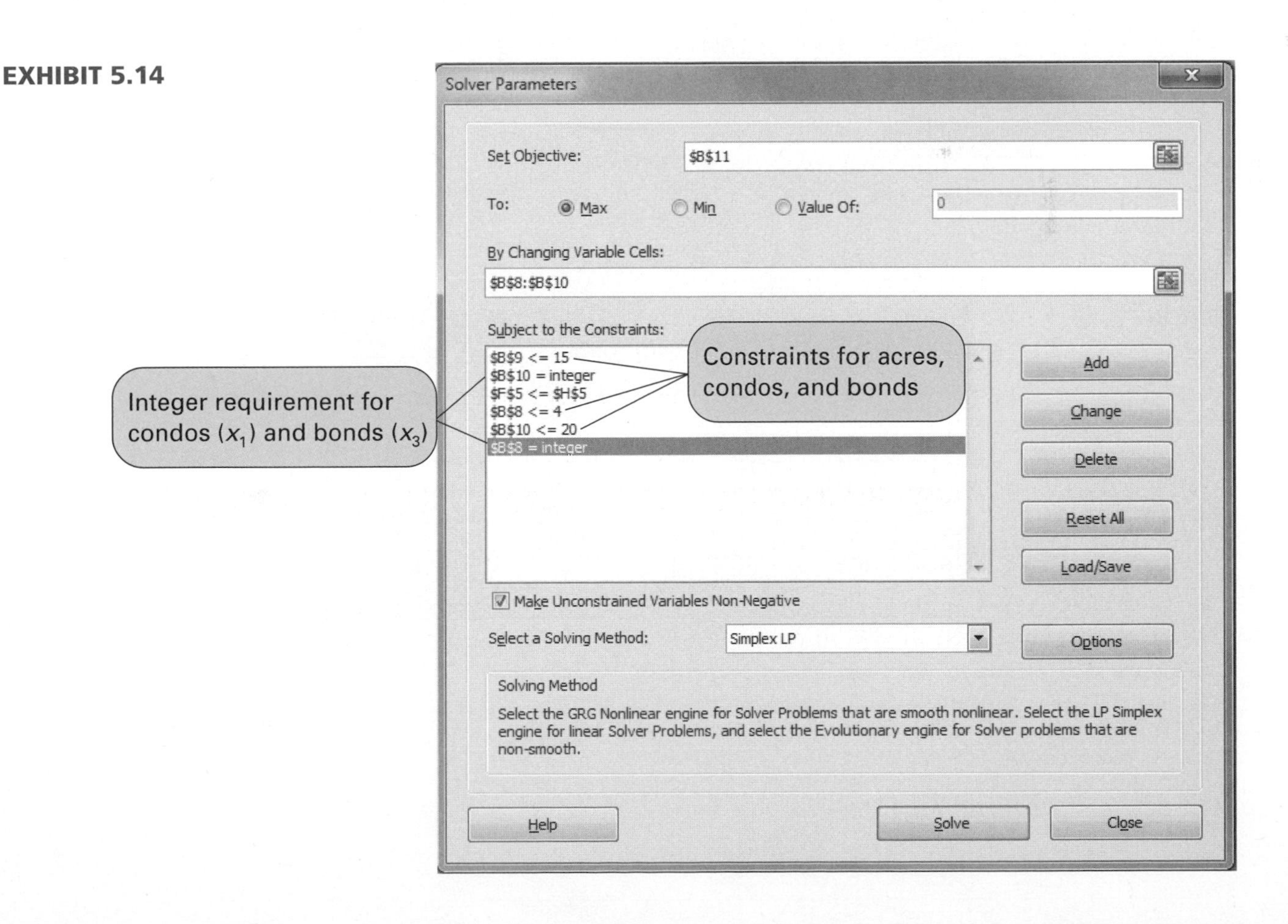

availability of each type of investment (i.e., 4 condos, 15 acres of land, and 20 bonds) in the spreadsheet setup. Instead, it was easier just to enter these constraints directly into Solver, as shown in Exhibit 5.14. Notice in Solver that we have designated B8 and B10 as integers, but we have not designated B9 as an integer, reflecting the fact that x_1 and x_3 are integers, whereas x_2 is a real variable.

Solution of the Mixed Integer Model with QM for Windows

The "Integer and Mixed Integer" module of QM for Windows is used to solve our investment example. When the problem data are input, we must designate the variable types for each decision variable. In this case, x_1 and x_3 are entered as integer variables, and x_2 is entered as a real variable. The QM for Windows input screen for our investment example is shown in Exhibit 5.15, and the solution screen is shown in Exhibit 5.16.

EXHIBIT 5.15

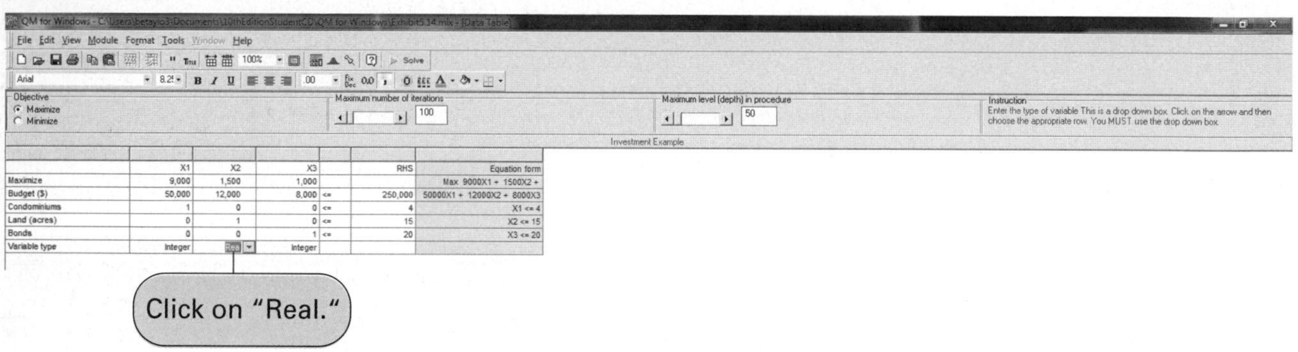

	X1	X2	X3		RHS	Equation form
Maximize	9,000	1,500	1,000			Max 9000X1 + 1500X2 +
Budget ($)	50,000	12,000	8,000	<=	250,000	50000X1 + 12000X2 + 8000X3
Condominiums	1	0	0	<=	4	X1 <= 4
Land (acres)	0	1	0	<=	15	X2 <= 15
Bonds	0	0	1	<=	20	X3 <= 20
Variable type	Integer	Real	Integer			

EXHIBIT 5.16

Integer & Mixed Integer Programming Resu...

Investment Example Solution

Variable	Type	Value
X1	Integer	4
X2	Real	4.17
X3	Integer	0
Solution value		42,250

Management Science Application

Forming Business Case Student Teams at Indiana University

The Kelley School of Business at Indiana University (Bloomington) has more than 4,000 students. The school has an "integrated core" program, which consists of four courses offered simultaneously in which business students must receive at least a C grade before they are allowed to move on to advanced courses in their area of specialization. A program requirement is that students must apply their knowledge of finance, marketing, operations, and strategy to a case study of a hypothetical business situation. Students are divided into teams of six or more for the cases, which account for 25% of the students' grades in the four program courses. About 800 students take the courses in the fall semester and about 300 in the fall and spring, resulting in between 50 and 130 teams. Assigning this number of students to teams in an equitable manner so that no team members perceive that they will be at a disadvantage because of their team assignment is a difficult and complex task. The program coordinator established conditions to ensure, as much as possible, that the teams be equitable, including equal academic performance, diversity of business functional areas, and an absence of a lone female or international student member. The need for equal academic performance among teams is evident, and since the case study is done in the last 10 days of the semester, the academic performance of each student to that point in time can be determined. Functional diversity was considered to be a desirable team characteristic; however, since the school has more finance and marketing majors, assigning students to teams arbitrarily would not necessarily achieve a functionally diverse team mix. Finally, past experience had shown that a single female or international student team member sometimes hindered team communication and did not always result in a positive experience for that one member. An integer programming model was developed to assign students to teams so that these conditions ensuring equitable teams would be met. The objective function minimizes the maximum deviation of a team from the average academic performance of the entire class. The model constraints ensure that all students are assigned to a team; teams consist of six students; no team has more than two finance or marketing majors; and that female and international students are assigned to teams in multiples of two (i.e., a team might not have any female or international students, but if it did there would be at least two). The integer programming modeling approach resulted in teams that were judged more cohesive, and that experienced less friction, than with the previous team assignment method, thus allowing the coordinator to conclude that the new method was more equitable from a student perspective.

Creatas Images/Creatas/Thinkstock

Source: Based on R. Cutshall, S. Gavirneni, and K. Schultz, "Indiana University's Kelley School of Business Uses Integer Programming to Form Equitable, Cohesive Student Teams," *Interfaces* 37, no. 3 (May–June 2007): 265–76.

0–1 Integer Programming Modeling Examples

Some of the most interesting and useful applications of integer programming involve 0–1 variables. In these applications the variables allow for the selection of an item (or activity) where a value of one indicates that the item is selected, and a value of zero indicates that it is not. For example, a decision variable might represent the purchase of a building; if the value of the variable is one, the building is purchased. If the value is zero, the building is not purchased. We will demonstrate three of the most popular applications of 0–1 integer programming—a capital budgeting problem, a fixed charge and facility location problem, and a set covering problem.

A Capital Budgeting Example

The University Bookstore at Tech is considering several expansion projects, including developing a store Web site for online retail and catalog purchases, buying an off-campus warehouse and its subsequent expansion, developing a clothing and gift department specializing in university logo apparel, opening a computer department carrying both hardware and software products, and creating a banking pavilion of three automated teller machines outside the store. Some of the projects will be developed over a 2-year period and some over a 3-year period, as funds permit. The net present value costs per year and the projected net present value of returns for a 5-year period for each of the projects are shown in the following table:

Project	NPV Return ($1,000s)	Project Costs/Year ($1,000s)		
		1	2	3
1. Web site	$120	$55	$40	$25
2. Warehouse	85	45	35	20
3. Clothing department	105	60	25	—
4. Computer department	140	50	35	30
5. ATMs	70	30	30	—
Available funds per year		$150	$110	$60

In addition, the store does not have enough space available to create both a computer department and a clothing department. The bookstore director wants to know which projects to select to maximize returns.

This problem requires a 0–1 integer programming model formulation where

$$
\begin{aligned}
x_1 &= \text{selection of Web site project}\\
x_2 &= \text{selection of warehouse project}\\
x_3 &= \text{selection of clothing department project}\\
x_4 &= \text{selection of computer department project}\\
x_5 &= \text{selection of ATM project}\\
x_i &= 1 \text{ if project } i \text{ is selected; 0 if project } i \text{ is not selected}
\end{aligned}
$$

$$
\begin{aligned}
&\text{maximize } Z = \$120x_1 + 85x_2 + 105x_3 + 140x_4 + 70x_5\\
&\text{subject to}\\
&55x_1 + 45x_2 + 60x_3 + 50x_4 + 30x_5 \le 150\\
&40x_1 + 35x_2 + 25x_3 + 35x_4 + 30x_5 \le 110\\
&25x_1 + 20x_2 + 30x_4 \le 60\\
&x_3 + x_4 \le 1\\
&x_i = 0 \text{ or } 1
\end{aligned}
$$

Exhibit 5.17 shows this capital budgeting example set up in Excel spreadsheet format. The decision variables for the projects are in cells **C7:C11**, and the objective function (Z) is in cell D17. The objective function is shown on the formula bar at the top of the spreadsheet. The model constraints are included in cells **E12:G12**. For example, cell E12 contains the budget constraint for year 1, **=SUMPRODUCT (C7:C11, E7:E11)**, and the available budget is in cell E14. Thus, this budget constraint will be included in Solver as **E12:G12≤E14:G14**. The mutually exclusive constraint, $x_3 + x_4 \le 1$, is included in cell E15 as **=C9+C10**.

Exhibit 5.18 shows the Solver Parameters screen for the capital budgeting example. The 0–1 variable restriction has been established by clicking on "bin" (meaning the variables are binary)

EXHIBIT 5.17

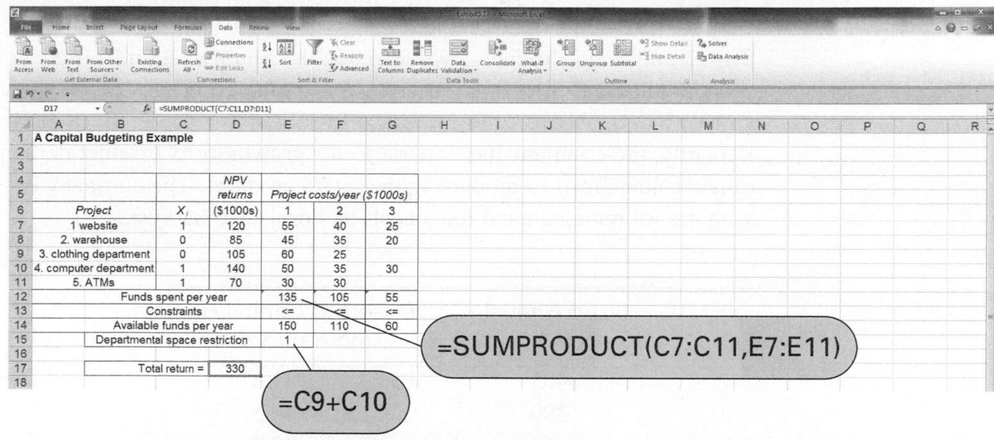

A Capital Budgeting Example

Project	X_i	NPV returns ($1000s)	Project costs/year ($1000s) 1	2	3
1 website	1	120	55	40	25
2. warehouse	0	85	45	35	20
3. clothing department	0	105	60	25	
4. computer department	1	140	50	35	30
5. ATMs	1	70	30	30	
Funds spent per year			135	105	55
Constraints			<=	<=	<=
Available funds per year			150	110	60
Departmental space restriction			1		
Total return =		330			

EXHIBIT 5.18

Solver Parameters

Set Objective: $D$17

To: Max Min Value Of: 0

By Changing Variable Cells: $C$7:$C$11

Subject to the Constraints:

$C$7:$C$11 = binary
$E$15 <= 1
$E$12:$G$12 <= $E$14:$G$14

Add Change Delete Reset All Load/Save

Make Unconstrained Variables Non-Negative

Select a Solving Method: Simplex LP Options

Solving Method

Select the GRG Nonlinear engine for Solver Problems that are smooth nonlinear. Select the LP Simplex engine for linear Solver Problems, and select the Evolutionary engine for Solver problems that are non-smooth.

Help Solve Close

0–1 integer restriction

Mutually exclusive constraint

in the "Add Constraint" window. (Exhibit 5.4 showed the "bin" option.) The mutually exclusive constraint is shown as **E15≤1**.

The model solution is

$$x_1 = 1 \text{ (Web site)}$$
$$x_4 = 1 \text{ (computer department)}$$
$$x_5 = 1 \text{ (ATMs)}$$
$$Z = \$330{,}000$$

A Fixed Charge and Facility Location Example

Frijo-Lane Food Products own farms in the Southwest and Midwest, where it grows and harvests potatoes. It then ships these potatoes to three processing plants in Atlanta, Baton Rouge, and Chicago, where different varieties of potato products, including chips, are produced. Recently, the company has experienced a growth in its product demand, so it wants to buy one or more new farms to produce more potato products. The company is considering six new farms with the following annual fixed costs and projected harvest:

Farm	Fixed Annual Costs ($1,000s)	Projected Annual Harvest (thousands of tons)
1	$405	11.2
2	390	10.5
3	450	12.8
4	368	9.3
5	520	10.8
6	465	9.6

The company currently has the following additional available production capacity (tons) at its three plants, which it wants to utilize:

Plant	Available Capacity (thousands of tons)
A	12
B	10
C	14

The shipping costs ($) per ton from the farms being considered for purchase to the plants are as follows:

	Plant (shipping costs, $/ton)		
Farm	A	B	C
1	18	15	12
2	13	10	17
3	16	14	18
4	19	15	16
5	17	19	12
6	14	16	12

The company wants to know which of the six farms it should purchase to meet available production capacity at the minimum total cost, including annual fixed costs and shipping costs.

This problem is formulated as a 0–1 integer programming model because the selection of the farms requires 0–1 decision variables:

y_i = 0 if farm i is not selected, and 1 if farm i is selected,
where

$$i = 1, 2, 3, 4, 5, 6$$

The variables for the amounts shipped from each prospective farm to each plant are non-integer, defined as follows:

$$x_{ij} = \text{potatoes (tons, 1,000s) shipped from farm } i \text{ to plant } j,$$
where
$$i = 1, 2, 3, 4, 5, 6 \text{ and } j = \text{A, B, C}$$

The objective function (Z) combines shipping costs and the annual fixed costs, where Z = \$1,000s:

$$\begin{aligned}\text{minimize } Z = {} & 18x_{1A} + 15x_{1B} + 12x_{1C} + 13x_{2A} + 10x_{2B} + 17x_{2C} + 16x_{3A} + 14x_{3B} + \\ & 18x_{3C} + 19x_{4A} + 15x_{4B} + 16x_{4C} + 17x_{5A} + 19x_{5B} + 12x_{5C} + 14x_{6A} + \\ & 16x_{6B} + 12x_{6C} + 405y_1 + 390y_2 + 450y_3 + 368y_4 + 520y_5 + 465y_6\end{aligned}$$

The constraints for production capacity require that potatoes be shipped (i.e., x_{ij} variables have positive values) only from a farm if it is selected; for example, for farm 1,

$$x_{1A} + x_{1B} + x_{1C} \le 11.2y_1$$

In this constraint, if $y_1 = 1$, then the 11.2 thousand tons of potatoes are available at farm 1 for shipment to one or more of the plants. Similar constraints are developed for each of the five other farms. Constraints are also necessary for the production capacity to be filled at each plant.

The complete model is formulated as

$$\begin{aligned}\text{minimize } Z = {} & 18x_{1A} + 15x_{1B} + 12x_{1C} + 13x_{2A} + 10x_{2B} + 17x_{2C} + 16x_{3A} + 14x_{3B} \\ & + 18x_{3C} + 19x_{4A} + 15x_{4B} + 16x_{4C} + 17x_{5A} + 19x_{5B} + 12x_{5C} + 14x_{6A} \\ & + 16x_{6B} + 12x_{6C} + 405y_1 + 390y_2 + 450y_3 + 368y_4 + 520y_5 + 465y_6\end{aligned}$$

subject to

$$\begin{aligned} x_{1A} + x_{1B} + x_{1C} - 11.2y_1 &\le 0 \\ x_{2A} + x_{2B} + x_{2C} - 10.5y_2 &\le 0 \\ x_{3A} + x_{3B} + x_{3C} - 12.8y_3 &\le 0 \\ x_{4A} + x_{4B} + x_{4C} - 9.3y_4 &\le 0 \\ x_{5A} + x_{5B} + x_{5C} - 10.8y_5 &\le 0 \\ x_{6A} + x_{6B} + x_{6C} - 9.6y_6 &\le 0 \\ x_{1A} + x_{2A} + x_{3A} + x_{4A} + x_{5A} + x_{6A} &= 12 \\ x_{1B} + x_{2B} + x_{3B} + x_{4B} + x_{5B} + x_{6B} &= 10 \\ x_{1C} + x_{2C} + x_{3C} + x_{4C} + x_{5C} + x_{6C} &= 14 \\ x_{ij} &\ge 0 \\ y_i &= 0 \text{ or } 1 \end{aligned}$$

Exhibit 5.19 shows this example model set up in Excel spreadsheet format. The decision variables for the shipments between farms and plants, x_{ij}, are in cells **C5:E10**. The 0–1 decision variables for farm selection, y_i, are in cells **C17:C22**. The objective function (Z) is embedded in cell C24 and is shown on the formula bar at the top of the spreadsheet. The model constraints for potato availability at the farms are embedded in cells **H5:H10**. For example, the total amount shipped from farm 1, $x_{1A} + x_{1B} + x_{1C}$, is in cell G5 as **=C5+D5+E5**. The constraint formulation is in cell H5 as **=G5–C17*F5**. The constraints for potatoes shipped from the plants are in cells **C12:E12**. For example, the constraint formula in C12 is **=SUM (C5:C10)**.

EXHIBIT 5.19

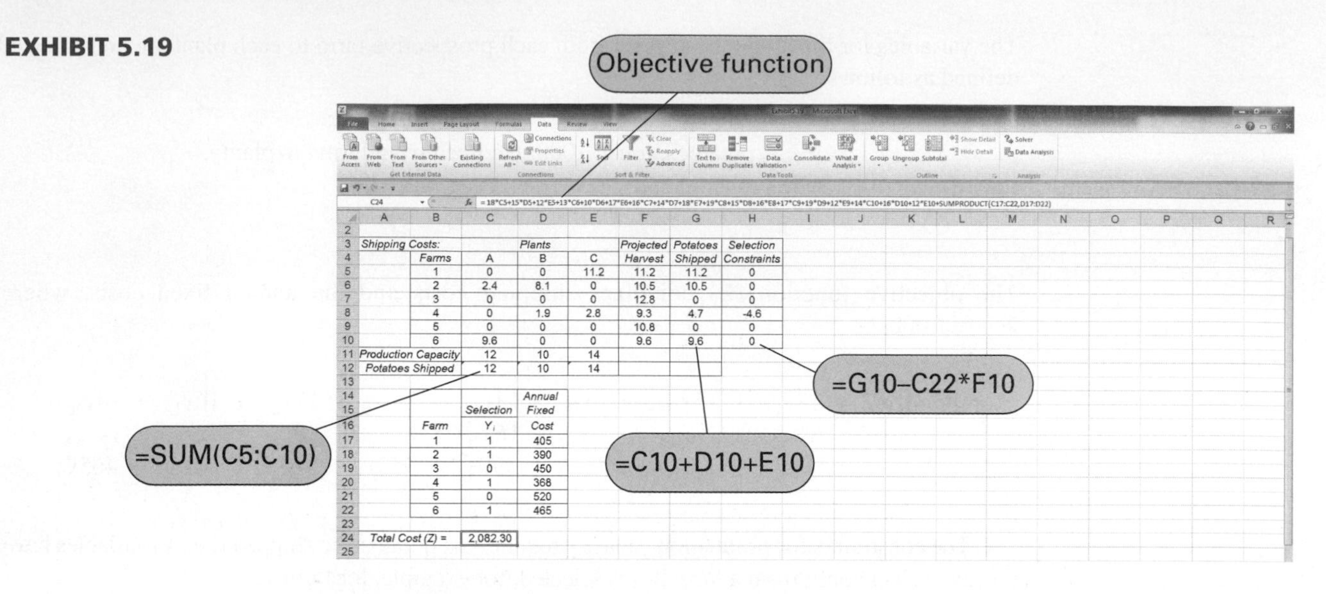

Shipping Costs:	Plants			Projected	Potatoes	Selection
Farms	A	B	C	Harvest	Shipped	Constraints
1	0	0	11.2	11.2	11.2	0
2	2.4	8.1	0	10.5	10.5	0
3	0	0	0	12.8	0	0
4	0	1.9	2.8	9.3	4.7	-4.6
5	0	0	0	10.8	0	0
6	9.6	0	0	9.6	9.6	0
Production Capacity	12	10	14			
Potatoes Shipped	12	10	14			

Farm	Selection Y_i	Annual Fixed Cost
1	1	405
2	1	390
3	0	450
4	1	368
5	0	520
6	1	465

Total Cost (Z) = 2,082.30

Exhibit 5.20 shows the Solver Parameters screen for this example. The 0–1 variable restriction for the y_i variables is established by the constraint **C17:C22=binary**. The farm harvest (potato availability) constraints are shown as **H5:H10≤0**, and the plant production capacity constraints are shown as **C12:E12=C11:E11**.

EXHIBIT 5.20

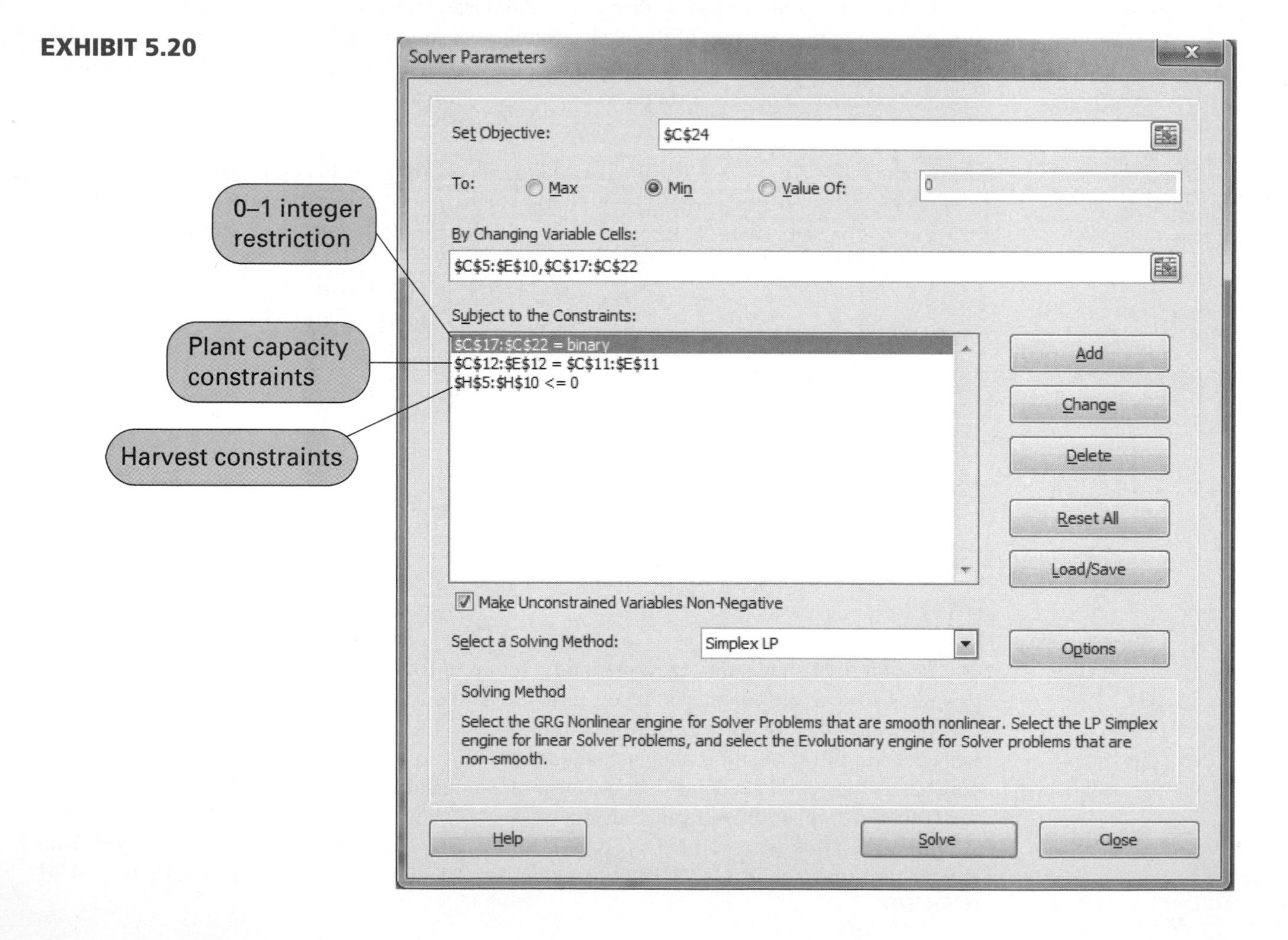

The model solution is

$$x_{1C} = 11{,}200 \text{ tons} \quad y_1 = 1 \text{ (farm 1)}$$
$$x_{2A} = 2{,}400 \text{ tons} \quad y_2 = 1 \text{ (farm 2)}$$
$$x_{2B} = 8{,}100 \text{ tons} \quad y_4 = 1 \text{ (farm 4)}$$
$$x_{4B} = 1{,}900 \text{ tons} \quad y_6 = 1 \text{ (farm 6)}$$
$$x_{4C} = 2{,}800 \text{ tons}$$
$$x_{6A} = 9{,}600 \text{ tons} \quad Z = \$2{,}082{,}300$$

A Set Covering Example

American Parcel Service (APS) has determined that it needs to add several new package distribution hubs to service cities east of the Mississippi River. The company wants to construct the minimum set of new hubs in the following 12 cities so that there is a hub within 300 miles of each city (i.e., every city is covered by a hub):

City	*Cities Within 300 Miles*
1. Atlanta	Atlanta, Charlotte, Nashville
2. Boston	Boston, New York
3. Charlotte	Atlanta, Charlotte, Richmond
4. Cincinnati	Cincinnati, Detroit, Indianapolis, Nashville, Pittsburgh
5. Detroit	Cincinnati, Detroit, Indianapolis, Milwaukee, Pittsburgh
6. Indianapolis	Cincinnati, Detroit, Indianapolis, Milwaukee, Nashville, St. Louis
7. Milwaukee	Detroit, Indianapolis, Milwaukee
8. Nashville	Atlanta, Cincinnati, Indianapolis, Nashville, St. Louis
9. New York	Boston, New York, Richmond
10. Pittsburgh	Cincinnati, Detroit, Pittsburgh, Richmond
11. Richmond	Charlotte, New York, Pittsburgh, Richmond
12. St. Louis	Indianapolis, Nashville, St. Louis

This problem requires a 0–1 integer programming model in which the decision variables are the available cities:

$x_i = $ city i, $i = 1$ to 12
where
$x_i = 0$, if city i is not selected as a hub and
$x_i = 1$, if city i is selected

The objective function (Z) is to minimize the number of hubs:

$$\text{minimize } Z = x_1 + x_2 + x_3 + x_4 + x_5 + x_6 + x_7 + x_8 + x_9 + x_{10} + x_{11} + x_{12}$$

The model constraints establish the set covering requirement (i.e., that each city be within 300 miles of a hub). For example, Atlanta covers itself, Charlotte, and Nashville:

$$\text{Atlanta: } x_1 + x_3 + x_8 \geq 1$$

In this constraint, at least one of the variables must equal 1 in order for Atlanta to be assured of a hub within 300 miles. Similar constraints are developed for the other 11 cities.

The complete 0–1 integer programming model is summarized as follows:

$$\text{minimize } Z = x_1 + x_2 + x_3 + x_4 + x_5 + x_6 + x_7 + x_8 + x_9 + x_{10} + x_{11} + x_{12}$$

subject to

$$
\begin{aligned}
&\text{Atlanta:} & x_1 + x_3 + x_8 &\geq 1\\
&\text{Boston:} & x_2 + x_9 &\geq 1\\
&\text{Charlotte:} & x_1 + x_3 + x_{11} &\geq 1\\
&\text{Cincinnati:} & x_4 + x_5 + x_6 + x_8 + x_{10} &\geq 1\\
&\text{Detroit:} & x_4 + x_5 + x_6 + x_7 + x_{10} &\geq 1\\
&\text{Indianapolis:} & x_4 + x_5 + x_6 + x_7 + x_8 + x_{12} &\geq 1\\
&\text{Milwaukee:} & x_5 + x_6 + x_7 &\geq 1\\
&\text{Nashville:} & x_1 + x_4 + x_6 + x_8 + x_{12} &\geq 1\\
&\text{New York:} & x_2 + x_9 + x_{11} &\geq 1\\
&\text{Pittsburgh:} & x_4 + x_5 + x_{10} + x_{11} &\geq 1\\
&\text{Richmond:} & x_3 + x_9 + x_{10} + x_{11} &\geq 1\\
&\text{St. Louis:} & x_6 + x_8 + x_{12} &\geq 1\\
& & x_i &= 0 \text{ or } 1
\end{aligned}
$$

The Excel spreadsheet for this model is shown in Exhibit 5.21. The decision variables for the cities and hub sites are in cells **B20:M20**, and the objective function (Z) is in cell B22. The formula for the objective function embedded in B22 is the sum of the cities selected as hubs, =**SUM (B20:M20)**. The model constraints are in cells **N7:N18**. Notice that the values of 1 in cells B7, D7, and I7 indicate the cities covered by a possible hub at Atlanta.

EXHIBIT 5.21

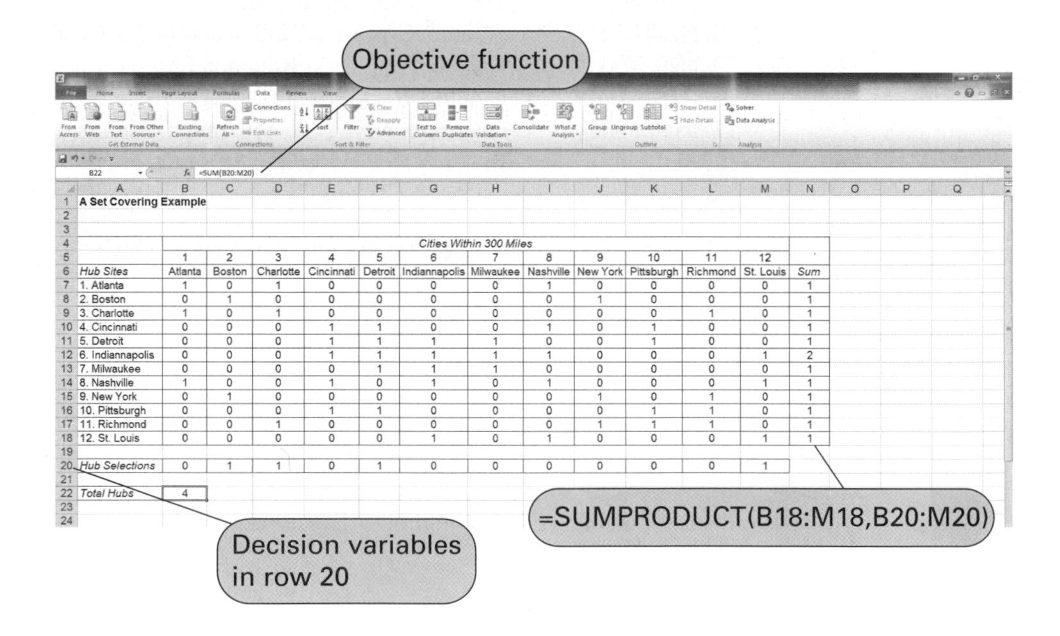

A Set Covering Example

	Cities Within 300 Miles												
	1	2	3	4	5	6	7	8	9	10	11	12	
Hub Sites	Atlanta	Boston	Charlotte	Cincinnati	Detroit	Indiannapolis	Milwaukee	Nashville	New York	Pittsburgh	Richmond	St. Louis	*Sum*
1. Atlanta	1	0	1	0	0	0	0	1	0	0	0	0	1
2. Boston	0	1	0	0	0	0	0	0	1	0	0	0	1
3. Charlotte	1	0	1	0	0	0	0	0	0	0	1	0	1
4. Cincinnati	0	0	0	1	1	0	0	1	0	1	0	0	1
5. Detroit	0	0	0	1	1	1	1	0	0	1	0	0	1
6. Indiannapolis	0	0	0	1	1	1	1	1	0	0	0	1	2
7. Milwaukee	0	0	0	0	1	1	1	0	0	0	0	0	1
8. Nashville	1	0	0	1	0	1	0	1	0	0	0	1	1
9. New York	0	1	0	0	0	0	0	0	1	0	1	0	1
10. Pittsburgh	0	0	0	1	1	0	0	0	0	1	1	0	1
11. Richmond	0	0	1	0	0	0	0	0	1	1	1	0	1
12. St. Louis	0	0	0	0	0	1	0	1	0	0	0	1	1
Hub Selections	0	1	1	0	1	0	0	0	0	0	0	1	
Total Hubs	4												

Exhibit 5.22 shows the Solver Parameters screen for this problem. The 0–1 variable restriction has been established by the constraints **B20:M20** = **binary** and Notice that the formulas in cells **N7:N18** are set ≥ 1 (i.e., **N7:N18** ≥ **1**) to complete the "set covering" constraints in the model.

EXHIBIT 5.22

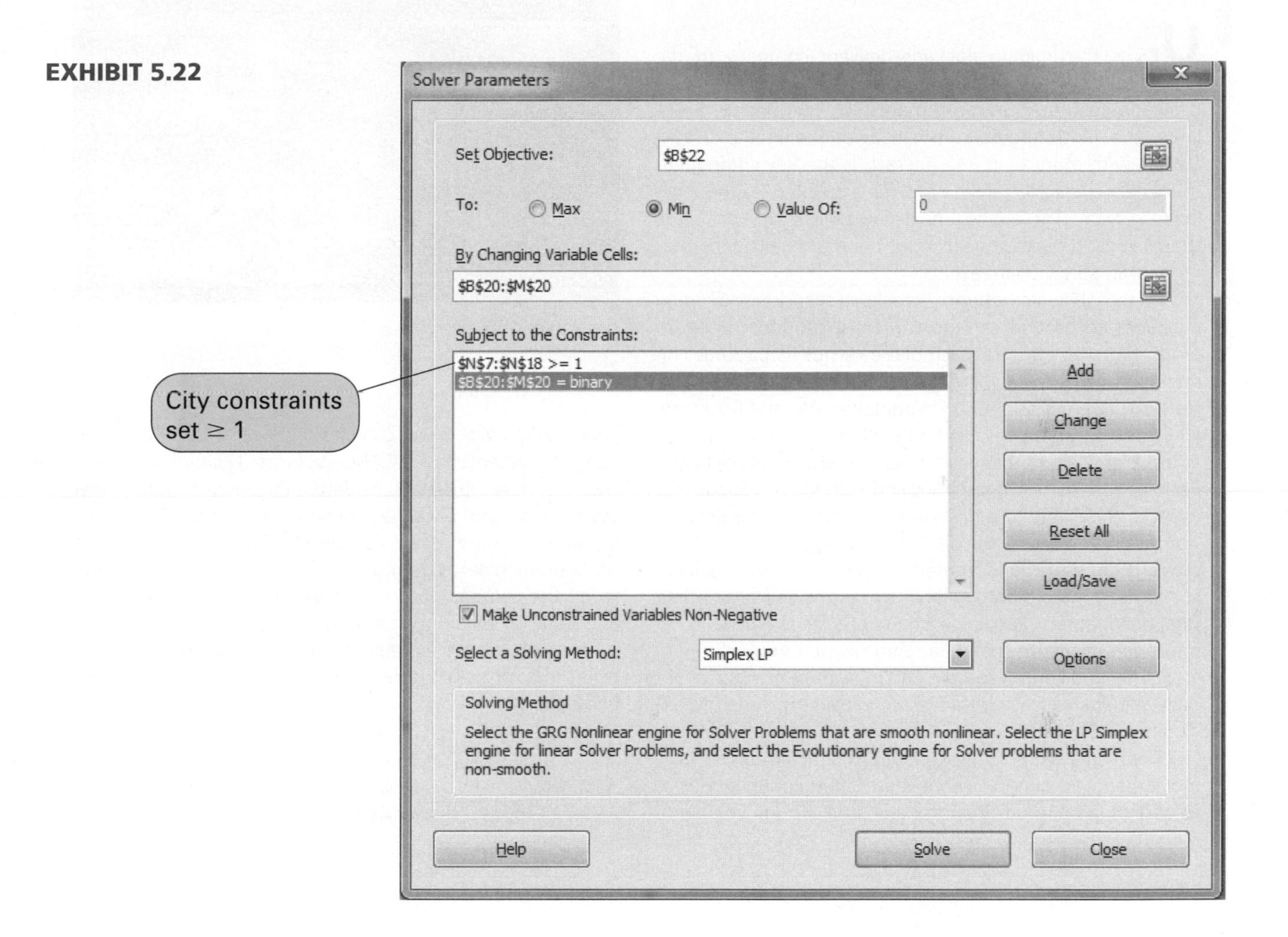

Notice that there is only one city covered by two hubs, Indianapolis (i.e., 2 in cell N12), which is overlapped by both Detroit and St. Louis. Also, there are multiple optimal solutions to this problem (i.e., different mixes of four hubs).

Management Science Application

Planning Next-Day Air Shipments at UPS

Thomas S. England/Bloomberg News

UPS is the leading package-delivery company in the world, each day delivering more than 13 million packages to 8 million customers in more than 200 countries. UPS Airlines, the ninth-largest commercial airline in the United States, with more than 345 aircraft, is the key component that enables the company to provide expedited (express) delivery service. The airline's next-day service delivers more than 1 million packages each night and annually generates more than $5 billion in revenue.

In the UPS air network, trucks transport originating packages to ground centers and from ground centers to more than 100 airports. Each plane carries its packages directly to one of six regional hubs (Columbia, SC, Dallas, TX, Hartford, CT, Ontario, CA, Philadelphia, PA, and Rockford, IL) or one all-points hub (in Louisville, KY), and it may stop at an airport along the way to pick up additional packages. Packages are sorted at the hubs and loaded onto aircraft for outbound delivery to a destination airport. At the destination airport, workers transfer the packages to trucks that move them to ground centers, where they are scanned, sorted, and loaded onto smaller trucks for delivery to their final destinations. Each type of aircraft has a maximum flying range, effective speed, landing restrictions at different locations, and cargo capacity. The maximum number of airports a plane can visit on a pickup or delivery route (not including a hub) is two.

Planning a shipping network of this size with seven hubs, more than 100 airports, more than 2,000 combinations of airport origin and destination pairs, and a huge volume of packages is a very large, complex problem. In 2000 the company implemented VOLCANO (Volume, Location, and Aircraft Network Optimizer) to determine aircraft routes, fleet assignments, and package routing to ensure overnight delivery at a minimum cost. VOLCANO is based on an integer programming set covering model formulation. This system saved the company more than $87 million between 2000 and 2002 by ensuring that it would use the most efficient routes with the fewest planes. UPS estimated that it would result in savings of more than $189 million over the next decade.

Source: Based on A. P. Armacost, C. Barnharrt, K. A. Ware, and A. M. Wilson, "UPS Based on Optimizes Its Air Network," *Interfaces* 34, no. 1 (January–February 2004): 15–25.

Summary

In this chapter, we found that simply rounding off non-integer simplex solution values for models requiring integer solutions is not always appropriate. Rounding can often lead to suboptimal results. Therefore, direct solution approaches using the computer are needed to solve the three forms of linear integer programming models—total integer models, 0–1 integer models, and mixed integer models.

Having analyzed integer problems and the computer techniques for solving them, we have now covered most of the basic forms of linear programming models and solution techniques. Two exceptions are the transportation model and a problem that contains more than one objective, goal programming. These special cases of linear programming are presented in the next two chapters.

Example Problem Solution

The following example problem demonstrates the model formulation and solution of a total integer programming problem.

The Problem Statement

A textbook publishing company has developed two new sales regions and is planning to transfer some of its existing sales force into these two regions. The company has 10 salespeople available for the transfer. Because of the different geographic configurations and the location of schools in each region, the average annual expenses for a salesperson differ in the two regions; the average is $10,000 per salesperson in region 1 and $7,000 per salesperson in region 2. The total annual expense budget for the new regions is $72,000. It is estimated that a salesperson in region 1 will generate an average of $85,000 in sales each year, and a salesperson in region 2 will generate $60,000 annually in sales. The company wants to know how many salespeople to transfer into each region to maximize increased sales.

Formulate this problem and solve it by using QM for Windows.

Solution

Step 1: Formulate the Integer Programming Model

$$\begin{aligned} \text{maximize } Z &= \$85{,}000x_1 + 60{,}000x_2 \\ \text{subject to} & \\ x_1 + x_2 &\leq 10 \text{ salespeople} \\ \$10{,}000x_1 + 7{,}000x_2 &\leq \$72{,}000 \text{ expense budget} \\ x_1, x_2 &\geq 0 \text{ and integer} \end{aligned}$$

Step 2: Solve the Model Using QM for Windows

Original Problem w/answers

Chapter5-Example Solution

	X1	X2		RHS
Maximize	85,000	60,000		
Salespeople	1	1	<=	10
Expense budget ($)	10,000	7,000	<=	72,000
Variable type	Integer	Integer		
Solution->	3	6	Optimal Z->	615,000

Problems

1. Consider the following linear programming model:

$$\begin{aligned} \text{maximize } Z &= 5x_1 + 4x_2 \\ \text{subject to} & \\ 3x_1 + 4x_2 &\leq 10 \\ x_1, x_2 &\geq 0 \text{ and integer} \end{aligned}$$

Demonstrate the graphical solution of this model.

2. Solve the following linear programming model by using the computer:

$$\text{minimize } Z = 3x_1 + 6x_2$$
subject to
$$7x_1 + 3x_2 \geq 40$$
$$x_1, x_2 \geq 0 \text{ and integer}$$

3. A tailor makes wool tweed sport coats and wool slacks. He is able to get a shipment of 150 square yards of wool cloth from Scotland each month to make coats and slacks, and he has 200 hours of his own labor to make them each month. A coat requires 3 square yards of wool and 10 hours to make, and a pair of slacks requires 5 square yards of wool and 4 hours to make. The tailor earns $50 in profit from each coat he makes and $40 from each pair of slacks. He wants to know how many coats and pairs of slacks to produce to maximize profit.
   a. Formulate an integer linear programming model for this problem.
   b. Determine the integer solution to this problem by using the computer. Compare this solution with the solution without integer restrictions and indicate whether the rounded-down solution would have been optimal.

4. A jeweler and her apprentice make silver pins and necklaces by hand. Each week they have 80 hours of labor and 36 ounces of silver available. It requires 8 hours of labor and 2 ounces of silver to make a pin and 10 hours of labor and 6 ounces of silver to make a necklace. Each pin also contains a small gem of some kind. The demand for pins is no more than six per week. A pin earns the jeweler $400 in profit, and a necklace earns $100. The jeweler wants to know how many of each item to make each week to maximize profit.
   a. Formulate an integer programming model for this problem.
   b. Solve this model by using the computer. Compare this solution with the solution without integer restrictions and indicate whether the rounded-down solution would have been optimal.

5. A glassblower makes glass decanters and glass trays on a weekly basis. Each item requires 1 pound of glass, and the glassblower has 15 pounds of glass available each week. A glass decanter requires 4 hours of labor, a glass tray requires only 1 hour of labor, and the glassblower works 25 hours a week. The profit from a decanter is $50, and the profit from a tray is $10. The glassblower wants to determine the total number of decanters (x_1) and trays (x_2) that he needs to produce in order to maximize his profit.
   a. Formulate an integer programming model for this problem.
   b. Solve this model by using the computer.

6. The Livewright Medical Supplies Company has a total of 12 salespeople it wants to assign to three regions—the South, the East, and the Midwest. A salesperson in the South earns $600 in profit per month for the company, a salesperson in the East earns $540, and a salesperson in the Midwest earns $375. The southern region can have a maximum assignment of 5 salespeople. The company has a total of $750 per day available for expenses for all 12 salespeople. A salesperson in the South has average expenses of $80 per day, a salesperson in the East has average expenses of $70 per day, and a salesperson in the Midwest has average daily expenses of $50. The company wants to determine the number of salespeople to assign to each region to maximize profit.
   a. Formulate an integer programming model for this problem.
   b. Solve this model by using the computer.

7. Helen Holmes makes pottery by hand in her basement. She has 20 hours available each week to make bowls and vases. A bowl requires 3 hours of labor, and a vase requires 2 hours of labor. It requires 2 pounds of special clay to make a bowl and 5 pounds to produce a vase; she is able to acquire 35 pounds of clay per week. Helen sells her bowls for $50 and her vases for $40. She wants to know how many of each item to make each week to maximize her revenue.
   a. Formulate an integer programming model for this problem.
   b. Solve this model by using the computer. Compare this solution with the solution without integer restrictions and indicate whether the rounded-down solution would have been optimal.

8. Lauren Moore has sold her business for $500,000 and wants to invest in condominium units (which she intends to rent) and land (which she will lease to a farmer). She estimates that she will receive an annual return of $8,000 for each condominium and $6,000 for each acre of land. A condominium unit costs $70,000, and land costs $30,000 per acre. A condominium will cost her $1,000 per unit, an acre of land will cost $2,000 for maintenance and upkeep, and $14,000 has been budgeted for these annual expenses. Lauren wants to know how much to invest in condominiums and land to maximize her annual return.
   a. Formulate a mixed integer programming model for this problem.
   b. Solve this model by using the computer.

9. The owner of the Consolidated Machine Shop has $10,000 available to purchase a lathe, a press, a grinder, or some combination thereof. The following 0–1 integer linear programming model has been developed to determine which of the three machines (lathe, x_1; press, x_2; or grinder, x_3) should be purchased in order to maximize annual profit:

$$\begin{aligned} &\text{maximize } Z = 1{,}000x_1 + 700x_2 + 800x_3 \text{ (profit, \$)} \\ &\text{subject to} \\ &\$5{,}000x_1 + 6{,}000x_2 + 4{,}000x_3 \leq 10{,}000 \text{ (cost, \$)} \\ &\qquad x_1, x_2, x_3 = 0 \text{ or } 1 \end{aligned}$$

   Solve this model by using the computer.

10. Solve the following mixed integer linear programming model by using the computer:

$$\begin{aligned} &\text{maximize } Z = 5x_1 + 6x_2 + 4x_3 \\ &\text{subject to} \\ &5x_1 + 3x_2 + 6x_3 \leq 20 \\ &\qquad x_1 + 3x_2 \leq 12 \\ &\qquad x_1, x_3 \geq 0 \\ &\qquad x_2 \geq 0 \text{ and integer} \end{aligned}$$

11. Northwoods Backpackers is a retail catalog store in Vermont that specializes in outdoor clothing and camping equipment. Phone orders are taken each day by a large pool of computer operators, some of whom are permanent and some temporary. A permanent operator can process an average of 76 orders per day, whereas a temporary operator can process an average of 53 orders per day. The company averages at least 600 orders per day. The store has 10 computer workstations. A permanent operator processes about 1.3 orders with errors each day, whereas a temporary operator averages 4.1 orders with errors daily. The store wants to limit errors to 24 per day. A permanent operator is paid $81 per day, including benefits, and a temporary operator is paid $50 per day. The company wants to know the number of permanent and temporary operators to hire to minimize costs.

    Formulate an integer programming model for this problem and solve it by using the computer.

12. Consider the following linear programming model:

$$\begin{aligned} &\text{maximize } Z = 20x_1 + 30x_2 + 10x_3 + 40x_4 \\ &\text{subject to} \\ &2x_1 + 4x_2 + 3x_3 + 7x_4 \leq 10 \\ &10x_1 + 7x_2 + 20x_3 + 15x_4 \leq 40 \\ &x_1 + 10x_2 + x_3 \leq 10 \\ &x_1, x_2, x_3, x_4 = 0 \text{ or } 1 \end{aligned}$$

    Solve this problem by using the computer.

13. In the example problem solution for this chapter on page 209, a textbook company was attempting to determine how many sales representatives to assign to each of two new regions. The company

has now decided that if any sales representatives are assigned to region 1, a sales office must be established there, at an annual cost of $18,000. This altered problem is an example of a type of integer programming problem known as a "fixed charge" problem.

a. Reformulate the integer programming model to reflect this new condition.
b. Solve this new problem by using the computer.

14. The Texas Consolidated Electronics Company is contemplating a research and development program encompassing eight research projects. The company is constrained from embarking on all projects by the number of available management scientists (40) and the budget available for R&D projects ($300,000). Further, if project 2 is selected, project 5 must also be selected (but not vice versa). Following are the resource requirements and the estimated profit for each project:

Project	Expense ($1,000s)	Management Scientists Required	Estimated Profit ($1,000,000s)
1	$ 60	7	$0.36
2	110	9	0.82
3	53	8	0.29
4	47	4	0.16
5	92	7	0.56
6	85	6	0.61
7	73	8	0.48
8	65	5	0.41

Formulate the integer programming model for this problem and solve it by using the computer.

15. Mazy's Department Store has decided to stay open for business on a 24-hour basis. The store manager has divided the 24-hour day into six 4-hour periods and has determined the following minimum personnel requirements for each period:

Time	Personnel Needed
Midnight–4:00 A.M.	90
4:00–8:00 A.M.	215
8:00 A.M.–Noon	250
Noon–4:00 P.M.	65
4:00–8:00 P.M.	300
8:00 P.M.–Midnight	125

Store personnel must report to work at the beginning of one of these time periods and must work for 8 consecutive hours. The store manager wants to know the minimum number of employees to assign to each 4-hour segment to minimize the total number of employees. Formulate and solve this problem.

16. The athletic boosters club for Beaconville has planned a 2-day fund-raising drive to purchase uniforms for all the local high schools and to improve facilities. Donations will be solicited during the day and night by telephone and personal contact. The boosters club has arranged for local college students to donate their time to solicit donations. The average donation from each type of contact and the time for a volunteer to solicit each type of donation are as follows:

	Average Donation ($)		Average Interview Time (min.)	
	Phone	Personal	Phone	Personal
Day	$16	$33	6	13
Night	17	37	7	19

The boosters club has gotten several businesses and car dealers to donate gasoline and cars for the college students to use to make a maximum of 575 personal contacts daily during the fund-raising drive. The college students will donate a total of 22 hours during the day and 43 hours at night during the drive.

The president of the boosters club wants to know how many different types of donor contacts to schedule during the drive to maximize total donations. Formulate and solve an integer programming model for this problem. What is the difference in the total maximum value of donations between the integer and non-integer rounded-down solutions to this problem?

17. The Metropolitan Arts Council (MAC) wants to advertise its upcoming season of plays, concerts, and ballets. A television commercial that costs $25,000 will supposedly reach 53,000 potential arts customers. The breakdown of the audience is as follows:

Age	Male	Female
≥35	12,000	20,000
<35	7,000	14,000

A newspaper ad costs $7,000, and the newspaper claims that its ads will reach an audience of 30,000 potential arts customers, broken down as follows:

Age	Male	Female
≥35	12,000	8,000
<35	6,000	4,000

A radio ad costs $9,000, and it is estimated to reach an audience of 41,000 arts customers, with the following audience breakdown:

Age	Male	Female
≥35	7,000	11,000
<35	10,000	13,000

The arts council has established several marketing guidelines. It wants to reach at least 200,000 potential arts customers. It believes older people are more likely to buy tickets than younger people, so it wants to reach at least 1.5 times as many people over 35 as people under 35. The council also believes women are more likely to instigate ticket purchases to the arts than men, so it wants its advertising audience to be at least 60% female.

a. Formulate and solve an integer programming model for MAC to determine the number of ads of each type it should use at the minimum cost.
b. Solve this model without integer restrictions and compare the results.

18. Juan Hernandez, a Cuban athlete who visits the United States and Europe frequently, is allowed to return with a limited number of consumer items not generally available in Cuba. The items, which are carried in a duffel bag, cannot exceed a weight of 5 pounds. Once Juan is in Cuba, he sells the items at highly inflated prices. The three most popular items in Cuba are denim jeans, CD players, and CDs of U.S. rock groups. The weight and profit (in U.S. dollars) of each item are as follows:

Item	Weight (lb.)	Profit
Denim jeans	2	$ 90
CD players	3	150
Compact discs	1	30

Juan wants to determine the combination of items he should pack in his duffel bag to maximize his profit. This problem is an example of a type of integer programming problem known as a "knapsack" problem. Formulate and solve this problem.

19. The Avalon Floor Cleaner Company is trying to determine the number of salespeople it should allocate to its three regions—the East, the Midwest, and the West. The company has 100 salespeople that it wants to assign to the three regions. The annual average unit sales volume achieved by a salesperson in each region is as follows:

Region	Units per Salesperson
East	25,000
Midwest	18,000
West	31,000

Because travel distances, costs of living, and other factors vary among the three regions, the annual cost of having a salesperson is $5,000 in the East, $11,000 in the Midwest, and $7,000 in the West. The company has $700,000 budgeted for expenses. To ensure nationwide exposure for its product, the company has decided that each region must have at least 10 salespeople. The company wants to know how many salespeople to allocate to each region to maximize total average units sold. Formulate an integer programming model for this problem and solve it by using the computer.

20. During the war with Iraq in 1991, the Terraco Motor Company produced a lightweight, all-terrain vehicle code-named "J99 Terra" for the military. The company is now planning to sell the Terra to the public. It has five plants that manufacture the vehicle and four regional distribution centers. The company is unsure of public demand for the Terra, so it is considering reducing its fixed operating costs by closing one or more plants, even though it would incur an increase in transportation costs. The relevant costs for the problem are provided in the following table. The transportation costs are per thousand vehicles shipped; for example, the cost of shipping 1,000 vehicles from plant 1 to warehouse C is $32,000.

From Plant	Transportation Costs ($1,000s) to Warehouse A	B	C	D	Annual Production Capacity	Annual Fixed Operating Costs
1	$56	$21	$32	$65	12,000	$2,100,000
2	18	46	7	35	18,000	850,000
3	12	71	41	52	14,000	1,800,000
4	30	24	61	28	10,000	1,100,000
5	45	50	26	31	16,000	900,000
Annual Demand	6,000	14,000	8,000	10,000		

Formulate and solve an integer programming model for this problem to assist the company in determining which plants should remain open and which should be closed and the number of vehicles that should be shipped from each plant to each warehouse to minimize total cost.

21. The Otter Creek Winery produces three kinds of table wine—a blush, a white, and a red. The winery has 30,000 pounds of grapes available to produce wine this season. A cask of blush requires 360 pounds of grapes, a cask of white requires 375 pounds, and a cask of red requires 410 pounds. The winery has enough storage space in its aging room to store 67 casks of wine. The winery has 2,200 hours of production capacity, and it requires 14 hours to produce a cask of blush, 10 hours to produce a cask of white, and 18 hours to produce a cask of red. From records of previous years'

sales, the winery knows it will sell at least twice as much blush as red and at least 1.5 times as much white as blush. The profit for a cask of blush is $12,100, the profit for a cask of white is $8,700, and the profit for a cask of red is $10,500. The winery wants to know the number of casks of each table wine to produce. Formulate and solve an integer programming model for this problem.

22. Brenda Last, an undergraduate business major at State University, is attempting to determine her course schedule for the fall semester. She is considering seven 3-credit-hour courses, which are shown in the following table. Also included are the average number of hours she expects to have to devote to each course each week (based on information from other students) and her *minimum* expected grade in each course, based on an analysis of the grading records of the teachers in each course:

Course	Average Hours per Week	Minimum Grade
Management I	5	B
Principles of Accounting	10	C
Corporate Finance	8	C
Quantitative Methods	12	D
Marketing Management	7	C
Java Programming	10	D
English Literature	8	B

An A in a course earns 4 quality credits per hour, a B earns 3 quality credits, a C earns 2 quality credits, a D earns 1 quality credit, and an F earns no quality credits per hour. Brenda wants to select a schedule that will provide at least a 2.0 grade point average. In order to remain a full-time student, which she must do to continue receiving financial aid, she must take at least 12 credit hours. Principles of Accounting, Corporate Finance, Quantitative Methods, and Java Programming all require a lot of computing and mathematics, and Brenda would like to take no more than two of these courses. To remain on schedule and meet prerequisites, she needs to take at least three of the following courses: Management I, Principles of Accounting, Java Programming, and English Literature. Brenda wants to develop a course schedule that will minimize the number of hours she has to work each week.

a. Formulate a 0–1 integer programming model for this problem.
b. Solve this model by using the computer. Indicate how many total hours Brenda should expect to work on these courses each week and her minimum grade point average.

23. The artisans at Jewelry Junction in Phoenix are preparing to make gold jewelry during a 2-month period for the Christmas season. They can make bracelets, necklaces, and pins. Each bracelet requires 6.3 ounces of gold and 17 hours of labor, each necklace requires 3.9 ounces of gold and 10 hours of labor, and each pin requires 3.1 ounces of gold and 7 hours of labor. Jewelry Junction has available 125 ounces of gold and 320 hours of labor. A bracelet sells for $1,650, a necklace for $850, and a pin for $790. The store wants to know how many of each item to produce to maximize revenue.

a. Formulate an integer programming model for this problem.
b. Solve this model by using the computer. Compare this solution with the solution with the integer restrictions relaxed and indicate whether the rounded-down solution would have been optimal.

24. Harry and Melissa Jacobson produce handcrafted furniture in a workshop on their farm. They have obtained a load of 600 board feet of birch from a neighbor and are planning to produce round kitchen tables and ladder-back chairs during the next 3 months. Each table will require 30 hours of labor, each chair will require 18 hours, and between them they have a total of 480 hours of labor available. A table requires 40 board feet of wood to make, and a chair requires 15 board feet. A table earns the couple $575 in profit, and a chair earns $120 in profit. Most people who buy

a table also want four chairs to go with it, so for every table that is produced, at least four chairs must also be made, although additional chairs can also be sold separately. Formulate and solve an integer programming model to determine the number of tables and chairs the Jacobsons should make to maximize profit.

25. The Jacobsons in Problem 24 have been approached by a home furnishings company that needs a wooden stool for its catalog. The company has asked the Jacobsons to produce a batch of 20 wooden stools, and the Jacobsons would realize a profit of $65 for each stool. A stool will require 4 board feet of wood and 5 hours to produce. Formulate and solve an integer programming model to help the Jacobsons determine whether they should also produce the stools.

26. The Skimmer Boat Company manufactures three kinds of molded fiberglass recreational boats—a bass fishing boat, a ski boat, and a speedboat. The profit for a bass boat is $20,500, the profit for a ski boat is $12,000, and the profit for a speedboat is $22,300. The company believes it will sell more bass boats than the other two boats combined but no more than twice as many. The ski boat is its standard production model, and bass boats and speedboats are modifications. The company has production capacity to manufacture 210 standard (ski-type) boats; however, a bass boat requires 1.3 times the standard production capacity, and a speedboat requires 1.5 times the normal production capacity. In addition, a bass boat uses one high-powered engine and a speedboat uses two, and only 160 high-powered engines are available. The company wants to know how many boats of each type to produce to maximize profit. Formulate and solve an integer programming model for this problem.

27. The Reliance Manufacturing Company produces an aircraft part. The company can produce the part entirely at a flexible work center with multiple computerized machines. The company has four work centers, all of which are different because they were purchased at different times. Each work center has a single operator; however, the company's operators have different skill levels, resulting in different levels of daily output and product quality. The following tables show the average daily output and average number of defects per day for each of the company's five operators who are capable of producing the aircraft part:

	Average Daily Output per Machine			
Operator	A	B	C	D
1	18	20	21	17
2	19	15	22	18
3	20	20	17	19
4	24	21	16	23
5	22	19	21	21

	Average Number of Defects Daily per Machine			
Operator	A	B	C	D
1	0.3	0.9	0.6	0.4
2	0.8	0.5	1.1	0.7
3	1.1	1.3	0.6	0.8
4	1.2	0.8	0.6	0.9
5	1.0	0.9	1.0	1.0

The company wants to determine which operator to assign to each machine to maximize daily output and keep the percentage of defects to less than 4%.

a. Formulate a 0–1 integer programming model for this problem.
b. Solve this model by using the computer.

28. Corsouth Mortgage Associates is a large home mortgage firm in the Southeast. It has a pool of permanent and temporary computer operators who process mortgage accounts, including posting payments and updating escrow accounts for insurance and taxes. A permanent operator can process 220 accounts per day, and a temporary operator can process 140 accounts per day. On average, the firm must process and update at least 6,300 accounts daily. The company has 32 computer workstations available. Permanent and temporary operators work 8 hours per day. A permanent operator averages about 0.4 error per day, whereas a temporary operator averages 0.9 error per day. The company wants to limit errors to 15 per day. A permanent operator is paid $120 per day, whereas a temporary operator is paid $75 per day. Corsouth wants to determine the number of permanent and temporary operators it needs to minimize cost. Formulate and solve an integer programming model for this problem and compare this solution to the non-integer solution.

29. In Problem 28, Corsouth Mortgage Associates is considering hiring some hourly, part-time computer operators in addition to its permanent and temporary operators. A part-time operator can process 12 accounts per hour, averages 0.16 error per hour, and is paid $4.50 per hour. Corsouth wants to know the number of permanent and temporary employees it should use, plus the number of part-time hours it should arrange for. Formulate and solve a mixed integer model for this problem.

30. Texmart is a locally owned "big-box" retail store chain in Texas with 75 stores, primarily located in the Dallas–Fort Worth area. In order to compete with national big-box store chains, Texmart is planning to undertake several "sustainability" (i.e., "green") projects at its stores. The national chains have been heavily publicizing their sustainability efforts, including the reduction of greenhouse gas (GHG) emissions, which has had a positive effect on their sales. They have also demonstrated that sustainability projects can have a positive impact on cost (especially energy) savings. The projects Texmart is considering include installing solar panels at some or all of its stores; installing small wind turbines; replacing some or all of its 165 trucks with more fuel-efficient hybrid trucks; reducing waste, including recycling; and reducing plastic bags in their stores. The costs for these projects, the resulting reduction in GHG emissions, the energy savings, and the annual costs savings are shown in the following table:

	Sustainable Projects				
	Solar Power	Wind Power	Shipping/ Vehicles	Waste/ Recycling	Plastic Bags
Media/public relations score	3	2	1	1	2
Cost	2,600,000	950,000	38,000	365,000	175,000
GHG reductions (metric tons per year)	17,500	8,600	25	1,700	900
Cost savings ($)	220,000	125,000	26,000	75,000	45,000
Energy savings (kWh)	400,000	150,000	34,000	1,200	55,000
Units	75	75	165	75	75

The media public/relations score in this table designates the importance of a particular project relative to the other projects in generating public awareness and publicity. For example, a score of 3 indicates that the solar power project will have the greatest public impact. However, Texmart believes if it undertakes a project, it will require a threshold number of projects to make an impact—specifically, a solar power installation at 1 store or more, wind power projects at least 3 stores, at least 10 new trucks, at least 2 waste/recycling store projects, and at least 6 stores with plastic bag reduction projects. Texmart has budgeted $30 million for sustainable projects, and it wants to achieve GHG emission reductions of at least 250,000 metric tons per year; it wants to achieve annual cost savings of at least $4 million; and, it wants to achieve annual energy savings of at least 5 million kilowatt hours (kWh), while maximizing the public relations impact of its sustainability program. Develop and solve a linear programming model to help Texmart determine how many projects of each type it should undertake.

31. In Problem 30, reformulate and solve the model so that Texmart maximizes its cost savings and does not consider the media/public relations impact at all.

32. The Office of Homeland Security and the Centers for Disease Control and Prevention have jointly mandated that cities and localities develop plans for establishing emergency command centers in multiple locations to dispense vaccines and antibiotics and provide security and police coordination in the event of major emergency events, such as terrorist attacks, bioterrorism, and natural pandemics. These command centers must be located such that one is within 15 minutes of every person in the general population. For cities with large populations and geographic areas, this could result in a large number of centers and create logistical problems in setting them up quickly and staffing them. As such, cities would like to create the minimum number of command centers possible. In order to determine where it should locate its command centers, a city has divided its geographic area into four quadrants—northeast, northwest, southeast, and southwest. Within each quadrant, it has established four sectors with potential sites for command centers—A, B, C, and D—usually where there is a sufficient facility capable of housing a center and not necessarily centrally located. The potential command center site in a sector is always within 15 minutes of the population in the other three sectors in the quadrant and also within 15 minutes of the population in some sectors in adjacent quadrants, as shown in the following table:

Sector	Adjacent Quadrant Sectors Within 15 Minutes
1. NE-A	NW-A, SE-A
2. NE-B	NW-A, NW-B
3. NE-C	SE-A, SE-C
4. NE-D	NW-A, NW-C, SE-B, SW-A
5. NW-A	NE-A, NE-B, NE-D
6. NW-B	NE-B
7. NW-C	NE-D, SE-B
8. NW-D	SW-B, SW-D
9. SE-A	NE-C
10. SE-B	NE-D, NW-C, SW-A, SW-C
11. SE-C	NE-C, SW-C
12. SE-D	SW-C, SW-D
13. SW-A	NE-D, SE-B
14. SW-B	NW-D
15. SW-C	SE-B, SE-C, SE-D
16. SW-D	NW-D, SE-D

Determine the location of the minimum number of command centers for the city.

33. Tech offers a number of off-campus courses in its professional and executive masters of information technology (MIT) programs at various locations around the state. The courses are taught by a group of regular Tech information technology faculty and adjuncts they hire from around the state. There are two main considerations in assigning individual faculty to courses: the travel distance and the average course teaching evaluation scores from past years. The college would like to minimize the mileage for teaching the courses not only for each faculty member's benefit but also to reduce program expenses. It would also like to have the faculty member who does the best job with a particular course teach that course. The following table shows the mileage for each faculty member to each course location and his or her average teaching evaluation scores (on a 5.00 scale, where "5" is outstanding and "1" is poor) for the fall semester:

Course (mileage/score)	Teacher							
	Abrahams	Bray	Clayton	Dennis	Evans	Farah	Gonzalez	Hampton
MIT 125	35/4.10	35/4.63	71/4.55	35/4.43	119/3.78	215/4.21	35/4.93	94/4.06
MIT 225	74/4.05	74/3.78	95/4.61	74/4.26	147/3.69	135/2.78	74/4.12	112/4.02
MIT 250	74/3.97	74/4.17	95/4.12	74/3.45	147/3.91	135/3.94	74/4.33	112/4.11
MIT 300	35/2.95	35/3.67	71/4.14	35/3.95	119/3.25	215/4.33	35/4.71	94/3.99
MIT 325	210/4.63	210/4.12	105/3.95	210/4.55	55/3.45	45/3.52	210/4.85	110/4.06
MIT 375	175/4.15	175/3.95	115/4.06	175/3.92	67/3.88	65/4.11	175/4.13	134/4.07
MIT 400	210/4.22	210/4.55	105/3.52	210/4.34	55/3.66	45/4.16	210/4.66	110/3.95
MIT 425	175/4.36	175/4.44	115/3.95	175/3.88	67/4.01	65/3.77	175/4.15	134/4.23
MIT 450	74/3.78	74/4.03	95/4.11	74/3.94	147/3.54	135/2.95	74/4.24	112/3.96
MIT 500	175/3.87	175/3.45	115/3.86	175/3.84	67/3.92	65/3.45	175/3.81	134/3.76

Formulate and solve a linear programming model that will assign all the faculty members such that all courses are covered and each faculty member teaches at least one, but no more than two courses. Explain your solution approach.

34. Rowntown Cab Company has 70 drivers that it must schedule in three 8-hour shifts. However, the demand for cabs in the metropolitan area varies dramatically according to the time of day. The slowest period is between midnight and 4:00 A.M. The dispatcher receives few calls, and the calls that are received have the smallest fares of the day. Very few people are going to the airport at that time of night or taking other long-distance trips. It is estimated that a driver will average $80 in fares during that period. The largest fares result from the airport runs in the morning. Thus, the drivers who start their shift during the period from 4:00 A.M. to 8:00 A.M. average $500 in total fares, and drivers who start at 8:00 A.M. average $420. Drivers who start at noon average $300, and drivers who start at 4:00 P.M. average $270. Drivers who start at the beginning of the 8:00 P.M. to midnight period earn an average of $210 in fares during their 8-hour shift.

    To retain customers and acquire new ones, Rowntown must maintain a high customer service level. To do so, it has determined the minimum number of drivers it needs working during every 4-hour time segment—10 from midnight to 4:00 A.M., 12 from 4:00 to 8:00 A.M., 20 from 8:00 A.M. to noon, 25 from noon to 4:00 P.M., 32 from 4:00 to 8:00 P.M., and 18 from 8:00 P.M. to midnight.

    a. Formulate and solve an integer programming model to help Rowntown Cab schedule its drivers.
    b. If Rowntown has a maximum of only 15 drivers who will work the late shift from midnight to 8:00 A.M., reformulate the model to reflect this complication and solve it.
    c. All the drivers like to work the day shift from 8:00 A.M. to 4:00 P.M., so the company has decided to limit the number of drivers who work this 8-hour shift to 20. Reformulate the model in (b) to reflect this restriction and solve it.

35. Globex Investment Capital Corporation owns six companies that have the following estimated returns (in millions of dollars) if sold in one of the next 3 years:

Company	Year Sold (estimated return, $1,000,000s)		
	1	2	3
1	$14	$18	$23
2	9	11	15
3	18	23	27
4	16	21	25
5	12	16	22
6	21	23	28

To operating funds, the company must sell at least $20 million worth of assets in year 1, $25 million in year 2, and $35 million in year 3. Globex wants to develop a plan for selling these companies during the next 3 years to maximize return.

Formulate an integer programming model for this problem and solve it by using the computer.

36. The Heartland Distribution Company is a food warehouse and distributor that has a contract with a grocery store chain in several Midwest and Southeast cities. The company wants to construct new warehouses/distribution centers in some of the cities it services to serve the stores in those cities plus all the other stores in the other cities that don't have distribution centers. A distribution center can effectively service all stores within a 300-mile radius. The company also wants to limit its fixed annual costs to under $900,000. The company wants to build the minimum number of distribution centers possible.

    The following table shows the cities within 300 miles of every city and the projected fixed annual charge for a distribution center in each city:

City	Annual Fixed Charge ($1,000s)	Cities Within 300 Miles
1. Atlanta	$276	Atlanta, Charlotte, Nashville
2. Charlotte	253	Atlanta, Charlotte, Richmond
3. Cincinnati	394	Cincinnati, Cleveland, Indianapolis, Louisville, Nashville, Pittsburgh
4. Cleveland	408	Cincinnati, Cleveland, Indianapolis, Pittsburgh
5. Indianapolis	282	Cincinnati, Cleveland, Indianapolis, Louisville, Nashville, St. Louis
6. Louisville	365	Cincinnati, Indianapolis, Louisville, Nashville, St. Louis
7. Nashville	268	Atlanta, Cincinnati, Indianapolis, Louisville, Nashville, St. Louis
8. Pittsburgh	323	Cincinnati, Cleveland, Pittsburgh, Richmond
9. Richmond	385	Charlotte, Pittsburgh, Richmond
10. St. Louis	298	Indianapolis, Louisville, Nashville, St. Louis

a. Formulate an integer programming model for this problem and solve it by using the computer.
b. What is the solution if the cost constraint is removed from the original model formulation? What is the difference in cost?

37. The Kreeger Grocery Store chain has bought out a competing grocery store chain. However, it now has too many stores in close proximity to each other in certain cities. In Roanoke the chain has 10 stores, and it does not want any stores closer than 2 miles to each other. Following are the monthly revenues ($1,000s) from each store and a map showing the general proximity of the stores. Stores within 2 miles of each other are circled.

Store	Monthly Revenue ($1,000s)
1	$127
2	83
3	165
4	96
5	112
6	88
7	135
8	141
9	117
10	94

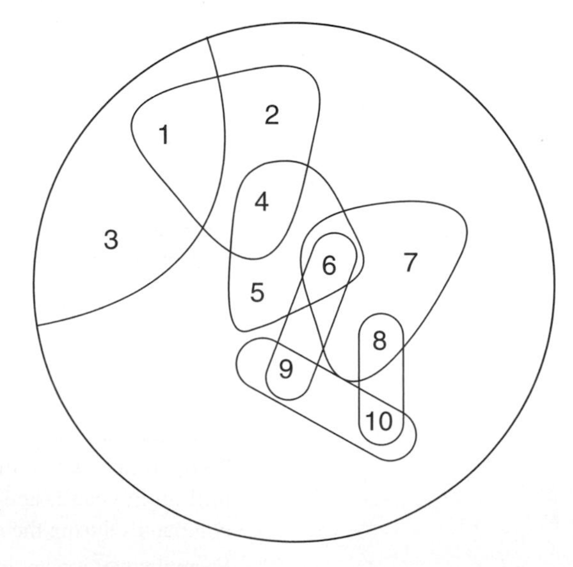

Develop and solve an integer programming model to determine which stores the Kreeger chain should keep open in Roanoke.

38. A youth soccer club has contracted with Holiday Helpers, a local travel agency, to broker hotel rooms for out-of-town teams that have entered the club's Labor Day weekend soccer tournament. The agency has 12 teams it needs to arrange rooms for at 8 possible hotels. The following tables show the number of rooms each team needs, the number of rooms available at each hotel, the room rate at each hotel, and the maximum room rate each team wants to pay:

Team	Max Rate	Rooms Needed
1. Arsenal	$ 70	15
2. United	75	18
3. Wildcats	60	20
4. Rage	80	12
5. Rapids	110	17
6. Storm	90	10
7. Tigers	70	18
8. Stars	80	18
9. Comets	80	20
10. Hurricanes	65	16
11. Strikers	90	20
12. Bees	100	14

Hotel	Room Rate	Rooms Available
A. Holiday	$ 90	41
B. Roadside	75	26
C. Bates	55	38
D. Hampson	95	25
E. Tilton	100	26
F. Marks	80	38
G. Bayside	70	35
H. Harriott	80	52

a. All of a team's rooms must be at the same hotel. Formulate a model and develop a solution for the agency to reserve rooms for as many teams as possible, according to their needs.

b. The travel agency has requested that each team indicate three hotels it would prefer to stay at, in order of priority, based on price, location, and facilities. The teams' preferences are shown in the following table:

Hotel Priority	Team 1	2	3	4	5	6	7	8	9	10	11	12
1	C	B	C	F	D	F	C	F	H	C	A	E
2	G	G	—	H	E	A	G	H	F	—	H	A
3	—	C	—	B	A	H	—	B	B	—	F	D

Determine a revised hotel room allocation to assign rooms to all teams while reflecting their preferences to the greatest possible extent.

39. The Tech Software School contracts to train employees of companies in the use of various software products and packages. The training programs are all delivered on three consecutive days but require different daily durations, depending on the software product. For a specific 7-day week (including weekends), the school has contracted with eight different companies for training sessions for different software products. The companies can send their employees to the Tech training facility only on certain days, as shown in the following table, with the hourly training requirements per day:

Company	Hours: First Day	Second Day	Third Day	Earliest Start Day	Latest Start Day
A	4	6	3	1	4
B	3	2	5	1	3
C	7	1	1	2	5
D	1	3	6	2	5
E	8	5	2	3	5
F	5	5	5	1	5
G	6	3	3	3	5
H	4	6	6	3	4

The school has 20 staff-hours per day available for these training sessions.

a. Develop a *feasible* training schedule for the eight companies.
b. Develop a *feasible* training schedule for the companies that will minimize the weekend training hours (where the weekend includes days 6 and 7).

40. Each day Seacoast Food Services makes deliveries to four restaurants it supplies in the metro Atlanta area. The service uses one truck that starts at its warehouse, makes a delivery to each restaurant, and then returns to the warehouse. The mileage between the warehouse, 1, and each of the restaurants, 2, 3, 4, and 5, is shown in the following table:

	Location				
Location	**1**	**2**	**3**	**4**	**5**
1	—	10	15	20	40
2	10	—	12	16	24
3	15	12	—	10	20
4	20	16	10	—	16
5	40	24	20	16	—

Formulate and solve an integer programming model to determine the route (or tour) the truck should take to start at the warehouse, visit each restaurant once, and return to the warehouse with the minimum total distance traveled.

41. HTM Realty has a storefront rental property in downtown Draperton, a small college town next to the Tech campus. The property encompasses four business spaces that open onto College Avenue. HTM has a long-term tenant in each space, and their leases are all running out at approximately the same time. HTM wants to significantly increase the rent for each space, and the current tenants have all indicated that they will stay on at a slightly increased rent but not at the rental price that HTM wants. HTM has shopped the spaces around and has gotten commitments from four new tenants at the higher rent. However, HTM realizes that there is a very high likelihood that the current tenants will remain in business and pay their rent for a new 3-year lease, whereas there is a much smaller likelihood that the new tenants will be able to stay in business and fulfill their lease payments. The following table shows the current and prospective new tenants for each space, their monthly payments for a 3-year lease, and the probability that each will remain in business:

Space	Current Tenant	Lease Payment	Probability	New Tenant	Lease Payment	Probability
A	Doughnut Shop	$1,200	.90	Sub Shop	$2,400	.50
B	Jewelry Store	800	.90	Dress Shop	1,200	.70
C	Music Store	1,000	.95	Restaurant	2,000	.40
D	Gift & Card Shop	1,100	.95	Tech Logo Store	1,800	.60

HTM wants to lease each space to businesses that will maximize their expected lease payments. It also wants a likelihood that at least three of the businesses will stay in business. Formulate and solve an integer programming model for this problem.

42. The *Valley Times* and *World News* publishes and delivers a morning newspaper 7 days a week. The bundled papers are delivered by trucks to a number of area communities where they are picked up by carriers at a central location in the community, who then deliver the individual newspapers to customers along various predetermined routes. The papers arrive at the central distribution location around 6 A.M., and they must be delivered by 8 A.M. The routes include homes, stores, restaurants, and newspaper machines. One community, Hannah Creek, has

10 routes, and six available paper carriers, who have requested a delivery contract. The following table shows the time (in minutes) required to deliver the newspapers along each route, the number of papers delivered on each route, and the vehicle capacity (in number of papers) of each carrier:

Carrier	A	B	C	D	E	F
Vehicle capacity	600	720	450	510	660	550

Route	1	2	3	4	5	6	7	8	9	10
Time (mins)	29	35	15	35	20	23	35	40	45	50
Papers delivered	200	240	110	90	65	135	80	170	150	270

Formulate and solve a linear integer programming model that will minimize the total number of carriers needed to make the deliveries and that will determine which carrier will be assigned each route and the total number of papers delivered by each of the selected carriers.

43. The BigRig Trucking Company participates in an Internet transportation exchange where customers advertise their shipments including load weight and volume, and trip origin and destination. BigRig then computes the cost and time of the trip and determines the bid it should make for the shipment to achieve a certain profit level. Twelve customers have posted shipments on the exchange, and BigRig has three trucks available for shipments. Each truck has a load capacity of 80,000 pounds and 5,500 cubic feet and available driving time of 90 hours. The following table shows the load parameters (i.e., weight in pounds and volume in cubic feet) for each customer shipment and the profit BigRig would realize from each shipment:

Customer	Profit ($)	Load (lb.)	Load (ft.^{3})	Time (hours)
1	20,000	44,000	1,600	51
2	17,000	39,000	2,100	22
3	15,000	24,000	3,200	45
4	7,000	33,000	3,700	36
5	18,000	18,000	4,400	110
6	12,000	21,000	2,900	105
7	5,000	15,000	1,100	44
8	4,600	19,000	1,600	56
9	11,000	23,000	800	60
10	6,200	36,000	1,800	25
11	14,000	55,000	3,700	37
12	9,000	45,000	2,900	41

Formulate and solve a linear programming model to determine which customer shipments BigRig should bid on in order to maximize profit.

44. Twobucks Coffee Company currently operates 12 coffee shops in downtown Nashville. The company has been losing money and wants to downsize by closing some stores. Its policy has been to saturate the downtown area with stores so that one is virtually always in the sight of a potential or current customer. However, the company's new policy is to have enough stores so that each is within 5 minutes' walking distance of another store. The company would also like to have annual operating costs of no more than $900,000.

The following table shows the coffee shops within 5 minutes' walking distance of another shop and the average annual cost of the existing stores:

Store Location	Annual Average Operating Cost ($)	Stores Within 5 Minutes' Walking Distance
1. 3rd Street	456	3rd Street, Rose Street
2. 10th Street	207	10th Street, South Street, Broad Street
3. South Street	139	South Street, Hill Street, 10th Street
4. Mulberry Avenue	246	Mulberry Avenue, Beamer Boulevard
5. Rose Street	177	Rose Street, 3rd Street, Church Street
6. Wisham Avenue	212	Wisham Avenue, Broad Street, 23rd Avenue
7. Richmond Road	195	Richmond Road, Broad Street
8. Hill Street	170	Hill Street, South Street
9. 23rd Avenue	184	23rd Avenue, Wisham Avenue
10. Broad Street	163	Broad Street, Wisham Avenue, Richmond Road, 10th Street
11. Church Street	225	Church Street, Rose Street, Beamer Boulevard
12. Beamer Boulevard	236	Beamer Boulevard, Mulberry Avenue, Church Street

Formulate and solve a linear programming model that will select the minimum number of stores the company will need to achieve its new policy objective.

45. The town of Hillsboro recently purchased a 55-acre tract of farm land, and it has $550,000 budgeted to develop recreational facilities. The impetus for the purchase was the need for more soccer fields to meet the increasing demand of youth soccer in the area. However, once the land was purchased, a number of other interest groups began to lobby the town council to develop other recreational facilities including rugby, football, softball, and baseball fields, plus walking and running trails, a children's playground, and a dog park. The following table shows the amount of acreage required by each project, the annual expected usage for each facility, and the cost to construct each facility. Also included is a priority designation determined by the town's recreation committee based on several public hearings and their perceptions of the critical need of each facility.

Facility	Annual Usage (people)	Acres	Cost ($)	Priority
Rugby fields	4,700	7	75,000	3
Football fields	12,500	12	180,000	2
Soccer fields	32,000	20	350,000	1
Dog park	7,500	6	45,000	3
Playground	41,000	3	120,000	2
Walking/running trails	47,000	25	80,000	1
Softball fields	23,000	5	115,000	2
Baseball fields	16,000	8	210,000	3

a. Formulate and solve a linear programming model that will maximize annual usage and achieve an average priority level of no more than 1.75.

b. Reformulate the model such that the objective is to achieve the minimum average priority level while achieving an annual usage of at least 120,000.

c. What combination of facilities will use the maximum acreage available without exceeding the budget and achieving an average priority level of no more than 1.75? What is the annual usage with these facilities?

Case Problem

PM Computer Services

Paulette Smith and Maureen Becker are seniors in engineering and business, respectively, at State University. They have set up a company, PM Computer Services, to assemble and sell their own brand of personal computers. They buy component parts on the open market from a variety of sources in the United States and overseas, and they assemble their computers, mostly at night, in their three-bedroom apartment. They sell their computers primarily to departments at State University and to other students. They hire other students to perform the assembly operations and to test and package the computers. In addition to managing the operations, Paulette and Maureen help with all other tasks, including sales and accounting.

They pay the students who work for them $8 per hour for a 40-hour week, or $1,280 per month. They hire students on a monthly basis, and their delivery schedule is also on a monthly (i.e., end-of-the-month) basis. PM currently has five employees. PM Computers has determined that each of its employees is able to produce 12.7 computers, on average, per month. When the monthly demand for its computers exceeds its regular production capacity, PM employs limited overtime. Each computer produced on an overtime basis adds $12 to the labor cost of a computer. A PM employee can produce 0.6 computer per month on an overtime basis.

Paulette and Maureen have received the following computer orders for the next 6 months:

Month *i*	Computer Orders
1	63
2	74
3	95
4	57
5	68
6	86

In the past, PM has met its demand strictly from regular and overtime production. To meet demand in some months when it did not have sufficient regular and overtime production, the company would plan ahead and produce computers in previous months with available capacity. However, Paulette and Maureen's apartment was completely filled with components and workspace, so they could not store completed computers. Instead, they leased warehouse space in town to store their completed computers for delivery in future months. They had to transport the computers across town to the warehouse and pay for all handling; also, the warehouse had to be climate controlled. The cost of holding a computer in storage at the warehouse is $15 per month.

Paulette and Maureen are considering an alternative production strategy wherein they would hire new workers on a monthly basis as needed and lay off workers when they are not needed. They estimate the cost of hiring new workers to be $200, primarily for related paperwork and training. The cost of laying off a worker is $320, or approximately 1 week's wages. They may want to rehire some of the workers they lay off at a later date, so they want them to leave with a good feeling about PM.

Determine a planning schedule for PM Computer Services, indicating the number of employees working each month, including the number hired and the number laid off, the number of computers produced each month in both regular time and overtime, and the number of computers carried over in inventory each month. There should be no inventory left over after month 6. Provide integer solution values for these different variables. Compare this solution with the one you would obtain without integer restrictions.

Case Problem

The Tennessee Pterodactyls

The Tennessee Pterodactyls is a new professional basketball franchise in Nashville. The team's general manager, Jerry East, and coach, Phil Riley, are trying to develop a roster of players. They drafted seven players from a pool to which the other teams in the league contributed two players each. However, the general manager and coach perceive these acquisitions to be no more than role players. They believe that the nucleus of their new team must come from the free agents who are currently available on the market. The team is well under the salary cap, and the owner has made $50 million per year available to them to sign players. The coach and general manager have put together the following list of 12 free agents, with important statistics for each, including their rumored asking price in terms of annual salary:

Player	Position	Per-Game Averages				Projected Annual Salary ($1,000,000s)
		Points	*Rebounds*	*Assists*	*Minutes*	
1. Mack Madonna	Back court	14.7	4.4	9.3	40.3	$ 8.2
2. Darrell Boards	Front court	12.6	10.6	2.1	34.5	6.5
3. Silk Curry	Back court	13.5	8.7	1.7	29.3	5.2
4. Ramon Dion	Back court	27.1	7.1	4.5	42.5	16.4
5. Joe Eastcoast	Back court	18.1	7.5	5.1	41.0	14.3
6. Abdul Famous	Front court	22.8	9.5	2.4	38.5	23.5
7. Hiram Grant	Front court	9.3	12.2	3.5	31.5	4.7
8. Antoine Roadman	Front court	10.2	12.6	1.8	44.4	7.1
9. Fred Westcoast	Front court	16.9	2.5	11.4	42.7	15.8
10. Magic Jordan	Back court	28.5	6.5	1.3	38.1	26.4
11. Barry Bird	Front court	24.8	8.6	6.9	42.6	19.5
12. Grant Hall	Front court	11.3	12.5	3.2	39.5	8.6

Jerry and Phil want to sign five free agents. They would like the group they sign to average at least 80 points per game (16 points per player), pull down an average of 40 rebounds per game (8 per player), dish out an average of 25 assists, and have averaged 190 minutes (38 minutes per player) per game in the past. Furthermore, they do not want to sign more than two front court and three back court players. Their immediate objective is to identify the players who as a group will meet their objectives at the minimum cost.

A. Formulate an integer linear programming model to help the general manager and coach determine which players they should sign and solve it by using the computer.
B. Is the money provided by the owner sufficient to sign the group of players identified in (A)? If not, reformulate the model so that the available funds are a constraint and the objective is to maximize the average points of the group.

Case Problem

New Offices at Atlantic Management Systems

Atlantic Management Systems is a consulting firm that specializes in developing computerized decision support systems for computer manufacturing companies. The firm currently has offices in Chicago, Charlotte, Pittsburgh, and Houston. It is considering opening new offices in one or more cities—including Atlanta, Boston, Washington, DC, St. Louis, Miami, Denver, and Detroit—and it has $14 million available for this purpose. Because of the highly specialized nature of its high-tech consulting work, the firm must necessarily staff any new offices with a minimum number of its employees from its existing offices. However, it has a limited number of employees available to transfer to any new offices. In addition, the cost of transferring employees depends on the city they are leaving and the city to which they might move.

Following are the costs for opening a new office in each of the prospective cities and the start-up staffing needs at each office:

Prospective Office	Setup Cost ($1,000,000s)	Staffing Needs (employees)
1. Atlanta	$1.7	9
2. Boston	3.6	14
3. Denver	2.1	8
4. Detroit	2.5	12
5. Miami	3.1	11
6. St. Louis	2.7	7
7. Washington, DC	4.1	18

The numbers of employees available for transfer from each of the current offices are as follows:

Existing Office	Available Employees
Chicago	24
Charlotte	19
Pittsburgh	16
Houston	21

The costs (in thousands of dollars) of transferring an employee from an existing office to a new office differ according to housing costs and moving expenses plus cost of living adjustments. They are as follows:

	Prospective New Offices (costs, $1,000s)						
Current Office	*1. Atlanta*	*2. Boston*	*3. Denver*	*4. Detroit*	*5. Miami*	*6. St. Louis*	*7. Washington, DC*
1. Chicago	$19	$32	$27	$14	$23	$14	$41
2. Charlotte	14	47	31	28	35	18	53
3. Pittsburgh	16	39	26	23	31	19	48
4. Houston	22	26	21	18	28	24	43

The firm has ranked the possible new offices according to their profit potential, with Washington, DC, being the best (i.e., greatest potential), as follows:

New Office	Rank
Washington, DC	1
Miami	2
Atlanta	3
Boston	4
Denver	5
St. Louis	6
Detroit	7

In addition, the firm wants at least one new office in the Midwest (i.e., Detroit and/or St. Louis) and one new office in the Southeast (i.e., Atlanta or Miami).

Formulate and solve an integer programming model to help Atlantic Systems determine how many new offices it should open, where they should be located, and how to transfer employees.

Case Problem

Scheduling Television Advertising Slots at the United Broadcast Network

The United Broadcast Network (UBN) sells to advertisers commercial advertising slots on its television shows. The network announces its new fall television schedule during the previous spring and shortly thereafter begins selling its inventory of advertising slots to its customers. Long-standing priority customers receive the first opportunities to purchase advertising slots. The new fall season begins in the third week of September. Advertising slots are mostly 15 seconds and 30 seconds in duration. The network develops a detailed sales plan for the year, typically on a monthly, bimonthly, 6-week, or quarterly basis. Advertisers generally have certain shows in mind that they prefer to advertise on in order to reach a certain demographic audience that they want to market their product to, as well as a specific advertising budget. The network sells advertising slots at rates based on a performance score related to the popularity of a show, primarily determined by demographics and audience size.

Nanocom, a business software development firm, wants to purchase 30- and 15-second advertising slots on several shows this coming fall. Nanocom wants to reach an older, more mature, upper-income audience that is likely to include many high-tech businesspeople. It has an advertising budget of $600,000, and it has informed the UBN advertising staff that it prefers to purchase ads on the following shows: *Bayside, Newsline, The Hour, Cops and Lawyers, The Judge, Friday Night Football*, and *ER Doctor. Newsline* and *The Hour* are news magazine shows, whereas the others are adult dramas, except for *Friday Night Football*, which is professional football. Nanocom wants at least 50% of the total number of advertising slots it purchases to be on *Newsline, The Hour*, and *Friday Night Football*. Nanocom would like UBN to develop a sales plan covering the 6-week period beginning with the third week in October through November (including the November sweeps).

As indicated, UBN bases its sales plans on performance scores for the different shows. The primary objective of both UBN and the advertiser is to develop a sales plan that will achieve the highest total performance score. The performance scores are based on several factors, including how well the show matches the desired audience demographics, the audience strength of the show, the historical ratings for the time slot, the competing shows in the same time slot, and the performance of adjacent shows. The network sets its advertising rates based on these performance scores. Both the performance scores and the rates are then multiplied by a weighting factor that is based on the week of the show and, because total viewers vary according to the week of the year, the total audience expected during that week. The following table shows the costs, performance scores, and available inventory of advertising slots for 15- and 30-second commercials for each show:

Show	Commercial Length (seconds)	Cost ($1,000s)	Performance Score	Available Inventory
Bayside	30	$50.0	115.2	3
	15	25.0	72.0	3
Newsline	30	41.0	160.0	4
	15	20.5	100.0	1
The Hour	30	36.0	57.6	3
	15	18.0	36.0	3
Cops and Lawyers	30	45.0	136.0	4
	15	27.5	85.0	2
The Judge	30	52.0	100.8	2
	15	26.0	63.0	2
Friday Night Football	30	25.0	60.8	4
	15	12.5	38.0	2
ER Doctor	30	46.0	129.6	3
	15	23.0	81.0	1

The network advises, and Nanocom agrees, that there should be a minimum and maximum number of advertising slots during each of the 6 weeks and that Nanocom should have at most only one ad slot (either 15 or 30 seconds) per show per week. The following table shows the weighting factor each week and the minimum and maximum numbers of slots per week:

	October		November			
Week	*3*	*4*	*1*	*2*	*3*	*4*
Weight	1.1	1.2	1.2	1.4	1.4	1.6
Minimum	1	1	2	2	2	3
Maximum	4	5	5	5	5	5

Assist UBN in developing a sales plan for the 6-week period to maximize the total performance score for Nanocom.

Case Problem

The Draperton Parks and Recreation Department's Formation of Girls' Basketball Teams

Each fall the Draperton Parks and Recreation Department holds a series of tryouts for its boys' and girls' youth basketball leagues. All leagues are formed by age group, each usually encompassing a couple of ages. One such grouping is the 12- to 13-year-old girls' league. This is a particularly competitive age group because the girls are all trying to improve their skills in hopes of making the high school junior varsity team in the immediate future and the high school varsity in several years. Some of the girls even hope to make the high school varsity next year, when they are freshmen. This is also the first age group in which not all girls who try out are placed on teams; some are cut. In the younger age groups, all players are assigned to teams. However, at this age group, four teams are formed, each with only seven players. This policy was agreed upon by a committee consisting of the Recreation Department director, a group of parents of past players, and the high school boys' and girls' basketball coaching staff. The small roster size allows each player to get a lot of playing time and makes the games more competitive. It also makes the teams more competitive in tournaments with teams from other towns and cities. This makes the girls very competitive, and it makes their parents even more competitive. Sandy Duncan, the Recreation Department's director of team sports, knows from past experience that parents do a lot of jockeying and politicking during tryouts. This has given rise to parental protests regarding the tryout process and about how players are selected and assigned to teams. Sandy and the player evaluators have frequently been accused of showing favoritism and stupidity.

This year, in an effort to blunt some of this criticism, Sandy has devised a tryout process that will hopefully remove her (and any other person) from direct involvement with the player selection and team formation process. First, she conducted tryouts for all 36 players who were trying out over two weekends that included four sessions each weekend: one on Friday evening, two on Saturday, and one on Sunday afternoon. She formed an evaluation team of several coaches from other towns and some of the current players and assistant coaches from the nearby State University women's basketball team. Each player was assigned a tryout number and evaluated for different skills and ability, and at the end of the tryouts, each player was given an overall score from 1 to 10, with 10 being the highest, as follows:

Player	Score	Player	Score	Player	Score
1	7	13	10	25	8
2	6	14	9	26	4
3	9	15	9	27	9
4	5	16	4	28	9
5	7	17	5	29	3
6	8	18	5	30	3
7	3	19	6	31	3
8	3	20	3	32	6
9	4	21	2	33	6
10	6	22	1	34	7
11	4	23	7	35	8
12	8	24	5	36	10

Sandy took a course in management science while she was a student at State University, and she wants to develop an integer linear programming model to use these evaluation scores to select the players and assign them to the four teams in a fair and equitable manner. She wants to select the best players, and in order to have competitive, equally matched teams, she wants each team to have an average overall player evaluation score between 6 and 7. As she sat down one evening to work on this problem, she quickly realized that it would require a relatively complex integer programming model. Help Sandy formulate and solve an integer programming model for this problem to select the best players, assign seven players to each team, and achieve the average team evaluation score Sandy wants for each team. Compare the teams formed by your model and determine whether you think they are competitive (i.e., whether the model has achieved Sandy's objectives). Also, determine whether any deserving players were unfairly cut by the model.

Case Problem

DEVELOPING PROJECT TEAMS FOR MANAGEMENT 4394

All students in the College of Business at State University are required to take a capstone course, Management 4394, the last semester of their senior year. This course consists primarily of a semester-long case project that pulls together all the business disciplines to test the students' knowledge in the major business areas of management, marketing, finance, accounting, quantitative methods, and operations. A difficult task for the course instructor is dividing the students in a class into equitable teams that will enhance the learning process and engender teamwork. The project teams should be relatively equal in terms of student ability and performance. Further, the teams should reflect different functional skills; that is, the teams should include a mix of different majors. Also, the college would like the teams to be diverse in terms of gender and nationalities in order to prepare the students for working with diverse individuals when they graduate and take a job. Consider one course section of Management 4394 with 18 students. The instructor wants to divide the class into six project teams of three members each. The instructor's main objective for the teams is that they be relatively equal in terms of academic capability. Since the only means of assessing capability is a student's grade point average (GPA), the instructor has determined that one way to achieve this objective is for the teams to have a minimum average GPA of 2.80 while also striving to achieve the maximum overall average team GPA possible. It is also important that the teams have a mix of members with different majors to enhance the teams' functional capabilities, but the instructor does not want there to be more than two of the same major on any single team. In order to achieve some degree of diversity among team members the instructor wants there to be at least one female and one international student on each team, but not more than two females or two international students on a team. Following is a table that shows the breakdown of GPAs, majors, gender, and internationality of each student in the class.

Student	GPA	Gender	International (yes or no)	Major
1	3.04	M	Yes	FIN
2	2.35	M	No	FIN/ACCT
3	2.26	M	Yes	MGT
4	2.15	M	Yes	MKTG
5	3.23	F	No	MGT
6	3.95	F	No	FIN
7	2.87	M	No	FIN
8	2.65	F	No	FIN
9	3.12	M	No	FIN
10	3.08	M	Yes	MKTG
11	3.35	M	Yes	MKTG
12	2.78	M	No	ACCT
13	2.56	F	Yes	MKTG
14	2.91	F	No	ACCT
15	3.40	F	Yes	MGT
16	3.12	M	Yes	FIN
17	2.75	M	Yes	ACCT
18	3.06	F	No	MGT

Note that only two students are both female and international, and that student 2 is a double major.

Formulate and solve an integer linear programming model for this problem to determine project teams that will meet the instructor's guidelines for team formation. How successfully do you think your model achieves the instructor's guidelines? In other words, do you consider the teams to be diverse and relatively equitable in terms of functional and academic capability? If not, how do you think you might need to change your model to better meet the team formation guidelines?

This case problem is based on R. Cutshall, S. Gavirneni, and K. Schultz, "Indiana University's Kelley School of Business Uses Integer Programming to Form Equitable, Cohesive Student Teams," *Interfaces* 37, no. 3 (May–June 2007): 265–76.

Case Problem

Scheduling the Lead Balloon's Summer Tour

The Lead Balloon is a popular '80s rock band that broke up in the 1990s but recently got back together to publish a new CD and go on tour. They want to schedule a summer tour that will encompass 16 possible cities in the eastern half of the United States. However, because of family and business commitments, plus the fact they are not so young and energetic as they used to be, the tour will encompass only 12 weeks, from mid-May to mid-August, when their children will be out of school, so that their families can join them on the tour most of the time. The band also wants to limit the number of concerts each week. Following is a list of the possible cities with the weeks that the local arenas in each city will be available (marked with an X) during the tour. Also included is the concert capacity of the arena in each city.

Based on its popularity and fame it is a foregone conclusion that the band will sell out everywhere it goes. Thus, the band wants to schedule the cities that will maximize the total attendance at its concerts, which basically means it wants to schedule the cities with the largest arenas, if possible. The band has also stipulated that it wants to perform at least one concert but no more than two concerts each week. Also, because of the band's high-tech equipment and gear, and sound and stage setup requirements, logistical constraints limit the travel between two cities in any one week to 500 miles.

Develop a summer tour schedule for the band that will maximize total attendance.

	May		June				July				August		
City	**3**	**4**	**1**	**2**	**3**	**4**	**1**	**2**	**3**	**4**	**1**	**2**	**Capacity**
Atlanta		X			X		X		X				19,500
Boston			X		X	X		X		X			18,500
Cincinnati	X	X		X					X		X	X	16,200
Charlotte	X	X		X		X		X			X	X	17,600
Cleveland			X			X	X			X			22,700
Detroit	X	X			X	X		X	X			X	24,200
DC		X					X				X	X	21,500
Philadelphia	X		X	X	X			X	X				21,300
Dallas					X	X				X		X	17,400
Miami	X	X		X			X				X		15,700
Orlando	X		X			X		X	X				16,300
Houston		X		X	X	X					X		18,200
Indianapolis			X	X				X	X		X	X	25,400
Nashville		X		X		X			X	X	X		16,500
Chicago			X		X		X	X				X	22,800
New York	X	X				X				X		X	20,700

Case Problem

Developing Kathleen Taylor's 401(k) Plan

Kathleen Taylor has been working for a government contractor, Summit Solutions, in Washington, DC, for over a year. She is now eligible to participate in the company's 401(k) retirement plan. The company has provided Kathleen with the information in the table on the following page about the various funds that are available for her to invest in.

	Evening Star Rating	5-Year Return	Fund Size ($1,000,000s)	Expense Ratio
International				
Parham International Value	3	12.58	2,066	1.67
Jarus Overseas	4	6.94	7,404	1.97
U.S. Fund EuroPacific	2	7.86	19,859	1.19
Admiral Foreign Investor	5	7.54	42,035	1.32
SPS Emerging Market	3	5.57	11,139	1.15
Small-Cap				
Maxam Small Cap Return	4	5.87	1,302	1.43
U.S. Small Cap Select	2	3.99	770	1.21
Veritas Small Cap Value	4	2.97	3,049	1.74
Oak Small Cap Growth	3	3.8	4,320	1.49
Mid-Cap				
Federal Mid Cap Growth	3	4.13	729	1.05
T. Row Price Growth	2	6.30	4,145	1.33
Draper Structured Mid Cap	2	7.73	1,467	1.31
Jarus Mid Cap Value	3	4.05	4,632	1.52
Maxam Mid Cap Opportunity	4	5.76	2,537	1.17
Large-Cap				
T. Row Price Bluechip Growth	3	6.01	6,374	1.23
Jarus Diversified Equity Income	2	3.22	962	1.05
Draper Strategic Growth	4	7.81	9,991	1.16
Centennial Common Stock Growth	3	2.16	20,194	0.99
Marius Stock Growth	2	9.88	19,648	1.12
Bond				
Maxam Global Bond	3	8.37	12,942	1.03
Parham High Yield	5	6.25	3,106	1.00
Madison Federated Bond Portfolio	3	5.66	8,402	1.14
Draper U.S. Government Bond	4	7.41	1,045	0.90
Federated High Income	3	5.85	11,211	0.82

International funds invest in global/overseas companies; small-cap funds invest in companies that have a market capitalization (i.e., the number of outstanding shares multiplied by the stock price per share), in general, between $300 million and $2 billion; mid-cap funds invest in companies with a market capitalization between $2 and $10 billion; and large-cap funds invest in companies over $10 billion. The *Evening Star* rating is developed by an independent investment analysis firm, and it rates funds based on their risk-adjusted performance over various time periods. "5" is best, "1" is worst; stocks trading close to their analysts' fair value estimates receive a "3", while stocks trading at large discounts compared to their analysts' fair value estimate receive a "4" or "5" rating. The expense ratio is the total percentage of fund assets used for operating expenses (i.e., administrative, management, advertising, etc.).

Since Kathleen is young and expects to build her 401(k) plan over a long period of time, she wants to employ a relatively aggressive investment strategy. She has read investment literature that suggests a relatively aggressive plan would invest 5% to 35% in international funds, 5% to 25% in small-cap funds, 5% to 35% in mid-cap funds, 20% to 50% in large-cap funds, and 5% to 10% in bond funds. Kathleen plans to contribute $900 of her salary each month to her plan, which the company will match. She has also developed a few of her own investment guidelines: to diversify, she wants to invest in five funds, one in each investment category; she wants to achieve an average *Evening Star* rating of at least 3.7; she wants to invest in funds that average at least $10,000 million in size; she wants to achieve an average expense ratio for her five funds of no more than 1.10; and she wants to maximize the average 5-year return of the five funds she selects, weighted by the amount she invests in each.

Develop an investment plan for Kathleen using linear programming. How would her investment plan change if she wanted to maximize the *Evening Star* rating?

CHAPTER

6

Transportation, Transshipment, and Assignment Problems

In this chapter, we examine three special types of linear programming model formulations—*transportation, transshipment,* and *assignment problems.* They are part of a larger class of linear programming problems known as *network flow problems.* We are considering these problems in a separate chapter because they represent a popular group of linear programming applications.

See Web site Module B for a chapter on the "Transportation and Assignment Solution Methods."

These problems have special mathematical characteristics that have enabled management scientists to develop very efficient, unique mathematical solution approaches to them. These solution approaches are variations of the traditional simplex solution procedure. Like the simplex method, we have placed these detailed manual, mathematical solution procedures—called the *transportation method* and *assignment method*—on the companion Web site that accompanies this text. As in previous chapters, we will focus on model formulation and solution by using the computer, specifically by using Excel and QM for Windows.

The Transportation Model

*In a **transportation problem,** items are allocated from sources to destinations at a minimum cost.*

The **transportation model** is formulated for a class of problems with the following unique characteristics: (1) A product is *transported* from a number of sources to a number of destinations at the minimum possible cost; and (2) each source is able to supply a fixed number of units of the product, and each destination has a fixed demand for the product. Although the general transportation model can be applied to a wide variety of problems, it is this particular application to the transportation of goods that is most familiar and from which the problem draws its name.

The following example demonstrates the formulation of the transportation model. Wheat is harvested in the Midwest and stored in grain elevators in three different cities—Kansas City, Omaha, and Des Moines. These grain elevators supply three flour mills, located in Chicago, St. Louis, and Cincinnati. Grain is shipped to the mills in railroad cars, each car capable of holding 1 ton of wheat. Each grain elevator is able to supply the following number of tons (i.e., railroad cars) of wheat to the mills on a monthly basis:

Grain Elevator	*Supply*
1. Kansas City	150
2. Omaha	175
3. Des Moines	275
Total	600 tons

Each mill demands the following number of tons of wheat per month:

Mill	*Demand*
A. Chicago	200
B. St. Louis	100
C. Cincinnati	300
Total	600 tons

The cost of transporting 1 ton of wheat from each grain elevator (source) to each mill (destination) differs, according to the distance and rail system. (For example, the cost of

EXHIBIT 6.11

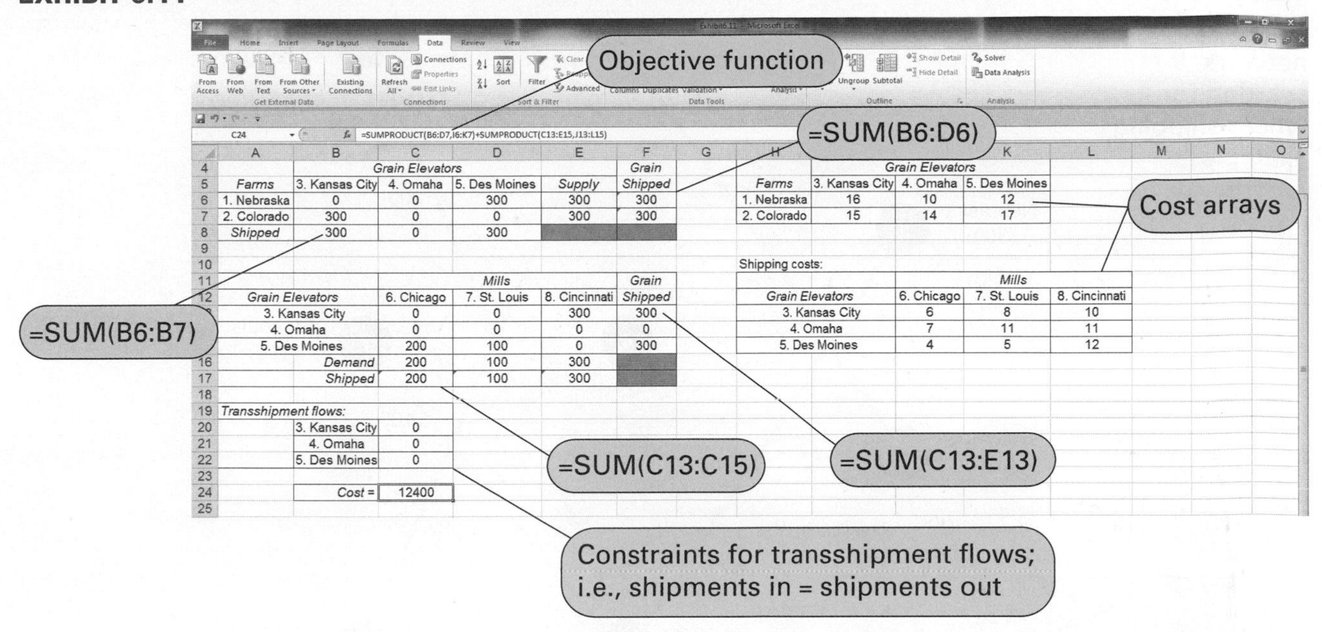

EXHIBIT 6.12

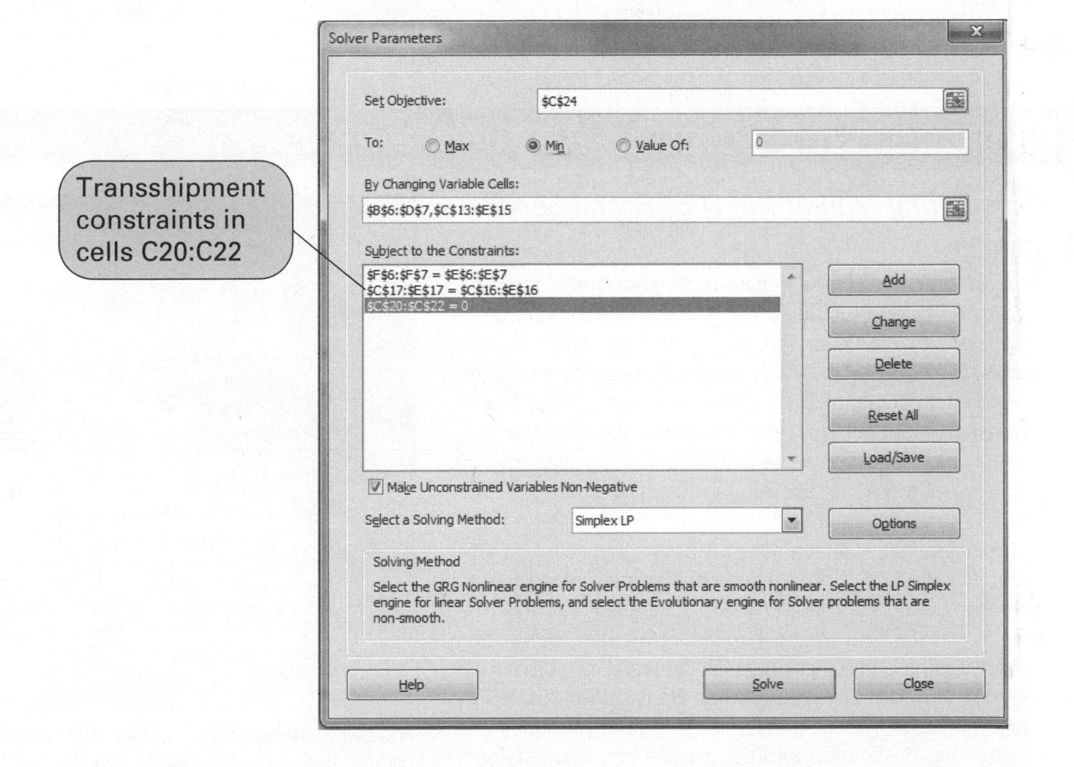

Two cost arrays have been developed for the shipping costs in cells **I6:K7** and cells **J13:L15**, which are then multiplied by the variables in cells **B6:D7** and **C13:E15** and added together. The objective function, **= SUMPRODUCT(B6:D7,I6:K7) + SUMPRODUCT (C13:E15,J13:L15)**, is shown on the toolbar at the top of Exhibit 6.11. Constructing the objective function with cost arrays like this is a little easier than typing in all the variables and costs in a single objective function when there are many variables and costs. A network diagram of the optimal solution is shown in Figure 6.4.

FIGURE 6.4
Transshipment network solution for wheat-shipping example

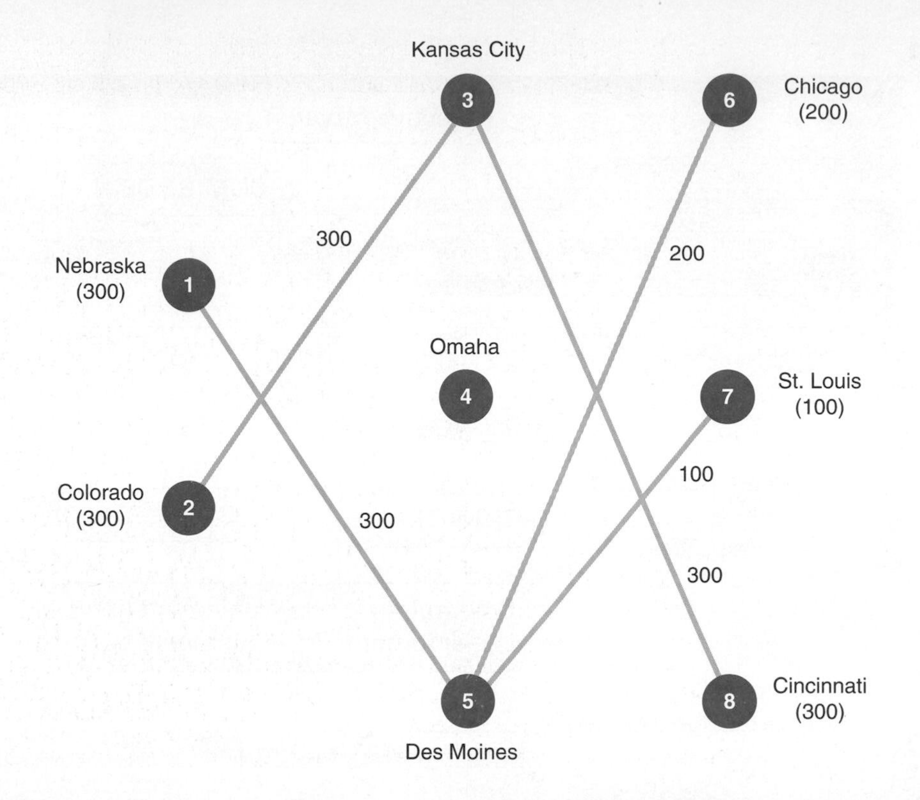

Management Science Application

Transporting Mail at the U.S. Postal Service

The United States Postal Service (USPS) has one of the most complex transportation networks in the world, annually delivering more than 200 billion pieces of mail. USPS delivers various types of mail in different shapes, weights, and sizes, through intermediate points and facilities, using various modes of transportation. On an average weekday the USPS highway transportation network alone will include more than 75,000 trips between more than 30,000 facilities. USPS transportation analysts use management science models to identify cost-savings opportunities in its highway transportation network. One model, the Highway Corridor Analytic Program (HCAP), optimizes the selection of trips between facilities (such as bulk mail centers, transfer and mail consolidation centers, processing and distribution centers, priority mail centers, and air mail centers) to minimize total transportation costs subject to delivery requirements. The models includes originating facilities and pickup times, destination facilities and delivery times, delivery volume, whether or not the delivery can be split, stops between the origins and destinations, transportation capacity, and the variable transportation costs (i.e., cost per unit of volume). The model identifies existing and potential new trips that can possibly transport the mail from an origin to a destination, including stops. Thus, the model includes 0–1 decision variables (that equal one if a trip is selected and zero if it is not) and real variables that indicate the proportion of a delivery's volume that is routed using a specific trip. The model solution identifies the optimal set of trips that satisfies the delivery requirements. The HCAP model has saved the USPS more than $5 million in annual transportation cost savings.

AP Photo/The Republic, Joel Philippsen

Source: Based on A. Pajunas, E. Matto, M. Trick, and L. Zuluaga, "Optimizing Highway Transportation at the United States Postal Service," *Interfaces* 37, no. 6 (November–December 2007): 515–25.

The Assignment Model

*An **assignment model** is for a special form of transportation problem in which all supply and demand values equal one.*

The **assignment model** is a special form of a linear programming model that is similar to the transportation model. There are differences, however. In the assignment model, the supply at each source and the demand at each destination are each limited to one unit.

The following example will demonstrate the assignment model. The Atlantic Coast Conference (ACC) has four basketball games on a particular night. The conference office wants to assign four teams of officials to the four games in a way that will minimize the total distance traveled by the officials. The supply is always one team of officials, and the demand is for only one team of officials at each game. The distances in miles for each team of officials to each game location are shown in the following table:

The travel distances to each game for each team of officials

	Game Sites			
Officials	Raleigh	Atlanta	Durham	Clemson
A	210	90	180	160
B	100	70	130	200
C	175	105	140	170
D	80	65	105	120

The linear programming formulation of the assignment model is similar to the formulation of the transportation model, except all the supply values for each source equal one, and all the demand values at each destination equal one. Thus, our example is formulated as follows:

$$\begin{aligned}\text{minimize } Z = {} & 210x_{AR} + 90x_{AA} + 180x_{AD} + 160x_{AC} + 100x_{BR} + 70x_{BA} + 130x_{BD} \\ & + 200x_{BC} + 175x_{CR} + 105x_{CA} + 140x_{CD} + 170x_{CC} + 80x_{DR} + 65x_{DA} \\ & + 105x_{DD} + 120x_{DC}\end{aligned}$$

subject to

$$\begin{aligned}x_{AR} + x_{AA} + x_{AD} + x_{AC} &= 1\\ x_{BR} + x_{BA} + x_{BD} + x_{BC} &= 1\\ x_{CR} + x_{CA} + x_{CD} + x_{CC} &= 1\\ x_{DR} + x_{DA} + x_{DD} + x_{DC} &= 1\\ x_{AR} + x_{BR} + x_{CR} + x_{DR} &= 1\\ x_{AA} + x_{BA} + x_{CA} + x_{DA} &= 1\\ x_{AD} + x_{BD} + x_{CD} + x_{DD} &= 1\\ x_{AC} + x_{BC} + x_{CC} + x_{DC} &= 1\\ x_{ij} &\geq 0\end{aligned}$$

This is a *balanced* assignment model. An *unbalanced* model exists when supply exceeds demand or demand exceeds supply.

Computer Solution of an Assignment Problem

The assignment problem can be solved using the assignment modules in QM for Windows and Excel QM, and with Excel spreadsheets. We will solve our example of assigning ACC officials to game sites first by using Excel, followed by Excel QM, and then by using QM for Windows.

Computer Solution with Excel

As is the case with transportation problems, Excel can be used to solve assignment problems, but only as linear programming models. Exhibit 6.13 shows an Excel spreadsheet for our ACC basketball officials example. The objective function in cell C11 was developed by creating a mileage array in cells **C16:F19** and multiplying it by the decision variables in cells **C5:F8**. The model constraints for available teams (supply) are contained in the cells in column H, and the constraints for teams of officials at the game sites (demand) are contained in the cells in row 10.

EXHIBIT 6.13

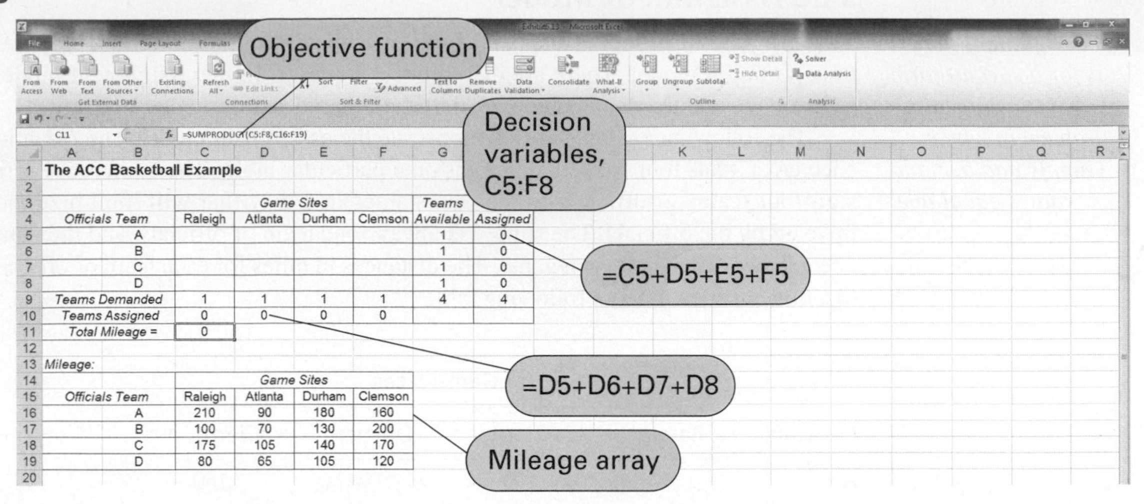

Exhibit 6.14 shows the Solver Parameters screen for our example. Before clicking on "Solve," remember to invoke "Simplex LP" to solve as a linear programming model. The optimal solution is shown in Exhibit 6.15. A network diagram of the optimal solution is shown in Figure 6.5.

EXHIBIT 6.14

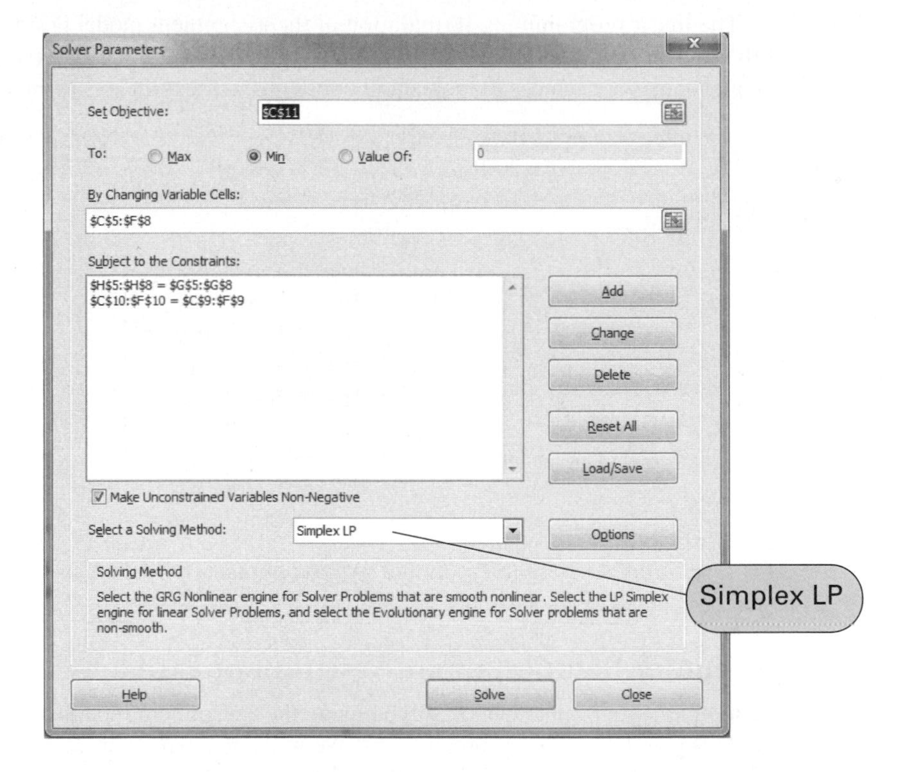

Computer Solution with Excel QM

Excel QM also includes a spreadsheet macro for assignment problems. It is very similar to the Excel QM macro demonstrated earlier in this chapter for the transportation problem, and the spreadsheet setup is very much like that of the Excel spreadsheet in Exhibit 6.15. The assignment macro is accessed (from the "Excel QM" menu) and applied similarly to the transportation problem macro. It already includes all the cell and constraint formulas necessary to solve the problem. The solution to our ACC basketball example with Excel QM is shown in Exhibit 6.16.

EXHIBIT 6.15

C11 =SUMPRODUCT(C5:F8,C16:F19)

The ACC Basketball Example

	Game Sites				Teams	Teams
Officials Team	Raleigh	Atlanta	Durham	Clemson	*Available*	*Assigned*
A	0	1	0	0	1	1
B	1	0	0	0	1	1
C	0	0	1	0	1	1
D	0	0	0	1	1	1
Teams Demanded	1	1	1	1	4	4
Teams Assigned	1	1	1	1		
Total Mileage =	450					

Mileage:

	Game Sites			
Officials Team	Raleigh	Atlanta	Durham	Clemson
A	210	90	180	160
B	100	70	130	200
C	175	105	140	170
D	80	65	105	120

FIGURE 6.5
Assignment network solution for ACC officials example

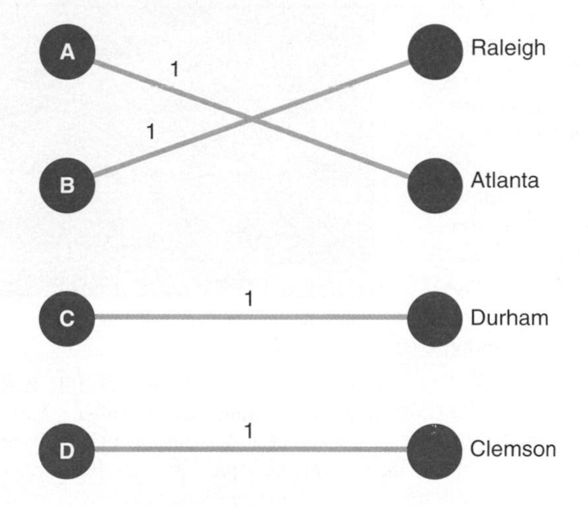

EXHIBIT 6.16

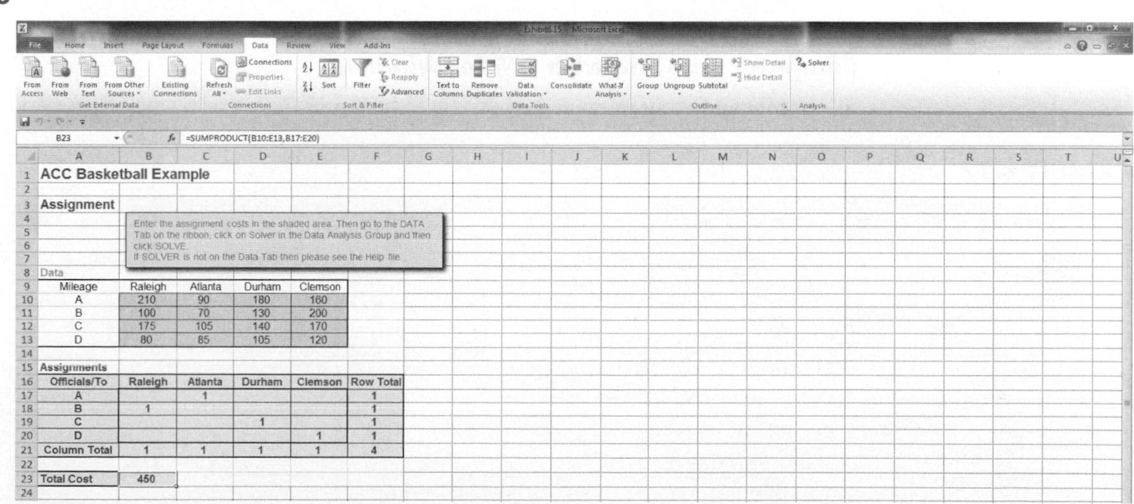

B23 =SUMPRODUCT(B10:E13,B17:E20)

ACC Basketball Example

Assignment

Enter the assignment costs in the shaded area. Then go to the DATA Tab on the ribbon, click on Solver in the Data Analysis Group and then click SOLVE.
If SOLVER is not on the Data Tab then please see the Help file

Data

Mileage	Raleigh	Atlanta	Durham	Clemson
A	210	90	180	160
B	100	70	130	200
C	175	105	140	170
D	80	85	105	120

Assignments

Officials/To	Raleigh	Atlanta	Durham	Clemson	Row Total
A		1			1
B	1				1
C			1		1
D				1	1
Column Total	1	1	1	1	4

Total Cost	450

Management Science Application

Supplying Empty Freight Cars at Union Pacific Railroad

Union Pacific (UP) Railroad, serving 23 states in the western two-thirds of the United States, is the largest railroad in North America, with more than 32,000 miles of track and an annual payroll of $37 billion. It has approximately 105,000 freight cars including automobile cars, boxcars, flat cars, and gondolas. UP delivers empty cars to its customers, who, in turn, use them to transport goods between locations in North America and Mexico. It also transports loaded cars from its customers to their clients, and cars loaded by shippers of other railroads whose clients are in the UP rail network. The assignment of empty cars to meet customer demand is a difficult problem for UP because of a variety of factors, including the fact that empty cars are scattered across the vast UP rail network, often at great distances from customer demand; the supply of available cars is frequently much smaller than demand; and car assignments must be made to customers according to UP's standards for excellent service at minimum cost. The company solved this problem by using a transportation model that minimizes transportation costs (including penalty costs, incentives, unmet demand costs, and holding costs) subject to supply-and-demand constraints. Decision variables include the number of cars of a certain type assigned according to a daily schedule to meet customer demand. UP uses the model to regularly solve its car assignment problem and to investigate new operating policies. The model has resulted in a 35% return on investment by reducing the manpower required in the process to meet demand and by reducing the operating costs of managing the movement of empty cars.

Susan E. Benson/Stock Connection

Source: Based on A. K. Narisetty, J.-P. P. Richard, D. Ramcharan, D. Murphy, G. Minks, and J. Fuller, "An Optimization Model for Empty Freight Car Assignment at Union Pacific Railroad," *Interfaces* 38, no. 2 (March–April 2008): 89–102.

Computer Solution with QM for Windows

The data input for our example is shown in Exhibit 6.17, followed by the solution in Exhibit 6.18.

EXHIBIT 6.17

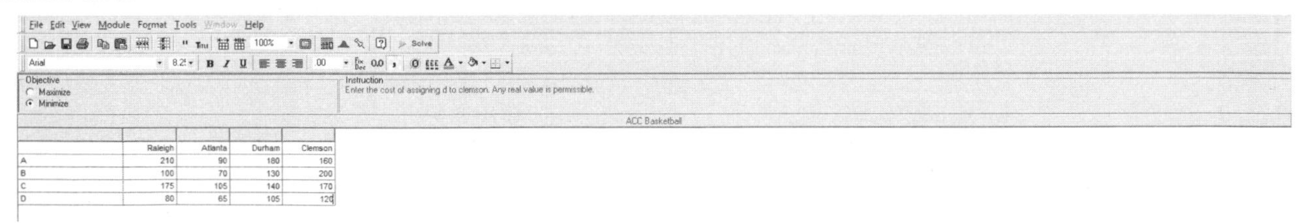

	Raleigh	Atlanta	Durham	Clemson
A	210	90	180	160
B	100	70	130	200
C	175	105	140	170
D	80	65	105	120

EXHIBIT 6.18

Assignments

ACC Basketball Solution

Optimal solution value = $450	Raleigh	Atlanta	Durham	Clemson
A	210	Assign 90	180	160
B	Assign 100	70	130	200
C	175	105	Assign 140	170
D	80	65	105	Assign 120

Management Science Application

Assigning Umpire Crews at Professional Tennis Tournaments

At tennis tournaments, as in all sporting events, the role of the officials who interpret and enforce the rules of play is critical. In professional tennis a set of umpires oversee matches. At a large tennis tournament such as the U.S. Open, up to 18 matches may be played simultaneously, and on each court up to 10 umpires, including a chair umpire and 9 line umpires, may be calling a match. Chief umpires assign umpire crews to matches according to an intricate set of guidelines based on experience, qualifications, and skills. Chair umpire assignments are based primarily on nationality, player histories, and experience, whereas line umpires are assigned according to skill and experience at a particular position (i.e., lines on the court). Also, since coaching is prohibited during matches, all players, male or female, must be accompanied by an umpire if they leave the court. As a result, line umpires also must be assigned to courts to ensure that at least two male and two female umpires are on each court at all times. Also, line umpires cannot stay on a court all day; they must be rotated to allow for breaks and meals. Umpire crews travel from court to court in a session as groups and typically rotate on and off courts every hour, thus requiring multiple crews to be assigned to groups of courts. Tournaments can run anywhere from 1 to 3 weeks, with increasing visibility (thus increasing umpire skill requirements) as the tournament progresses. In 2002 while attending the U.S. Open, one of the authors noticed the difficulty the chief umpires were having trying to reschedule crews following extensive rain delays. He and his coauthors subsequently developed a linear programming model for assigning umpire crews to matches during a tournament. The model is based on a skill rating system that rates umpires from 1 (the best) to 7 according to their historical performance. Each court is assigned a priority, and position (line) assignments on each court are based on a required maximum and minimum skill rating. The overall objective (function) of the model minimizes the weighted deviation of the skill ratings of the umpires assigned to each position on the court from target skill ratings for each court position (i.e., chair and lines). The model includes several constraints. First, each available umpire can be assigned only once during a scheduling session (i.e., time period); second, each crew must have the correct number of umpires, and each umpire must have a skill rating consistent with the court position for the assignment; and third, each crew must have the appropriate minimal gender makeup. In 2004 the United States Tennis Association spent 2,513 hours scheduling umpire crews for all U.S. tournaments. The model developed by the authors reduced this time by 68% to 91%, enabling chief umpires to spend about 75% of the time they previously used for umpire crew scheduling for managing the tournament instead.

AP Photo/Sang Tan

Source: Based on A. Farmer, J. Smith, and L. Miller, "Scheduling Umpire Crews for Professional Tennis Tournaments," *Interfaces* 37, no. 2 (March–April 2007): 187–96.

Summary

In this chapter, three special types of linear programming problems were presented—the transportation problem, the transshipment problem, and the assignment problem. As mentioned at the beginning of the chapter, they are part of a larger class of linear programming problems, known as network flow problems. In the next chapter, we examine several additional examples of network flow problems, including the shortest route problem, the minimal spanning tree problem, and the maximal flow problem. Although they have different objectives, they share the same general characteristics as the transportation and assignment problems—that is, the flow of items from sources to destinations.

Example Problem Solution

This example will demonstrate the procedure for solving a transportation problem.

Problem Statement

A concrete company transports concrete from three plants to three construction sites. The supply capacities of the three plants, the demand requirements at the three sites, and the transportation costs per ton are as follows:

	Construction Site			
Plant	A	B	C	**Supply (tons)**
1	$ 8	$ 5	$ 6	120
2	15	10	12	80
3	3	9	10	80
Demand (tons)	150	70	100	280

Determine the linear programming model formulation for this problem and solve it by using Excel.

Solution

Step 1: The Linear Programming Model Formulation

$$\text{minimize } Z = \$8x_{1A} + 5x_{1B} + 6x_{1C} + 15x_{2A} + 10x_{2B} + 12x_{2C} + 3x_{3A} + 9x_{3B} + 10x_{3C}$$

subject to

$$x_{1A} + x_{1B} + x_{1C} = 120$$
$$x_{2A} + x_{2B} + x_{2C} = 80$$
$$x_{3A} + x_{3B} + x_{3C} = 80$$
$$x_{1A} + x_{2A} + x_{3A} \leq 150$$
$$x_{1B} + x_{2B} + x_{3B} \leq 70$$
$$x_{1C} + x_{2C} + x_{3C} \leq 100$$
$$x_{ij} \geq 0$$

Step 2: Excel Solution

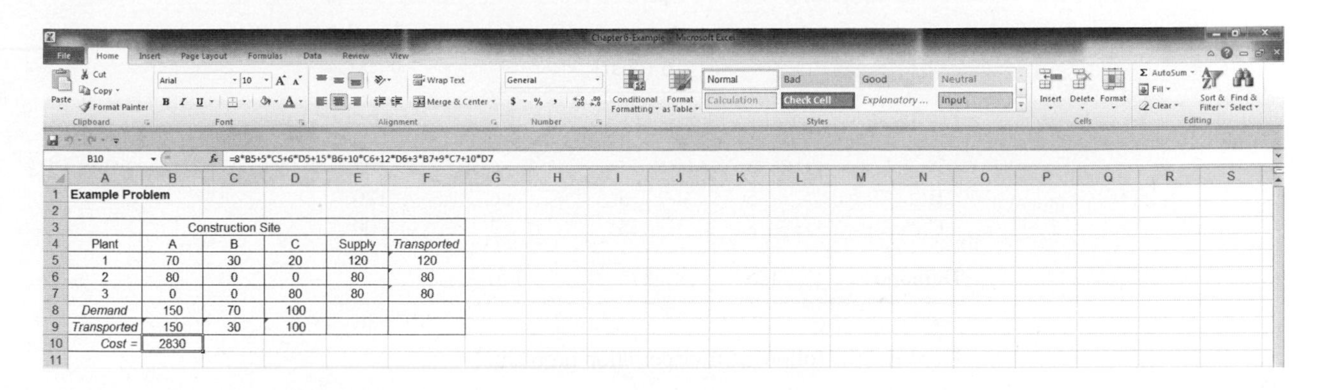

B10 f_x =8*B5+5*C5+6*D5+15*B6+10*C6+12*D6+3*B7+9*C7+10*D7

	A	B	C	D	E	F
1	Example Problem					
2						
3		Construction Site				
4	Plant	A	B	C	Supply	Transported
5	1	70	30	20	120	120
6	2	80	0	0	80	80
7	3	0	0	80	80	80
8	Demand	150	70	100		
9	Transported	150	30	100		
10	Cost =	2830				
11						

Problems

1. Green Valley Mills produces carpet at plants in St. Louis and Richmond. The carpet is then shipped to two outlets, located in Chicago and Atlanta. The cost per ton of shipping carpet from each of the two plants to the two warehouses is as follows:

	To (cost)	
From	Chicago	Atlanta
St. Louis	$40	$65
Richmond	70	30

The plant at St. Louis can supply 250 tons of carpet per week; the plant at Richmond can supply 400 tons per week. The Chicago outlet has a demand of 300 tons per week, and the outlet at Atlanta demands 350 tons per week. The company wants to know the number of tons of carpet to ship from each plant to each outlet in order to minimize the total shipping cost. Solve this transportation problem.

2. A transportation problem involves the following costs, supply, and demand:

	To (cost)				
From	1	2	3	4	**Supply**
1	$500	$750	$300	$450	12
2	650	800	400	600	17
3	400	700	500	550	11
Demand	10	10	10	10	

Solve this problem by using the computer.

3. Given a transportation problem with the following costs, supply, and demand, find the optimal solution by using the computer:

	To (cost)			
From	**1**	**2**	**3**	**Supply**
A	$ 6	$ 7	$ 4	100
B	5	3	6	180
C	8	5	7	200
Demand	135	175	170	

4. Consider the following transportation problem:

	To (cost)			
From	**1**	**2**	**3**	**Supply**
A	$ 6	$ 9	$100	130
B	12	3	5	70
C	4	8	11	100
Demand	80	110	60	

Formulate this problem as a linear programming model and solve it by using the computer.

5. Solve the following linear programming problem:

minimize $Z = 3x_{11} + 12x_{12} + 8x_{13} + 10x_{21} + 5x_{22} + 6x_{23} + 6x_{31} + 7x_{32} + 10x_{33}$
subject to

$$
\begin{aligned}
x_{11} + x_{12} + x_{13} &= 90 \\
x_{21} + x_{22} + x_{23} &= 30 \\
x_{31} + x_{32} + x_{33} &= 100 \\
x_{11} + x_{21} + x_{31} &\leq 70 \\
x_{12} + x_{22} + x_{32} &\leq 110 \\
x_{13} + x_{23} + x_{33} &\leq 80 \\
x_{ij} &\geq 0
\end{aligned}
$$

6. Consider the following transportation problem:

	To (cost)			
From	**1**	**2**	**3**	**Supply**
A	$ 6	$ 9	$ 7	130
B	12	3	5	70
C	4	8	11	100
Demand	80	110	60	

Solve it by using the computer.

7. Steel mills in three cities produce the following amounts of steel:

Location	*Weekly Production (tons)*
A. Bethlehem	150
B. Birmingham	210
C. Gary	320
	680

These mills supply steel to four cities, where manufacturing plants have the following demand:

Location	*Weekly Demand (tons)*
1. Detroit	130
2. St. Louis	70
3. Chicago	180
4. Norfolk	240
	620

Shipping costs per ton of steel are as follows:

	To (cost)			
From	**1**	**2**	**3**	**4**
A	$14	$ 9	$16	$18
B	11	8	7	16
C	16	12	10	22

Because of a truckers' strike, shipments are prohibited from Birmingham to Chicago. Formulate this problem as a linear programming model and solve it by using the computer.

8. In Problem 7, what would be the effect on the optimal solution of a reduction in production capacity at the Gary mill from 320 tons to 290 tons per week?

9. Coal is mined and processed at the following four mines in Kentucky, West Virginia, and Virginia:

Location	*Capacity (tons)*
A. Cabin Creek	90
B. Surry	50
C. Old Fort	80
D. McCoy	60
	280

These mines supply the following amount of coal to utility power plants in three cities:

Plant	*Demand (tons)*
1. Richmond	120
2. Winston-Salem	100
3. Durham	110
	330

The railroad shipping costs (in thousands of dollars) per ton of coal are shown in the following table. Because of railroad construction, shipments are prohibited from Cabin Creek to Richmond:

	To (cost, in $1,000s)		
From	1	2	3
A	$ 7	$10	$ 5
B	12	9	4
C	7	3	11
D	9	5	7

Formulate this problem as a linear programming model and solve it by using the computer.

10. Oranges are grown, picked, and then stored in warehouses in Tampa, Miami, and Fresno. These warehouses supply oranges to markets in New York, Philadelphia, Chicago, and Boston. The following table shows the shipping costs per truckload (in hundreds of dollars), supply, and demand. Because of an agreement between distributors, shipments are prohibited from Miami to Chicago:

	To (cost, in $100s)				
From	New York	Philadelphia	Chicago	Boston	**Supply**
Tampa	$ 9	$ 14	$ 12	$ 17	200
Miami	11	10	6	10	200
Fresno	12	8	15	7	200
Demand	130	170	100	150	

Formulate this problem as a linear programming model and solve it by using the computer.

11. A manufacturing firm produces diesel engines in four cities—Phoenix, Seattle, St. Louis, and Detroit. The company is able to produce the following numbers of engines per month:

Plant	*Production*
1. Phoenix	5
2. Seattle	25
3. St. Louis	20
4. Detroit	25

Three trucking firms purchase the following numbers of engines for their plants in three cities:

Firm	*Demand*
A. Greensboro	10
B. Charlotte	20
C. Louisville	15

The transportation costs per engine (in hundreds of dollars) from sources to destinations are shown in the following table. However, the Charlotte firm will not accept engines made in

Seattle, and the Louisville firm will not accept engines from Detroit; therefore, those routes are prohibited:

	To (cost, in $100s)		
From	A	B	C
1	$ 7	$ 8	$ 5
2	6	10	6
3	10	4	5
4	3	9	11

Formulate this problem as a linear programming model and solve it by using the computer.

12. The Interstate Truck Rental firm has accumulated extra trucks at three of its truck leasing outlets, as shown in the following table:

Leasing Outlet	Extra Trucks
1. Atlanta	70
2. St. Louis	115
3. Greensboro	60
Total	245

The firm also has four outlets with shortages of rental trucks, as follows:

Leasing Outlet	Truck Shortage
A. New Orleans	80
B. Cincinnati	50
C. Louisville	90
D. Pittsburgh	25
Total	245

The firm wants to transfer trucks from those outlets with extras to those with shortages at the minimum total cost. The following costs of transporting these trucks from city to city have been determined:

	To (cost)			
From	A	B	C	D
1	$ 70	$80	$45	$90
2	120	40	30	75
3	110	60	70	80

Solve this problem by using the computer.

13. The Shotz Beer Company has breweries in two cities; the breweries can supply the following numbers of barrels of draft beer to the company's distributors each month:

Brewery	Monthly Supply (bbl)
A. Tampa	3,500
B. St. Louis	5,000
Total	8,500

The distributors, which are spread throughout six states, have the following total monthly demand:

Distributor	Monthly Demand (bbl)
1. Tennessee	1,600
2. Georgia	1,800
3. North Carolina	1,500
4. South Carolina	950
5. Kentucky	1,250
6. Virginia	1,400
Total	8,500

The company must pay the following shipping costs per barrel:

	To (cost)					
From	1	2	3	4	5	6
A	$0.50	$0.35	$0.60	$0.45	$0.80	$0.75
B	0.25	0.65	0.40	0.55	0.20	0.65

Solve this problem by using the computer.

14. In Problem 13, the Shotz Beer Company management negotiated a new shipping contract with a trucking firm between its Tampa brewery and its distributor in Kentucky. This contract reduces the shipping cost per barrel from $0.80 per barrel to $0.65 per barrel. How will this cost change affect the optimal solution?

15. Computers Unlimited sells microcomputers to universities and colleges on the East Coast and ships them from three distribution warehouses. The firm is able to supply the following numbers of microcomputers to the universities by the beginning of the academic year:

Distribution Warehouse	Supply (microcomputers)
1. Richmond	420
2. Atlanta	610
3. Washington, DC	340
Total	1,370

Four universities have ordered microcomputers that must be delivered and installed by the beginning of the academic year:

University	Demand (microcomputers)
A. Tech	520
B. A & M	250
C. State	400
D. Central	380
Total	1,550

The shipping and installation costs per microcomputer from each distributor to each university are as follows:

From	To (cost) A	B	C	D
1	$22	$17	$30	$18
2	15	35	20	25
3	28	21	16	14

Solve this problem by using the computer.

16. In Problem 15, Computers Unlimited wants to better meet demand at the four universities it supplies. It is considering two alternatives: (1) expand its warehouse at Richmond to a capacity of 600, at a cost equivalent to an additional $6 in handling and shipping per unit; or (2) purchase a new warehouse in Charlotte that can supply 300 units with shipping costs of $19 to Tech, $26 to A & M, $22 to State, and $16 to Central. Which alternative should management select, based solely on transportation costs (i.e., no capital costs)?

17. Computers Unlimited in Problem 15 has determined that when it is unable to meet the demand for microcomputers at the universities it supplies, the universities tend to purchase microcomputers elsewhere in the future. Thus, the firm has estimated a shortage cost for each microcomputer demanded but not supplied that reflects the loss of future sales and goodwill. These costs for each university are as follows:

University	Cost/Microcomputer
A. Tech	$40
B. A & M	65
C. State	25
D. Central	50

Solve Problem 15 with these shortage costs included. Compute the total transportation cost and the total shortage cost.

18. A severe winter ice storm has swept across North Carolina and Virginia, followed by over a foot of snow and frigid, single-digit temperatures. These weather conditions have resulted in numerous downed power lines and power outages, causing dangerous conditions for much of the population. Local utility companies have been overwhelmed and have requested assistance from unaffected utility companies across the Southeast. The following table shows the number of utility trucks with crews available from five different companies in Georgia, South Carolina, and Florida; the demand for crews in seven different areas that local companies cannot get to; and the weekly cost (in thousands of dollars) of a crew going to a specific area (based on the visiting company's normal charges, the distance the crew has to come, and living expenses in an area):

	Area (cost, in $1,000s)							
Crew	**NC-E**	**NC-SW**	**NC-P**	**NC-W**	**VA-SW**	**VA-C**	**VA-T**	**Crews Available**
GA-1	15.2	14.3	13.9	13.5	14.7	16.5	18.7	12
GA-2	12.8	11.3	10.6	12.0	12.7	13.2	15.6	10
SC-1	12.4	10.8	9.4	11.3	13.1	12.8	14.5	14
FL-1	18.2	19.4	18.2	17.9	20.5	20.7	22.7	15
FL-2	19.3	20.2	19.5	20.2	21.2	21.3	23.5	12
Crews Needed	9	7	6	8	10	9	7	

Determine the number of crews that should be sent from each utility to each affected area to minimize total costs.

19. A large manufacturing company is closing three of its existing plants and intends to transfer some of its more skilled employees to three plants that will remain open. The number of employees available for transfer from each closing plant is as follows:

Closing Plant	Transferable Employees
1	60
2	105
3	70
Total	235

The following number of employees can be accommodated at the three plants remaining open:

Open Plants	Employees Demanded
A	45
B	90
C	35
Total	170

Each transferred employee will increase product output per day at each plant, as shown in the following table:

	To (employees)		
From	A	B	C
1	5	8	6
2	10	9	12
3	7	6	8

The company wants to transfer employees to ensure the maximum increase in product output. Solve this problem by using the computer.

20. The Sav-Us Rental Car Agency has six lots in Nashville, and it wants to have a certain number of cars available at each lot at the beginning of each day for local rental. The agency would like a model it could quickly solve at the end of each day that would tell it how to redistribute the cars among the six lots in the minimum total time. The times required to travel between the six lots are as follows:

	To (min.)					
From	1	2	3	4	5	6
1	—	12	17	18	10	20
2	14	—	10	19	16	15
3	14	10	—	12	8	9
4	8	16	14	—	12	15
5	11	21	16	18	—	10
6	24	12	9	17	15	—

The agency would like the following number of cars at each lot at the end of the day. Also shown is the number of available cars at each lot at the end of a particular day. Determine the optimal reallocation of rental cars by using any initial solution approach and any solution method:

	Lot (cars)					
Cars	1	2	3	4	5	6
Available	37	20	14	26	40	28
Desired	30	25	20	40	30	20

21. Bayville has built a new elementary school, increasing the town's total to four schools—Addison, Beeks, Canfield, and Daley. Each has a capacity of 400 students. The school board wants to assign children to schools so that their travel time by bus is as short as possible. The school board has partitioned the town into five districts conforming to population density—north, south, east, west, and central. The average bus travel time from each district to each school is shown as follows:

	Travel Time (min.)				
District	Addison	Beeks	Canfield	Daley	**Student Population**
North	12	23	35	17	250
South	26	15	21	27	340
East	18	20	22	31	310
West	29	24	35	10	210
Central	15	10	23	16	290

Determine the number of children that should be assigned from each district to each school to minimize total student travel time.

22. In Problem 21, the school board determined that it does not want any of the schools to be overly crowded compared with the other schools. It would like to assign students from each district to each school so that enrollments are evenly balanced between the four schools. However, the school board is concerned that this might significantly increase travel time. Determine the number of students to be assigned from each district to each school such that school enrollments are evenly balanced. Does this new solution appear to significantly increase travel time per student?

23. The Easy Time Grocery chain operates in major metropolitan areas on the East Coast. The stores have a "no-frills" approach, with low overhead and high volume. They generally buy their stock in volume at low prices. However, in some cases they actually buy stock at stores in other areas and ship it in. They can do this because of high prices in the cities they operate in compared with costs in other locations. One example is baby food. Easy Time purchases baby food at stores in Albany, Binghamton, Claremont, Dover, and Edison and then trucks it to six stores in and around New York City. The stores in the outlying areas know what Easy Time is up to, so they limit the number of cases of baby food Easy Time can purchase. The following table shows the profit Easy Time makes per case of baby food, based on where the chain purchases it and at

which store it is sold, plus the available baby food per week at purchase locations and the shelf space available at each Easy Time store per week:

Purchase Location	Easy Time Store (profit/case) 1	2	3	4	5	6	Supply
Albany	$ 9	$ 8	$11	$12	$7	$ 8	26
Binghamton	10	10	8	6	9	7	40
Claremont	8	6	6	5	7	4	20
Dover	4	6	9	5	8	10	40
Edison	12	10	8	9	6	7	45
Demand	25	15	30	18	27	35	

Determine where Easy Time should purchase baby food and how the food should be distributed to maximize profit.

24. Suppose that in Problem 23 Easy Time can purchase all the baby food it needs from a New York City distributor at a price that will result in a profit of $9 per case at stores 1, 3, and 4; $8 per case at stores 2 and 6; and $7 per case at store 5. Should Easy Time purchase all, none, or some of its baby food from the distributor rather than purchase it at other stores and truck it in?

25. In Problem 23, if Easy Time could arrange to purchase more baby food from one of the outlying locations, which should it be, how many additional cases could be purchased, and how much would this increase profit?

26. The Roadnet Transport Company expanded its shipping capacity by purchasing 90 trailer trucks from a competitor that went bankrupt. The company subsequently located 30 of the purchased trucks at each of its shipping warehouses in Charlotte, Memphis, and Louisville. The company makes shipments from each of these warehouses to terminals in St. Louis, Atlanta, and New York. Each truck is capable of making one shipment per week. The terminal managers have indicated their capacity of extra shipments. The manager at St. Louis can accommodate 40 additional trucks per week, the manager at Atlanta can accommodate 60 additional trucks, and the manager at New York can accommodate 50 additional trucks. The company makes the following profit per truckload shipment from each warehouse to each terminal. The profits differ as a result of differences in products shipped, shipping costs, and transport rates:

Warehouse	Terminal (profit) St. Louis	Atlanta	New York
Charlotte	$1,800	$2,100	$1,600
Memphis	1,000	700	900
Louisville	1,400	800	2,200

Determine how many trucks to assign to each route (i.e., warehouse to terminal) in order to maximize profit.

27. During the Gulf War in Iraq, large amounts of military matériel and supplies had to be shipped daily from supply depots in the United States to bases in the Middle East. The critical factor in the movement of these supplies was speed. The following table shows the number of planeloads of supplies available each day from each of six supply depots and the number of daily loads demanded at each

of five bases. (Each planeload is approximately equal in tonnage.) Also included are the transport hours per plane, including loading and fueling, actual flight time, and unloading and refueling:

Supply Depot	Military Base A	B	C	D	E	Supply
1	36	40	32	43	29	7
2	28	27	29	40	38	10
3	34	35	41	29	31	8
4	41	42	35	27	36	8
5	25	28	40	34	38	9
6	31	30	43	38	40	6
Demand	9	6	12	8	10	

Determine the optimal daily flight schedule that will minimize total transport time.

28. PM Computer Services produces personal computers from component parts it buys on the open market. The company can produce a maximum of 300 personal computers per month. PM wants to determine its production schedule for the first 6 months of the new year. The cost to produce a personal computer in January will be $1,200. However, PM knows the cost of component parts will decline each month so that the overall cost to produce a PC will be 5% less each month. The cost of holding a computer in inventory is $15 per unit per month. Following is the demand for the company's computers each month:

Month	Demand	Month	Demand
January	180	April	210
February	260	May	400
March	340	June	320

Determine a production schedule for PM that will minimize total cost.

29. In Problem 28, suppose that the demand for personal computers increased each month, as follows:

Month	Demand	Month	Demand
January	410	April	620
February	320	May	430
March	500	June	380

In addition to the regular production capacity of 300 units per month, PM Computer Services can also produce an additional 200 computers per month by using overtime. Overtime production adds 20% to the cost of a personal computer.

Determine a production schedule for PM that will minimize total cost.

30. National Foods Company has five plants where it processes and packages fruits and vegetables. It has suppliers in six cities in California, Texas, Alabama, and Florida. The company owns and operates its own trucking system for transporting fruits and vegetables from its suppliers to its plants. However, it is now considering outsourcing all its shipping to outside trucking firms and getting rid of its own trucks. It currently spends $245,000 per month to operate its own trucking system. It has determined

monthly shipping costs (in thousands of dollars per ton) of using outside shippers from each of its suppliers to each of its plants, as shown in the following table:

	Processing Plants ($1,000s/ton)					
Suppliers	Denver	St. Paul	Louisville	Akron	Topeka	**Supply (tons)**
Sacramento	$3.7	$4.6	$4.9	$5.5	$4.3	18
Bakersfield	3.4	5.1	4.4	5.9	5.2	15
San Antonio	3.3	4.1	3.7	2.9	2.6	10
Montgomery	1.9	4.2	2.7	5.4	3.9	12
Jacksonville	6.1	5.1	3.8	2.5	4.1	20
Ocala	6.6	4.8	3.5	3.6	4.5	15
Demand (tons)	20	15	15	15	20	90

Should National Foods continue to operate its own shipping network or sell its trucks and outsource its shipping to independent trucking firms?

31. In Problem 30, National Foods would like to know what the effect would be on the optimal solution and the company's decision regarding its shipping if it negotiates with its suppliers in Sacramento, Jacksonville, and Ocala to increase their capacity to 25 tons per month. What would be the effect of negotiating instead with its suppliers at San Antonio and Montgomery to increase their capacity to 25 tons each?

32. Orient Express is a global distribution company that transports its clients' products to customers in Hong Kong, Singapore, and Taipei. All the products Orient Express ships are stored at three distribution centers—one in Los Angeles, one in Savannah, and one in Galveston. For the coming month the company has 450 containers of computer components available at the Los Angeles center, 600 containers available at Savannah, and 350 containers available at Galveston. The company has orders for 600 containers from Hong Kong, 500 containers from Singapore, and 500 containers from Taipei. The shipping costs per container from each U.S. port to each of the overseas ports are shown in the following table:

U.S. Distribution Center	Overseas Port (cost/container)		
	Hong Kong	Singapore	Taipei
Los Angeles	$300	$210	$340
Savannah	490	520	610
Galveston	360	320	500

Orient Express, as the overseas broker for its U.S. customers, is responsible for unfulfilled orders, and it incurs stiff penalty costs from overseas customers if it does not meet an order. The Hong Kong customers charge a penalty cost of $800 per container for unfulfilled demand, Singapore customers charge a penalty cost of $920 per container, and Taipei customers charge $1,100 per container. Formulate and solve a transportation model to determine the shipments from each U.S. distribution center to each overseas port that will minimize shipping costs. Indicate what portion of the total cost is a result of penalties.

33. Binford Tools manufactures garden tools. It uses inventory, overtime, and subcontracting to absorb demand fluctuations. Expected demand, regular and overtime production capacity, and subcontracting capacity are provided in the following table for the next four quarters for its basic line of steel garden tools:

Quarter	Demand	Regular Capacity	Overtime Capacity	Subcontracting Capacity
1	9,000	9,000	1,000	3,000
2	12,000	10,000	1,500	3,000
3	16,000	12,000	2,000	3,000
4	19,000	12,000	2,000	3,000

The regular production cost per unit is $20, the overtime cost per unit is $25, the cost to subcontract a unit is $27, and the inventory carrying cost is $2 per unit. The company has 300 units in inventory at the beginning of the year.

Determine the optimal production schedule for the four quarters to minimize total costs.

34. The National Western Railroad's rail network covers most of the U.S. West and Midwest. On a daily basis it sends empty freight cars from various locations in its rail network to its customers for their use. Sometimes there are not enough freight cars to meet customer demand. The transportation costs for shipping empty freight cars, shown as follows, are directly related to distance traveled and the number of rail centers that must handle the car movement.

	Customer Location									
Freight Car Location	A. Milwaukee	B. Omaha	C. Topeka	D. Tucson	E. Denver	F. Wichita	G. Minneapolis	H. Memphis	I. Kansas City	Supply
1. Portland	27	23	23	26	21	29	40	45	23	1,100
2. Fresno	31	26	25	22	20	34	47	43	26	720
3. Long Beach	38	31	32	18	24	27	51	48	34	1,450
4. Salt Lake City	28	18	17	24	9	20	32	35	19	980
5. El Paso	41	27	24	11	18	22	46	30	25	650
6. Houston	38	24	22	16	27	25	41	28	23	1,025
7. St. Louis	15	14	10	27	23	12	19	10	9	1,330
8. Chicago	12	13	15	31	26	17	14	15	14	1,275
Demand	974	1,225	1,690	710	1,261	663	301	479	1,227	

Determine the number of empty freight cars that should be sent from each rail network location to customers to meet demand at the minimum total cost.

35. Al, Barbara, Carol, and Dave have joined together to purchase two season tickets to the Giants' home football games. Because there are eight home games, each person will get tickets to two games. Each person has ranked the games they prefer from 1 to 8, with 1 being most preferred and 8 least preferred, as follows:

	Person			
Game	Al	Barbara	Carol	Dave
1. Cowboys	1	2	1	4
2. Redskins	3	4	4	1
3. Cardinals	7	8	8	7
4. Eagles	2	7	5	3
5. Bengals	5	6	6	8
6. Packers	6	3	2	5
7. Saints	8	5	7	6
8. Jets	4	1	3	2

Determine the two games each person should get tickets for that will result in the groups' greatest degree of satisfaction. Do you think the participants will think your allocation is fair?

36. World Foods, Inc., imports food products such as meats, cheese, and pastries to the United States from warehouses at ports in Hamburg, Marseilles, and Liverpool. Ships from these ports deliver the products to Norfolk, New York, and Savannah, where they are stored in company warehouses before being shipped to distribution centers in Dallas, St. Louis, and Chicago. The products are then distributed to specialty food stores and sold through catalogs. The shipping costs ($/1,000 lb.) from the European ports to the U.S. cities and the available supplies (1,000 lb.) at the European ports are provided in the following table:

	U.S. City			
European Port	**4. Norfolk**	**5. New York**	**6. Savannah**	**Supply**
1. Hamburg	$420	$390	$610	55
2. Marseilles	510	590	470	78
3. Liverpool	450	360	480	37

The transportation costs ($/1,000 lb.) from each U.S. city of the three distribution centers and the demands (1,000 lb.) at the distribution centers are as follows:

	Distribution Center		
Warehouse	**7. Dallas**	**8. St. Louis**	**9. Chicago**
4. Norfolk	$ 75	$ 63	$81
5. New York	125	110	95
6. Savannah	68	82	95
Demand	60	45	50

Determine the optimal shipments between the European ports and the warehouses and the distribution centers to minimize total transportation costs.

37. A sports apparel company has received an order for a college basketball team's national championship T-shirt. The company can purchase the T-shirts from textile factories in Mexico, Puerto Rico, and Haiti. The shirts are shipped from the factories to companies in the United States that silk-screen the shirts before they are shipped to distribution centers. Following are the production and transportation costs ($/shirt) from the T-shirt factories to the silk-screen companies to the distribution centers, plus the supply of T-shirts at the factories and demand for the shirts at the distribution centers:

	Silk-Screen Company			
T-shirt Factory	**4. Miami**	**5. Atlanta**	**6. Houston**	**Supply (1,000s)**
1. Mexico	$4	$6	$3	18
2. Puerto Rico	3	5	5	15
3. Haiti	2	4	4	23

	Distribution Center		
Silk-Screen Company	**7. New York**	**8. St. Louis**	**9. Los Angeles**
4. Miami	$ 5	$ 7	$ 9
5. Atlanta	7	6	10
6. Houston	8	6	8
Demand (1,000s)	20	12	20

Determine the optimal shipments to minimize total production and transportation costs for the apparel company.

38. Walsh's Fruit Company contracts with growers in Ohio, Pennsylvania, and New York to purchase grapes. The grapes are processed into juice at the farms and stored in refrigerated vats. Then the juice is shipped to two plants, where it is processed into bottled grape juice and frozen concentrate. The juice and concentrate are then transported to three food warehouses/distribution centers. The transportation costs per ton from the farms to the plants and from the plants to the distributors, and the supply at the farms and demand at the distribution centers are summarized in the following tables:

	Plant		
Farm	**4. Indiana**	**5. Georgia**	**Supply (1,000 tons)**
1. Ohio	$16	21	72
2. Pennsylvania	18	16	105
3. New York	22	25	83

	Distribution Center		
Plant	**6. Virginia**	**7. Kentucky**	**8. Louisiana**
4. Indiana	$23	$15	$29
5. Georgia	20	17	24
Demand (1,000 tons)	90	80	120

a. Determine the optimal shipments from farms to plants to distribution centers to minimize total transportation costs.
b. What would be the effect on the solution if the capacity at each plant were 140,000 tons?

39. A national catalog and Internet retailer has three warehouses and three major distribution centers located around the country. Normally, items are shipped directly from the warehouses to the distribution centers; however, each of the distribution centers can also be used as an intermediate transshipment point. The transportation costs ($/unit) between warehouses and distribution centers, the supply at the warehouses (100 units), and the demand at the distribution centers (100 units) for a specific week are shown in the following table:

	Distribution Center			
Warehouse	**A**	**B**	**C**	**Supply**
1	$12	$ 11	$ 7	70
2	8	6	14	80
3	9	10	12	50
Demand	60	100	40	

The transportation costs ($/unit) between the distribution centers are

	Distribution Center		
Distribution Center	**A**	**B**	**C**
A	$—	$8	$3
B	1	—	2
C	7	2	—

Determine the optimal shipments between warehouses and distribution centers to minimize total transportation costs.

40. Horizon Computers manufactures laptops in Germany, Belgium, and Italy. Because of high tariffs between international trade groups, it is sometimes cheaper to ship partially completed laptops to factories in Puerto Rico, Mexico, and Panama and have them completed before final shipment to U.S. distributors in Texas, Virginia, and Ohio. The cost ($/unit) of the completed laptops plus tariffs and shipment costs from the European plants directly to the United States and supply and demand are as follows:

	U.S. Distributor			
European Plant	7. Texas	8. Virginia	9. Ohio	**Supply (1,000s)**
1. Germany	$2,600	$1,900	$2,300	5.2
2. Belgium	2,200	2,100	2,600	6.3
3. Italy	1,800	2,200	2,500	4.5
Demand (1,000s)	2.1	3.7	7.8	

Alternatively, the unit costs of shipping partially completed laptops to plants for finishing before sending them to the United States are as follows:

	Factory		
European Plant	4. Puerto Rico	5. Mexico	6. Panama
1. Germany	$1,400	$1,200	$1,100
2. Belgium	1,600	1,100	900
3. Italy	1,500	1,400	1,200

	U.S. Distributor		
Factory	7. Texas	8. Virginia	9. Ohio
4. Puerto Rico	$800	$700	$ 900
5. Mexico	600	800	1,100
6. Panama	900	700	1,200

Determine the optimal shipments of laptops that will meet demand at the U.S. distributors at the minimum total cost.

41. The Midlands Field Produce Company contracts with potato farmers in Colorado, Minnesota, North Dakota, and Wisconsin for monthly potato shipments. Midlands picks up the potatoes at the farms and ships mostly by truck (and sometimes by rail) to its sorting and distribution centers in Ohio, Missouri, and Iowa. At these centers the potatoes are cleaned, rejects are discarded, and the potatoes are sorted according to size and quality. They are then shipped to combination plants and distribution centers in Virginia, Pennsylvania, Georgia, and Texas, where the company produces a variety of potato products and distributes bags of potatoes to stores. Exceptions are the Ohio distribution center, which will accept potatoes only from farms in Minnesota, North Dakota, and Wisconsin, and the Texas plant, which won't accept shipments from Ohio because of disagreements over delivery schedules and quality issues. Following are summaries of the shipping costs from the farms to the distribution centers and the processing and shipping costs from the distribution centers to the plants, as well as the available monthly supply at each farm, the processing capacity at the distribution centers, and the final demand at the plants (in bushels):

Farm	Distribution Center ($/bushel) 5. Ohio	6. Missouri	7. Iowa	Supply (bushels)
1. Colorado	$—	$1.09	$1.26	1,600
2. Minnesota	0.89	1.32	1.17	1,100
3. North Dakota	0.78	1.22	1.36	1,400
4. Wisconsin	1.19	1.25	1.42	1,900
Processing Capacity (bushels)	1,800	2,200	1,600	

Distribution Center	Plant ($/bushel) 8. Virginia	9. Pennsylvania	10. Georgia	11. Texas
5. Ohio	$4.56	$3.98	$4.94	$—
6. Missouri	3.43	5.74	4.65	5.01
7. Iowa	5.39	6.35	5.70	4.87
Demand (bushels)	1,200	900	1,100	1,500

Formulate and solve a linear programming model to determine the optimal monthly shipments from the farms to the distribution centers and from the distribution centers to the plants to minimize total shipping and processing costs.

42. KanTech Corporation is a global distributor of electrical parts and components. Its customers are electronics companies in the United States, including computer manufacturers and audio/visual product manufacturers. The company contracts to purchase components and parts from manufacturers in Russia, Eastern and Western Europe, and the Mediterranean, and it has them delivered to warehouses in three European ports, Gdansk, Hamburg, and Lisbon. The various components and parts are loaded into containers based on demand from U.S. customers. Each port has a limited fixed number of containers available each month. The containers are then shipped overseas by container ships to the ports of Norfolk, Jacksonville, New Orleans, and Galveston. From these seaports, the containers are typically coupled with trucks and hauled to inland ports in Front Royal (Virginia), Kansas City, and Dallas. There are a fixed number of freight haulers available at each port each month. These inland ports are sometimes called "freight villages," or intermodal junctions, where the containers are collected and transferred from one transport mode to another (i.e., from truck to rail or vice versa). From the inland ports, the containers are transported to KanTech's distribution centers in Tucson, Pittsburgh, Denver, Nashville, and Cleveland. Following are the handling and shipping costs ($/container) between each of the embarkation and destination points along this overseas supply chain and the available containers at each port:

European Port	U.S. Port 4. Norfolk	5. Jacksonville	6. New Orleans	7. Galveston	Available Containers
1. Gdansk	$1,725	$1,800	$2,345	$2,700	125
2. Hamburg	1,825	1,750	1,945	2,320	210
3. Lisbon	2,060	2,175	2,050	2,475	160

	Inland Port			
U.S. Port	8. Dallas	9. Kansas City	10. Front Royal	Intermodal Capacity (containers)
4. Norfolk	$825	$545	$ 320	85
5. Jacksonville	750	675	450	110
6. New Orleans	325	605	690	100
7. Galveston	270	510	1,050	130
Intermodal Capacity (containers)	170	240	140	

	Distribution Center				
Inland Port	11. Tucson	12. Denver	13. Pittsburgh	14. Nashville	15. Cleveland
8. Dallas	$ 450	$830	$ 565	$420	$960
9. Kansas City	880	520	450	380	660
10. Front Royal	1,350	390	1,200	450	310
Demand	85	60	105	50	120

Formulate and solve a linear programming model to determine the optimal shipments from each point of embarkation to each destination along this supply chain that will result in the minimum total shipping cost.

43. In Problem 42, KanTech Corporation is just as concerned that its U.S. distributors receive shipments in the minimum amount of time as they are about minimizing their shipping costs. Suppose that each U.S. distributor receives one major container shipment each month. Following are summaries of the shipping times (in days) between each of the embarkation and destination points along KanTech's overseas supply chain. These times encompass not only travel time but also processing, loading, and unloading times at each port:

	U.S. Port			
European Port	4. Norfolk	5. Jacksonville	6. New Orleans	7. Galveston
1. Gdansk	22	24	27	30
2. Hamburg	17	20	23	26
3. Lisbon	25	21	24	26

	Inland Port		
U.S. Port	8. Dallas	9. Kansas City	10. Front Royal
4. Norfolk	10	8	6
5. Jacksonville	12	9	8
6. New Orleans	8	7	10
7. Galveston	12	6	8

	Distribution Center				
Inland Port	11. Tucson	12. Denver	13. Pittsburgh	14. Nashville	15. Cleveland
8. Dallas	5	6	5	7	8
9. Kansas City	6	4	4	5	7
10. Front Royal	10	5	7	4	6

a. Formulate and solve a linear programming model to determine the optimal shipping route for each distribution center along this supply chain that will result in the minimum total shipping time. Determine the shipping route and time for each U.S. distributor.
b. Suppose the European ports can accommodate only three shipments each. How will this affect the solution in part (a)?

44. Blue Mountain Coffee Company produces various blends of Free Trade, organic specialty coffees that it sells to wholesale customers. The company imports 25 million pounds of coffee beans annually from coffee plantations in Brazil, Indonesia, Kenya, Colombia, Côte d'Ivoire, and Guatemala. The beans are shipped from these countries to U.S. ports in Galveston, New Orleans, Savannah, and Jacksonville, where they are loaded onto container trucks and shipped to the company's plant in Vermont. The shipping costs (in dollars per million pounds) from the countries to the U.S. ports, the amount of beans (in millions of pounds) contracted from the growers in each country, and the port capacities are shown in the following table:

	U.S. Port				
Grower Country	Galveston	New Orleans	Savannah	Jacksonville	**Supply**
1. Brazil	$30,000	$36,000	$29,000	$41,000	6.6
2. Colombia	19,000	23,000	28,000	35,000	3.2
3. Indonesia	53,000	47,000	45,000	39,000	4.1
4. Kenya	45,000	54,000	48,000	41,000	5.8
5. Côte d' Ivoire	35,000	33,000	27,000	29,000	1.7
6. Guatemala	14,000	17,000	24,000	28,000	3.6
Capacity	7.8	9	8.1	6.7	

The shipping costs from each port to the plant in Vermont are shown in the following table:

U.S. Port	Vermont
7. Galveston	$61,000
8. New Orleans	55,000
9. Savannah	38,000
10. Jacksonville	43,000

Determine the optimal shipments from the grower countries to the plant in Vermont that will minimize shipping costs.

45. The Pinnacle Company is a U.S.–based manufacturer of furniture and appliances that offshored all of its actual manufacturing operations to Asia about a decade ago. It then set up distribution centers at various locations on the East Coast, near ports where its items were imported on container ships. In many cases, Pinnacle's appliances and furniture arrive partially assembled, and the company completes the assembly at its distribution centers before sending the finished products to retailers. For example, appliance motors, electric controls, housings, and furniture pieces might arrive from different Asian manufacturers in separate containers. Recently Pinnacle began exporting its products to various locations in Europe, and demand steadily increased. As a result, the company determined that shipping items to the United States, assembling the products, and then turning around and shipping them to Europe was inefficient and not cost effective. The company now plans to open three new distribution centers near ports in Europe, and it will ship its items from Asian ports to distribution centers at the European ports, offload some of the items for final product assembly, and then ship the partially filled containers on to the U.S. distribution centers. The following table shows the seven possible locations near container ports in Europe, and their container capacity that Pinnacle has identified to construct its proposed three

distribution centers; the container shipments from each of its Asian ports; and the container shipping cost from each of the Asian ports to each possible distribution center location.

	Proposed Distribution Center							
Asian Port	**F. Rotterdam**	**G. Hamburg**	**H. Antwerp**	**I. Bremen**	**J. Valencia**	**K. Lisbon**	**L. Le Havre**	
Center Cost	**$16,725,000**	**$19,351,000**	**$13,766,000**	**$15,463,000**	**$12,542,000**	**$13,811,000**	**$22,365,000**	**Container Shipments**
A. Hong Kong	$3,466	$3,560	$3,125	$3,345	$3,060	$3,120	$3,658	235
B. Shanghai	$3,190	$3,020	$3,278	$3,269	$2,987	$2,864	$3,725	170
C. Busan	$2,815	$2,700	$2,890	$3,005	$2,465	$2,321	$3,145	165
D. Mumbai	$2,412	$2,560	$2,515	$2,875	$2,325	$2,133	$2,758	325
E. Kaoshiung	$2,600	$2,800	$2,735	$2,755	$2,473	$2,410	$2,925	405
Capacity	565	485	520	490	310	410	605	

The following table shows the demand from each of the U.S. ports and the cost for container shipments from each of the possible distribution center locations to each of the U.S. ports:

	U.S. Port			
Proposed Distribution Centers	**M. New York**	**N. Savannah**	**O. Miami**	**P. New Orleans**
F. Rotterdam	$2,045	$1,875	$1,675	$2,320
G. Hamburg	$2,875	$2,130	$1,856	$2,415
H. Antwerp	$2,415	$2,056	$1,956	$2,228
I. Bremen	$2,225	$1,875	$2,075	$2,652
J. Valencia	$1,865	$1,725	$1,548	$1,815
K. Lisbon	$1,750	$1,555	$1,420	$1,475
L. Le Havre	$3,056	$2,280	$2,065	$2,425
Demand	440	305	190	365

Formulate and solve a linear programming model to determine which three distribution center locations in Europe Pinnacle should select, and the shipments from each of the Asian ports to these selected distribution centers and from the European distribution centers to the U.S. ports.

46. In Problem 45, reformulate the model so that Pinnacle minimizes its shipping costs while selecting its three new distribution centers within a budget of $45 million. What is the difference in this solution, if any?

47. Solve the following linear programming problem:

$$\begin{aligned}\text{minimize } Z = {} & 18x_{11} + 30x_{12} + 20x_{13} + 18x_{14} + 25x_{21} + 27x_{22} + 22x_{23} \\ & + 16x_{24} + 30x_{31} + 26x_{32} + 19x_{33} + 32x_{34} + 40x_{41} + 36x_{42} \\ & + 27x_{43} + 29x_{44} + 30x_{51} + 26x_{52} + 18x_{53} + 24x_{54}\end{aligned}$$

subject to

$$\begin{aligned} x_{11} + x_{12} + x_{13} + x_{14} &\le 1 \\ x_{21} + x_{22} + x_{23} + x_{24} &\le 1 \\ x_{31} + x_{32} + x_{33} + x_{34} &\le 1 \\ x_{41} + x_{42} + x_{43} + x_{44} &\le 1 \\ x_{51} + x_{52} + x_{53} + x_{54} &\le 1 \\ x_{11} + x_{21} + x_{31} + x_{41} + x_{51} &= 1 \\ x_{12} + x_{22} + x_{32} + x_{42} + x_{52} &= 1 \\ x_{13} + x_{23} + x_{33} + x_{43} + x_{53} &= 1 \\ x_{14} + x_{24} + x_{34} + x_{44} + x_{54} &= 1 \\ x_{ij} &\ge 0 \end{aligned}$$

48. A plant has four operators to be assigned to four machines. The time (minutes) required by each worker to produce a product on each machine is shown in the following table:

	Machine (min.)			
Operator	A	B	C	D
1	10	12	9	11
2	5	10	7	8
3	12	14	13	11
4	8	15	11	9

Determine the optimal assignment and compute total minimum time.

49. A shop has four machinists to be assigned to four machines. The hourly cost of having each machine operated by each machinist is as follows:

	Machine (cost/hr.)			
Machinist	A	B	C	D
1	$12	$11	$ 8	$14
2	10	9	10	8
3	14	8	7	11
4	6	8	10	9

However, because he does not have enough experience, machinist 3 cannot operate machine B.

a. Determine the optimal assignment and compute total minimum cost.
b. Formulate this problem as a general linear programming model.

50. The Omega pharmaceutical firm has five salespersons, whom the firm wants to assign to five sales regions. Given their various previous contacts, the salespersons are able to cover the regions in different amounts of time. The amount of time (days) required by each salesperson to cover each city is shown in the following table:

	Region (days)				
Salesperson	A	B	C	D	E
1	17	10	15	16	20
2	12	9	16	9	14
3	11	16	14	15	12
4	14	10	10	18	17
5	13	12	9	15	11

Which salesperson should be assigned to each region to minimize total time? Identify the optimal assignments and compute total minimum time.

51. The Bunker Manufacturing firm has five employees and six machines and wants to assign the employees to the machines to minimize cost. A cost table showing the cost incurred by each employee on each machine follows:

	Machine					
Employee	A	B	C	D	E	F
1	$12	$ 7	$20	$14	$ 8	$10
2	10	14	13	20	9	11
3	5	3	6	9	7	10
4	9	11	7	16	9	10
5	10	6	14	8	10	12

Because of union rules regarding departmental transfers, employee 3 cannot be assigned to machine E, and employee 4 cannot be assigned to machine B. Solve this problem, indicate the optimal assignment, and compute total minimum cost.

52. Given the following cost table for an assignment problem, determine the optimal assignment and compute total minimum cost:

	Machine			
Operator	A	B	C	D
1	$10	$2	$ 8	$ 6
2	9	5	11	9
3	12	7	14	14
4	3	1	4	2

53. An electronics firm produces electronic components, which it supplies to various electrical manufacturers. Quality control records indicate that different employees produce different numbers of defective items. The average number of defects produced by each employee for each of six components is given in the following table:

	Component					
Employee	A	B	C	D	E	F
1	30	24	16	26	30	22
2	22	28	14	30	20	13
3	18	16	25	14	12	22
4	14	22	18	23	21	30
5	25	18	14	16	16	28
6	32	14	10	14	18	20

Determine the optimal assignment that will minimize the total average number of defects produced by the firm per month.

54. A dispatcher for Citywide Taxi Company has six taxicabs at different locations and five customers who have called for service. The mileage from each taxi's present location to each customer is shown in the following table:

	Customer				
Cab	1	2	3	4	5
A	7	2	4	10	7
B	5	1	5	6	6
C	8	7	6	5	5
D	2	5	2	4	5
E	3	3	5	8	4
F	6	2	4	3	4

Determine the optimal assignment(s) that will minimize the total mileage traveled.

55. The Southeastern Conference has nine basketball officials who must be assigned to three conference games, three to each game. The conference office wants to assign the officials so that the total distance they travel will be minimized. The distance (in miles) each official would travel to each game is given in the following table:

	Game		
Official	Athens	Columbia	Nashville
1	165	90	130
2	75	210	320
3	180	170	140
4	220	80	60
5	410	140	80
6	150	170	190
7	170	110	150
8	105	125	160
9	240	200	155

Determine the optimal assignment(s) to minimize the total distance traveled by the officials.

56. In Problem 55, officials 2 and 8 recently had a confrontation with one of the coaches in the game in Athens. They were forced to eject the coach after several technical fouls. The conference office decided that it would not be a good idea to have these two officials work the Athens game so soon after this confrontation, so they decided that officials 2 and 8 will not be assigned to the Athens game. How will this affect the optimal solution to this problem?

57. State University has planned six special catered events for the Saturday of its homecoming football game. The events include an alumni brunch, a parents' brunch, a booster club luncheon, a postgame party for season ticket holders, a lettermen's dinner, and a fund-raising dinner for major contributors. The university wants to use local catering firms as well as the university catering service to cater these events, and it has asked the caterers to bid on each event. The bids

(in thousands of dollars) based on menu guidelines for the events prepared by the university are shown in the following table:

	Event					
Caterer	Alumni Brunch	Parents' Brunch	Booster Club Lunch	Postgame Party	Lettermen's Dinner	Contributors' Dinner
Al's	$12.6	$10.3	$14.0	$19.5	$25.0	$30.0
Bon Apetít	14.5	13.0	16.5	17.0	22.5	32.0
Custom	13.0	14.0	17.6	21.5	23.0	35.0
Divine	11.5	12.6	13.0	18.7	26.2	33.5
Epicurean	10.8	11.9	12.9	17.5	21.9	28.5
Fouchéss	13.5	13.5	15.5	22.3	24.5	36.0
University	12.5	14.3	16.0	22.0	26.7	34.0

The Bon Apetít, Custom, and University caterers can handle two events, whereas each of the other four caterers can handle only one. The university is confident that all the caterers will do a high-quality job, so it wants to select the caterers for the events that will result in the lowest total cost.

Determine the optimal selection of caterers to minimize total cost.

58. A university department head has five instructors to be assigned to four different courses. All the instructors have taught the courses in the past and have been evaluated by the students. The rating for each instructor for each course is given in the following table (a perfect score is 100):

	Course			
Instructor	A	B	C	D
1	80	75	90	85
2	95	90	90	97
3	85	95	88	91
4	93	91	80	84
5	91	92	93	88

The department head wants to know the optimal assignment of instructors to courses to maximize the overall average evaluation. The instructor who is not assigned to teach a course will be assigned to grade exams.

59. The coach of the women's swim team at State University is preparing for the conference swim meet and must choose the four swimmers she will assign to the 800-meter medley relay team. The medley relay consists of four strokes—backstroke, breaststroke, butterfly, and freestyle. The coach has computed the average times (in minutes) each of her top six swimmers has achieved in each of the four strokes for 200 meters in previous swim meets during the season, as follows:

	Stroke (min.)			
Swimmer	Backstroke	Breaststroke	Butterfly	Freestyle
Annie	2.56	3.07	2.90	2.26
Beth	2.63	3.01	3.12	2.35
Carla	2.71	2.95	2.96	2.29
Debbie	2.60	2.87	3.08	2.41
Erin	2.68	2.97	3.16	2.25
Fay	2.75	3.10	2.93	2.38

Determine for the coach the medley relay team and its total expected relay time.

60. Biggio's Department Store has six employees available to assign to four departments in the store—home furnishings, china, appliances, and jewelry. Most of the six employees have worked in each of the four departments on several occasions in the past and have demonstrated that they perform better in some departments than in others. The average daily sales for each of the six employees in each of the four departments are shown in the following table:

	Department Sales			
Employee	**Home Furnishings**	**China**	**Appliances**	**Jewelry**
1	$340	$160	$610	$290
2	560	370	520	450
3	270	—	350	420
4	360	220	630	150
5	450	190	570	310
6	280	320	490	360

Employee 3 has not worked in the china department before, so the manager does not want to assign this employee to china.

Determine which employee to assign to each department and indicate the total expected daily sales.

61. The Vanguard Publishing Company wants to hire seven of the eight college students who have applied as salespeople to sell encyclopedias during the summer. The company desires to allocate them to three sales territories. Territory 1 requires three salespeople, and territories 2 and 3 require two salespeople each. It is estimated that each salesperson will be able to generate the amounts of dollar sales per day in each of the three territories as given in the following table:

	Territory		
Salesperson	**1**	**2**	**3**
A	$110	$150	$130
B	90	120	80
C	205	160	175
D	125	100	115
E	140	105	150
F	100	140	120
G	180	210	160
H	110	120	90

Help the company allocate the salespeople to the three territories so that sales will be maximized.

62. Carolina Airlines, a small commuter airline in North Carolina, has six flight attendants that it wants to assign to six monthly flight schedules in a way that will minimize the number of nights they will be away from their homes. The numbers of nights each attendant must be away from home with each schedule are given in the following table:

	Schedule					
Attendant	**A**	**B**	**C**	**D**	**E**	**F**
1	7	4	6	10	5	8
2	4	5	5	12	7	6
3	9	9	11	7	10	8
4	11	6	8	5	9	10
5	5	8	6	10	7	6
6	10	12	11	9	9	10

Identify the optimal assignments that will minimize the total number of nights the attendants will be away from home.

63. The football coaching staff at Tech focuses its recruiting on several key states, including Georgia, Florida, Virginia, Pennsylvania, New York, and New Jersey. The staff includes seven assistant coaches, two of whom are responsible for Florida, a high school talent–rich state, whereas one coach is assigned to each of the other five states. The staff has been together for a long time, and at one time or another all the coaches have recruited in all the states. The head coach has accumulated some data on the past success rate (i.e., percentage of targeted recruits signed) for each coach in each state, as shown in the following table:

	State					
Coach	GA	FL	VA	PA	NY	NJ
Allen	62	56	65	71	55	63
Bush	65	70	63	81	75	72
Crumb	46	53	62	55	64	50
Doyle	58	66	70	67	71	49
Evans	77	73	69	80	80	74
Fouch	68	73	72	80	78	57
Goins	72	60	74	72	62	61

Determine the optimal assignment of coaches to recruiting regions that will maximize the overall success rate and indicate the average percentage success rate for the staff with this assignment.

64. Kathleen Taylor is a freshman at Roanoke College, and she wants to develop her schedule for the spring semester. Courses are offered with class periods either on Monday and Wednesday or Tuesday and Thursday for 1 hour and 15 minutes duration, with 15 minutes between class periods. For example, a course designated as 8M meets on Monday and Wednesday from 8:00 A.M. to 9:15 A.M.; the next class on Monday and Wednesday (9M) meets from 9:30 to 10:45; the next class (11M) is from 11:00 A.M. to 12:15 P.M.; and so on. Kathleen wants to take the following six freshman courses, with the available sections shown in order of her preference, based on the professor who's teaching the course and the time:

Course	Sections Available
Math	11T, 12T, 9T, 11M, 12M, 9M, 8T, 8M
History	11T, 11M, 14T, 14M, 8T, 8M
English	9T, 11T, 14T, 11M, 12T, 14M, 12M, 9M
Biology	14T, 11M, 12M, 14M, 9M, 8T, 8M
Spanish	9T, 11M, 12M, 8T
Psychology	14T, 11T, 12T, 9T, 14M, 8M

For example, there are eight sections of math offered, and Kathleen's first preference is the 11T section, her second choice is the 12T section, and so forth.

a. Determine a class schedule for Kathleen that most closely meets her preferences.
b. Determine a class schedule for Kathleen if she wants to leave 11:00 A.M. to noon open for lunch every day.
c. Suppose Kathleen wants all her classes on two days, either Monday and Wednesday or Tuesday and Thursday. Determine schedules for each and indicate which most closely matches her preferences.

65. CareMed, an HMO health care provider, operates a 24-hour outpatient clinic in Draperton, near the Tech campus. The facility has a medical staff with doctors and nurses who see regular local patients according to a daily appointment schedule. However, the clinic sees a number of Tech

students who visit the clinic each day and evening without appointments because their families are part of the CareMed network. The clinic has 12 nurses who work according to three 8-hour shifts. Five nurses are needed from 8:00 A.M. to 4:00 P.M., four nurses work from 4:00 P.M. to midnight, and 3 nurses work overnight from midnight to 8:00 A.M. The clinic administrator wants to assign nurses to a shift according to their preferences and seniority (i.e., when the number of nurses who prefer a shift exceeds the shift demand, the nurses are assigned according to seniority). While the majority of nurses prefer the day shift, some prefer other shifts because of the work and school schedules of their spouses and families. Following are the nurses' shift preferences (where 1 is most preferred) and their years working at the clinic:

	Shift			
Nurse	8 A.M. to 4 P.M.	4 P.M. to Midnight	Midnight to 8 A.M.	Years' Experience
Adams	1	2	3	2
Baxter	1	3	2	5
Collins	1	2	3	7
Davis	3	1	2	1
Evans	1	3	2	3
Forrest	1	2	3	4
Gomez	2	1	3	1
Huang	3	2	1	1
Inchio	1	3	2	2
Jones	2	1	3	3
King	1	3	2	5
Lopez	2	3	1	2

Formulate and solve a linear programming model to assign the nurses to shifts according to their preferences and seniority.

66. The MidSouth Trucking Company based in Nashville has eight trucks located throughout the East and Midwest that have delivered their loads and are available for shipments. Through their Internet logistics site MidSouth has received shipping requests from 12 customers. The following table shows the mileage for a truck to travel to a customer location, pick up the load, and deliver it.

Truck	A	B	C	D	E	F	G	H	I	J	K	L
1	500	730	620	410	550	600	390	480	670	710	440	590
2	900	570	820	770	910	660	650	780	840	950	590	670
3	630	660	750	540	680	750	810	560	710	1200	490	650
4	870	1200	810	670	710	820	1200	630	700	900	540	620
5	950	910	740	810	630	590	930	650	840	930	460	560
6	1100	860	800	590	570	550	780	610	1300	840	550	790
7	610	710	910	550	810	730	910	720	850	760	580	630
8	560	690	660	640	720	670	830	690	880	1000	710	680

Determine the optimal assignment of trucks to customers that will minimize the total mileage.

67. In Problem 66, assume that the customers have the following truck capacity percentage loads:

	Customer											
	A	B	C	D	E	F	G	H	I	J	K	L
Capacity	89	78	94	82	90	83	88	79	71	96	78	85

Determine the optimal assignment of trucks to customers that will minimize total mileage while also achieving at least an average truck load capacity of 85%. Does this load capacity requirement significantly increase the total mileage?

68. The Hilton Island Tennis Club is hosting its annual professional tennis tournament. One of the tournament committee's scheduling activities is to assign chair umpires to the various matches. The tennis association rates chair umpires from 1 (best) to worst (4) based on experience, consistency, and player and coach evaluations. On a particular afternoon 12 matches start at 1:00 P.M. Fifteen chair umpires are available to officiate the matches. The matches are prioritized from 1 to 12 in order of importance, based on the rankings and seedings of the players involved, and whether it is a singles or a doubles match. The committee wants to assign the best umpires to the highest priority matches. The following tables show the match priorities and the umpire ratings.

Umpire	1	2	3	4	5	6	7	8	9	10	11	12	13	14	15
Rating	**3**	**4**	**3**	**1**	**2**	**2**	**3**	**2**	**4**	**4**	**2**	**1**	**4**	**3**	**3**

Match	1	2	3	4	5	6	7	8	9	10	11	12
Priority	2	11	6	4	9	10	12	3	1	5	7	8

Determine the optimal assignment of umpires to matches.

69. The National Collegiate Lacrosse Association is planning its annual national championship tournament. It selects 16 teams from conference champions and the highest ranked at-large teams to play in the single-elimination tournament. The teams are ranked from 1 (best) to 16 (worst), and in the first round of the tournament, the association wants to pair the teams so that high-ranked teams play low-ranked teams (i.e., seed them so that 1 plays 16, 2 plays 15, etc.). The eight first-round game sites are predetermined and have been selected based on stadium size and conditions, as well as historical local fan interest in lacrosse. Because of limited school budgets for lacrosse and a desire to boost game attendance, the association wants to assign teams to game sites so that all schools will have to travel the least amount possible. The following table shows the 16 teams in order of their ranking and the distance (in miles) for each of the teams to each of the 8 game sites.

		Game Site							
Team	**Rank**	**1**	**2**	**3**	**4**	**5**	**6**	**7**	**8**
Jackets	1	146	207	361	215	244	192	187	467
Big Red	2	213	0	193	166	312	233	166	631
Knights	3	95	176	348	388	377	245	302	346
Tigers	4	112	243	577	0	179	412	276	489
Bulldogs	5	375	598	112	203	263	307	422	340
Wasps	6	199	156	196	257	389	388	360	288
Blue Jays	7	345	231	207	326	456	276	418	374
Blue Devils	8	417	174	178	442	0	308	541	462
Cavaliers	9	192	706	401	194	523	233	244	446
Rams	10	167	157	233	294	421	272	367	521
Eagles	11	328	428	175	236	278	266	409	239
Beavers	12	405	310	282	278	344	317	256	328
Bears	13	226	268	631	322	393	338	197	297
Hawks	14	284	161	176	267	216	281	0	349
Lions	15	522	209	218	506	667	408	270	501
Panthers	16	197	177	423	183	161	510	344	276

Formulate and solve a linear programming model that will assign the teams to the game sites according to the association's guidelines.

70. Suppose that in Problem 69, the association wants to consider allowing some flexibility in pairing the teams according to their rankings (i.e., seedings)—for example 1 might play 14 or 2 might play 12—in order to see to what extent overall team travel might be reduced. Reformulate and solve the model in order to see what effect this might have.

Case Problem

The Department of Management Science and Information Technology at Tech

The management science and information technology department at Tech offers between 36 and forty 3-hour course sections each semester. Some of the courses are taught by graduate student instructors, whereas 20 of the course sections are taught by the 10 regular, tenured faculty in the department. Before the beginning of each year, the department head sends the faculty a questionnaire, asking them to rate their preference for each course using a scale from 1 to 5, where 1 is "strongly preferred," 2 is "preferred but not as strongly as 1," 3 is "neutral," 4 is "prefer not to teach but not strongly," and 5 is "strongly prefer not to teach this course." The faculty have returned their preferences, as follows:

For the fall semester the department will offer two sections each of 3424 and 4464; three sections of 3434, 3444, 4434, 4444, and 4454; and one section of 3454.

The normal semester teaching load for a regular faculty member is two sections. (Once the department head determines the courses, he will assign the faculty he schedules the course times so they will not conflict.) Help the department head determine a teaching schedule that will satisfy faculty teaching preferences to the greatest degree possible.

	Course							
Faculty Member	*3424*	*3434*	*3444*	*3454*	*4434*	*4444*	*4454*	*4464*
Clayton	2	4	1	3	2	5	5	5
Houck	3	3	4	1	2	5	5	4
Huang	2	3	2	1	3	4	4	4
Major	1	4	2	5	1	3	2	2
Moore	1	1	4	4	2	3	3	5
Ragsdale	1	3	1	5	4	1	1	2
Rakes	3	1	2	5	3	1	1	1
Rees	3	4	3	5	5	1	1	3
Russell	4	1	3	2	2	5	5	5
Sumichrast	4	3	1	5	2	3	3	1

Case Problem

Stateline Shipping and Transport Company

Rachel Sundusky is the manager of the South-Atlantic office of the Stateline Shipping and Transport Company. She is in the process of negotiating a new shipping contract with Polychem, a company that manufactures chemicals for industrial use. Polychem wants Stateline to pick up and transport waste products from its six plants to three waste disposal sites. Rachel is very concerned about this proposed arrangement. The chemical wastes that will be hauled can be hazardous to humans and the environment if they leak. In addition, a number of towns and communities in the region where the plants are located prohibit hazardous materials from being shipped through their municipal limits. Thus, not only will the shipments have to be handled carefully and transported at reduced speeds, they will also have to traverse circuitous routes in many cases.

Rachel has estimated the cost of shipping a barrel of waste from each of the six plants to each of the three waste disposal sites as shown in the following table:

	Waste Disposal Site		
Plant	*Whitewater*	*Los Canos*	*Duras*
Kingsport	$12	$15	$17
Danville	14	9	10
Macon	13	20	11
Selma	17	16	19
Columbus	7	14	12
Allentown	22	16	18

The plants generate the following amounts of waste products each week:

Plant	Waste per Week (bbl)
Kingsport	35
Danville	26
Macon	42
Selma	53
Columbus	29
Allentown	38

The three waste disposal sites at Whitewater, Los Canos, and Duras can accommodate a maximum of 65, 80, and 105 barrels per week, respectively.

In addition to shipping directly from each of the six plants to one of the three waste disposal sites, Rachel is also considering using each of the plants and waste disposal sites as intermediate shipping points. Trucks would be able to drop a load at a plant or disposal site to be picked up and carried on to the final destination by another truck, and vice versa. Stateline would not incur any handling costs because Polychem has agreed to take care of all local handling of the waste materials at the plants and the waste disposal sites. In other words, the only cost Stateline incurs is the actual transportation cost. So Rachel wants to be able to consider the possibility that it may be cheaper to drop and pick up loads at intermediate points rather than ship them directly.

Rachel estimates the shipping costs per barrel between each of the six plants to be as follows:

	Plant					
Plant	Kingsport	Danville	Macon	Selma	Columbus	Allentown
Kingsport	$—	$ 6	$ 4	$ 9	$ 7	$ 8
Danville	6	—	11	10	12	7
Macon	5	11	—	3	7	15
Selma	9	10	3	—	3	16
Columbus	7	12	7	3	—	14
Allentown	8	7	15	16	14	—

The estimated shipping cost per barrel between each of the three waste disposal sites is as follows.

	Waste Disposal Site		
Waste Disposal Site	Whitewater	Los Canos	Duras
Whitewater	$—	$12	$10
Los Canos	12	—	15
Duras	10	15	—

Rachel wants to determine the shipping routes that will minimize Stateline's total cost in order to develop a contract proposal to submit to Polychem for waste disposal. She particularly wants to know if it would be cheaper to ship directly from the plants to the waste sites or if she should drop and pick up some loads at the various plants and waste sites. Develop a model to assist Rachel and solve the model to determine the optimal routes.

Case Problem

BURLINGHAM TEXTILE COMPANY

Brenda Last is the personnel director at the Burlingham Textile Company. The company's plant is expanding, and Brenda must fill five new supervisory positions in carding, spinning, weaving, inspection, and shipping. Applicants for the positions are required to take a written psychological and aptitude test. The test has different modules that indicate an applicant's aptitude and suitability for a specific area and position. For example, one module tests the psychological traits and intellectual skills that are best suited for the inspection department, which are different from the traits and skills required in shipping. Brenda has had 10 applicants for the five positions and has compiled the results from the test. The test scores for each position module for each applicant (where the higher the score, the better) are as follows:

Brenda wants to offer the vacant positions to the five most qualified candidates. Determine an optimal assignment for Brenda.

There is a possibility that one or more of the successful applicants will turn down a position offer, and Brenda wants to be able to hire the next-best person into a position if someone rejects a job offer. If the applicant selected for the carding job turns it down, whom should Brenda offer this job to next? If the applicants for both the carding and spinning jobs turn them down, which of the remaining applicants should Brenda offer each job to? How would a third applicant be selected if three of the job offers were declined?

Brenda believes this is a particularly good group of applicants. She would like to retain a few of the people for several more supervisory positions that she believes will open up soon. She has two vacant clerical positions that she can offer to the two best applicants not selected for the five original supervisory positions. Then when the supervisory positions open, she can move these people into them. How should Brenda identify these two people?

	Test Module Scores				
Applicant	*Carding*	*Spinning*	*Weaving*	*Inspection*	*Shipping*
Roger Acuff	68	75	72	86	78
Melissa Ball	73	82	66	78	85
Angela Coe	92	101	90	79	74
Maureen Davis	87	98	75	90	92
Fred Evans	58	62	93	81	75
Bob Frank	93	79	94	92	96
Ellen Gantry	77	92	90	81	93
David Harper	79	66	90	85	86
Mary Inchavelia	91	102	95	90	88
Marilu Jones	72	75	67	93	93

Case Problem

THE GRAPHIC PALETTE

The Graphic Palette is a firm in Charleston, South Carolina, that does graphic artwork and produces color and black-and-white posters, lithographs, and banners. The firm's owners, Kathleen and Lindsey Taylor, have been approached by a client to produce a spectacularly colored poster for an upcoming arts festival. The poster is more complex than anything Kathleen and Lindsey have previously worked on. It requires color screening in three stages, and the processing must proceed rapidly to produce the desired color effect.

By suspending all their other jobs, they can devote three machines to the first stage, four to the second stage, and two to the last stage of the process. Posters that come off the machines at each stage can be processed on any of the machines at the next stage. However, all the machines are of different models and of varying ages, so they cannot process the same number of posters in the specified time frame necessary to complete the job. The different machine capacities at each stage are as follows:

Stage 1	Stage 2	Stage 3
Machine 1 = 750	Machine 4 = 530	Machine 8 = 620
Machine 2 = 900	Machine 5 = 320	Machine 9 = 750
Machine 3 = 670	Machine 6 = 450	
	Machine 7 = 250	

Because the machines are of different ages and types, the costs of producing posters on them differ. For example, a poster that starts on machine 1 and then proceeds to machine 4 costs $18. If this poster at machine 4 is then processed on machine 8, it costs an additional $36. The processing costs for each combination of machines for stages 1, 2, and 3 are as follows:

	Machine (cost)			
Machine	4	5	6	7
1	$18	$23	$25	$21
2	20	26	24	19
3	24	24	22	23

	Machine (cost)	
Machine	8	9
4	$36	$41
5	40	52
6	42	46
7	33	49

Kathleen and Lindsey are unsure how to route the posters from one stage to the next to make as many posters as they possibly can at the lowest cost. Determine how to route the posters through the various stages for Graphic Palette to minimize costs.

Case Problem

SCHEDULING AT HAWK SYSTEMS, INC.

Jim Huang and Roderick Wheeler were sales representatives in a computer store at a shopping mall in Arlington, Virginia, when they got the idea of going into business in the burgeoning and highly competitive microcomputer market. Jim went to Taiwan over the summer to visit relatives and made a contact with a new firm producing display monitors for microcomputers, which was looking for an East Coast distributor in America. Jim made a tentative deal with the firm to supply a maximum of 500 monitors per month and called Rod to see if he could find a building they could operate out of and some potential customers.

Rod went to work. The first thing he did was send bids to several universities in Maryland, Virginia, and Pennsylvania for contracts as an authorized vendor for monitors at the schools. Next, he started looking for a facility to operate from. Jim and his operation would provide minor physical modifications to the monitors, including some labeling, testing, packaging, and then storage in preparation for shipping. He knew he needed a building with good security, air-conditioning, and a loading dock. However, his search proved to be more difficult than he anticipated. Building space of the type and size he needed was very limited in the area and very expensive. Rod began to worry that he would not be able to find a suitable facility at all. He decided to look for space in the Virginia and Maryland suburbs and countryside; and although he found some good locations, the shipping costs out to those locations were extremely high.

Disheartened by his lack of success, Rod sought help from his sister-in-law Miriam, a local real estate agent. Rod poured out the details of his plight to Miriam over dinner at Rod's mother's house, and she was sympathetic. She told Rod that she owned a building in Arlington that might be just what he was looking for, and she would show it to him the next day. As promised, she showed him the ground floor of the building, and it was perfect. It had plenty of space, good security, and a nice office; furthermore, it was in an upscale shopping area with lots of good restaurants. Rod was elated; it was just the type of environment he had envisioned for them to set up their business in. However, his joy soured when he asked Miriam what the rent was. She said she had not worked out the details, but the rent would be around $100,000 per year. Rod was shocked, so Miriam said she would offer him an alternative: a storage fee of $10 per monitor for every monitor purchased and in stock the first month of operation, with an increase of $2 per month per unit for the remainder of the year. Miriam explained that based on what he told her about the business, they would not have any sales until the universities opened around the end of August or the first of September, and that their sales would fall off to nothing in May or June. She said her offer meant that she would share in their success or failure. If they ended up with some university contracts, she would reap a reward along with them; if they did not sell many monitors, she would lose on the deal. But in the summer months after school ended, if they had no monitors in stock, they would pay her nothing.

Rod mulled this over, and it sounded fair. He loved the building. Also, he liked the idea that they would not be indebted for a flat lease payment and that the rent was essentially on a per-unit basis. If they failed, at least they would not be stuck with a huge lease. So he agreed to Miriam's offer.

When Jim returned from Taiwan, he was skeptical about Rod's lease arrangement with Miriam. He was chagrined that Rod hadn't performed a more thorough analysis of the costs, but Rod explained that it was pretty hard to do an analysis when he did not know their costs, potential sales, or selling price. Jim said he had a point, and his concern was somewhat offset by the fact that Rod had gotten contracts with five universities as an authorized vendor for monitors at a selling price of $180 per unit. So the two sat down to begin planning their operation.

First, Jim said he had thought of a name for their enterprise, Hawk Systems, Inc., which he said stood for Huang and Wheeler Computers. When Rod asked how Jim got a *k* out of *computers,* Jim cited poetic license.

Jim said that he had figured that the total cost of the units for them—including the purchase of the units, shipping, and their own material, labor, and administrative costs—would be $100 per unit during the first 4 months but would then drop to $90 per month for the following 4 months and, finally, to $85 per month for the remainder of the year. Jim said that the Taiwan firm was anticipating being able to lower the purchase price because its production costs would go down as it gained experience.

Jim thought their own costs would go down, too. He also explained that they would not be able to return any items, so it was important that they develop a good order plan that would minimize costs. This was now much more important than Jim had originally thought because of their peculiar lease arrangement based on their inventory level. Rod said that he had done some research on past computer sales at the universities they had contracted with and had come up with the following sales forecast for the next 9 months of the academic year (from September through May):

September	340
October	650
November	420
December	200
January	660
February	550
March	390
April	580
May	120

Rod explained to Jim that computer equipment purchases at universities go up in the fall, then drop until January, and then peak again in April, just before university budgets are exhausted at the end of the academic year.

Jim then asked Rod what kind of monthly ordering schedule from Taiwan they should develop to meet demand while minimizing their costs. Rod said that it was a difficult question, but he remembered that when he was in college in a management science course, he had seen a production schedule developed using a transportation model. Jim suggested he get out his old textbook and get busy, or they would be turning over all their profits to Miriam.

However, before Rod was able to develop a schedule, Jim got a call from the Taiwan firm, saying that it had gotten some more business later in the year and it could no longer supply up to 500 units per month. Instead, it could supply 700 monitors for the first 4 months and 300 for the next 5. Jim and Rod worried about what this would do to their inventory costs.

A. Formulate and solve a transportation model to determine an optimal monthly ordering and distribution schedule for Hawk Systems that will minimize costs.

B. If Hawk Systems has to borrow approximately $200,000 to start up the business, will it end up making anything the first year?

C. What will the change in the supply pattern from the Taiwan firm cost Hawk Systems?

D. How did Miriam fare with her alternative lease arrangement? Would she have been better off with a flat $100,000 lease payment?

Case Problem

Global Shipping at Erken Apparel International

Erken Apparel International manufactures clothing items around the world. It has currently contracted with a U.S. retail clothing wholesale distributor for men's goatskin and lambskin leather jackets for the next Christmas season. The distributor has distribution centers in Indiana, North Carolina, and Pennsylvania. The distributor supplies the leather jackets to a discount retail chain, a chain of mall boutique stores, and a department store chain. The jackets arrive at the distribution centers unfinished, and at the centers the distributor adds a unique lining and label specific to each of its customers. The distributor has contracted with Erken to deliver the following number of leather jackets to its distribution centers in late fall:

Distribution Center	Goatskin Jackets	Lambskin Jackets
Indiana	1,000	780
North Carolina	1,400	950
Pennsylvania	1,600	1,150

Erken has tanning factories and clothing manufacturing plants to produce leather jackets in Spain, France, Italy, Venezuela, and Brazil. Its tanning facilities are in Mende in France, Foggia in Italy, Saragosa in Spain, Feira in Brazil, and El Tigre in Venezuela. Its manufacturing plants are in Limoges, Naples, and Madrid in Europe and in Sao Paulo and Caracas in South America. Following are the supplies of available leather from each tanning facility and the processing capacity at each plant (in pounds) for this particular order of leather jackets:

Tanning Factory	Goatskin Supply (lb.)	Lambskin Supply (lb.)
Mende	4,000	4,400
Foggia	3,700	5,300
Saragosa	6,500	4,650
Feira	5,100	6,850
El Tigre	3,600	5,700

Plant	Production Capacity (lb.)
Madrid	7,800
Naples	5,700
Limoges	8,200
Sao Paulo	7,600
Caracas	6,800

In the production of jackets at the plants, 37.5% of the goatskin leather and 50% of the lambskin leather is waste (i.e., it is discarded during the production process and sold for other byproducts). After production, a goatskin jacket weighs approximately 3 pounds, and a lambskin jacket weighs approximately 2.5 pounds (neither with linings, which are added in the United States).

Following are the costs per pound, in U.S. dollars, for tanning the uncut leather, shipping it, and producing the leather jackets at each plant:

Tanning Factory	Plant ($/lb.)				
	Madrid	Naples	Limoges	Sao Paulo	Caracas
Mende	$24	$22	$16	$21	$23
Foggia	31	17	22	19	22
Saragosa	18	25	28	23	25
Feira	—	—	—	16	18
El Tigre	—	—	—	14	15

Note that the cost of jacket production is the same for goatskin and lambskin. Also, leather can be tanned in France, Spain, and Italy and shipped directly to the South American plants for jacket production, but the opposite is not possible due to high tariff restrictions (i.e., tanned leather is not shipped to Europe for production).

Once the leather jackets are produced at the plants in Europe and South America, Erken transports them to ports in Lisbon, Marseilles, and Caracas and then from these ports to U.S. ports in New Orleans, Jacksonville, and Savannah. The available shipping capacity at each port and the transportation costs from the plants to the ports are as follows:

Plant	Port ($/lb.)		
	Lisbon	Marseilles	Caracas
Madrid	0.75	1.05	—
Naples	3.45	1.35	—
Limoges	2.25	0.60	—
Sao Paulo	—	—	1.15
Caracas	—	—	0.20
Capacity (lb.)	8,000	5,500	9,000

The shipping costs ($/lb.) from each port in Europe and South America to the U.S. ports and the available truck and rail capacity for transport at the U.S. ports are as follows:

Port	U.S. Port ($/lb.)		
	New Orleans	Jacksonville	Savannah
Lisbon	$2.35	$1.90	$1.80
Marseilles	3.10	2.40	2.00
Caracas	1.95	2.15	2.40
Capacity (lb.)	8,000	5,200	7,500

The transportation costs ($/lb.) from the U.S. ports to the three distribution centers are as follows:

U.S. Port	Distribution Center ($/lb.)		
	Indiana	North Carolina	Pennsylvania
New Orleans	$0.65	$0.52	$0.87
Jacksonville	0.43	0.41	0.65
Savannah	0.38	0.34	0.50

Erken wants to determine the least costly shipments of material and jackets that will meet the demand at the U.S. distribution centers. Develop a transshipment model for Erken that will result in a minimum cost solution.

Case Problem

WeeMow Lawn Service

WeeMow Lawn Service provides lawn service, including mowing lawns and lawn care, landscaping, and lawn maintenance, to residential and commercial customers in the Draper town community. In the summer WeeMow has three teams that it schedules daily for jobs. Team 1 has five members, team 2 has four members, and team 3 has three members. On a normal summer day WeeMow will have approximately 14 jobs. Each team works 10 hours a day, but because of the heat plus work breaks, each team actually works only 45 minutes out of every hour. The following table shows the times (in minutes) and costs (in dollars) required for each team to complete the 14 jobs for a specific day.

	Team 1		Team 2		Team 3	
Job	Time (min.)	Cost ($)	Time (min.)	Cost ($)	Time (min.)	Cost ($)
A	45	48	65	55	78	50
B	67	70	72	60	85	55
C	90	94	105	84	125	75
D	61	65	78	65	97	60
E	75	80	93	75	107	66
F	48	55	70	65	95	60
G	65	70	83	70	110	67
H	67	72	84	74	100	65
I	95	100	110	90	130	80
J	60	65	78	65	95	62
K	47	55	64	57	84	56
L	114	118	135	110	155	95
M	85	90	107	98	125	75
N	63	67	81	72	102	68

The WeeMow manager wants to develop a schedule of team assignments to the jobs for this day. Formulate and solve a linear programming model to determine the assignments of teams to jobs that will minimize total job time for the day, given a daily budget of $1,000. Indicate how many minutes each crew will work during the day. Next, reformulate and solve this model if cost minimization is the objective. Of these two models, which one do you think the manger should select and why?

CHAPTER

7

Network Flow Models

*A **network** is an arrangement of paths connected at various points, through which items move.*

A **network** is an arrangement of paths connected at various points, through which one or more items move from one point to another. Everyone is familiar with such networks as highway systems, telephone networks, railroad systems, and television networks. For example, a railroad network consists of a number of fixed rail routes (paths) connected by terminals at various junctions of the rail routes.

Networks are popular because they provide a picture of a system and because a large number of systems can be easily modeled as networks.

In recent years, using network models has become a very popular management science technique for a couple of very important reasons. First, a network is drawn as a diagram, which literally provides a *picture* of the system under analysis. This enables a manager to visually interpret the system and thus enhances the manager's understanding. Second, a large number of real-life systems can be modeled as networks, which are relatively easy to conceive and construct.

Network flow models describe the flow of items through a system.

In this and the next chapter, we will look at several different types of network models. In this chapter we will present a class of network models directed at the *flow of items* through a system. As such, these models are referred to as **network flow models**. We will discuss the use of network flow models to analyze three types of problems: the shortest route problem, the minimal spanning tree problem, and the maximal flow problem. In Chapter 8, we will present the network techniques PERT and CPM, which are used extensively for project analysis.

Network Components

Nodes, denoted by circles, represent junction points connecting branches.

Branches, represented as lines, connect nodes and show flow from one point to another.

Networks are illustrated as diagrams consisting of two main components: nodes and branches. **Nodes** represent junction points—for example, an intersection of several streets. **Branches** connect the nodes and reflect the flow from one point in the network to another. Nodes are denoted in the network diagram by *circles*, and branches are represented by *lines* connecting the nodes. Nodes typically represent localities, such as cities, intersections, or air or railroad terminals; branches are the paths connecting the nodes, such as roads connecting cities and intersections or railroad tracks or air routes connecting terminals. For example, the different railroad routes between Atlanta, Georgia, and St. Louis, Missouri, and the intermediate terminals are shown in Figure 7.1.

FIGURE 7.1
Network of railroad routes

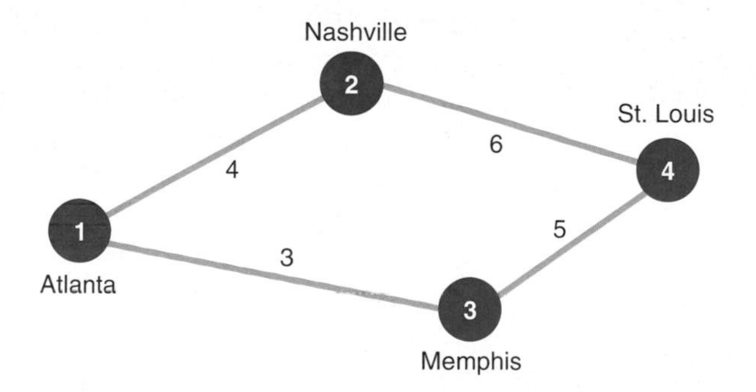

The network shown in Figure 7.1 has four nodes and four branches. The node representing Atlanta is referred to as the *origin*, and any of the three remaining nodes could be the *destination*, depending on what we are trying to determine from the network. Notice that a number has been assigned to each node. Numbers provide a more convenient means of identifying the nodes and branches than do names. For example, we can refer to the origin (Atlanta) as node 1 and the branch from Atlanta to Nashville as branch 1–2.

Typically, a value that represents a distance, length of time, or cost is assigned to each branch. Thus, the purpose of the network is to determine the shortest distance, shortest length of

The values assigned to branches typically represent distance, time, or cost.

time, or lowest cost between points in the network. In Figure 7.1, the values 4, 6, 3, and 5, corresponding to the four branches, represent the lengths of time, in hours, between the attached nodes. Thus, a traveler can see that the route to St. Louis through Nashville requires 10 hours and the route to St. Louis through Memphis requires 8 hours.

The Shortest Route Problem

*The **shortest route problem** is to find the shortest distance between an origin and various destination points.*

The **shortest route problem** is to determine the shortest distance between an originating point and several destination points. For example, the Stagecoach Shipping Company transports oranges by six trucks from Los Angeles to six cities in the West and Midwest. The different routes between Los Angeles and the destination cities and the length of time, in hours, required by a truck to travel each route are shown in Figure 7.2.

FIGURE 7.2
Shipping routes from Los Angeles

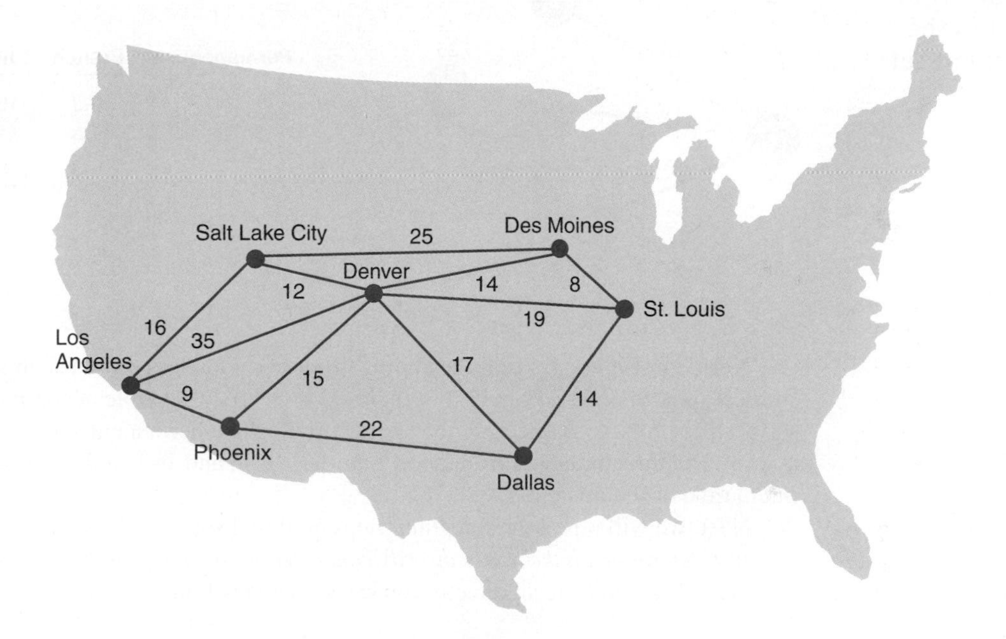

The shipping company manager wants to determine the best routes (in terms of the minimum travel time) for the trucks to take to reach their destinations. This problem can be solved by using the shortest route solution technique. In applying this technique, it is convenient to represent the system of truck routes as a network, as shown in Figure 7.3.

FIGURE 7.3
Network of shipping routes

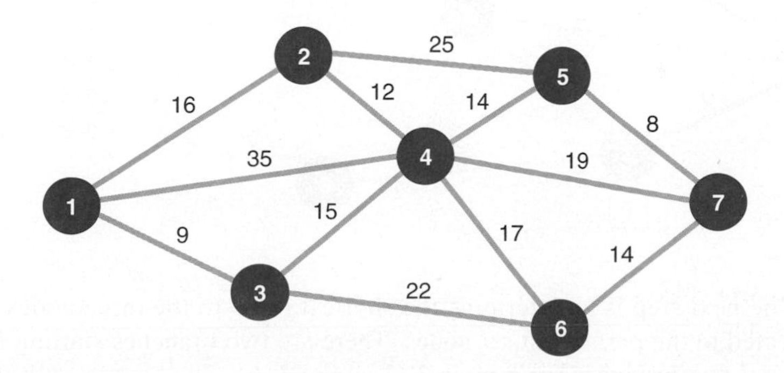

To repeat our objective as it relates to Figure 7.3, we want to determine the shortest routes from the origin (node 1) to the six destinations (nodes 2 through 7).

Determine the initial shortest route from the origin (node 1) to the closest node (3).

The Shortest Route Solution Approach

We begin the shortest route solution technique by starting at node 1 (the origin) and determining the shortest time required to get to a directly connected (i.e., adjacent) node. The three nodes directly connected to node 1 are 2, 3, and 4, as shown in Figure 7.4. Of these three nodes, the shortest time is 9 hours to node 3. Thus, we have determined our first shortest route from nodes 1 to 3 (i.e., from Los Angeles to Phoenix). We will now refer to nodes 1 and 3 as the **permanent** set to indicate that we have found the shortest route to these nodes. (Because node 1 has no route to it, it is automatically in the permanent set.)

*The **permanent set** indicates the nodes for which the shortest route to has been found.*

FIGURE 7.4
Network with node 1 in the permanent set

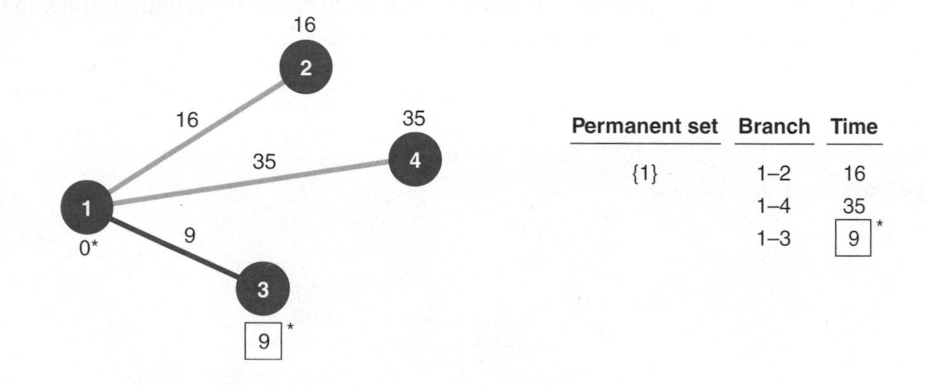

Permanent set	Branch	Time
{1}	1–2	16
	1–4	35
	1–3	9*

Notice in Figure 7.4 that the shortest route to node 3 is drawn with a heavy line, and the shortest time to node 3 (9 hours) is enclosed by a box. The table accompanying Figure 7.4 describes the process of selecting the shortest route. The permanent set is shown to contain only node 1. The three branches from node 1 are 1–2, 1–4, and 1–3, and this last branch has the minimum time of 9 hours.

Determine all nodes directly connected to the permanent set.

Next, we will repeat the foregoing steps used to determine the shortest route to node 3. First, we must *determine all the nodes directly connected to the nodes in the permanent set* (nodes 1 and 3). Nodes 2, 4, and 6 are all directly connected to nodes 1 and 3, as shown in Figure 7.5.

FIGURE 7.5
Network with nodes 1 and 3 in the permanent set

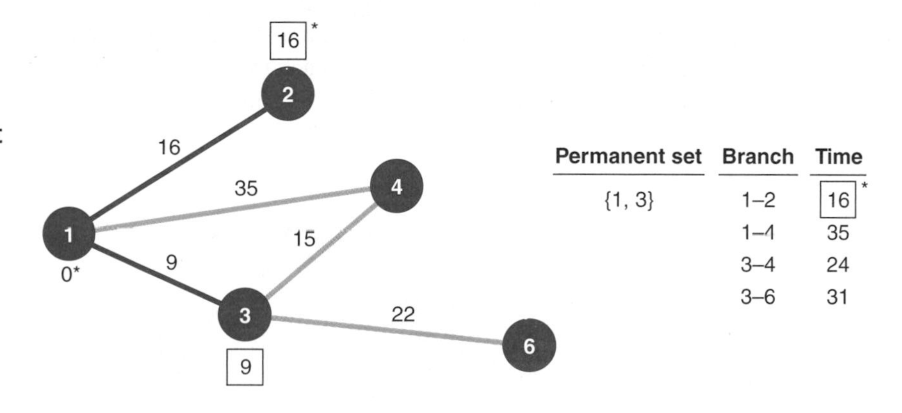

Permanent set	Branch	Time
{1, 3}	1–2	16*
	1–4	35
	3–4	24
	3–6	31

The next step is to determine the shortest route to the three nodes (2, 4, and 6) directly connected to the permanent set nodes. There are two branches starting from node 1 (1–2 and 1–4) and two branches from node 3 (3–4 and 3–6). The branch with the shortest time is to node 2, with a time of 16 hours. Thus, node 2 becomes part of the permanent set. Notice in our computations accompanying Figure 7.5 that the time to node 6 (branch 3–6) is 31 hours, which is determined by adding the branch 3–6 time of 22 hours to the shortest route time of 9 hours at node 3.

Redefine the permanent set.

As we move to the next step, the permanent set consists of nodes 1, 2, and 3. This indicates that we have found the shortest route to nodes 1, 2, and 3. We must now determine which nodes are directly connected to the permanent set nodes. Node 5 is the only *adjacent* node not currently connected to the permanent set, so it is connected directly to node 2. In addition, node 4 is now connected directly to node 2 (because node 2 has joined the permanent set). These additions are shown in Figure 7.6.

FIGURE 7.6
Network with nodes 1, 2, and 3 in the permanent set

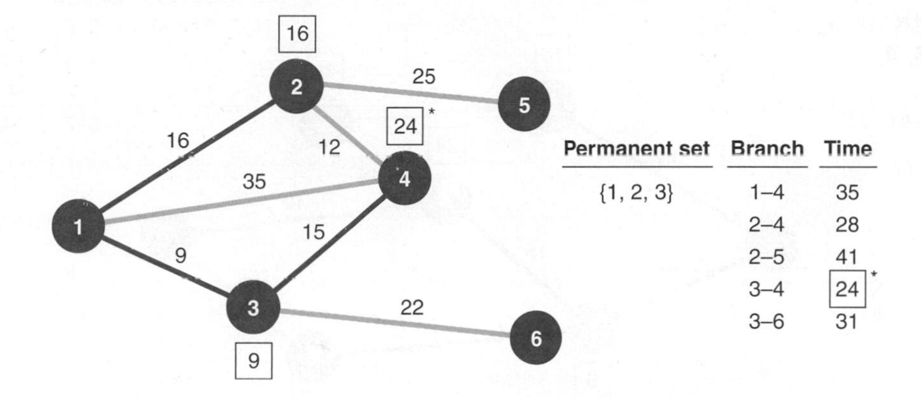

Permanent set	Branch	Time
{1, 2, 3}	1–4	35
	2–4	28
	2–5	41
	3–4	24*
	3–6	31

Five branches lead from the permanent set nodes (1, 2, and 3) to their directly connected nodes, as shown in the table accompanying Figure 7.6. The branch representing the route with the shortest time is 3–4, with a time of 24 hours. Thus, we have determined the shortest route to node 4, and it joins the permanent set. Notice that the shortest time to node 4 (24 hours) is the route from node 1 through node 3. The other routes to node 4 from node 1 through node 2 are longer; therefore, we will not consider them any further as possible routes to node 4.

To summarize, the shortest routes to nodes 1, 2, 3, and 4 have all been determined, and these nodes now form the permanent set.

Next, we repeat the process of determining the nodes directly connected to the permanent set nodes. These directly connected nodes are 5, 6, and 7, as shown in Figure 7.7. Notice in Figure 7.7 that we have eliminated the branches from nodes 1 and 2 to node 4 because we determined that the route with the shortest time to node 4 does not include these branches.

FIGURE 7.7
Network with nodes 1, 2, 3, and 4 in the permanent set

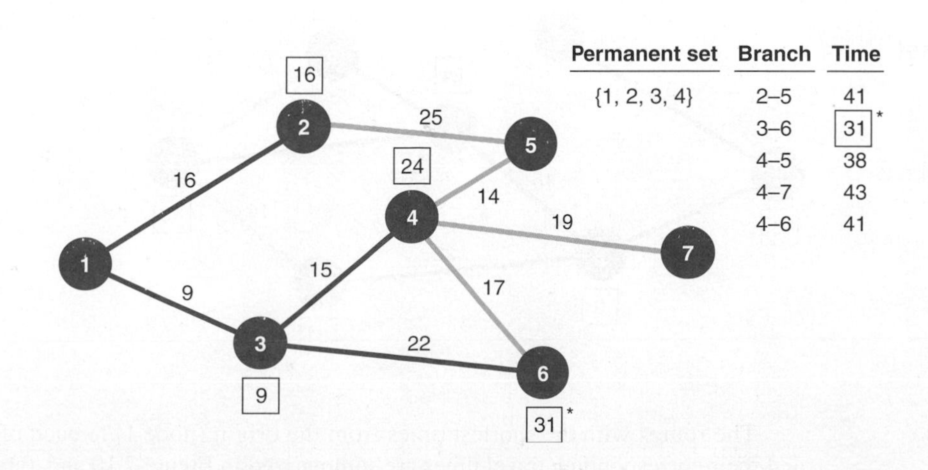

Permanent set	Branch	Time
{1, 2, 3, 4}	2–5	41
	3–6	31*
	4–5	38
	4–7	43
	4–6	41

From the table accompanying Figure 7.7 we can see that of the branches leading to nodes 5, 6, and 7, branch 3–6 has the shortest *cumulative* time, of 31 hours. Thus, node 6 is added to our permanent set. This means that we have now found the shortest routes to nodes 1, 2, 3, 4, and 6.

Repeating the process, we observe that the nodes directly connected (adjacent) to our permanent set are nodes 5 and 7, as shown in Figure 7.8. (Notice that branch 4–6 has been eliminated because the best route to node 6 goes through node 3 instead of node 4.)

FIGURE 7.8
Network with nodes 1, 2, 3, 4, and 6 in the permanent set

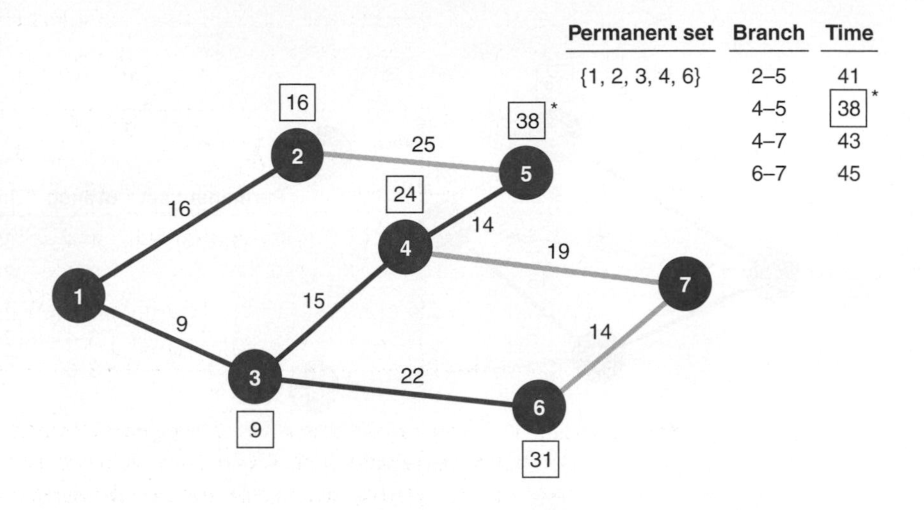

Of the branches leading from the permanent set nodes to nodes 5 and 7, branch 4–5 has the shortest cumulative time, of 38 hours. Thus, node 5 joins the permanent set. We have now determined the routes with the shortest times to nodes 1, 2, 3, 4, 5, and 6 (as denoted by the heavy branches in Figure 7.8).

The only remaining node directly connected to the permanent set is node 7, as shown in Figure 7.9. Of the three branches connecting node 7 to the permanent set, branch 4–7 has the shortest time, of 43 hours. Therefore, node 7 joins the permanent set.

FIGURE 7.9
Network with nodes 1, 2, 3, 4, 5, and 6 in the permanent set

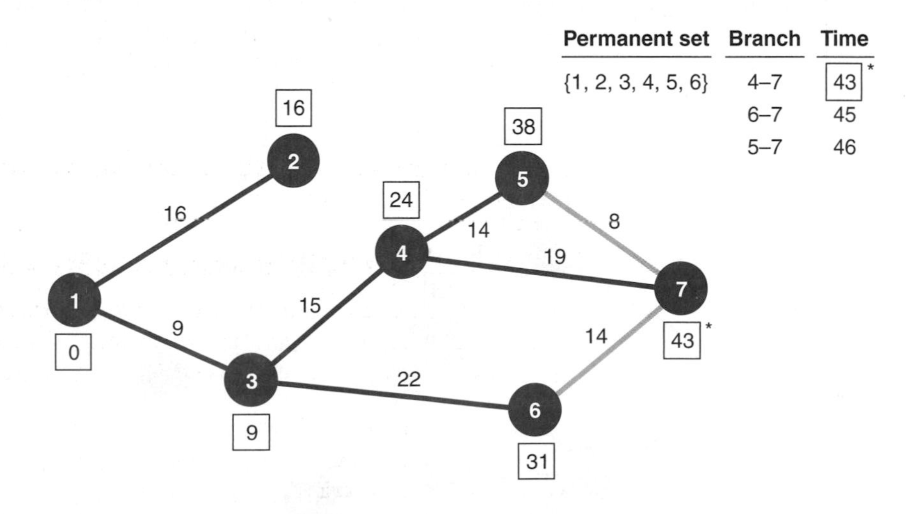

The routes with the shortest times from the origin (node 1) to each of the other six nodes and their corresponding travel times are summarized in Figure 7.10 and Table 7.1.

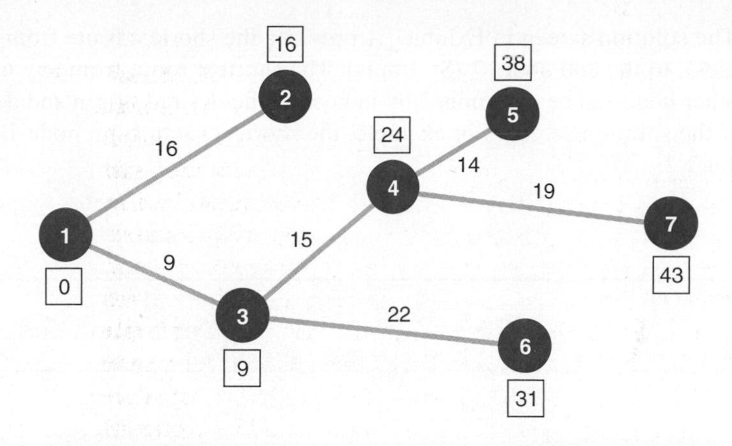

FIGURE 7.10
Network with optimal routes from Los Angeles to all destinations

TABLE 7.1
Shortest travel time from origin to each destination

From Los Angeles to:	Route	Total Hours
Salt Lake City (node 2)	1–2	16
Phoenix (node 3)	1–3	9
Denver (node 4)	1–3–4	24
Des Moines (node 5)	1–3–4–5	38
Dallas (node 6)	1–3–6	31
St. Louis (node 7)	1–3–4–7	43

Steps of the shortest route solution method.

In summary, the steps of the shortest route solution method are as follows:

1. Select the node with the shortest direct route from the origin.
2. Establish a *permanent set* with the origin node and the node that was selected in step 1.
3. Determine all nodes directly connected to the permanent set nodes.
4. Select the node with the shortest route (branch) from the group of nodes directly connected to the permanent set nodes.
5. Repeat steps 3 and 4 until all nodes have joined the permanent set.

Computer Solution of the Shortest Route Problem with QM for Windows

QM for Windows includes modules for each of the three network flow models that will be presented in this chapter—shortest route, minimal spanning tree, and maximal flow. The QM for Windows solution for the shortest route from node 1 to node 7 for the Stagecoach Shipping Company example is shown in Exhibit 7.1.

EXHIBIT 7.1

Networks Results

Stagecoach Shipping Company Solution

Total distance = 43	Start node	End node	Distance	Cumulative Distance
Los Angeles to Phoenix	1	3	9	9
Phoenix to Denver	3	4	15	24
Denver to St. Louis	4	7	19	43

The solution screen in Exhibit 7.1 provides the shortest route from the start node, 1 (Los Angeles), to the end node, 7 (St. Louis). The shortest route from any node in the network to any other node can be determined by indicating the desired origin and destination nodes at the top of the solution screen. For example, the shortest route from node 1 to node 5 is shown in Exhibit 7.2.

EXHIBIT 7.2

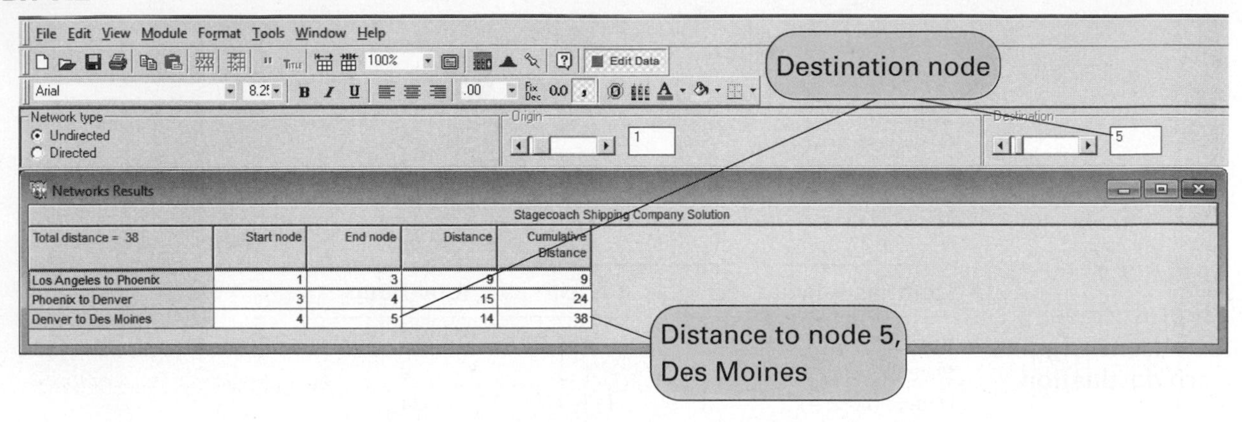

Computer Solution of the Shortest Route Problem with Excel

We can also solve the shortest route problem with Excel spreadsheets by formulating and solving the shortest route network as a 0–1 integer linear programming problem. To formulate the linear programming model, we first define a decision variable for each branch in the network, as follows:

x_{ij} = 0 if branch i–j is not selected as part of the shortest route and 1 if branch i–j is selected

To reduce the complexity and size of our model formulation, we will also assume that items can flow only from a lower node number to a higher node number (e.g., 3 to 4 but not 4 to 3). This greatly reduces the number of variables and branches to consider.

The objective function is formulated by minimizing the sum of the multiplication of each branch value and the travel time for each branch:

$$\text{minimize } Z = 16x_{12} + 9x_{13} + 35x_{14} + 12x_{24} + 25x_{25} + 15x_{34} + 22x_{36} + 14x_{45} + 17x_{46} + 19x_{47} + 8x_{57} + 14x_{67}$$

There is one constraint for each node, indicating that whatever comes into a node must also go out. This is referred to as *conservation of flow*. This means that one "truck" leaves node 1 (Los Angeles) either through branch 1–2, branch 1–3, or branch 1–4. This constraint for node 1 is formulated as

$$x_{12} + x_{13} + x_{14} = 1$$

At node 2, a truck would come in through branch 1–2 and depart through 2–4 or 2–5, as follows:

$$x_{12} = x_{24} + x_{25}$$

This constraint can be rewritten as

$$x_{12} - x_{24} - x_{25} = 0$$

Constraints at nodes 3, 4, 5, 6, and 7 are constructed similarly, resulting in the complete linear programming model, summarized as follows:

$$\text{minimize } Z = 16x_{12} + 9x_{13} + 35x_{14} + 12x_{24} + 25x_{25} + 15x_{34} + 22x_{36} + 14x_{45} + 17x_{46} + 19x_{47} + 8x_{57} + 14x_{67}$$

subject to

$$\begin{aligned} x_{12} + x_{13} + x_{14} &= 1 \\ x_{12} - x_{24} - x_{25} &= 0 \\ x_{13} - x_{34} - x_{36} &= 0 \\ x_{14} + x_{24} + x_{34} - x_{45} - x_{46} - x_{47} &= 0 \\ x_{25} + x_{45} - x_{57} &= 0 \\ x_{36} + x_{46} - x_{67} &= 0 \\ x_{47} + x_{57} + x_{67} &= 1 \\ x_{ij} &= 0 \text{ or } 1 \end{aligned}$$

This model is solved in Excel much the same way as any other linear programming problem, using "Solver". Exhibit 7.3 shows an Excel spreadsheet set up to solve the Stagecoach Shipping Company example. The decision variables, x_{ij}, are represented by cells **A6:A17**. Thus, a value of 1 in one of these cells means that branch has been selected as part of the shortest route. Cells **F6:F17** contain the travel times (in hours) for each branch, and the objective function formula is contained in cell F18, shown on the formula bar at the top of the screen. The model constraints reflecting the flow through each node are included in the box on the right side of the spreadsheet. For example, cell I6 contains the constraint formula for node 1, **=A6+A7+A8,** and cell I7 contains the constraint formula for node 2, **=A6−A9−A10**.

EXHIBIT 7.3

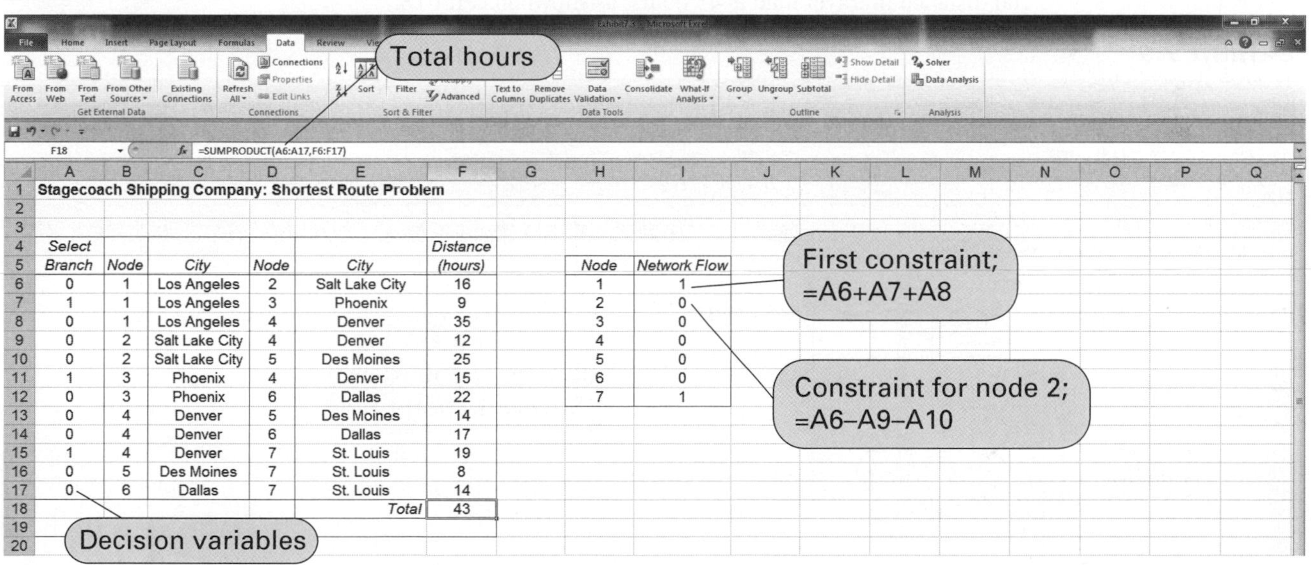

Select Branch	Node	City	Node	City	Distance (hours)
0	1	Los Angeles	2	Salt Lake City	16
1	1	Los Angeles	3	Phoenix	9
0	1	Los Angeles	4	Denver	35
0	2	Salt Lake City	4	Denver	12
0	2	Salt Lake City	5	Des Moines	25
1	3	Phoenix	4	Denver	15
0	3	Phoenix	6	Dallas	22
0	4	Denver	5	Des Moines	14
0	4	Denver	6	Dallas	17
1	4	Denver	7	St. Louis	19
0	5	Des Moines	7	St. Louis	8
0	6	Dallas	7	St. Louis	14
				Total	43

Node	Network Flow
1	1
2	0
3	0
4	0
5	0
6	0
7	1

Exhibit 7.4 shows the Solver Parameters screen, which is accessed from the "Data" tab at the top of the screen. Notice that the constraint for node 1 embedded in cell I6 equals 1, indicating that one truck is leaving node 1, and the constraint for node 7, embedded in cell I12, also equals 1, indicating that one truck will end up at node 7. This is an integer programming model, so be sure to click on "Options" and deactivate the "Ignore Integer Constraints" button.

EXHIBIT 7.4

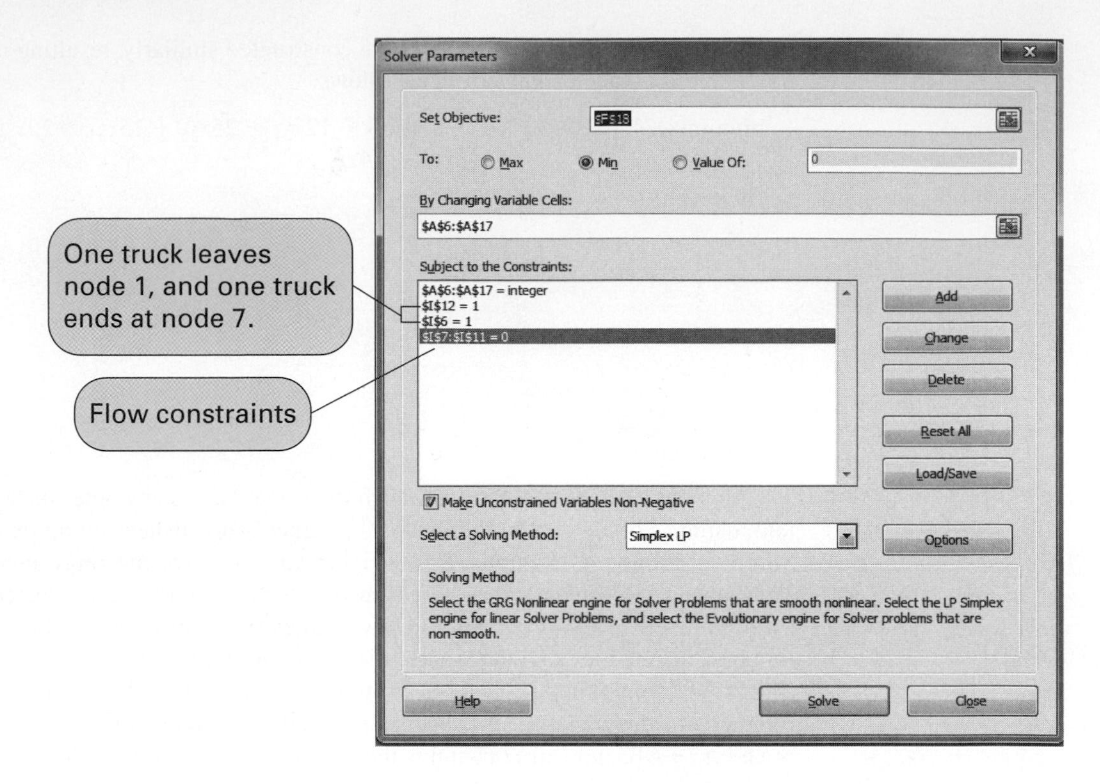

The Excel solution is shown in Exhibit 7.5. Notice that cells A7, A11, and A15 have values of 1 in them, indicating that these branches, 1–3, 3–4, and 4–7, are on the shortest route. The total distance in travel time is 43 hours, as shown in cell F18.

EXHIBIT 7.5

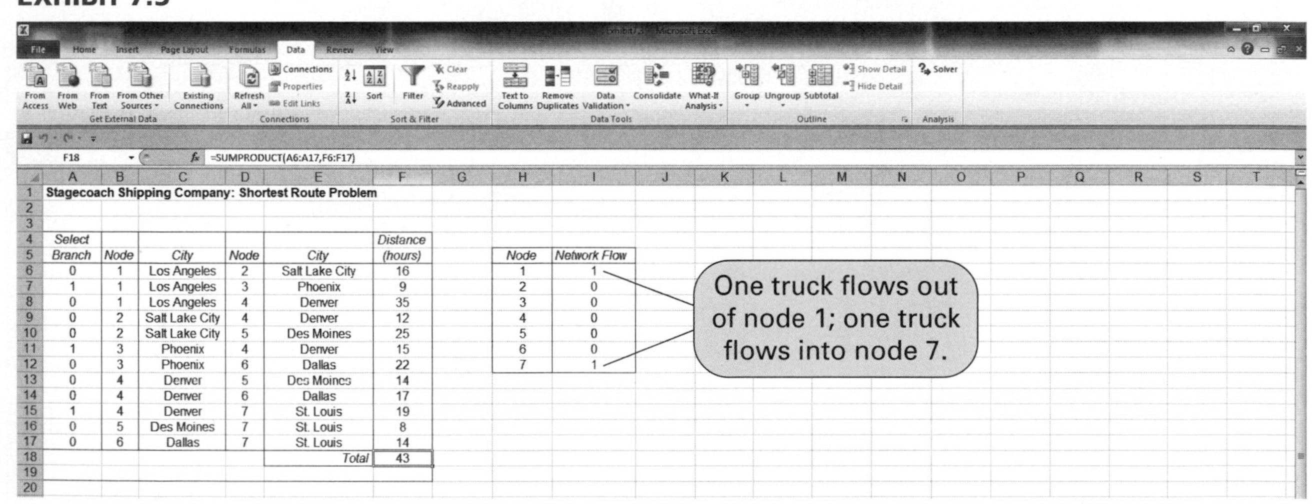

Stagecoach Shipping Company: Shortest Route Problem

Select Branch	Node	City	Node	City	Distance (hours)
0	1	Los Angeles	2	Salt Lake City	16
1	1	Los Angeles	3	Phoenix	9
0	1	Los Angeles	4	Denver	35
0	2	Salt Lake City	4	Denver	12
0	2	Salt Lake City	5	Des Moines	25
1	3	Phoenix	4	Denver	15
0	3	Phoenix	6	Dallas	22
0	4	Denver	5	Des Moines	14
0	4	Denver	6	Dallas	17
1	4	Denver	7	St. Louis	19
0	5	Des Moines	7	St. Louis	8
0	6	Dallas	7	St. Louis	14
				Total	43

Node	Network Flow
1	1
2	0
3	0
4	0
5	0
6	0
7	1

The Minimal Spanning Tree Problem

*The **minimal spanning tree problem** is to connect all nodes in a network so that the total branch lengths are minimized.*

In the shortest route problem presented in the previous section, the objective was to determine the shortest routes between the origin and the destination nodes in the network. In our example, we determined the best route from Los Angeles to each of the six destination cities. The **minimal spanning tree problem** is similar to the shortest route problem, except that the objective is to connect all the nodes in the network so that the total branch lengths are minimized.

The resulting network *spans* (connects) all the points in the network at a minimum total distance (or length).

To demonstrate the minimal spanning tree problem, we will consider the following example. The Metro Cable Television Company is to install a television cable system in a community consisting of seven suburbs. Each of the suburbs must be connected to the main cable system. The cable television company wants to lay out the main cable network in a way that will minimize the total length of cable that must be installed. The possible paths available to the cable television company (by consent of the town council) and the feet of cable (in thousands of feet) required for each path are shown in Figure 7.11.

FIGURE 7.11
Network of possible cable TV paths

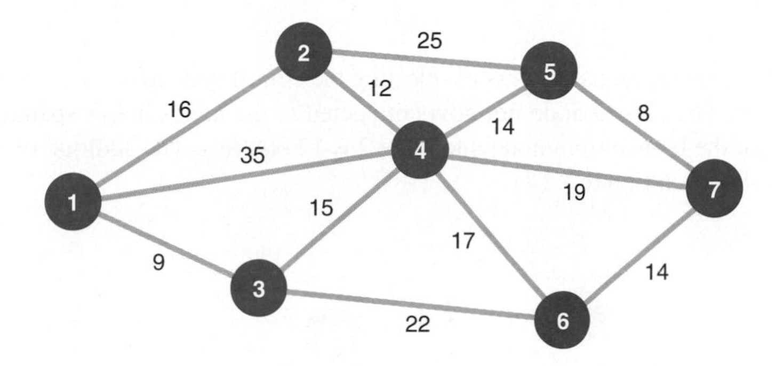

In Figure 7.11, the branch from node 1 to node 2 represents the available cable path between suburbs 1 and 2. The branch requires 16,000 feet of cable. Notice that the network shown in Figure 7.11 is identical to the network in Figure 7.2 that we used to demonstrate the shortest route problem. The networks were intentionally made identical to demonstrate the difference between the results of the two types of network models.

Start with any node in the network and select the closest node to join the spanning tree. Select the closest node to any node in the spanning area.

The Minimal Spanning Tree Solution Approach

The solution approach to the minimal spanning tree problem is actually easier than the shortest route solution method. In the minimal spanning tree solution approach, we can start at any node in the network. However, the conventional approach is to start with node 1. Beginning at node 1, we select the closest node (i.e., the shortest branch) to join our spanning tree. The shortest branch from node 1 is to node 3, with a length of 9 (thousand feet). This branch is indicated with a heavy line in Figure 7.12. Now we have a *spanning tree* consisting of two nodes: 1 and 3.

FIGURE 7.12
Spanning tree with nodes 1 and 3

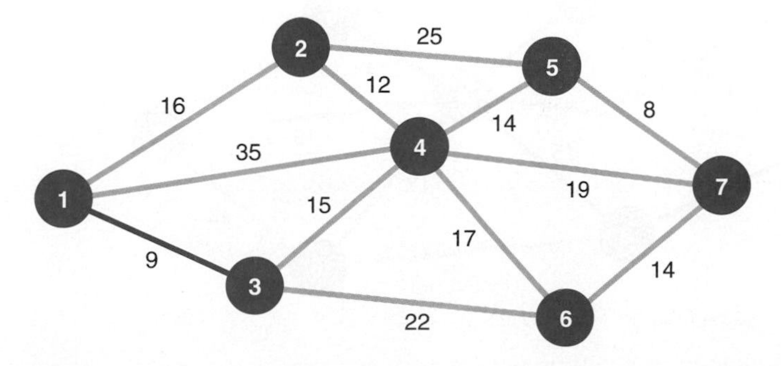

The next step is to select the closest node not currently in the spanning tree. The node closest to either node 1 or node 3 (the nodes in our present spanning tree) is node 4, with a branch length of 15,000 feet. The addition of node 4 to our spanning tree is shown in Figure 7.13.

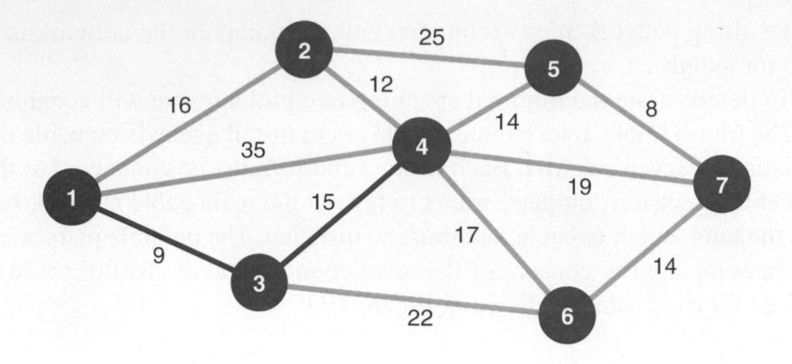

FIGURE 7.13
Spanning tree with nodes 1, 3, and 4

Next, we repeat the process of selecting the closest node to our present spanning tree (nodes 1, 3, and 4). The closest node not now connected to the nodes in our spanning tree is node 2. The length of the branch from node 4 to node 2 is 12,000 feet. The addition of node 2 to the spanning tree is shown in Figure 7.14.

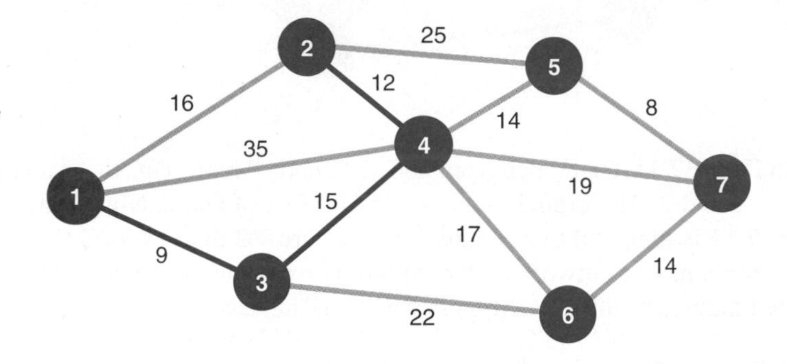

FIGURE 7.14
Spanning tree with nodes 1, 2, 3, and 4

Our spanning tree now consists of nodes 1, 2, 3, and 4. The node closest to this spanning tree is node 5, with a branch length of 14,000 feet to node 4. Thus, node 5 joins our spanning tree, as shown in Figure 7.15.

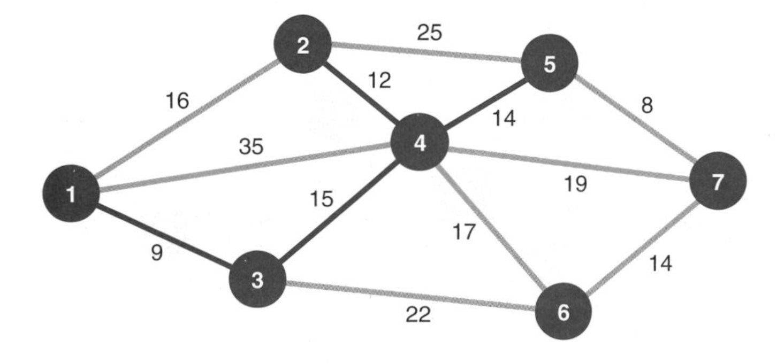

FIGURE 7.15
Spanning tree with nodes 1, 2, 3, 4, and 5

The spanning tree now contains nodes 1, 2, 3, 4, and 5. The closest node not currently connected to the spanning tree is node 7. The branch connecting node 7 to node 5 has a length of 8,000 feet. Figure 7.16 shows the addition of node 7 to the spanning tree.

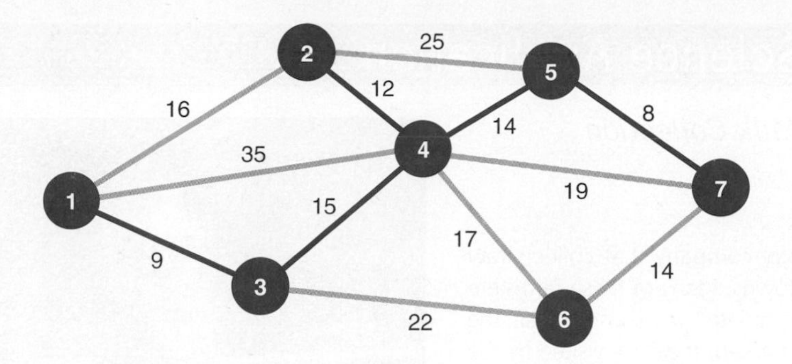

FIGURE 7.16
Spanning tree with nodes 1, 2, 3, 4, 5, and 7

Now our spanning tree includes nodes 1, 2, 3, 4, 5, and 7. The only remaining node not connected to the spanning tree is node 6. The node in the spanning tree closest to node 6 is node 7, with a branch length of 14,000 feet. The complete spanning tree, which now includes all seven nodes, is shown in Figure 7.17.

The spanning tree shown in Figure 7.17 requires the minimum amount of television cable to connect the seven suburbs—72,000 feet. This same minimal spanning tree could have been obtained by starting at any of the six nodes other than node 1.

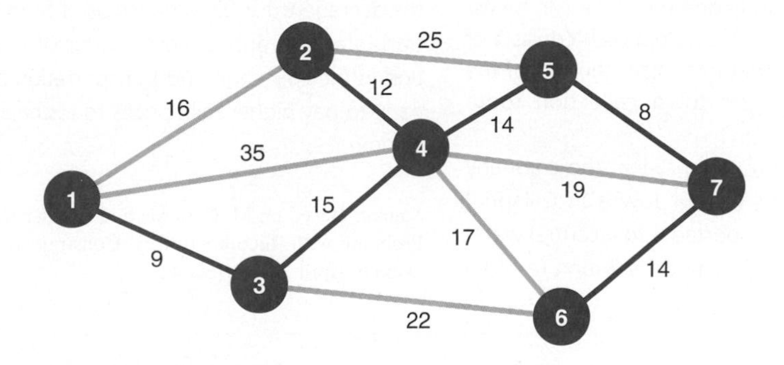

FIGURE 7.17
Minimal spanning tree for cable TV network

Notice the difference between the minimal spanning tree network shown in Figure 7.17 and the shortest route network shown in Figure 7.10. The shortest route network represents the shortest paths between the origin and each of the destination nodes (i.e., six different routes). In contrast, the minimal spanning tree network shows how to connect all seven nodes so that the total distance (length) is minimized.

Steps of the minimal spanning tree solution method.

In summary, the steps of the minimal spanning tree solution method are as follows:

1. Select any starting node (conventionally, node 1 is selected).
2. Select the node closest to the starting node to join the spanning tree.
3. Select the closest node not currently in the spanning tree.
4. Repeat step 3 until all nodes have joined the spanning tree.

Computer Solution of the Minimal Spanning Tree Problem with QM for Windows

The minimal spanning tree solution for our Metro Cable Television Company example using QM for Windows is shown in Exhibit 7.6.

Management Science Application

Determining Optimal Milk Collection Routes in Italy

blojfo/Shutterstock

ASSO.LA.C. is an Italian dairy company that collects raw milk from farmers in different towns. Many of these farms are small and inaccessible by a "complete" truck and trailer. The trailer must be uncoupled and parked, the farm visited by the truck and the milk collected, and then the truck and trailer must be coupled again. The milk is collected from the farms by a fleet of tank trucks that transport the milk to a central warehouse. The trucks are compartmentalized to collect milk of different types, and when milk of a particular type has been collected in a truck compartment, it cannot be used to hold a different type of milk.

The problem has the following constraints: Each node in the network (of farms) can be a loading point or parking area (for decoupling), the tank truck cannot exceed its capacity, multiple trucks can collect milk from a particular farmer, the time required for collections along a route cannot exceed the work shift, and only one milk type can be assigned to a compartment. The solution approach consists of two mathematical models: one that minimizes the size of the fleet and one that minimizes the town collection route length.

In the test case for the solution approach, the company collected milk from 158 farmers in four towns in Calabria, a region in southern Italy, and transported it to a central warehouse in the city of Castrovillari. The model solution resulted in a reduction in the total truck travel distance of 14.4%, and the average truck fill rate increased from 85% to 95%, which translates into annual cost savings of € 166,000. This reduction in the collection and transportation costs allows the company to pay higher milk prices to farmers, thus inviting higher volume.

Source: Based on M. Caramia and F. Guerriero, "A Milk Collection Problem with Incompatibility Constraints," *Interfaces* 40, no. 2 (March–April 2010): 130–43.

EXHIBIT 7.6

Networks Results

Metro Cable Television Company Solution

Branch name	Start node	End node	Cost	Include	Cost
1	0	2	16		
2	1	3	9	Y	9
3	1	4	35		
4	2	4	12	Y	12
5	2	5	25		
6	3	4	15	Y	15
7	3	6	22		
8	4	5	14	Y	14
9	4	6	17		
10	4	7	19		
11	5	7	8	Y	8
12	6	7	14	Y	14
Total					72

The Maximal Flow Problem

In the shortest route problem, we determined the shortest truck route from the origin (Los Angeles) to six destinations. In the minimal spanning tree problem, we found the shortest connected network for television cable. In neither of these problems was the capacity of a branch limited to a specific number of items. However, there are network problems in which the branches of the network have limited flow capacities. The objective of these networks is to maximize the total amount of flow from an origin to a destination. These problems are referred to as **maximal flow problems**.

*The **maximal flow problem** is to maximize the amount of flow of items from an origin to a destination.*

Maximal flow problems can involve the flow of water, gas, or oil through a network of pipelines; the flow of forms through a paper processing system (such as a government agency); the flow of traffic through a road network; or the flow of products through a production line system. In each of these examples, the branches of the network have limited and often different flow capacities. Given these conditions, the decision maker wants to determine the maximum flow that can be obtained through the system.

An example of a maximal flow problem is illustrated by the network of a railway system between Omaha and St. Louis shown in Figure 7.18. The Scott Tractor Company ships tractor parts from Omaha to St. Louis by railroad. However, a contract limits the number of railroad cars the company can secure on each branch during a week.

FIGURE 7.18
Network of railway system

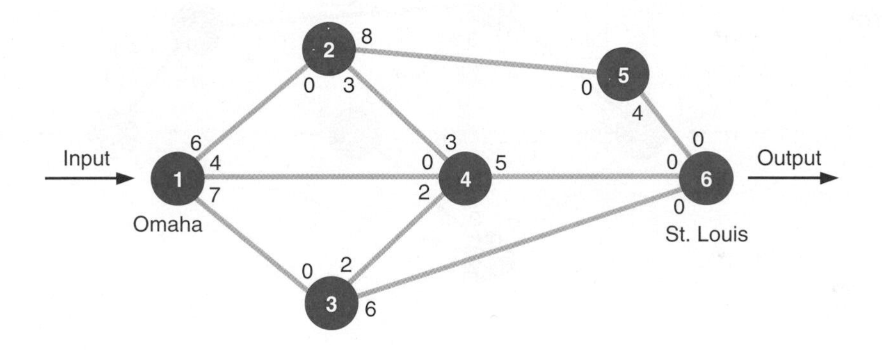

For a directed branch, flow is possible in only one direction.

Given these limiting conditions, the company wants to know the maximum number of railroad cars containing tractor parts that can be shipped from Omaha to St. Louis during a week. The number of railroad cars available to the tractor company on each rail branch is indicated by the number on the branch to the *immediate right of each node* (which represents a rail junction). For example, six cars are available from node 1 (Omaha) to node 2, eight cars are available from node 2 to node 5, five cars are available from node 4 to node 6 (St. Louis), and so forth. The number on each branch to the *immediate* left of each node is the number of cars available for shipping in the opposite direction. For example, no cars are available from node 2 to node 1. The branch from node 1 to node 2 is referred to as a *directed* branch because flow is possible in only one direction (from node 1 to node 2, but not from 2 to 1). Notice that flow is possible in both directions on the branches between nodes 2 and 4 and nodes 3 and 4. These are referred to as *undirected branches*.

The Maximal Flow Solution Approach

The first step in determining the maximum possible flow of railroad cars through the rail system is to *choose any path arbitrarily from origin to destination* and ship as much as possible on that path. In Figure 7.19, we will arbitrarily select the path 1–2–5–6. The maximum number of

Time Out for E. W. Dijkstra, L. R. Ford, Jr., and D. R. Fulkerson

In 1959, E. W. Dijkstra of the Netherlands proposed the solution procedures (shown in this chapter) for both the minimal spanning tree problem and the shortest route problem. Earlier, in 1955, L. R. Ford, Jr., and D. R. Fulkerson (colleagues of George Dantzig's) of the RAND Corporation introduced the procedure for solving the maximal flow network problem, which evolved from the study of transportation problems. The original problem analyzed was of a rail network connecting cities in which each rail link had capacity limitations.

Arbitrarily choose any path through the network from origin to destination and ship as much as possible.

railroad cars that can be sent through this route is four. We are limited to four cars because that is the maximum amount available on the branch between nodes 5 and 6. This path is shown in Figure 7.19.

FIGURE 7.19
Maximal flow for path 1–2–5–6

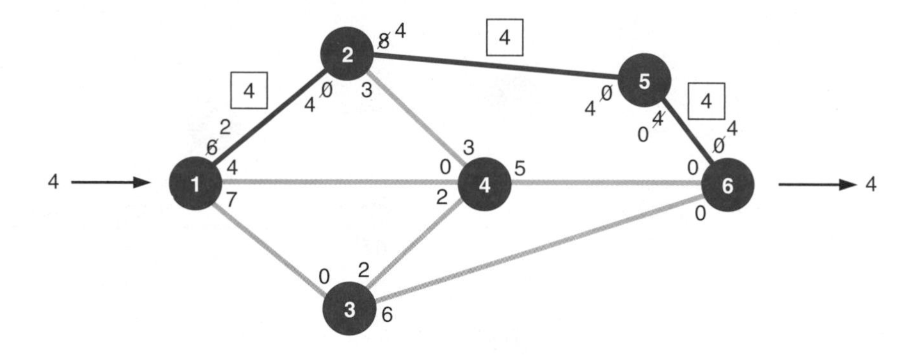

Recompute branch flow in both directions.

Notice that the remaining capacities of the branches from node 1 to node 2 and from node 2 to node 5 are now two and four cars, respectively, and that no cars are available from node 5 to node 6. These values were computed by subtracting the flow of four cars from the original number available. The actual flow of four cars along each branch is shown enclosed in a box. Notice that the present input of four cars into node 1 and output of four cars from node 6 are also designated.

The final adjustment on this path is to add the designated flow of four cars to the values at the immediate left of each node on our path, 1–2–5–6. These are the flows in the opposite direction. Thus, the value 4 is added to the zeros at nodes 2, 5, and 6. It may seem incongruous to designate flow in a direction that is not possible; however, it is the means used in this solution approach to compute the *net flow* along a branch. (If, for example, a later iteration showed a flow of one car from node 5 to node 2, then the net flow in the correct direction would be computed by subtracting this flow of one in the wrong direction from the previous flow of four in the correct direction. The result would be a net flow of three in the correct direction.)

Select other feasible paths arbitrarily and determine maximum flow along the paths until flow is no longer possible.

We have now completed one iteration of the solution process and must repeat the preceding steps. Again, we arbitrarily select a path. This time we will select path 1–4–6, as shown in Figure 7.20. The maximum flow along this path is four cars, which is subtracted at each of the nodes. This increases the total flow through the network to eight cars (because the flow of four along path 1–4–6 is added to the flow previously determined in Figure 7.19).

As a final step, the flow of four cars is added to the flow along the path in the opposite direction at nodes 4 and 6.

FIGURE 7.20
Maximal flow for path 1–4–6

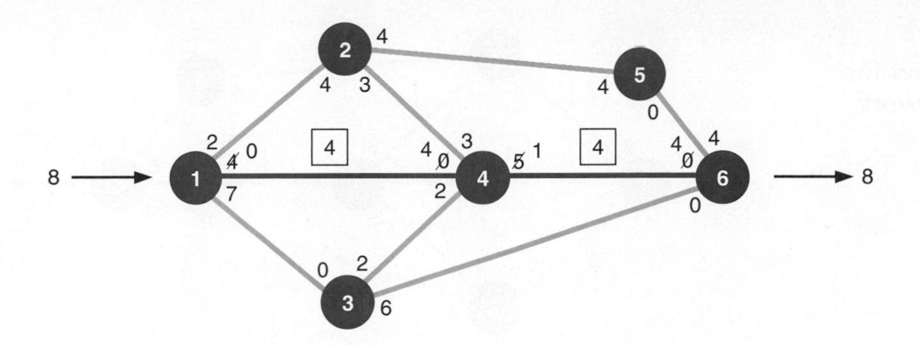

Now, we arbitrarily select another path. This time, we will choose path 1–3–6, with a maximum possible flow of six cars. This flow of six is subtracted from the branches along path 1–3–6 and added to the branches in the opposite direction, as shown in Figure 7.21.

FIGURE 7.21
Maximal flow for path 1–3–6

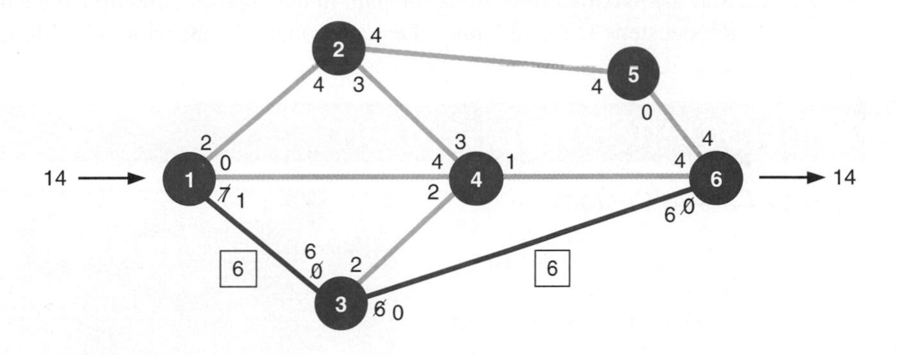

The flow of 6 for this path is added to the previous flow of 8, which results in a total flow of 14 railroad cars.

Notice that at this point the number of paths we can take is restricted. For example, we cannot take the branch from node 3 to node 6 because zero flow capacity is available. Likewise, no path that includes the branch from node 1 to node 4 is possible.

The available flow capacity along the path 1–3–4–6 is one car, as shown in Figure 7.22. This increases the total flow from 14 cars to 15 cars. The resulting network is shown in Figure 7.23. Close observation of the network in Figure 7.23 shows that there are no more paths with available flow capacity. All paths out of nodes 3, 4, and 5 show zero available capacity, which prohibits any further paths through the network.

Steps of the maximal flow solution method.

This completes the maximal flow solution for our example problem. The maximum flow is 15 railroad cars. The flows that will occur along each branch appear in boxes in Figure 7.23.

FIGURE 7.22
Maximal flow for path 1–3–4–6

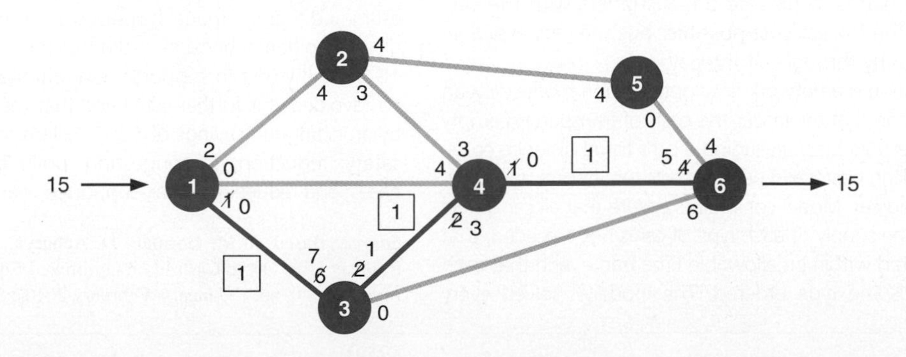

FIGURE 7.23
Maximal flow for railway network

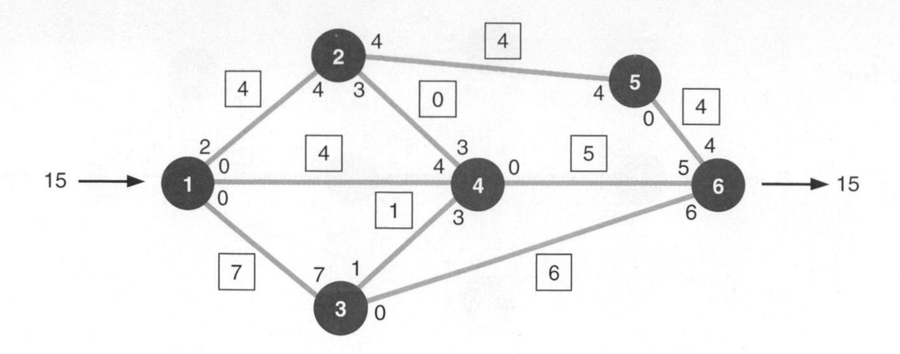

In summary, the steps of the maximal flow solution method are as follows:

1. Arbitrarily select any path in the network from origin to destination.
2. Adjust the capacities at each node by subtracting the maximal flow for the path selected in step 1.
3. Add the maximal flow along the path in the opposite direction at each node.
4. Repeat steps 1, 2, and 3 until there are no more paths with available flow capacity.

Management Science Application

Distributing Railway Cars to Customers at CSX

CSX Transportation, Inc., is one of the nation's major freight railroad companies, with a 21,000-mile rail network and 35,000 employees. CSX serves 23 states and 70 ports in the eastern part of the United States and Canada, and it has annual revenues of $11 billion. Every day CSX distributes hundreds of empty railroad cars (i.e., boxcars, gondolas, hoppers, tank cars) to its customers. CSX delivers an empty car to the location where a customer is loading, the car is loaded, and then it is transported to the customer's destination. At the destination, the car is unloaded and becomes available for another customer order. The empty car distribution problem is large and complex and has significant cost implications.

CSX maintains a fleet of 90,000 railroad cars that it relocates in a rail network with thousands of locations every day. Hauling empty cars hundreds of thousands of miles every day wears out tracks and cars, requires additional trains to haul the cars, and congests rail yards while generating no revenue. The company wants to provide the customers with the cars they need at the lowest cost possible, but the cars available change constantly throughout the day.

CSX's solution is a network flow optimization model with an objective function that minimizes the costs of transporting empty cars (the model variables), including a car's travel distance costs, time and handling costs, and penalty costs for lateness and delivering the wrong car. Model constraints ensure that all car orders are met, that the supply of each type of car is not exceeded, that cars are delivered within an allowable time frame, and that each car delivered is the type ordered. The model is solved every 15 minutes throughout the day in order to respond to rapidly changing conditions, such as supply and demand changes.

CSX estimates that the distribution system saves the company $51 million annually and has saved the company $561 million since its implementation several years ago. It is also estimated that the model has saved $1.4 billion in capital expenditure avoidance because it did not have to buy an estimated 18,000 new cars to support the additional miles the company has avoided. It is further estimated that the public has benefited by an estimated savings of $600 million for improved highway safety; reductions in congestion, pollution, and greenhouse gases; and reductions in tax-supported road maintenance.

Source: Based on M. Gorman, D. Acharya, and D. Sellers, "CSX Railway Uses OR to Cash In on Optimized Equipment Distribution," *Interfaces* 40, no. 1 (January–February 2010): 5–16.

Computer Solution of the Maximal Flow Problem with QM for Windows

The program input and maximal flow solution for the Scott Tractor Company example using QM for Windows is shown in Exhibit 7.7. (Note that this problem has multiple optimal solutions, and the branch flows are slightly different for this solution.)

EXHIBIT 7.7

Networks Results

Scott Tractor Company Solution

Branch name	Start node	End node	Capacity	Reverse capacity	Flow
Maximal Network Flow	15				
1	1	2	6	0	5
2	1	3	7	0	6
3	1	4	4	4	4
4	2	4	3	3	1
5	2	5	8	0	4
6	3	4	2	2	0
7	3	6	6	0	6
8	4	6	5	0	5
9	5	6	4	0	4

Computer Solution of the Maximal Flow Problem with Excel

The maximal flow problem can also be solved with Excel, much the same way as we solved the shortest route problem, by formulating it as an integer linear programming model and solving it by using "Solver".

The decision variables represent the flow along each branch, as follows:

$$x_{ij} = \text{flow along branch } i\text{–}j \text{ and integer}$$

To reduce the size and complexity of the model formulation, we will also eliminate flow along a branch in the opposite direction (e.g., flow from 4 to 2 is zero).

Before proceeding with the model formulation and developing the objective function, we must alter the network slightly to be able to solve it as a linear programming problem. We will create a new branch from node 6 back to node 1, x_{61}. The flow along this branch from the end of the network back to the start corresponds to the maximum amount that can be shipped from node 6 to node 1 and then back through the network to node 6. In effect, we are creating a continual flow through the network so that the most that goes through and comes out at node 6 is the most that can go through the network beginning at node 1. As a result, the objective is to maximize the amount that flows from node 6 back to node 1:

$$\text{maximize } Z = x_{61}$$

The constraints follow the same premise as the shortest route problem; that is, whatever flows into a node must flow out. Thus, at node 1, we have the following constraint, showing that the amount flowing from 6–1 must flow out through branches 1–2, 1–3, and 1–4:

$$x_{61} = x_{12} + x_{13} + x_{14}$$

This constraint can be rewritten as

$$x_{61} - x_{12} - x_{13} - x_{14} = 0$$

Similarly, the constraint at node 2 is written as

$$x_{12} - x_{24} - x_{25} = 0$$

We must also develop a set of constraints reflecting the capacities along each branch, as follows:

$$\begin{array}{ll} x_{12} \leq 6 & x_{34} \leq 2 \\ x_{13} \leq 7 & x_{36} \leq 6 \\ x_{14} \leq 4 & x_{46} \leq 5 \\ x_{24} \leq 3 & x_{56} \leq 4 \\ x_{25} \leq 8 & x_{61} \leq 17 \end{array}$$

The capacity for x_{61} can be any relatively large number (compared to the other branch capacities); we set it at the sum of the capacities on the branches leaving node 1.

The complete linear programming model for our Scott Tractor Company example is summarized as follows:

$$\begin{aligned} &\text{maximize } Z = x_{61} \\ &\text{subject to} \\ &x_{61} - x_{12} - x_{13} - x_{14} = 0 \\ &x_{12} - x_{24} - x_{25} = 0 \\ &x_{13} - x_{34} - x_{36} = 0 \\ &x_{14} + x_{24} + x_{34} - x_{46} = 0 \\ &x_{25} - x_{56} = 0 \\ &x_{36} + x_{46} + x_{56} - x_{61} = 0 \\ &x_{12} \leq 6 \\ &x_{13} \leq 7 \\ &x_{14} \leq 4 \\ &x_{24} \leq 3 \\ &x_{25} \leq 8 \\ &x_{34} \leq 2 \\ &x_{36} \leq 6 \\ &x_{46} \leq 5 \\ &x_{56} \leq 4 \\ &x_{61} \leq 17 \\ &x_{ij} \geq 0 \text{ and integer} \end{aligned}$$

The Excel spreadsheet set up to solve this integer linear programming model for the Scott Tractor Company example is shown in Exhibit 7.8. The decision variables, x_{ij}, are represented by cells **C6:C15**. Cells **D6:D15** include the branch capacities. The objective function, which is to maximize the flow along branch 6–1, is included in cell C16. The model constraints reflecting the flow through each node are included in the box on the right side of the spreadsheet. For example, cell G6 contains the constraint formula at node 1, **=C15−C6−C7−C8,** and cell G7 contains the constraint formula for node 2, **=C6−C9−C10.** The constraints for the branch capacities are obtained by adding the formula **C6:C15 ≤ D6:D15** in Solver in Exhibit 7.9.

EXHIBIT 7.8

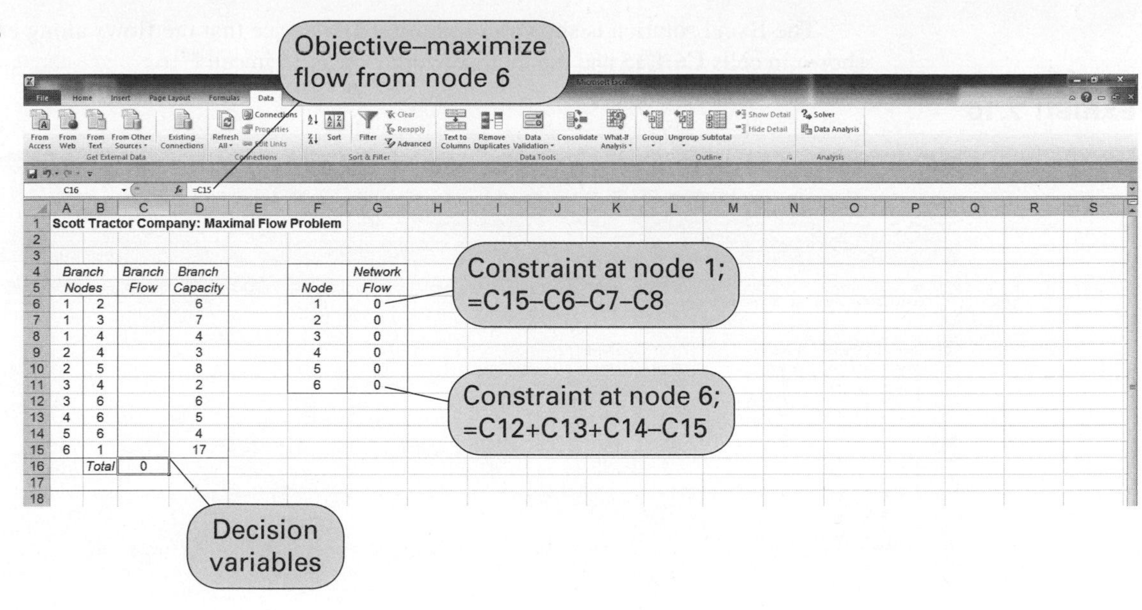

Branch Nodes		Branch Flow	Branch Capacity
1	2		6
1	3		7
1	4		4
2	4		3
2	5		8
3	4		2
3	6		6
4	6		5
5	6		4
6	1		17
	Total	0	

Node	Network Flow
1	0
2	0
3	0
4	0
5	0
6	0

Exhibit 7.9 shows the Solver Parameters screen for this model.

EXHIBIT 7.9

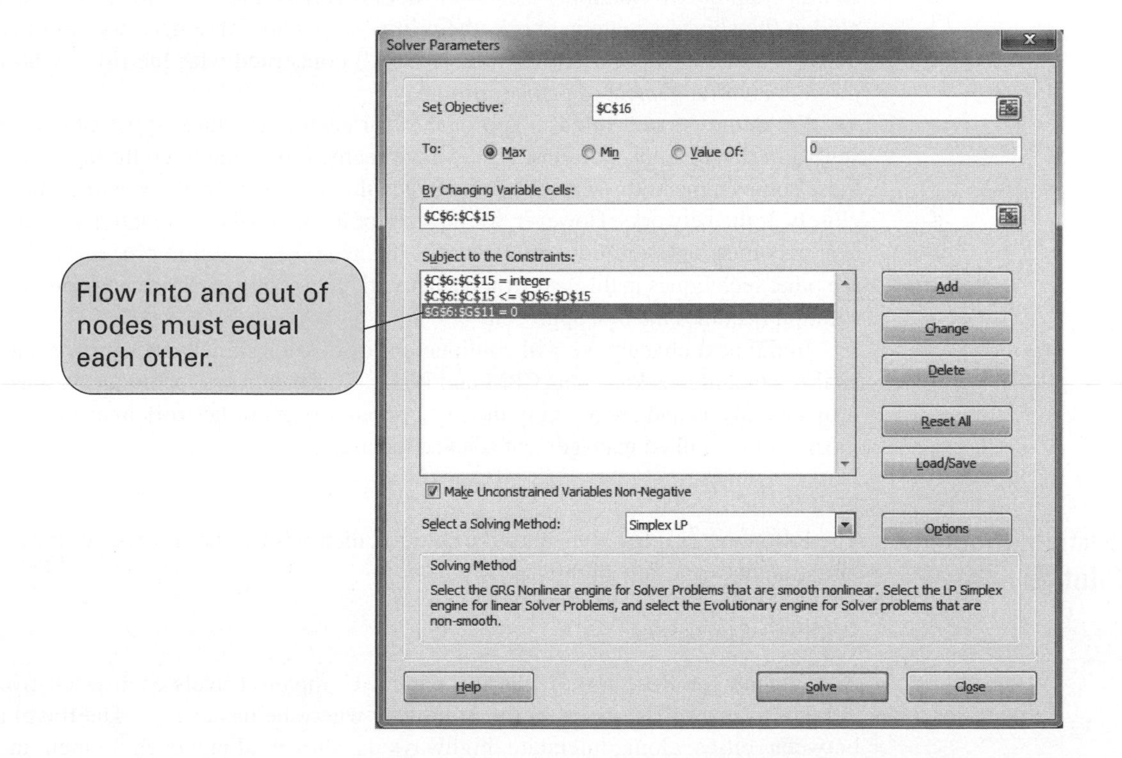

The Excel solution is shown in Exhibit 7.10. Notice that the flows along each branch are shown in cells **C6:C15** and the total network flow is 15 in cell C16.

EXHIBIT 7.10

G11 =C12+C13+C14-C15

Scott Tractor Company: Maximal Flow Problem

Branch Nodes		Branch Flow	Branch Capacity
1	2	4	6
1	3	7	7
1	4	4	4
2	4	0	3
2	5	4	8
3	4	1	2
3	6	6	6
4	6	5	5
5	6	4	4
6	1	15	17
	Total	15	

Node	Network Flow
1	0
2	0
3	0
4	0
5	0
6	0

Summary

In this chapter, we examined a class of models referred to as network flow models. These included the shortest route network, the minimal spanning tree network, and the maximal flow network model. These network models are all concerned with the flow of an item (or items) through an arrangement of paths (or routes).

We demonstrated solution approaches for each of the three types of network models presented in this chapter. At times, it may have seemed tiresome to go through the various steps of these solution methods, when the solutions could have more easily been found by simply looking closely at the networks. However, as the sizes of networks increase, intuitive solution by observation becomes more difficult, thus creating the need for a solution procedure. Of course, as with the other techniques in this text, when a network gets extremely large and complex, computerized solution becomes the best approach.

In the next chapter, we will continue our discussion of networks by presenting the network analysis techniques known as CPM and PERT. These network techniques are used primarily for project analysis and are not only the most popular types of network analysis but also two of the most widely applied management science techniques.

Example Problem Solution

The following example illustrates the solution methods for the shortest route and minimal spanning tree network flow problems.

Problem Statement

A salesman for Healthproof Pharmaceutical Company travels each week from his office in Atlanta to one of five cities in the Southeast where he has clients. The travel time (in hours) between cities along interstate highways is shown along each branch in the following network:

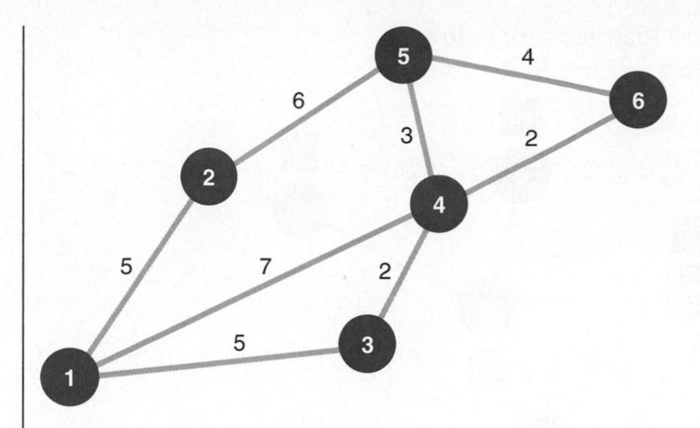

A. Determine the shortest route from Atlanta to each of the other five cities in the network.
B. Assume that the network now represents six different communities in a city and that the local transportation authority wants to design a rail system that will connect all six communities with the minimum amount of track. The miles between each community are shown on each branch. Develop a minimal spanning tree for this problem.

Solution

Step 1 (Part A): Determine the Shortest Route Solution

	Branch	Time
1. Permanent set {1}	1–2	[5]
	1–3	5
	1–4	7
2. Permanent set {1, 2}	1–3	[5]
	1–4	7
	2–5	11
3. Permanent set {1, 2, 3}	1–4	[7]
	2–5	11
	3–4	7
4. Permanent set {1, 2, 3, 4}	4–5	10
	4–6	[9]
5. Permanent set {1, 2, 3, 4, 6}	4–5	[10]
	6–5	13
6. Permanent set {1, 2, 3, 4, 5, 6}		

The shortest route network follows:

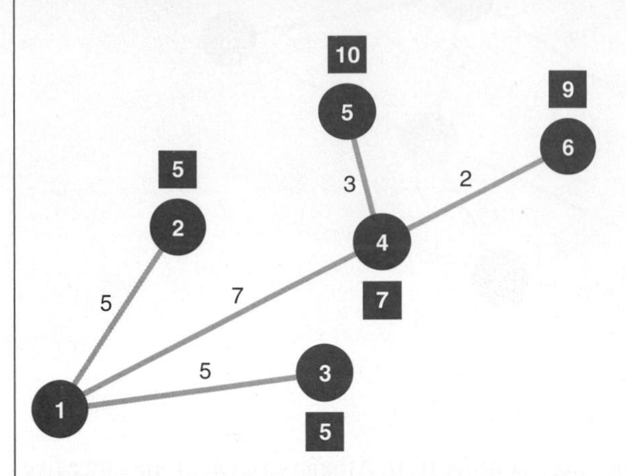

Step 2 (Part B): Determine the Minimal Spanning Tree

1. The closest unconnected node to node 1 is node 2.
2. The closest unconnected node to 1 and 2 is node 3.
3. The closest unconnected node to 1, 2, and 3 is node 4.
4. The closest unconnected node to 1, 2, 3, and 4 is node 6.
5. The closest unconnected node to 1, 2, 3, 4, and 6 is node 5.

The minimal spanning tree follows; the shortest total distance is 17 miles:

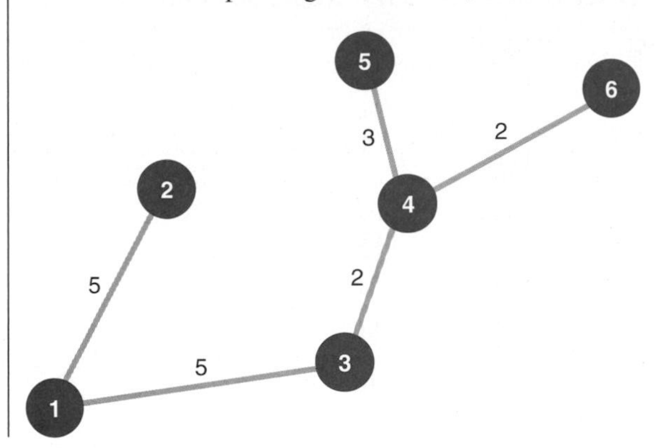

Problems

1. Given the following network with the indicated distances between nodes (in miles), determine the shortest route from node 1 to each of the other six nodes (2, 3, 4, 5, 6, and 7):

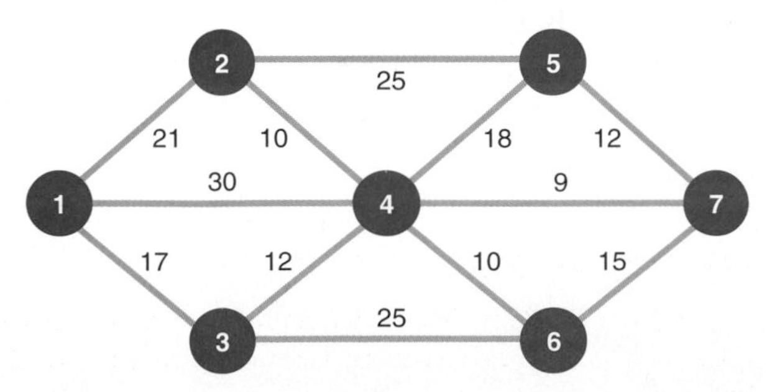

2. Frieda Millstone and her family live in Roanoke, Virginia, and they are planning an auto vacation across Virginia, their ultimate destination being Washington, DC. The family has developed the following network of possible routes and cities to visit on their trip:

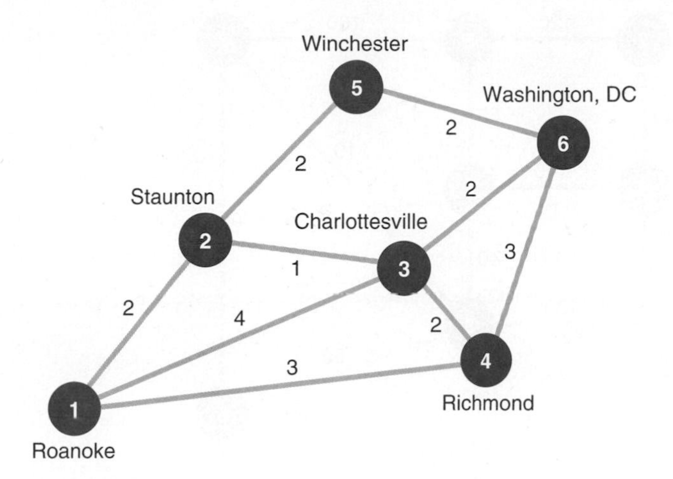

The time, in hours, between cities (which is affected by the type of road and the number of intermediate towns) is shown along each branch. Determine the shortest route that the Millstone family can travel from Roanoke to Washington, DC.

3. The Roanoke, Virginia, distributor of Rainwater Beer delivers beer by truck to stores in six other Virginia cities, as shown in the following network:

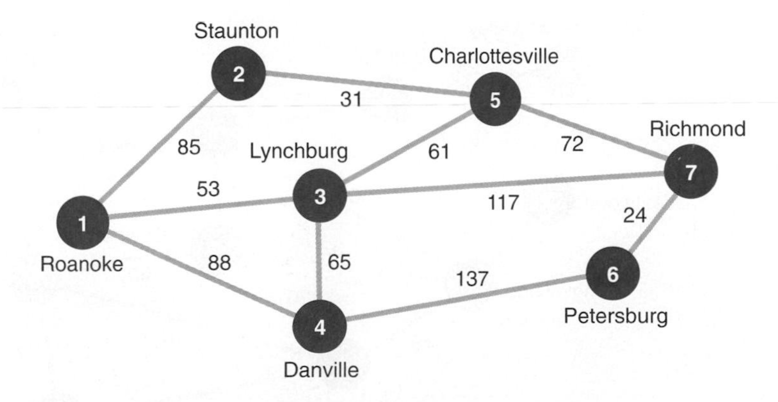

The mileage between cities is shown along each branch. Determine the shortest truck route from Roanoke to each of the other six cities in the network.

4. The plant engineer for the Bitco manufacturing plant is designing an overhead conveyor system that will connect the distribution/inventory center to all areas of the plant. The network

of possible conveyor routes through the plant, with the length (in feet) along each branch, follows:

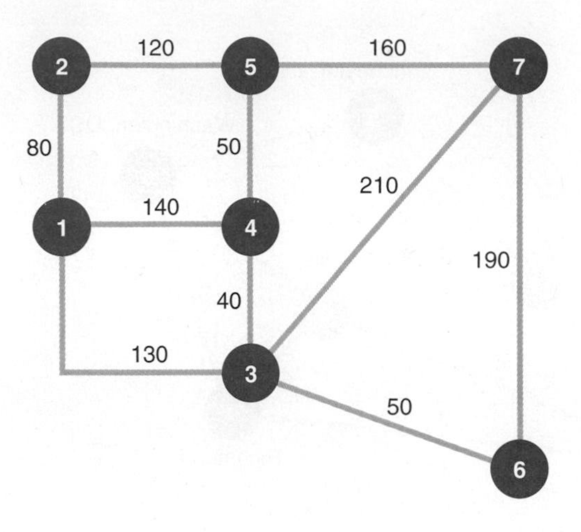

Determine the shortest conveyor route from the distribution/inventory center at node 1 to each of the other six areas of the plant.

5. The Burger Doodle restaurant franchises in Los Angeles are supplied from a central warehouse in Inglewood. The location of the warehouse and its proximity, in minutes of travel time, to the franchises are shown in the following network:

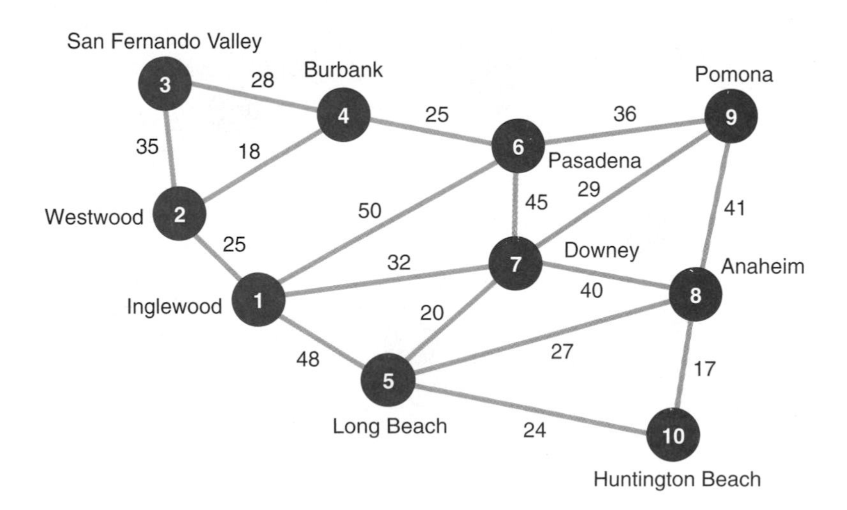

Trucks supply each franchise on a daily basis. Determine the shortest route from the warehouse at Inglewood to each of the nine franchises.

6. The Petroco gasoline distributor in Jackson, Mississippi, supplies service stations in nine other southeastern cities, as shown in the following network:

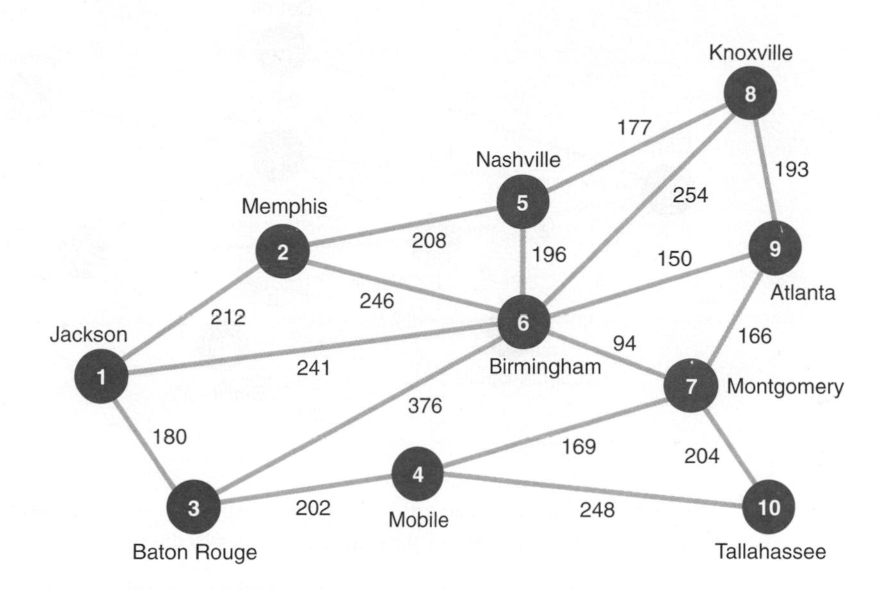

The distance, in miles, is shown on each branch. Determine the shortest route from Jackson to the nine other cities in the network.

7. The Hylton Hotel has a limousine van that transports guests to various business and tourist locations around the city. The following network indicates the different routes the limousine could follow from the hotel at node 1 to the nine locations (nodes 2 through 10):

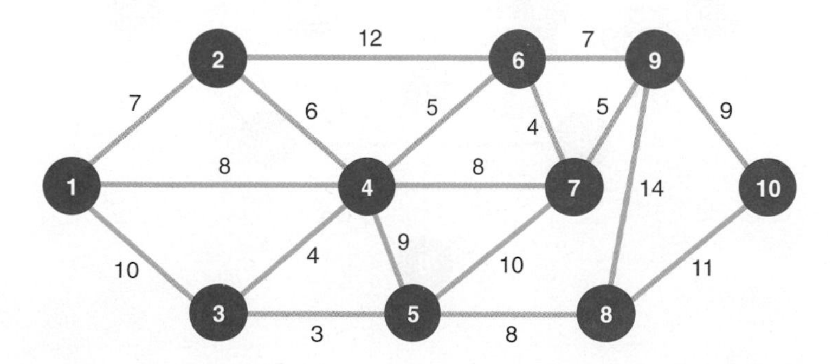

The values on each branch in the network are the distances, in miles, between the locations. Determine the shortest route from the hotel to each of the nine destinations and indicate the distance for each route.

8. A steel mill in Gary supplies steel to manufacturers in eight other midwestern cities by truck, as shown in the following network:

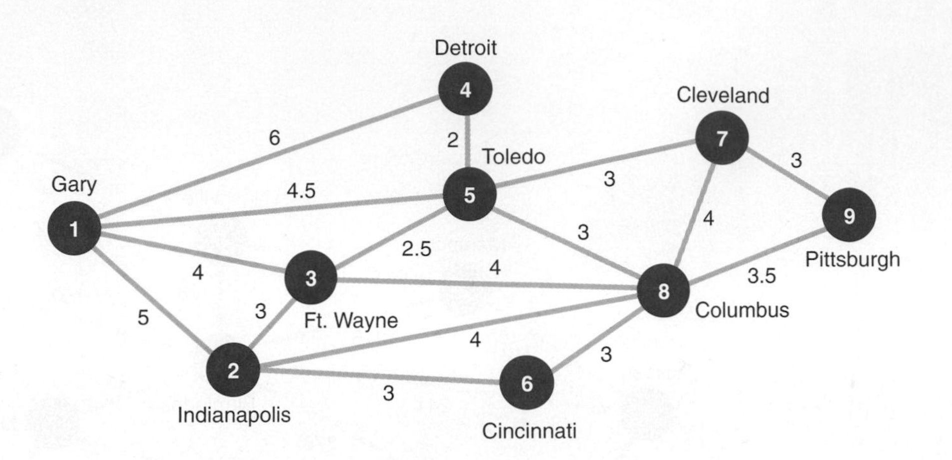

The travel time between cities, in hours, is shown along each branch. Determine the shortest route from Gary to each of the other eight cities in the network.

9. Determine the shortest route from node 1 (origin) to node 12 (destination) for the following network. Distances are given along the network branches:

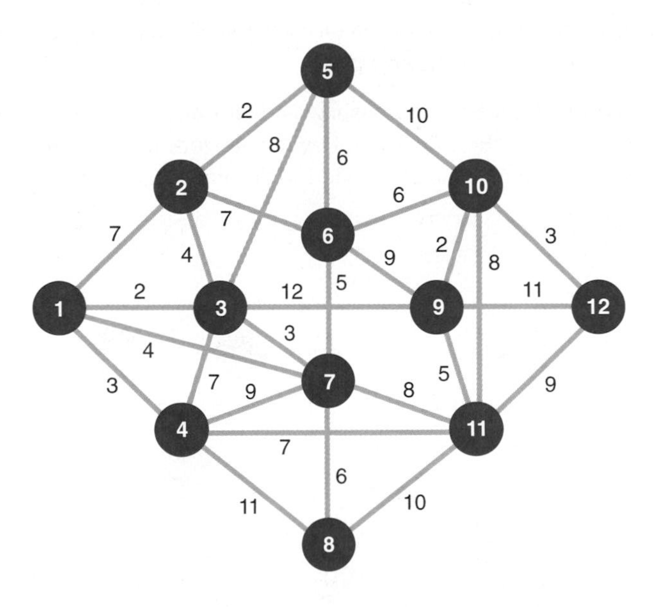

10. The members of the Vistas Club are planning a month long hiking and camping trip through the Blue Ridge Mountains, from north Georgia to Virginia. There are a number of trails through the mountains that connect at various campgrounds, crossings, and cabins. The following network shows the different connecting trails and the distance of each connecting branch, in miles:

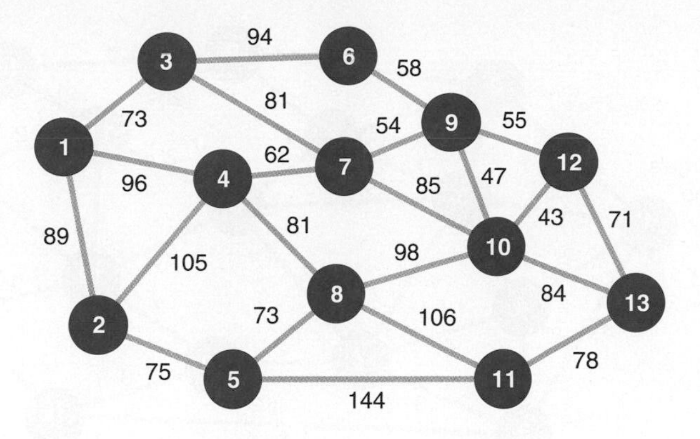

Determine the shortest path from node 1 in north Georgia to node 13 in Virginia and indicate the total mileage of the trail.

11. Hard Rock Concrete Supply makes concrete at its plant in Centerville, Virginia, and delivers it to construction sites throughout the metropolitan Washington, DC, area. The following network shows the possible routes and distances (in miles) from the concrete plant to seven construction sites:

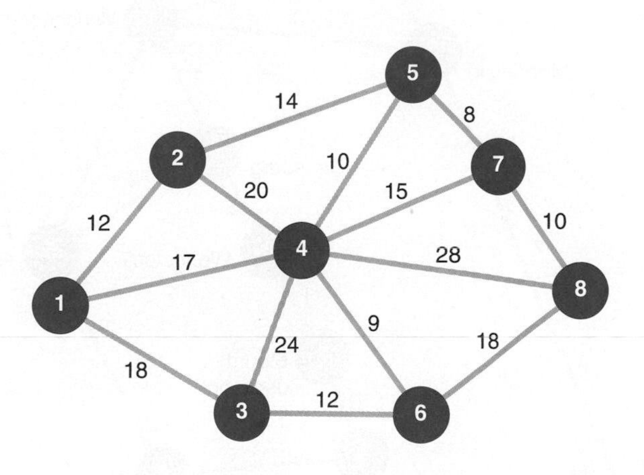

Determine the shortest route a concrete truck would take from the plant at node 1 to node 8 and the total distance for this route, using the shortest route method.

12. George is camped deep in the jungle, and he wants to make his way back to the coast and civilization. Each of the paths he can take through the jungle has obstacles that can delay him, including hostile natives, wild animals, dense forests and vegetation, swamps, rivers and streams, snakes, insects, and mountains. Following is the network of paths that George can take, with the time (in days) to travel each branch:

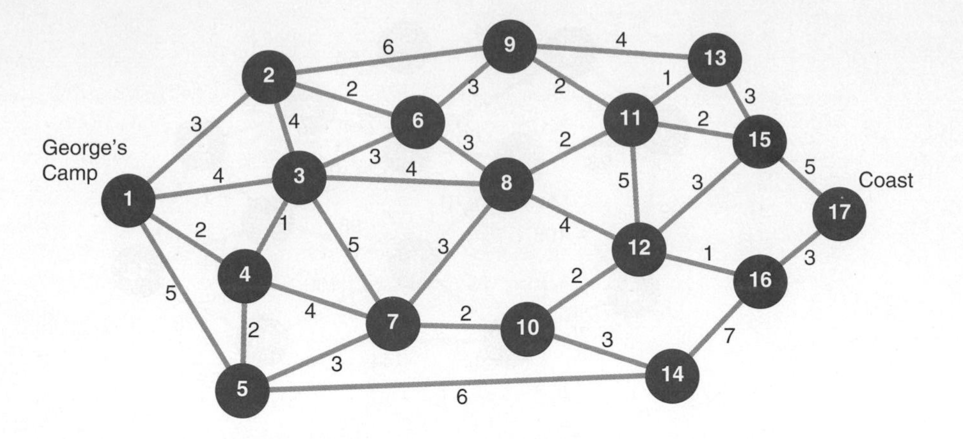

Determine the shortest (time) path for George to take and indicate the total time.

13. In 1862, during the second year of the Civil War, General Thomas J. "Stonewall" Jackson fought a brilliant military campaign in the Shenandoah Valley in Virginia. One of his victories was at the Battle of McDowell. Using the following figure and your imagination, determine the shortest path and how long it will take (in days) for General Jackson to move his army from Winchester to McDowell to fight the battle:

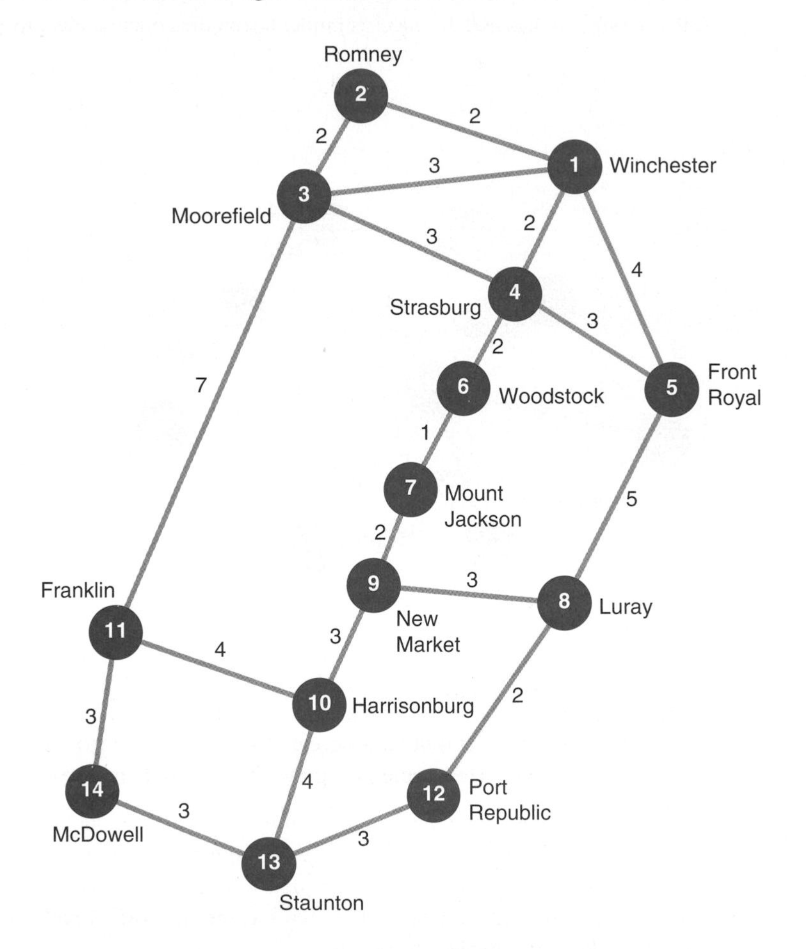

14. The *Voyager* spacecraft has been transported by an alien being to the Delta Quadrant of space, millions of light-years from Earth. Captain Janeway and her crew are attempting to plot the shortest course home. Following is a network of the possible routes through the Delta Quadrant, where the nodes are different worlds, planetary systems, and space anomalies, and the values on the branches are the actual times, in years, between nodes. Determine the shortest route from the *Voyager*'s present location (at node 1) to Earth.

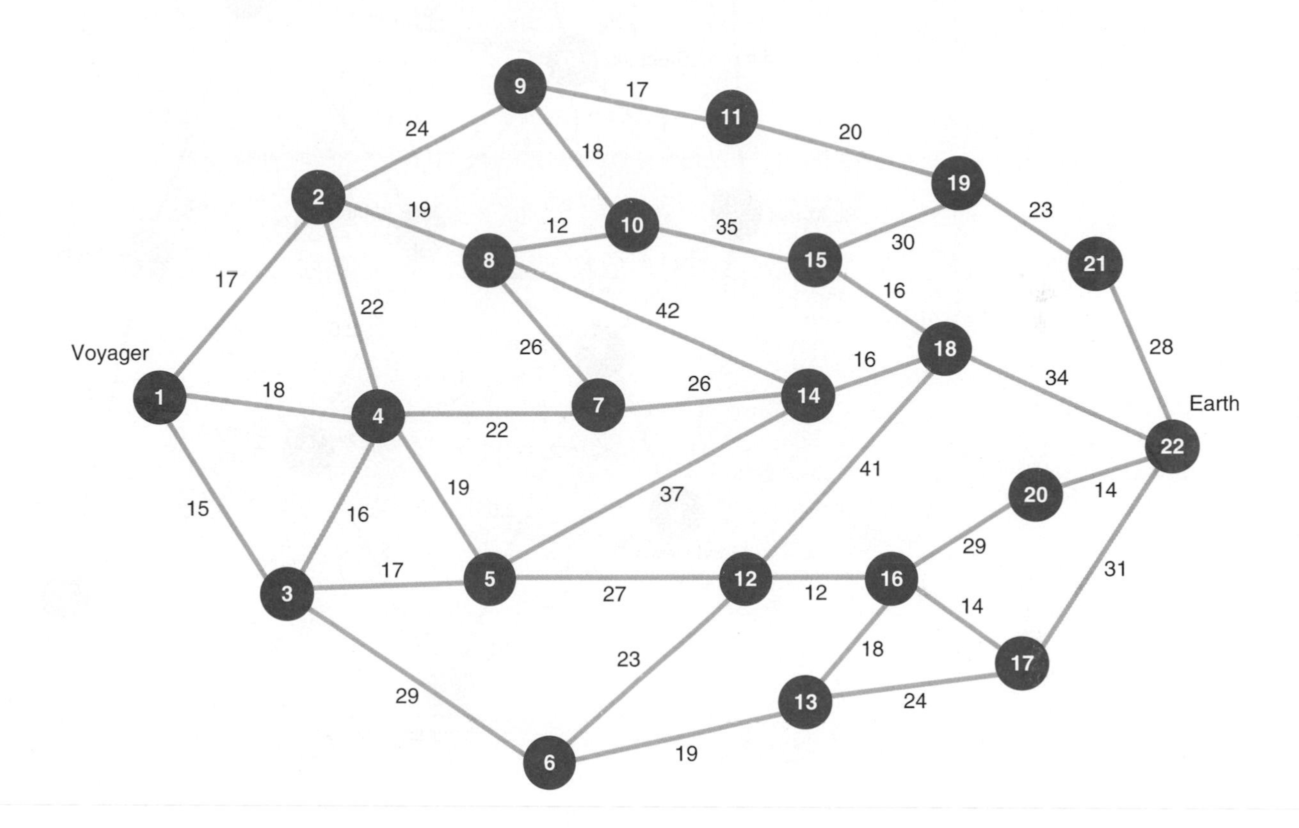

15. John Clooney, a bush pilot in Alaska, makes regular charter flights in his floatplane to various towns and cities in western Alaska. His passengers include hunters, fishermen, backpackers and campers, and tradespeople hired for jobs in the different localities. He also carries some cargo for delivery. The following network shows the possible air routes between various towns and cities John might take (with the times, in hours). For safety reasons, he flies point-to-point, flying over at least one town along a route, even though he might not land there. In the upcoming week John has scheduled charter flights for Kotzebue, Nome, and Stebbins. Determine the shortest route between John's home base in Anchorage and each of these destinations.

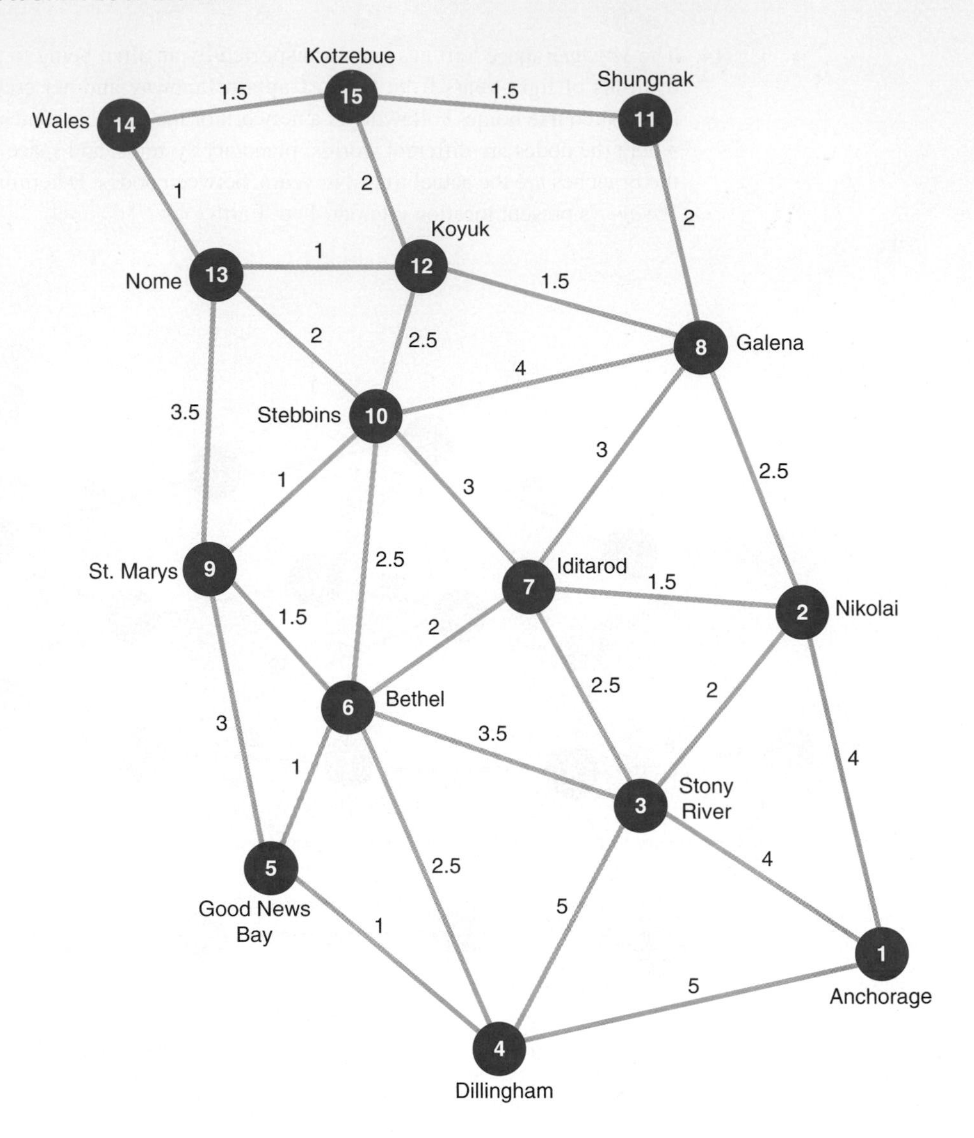

16. From 1840 to 1850, more than 12,000 pioneers migrated west in wagon trains. It was typically about a 2,000-mile journey, and pioneers averaged about 10 miles per day. One pioneer family, the Smiths, is planning to join a wagon train traveling west to Oregon. The destination is Fort Vancouver, which is near present-day Portland. The family plans to join a wagon train in St. Louis. However, the trains follow various trails west that are mostly determined by the location of forts and trading posts along the way. The Smith family wants to choose a wagon train that will get them to Oregon in the shortest amount of time. They have checked around with the different wagon train leaders plus immigrants, soldiers, fur traders, and scouts who have previously made the trip west, and from the information they have gathered, they have developed the following network, with estimated times (in days) along each branch:

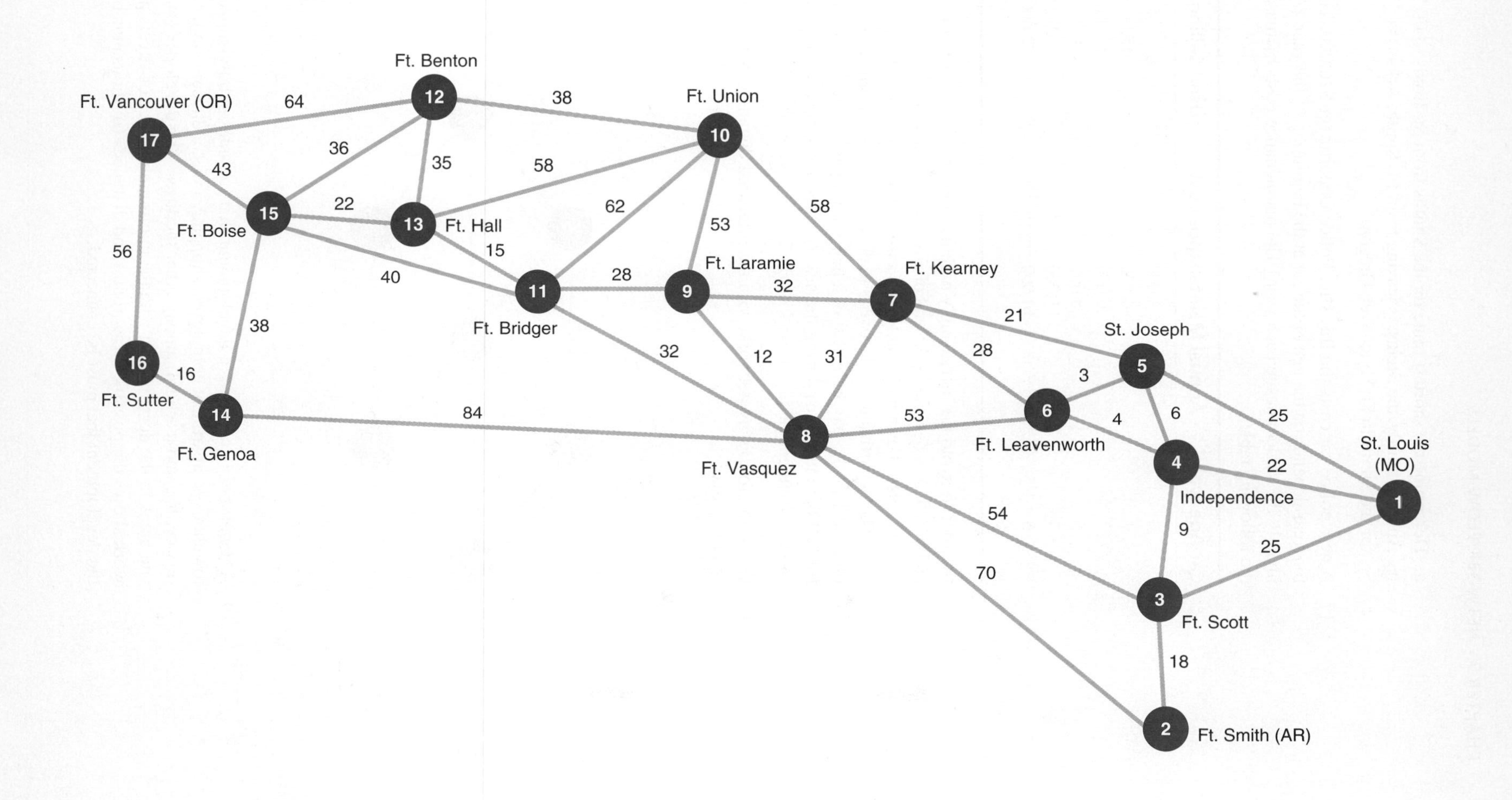

Ft. Benton
12
Ft. Vancouver (OR)
17
Ft. Union
10
15
Ft. Boise
13
Ft. Hall
Ft. Laramie
9
11
Ft. Bridger
Ft. Kearney
7
St. Joseph
5
16
Ft. Sutter
14
Ft. Genoa
8
Ft. Vasquez
6
Ft. Leavenworth
4
Independence
St. Louis
(MO)
1
3
Ft. Scott
2
Ft. Smith (AR)
64
38
36
35
58
43
22
62
58
53
56
15
40
28
32
38
21
32
12
31
28
3
16
84
53
4
6
25
22
54
9
25
70
18

a. Determine the shortest route for the Smiths from St. Louis to Ft. Vancouver.
b. Another family, the Steins, is leaving from Ft. Smith, Arkansas. Determine the shortest route for the Steins to Ft. Vancouver, Oregon.

17. A new police car costs the Bay City Police Department $26,000. The annual maintenance cost for a car depends on the age of the car at the beginning of the year. (All cars accumulate approximately the same mileage each year.) The maintenance costs increase as a car ages, as shown in the following table:

Car Age (yr.)	Annual Maintenance Cost	Used Selling Price
0	$ 3,000	$ —
1	4,500	15,000
2	6,000	12,000
3	8,000	8,000
4	11,000	4,000
5	14,000	2,000
6		0

In order to avoid the increasingly high maintenance costs, the police department can sell a car and purchase a new car at the end of any year. The selling price for a car at the end of each year of use is also shown in the table. It is assumed that the price of a new car will increase $500 each year. The department's objective is to develop a car replacement schedule that will minimize total cost (i.e., the purchase cost of a new car plus maintenance costs minus the money received for selling a used car) during the next 6 years. Develop a car replacement schedule, using the shortest route method.

18. Given the following network, with the indicated distances between nodes, develop a minimal spanning tree:

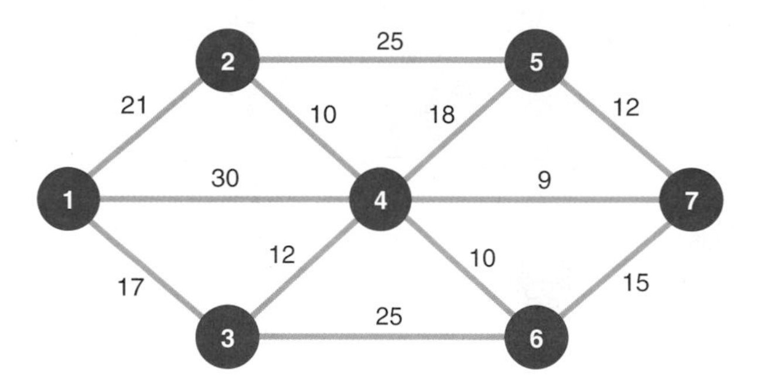

19. A developer is planning a development that includes subdivisions of houses, cluster houses, townhouses, apartment complexes, shopping areas, a daycare center and playground, a community center, and a school, among other facilities. The developer wants to connect the facilities and areas in the development with the minimum number of streets possible. The following network shows the possible street routes and distances (in thousands of feet) between 10 areas in the development that must be connected:

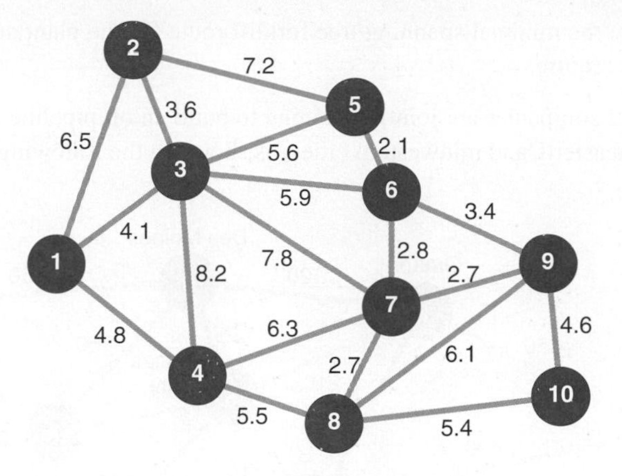

Determine a minimal spanning tree network of streets to connect the 10 areas, and indicate the total streets (in thousands of feet) needed.

20. One of the opposing forces in a simulated army battle wishes to set up a communications system that will connect the eight camps in its command. The following network indicates the distances (in hundreds of yards) between the camps and the different paths over which a communications line can be constructed:

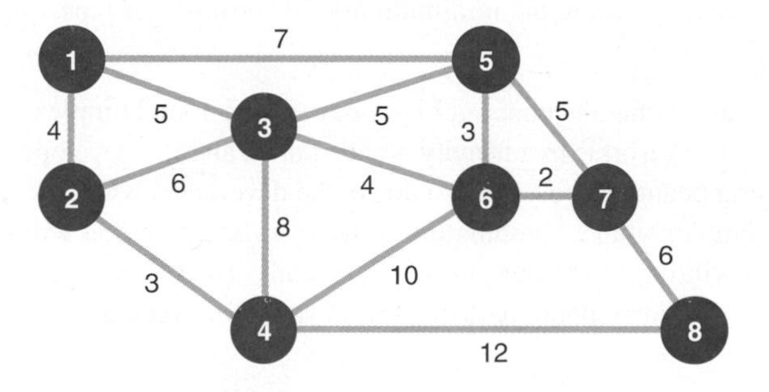

Using the minimal spanning tree approach, determine the minimum distance communication system that will connect all eight camps.

21. The management of the Dynaco manufacturing plant wants to connect the eight major manufacturing areas of its plant with a forklift route. Because the construction of such a route will take a considerable amount of plant space and disrupt normal activities, management wants to minimize the total length of the route. The following network shows the distance, in yards, between the manufacturing areas (denoted by nodes 1 through 8):

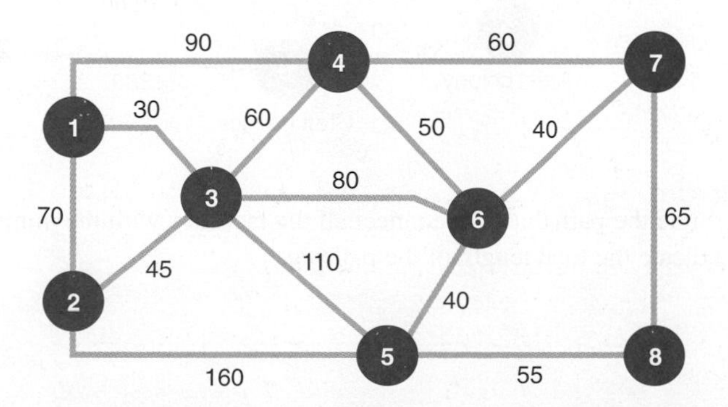

Determine the minimal spanning tree forklift route for the plant and indicate the total yards the route will require.

22. Several oil companies are jointly planning to build an oil pipeline to connect several southwestern, southeastern, and midwestern cities, as shown in the following network:

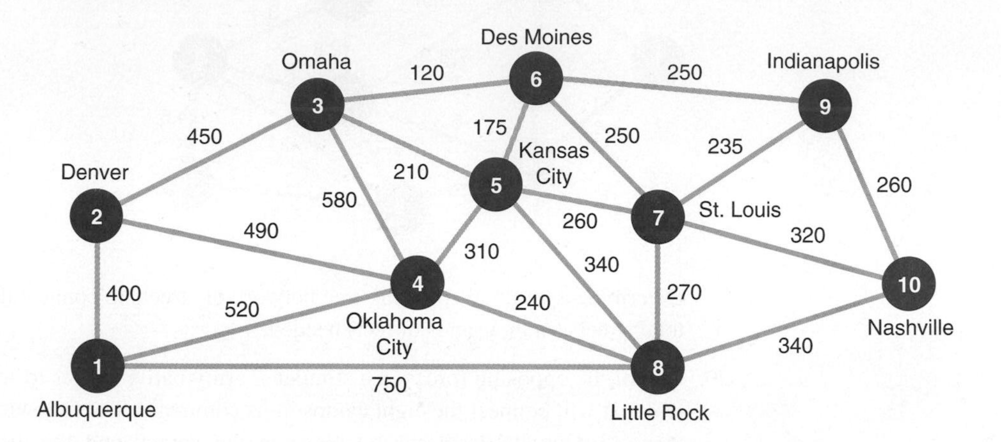

The miles between cities are shown on each branch. Determine a pipeline system that will connect all 10 cities, using the minimum number of miles of pipe, and indicate how many miles of pipe will be used.

23. A major hotel chain is constructing a new resort hotel complex in Greenbranch Springs, West Virginia. The resort is in a heavily wooded area, and the developers want to preserve as much of the natural beauty as possible. To do so, the developers want to connect all the various facilities in the complex with a combination walking–riding path that will minimize the amount of pathway that will have to be cut through the woods. The following network shows possible connecting paths and corresponding distances (in yards) between the facilities:

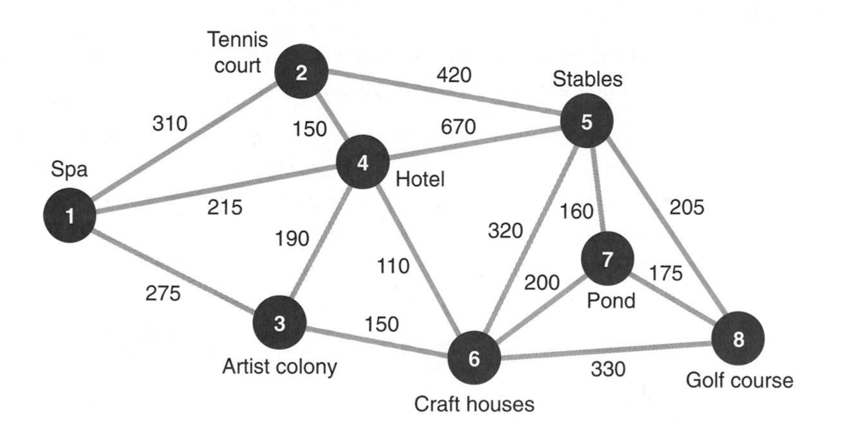

Determine the path that will connect all the facilities with the minimum amount of construction and indicate the total length of the pathway.

24. The Barrett Textile Mill is remodeling its plant and installing a new ventilation system. The possible ducts connecting the different rooms and buildings at the plant, with the length (in feet) along each branch, are shown in the following network:

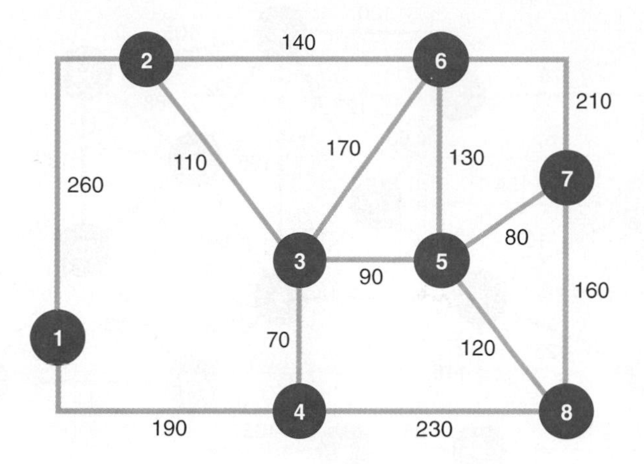

Determine the ventilation system that will connect the various rooms and buildings of the plant, using the minimum number of feet of ductwork, and indicate how many feet of ductwork will be used.

25. The town council of Whitesville has decided to construct a bicycle path that will connect the various suburbs of the town with the shopping center, the downtown area, and the local college. The council hopes the local citizenry will use the bike path, thus conserving energy and decreasing traffic congestion. The various paths that can be constructed and their lengths (in miles) are shown in the following network:

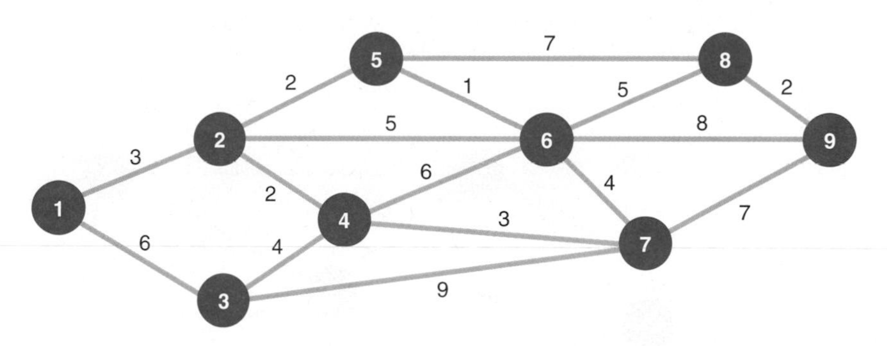

Determine the bicycle pathway that will require the minimum amount of construction to connect all the areas of the town. Indicate the total length of the path.

26. Determine the minimal spanning tree for the network in Problem 9.

27. State University has decided to reconstruct the sidewalks throughout the east side of its campus to provide wheelchair access. However, upgrading sidewalks is a very expensive undertaking, so for the first phase of this project, university administrators want to make sure they connect all buildings with wheelchair access with the minimum number of refurbished sidewalks possible.

Following is a network of the existing sidewalks on the east side of campus, with the feet between each building shown on the branches:

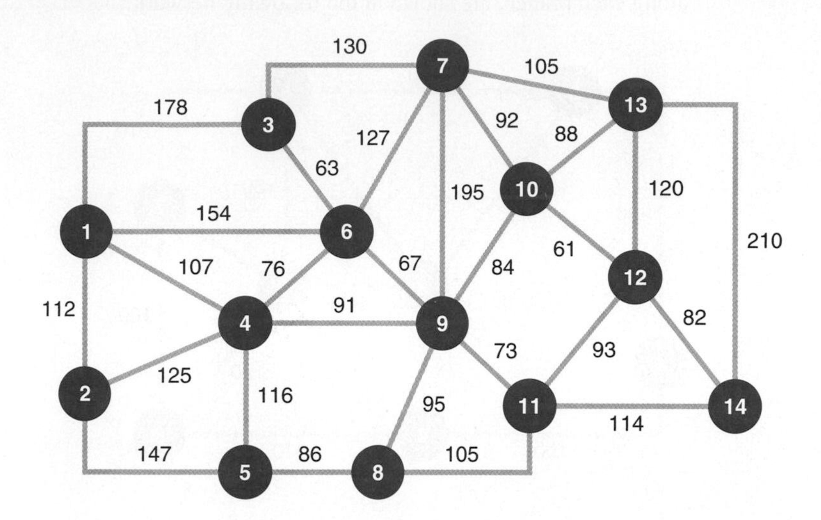

Determine a minimal spanning tree network that will connect all the buildings on campus with wheelchair access sidewalks and indicate the number of feet of sidewalk.

28. Tech wants to develop an area network that will connect its server at its computer and satellite center with the main campus buildings to improve Internet service. The cable will be laid primarily through existing electrical tunnels, although some cable will have to be buried underground. The following network shows the possible cable connections between the computer center at node 1 and the various buildings, with the distances, in feet, along the branches:

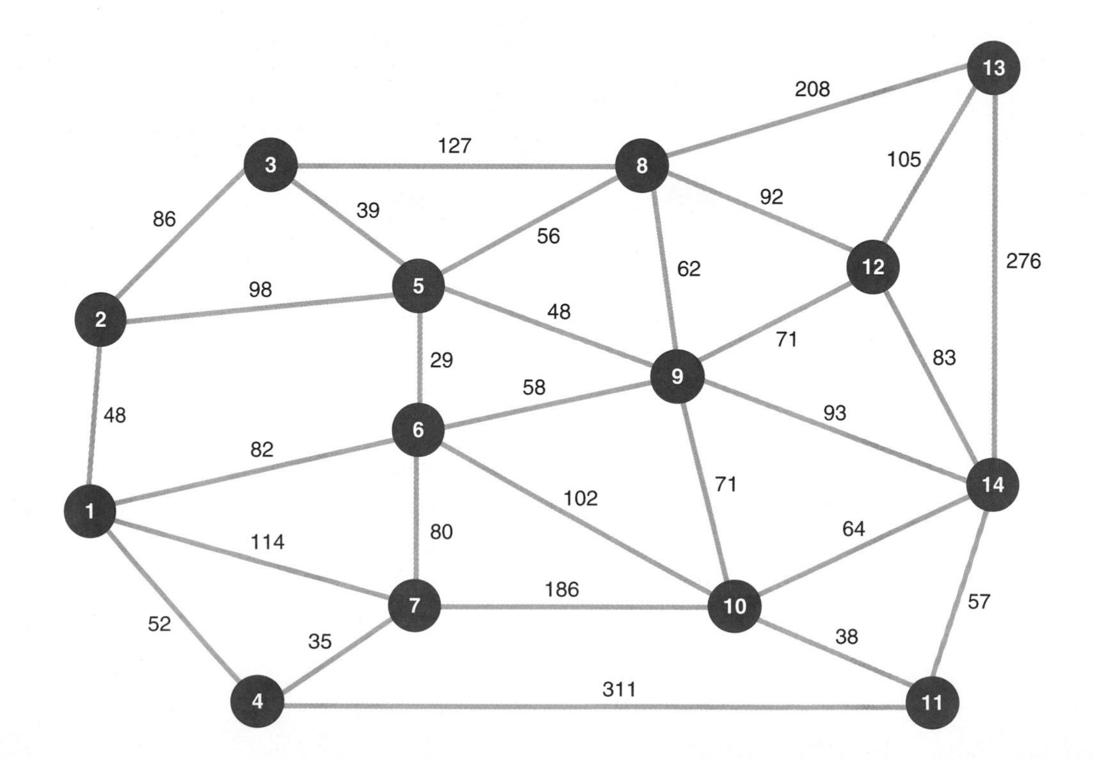

Determine a minimal spanning tree network that will connect all the buildings and indicate the total amount of cable that will be needed to do so.

29. Given the following network, with the indicated flow capacities of each branch, determine the maximum flow from source node 1 to destination node 6 and the flow along each branch:

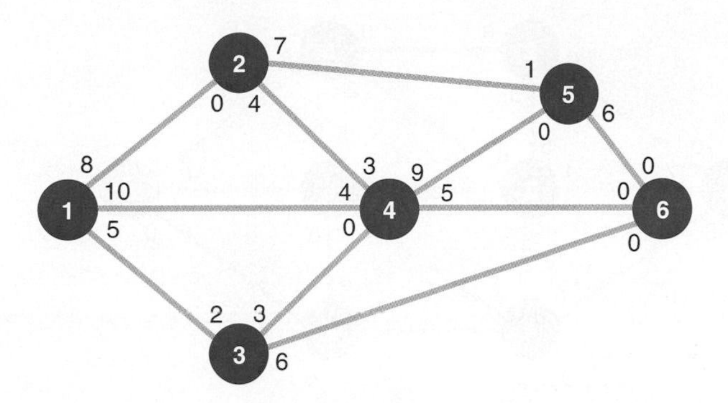

30. Given the following network, with the indicated flow capacities along each branch, determine the maximum flow from source node 1 to destination node 7 and the flow along each branch:

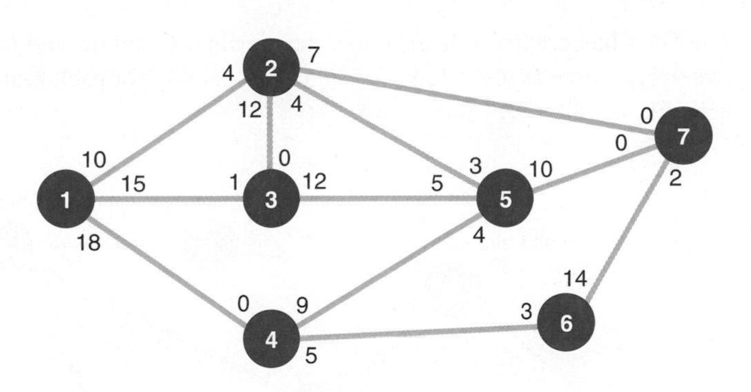

31. Given the following network, with the indicated flow capacities along each branch, determine the maximum flow from source node 1 to destination node 6 and the flow along each branch:

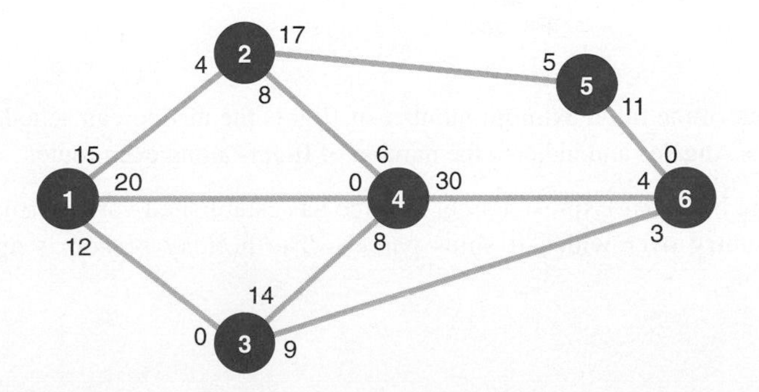

32. A new stadium complex is being planned for Denver, and the Denver traffic engineer is attempting to determine whether the city streets between the stadium complex and the interstate highway can accommodate the expected flow of 21,000 cars after each game. The various traffic arteries between the stadium (node 1) and the interstate (node 8) are shown in the following network:

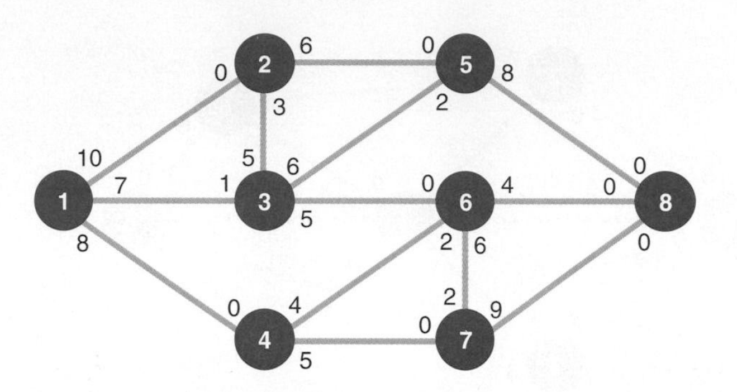

The flow capacities on each street are determined by the number of available lanes, the use of traffic police and lights, and whether any lanes can be opened or closed in either direction. The flow capacities are given in thousands of cars. Determine the maximum traffic flow the streets can accommodate and the amount of traffic along each street. Will the streets be able to handle the expected flow after a game?

33. The FAA has granted a license to a new airline, Omniair, and awarded it several routes between Los Angeles and Chicago. The flights per day for each route are shown in the following network:

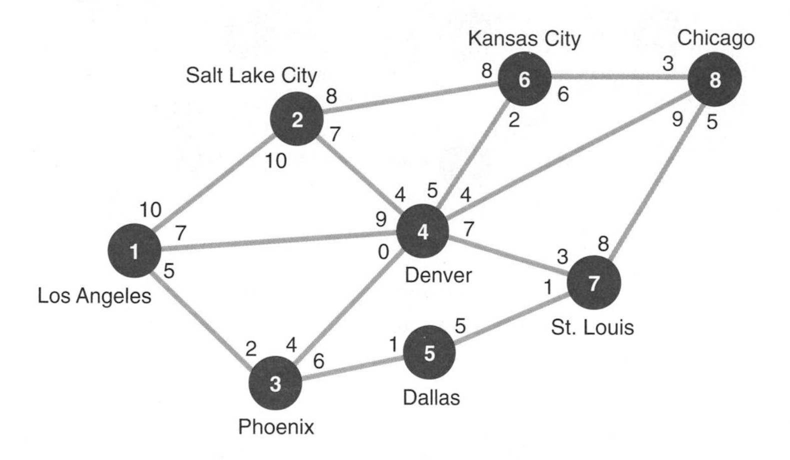

Determine the maximum number of flights the airline can schedule per day from Chicago to Los Angeles and indicate the number of flights along each route.

34. The National Express Parcel Service has established various truck and air routes around the country over which it ships parcels. The holiday season is approaching, which means a

dramatic increase in the number of packages that will be sent. The service wants to know the maximum flow of packages it can accommodate (in tons) from station 1 to station 7. The network of routes and the flow capacities (in tons of packages per day) along each route are shown in the following network:

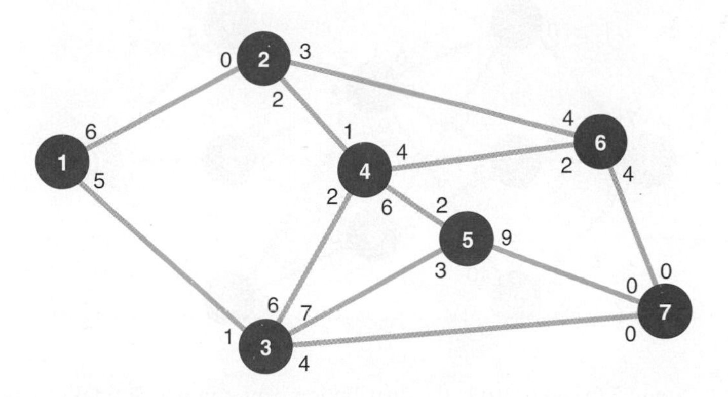

Determine the maximum tonnage of packages that can be transported per day from station 1 to station 7 and indicate the flow along each branch.

35. The traffic management office in Richmond is attempting to analyze the potential traffic flow from a new office complex under construction to an interstate highway interchange during the evening rush period. Cars leave the office complex via one of three exits, and then they travel through the city streets until they arrive at the interstate interchange. The following network shows the various street routes (branches) from the office complex (node 1) to the interstate interchange (node 9):

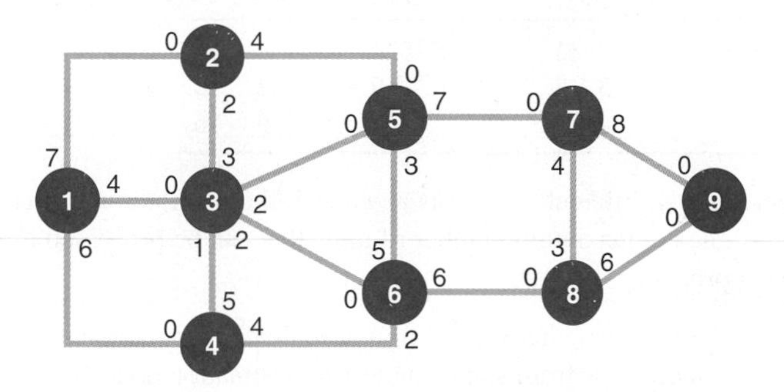

All intermediate nodes represent street intersections, and the values accompanying the branches emanating from the nodes represent the traffic capacities of each street, expressed in thousands of cars per hour. Determine the maximum flow of cars that the street system can absorb during the evening rush hour.

36. The Dynaco Company manufactures a product in five stages. Each stage of the manufacturing process is conducted at a different plant. The following network shows the five different stages

and the routes over which the partially completed products are shipped to the various plants at the different stages:

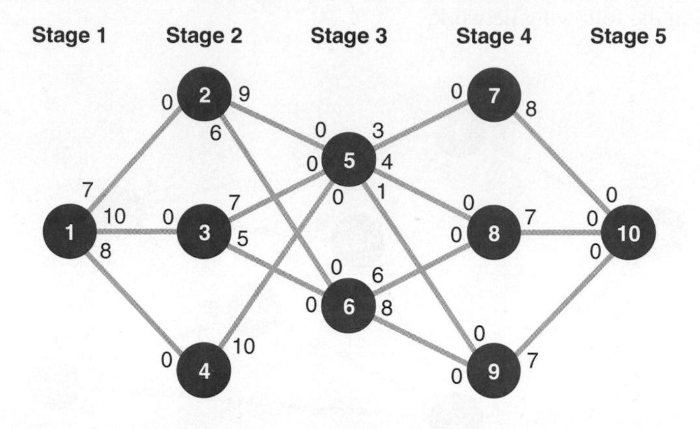

Stage 5 (at node 10) is the distribution center in which final products are stored. Although each node represents a different plant, plants at the same stage perform the same operation. (For example, at stage 2 of the manufacturing process, plants 2, 3, and 4 all perform the same manufacturing operation.) The values accompanying the branches emanating from each node indicate the maximum number of units (in thousands) that a particular plant can produce and ship to another plant at the next stage. (For example, plant 3 has the capacity to process and ship 7,000 units of the product to plant 5.) Determine the maximum number of units that can be processed through the five-stage manufacturing process and the number of units processed at each plant.

37. Suppose in Problem 36 that the processing cost per unit at each plant is different because of different machinery, workers' abilities, overhead, and so on, according to the following table:

Stage 1	Stage 2	Stage 3	Stage 4
1. $3	2. $5	5. $22	7. $12
	3. 7	6. 19	8. 14
	4. 4		9. 16

There are no differentiated costs at stage 5, the distribution center. Given a budget of $700,000, determine the maximum number of units that can be processed through the five-stage manufacturing process.

38. Given the following network, with the indicated flow capacities along each branch, determine the maximum flow from source node 1 to destination node 10 and the flow along each path:

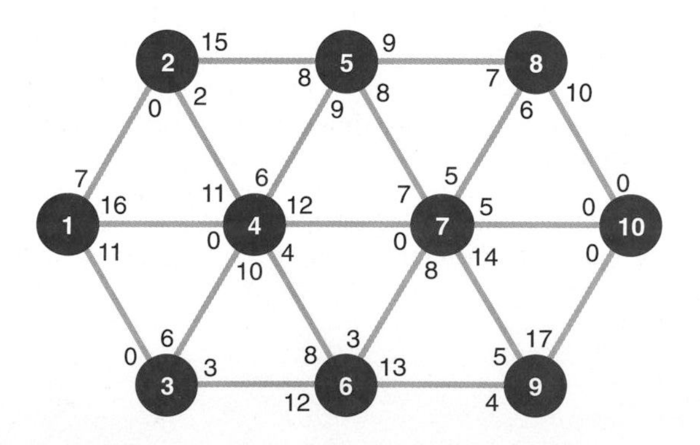

39. A manufacturing company produces different variations of a product at different work centers in its plant on a daily basis. Following is a network showing the various work centers in the plant, the daily capacities at each work center, and the flow of the partially completed products between work centers:

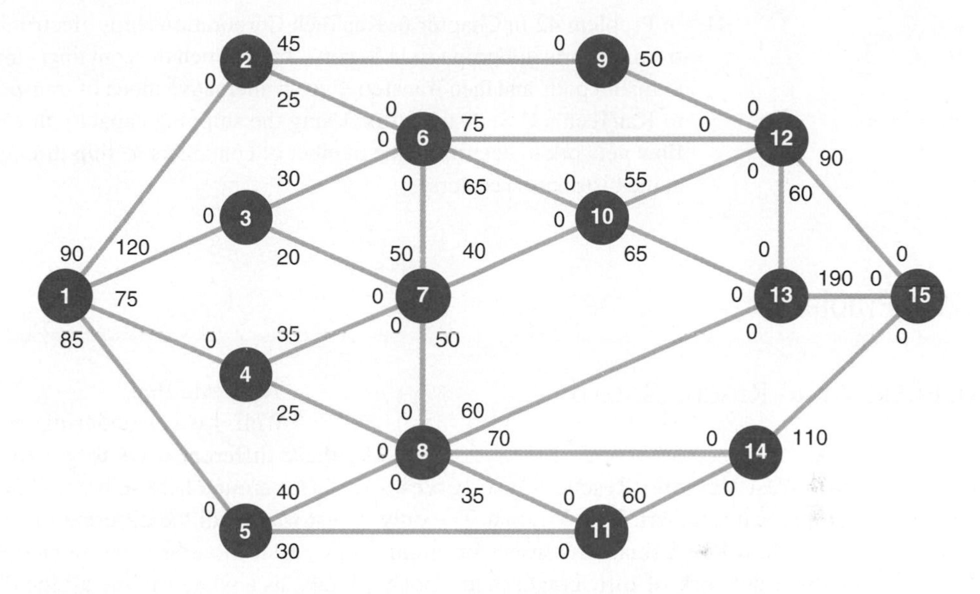

Node 1 represents the point where raw materials enter the process, and node 15 is the packaging and distribution center. Determine the maximum number of units that can be completed each day and the number of units processed at each work center.

40. The following network shows the major roads that would be part of the hurricane evacuation routes for Charleston, South Carolina:

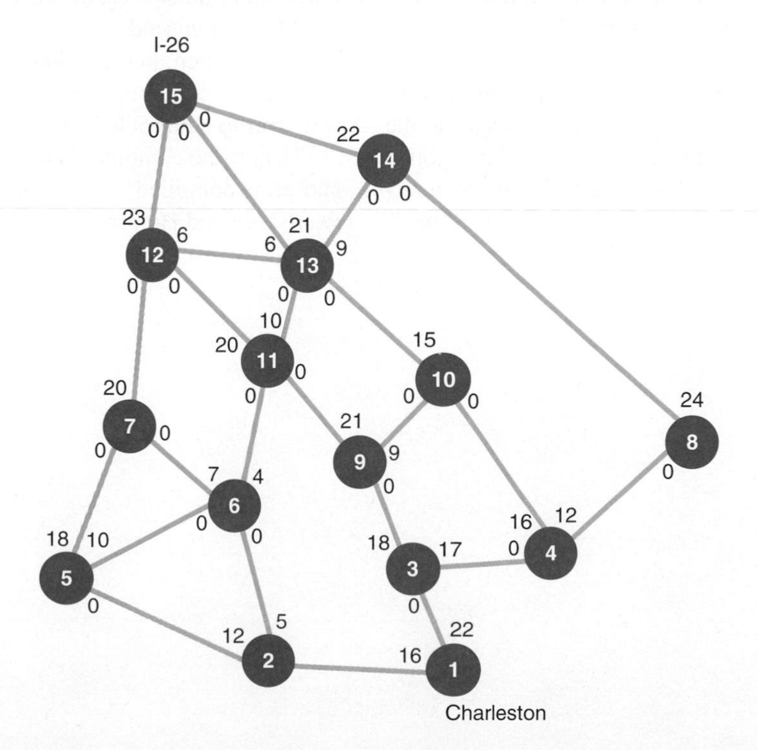

The destination is the interchange of Interstate 26 and Interstate 526, north of Charleston. The nodes represent road intersections, and the values at the nodes are the traffic capacities, in thousands of cars. Determine the maximum flow of cars that can leave Charleston during a hurricane evacuation and the optimal flow along each road segment (branch 1).

41. In Problem 42 in Chapter 6, KanTech Corporation ships electronic components in containers from seaports in Europe to U.S. ports, from which the containers are transported by truck or rail to inland ports and then transferred to an alternative mode of transportation before being shipped to KanTech's U.S. distributors. Using the shipping capacity at each port, develop a maximal flow network to determine the number of containers to ship through each port to meet demand at the distribution centers.

Case Problem

The Pearlsburg Rescue Squad

The Pearlsburg (West Virginia) Rescue Squad serves a mountainous, rural area in southern West Virginia. The only access to the homes, farms, and small crossroad communities and villages is a network of dirt, gravel, and poorly paved roads. The rescue squad had just returned from an emergency at Blake's Crossing, and two of the squad members, Melanie Hart and Ben Cross, were cleaning up and getting the truck back in proper order for the next call.

"You know, Ben," said Melanie, "that was close. We could have lost little Randy if we had been a few minutes later in getting to their farm."

"Yep," said Ben.

After a moment Melanie continued, "I was sort of wondering about the route Dave took to get there. It seems to me if we had gone around by Cedar Creek, we could have gotten there a few minutes sooner. And as close as things got, a few minutes could have made a big difference, don't you think?"

"Yep," said Ben.

"Well, I was wondering, Ben, why don't we study all these different ways to get to the little communities and farms around here so we will always know what the quickest way to all the different places is?"

Ben thought a moment before answering. "It would take us a while to time all the different ways you could get from here to everywhere we go."

"That's true," Melanie answered, "but I've been studying a way to work out this sort of problem in one of my college courses. All we have to do is get the times it takes to travel each piece of road between all these little communities, and I think I can do the rest. Will you help me?"

"Sure," Ben said.

"Okay, then, here's what we'll do. I'll write down all the routes you should time, and I'll time the rest all the way over to Holbrook."

Here is the combined list of times (in minutes) Melanie and Ben compiled for all the route segments between Pearlsburg and Holbrook:

Segment	*Time*	*Segment*	*Time*
Pearlsburg to Kitchen Corner	10	Cedar Creek to Blake's Crossing	10
Pearlsburg to Quarry	15	Cedar Creek to Wellis Farm	17
Pearlsburg to Morgan Creek	12	Cedar Creek to Homer	5
Kitchen Corner to Cutter's Store	20	Cutter's Store to Blake's Crossing	12
Kitchen Corner to Stone House	14	Cutter's Store to Bottom Town	14
Kitchen Corner to Quarry	8	Blake's Crossing to Bottom Town	6
Quarry to Blake's Crossing	18	Blake's Crossing to Holbrook	15
Quarry to Cedar Creek	9	Blake's Crossing to Wellis Farm	9
Morgan Creek to Quarry	16	Homer to Wellis Farm	11
Morgan Creek to Cedar Creek	7	Homer to McKinney Farm	8
Morgan Creek to Homer	18	McKinney Farm to Wellis Farm	21
Morgan Creek to McKinney Farm	11	Wellis Farm to Holbrook	10
Stone House to Cutter's Store	10	Bottom Town to Holbrook	12
Stone House to Blake's Crossing	6		

Determine the shortest routes from Pearlsburg to all the different communities and farms visited by the rescue squad.

Case Problem

Around the World in 80 Days

In the novel *Around the World in 80 Days* by Jules Verne, Phileas T. Fogg wagered four of his fellow members of the Reform Club in London £5,000 apiece that he could travel around the world in 80 days. This would have been an astounding feat in 1872, the year in which the novel is set. Then, the fastest modes of travel were rail and ship; however, much of the world still traveled by coach, wagon, horse, elephant, or donkey, or on foot. Phileas Fogg was not a frivolous man. He would not have undertaken such a large wager if he had not carefully researched the feasibility of such a trip and been confident of his chances. Although he undoubtedly analyzed various routes to circumnavigate the globe, he did not have knowledge of techniques such as the shortest route method, nor did he have a computer to help him select an optimal route. If he had, he might have chosen a route that would have completed his trip in less than 80 days or possibly would not have been so quick to make his wager. On the next page is a network of the various routes of the day for traveling around the world from London. Fogg traveled eastward. The travel time, in days, is shown on each branch. Travel times are based as much on available modes of transportation at the time as on distance. Using the shortest route method (and the computer), select the best route for Mr. Fogg. Would the route you determined have won him his wager?

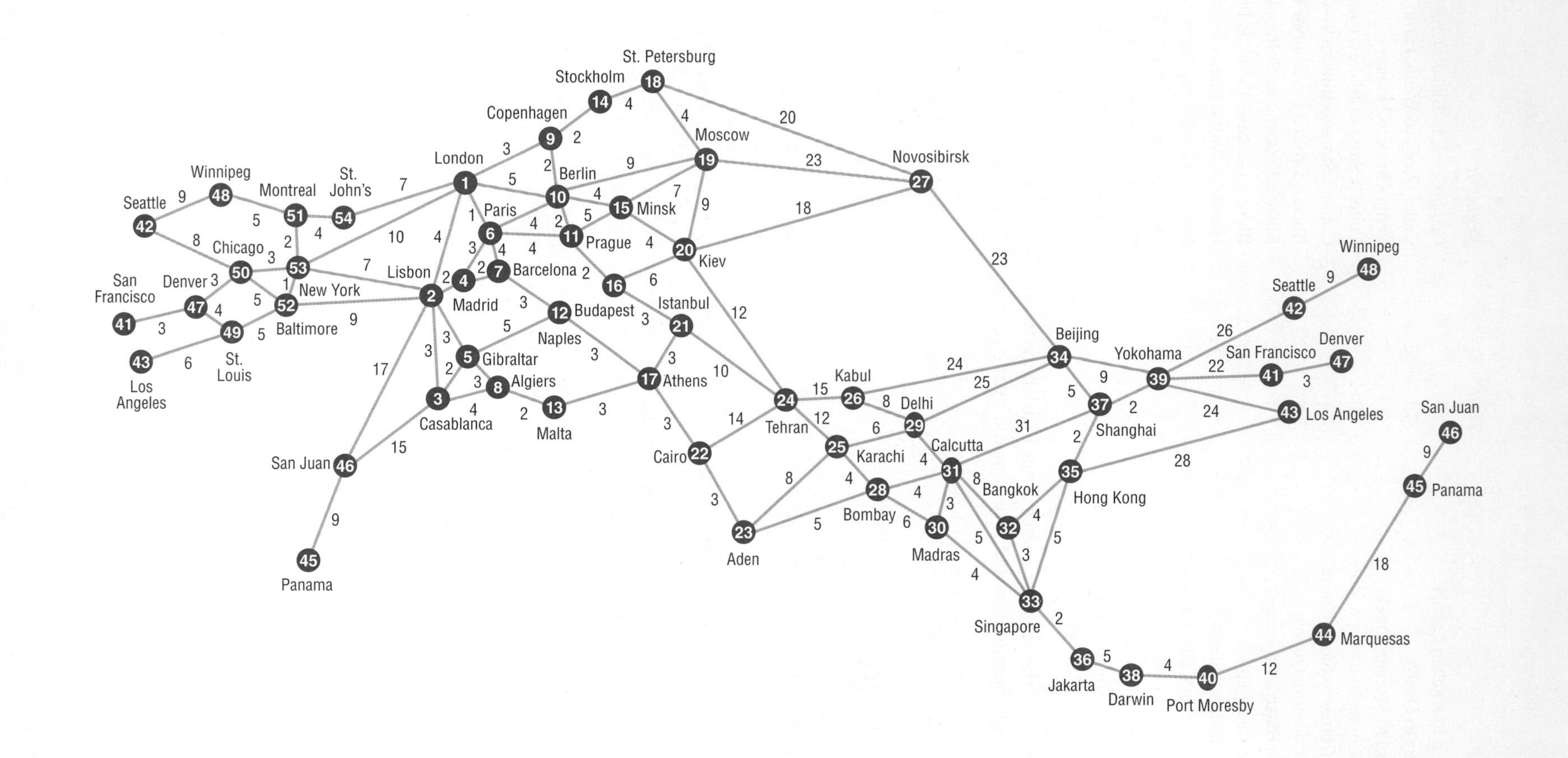

St. Petersburg
Stockholm
Copenhagen
Moscow
Novosibirsk
London
Berlin
Winnipeg
Montreal
St. John's
Seattle
Paris
Minsk
Chicago
Prague
Kiev
Lisbon
Barcelona
San Francisco
Denver
New York
Madrid
Istanbul
Budapest
Baltimore
Naples
St. Louis
Gibraltar
Algiers
Athens
Los Angeles
Casablanca
Malta
San Juan
Cairo
Tehran
Kabul
Delhi
Karachi
Calcutta
Bombay
Bangkok
Aden
Madras
Panama
Beijing
Yokohama
Shanghai
Hong Kong
Singapore
Jakarta
Darwin
Port Moresby
Marquesas

Case Problem

The Battle of the Bulge

On December 16, 1944, in the last year of World War II, two German panzer armies, supported by a third army of infantry, together totaling more than 250,000 men, staged a massive counteroffensive in northern France, overwhelming the American First Army in the Ardennes. The offensive emanated from the German defensive line along the Our River, north of the city of Luxembourg, and was directed almost due west toward Namur and Liége in Belgium. The result, after several days of fighting, was a huge "bulge" in the Allied line and, therefore, this last major ground battle of World War II became known as the Battle of the Bulge.

On December 20, General Dwight D. Eisenhower, Supreme Allied Commander, called on General George Patton to attack the German offensive with his Third Army, which was then situated near Verdun, approximately 100 miles due south of the German left flank. Patton's immediate objective was to relieve the 101st Airborne and elements of Patton's own 9th and 10th Armored Divisions surrounded at Bastogne. Within 48 hours, on December 22, Patton was able to begin his counteroffensive, with three divisions totaling approximately 62,000 men.

The winter weather was cold with snow and fog, and the roads were icy, making the movement of troops, tanks, supplies, and equipment a logistical nightmare. Nevertheless, on December 26 Bastogne was relieved, and on January 12, 1945, the Battle of the Bulge effectively ended in one of the great Allied victories of the war.

General Patton's staff did not have knowledge of the maximal flow technique nor access to computers to help plan the Third Army's troop movements during the Battle of the Bulge. However, the figure on the following page shows the road network between Verdun and Bastogne, with (imagined) troop capacities (in thousands) along each road branch between towns. Using the maximal flow technique (and your imagination), determine the number of troops that should be sent along each road in order to get the maximum number of troops to Bastogne. Also indicate the total number of troops that would be able to get to Bastogne.

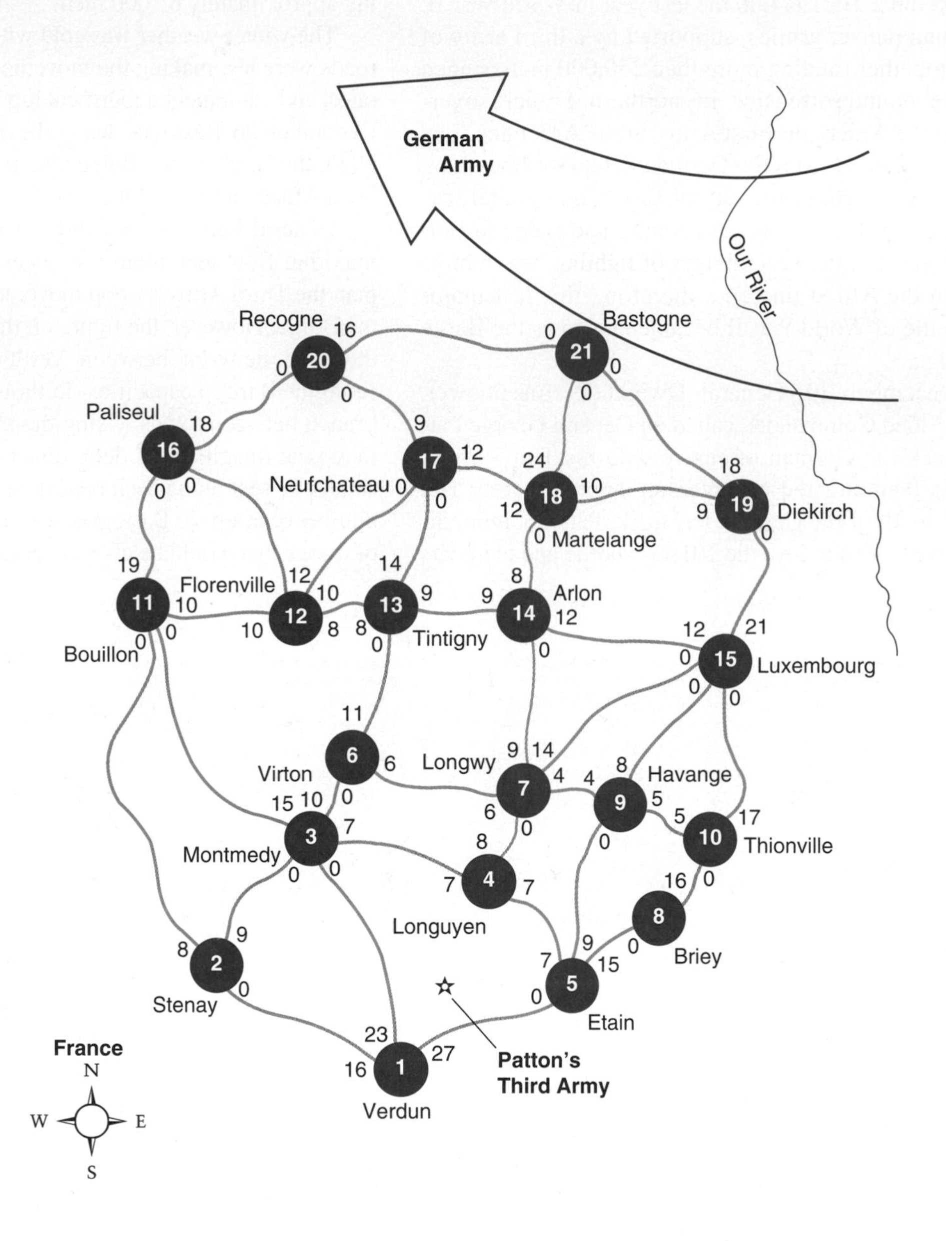

German Army
Our River
Recogne
Bastogne
Paliseul
Neufchateau
Martelange
Diekirch
Florenville
Bouillon
Tintigny
Arlon
Luxembourg
Virton
Longwy
Havange
Montmedy
Thionville
Longuyen
Briey
Stenay
Etain
Verdun
Patton's Third Army
France
N
W
E
S

Case Problem

Nuclear Waste Disposal at PAWV Power and Light

PAWV Power and Light has contracted with a waste disposal firm to have nuclear waste from its nuclear power plants in Pennsylvania disposed of at a government-operated nuclear waste disposal site in Nevada. The waste must be shipped in reinforced container trucks across the country, and all travel must be confined to the interstate highway system. The government insists that the waste transport must be completed within 42 hours and that the trucks travel through the least populated areas possible. The following network shows the various interstate segments the trucks might use from Pittsburgh to the Nevada waste site and the travel time (in hours) estimated for each road segment.

The approximate population (in millions) for the metropolitan areas the trucks might travel through are as follows:

City	Population (1,000,000s)	City	Population (1,000,000s)
Akron	0.50	Las Vegas	1.60
Albuquerque	1.00	Lexington	0.50
Amarillo	0.30	Little Rock	0.60
Charleston	1.30	Louisville	0.93
Cheyenne	0.16	Memphis	1.47
Chicago	10.00	Nashville	1.00
Cincinnati	1.20	Oklahoma City	1.30
Cleveland	1.80	Omaha	1.40
Columbus	0.75	Salt Lake City	1.20
Davenport/Moline/ Rock Island	1.00	Springfield	0.36
Denver	2.20	St. Louis	2.00
Des Moines	0.56	Toledo	0.76
Evansville	0.30	Topeka	0.30
Indianapolis	1.60	Tulsa	1.00
Kansas City	2.10	Wichita	0.73
Knoxville	0.54		

Determine the optimal route the trucks should take from Pittsburgh to the Nevada site to complete the trip within 42 hours and expose the trucks to the least number of people possible.

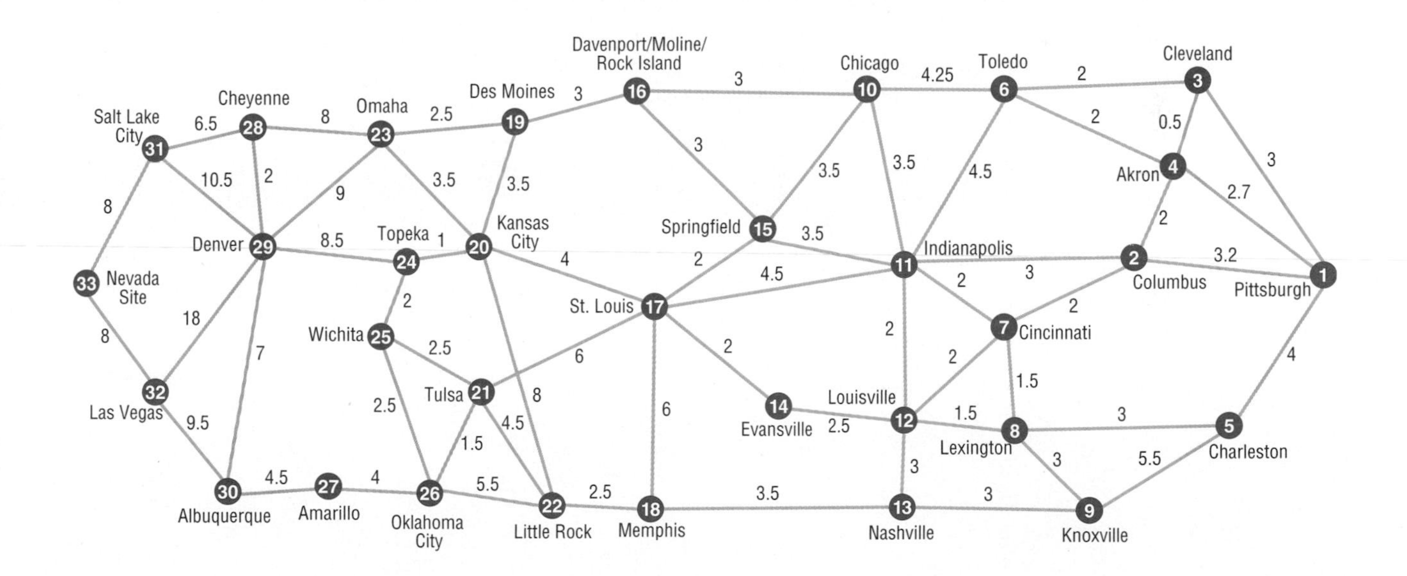

Case Problem

A Day in Paris

Kathleen Taylor, a student at Tech, is planning to visit her sister, Lindsey, who is living in Toulouse, France, over the summer break. She is going to fly from Dulles Airport in Washington to Charles de Gaulle Airport in Paris, and because of the time changes, this travel will take a full day. In Paris, Kathleen is going to spend 2 nights and 1 full day before taking the train to Toulouse. Kathleen has never been to Paris, so she wants to spend her 1 day there seeing as many of the famous attractions as she can, including the Eiffel Tower, the Louvre, Notre Dame Cathedral, the Arch de Triumph, the Pantheon, and the Palace at Versailles. She plans to stay at a youth hostel very near Sacre Coeur in Montmartre and from there use the Paris Metro to visit as many of the sites as she can in a day. She has downloaded a detailed Metro map from the French rail Web site at www.ratp.fr and has discovered that the Metro system is huge, with almost 250 stations and 14 lines throughout Paris.

There's no question that Kathleen can get to all the sites by the Metro, but she is concerned about her limited time frame and her ability to get from location to location quickly. She has determined the following information regarding the average times (in minutes) between stations for each line:

Line	Time	Line	Time
1	2.1	8	1.8
2	1.5	9	1.6
3	1.5	10	1.6
4	1.3	11	1.6
5	2.1	12	1.6
6	1.5	13	2.3
7	1.3	14	3.4

She's also guessing that the subway stops about 1 minute at each station. If she has to change lines, she assumes it will take her at least 5 minutes. She plans to leave early in the morning, when the sites open, and she has no specific time she must be back to the hostel.

Kathleen, a business student, would like to use some of kind of logical, systematic approach to help her plan her movement using the Metro around the city to the different sites. Help Kathleen develop a route around the city to each of the sites she wants to see, starting from her youth hostel in Montmartre, for the day she'll be in Paris. Do you think she'll be able to see all the sites she wants to see?

CHAPTER

8

Project Management

One of the most popular uses of networks is for project analysis. Such projects as the construction of a building, the development of a drug, or the installation of a computer system can be represented as networks. These networks illustrate the way in which the parts of the project are organized, and they can be used to determine the time duration of the projects. The network techniques that are used for project analysis are CPM and PERT. CPM stands for *critical path method*, and PERT is an acronym for *project evaluation and review technique*. These two techniques are very similar.

There were originally two primary differences between CPM and PERT. With CPM, a single, or deterministic, estimate for activity time was used, whereas with PERT probabilistic time estimates were employed. The other difference was related to the mechanics of drawing the project network. In PERT, activities were represented as arcs, or arrowed lines, between two nodes, or circles, whereas in CPM, activities were represented as the nodes or circles. However, these were minor differences, and over time CPM and PERT have been effectively merged into a single technique, conventionally referred to as simply CPM/PERT.

CPM and PERT were developed at approximately the same time (although independently) during the late 1950s. The fact that they have already been so frequently and widely applied attests to their value as management science techniques.

The Elements of Project Management

Management is generally perceived to be concerned with the planning, organization, and control of an ongoing process or activity such as the production of a product or delivery of a service. *Project* management is different in that it reflects a commitment of resources and people to a typically important activity for a relatively short time frame, after which the management effort is dissolved. Projects do not have the continuity of supervision that is typical in the management of a production process. As such, the features and characteristics of project management tend to be somewhat unique.

Figure 8.1 provides an overview of project management, which encompasses three major processes—planning, scheduling, and control. It also includes a number of the more prominent elements of these processes. In the remainder of this section we will discuss some of the features of the project planning process.

Project Planning

Project plans generally include the following basic elements:

- *Objectives*—A detailed statement of what is to be accomplished by the project, how it will achieve the company's goals and meet the strategic plan, and an estimate of when it needs to be completed, the cost, and the return.
- *Project scope*—A discussion of how to approach the project, the technological and resource feasibility, the major tasks involved, and a preliminary schedule; it includes a justification of the project and what constitutes project success.
- *Contract requirements*—A general structure of managerial, reporting, and performance responsibilities, including a detailed list of staff, suppliers, subcontractors, managerial requirements and agreements, reporting requirements, and a projected organizational structure.
- *Schedules*—A list of all major events, tasks, and subschedules, from which a master schedule is developed.
- *Resources*—The overall project budget for all resource requirements and procedures for budgetary control.
- *Personnel*—Identification and recruitment of personnel required for the project team, including special skills and training.
- *Control*—Procedures for monitoring and evaluating progress and performance, including schedules and cost.
- *Risk and problem analysis*—Anticipation and assessment of uncertainties, problems, and potential difficulties that might increase the risk of project delays and/or failure and threaten project success.

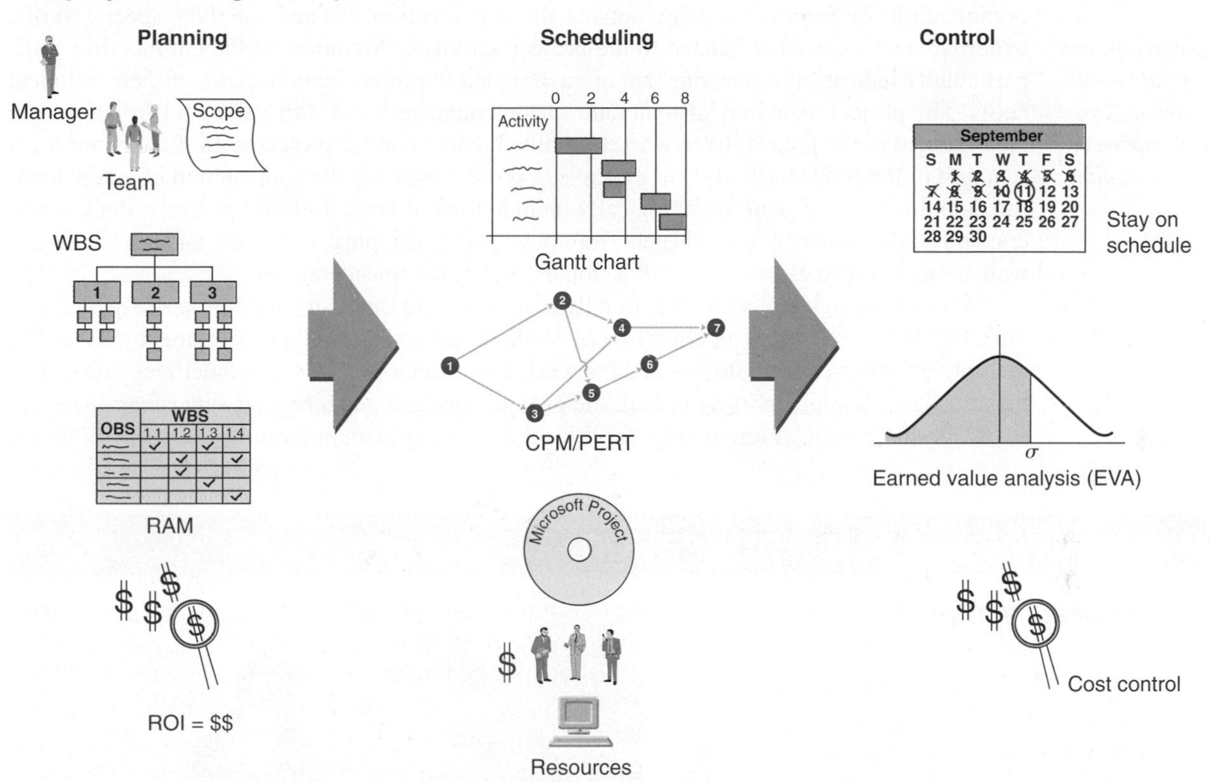

FIGURE 8.1
The project management process

Project Return

Return on investment is a measure used to evaluate projects calculated by dividing the dollar gain minus the dollar cost of a project by the cost.

In order for a project to be selected to be undertaken, it typically has to offer some kind of positive gain or benefit for the organization that is considering it. In a business, one of the most popular measures of benefit is **return on investment (ROI)**. ROI is a performance measure that is often used to evaluate the expected outcome of a project or to compare a number of different projects. To calculate ROI, the benefit (return) of a project is divided by the cost of the project; the result is expressed as a percentage or a ratio:

$$\text{ROI} = \frac{(\text{gain from project} - \text{cost of project})}{\text{cost of project}}$$

If a project does not have a positive ROI, or if another project has a higher ROI, then the project might not be undertaken. ROI is a very popular metric for project planning because of its versatility and simplicity.

However, projects sometimes have benefits that cannot be measured in a tangible way with something like an ROI; these benefits are referred to as "soft" returns. For example, a project that has raising employee satisfaction as its goal can result in real benefits—increased productivity, improved quality, and lower costs—that are difficult to measure monetarily in the short run. A project by an online retailer to install backup power generators to keep orders coming in and customers happy during a power outage is like insurance for something that may never happen, making an ROI difficult to determine. A "green" project may not have a tangible dollar ROI, but it can protect a company against regulatory infractions and improve the company's public image. Projects undertaken by government agencies and not-for-profit organizations typically do not have an ROI-type benefit; they are undertaken to benefit the "public good." In general, it may be more appropriate to measure a project's benefit not just in terms of financial return but also in terms of the positive impact it may have on a company's employees and customers (i.e., quality improvement).

The Project Team

Project teams are made up of individuals from various areas and departments within a company.

A project team typically consists of a group of individuals selected from other areas in the organization, or from consultants outside the organization, because of their special skills, expertise, and experience related to the project activities. Members of the engineering staff, particularly industrial engineering, are often assigned to project work because of their technical skills. The project team may also include various managers and staff personnel from specific areas related to the project. Even workers can be involved on the project team if their jobs are a function of the project activity. For example, a project team for the construction of a new loading dock facility at a plant might logically include truck drivers; forklift operators; dock workers; and staff personnel and managers from purchasing, shipping, receiving, and packaging; as well as engineers to assess vehicle flow, routes, and space considerations.

Assignment to a project team is usually temporary and thus can have both positive and negative repercussions. The temporary loss of workers and staff from their permanent jobs can be disruptive for both the employees and the work area. An employee must sometimes "serve two masters," in a sense, reporting to both the project manager and a regular supervisor. Alternatively, because projects are usually "exciting," they provide an opportunity to do work that is

Management Science Application

Managing Projects in China

China is the most likely country in the world where U.S. companies will be doing global business in the future. Business between U.S. and Chinese companies includes projects in which diversity and teamwork will be critical factors in determining project success; however, the cultural differences between the United States and China are probably as great as they could possibly be, thus making project teamwork a potential problem that requires special attention. That doesn't mean that the way things are done in the United States (or in China) is better or worse, just different. Following are some examples of these cultural differences that should be addressed to achieve project success.

Jupiterimages/Comstock/Thinkstock

First, whereas U.S. companies value individualism, China, reflecting 2,000 years of traditional Confucian values, does not. Chinese managers can be good, but their leadership style is typically different from that of U.S. project managers, being more collaborative rather than direct. Relationships are very important in China. The Chinese word *guanxi* (gwan-shee) refers to the strong reliance of business on social connections. A Chinese business partner with strong, well-placed relationships can enhance project success; however, such relationships are developed slowly. The Chinese focus on the long-term rather than the short-term view prevalent in many Western companies. "Face" or *mianzi* (me-ahn-zee) is very important in China. Every conversation, meeting, meal, and social or business engagement is an opportunity for an individual to gain or lose stature. Rudeness and anger have very negative consequences; raising your voice to a Chinese manager will cause him or her shame and embarrassment and brand the offending person as a barbarian. Rank is important in China; people of high rank are afforded an extraordinary degree of deference. Consequently, instructions from team leaders are respected and followed, and advice from junior team members is not generally expected or valued. Humility helps—U.S. food is not better than Chinese food, and the U.S. way of doing business is not better than the Chinese way; they're just different. Gifts are important in China, not so much for their intrinsic value as for their symbolic value (of respect). They should reflect a degree of thoughtfulness and effort, for example, not just a logo T-shirt. Patience is a respected virtue in China, where time is used as a competitive tool. The little things can be important; learn the Chinese way of doing things like smiling, being friendly, and paying attention, which can pay off in successful teamwork.

Source: Based on Bud Baker, "When in China . . . ," *PM Network* 20, no. 6 (June 2006): 24–25.

new and innovative, and the employee may be reluctant to report back to a more mundane, regular job after the project is completed.

The project manager is often under great pressure.

The most important member of a project team is the *project manager*. The job of managing a project is subject to a great deal of uncertainty and the distinct possibility of failure. Because each project is unique and usually has not been attempted previously, the outcome is not as certain as the outcome of an ongoing process would be. A degree of security is attained in the supervision of a continuing process that is not present in project management. The project team members are often from diverse areas of the organization and possess different skills, which must be coordinated into a single, focused effort to successfully complete the project. In addition, the project is invariably subject to time and budgetary constraints that are not the same as normal work schedules and resource consumption in an ongoing process. Overall, there is usually more perceived and real pressure associated with project management than in a normal management position. However, there are potential rewards, including the ability to demonstrate one's management abilities in a difficult situation, the challenge of working on a unique project, and the excitement of doing something new.

Scope Statement

*A **scope statement** includes a project justification and the expected results.*

A **scope statement** is a document that provides a common understanding of a project. It includes a justification for the project that describes what factors have created a need within the company for the project. It also includes an indication of what the expected results of the project will be and what will constitute project success. Further, the scope statement might include a list of the types of planning reports and documents that are part of the project management process.

A similar planning document is the *statement of work (SOW)*. In a large project, the SOW is often prepared for individual team members, groups, departments, subcontractors, and suppliers. This statement describes the work in sufficient detail so that the team member responsible for it knows what is required and whether he or she has sufficient resources to accomplish the work successfully and on time. For suppliers and subcontractors, it is often the basis for determining whether they can perform the work and for bidding on it. Some companies require that an SOW be part of an official contract with a supplier or subcontractor.

Work Breakdown Structure

*A **work breakdown structure** is an organizational chart that breaks down the project into modules for planning.*

The **work breakdown structure** (WBS) is an organizational chart used for project planning. It organizes the work to be done on a project by breaking down the project into its major components, referred to as *modules*. These components are then subdivided into more detailed subcomponents, which are further broken down into activities, and, finally, into individual tasks. The end result is an organizational structure of the project made up of different levels, with the overall project at the top level and the individual tasks at the bottom. A WBS helps identify activities and determine individual tasks, project workloads, and the resources required. It also helps to identify the relationships between modules and activities and avoid unnecessary duplication of activities. A WBS provides the basis for developing and managing the project schedule, resources, and modifications.

There is no specific model for a WBS, although it is most often in the form of a chart or a table. In general, there are two good ways to develop a WBS. One way is to start at the top and work your way down, asking "What components constitute this level?" until the WBS is developed in sufficient detail. Another way is to brainstorm the entire project, writing down each item on a sticky note and then organizing the sticky notes into a WBS. The upper levels of the WBS tend to contain the summary activities, major components or functional areas involved in the project that indicate what is to be done. The lower levels tend to describe the detailed work activities of the project within the major components or modules. They typically indicate how things are done.

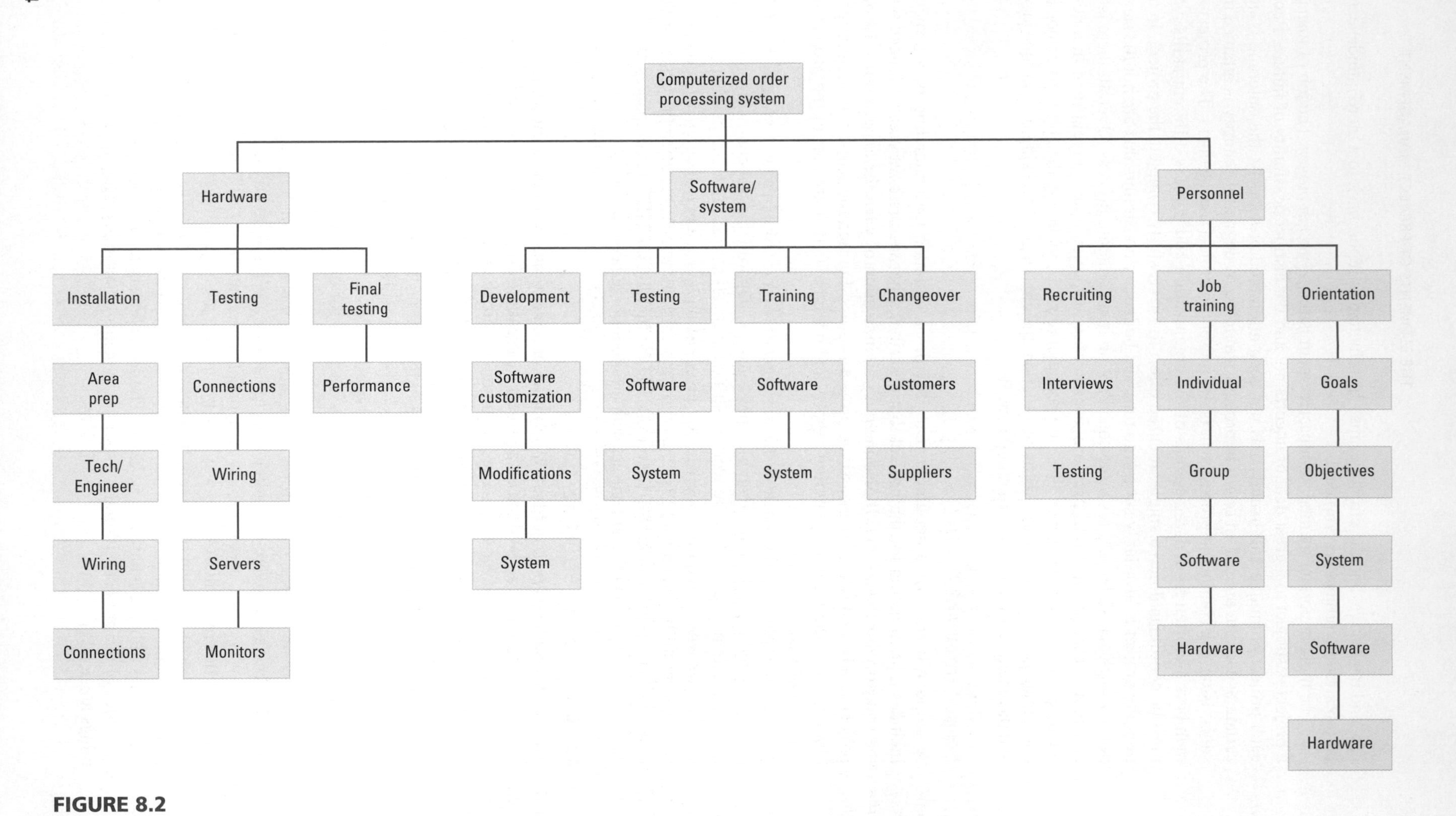

FIGURE 8.2
Work breakdown structure for computer order-processing system project

Figure 8.2 shows a WBS for a project for installing a new computerized order processing system for a manufacturing company that links customers, the manufacturer, and the manufacturer's suppliers. The WBS is organized according to the three major project categories for the development of the system: hardware, software/system, and personnel. Within each of these categories, the major tasks and activities under those tasks are detailed. For example, under hardware, a major task is installation, and activities required in installation include area preparation, technical/engineering layouts and configurations, wiring, and electrical connections.

Responsibility Assignment Matrix

After the WBS is developed, to organize the project work into smaller, manageable elements, the project manager assigns the work elements to organizational units—departments, groups, individuals, or subcontractors—by using an *organizational breakdown structure (OBS)*. An OBS is a table or chart that shows which organizational units are responsible for work items. After the OBS is developed, the project manager can then develop a **responsibility assignment matrix** *(RAM)*. A RAM shows who in the organization is responsible for doing the work in the project. Figure 8.3 shows a RAM for the "Hardware/Installation" category from the WBS for the computerized order processing project shown in Figure 8.2. Notice that there are three levels of work assignments in the matrix, reflecting who is responsible for the work, who actually performs the work, and who performs support activities. As with the WBS, there are many different forms both the OBS and RAM can take, depending on the needs and preferences of the company, project team, and project manager.

*A **responsibility assignment matrix** is a table or chart that shows who is responsible for project work.*

FIGURE 8.3
A responsibility assignment matrix

	WBS Activities—Hardware/Installation			
OBS Units	1.1.1 Area prep	1.1.2 Tech/Engineer	1.1.3 Wiring	1.1.4 Connections
Hardware engineering	3	1	1	1
Systems engineering		3		3
Software engineering		3		
Technical support	1	2		2
Electrical staff	2		2	2
Hardware vendor	3	3	3	3
Quality manager				3
Customer/supplier liaison				3

Level of responsibility: 1 = Overall responsibility 2 = Performance responsibility 3 = Support

Project Scheduling

A project schedule evolves from the planning documents discussed previously. It is typically the most critical element in the project management process, especially during the implementation phase (i.e., the actual project work), and it is the source of most conflict and problems. One reason is that frequently the single most important criterion for the success of a project is that it be finished on time. If a stadium is supposed to be finished in time for the first game of the season and it's not, there will be a lot of angry ticket holders; if a school building is not completed by the time the school year starts, there will be a lot of angry parents; if a shopping mall is not completed on time, there will be a lot of angry tenants; if a new product is not completed by the scheduled launch date, millions of dollars can be lost; and if a new military weapon is not

Time Out for Henry Gantt

Both CPM and PERT are based on the network diagram, which is an outgrowth of the bar, or Gantt, chart that was designed to control the time element of a program. Henry Gantt, a pioneer in the field of industrial engineering, first employed the Gantt chart in the artillery ammunition shops at the Frankford Arsenal in 1914, when World War I was declared. The chart graphically depicted actual versus estimated production time.

completed on time, it could affect national security. Also, time is a measure of progress that is very visible. It is an absolute with little flexibility; you can spend less money or use fewer people, but you cannot slow down or stop the passage of time.

Developing a schedule encompasses four basic steps. First, *define the activities* that must be performed to complete the project; second, *sequence the activities* in the order in which they must be completed; third, *estimate the time* required to complete each activity; and fourth, *develop the schedule* based on the sequencing and time estimates of the activities.

Because scheduling involves a quantifiable measure, time, there are several quantitative techniques available that can be used to develop a project schedule, including the Gantt chart and CPM/PERT networks. There are also various computer software packages that can be used to schedule projects, including the popular Microsoft Project. Later in this chapter we will discuss CPM/PERT and Microsoft Project in greater detail. For now, we will describe one of the oldest and most widely used scheduling techniques, the Gantt chart.

The Gantt Chart

A ***Gantt chart*** *is a graph or bar chart with a bar for each project activity that shows the passage of time.*

Using a **Gantt chart** is a traditional management technique for scheduling and planning small projects that have relatively few activities and precedence relationships. This scheduling technique (also called a *bar chart*) was developed by Henry Gantt, a pioneer in the field of industrial engineering at the artillery ammunition shops of the Frankford Arsenal in 1914. The Gantt chart has been a popular project scheduling tool since its inception and is still widely used today. It is the direct precursor of the CPM/PERT technique, which we will discuss later.

The Gantt chart is a graph with a bar representing time for each activity in the project being analyzed. Figure 8.4 illustrates a Gantt chart of a simplified project description for building a house. The project contains only seven general activities, such as designing the house, laying the foundation, ordering materials, and so forth. The first activity is "design house and obtain financing," and it requires 3 months to complete, shown by the bar from left to right across the chart. After the first activity is finished, the next two activities, "lay foundation" and "order and receive materials," can start simultaneously. This set of activities demonstrates how a precedence relationship works; the design of the house and the financing must precede the next two activities.

Slack *is the amount of time an activity can be delayed without delaying the project.*

The activity "lay foundation" requires 2 months to complete, so it will be finished, at the earliest, at the end of month 5. "Order and receive materials" requires 1 month to complete, and it could be finished after month 4. However, observe that it is possible to delay the start of this activity 1 month, until month 4. This delay would still enable the activity to be completed by the end of month 5, when the next activity, "build house," is scheduled to start. This extra time for the activity "order and receive materials" is called **slack**. Slack is the amount of time by which an activity can be delayed without delaying any of the activities that follow it or the project as a whole. The remainder of the Gantt chart is constructed in a similar manner, and the project is scheduled to be completed at the end of month 9.

FIGURE 8.4
A Gantt chart

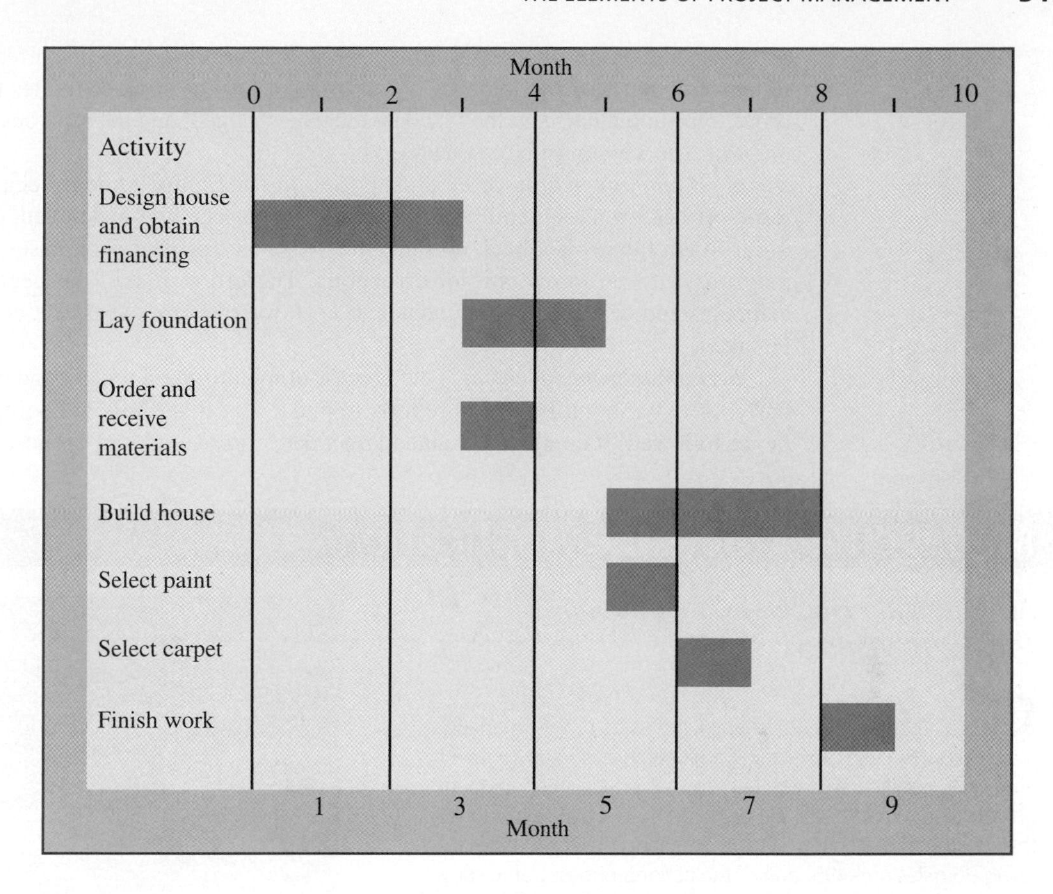

A Gantt chart provides a visual display of a project schedule, indicating when activities are scheduled to start and to finish and where extra time is available and activities can be delayed. A project manager can use a Gantt chart to monitor the progress of activities and see which ones are ahead of schedule and which ones are behind schedule. A Gantt chart also indicates the precedence relationships between activities; however, these relationships are not always easily discernible. This problem is one of the disadvantages of the Gantt chart method, and it limits the chart's use to smaller projects with relatively few activities. The CPM/PERT network technique, which we will talk about later, does not suffer this disadvantage.

Project Control

Project control is the process of making sure a project progresses toward successful completion. It requires that the project be monitored and progress measured so that any deviations from the project plan, and particularly the project schedule, are minimized. If the project is found to be deviating from the plan (i.e., it is not on schedule, cost overruns are occurring, activity results are not as expected), corrective action must be taken. In the rest of this section we will describe several key elements of project control, including time and cost management and performance monitoring.

Time management is the process of making sure a project schedule does not slip and that a project is on time. This requires monitoring of individual activity schedules and frequent updates. If the schedule is being delayed to an extent that jeopardizes the project success, it may be necessary for the project manager to shift resources to accelerate critical activities. Some activities may have slack time, so resources can be shifted from them to activities that are not on schedule. This is referred to as *time–cost trade-off*. However, this can also push the project cost above the budget. In some cases it may be that the work needs to be corrected or made more efficient. In other cases, it may occur that original activity time estimates upon implementation prove to be unrealistic and the schedule must be changed, and the repercussions of such changes on project success must be evaluated.

Cost management is often closely tied to time management because of the time–cost trade-off occurrences mentioned previously. If the schedule is delayed, costs tend to go up in order to get the project back on schedule. Also, as a project progresses, some cost estimates may prove to be unrealistic or erroneous. Therefore, it may be necessary to revise cost estimates and develop budget updates. If cost overruns are excessive, corrective actions must be taken.

Performance management is the process of monitoring a project and developing timed (i.e., daily, weekly, monthly) status reports to make sure that goals are being met and the plan is being followed. It compares planned target dates for events, milestones, and work completion

Management Science Application

An Interstate Highway Construction Project in Virginia

One of the most frequent applications of project management is construction projects. Many companies and government agencies contractually require a formal project management process plan as part of the bid process. One such project involves the construction of four new high occupancy toll (HOT) lanes totaling 96 miles on I-495, one of the nation's busiest traffic corridors, near Washington, DC. The cost of the project is estimated at $1.4 billion.

HOT lanes are toll lanes that operate alongside existing highways to provide drivers with a faster and more reliable travel option—for a fee. This Virginia Department of Transportation (VDOT) 5-year project, scheduled to be completed in 2013, also includes replacing more than 50 aging bridges and overpasses, upgrading 10 interchanges, improving bike and pedestrian access, and improving sound protection for local neighborhoods. The project employs as many as 500 skilled workers on-site who work under extreme safety conditions. As many as a 250,000 vehicles pass through this corridor daily, so all work is done under heavy traffic conditions, and much of it is done at night, during off-hours and weekends, which adds costs for premium-time and late-shift pay. All these factors make scheduling complex and difficult.

Dennis Brack/Newscom

A dimension that has added complexity to the HOT project is that it requires a public–private partnership between VDOT and private contractors. While VDOT as a government agency has a responsibility to review and evaluate every contract package carefully (and sometimes slowly), the private contractors want to accelerate the contract biding and award process in order to maintain schedules. When designs have not been approved and contracts have not been bid, the contractors may not be able to keep workers busy, and work will slow down and lag. Thus, the project management process requires a high degree of communication and coordination between team members, as well as focused administration by the project leaders.

Sources: Based on S. Gale, "A Closer Look: Virginia Department of Transportation," *PM Network* 23, no. 4 (April 2009): 48–51; and the Virginia Department of Transportation Web site, http://virginiadot.org.

with dates actually achieved to determine whether the project is on schedule or behind schedule. Key measures of performance include deviation from the schedule, resource usage, and cost overruns. The project manager and individuals and organizational units with performance responsibility develop these status reports.

Earned value analysis (EVA) is a specific system for performance management. Activities "earn value" as they are completed. EVA is a recognized standard procedure for numerically measuring a project's progress, forecasting its completion date and final cost, and providing measures of schedule and budget variation as activities are completed. For example, an EVA metric such as *schedule variance* compares the work performed during a time period with the work that was scheduled to be performed; a negative variance means the project is behind schedule. *Cost variance* is the budgeted cost of work performed minus the actual cost of the work; a negative variance means the project is over budget. EVA works best when it is used in conjunction with a WBS that compartmentalizes project work into small packages that are easier to measure. The drawbacks of EVA are that it is sometimes difficult to measure work progress, and the time required for data measurement can be considerable.

CPM/PERT

CPM used a single activity time estimate, and in the network, activities were nodes.

PERT used multiple activity time estimates, and in the network, activities were lines between nodes.

Critical path method (CPM) and **project evaluation and review technique (PERT)** were originally developed as separate techniques. (See the TIME OUT box on page 350.) Both are derivatives of the Gantt chart and, as a result, are very similar. There were originally two primary differences between CPM and PERT. With CPM, a single estimate for activity time was used that did not allow for variation; activity times were treated as if they were known with certainty. With PERT, multiple time estimates were used for each activity that reflected variation; activity times were treated as probabilistic. The other difference was in the mechanics of drawing a network. In PERT, activities were represented as arcs, or lines with arrows, between circles called *nodes*, whereas in CPM activities were represented by the nodes, and the arrows between them showed precedence relationships (i.e., which activity came before another). However, over time CPM and PERT have effectively merged into a single technique conventionally known as CPM/PERT.

The advantage of CPM/PERT over the Gantt chart is in the use of a network (instead of a graph) to show the precedence relationships between activities. A Gantt chart does not clearly show precedence relationships, especially for larger networks. Put simply, a CPM/PERT network provides a better picture; it is visually easier to use, which makes using CPM/PERT a popular technique for project planners and managers.

AOA Networks

A CPM/PERT network is drawn with branches and nodes, as shown in Figure 8.5. As previously mentioned, when CPM and PERT were first developed, they employed different conventions for drawing a network. With CPM, the nodes in Figure 8.5 represent the project activities. The branches with arrows in between the nodes indicate the precedence relationships between activities. For example, in Figure 8.5, activity 1, represented by node 1, precedes activity 2, and 2 must be finished before 3 can start. This approach to constructing a network is called *activity-on-node (AON)*. Alternatively, with PERT, the branches in between the nodes represent activities, and the nodes reflect events or points in time such as the end of one activity and the start of another. This approach is called *activity-on-arrow (AOA)*, and the activities are identified by the node numbers at the start and end of an activity (for example, activity 1 $\rightarrow$ 2, which precedes activity 2 $\rightarrow$ 3 in Figure 8.5). In this chapter we will focus on the AON convention, but we will first provide an overview of the AOA network.

FIGURE 8.5
Nodes and branches

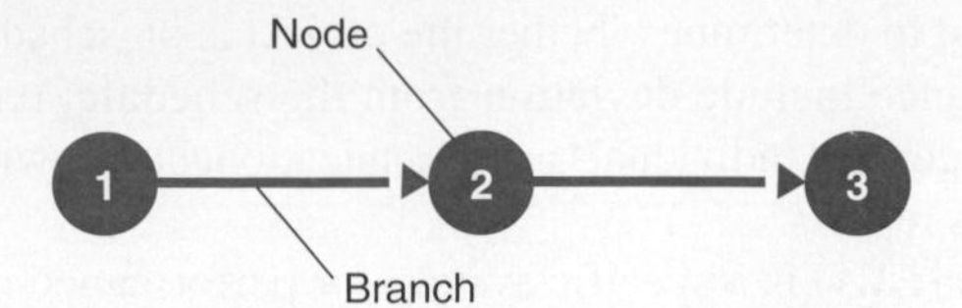

To demonstrate how an AOA network is drawn, we will revisit the example of building a house that we used as a Gantt chart example in Figure 8.4. The comparable CPM/PERT network for this project is shown in Figure 8.6. The precedence relationships are reflected in this network by the arrangement of the arrowed (or directed) branches. The first activity in the project is to design the house and obtain financing. This activity must be completed before any subsequent activities can begin. Thus, activity 2 ⟶ 3, laying the foundation, and activity 2 ⟶ 4, ordering and receiving materials, can start only when node 2 is *realized*, indicating that activity 1 ⟶ 2 has been completed. The number 3 above this branch signifies a time estimate of 3 months for the completion of this activity. Activity 2 ⟶ 3 and activity 2 ⟶ 4 can occur concurrently; neither depends on the other, and both depend only on the completion of activity 1 ⟶ 2.

FIGURE 8.6
Expanded network for building a house, showing concurrent activities

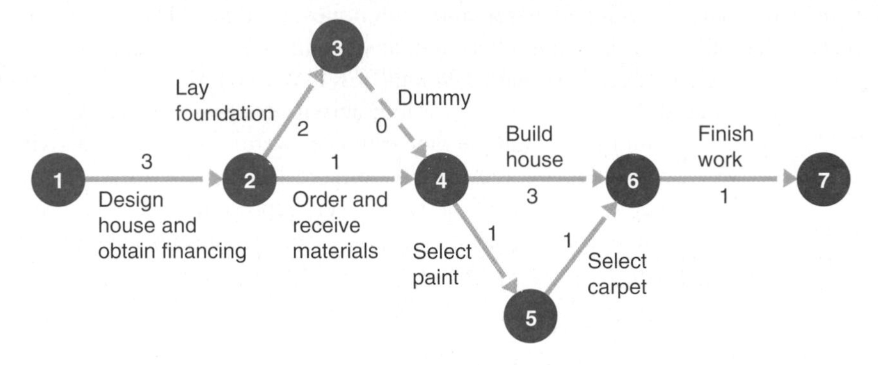

A dummy activity shows a precedence relationship but no passage of time.

When the activities of laying the foundation (2 ⟶ 3) and ordering and receiving materials (2 ⟶ 4) are completed, activities 4 ⟶ 5 and 4 ⟶ 6 can begin simultaneously. However, notice activity 3 ⟶ 4, which is referred to as a **dummy activity**. A dummy activity is used in an AOA network to show a precedence relationship, but it does not represent any actual passage of time. These activities have the precedence relationship shown in Figure 8.7(a). However, in an AOA network, two or more activities are not allowed to share the same start and ending nodes because

Time Out for Morgan R. Walker, James E. Kelley, Jr., and D. G. Malcolm

In 1956, a research team at E. I. du Pont de Nemours & Company, Inc., led by a du Pont engineer, Morgan R. Walker, and a Remington-Rand computer specialist, James E. Kelley, Jr., initiated a project to develop a computerized system to improve the planning, scheduling, and reporting of the company's engineering programs (including plant maintenance and construction projects). The resulting network approach is known as the critical path method (CPM). At virtually the same time, the U.S. Navy established a research team composed of members of the Navy Special Projects Office, Lockheed (the prime contractor), and the consulting firm of Booz, Allen, Hamilton, led by D. G. Malcolm, that developed PERT for the design of a management control system for the Polaris Missile Project (a ballistic missile–firing nuclear submarine). The Polaris project eventually included 23 PERT networks encompassing 2,000 events and 3,000 activities.

FIGURE 8.7
A dummy activity

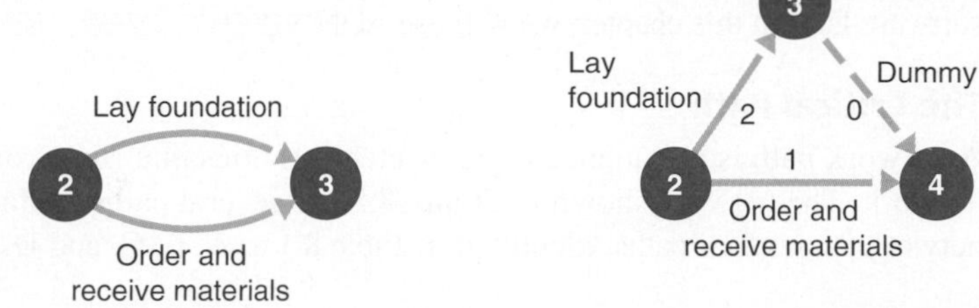

that would give them the same name designation (i.e., 2 → 3). Activity 3 → 4 is inserted to give the two activities separate end nodes and thus separate identities, as shown in Figure 8.7(b).

Returning to the network shown in Figure 8.6, we see that two activities start at node 4. Activity 4 → 6 is the actual building of the house, and activity 4 → 5 is the search for and selection of the paint for the exterior and interior of the house. Activity 4 → 6 and activity 4 → 5 can begin simultaneously and take place concurrently. Following the selection of the paint (activity 4 → 5) and the realization of node 5, the carpet can be selected (activity 5 → 6) because the carpet color is dependent on the paint color. This activity can also occur concurrently with the building of the house (activity 4 → 6). When the building is completed and the paint and carpet are selected, the house can be finished (activity 6 → 7).

AON Networks

Figure 8.8 shows the comparable AON network to the AOA network in Figure 8.6 for our house-building project. Notice that the activities and activity times are on the nodes and not on the activities, as with the AOA network. The branches or arrows simply show the precedence relationships between the activities. Also, notice that there is no dummy activity; dummy activities are not required in an AON network because no two activities have the same start and ending nodes, so they will never be confused. This is one of the advantages of the AON convention, although both AOA and AON have minor advantages and disadvantages. In general, the two methods both accomplish the same thing, and the one that is used is usually a matter of individual preference. However, for our purposes, the AON network has one distinct advantage: It is the convention used in

FIGURE 8.8
AON network for house-building project

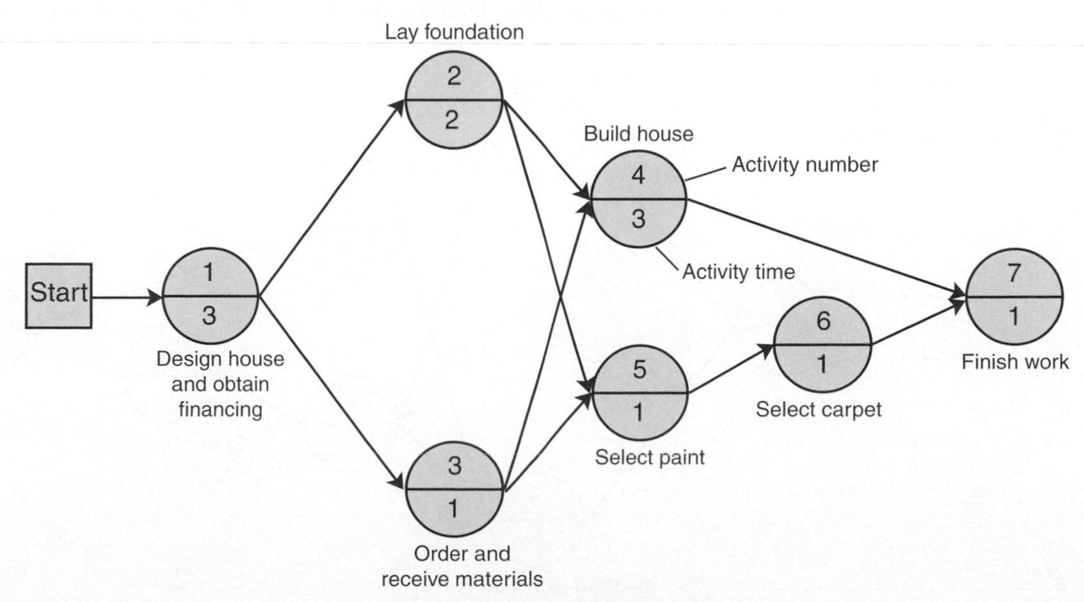

the popular Microsoft Project software package. Because we will demonstrate how to use that software later in this chapter, we will use AON.

The Critical Path

A network path is a sequence of connected activities that runs from the start to the end of the network. The network shown in Figure 8.8 has several paths. In fact, close observation of this network shows four paths, identified in Table 8.1 as A, B, C, and D.

TABLE 8.1
Paths through the house-building network

Path	Events
A	$1 \rightarrow 2 \rightarrow 4 \rightarrow 7$
B	$1 \rightarrow 2 \rightarrow 5 \rightarrow 6 \rightarrow 7$
C	$1 \rightarrow 3 \rightarrow 4 \rightarrow 7$
D	$1 \rightarrow 3 \rightarrow 5 \rightarrow 6 \rightarrow 7$

The project cannot be completed (i.e., the house cannot be built) until the longest path in the network is completed. This is the minimum time in which the project can be completed. The longest path is referred to as the **critical path**. To better understand the relationship between the minimum project time and the critical path, we will determine the length of each of the four paths shown in Figure 8.8.

*The **critical path** is the longest path through the network; it is the minimum time in which the network can be completed.*

By summing the activity times (shown in Figure 8.8) along each of the four paths, we can compute the length of each path, as follows:

$$\text{path A: } 1 \rightarrow 2 \rightarrow 4 \rightarrow 7$$
$$3 + 2 + 3 + 1 = 9 \text{ months}$$
$$\text{path B: } 1 \rightarrow 2 \rightarrow 5 \rightarrow 6 \rightarrow 7$$
$$3 + 2 + 1 + 1 + 1 = 8 \text{ months}$$
$$\text{path C: } 1 \rightarrow 3 \rightarrow 4 \rightarrow 7$$
$$3 + 1 + 3 + 1 = 8 \text{ months}$$
$$\text{path D: } 1 \rightarrow 3 \rightarrow 5 \rightarrow 6 \rightarrow 7$$
$$3 + 1 + 1 + 1 + 1 = 7 \text{ months}$$

Because path A is the longest, it is the critical path; thus, the minimum completion time for the project is 9 months. Now let us analyze the critical path more closely. From Figure 8.9 we can

FIGURE 8.9
Activity start time

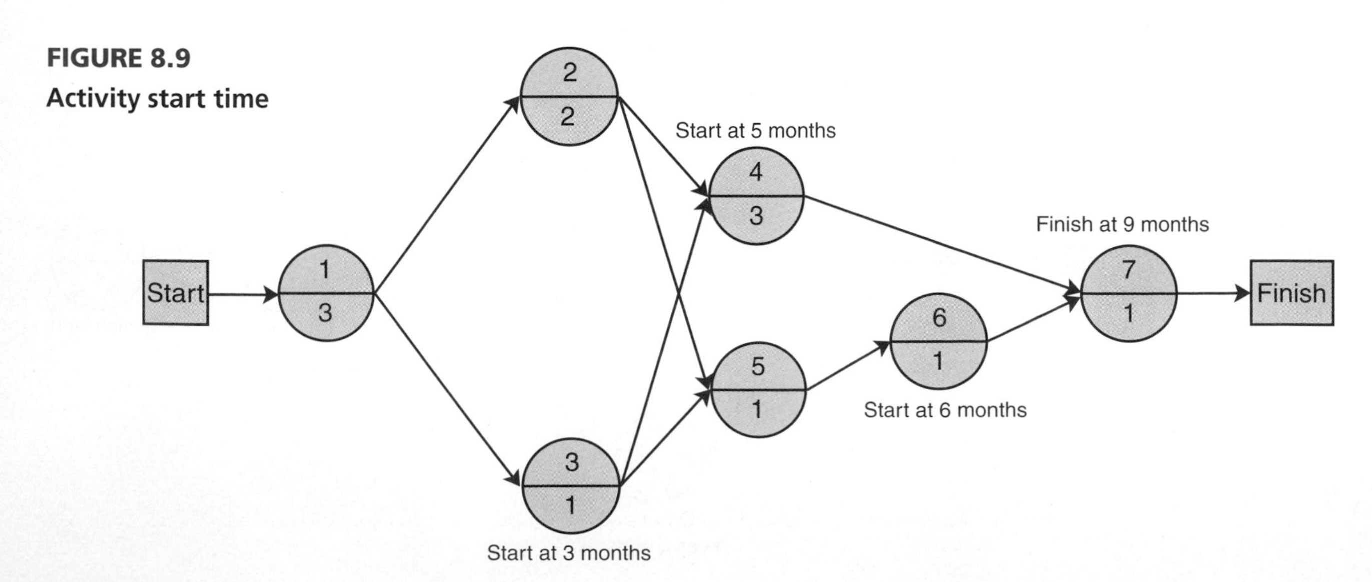

see that activity 3 cannot start until 3 months have passed. It is also easy to see that activity 4 will not start until 5 months have passed (i.e., the sum of activity 1 and 2 times). The start of activity 4 is dependent on the two activities leading into node 4. Activity 2 is completed after 5 months, but activity 3 is completed at the end of 4 months. Thus, we have two possible start times for activity 4: 5 months and 4 months. However, because the activity on node 4 cannot start until all preceding activities have been finished, the soonest node 4 can be started is 5 months.

Now, let us consider the activity following node 4. Using the same logic as before, we can see that activity 7 cannot start until after 8 months (5 months at node 4 plus the 3 months required by activity 4), or after 7 months (on path 1-3-5-6-7). Because all activities preceding node 7 must be completed before activity 7 can start, the soonest this can occur is 8 months. Adding 1 month for activity 7 to the time at node 7 gives a project duration of 9 months. Recall that this is the time of the longest path in the network, or the critical path.

This brief analysis demonstrates the concept of a critical path and the determination of the minimum completion time of a project. However, this is a cumbersome method for determining a critical path. Next, we will discuss a mathematical approach for scheduling the project activities and determining the critical path.

Activity Scheduling

In our analysis of the critical path, we determined the earliest time that each activity could be finished. For example, we found that the earliest time activity 4 could start was at 5 months. This time is referred to as the *earliest start time*, and it is expressed symbolically as *ES*.

ES is the earliest time an activity can start.

In order to show the earliest start time on the network, as well as some other activity times we will develop in the scheduling process, we will alter our node structure a little. Figure 8.10 shows the structure for node 1, the first activity in our example network for designing a house and obtaining financing.

FIGURE 8.10
Activity-on-node configuration

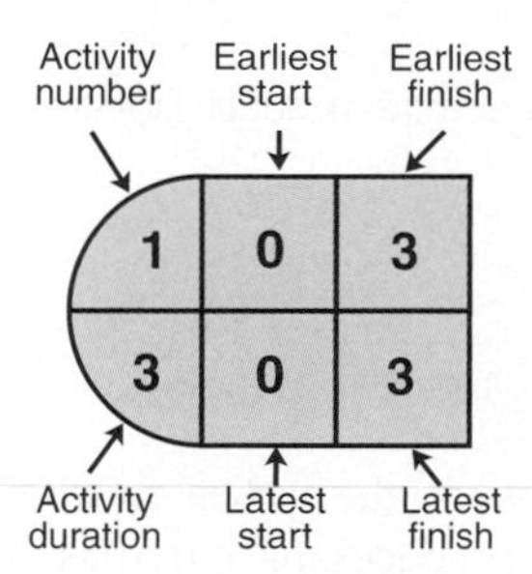

To determine the earliest start time for every activity, we make a *forward pass* through the network. That is, we start at the first node and move forward through the network. The earliest time for an activity is the maximum time for all preceding activities that have been completed—the time when the activity start node is realized.

EF is the earliest start time plus the activity time estimate.

The *earliest finish time, EF*, for an activity is the earliest start time plus the activity time estimate. For example, if the earliest start time for activity 1 is at time 0, then the earliest finish time is 3 months. In general, the earliest start and finish times for an activity are computed according to the following mathematical formulas:

$$ES = \text{Maximum } (EF \text{ immediate predecessors})$$
$$EF = ES + t$$

The earliest start and earliest finish times for all the activities in our project network are shown in Figure 8.11.

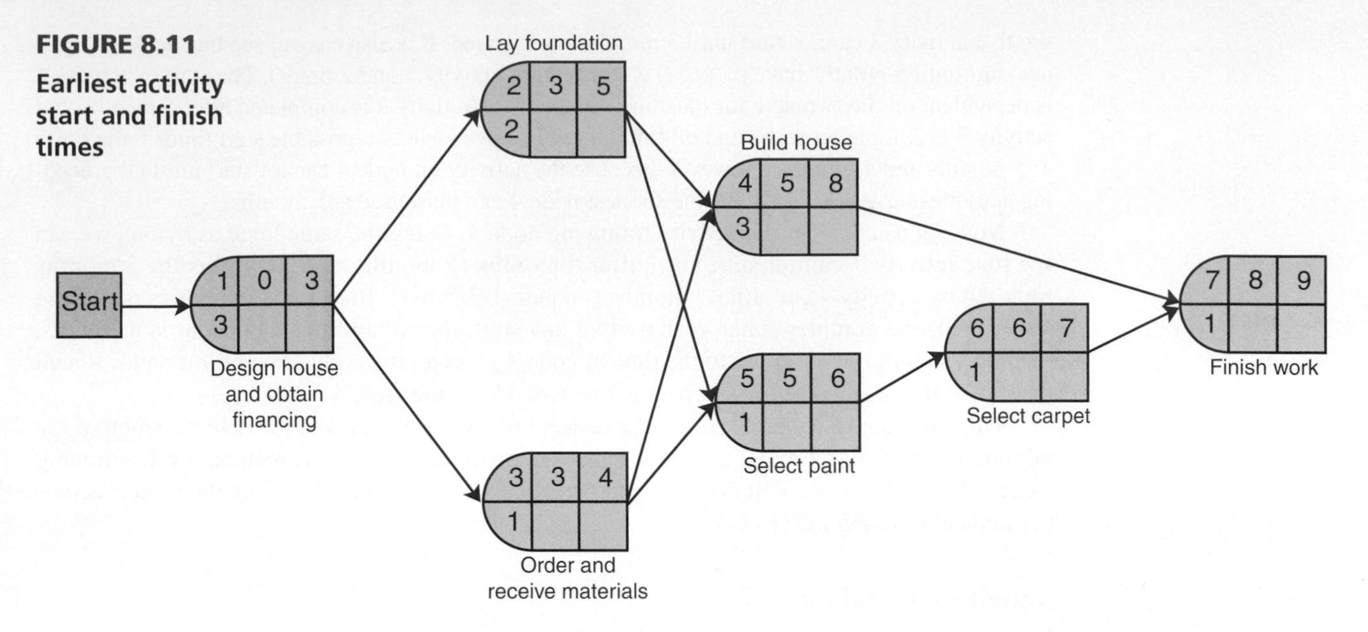

FIGURE 8.11
Earliest activity start and finish times

The earliest start time for the first activity in the network (for which there are no predecessor activities) is always zero, or $ES = 0$. This enables us to compute the earliest finish time for activity 1 as

$$EF = ES + t = 0 + 3 = 3 \text{ months}$$

The earliest start time for activity 2 is

$$\begin{aligned} ES &= \text{Max}\,(EF \text{ immediate predecessors}) \\ &= 3 \text{ months} \end{aligned}$$

The corresponding earliest finish time is

$$EF = ES + t = 3 + 2 = 5 \text{ months}$$

For activity 3 the earliest start time is 3 months, and the earliest finish time is 4 months.

Now consider activity 4, which has two predecessor activities. The earliest start time is computed as

$$\begin{aligned} ES &= \text{Max}\,(EF \text{ immediate predecessors}) \\ &= \text{Max}\,(5, 4) = 5 \text{ months} \end{aligned}$$

and the earliest finish time is

$$EF = ES + t = 5 + 3 = 8 \text{ months}$$

All the remaining earliest start and finish times are computed similarly. Notice in Figure 8.11 that the earliest finish time for activity 7, the last activity in the network, is 9 months, which is the total project duration, or critical path time.

LS is the latest time an activity can start without delaying the critical path time.

Companions to the earliest start and finish are the *latest start* (*LS*) and *finish* (*LF*) times. The latest start time is the latest time an activity can start without delaying the completion of the project beyond the project's critical path time. For our example, the project completion time (and earliest finish time) at node 7 is 9 months. Thus, the objective of determining latest times is to see how long each activity can be delayed without the project exceeding 9 months.

In general, the latest start and finish times for an activity are computed according to the following formulas:

$$LS = LF - t$$
$$LF = \text{Minimum } (LS \text{ immediately following predecessors})$$

A forward pass is used to determine earliest times.

A backward pass is used to determine latest times.

Whereas a **forward pass** through the network is made to determine the earliest times, the latest times are computed using a **backward pass**. We start at the end of the network at node 7 and work backward, computing the latest time for each activity. Because we want to determine how long each activity in the network can be delayed without extending the project time, the latest finish time at node 7 cannot exceed the earliest finish time. Therefore, the latest finish time at node 7 is 9 months. This and all other latest times are shown in Figure 8.12.

FIGURE 8.12
Latest activity start and finish times

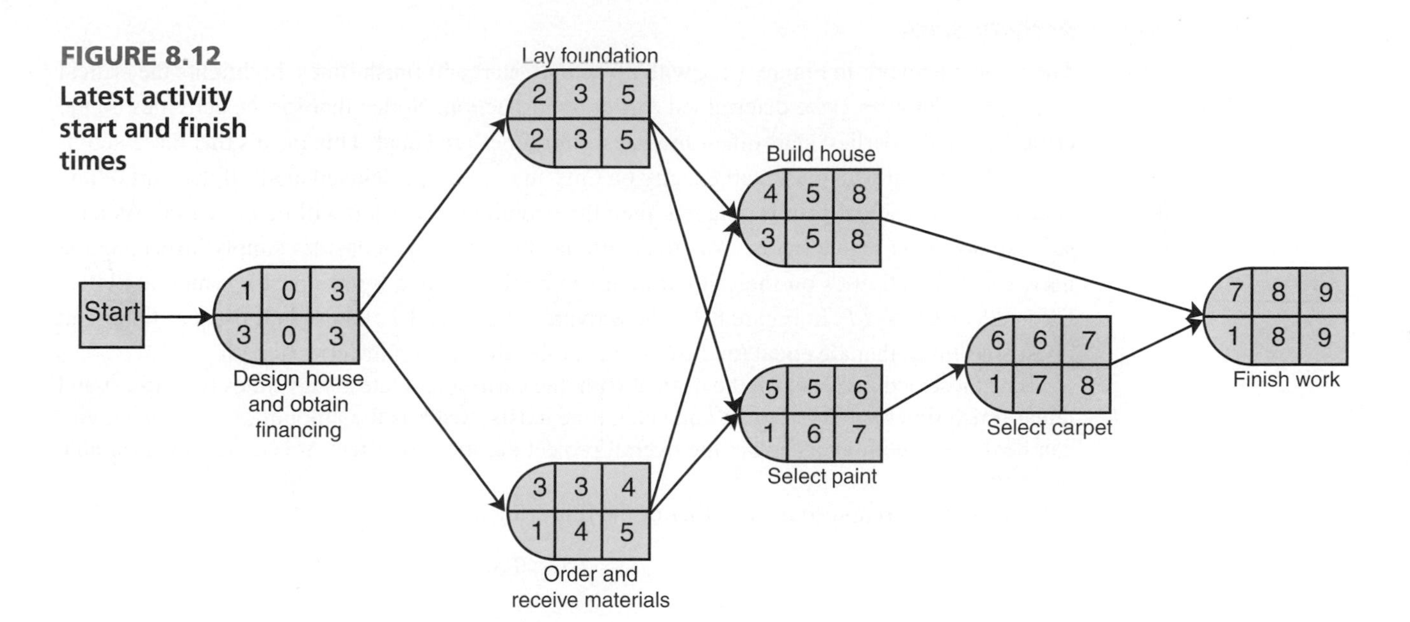

Starting at the end of the network, the critical path time, which is also equal to the earliest finish time of activity 7, is 9 months. This automatically becomes the latest finish time for activity 7, or

$$LF = 9 \text{ months}$$

Using this value, the latest start time for activity 7 can be computed:

$$LS = LF - t = 9 - 1 = 8 \text{ months}$$

The latest finish time for activity 6 is the minimum of the latest start times for the activities following node 6. Because activity 7 follows node 6, the latest start time is computed as follows:

$$LF = \text{Min } (LS \text{ following activities})$$
$$= 8 \text{ months}$$

The latest start time for activity 6 is

$$LS = LF - t = 8 - 1 = 7 \text{ months}$$

For activity 4, the latest finish time is 8 months, and the latest start time is 5 months; for activity 5, the latest finish time is 7 months, and the latest start time is 6 months.

Now consider activity 3, which has two activities following it. The latest finish time is computed as follows:

$$LF = \text{Min } (LS \text{ following activities})$$
$$= \text{Min } (5, 6) = 5 \text{ months}$$

The latest start time is

$$LS = LF - t = 5 - 1 = 4 \text{ months}$$

All the remaining latest start and latest finish times are computed similarly. Figure 8.12 includes the earliest and latest start times and earliest and latest finish times for all activities.

Activity Slack

The project network in Figure 8.12, with all activity start and finish times, highlights the critical path ($1 \rightarrow 2 \rightarrow 4 \rightarrow 7$) we determined earlier by inspection. Notice that for the activities on the critical path, the earliest start times and latest start times are equal. This means that these activities on the critical path must start exactly on time and cannot be delayed at all. If the start of any activity on the critical path is delayed, then the overall project time will be increased. As a result, we now have an alternative way to determine the critical path besides simply inspecting the network. The activities on the critical path can be determined by seeing for which activities $ES = LS$ or $EF = LF$. In Figure 8.12, the activities 1, 2, 4, and 7 all have earliest start times and latest start times that are equal (and $EF = LF$); thus, they are on the critical path.

For those activities not on the critical path, the earliest and latest start times (or earliest and latest finish times) are not equal, and slack time exists. Slack is the amount of time an activity can be delayed without affecting the overall project duration. In effect, it is extra time available for completing an activity.

Slack, S, is computed using either of the following formulas:

$$S = LS - ES$$

or

$$S = LF - EF$$

For example, the slack for activity 3 is computed as follows:

$$S = LS - ES = 4 - 3 = 1 \text{ month}$$

If the start of activity 3 were delayed for 1 month, the activity could still be completed by month 5 without delaying the project completion time. The slack for each activity in our example project network is shown in Table 8.2 and in Figure 8.13.

TABLE 8.2
Activity slack

Activity	*LS*	*ES*	*LF*	*EF*	Slack, *S*
*1	0	0	3	3	0
*2	3	3	5	5	0
3	4	3	5	4	1
*4	5	5	8	8	0
5	6	5	7	6	1
6	7	6	8	7	1
*7	8	8	9	9	0

*Critical path activities

FIGURE 8.13
Activity slack

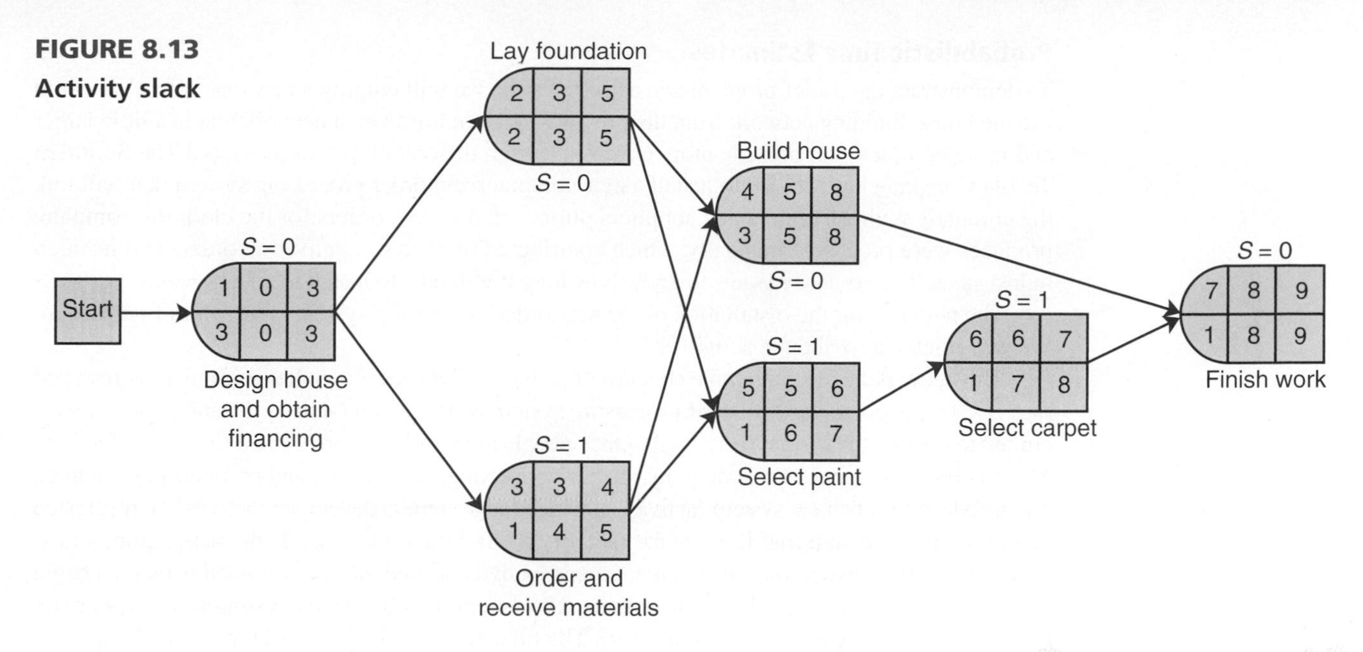

Notice in Figure 8.13 that activity 3 can be delayed 1 month and activity 5, which follows it, can be delayed 1 more month, but then activity 6 cannot be delayed at all, even though it has 1 month of slack. If activity 3 starts late, at month 4 instead of month 3, then it will be completed at month 5, which will not allow activity 5 to start until month 5. If the start of activity 5 is delayed 1 month, then it will be completed at month 7, and activity 6 cannot be delayed at all without exceeding the critical path time. The slack on these three activities is called **shared slack**. This means that the sequence of activities 3 → 5 → 6 can be delayed 2 months jointly without delaying the project, but not 3 months.

Shared slack is total slack available for combination of activities.

Slack is obviously beneficial to a project manager because it enables resources to be temporarily pulled away from activities with slack and used for other activities that might be delayed for various reasons or for which the time estimate has proven to be inaccurate.

The times for the network activities are simply estimates for which there is usually not a lot of historical basis (because projects tend to be unique undertakings). As such, activity time estimates are subject to quite a bit of uncertainty. However, the uncertainty inherent in activity time estimates can be reflected to a certain extent by using probabilistic time estimates instead of the single, deterministic estimates we have used so far.

Probabilistic Activity Times

In the project network for building a house presented in the previous section, all the activity time estimates were single values. By using only a single activity time estimate, we in effect assume that activity times are known with certainty (i.e., they are deterministic). For example, in Figure 8.8, the time estimate for activity 2 (laying the foundation) is shown to be 2 months. Because only this one value is given, we must assume that the activity time does not vary (or varies very little) from 2 months. In reality, however, it is rare that activity time estimates can be made with certainty. Project activities are likely to be unique, and thus there is little historical evidence that can be used as a basis to predict actual times. However, we earlier indicated that one of the primary differences between CPM and PERT was that PERT used probabilistic activity times. It is this approach to estimating activity times for a project network that we discuss in this section.

Probabilistic Time Estimates

To demonstrate the use of probabilistic activity times, we will employ a new example. (We could use the house-building network from the previous section; however, a network that is a little larger and more complex will provide more experience with different types of projects.) The Southern Textile Company has decided to install a new computerized order processing system that will link the company with customers and suppliers online. In the past, orders for the cloth the company produces were processed manually, which contributed to delays in delivering orders and resulted in lost sales. The company wants to know how long it will take to install the new system.

The network for the installation of the new order processing system is shown in Figure 8.14. We will briefly describe the activities.

The network begins with three concurrent activities: The new computer equipment is installed (activity 1); the computerized order processing system is developed (activity 2); and people are recruited to operate the system (activity 3). Once people are hired, they are trained for the job (activity 6), and other personnel in the company, such as marketing, accounting, and production personnel, are introduced to the new system (activity 7). Once the system is developed (activity 2), it is tested manually to make sure that it is logical (activity 5). Following activity 1, the new equipment is tested, and any necessary modifications are made (activity 4), and the newly trained personnel begin training on the computerized system (activity 8). Also, node 9 begins the testing of the system on the computer to check for errors (activity 9). The final activities include a trial run and changeover to the system (activity 11) and final debugging of the computer system (activity 10).

*A **beta distribution** is a probability distribution whose shape is based on 3 time estimates.*

At this stage in a project network, we previously assigned a single time estimate to each network activity. In a PERT project network, however, we determine *three time estimates* for each activity, which will enable us to estimate the mean and variance for a **beta distribution** of the activity times. We assume that the activity times can be described by a beta distribution for several reasons. First, the beta distribution mean and variance can be approximated with three estimates. Second, the beta distribution is continuous, but it has no predetermined shape (such as the bell shape of the normal curve). It will take on the shape indicated—that is, be skewed—by the time estimates given. This is beneficial because typically we have no prior knowledge of the shapes of the distributions of activity times in a unique project network. Third, although other types of distributions have been shown to be no more or less accurate than the beta, it has become traditional to use the beta distribution for probabilistic network analysis.

*Three time estimates for each activity—a **most likely**, an **optimistic**, and a **pessimistic**—provide an estimate of the mean and variance of a beta distribution.*

The three time estimates for each activity are the most likely time, the optimistic time, and the pessimistic time. The **most likely time** is the time that would most frequently occur if the activity were repeated many times. The **optimistic time** is the shortest possible time within which the activity could be completed if everything went right. The **pessimistic time** is the longest possible time the activity would require to be completed, assuming that everything went wrong. In general, the person most familiar with an activity makes these estimates to the best of his or her knowledge and ability. In other words, the estimate is subjective. (See Chapter 11 for subjective probability.)

These three time estimates can subsequently be used to estimate the mean and variance of a beta distribution. If

$$a = \text{optimistic time estimate}$$
$$m = \text{most likely time estimate}$$
$$b = \text{pessimistic time estimate}$$

the mean and variance are computed as follows:

$$\text{mean (expected time)}: t = \frac{a + 4m + b}{6}$$

$$\text{variance}: v = \left(\frac{b - a}{6}\right)^2$$

These formulas provide a reasonable estimate of the mean and variance of the beta distribution, a distribution that is continuous and can take on various shapes—that is, exhibit skewness.

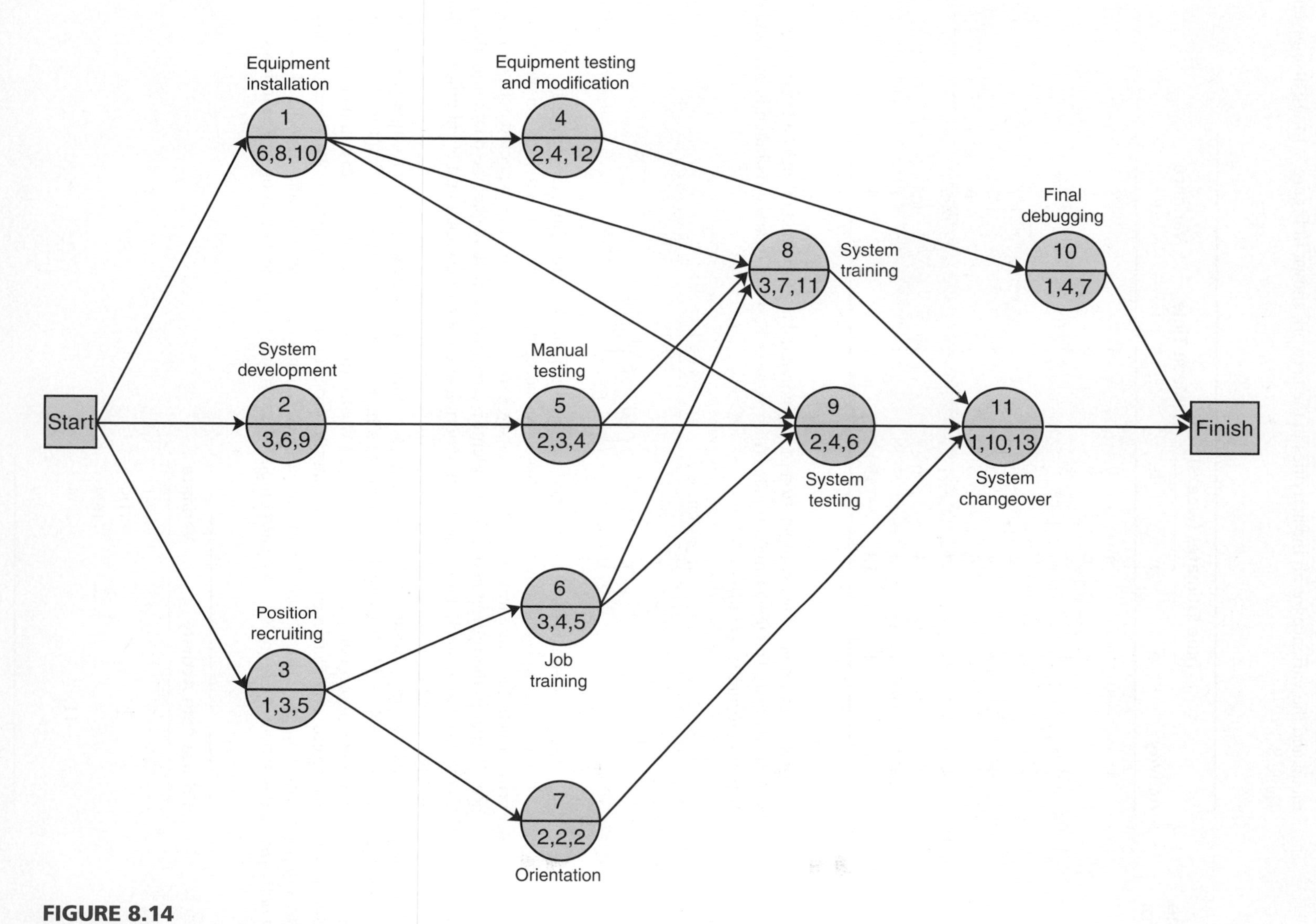

FIGURE 8.14
Network for order processing system installation

The three time estimates for each activity are shown in Figure 8.14 and in Table 8.3. The mean and the variance for all the activities in the network shown in Figure 8.14 are also given in Table 8.3.

TABLE 8.3
Activity time estimates for Figure 8.14

	Time Estimates (weeks)			Mean Time	Variance
Activity	*a*	*m*	*b*	*t*	*v*
1	6	8	10	8	4/9
2	3	6	9	6	1
3	1	3	5	3	4/9
4	2	4	12	5	25/9
5	2	3	4	3	1/9
6	3	4	5	4	1/9
7	2	2	2	2	0
8	3	7	11	7	16/9
9	2	4	6	4	4/9
10	1	4	7	4	1
11	1	10	13	9	4

As an example of the computation of the individual activity mean times and variances, consider activity 1. The three time estimates ($a = 6, m = 8, b = 10$) are substituted in our beta distribution formulas as follows:

$$t = \frac{a + 4m + b}{6} = \frac{6 + 4(8) + 10}{6} = 8 \text{ weeks}$$

$$v = \left(\frac{b - a}{6}\right)^2 = \left(\frac{10 - 6}{6}\right)^2 = \frac{4}{9} \text{ week}$$

The other values for the mean and variance in Table 8.3 are computed similarly.

Once the expected activity times have been computed for each activity, we can determine the critical path the same way we did previously, except that we use the expected activity times, *t*. Recall that in the project network, we identified the critical path as the one containing those activities with zero slack. This requires the determination of earliest and latest event times, as shown in Figure 8.15.

The critical path has no slack.

The project variance is the sum of the variances of the critical path activities

Observing Figure 8.15, we can see that the critical path encompasses activities $2 \rightarrow 5 \rightarrow 8 \rightarrow 11$, because these activities have no available slack. We can also see that the *expected* project completion time (t_p) is 25 weeks. However, it is possible to compute the variance for project completion time. To determine the project variance, we *sum the variances for those activities on the critical path*. Using the variances computed in Table 8.3 and the critical path activities shown in Figure 8.15, we can compute the variance for project duration (v_p) as follows:

Critical Path Activity	Variance
2	1
5	1/9
8	16/9
11	4
	62/9

$$v_p = 62/9 = 6.9 \text{ weeks}$$

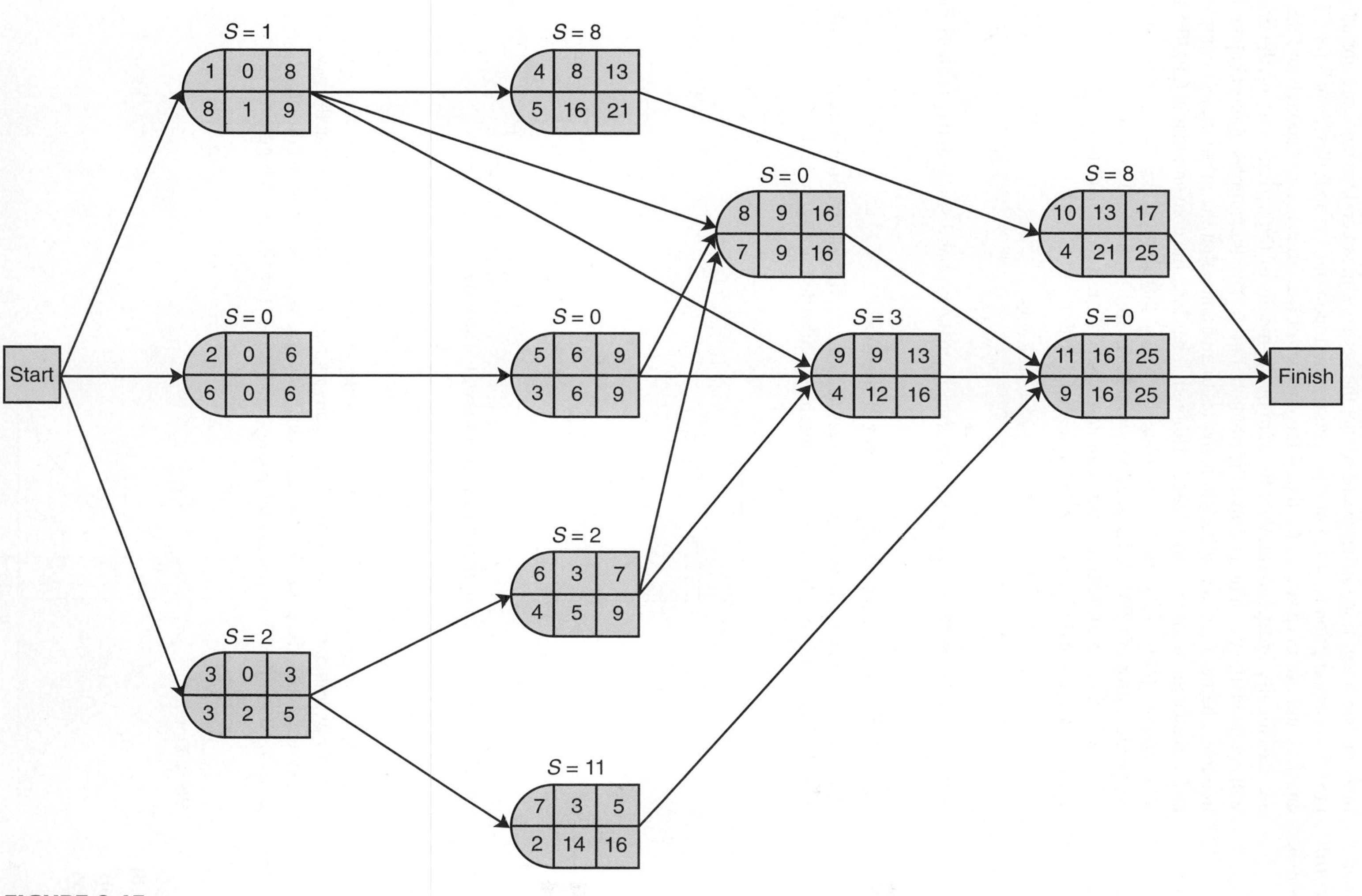

FIGURE 8.15
Earliest and latest activity times

The expected project time is assumed to be normally distributed based on the central limit theorem.

The CPM/PERT method assumes that the activity times are statistically independent, which allows us to sum the individual expected activity times and variances to get an expected *project* time and variance. It is further assumed that the network mean and variance are normally distributed. This assumption is based on the central limit theorem of probability, which for CPM/PERT analysis and our purposes states that if the number of activities is large enough and the activities are statistically independent, then the sum of the means of the activities along the critical path will approach the mean of a normal distribution. For the small examples in this chapter, it is questionable whether there are sufficient activities to guarantee that the mean project completion time and variance are normally distributed. Although it has become conventional in CPM/PERT analysis to employ probability analysis by using the normal distribution, regardless of the network size, the prudent user of CPM/PERT analysis should bear this limitation in mind.

Given these assumptions, we can interpret the expected project time (t_p) and variance (v_p) as the mean (μ) and variance (σ^2) of a normal distribution:

$$\mu = 25 \text{ weeks}$$
$$\sigma^2 = 6.9 \text{ weeks}$$

In turn, we can use these statistical parameters to make probabilistic statements about the project.

Probability Analysis of the Project Network

Using the normal distribution, probabilities can be determined by computing the number of standard deviations (Z) a value is from the mean, as illustrated in Figure 8.16.

FIGURE 8.16 Normal distribution of network duration

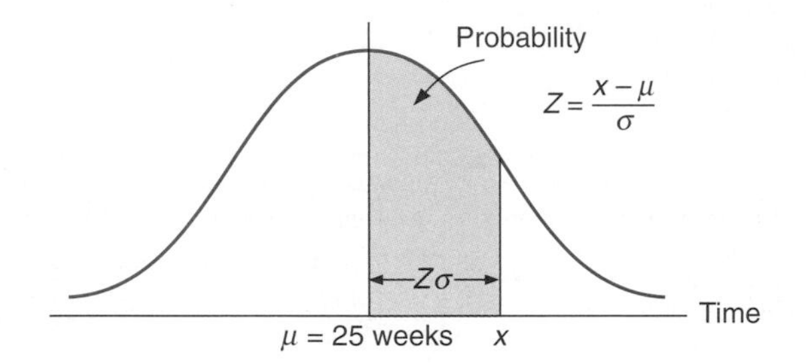

The Z value is computed using the following formula:

$$Z = \frac{x - \mu}{\sigma}$$

This value is then used to find the corresponding probability in Table A.1 of Appendix A.

For example, suppose the textile company manager told customers that the new order processing system would be completely installed in 30 weeks. What is the probability that it will, in fact, be ready by that time? This probability is illustrated as the shaded area in Figure 8.17.

FIGURE 8.17 Probability that the network will be completed in 30 weeks or less

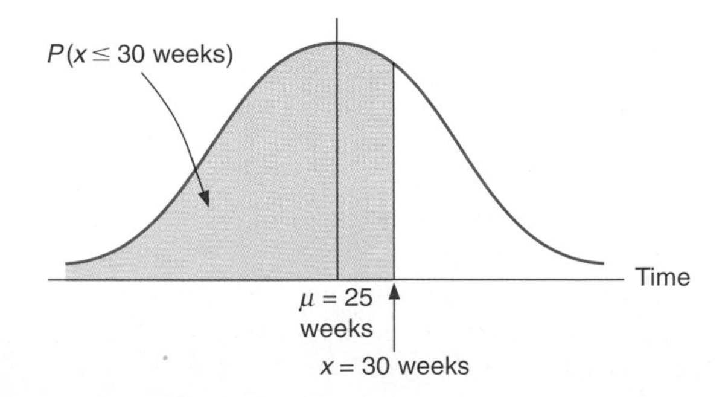

To compute the Z value for a time of 30 weeks, we must first compute the standard deviation (σ) from the variance (σ^2):

$$\sigma^2 = 6.9$$
$$\sigma = \sqrt{6.9} = 2.63$$

Next, we substitute this value for the standard deviation, along with the value for the mean and our proposed project completion time (30 weeks), into the following formula:

$$Z = \frac{x - \mu}{\sigma} = \frac{30 - 25}{2.63} = 1.90$$

A Z value of 1.90 corresponds to a probability of .4713 in Table A.1 in Appendix A. This means that there is a .9713 $(.5000 + .4713)$ probability of completing the project in 30 weeks or less.

Suppose one customer, frustrated with delayed orders, has told the textile company that if the new ordering system is not working within 22 weeks, she will trade elsewhere. The probability of the project's being completed within 22 weeks is computed as follows:

$$Z = \frac{22 - 25}{2.63} = \frac{-3}{2.63} = -1.14$$

A Z value of 1.14 (the negative indicates the area is to the left of the mean) corresponds to a probability of .3729 in Table A.1 of Appendix A. Thus, there is only a .1271 (.5000 − .3729) probability that the customer will be retained, as illustrated in Figure 8.18.

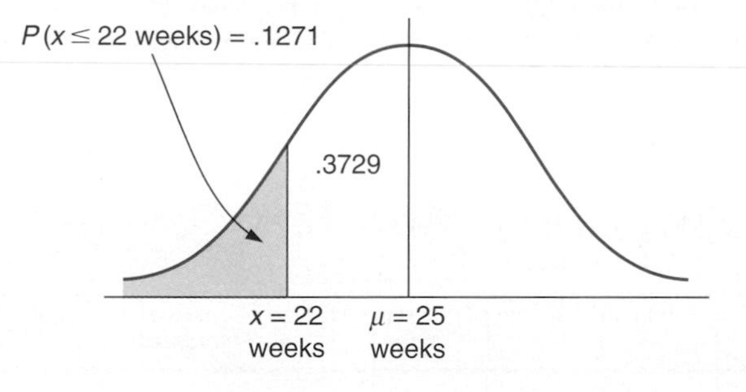

FIGURE 8.18
Probability that the network will be completed in 22 weeks or less

CPM/PERT Analysis with QM for Windows and Excel QM

The capability to perform CPM/PERT network analysis is a standard feature of most management science software packages for the personal computer. To illustrate the application of QM for Windows and Excel QM, we will use our example of installing an order processing system at Southern Textile Company. The QM for Windows solution output is shown in Exhibit 8.1, and the Excel QM solution is shown in Exhibit 8.2.

Management Science Application

The T5 Project at Heathrow Airport

British Airport Authority's terminal five (T5) at Heathrow Airport in London was completed in March 2008, for a cost of £4.3 billion. It was one of Europe's largest construction projects, taking 5.5 years to complete the construction, with the help of 60,000 people. Located between two runways in a space equal in size to Hyde Park in London, T5 has the largest single-span roof in Europe, made up of six sections and requiring 10 months to lift into place. T5 also boasts 11 miles of baggage conveyor belt. The facility has provided Heathrow (and British Airways) with 47 additional aircraft stands and increased Heathrow's capacity by 35 million passengers per year.

For the first time in a project of this size and complexity, off-site prefabrication was used extensively. This involved assembling components in off-site modules (2,800 in all) and then transporting them to the building site, where they were bolted together, thus reducing on-site construction time and disruption and construction traffic at the airport. Key clients worked collaboratively through integrated teams. A Web-based system was created to provide an effective communication vehicle for all project participants so that they could collaborate effectively. The Internet enabled the diverse off-site project module manufacturers in Dover and Scotland to be coordinated though a database system, or "virtual factory." The supply of material, equipment, and workflows could all be monitored through one integrated system. All project participants could see when modules were in production, completed, delivered, and stored. The database system included a "lessons learned" component, where lessons learned were recorded for all participants to see and learn from. The system was especially important for planning and coordinating deliveries between suppliers because of the high volume of deliveries and the limited on-site space involved.

e30/ZUMA Press/Newscom

The modules in particular required wide load deliveries, and roads had to be closed for delivery.

The collaborative nature of this project, combined with the "virtual factory" web-based computer system facilitated the performance of the integrated project teams, resulting in a reduction in construction times, a safer working environment, and better quality.

Sources: Based on J. Summers, "The Virtual Factory," *Quality World* 31, no. 10 (October 2005): 24–28; and P. Curmi, "Terminal Velocity," *Quality World* 33, no. 8 (August 2007): 17–21.

EXHIBIT 8.1

Project Management (PERT/CPM) Results

Southern Textile Company Solution

Activity	Activity time	Early Start	Early Finish	Late Start	Late Finish	Slack	Standard Deviation
Project	25						2.62
1	8	0	8	1	9	1	.67
2	6	0	6	0	6	0	1
3	3	0	3	2	5	2	.67
4	5	8	13	16	21	8	1.67
5	3	6	9	6	9	0	.33
6	4	3	7	5	9	2	.33
7	2	3	5	14	16	11	0
8	7	9	16	9	16	0	1.33
9	4	9	13	12	16	3	.67
10	4	13	17	21	25	8	1
11	9	16	25	16	25	0	2

EXHIBIT 8.2

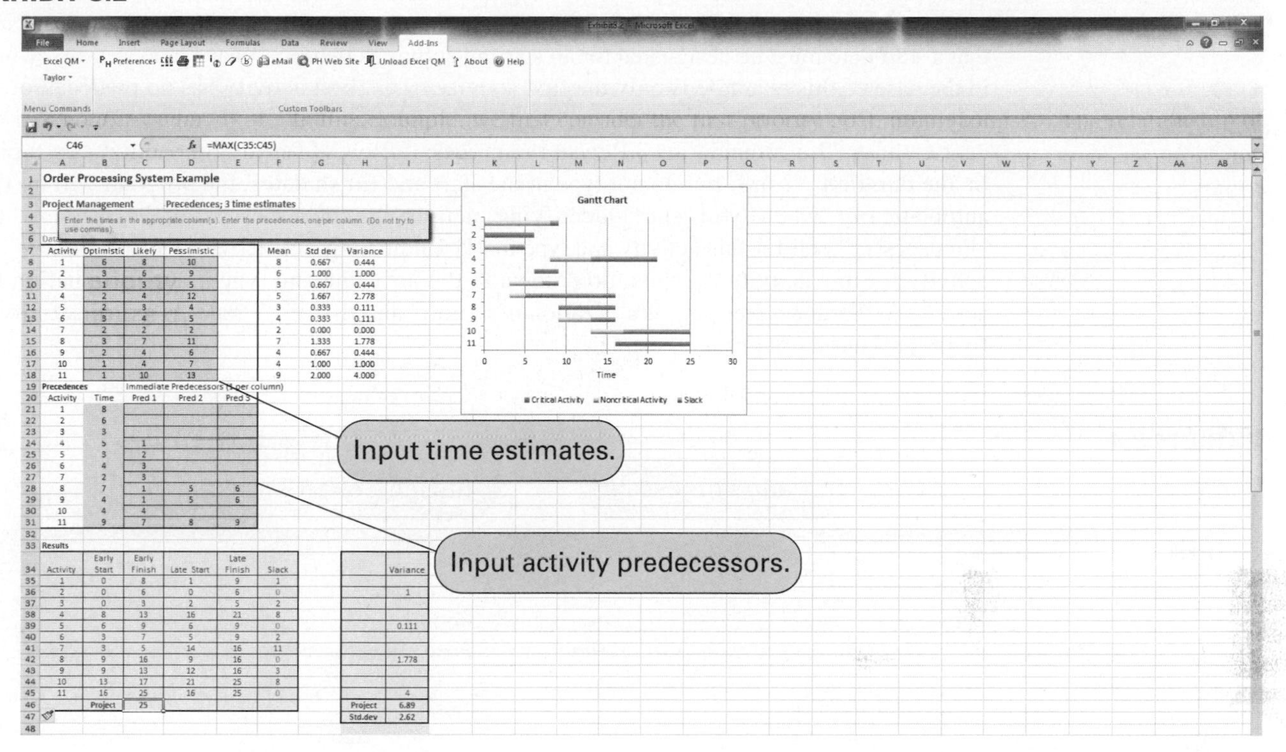

Microsoft Project

Microsoft Project is a popular and widely used software package for project management and CPM/PERT analysis. It is also relatively easy to use. We will demonstrate how to use Microsoft Project, using the project network for building a house shown in Figure 8.13. (Note that the Microsoft Project file for this example can be downloaded from the text Web site.)

When you open Microsoft Project, a screen comes up for a new project, as shown in Exhibit 8.3. Notice that the "Gantt Chart Tools" tab on the toolbar ribbon at the top of the screen is highlighted. The initial step is to set the project up in this window. First, type in the activity name, "Design and finance" under the "Task Name" column, and then type in the activity duration in the "Duration" column, which is 3 months for this activity. This only requires

EXHIBIT 8.3

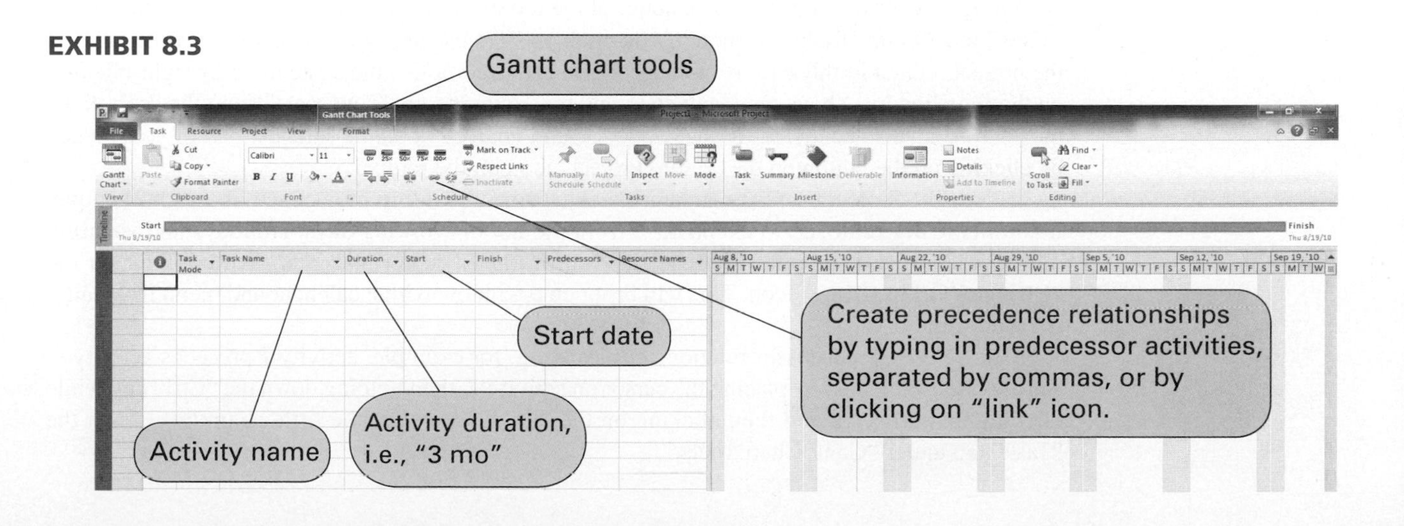

that you type "3 mo" and Project will recognize it as 3 months. Next, type in a "Start" date, such as "May 11, 2011." The start date can also be selected from the drop-down calendar. (Note that a start date must be designated for all starting activities that don't have predecessors, which in the case of this example is only the first activity.) This first activity has no predecessor, so leave the "Predecessor" cell for this activity blank and drop down to the next line to enter the next activity, "Lay foundation." Repeat the process followed for the first activity by typing in the duration, "2 mo" but do not type in the start and finish dates; the program will do this automatically when you start identifying activity predecessors. Toggle over to the "Predecessor" cell for this activity and type in "1" (indicating that activity 1 is the predecessor activity for this activity 2). Next, drop down to the next line and type in the information for activity 3, "Order materials," the duration of "1 mo" and that its "Predecessor" is activity 1. Exhibit 8.4 shows our progress so far.

EXHIBIT 8.4

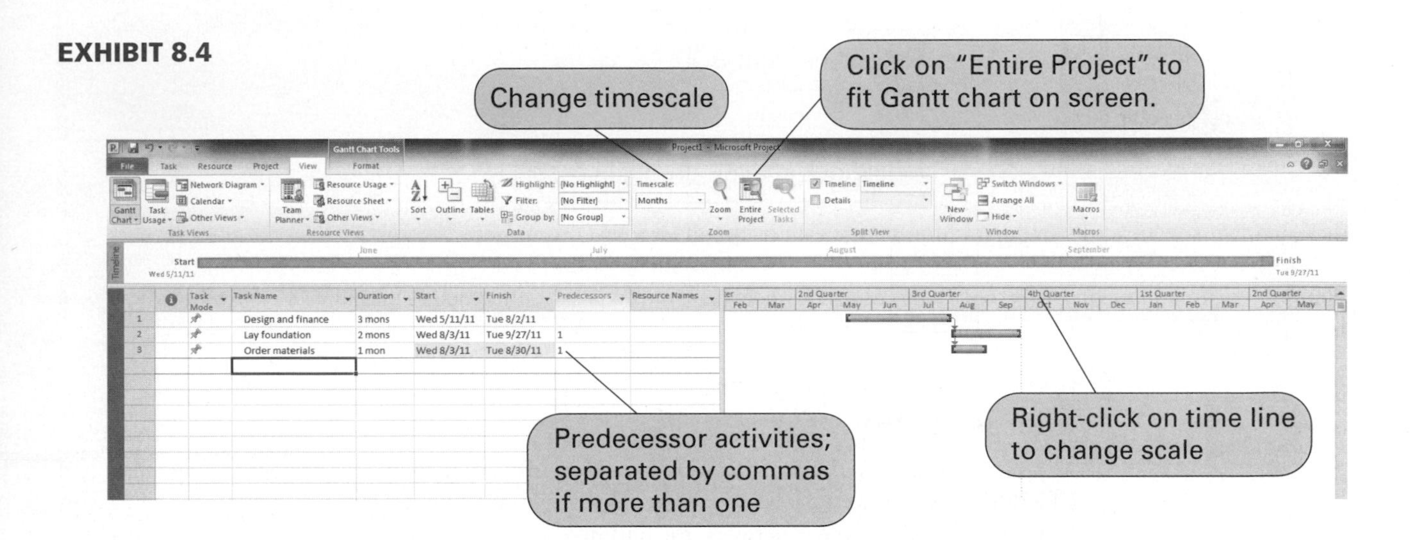

Before proceeding to finishing inputting the project data, we'll make a few comments. First, note the emerging Gantt chart forming on the calendar part of your window (on the right-hand part of the screen) for the activities we've entered. This chart might not be on your screen when you start because the timescale on the calendar does not conform to your start date. If the Gantt chart is not there, you can make it appear by moving the tab at the bottom of the calendar part of the screen ahead (i.e., to the right) until your start date appears, and this view should contain the beginning of your Gantt chart. Also, notice at the top of the window that we have switched to the "View" tab. On this toolbar ribbon, if you click on "Timescale," you can insert the time units of the project, which in this case is months, or you can accomplish the same thing by right-clicking on the time line just above the Gantt chart and then clicking on "Timescale" from the drop-down menu. You can also condense the size of your Gantt chart to fit on the calendar part of your screen by clicking on the button on the "Entire Project" icon on the toolbar.

Microsoft Project uses a "standard" calendar for scheduling project activities; for example, it automatically removes weekend days from the list of working days. Holidays and vacation days can also be designated by first clicking on the "Project" tab and then clicking on the "Change Working Time" icon. This will bring up a window with a calendar and menu for changing work times.

The activity predecessor relationships showing, for example, activity 1 precedes activity 2, can also be established by placing the cursor on activity 1, then holding down the "Ctrl" key while clicking on activity 2 and then clicking on the "link" icon (i.e., the little chain link) from the "Task" tab under "Gantt Chart Tools."

Next, we will finish typing in the rest of the information for our project, as shown in Exhibit 8.5. Notice on this screen that we have switched back to the "Format" tab under "Gantt Chart Tools," which allows us to select the "Critical Tasks" box to highlight the critical path on the Gantt Chart in red. Note that the "pins" in the cells next to each activity in the "Task Mode" column on our data window indicate that we have manually scheduled activities (or tasks); the other "Task Mode" option is to allow Microsoft Project to automatically schedule the activities. For our purposes, manual scheduling is sufficient.

EXHIBIT 8.5

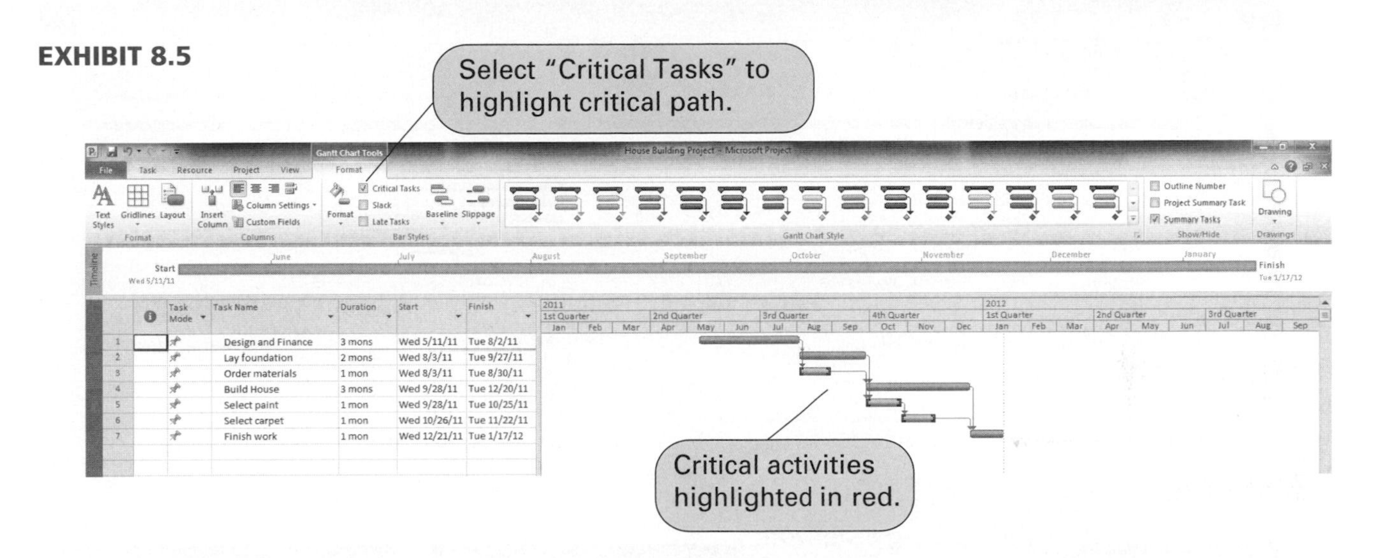

In order to show the network diagram of our project, click on "View" on the toolbar and then on "Network Diagram," which results in the screen shown in Exhibit 8.6. (We also used the "Zoom" icon to make the network larger.) Notice that the critical path is highlighted in red. Clicking on the "Gantt Chart" icon on the far left of the toolbar ribbon will return you to the Gantt Chart screen, where we typed in our project data.

EXHIBIT 8.6

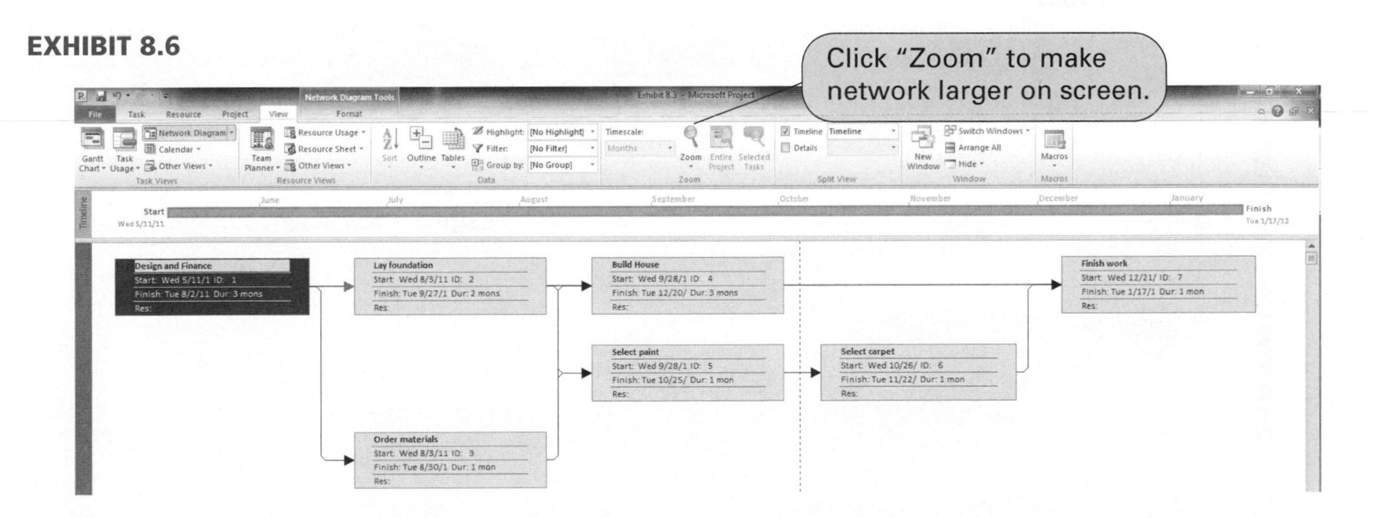

The latest version of Microsoft Project does not have PERT (three time estimate) capabilities, thus to use Microsoft Project when you have three time estimates, you must first manually (or using Excel) compute the mean activity times and use these as single time estimates in Microsoft Project. Exhibit 8.7 shows the Microsoft Project "Gantt Chart" window for our

"Order Processing System" project example, where the mean activity time estimates are used. Exhibit 8.8 shows the network diagram. Notice that we gave this project a start date of May 11, 2011, which had to be designated as the start date for all three of our starting activities (that do not have predecessors): 1, 2, and 3.

EXHIBIT 8.7

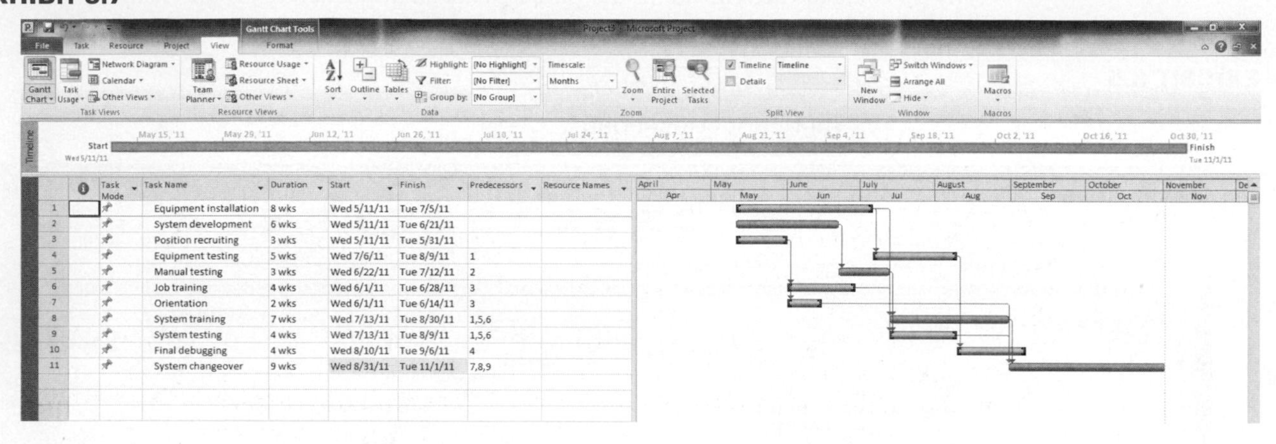

	Task Name	Duration	Start	Finish	Predecessors
1	Equipment installation	8 wks	Wed 5/11/11	Tue 7/5/11	
2	System development	6 wks	Wed 5/11/11	Tue 6/21/11	
3	Position recruiting	3 wks	Wed 5/11/11	Tue 5/31/11	
4	Equipment testing	5 wks	Wed 7/6/11	Tue 8/9/11	1
5	Manual testing	3 wks	Wed 6/22/11	Tue 7/12/11	2
6	Job training	4 wks	Wed 6/1/11	Tue 6/28/11	3
7	Orientation	2 wks	Wed 6/1/11	Tue 6/14/11	3
8	System training	7 wks	Wed 7/13/11	Tue 8/30/11	1,5,6
9	System testing	4 wks	Wed 7/13/11	Tue 8/9/11	1,5,6
10	Final debugging	4 wks	Wed 8/10/11	Tue 9/6/11	4
11	System changeover	9 wks	Wed 8/31/11	Tue 11/1/11	7,8,9

EXHIBIT 8.8

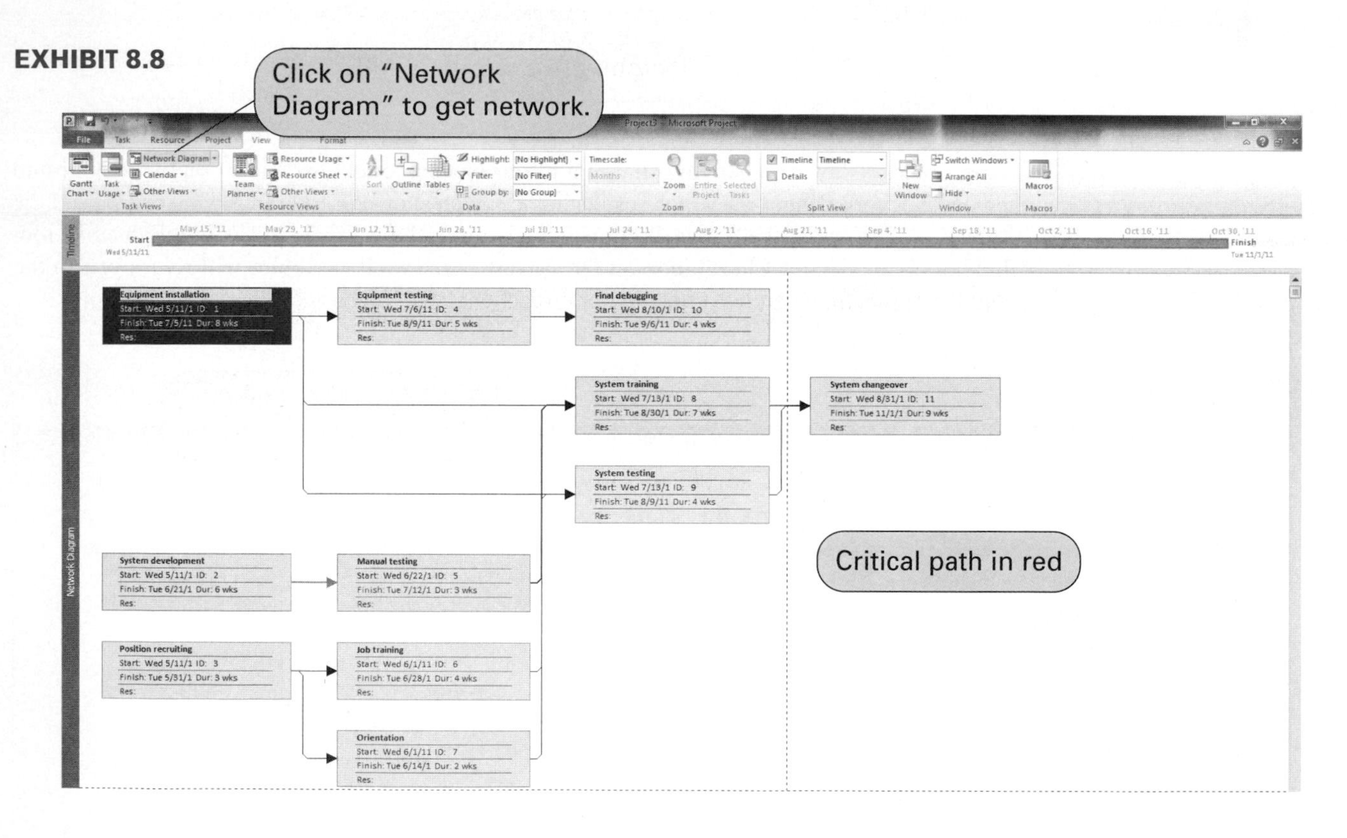

Microsoft Project also has many additional tools and features, including project updating and activity completion, resource management, work stoppages, development of work breakdown structures (WBS), and the ability to change work times, including shortened days, holidays, and vacation time. To access information about these features, access the various "Help" screens from the different Microsoft Project windows or press the F1 key.

Project Crashing and Time–Cost Trade-Off

To this point, we have demonstrated the use of CPM and PERT network analysis for determining project time schedules. This in itself is valuable to a manager planning a project. However, in addition to scheduling projects, a project manager is frequently confronted with the problem of having to reduce the scheduled completion time of a project to meet a deadline. In other words, the manager must finish the project sooner than indicated by the CPM or PERT network analysis.

Project duration can be reduced by assigning more labor to project activities, often in the form of overtime, and by assigning more resources (material, equipment, etc). However, additional labor and resources cost money and hence increase the overall project cost. Thus, the decision to reduce the project duration must be based on an analysis of the *trade-off* between time and cost.

Project crashing shortens the project time by reducing critical activity times at a cost.

Project crashing is a method for shortening project duration by reducing the time of one or more of the critical project activities to a time that is less than the normal activity time. This reduction in the normal activity times is referred to as *crashing*. Crashing is achieved by devoting more resources, measured in terms of dollars, to the activities to be crashed.

To demonstrate how project crashing works, we will employ the original network for constructing a house shown in Figure 8.8. This network is repeated in Figure 8.19, except that the activity times previously shown as months have been converted to weeks. Although this example network encompasses only single-activity time estimates, the project crashing procedure can be applied in the same manner to PERT networks with probabilistic activity time estimates.

FIGURE 8.19
The project network for building a house

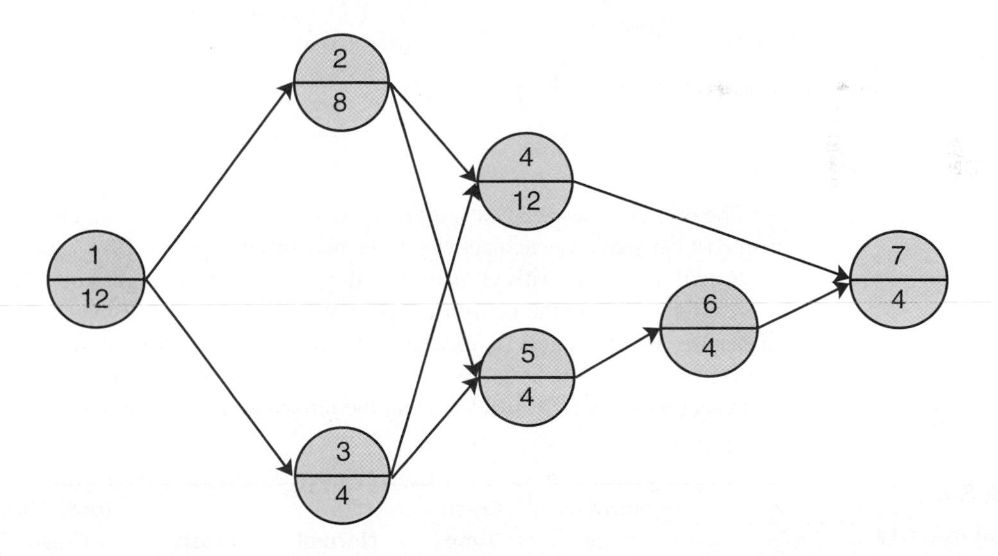

In Figure 8.19, we will assume that the times (in weeks) shown on the network activities are the *normal activity times*. For example, normally 12 weeks are required to complete activity 1. Furthermore, we will assume that the cost required to complete this activity in the time indicated is $3,000. This cost is referred to as the *normal activity cost*. Next, we will assume that the building contractor has estimated that activity 1 can be completed in 7 weeks, but it will cost $5,000 to complete the activity instead of $3,000. This new estimated activity time is known as the *crash time*, and the revised cost is referred to as the *crash cost*.

Activity 1 can be crashed a total of 5 weeks (normal time − crash time = 12 − 7 = 5 weeks), at a total crash cost of $2,000 (crash cost − normal cost = $5,000 − 3,000 = $2,000). Dividing the total crash cost by the total allowable crash time yields the crash cost per week:

$$\frac{\text{total crash cost}}{\text{total crash time}} = \frac{\$2{,}000}{5} = \$400 \text{ per week}$$

Crash cost and crash time have a linear relationship.

If we assume that the relationship between crash cost and crash time is linear, then activity 1 can be crashed by any amount of time (not exceeding the maximum allowable crash time) at a rate of $400 per week. For example, if the contractor decided to crash activity 1 by only 2 weeks (for an activity time of 10 weeks), the crash cost would be $800 ($400 per week × 2 weeks). The linear relationships between crash cost and crash time and between normal cost and normal time are illustrated in Figure 8.20.

FIGURE 8.20
Time–cost relationship for crashing activity 1

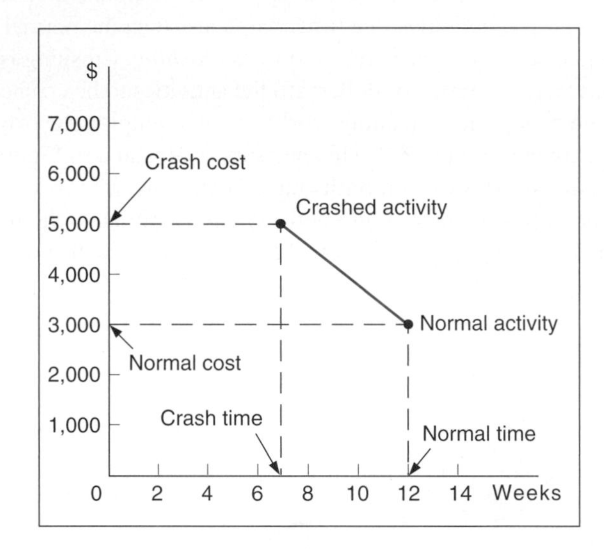

The normal times and costs, the crash times and costs, the total allowable crash times, and the crash cost per week for each activity in the network in Figure 8.19 are summarized in Table 8.4.

Recall that the critical path for the house-building network encompassed activities $1 \rightarrow 2 \rightarrow 4 \rightarrow 7$, and the project duration was 9 months, or 36 weeks. Suppose that the home builder needed the house in 30 weeks and wanted to know how much extra cost would be incurred to complete the house in this time. To analyze this situation, the contractor would crash the project network to 30 weeks, using the information in Table 8.4.

TABLE 8.4
Normal activity and crash data for the network in Figure 8.19

Activity	Normal Time (weeks)	Crash Time (weeks)	Normal Cost	Crash Cost	Total Allowable Crash Time (weeks)	Crash Cost per Week
1	12	7	$ 3,000	$ 5,000	5	$ 400
2	8	5	2,000	3,500	3	500
3	4	3	4,000	7,000	1	3,000
4	12	9	50,000	71,000	3	7,000
5	4	1	500	1,100	3	200
6	4	1	500	1,100	3	200
7	4	3	15,000	22,000	1	7,000
			$75,000	$110,700		

As activities are crashed, the critical path may change, and several paths may become critical.

The objective of project crashing is to reduce the project duration while minimizing the cost of crashing. Because the project completion time can be shortened only by crashing activities on the critical path, it may turn out that not all activities have to be crashed. However, as activities are crashed, the critical path may change, requiring crashing of previously noncritical activities to further reduce the project completion time.

We start the crashing process by looking at the critical path and seeing which activity has the minimum crash cost per week. Observing Table 8.4 and Figure 8.21, we see that on the critical path activity 1 has the minimum crash cost of $400. Thus, activity 1 will be reduced as much as possible. Table 8.4 shows that the maximum allowable reduction for activity 1 is 5 weeks, *but we can reduce activity 1 only to the point where another path becomes critical.* When two paths simultaneously become critical, activities on both must be reduced by the same amount. (If we reduce the activity time beyond the point where another path becomes critical, we may incur an unnecessary cost.) This last stipulation means that we must keep up with all the network paths as we reduce individual activities, a condition that makes manual crashing very cumbersome. Later we will demonstrate an alternative method for project crashing, using linear programming; however, for the moment we will pursue this example in order to demonstrate the logic of project crashing.

FIGURE 8.21
Network with normal activity times and weekly activity crashing costs

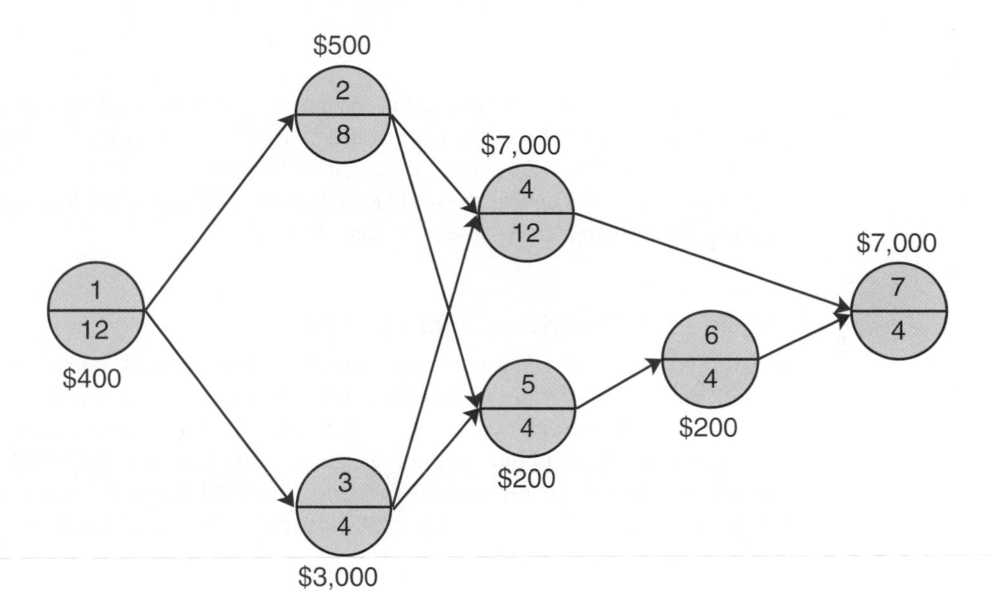

It turns out that activity 1 can be crashed by the total amount of 5 weeks without another path's becoming critical because activity 1 is included in all four paths in the network. Crashing this activity results in a revised project duration of 31 weeks, at a crash cost of $2,000. The revised network is shown in Figure 8.22.

This process must now be repeated. The critical path in Figure 8.22 remains the same, and the new minimum activity crash cost on the critical path is $500 for activity 2. Activity 2 can be crashed a total of 3 weeks, but because the contractor desires to crash the network to only 30 weeks, we need to crash activity 2 by only 1 week. Crashing activity 2 by 1 week does not result in any other path's becoming critical, so we can safely make this reduction. Crashing activity 2 to 7 weeks (i.e., a 1-week reduction) costs $500 and reduces the project duration to 30 weeks.

The extra cost of crashing the project to 30 weeks is $2,500. Thus, the contractor could inform the customer that an additional cost of only $2,500 would be incurred to finish the house in 30 weeks.

FIGURE 8.22
Revised network with activity 1 crashed

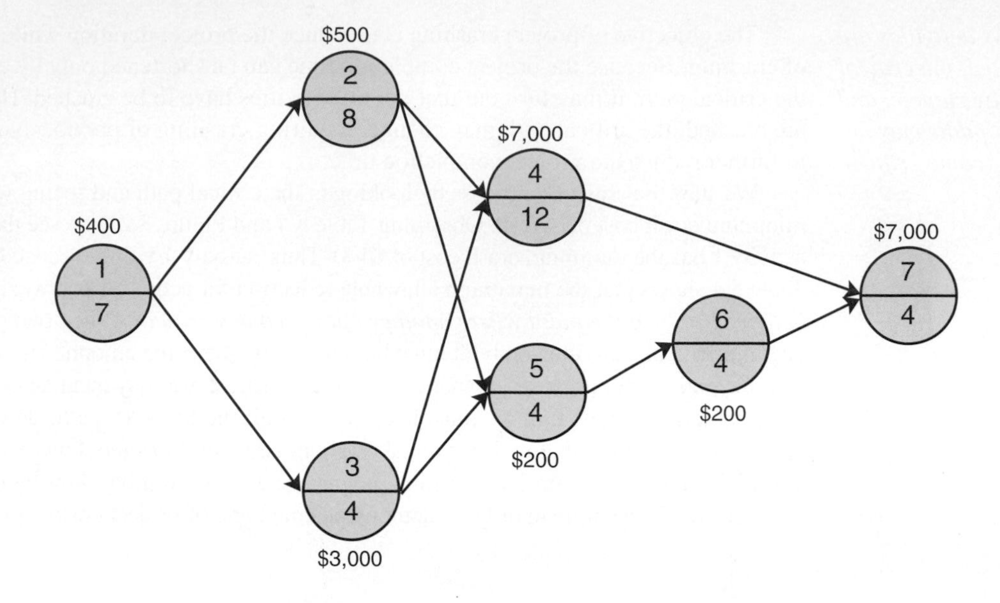

As indicated earlier, the manual procedure for crashing a network is very cumbersome and generally unacceptable for project crashing. It is basically a trial-and-error approach that is useful for demonstrating the logic of crashing; however, it quickly becomes unmanageable for larger networks. This approach would have become difficult if we had pursued even the house-building example to a crash time more than 30 weeks.

Project Crashing with QM for Windows

QM for Windows also has the capability to crash a network *completely*. In other words, it crashes the network by the maximum amount possible. In our house-building example in the previous section, we crashed the network to only 30 weeks, and we did not consider by how much the network could have actually been crashed. Alternatively, QM for Windows crashes a network by the maximum amount possible. The QM for Windows solution for our house-building example is shown in Exhibit 8.9. Notice that the network has been crashed to 24 weeks, at a total crash cost of $31,500.

EXHIBIT 8.9

Project Management (PERT/CPM) Results

House Building Example Solution

Activity	Normal time	Crash time	Normal Cost	Crash Cost	Crash cost/pd	Crash by	Crashing cost
Project	36	24					
1	12	7	3000	5000	400	5	2000
2	8	5	2000	3500	500	3	1500
3	4	3	4000	7000	3000	0	0
4	12	9	50000	71000	7000	3	21000
5	4	1	500	1100	200	0	0
6	4	1	500	1100	200	0	0
7	4	3	15000	22000	7000	1	7000
TOTALS			75000				31500

The General Relationship of Time and Cost

In our discussion of project crashing, we demonstrated how the project critical path time could be reduced by increasing expenditures for labor and direct resources. The implicit objective of crashing was to reduce the scheduled completion time for its own sake—that is, to reap the results of the project sooner. However, there may be other important reasons for reducing project time. As projects continue over time, they consume various *indirect costs*, including the cost of facilities, equipment, and machinery; interest on investment; utilities, labor, and personnel costs; and the loss of skills and labor from members of the project team who are not working at their regular jobs. There also may be direct financial penalties for not completing a project on time. For example, many construction contracts and government contracts have penalty clauses for exceeding the project completion date.

Crash costs increase as project time decreases; indirect costs increase as project time increases.

In general, project crash costs and indirect costs have an inverse relationship; crash costs are highest when the project is shortened, whereas indirect costs increase as the project duration increases. This time–cost relationship is illustrated in Figure 8.23. The best, or optimal, project time is at the minimum point on the total cost curve.

Management Science Application

Reconstructing the Pentagon After 9/11

On September 11, 2001, at 9:37 A.M., American Airlines flight 77, which had been hijacked by terrorists, was flown into the west face of the Pentagon in Arlington, Virginia. More than 400,000 square feet of office space was destroyed, and an additional 1.6 million square feet was damaged. Almost immediately, the "Phoenix Project" to restore the Pentagon was initiated. A deadline of 1 year was established for project completion, which required the demolition and removal of the destroyed portion of the building, followed by the building restoration, including restoration of the limestone facade. The Pentagon consists of five rings of offices (housing 25,000 employees) that emanate from the center of the building; ring A is the innermost ring, and ring E is the outermost. Ten corridors radiate out from the building's hub, bisecting the ring, and forming the Pentagon's five distinctive wedges. At the time of the attack, the Pentagon was undergoing a 20-year, $1.2 billion renovation program, and the renovation of Wedge 1, which was demolished in the 9/11 attack, was nearing completion. As a result, the Phoenix Project leaders were able to use the Wedge 1 renovation project structure and plans as a basis for their own reconstruction plan and schedule, saving much time in the process. Project leaders were in place and able to assign resources on the very day of the attack. The project included more than 30,000 activities and a 3,000-member project team and required 3 million person-hours of work. Over 56,000 tons of contaminated debris were removed from the site, 2.5 million pounds of limestone were used to reconstruct the facade (using the original drawings from 1941), 21,000 cubic yards of concrete were poured, and 3,800 tons of reinforcing steel were placed. The Phoenix Project was completed almost a month ahead of schedule and nearly $194 million under the original budget estimate of $700 million.

AP Photo/Dept. of Defense, Grant Greenwalt

Source: Based on N. Bauer, "Rising from the Ashes," *PM Network* 18, no. 5 (May 2004): 24–32.

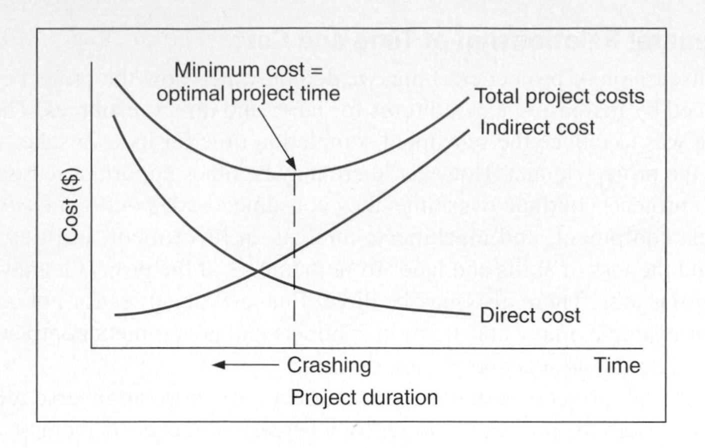

FIGURE 8.23
The time–cost trade-off

Formulating the CPM/PERT Network as a Linear Programming Model

First we will look at the linear programming formulation of the general CPM/PERT network model and then at the formulation of the project crashing network.

We will formulate the linear programming model of a CPM/PERT network, using the AOA convention. As the first step in formulating the linear programming model, we will define the decision variables. In our discussion of CPM/PERT networks using the AOA convention, we designated an activity by its start and ending node numbers. Thus, an activity starting at node 1 and ending at node 2 was referred to as activity $1 \rightarrow 2$. We will use a similar designation to define the decision variables of our linear programming model.

The objective function for the linear programming model is to minimize project duration.

We will use a different scheduling convention. Instead of determining the earliest activity start time for each activity, we will use the *earliest event time* at each node. This is the earliest time that a node (i or j) can be realized. In other words, it is the earliest time that the event the node represents, either the completion of all the activities leading into it or the start of all activities leaving it, can occur. Thus, for an activity $i \rightarrow j$, the earliest event time of node i will be x_i, and the earliest event time of node j will be x_j.

The objective of the project network is to determine the earliest time the project can be completed (i.e., the critical path time). We have already determined from our discussion of CPM/PERT network analysis that the earliest event time of the last node in the network equals the critical path time. If we let x_i equal the earliest event times of the nodes in the network, then the objective function can be expressed as

$$\text{minimize } Z = \sum_i x_i$$

Because the value of Z is the sum of all the earliest event times, it has no real meaning; however, it will ensure the earliest event time at each node.

Next, we must develop the model constraints. We will define the time for activity $i \rightarrow j$ as t_{ij} (as we did earlier in this chapter). From our previous discussion of CPM/PERT network analysis, we know that the difference between the earliest event time at node j and the earliest event time at node i must be at least as great as the activity time t_{ij}. A set of constraints that expresses this condition is defined as

$$x_j - x_i \geq t_{ij}$$

The general linear programming model of formulation of a CPM/PERT network can be summarized as

$$\text{minimize } Z = \sum_i x_i$$

subject to

$$x_j - x_i \geq t_{ij}, \text{ for all activities } i \rightarrow j$$
$$x_i, x_j \geq 0$$

where

x_i = earliest event time of node i
x_j = earliest event time of node j
t_{ij} = time of activity $i \rightarrow j$

The solution of this linear programming model will indicate the earliest event time of each node in the network and the project duration.

As an example of the linear programming model formulation and solution of a project network, we will use the house-building network from Figure 8.6 with times converted to weeks. This network, with activity times in weeks and earliest event times, is shown in Figure 8.24.

FIGURE 8.24
CPM/PERT network for the house-building project, with earliest event times

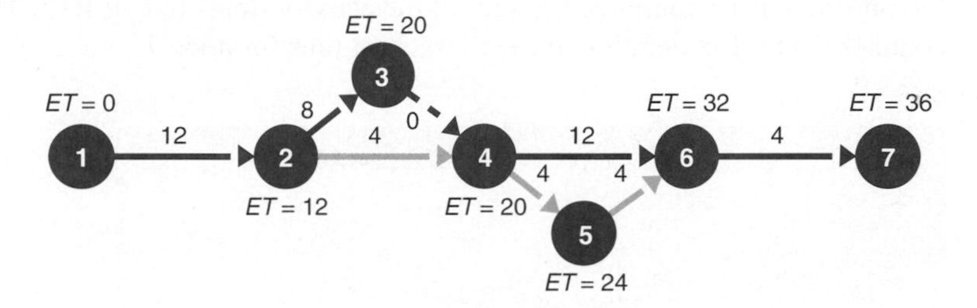

The linear programming model for the network in Figure 8.24 is

$$\text{minimize } Z = x_1 + x_2 + x_3 + x_4 + x_5 + x_6 + x_7$$

subject to

$$x_2 - x_1 \geq 12$$
$$x_3 - x_2 \geq 8$$
$$x_4 - x_2 \geq 4$$
$$x_4 - x_3 \geq 0$$
$$x_5 - x_4 \geq 4$$
$$x_6 - x_4 \geq 12$$
$$x_6 - x_5 \geq 4$$
$$x_7 - x_6 \geq 4$$
$$x_i, x_j \geq 0$$

Notice in this model that there is a constraint for each activity in the network.

Solution of the CPM/PERT Linear Programming Model with Excel

The linear programming model of the CPM/PERT network in the preceding section enables us to use Excel to schedule the project. Exhibit 8.10 shows an Excel spreadsheet set up to determine the earliest event times for each node—that is, the x_i and x_j values in our linear programming model for our house-building example. The earliest start times, our decision variables, are in cells **B6:B12**. Cells **F6:F13** contain the model constraints. For example, cell F6 contains the constraint formula **=B7−B6**, and cell F7 contains the formula **=B8−B7**. These constraint formulas will be set ≥ the activity times in column G when we access Solver. (Also, because activity 3 → 4 is a dummy, a constraint for **F9=0** must be added to Solver.)

EXHIBIT 8.10

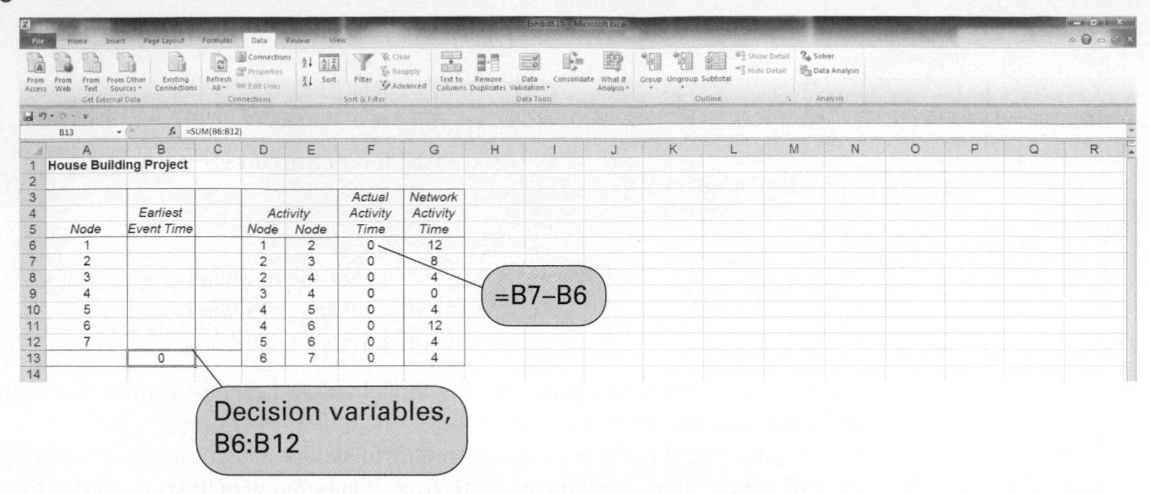

	A	B	C	D	E	F	G
1	House Building Project						
2							
3						Actual	Network
4		Earliest		Activity		Activity	Activity
5	Node	Event Time		Node	Node	Time	Time
6	1			1	2	0	12
7	2			2	3	0	8
8	3			2	4	0	4
9	4			3	4	0	0
10	5			4	5	0	4
11	6			4	6	0	12
12	7			5	6	0	4
13		0		6	7	0	4
14							

The Solver Parameters window with the model data is shown in Exhibit 8.11. Notice that the objective is to minimize the sum of the activity times in cell B13. Thus, cell B12 actually contains the project duration, the earliest event time for node 7.

EXHIBIT 8.11

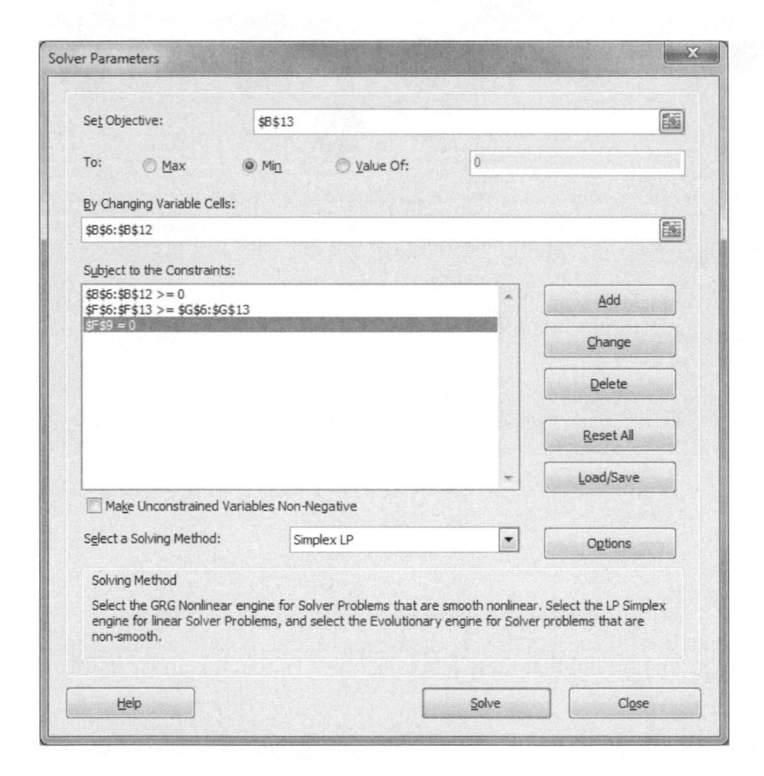

The solution is shown in Exhibit 8.12. Notice that the earliest time at each node is given in cells **B6:B12**, and the total project duration is 36 weeks. However, this output does not indicate the critical path. The critical path can be determined by accessing the sensitivity report for this problem. Recall that when you click on "Solve" from Solver, a screen comes up, indicating that Solver has reached a solution. This screen also provides the opportunity to generate several different kinds of reports, including an answer report and a sensitivity report. When we click on "Sensitivity" under the report options, the information shown in Exhibit 8.13 is provided.

EXHIBIT 8.12

B13 =SUM(B6:B12)

House Building Project

Node	Earliest Event Time		Activity Node	Activity Node	Actual Activity Time	Network Activity Time
1	0		1	2	12	12
2	12		2	3	8	8
3	20		2	4	8	4
4	20		3	4	0	0
5	24		4	5	4	4
6	32		4	6	12	12
7	36		5	6	8	4
	144		6	7	4	4

EXHIBIT 8.13

Changing Cells

Cell	Name	Final Value	Reduced Cost	Objective Coefficient	Allowable Increase	Allowable Decrease
$B$6	Start Time	0	1	0	1E+30	1
$B$7	Start Time	12	0	0	1E+30	1
$B$8	Start Time	20	0	0	1E+30	1
$B$9	Start Time	20	0	0	1E+30	1
$B$10	Start Time	28	0	0	0	1
$B$11	Start Time	32	0	0	1E+30	1
$B$12	Start Time	36	0	1	1E+30	1

Constraints

Cell	Name	Final Value	Shadow Price	Constraint R.H. Side	Allowable Increase	Allowable Decrease
$F$6	Time	12	1	12	1E+30	12
$F$7	Time	8	1	8	1E+30	4
$F$8	Time	8	0	4	4	1E+30
$F$9	Time	0	1	0	1E+30	4
$F$10	Time	8	0	4	4	1E+30
$F$11	Time	12	1	12	1E+30	4
$F$12	Time	4	0	4	4	1E+30
$F$13	weeks Time	4	1	4	1E+30	36

A shadow price of 1 indicates critical path.

The information in which we are interested is the shadow price for each of the activity constraints. The shadow price for each activity will be either 1 or 0. A positive shadow price of 1 for an activity means that you can reduce the overall project duration by an amount with a corresponding decrease by the same amount in the activity duration. Alternatively, a shadow price of 0 means that the project duration will not change, even if you change the activity duration by some amount. This means that those activities with a shadow price of 1 are on the critical path. Cells F6, F7, F9, F11, and F13 have shadow prices of 1, and referring back to Exhibit 8.12, we see that these cells correspond to activities $1 \rightarrow 2$, $2 \rightarrow 3$, $3 \rightarrow 4$, $4 \rightarrow 6$, and $6 \rightarrow 7$, which are the activities on the critical path.

Project Crashing with Linear Programming

The linear programming model required to perform project crashing analysis differs from the linear programming model formulated for the general CPM/PERT network in the previous section. The linear programming model for project crashing is somewhat longer and more complex.

The objective of the project crashing model is to minimize the cost of crashing.

The objective for our general linear programming model was to minimize project duration; the objective of project crashing is to minimize the cost of crashing, given the limits on how much individual activities can be crashed. As a result, the general linear programming model formulation must be expanded to include crash times and cost. We will continue to define the earliest event times for activity $i \rightarrow j$ as x_i and x_j. In addition, we will define the amount of time each activity $i \rightarrow j$ is crashed as y_{ij}. Thus, the decision variables are defined as

$$x_i = \text{earliest event time of node } i$$
$$x_j = \text{earliest event time of node } j$$
$$y_{ij} = \text{amount of time by which activity } i \rightarrow j \text{ is crashed (i.e., reduced)}$$

The objective of project crashing is to reduce the project duration at the minimum possible crash cost. For our house-building network, the objective function is written as

$$\text{minimize } Z = \$400y_{12} + 500y_{23} + 3{,}000y_{24} + 200y_{45} + 7{,}000y_{46} + 200y_{56} + 7{,}000y_{67}$$

The objective function coefficients are the activity crash costs per week from Table 8.4; the variables y_{ij} indicate the number of weeks each activity will be reduced. For example, if activity $1 \rightarrow 2$ is crashed by 2 weeks, then $y_{12} = 2$, and a cost of \$800 is incurred.

The model constraints must specify the limits on the amount of time each activity can be crashed. Using the allowable crash times for each activity from Table 8.4 enables us to develop the following set of constraints:

$$y_{12} \leq 5$$
$$y_{23} \leq 3$$
$$y_{24} \leq 1$$
$$y_{34} \leq 0$$
$$y_{45} \leq 3$$
$$y_{46} \leq 3$$
$$y_{56} \leq 3$$
$$y_{67} \leq 1$$

For example, the first constraint, $y_{12} \leq 5$, specifies that the amount of time by which activity $1 \rightarrow 2$ is reduced cannot exceed 5 weeks.

The next group of constraints must mathematically represent the relationship between earliest event times for each activity in the network, as the constraint $x_j - x_i \geq t_{ij}$ did in our original linear programming model. However, we must now reflect the fact that activity times can be crashed by an amount y_{ij}. Recall the formulation of the activity $1 \rightarrow 2$ constraint for the general linear programming model formulation in the previous section:

$$x_2 - x_1 \geq 12$$

This constraint can also be written as

$$x_1 + 12 \leq x_2$$

This latter constraint indicates that the earliest event time at node 1 (x_1), plus the normal activity time (12 weeks) cannot exceed the earliest event time at node 2 (x_2). To reflect the fact that this activity can be crashed, it is necessary only to subtract the amount by which it can be crashed from the left-hand side of the preceding constraint:

amount activity $1 \rightarrow 2$ can be crashed

$$x_1 + 12 - y_{12} \leq x_2$$

This revised constraint now indicates that the earliest event time at node 2 (x_2) is determined not only by the earliest event time at node 1 plus the activity time but also by the amount the activity is crashed. Each activity in the network must have a similar constraint, as follows:

$$x_1 + 12 - y_{12} \leq x_2$$
$$x_2 + 8 - y_{23} \leq x_3$$
$$x_2 + 4 - y_{24} \leq x_4$$
$$x_3 + 0 - y_{34} \leq x_4$$
$$x_4 + 4 - y_{45} \leq x_5$$
$$x_4 + 12 - y_{46} \leq x_6$$
$$x_5 + 4 - y_{56} \leq x_6$$
$$x_6 + 4 - y_{67} \leq x_7$$

Finally, we must indicate the project duration we are seeking (i.e., the crashed project time). Because the housing contractor wants to crash the project from the 36-week normal critical path time to 30 weeks, our final model constraint specifies that the earliest event time at node 7 should not exceed 30 weeks:

$$x_7 \leq 30$$

The complete linear programming model formulation is summarized as follows:

minimize $Z = \$400y_{12} + 500y_{23} + 3{,}000y_{24} + 200y_{45} + 7{,}000y_{46} + 200y_{56} + 7{,}000y_{67}$
subject to

$$\begin{aligned}
y_{12} &\leq 5 \\
y_{23} &\leq 3 \\
y_{24} &\leq 1 \\
y_{34} &\leq 0 \\
y_{45} &\leq 3 \\
y_{46} &\leq 3 \\
y_{56} &\leq 3 \\
y_{67} &\leq 1 \\
y_{12} + x_2 - x_1 &\geq 12 \\
y_{23} + x_3 - x_2 &\geq 8 \\
y_{24} + x_4 - x_2 &\geq 4 \\
y_{34} + x_4 - x_3 &\geq 0 \\
y_{45} + x_5 - x_4 &\geq 4 \\
y_{46} + x_6 - x_4 &\geq 12 \\
y_{56} + x_6 - x_5 &\geq 4 \\
y_{67} + x_7 - x_6 &\geq 4 \\
x_7 &\leq 30 \\
x_i, y_{ij} &\geq 0
\end{aligned}$$

Project Crashing with Excel

Because we have been able to develop a linear programming model for project crashing, we can also solve this model by using Excel. Exhibit 8.14 shows a modified version of the Excel spreadsheet we developed earlier, in Exhibit 8.10, to determine the earliest event times for our CPM/PERT network for the house-building project. We have added columns H, I, and J for the activity crash costs, the activity crash times, and the actual activity crash times. Cells **J6:J13** correspond to the y_{ij} variables in the linear programming model. The constraint formulas for each activity are included in cells **F6:F13**. For example, cell F6 contains the formula **=J6+B7−B6**, and cell F7 contains the formula **=J7+B8−B7**. These constraints and the others in column F must be set ≥ the activity times in column G. The objective function formula in cell B16 is shown on the formula bar at the top of the spreadsheet. The crashing goal of 30 weeks is included in cell B15.

The problem in Exhibit 8.14 is solved by using Solver, as shown in Exhibit 8.15. Notice that there are two sets of decision variables, in cells **B6:B12** and in cells **J6:J13**. The project crashing solution is shown in Exhibit 8.16.

EXHIBIT 8.14

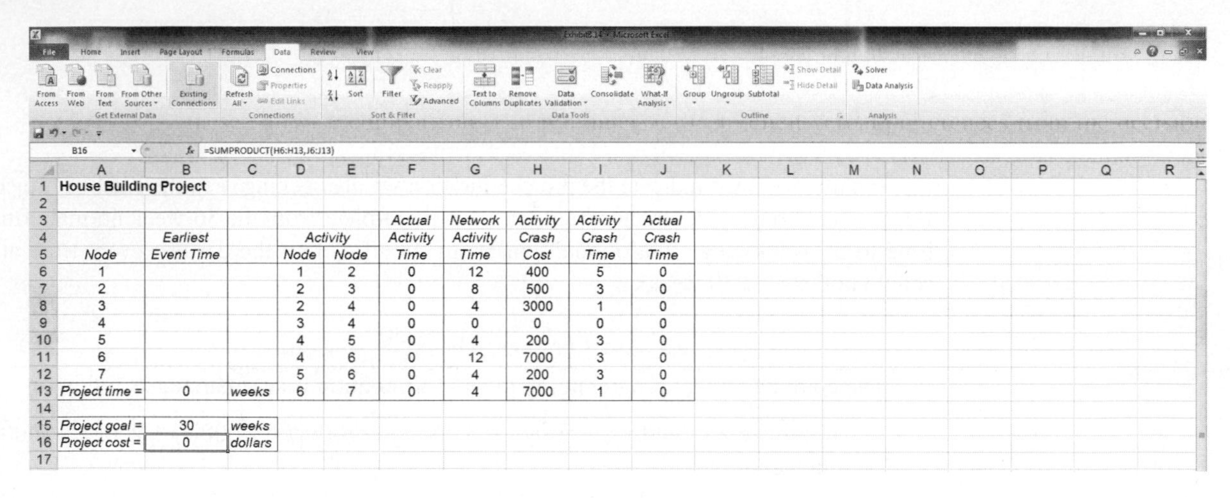

B16 =SUMPRODUCT(H6:H13,J6:J13)

House Building Project

Node	Earliest Event Time		Activity Node	Activity Node	Actual Activity Time	Network Activity Time	Activity Crash Cost	Activity Crash Time	Actual Crash Time
1			1	2	0	12	400	5	0
2			2	3	0	8	500	3	0
3			2	4	0	4	3000	1	0
4			3	4	0	0	0	0	0
5			4	5	0	4	200	3	0
6			4	6	0	12	7000	3	0
7			5	6	0	4	200	3	0
Project time =	0	weeks	6	7	0	4	7000	1	0
Project goal =	30	weeks							
Project cost =	0	dollars							

EXHIBIT 8.15

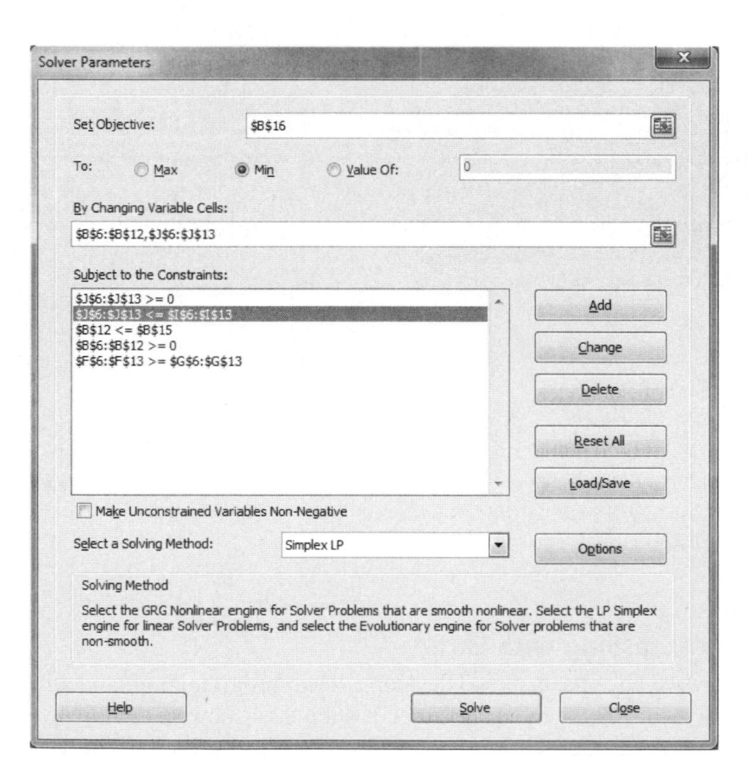

EXHIBIT 8.16

B16 =SUMPRODUCT(H6:H13,J6:J13)

	A	B	C	D	E	F	G	H	I	J
1	**House Building Project**									
2										
3						*Actual*	*Network*	*Activity*	*Activity*	*Actual*
4		*Earliest*		*Activity*		*Activity*	*Activity*	*Crash*	*Crash*	*Crash*
5	*Node*	*Event Time*		*Node*	*Node*	*Time*	*Time*	*Cost*	*Time*	*Time*
6	1	0		1	2	12	12	400	5	5
7	2	7		2	3	8	8	500	3	1
8	3	14		2	4	7	4	3000	1	0
9	4	14		3	4	0	0	0	0	0
10	5	22		4	5	8	4	200	3	0
11	6	26		4	6	12	12	7000	3	0
12	7	30		5	6	4	4	200	3	0
13	*Project time =*	30	*weeks*	6	7	4	4	7000	1	0
14										
15	*Project goal =*	30	*weeks*							
16	*Project cost =*	2500	*dollars*							
17										
18										

Summary

In this chapter, we discussed two of the most popular management science techniques—CPM and PERT networks. Their popularity is due primarily to the fact that a network forms a picture of the system under analysis that is easy for a manager to interpret. Sometimes, it is difficult to explain a set of mathematical equations to a manager, but a network often can be easily explained. CPM/PERT has been applied in a variety of government agencies concerned with project control, including various military agencies, the National Aeronautics and Space Administration (NASA), the Federal Aviation Administration (FAA), and the General Services Administration (GSA). These agencies are frequently involved in large-scale projects involving millions of dollars and many subcontractors. Examples of such governmental projects include the development of weapons systems, aircraft, and NASA space exploration projects. It has become a common practice for these agencies to require subcontractors to develop and use a CPM/PERT analysis to maintain management control of the myriad project components and subprojects.

CPM/PERT has also been widely applied in the private sector. Two of the major areas of application of CPM/PERT in the private sector have been research and development (R&D) and construction. CPM/PERT has been applied to various R&D projects, such as developing new drugs, planning and introducing new products, and developing new and more powerful computer systems. CPM/PERT analysis has been particularly applicable to construction projects. Almost every type of construction project—from building a house to constructing a major sports stadium to building a ship to constructing an oil pipeline—has been subjected to network analysis.

Network analysis is also applicable to the planning and scheduling of major events, such as summit conferences, sports festivals, basketball tournaments, football bowl games, parades, political conventions, school registrations, and rock concerts. The availability of powerful, user-friendly project management software packages for the personal computer has also served to increase the use of this technique.

Example Problem Solution

The following example will illustrate CPM/PERT network analysis and probability analysis.

Problem Statement

Given the following AON network and activity time estimates, determine the expected project completion time and variance, as well as the probability that the project will be completed in 28 days or less:

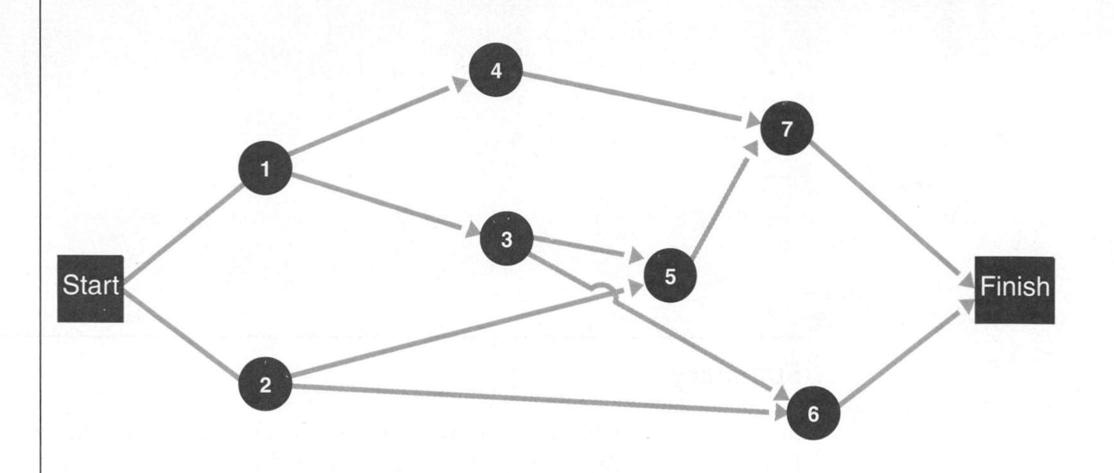

Activity	Time Estimates (weeks)		
	a	*m*	*b*
1	5	8	17
2	7	10	13
3	3	5	7
4	1	3	5
5	4	6	8
6	3	3	3
7	3	4	5

Solution

Step 1: Compute the Expected Activity Times and Variances

Using the following formulas, compute the expected time and variance for each activity:

$$t = \frac{a + 4m + b}{6}$$

$$v = \left(\frac{b - a}{6}\right)^2$$

For example, the expected time and variance for activity 1 is

$$t = \frac{5 + 4(8) + 17}{6}$$
$$= 9$$
$$v = \left(\frac{17-5}{6}\right)^2$$
$$= 4$$

These values and the remaining expected times and variances for each activity follow:

Activity	t	v
1	9	4
2	10	1
3	5	4/9
4	3	4/9
5	6	4/9
6	3	0
7	4	1/9

Step 2: Determine the Earliest and Latest Times at Each Node

The earliest and latest activity times and the activity slack are shown on the following network:

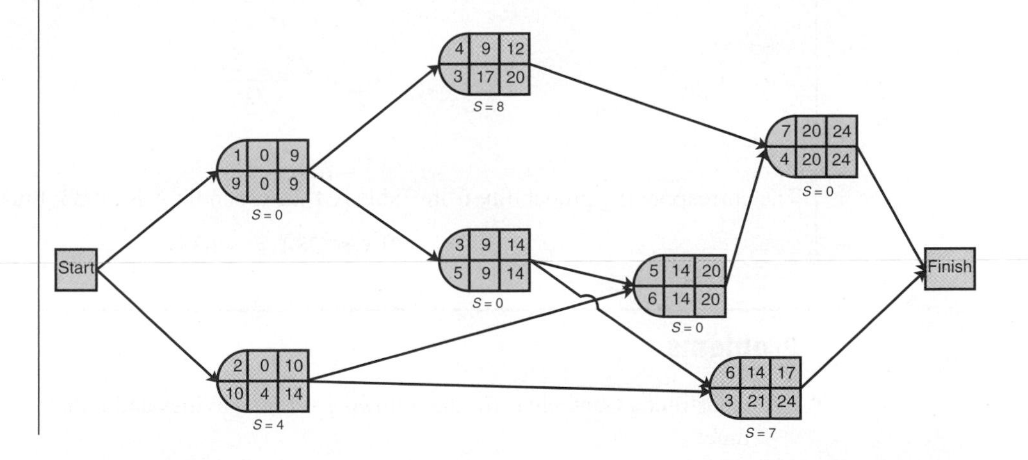

Step 3: Identify the Critical Path and Compute Expected Project Completion Time and Variance

After observing the foregoing network and those activities with no slack (i.e., $S = 0$), we can identify the critical path as $1 \rightarrow 3 \rightarrow 5 \rightarrow 7$. The expected project completion time (t_p) is 24 days. The variance is computed by summing the variances for the activities in the critical path:

$$v_p = 4 + 4/9 + 4/9 + 1/9$$
$$= 5 \text{ days}$$

Step 4: Determine the Probability That the Project Will Be Completed in 28 Days or Less

The following normal probability distribution describes the probability analysis:

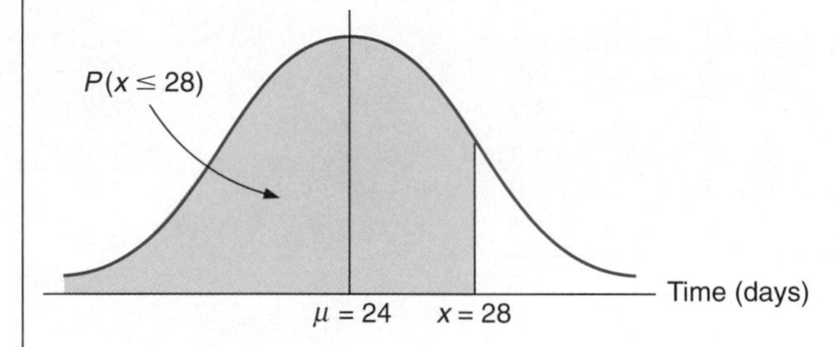

Compute Z by using the following formula:

$$Z = \frac{x - \mu}{\sigma}$$
$$= \frac{28 - 24}{\sqrt{5}}$$
$$= 1.79$$

The corresponding probability from Table A.1 in Appendix A is .4633; thus,

$$P(x \leq 28) = .9633$$

Problems

1. Construct a Gantt chart for the following set of activities and indicate the project completion time:

Activity	Activity Predecessor	Time (weeks)
1	—	5
2	—	4
3	1	3
4	2	6

2. Construct a Gantt chart for the following set of activities and indicate the project completion time and slack for each activity:

Activity	Activity Predecessor	Time (weeks)
1	—	3
2	—	7
3	1	2
4	2	5
5	2	6
6	4	1
7	5	4

3. Construct a Gantt chart and project network for the following set of activities, compute the length of each path in the network, and indicate the critical path:

Activity	Activity Predecessor	Time (months)
1	—	4
2	—	7
3	1	8
4	1	3
5	2	9
6	3	5
7	3	2
8	4, 5, 6	6
9	2	5

4. Identify all the paths in the following network, compute the length of each, and indicate the critical path (activity times are in weeks):

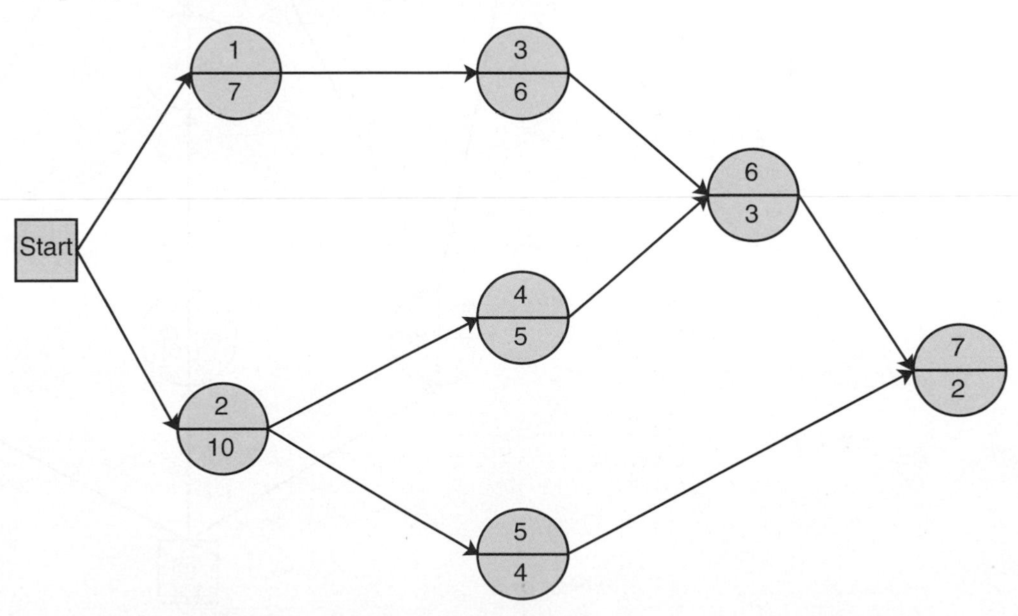

5. For the network in Problem 4, determine the earliest and latest activity times and the slack for each activity. Indicate how the critical path would be determined from this information.

6. Given the following network, with activity times in months, determine the earliest and latest activity times and slack for each activity. Indicate the critical path and the project duration.

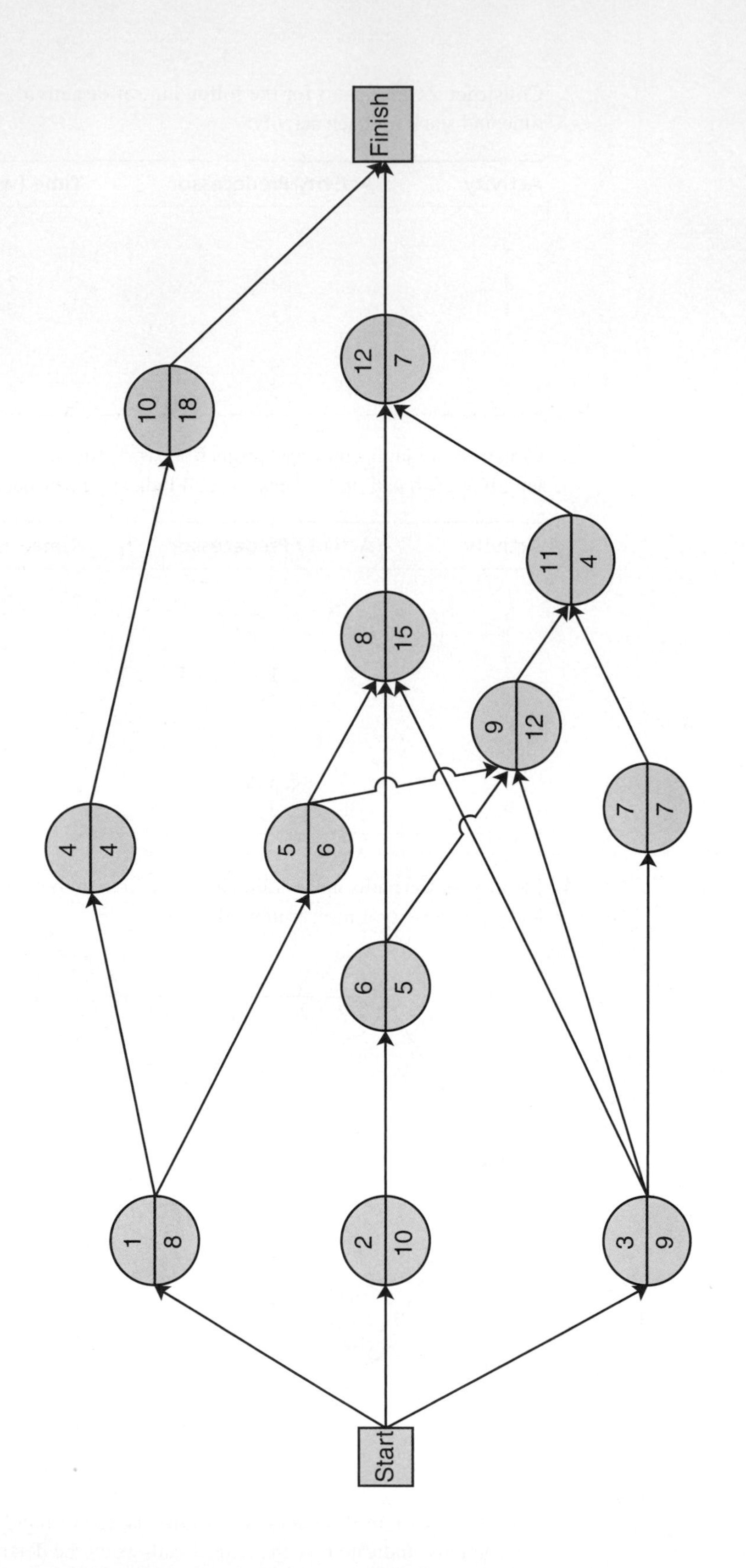

Start
1
8
2
10
3
9
4
4
5
6
6
5
7
7
8
15
9
12
10
18
11
4
12
7
Finish

7. Given the following network, with activity times in weeks, determine the earliest and latest activity times and the slack for each activity. Indicate the critical path and the project duration:

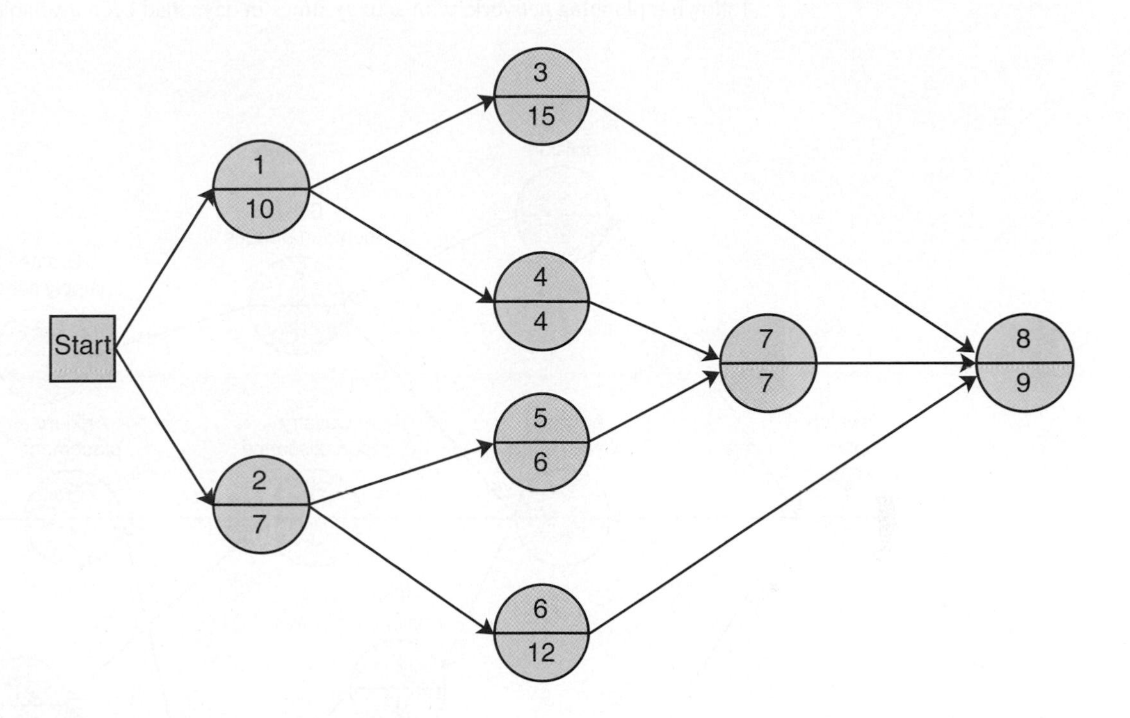

8. In one of the little-known battles of the Civil War, General Tecumseh Beauregard lost the Third Battle of Bull Run because his preparations were not complete when the enemy attacked. If the critical path method had been available, the general could have planned better. Suppose that the following planning network, with activity times in days, had been available:

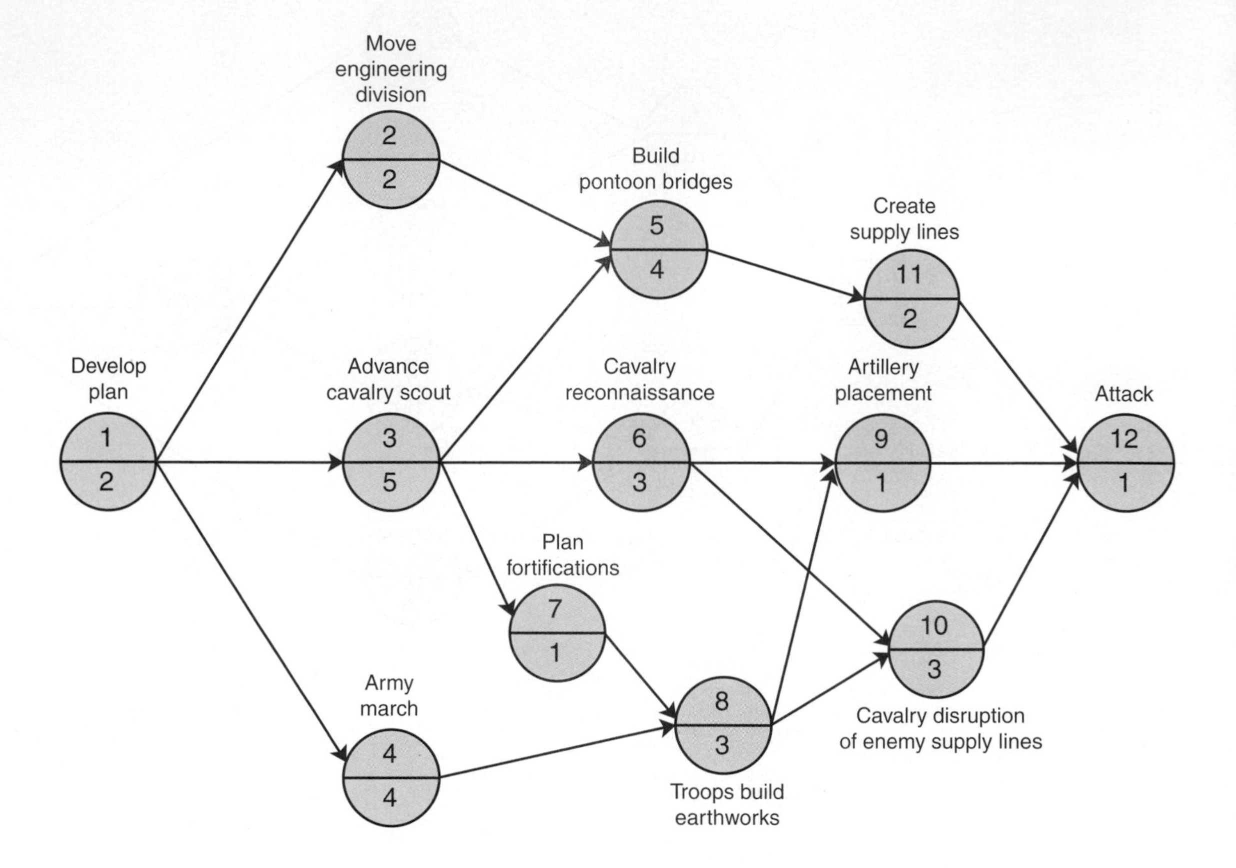

Determine the earliest and latest activity times and the activity slack for the network. Indicate the critical path and the time between the general's receipt of battle orders and the onset of battle.

9. A group of developers is building a new shopping center. A consultant for the developers has constructed the following project network and assigned activity times, in weeks.

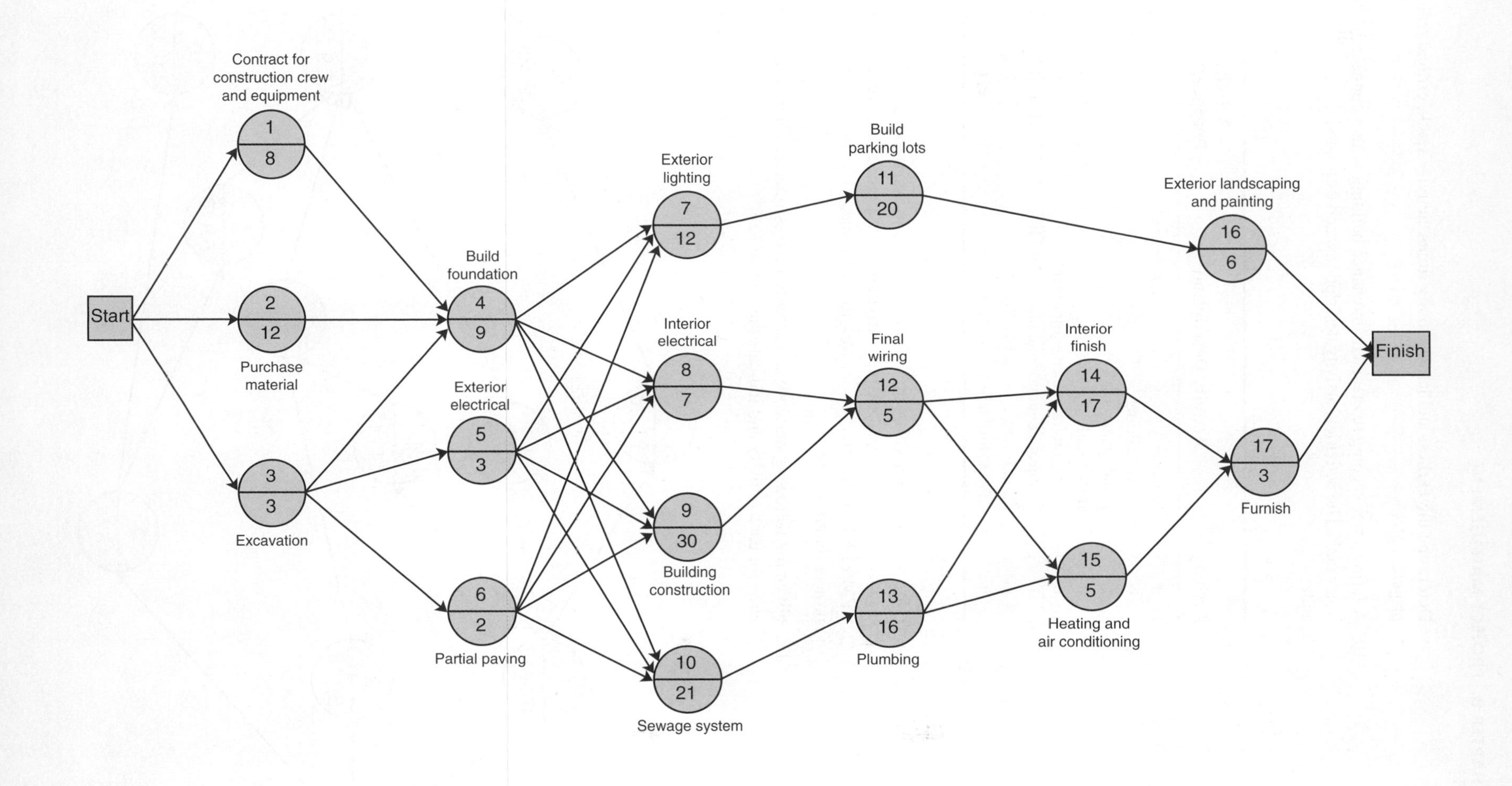

Start
Contract for construction crew and equipment
1
8
2
12
Purchase material
3
3
Excavation
Build foundation
4
9
Exterior electrical
5
3
6
2
Partial paving
Exterior lighting
7
12
Interior electrical
8
7
9
30
Building construction
10
21
Sewage system
Build parking lots
11
20
Final wiring
12
5
13
16
Plumbing
Interior finish
14
17
15
5
Heating and air conditioning
Exterior landscaping and painting
16
6
17
3
Furnish
Finish

Determine the earliest and latest activity times, activity slack, critical path, and duration for the project.

10. A farm owner is going to erect a maintenance building with a connecting electrical generator and water tank. The activities, activity descriptions, and estimated durations are given in the following table:

Activity	Activity Description	Activity Predecessor	Activity Duration (weeks)
a	Excavate	—	2
b	Erect building	a	6
c	Install generator	a	4
d	Install tank	a	2
e	Install maintenance equipment	b	4
f	Connect generator and tank to building	b, c, d	5
g	Paint finish	b	3
h	Check out facility	e, f	2

(Notice that the activities are defined not by node numbers but by activity descriptions. This alternative form of expressing activities and precedence relationships is often used in CPM/PERT.) Construct the network for this project, identify the critical path, and determine the project duration time.

11. Given the following network and activity time estimates, determine the expected time and variance for each activity and indicate the critical path:

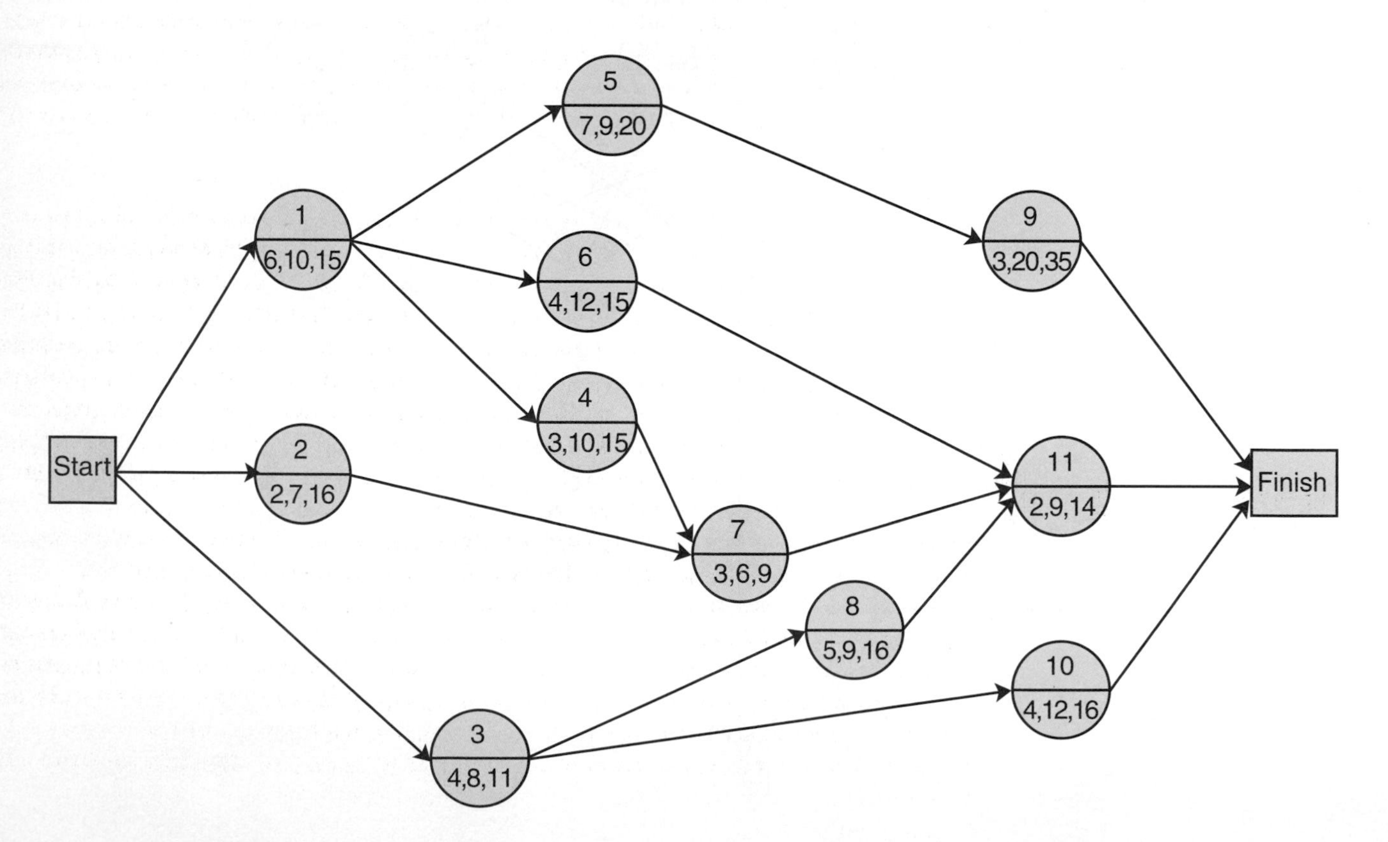

12. The Farmer's American Bank of Leesburg is planning to install a new computerized accounts system. Bank management has determined the activities required to complete the project, the precedence relationships of the activities, and activity time estimates, as shown in the following table:

Activity	Activity Description	Activity Predecessor	Time Estimates (weeks)		
			a	*m*	*b*
a	Position recruiting	—	5	8	17
b	System development	—	3	12	15
c	System training	a	4	7	10
d	Equipment training	a	5	8	23
e	Manual system test	b, c	1	1	1
f	Preliminary system changeover	b, c	1	4	13
g	Computer–personnel interface	d, e	3	6	9
h	Equipment modification	d, e	1	2.5	7
i	Equipment testing	h	1	1	1
j	System debugging and installation	f, g	2	2	2
k	Equipment changeover	g, i	5	8	11

Determine the expected project completion time and variance and determine the probability that the project will be completed in 40 weeks or less.

13. Consider the network in Problem 6, but with the following new time estimates for each activity:

Activity	Time Estimates (mo.)		
	a	*m*	*b*
1	4	8	12
2	6	10	15
3	2	10	14
4	1	4	13
5	3	6	9
6	3	6	18
7	2	8	12
8	9	15	22
9	5	12	21
10	7	20	25
11	5	6	12
12	3	8	20

Determine the following:

a. Expected activity times
b. Earliest activity times
c. Latest activity times
d. Activity slack
e. Critical path
f. Expected project duration and variance

14. The Virginia Department of Transportation is undertaking a construction project to widen a large section of interstate highway near Washington, DC. The project includes the construction of a number of new bridges, interchanges, and overpasses. The first step in the project is to appoint a project manager and develop a project team, whose members will be departmental employees and private consultants and operatives. The team for a project this size also requires a large support

staff with a variety of technical skills. Once the team is selected and the staff is in place, the next task is to select a primary contractor to manage and oversee the construction. The following activities are standard for the planning and scheduling of this process:

Activity	Activity Description	Activity Predecessor	Time Estimates (weeks)		
			a	*m*	*b*
a	Select project manager	—	1	2	2
b	Develop team recruitment plan	a	1	2	3
c	Implement team recruitment process	b	3	6	8
d	Staff needs assessment	b	1	2	3
e	Staff recruitment advertising	d	1	1	1
f	Staff selection evaluation procedures	b	1	2	4
g	Select project team members	c	2	4	6
h	Develop contractor bid requirements	a	1	4	6
i	Develop contractor bid evaluation procedure	g, h	2	3	4
j	Staff applicant evaluation and selection	e, f, g	2	4	5
k	Solicit contractor bids	i, j	4	6	10
l	Contractor bid evaluation	k	3	6	8
m	Select contractor and complete negotiations	l	2	3	5

Construct the CPM/PERT network for this project and determine the project schedule. Identify the critical path and determine the expected project duration time and variance. What is the probability that the team and contractor will be in place and the project started within 6 months? within 1 year?

15. The Matsusaki Company near Nashville, Tennessee, is a direct auto parts supplier for the nearby Neptune auto manufacturing plant. In order to gain a competitive advantage and meet quality requirements of its customers, Matsusaki has undertaken a project to achieve ISO certification. The goal of the international ISO organization and the certification process is the development of a quality management system that provides for continual improvement emphasizing defect prevention and the reduction of process variation and waste. The project management team has developed the following list of required activities necessary to achieve ISO certification:

Activity	Activity Description	Activity Predecessor	Time Estimates (weeks)		
			a	*m*	*b*
a	Appointment of company task force	—	1	2	3
b	Development of feasibility plan	a	2	3	4
c	Selection of third party registrar and auditor	b	1	2	3
d	Development of quality manual, procedure, and instruction documents	b	10	14	20
e	Quality system training for all employees	b	4	7	10
f	Outside training for company quality managers and internal plant auditors	b	4	7	10
g	Plant preparation and organization to meet ISO 9001 standards including use of SPC charts to identify non-conforming processes	e	18	24	32
h	Preparation of ISO application documents with outside auditor; set schedule for submissions to registrar	c	4	6	10
i	Internal audit of plant	d, f, h, g	10	12	16

Activity	Activity Description	Activity Predecessor	Time Estimates (weeks)		
			a	*m*	*b*
j	Corrective action plans for necessary process improvements	i	8	12	20
k	Outside auditor visit and audit	j	1	2	3
l	Corrective action plans developed based on auditor report and implemented	k	2	4	5
m	Follow up visit and re-audit by auditor	l	1	2	4
n	Auditor recommendation to registrar for ISO certification	m	1	2	2

Construct the network for this project and determine the project schedule. Identify the critical path and determine the expected project duration time and variance. What is the probability that the certification project can be completed within 1 year? within 18 months?

16. Consider the network in Problem 8, but with the following new time estimates for each activity:

Activity	Time Estimates (days)		
	a	*m*	*b*
1	1	2	6
2	1	3	5
3	3	5	10
4	3	6	14
5	2	4	9
6	2	3	7
7	1	1.5	2
8	1	3	5
9	1	1	5
10	2	4	9
11	1	2	3
12	1	1	1

Determine the following:

a. Expected activity times
b. Earliest activity times
c. Latest activity times
d. Activity slack
e. Critical path
f. Expected project duration and variance

17. For the CPM/PERT network in Problem 13, determine the probability that the network duration will exceed 50 months.

18. The Center for Information Technology at State University has outgrown its office in Bates (B) Hall and is moving to Allen (A) Hall, which has more space. The move will take place during the 3-week break between the end of summer semester and the beginning of fall semester. Movers will be hired from the university's physical plant to move the furniture, boxes of books, and files that the faculty will pack. The center has hired a local retail computer firm to move its

office computers so they won't be damaged. Following is a list of activities, their precedence relationships, and probabilistic time estimates for this project:

Activity	Activity Description	Activity Predecessor	Time Estimates (days)		
			a	*m*	*b*
a	Pack A offices	—	1	3	5
b	Network A offices	—	2	3	5
c	Pack B offices	—	2	4	7
d	Movers move A offices	a	1	3	4
e	Paint and clean A offices	d	2	5	8
f	Move computers	b, e	1	2	2
g	Movers move B offices	b, c, e	3	6	8
h	Computer installation	f	2	4	5
i	Faculty move and unpack	g	3	4	6
j	Faculty set up computers and offices	h, i	1	2	4

Determine the earliest and latest start and finish times, the critical path, and the expected project duration. What is the probability that the center will complete its move before the start of the fall semester?

19. Jane and Jim Smith are going to give a dinner party on Friday evening at 7:00 P.M. Their two children, Jerry and Judy, are going to help them get ready. The Smiths will all get home from work and school at 4:00 P.M. Jane and Jim have developed a project network to help them schedule their dinner preparations. Following is a list of the activities, the precedence relationships, and the activity times involved in the project:

Activity	Activity Description	Activity Predecessor	Time Estimates (min.)		
			a	*m*	*b*
a	Prepare salad	—	18	25	31
b	Prepare appetizer	—	15	23	30
c	Dust/clean	—	25	40	56
d	Vacuum	c	20	30	45
e	Prepare dessert	a, b	12	21	30
f	Set table	d	10	17	25
g	Get ice from market	f	5	12	18
h	Prepare/start tenderloin	e	10	20	25
i	Cut and prepare vegetables	e	9	15	22
j	Jim shower and dress	g	8	10	15
k	Jane shower and dress	h, i, j	20	27	40
l	Uncork wine/water carafes	h, i, j	6	10	15
m	Prepare bread	k, l	4	7	10
n	Prepare and set out dishes	k, l	10	14	20
o	Cut meat	k, l	7	15	20
p	Heat vegetable dish	k, l	3	4	6
q	Put out appetizers	m	4	6	7
r	Pour champagne	o	5	8	10

Develop a project network and determine the earliest and latest start and finish times, activity slack, and critical path. Compute the probability that they will be ready by 7:00 P.M.

20. The Stone River Textile Mill was inspected by OSHA and found to be in violation of a number of safety regulations. The OSHA inspectors ordered the mill to alter some existing machinery to make it safer (add safety guards, etc.); purchase some new machinery to replace older, dangerous machinery; and relocate some machinery to make safer passages and unobstructed entrances and exits. OSHA gave the mill only 35 weeks to make the changes; if the changes were not made by then, the mill would be fined $300,000.

The mill determined the activities in a CPM/PERT network that would have to be completed and then estimated the indicated activity times, as shown in the following table:

Activity	Activity Description	Activity Predecessor	Time Estimates (weeks)		
			a	*m*	*b*
a	Order new machinery	—	1	2	3
b	Plan new physical layout	—	2	5	8
c	Determine safety changes in existing machinery	—	1	3	5
d	Receive equipment	a	4	10	25
e	Hire new employees	a	3	7	12
f	Make plant alterations	b	10	15	25
g	Make changes in existing machinery	c	5	9	14
h	Train new employees	d, e	2	3	7
i	Install new machinery	d, e, f	1	4	6
j	Relocate old machinery	d, e, f, g	2	5	10
k	Conduct employee safety orientation	h, i, j	2	2	2

Construct the project network for this project and determine the following:

a. Expected activity times
b. Earliest and latest activity times and activity slack
c. Critical path
d. Expected project duration and variance
e. The probability that the mill will be fined $300,000

21. In the Third Battle of Bull Run, for which a CPM/PERT network was developed in Problem 16, General Beauregard would have won if his preparations had been completed in 15 days. What would the probability of General Beauregard's winning the battle have been?

22. On May 21, 1927, Charles Lindbergh landed at Le Bourget Field in Paris, completing his famous transatlantic solo flight. The preparation period prior to his flight was quite hectic, and time was very critical because several other famous pilots of the day were also planning transatlantic flights. Once Ryan Aircraft was contracted to build the *Spirit of St. Louis*, it took only a little over 2.5 months to construct the plane and fly it to New York for the takeoff. If CPM/PERT had been available to Charles Lindbergh, it no doubt would have been useful in helping him plan this project. Use your imagination and assume that a CPM/PERT network with the following estimated activity times was developed for the flight.

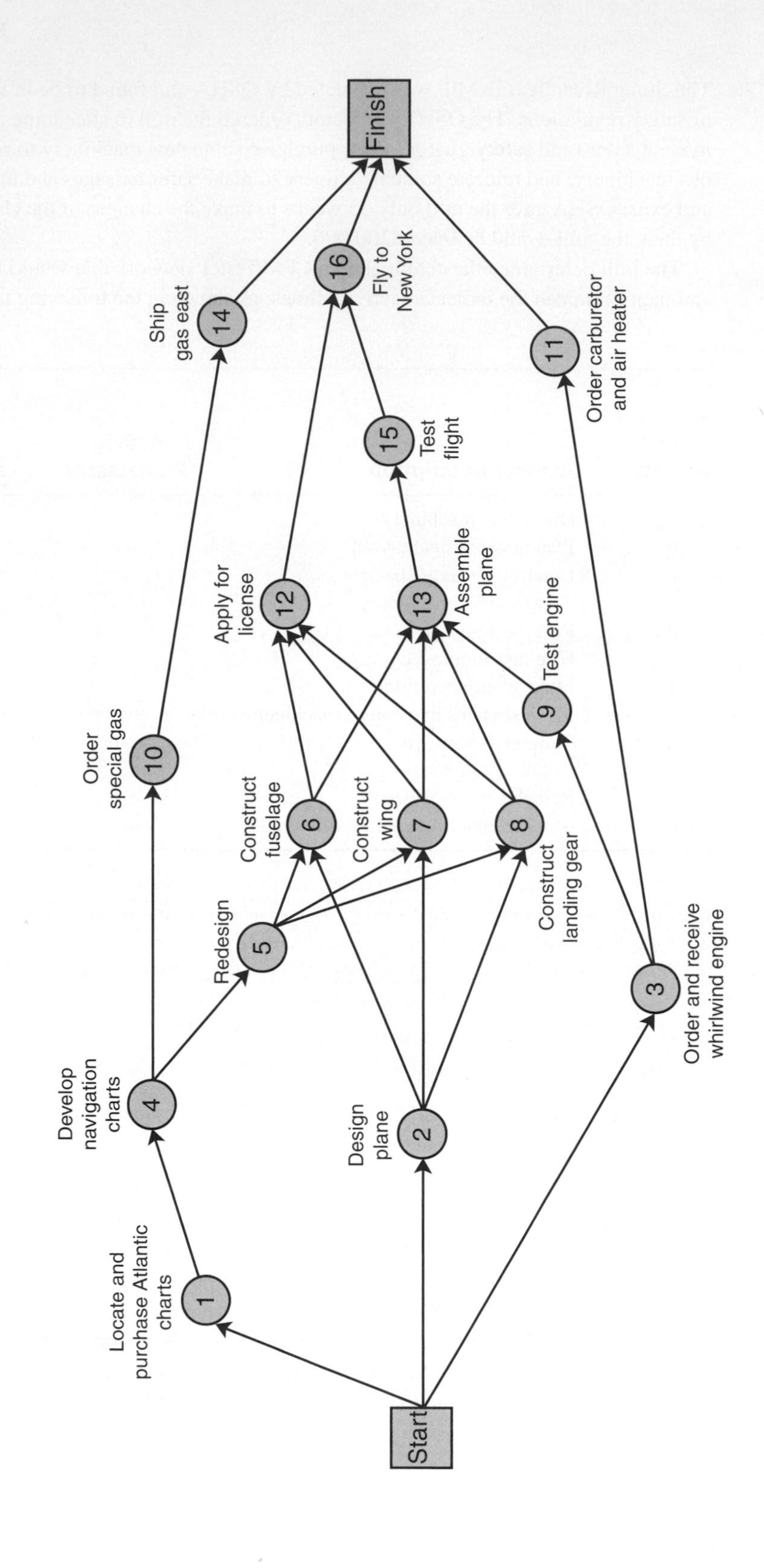

Start
Finish
1
Locate and purchase Atlantic charts
2
Design plane
3
Order and receive whirlwind engine
4
Develop navigation charts
5
Redesign
6
Construct fuselage
7
Construct wing
8
Construct landing gear
9
Test engine
10
Order special gas
11
Order carburetor and air heater
12
Apply for license
13
Assemble plane
14
Ship gas east
15
Test flight
16
Fly to New York

Activity	Time Estimates (days)		
	a	*m*	*b*
1	1	3	5
2	4	6	10
3	20	35	50
4	4	7	12
5	2	3	5
6	8	12	25
7	10	16	21
8	5	9	15
9	1	2	2
10	6	8	14
11	5	8	12
12	5	10	15
13	4	7	10
14	5	7	12
15	5	9	20
16	1	3	7

Determine the expected project duration and variance and the probability of completing the project in 67 days.

23. RusTech Tooling is a large job shop operation that builds machine tools and dies to manufacture parts for specialized items. The company primarily bids on government-related contracts to produce parts for such items as military aircraft and weapons systems, as well as the space program. The company is bidding on a contract to produce a component part for the fuselage assembly in a new space shuttle. A major criterion for selecting the winning bid, besides low cost, is the time required to produce the part. However, if the company is awarded the contract, it will be strictly held to the completion date specified in the bid, and any delays will result in severe financial penalties. To determine the project completion time to put in its bid, the company has identified the project activities, precedence relationships, and activity times shown in the following table:

Activity	Activity Predecessor	Time Estimates (weeks)		
		a	*m*	*b*
a	—	3	5	9
b	a	2	5	8
c	a	1	4	6
d	a	4	6	10
e	b	2	8	11
f	b	5	9	16
g	c	4	12	20
h	c	6	9	13
i	d	3	7	14
j	d	8	14	22
k	f, g	9	12	20
l	h, i	6	11	15
m	e	4	7	12
n	j	3	8	16
o	n	5	10	18

If RusTech wants to be 90% certain that it can deliver the part without incurring a penalty, what time frame should it specify in the bid?

24. PM Computers is an international manufacturer of computer equipment and software. It is going to introduce a number of new products in the coming year, and it wants to develop marketing programs to accompany the product introductions. The marketing program includes the preparation of printed materials distributed directly by the company and used by the company's marketing personnel, vendors, and representatives; print advertising in regular magazines, trade journals, and newspapers; and television commercials. The program also includes extensive training programs for marketing personnel, vendors, and representatives about the new products. A project management team with members from the marketing department and manufacturing areas has developed the following list of activities for the development of the marketing program:

Activity	Activity Description	Activity Predecessor	Time Estimates (days)		
			a	*m*	*b*
a	Preliminary budget and plan approval	—	10	15	20
b	Select marketing personnel for training	—	5	9	12
c	Develop overall media plan	a	15	25	30
d	Prepare separate media plans	c	12	20	25
e	Develop training plan	c	5	8	12
f	Design training course	e	6	14	20
g	Train marketing personnel	b, f	16	20	25
h	Plan TV commercials with agency	d	15	25	35
i	Draft in-house print materials	d	8	15	20
j	Develop print advertising layouts with agency	d	16	23	30
k	Review print advertising layouts	j	4	9	12
l	Review TV commercials	h	3	7	12
m	Review and print in-house materials	i	3	5	7
n	Release advertising to print media	g, i, k	2	4	8
o	Release TV commercials to networks	l	4	7	10
p	Final marketing personnel review	g, i, k	4	5	9
q	Run media advertising, mailings	m, n, o	15	20	30

Construct the network for this project and determine the activity schedule. Identify the critical path and determine the expected project duration time and variance. What is the probability that the program can be completed within 4 months?

25. A marketing firm is planning to conduct a survey of a segment of the potential product audience for one of its customers. The planning process for preparing to conduct the survey consists of six activities, with precedence relationships and activity time estimates as follows:

Activity	Activity Description	Activity Predecessor	Time Estimate (days)
a	Determine survey objectives	—	3
b	Select and hire personnel	a	3
c	Design questionnaire	a	5
d	Train personnel	b, c	4
e	Select target audience	c	3
f	Make personnel assignments	d, e	2

a. Determine all paths through the network from node a to node f and the duration of each. Also indicate the critical path.
b. Determine the earliest and latest activity start and finish times.
c. Determine the slack for each activity.

26. Lakeland-Bering Aircraft Company is preparing a contract proposal to submit to the defense department for a new military aircraft, the X-300J jet fighter. Part of the proposal is a development and production schedule for completion of the first aircraft. The project consists of three primary categories: engine design and development, development and production of the airframe (e.g., the aircraft body), and design and development of the aircraft avionics (e.g., the electronic systems, equipment, and other devices used to operate and control the aircraft). Following is a list of the project activities, with time estimates (in months):

Activity	Activity Description	Activity Predecessor	Time Estimates (mo.)		
			a	*m*	*b*
1	General design	—	5	10	22
2	Engine design	—	20	33	57
3	Airframe design	1	6	17	33
4	Avionics design	1	2	8	20
5	Develop test engine	2	7	10	12
6	Develop test airframe	3	8	11	18
7	Develop interim avionics	4	7	15	28
8	Develop engine	2	19	24	41
9	Assemble test aircraft	5, 6, 7	4	6	10
10	Test avionics	4	5	10	16
11	Conduct engine/airframe flight trials	9	5	9	12
12	Conduct avionics flight trials	10	5	9	24
13	Produce engine	8	11	14	15
14	Produce airframe	11	12	15	20
15	Produce avionics	11, 12	14	16	25
16	Final assembly/finish	13, 14, 15	2	3	5

Develop the project network and determine the critical path, the expected project duration, and the variance. What is the probability that the project will be completed within 8 years?

27. The Valley United Soccer Club is planning a soccer tournament for the weekend of April 29 and 30. The club's officers know that by March 30 they must send out acceptances to teams that have applied to enter and that by April 15 they must send out the tournament game schedule to teams that have been selected to play. Their tentative plan is to begin the initial activities for tournament preparation, including sending out the application forms to prospective teams, on January 20. Following is a list of tournament activities, their precedence relationships, and estimates of their duration, in days:

Activity	Activity Description	Activity Predecessor	Time Estimates (days)		
			a	*m*	*b*
a	Send application forms	—	5	7	10
b	Get volunteer workers	—	10	18	26
c	Line up referees	—	7	10	14
d	Reserve fields	—	14	21	35
e	Receive and process forms	a	30	30	30
f	Determine age divisions	b, c, d, e	4	9	12
g	Assign fields to divisions	f	4	7	10
h	Sell program ads	b	14	21	30
i	Acquire donated items for team gift bags	b	15	20	26
j	Schedule games	g	6	14	18
k	Design ads	h	5	8	10
l	Fill gift bags	i	9	12	17
m	Process team T-shirt orders	e	7	10	14
n	Send acceptance letters	f	4	7	12
o	Design and print programs	j, k, l, n	14	18	24
p	Put together registration boxes (programs, gifts, etc.)	o	5	7	10
q	Send out game schedules	j, k, l, n	5	8	12
r	Assign referees to games	j, k, l, n	4	7	10
s	Get trophies	f	20	28	35
t	Silk-screen T-shirts	m	12	17	25
u	Package team T-shirt orders	t	5	8	12

Develop a project network for the club's tournament preparation process and determine the likelihood that it will meet its schedule milestones and complete the process according to the scheduled tournament date of April 29.

28. During a violent thunderstorm with very high wind gusts in the third week of March, the broadcast tower for the public radio station WVPR, atop Poor Mountain in Roanoke, collapsed. This greatly reduced the strength of the station's signal in the area. The station management immediately began plans to construct a new tower. Following is a list of the required activities for building the new tower with optimistic (*a*), most likely (*m*), and pessimistic (*b*) time estimates (in days); however, the sequence of the activities has not been designated:

Activity	Activity Description	Time Estimates (days)		
		a	*m*	*b*
a	Removal of debris	5	8	12
b	Survey new tower site	3	6	8
c	Procure structural steel	15	21	30
d	Procure electrical/broadcasting equipment	18	32	40
e	Grade tower site	4	7	10
f	Pour concrete footings and anchors	10	18	22
g	Deliver and unload steel	3	5	9
h	Deliver and unload electrical/broadcast equipment	1	2	4
i	Erect tower	25	35	50
j	Connect electrical cables between tower and building	4	6	10
k	Construct storm drains and tiles	10	15	21
l	Backfill and grade tower site	4	7	9
m	Clean up	3	6	10
n	Obtain inspection approval	1	4	7

Using your best judgment, develop a CPM/PERT network for this project and determine the expected project completion time. Also determine the probability that the station signal will be back at full strength within 3 months.

29. The following table contains the activities for planning a wedding and the activity time estimates; however, the precedence relationships between activities are not included:

Activity	Activity Description	Time (days)		
		a	*m*	*b*
a	Determine date	1	10	15
b	Obtain marriage license	1	5	8
c	Select bridal attendants	3	5	7
d	Order dresses	10	14	21
e	Fit dresses	5	10	12
f	Select groomsmen	1	2	4
g	Order tuxedos	3	5	7
h	Find and rent church	6	14	20
i	Hire florist	3	6	10
j	Develop/print programs	15	22	30
k	Hire photographer	3	10	15
l	Develop guest list	14	25	40
m	Order invitations	7	12	20
n	Address and mail invitations	10	15	25
o	Compile RSVP list	30	45	60
p	Reserve reception hall	3	7	10
q	Hire caterer	2	5	8
r	Determine reception menu	10	12	16
s	Make final order	2	4	7
t	Hire band	10	18	21
u	Decorate reception hall	1	2	3
v	Wedding ceremony	0.5	0.5	0.5
w	Wedding reception	0.5	0.5	0.5

Using your best judgment, determine the project network, critical path, and expected project duration. If it is January 1 and a couple is planning a June 1 wedding, what is the probability that it can be done on time?

30. The following table provides the information necessary to construct a project network and project crash data:

Activity	(*i*, *j*)	Activity Predecessor	Activity Time (weeks)		Activity Cost	
			Normal	*Crash*	*Normal*	*Crash*
a	(1, 2)	—	20	8	$1,000	$1,480
b	(1, 4)	—	24	20	1,200	1,400
c	(1, 3)	—	14	7	700	1,190
d	(2, 4)	a	10	6	500	820
e	(3, 4)	c	11	5	550	730

a. Construct the project network.

b. Compute the total allowable crash time per activity and the crash cost per week for each activity.

c. Determine the maximum possible crash time for the network and manually crash the network the maximum amount.
d. Compute the normal project cost and the cost of the crashed project.
e. Formulate the general linear programming model for this network and solve it.
f. Formulate the linear programming crashing model that would crash this network by the maximum amount and solve it.

31. The following table provides the information necessary to construct a project network and project crash data:

Activity	(i, j)	Activity Predecessor	Activity Time (weeks)		Activity Cost	
			Normal	*Crash*	*Normal*	*Crash*
a	(1, 2)	—	16	8	$2,000	$4,400
b	(1, 3)	—	14	9	1,000	1,800
c	(2, 4)	a	8	6	500	700
d	(2, 5)	a	5	4	600	1,300
e	(3, 5)	b	4	2	1,500	3,000
f	(3, 6)	b	6	4	800	1,600
g	(4, 6)	c	10	7	3,000	4,500
h	(5, 6)	d, e	15	10	5,000	8,000

a. Construct the project network.
b. Manually crash the network to 28 weeks.
c. Formulate the general linear programming model for this network.
d. Formulate the linear programming crashing model that would crash this model by the maximum amount.

32. Formulate the general linear programming model for Problem 4, and solve it.

33. Formulate the general linear programming model for the project network for installing an order processing system shown in Figure 8.14 and solve it.

34. Reconstruct the example problem at the end of this chapter as an AOA network. Assume that the most likely times (m) are the normal activity times and that the optimistic times (a) are the activity crash times. Further assume that the activities have the following normal and crash costs:

Activity	(i, j)	Costs (normal cost, crash cost)
1	(1, 2)	($100, $400)
2	(1, 3)	($250, $400)
3	(2, 3)	($400, $800)
4	(2, 4)	($200, $400)
5	(3, 4)	($150, $300)
6	(3, 5)	($100, $100)
7	(4, 5)	($300, $500)

a. Formulate the general linear programming model for this project network, using most likely activity times (m).

b. Formulate the linear programming crashing model that would crash this network by the maximum amount.

c. Solve this model by using the computer.

35. The following table provides the crash data for the network project described in Problem 12:

	Activity Time (weeks)		Activity Cost	
Activity	*Normal*	*Crash*	*Normal*	*Crash*
a	9	7	$4,800	$ 6,300
b	11	9	9,100	15,500
c	7	5	3,000	4,000
d	10	8	3,600	5,000
e	1	1	0	0
f	5	3	1,500	2,000
g	6	5	1,800	2,000
h	3	3	0	0
i	1	1	0	0
j	2	2	0	0
k	8	6	5,000	7,000

The normal activity times are considered to be deterministic and not probabilistic. Using the computer, crash the network to 26 weeks. Indicate how much it would cost the bank and then indicate the critical path.

36. The following table provides the crash data for the network project described in Problem 6:

	Activity Time (months)		Activity Cost ($1,000s)	
Activity	*Normal*	*Crash*	*Normal*	*Crash*
1	8	5	$ 700	$1,200
2	10	9	1,600	2,000
3	9	7	900	1,500
4	4	2	500	700
5	6	3	500	900
6	5	4	500	800
7	7	5	700	1,000
8	15	12	1,400	2,000
9	12	10	1,800	2,300
10	18	14	1,400	3,200
11	4	3	500	800
12	7	6	800	1,400

Using the computer, crash the network to 32 months. Indicate the first critical path activities and then the cost of crashing the network.

Case Problem

THE BLOODLESS COUP CONCERT

John Aaron called the meeting of the Programs and Arts Committee of the Student Government Association to order.

"Okay, okay, everybody, quiet down. I have an important announcement to make," he shouted above the noise. The room got quiet, and John started again, "Well, you guys, we can have the Coup."

His audience looked puzzled, and Randy Jones asked, "What coup have we scored this time, John?"

"The Coup, the Coup! You know, the rock group, the Bloodless Coup!"

Everyone in the room cheered and started talking excitedly. John stood up and waved his arms and shouted, "Hey, calm down, everybody, and listen up." The room quieted again, and everyone focused on John. "The good news is that they can come." He paused a moment. "The bad news is that they will be here in 18 days."

The students groaned and seemed to share Jim Hastings' feelings: "No way, man. It can't be done. Why can't we put it off for a couple of weeks?"

John answered, "They're just starting their new tour and are looking for some warm-up concerts. They'll be traveling near here for their first concert date in DC and saw they had a letter from us, so they said they could come now. But that's it—now or never." He looked around the room at the solemn faces. "Look, you guys, we can handle this. Let's think of what we have to do. Come on, perk up. Let's make a list of everything we have to do to get ready and figure out how long it will take. So somebody tell me what we have to do first!"

Anna Mendoza shouted from the back of the room, "We have to find a place; you know, get an auditorium somewhere. I've done that before, and it should take anywhere from 2 days up to 7 days, most likely about 4 days."

"Okay, that's great," John said, as he wrote down the activity "Secure auditorium" on the blackboard, with the times out to the side. "What's next?"

"We need to print tickets—and quick," Tracey Shea called. "It could only take 1 day if the printer isn't busy, but it could take up to 4 days if he is. It should probably take about 2 days."

"But we can't print tickets until we know where the concert will be because of the security arrangement," Andy Taylor noted.

"Right," said John. "Get the auditorium first; then print the tickets. What else?"

"We need to make hotel and transportation arrangements for the Coup and their entourage while they're here," Jim Hastings proposed. "But we better not do that until we get the auditorium. If we can't find a place for the concert, everything falls through."

"How long do you think it will take to make the arrangements?" John asked.

"Oh, between 3 and 10 days, probably about 5, most likely," Jim answered.

"We also have to negotiate with the local union for concert employees, stagehands, and whoever else we need to hire," Reggie Wilkes interjected. "That could take 1 day or up to 8 days, but 3 days would be my best guess."

"We should probably also hold off on talking to the union until we get the auditorium," John added. "That will probably be a factor in the negotiations."

"After we work things out with the union, we can hire some stagehands," Reggie continued. "That could take as few as 2 days but as long as 7. I imagine it'll take about 4 days. We should also be able to get some student ushers at the same time, once we get union approval. That could take only 1 day, but it has taken 5 days in the past; 3 days is probably the most likely."

"We need to arrange a press conference," said Art Cohen, who was leaning against a wall. "This is a heavy group, big time."

"But doesn't a press conference usually take place at the hotel?" John asked.

"Yeah, that's right," Art answered. "We can't make arrangements for the press conference until we work things out with the hotel. When we do that, it should take about 3 days to set up a press conference—2 days if we're lucky, and 4 at the most."

The room got quiet as everyone thought.

"What else?" John asked.

"Hey, I know," Annie Roark spoke up. "Once we hire the stagehands, they have to set up the stage. I think that could be done in a couple of days, but it could take up to 6 days, with 3 most likely." She paused for a moment before adding, "And we can also assign the ushers to their jobs once we hire them. That shouldn't take long, maybe only 1 day—3 days worst, probably 2 days would be a good time to put down."

"We also have to do some advertising and promotion if we want people to show up for this thing," mentioned Art nonchalantly. "I guess we need to wait until we print the tickets, so we'll have something to sell. That depends on the media, the paper, and radio stations. I've worked with this before. It could get done really quick, like 2 days, if we can make the right contacts. But it could take a lot longer, like 12 days, if we hit any snags. We probably ought to count on 6 days as our best estimate."

"Hey, if we're going to promote this, shouldn't we also have a preliminary act, some other group?" Annie asked.

"Wow, I forgot all about that!" John exclaimed. "Hiring another act will take me between 4 and 8 days; I can probably do it in 5. I can start on that right away, at the same time you guys are arranging for an auditorium." He thought for a moment. "But we really can't begin to work on the promotion until I get the lead-in group. So what's left?"

"Sell the tickets," shouted several people at once.

"Right," said John. "We have to wait until they're printed; but I don't think we have to wait for the advertising and promotion to start, do we?"

"No," Jim replied. "But we should hire the preliminary act first so people will know what they're buying a ticket for."

"Agreed," said John. "The tickets could go quick; I suppose in the first day."

"Or," interrupted Mike Eggleston, "it could take longer. I remember 2 years ago, it took 12 days to sell out for the Cosmic Modem."

"Okay, so it's between 1 and 12 days to sell the tickets," said John, "but I think about 5 days is more likely. Everybody agree?"

The group nodded in unison, and they all turned at once to the list of activities and times John had written on the blackboard.

Use PERT analysis to determine the probability that the concert preparations will be completed in time.

Case Problem

Moore Housing Contractors

Moore Housing Contractors is negotiating a deal with Countryside Realtors to build six houses in a new development. Countryside wants Moore Contractors to start in late winter or early spring, when the weather begins to moderate, and build on through the summer and into the fall. The summer months are an especially busy time for the realty company, and it believes it can sell the houses almost as soon as they are ready, and maybe even before. The houses all have similar floor plans and are of approximately equal size; only the exteriors are noticeably different. The completion time is so critical for Countryside Realtors that it is insisting that a project management network accompany the contractor's bid for the job, with an estimate of the completion time for a house. The realtor also needs to be able to plan its offerings and marketing for the summer. It wants each house to be completed within 45 days after it is started. If a house is not completed within this time frame, it wants to be able to charge the contractor a penalty. Mary and Sandy Moore, the president and vice president, respectively, of Moore Contractors, are concerned about the prospect of a penalty charge. They want to be very confident that they can meet the deadline for a house before they enter into any kind of agreement with a penalty involved. (If there is a reasonable likelihood that they cannot finish a house within 45 days, they want to increase their bid to cover potential penalty charges.)

The Moores are experienced home builders, so it was not difficult for them to list the activities involved in building a house or to estimate activity times. However, they made their estimates conservatively and tended to increase their pessimistic estimates to compensate for the possibility of bad weather and variations in their workforce. A list of the activities involved in building a house and the activity time estimates follow on the next page.

Activity	Activity Description	Activity Predecessor	Time (days)		
			a	*m*	*b*
a	Excavation, pour footers	—	3	4	6
b	Lay foundation	a	2	3	5
c	Frame and roof	b	2	4	5
d	Lay drain tiles	b	1	2	4
e	Sewer (floor) drains	b	1	2	3
f	Install insulation	c	2	4	5
g	Pour basement floor	e	2	3	5
h	Rough plumbing, pipes	e	2	4	7
i	Install windows	f	1	3	4
j	Rough electrical wiring	f	1	2	4
k	Install furnace, air conditioner	c, g	3	5	8
l	Exterior brickwork	i	5	6	10
m	Install plasterboard, mud, plaster	j, h, k	6	8	12
n	Roof shingles, flashing	l	2	3	6
o	Attach gutter, downspouts	n	1	2	5
p	Grading	d, o	2	3	7
q	Lay subflooring	m	3	4	6
r	Lay driveway, walks, landscape	p	4	6	10
s	Finish carpentry	q	3	5	12
t	Kitchen cabinetry, sink, and appliances	q	2	4	8
u	Bathroom cabinetry, fixtures	q	2	3	6
v	Painting (interior and exterior)	t, u	4	6	10
w	Finish wood floors, lay carpet	v, s	2	5	8
x	Final electrical, light fixtures	v	1	3	4

a. Develop a CPM/PERT network for Moore Housing Contractors and determine the probability that the company can complete a house within 45 days. Does it appear that the Moores might need to increase their bid to compensate for potential penalties?

b. Indicate which project activities Moore Housing Contractors should be particularly diligent to keep on schedule by making sure workers and materials are always available. Also indicate which activities the company might shift workers from as the need arises.

CHAPTER

9

Multicriteria Decision Making

In all the linear programming models presented in Chapters 2 through 8, a single objective was either maximized or minimized. However, a company or an organization often has more than one objective, which may relate to something other than profit or cost. In fact, a company may have several criteria, that is, *multiple criteria*, that it will consider in making a decision instead of just a single objective. For example, in addition to maximizing profit, a company in danger of a labor strike might want to avoid employee layoffs, or a company about to be fined for pollution infractions might want to minimize the emission of pollutants. A company deciding between several potential research and development projects might want to consider the probability of success of each of the projects, the cost and time required for each, and potential profitability in making a selection.

In this chapter, we discuss three techniques that can be used to solve problems when they have multiple objectives: *goal programming*, the *analytical hierarchy process*, and *scoring models*. Goal programming is a variation of linear programming in that it considers more than one objective (called goals) in the objective function. Goal programming models are set up in the same general format as linear programming models, with an objective function and linear constraints. The model solutions are very much like the solutions to linear programming models. The format for the analytical hierarchy process and scoring models, however, is quite different from that of linear programming. These methods are based on a comparison of decision alternatives for different criteria that reflects the decision maker's preferences. The result is a mathematical "score" for each alternative that helps the decision maker rank the alternatives in terms of preferability.

Goal Programming

***Goal programming** is a form of linear programming that includes multiple goals instead of a single objective.*

As we mentioned, a **goal programming** model is very similar to a linear programming model, with an objective function, decision variables, and constraints. Like linear programming, goal programming models with two decision variables can be solved graphically and by using QM for Windows and Excel. We begin our presentation of goal programming as we did linear programming, demonstrating through an example of how to formulate a model. This will illustrate the main differences between goal and linear programming.

Model Formulation

We will use the Beaver Creek Pottery Company example again to illustrate the way a goal programming model is formulated and the differences between a linear programming model and a goal programming model. Recall that this model was originally formulated in Chapter 2 as follows:

$$\begin{aligned}
&\text{maximize } Z = \$40x_1 + 50x_2\\
&\text{subject to}\\
&x_1 + 2x_2 \leq 40 \text{ hr. of labor}\\
&4x_1 + 3x_2 \leq 120 \text{ lb. of clay}\\
&x_1, x_2 \geq 0
\end{aligned}$$

where

$$\begin{aligned}
x_1 &= \text{number of bowls produced}\\
x_2 &= \text{number of mugs produced}
\end{aligned}$$

The objective function, Z, represents the total profit to be made from bowls and mugs, given that \$40 is the profit per bowl and \$50 is the profit per mug. The first constraint is for available labor. It shows that a bowl requires 1 hour of labor, a mug requires 2 hours, and 40 hours of labor are available daily. The second constraint is for clay, and it shows that each bowl requires 4 pounds of clay, each mug requires 3 pounds, and the daily limit of clay is 120 pounds.

This is a standard linear programming model; as such, it has a single objective function for profit. However, let us suppose that instead of having one objective, the pottery company has several objectives, listed here *in order of importance*:

1. To avoid layoffs, the company does not want to use fewer than 40 hours of labor per day.
2. The company would like to achieve a *satisfactory* profit level of $1,600 per day.
3. Because the clay must be stored in a special place so that it does not dry out, the company prefers not to keep more than 120 pounds on hand each day.
4. Because high overhead costs result when the plant is kept open past normal hours, the company would like to minimize the amount of overtime.

The different objectives in a goal programming problem are referred to as goals.

These different objectives are referred to as *goals* in the context of the goal programming technique. The company would, naturally, like to come as close as possible to achieving each of these goals. Because the regular form of the linear programming model presented in previous chapters considers only one objective, we must develop an alternative form of the model to reflect these multiple goals. Our first step in formulating a goal programming model is to transform the linear programming model constraints into goals.

Labor Goal

The first goal of the pottery company is to avoid *underutilization* of labor—that is, using fewer than 40 hours of labor each day. To represent the possibility of underutilizing labor, the linear programming constraint for labor, $x_1 + 2x_2 \leq 40$ hours of labor, is reformulated as

$$x_1 + 2x_2 + d_1^- - d_1^+ = 40 \text{ hr.}$$

All ***goal constraints*** *are equalities that include* ***deviational variables****, d^- and d^+.*

This reformulated equation is referred to as a **goal constraint**. The two new variables, d_1^- and d_1^+, are called **deviational variables**. They represent the number of labor hours less than 40 (d_1^-) and the number of labor hours exceeding 40 (d_1^+). More specifically, d_1^- represents labor underutilization, and d_1^+ represents *overtime*. For example, if $x_1 = 5$ bowls and $x_2 = 10$ mugs, then a total of 25 hours of labor have been expended. Substituting these values into our goal constraint gives

$$(5) + 2(10) + d_1^- - d_1^+ = 40$$
$$25 + d_1^- - d_1^+ = 40$$

A positive deviational variable (d^+) is the amount by which a goal level is exceeded.

A negative deviational variable (d^-) is the amount by which a goal level is underachieved.

Because only 25 hours were used in production, labor was underutilized by 15 hours ($40 - 25 = 15$). Thus, if we let $d_1^- = 15$ hours and $d_1^+ = 0$ (because there is obviously no overtime), we have

$$25 + d_1^- - d_1^+ = 40$$
$$25 + 15 - 0 = 40$$
$$40 = 40$$

Now consider the case where $x_1 = 10$ bowls and $x_2 = 20$ mugs. This means that a total of 50 hours have been used for production, or 10 hours above the goal level of 40 hours. This extra 10 hours is overtime. Thus, $d_1^- = 0$ (because there is no underutilization) and $d_1^+ = 10$ hours.

In each of these two brief examples, at least one of the deviational variables equaled zero. In the first example, $d_1^+ = 0$, and in the second example, $d_1^- = 0$. This is because it is impossible to use fewer than 40 hours of labor and more than 40 hours of labor *at the same time*. Of course, both deviational variables, d_1^- and d_1^+, could have equaled zero if exactly 40 hours were used in production. These examples illustrate one of the fundamental characteristics of goal programming: At least one or both of the deviational variables in a goal constraint must equal zero.

At least one or both deviational variables in a goal constraint must equal zero.

The next step in formulating our goal programming model is to represent the goal of not using fewer than 40 hours of labor. We do this by creating a new form of objective function:

$$\text{minimize } P_1 d_1^-$$

The objective function in a goal programming model seeks to minimize the deviation from goals in order of the goal priorities.

The objective function in all goal programming models is to *minimize* deviation from the goal constraint levels. In this objective function, the goal is to minimize d_1^-, the underutilization of labor. If d_1^- equaled zero, then we would not be using fewer than 40 hours of labor. Thus, it is our objective to make d_1^- equal zero or the minimum amount possible. The symbol P_1 in the objective function designates the minimization of d_1^- as the *first-priority* goal. This means that when this model is solved, the first step will be to minimize the value of d_1^- before any other goal is addressed.

The fourth-priority goal in this problem is also associated with the labor constraint. The fourth goal, P_4, reflects a desire to minimize overtime. Recall that hours of overtime are represented by d_1^+; the objective function, therefore, becomes

$$\text{minimize } P_1d_1^-, P_4d_1^+$$

As before, the objective is to minimize the deviational variable d_1^+. In other words, if d_1^+ equaled zero, there would be no overtime at all. In solving this model, the achievement of this fourth-ranked goal will not be attempted until goals one, two, and three have been considered.

Profit Goal

The second goal in our goal programming model is to achieve a daily profit of $1,600. Recall that the original linear programming objective function was

$$Z = 40x_1 + 50x_2$$

Now we reformulate this objective function as a goal constraint with the following goal level:

$$40x_1 + 50x_2 + d_2^- - d_2^+ = \$1{,}600$$

The deviational variables d_2^- and d_2^+ represent the amount of profit less than $1,600 ($d_2^-$) and the amount of profit exceeding $1,600 ($d_2^+$). The pottery company's goal of achieving $1,600 in profit is represented in the objective function as

$$\text{minimize } P_1d_1^-, P_2d_2^-, P_4d_1^+$$

Notice that only d_2^- is being minimized, not d_2^+, because it is logical to assume that the pottery company would be willing to accept all profits in excess of $1,600 (i.e., it does not desire to minimize d_2^+, excess profit). By minimizing d_2^- at the second-priority level, the pottery company hopes that d_2^- will equal zero, which will result in at least $1,600 in profit.

Material Goal

The third goal of the company is to avoid keeping more than 120 pounds of clay on hand each day. The goal constraint is

$$4x_1 + 3x_2 + d_3^- - d_3^+ = 120 \text{ lb.}$$

Because the deviational variable d_3^- represents the amount of clay less than 120 pounds, and d_3^+ represents the amount in excess of 120 pounds, this goal can be reflected in the objective function as

$$\text{minimize } P_1d_1^-, P_2d_2^-, P_3d_3^+, P_4d_1^+$$

The term $P_3d_3^+$ represents the company's desire to minimize d_3^+, the amount of clay in excess of 120 pounds. The P_3 designation indicates that it is the pottery company's third most important goal.

The complete goal programming model can now be summarized as follows:

$$\begin{aligned}&\text{minimize } P_1d_1^-, P_2d_2^-, P_3d_3^+, P_4d_1^+\\&\text{subject to}\\&\quad x_1 + 2x_2 + d_1^- - d_1^+ = 40\\&\quad 40x_1 + 50x_2 + d_2^- - d_2^+ = 1{,}600\\&\quad 4x_1 + 3x_2 + d_3^- - d_3^+ = 120\\&\quad x_1, x_2, d_1^-, d_1^+, d_2^-, d_2^+, d_3^-, d_3^+ \geq 0\end{aligned}$$

Terms are not summed in the objective function because the deviational variables often have different units of measure.

The one basic difference between this model and the standard linear programming model is that the objective function terms *are not summed* to equal a total value, Z. This is because the deviational variables in the objective function represent different units of measure. For example, d_1^- and d_1^+ represent hours of labor, d_2^- represents dollars, and d_3^+ represents pounds of clay. It would be illogical to sum hours, dollars, and pounds. The objective function in a goal programming model specifies only that the deviations from the goals represented in the objective function be minimized *individually*, in order of their priority.

Alternative Forms of Goal Constraints

Let us now alter the preceding goal programming model so that our fourth-priority goal limits overtime to 10 hours instead of minimizing overtime. Recall that the goal constraint for labor is

$$x_1 + 2x_2 + d_1^- - d_1^+ = 40$$

In this goal constraint, d_1^+ represents overtime. Because the new fourth-priority goal is to limit overtime to 10 hours, the following goal constraint is developed:

$$d_1^+ + d_4^- - d_4^+ = 10$$

Goal constraints can include all deviational variables.

Although this goal constraint looks unusual, it is acceptable in goal programming to have an equation with *all deviational variables.* In this equation, d_4^- represents the amount of overtime less than 10 hours, and d_4^+ represents the amount of overtime greater than 10 hours. Because the company desires to *limit overtime* to 10 hours, d_4^+ is minimized in the objective function:

$$\text{minimize } P_1d_1^-, P_2d_2^-, P_3d_3^+, P_4d_4^+$$

Now let us consider the addition of a fifth-priority goal to this example. Assume that the pottery company has limited warehouse space, so it can produce no more than 30 bowls and 20 mugs daily. If possible, the company would like to produce these numbers. However, because the profit for mugs is greater than the profit for bowls (i.e., \$50 rather than \$40), it is more important to achieve the goal for mugs. This fifth goal requires that two new goal constraints be formulated, as follows:

$$\begin{aligned}x_1 + d_5^- &= 30 \text{ bowls}\\x_2 + d_6^- &= 20 \text{ mugs}\end{aligned}$$

Notice that the positive deviational variables d_5^+ and d_6^+ have been deleted from these goal constraints. This is because the statement of the fifth goal specifies that "no more than 30 bowls and 20 mugs" can be produced. In other words, positive deviation, or overproduction, is not possible.

Because the actual goal of the company is to achieve the levels of production shown in these two goal constraints, the negative deviational variables d_5^- and d_6^- are minimized in the objective function. However, recall that it is *more important* to the company to achieve the goal for mugs because mugs generate greater profit. This condition is reflected in the objective function, as follows:

$$\text{minimize } P_1d_1^-, P_2d_2^-, P_3d_3^+, P_4d_4^+, 4P_5d_5^- + 5P_5d_6^-$$

Because the goal for mugs is more important than the goal for bowls, the *degree of importance* should be in proportion to the amount of profit (i.e., \$50 for each mug and \$40 for each bowl). Thus, the goal for mugs is more important than the goal for bowls by a ratio of 5 to 4.

The coefficients of 5 for $P_5d_6^-$ and 4 for $P_5d_5^-$ are referred to as *weights*. In other words, the minimization of d_6^- is "weighted" higher than the minimization of d_5^- at the fifth priority level.

Two or more goals at the same priority level can be assigned weights to indicate their relative importance.

When this model is solved, the achievement of the goal for minimizing (bowls) is more important, even though both goals are at the same priority level.

Notice, however, that these two weighted goals have been summed because they are at the same priority level. Their sum represents achievement of the desired goal at this particular priority level. The complete goal programming model, with the new goals for both overtime and production, is

$$\text{minimize } P_1d_1^-, P_2d_2^-, P_3d_3^+, P_4d_4^+, 4P_5d_5^- + 5P_5d_6^-$$

subject to

$$\begin{aligned} x_1 + 2x_2 + d_1^- - d_1^+ &= 40 \\ 40x_1 + 50x_2 + d_2^- - d_2^+ &= 1{,}600 \\ 4x_1 + 3x_2 + d_3^- - d_3^+ &= 120 \\ d_1^+ + d_4^- - d_4^+ &= 10 \\ x_1 + d_5^- &= 30 \\ x_2 + d_6^- &= 20 \\ x_1, x_2, d_1^-, d_1^+, d_2^-, d_2^+, d_3^-, d_3^+, d_4^-, d_4^+, d_5^-, d_6^- &\geq 0 \end{aligned}$$

Graphical Interpretation of Goal Programming

In Chapter 2, we analyzed the solution of linear programming models by using graphical analysis. Because goal programming models are linear, they can also be analyzed graphically. The original goal programming model for Beaver Creek Pottery Company, formulated at the beginning of this chapter, will be used as an example:

$$\text{minimize } P_1d_1^-, P_2d_2^-, P_3d_3^+, P_4d_1^+$$

subject to

$$\begin{aligned} x_1 + 2x_2 + d_1^- - d_1^+ &= 40 \\ 40x_1 + 50x_2 + d_2^- - d_2^+ &= 1{,}600 \\ 4x_1 + 3x_2 + d_3^- - d_3^+ &= 120 \\ x_1, x_2, d_1^-, d_1^+, d_2^-, d_2^+, d_3^-, d_3^+ &\geq 0 \end{aligned}$$

To graph this model, the deviational variables in each goal constraint are set equal to zero, and we graph each subsequent equation on a set of coordinates, just as in Chapter 2. Figure 9.1 is a graph of the three goal constraints for this model.

Notice that in Figure 9.1 there is no feasible solution space indicated, as in a regular linear programming model. This is because all three goal constraints are *equations*; thus, all solution points are on the constraint lines.

FIGURE 9.1
Goal constraints

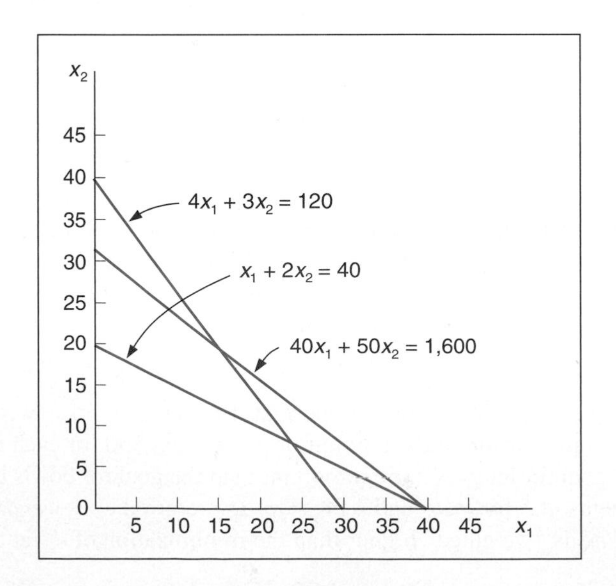

Graphical solution illustrates the goal programming solution logic—seeking to achieve goals by minimizing deviation in order of their priority.

The solution logic in a goal programming model is to attempt to achieve the goals in the objective function, in order of their priorities. As a goal is achieved, the next highest-ranked goal is then considered. However, a higher-ranked goal that has been achieved is never given up in order to achieve a lower-ranked goal.

In this example, we first consider the first-priority goal, minimizing d_1^-. The relationship of d_1^- and d_1^+ to the goal constraint is shown in Figure 9.2. The area below the goal constraint line $x_1 + 2x_2 = 40$ represents possible values for d_1^-, and the area above the line represents values

FIGURE 9.2
The first-priority goal: minimize d_1^-

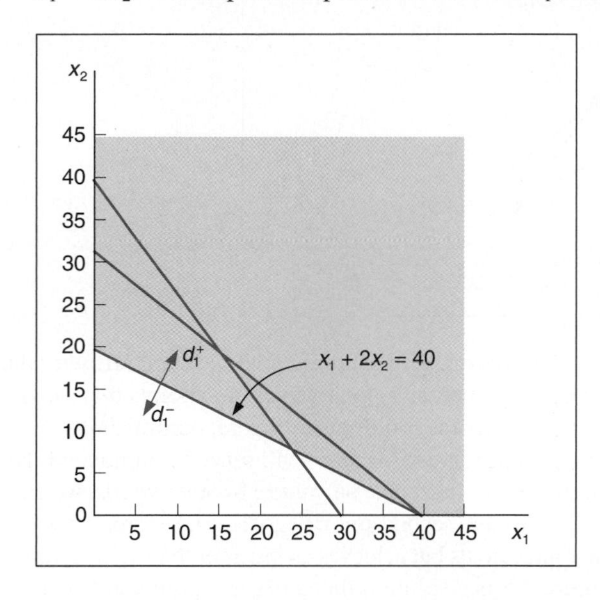

for d_1^+. In order to achieve the goal of minimizing d_1^-, the area below the constraint line corresponding to d_1^- is eliminated, leaving the shaded area as a possible solution area.

One goal is never achieved at the expense of another higher-priority goal.

Next, we consider the second-priority goal, minimizing d_2^-. In Figure 9.3, the area below the constraint line $40x_1 + 50x_2 = 1{,}600$ represents the values for d_2^-, and the area above the line represents the values for d_2^+. To minimize d_2^-, the area below the constraint line corresponding to d_2^- is eliminated. Notice that by eliminating the area for d_2^-, we do not affect the first-priority goal of minimizing d_1^-.

FIGURE 9.3
The second-priority goal: minimize d_2^-

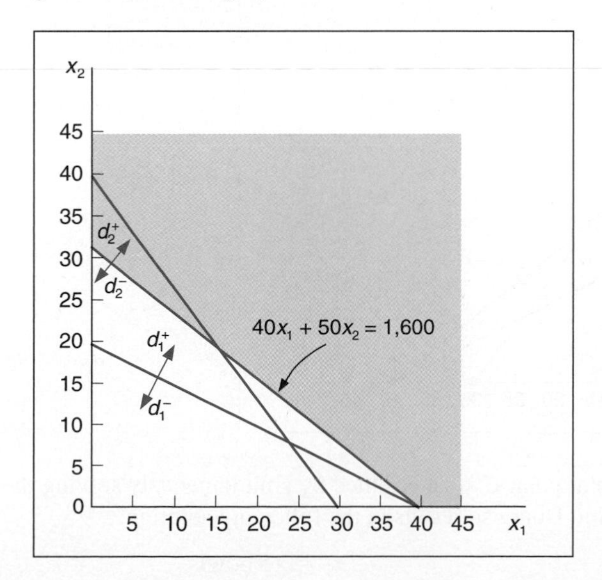

Next, the third-priority goal, minimizing d_3^+, is considered. Figure 9.4 shows the areas corresponding to d_3^- and d_3^+. To minimize d_3^+, the area above the constraint line $4x_1 + 3x_2 = 120$

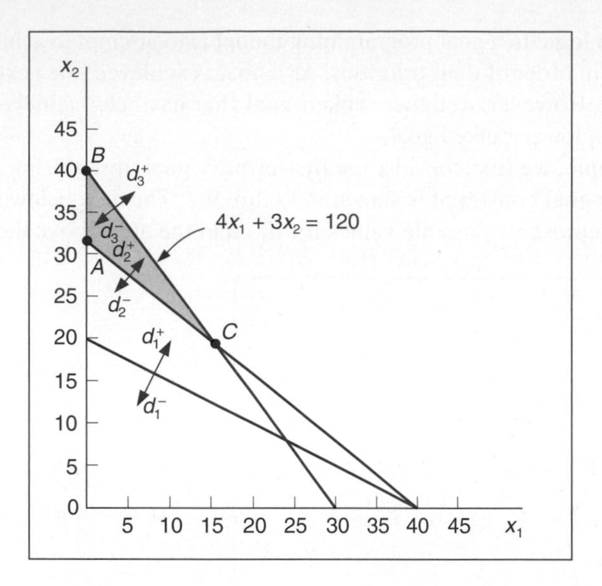

FIGURE 9.4
The third-priority goal: minimize d_3^+

is eliminated. After considering the first three goals, we are left with the area between the line segments *AC* and *BC*, which contains possible solution points that satisfy the first three goals.

Finally, we must consider the fourth-priority goal, minimizing d_1^+. To achieve this final goal, the area above the constraint line $x_1 + 2x_2 = 40$ must be eliminated. However, if we eliminate this area, then both d_2^- and d_3^- must take on values. In other words, we cannot minimize d_1^+ totally without violating the first- and second-priority goals. Therefore, we want to find a solution point that satisfies the first three goals but achieves as much of the fourth-priority goal as possible.

Point *C* in Figure 9.5 is a solution that *satisfies* these conditions. Notice that if we move down the goal constraint line $4x_1 + 3x_2 = 120$ toward point *D*, d_1^+ is further minimized; however, d_2^- takes on a value as we move past point *C*. Thus, the minimization of d_1^+ would be accomplished only at the expense of a higher-ranked goal.

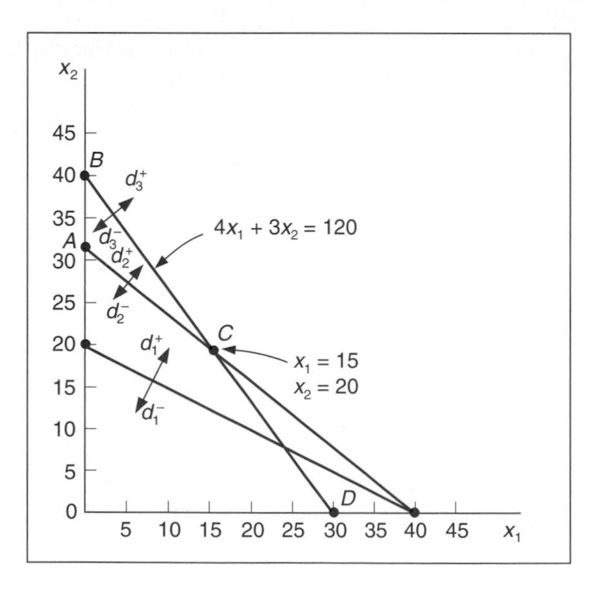

FIGURE 9.5
The fourth-priority goal (minimize d_1^+) and the solution

The solution at point *C* is determined by simultaneously solving the two equations that intersect at this point. Doing so results in the following solution:

$$x_1 = 15 \text{ bowls}$$
$$x_2 = 20 \text{ mugs}$$
$$d_1^+ = 15 \text{ hours}$$

Goal programming solutions do not always achieve all goals, and they are not optimal; they achieve the best or most satisfactory solution possible.

Because the deviational variables d_1^-, d_2^-, and d_3^+ all equal zero, they have been minimized, and the first three goals have been achieved. Because $d_1^+ = 15$ hours of overtime, the fourth-priority goal has not been achieved. The solution to a goal programming model such as this one is referred to as the *most satisfactory* solution rather than the optimal solution because it satisfies the specified goals as well as possible.

Computer Solution of Goal Programming Problems with QM for Windows and Excel

Goal programming problems can be solved using QM for Windows and Excel spreadsheets. In this section, we will demonstrate how to use both computer alternatives to solve the Beaver Creek Pottery Company example we solved graphically in the previous section.

Management Science Application

Developing Television Advertising Sales Plans at NBC

The National Broadcasting Company (NBC), a subsidiary of General Electric, generated more than $4 billion in revenue in 2000 from its television network. In May the network announces its programming schedule for the broadcast year, which begins in the third week in September. Soon after that initial announcement, the network begins the sale of its inventory of advertising slots on its TV shows to advertisers. During this time advertising agencies approach the network with requests to purchase time for their clients for the year. The requests generally include a budget, the demographic audience in which the client is interested, and a desired program mix. NBC subsequently develops a sales plan consisting of a schedule of television commercials that will meet the client's requirements. The most popular commercial durations are 30 and 15 seconds, although 60- and 120-second slots are not uncommon.

Diver721/Dreamstime.com

After the network finalizes its programming schedule, it develops a rating forecast that projects the audience size for several demographic groupings for each airing of a show. These ratings projections are based on such factors as the strength of the show, historical ratings for a time slot, competing shows on other networks, and the performance of adjacent shows. The network then develops advertising rate cards that set the prices of advertising slots for each airing of a show. The prices vary according to the time of the year the show airs. For example, prices are high during the sweeps months of November, February, and May and are lower in January and the summer; thus prices are weighted for different weeks of the year. The sales management staff prioritizes sales requests based on the client's importance to NBC. NBC also desires to make the most profitable use of its limited inventory. It is optimal for NBC to use as little premium inventory as possible in meeting each client's request.

NBC uses a 0–1 integer goal programming model to develop annual advertising sales plans for individual clients. The model minimizes the amount of premium inventory assigned to a sales plan while also minimizing the penalty incurred in not meeting other goals. Goals are for inventory (i.e., airtime availability), product conflict (i.e., no two similar products should advertise on the same show), client budget constraints, show–mix goals, unit–mix (i.e., commercial length) goals, and weekly weighting constraints. The decision variables are 0–1 and indicate the number of commercials of each length aired on shows during weeks included in the sales plan. The model has enabled NBC to save millions of dollars of premium inventory while meeting customer requirements, and it has reduced the time required to develop a sales plan from 3 to 4 hours to 20 minutes. It is estimated that this model and associated planning systems increase NBC's revenue by at least $50 million per year.

Source: Based on S. Bollapragada, et al., "NBC Optimization Systems Increase Revenues and Productivity," *Interfaces* 32, no. 1 (January–February 2002): 47–60.

QM for Windows

We will demonstrate how to solve a goal programming model by using our Beaver Creek Pottery Company example, which was formulated as follows:

$$\text{minimize } P_1d_1^-, P_2d_2^-, P_3d_3^+, P_4d_1^+$$
$$\text{subject to}$$
$$x_1 + 2x_2 + d_1^- - d_1^+ = 40$$
$$40x_1 + 50x_2 + d_2^- - d_2^+ = 1{,}600$$
$$4x_1 + 3x_2 + d_3^- - d_3^+ = 120$$
$$x_1, x_2, d_1^-, d_1^+, d_2^-, d_2^+, d_3^-, d_3^+ \geq 0$$

QM for Windows includes a goal programming module that can be accessed by clicking on the "Module" button at the top of the screen. The model parameters are entered onto the data input screen as shown in Exhibit 9.1. Notice that each prioritized deviational variable must be assigned a weight, which for this problem is always one. The solution is obtained by clicking on the "Solve" button at the top of the screen. The solution summary for our model is shown in Exhibit 9.2.

EXHIBIT 9.1

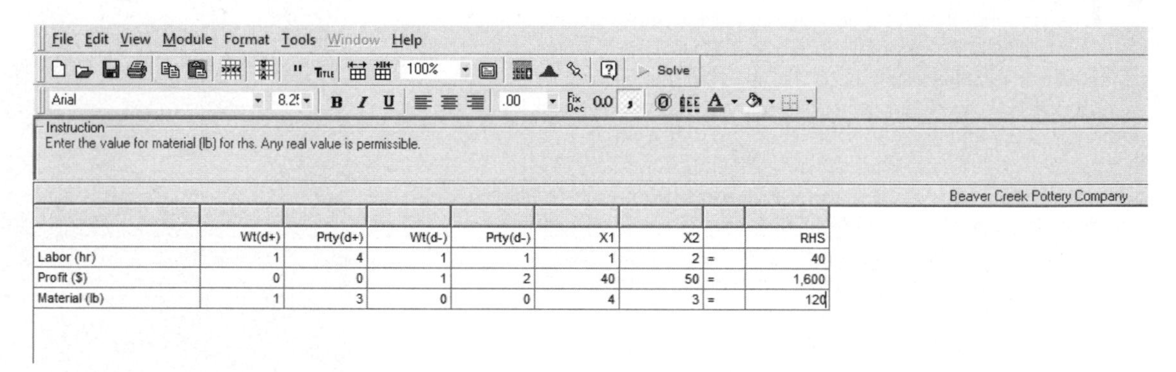

	Wt(d+)	Prty(d+)	Wt(d-)	Prty(d-)	X1	X2		RHS
Labor (hr)	1	4	1	1	1	2	=	40
Profit ($)	0	0	1	2	40	50	=	1,600
Material (lb)	1	3	0	0	4	3	=	120

EXHIBIT 9.2

Summary

Beaver Creek Pottery Company Solution

Item			
Decision variable analysis	Value		
X1	15		
X2	20		
Priority analysis	Nonachievement		
Priority 1	0		
Priority 2	0		
Priority 3	0		
Priority 4	15		
Constraint Analysis	RHS	d+ (row i)	d- (row i)
Labor (hr)	40	15	0
Profit ($)	1,600	0	0
Material (lb)	120	0	0

QM for Windows will also provide a graphical analysis of a goal programming model. We click on the "Windows" button and then select "Graph" from the menu. The graph for our example is shown in Exhibit 9.3.

EXHIBIT 9.3

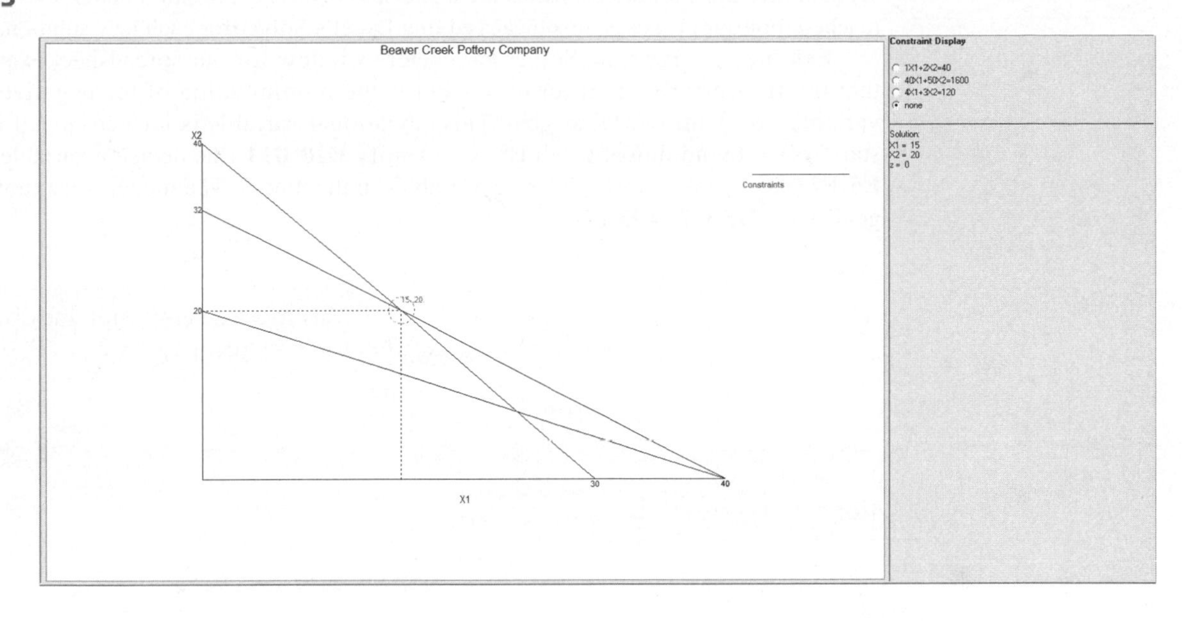

Excel Spreadsheets

Solving a goal programming problem by using Excel is similar to solving a linear programming model, although not quite as straightforward. Exhibit 9.4 shows the spreadsheet format for our Beaver Creek Pottery Company example. Cells G5, G6, and G7, under the heading "Constraint Total," contain the formulas for our goal constraints, including deviational variables. The formula for the labor constraint is shown on the formula bar at the top of the screen in Exhibit 9.4. The model decision variables are in cells B10 and B11, and the deviational variables are in cells **E5:F7**. The goals are established by setting the constraint formulas in G5 to G7 equal to the goal levels in cells I5 to I7.

EXHIBIT 9.4

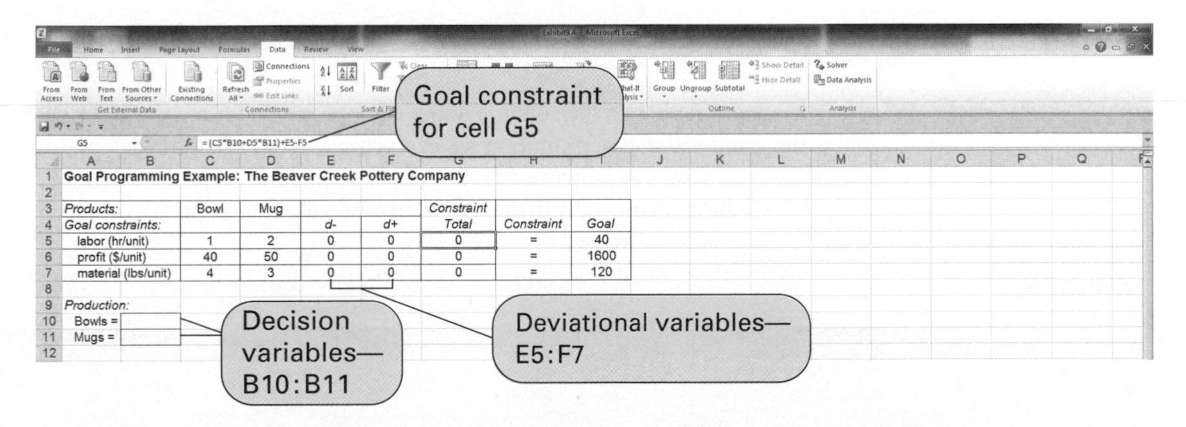

When using a spreadsheet (or any regular linear programming program) to solve a goal programming problem, it must be solved *sequentially*. In this procedure, a new problem is formulated and solved for each priority goal in the objective function, beginning with the highest priority. In other words, the minimization of the deviational variable at the highest priority is the initial objective. Once a solution for this formulation is achieved, the value of the deviational variable that is the objective is added to the model as a constraint, and the second-priority deviational variable becomes the new objective. A new solution is achieved for each new objective sequentially until all the priorities are exhausted or it is clear that a better solution cannot be reached. For our purposes, this means editing Excel's Solver for each new solution.

Exhibit 9.5 shows the Solver Parameters window for our spreadsheet example. Recall that the first-priority goal for our model is the minimization of the negative deviational variable (d_1^-) for our labor goal. This deviational variable is located in cell E5; thus, we start Solver by minimizing cell E5. We identify **B10:B11** (the decision variables) as well as **E5:F7** (the deviational variables) as variables in the model. The model constraints are for our goals (i.e., **G5:G7 = I5:I7**).

EXHIBIT 9.5

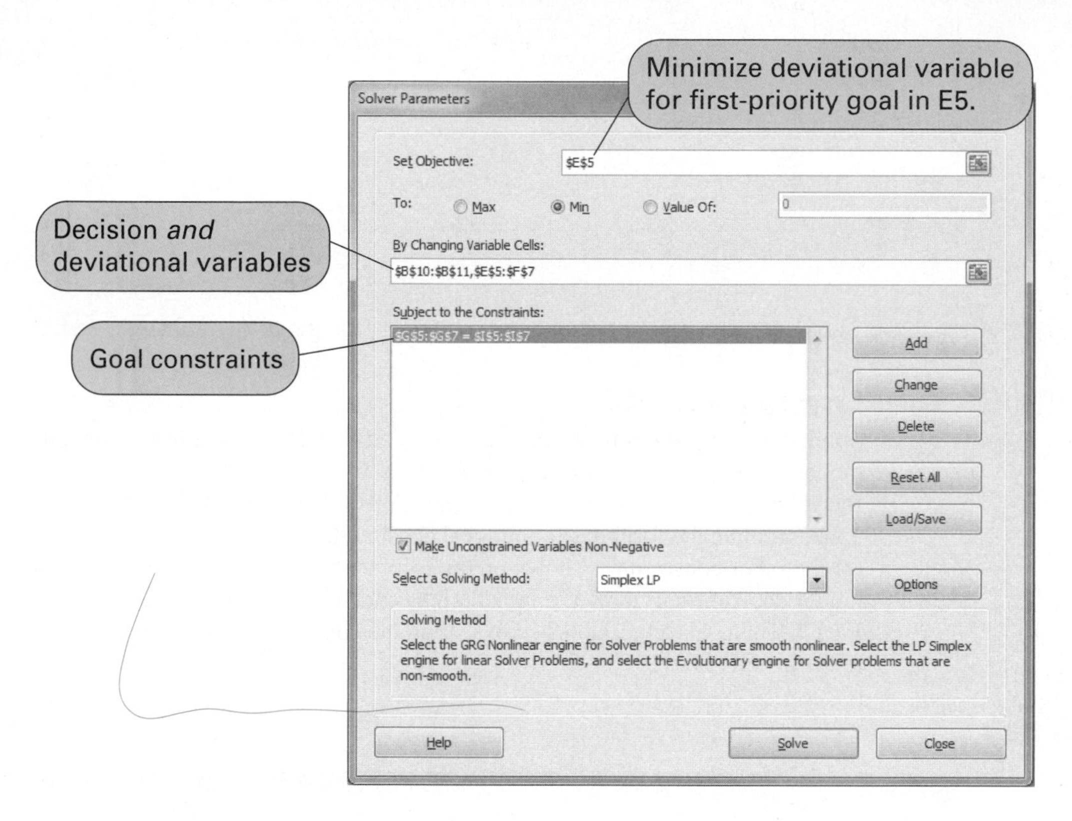

The spreadsheet with the solution to our example problem is shown in Exhibit 9.6. We know this is the most satisfactory solution we can achieve from the QM for Windows solution, so it will not be necessary to perform any additional sequential steps. From the spreadsheet we can see that we have achieved all the goals except the fourth-priority goal, to minimize d_1^+.

EXHIBIT 9.6

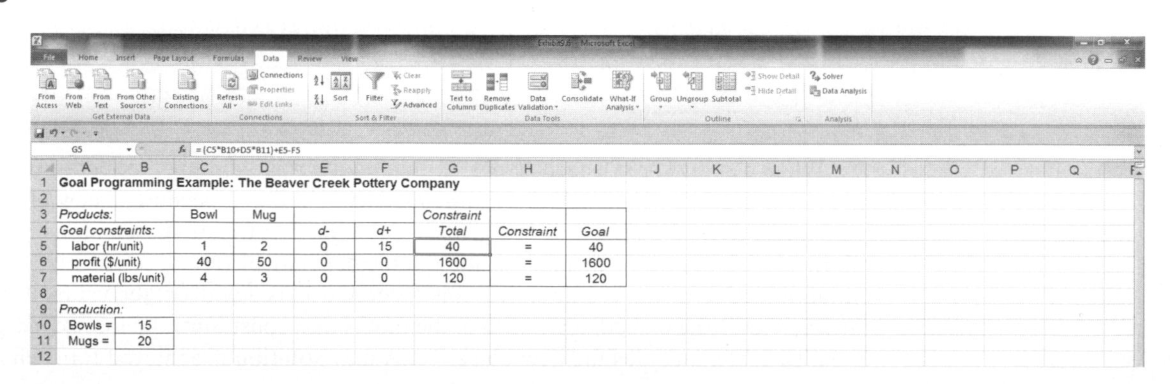

G5 = (C5*B10+D5*B11)+E5-F5

Goal Programming Example: The Beaver Creek Pottery Company

Products:	Bowl	Mug			Constraint		
Goal constraints:			d-	d+	Total	Constraint	Goal
labor (hr/unit)	1	2	0	15	40	=	40
profit ($/unit)	40	50	0	0	1600	=	1600
material (lbs/unit)	4	3	0	0	120	=	120

Production:

Bowls =	15
Mugs =	20

Time Out for Abraham Charnes and William W. Cooper

The concept of goal programming was first introduced in 1955 by Abraham Charnes and William Cooper of the Carnegie Institute of Technology, with R. O. Ferguson, a consultant for the Methods Engineering Council. They developed a model for a plan for executive compensation for a division of General Electric (using competing salary offers for executives as multiple goals), and they solved it by reducing it to an equivalent linear programming formulation. However, it was not until 1961 that Charnes and Cooper coined the name *goal programming*. Yuji Ijiri of Stanford (and a former doctoral student of Cooper's at Carnegie Tech) in 1965 developed the concept of preemptive priorities for treating multiple goals and the general solution approach.

However, if we had not achieved all our top goals so readily, the next step would be to include **E5=0** as a constraint in Solver and then minimize our next-priority goal, which would be E6 (i.e., d_2^-).

Next, we will use Excel to solve a slightly more complicated goal programming model—that is, the altered Beaver Creek Pottery Company example we developed at the beginning of this chapter, with goals for overtime and maximum storage levels for bowls and mugs:

$$\text{minimize } P_1d_1^-, P_2d_2^-, P_3d_3^+, P_4d_4^+, 4P_5d_5^- + 5P_5d_6^-$$

subject to

$$\begin{aligned}
x_1 + 2x_2 + d_1^- - d_1^+ &= 40\\
40x_1 + 50x_2 + d_2^- - d_2^+ &= 1{,}600\\
4x_1 + 3x_2 + d_3^- - d_3^+ &= 120\\
d_1^+ + d_4^- - d_4^+ &= 10\\
x_1 + d_5^- &= 30\\
x_2 + d_6^- &= 20\\
x_1, x_2, d_1^-, d_1^+, d_2^-, d_2^+, d_3^-, d_3^+, d_4^-, d_4^+, d_5^-, d_6^- &\geq 0
\end{aligned}$$

The spreadsheet for this modified version of our example is shown in Exhibit 9.7. The spreadsheet is set up much the same as the original version of this example, with the exception of the goal constraint for overtime. The formula for this goal constraint is included in cell G8 as **=F5+E8−F8**. In addition, the positive deviational variables for the last two goal constraints are now included in the formulas embedded in cells G9 and G10. For example, in cell G9 the constraint formula is **=C9*B13+E9**.

EXHIBIT 9.7

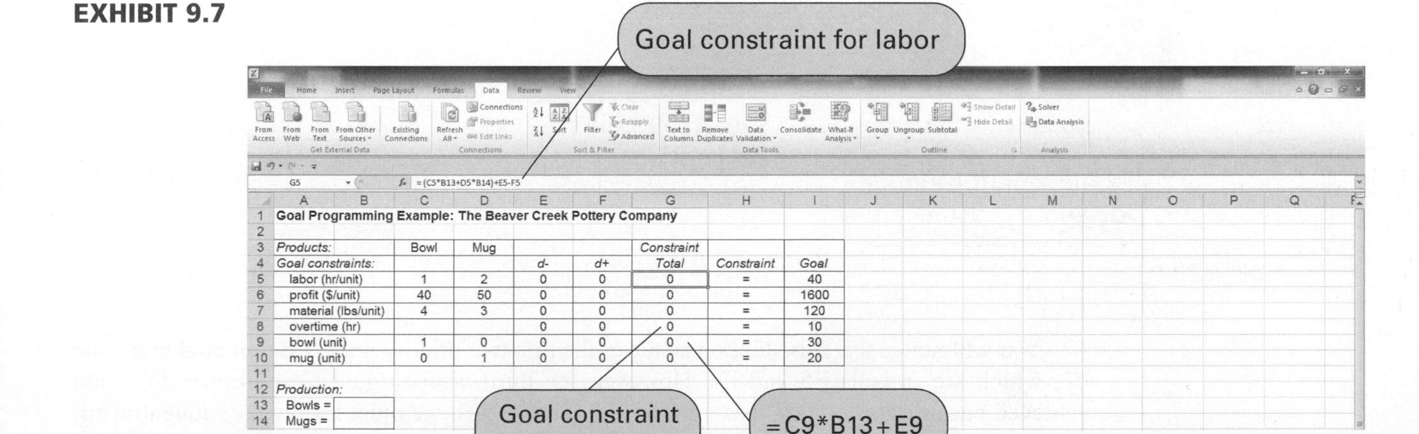

	A	C	D	E	F	G	H	I
1	Goal Programming Example: The Beaver Creek Pottery Company							
3	Products:	Bowl	Mug			Constraint		
4	Goal constraints:			d-	d+	Total	Constraint	Goal
5	labor (hr/unit)	1	2	0	0	0	=	40
6	profit ($/unit)	40	50	0	0	0	=	1600
7	material (lbs/unit)	4	3	0	0	0	=	120
8	overtime (hr)			0	0	0	=	10
9	bowl (unit)	1	0	0	0	0	=	30
10	mug (unit)	0	1	0	0	0	=	20
12	Production:							
13	Bowls =							
14	Mugs =							

The Solver Parameters window for this spreadsheet is shown in Exhibit 9.8, and the resulting solution is shown in Exhibit 9.9.

EXHIBIT 9.8

EXHIBIT 9.9

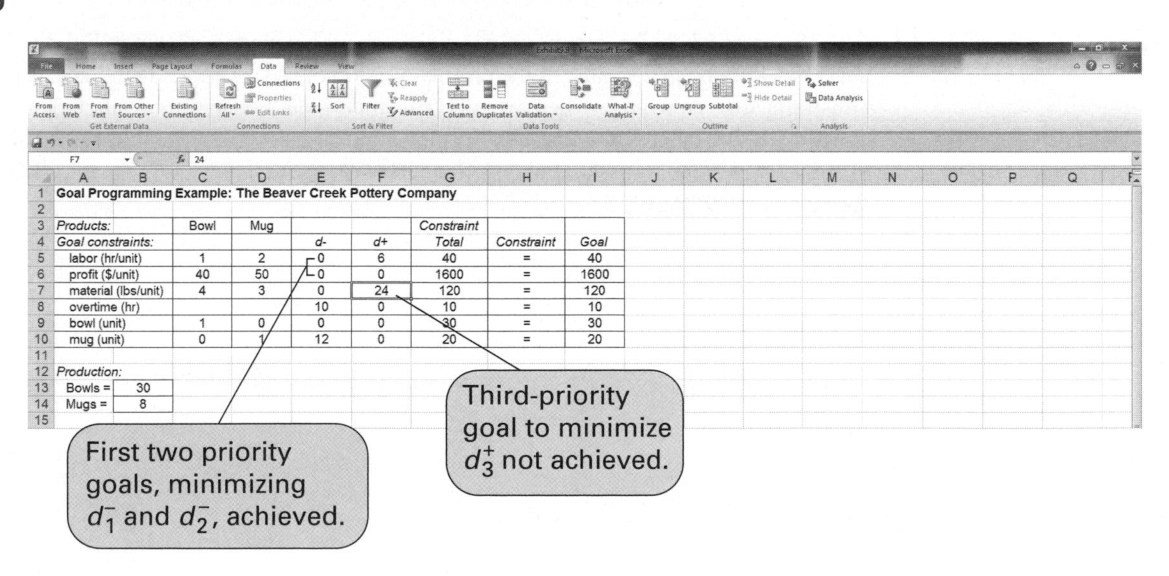

Goal Programming Example: The Beaver Creek Pottery Company

	A/B	C	D	E	F	G	H	I
3	Products:	Bowl	Mug			Constraint		
4	Goal constraints:			d-	d+	Total	Constraint	Goal
5	labor (hr/unit)	1	2	0	6	40	=	40
6	profit ($/unit)	40	50	0	0	1600	=	1600
7	material (lbs/unit)	4	3	0	24	120	=	120
8	overtime (hr)			10	0	10	=	10
9	bowl (unit)	1	0	0	0	30	=	30
10	mug (unit)	0	1	12	0	20	=	20
12	Production:							
13	Bowls =	30						
14	Mugs =	8						

You will notice that this solution achieves the first two priority goals for minimizing d_1^- and d_2^-, which are in cells E5 and E6. However, the third-priority goal to minimize d_3^+ is not achieved because its cell (F7) has a value of 24 in it. Thus, we must follow the sequential approach to attempt to obtain a better solution. We accomplish this by including **E5=0** (the achievement of our first goal) as a constraint in Solver, as well as our second-priority goal that this solution achieved, **E6=0**, and then minimizing F7. This Solver Parameters screen is shown in Exhibit 9.10.

EXHIBIT 9.10

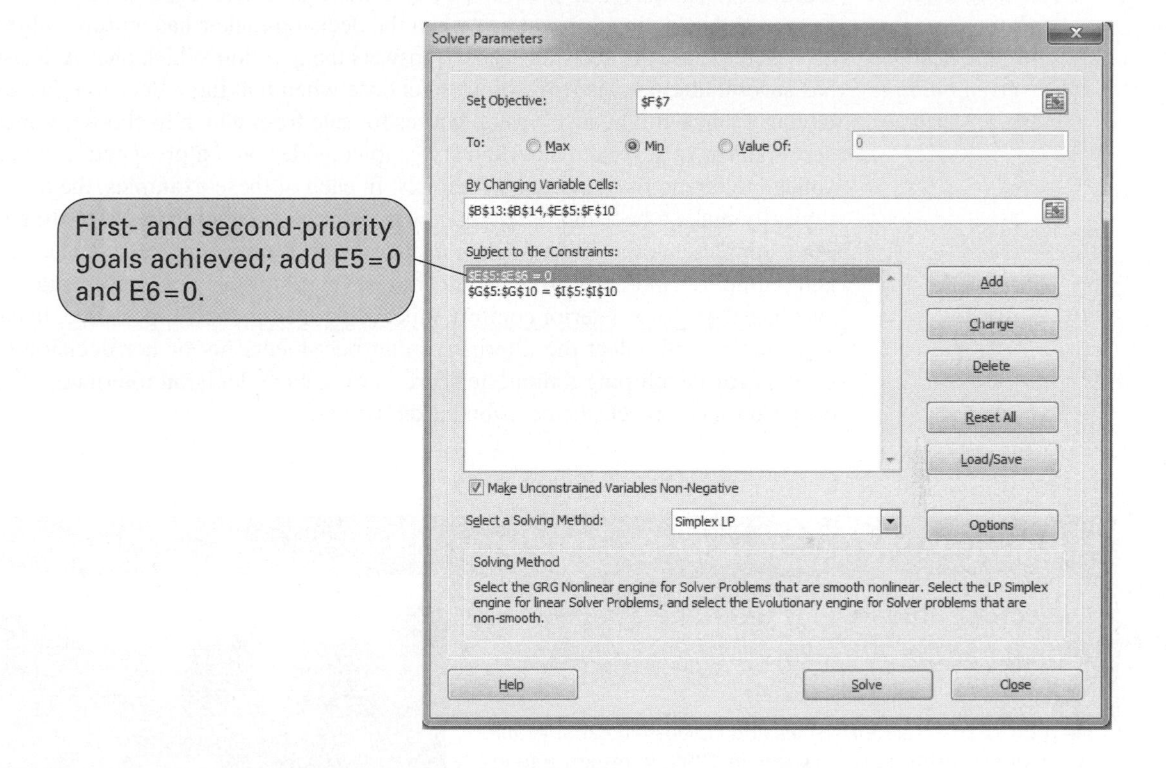

The solution is shown in Exhibit 9.11. This is the same solution that we obtained for our original version of the model, without the alterations for overtime and production levels. This solution achieves the first three priority goals. We could continue to attempt to achieve the fourth-priority goal by including **F7=0** as a constraint in Solver and minimizing cell F5 (i.e., d_1^+); however, doing so will not result in a better solution without sacrificing the goal achievement at the higher-priority levels. Thus, this is the best solution we can achieve.

EXHIBIT 9.11

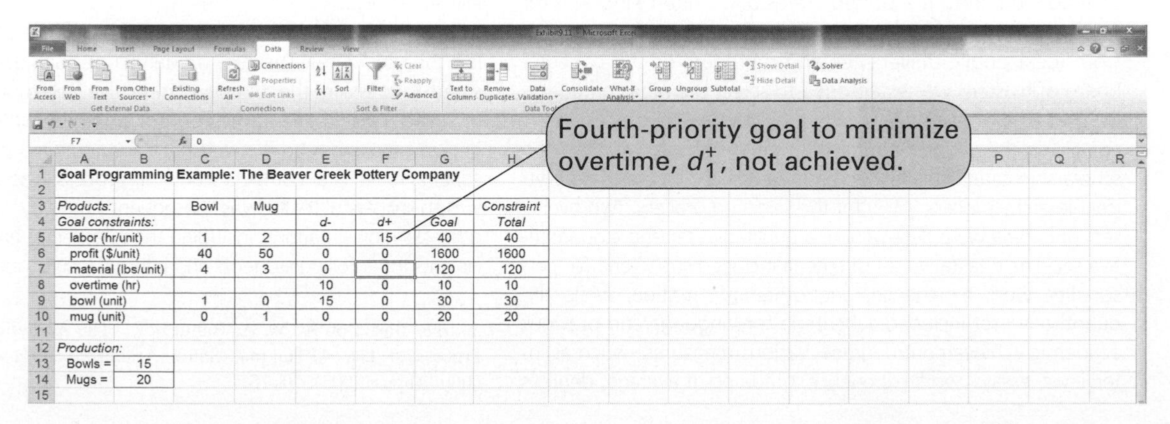

Products:	Bowl	Mug				Constraint
Goal constraints:			d-	d+	Goal	Total
labor (hr/unit)	1	2	0	15	40	40
profit ($/unit)	40	50	0	0	1600	1600
material (lbs/unit)	4	3	0	0	120	120
overtime (hr)			10	0	10	10
bowl (unit)	1	0	15	0	30	30
mug (unit)	0	1	0	0	20	20

Production:	
Bowls =	15
Mugs =	20

The Analytical Hierarchy Process

AHP is a method for ranking decision alternatives and selecting the best one given multiple criteria.

Goal programming is a method that provides us with a mathematical "quantity" for the decision variables that best achieves a set of goals. It answers the question "How much?" The **analytical hierarchy process (AHP)**, developed by Thomas Saaty, is a method for ranking decision alternatives and selecting the best one when the decision maker has multiple objectives, or criteria, on which to base the decision. Thus, it answers the question Which one? A decision maker usually has several alternatives from which to choose when making a decision. For example, someone buying a house might have several houses for sale from which to choose; someone buying a new car might have several makes and styles to consider; and a prospective student might select a college to attend from a group of schools. In each of these examples, the decision maker would typically make a decision based on how the alternatives compare, according to several criteria. For example, a home buyer might consider the cost, proximity of schools, trees, neighborhood, and public transportation when comparing several houses; a car buyer might compare different cars based on price, interior comfort, mileage per gallon, and appearance. In each case, the decision maker will select the alternative that best meets his or her decision criteria. AHP is a process for developing a numeric score to rank each decision alternative, based on how well each alternative meets the decision maker's criteria.

Management Science Application

Selecting Students for Graduate Studies Abroad at Dar Al-Hekma Women's College

Prior to 1999, all universities and colleges in Saudi Arabia were public institutions; however, in 1999, in response to increased demand for higher education, the Ministry of Higher Education in Saudi Arabia began allowing nonprofit, private colleges. Dar Al-Hekma, located in Jeddah, was one of the first private colleges for women instituted under this new policy.

Dar Al-Hekma offers a 4-year degree program with degrees in business information systems, interior design, and special education. One of the initial problems the new college faced was the lack of qualified women to staff the three departments. To help alleviate this problem, the Ministry of Higher Education provided the college with nine scholarships to send women to the United States and United Kingdom for graduate studies. A committee consisting of faculty members from local public universities and members of the college's board of trustees was established to select the nine candidates in what was expected to be a highly competitive process.

Given the various criteria that could be used to select the scholarship candidates, the committee decided to use the analytical hierarchy process (AHP) for the selection process. Two overall sets of criteria were determined: a qualitative set and a quantitative set. The qualitative set of criteria included subcriteria for personality, work experience, and other information, while the quantitative set included subcriteria for language and previous academic achievement/degrees. Various measures—such as interviews, essays, work experience, grade point average, degrees awarded, and TOEFL scores—were used to rank candidates in each category.

AP Photo/Sakchai Lalit

The AHP model was successful in helping the committee select the candidates. The AHP model reduced subjectivity and ensured fairness in the highly competitive selection process; it provided a systematic framework for committee interaction and decision making, and it saved time and removed potential personal and political conflicts from the process. The AHP process was so successful that it was subsequently used by the college for other decision situations, including the selection of financial aid recipients for enrolling students and internal faculty grant recipients.

Source: Based on A. M. A. Bahurmoz, "The Analytical Hierarchy Process at Dar Al-Hekma, Saudi Arabia," *Interfaces* 33, no. 4 (July–August 2003): 70–78.

We will demonstrate AHP by using an example. Southcorp Development builds and operates shopping malls around the country. The company has identified three potential sites for its latest project, near Atlanta, Birmingham, and Charlotte. The company has identified four primary criteria on which it will compare the sites—(1) the customer market base (including overall market size and population at different age levels); (2) income level; (3) infrastructure (including utilities and roads); and (4) transportation (i.e., proximity to interstate highways for supplier deliveries and customer access).

The overall objective of the company is to select the best site. This goal is at the top of the *hierarchy* of the problem. At the next (second) level of the hierarchy, we determine how the four *criteria* contribute to achieving the objective. At the level of the *problem hierarchy*, we determine how each of the three alternatives (Atlanta, Birmingham, and Charlotte) contributes to each of the four criteria.

The general mathematical process involved in AHP is to establish preferences at each of these hierarchical levels. First, we mathematically determine our preferences for each site for each criterion. For example, we first determine our site preferences for the customer market base. We might decide that Atlanta has a better market base than the other two cities; that is, we prefer Atlanta for the customer market criterion. Then we determine our site preferences for income level, and so on. Next, we mathematically determine our preferences for the criteria—that is, which of the four criteria is most important, which is the next most important, and so on. For example, we might decide that the customer market is a more important criterion than the others. Finally, we combine these two sets of preferences—for sites within each criterion and for the four criteria—to mathematically derive a score for each site, with the highest score being the best.

In the following sections, we describe these general steps in greater detail.

Pairwise Comparisons

In a pairwise comparison, two alternatives are compared according to a criterion, and one is preferred.

A preference scale assigns numeric values to different levels of preference.

In AHP the decision maker determines how well each alternative "scores" on a criterion by using **pairwise comparisons**. In a pairwise comparison, the decision maker compares two alternatives (i.e., a pair) according to one criterion and indicates a preference. For example, Southcorp might compare the Atlanta (A) site with the Birmingham (B) site and decide which one it prefers according to the customer market criterion. These comparisons are made by using a **preference scale**, which assigns numeric values to different levels of preference.

The standard preference scale used for AHP is shown in Table 9.1. This scale has been determined by experienced researchers in AHP to be a reasonable basis for comparing two items or alternatives. Each rating on the scale is based on a comparison of two items. For example, if the Atlanta site is "moderately preferred" to the Birmingham site, then a value of 3 is assigned to this particular comparison. The rating of 3 is a measure of the decision maker's preference for one of the alternatives over the other.

If Southcorp compares Atlanta to Birmingham and moderately prefers Atlanta, resulting in a comparison value of 3 for the customer market criterion, it is not necessary for Southcorp to compare Birmingham to Atlanta to determine a separate preference value for this

TABLE 9.1
Preference scale for pairwise comparisons

Preference Level	Numeric Value
Equally preferred	1
Equally to moderately preferred	2
Moderately preferred	3
Moderately to strongly preferred	4
Strongly preferred	5
Strongly to very strongly preferred	6
Very strongly preferred	7
Very strongly to extremely preferred	8
Extremely preferred	9

"opposite" comparison. The preference value of B for A is simply the reciprocal or inverse of the preference of A for B. Thus, in our example if the preference value of Atlanta for Birmingham is 3, the preference value of Birmingham for Atlanta is 1/3.

A pairwise comparison matrix summarizes the pairwise comparisons for a criterion.

Southcorp's pairwise comparison ratings for each of the three sites for the customer market criterion are summarized in a matrix, a rectangular array of numbers. This **pairwise comparison matrix** will have a number of rows and columns equal to the decision alternatives:

	Customer Market		
Site	*A*	*B*	*C*
A	1	3	2
B	1/3	1	1/5
C	1/2	5	1

This matrix shows that the customer market in Atlanta (A) is equally to moderately preferred (2) over the Charlotte (C) customer market, but Charlotte (C) is strongly preferred (5) over Birmingham (B). Notice that any site compared against itself, such as A compared to A, must be "equally preferred," with a preference value of 1. Thus, the values along the diagonal of our matrix must be 1s.

The remaining pairwise comparison matrices for the other three criteria—income level, infrastructure, and transportation—have been developed by Southcorp as follows:

$$\textit{Income Level}\quad \begin{matrix}A\\B\\C\end{matrix}\begin{bmatrix}1 & 6 & 1/3\\ 1/6 & 1 & 1/9\\ 3 & 9 & 1\end{bmatrix}\qquad \textit{Infrastructure}\quad \begin{matrix}A\\B\\C\end{matrix}\begin{bmatrix}1 & 1/3 & 1\\ 3 & 1 & 7\\ 1 & 1/7 & 1\end{bmatrix}\qquad \textit{Transportation}\quad \begin{matrix}A\\B\\C\end{matrix}\begin{bmatrix}1 & 1/3 & 1/2\\ 3 & 1 & 4\\ 2 & 1/4 & 1\end{bmatrix}$$

Developing Preferences Within Criteria

In synthesization, decision alternatives are prioritized within each criterion.

The next step in AHP is to prioritize the decision alternatives within each criterion. For our site selection example, this means that we want to determine which site is the most preferred, the second most preferred, and the third most preferred for each of the four criteria. This step in AHP is referred to as **synthesization**. The mathematical procedure for synthesization is very complex and beyond the scope of this text. Instead, we will use an approximation method for synthesization that provides a reasonably good estimate of preference scores for each decision in each criterion.

The first step in developing preference scores is to sum the values in each column of the pairwise comparison matrices. The column sums for our customer market matrix are shown as follows:

	Customer Market		
Site	*A*	*B*	*C*
A	1	3	2
B	1/3	1	1/5
C	1/2	5	1
	11/6	9	16/5

Next, we divide each value in a column by its corresponding column sum. This results in a *normalized* matrix, as follows:

	Customer Market		
Site	*A*	*B*	*C*
A	6/11	3/9	5/8
B	2/11	1/9	1/16
C	3/11	5/9	5/16

Notice that the values in each column sum to 1. The next step is to average the values in each *row*. At this point, we convert the fractional values in the matrix to decimals, as shown in Table 9.2. The row average for each site is also shown in Table 9.2.

TABLE 9.2
The normalized matrix with row averages

	Customer Market			
Site	*A*	*B*	*C*	**Row Average**
A	0.5455	0.3333	0.6250	0.5012
B	0.1818	0.1111	0.0625	0.1185
C	0.2727	0.5556	0.3125	0.3803
				1.0000

The row averages in Table 9.2 provide us with Southcorp's preferences for the three sites for the customer market criterion. The most preferred site is Atlanta, followed by Charlotte; the least preferred site (for this criterion) is Birmingham. We can write these preferences as a matrix with one column, which we will refer to as a *vector*:

Customer Market

$$\begin{matrix} A \\ B \\ C \end{matrix} \begin{bmatrix} 0.5012 \\ 0.1185 \\ 0.3803 \end{bmatrix}$$
$$1.0000$$

The preference vectors for the other decision criteria are computed similarly:

$$\underset{\textit{Income Level}}{\begin{matrix} A \\ B \\ C \end{matrix} \begin{bmatrix} 0.2819 \\ 0.0598 \\ 0.6583 \end{bmatrix}} \quad \underset{\textit{Infrastructure}}{\begin{matrix} A \\ B \\ C \end{matrix} \begin{bmatrix} 0.1790 \\ 0.6850 \\ 0.1360 \end{bmatrix}} \quad \underset{\textit{Transportation}}{\begin{matrix} A \\ B \\ C \end{matrix} \begin{bmatrix} 0.1561 \\ 0.6196 \\ 0.2243 \end{bmatrix}}$$

These four preference vectors for the four criteria are summarized in a single preference matrix, shown in Table 9.3.

TABLE 9.3
Criteria preference matrix

	Criterion			
Site	Market	Income Level	Infrastructure	Transportation
A	0.5012	0.2819	0.1790	0.1561
B	0.1185	0.0598	0.6850	0.6196
C	0.3803	0.6583	0.1360	0.2243

Ranking the Criteria

The next step in AHP is to determine the relative importance, or weight, of the criteria—that is, to rank the criteria from most important to least important. This is accomplished the same way we ranked the sites within each criterion: using pairwise comparisons. The following pairwise comparison matrix for the four criteria in our example was developed by using the preference scale in Table 9.1:

Criterion	Market	Income	Infrastructure	Transportation
Market	1	1/5	3	4
Income	5	1	9	7
Infrastructure	1/3	1/9	1	2
Transportation	1/4	1/7	1/2	1

The normalized matrix converted to decimals, with the row averages for each criterion, is shown in Table 9.4.

TABLE 9.4
Normalized matrix for criteria, with row averages

Criterion	Market	Income	Infrastructure	Transportation	Row Averages
Market	0.1519	0.1375	0.2222	0.2857	0.1993
Income	0.7595	0.6878	0.6667	0.5000	0.6535
Infrastructure	0.0506	0.0764	0.0741	0.1429	0.0860
Transportation	0.0380	0.0983	0.0370	0.0714	0.0612
					1.0000

The preference vector, computed from the normalized matrix by computing the row averages in Table 9.4, is as follows:

$$\begin{array}{r} \textit{Criterion} \\ \begin{array}{r} \text{Market} \\ \text{Income} \\ \text{Infrastructure} \\ \text{Transportation} \end{array} \begin{bmatrix} 0.1993 \\ 0.6535 \\ 0.0860 \\ 0.0612 \end{bmatrix} \end{array}$$

Clearly, income level is the highest-priority criterion, and customer market is second. Infrastructure and transportation appear to be relatively unimportant third- and fourth-ranked priorities in terms of the overall objective of determining the best site for the new shopping mall. The next step in AHP is to combine the preference matrices we developed for the sites for each criterion in Table 9.3 with the preceding preference vector for the four criteria.

Developing an Overall Ranking

Recall that earlier we summarized Southcorp's preferences for each site for each criterion in a preference matrix shown in Table 9.3 and repeated as follows:

$$\textit{Site}\ \begin{array}{c} \\ A \\ B \\ C \end{array} \overset{\textbf{\textit{Criterion}}}{\begin{array}{cccc} \text{Market} & \text{Income} & \text{Infrastructure} & \text{Transportation} \\ \end{array}} \begin{bmatrix} 0.5012 & 0.2819 & 0.1790 & 0.1561 \\ 0.1185 & 0.0598 & 0.6850 & 0.6196 \\ 0.3803 & 0.6583 & 0.1360 & 0.2243 \end{bmatrix}$$

In the previous section, we used pairwise comparisons to develop a preference vector for the four criteria in our example:

$$\textbf{\textit{Criterion}} \quad \begin{array}{r} \text{Market} \\ \text{Income} \\ \text{Infrastructure} \\ \text{Transportation} \end{array} \begin{bmatrix} 0.1993 \\ 0.6535 \\ 0.0860 \\ 0.0612 \end{bmatrix}$$

An overall score for each site is computed by multiplying the values in the criteria preference vector by the preceding criteria matrix and summing the products, as follows:

$$\text{site A score} = 0.1993(0.5012) + 0.6535(0.2819) + 0.0860(0.1790) + 0.0612(0.1561) = 0.3091$$
$$\text{site B score} = 0.1993(0.1185) + 0.6535(0.0598) + 0.0860(0.6850) + 0.0612(0.6196) = 0.1595$$
$$\text{site C score} = 0.1993(0.3803) + 0.6535(0.6583) + 0.0860(0.1360) + 0.0612(0.2243) = 0.5314$$

The three sites, in order of the magnitude of their scores, result in the following AHP ranking:

Site	Score
Charlotte	0.5314
Atlanta	0.3091
Birmingham	0.1595
	1.0000

Based on these scores developed by AHP, Charlotte should be selected as the site for the new shopping mall, with Atlanta ranked second and Birmingham third. To rely on this result to make its site selection decision, Southcorp must have confidence in the judgments it made in the pairwise comparisons, and it must also have confidence in AHP. However, whether the AHP-recommended decision is the one made by Southcorp or not, following this process can help identify and prioritize criteria and enlighten the company about how it makes decisions.

Following is a summary of the mathematical steps used to arrive at the AHP-recommended decision:

1. Develop a pairwise comparison matrix for each decision alternative (site) for each criterion.
2. *Synthesization*:
   a. Sum the values in each column of the pairwise comparison matrices.
   b. Divide each value in each column of the pairwise comparison matrices by the corresponding column sum—these are the *normalized* matrices.
   c. Average the values in each row of the normalized matrices—these are the *preference vectors*.
   d. Combine the vectors of preferences for each criterion (from step 2c) into one preference matrix that shows the preference for each site for each criterion.

3. Develop a pairwise comparison matrix for the criteria.
4. Compute the normalized matrix by dividing each value in each column of the matrix by the corresponding column sum.
5. Develop the preference vector by computing the row averages for the normalized matrix.
6. Compute an overall score for each decision alternative by multiplying the criteria preference vector (from step 5) by the criteria matrix (from step 2d).
7. Rank the decision alternatives, based on the magnitude of their scores computed in step 6.

AHP Consistency

AHP is based primarily on the pairwise comparisons a decision maker uses to establish preferences between decision alternatives for different criteria. The normal procedure in AHP for developing these pairwise comparisons is for an interviewer to elicit verbal preferences from the decision maker, using the preference scale in Table 9.1. However, when a decision maker has to make a lot of comparisons (i.e., three or more), he or she can lose track of previous responses. Because AHP is based on these responses, it is important that they be in some sense valid and especially that the

Management Science Application

Analyzing Advanced-Technology Projects at NASA

Evaluating advanced-technology projects at the National Aeronautics and Space Administration (NASA) is a critical multicriteria decision-making process. During the past two decades, technological advances have increased the number of possible projects that NASA might undertake. At the same time, the government and public have become more demanding about cost accountability. The project evaluation and selection process is very complex, requiring the consideration and analysis of vast amounts of information, concerning such things as safety, systems engineering, cost, reliability, implementation, and project success without a formal, structured decision-making model.

At the Shuttle Project Engineering Office at the Kennedy Space Center, a multicriteria group decision-making model called CROSS (consensus-ranking organizational-support system) was developed for the project evaluation process, based in part on the analytical hierarchy process (AHP).

Prior to the implementation of CROSS, the Shuttle Project Engineering Office used a 15-member working committee to evaluate advanced-technology projects, which are independent requests for engineering changes to the space shuttle program. Each committee member represents a specific department at the Kennedy Space Center. Committee members assigned each an assessment score from 1 to 10, and these scores were averaged to produce an overall score. Projects were then sorted into three categories, based on these scores: superior, borderline, and inferior. This process was considered to be very intuitive and subjective—and thus subject to inconsistency.

CROSS guides the committee through a systematic evaluation process, using AHP to weight the stakeholder departments that will be responsible for implementation of selected projects. The stakeholders then use AHP to identify and weight the criteria that should be used to evaluate the projects. For example, a typical evaluation scenario might include a budget of $6 million, with 10 possible projects (costing a total of $15 million) encompassing six stakeholder departments and 40 criteria. Individual project scores are calculated by using an Excel program. AHP is also used to evaluate the decision makers' consistency. CROSS helped NASA decision makers to think systematically about the project evaluation and selection process and improve the quality of their decisions. It also brought consistency to the decision-making process.

NASA/Jack Pfaller

Source: Based on M. Tavana, "CROSS: A Multicriteria Group-Decision-Making Model for Evaluating and Prioritizing Advanced-Technology Projects at NASA," *Interfaces* 33, no. 3 (May–June 2003): 40–56.

responses be consistent. That is, a preference indicated for one set of pairwise comparisons needs to be consistent with another set of comparisons.

Using our site selection example, suppose for the income level criterion Southcorp indicates that A is "very strongly preferred" to B and that A is "moderately preferred" to C. That's fine, but then suppose Southcorp says that C is "equally preferred" to B for the same criterion. That comparison is not entirely consistent with the previous two pairwise comparisons. To say that A is strongly preferred over B and moderately preferred over C and then turn around and say C is equally preferred to B does not reflect a consistent preference rating for the three sets of pairwise comparisons. A more logical comparison would be that C is preferred to B to some degree. This kind of inconsistency can creep into AHP when the decision maker is asked to make verbal responses for a lot of pairwise comparisons. In general, they are usually not a serious problem; some degree of slight inconsistency is expected. However, a consistency index can be computed that measures the degree of inconsistency in the pairwise comparisons.

A consistency index measures the degree of inconsistency in pairwise comparisons.

To demonstrate how to compute the consistency index (*CI*), we will check the consistency of the pairwise comparisons for the four site selection criteria. This matrix, shown as follows, is multiplied by the preference vector for the criteria:

$$\begin{array}{r} \\ \text{Market} \\ \text{Income} \\ \text{Infrastructure} \\ \text{Transportation} \end{array} \begin{array}{c} \begin{array}{cccc} \textbf{\textit{Market}} & \textbf{\textit{Income}} & \textbf{\textit{Infrastructure}} & \textbf{\textit{Transportation}} \end{array} \\ \begin{bmatrix} 1 & 1/5 & 3 & 4 \\ 5 & 1 & 9 & 7 \\ 1/3 & 1/9 & 1 & 2 \\ 1/4 & 1/7 & 1/2 & 1 \end{bmatrix} \end{array} \times \begin{array}{c} \textbf{\textit{Criteria}} \\ \begin{bmatrix} 0.1993 \\ 0.6535 \\ 0.0860 \\ 0.0612 \end{bmatrix} \end{array}$$

The product of the multiplication of this matrix and vector is computed as follows:

$$\begin{aligned}
(1)(0.1993) + (1/5)(0.6535) + (3)(0.0860) + (4)(0.0612) &= 0.8328 \\
(5)(0.1993) + (1)(0.6535) + (9)(0.0860) + (7)(0.0612) &= 2.8524 \\
(1/3)(0.1993) + (1/9)(0.6535) + (1)(0.0860) + (2)(0.0612) &= 0.3474 \\
(1/4)(0.1993) + (1/7)(0.6535) + (1/2)(0.0860) + (1)(0.0612) &= 0.2473
\end{aligned}$$

Next, we divide each of these values by the corresponding weights from the criteria preference vector:

$$\begin{aligned}
0.8328/0.1993 &= 4.1786 \\
2.8524/0.6535 &= 4.3648 \\
0.3474/0.0860 &= 4.0401 \\
0.2474/0.0612 &= \underline{4.0422} \\
& 16.6257
\end{aligned}$$

If the decision maker, Southcorp, were a perfectly consistent decision maker, then each of these ratios would be exactly 4, the number of items we are comparing—in this case, four criteria. Next, we average these values by summing them and dividing by 4:

$$\frac{16.6257}{4} = 4.1564$$

The consistency index, CI, is computed using the following formula:

$$CI = \frac{4.1564 - n}{n - 1}$$

where

$$n = \text{the number of items being compared}$$
$$4.1564 = \text{the average we computed previously}$$
$$CI = \frac{4.1564 - 4}{3}$$
$$= 0.0521$$

If $CI = 0$, then Southcorp would be a perfectly consistent decision maker. Because Southcorp is not perfectly consistent, the next question is the degree of inconsistency that is acceptable. An acceptable level of consistency is determined by comparing the CI to a *random index, RI*, which is the consistency index of a randomly generated pairwise comparison matrix. The RI has the values shown in Table 9.5, depending on the number of items, n, being compared. In our example, $n = 4$ because we are comparing four criteria.

TABLE 9.5
***RI* values for *n* items being compared**

n	2	3	4	5	6	7	8	9	10
RI	0	0.58	0.90	1.12	1.24	1.32	1.41	1.45	1.51

Management Science Application

Ranking Twentieth-Century Army Generals

A group of 10 military historians participated in a study to rank the greatest U.S. Army generals of the twentieth century using the analytical hierarchy process. An initial set of 21 generals was pared to a group of 7, which included Omar Bradley, Dwight Eisenhower, Douglas MacArthur, George Marshall, George Patton, John Pershing, and Matthew Ridgeway. The group used two main criteria—skills and actions. The skills criteria were subsequently broken into four subcriteria including conceptual skills to handle ideas, process information, and deal with uncertainty; interpersonal skills to communicate and work with others; tactical skills to attain proficiency; and technical skills to assemble and organize equipment, personnel, schedules, budgets, and facilities to accomplish their missions. The actions criteria were broken down into contribution to conflict (i.e., the general's impact on the wartime conflict he conducted and his contributions to the outcome); responsibility for the size and location of his forces and the importance of his decisions; success; and total wartime service. Each general's performance criteria were evaluated according to a set of ratings—superior, very good, good, and poor. The rankings fell into three groups. The top-ranked general was George Marshall followed by Dwight Eisenhower. The second group included Matthew Ridgeway, Douglas MacArthur, and John Pershing, and the final grouping included George Patton and Omar Bradley.

Source: Based on T. Retchless, B. Golden, and E. Wasil, "Ranking U.S. Army Generals of the 20th Century: A Group Decision-Making Application of the Analytical Hierarchy Process," *Interfaces* 37, no. 2 (March–April 2007): 163–75.

The degree of consistency for the pairwise comparisons in the decision criteria matrix is determined by computing the ratio of CI to RI:

$$\frac{CI}{RI} = \frac{0.0521}{0.90} = 0.0580$$

In general, the degree of consistency is satisfactory if $CI/RI < 0.10$, and in this case, it is. If $CI/RI > 0.10$, then there are probably serious inconsistencies, and the AHP results may not be meaningful.

Remember that in this instance, we have evaluated the degree of consistency only for the pairwise comparisons in the decision criteria preference matrix. This does not mean that we have verified the consistency for the entire AHP. We would still have to evaluate the pairwise comparisons for each of the four individual criterion matrices before we could be sure the entire AHP for this problem was consistent.

AHP with Excel Spreadsheets

The various computational steps involved in AHP can be accomplished with Excel spreadsheets. Exhibit 9.12 shows a spreadsheet formatted for our Southcorp site selection example. The spreadsheet includes the pairwise comparison matrix for our customer market criterion. Notice that the pairwise comparisons have been entered as fractions. We can enter fractions in Excel by dragging the cursor over the object cells, in this case **B5:D7**, and then clicking on "Format" at the top of the screen. When the "Format" menu comes down, we click on "Cells" and then select "Fraction" from the menu.

EXHIBIT 9.12

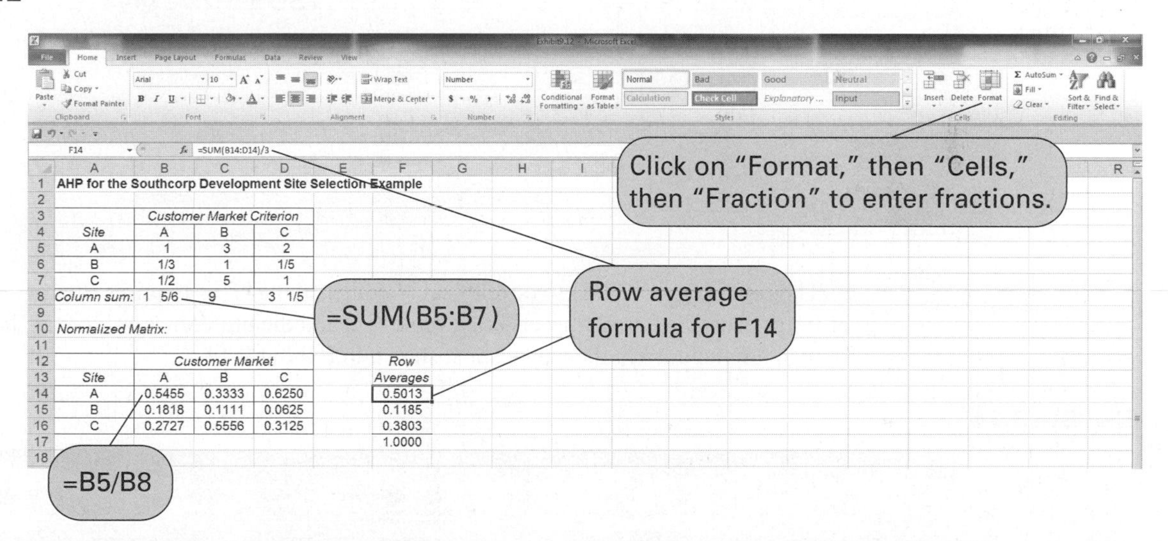

The sum for column A in the pairwise comparison matrix is obtained by inserting the formula **=SUM(B5:B7)** in cell B8. The sums for columns B and C can then be obtained by putting the cursor on cell B8, clicking on the right mouse button, and clicking on "Copy." Then we drag the cursor over cells **B8:D8** and press the "Enter" key. This will compute the sum for the other two columns.

The cell values for the normalized matrix at the bottom of the spreadsheet in Exhibit 9.12 can be computed as follows. First, make sure you convert the cell values to decimal numbers. To start, divide the value in cell B5 by the column sum in cell B8 by typing **=B5/B8** in cell B14. This results in the value 0.5455 in cell B14. Next, with the cursor on B14, click the right mouse button and then click on "Copy." Drag the cursor across cells **B14:D14** and press the "Enter" key. This will compute the values for cells C14 and D14. Repeat this process for each row to complete the normalized matrix.

To compute the row averages from the normalized matrix, first type the formula **=SUM(B14:D14)/3** in cell F14. This formula is displayed on the formula bar at the top of the spreadsheet. This will compute the value 0.5013 in cell F14. Next, with the cursor on cell F14, click the right mouse button and copy the formula in cell F14 to F15 and F16 by dragging the cursor across cells **F14:F16**.

Recall that these row averages represent the preference vector for the customer market criterion. To complete this stage in AHP, we would need to compute the preference vectors for the other three criteria for income level, infrastructure, and transportation.

The next step in AHP is to develop the pairwise comparison matrix and normalized matrix for our four criteria. The spreadsheet to accomplish this step is shown in Exhibit 9.13. These two matrices are constructed in the same way as the two similar matrices in Exhibit 9.12.

EXHIBIT 9.13

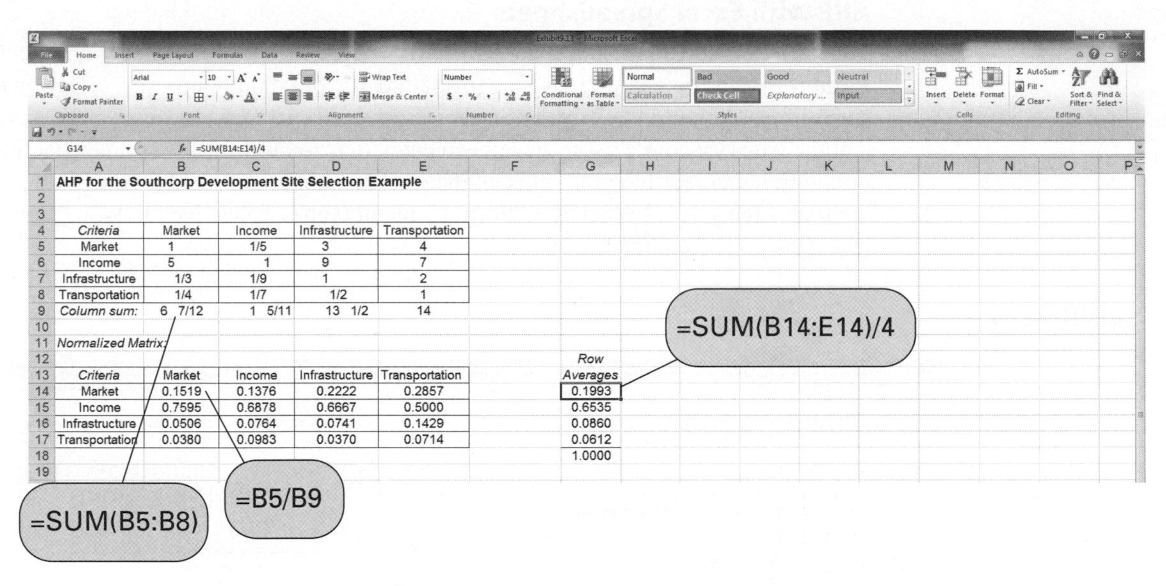

AHP for the Southcorp Development Site Selection Example

Criteria	Market	Income	Infrastructure	Transportation
Market	1	1/5	3	4
Income	5	1	9	7
Infrastructure	1/3	1/9	1	2
Transportation	1/4	1/7	1/2	1
Column sum:	6 7/12	1 5/11	13 1/2	14

Normalized Matrix:

Criteria	Market	Income	Infrastructure	Transportation	Row Averages
Market	0.1519	0.1376	0.2222	0.2857	0.1993
Income	0.7595	0.6878	0.6667	0.5000	0.6535
Infrastructure	0.0506	0.0764	0.0741	0.1429	0.0860
Transportation	0.0380	0.0983	0.0370	0.0714	0.0612
					1.0000

The final step in AHP is to develop an overall ranking. This requires Southcorp's preference matrix for each site for all four criteria (Table 9.3) and the preference vector for the criteria, both of which are included in the spreadsheet shown in Exhibit 9.14.

EXHIBIT 9.14

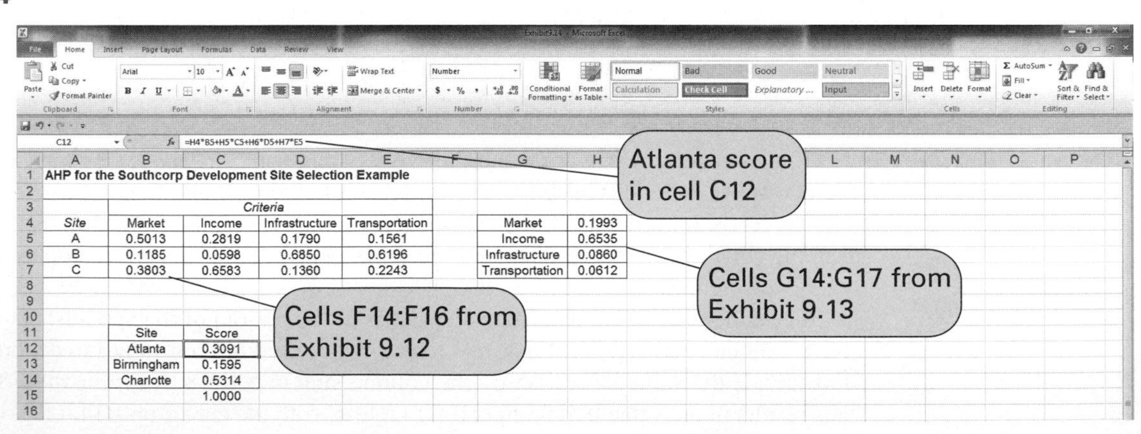

AHP for the Southcorp Development Site Selection Example

Site	Market	Income	Infrastructure	Transportation
	Criteria			
A	0.5013	0.2819	0.1790	0.1561
B	0.1185	0.0598	0.6850	0.6196
C	0.3803	0.6583	0.1360	0.2243

Market	0.1993
Income	0.6535
Infrastructure	0.0860
Transportation	0.0612

Site	Score
Atlanta	0.3091
Birmingham	0.1595
Charlotte	0.5314
	1.0000

All the values in the two matrices in Exhibit 9.14 were entered as input from previous AHP steps (i.e., Exhibits 9.12 and 9.13). The score for Atlanta was computed using the formula embedded in cell C12 and shown on the formula bar. Similar formulas were used for the other two site scores in cells C13 and C14.

Exhibit 9.15 shows the spreadsheet for evaluating the degree of consistency for the pairwise comparisons in the preference matrix for the four criteria. The spreadsheet is broken down into steps that are self-explanatory.

EXHIBIT 9.15

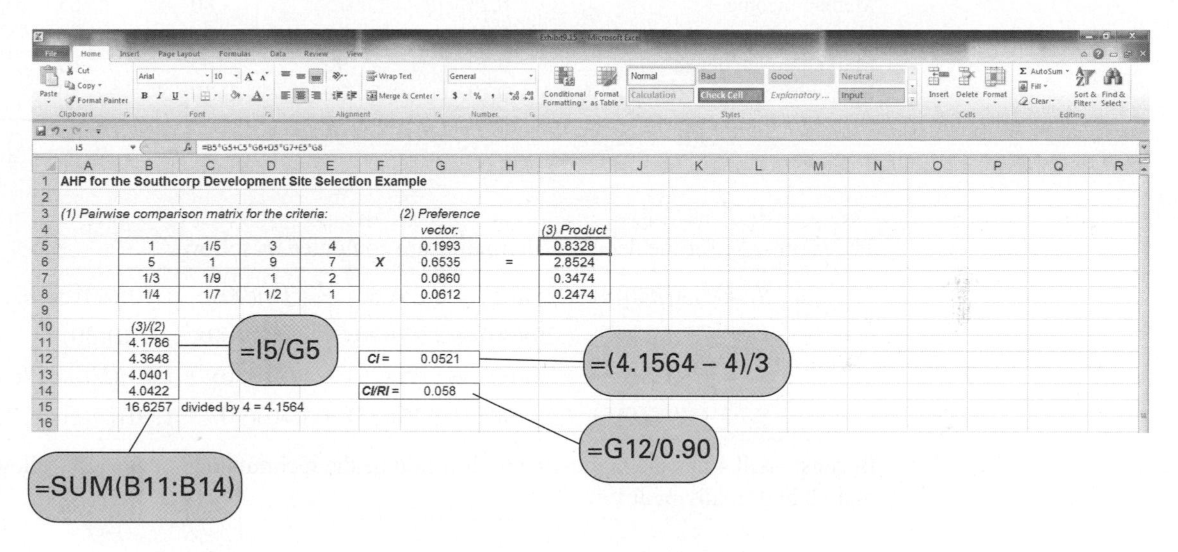

Scoring Models

For selecting among several alternatives according to various criteria, a *scoring model* is a method similar to AHP, but it is mathematically simpler. There are several versions of scoring models. In the scoring model that we will use, the decision criteria are weighted in terms of their relative importance, and each decision alternative is graded in terms of how well it satisfies the criteria, according to the following formula:

$$S_i = \sum g_{ij} w_j$$

where

w_j = a weight between 0 and 1.00 assigned to criterion j, indicating its relative importance, where 1.00 is extremely important and 0 is not important at all. The sum of the total weights equals 1.00.

g_{ij} = a grade between 0 and 100 indicating how well the decision alternative i satisfies criterion j, where 100 indicates extremely high satisfaction and 0 indicates virtually no satisfaction.

S_i = the total "score" for decision alternative i, where the higher the score, the better.

To demonstrate the scoring model, we will use an example. Sweats and Sweaters is a chain of stores specializing in cotton apparel. The company wants to open a new store in one of four malls around the Atlanta metropolitan area. The company has indicated five criteria that are important in its decision about where to locate: proximity of schools and colleges, area median income, mall vehicle traffic flow and parking, quality and size (in terms of number of stores in the mall), and proximity of other malls or shopping areas. The company has weighted each of these criteria in terms of its relative importance in the decision-making process, and it has analyzed each potential mall location and graded them according to each criterion as shown in the following table:

Decision Criteria	Weight (0 to 1.00)	Grades for Alternative (0 to 100)			
		Mall 1	Mall 2	Mall 3	Mall 4
School proximity	0.30	40	60	90	60
Median income	0.25	75	80	65	90
Vehicle traffic	0.25	60	90	79	85
Mall quality and size	0.10	90	100	80	90
Proximity of other shopping	0.10	80	30	50	70

The scores, S_i, for the decision alternative are computed as follows:

$$S_1 = (.30)(40) + (.25)(75) + (.25)(60) + (.10)(90) + (.10)(80) = 62.75$$
$$S_2 = (.30)(60) + (.25)(80) + (.25)(90) + (.10)(100) + (.10)(30) = 73.50$$
$$S_3 = (.30)(90) + (.25)(65) + (.25)(79) + (.10)(80) + (.10)(50) = 76.00$$
$$S_4 = (.30)(60) + (.25)(90) + (.25)(85) + (.10)(90) + (.10)(70) = 77.75$$

Because mall 4 has the highest score, it would be the recommended decision, followed by mall 3, mall 2, and finally mall 1.

Scoring Model with Excel Solution

The steps included in computing the scores of the decisions in the scoring model are relatively simple and straightforward. Exhibit 9.16 shows a spreadsheet set up for the Sweats and Sweaters store location example. The formula that computes the score for mall 1 in cell D10, **=SUMPRODUCT (C5:C9, D5:D9)**, is shown on the formula bar at the top of the spreadsheet. This formula is copied and pasted using the right mouse button into cells E10, F10, and G10.

EXHIBIT 9.16

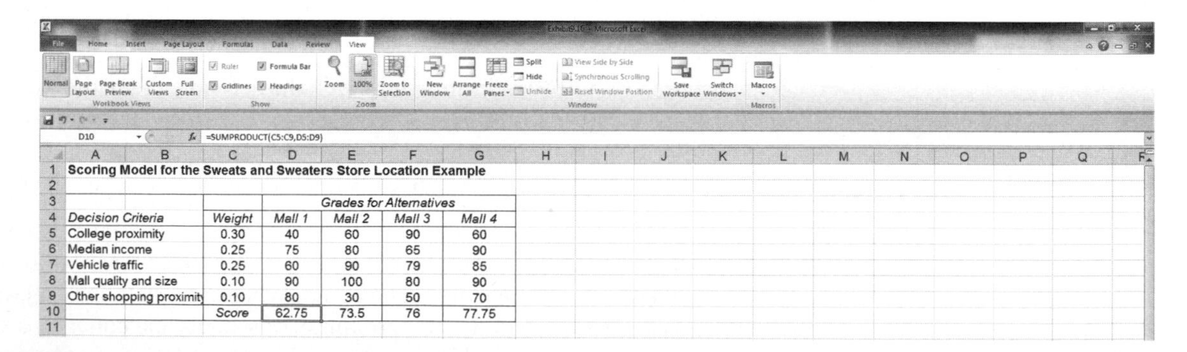

D10 =SUMPRODUCT(C5:C9,D5:D9)

Scoring Model for the Sweats and Sweaters Store Location Example

Decision Criteria	Weight	Grades for Alternatives			
		Mall 1	Mall 2	Mall 3	Mall 4
College proximity	0.30	40	60	90	60
Median income	0.25	75	80	65	90
Vehicle traffic	0.25	60	90	79	85
Mall quality and size	0.10	90	100	80	90
Other shopping proximity	0.10	80	30	50	70
	Score	62.75	73.5	76	77.75

Management Science Application

A Scoring Model for Determining U.S. Army Installation Regions

Jay Mallin/Newscom

Before 2002 the U.S. Army allowed five major commands to manage army installations, which included about 100 major army bases in the United States. However, senior military leaders were concerned that this level of decentralization did not result in standard services and equitable resource allocation among the installations. Furthermore, decentralization unnecessarily increased overhead costs and resulted in redundant support activities. As a result, the Installation Management Agency (IMA), headquartered in Washington, DC, was created to manage all army installations worldwide according to a regional organization plan, with four regions in the United States. The IMA was created to achieve three main functions: centralize command and control of installations; ensure operational capability (i.e., that the installations have the facilities and equipment to train, deploy, sustain, and care for the units, soldiers, and families); and improve and ensure the quality of the resource allocation process for analyzing and prioritizing resource needs. Partly because of political pressures, the United States Military Academy was asked to analyze the four-region management structure to justify its creation and to recommend possibly better alternative regional organizations in terms of number, location, and manpower. Eight regional organizational alternatives were identified for analysis, from a single region up to eight regions. The primary dimensions of the regional alternatives were the IMA functions mentioned above, plus the manpower assigned to a region, the number of regions, the regional boundaries, and the location of the headquarters. A scoring model was developed that measured how well each regional alternative would perform the IMA functions according to a number of measures related to the dimensions of the analysis. Weights were assigned to each measure based on their importance. The scoring model analysis showed that four regions managed installations effectively. Reducing the number of regions below four reduced the functional effectiveness, whereas increasing the number of regions to more than four provided little additional benefit.

Source: Based on T. Trainor, G. Parnell, B. Kwinn, J. Brence, E. Tollefson, and P. Downes, "The U.S. Army Uses Decision Analysis in Designing Its U.S. Installation Regions," *Interfaces* 37, no. 3 (May–June 2007): 253–64.

Summary

This chapter introduced the concept of making decisions when there is more than one objective or criterion to consider. Three specific modeling techniques were presented to solve decision-making problems with multiple criteria: goal programming, the analytical hierarchy process (AHP), and scoring methods. These techniques can be applied to a wide variety of decision-making situations when there are objectives besides just profit or cost. They are often applicable to decision-making problems in public or governmental organizations in which the levels of service or efficiency in carrying out numerous goals are more important than are profit or cost.

The presentation of goal programming completes the coverage of linear programming. One of the implicit assumptions of linear programming has been that all parameters and values in the model were known with certainty. Upcoming chapters will consider techniques that are probabilistic and are used when a situation includes uncertainty.

Example Problem Solutions

As a prelude to the problems section, the following examples demonstrate the formulations and solutions of a goal programming problem, and an AHP problem.

Problem Statement

Rucklehouse Public Relations has contracted to do a survey following an election primary in New Hampshire. The firm must assign interviewers to carry out the survey. The interviews are conducted by telephone and in person. One person can conduct 80 telephone interviews or 40 personal interviews in a day. It costs \$50 per day for a telephone interviewer and \$70 per day for a personal interviewer. The following three goals, which are listed in order of their priority, have been established by the firm to ensure a representative survey:

a. At least 3,000 total interviews should be conducted.
b. An interviewer should conduct only one type of interview each day. The firm wants to maintain its daily budget of \$2,500.
c. At least 1,000 interviews should be by telephone.

Formulate a goal programming model to determine the number of interviewers to hire to satisfy these goals and then solve the model.

Solution

Step 1: Model Formulation

$$\text{minimize } P_1 d_1^-, P_2 d_2^+, P_3 d_3^-$$

subject to

$$\begin{aligned} 80x_1 + 40x_2 + d_1^- - d_1^+ &= 3{,}000 \text{ interviews} \\ 50x_1 + 70x_2 + d_2^- - d_2^+ &= \$2{,}500 \text{ budget} \\ 80x_1 + d_3^- - d_3^+ &= 1{,}000 \text{ telephone interviews} \end{aligned}$$

where

x_1 = number of telephone interviewers
x_2 = number of personal interviewers

Step 2: The QM for Windows Solution

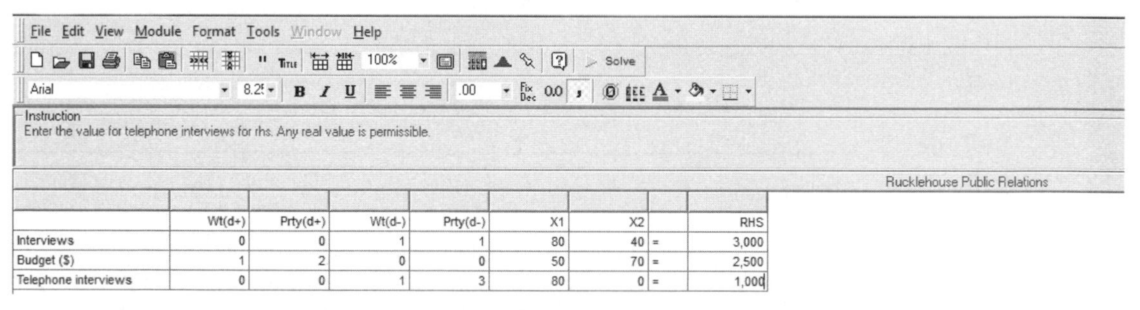

	Wt(d+)	Prty(d+)	Wt(d-)	Prty(d-)	X1	X2		RHS
Interviews	0	0	1	1	80	40	=	3,000
Budget ($)	1	2	0	0	50	70	=	2,500
Telephone interviews	0	0	1	3	80	0	=	1,000

Summary

Rucklehouse Public Relations Solution

Item			
Decision variable analysis	Value		
X1	30.56		
X2	13.89		
Priority analysis	Nonachievement		
Priority 1	0		
Priority 2	0		
Priority 3	0		
Constraint Analysis	RHS	d+ (row i)	d- (row i)
Interviews	3,000	0	0
Budget ($)	2,500	0	0
Telephone interviews	1,000	1,444.44	0

Problem Statement

Grace LeMans wants to purchase a new mountain bike, and she is considering three models—the Xandu Mark III, the Yellow Hawk Z9, and the Zodiak MB5. Grace has identified three criteria for selection on which she will base her decision: purchase price, gear action, and weight/durability. Grace has developed the following pairwise comparison matrices for the three criteria:

	Price		
Bike	*X*	*Y*	*Z*
X	1	3	6
Y	1/3	1	2
Z	1/6	1/2	1

	Gear Action		
Bike	*X*	*Y*	*Z*
X	1	1/3	1/7
Y	3	1	1/4
Z	7	4	1

	Weight/Durability		
Bike	*X*	*Y*	*Z*
X	1	3	1
Y	1/3	1	1/2
Z	1	2	1

Grace has prioritized her decision criteria according to the following pairwise comparisons:

Criteria	Price	Gears	Weight
Price	1	3	5
Gears	1/3	1	2
Weight	1/5	1/2	1

Using AHP, develop an overall ranking of the three bikes Grace is considering.

Solution

Step 1: Develop Normalized Matrices and Preference Vectors for All the Pairwise Comparison Matrices for Criteria

	Price			
Bike	*X*	*Y*	*Z*	**Row Averages**
X	0.6667	0.6667	0.6667	0.6667
Y	0.2222	0.2222	0.2222	0.2222
Z	0.1111	0.1111	0.1111	0.1111
				1.0000

	Gear Action			
Bike	*X*	*Y*	*Z*	Row Averages
X	0.0909	0.0625	0.1026	0.0853
Y	0.2727	0.1875	0.1795	0.2132
Z	0.6364	0.7500	0.7179	0.7014
				1.0000

	Weight/Durability			
Bike	*X*	*Y*	*Z*	Row Averages
X	0.4286	0.5000	0.4000	0.4429
Y	0.1429	0.1667	0.2000	0.1698
Z	0.4286	0.3333	0.4000	0.3873
				1.0000

The preference vectors are summarized in the following matrix:

	Criteria		
Bike	*Price*	*Gears*	*Weight*
X	0.6667	0.0853	0.4429
Y	0.2222	0.2132	0.1698
Z	0.1111	0.7014	0.3873

Step 2: Rank the Criteria

Criteria	Price	Gears	Weight	Row Averages
Price	0.6522	0.6667	0.6250	0.6479
Gears	0.2174	0.2222	0.2500	0.2299
Weight	0.1304	0.1111	0.1250	0.1222
				1.0000

The preference vector for the criteria is

$$\begin{array}{l} \\ \text{Price} \\ \text{Gears} \\ \text{Weight} \end{array} \begin{array}{c} \textbf{\textit{Criteria}} \\ \begin{bmatrix} 0.6479 \\ 0.2299 \\ 0.1222 \end{bmatrix} \end{array}$$

Step 3: Develop an Overall Ranking

$$\begin{array}{c} \\ X \\ Y \\ Z \end{array} \begin{array}{c} \textbf{\textit{Price}} \quad \textbf{\textit{Gears}} \quad \textbf{\textit{Weight}} \\ \begin{bmatrix} 0.6667 & 0.0853 & 0.4429 \\ 0.2222 & 0.2132 & 0.1698 \\ 0.1111 & 0.7014 & 0.3873 \end{bmatrix} \end{array} \times \begin{array}{l} \text{Price} \\ \text{Gears} \\ \text{Weight} \end{array} \begin{bmatrix} 0.6479 \\ 0.2299 \\ 0.1222 \end{bmatrix}$$

$$\text{bike X score} = 0.6667(0.6479) + 0.0853(0.2299) + 0.4429(0.1222) = 0.5057$$

$$\text{bike Y score} = 0.2222(0.6479) + 0.2132(0.2299) + 0.1698(0.1222) = 0.2138$$

$$\text{bike Z score} = 0.1111(0.6479) + 0.7014(0.2299) + 0.3873(0.1222) = 0.2806$$

The ranking of the three bikes, in order of the magnitude of their scores, is

Bike	Score
Xandu	0.5057
Zodiak	0.2806
Yellow Hawk	0.2138
	1.0000

Problems

1. A manufacturing company produces products 1, 2, and 3. The three products have the following resource requirements and produce the following profit:

Product	Labor (hr./unit)	Material (lb./unit)	Profit ($/unit)
1	5	4	$3
2	2	6	5
3	4	3	2

At present, the firm has a daily labor capacity of 240 available hours and a daily supply of 400 pounds of material. The general linear programming formulation for this problem is as follows:

$$\begin{aligned} &\text{maximize } Z = 3x_1 + 5x_2 + 2x_3 \\ &\text{subject to} \\ &\quad 5x_1 + 2x_2 + 4x_3 \le 240 \\ &\quad 4x_1 + 6x_2 + 3x_3 \le 400 \\ &\quad x_1, x_2, x_3 \ge 0 \end{aligned}$$

Management has developed the following set of goals, arranged in order of their importance to the firm:

(1) Because of recent labor relations difficulties, management wants to avoid underutilization of normal production capacity.
(2) Management has established a satisfactory profit level of $500 per day.
(3) Overtime is to be minimized as much as possible.
(4) Management wants to minimize the purchase of additional materials to avoid handling and storage problems.

Formulate a goal programming model to determine the number of each product to produce to best satisfy the goals.

2. In Problem 15 in Chapter 6, Computers Unlimited sells microcomputers and distributes them from three warehouses to four universities. The available supply at the three warehouses, demand at the four universities, and shipping costs are shown in the following table:

	University				
Warehouse	Tech	A&M	State	Central	**Supply**
Richmond	$22	17	30	18	420
Atlanta	15	35	20	25	610
Washington	28	21	16	14	340
Demand	520	250	400	380	

Instead of its original objective of cost minimization, Computers Unlimited has indicated the following goals, arranged in order of their importance:

(1) A&M has been one of its better long-term customers, so Computers Unlimited wants to meet all of A&M's demands.

(2) Because of recent problems with a trucking union, it wants to ship at least 80 units from the Washington warehouse to Central University.

(3) To maintain the best possible relations with all its customers, Computers Unlimited would like to meet no less than 80% of each customer's demand.

(4) It would like to keep total transportation costs to no more than 110% of the $22,470 total cost achieved with the optimal allocation, using the transportation solution method.

(5) Because of dissatisfaction with the trucking firm it uses for the Atlanta-to-State deliveries, it would like to minimize the number of units shipped over this route.

a. Formulate a goal programming model for this problem to determine the number of microcomputers to ship on each route to achieve the goals.

b. Solve this model by using the computer.

3. The Bay City Parks and Recreation Department has received a federal grant of $600,000 to expand its public recreation facilities. City council representatives have demanded four different types of facilities—gymnasiums, athletic fields, tennis courts, and swimming pools. In fact, the demand by various communities in the city has been for 7 gyms, 10 athletic fields, 8 tennis courts, and 12 swimming pools. Each facility costs a certain amount, requires a certain number of acres, and is expected to be used a certain amount, as follows:

Facility	Cost	Required Acres	Expected Usage (people/week)
Gymnasium	$80,000	4	1,500
Athletic field	24,000	8	3,000
Tennis court	15,000	3	500
Swimming pool	40,000	5	1,000

The Parks and Recreation Department has located 50 acres of land for construction (although more land could be located, if necessary). The department has established the following goals, listed in order of their priority:

(1) The department wants to spend the total grant because any amount not spent must be returned to the government.

(2) The department wants the facilities to be used by a total of at least 20,000 people each week.

(3) The department wants to avoid having to secure more than the 50 acres of land already located.

(4) The department would like to meet the demands of the city council for new facilities. However, this goal should be weighted according to the number of people expected to use each facility.

a. Formulate a goal programming model to determine how many of each type of facility should be constructed to best achieve the city's goals.

b. Solve this model by using the computer so that the solution values are integers.

4. A farmer in the Midwest has 1,000 acres of land on which she intends to plant corn, wheat, and soybeans. Each acre of corn costs $100 for preparation, requires 7 worker-days of labor, and yields a profit of $30. An acre of wheat costs $120 to prepare, requires 10 worker-days of labor, and yields $40 profit. An acre of soybeans costs $70 to prepare, requires 8 worker-days, and yields $20 profit. The farmer has taken out a loan of $80,000 for crop preparation and has contracted with a union for 6,000 worker-days of labor. A midwestern granary has agreed to purchase 200 acres of corn, 500 acres of wheat, and 300 acres of soybeans. The farmer has established the following goals, in order of their importance:

(1) To maintain good relations with the union, the labor contract must be honored; that is, the full 6,000 worker-days of labor contracted for must be used.

(2) Preparation costs should not exceed the loan amount so that additional loans will not have to be secured.

(3) The farmer desires a profit of at least $105,000 to remain in good financial condition.

(4) Contracting for excess labor should be avoided.

(5) The farmer would like to use as much of the available acreage as possible.

(6) The farmer would like to meet the sales agreement with the granary. However, the goal should be weighted according to the profit returned by each crop.

a. Formulate a goal programming model to determine the number of acres of each crop the farmer should plant to satisfy the goals in the best possible way.

b. Solve this model by using the computer.

5. The Growall Fertilizer Company produces three types of fertilizer—Supergro, Dynaplant, and Soilsaver. The company has the capacity to produce a maximum of 2,000 tons of fertilizer in a week. It costs $800 to produce a ton of Supergro, $1,500 for Dynaplant, and $500 for Soilsaver. The production process requires 10 hours of labor for a ton of Supergro, 12 hours for a ton of Dynaplant, and 18 hours for a ton of Soilsaver. The company has 800 hours of normal production labor available each week. Each week the company can expect a demand for 800 tons of Supergro, 900 tons of Dynaplant, and 1,100 tons of Soilsaver. The company has established the following goals, in order of their priority:

(1) The company does not want to spend over $20,000 per week on production, if possible.

(2) The company would like to limit overtime to 100 hours per week.

(3) The company wants to meet demand for all three fertilizers; however, it is twice as important to meet the demand for Supergro as it is to meet the demand for Dynaplant, and it is twice as important to meet the demand for Dynaplant as it is to meet the demand for Soilsaver.

(4) It is desirable to avoid producing under capacity, if possible.

(5) Because of union agreements, the company wants to avoid underutilization of labor.

a. Formulate a goal programming model to determine the number of tons of each brand of fertilizer to produce to satisfy the goals.

b. Solve this model by using the computer.

6. The Barrett Textile Mill was checked by inspectors enforcing Occupational Safety and Health Administration (OSHA) codes. The inspectors found violations in four categories: hazardous materials, fire protection, hand-powered tools, and machine guarding. In each category the mill was not in 100% compliance. Each percentage point of increase in the compliance level in each category will reduce the frequency of accidents, decrease the accident cost per worker,

and constitute progress toward satisfying the OSHA compliance level. However, achieving compliance does cost the mill money. The following table shows the benefits (in accident frequency and accident cost per worker) and the costs of a percentage point increase in compliance in each category:

Category	Accident Frequency Reduction (accidents/ 10^5 hr. of exposure)	Accident Cost/ Worker Reduction	Cost/Percentage Point Compliance
1. Hazardous materials	0.18	$1.21	$135
2. Fire protection	0.11	0.48	87
3. Hand-powered tools	0.17	0.54	58
4. Machine guarding	0.21	1.04	160

To achieve 100% compliance in all four categories, the mill will have to increase compliance in hazardous materials by 60 percentage points (i.e., it is now at 40% compliance), in fire protection by 28 percentage points, in hand-powered tools by 35 percentage points, and in machine guarding by 17 percentage points. However, the management of the mill faces a dilemma, in that only $52,000 is available to spend on safety. Any larger expenditure could jeopardize the financial standing of the mill. Thus, management hopes to achieve a level of accident reduction and compliance that is within the company's budget limitation and that will satisfy OSHA authorities enough to temporarily delay punitive action. Therefore, management has established four goals, listed here in order of importance:

(1) Do not exceed the budget constraint of $52,000.
(2) Achieve the percentage increases in compliance necessary to achieve 100% compliance in each category.
(3) Achieve total accident frequency reduction of 20 accidents/10^5 hours of exposure. (This goal denotes management's desire to minimize the frequency of accidents even if 100% compliance cannot be achieved in all categories.)
(4) Reduce the total accident cost per worker by $115.

a. Formulate a goal programming model to determine the percentage points of compliance needed in each category to satisfy the goals.
b. Solve this model by using the computer.

7. Solve the following goal programming model graphically and by using the computer:

$$\begin{aligned} &\text{minimize } P_1d_1^+, P_2d_2^-, P_3d_3^- \\ &\text{subject to} \\ &4x_1 + 2x_2 + d_1^- - d_1^+ = 80 \\ &x_1 + d_2^- - d_2^+ = 30 \\ &x_2 + d_3^- - d_3^+ = 50 \\ &x_j, d_i^-, d_i^+ \geq 0 \end{aligned}$$

8. Solve the following goal programming model graphically and by using the computer:

$$\begin{aligned} &\text{minimize } P_1d_1^- + P_1d_1^+, P_2d_2^-, P_3d_3^-, 3P_4d_2^+ + 5P_4d_3^+ \\ &\text{subject to} \\ &x_1 + x_2 + d_1^- - d_1^+ = 800 \\ &5x_1 + d_2^- - d_2^+ = 2{,}500 \\ &3x_2 + d_3^- - d_3^+ = 1{,}400 \\ &x_j, d_i^-, d_i^+ \geq 0 \end{aligned}$$

9. Solve the following goal programming model by using the computer:

$$\begin{aligned}
&\text{minimize } P_1 d_3^-, P_2 d_2^-, P_3 d_1^+, P_4 d_2^+ \\
&\text{subject to} \\
&4x_1 + 6x_2 + d_1^- - d_1^+ = 48 \\
&2x_1 + x_2 + d_2^- - d_2^+ = 20 \\
&d_2^+ + d_3^- - d_3^+ = 10 \\
&x_2 + d_4^- = 6 \\
&x_j, d_i^-, d_i^+ \geq 0
\end{aligned}$$

10. The Wearever Carpet Company manufactures two brands of carpet—shag and sculptured—in 100-yard lots. It requires 8 hours to produce one lot of shag carpet and 6 hours to produce one lot of sculptured carpet. The company has the following production goals, in prioritized order:
    (1) Do not underutilize production capacity, which is 480 hours.
    (2) Achieve product demand of 40 (100-yard) lots for shag and 50 (100-yard) lots for sculptured carpet. Meeting demand for shag is more important than meeting demand for sculptured, by a ratio of 5 to 2.
    (3) Limit production overtime to 20 hours.
    a. Formulate a goal programming model to determine the amount of shag and sculptured carpet to produce to best meet the company's goals.
    b. Solve this model by using the computer.

11. The East Midvale Textile Company produces denim and brushed-cotton cloth. The average production rate for both types of cloth is 1,000 yards per hour, and the normal weekly production capacity (running two shifts) is 80 hours. The marketing department estimates that the maximum weekly demand is for 60,000 yards of denim and 35,000 yards of brushed cotton. The profit is $3.00 per yard for denim and $2.00 per yard for brushed cotton. The company has established the following four goals, listed in order of importance:
    (1) Eliminate underutilization of production capacity to maintain stable employment levels.
    (2) Limit overtime to 10 hours.
    (3) Meet demand for denim and brushed cotton weighted according to profit for each.
    (4) Minimize overtime as much as possible.
    a. Formulate a goal programming model to determine the number of yards (in 1,000-yard lots) to produce to satisfy the goals.
    b. Solve this model by using the computer.

12. The Oregon Atlantic Company produces two kinds of paper—newsprint and white wrapping paper (butcher paper). It requires 5 minutes to produce a yard of newsprint and 8 minutes to produce a yard of wrapping paper. The company has 4,800 minutes of normal production capacity available each week. The profit is $0.20 for a yard of newsprint and $0.25 for a yard of wrapping paper. The weekly demand is for 500 yards of newsprint and 400 yards of wrapping paper. The company has established the following goals, in order of priority:
    (1) Limit overtime to 480 minutes.
    (2) Achieve a profit of $300 each week.
    (3) Fulfill the demand for the products in order of magnitude of their profits.
    (4) Avoid underutilization of production capacity.
    a. Formulate a goal programming model to determine the number of yards of each type of paper to produce weekly to satisfy the various goals.
    b. Solve the goal programming model by using the computer.

13. A rural clinic hires its staff from nearby cities and towns on a part-time basis. The clinic attempts to have a general practitioner (GP), a nurse, and an internist on duty during at least a portion of each week. The clinic has a weekly budget of $1,200. A GP charges the clinic $40 per hour,

a nurse charges $20 per hour, and an internist charges $150 per hour. The clinic has established the following goals, in order of priority:

(1) A nurse should be available at least 30 hours per week.
(2) The weekly budget of $1,200 should not be exceeded.
(3) A GP or an internist should be available at least 20 hours per week.
(4) An internist should be available at least 6 hours per week.

a. Formulate a goal programming model to determine the number of hours to hire each staff member to satisfy the various goals.
b. Solve the model by using the computer.

14. The Eaststate Manufacturing Company produces four different airplane parts from fabricated sheet metal for several major aircraft companies. The manufacturing process consists of four operations—stamping, assembly, finishing, and packaging. The processing times per unit for each operation and total available hours per year to produce these parts are as follows:

	Part (hr./unit)				
Operation	**1**	**2**	**3**	**4**	**Total Hours/Year**
Stamping	0.06	0.17	0.10	0.14	700
Assembly	0.18	0.20	—	0.14	700
Finishing	0.07	0.20	0.08	0.12	800
Packaging	0.09	0.12	0.07	0.15	600

The sheet metal required for each part, the estimated annual demand, and the profit per part are as follows:

Part	Sheet Metal (ft.^{2})	Estimated Annual Demand	Profits
1	2.6	2,600	$ 90
2	1.4	1,800	100
3	2.5	4,100	80
4	3.2	1,200	120

The company has 15,000 square feet of fabricated metal delivered each month. The company has the following prioritized production goals:

(1) Avoid overtime, which would erode profit levels.
(2) Meet parts demand.
(3) Achieve an annual profit of $700,000.
(4) Avoid ordering more material because a surcharge is required by the supplier for changing the standard monthly order.

a. Formulate a goal programming model to determine the amount of each part to produce to achieve the company's objectives.
b. Solve this model by using the computer.
c. How would the solution be affected if the first two priorities were reversed?

15. Mac's Warehouse is a large discount store that operates 7 days per week. The store needs the following number of full-time employees working each day of the week:

Day	Number of Employees	Day	Number of Employees
Sunday	47	Thursday	34
Monday	22	Friday	43
Tuesday	28	Saturday	53
Wednesday	35		

Each employee must work 5 consecutive days each week and then have 2 days off. For example, any employee who works Sunday through Thursday has Friday and Saturday off. The store currently has a total of 60 employees available to work. Mac's has developed the following set of prioritized goals for employee scheduling:

(1) The store would like to avoid hiring any additional employees.
(2) The most important days for the store to be fully staffed are Saturday and Sunday.
(3) The next most important day to be fully staffed is Friday.
(4) The store would like to be fully staffed the remaining 4 days in the week.

a. Formulate a goal programming model to determine the number of employees who should begin their 5-day workweek each day of the week to achieve the store's objectives.
b. Solve this model by using the computer.

16. Infocomp Systems Lab is a research and development (R&D) company that develops computer systems and software primarily for the medical industry. The lab has proposals from its own researchers for eight new projects. Each of the proposed research projects requires limited resources, and it is not possible to undertake all of them. The following table shows the developmental budget, the number of researchers, and the expected annual sales from each project if successfully developed and implemented:

Project	Developmental Budget ($1,000,000s)	Number of Research Personnel	Expected Annual Sales ($1,000,000s)
1	$0.675	6	$0.82
2	1.050	5	1.75
3	0.725	7	1.60
4	0.430	8	1.90
5	1.240	10	0.93
6	0.890	6	1.70
7	1.620	7	1.30
8	1.200	6	1.80

The lab has developed the following set of prioritized goals for selecting which projects to initiate:

(1) The company would like to remain within a total developmental budget of $5,000,000.
(2) The number of available research personnel is 27, and Infocomp would like to avoid obtaining extra researchers.
(3) The company would like the expected future annual sales from the implemented projects to be at least $6,500,000.
(4) Projects 1, 3, 4, and 6 are considered offensive in that they represent new product initiatives, while projects 2, 5, 7, and 8 are existing product upgrades and thus defensive in nature. The lab would like to select at least two projects from each group.
(5) Projects 2, 3, 5, 6, and 7 are considered the most risky of the projects, and the company would prefer not to select any more than three of these projects.
(6) The lab's owner has indicated that she would like to see projects 5 and 6 initiated if doing so would not interfere with the achievement of any of the more important goals determined by the lab's top management.

a. Formulate a goal programming model to determine which projects Infocomp Systems Lab should select to best achieve its goals.
b. Solve this model by using the computer. (Note that the solution requires 0–1 integer values for the variables in the model.)

17. Texmart is a locally owned "big-box" retail store chain in Texas with 75 stores, primarily located in the Dallas–Fort Worth area. In order to compete with national big-box store chains,

Texmart is planning to undertake several "sustainability" (i.e., "green") projects at its stores. These national chains have been heavily publicizing their sustainability efforts, including the reduction of greenhouse gas (GHG) emissions, which has had a positive effect on their sales. They have also demonstrated that sustainability projects can have a positive impact on cost (especially energy) savings. The projects Texmart is considering include installing solar panels at some or all of its stores, installing small wind turbines, replacing some or all of its 165 trucks with more fuel-efficient hybrid trucks, and implementing waste reduction programs at its stores, including recycling and reducing the use of plastic bags in stores. The costs for these projects (per store or truck), the resulting reduction in GHG emissions, the energy savings, and the annual costs savings are shown in the following table:

	Sustainable Projects				
	Solar Power	Wind Power	Shipping/ Vehicles	Waste/ Recycling	Plastic Bags
Cost ($)	$2,600,000	950,000	38,000	365,000	175,000
GHG reductions (metric tons per year)	17,500	8,600	25	1,700	900
Cost savings ($)	220,000	125,000	26,000	75,000	45,000
Energy savings (kWh)	400,000	150,000	34,000	1,200	55,000
Units	75	75	165	75	75

Texmart has established the following four goals, in order of importance:

(1) It doesn't want to exceed is program budget of $30 million for sustainable projects.

(2) It wants to achieve GHG emission reductions of at least 250,000 metric tons per year.

(3) It wants to achieve annual cost savings of at least $4 million.

(4) It wants to achieve annual energy savings of at least 5 million kilowatt hours (kWh).

a. Formulate a goal programming model to determine the number of each type of project to undertake to satisfy the goals.

b. Solve the model by using the computer.

18. Dampier Associates is a holding company that specializes in buying out small to medium-sized textile companies. The company is currently considering purchasing three companies in the Carolinas: Alton Textiles, Bonham Mills, and Core Textiles. The main criteria the company uses to determine which companies it will purchase are current profitability and growth potential. Dampier moderately prefers growth potential over profitability in making a purchase decision. Dampier's pairwise comparisons for the three target textile companies it is considering are as follows:

	Profitability		
Company	A	B	C
A	1	1/3	7
B	3	1	9
C	1/7	1/9	1

	Growth Potential		
Company	A	B	C
A	1	1/2	1/5
B	2	1	1/3
C	5	3	1

Develop an overall ranking of the three companies for Dampier Associates by using AHP.

19. Check the pairwise comparisons for Dampier Associates in Problem 18 for consistency.

20. Bernard Mee, the head of the department of management science at Tech, is evaluating faculty for raises at the end of the academic year. He is considering three faculty members for raises: John Abbott, Megan Bates, and Debbie Cook. Faculty evaluations are based on three criteria—teaching, research, and service. Professor Mee's pairwise comparisons for each of the three faculty members for each criterion and his pairwise comparison matrix for the three criteria are as follows:

	Teaching		
Faculty Member	A	B	C
A	1	2	1/3
B	1/2	1	1/5
C	3	5	1

	Research		
Faculty Member	A	B	C
A	1	3	1/2
B	1/3	1	1
C	2	1	1

	Service		
Faculty Member	A	B	C
A	1	3	6
B	1/3	1	2
C	1/6	1/2	1

Criterion	Teaching	Research	Service
Teaching	1	3	5
Research	1/3	1	2
Service	1/5	1/2	1

Determine an overall ranking of the three faculty members by using AHP.

21. Check the pairwise comparisons for the criteria in Problem 20 for consistency and indicate whether the level of consistency is acceptable.

22. Megan Moppett is a sales representative for Technical Software Systems (TSS), and she receives a commission for every new system installation she sells to a client. Her earnings during the past few years have been very high, and she wants to invest in a mutual fund. She is considering three funds: the Temple Global Fund, the Alliance Blue Chip Fund, and the Madison Bond Fund. She has three criteria for selection—potential return (based on historical trends and forecasts), risk, and the

fund's load factor. Megan's pairwise comparisons for the funds for each of their criteria and her pairwise comparison of the three criteria are as follows:

	Return		
Fund	**Global**	**Blue Chip**	**Bond**
Global	1	1/4	2
Blue Chip	4	1	6
Bond	1/2	1/6	1

	Risk		
Fund	**Global**	**Blue Chip**	**Bond**
Global	1	2	1/3
Blue Chip	1/2	1	1/5
Bond	3	5	1

	Load		
Fund	**Global**	**Blue Chip**	**Bond**
Global	1	1	1/3
Blue Chip	1	1	1/3
Bond	3	3	1

Criterion	Return	Risk	Load
Return	1	3	5
Risk	1/3	1	2
Load	1/5	1/2	1

Determine the fund in which Megan should invest.

23. In Problem 22, if Megan Moppett has $85,000 to invest and she wants to diversify by investing in all three funds, how much should she invest in each?

24. Alex Wall is shopping for a new four-wheel-drive utility vehicle and has identified three models from which she will choose—an Explorer, a Trooper, and a Passport. She will make her selection based on *Consumer Digest* ratings, price, and each vehicle's appearance. Following are Alex's pairwise comparisons for the vehicles for each of her criteria and her criteria preferences:

	Consumer Digest Rating		
Vehicle	**Explorer**	**Trooper**	**Passport**
Explorer	1	4	3
Trooper	1/4	1	1/2
Passport	1/3	2	1

	Price		
Vehicle	**Explorer**	**Trooper**	**Passport**
Explorer	1	1/4	1/6
Trooper	4	1	2
Passport	6	1/2	1

	Appearance		
Vehicle	**Explorer**	**Trooper**	**Passport**
Explorer	1	4	3
Trooper	1/4	1	1/2
Passport	1/3	2	1

Criterion	*Consumer Digest* Rating	Price	Appearance
Consumer Digest Rating	1	2	4
Price	1/2	1	3
Appearance	1/4	1/3	1

Using AHP, determine which vehicle Alex should purchase.

25. Station WRCH in Richmond, Virginia, is interviewing candidates for the job of news anchor on its 6:00 P.M. *Eyewitness News* show. There are three final candidates for the job—June Pawlie, Kellie Cooric, and Tim Brokenaw. The criteria the station manager will use to make the selection are camera appearance, intelligence, and speaking ability (or speech). The station manager's pairwise comparisons for the job for each of these candidates and for the three criteria are as follows:

	Appearance		
Anchor	**Pawlie**	**Cooric**	**Brokenaw**
Pawlie	1	2	7
Cooric	1/2	1	5
Brokenaw	1/7	1/5	1

	Intelligence		
Anchor	**Pawlie**	**Cooric**	**Brokenaw**
Pawlie	1	1/3	1/4
Cooric	3	1	1/2
Brokenaw	4	2	1

	Speech		
Anchor	**Pawlie**	**Cooric**	**Brokenaw**
Pawlie	1	1/3	2
Cooric	3	1	6
Brokenaw	1/2	1/6	1

Criterion	Appearance	Intelligence	Speech
Appearance	1	8	3
Intelligence	1/8	1	1/5
Speech	1/3	5	1

Using AHP, determine which candidate the station manager should hire as news anchor.

26. Carol Latta is visiting hotels in Los Angeles to decide where to hold a convention for a national organization of college business school teachers she represents. There are three hotels from which to choose—the Cheraton, the Milton, and the Harriott. The criteria she is to use to make her selection are ambiance, location (based on safety and walking distance to attractions and restaurants), and cost to the organization. Following are the pairwise comparisons she has developed that indicate her preference for each hotel for each criterion and her pairwise comparisons for the criteria:

	Ambiance		
Hotel	Cheraton	Milton	Harriott
Cheraton	1	1/2	1/5
Milton	2	1	1/3
Harriott	5	3	1

	Location		
Hotel	Cheraton	Milton	Harriott
Cheraton	1	5	3
Milton	1/5	1	1/4
Harriott	1/3	4	1

	Cost		
Hotel	Cheraton	Milton	Harriott
Cheraton	1	2	5
Milton	1/2	1	2
Harriott	1/5	1/2	1

Criterion	Ambiance	Location	Cost
Ambiance	1	2	4
Location	1/2	1	3
Cost	1/4	1/3	1

Develop an overall ranking of the three hotels, using AHP, to help Carol Latta decide where to hold the meeting.

27. Aaron Zeitel is a high school senior deciding which college to attend in the fall. He has narrowed his choices to three liberal arts schools: Arrington, Barton, and Claiborne. His criteria for selection are the school's academic reputation, location (and especially proximity to his home), the cost of tuition and room and board, and the social and cultural opportunities available. Following

are Aaron's pairwise comparisons of the schools for each of the four criteria and his pairwise comparisons for the criteria:

	Academic		
College	A	B	C
A	1	1/2	3
B	2	1	4
C	1/3	1/4	1

	Location		
College	A	B	C
A	1	1	3
B	1	1	5
C	1/3	1/5	1

	Cost		
College	A	B	C
A	1	1/2	1/4
B	2	1	1/2
C	4	2	1

	Social		
College	A	B	C
A	1	3	6
B	1/3	1	3
C	1/6	1/3	1

Criterion	Academic	Location	Cost	Social
Academic	1	4	1/2	3
Location	1/4	1	1/5	1/2
Cost	2	5	1	3
Social	1/3	2	1/3	1

Using AHP, determine which college Aaron should select and check the consistency of the pairwise comparison matrix for the criteria.

28. Whitney Eggleston operates a computerized dating service for students at Tech. She uses AHP to help match her clients. Whitney is attempting to match Chris with either Robin, Terry, or Kelly. She evaluates her clients according to three criteria—physical attractiveness, intelligence, and personality, and she had Chris do pairwise comparisons on this set of criteria, as follows:

Criterion	Attractiveness	Intelligence	Personality
Attractiveness	1	1	3
Intelligence	1	1	2
Personality	1/3	1/2	1

Whitney herself did the pairwise comparisons for Robin, Terry, and Kelly, based on their data sheets and a personal interview with each:

	Attractiveness		
Client	Robin	Terry	Kelly
Robin	1	3	5
Terry	1/3	1	2
Kelly	1/5	1/2	1

	Intelligence		
Client	Robin	Terry	Kelly
Robin	1	2	1/2
Terry	1/2	1	1/4
Kelly	2	4	1

	Personality		
Client	Robin	Terry	Kelly
Robin	1	2	1/3
Terry	1/2	1	1/2
Kelly	3	2	1

Who is the best match for Chris, according to Whitney's AHP analysis?

29. Rockingham Systems is considering three R&D projects it has identified as A, B, and C. Rockingham is not sure it will undertake all three projects, so it wants to rank them in terms of preferability. Rockingham will use three criteria to rank the projects—profit potential, probability of success, and cost. Following are Rockingham's pairwise comparisons for the projects for each of the three criteria and for the criteria:

	Profit		
Project	A	B	C
A	1	4	6
B	1/4	1	2
C	1/6	1/2	1

	P(Success)		
Project	A	B	C
A	1	2	1/3
B	1/2	1	1/6
C	3	6	1

Project	Cost		
	A	B	C
A	1	1/3	1/4
B	3	1	1/2
C	4	2	1

Criteria	Profit	*P*(Success)	Cost
Profit	1	2	6
P(Success)	1/2	1	4
Cost	1/6	1/4	1

Rank the projects for Rockingham Systems by using AHP.

30. Professor Rakes is selecting a new graduate assistant from a pool of second-year MBA students. He will make his selection based on the student's grade point average (GPA) to date, the overall GMAT (entrance exam) score, and the undergraduate degree discipline. Professor Rakes has developed the following pairwise comparisons for these criteria:

Criteria	GPA	GMAT	Degree
GPA	1	5	1/3
GMAT	1/5	1	1/6
Degree	3	6	1

The files of the three students, Adrian, Bon, and Corey, provide the following data:

Student	GPA	GMAT	Degree
Adrian	3.6	560	Business
Bon	3.8	670	English
Corey	3.0	610	Engineering

Use AHP to help Professor Rakes select a graduate assistant.

31. The Bay City Parks and Recreation Department is considering building several new facilities, including a gym, an athletic field, a tennis pavilion, and a pool. It will base its decision on which facilities to build depending on projected usage (from surveys) and construction and operating costs. The department strongly prefers usage to cost as a criterion for selection. Following are the department's pairwise comparisons, reflecting its preferences for each facility for the two criteria:

Facility	Usage			
	Gym	Field	Tennis	Pool
Gym	1	1/3	3	2
Field	3	1	5	4
Tennis	1/3	1/5	1	1/3
Pool	1/2	1/4	3	1

	Cost			
Facility	Gym	Field	Tennis	Pool
Gym	1	1/4	1/2	3
Field	4	1	3	7
Tennis	2	1/3	1	4
Pool	1/3	1/7	1/4	1

Rank the facilities, using AHP, and check the pairwise comparisons for consistency.

32. Students at a college in Ohio are planning a spring break vacation to one of three locations: Myrtle Beach (MB), Daytona Beach (DB), or Ft. Lauderdale (FL). They are to base their decision on three criteria—weather, cost, and potential fun (based on an Internet survey of friends and acquaintances at other colleges). The students have developed the following pairwise comparisons for the locations for each criterion and for the three criteria:

	Weather		
Location	MB	DB	FL
MB	1	1/3	1/3
DB	3	1	1
FL	3	1	1

	Cost		
Location	MB	DB	FL
MB	1	3	5
DB	1/3	1	2
FL	1/5	1/2	1

	Fun		
Location	MB	DB	FL
MB	1	1/2	5
DB	2	1	3
FL	1/5	1/3	1

Criteria	Weather	Cost	Fun
Weather	1	4	1/4
Cost	1/4	1	1/5
Fun	4	5	1

If the students use AHP to help make a decision, which location will they select for their spring break vacation?

33. Check the pairwise comparisons in Problem 32 for consistency.

34. The management science and information technology majors at Tech select one of two available options within the major—decision support systems (DSS) or operations management (OM).

Student advisers use AHP with the students to determine which option they should select. The criteria used by the advisers are student aptitude and interests, faculty who teach in the options, and potential job availability. An adviser has helped one major develop the following pairwise comparisons:

	Aptitude	
Option	**DSS**	**OM**
DSS	1	3
OM	1/3	1

	Faculty	
Option	**DSS**	**OM**
DSS	1	1/5
OM	5	1

	Jobs	
Option	**DSS**	**OM**
DSS	1	4
OM	1/4	1

Criteria	Aptitude	Faculty	Jobs
Aptitude	1	1/2	1/4
Faculty	2	1	1/3
Jobs	4	3	1

Which option should the student select?

35. The Des Moines Twisters, a women's professional basketball team, needs a new point guard and is considering signing one of three college players—Keisha Jones, Natasha Franklin, and Kathleen Taylor. The team will base its decision of which player to sign on four criteria: the player's shooting ability, rebounding ability, ball handling skills, and attitude. Following is a survey of the three players' college game statistics for their senior year:

	Per-Game Statistics						
Player	**Points**	**Field Goal (%)**	**Free Throw (%)**	**Rebounds**	**Assists**	**Turnovers**	**Minutes**
Keisha Jones	18.6	47.5	75.7	10.3	4.1	3.2	33.9
Natasha Franklin	22.9	38.5	64.2	12.4	3.6	1.8	32.5
Kathleen Taylor	16.5	62.1	88.3	7.6	6.7	0.6	37.4

Although the coach can base some of her pairwise comparisons on these statistics, it is more difficult to assess attitude. She has interviewed each player, and she moderately prefers Keisha Jones's attitude to Kathleen Taylor's and strongly prefers Keisha Jones's attitude to Natasha Franklin's.

Assume the coach's role and develop pairwise comparisons for all four of the criteria and between the criteria. Use AHP to determine which player you would select.

36. The town of Blacksburg needs a larger modern middle school. The current middle school is in the center of town and is over 40 years old. There are two proposals for a new school—renovate and expand the current facility and keep it in town or build a new school on the outskirts of town. Different groups in town have strong feelings about the proposals. Some citizens want to retain the sense of tradition of the old school and like it in town, where it helps engender a sense of community. Others view the old school as antiquated and beyond saving and believe keeping the school in town near bars, traffic, and college students to be negative. The county school board will make the final decision. The school board has asked several management science professors from the local college to use AHP to help evaluate the proposals. The school board has identified four groups from which it wants to solicit input regarding their preferences: the middle school PTA, the middle school teachers, current and former middle school students, and the town council. The management science professors have developed the following pairwise comparison matrices for each of these groups:

	PTA	
Proposal	**Renovate**	**New**
Renovate	1	1/3
New	3	1

	Teachers	
Proposal	**Renovate**	**New**
Renovate	1	1/9
New	9	1

	Students	
Proposal	**Renovate**	**New**
Renovate	1	2
New	1/2	1

	Town Council	
Proposal	**Renovate**	**New**
Renovate	1	5
New	1/5	1

The school board's pairwise comparison of the four groups from which it is soliciting preferences is as follows:

Group	PTA	Teachers	Students	Town Council
PTA	1	5	2	1/4
Teachers	1/5	1	1/4	1/7
Students	1/2	4	1	1/5
Town council	4	7	5	1

a. Based on the AHP analysis conducted by the management science professors, which proposal should the school board select?
b. Check the school board's pairwise comparison of the criteria for consistency.

37. Given the following pairwise comparisons, indicate your preferences according to the preference scale in Table 9.1.
a. Steak to chicken
b. Hot dogs to hamburgers
c. Republicans to Democrats
d. Soccer to football
e. College basketball to professional basketball
f. Management science to management
g. Domino's to Pizza Hut
h. McDonald's to Wendy's
i. Ford to Honda
j. Dickens to Faulkner
k. Beatles to Beethoven
l. New York to Los Angeles
m. Chicago to Atlanta
n. American League to National League

38. Federated Health Care has contracted to be Tech's primary health care provider for faculty and staff. There are three major hospitals in the area (within 35 miles)—County, Memorial, and General—that have full-service emergency rooms. Federated wants to designate one of the hospitals as its primary care emergency room for its members. The company's criteria for selection are quality of medical care, as determined by a patient survey; distance to the emergency room by the majority of its members; speed of medical attention at the emergency room; and cost. Following are the pairwise comparisons of the emergency rooms for each of the four criteria and the pairwise comparisons for the criteria:

	Medical Care		
Hospital	County	Memorial	General
County	1	1/6	1/3
Memorial	6	1	3
General	3	1/3	1

	Distance		
Hospital	County	Memorial	General
County	1	7	4
Memorial	1/7	1	2
General	1/4	1/2	1

	Speed of Attention		
Hospital	County	Memorial	General
County	1	1/2	3
Memorial	2	1	4
General	1/3	1/4	1

Hospital	Cost		
	County	Memorial	General
County	1	6	4
Memorial	1/6	1	1/2
General	1/4	2	1

Criterion	Medical Care	Distance	Speed of Attention	Cost
Medical care	1	8	6	3
Distance	1/8	1	1/2	1/6
Speed of attention	1/6	2	1	1/4
Cost	1/3	6	4	1

Using AHP, determine which hospital emergency room Federated Health Care should designate as its primary care provider.

39. The department of management science at Tech offers four sections of introductory quantitative methods in the fall semester, each taught by a different teacher. A group of students have developed the following criteria to rank the sections: time and day, grading history of the instructor, classroom atmosphere (i.e., relaxed or formal), amount of homework, and the teacher's sense of humor. Following are the students' preferences for these criteria:

Criterion	Time/Day	Grading	Atmosphere	Homework	Jokes
Time/day	1	2	7	3	8
Grading	1/2	1	6	3	9
Atmosphere	1/7	1/6	1	1/4	2
Homework	1/3	1/3	4	1	7
Jokes	1/8	1/9	1/2	1/7	1

Following are the students' pairwise comparisons of the sections for the criteria:

Section	Time/Day			
	1	2	3	4
1	1	3	5	7
2	1/3	1	2	5
3	1/5	1/2	1	3
4	1/7	1/5	1/3	1

Section	Grading			
	1	2	3	4
1	1	1/7	1/8	2
2	7	1	2	6
3	8	1/2	1	5
4	1/2	1/6	1/5	1

	Atmosphere			
Section	**1**	**2**	**3**	**4**
1	1	6	3	3
2	1/6	1	3	2
3	1/3	1/3	1	1
4	1/3	1/2	1	1

	Homework			
Section	**1**	**2**	**3**	**4**
1	1	1/4	1/8	1/2
2	4	1	1/3	4
3	8	3	1	5
4	2	1/4	1/5	1

	Jokes			
Section	**1**	**2**	**3**	**4**
1	1	1/5	3	4
2	5	1	6	7
3	1/3	1/6	1	3
4	1/4	1/7	1/3	1

a. Using AHP, rank the courses for the students.
b. Using this framework and these criteria, but with your own preferences for a course with multiple sections that you are familiar with at your school, develop a ranking of the sections.

40. A faculty committee in the department of management science at Tech is evaluating three new textbooks for its introductory management science course, which all business students are required to take. The texts, identified by the authors, are Adams/Jones, Barnes, and Cook/Smith. The committee's selection criteria are topical coverage, readability, cost, and the available supplements. Following are the committee's pairwise comparisons of the three textbooks for each of the four criteria and the committee's pairwise comparisons for the criteria:

	Coverage		
Textbook	A	B	C
A	1	1/5	1/4
B	5	1	3
C	4	1/3	1

	Readability		
Textbook	A	B	C
A	1	2	3
B	1/2	1	3
C	1/3	1/3	1

Textbook	Cost		
	A	B	C
A	1	1/2	1/5
B	2	1	1/3
C	5	3	1

Textbook	Supplements		
	A	B	C
A	1	4	7
B	1/4	1	3
C	1/7	1/3	1

Criterion	Coverage	Readability	Cost	Supplements
Coverage	1	1/2	1/4	2
Readability	2	1	1/3	5
Cost	4	3	1	3
Supplements	1/2	1/5	1/3	1

Using AHP, determine which textbook the committee should select. Check the consistency of the pairwise comparison matrix for the criteria.

41. On the day of the professional football draft, the owner, general manager, and coaching staff of the New York Gladiators are attempting to decide which of the available players they should select as their pick approaches. They have three players on their big board whom they would like: Al Stonecrusher, a defensive lineman, Bruce Kowslaski, a tight end, and Charlie Speedman, a running back. They evaluate players according to four criteria—the player's projected salary demands, speed, and size (i.e., height and weight), and the team's position needs. Following are the group's pairwise comparisons for the players for each of the four criteria and the pairwise comparisons for the criteria:

Player	Salary		
	A	B	C
A	1	5	3
B	1/5	1	1/2
C	1/3	2	1

Player	Speed		
	A	B	C
A	1	1/5	3
B	5	1	6
C	1/3	1/6	1

	Position		
Player	A	B	C
A	1	1/2	1/5
B	2	1	1/3
C	5	3	1

	Size		
Player	A	B	C
A	1	4	2
B	1/4	1	1/3
C	1/2	3	1

Criterion	**Salary**	**Speed**	**Position**	**Size**
Salary	1	1/5	1/3	1/2
Speed	5	1	3	2
Position	3	1/3	1	2
Size	2	1/2	1/2	1

Using AHP, determine which player the Gladiators should select. Check the consistency of the pairwise comparison matrix for the criteria.

42. The four most famous generals during the American Civil War were Ulysses S. Grant, William Tecumseh Sherman, Robert E. Lee, and Stonewall Jackson. Historians have always debated which was the greatest general from among these four. Following are the pairwise comparisons for five criteria that might be used to evaluate the generals: leadership, including interpersonal skills and the ability to communicate with others and delegate responsibility; tactical and technical skills in planning and directing their armies, including the efficient use of their resources; their direct impact on the battles they were involved in; their decision-making skills both between and during battles; and their overall success in the war.

	Interpersonal Skills/Leadership			
General	Grant	Lee	Jackson	Sherman
Grant	1	1/4	1/3	2
Lee	4	1	2	8
Jackson	3	1/2	1	5
Sherman	1/2	1/8	1/5	1

	Tactical Skills			
General	Grant	Lee	Jackson	Sherman
Grant	1	1/4	1/5	2
Lee	4	1	1/3	6
Jackson	5	3	1	8
Sherman	1/2	1/6	1/8	1

	Impact on Battle			
General	**Grant**	**Lee**	**Jackson**	**Sherman**
Grant	1	1/4	1/7	2
Lee	4	1	1/3	3
Jackson	7	3	1	9
Sherman	1/2	1/3	1/9	1

	Decision Making			
General	**Grant**	**Lee**	**Jackson**	**Sherman**
Grant	1	1/2	3	3
Lee	2	1	3	5
Jackson	1/3	1/3	1	3
Sherman	1/3	1/5	1/3	1

	Overall Success			
General	**Grant**	**Lee**	**Jackson**	**Sherman**
Grant	1	5	7	6
Lee	1/5	1	2	3
Jackson	1/7	1/2	1	2
Sherman	1/6	1/3	1/2	1

Criterion	**Interpersonal Skills/Leadership**	**Tactical Skills**	**Impact on Battles**	**Decision Making**	**Overall Success**
Interpersonal skills/leadership	1	1/4	1/3	1/5	1/9
Tactical skills	4	1	1/2	1/3	1/6
Impact on battles	3	2	1	1/2	1/5
Decision making	5	3	2	1	1/4
Overall success	9	6	5	4	1

Using AHP, rank these four Civil War generals.

43. For Problem 42, develop your own pairwise comparisons for and among the criteria using your own knowledge of Civil War history and any Internet research you might perform, and develop your own AHP ranking of the four generals.

44. The owners of the Blitz professional soccer team are disappointed in their team's home attendance, so they are considering moving to one of four cities: Atlanta, Birmingham, Charlotte, or Durham. The criteria the owners are using to make their decision are population in the 15–40 age group, soccer interest (measured by youth, high school, and college soccer game participation and attendance), media market, entertainment and sports competition,

and the proposed playing facility. Following are the owners' pairwise comparisons for the four potential cities for each of the five criteria, and the owners' pairwise comparisons for the criteria:

	Population			
City	A	B	C	D
A	1	5	4	9
B	1/5	1	1/3	4
C	1/4	3	1	5
D	1/9	1/4	1/5	1

	Soccer Interest			
City	A	B	C	D
A	1	3	1	1/4
B	1/3	1	1/3	1/6
C	1	3	1	1/2
D	4	6	2	1

	Media Market			
City	A	B	C	D
A	1	5	3	7
B	1/5	1	1/4	2
C	1/3	4	1	3
D	1/7	1/2	1/3	1

	Competition			
City	A	B	C	D
A	1	2	1/3	1/7
B	1/2	1	1/4	1/6
C	3	4	1	1/4
D	7	6	4	1

	Playing Facility			
City	A	B	C	D
A	1	1/2	1/4	1/7
B	2	1	1/3	1/5
C	4	3	1	1/4
D	7	5	4	1

Criterion	Population	Soccer Interest	Media Market	Competition	Playing Facility
Population	1	1/5	1/2	2	1/4
Soccer interest	5	1	3	6	4
Media market	2	1/3	1	3	1/2
Competition	1/2	1/6	1/3	1	1/2
Playing facility	4	1/4	2	2	1

Using AHP determine which city the owners should select to relocate to.

45. Check the pairwise comparisons for the five criteria in Problem 44 for consistency.

46. Texmart is a locally owned "big-box" retail store chain in Texas, with stores primarily located in the Dallas–Fort Worth area. In order to compete with national big-box store chains, it is planning to undertake several "sustainability" (i.e., "green") projects at its stores. These national chains have been heavily publicizing their sustainability efforts, including the reduction of greenhouse gas (GHG) emissions (in metric tons per year) and energy savings (in kilowatt hours), which have had a positive effect on their sales. They have also demonstrated that sustainability projects can have a positive impact on cost (especially energy) savings. The projects Texmart is considering include installing solar panels at its stores, installing small wind turbines at its stores, replacing their trucks with more fuel-efficient hybrid trucks, and implementing waste reduction programs at its stores, including recycling and reducing the use of plastic bags. The criteria Texmart wants to use to evaluate the projects include the media and public response to the projects they undertake, the cost of the projects, the amount of GHG emissions reduction, and the amount of energy savings. Following are the pairwise comparisons for the projects for these four criteria, and the pairwise comparisons for the criteria:

	Media/Public Response			
Project	**Solar Power**	**Wind Power**	**Shipping/Vehicles**	**Waste/Recycling**
Solar power	1	3	5	7
Wind power	1/3	1	3	5
Shipping/vehicles	1/5	1/3	1	3
Waste/recycling	1/7	1/5	1/3	1

	Cost ($)			
Project	**Solar Power**	**Wind Power**	**Shipping/Vehicles**	**Waste/Recycling**
Solar power	1	1/3	1/5	1/7
Wind power	3	1	1/2	1/5
Shipping/vehicles	5	2	1	1/3
Waste/recycling	7	5	3	1

	GHG Reductions			
Project	**Solar Power**	**Wind Power**	**Shipping/Vehicles**	**Waste/Recycling**
Solar power	1	3	2	4
Wind power	1/3	1	2	3
Shipping/vehicles	1/2	1/2	1	3
Waste/recycling	1/4	1/3	1/3	1

Project	Energy Savings			
	Solar Power	Wind Power	Shipping/Vehicles	Waste/Recycling
Solar power	1	2	4	8
Wind power	1/2	1	3	5
Shipping/vehicles	1/4	1/3	1	3
Waste/recycling	1/8	1/5	1/3	1

Criterion	Media/Public Response	Cost ($)	GHG Reduction	Energy Savings
Media/public response	1	2	3	5
Cost ($)	1/2	1	2	3
GHG reduction	1/3	1/2	1	2
Energy savings	1/5	1/3	1/2	1

Using AHP, rank the projects for Texmart.

47. Labran Jones has played for the Cleveland professional basketball team for the past eight seasons and has established himself as one of the top players in the league. He has recently become a free agent, meaning he can sign a new contract with Cleveland or with any other team in the league. While he has enjoyed playing for Cleveland, which is near his hometown, the team has never seriously contended for a championship, so Labran is strongly considering moving to one of three other teams that he thinks have more championship potential. Other factors he is considering are salary (although Cleveland is offering him more money than the other teams), the possible media attention and endorsements he might receive by playing in another city, and the city itself and lifestyle where he would be playing. Following are the pairwise comparisons for the four teams for these four criteria and the pairwise comparisons for the criteria:

City	Salary ($)			
	Cleveland	Miami	New York	Chicago
Cleveland	1	5	4	3
Miami	1/5	1	1/3	1/2
New York	1/4	3	1	2
Chicago	1/3	2	1/2	1

City	Media Exposure/Endorsements			
	Cleveland	Miami	New York	Chicago
Cleveland	1	1	1/4	1/2
Miami	1	1	1/4	1/2
New York	4	4	1	2
Chicago	2	2	1/2	1

City	City/Lifestyle			
	Cleveland	Miami	New York	Chicago
Cleveland	1	1/4	3	2
Miami	4	1	5	4
New York	1/3	1/5	1	1/2
Chicago	1/2	1/4	2	1

City	Championship Potential			
	Cleveland	Miami	New York	Chicago
Cleveland	1	1/7	1/2	1/4
Miami	7	1	5	3
New York	2	1/5	1	1/2
Chicago	4	1/3	2	1

Salary	Criterion			
	Salary ($)	Media Exposure	City/Lifestyle	Championship Potential
Salary ($)	1	1/4	1/3	1/5
Media exposure	4	1	1/2	1/3
City/lifestyle	3	2	1	1/2
Championship potential	5	3	2	1

Using AHP, determine which team Labran should select to sign a new contract with.

48. For Problem 47, check the consistency of the pairwise comparison matrices for all four criteria and the criteria.

49. Balston Healthcare operates three hospitals and a number of clinics in its citywide network. It is planning to open a new wellness center and clinic facility that focuses on geriatric clients in one of four suburbs. The following table shows the weighted criteria for each location:

Location Factors	Weight	Scores (0 to 100)			
		Ashcroft	Brainerd	Crabtree	Dowling
Elderly population	0.55	75	80	65	75
Income level	0.15	65	75	90	85
Land availability	0.10	90	70	90	80
Average age	0.10	80	70	80	75
Public transportation	0.05	95	55	75	95
Crime rate	0.05	95	70	85	90

Recommend a site for the new Balston healthcare facility, based on these weighted location factors and scores.

50. The owners of the Blitz professional soccer team in Problem 44 have decided to also use a scoring model, with an expanded set of criteria, to help them decide in which city—Atlanta, Birmingham, Charlotte, or Durham—they should relocate their team. They have graded the possible cities according to the following weighted criteria:

Decision Criteria	Weight	Relocation City			
		Atlanta	Birmingham	Charlotte	Durham
Soccer interest	0.25	70	40	75	90
Entertainment competition	0.18	33	45	60	95
Playing facility	0.12	50	65	70	85
Population (ages 15–40)	0.10	100	70	90	25
Media market	0.07	100	65	95	40
Income level (ages 15–40)	0.05	80	70	80	60
Tax incentives	0.05	20	40	75	60
Airline transportation	0.05	100	70	95	65
Cultural diversity	0.05	100	80	90	75
General sports interest	0.03	75	95	85	65
Local government support	0.03	30	60	75	90
Community support	0.02	20	35	50	75

Develop a scoring model to help the owners decide which city to select to relocate to. Based on these scoring model results and the results of the AHP model in Problem 44, which city would you recommend?

51. Visit one or more local car dealers and select four new models of cars you might like to purchase. Using AHP and your own preferences, rank your selections according to the following criteria: price, style/appearance, reliability/maintenance, engine size, gas mileage, safety, and features/options. (You may need to access some additional references, such as *Consumer Reports*, to facilitate your judgments.) After you have ranked your car selections using AHP, develop a scoring model to perform the same analysis. Then compare the results of the two models and discuss which method you prefer.

52. Arsenal Electronics is to construct a new $1.2 billion semiconductor plant and has selected four small towns in the Midwest as potential sites. The important decision criteria and grades for each town are as follows:

Decision Criterion	Weight	Town			
		Abbeton	Bayside	Cane Creek	Dunnville
Work ethic	0.18	80	90	70	75
Quality of life	0.16	75	85	95	90
Labor laws/unionization	0.12	90	90	60	70
Infrastructure	0.10	60	50	60	70
Education	0.08	80	90	85	95
Labor skill and education	0.07	75	65	70	80
Cost of living	0.06	70	80	85	75
Taxes	0.05	65	70	55	60
Incentive package	0.05	90	95	70	80
Government regulations	0.03	40	50	65	55
Environmental regulations	0.03	65	60	70	80
Transportation	0.03	90	80	95	80
Space for expansion	0.02	90	95	90	90
Urban proximity	0.02	60	90	70	80

Develop a scoring model to determine in which town the plant should be built.

53. The Dynaco Manufacturing Company is to build a new plant to make ring bearings (used in automobiles and trucks). The site selection team is evaluating three sites, and it has graded the important weighted criteria for each as follows:

		Building Site		
Decision Criterion	**Weight**	1	2	3
Labor pool and climate	0.30	80	65	90
Proximity to suppliers	0.20	100	91	75
Wage rates	0.15	60	95	72
Community environment	0.15	75	80	80
Proximity to customers	0.10	65	90	95
Shipping modes	0.05	85	92	65
Air service	0.05	50	65	90

Develop a scoring model to determine which site the selection team should recommend.

54. Exotech Computers manufactures computer components such as chips, circuit boards, motherboards, keyboards, LCD panels, and the like, and sells them around the world. It wants to construct a new warehouse and distribution center in Asia to serve emerging markets there. It has identified sites in Shanghai, Hong Kong, and Singapore and has graded the important weighted decision criteria for each site as follows:

		City		
Decision Criterion	**Weight**	Shanghai	Hong Kong	Singapore
Political stability	0.25	50	60	90
Economic growth	0.18	90	70	75
Port facilities	0.15	60	95	90
Container support	0.10	50	80	90
Land construction cost	0.08	90	20	30
Transportation/distribution	0.08	50	80	70
Duties and tariffs	0.07	70	90	90
Trade regulations	0.05	70	95	95
Airline service	0.02	60	80	70
Area roads	0.02	60	70	80

Recommend a site based on these decision criteria and grades.

55. State University is to construct a new student center and athletic complex that will include a bookstore, a post office, theaters, a market, a minimall, meeting rooms, a swimming pool, and weight and exercise rooms. The university administration has hired a site selection specialist to

identify the best potential sites on campus for the new facility. The site specialist has identified four sites on campus and has graded the important weighted decision criteria for each site as follows:

		Campus Site			
Decision Criterion	**Weight**	South	West A	West B	East
Proximity to housing	0.23	70	90	65	85
Student traffic	0.22	75	80	60	85
Parking availability	0.16	90	60	80	70
Plot size, terrain	0.12	80	70	90	75
Infrastructure	0.10	50	60	40	60
Off-campus accessibility	0.06	90	70	70	70
Proximity to dining facilities	0.05	60	80	70	90
Visitor traffic	0.04	70	80	65	55
Landscape and aesthetics	0.02	50	40	60	70

Which site should the specialist recommend?

56. Following an all-star school soccer career, Kelly Williams has been offered scholarships to five universities. She has had a difficult time making a decision and has decided to use a scoring model to help evaluate the different offers. The following table includes a weighted list of criteria she has developed and a grade showing how well each school satisfies the criteria:

		University				
Decision Criterion	**Weight**	Tech	State	A&M	Central	Western
Playing time	0.29	61	52	93	83	77
Coach	0.20	87	92	66	55	80
Conference affiliation	0.17	42	57	80	92	63
School prestige	0.13	98	72	65	81	83
Program status	0.08	78	82	59	62	75
Degree program/major	0.08	72	80	75	93	89
Dollar value of scholarship	0.05	93	88	75	82	74

Rank the universities according to their scores to assist Kelly in making her decision.

57. The Carter family wants to purchase a time-share condominium for 1 week during the summer in Hilton Head, South Carolina. There are many resort complexes from which to choose, and the

Carters have narrowed their list to five. They have graded their choices according to the following weighted criteria:

		Resort Location				
Decision Criterion	**Weight**	**Albermarle**	**Beachfront**	**Calypso**	**Dafuskie**	**Edenisle**
Specific summer week	.40	80	70	70	90	60
Cost	.20	50	70	90	60	90
Proximity to beach	.15	70	60	70	50	80
Condominium quality	.05	90	80	90	60	80
Pool size	.05	40	60	70	80	100
Cleanliness	.05	70	90	90	80	90
Crowdedness of beach	.05	30	80	50	70	60
Condominium size	.05	100	70	90	80	90

Use a scoring model to determine a recommended resort for the Carters' time-share purchase.

58. Robin Dillon has recently accepted a new job in the Washington, DC, area and has been hunting for a condominium to purchase. From friends and coworkers she has compiled a list of five possible condominium complexes that she might move into. The following table indicates the weighted criteria that Robin intends to use in her decision-making process and a grade indicating how well each complex satisfies each criterion:

		Condominium Complex				
Decision Criterion	**Weight**	**Fairfax Forest**	**Dupont Gardens**	**Tysons Terrace**	**Alexandria Commons**	**Manassas Farms**
Purchase price	.30	92	85	75	62	79
Neighborhood location	.18	76	63	95	90	80
Proximity to Metro train	.12	78	75	76	85	60
Shopping	.10	65	80	98	92	75
Security	.10	75	78	90	95	83
Recreational facilities	.05	96	90	82	81	93
Distance to job	.05	85	67	95	75	65
Condo floor plan	.05	80	78	86	92	90
Complex size	.05	65	60	92	89	70

Use a scoring model to help Robin determine which condominium she should purchase.

59. The New River Rapids Under-15 Girls' Travel Soccer Team is determining which tournaments it should enter during the fall season. The manager has identified six possible tournaments, and the team wants to enter three of them. The team manager first surveyed the parents and players and determined a list of decision criteria and their relative weights. The manager then asked the parents and players to grade the tournaments on how well they satisfied the criteria. The criteria, weights, and average grades are summarized as follows:

Decision Criterion	Weight	Tournament					
		Roanoke	Greensboro	Bristol	Richmond	Knoxville	Charlotte
Distance	0.24	81	73	85	92	66	84
Tournament dates	0.21	91	88	79	83	75	69
Hotel cost	0.16	86	75	92	67	73	71
Level of competition	0.14	88	93	63	79	89	94
Field quality	0.10	84	91	70	87	77	81
Shopping	0.07	81	83	77	92	90	97
Attractions	0.06	73	82	80	91	96	87
Restaurants	0.02	83	88	65	95	72	91

To what three tournaments should the manager recommend the team apply, given the preferences indicated by the parents and players?

60. As part of an aggressive expansion plan, StarTrack Coffee is planning to open three new retail stores in the city. The following table shows the location factors it considers to be important indicators of future profitability and how management has graded each location according to each of these factors:

Location	Location Factor				
	Business Density	Shopping Density	Vehicle Traffic	Pedestrian Traffic	Land and Construction Cost
1	75	81	55	75	63
2	62	56	83	67	82
3	73	45	71	70	74
4	81	69	77	65	66
5	77	86	75	65	82
6	64	75	65	80	91
7	89	86	67	73	67
8	91	90	64	80	65
9	56	64	77	69	82
10	66	68	81	72	87
11	67	81	75	66	85
12	83	73	77	70	90

Use your own judgment to determine weights for each of the location factors and recommend the three new store sites. Are there other location factors that you think might be important?

61. Federated Electronics, Ltd., manufactures display screens and monitors for computers and televisions, which it sells to companies around the world. It wants to construct a new warehouse and distribution center in Asia to serve emerging markets there. It has identified potential sites in the

port cities Shanghai, Singapore, Pusan, Kaohsiung, and Hong Kong. The following table shows the factors in the location decision and the grade of each location for each factor:

	Port Score (0 to 100)				
Location Factor	Shanghai	Singapore	Pusan	Kaohsiung	Hong Kong
Facility cost	65	75	80	90	55
Labor rates	75	70	85	95	60
Labor availability	70	65	85	80	70
Infrastructure	80	80	65	70	95
Transportation	75	65	75	75	90
Container availability	70	80	65	75	85
Expansion/modernization	80	75	90	80	95
Political stability	65	70	85	80	90
Duties, tariffs, and fees	75	80	80	90	70
Trade regulations	65	75	80	80	75

The weights indicating the importance of each location factor are not included. Determine what you think these weights should be and recommend the best location for the new distribution center.

62. Select four fast-food restaurants (e.g., McDonald's, Burger King, Wendy's, Domino's) in your area and develop a scoring model that includes decision criteria, weights, and grades to rank the restaurants from best to worst.

Case Problem

Oakdale County School Busing

The Oakdale County School Board was meeting in special session. A federal judge had ordered the board to present an acceptable busing plan for racially balancing the four high schools in Oakdale County within a week. The judge had previously given the school board several opportunities to informally present a plan, but the members had been unable to agree among themselves. Every time they met and started to develop a plan to bus students from one high school district to another, an argument would arise before they got past the first busing move, and they would adjourn the meeting. This time, however, they knew the judge had lost patience and they had to agree on something.

Of the four schools, only West High School was racially balanced, with 500 white students and 500 black students. North High School had 1,000 white students and only 300 black students; East High School was almost as bad, with 1,050 white students and 400 black students. South High School was predominantly black, with 800 black students and 450 white students. Overall, of the 5,000 students in Oakdale County, 60% were white and 40% black.

"Look," said John Connor, a school board member from the West district, "rather than starting off by trying to shift students from one district to another, why don't we try to establish what we want to accomplish—you know, what our goals are?"

Several of the other members nodded in agreement, and Fred Harvey, the board chairman, said, "Good idea, John."

"Okay, the first goal seems pretty evident to me," John said. "Sixty percent of our students are white, and 40% are black, so that's what we need our schools to be, 60% and 40%."

"That's okay for you to say, John," Betty Philips argued, "because your district has those proportions already—so you won't have any busing. But my district in the North is a long way from that ratio, and we would have to bus a lot of our students to achieve a 60%/40% ratio."

"*I'm* not saying it, Betty," said John. "That is basically what Judge Barry has been saying for 6 months."

"John's right, Betty, and we're not busing students yet; we're just putting down our objectives," said Fred. "I think that has to be our highest-priority objective. How about the rest of you?"

They all nodded their agreement, even Betty Philips, reluctantly.

"Since we know we're going to have to bus students to achieve this ratio at each school, I think we ought to try to minimize the amount of traveling the students will have to do," suggested Mickey Gibboney, a member from the South district.

Fred Harvey noted that page 10 of their handout had a chart showing the average mileage a student in one district would have to travel on a bus to the high school in each of the other districts. The chart looked like this:

	Distance (mi.)			
District/School	*North*	*South*	*East*	*West*
North	—	30	12	20
South	30	—	18	26
East	12	18	—	24
West	20	26	24	—

"Why don't we try to set some reasonable objectives for total busing miles, for the students' sake and for budgeting reasons?" Cassandra Watkins asked. "I would suggest about 30,000 miles per day, based on the miles we bus students now. If we get much higher than that, we're not going to have the money to pay for it, and it means we'll be busing students all over the place."

The other members nodded and agreed.

"Okay," said Fred Harvey, "that'll be our number-two goal."

Betty Philips spoke up again. "I'll tell you another thing I don't want to see happen, and that's any more overcrowding at North High School. We have 100 students more than capacity now."

"You think you have problems!" Bob Wilson exclaimed. "In East we have 1,450 students and capacity for 1,000. I think no overcrowding is a great idea!"

"I agree," said Mickey Gibboney. "We're 250 over our capacity at South High School."

"That's a nice idea," John Connor responded, "and I realize that we have 200 students less than our capacity at West High School. However, let's face it, in the county we have capacity for 4,400, not 5,000, students, so there's going to be some overcrowding. I think our objective should be that all four schools share in the overcrowding proportionally."

"That sounds reasonable to me," said Fred Harvey. "How about the rest of you? Okay to say our number-three goal is to be as close to capacity at each school as possible but share proportionally in the overcrowding?"

They all voiced their approval.

"Well," John Connor concluded, "I think we have identified the things we want to accomplish in our plan. Now if we could just use some magic trick to find a plan for busing students between the districts that would achieve all these goals."

The others nodded and frowned.

A. Formulate a goal programming model to help the board with its dilemma.
B. Solve the goal programming model by using the computer.

Case Problem

Catawba Valley Highway Patrol

Broderick Crawford is the district commander for the Catawba Valley highway patrol district in western Pennsylvania. He is attempting to assign highway patrol cars to different road segments in his district. The primary function of the highway patrol force is to patrol roads outside incorporated city and town limits in the district to deter traffic violators and accidents. This objective is typically achieved by maintaining a visible presence—letting motorists see patrol units on a regular basis and giving out warnings, citations, and so forth. Secondary activities of a patrol unit include providing assistance to motorists, answering distress calls, handling emergencies and accidents when called to the scene, and occasionally apprehending criminals.

Commander Crawford has 23 patrol cars that he wants to assign to the following six major road segment areas:

road segment 1, interstate, north
road segment 2, urban area, north
road segment 3, four-lane highway, east
road segment 4, two-lane highway, west
road segment 5, interstate/four-lane highway, south
road segment 6, two-lane highway (heavy truck traffic), south

Each of these road segments includes the primary arteries, as indicated earlier, plus adjoining roads. All the road segments have different levels of traffic density and accident rates, which are key factors in determining how many patrol units to assign. However, these factors do not always coincide. For example, interstate highway segments typically have high traffic density but low accident rates, whereas some two-lane highways have low traffic density but high accident rates. Differences often occur because of variations in road conditions (such as sharp curves, visibility, and width). Other conditions, such as heavy truck traffic (as on segment 6), also contribute to high accident rates.

Each segment requires different operating costs, including maintenance and repair, fuel, and so on because of different operating conditions. The commander's most pressing objective is to limit daily operating costs to $450. The daily operating costs per road segment are as follows:

Road Segment	Operating Cost
1	$20
2	18
3	22
4	24
5	17
6	19

The commander would like to reduce the accident rate for the district as well as increase both physical and sight contacts, which are deterrents to potential traffic violators. The commander would also like to achieve a reasonable average response time for a patrol unit to respond to a call for each road segment. The average accident rate reduction (per million miles traveled) and physical contacts and sight contacts per car for each road segment are shown in the following table:

	Patrol Unit		
	Physical	*Sight*	
Road	*Accident Reduction*	*Contacts*	*Contacts*
Segment	*(per million mi. traveled)*	*(per day)*	*(per day)*
1	0.27	18	1,700
2	0.21	26	900
3	0.28	10	650
4	0.19	34	230
5	0.23	25	1,600
6	0.33	17	520

The commander's second-most-important goal is to reduce the average accident rate for the district by five accidents per million miles traveled. The commander's next goals (in order) are to achieve 350 physical contacts and 30,000 sight contacts per day in the district.

If no patrol units are assigned in the district, the average time to respond to a distress call anywhere in the district from the main district headquarters and motor pool is 28 minutes. Each car assigned to a road segment reduces the overall average response time in the district by the following amounts:

Road Segment	Reduction in Average Response Time (min.)
1	0.32
2	0.65
3	0.43
4	0.87
5	0.55
6	0.49

The commander's last objective is to achieve an average response time to distress calls of 15 minutes. Because of local and political pressure, the commander has to assign at least two patrol units to each road segment. In addition, the commander believes that a maximum of five patrol units is sufficient for any particular road segment.

Formulate and solve a goal programming model to determine the number of patrol units to assign to each road segment to achieve the commander's goals.

Case Problem

KATHERINE MILLER'S JOB SELECTION

Katherine Miller is a senior in the department of information technology at Tech. For the past few months she has been involved in the job search process. She has an excellent résumé, with a high grade point average and a strong record of campus participation in clubs and activities. As a result, she has had a number of good interviews with various companies. She now has job offers from five companies—American Systems Developers, Anderssun Consulting, National Computing Software Systems (NCSS), the Gulf-South Company, and Electronic Village.

American Systems Developers and Anderssun Consulting are both large national consulting firms with offices in several major cities. If Katherine accepted the offer of either of these firms, she would primarily work on project teams assigned to develop decision support and information systems for corporate clients around the country. If she went with American Systems Developers, her home base would be in Atlanta, and if she accepted Anderssun's offer, she would be located in Washington, DC. However, in both cases she would be traveling a great deal and could sometimes be on the road at a client location for as much as 6 to 9 months. NCSS is a software and computer systems development company with a campus-like location in Chicago. Although her job with NCSS would involve some traveling, it would never be more than several weeks at any one time. Gulf-South is a bank holding company that operates eight different banks and its various branches in six southeastern states. If Katherine accepted Gulf-South's offer, she would be located in Tampa, where she would work in operations systems. She would be involved in developing information and support systems for bank operations, and she would have minimal travel. Electronic Village is a national chain of discount stores specializing in electronic products, such as televisions, DVRs, CD players, MP3 players, and computers. Her job with Electronic Village would be at its corporate headquarters in Nashville, Tennessee, where she would develop and maintain computer systems to be used for inventory control at the hundreds of Electronic Village stores across the country. She would be required to travel very little.

American Systems Developers has offered Katherine a starting salary of $38,000 annually, and Anderssun Consulting has offered her $41,000 per year. NCSS has offered her an annual salary of $46,000, whereas Gulf-South has offered her $35,000 per year, and Electronic Village has offered her a salary of $32,000 per year.

Katherine is having a difficult time making her decision. All the companies have excellent reputations, are financially healthy, and have good opportunities for advancement. All are demanding in terms of the workload they require. All five companies have given Katherine only a few weeks to make a decision regarding their offers.

Katherine has decided to use the analytical hierarchy process (AHP) to help decide which job offer she should accept. She has developed a list of criteria that are important to her in deciding which job to take. The criteria, in no particular order, are (1) salary; (2) cost of living in the city where she would be located; (3) amount of travel associated with her job; (4) climate (weather) where she would be located; (5) entertainment and cultural opportunities, including sports, theater, museums, parks, and so on; (6) universities where she can work on an MBA degree part time and at night; (7) the crime rate in the city where she would live; (8) the nature of the job and what she would be doing; and (9) her proximity to friends and relatives. Katherine realizes that she has very limited information on which to compare the five jobs, based on most of these criteria, so she knows she needs to go to the library and do some research, especially on the five different job locations.

Put yourself in Katherine's shoes and, using all or some of her criteria and your own preferences and knowledge, develop an overall ranking of the jobs, using AHP.

Case Problem

SELECTING NATIONAL BASEBALL HALL OF FAME MEMBERS

Each November the Baseball Writers' Association of America (BBWAA) releases its list of that year's nominees for induction into the National Baseball Hall of Fame in Cooperstown, New York. There are 575 voting members of the BBWAA, and candidates must be named on 75% of the ballots cast to gain entry into the National Baseball Hall of Fame. There is no limit to the number of players who can be nominated or who can be elected. Go to the BBWAA Web site at http://bbwaa.com/ and from the list of this year's nominees select the five or six "field players" (as opposed to pitchers) who you think are the top candidates for selection, and develop an AHP model using your own criteria to rank these candidates. Criteria might include, among other things, hits, lifetime batting average, home runs, runs batted in, runs, stolen bases, awards, years played, fielding percentage, number of World Series participated in, and a comparison with Hall of Famers overall and at their field position (e.g., catcher, first base, outfield). Player bios including their playing statistics can be obtained by searching the Internet.

CHAPTER

10

Nonlinear Programming

More attention has been devoted to linear programming in this text than to any other single topic. It is a very versatile technique that can be and has been applied to a wide variety of problems. Besides chapters devoted specifically to linear programming models and applications, we have also presented several variations of linear programming, integer and goal programming, and unique applications of linear programming for transportation and assignment problems. In all these cases, all the objective functions and constraints were linear; that is, they formed a line or plane in space. However, many realistic business problems have relationships that can be modeled only with nonlinear functions. When problems fit the general linear programming format but include nonlinear functions, they are referred to as **nonlinear programming** problems.

***Nonlinear programming** has the same format as a linear programming model, but the objective function or constraints, or both, are nonlinear functions.*

Nonlinear programming problems are given a separate name because they are solved in a different manner than are linear programming problems. In fact, their solution is considerably more complex than that of linear programming problems, and it is often difficult, if not impossible, to determine an optimal solution, even for a relatively small problem. In linear programming problems, solutions are found at the intersections of lines or planes, and though there may be a very large number of possible solution points, the number is finite, and a solution can eventually be found. However, in nonlinear programming there may be no intersection or corner points; instead, the solution space can be an undulating line or surface, which includes virtually an infinite number of points. For a realistic problem, the solution space may be like a mountain range, with many peaks and valleys, and the maximum or minimum solution point could be at the top of any peak or at the bottom of any valley. What is difficult in nonlinear programming is determining if the point at the top of a peak is just the highest point in the immediate area (called a *local optimal*, in calculus terms) or the highest point of all (called the *global optimal*).

The solution techniques for nonlinear programming problems generally involve searching the solution surface for peaks or valleys—that is, high points or low points. The problem encountered by these methods is that they sometimes have trouble determining whether the high point they have identified is just a local optimal solution or the global optimal solution. Thus, finding a solution is often difficult and can involve very complex mathematics that are beyond the scope of this text.

In this chapter, we present the basic structure of nonlinear programming problems and use Excel to solve simple models.

Nonlinear Profit Analysis

We begin our presentation of nonlinear programming models by determining the optimal value for a single nonlinear function. To demonstrate the solution procedure, we will use a profit function based on break-even analysis. In Chapter 1 we used break-even analysis to begin our study of model building, so it seems appropriate that we return to this basic model to complete our study of model building.

Recall that in break-even analysis the profit function, Z, is formulated as

$$Z = vp - c_f - vc_v$$

where

v = sales volume (i.e., demand)
p = price
c_f = fixed cost
c_v = variable cost

One important but somewhat unrealistic assumption of this break-even model is that volume, or demand, is independent of price (i.e., volume remains constant, regardless of the price of the product). It would be more realistic for the demand to vary as price increased or decreased. For our Western Clothing Company example from Chapter 1, let us suppose that the dependency of demand on price is defined by the following linear function:

$$v = 1{,}500 - 24.6p$$

This linear relationship is illustrated in Figure 10.1. The figure illustrates the fact that as price increases, demand decreases, up to a particular price level ($60.98) that will result in no sales volume.

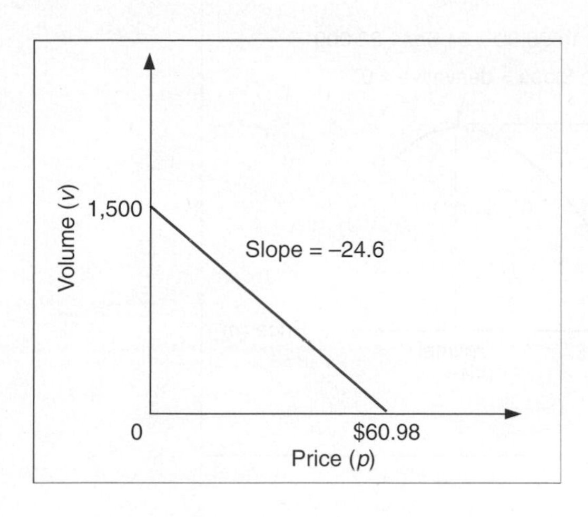

FIGURE 10.1
Linear relationship of volume to price

Now we will insert our new relationship for volume (v) into our original profit equation:

$$\begin{aligned} Z &= vp - c_f - vc_v \\ &= (1{,}500 - 24.6p)p - c_f - (1{,}500 - 24.6p)\,c_v \\ &= 1{,}500p - 24.6p^2 - c_f - 1{,}500c_v + 24.6pc_v \end{aligned}$$

Substituting values for fixed cost ($c_f = \$10{,}000$) and variable cost ($c_v = \$8$) into this new profit function results in the following equation:

$$\begin{aligned} Z &= 1{,}500p - 24.6p^2 - 10{,}000 - 1{,}500(8) + 24.6p(8) \\ &= 1{,}696.8p - 24.6p^2 - 22{,}000 \end{aligned}$$

Because of the squared term, this equation for profit is now a nonlinear, or quadratic, function that relates profit to price, as shown in Figure 10.2.

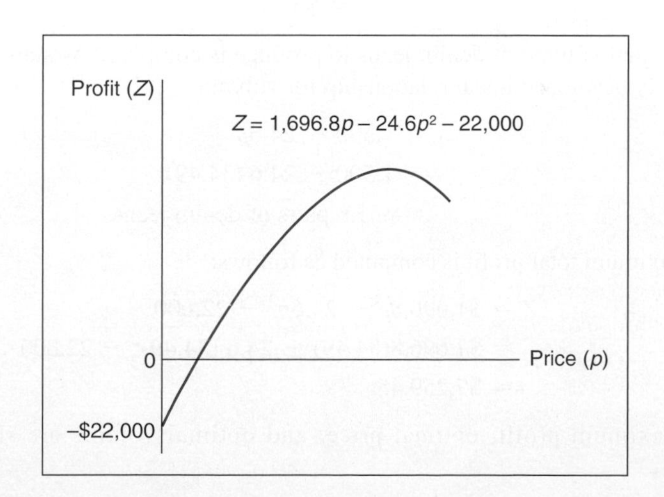

FIGURE 10.2
The nonlinear profit function

In Figure 10.2, the greatest profit will occur at the point where the profit curve is at its highest. At that point the slope of the curve will equal zero, as shown in Figure 10.3.

FIGURE 10.3
Maximum profit for the profit function

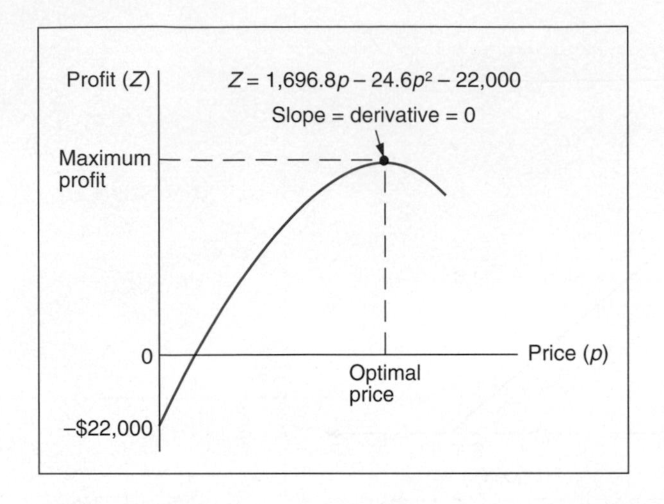

The slope of a curve at any point is equal to the derivative of the curve's function.

In calculus, the slope of a curve at any point is equal to the derivative of the mathematical function that defines the curve. The derivative of our profit function is determined as follows:

$$Z = 1{,}696.8p - 24.6p^2 - 22{,}000$$

$$\frac{\partial Z}{\partial p} = 1.696.8 - 49.2p$$

The slope of a curve at its highest point equals zero.

Given this derivative, the slope of the profit curve at its highest point is defined by the following relationship:

$$0 = 1{,}696.8 - 49.2p$$

Now we can solve this relationship for the optimal price, p, which will maximize total profit:

$$\begin{aligned} 0 &= 1{,}696.8 - 49.2p \\ 49.2p &= 1{,}696.8 \\ p &= 1{,}696.8/49.2 \\ &= \$34.49 \end{aligned}$$

The optimal volume of denim jeans to produce is computed by substituting this price into our previously developed linear relationship for volume:

$$\begin{aligned} v &= 1{,}500 - 24.6p \\ &= 1{,}500 - 24.6\,(34.49) \\ &= 651.6 \text{ pairs of denim jeans} \end{aligned}$$

The maximum total profit is computed as follows:

$$\begin{aligned} Z &= \$1{,}696.8p - 24.6p^2 - 22{,}000 \\ &= \$1{,}696.8(34.49) - 24.6\,(34.49)^2 - 22{,}000 \\ &= \$7{,}259.45 \end{aligned}$$

The maximum profit, optimal price, and optimal volume are shown graphically in Figure 10.4.

FIGURE 10.4
Maximum profit, optimal price, and optimal volume

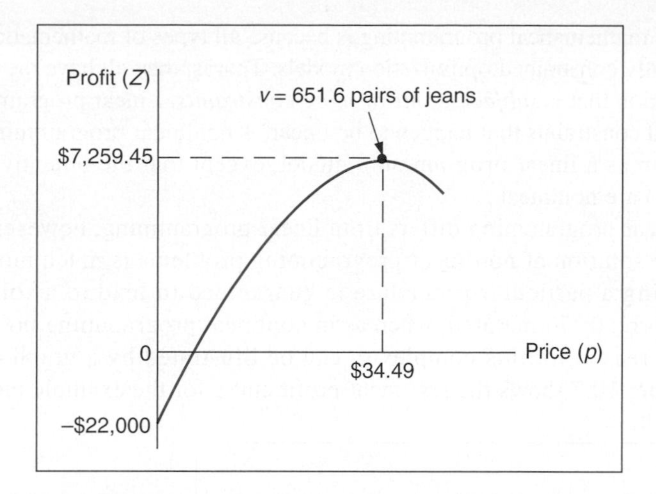

An important concept we have yet to mention is that by extending the break-even model this way, we have converted it into an *optimization* model. In other words, we are now able to maximize an objective function (profit) by determining the optimal value of a variable (price). This is exactly what we did in linear programming when we determined the values of decision variables that optimized an objective function. The use of calculus to find optimal values for variables is often referred to as **classical optimization**.

***Classical optimization** is the use of calculus to determine the optimal value of a variable.*

Constrained Optimization

In the preceding section, the profit analysis model was developed as an extension of the break-even model. Recall that the total profit function was

$$Z = vp - c_f - vc_v$$

where

v = volume
p = price
c_f = fixed cost
c_v = variable cost

and the demand function (i.e., volume as a function of price) was

$$v = 1{,}500 - 24.6p$$

By substituting this demand function into our total profit equation, we developed a nonlinear function:

$$Z = 1{,}500p - 24.6p^2 - c_f - 1{,}500c_v + 24.6pc_v$$

Then, by substituting values for c_f ($10,000) and c_v ($8) into this function, we obtained

$$Z = 1{,}696.8p - 24.6p^2 - 22{,}000$$

We then differentiated this function, set it equal to zero, and solved for the value of p ($34.49), which corresponded to the maximum point on the profit curve (where the slope equaled zero).

*An **unconstrained optimization model** includes a nonlinear objective function and no constraints.*

*A **constrained optimization model** includes a nonlinear objective function and one or more constraints.*

This type of model is referred to as an **unconstrained optimization model**. It consists of a single nonlinear objective function and *no* constraints. If we add one or more constraints to this model, it becomes a **constrained optimization model**. A constrained optimization model is more commonly referred to as a *nonlinear programming model*. The reason this type of model is designated as a form of mathematical programming is because all types of mathematical programming models are actually constrained optimization models. That is, they all have the general form of an objective function that is *subject to one or more constraints*. Linear programming has an objective function and constraints that happen to be linear. A nonlinear programming model has the same general form as a linear programming model, except that the objective function *and/or* the constraint(s) are nonlinear.

Nonlinear programming differs from linear programming, however, in one other critical aspect: The solution of nonlinear programming problems is much more complex. In linear programming a particular procedure is guaranteed to lead to a solution if the problem has been correctly formulated, whereas in nonlinear programming no guaranteed procedure exists. The reason for this complexity can be illustrated by a graph of our profit analysis model. Figure 10.5 shows the nonlinear profit curve for the example model.

FIGURE 10.5
Nonlinear profit curve for the profit analysis model

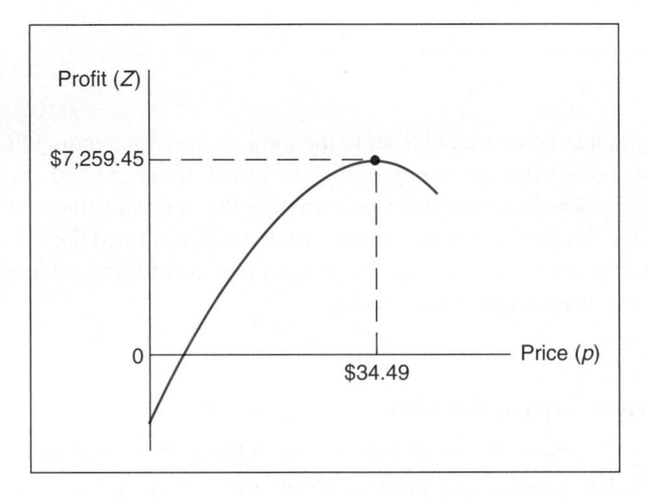

As stated previously, the solution is found by taking the derivative of the profit function, setting it equal to zero, and solving for p (price). This results in an optimal value for p that corresponds to the maximum profit, as shown in Figure 10.5.

Now we will transform this unconstrained optimization model into a nonlinear programming model by adding the constraint

$$p \leq \$20$$

In other words, because of market conditions, we are restricting the price to a maximum of $20. This constraint results in a feasible solution space, as shown in Figure 10.6.

FIGURE 10.6
A constrained optimization model

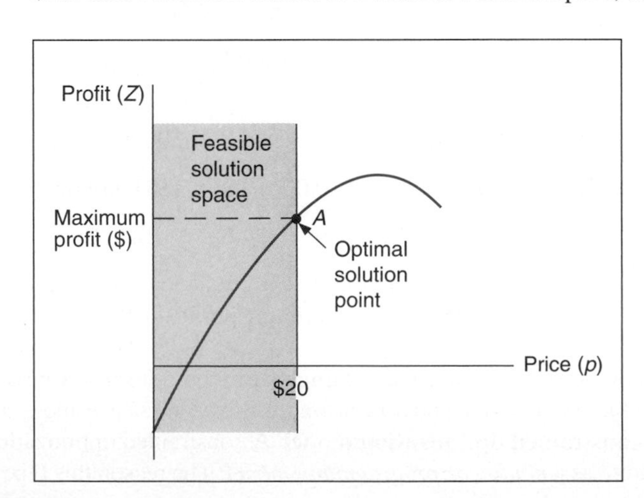

As in a linear programming problem, the solution is on the boundary of the feasible solution space formed by the constraint. Thus, in Figure 10.6, point A is the optimal solution. It corresponds to the maximum value of the portion of the objective function that is still feasible. However, the difficulty with nonlinear programming is that the solution is not always on the boundary of the feasible solution space formed by the constraint. For example, consider the addition of the following constraint to our original nonlinear objective function:

$$p \leq \$40$$

This constraint also creates a feasible solution space, as shown in Figure 10.7.

FIGURE 10.7
A constrained optimization model with a solution point not on the constraint boundary

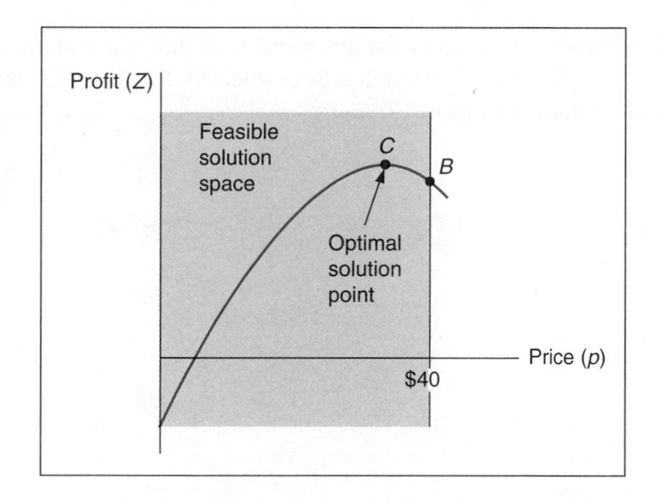

But notice in Figure 10.7 that the solution is no longer on the boundary of the feasible solution space, as it would be in linear programming. Point C represents a greater profit than point B, and it is also in the feasible solution space. This means that we cannot simply look at points on the solution space boundary to find the solution; instead, we must also consider other points on the surface of the objective function. This greatly complicates the process of finding a solution to a nonlinear programming problem, and this difficulty is aggravated by an increased number of variables and constraints and by nonlinear functions of a higher order. You can imagine the difficulties of finding a solution if you contemplate a model in space that is made up of intersecting cones, ellipses, and undulating surfaces, as well as planes, and a solution that is not even on the boundary of the solution space.

See Web site Module D for a chapter on "Nonlinear Programming Solution Techniques."

A number of different solution approaches to nonlinear programming problems are available. As we have already indicated, they typically represent a convergence of the principles of calculus and mathematical programming. However, as noted earlier, the solution techniques can be very complex. Thus, we will confine our discussion of nonlinear programming solution methods to several different model applications and the Excel solution.

Solution of Nonlinear Programming Problems with Excel

Excel can solve nonlinear programming problems by using "Solver" that we used previously in this text to solve linear programming problems. Exhibit 10.1 shows an Excel spreadsheet set up to solve our initial Western Clothing Company example. The demand function contained in cell C4 is **=1500−24.6*C5**. The formula for profit is contained in cell C3 and is shown on the formula bar at the top of the spreadsheet.

EXHIBIT 10.1

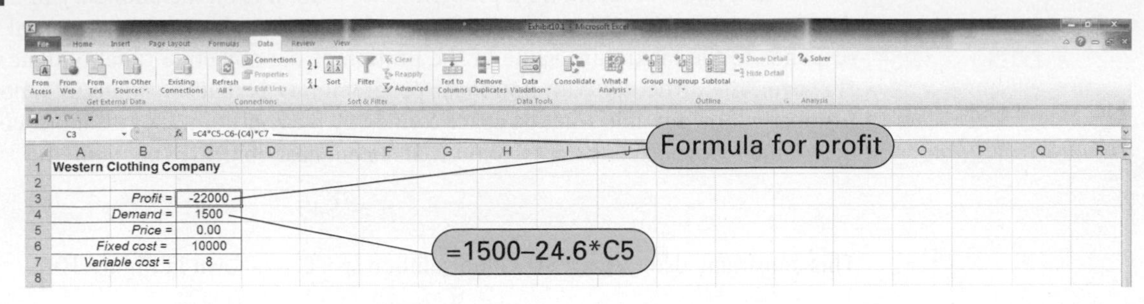

The Solver Parameters window for this problem is shown in Exhibit 10.2. It is important to make sure that the "Nonlinear" option has been selected. Exhibit 10.3 shows the Excel solution for the Western Clothing Company example.

EXHIBIT 10.2

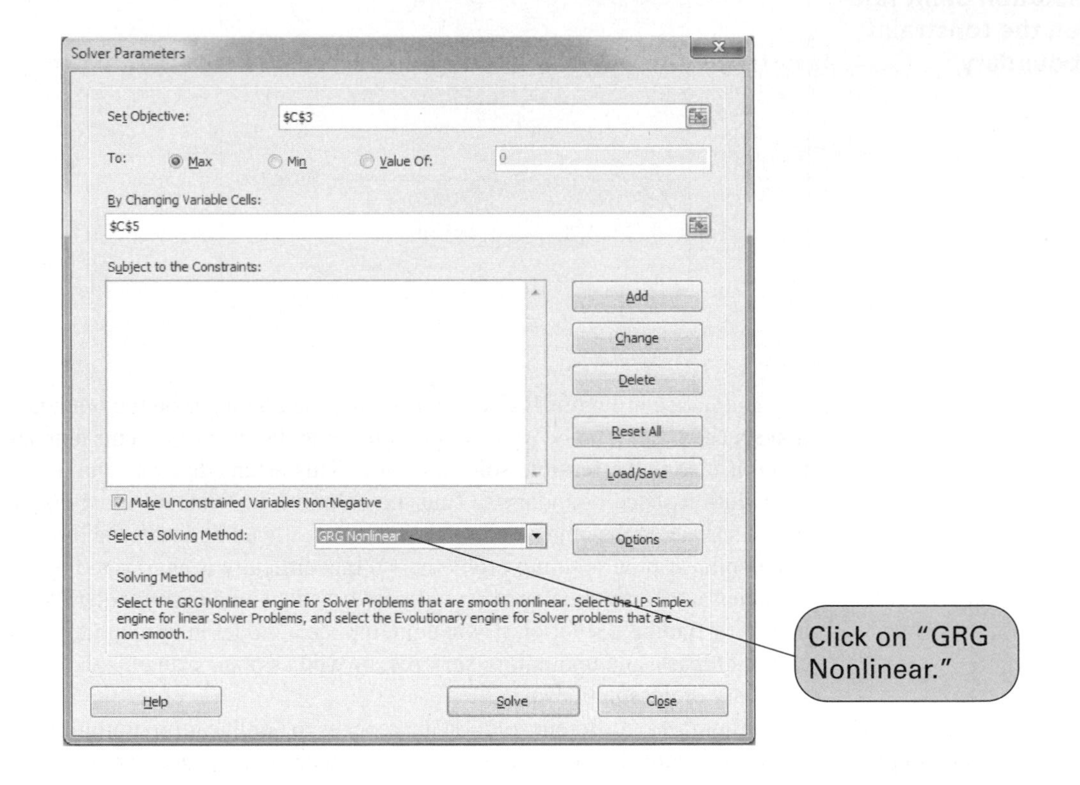

EXHIBIT 10.3

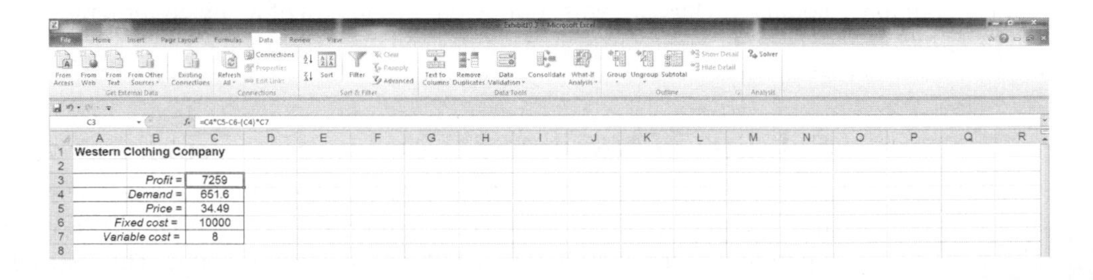

Exhibit 10.4 shows an Excel spreadsheet set up to solve a nonlinear version of the Beaver Creek Pottery Company example from Chapter 2 that is formulated as

$$\text{maximize } Z = \$(4 - 0.1x_1)x_1 + (5 - 0.2x_2)x_2$$

subject to

$$x_1 + 2x_2 = 40 \text{ hours (labor)}$$

where

$$x_1 = \text{number of bowls produced}$$
$$x_2 = \text{number of mugs produced}$$
$$4 - 0.1x_1 = \text{profit (\$) per bowl}$$
$$5 - 0.2x_2 = \text{profit (\$) per mug}$$

EXHIBIT 10.4

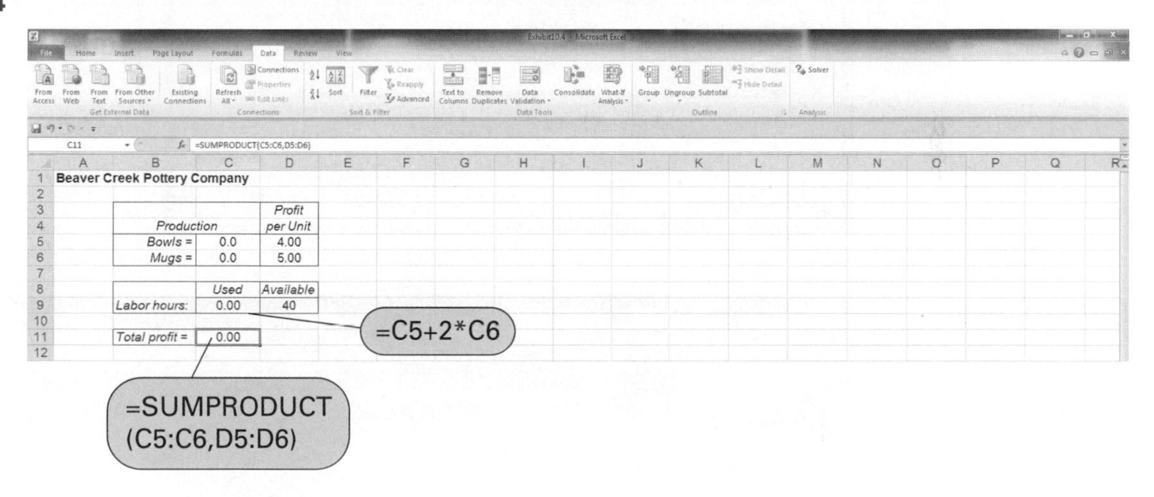

The numbers of bowls and mugs produced are contained in cells C5 and C6, respectively. The profit formula for a bowl is contained in cell D5. The total profit contained in cell C11 is computed using the formula **=SUMPRODUCT(C5:C6,D5:D6)** and is shown on the formula bar at the top of the spreadsheet. The formula for labor, **=C5+2*C6,** is contained in cell C9. The Solver Parameters window for this problem is shown in Exhibit 10.5, and the final solution is shown in Exhibit 10.6.

Excel will also provide the value of the *Lagrange multiplier*, which provides the dual value of the labor resource. To derive the Lagrange multiplier, after you click on "Solve" in Solver, the Solver Results screen shown in Exhibit 10.7 will appear. On this screen, under "Reports," select "Sensitivity." This will generate the sensitivity report shown in Exhibit 10.8. Note that in addition to the problem solution, the value of the Lagrange multiplier is also provided for the labor constraint.

The Lagrange multiplier value of .33 is analogous to the dual value in a linear programming problem. It reflects the approximate change in the objective function resulting from a unit change in the quantity (right-hand-side) value of the constraint equation. For this example, if the quantity of labor hours is increased from 40 to 41 hours, the value of Z will increase by \$0.33—from \$70.42 to \$70.75.

EXHIBIT 10.5

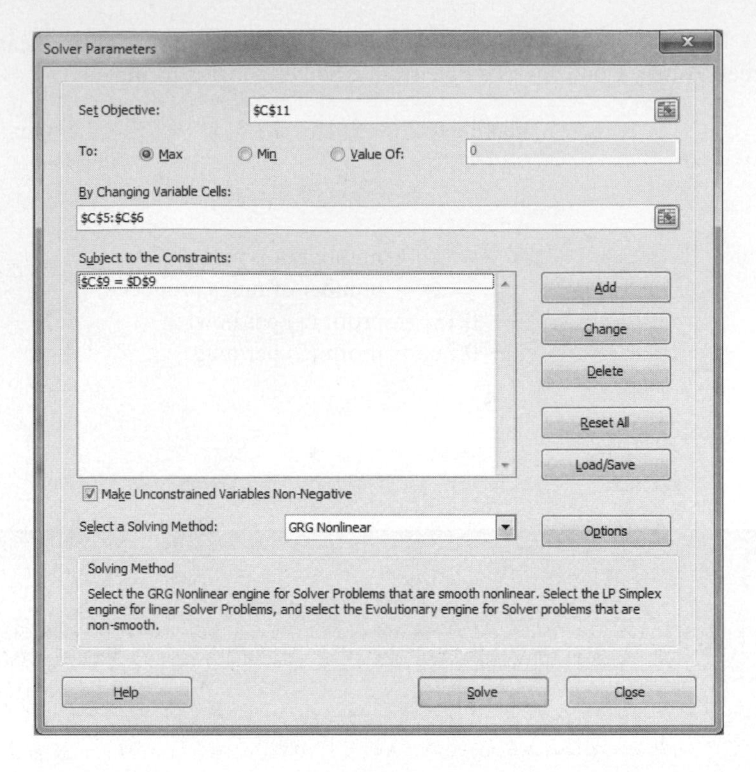

EXHIBIT 10.6

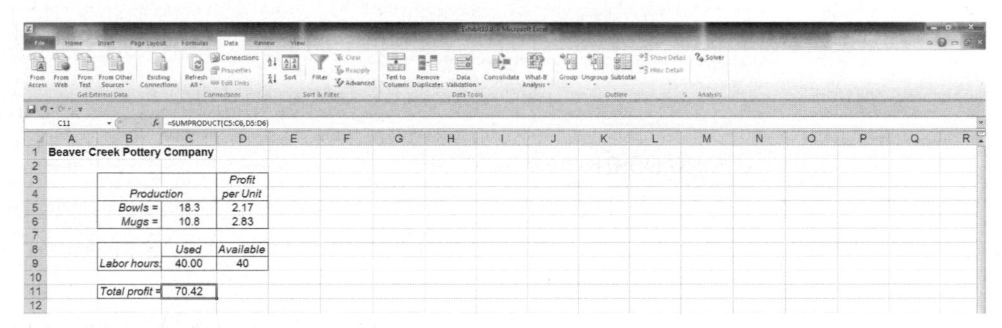

EXHIBIT 10.7

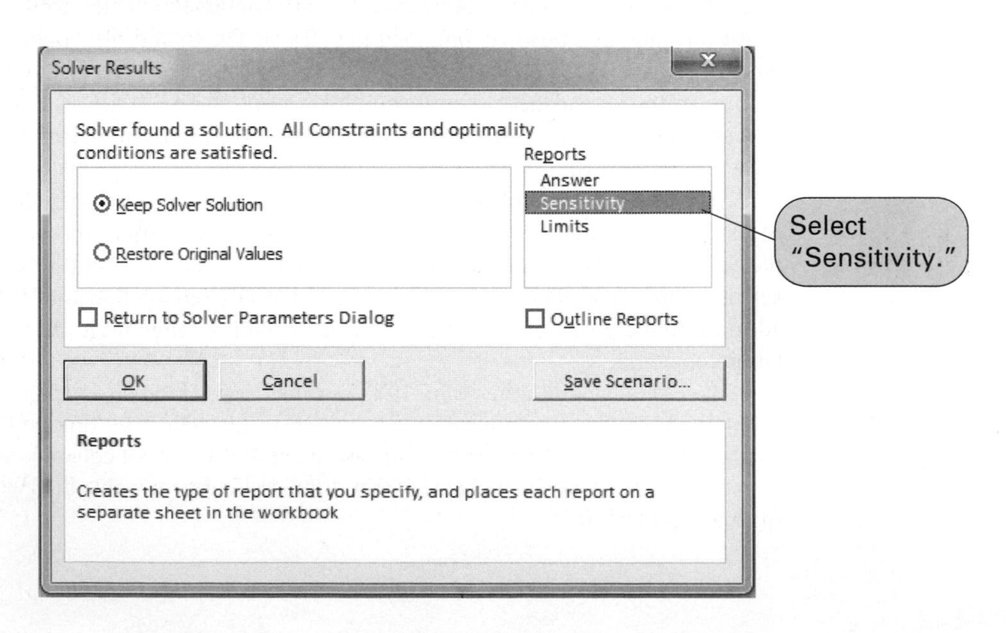

EXHIBIT 10.8

Changing Cells

Cell	Name	Final Value	Reduced Gradient
$C$5	Bowls = Producrtion	18.3	0.0
$C$6	Mugs = Producrtion	10.8	0.0

Constraints

Cell	Name	Final Value	Lagrange Multiplier
$C$9	Labor hours: Used	40.00	0.33

Lagrange multiplier for labor

Management Science Application

Making Supplier Decisions at Ford with Nonlinear Programming

In 2005, Ford took back a number of supplier component operations (under the name Visteon) it spun off in 2000 in order to sustain the viability of its supply chain. It created a holding company for these businesses, called Automotive Components Holdings, LLC (ACH), with a commitment to sell or close all ACH businesses by 2009.

A major ACH business with $1.5 billion in annual revenue was Interiors, which produced instrument panels, consoles, door-trim panels, and cockpit modules for Ford assembly plants. Interiors manufactured these components at two underutilized and unprofitable plants in Saline and Utica in southeast Michigan; these plants consumed substantial amounts of cash. However, despite improvements to these plants and a concerted marketing effort, Ford was unable to sell the Interiors business, and it needed to decide how to best resolve the problem. Ford identified the following solution alternatives: consolidate the entire ACH Interiors business into one plant, thereby improving its utilization and enhancing its attractiveness to a potential buyer; shut down both plants and outsource all of Ford's Interiors business taking advantage of unused capacity at outside suppliers; or use some combination of plant consolidation and outsourcing.

The problem Ford addressed with its management science application was which of the three alternatives to select—that is, which plants to close and where to outsource the components that would minimize the overall production, distribution, and restructuring costs. Key to this problem for Ford was identifying an optimal sourcing pattern for the interiors components it needed, where the best sourcing pattern would result in the minimum present value of the 5-year cost to Ford. The model created by these sourcing patterns related component parts pricing, fixed costs, and semi-fixed costs to capacity utilization as nonlinear operational constraints and a nonlinear objective function, thus requiring a nonlinear programming model for solution. The model had to reflect millions of possible product-to-process-to-suppliers allocations.

Monty Rakusen/Cultura/Newscom

The model's solution recommended outsourcing several product lines to third-party suppliers, closing the plant at Utica, and consolidating the remaining ACH business at the Saline plant. This solution resulted in 5-year cost savings of more than $50 million and up-front investment savings of $45 million.

Source: Based on W. Klampfl, Y. Fradkin, C. McDaniel, and M. Wolcott, "Ford Uses OR to Make Urgent Sourcing Decisions in a Distressed Supplier Environment," *Interfaces* 39, no. 5 (September–October 2009): 428–42.

A Nonlinear Programming Model with Multiple Constraints

Now that we have shown how Excel can be used to solve nonlinear problems, we can look at more complex problems—for example, a problem with more than one constraint. Consider the Western Clothing Company example presented earlier, except now the company produces two kinds of jeans, designer and straight-leg, and production is subject to resource constraints for denim cloth, cutting time, and sewing time. The company sells its jeans to several upscale clothing store chains, and sales demand is dependent on the price at which the company sells the jeans. The demand for designer jeans (x_1) and the demand for straight-leg jeans (x_2) are defined by the following relationships:

$$x_1 = 1{,}500 - 24.6p_1$$
$$x_2 = 2{,}700 - 63.8p_2$$

where

p_1 = price of designer jeans
p_2 = price of straight-leg jeans

The cost of producing designer jeans is \$12 per pair, and the cost of producing straight-leg jeans is \$9 per pair; thus, the objective function for profit is

$$\text{maximize } Z = (p_1 - 12)x_1 + (p_2 - 9)x_2$$

The production of jeans is subject to the following resource constraints for available cloth, cutting time, and sewing time:

$$\text{cloth: } 2x_1 + 2.7x_2 \le 6{,}000 \text{ yards}$$
$$\text{cutting: } 3.6x_1 + 2.9x_2 \le 8{,}500 \text{ minutes}$$
$$\text{sewing: } 7.2x_1 + 8.5x_2 \le 15{,}000 \text{ minutes}$$

The complete nonlinear programming model formulation is summarized as follows:

$$\text{maximize } Z = (p_1 - 12)x_1 + (p_2 - 9)x_2$$

subject to

$$2x_1 + 2.7x_2 \le 6{,}000$$
$$3.6x_1 + 2.9x_2 \le 8{,}500$$
$$7.2x_1 + 8.5x_2 \le 15{,}000$$

where

$$x_1 = 1{,}500 - 24.6p_1$$
$$x_2 = 2{,}700 - 63.8p_2$$

p_1 = price of designer jeans
p_2 = price of straight-leg jeans

Notice that the decision variables for this problem are p_1 and p_2, not x_1 and x_2. The demand variables, x_1 and x_2, are functions of price and, thus, are dependent variables. Also notice that we did not substitute the functional relationships for x_1 and x_2 into the objective function, creating a quadratic function. The model is still nonlinear, but we can solve it directly with Excel in its present form.

Exhibit 10.9 shows an Excel spreadsheet set up to solve our Western Clothing Company example with constraints. The decision variables for price are contained in cells **D5:D6**. The formula for demand for designer jeans is contained in cell C5 and is shown on the formula bar at the top of the screen. The formula for total profit,

=SUMPRODUCT(C5:C6,E5:E6), is contained in cell E7. The formulas for the resource constraints are contained in cells C11, C12, and C13, and the resource availabilities for each constraint are in cells **D11:D13**. The Solver Parameters window and solution are shown in Exhibit 10.10 and Exhibit 10.11, respectively.

EXHIBIT 10.9

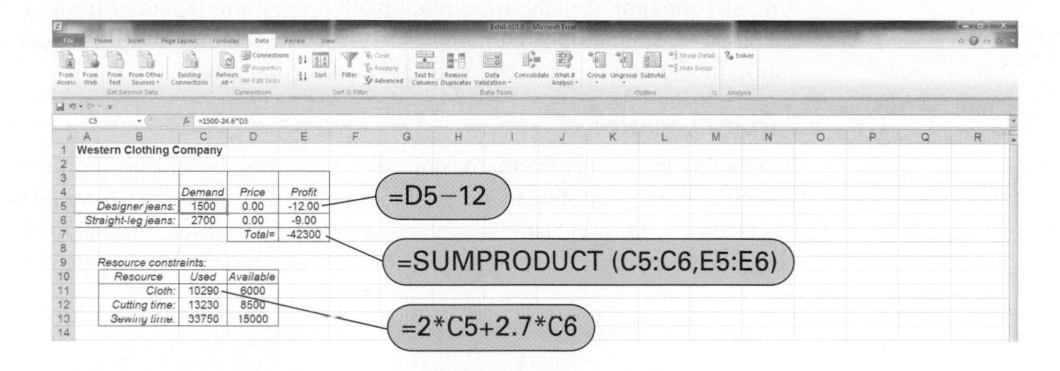

EXHIBIT 10.10

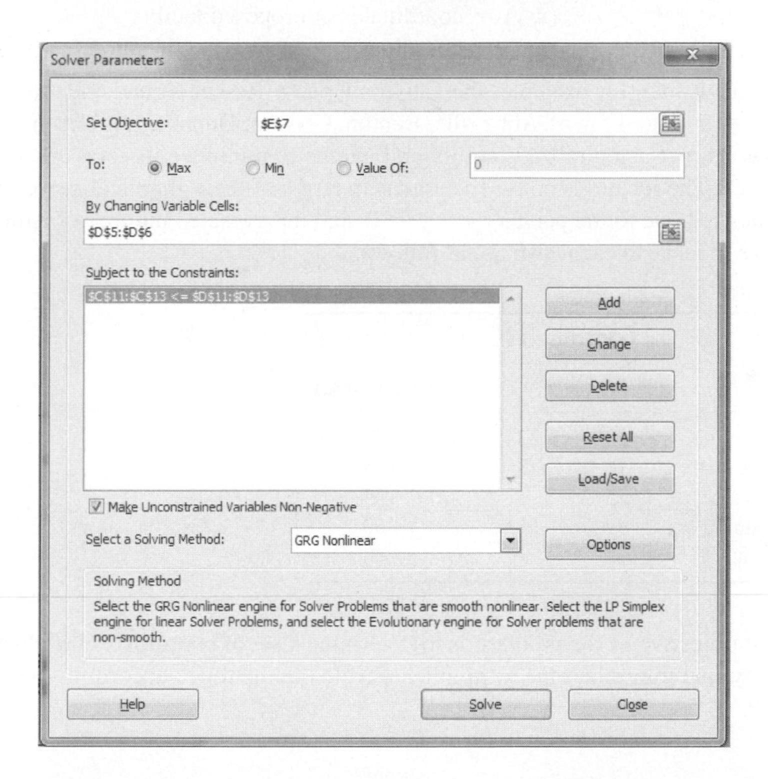

EXHIBIT 10.11

Nonlinear Model Examples

In our previous profit analysis examples, the models followed the traditional mathematical (i.e., linear) programming formulation, with an objective function, decision variables, and constraints. However, many types of problems employ well-known nonlinear functions but are not typically thought of in the traditional math programming model framework. In this section we consider several additional examples that include familiar nonlinear functions—the facility location model and the investment portfolio selection model.

Facility Location

In a facility location problem, the objective, in general, is to locate a centralized facility that serves several customers or other facilities to minimize distance or miles traveled between the facility and its customers. This problem uses as a measure of distance the formula for the straight-line distance between two points on a set of x, y coordinates, which is also the hypotenuse of a right triangle:

$$d = \sqrt{(x_i - x)^2 + (y_i - y)^2}$$

where

(x, y) = coordinates of proposed facility
(x_i, y_i) = coordinates of customer or location facility, i

Consider, for example, the Clayton County Rescue Squad and Ambulance Service, which serves five rural towns, Abbeville, Benton, Clayton, Dunning, and Eden. The rescue squad wants to construct a centralized facility and garage to minimize its total annual travel mileage to the towns. The locations of the five towns in terms of their graphical x, y coordinates, measured in miles relative to the point $x = 0$, $y = 0$, and the expected number of annual trips the squad will have to make to each town are as follows:

	Coordinates		
Town	*x*	*y*	**Annual Trips**
Abbeville	20	20	75
Benton	10	35	105
Clayton	25	9	135
Dunning	32	15	60
Eden	10	8	90

The objective of the problem is to determine a set of coordinates (x, y) for the rescue squad facility that minimizes the total miles traveled to the town, according to the following function:

$$\sum_i d_i t_i$$

where

d_i = distance to town i
t_i = annual trips to town i

The Excel solution to this problem is shown in Exhibit 10.12. The coordinates (x, y) for the new rescue squad facility are in cells **C14:C15**. The formulas for the distances between the rescue squad facility and each of the towns (d_i) are in cells **E6:E10**. For example, the formula for the distance, d_A, between the rescue squad facility and Abbeville in cell E6 is **=SQRT((B6–C14)^2+(C6–C15)^2)**. The formula for the total annual distance in cell D18 is **=SUMPRODUCT(E6:E10,D6:D10)**, which is also shown on the formula bar at the top of the spreadsheet. In the Solver Parameters window for this problem, D18 is the target cell that is minimized, and the changing cells (i.e., decision variables) are **C14:C15**. There are no constraints.

EXHIBIT 10.12

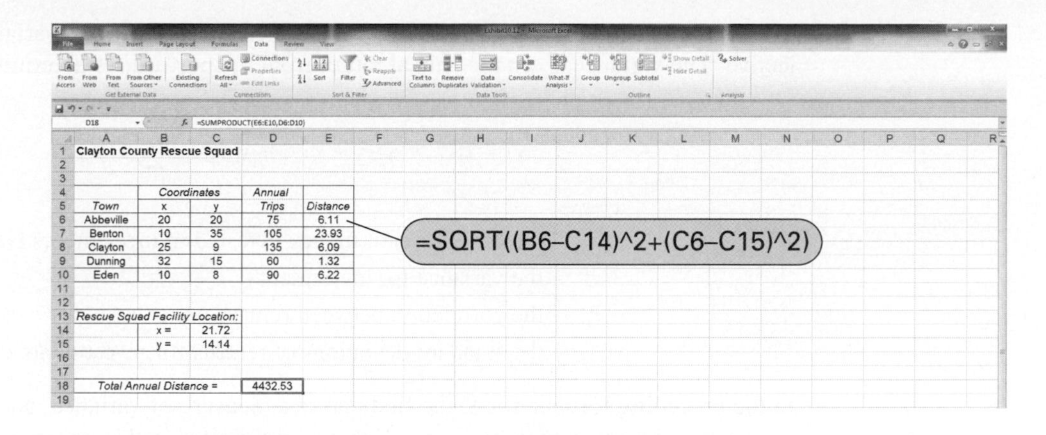

D18 =SUMPRODUCT(E6:E10,D6:D10)

Clayton County Rescue Squad

Town	Coordinates x	Coordinates y	Annual Trips	Distance
Abbeville	20	20	75	6.11
Benton	10	35	105	23.93
Clayton	25	9	135	6.09
Dunning	32	15	60	1.32
Eden	10	8	90	6.22

Rescue Squad Facility Location:	
x =	21.72
y =	14.14

Total Annual Distance =	4432.53

=SQRT((B6–C14)^2+(C6–C15)^2)

The location of the rescue squad facility is shown graphically in Figure 10.8.

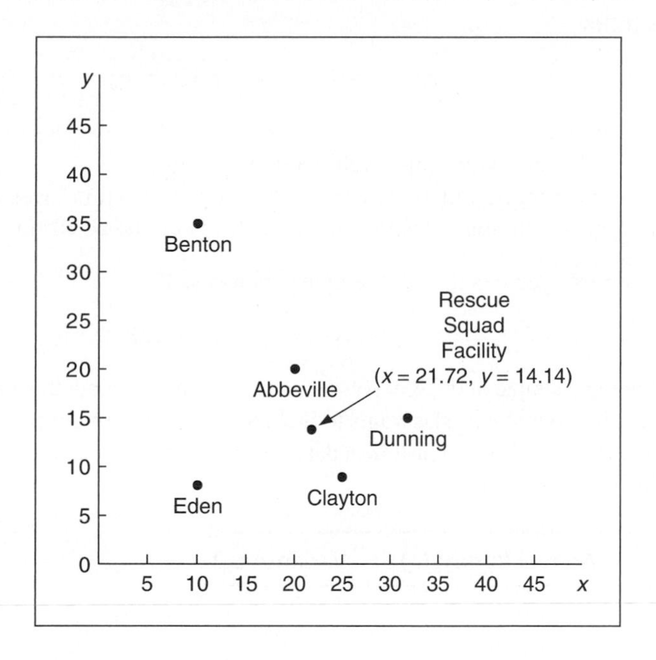

FIGURE 10.8
Rescue squad facility location

Investment Portfolio Selection

A classic example of nonlinear programming is the investment portfolio selection model developed by Harry Markowitz in 1959. This model is based on the assumption that most investors are concerned with two factors—return on investment and risk. Thus, the objective of the portfolio selection model is to minimize some measure of portfolio risk while achieving a minimum return on the total portfolio investment. Risk is reflected by the variability in the value of the investment, and in this model, *variance* in the return on investment is the measure of risk.

Covariance, which is a measure of correlation, is also used to reflect risk. Individual investment returns within a portfolio typically exhibit statistical dependence. Over time, the returns of any two stocks may exhibit positive or negative correlation; that is, two stocks of the same general type will go up or down together. To adjust for this possible correlation, investors often attempt to diversify their portfolios. To reflect the risk associated with *not* diversifying, the model includes covariance.

The minimization of portfolio risk, as measured by the portfolio variance, is the model objective. The variance, S, on the annual return from the portfolio is determined by the following formula:

$$S = x_1^2 s_1^2 + x_2^2 s_2^2 + \cdots + x_n^2 s_n^2 + \sum_{i \neq j} x_i x_j r_{ij} s_i s_j$$

where

x_i, x_j = the proportion of money invested in investments i and j
s_i^2 = the variance for investment i
r_{ij} = the correlation between returns on investments i and j
s_i, s_j = the standard deviations of returns for investments i and j

In the preceding formula for S, the first part is a measure of variance, and the second part is a measure of the covariance. In many cases, the parameters in this equation may be estimates or sample values (the sample variance, sample covariance, etc.).

The general formulation of the portfolio selection model is as follows. The investor desires to achieve a minimum expected annual return from the portfolio, which is formulated as a model constraint as follows:

$$r_i x_i + r_2 x_2 + \cdots + r_n x_n \geq r_m$$

where

r_i = expected annual return on investment i
x_i = proportion (or fraction) of money invested in investment i
r_m = minimum desired annual return from the portfolio

A second constraint specifies that all the money is invested:

$$x_1 + x_2 + \cdots + x_n = 1.0$$

The following example and Excel solution will demonstrate how this model is applied. Jessica Todd has identified four stocks she wants to include in her investment portfolio. She wants a total annual return of at least .11. From historical data, she has estimated the average returns and variances for the four investments as follows:

Stock (x_i)	Annual Return (r_i)	Variance (s_i)
1. Altacam	.08	.009
2. Bestco	.09	.015
3. Com.com	.16	.040
4. Delphi	.12	.023

She has also estimated the covariances between the stocks, as follows:

Stock Combination (i, j)	Covariances (r_{ij})
A,B (1,2)	.4
A,C (1,3)	.3
A,D (1,4)	.6
B,C (2,3)	.2
B,D (2,4)	.7
C,D (3,4)	.4

The nonlinear programming model formulation for this problem is as follows:

$$\begin{aligned}
\text{minimize } Z = S = {} & x_1^2(.009) + x_2^2(.015) + x_3^2(.040) + x_4^2(.230) \\
& + x_1x_2(.4)(.009)^{1/2}(.015)^{1/2} + x_1x_3(.3)(.009)^{1/2}(.040)^{1/2} \\
& + x_1x_4(.6)(.009)^{1/2}(.023)^{1/2} + x_2x_3(.2)(.015)^{1/2}(.040)^{1/2} \\
& + x_2x_4(.7)(.015)^{1/2}(.023)^{1/2} + x_3x_4(.4)(.040)^{1/2}(.023)^{1/2} \\
& + x_2x_1(.4)(.015)^{1/2}(.009)^{1/2} + x_3x_1(.3)(.040)^{1/2}(.009)^{1/2} \\
& + x_4x_1(.6)(.023)^{1/2}(.009)^{1/2} + x_3x_2(.2)(.040)^{1/2}(.015)^{1/2} \\
& + x_4x_2(.7)(.023)^{1/2}(.015)^{1/2} + x_4x_3(.4)(.023)^{1/2}(.040)^{1/2}
\end{aligned}$$

subject to

$$\begin{aligned}
.08x_1 + .09x_2 + .16x_3 + .12x_4 &\geq 0.11 \\
x_1 + x_2 + x_3 + x_4 &= 1.00 \\
x_i &\geq 0
\end{aligned}$$

The Excel spreadsheet for Jessica Todd's portfolio is shown in Exhibit 10.13. The decision variables representing the proportion of each stock in the portfolio (x_i) are in cells **E6:E9**. The formula for total return, **=SUMPRODUCT(B6:B9, E6:E9)**, is in cell B19. The objective is to minimize the total portfolio variance formula, which is shown on the formula bar at the top of the spreadsheet. This is a complicated formula and difficult to type in. It has been shortened slightly by doubling the covariance terms in the last part of the equation to reflect all the different investment pairs that are included in **B14:E17** (i.e., 2 times AB, AC, AD, BC, BD, and CD will include BA, CA, DA, CB, DB, and DC). In other words, all the covariance terms to the right of the diagonal in the matrix **B14:E17** are doubled to include the same values to the left of the diagonal.

EXHIBIT 10.13

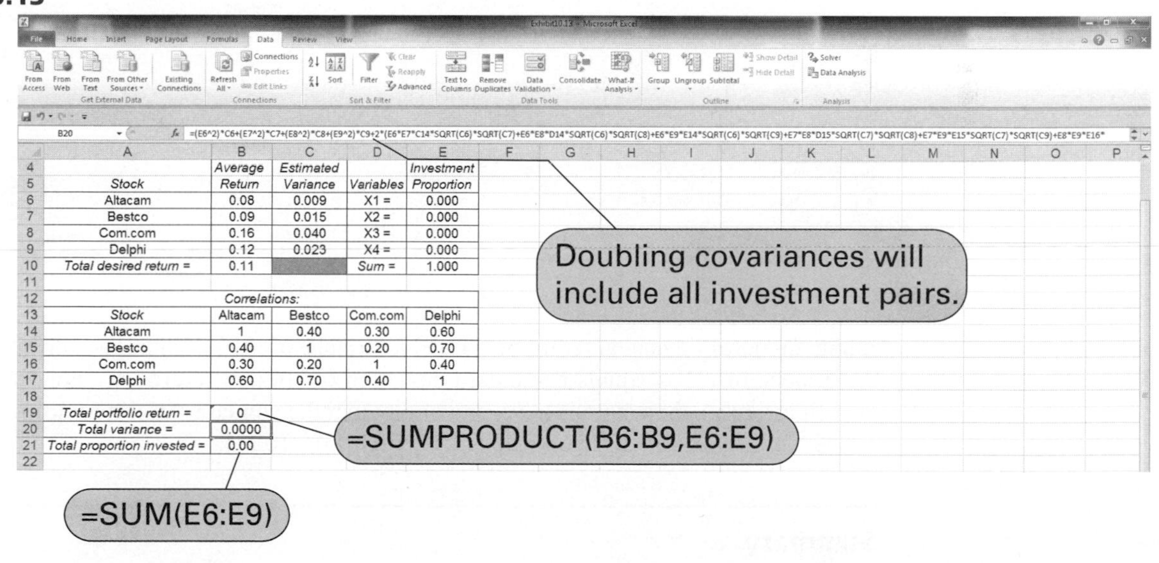

	A	B	C	D	E
4		Average	Estimated		Investment
5	Stock	Return	Variance	Variables	Proportion
6	Altacam	0.08	0.009	X1 =	0.000
7	Bestco	0.09	0.015	X2 =	0.000
8	Com.com	0.16	0.040	X3 =	0.000
9	Delphi	0.12	0.023	X4 =	0.000
10	Total desired return =	0.11		Sum =	1.000
11					
12		Correlations:			
13	Stock	Altacam	Bestco	Com.com	Delphi
14	Altacam	1	0.40	0.30	0.60
15	Bestco	0.40	1	0.20	0.70
16	Com.com	0.30	0.20	1	0.40
17	Delphi	0.60	0.70	0.40	1
18					
19	Total portfolio return =	0			
20	Total variance =	0.0000			
21	Total proportion invested =	0.00			
22					

The Solver Parameters window is shown in Exhibit 10.14. Notice that nonnegativity constraints for the variables are included—by clicking on "Options." The model solution is shown in Exhibit 10.15. It indicates that Jessica should invest 36.0% of her funds in Altacam (x_1), 27.2% in Bestco (x_2), 31.5% in Com.com (x_3), and 5.3% in Delphi (x_4), This will result in her desired

EXHIBIT 10.14

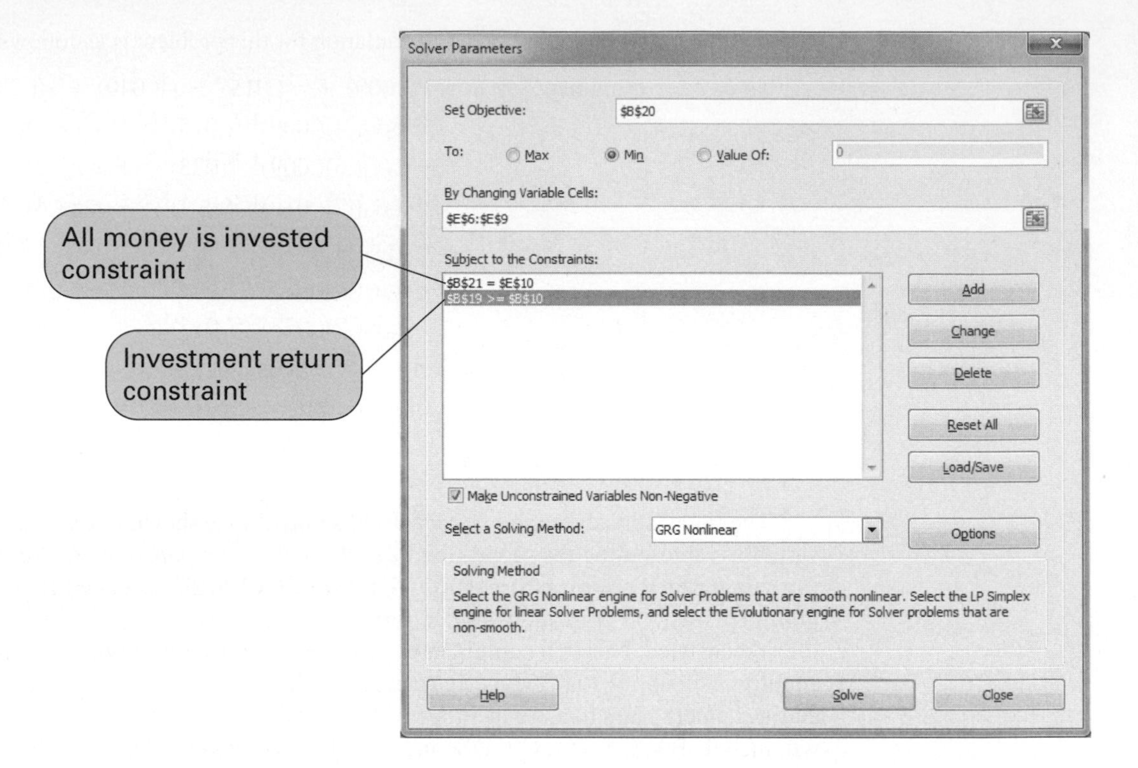

EXHIBIT 10.15

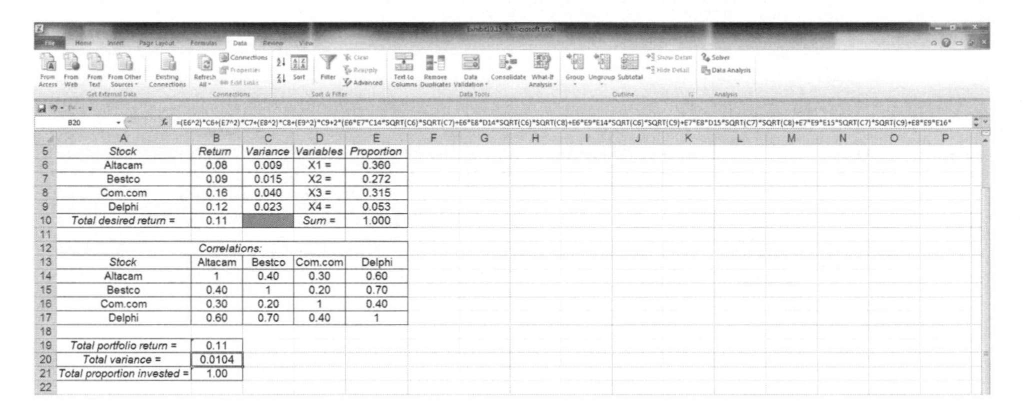

	A	B	C	D	E
5	Stock	Return	Variance	Variables	Proportion
6	Altacam	0.08	0.009	X1 =	0.360
7	Bestco	0.09	0.015	X2 =	0.272
8	Com.com	0.16	0.040	X3 =	0.315
9	Delphi	0.12	0.023	X4 =	0.053
10	Total desired return =	0.11		Sum =	1.000
11					
12		Correlations:			
13	Stock	Altacam	Bestco	Com.com	Delphi
14	Altacam	1	0.40	0.30	0.60
15	Bestco	0.40	1	0.20	0.70
16	Com.com	0.30	0.20	1	0.40
17	Delphi	0.60	0.70	0.40	1
18					
19	Total portfolio return =	0.11			
20	Total variance =	0.0104			
21	Total proportion invested =	1.00			
22					

minimum portfolio annual return of .11 and a variance of .0104. What does this variance mean? A variance of .0104 equals a standard deviation of .102. If returns are normally distributed (which is likely), then there is a .95 probability that the annual portfolio return will be between −.0899 [.11 − (1.96)(.102)] and .3099 [.11 + (1.96)(.102)].

Summary

In this chapter, we presented an overview of nonlinear programming problems and solutions. We have used relatively simple examples because of the complexity involved in formulating and solving larger problems. A more in-depth understanding of this topic requires significant knowledge of calculus and higher mathematics. Nevertheless, this chapter has provided a general understanding of the types of problems that require nonlinear solutions and the general format for modeling nonlinear problems. It has also demonstrated the capability of Excel for solving nonlinear programming problems.

Example Problem Solution

The following example illustrates the use of the substitution method and the method of Lagrange multipliers for solving a nonlinear programming problem.

Problem Statement

The Hickory Cabinet and Furniture Company makes chairs and tables. The company has developed the following nonlinear programming model to determine the optimal number of chairs (x_1) and tables (x_2) to produce each day to maximize profit, given a constraint for available mahogany wood:

$$\text{maximize } Z = \$280x_1 - 6x_1^2 + 160x_2 - 3x_2^2$$
$$\text{subject to}$$
$$20x_1 + 10x_2 = 800 \text{ bd. ft.}$$

Determine the Excel solution to this model.

Solution

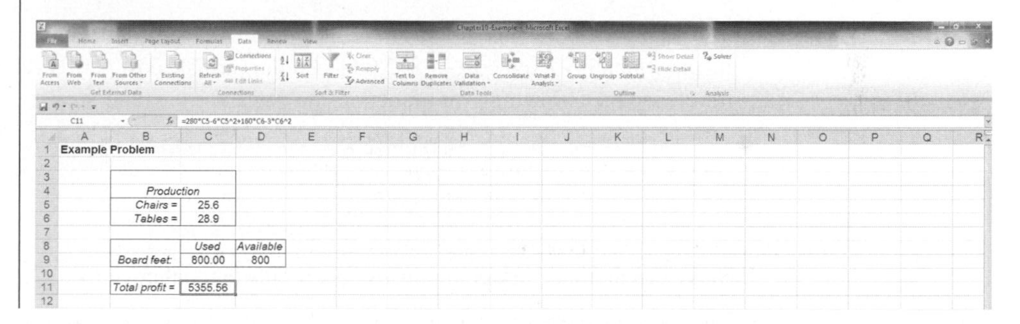

Problems

1. The Hickory Cabinet and Furniture Company makes chairs. The fixed cost per month of making chairs is \$7,500, and the variable cost per chair is \$40. Price is related to demand, according to the following linear equation:

$$v = 400 - 1.2p$$

Develop the nonlinear profit function for this company and determine the price that will maximize profit, the optimal volume, and the maximum profit per month.

2. Graphically illustrate the profit curve developed in Problem 1. Indicate the optimal price and the maximum profit per month.

3. The Rainwater Brewery produces beer. The annual fixed cost is \$150,000, and the variable cost per barrel is \$16. Price is related to demand, according to the following linear equation:

$$v = 75{,}000 - 1{,}153.8p$$

Develop the nonlinear profit function for the brewery and determine the price that will maximize profit, the optimal volume, and the maximum profit per year.

4. The Rolling Creek Textile Mill makes denim. The monthly fixed cost is \$8,000, and the variable cost per yard of denim is \$0.35. Price is related to demand, according to the following linear equation:

$$v = 17{,}000 - 5{,}666p$$

Develop the nonlinear profit function for the textile mill and determine the optimal price, the optimal volume, and the maximum profit per month.

5. The Grady Tire Company recaps tires. The weekly fixed cost is $2,500, and the variable cost per tire is $9. Price is related to demand, according to the following linear equation:

$$v = 200 - 4.75p$$

Develop the nonlinear profit function for the tire company and determine the optimal price, the optimal volume, and the maximum profit per week.

6. Andy Mendoza makes handcrafted dolls, which he sells at craft fairs. He is considering mass-producing the dolls to sell in stores. He estimates that the initial investment for plant and equipment will be $25,000, while labor, materials, packaging, and shipping will be about $10 per doll. He has determined that sales volume is related to price, according to the following linear equation:

$$v = 4{,}000 - 80p$$

Develop the nonlinear profit function for Andy and determine the price that will maximize profit, the optimal volume, and the maximum profit per month.

7. The Rainwater Brewery produces beer, which it sells to distributors in barrels. The brewery incurs a monthly fixed cost of $12,000, and the variable cost per barrel is $17. The brewery has developed the following profit function and demand constraint:

$$\text{maximize } Z = vp - \$12{,}000 - 17v$$

subject to

$$v = 1{,}800 - 15p$$

Solve this nonlinear programming model for the optimal price (p).

8. The Beaver Creek Pottery Company has developed the following nonlinear programming model to determine the optimal number of bowls (x_1) and mugs (x_2) to produce each day:

$$\text{maximize } Z = \$7x_1 - 0.3x_1^2 + 8x_2 - 0.4x_2^2$$

subject to

$$4x_1 + 5x_2 = 100 \text{ hr.}$$

Determine the optimal solution to this nonlinear programming model.

9. The Evergreen Fertilizer Company produces two types of fertilizers, Fastgro and Super Two. The company has developed the following nonlinear programming model to determine the optimal number of bags of Fastgro (x_1) and Super Two (x_2) that it must produce each day to maximize profit, given a constraint for available potassium:

$$\text{maximize } Z = \$30x_1 - 2x_1^2 + 25x_2 - 0.5x_2^2$$

subject to

$$3x_1 + 6x_2 = 300 \text{ lb.}$$

Determine the optimal solution to this nonlinear programming model.

10. The Rolling Creek Textile Mill produces denim and brushed cotton cloth. The company has developed the following nonlinear programming model to determine the optimal number of yards of denim (x_1) and brushed cotton (x_2) to produce each day in order to maximize profit, subject to a labor constraint:

$$\text{maximize } Z = \$10x_1 - 0.02x_1^2 + 12x_2 - 0.03x_2^2$$

subject to

$$0.2x_1 + 0.1x_2 = 40 \text{ hr.}$$

Determine the optimal solution to this nonlinear programming model.

11. In the investment portfolio selection model in this chapter, what if Jessica Todd decided that she wanted to maximize her return while incurring a portfolio variance of no more than 0.020? Analyze this scenario by using the Excel spreadsheet in Exhibit 10.13 to determine what Jessica's portfolio and return would be.

12. The Riverwood Paneling Company makes two kinds of wood paneling: Colonial and Western. The company has developed the following nonlinear programming model to determine the optimal number of sheets of Colonial paneling (x_1) and Western paneling (x_2) to produce to maximize profit, subject to a labor constraint:

$$\text{maximize } Z = \$25x_1 - 0.8x_1^2 + 30x_2 - 1.2x_2^2$$

subject to

$$x_1 + 2x_2 = 40 \text{ hr.}$$

Determine the optimal solution to this nonlinear programming model.

13. Interpret the meaning of the Lagrange multiplier in Problem 12.

14. Metro Telcom Systems develops, sells, and installs computer systems. The company has divided its customer base into five regions, and it has 15 representatives who sell and install the company's systems. The company wants to allocate salespeople to regions so that they maximize daily sales revenue. However, whereas the sales increase as the number of salespeople allocated to a region increases, they do so at a declining rate, according to the following nonlinear formula:

$$\text{total sales} = a - (b/x)$$

Following are the a and b parameters for daily sales in each region:

	Region				
	1	2	3	4	5
a	\$15,000	\$24,000	\$8,100	\$12,000	\$21,000
b	9,000	15,000	5,300	7,600	12,500

Because some of the regions are in urban areas and some are not, the representatives' daily expenses will differ among regions. The company has a daily expense budget of \$6,500, and the daily expenses (including travel costs) per representative for each region average \$355 for region 1, \$540 for region 2, \$290 for region 3, \$275 for region 4, and \$490 for region 5. Formulate and solve a nonlinear programming model for this problem to determine the number of representatives to allocate to each region to maximize daily sales.

15. Blue Ridge Power Company has a maximum capacity of 2.5 million kilowatt-hours (kWh) of electric power available on a daily basis. The demand (in millions of kWh) for power from its customers for high-demand (peak) hours and low-demand (off-peak) hours is determined by the following formulas:

$$\text{high demand: } 5.8 - 0.06p_h + 0.005p_l$$
$$\text{low demand: } 3.0 - 0.11p_l + 0.008p_h$$

The variable p_l equals the price per kilowatt-hour during low-demand hours, and p_h is the price per kilowatt-hour during high-demand (peak) hours.

Formulate and solve a nonlinear programming model to determine the price structure (per kWh) that will maximize revenue.

16. The Metro Police Department has partitioned the city into four quadrants. The department has 20 police patrol cars available for each shift per day to assign to the quadrants. The department wants to assign patrol cars to the quadrants during a shift so that the crime rate is minimized while providing average response times to calls of less than 10 minutes. As patrol cars

are assigned to a quadrant, the crime rate and response time decrease, but at a decreasing rate according to the following formula:

$$y = a + (b/x)$$

The a and b parameters for daily crime rate (expressed as crimes per 1,000 population) and response time (in minutes) are provided in the following table:

	Crime Rate		Response Time (min.)	
Quadrant	*a*	*b*	*a*	*b*
1	0.24	0.15	4	11
2	0.37	0.21	8	8
3	0.21	0.12	6	10
4	0.48	0.30	3	9

Formulate and solve a nonlinear programming model to determine the number of police patrol cars to assign to each quadrant that will result in the minimum overall crime rate.

17. The Burger Doodle restaurant chain purchases ingredients from four different food suppliers. The company wants to construct a new central distribution center to process and package the ingredients it uses in its menu items before shipping them to their various restaurants. The suppliers transport the food items in 40-foot tractor-trailer trucks. The coordinates of the four suppliers and the annual number of truckloads that will be transported to the distribution center are as follows:

	Coordinates		
Supplier	*x*	*y*	**Annual Truckloads**
A	200	200	65
B	100	500	120
C	250	600	90
D	500	300	75

Determine the set of coordinates for the new distribution center that will minimize the total miles traveled from the suppliers.

18. Home-Base, a home improvement and building supply chain, is going to build a new warehouse facility to serve its stores in six North Carolina cities—Charlotte, Winston-Salem, Greensboro, Durham, Raleigh, and Wilmington. The coordinates of these cities (in miles), using Columbia, South Carolina, as the graphical origin (0, 0), and the annual truckload trips that supply each store are as follows:

	Coordinates		
Store	*x*	*y*	**Truckload**
Charlotte	15	85	160
Winston-Salem	42	145	90
Greensboro	88	145	105
Durham	125	140	35
Raleigh	135	125	60
Wilmington	180	18	75

Determine the set of coordinates for the new warehouse that will minimize the total miles traveled to the stores and identify on a map the closest town to these coordinates.

19. An investment adviser is helping a couple plan a retirement portfolio. The adviser has recommended three stocks—Allied Electronics, Bank United, and Consolidated Computers. Following are the annual return and variance for each stock and the covariance between stocks:

Stock	Annual Return	Covariance
Allied Electronics	.14	.10
Bank United	.10	.04
Consolidated Computers	.12	.08

Stock Combination (*i*, *j*)	Covariance
A, B	.4
A, C	.7
B, C	.3

The couple wants a total portfolio return of at least .11. Determine the proportion of each stock to include in the portfolio to minimize the overall risk.

20. Mark Decker has identified four stocks for his portfolio, and he wants to determine the percentage of his total available funds he should invest in each stock. The alternative stocks include an Internet company, a computer software company, a computer manufacturer, and an entertainment conglomerate. He wants a total annual return of .12. From historical data, he has determined the average annual return and variance for each of the funds, as follows:

Stock	Annual Return	Variance
1. Internet	.18	.112
2. Software	.12	.061
3. Computer	.10	.045
4. Entertainment	.15	.088

He has also estimated the covariances between stocks, as follows:

Stock Combination (*i*, *j*)	Covariances
1, 2	.9
1, 3	.7
1, 4	.3
2, 3	.8
2, 4	.4
3, 4	.2

Determine the percentage of Mark's total funds that he should invest in each stock to minimize his overall risk.

21. The Accentuate Consulting Firm has eight projects for which it has contracted with clients to develop computer systems and software. The firm has 35 team members it can assign to the

projects. The firm has developed the following profit functions for each project, where x_i equals the number of team members assigned to a project:

Project	Profit
1	$\$250{,}000x_1^{0.50}$
2	$370{,}000x_2^{0.30}$
3	$140{,}000x_3^{0.70}$
4	$500{,}000x_4^{0.25}$
5	$230{,}000x_5^{0.33}$
6	$170{,}000x_6^{0.60}$
7	$280{,}000x_7^{0.45}$
8	$315{,}000x_8^{0.40}$

These functions take into account the project completion time, cost, and probability of successful completion based on the number of team members assigned. The functions reflect the fact that profit increases but at a decreasing rate as team members are assigned. Thus, as additional team members are assigned to a project, the marginal effect decreases. Determine the optimal number of team members to assign to each project in order to maximize profit.

Case Problem

Admissions at State University

State University has increased its tuition for in-state and out-of-state students in each of the past 5 years to offset cuts in its budget allocation from the state legislature. The university administration always thought that the number of applications received was independent of tuition; however, drops in applications and enrollments the past 2 years have proved this theory to be wrong. University admissions officials have developed the following relationships between the number of applicants who accept admission and enter State (x_i) and the cost of tuition per semester (t_i):

$$x_1 = 21{,}000 - 12t_1 \text{ (in-state)}$$
$$x_2 = 35{,}000 - 6t_2 \text{ (out-of-state)}$$

The university would like to develop a planning model that will indicate the in-state and out-of-state tuitions, as well as the number of students that could be expected to enroll in the freshman class. The university doesn't have enough classroom space for more than 1,400 freshmen, and it needs at least 700 freshmen to meet all its class-size objectives. The university knows from historical data that approximately 55% of all in-state freshmen will want to live in the school dormitories, 72% of all out-of-state students will want to live in the dormitories, and there will be at most 800 dormitory rooms available for freshmen.

The university also wants to make sure that it maintains high academic standards in its admissions decisions. It knows from historical data that the average SAT score for an in-state student is 960 and the average SAT score for an out-of-state student is 1,150. The university wants the entering freshman class to have an average SAT score of at least 1,000.

State University is a state-supported institution, so the state legislature wants to make sure that the university doesn't enroll just out-of-state students because they pay more tuition and they have better SAT scores. Thus, the government has instituted a policy that no more than 55% of the entering freshman class can be out-of-state students.

Develop and solve a nonlinear programming model for State University to indicate the tuition the university should charge, the total tuition, and the number of in-state and out-of-state students it can expect with these tuition rates.

Case Problem

Selecting a European Distribution Center Site for American International Automotive Industries

American International Automotive Industries (AIAI) manufactures engine, transmission, and chassis parts for manufacturers and repair companies in the United States, South America, Canada, Mexico, Asia, and Europe. The company transports parts to its foreign markets by container ships. To serve its customers in South America and Asia, AIAI has large warehouse and distribution centers. In Europe, it ships into Hamburg and Gdansk, where it has contracted with independent distribution companies to deliver its products to customers throughout Europe. However, AIAI has been displeased with a recent history of late deliveries and rough handling of its products. For a time AIAI was not overly concerned because its European market wasn't too big, and its European customers didn't complain. Plus, it had more pressing problems elsewhere. However, in the past few years, trade barriers have fallen in Europe, and Eastern European markets have opened up. AIAI's European business has expanded, as has new competition, and its customers have become more demanding and quality conscious. As a result, AIAI has initiated the selection process for the site of a new European warehouse and distribution center. Although it provides parts to a number of smaller truck and auto maintenance and service centers in Europe, it has seven major customers—auto and truck manufacturers in Vienna, Leipzig, Budapest, Prague, Krakow, Munich, and Frankfurt. Its customers have adopted manufacturing processes requiring continuous replenishment of parts and materials.

AIAI's European headquarters is in Hamburg. The vice president for construction and development in Dayton, Ohio, has asked the Hamburg office to do a preliminary site search based on proximity to customers and mileage. Each container is converted to a single tractor-trailer truckload. The number of containers shipped annually to each customer is as follows: Vienna, 160; Leipzig, 100; Budapest, 180; Prague, 210; Krakow, 90; Munich, 120; and Frankfurt, 50. When the vice president of construction in Dayton received this information, he pulled out his map of Europe and began to look for possible sites.

Assist AIAI with its site selection process in Europe. Recommend a city site that minimizes the total annual delivery mileage.

CHAPTER

11

Probability and Statistics

Deterministic techniques assume that no uncertainty exists in model parameters.

The techniques presented in Chapters 2 through 10 are typically thought of as **deterministic**; that is, they are not subject to uncertainty or variation. With deterministic techniques, the assumption is that conditions of complete certainty and perfect knowledge of the future exist. In the linear programming models presented in previous chapters, the various parameters of the models and the model results were assumed to be known with certainty. In the model constraints, we did not say that a bowl would require 4 pounds of clay "70% of the time." We specifically stated that each bowl would require exactly 4 pounds of clay (i.e., there was no uncertainty in our problem statement). Similarly, the solutions we derived for the linear programming models contained no variation or uncertainty. It was assumed that the results of the model would occur in the future, without any degree of doubt or chance.

Probabilistic techniques include uncertainty and assume that there can be more than one model solution.

In contrast, many of the techniques in management science do reflect *uncertain* information and result in *uncertain* solutions. These techniques are said to be **probabilistic**. This means that there can be more than one outcome or result to a model and that there is some doubt about which outcome will occur. The solutions generated by these techniques have a probability of occurrence. They may be in the form of *averages*; thc actual valucs that occur will vary over time.

Many of the upcoming chapters in this text present probabilistic techniques. The presentation of these techniques requires that the reader have a fundamental understanding of probability. Thus, the purpose of this chapter is to provide an overview of the fundamentals, properties, and terminology of probability and statistics.

Types of Probability

Two basic types of probability can be defined: *objective probability* and *subjective probability*. First, we will consider what constitutes objective probability.

Objective Probability

Consider a referee's flipping a coin before a football game to determine which team will kick off and which team will receive. Before the referee tosses the coin, both team captains know that they have a .50 (or 50%) probability (or chance) of winning the toss. None of the onlookers in the stands or anywhere else would argue that the probability of a head or a tail was not .50. In this example, the probability of .50 that either a head or a tail will occur when a coin is tossed is called an **objective probability**. More specifically, it is referred to as a **classical** or **a priori** (prior to the occurrence) **probability**, one of the two types of objective probabilities.

Objective probabilities that can be stated prior to the occurrence of an event are classical, or a priori, probabilities.

A classical, or a priori, probability can be defined as follows: Given a set of outcomes for an activity (such as a head or a tail when a coin is tossed), the probability of a specific (desired) outcome (such as a head) is the ratio of the number of specific outcomes to the total number of outcomes. For example, in our coin-tossing example, the probability of a head is the ratio of the number of specific outcomes (a head) to the total number of outcomes (a head and a tail), or 1/2. Similarly, the probability of drawing an ace from a deck of 52 cards would be found by dividing 4 (the number of aces) by 52 (the total number of cards in a deck) to get 1/13. If we spin a roulette wheel with 50 red numbers and 50 black numbers, the probability of the wheel's landing on a red number is 50 divided by 100, or 1/2.

These examples are referred to as a priori probabilities because we can state the probabilities *prior to* the actual occurrence of the activity (i.e., ahead of time). This is because we know (or assume we know) the number of specific outcomes and total outcomes prior to the occurrence of the activity. For example, we know that a deck of cards consists of 4 aces and 52 total cards before we draw a card from the deck and that a coin contains one head and one tail before we toss it. These probabilities are also known as *classical* probabilities because some of the earliest references in history to probabilities were related to games of chance, to which (as the preceding examples show) these probabilities readily apply.

Objective probabilities that are stated after the outcomes of an event have been observed are relative frequency probabilities.

The second type of objective probability is referred to as **relative frequency probability**. This type of objective probability indicates the relative frequency with which a specific outcome has been observed to occur in the long run. It is based on the observation of past occurrences. For example, suppose that over the past 4 years, 3,000 business students have taken the introductory management science course at State University, and 300 of them have made an A in the course. The relative frequency probability of making an A in management science would be 300/3,000 or .10. Whereas in the case of a classical probability we indicate a probability before an activity (such as tossing a coin) takes place, in the case of a relative frequency we determine the probability after observing, for example, what 3,000 students have done in the past.

Relative frequency is the more widely used definition of objective probability.

The relative frequency definition of objective probability is more general and widely accepted than the classical definition. Actually, the relative frequency definition can encompass the classical case. For example, if we flip a coin many times, in the long run the relative frequency of a head's occurring will be .50. If, however, you tossed a coin 10 times, it is conceivable that you would get 10 consecutive heads. Thus, the relative frequency (probability) of a head would be 1.0. This illustrates one of the key characteristics of a relative frequency probability: The relative frequency probability becomes more accurate as the total number of observations of the activity increases. If a coin were tossed about 350 times, the relative frequency would approach 1/2 (assuming a fair coin).

Subjective Probability

When relative frequencies are not available, a probability is often determined anyway. In these cases a person must rely on personal belief, experience, and knowledge of the situation to develop a probability estimate. A probability estimate that is not based on prior or past evidence is a **subjective probability**. For example, when a meteorologist forecasts a "60% chance of rain tomorrow," the .60 probability is usually based on the meteorologist's experience and expert analysis of the weather conditions. In other words, the meteorologist is not saying that these exact weather conditions have occurred 1,000 times in the past and on 600 occasions it has rained, thus there is a 60% probability of rain. Likewise, when a sportswriter says that a football team has an 80% chance of winning, it is usually not because the team has won 8 of its 10 previous games. The prediction is judgmental, based on the sportswriter's knowledge of the teams involved, the playing conditions, and so forth. If the sportswriter had based the probability estimate solely on the team's relative frequency of winning, then it would have been an objective probability. However, once the relative frequency probability becomes colored by personal belief, it is subjective.

***Subjective probability** is an estimate based on personal belief, experience, or knowledge of a situation.*

Subjective probability estimates are frequently used in making business decisions. For example, suppose the manager of Beaver Creek Pottery Company (referred to in Chapters 2 and 3) is thinking about producing plates in addition to the bowls and mugs it already produces. In making this decision, the manager will determine the chances of the new product's being successful and returning a profit. Although the manager can use personal knowledge of the market and judgment to determine a *probability of success*, direct relative frequency evidence is not generally available. The manager cannot observe the frequency with which the introduction of this new product was successful in the past. Thus, the manager must make a subjective probability estimate.

This type of subjective probability analysis is common in the business world. Decision makers frequently must consider their chances for success or failure, the probability of achieving a certain market share or profit, the probability of a level of demand, and the like without the benefit of relative frequency probabilities based on past observations. Although there may not be a consensus as to the accuracy of a subjective estimate, as there is with an objective probability (e.g., everyone is sure there is a .50 probability of getting a head when a coin is tossed), subjective probability analysis is often the only means available for making probabilistic estimates, and it is frequently used.

A brief note of caution must be made regarding the use of subjective probabilities. Different people will often arrive at different subjective probabilities, whereas everyone should arrive at the same objective probability, given the same numbers and correct calculations. Therefore, when a probabilistic analysis is to be made of a situation, the use of objective probability will provide more consistent results. In the material on probability in the remainder of this chapter, we will use objective probabilities unless the text indicates otherwise.

Fundamentals of Probability

*An **experiment** is an activity that results in one of several possible outcomes.*

Let us return to our example of a referee's tossing a coin prior to a football game. In the terminology of probability, the coin toss is referred to as an **experiment**. An experiment is an activity (such as tossing a coin) that results in one of the several possible outcomes. Our coin-tossing experiment can result in either one of two outcomes, which are referred to as **events**: a head or a tail. The probabilities associated with each event in our experiment follow:

Event	***Probability***
Head	.50
Tail	.50
	1.00

Two fundamentals of probability: $0 \leq P$ (events) ≤ 1.0, and the probabilities of all events in an experiment sum to one.

This simple example highlights two of the fundamental characteristics of probability. First, *the probability of an event is always greater than or equal to zero and less than or equal to one* [i.e., $0 \leq P$ (event) ≤ 1.0]. In our coin-tossing example, each event has a probability of .50, which is in the range of 0 to 1.0. Second, *the probability of all the events included in an experiment must sum to one*. Notice in our example that the probability of each of the two events is .50, and the sum of these two probabilities is 1.0.

*The events in an experiment are **mutually exclusive** if only one can occur at a time.*

The specific example of tossing a coin also exhibits a third characteristic: The events in a set of events are **mutually exclusive**. The events in an experiment are mutually exclusive if only one of them can occur at a time. In the context of our experiment, the term *mutually exclusive* means that any time the coin is tossed, *only one* of the two events can take place—either a head or a tail can occur, but not both. Consider a customer who enters a store to shop for shoes. The store manager estimates that there is a .60 probability that the customer will buy a pair of shoes and a .40 probability that the customer will not buy a pair of shoes. These two events are mutually exclusive because it is impossible to buy shoes and not buy shoes at the same time. In general, events are mutually exclusive if only one of the events can occur, but not both.

The probabilities of mutually exclusive events sum to one.

Because the events in our example of obtaining a head or tail are mutually exclusive, we can infer that the probabilities of mutually exclusive events *sum to 1.0*. Also, the probabilities of mutually exclusive events can be added. The following example will demonstrate these fundamental characteristics of probability.

The staff of the dean of the business school at State University has analyzed the records of the 3,000 students who received a grade in management science during the past 4 years. The dean wants to know the number of students who made each grade (A, B, C, D, or F) in the course. The dean's staff has developed the following table of information:

Event Grade	Number of Students	Relative Frequency	Probability
A	300	300/3,000	.10
B	600	600/3,000	.20
C	1,500	1,500/3,000	.50
D	450	450/3,000	.15
F	150	150/3,000	.05
	3,000		1.00

*A **frequency distribution** is an organization of numeric data about the events in an experiment.*

This example demonstrates several of the characteristics of probability. First, the *data* (numerical information) in the second column show how the students are distributed across the different grades (events). The third column shows the relative frequency with which each event occurred for the 3,000 observations. In other words, the relative frequency of a student's making a C is 1,500/3,000, which also means that the probability of selecting a student who had obtained a C at random from those students who took management science in the past 4 years is .50.

*A **probability distribution** shows the probability of occurrence of all events in an experiment.*

This information, organized according to the events in the experiment, is called a **frequency distribution**. The list of the corresponding probabilities for each event in the last column is referred to as a **probability distribution**.

All the events in this example are mutually exclusive; it is not possible for two or more of these events to occur at the same time. A student can make only one grade in the course, not two or more grades. As indicated previously, mutually exclusive probabilities of an experiment can be summed to equal one. There are five mutually exclusive events in this experiment, the probabilities of which (.10, .20, .50, .15, and .05) sum to one.

This example exhibits another characteristic of probability: Because the five events in the example are all that can occur (i.e., no other grade in the course is possible), the experiment is said to be **collectively exhaustive**. Likewise, the coin-tossing experiment is collectively exhaustive because the only two events that can occur are a head and a tail. In general, when a *set of events* includes all the events that can possibly occur, the set is said to be collectively exhaustive.

*A set of events is **collectively exhaustive** when it includes all the events that can occur in an experiment.*

The probability of a single event occurring, such as a student receiving an A in a course, is represented symbolically as $P(\text{A})$. This probability is called the **marginal probability** in the terminology of probability. For our example, the marginal probability of a student's getting an A in management science is

*A **marginal probability** is the probability of a single event occurring.*

$$P(\text{A}) = .10$$

For mutually exclusive events, it is possible to determine the probability that one or the other of several events will occur. This is done by summing the individual marginal probabilities of the events. For example, the probability of a student receiving an A or a B is determined as follows:

$$P(\text{A or B}) = P(\text{A}) + P(\text{B}) = .10 + .20 = .30$$

In other words, 300 students received an A and 600 students received a B; thus, the number of students who received an A or a B is 900. Dividing 900 students who received an A or a B by the total number of students, 3,000, yields the probability of a student's receiving an A or a B [i.e., $P(\text{A or B}) = .30$].

Mutually exclusive events can be shown pictorially with a **Venn diagram**. Figure 11.1 shows a Venn diagram for the mutually exclusive events A and B in our example.

*A **Venn diagram** visually displays mutually exclusive and non–mutually exclusive events.*

FIGURE 11.1
Venn diagram for mutually exclusive events

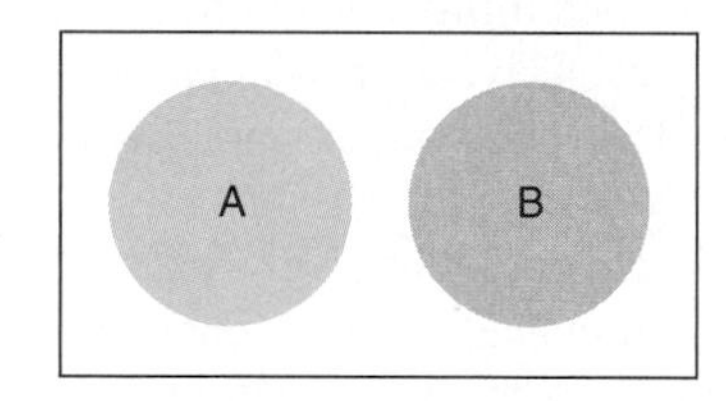

Now let us consider a case in which two events are *not* mutually exclusive. In this case the probability that A or B or *both* will occur is expressed as

$$P(\text{A or B}) = P(\text{A}) + P(\text{B}) - P(\text{AB})$$

where the term $P(\text{AB})$, referred to as the **joint probability** of A and B, is the probability that *both* A and B will occur. For mutually exclusive events, this term has to equal zero because both events cannot occur together. Thus, for mutually exclusive events our formula becomes

*A **joint probability** is the probability that two or more events that are not mutually exclusive can occur simultaneously.*

$$\begin{aligned} P(\text{A or B}) &= P(\text{A}) + P(\text{B}) - P(\text{AB}) \\ &= P(\text{A}) + P(\text{B}) - 0 \\ &= P(\text{A}) + P(\text{B}) \end{aligned}$$

which is the same formula we developed previously for mutually exclusive events.

The following example will illustrate the case in which two events are *not* mutually exclusive. Suppose it has been determined that 40% of all students in the school of business are at present taking management, and 30% of all the students are taking finance. Also, it has been determined that 10% take both subjects. Thus, our probabilities are

$$\begin{aligned} P(\text{M}) &= .40 \\ P(\text{F}) &= .30 \\ P(\text{MF}) &= .10 \end{aligned}$$

Management Science Application

Treasure Hunting with Probability and Statistics

Library of Congress, LC-USZ62-92066

In 1857, the SS *Central America,* a wooden-hulled steamship, sank during a hurricane approximately 20 miles off the coast of Charleston, South Carolina, in the Atlantic Ocean. A total of 425 passengers and crew members lost their lives, and approximately $400 million in gold bars and coins sank to the ocean bottom approximately 8,000 feet (i.e., 1.5 miles) below the surface. This shipwreck was the most famous of its time, causing a financial panic in the United States and a recommendation from insurance investigators that new ships be built with iron hulls and watertight compartments.

In 1985, the Columbus-America Discovery Group was formed to locate and recover the remains of the SS *Central America*. A search plan was developed, using statistical techniques and Monte Carlo simulation. From the available historical information about the wreck, three scenarios were constructed, describing different search areas based on different accounts from nearby ships and witnesses. Using historical statistical data on weather, winds and currents in the area, and estimates of uncertainties in celestial navigation, probability distributions were developed for different possible sites where the ship might have sunk. Using Monte Carlo simulation, a map was configured for each scenario, consisting of cells representing locations. Each cell was assigned a probability that the ship sank in that location. The probability maps for the three scenarios were combined into a composite probability map, which provided a guide for searching the ocean bottom using sonar. The search consisted of long, straight paths that concentrated on the high-probability areas. Based on the results of the sonar search of 1989, the discovery group recovered 1 ton of gold bars and coins from the wreck. In 1993 the total gold recovered from the wreck was valued at $21 million. However, because of the historical significance of all recovered items, they were expected to bring a much higher return when sold.

Source: Based on L. D. Stone, "Search for the SS *Central America:* Mathematical Treasure Hunting," *Interfaces* 22, no. 1 (January–February 1992): 32–54.

The probability of a student's taking one or the other or both of the courses is determined as follows:

$$P(\text{M or F}) = P(\text{M}) + P(\text{F}) - P(\text{MF}) = .40 + .30 - .10 = .60$$

Observing this formulation closely, we can see why the joint probability, $P(\text{MF})$, was subtracted out. The 40% of the students who were taking management also included those students taking both courses. Likewise, the 30% of the students taking finance also included those students taking both courses. Thus, if we add the two marginal probabilities, we are *double-counting* the percentage of students taking both courses. By subtracting out one of these probabilities (that we added in twice), we derive the correct probability.

Figure 11.2 contains a Venn diagram that shows the two events, M and F, that are not mutually exclusive, and the joint event, MF.

FIGURE 11.2
Venn diagram for non–mutually exclusive events and the joint event

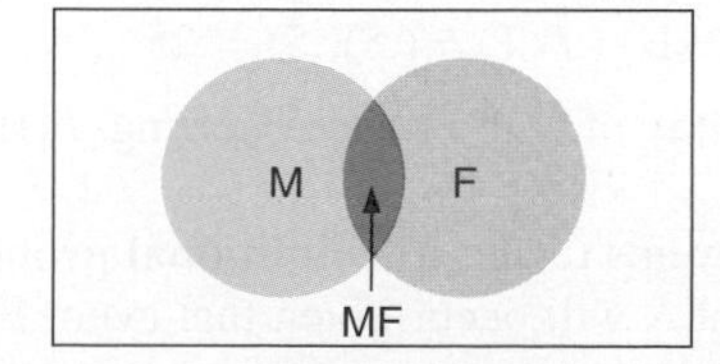

An alternative way to construct a probability distribution is to add the probability of an event to the sum of all previously listed probabilities in a probability distribution. Such a list is

In a ***cumulative probability distribution*** *the probability of an event is added to the sum of all previously listed probabilities in a distribution.*

referred to as a **cumulative probability distribution**. The cumulative probability distribution for our management science grade example is as follows:

Event Grade	Probability	Cumulative Probability
A	.10	.10
B	.20	.30
C	.50	.80
D	.15	.95
F	.05	1.00
	1.00	

The value of a cumulative probability distribution is that it organizes the event probabilities in a way that makes it easier to answer certain questions about the probabilities. For example, if we want to know the probability that a student will get a grade of C *or higher,* we can add the probabilities of the mutually exclusive events A, B, and C:

$$P(\text{A or B or C}) = P(\text{A}) + P(\text{B}) + P(\text{C}) = .10 + .20 + .50 = .80$$

Or we can look directly at the cumulative probability distribution and see that the probability of a C and the events preceding it in the distribution (A and B) equals .80. Alternatively, if we want to know the probability of a grade lower than C, we can subtract the cumulative probability of a C from 1.00 (i.e., 1.00 − .80 = .20).

Statistical Independence and Dependence

Statistically, events are either independent or dependent. If the occurrence of one event does not affect the probability of the occurrence of another event, the events are *independent*. Conversely, if the occurrence of one event affects the probability of the occurrence of another event, the events are *dependent*. We will first turn our attention to a discussion of independent events.

Independent Events

A succession of events that do not affect each other are ***independent events.***

When we toss a coin, the two events—getting a head and getting a tail—are independent. If we get a head on the first toss, this result has absolutely no effect on the probability of getting a head or a tail on the next toss. The probability of getting either a head or a tail will still be .50, regardless of the outcomes of previous tosses. In other words, the two events are **independent events**.

The probability that independent events will occur in succession is computed by multiplying the probabilities of each event.

When events are independent, it is possible to determine the probability that both events will occur in succession by multiplying the probabilities of each event. For example, what is the probability of getting a head on the first toss and a tail on the second toss? The answer is

$$P(\text{HT}) = P(\text{H}) \cdot P(\text{T})$$

where

$P(\text{H})$ = probability of a head
$P(\text{T})$ = probability of a tail
$P(\text{HT})$ = joint probability of a head and a tail

Therefore,

$$P(\text{HT}) = P(\text{H}) \cdot P(\text{T}) = (.5)(.5) = .25$$

A ***conditional probability*** *is the probability that an event will occur, given that another event has already occurred.*

As we indicated previously, the probability of both events occurring, $P(\text{HT})$, is referred to as the *joint probability*.

Another property of independent events relates to **conditional probabilities**. A conditional probability is the probability that event A will occur given that event B has already occurred. This relationship is expressed symbolically as

$$P(\text{A}|\text{B})$$

The term in parentheses, "A slash B," means "A, given the occurrence of B." Thus, the entire term $P(\text{A}|\text{B})$ is interpreted as the probability that A will occur, given that B has already occurred. If A and B are independent events, then

$$P(\text{A}|\text{B}) = P(\text{A})$$

In words, this result says that if A and B are independent, then the probability of A, given the occurrence of event B, is simply equal to the probability of A. Because the events are independent of each other, the occurrence of event B will have no effect on the occurrence of A. Therefore, the probability of A is in no way dependent on the occurrence of B.

In summary, if events A and B are independent, the following two properties hold:

1. $P(\text{AB}) = P(\text{A}) \cdot P(\text{B})$
2. $P(\text{A}|\text{B}) = P(\text{A})$

A probability tree is a diagram that shows the joint probabilities of the outcomes of an experiment.

Probability Trees

Consider an example in which a coin is tossed three consecutive times. The possible outcomes of this example can be illustrated by using a **probability tree**, as shown in Figure 11.3.

FIGURE 11.3
Probability tree for coin-tossing example

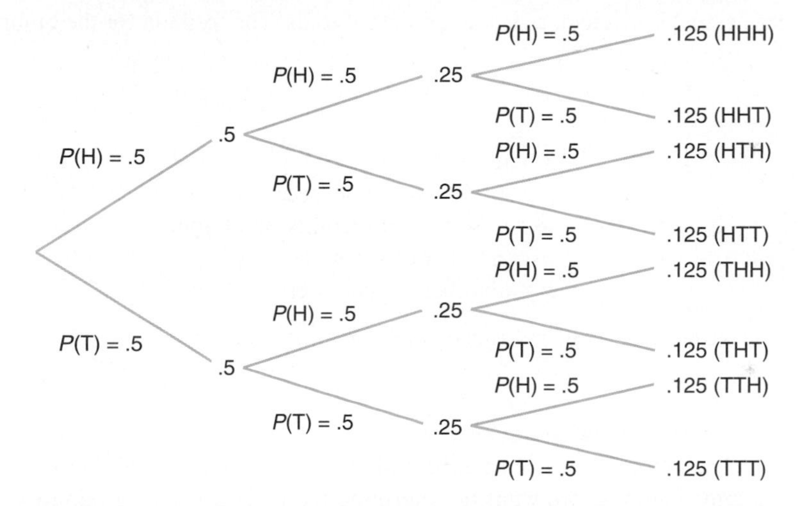

The probability tree in Figure 11.3 demonstrates the probabilities of the various occurrences, given three tosses of a coin. Notice that at each toss, the probability of either event's occurring remains the same: $P(\text{H}) = P(\text{T}) = .5$. Thus, the events are independent. Next, the joint probabilities that events will occur in succession are computed by multiplying the probabilities of all the events. For example, the probability of getting a head on the first toss, a tail on the second, and a tail on the third is .125:

$$P(\text{HTT}) = P(\text{H}) \cdot P(\text{T}) \cdot P(\text{T}) = (.5)(.5)(.5) = .125$$

However, do not confuse the results in the probability tree with conditional probabilities. The probability that a head and then two tails will occur on three consecutive tosses is computed prior to any tosses. If the first two tosses have already occurred, the probability of getting a tail on the third toss is still .5:

$$P(\text{T}|\text{HT}) = P(\text{T}) = .5$$

The Binomial Distribution

Some additional information can be drawn from the probability tree of our example. For instance, what is the probability of achieving exactly two tails on three tosses? The answer can be found by observing the instances in which two tails occurred. It can be seen that two tails in three tosses occurred three times, each time with a probability of .125. Thus, the probability of getting exactly two tails in three tosses is the sum of these three probabilities, or .375. The use of a probability tree can become very cumbersome, especially if we are considering an example with 20 tosses. However, the example of tossing a coin exhibits certain properties that enable us to define it as a **Bernoulli process**. The properties of a Bernoulli process follow:

Properties of a ***Bernoulli process.***

1. There are two possible outcomes for each trial (i.e., each toss of a coin). Outcomes can be success or failure, yes or no, heads or tails, good or bad, and so on.
2. The probability of the outcomes remains constant over time. In other words, the probability of getting a head on a coin toss remains the same, regardless of the number of tosses.
3. The outcomes of the trials are independent. The fact that we get a head on the first toss does not affect the probabilities on subsequent tosses.
4. The number of trials is *discrete* and an integer. The term *discrete* indicates values that are *countable* and, thus, usually an integer—for example, 1 car or 2 people rather than 1.34 cars or 2.51 people. There are 1, 2, 3, 4, 5, . . . tosses of the coin, not 3.36 tosses.

A ***binomial distribution*** *indicates the probability of* r *successes in* n *trials.*

Given the properties of a Bernoulli process, a **binomial distribution** function can be used to determine the probability of a number of successes in n trials. The binomial distribution is an example of a **discrete distribution** because the value of the distribution (the number of successes) is discrete, as is the number of trials. The formula for the binomial distribution is

$$P(r) = \frac{n!}{r!(n-r)!}\, p^r q^{n-r}$$

where

p = probability of a success
$q = 1 - p$ = probability of a failure
n = number of trials
r = number of successes in n trials

The terms $n!$, $(n - r)!$, and $r!$ are called *factorials*. Factorials are computed using the formula

$$m! = m(m-1)(m-2)(m-3)\ldots(2)(1)$$

The factorial 0! always equals one.

The binomial distribution formula may look complicated, but using it is not difficult. For example, suppose we want to determine the probability of getting exactly two tails in three tosses of a coin. For this example, getting a tail is a success because it is the object of the analysis. The probability of a tail, p, equals .5; therefore, $q = 1 - .5 = .5$. The number of tosses, n, is 3, and the number of tails, r, is 2. Substituting these values into the binomial formula will result in the probability of two tails in three coin tosses:

$$\begin{aligned} P(2\text{ tails}) &= P(r=2) = \frac{3!}{2!(3-2)!}(.5)^2(.5)^{3-2} \\ &= \frac{(3\cdot2\cdot1)}{(2\cdot1)(1)}(.25)(.5) \\ &= \frac{6}{2}(.125) \\ P(r=2) &= .375 \end{aligned}$$

Notice that this is the same result achieved by using a probability tree in the previous section.

Now let us consider an example of more practical interest. An electrical manufacturer produces microchips. The microchips are inspected at the end of the production process at a quality control station. Out of every batch of microchips, four are randomly selected and tested for defects. Given that 20% of all transistors are defective, what is the probability that each batch of microchips will contain exactly two defective microchips?

The two possible outcomes in this example are a good microchip and a defective microchip. Because defective microchips are the object of our analysis, a defective item is a success. The probability of a success is the probability of a defective microchip, or $p = .2$. The number of trials, n, equals 4. Now let us substitute these values into the binomial formula:

$$\begin{aligned} P(r = \text{defectives}) &= \frac{4!}{2!(4-2)!}(.2)^2(.8)^2 \\ &= \frac{(4\cdot 3\cdot 2\cdot 1)}{(2\cdot 1)(2\cdot 1)}(.04)(.64) \\ &= \frac{24}{4}(.0256) \\ &= .1536 \end{aligned}$$

Thus, the probability of getting exactly two defective items out of four microchips is .1536.

Now, let us alter this problem to make it even more realistic. The manager has determined that four microchips from every large batch should be tested for quality. If *two or more* defective microchips are found, the whole batch will be rejected. The manager wants to know the probability of rejecting an entire batch of microchips, if, in fact, the batch has 20% defective items.

From our previous use of the binomial distribution, we know that it gives us the probability of an *exact* number of *integer* successes. Thus, if we want the probability of two or more defective items, it is necessary to compute the probability of two, three, and four defective items:

$$P(r \geq 2) = P(r = 2) + P(r = 3) + P(r = 4)$$

Substituting the values $p = .2$, $n = 4$, $q = .8$, and $r = 2, 3,$ and 4 into the binomial distribution results in the probability of two or more defective items:

$$\begin{aligned} P(r \geq 2) &= \frac{4!}{2!(4-2)!}(.2)^2(.8)^2 + \frac{4!}{3!(4-3)!}(.2)^3(.8)^1 + \frac{4!}{4!(4-4)!}(.2)^4(.8)^0 \\ &= .1536 + .0256 + .0016 \\ &= .1808 \end{aligned}$$

Thus, the probability that a batch of microchips will be rejected due to poor quality is .1808.

Notice that the *collectively exhaustive* set of events for this example is 0, 1, 2, 3, and 4 defective transistors. Because the sum of the probabilities of a collectively exhaustive set of events equals 1.0,

$$P(r = 0, 1, 2, 3, 4) = P(r = 0) + P(r = 1) + P(r = 2) + P(r = 3) + P(r = 4) = 1.0$$

Recall that the results of the immediately preceding example show that

$$P(r = 2) + P(r = 3) + P(r = 4) = .1808$$

Given this result, we can compute the probability of "less than two defectives" as follows:

$$\begin{aligned} P(r < 2) &= P(r = 0) + P(r = 1) \\ &= 1.0 - [P(r = 2) + P(r = 3) + P(r = 4)] \\ &= 1.0 - .1808 \\ &= .8192 \end{aligned}$$

Although our examples included very small values for n and r that enabled us to work out the examples by hand, problems containing larger values for n and r can be solved easily by using an electronic calculator.

Dependent Events

As stated earlier, if the occurrence of one event affects the probability of the occurrence of another event, the events are **dependent events**. The following example illustrates dependent events.

Two buckets contain a number of colored balls each. Bucket 1 contains two red balls and four white balls, and bucket 2 contains one blue ball and five red balls. A coin is tossed. If a head results, a ball is drawn out of bucket 1. If a tail results, a ball is drawn from bucket 2. These events are illustrated in Figure 11.4.

FIGURE 11.4
Dependent events

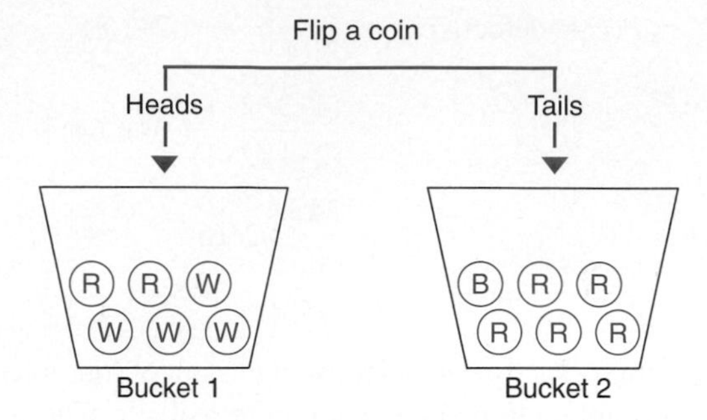

In this example, the probability of drawing a *blue* ball is clearly dependent on whether a head or a tail occurs on the coin toss. If a tail occurs, there is a 1/6 chance of drawing a blue ball from bucket 2. However, if a head results, there is no possibility of drawing a blue ball from bucket 1. In other words, the probability of the event "drawing a blue ball" is dependent on the event "flipping a coin."

Like statistically independent events, dependent events exhibit certain defining properties. In order to describe these properties, we will alter our previous example slightly, so that bucket 2 contains one white ball and five red balls. Our new example is shown in Figure 11.5. The outcomes that can result from the events illustrated in Figure 11.5 are shown in Figure 11.6. When the coin is flipped, one of two outcomes is possible, a head or a tail. The probability of getting a head is .50, and the probability of getting a tail is .50:

$$P(\text{H}) = .50$$
$$P(\text{T}) = .50$$

FIGURE 11.5
Another set of dependent events

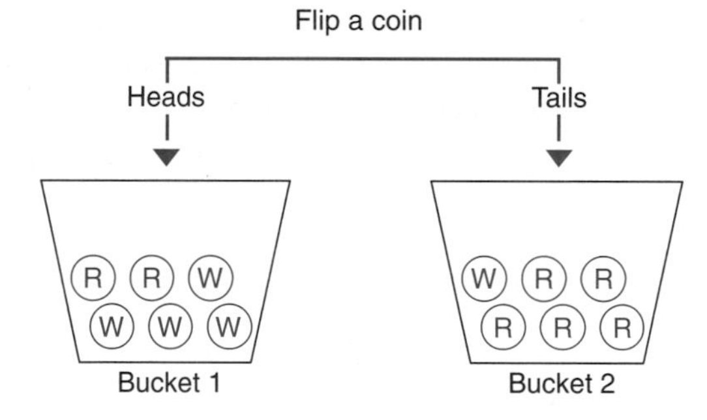

FIGURE 11.6
Probability tree for dependent events

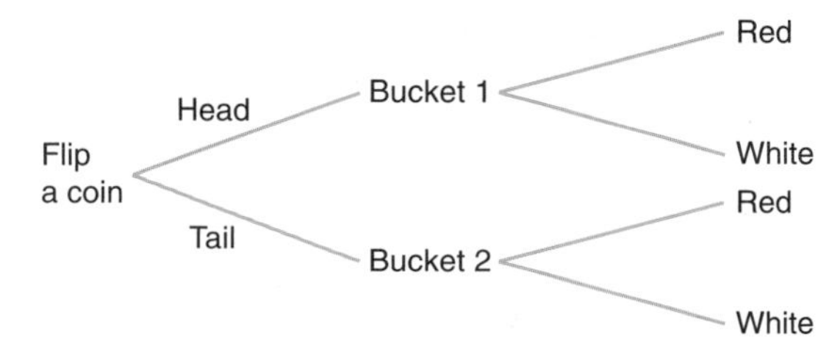

As indicated previously, these probabilities are referred to as *marginal* probabilities. They are also *unconditional* probabilities because they are the probabilities of the occurrence of a single event and are not conditional on the occurrence of any other event(s). They are the same as the probabilities of independent events defined earlier, and like those of independent events, the marginal probabilities of a collectively exhaustive set of events *sum to one*.

Once the coin has been tossed and a head or tail has resulted, a ball is drawn from one of the buckets. If a head results, a ball is drawn from bucket 1; there is a 2/6, or .33, probability of drawing a red ball and a 4/6, or .67, probability of drawing a white ball. If a tail resulted, a ball is drawn from bucket 2; there is a 5/6, or .83, probability of drawing a red ball and a 1/6, or .17, probability of drawing a white ball. These probabilities of drawing red or white balls are called *conditional* probabilities because they are conditional on the outcome of the event of tossing a coin. Symbolically, these conditional probabilities are expressed as follows:

$$P(\mathrm{R} \mid \mathrm{H}) = .33$$
$$P(\mathrm{W} \mid \mathrm{H}) = .67$$
$$P(\mathrm{R} \mid \mathrm{T}) = .83$$
$$P(\mathrm{W} \mid \mathrm{T}) = .17$$

The first term, which can be expressed verbally as "the probability of drawing a red ball, given that a head results from the coin toss," equals .33. The other conditional probabilities are expressed similarly.

Conditional probabilities can also be defined by the following mathematical relationship. Given two dependent events A and B,

$$P(\mathrm{A} \mid \mathrm{B}) = \frac{P(\mathrm{AB})}{P(\mathrm{B})}$$

the term $P(\mathrm{AB})$ is the joint probability of the two events, as noted previously. This relationship can be manipulated by multiplying both sides by $P(\mathrm{B})$, to yield

$$P(\mathrm{A} \mid \mathrm{B}) \cdot P(\mathrm{B}) = P(\mathrm{AB})$$

Thus, the joint probability can be determined by multiplying the conditional probability of A by the marginal probability of B.

Recall from our previous discussion of independent events that

$$P(\mathrm{AB}) = P(\mathrm{A}) \cdot P(\mathrm{B})$$

Substituting this result into the relationship for a conditional probability yields

$$P(\mathrm{A} \mid \mathrm{B}) = \frac{P(\mathrm{A}) \cdot (\mathrm{B})}{P(\mathrm{B})}$$
$$= P(\mathrm{A})$$

which is consistent with the property for independent events.

Returning to our example, the joint events are the occurrence of a head and a red ball, a head and a white ball, a tail and a red ball, and a tail and a white ball. The probabilities of these joint events are as follows:

$$P(\mathrm{RH}) = P(\mathrm{R} \mid \mathrm{H}) \cdot P(\mathrm{H}) = (.33)(.5) = .165$$
$$P(\mathrm{WH}) = P(\mathrm{W} \mid \mathrm{H}) \cdot P(\mathrm{H}) = (.67)(.5) = .335$$
$$P(\mathrm{RT}) = P(\mathrm{R} \mid \mathrm{T}) \cdot P(\mathrm{T}) = (.83)(.5) = .415$$
$$P(\mathrm{WT}) = P(\mathrm{W} \mid \mathrm{T}) \cdot P(\mathrm{T}) = (.17)(.5) = .085$$

The marginal, conditional, and joint probabilities for this example are summarized in Figure 11.7. Table 11.1 is a *joint probability table*, which summarizes the joint probabilities for the example.

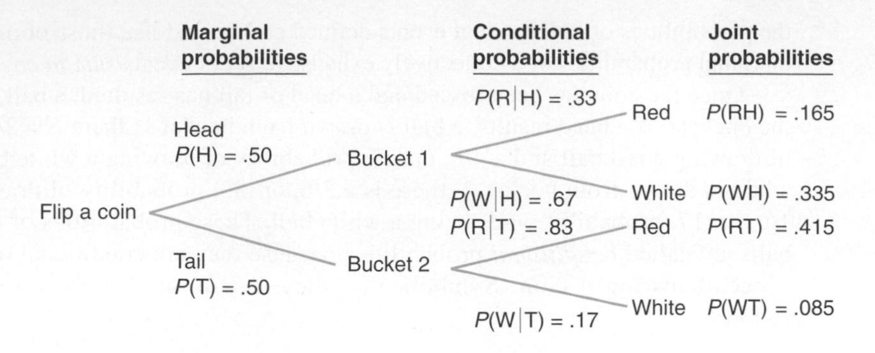

FIGURE 11.7 Probability tree with marginal, conditional, and joint probabilities

TABLE 11.1 Joint probability table

Flip a Coin	Draw a Ball: Red	Draw a Ball: White	Marginal Probabilities
Head	$P(RH) = .165$	$P(WH) = .335$	$P(H) = .50$
Tail	$P(RT) = .415$	$P(WT) = .085$	$P(T) = .50$
Marginal probabilities	$P(R) = .580$	$P(W) = .420$	1.00

Bayesian Analysis

In Bayesian analysis, additional information is used to alter the marginal probability of the occurrence of an event.

A ***posterior*** *probability is the altered marginal probability of an event, based on additional information.*

The concept of conditional probability given statistical dependence forms the necessary foundation for an area of probability known as *Bayesian analysis*. The technique is named after Thomas Bayes, an eighteenth-century clergyman who pioneered this area of analysis.

The basic principle of Bayesian analysis is that additional information (if available) can sometimes enable one to alter (improve) the marginal probabilities of the occurrence of an event. The altered probabilities are referred to as *revised*, or **posterior**, probabilities.

The concept of posterior probabilities will be illustrated using an example. A production manager for a manufacturing firm is supervising the machine setup for the production of a product. The machine operator sets up the machine. If the machine is set up correctly, there is a 10% chance that an item produced on the machine will be defective; if the machine is set up incorrectly, there is a 40% chance that an item will be defective. The production manager knows from past experience that there is a .50 probability that a machine will be set up correctly or incorrectly by an operator. In order to reduce the chance that an item produced on the machine will be defective, the manager has decided that the operator should produce a sample item. The manager wants to know the probability that the machine has been set up incorrectly if the sample item turns out to be defective.

The probabilities given in this problem statement can be summarized as follows:

$$P(C) = .50 \qquad P(D|C) = .10$$
$$P(IC) = .50 \qquad P(D|IC) = .40$$

where

C = correct
IC = incorrect
D = defective

The posterior probability for our example is the conditional probability that the machine has been set up incorrectly, given that the sample item proves to be defective, or $P(IC|D)$.

Bayes' rule is a formula for computing the posterior probability given marginal and conditional probabilities.

In Bayesian analysis, once we are given the initial marginal and conditional probabilities, we can compute the posterior probability by using **Bayes' rule**, as follows:

$$P(\text{IC}|\text{D}) = \frac{P(\text{D}|\text{IC})P(\text{IC})}{P(\text{D}|\text{IC})P(\text{IC}) + P(\text{D}|\text{C})P(\text{C})}$$
$$= \frac{(.40)(.50)}{(.40)(.50) + (.10)(.50)}$$
$$= .80$$

Previously, the manager knew that there was a 50% chance that the machine was set up incorrectly. Now, after producing and testing a sample item, the manager knows that if it is defective, there is an .80 probability that the machine was set up incorrectly. Thus, by gathering some additional information, the manager can revise the estimate of the probability that the machine was set up correctly. This will obviously improve decision making by allowing the manager to make a more informed decision about whether to have the machine set up again.

In general, given two events, A and B, and a third event, C, that is conditionally dependent on A and B, Bayes' rule can be written as

$$P(\text{A}|\text{C}) = \frac{P(\text{C}|\text{A})P(\text{A})}{P(\text{C}|\text{A})P(\text{A}) + P(\text{C}|\text{B})P(\text{B})}$$

Expected Value

It is often possible to assign numeric values to the various outcomes that can result from an experiment. When the values of the variables occur in no particular order or sequence, the variables are referred to as **random variables**. Every possible value of a variable has a probability of occurrence associated with it. For example, if a coin is tossed three times, the number of heads obtained is a random variable. The possible values of the random variable are 0, 1, 2, and 3 heads. The values of the variable are random because there is no way of predicting which value (0, 1, 2, or 3) will result when the coin is tossed three times. If three tosses are made several times, the values (i.e., numbers of heads) that will result will have no sequence or pattern; they will be random.

Like the variables defined in previous chapters in this text, random variables are typically represented symbolically by a letter, such as x, y, or z. Consider a vendor who sells hot dogs outside a building every day. If the number of hot dogs the vendor sells is defined as the random variable x, then x will equal 0, 1, 2, 3, 4, . . . hot dogs sold daily.

Although the exact values of the random variables in the foregoing examples are not known prior to the event, it is possible to assign a probability to the occurrence of the possible values that can result. Consider a production operation in which a machine breaks down periodically. From experience it has been determined that the machine will break down 0, 1, 2, 3, or 4 times per month. Although managers do not know the exact number of breakdowns (x) that will occur each month, they can determine the relative frequency probability of each number of breakdowns $P(x)$. These probabilities are as follows:

x	$P(x)$
0	.10
1	.20
2	.30
3	.25
4	.15
	1.00

These probability values taken together form a *probability distribution*. That is, the probabilities are distributed over the range of possible values of the random variable x.

*The **expected value** of a random variable is computed by multiplying each possible value of the variable by its probability and summing these products.*

The **expected value** of the random variable (the number of breakdowns in any given month) is computed by multiplying each value of the random variable by its probability of occurrence and summing these products.

For our example, the expected number of breakdowns per month is computed as follows:

$$\begin{aligned} E(x) &= (0)(.10) + (1)(.20) + (2)(.30) + (3)(.25) + (4)(.15) \\ &= 0 + .20 + .60 + .75 + .60 \\ &= 2.15 \text{ breakdowns} \end{aligned}$$

This means that, on average, management can expect 2.15 breakdowns every month.

The expected value is the mean of the probability distribution of the random variable.

The expected value is often referred to as the weighted average, or *mean*, of the probability distribution and is a measure of central tendency of the distribution. In addition to knowing the mean, it is often desirable to know how the values are dispersed (or scattered) around the mean. A measure of dispersion is the **variance**, which is computed as follows:

__Variance__ is a measure of the dispersion of random variable values around the expected value, or mean.

1. Square the difference between each value and the expected value.
2. Multiply these resulting amounts by the probability of each value.
3. Sum the values compiled in step 2.

The general formula for computing the variance, which we will designate as σ^2, is

$$\sigma^2 = \sum_{i=1}^{n} [x_i - E(x_i)]^2 P(x_i)$$

Management Science Application

A Probability Model for Analyzing Coast Guard Patrol Effectiveness

A primary responsibility of the U.S. Coast Guard is to monitor and protect the maritime borders of the United States. One measure of Coast Guard effectiveness in meeting this responsibility is determined by a probability model that indicates the probability that the Coast Guard will intercept a randomly selected target vessel attempting to penetrate a Coast Guard patrol perimeter. The probability $P(\text{I})$, the probability of intercepting the target vessel, is computed according to the formula

$$P(\text{I}) = P(\text{I}|\text{D})P(\text{D}|\text{A})P(\text{A}|\text{O})P(\text{O})$$

$P(\text{I}|\text{D})$ is the probability that a target vessel will be intercepted (I), given that it is detected (D). It is a constant value estimated from Coast Guard historical data. $P(\text{D}|\text{A})$ is the probability that a target vessel is detected, given that a Coast Guard vessel is available (A). This value is computed using a submodel based on the patrol area and speed of the Coast Guard vessel and the speed of the target vessel. $P(\text{A}|\text{O})$ is the probability that the Coast Guard vessel is available, given that it is on the scene (O) in the patrol area. $P(\text{O})$ is the probability that a Coast Guard vessel is on the scene in the patrol area, and currently it is assigned a constant value of one. The Coast Guard used this model to perform sensitivity analysis and analyze the effect of different parameter changes on effectiveness. An example might be testing the impact of a change in the patrol area or vessel speed.

Simon Alvinge/iStockphoto

Source: Based on S. O. Kimbrough, J. R. Oliver, and C. W. Pritchett, "On Post-Evaluation Analysis: Candle-Lighting and Surrogate Models," *Interfaces* 23, no. 3 (May–June 1993): 17–28.

The variance (σ^2) for the machine breakdown example is computed as follows:

x_i	$P(x_i)$	$x_i - E(x)$	$[x_i - E(x)]^2$	$[x_i - E(x)]^2 \cdot P(x_i)$
0	.10	−2.15	4.62	.462
1	.20	−1.15	1.32	.264
2	.30	−0.15	.02	.006
3	.25	0.85	.72	.180
4	.15	1.85	3.42	.513
	1.00			1.425

$$\sigma^2 = 1.425 \text{ breakdowns per month}$$

The ***standard deviation*** *is computed by taking the square root of the variance.*

The **standard deviation** is another widely recognized measure of dispersion. It is designated symbolically as σ and is computed by taking the square root of the variance, as follows:

$$\sigma = \sqrt{1.425}$$
$$= 1.19 \text{ breakdowns per month}$$

A small standard deviation or variance, relative to the expected value, or mean, indicates that most of the values of the random variable distribution are bunched close to the expected value. Conversely, a large relative value for the measures of dispersion indicates that the values of the random variable are widely dispersed from the expected value.

The Normal Distribution

Previously, a *discrete* value was defined as a value that is countable (and usually an integer). A random variable is discrete if the values it can equal are finite and countable. The probability distributions we have encountered thus far have been discrete distributions. In other words, the values of the random variables that made up these discrete distributions were always finite (for example, in the preceding example, there were five possible values of the random variable breakdowns per month). Because every value of the random variable had a unique probability of occurrence, the discrete probability distribution consisted of all the (finite) values of a random variable and their associated probabilities.

A ***continuous random variable*** *can take on an infinite number of values within an interval, or range.*

In contrast, a **continuous random variable** can take on an infinite number of values within some interval. This is because continuous random variables have values that are not specifically countable and are often fractional. The distinction between discrete and continuous random variables is sometimes made by saying that *discrete* relates to things that can be counted and *continuous* relates to things that are measured. For example, a load of oil being transported by tanker may consist of not just 1 million or 2 million barrels but 1.35 million barrels. If the range of the random variable is between 1 million and 2 million barrels, then there is an infinite number of possible (fractional) values between 1 million and 2 million barrels, even though the value 1.35 million corresponds to a discrete value of 1,350,000 barrels of oil. No matter how small an interval exists between two values in the distribution, there is always at least one value—and, in fact, an infinite number of values—between the two values.

In a ***continuous distribution****, the random variables can equal an infinite number of values within an interval.*

Because a continuous random variable can take on an extremely large or infinite number of values, assigning a unique probability to every value of the random variable would require an infinite (or very large) number of probabilities, each of which would be infinitely small. Therefore, we cannot assign a unique probability to each value of the continuous random variable, as we did in a discrete probability distribution. In a **continuous distribution**, we can refer only to the probability that a value of the random variable is within some *range*. For example, we can determine the probability that between 1.35 million and 1.40 million barrels of oil are transported, or the probability that fewer than or more than 1.35 million barrels are shipped, but we cannot determine the probability that exactly 1.35 million barrels of oil are transported.

The ***normal distribution*** *is a continuous probability distribution that is symmetrical on both sides of the mean—that is, it is shaped like a bell.*

One of the most frequently used continuous probability distributions is the **normal distribution**, which is a continuous curve in the shape of a bell (i.e., it is symmetrical).

The normal distribution is a popular continuous distribution because it has certain mathematical properties that make it easy to work with, and it is a reasonable approximation of the continuous probability distributions of a number of natural phenomena. Figure 11.8 is an illustration of the normal distribution.

FIGURE 11.8
The normal curve

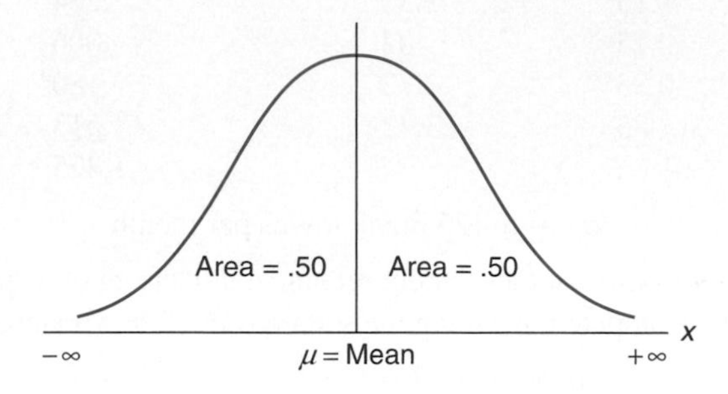

The center of a normal distribution is its mean, μ.

The fact that the normal distribution is a continuous curve reflects the fact that it consists of an infinite or extremely large number of points (on the curve). The bell-shaped curve can be flatter or taller, depending on the degree to which the values of the random variable are dispersed from the center of the distribution. The center of the normal distribution is referred to as the *mean* (μ), and it is analogous to the average of the distribution.

The area under the normal curve represents probability, *and the total area under the curve sums to one.*

Notice that the two ends (or tails) of the distribution in Figure 11.8 extend from $-\infty$ to $+\infty$. In reality, random variables do not often take on values over an infinite range. Therefore, when the normal distribution is applied, it actually approximates the distribution of a random variable with finite limits. The *area* under the normal curve represents *probability*. The entire area under the curve equals 1.0 because the sum of the probabilities of all values of a random variable in a probability distribution must equal 1.0. Fifty percent of the curve lies to the right of the mean, and 50% lies to the left. Thus, the probability that a random variable x will have a value greater (or less) than the mean is .50.

As an example of the application of the normal distribution, consider the Armor Carpet Store, which sells Super Shag carpet. From several years of sales records, store management has determined that the mean number of yards of Super Shag demanded by customers during a week is 4,200 yards, and the standard deviation is 1,400 yards. It is necessary to know both the mean and the standard deviation to perform a probabilistic analysis using the normal distribution. The store management assumes that the continuous random variable, yards of carpet demanded per week, is normally distributed (i.e., the values of the random variable have approximately the shape of the normal curve). The mean of the normal distribution is represented by the symbol μ, and the standard deviation is represented by the symbol σ:

$$\mu = 4{,}200 \text{ yd.}$$
$$\sigma = 1{,}400 \text{ yd.}$$

The store manager wants to know the probability that the demand for Super Shag in the upcoming week will exceed 6,000 yards. The normal curve for this example is shown in Figure 11.9. The probability that x (the number of yards of carpet) will be equal to or greater than 6,000 is expressed as

The area under a normal curve is measured by determining the number of standard deviations the value of a random variable is from the mean.

$$P(x \geq 6{,}000)$$

which corresponds to the area under the normal curve to the right of the value 6,000 because the area under the curve (in Figure 11.9) represents probability. In a normal distribution, area or probability is measured by determining the *number of standard deviations the value of the random variable* x *is from the mean*. The number of standard deviations a value is from the mean is represented by Z and is computed using the formula

$$Z = \frac{x - \mu}{\sigma}$$

FIGURE 11.9
The normal distribution for carpet demand

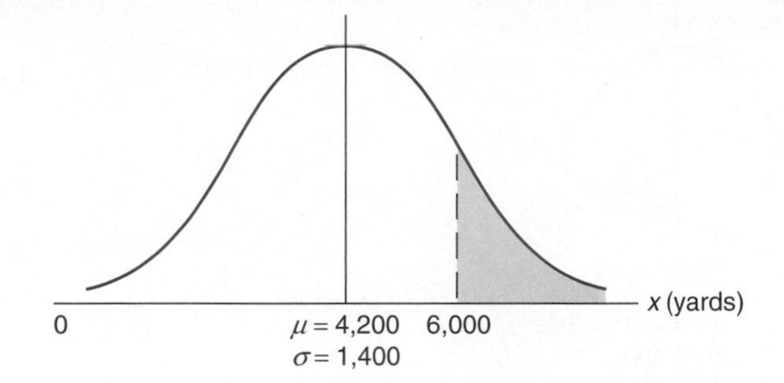

The number of standard deviations a value is from the mean gives us a consistent *standard* of measure for all normal distributions. In our example, the units of measure are yards; in other problems, the units of measure may be pounds, hours, feet, or tons. By converting these various units of measure into a common measure (number of standard deviations), we create a standard that is the same for all normal distributions.

Actually, the standard form of the normal distribution has a mean of zero ($\mu = 0$) and a standard deviation of one ($\sigma = 1$). The value Z enables us to convert this scale of measure into whatever scale our problem requires.

Figure 11.10 shows the *standard normal distribution*, with our example distribution of carpet demand above it. This illustrates the conversion of the scale of measure along the horizontal axis from yards to number of standard deviations.

FIGURE 11.10
The standard normal distribution

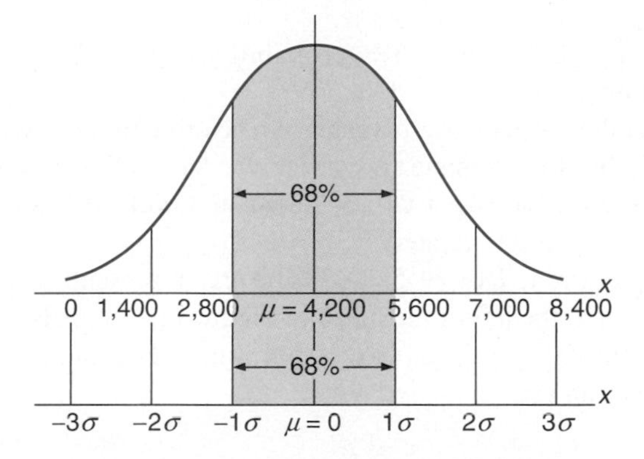

The horizontal axis along the bottom of Figure 11.10 corresponds to the standard normal distribution. Notice that the area under the normal curve between -1σ and 1σ represents 68% of the total area under the normal curve, or a probability of .68. Now look at the horizontal axis corresponding to yards in our example. Given that the standard deviation is 1,400 yards, the area between -1σ (2,800 yards) and 1σ (5,600 yards) is also 68% of the total area under the curve. Thus, if we measure distance along the horizontal axis in terms of the number of standard deviations, we will determine the same probability, no matter what the units of measure are. The formula for Z makes this conversion for us.

Returning to our example, recall that the manager of the carpet store wants to know the probability that the demand for Super Shag in the upcoming week will be 6,000 yards or more. Substituting the values $x = 6{,}000$, $\mu = 4{,}200$, and $\sigma = 1{,}400$ yards into our formula for Z, we can determine the number of standard deviations the value 6,000 is from the mean:

$$\begin{aligned} Z &= \frac{x - \mu}{\sigma} \\ &= \frac{6{,}000 - 4{,}200}{1{,}400} \\ &= 1.29 \text{ standard deviations} \end{aligned}$$

FIGURE 11.11
Determination of the Z value

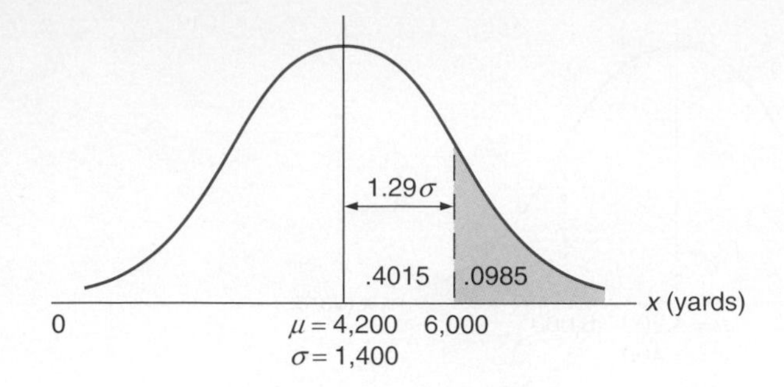

The value $x = 6{,}000$ is 1.29 standard deviations from the mean, as shown in Figure 11.11.

The area under the standard normal curve for values of Z has been computed and is displayed in easily accessible *normal tables*. Table A.1 in Appendix A is such a table. It shows that $Z = 1.29$ standard deviations corresponds to an area, or probability, of .4015. However, this is the area between $\mu = 4{,}200$ and $x = 6{,}000$ because what was measured was the area within 1.29 standard deviations of the mean. Recall, though, that 50% of the area under the curve lies to the right of the mean. Thus, we can subtract .4015 from .5000 to get the area to the right of $x = 6{,}000$:

$$\begin{aligned} P(x \geq 6{,}000) &= .5000 - .4015 \\ &= .0985 \end{aligned}$$

This means that there is a .0985 (or 9.85%) probability that the demand for carpet next week will be 6,000 yards or more.

Now suppose that the carpet store manager wishes to consider two additional questions: (1) What is the probability that demand for carpet will be 5,000 yards or less? (2) What is the probability that the demand for carpet will be between 3,000 yards and 5,000 yards? We will consider each of these questions separately.

First, we want to determine $P(x \leq 5{,}000)$. The area representing this probability is shown in Figure 11.12. The area to the left of the mean in Figure 11.12 equals .50. That leaves only the area between $\mu = 4{,}200$ and $x = 5{,}000$ to be determined. The number of standard deviations $x = 5{,}000$ is from the mean is

$$\begin{aligned} Z &= \frac{x - \mu}{\sigma} \\ &= \frac{5{,}000 - 4{,}200}{1{,}400} \\ &= \frac{800}{1{,}400} \\ &= .57 \text{ standard deviation} \end{aligned}$$

FIGURE 11.12
Normal distribution for $P(x \leq 5{,}000$ yards) yards)

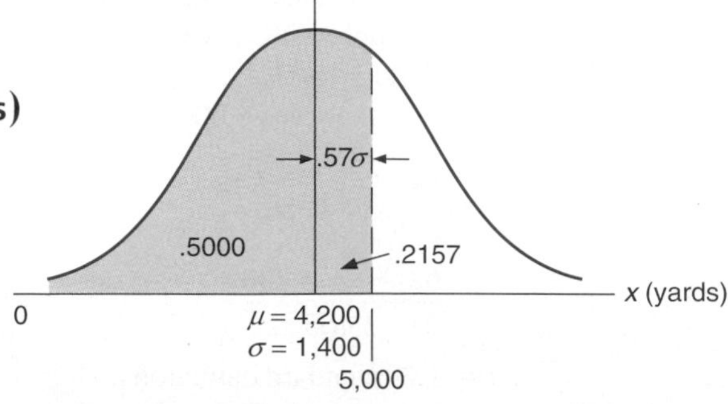

The value $Z = .57$ corresponds to a probability of .2157 in Table A.1 in Appendix A. Thus, the area between 4,200 and 5,000 in Figure 11.12 is .2157. To find our desired probability, we simply add this amount to .5000:

$$P(x \leq 5{,}000) = .5000 + .2157 = .7157$$

Next, we want to determine $P(3{,}000 \leq x \leq 5{,}000)$. The area representing this probability is shown in Figure 11.13. The shaded area in Figure 11.13 is computed by finding two areas—the area between $x = 3{,}000$ and $\mu = 4{,}200$ and the area between $\mu = 4{,}200$ and $x = 5{,}000$—and summing them. We already computed the area between 4,200 and 5,000 in the previous example and found it to be .2157. The area between $x = 3{,}000$ and $\mu = 4{,}200$ is found by determining how many standard deviations $x = 3{,}000$ is from the mean:

$$\begin{aligned} Z &= \frac{3{,}000 - 4{,}200}{1{,}400} \\ &= \frac{-1{,}200}{1{,}400} \\ &= -.86 \end{aligned}$$

FIGURE 11.13
Normal distribution with $P(3{,}000 \text{ yards} \leq x \leq 5{,}000 \text{ yards})$

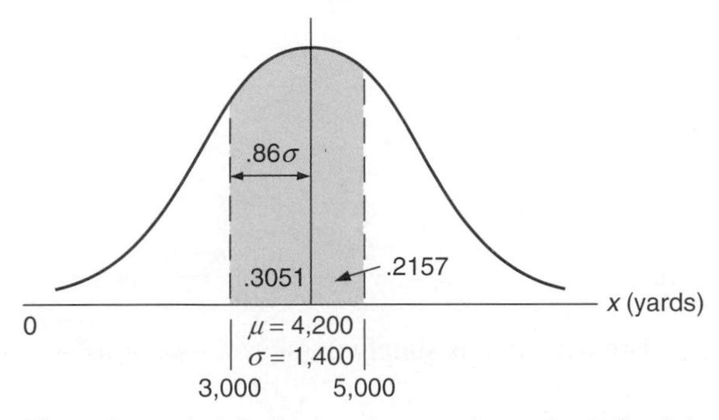

The minus sign indicates the area is to the left of the mean when we find the area corresponding to the Z value of .86 in Table A.1. This value is .3051. We now find our probability by summing .2157 and .3051:

$$\begin{aligned} P(3{,}000 \leq x \leq 5{,}000) &= .2157 + .3051 \\ &= .5208 \end{aligned}$$

The normal distribution, although applied frequently in probability analysis, is just one of a number of continuous probability distributions. In subsequent chapters, other continuous distributions will be identified. Being acquainted with the normal distribution will make the use of these other distributions much easier.

Sample Mean and Variance

*The **population mean** and variance are for the entire set of data being analyzed.*

*A **sample mean** and **variance** are derived from a subset of the population data and are used to make inferences about the population.*

The mean and variance we described in the previous section are actually more correctly referred to as the **population mean** (μ) and **variance** (σ^2). The population mean is the mean of the entire set of possible measurements or set of data being analyzed, and the population variance is defined similarly. Although we provided numeric values of the population mean and variance (or standard deviation) in our examples in the previous section, it is more likely that a **sample mean**, $\bar{x}$, would be used to estimate the population mean, and a **sample variance**, s^2, would be used to estimate the population variance. The computation of a true population mean and variance is usually too time-consuming and costly, or the entire population of data may not be available. Instead, a sample, which is a smaller subset of the population, is used. The mean and variance of this sample can then be used as an estimate of the population mean and variance if the population is assumed to be normally distributed. In other words, a smaller set of sample data is used to make generalizations or estimates, referred to as *inferences*, about the whole (population) set of data.

For example, in the previous section we indicated that the mean weekly demand for Super Shag carpet was 4,200 yards and the standard deviation was 1,400; thus $\mu = 4{,}200$ and $\sigma = 1{,}400$. However, we noted that the management of the Armor Carpet Store determined the mean and standard deviation from "several years of sales records." Thus, the mean and standard deviation were actually a sample mean and standard deviation, based on several years of sample data from a normal distribution. In other words, we used a sample mean and sample standard deviation to estimate the population mean and standard deviation. The population mean and standard deviation would have actually been computed from all weekly demand since the company started, for some extended period of time.

The sample mean, $\bar{x}$, is computed using the following formula:

$$\bar{x} = \frac{\sum_{i=1}^{n} x_i}{n}$$

The sample variance, s^2, is computed as follows:

$$s^2 = \frac{\sum_{i=1}^{n} (x_i - \bar{x})^2}{n - 1}$$

or, in shortcut form:

$$s^2 = \frac{\sum_{i=1}^{n} x_i^2 - \frac{\left(\sum_{i=1}^{n} x_i\right)^2}{n}}{n - 1}$$

The sample standard deviation is simply the square root of the variance:

$$s = \sqrt{s^2}$$

Let us consider our example for Armor Carpet Store again, except now we will use a sample of 10 weeks' demand data for Super Shag carpet, as follows:

Week *i*	**Demand x_i**
1	2,900
2	5,400
3	3,100
4	4,700
5	3,800
6	4,300
7	6,800
8	2,900
9	3,600
10	4,500
	$\Sigma = 42{,}000$

The sample mean is computed as

$$\bar{x} = \frac{\sum_{i=1}^{n} x_i}{n}$$

$$= \frac{42{,}000}{10} = 4{,}200 \text{ yd.}$$

The sample variance is computed as

$$s^2 = \frac{\sum_{i=1}^{n} x_i^2 - \frac{\left(\sum_{i=1}^{n} x_i\right)^2}{n}}{n - 1}$$

$$= \frac{(190{,}060{,}000) - \frac{(1{,}764{,}000{,}000)}{10}}{9}$$

$$s^2 = 1{,}517{,}777$$

The sample standard deviation, s, is

$$s = \sqrt{s^2}$$
$$= \sqrt{1{,}517{,}777}$$
$$= 1{,}232 \text{ yd.}$$

These values are very close to the mean and standard deviation we originally estimated in this example in the previous section (i.e., $\mu = 4{,}200$ yd., $\sigma = 1{,}400$ yd.). In general, the accuracy of the sample depends on two factors: the sample size (n) and the variation in the data. The larger the sample, the more accurate the sample statistics will be in estimating the population statistics. Also, the more variable the data are, the less accurate the sample statistics will be as estimates of the population statistics.

The Chi-Square Test for Normality

Several of the quantitative techniques that are presented in the remainder of this text include probabilistic data and parameters and statistical analysis. In many cases the problem data are assumed (or stated) to be normally distributed, with a mean and standard deviation, which enable statistical analysis to be performed based on the normal distribution. However, in reality it can never be simply assumed that data are normally distributed or in fact reflect any probability distribution. Frequently, a statistical test must be performed to determine the exact distribution (if any) to which the data conform.

The chi-square (χ^2) test is one such statistical test to see if observed data fit a particular probability distribution, including the normal distribution. The chi-square test compares the actual frequency distribution for a set of data with a theoretical frequency distribution that would be expected to occur for a specific distribution. This is also referred to as testing the *goodness-of-fit* of a set of data to a specific probability distribution.

To perform a chi-square goodness-of-fit test, the actual number of frequencies in each class or the range of a frequency distribution is compared to the theoretical frequencies that should occur in each class if the data followed a particular distribution. These numeric differences between the actual and theoretical values in each class are used in a formula to compute a χ^2 test statistic, or number, which is then compared to a number from a chi-square table called a *critical value.* If the computed χ^2 test statistic is greater than the tabular critical value, then the data do not follow the distribution being tested; if the critical value is greater than the computed test statistic, the distribution is presumed to exist. In statistical terminology this is referred to as testing the hypothesis that the data are, for example, normally distributed. If the χ^2 test statistic is greater than the tabular critical value, the hypothesis (H_o) that the data fits the hypothesized distribution is rejected, and the distribution is presumed to not exist; otherwise, it is accepted as existing.

To demonstrate how to apply the chi-square test for a normal distribution, we will use an expanded version of our Armor Carpet Store example used in the previous section. To test the distribution, we need a lot more data than the 10 weeks of demand we used previously. Instead, we will now assume that we have collected a sample of 200 weeks of demand for

Super Shag carpet and that these data have been grouped according to the following frequency distribution:

Range, Weekly Demand (yd.)	Frequency (weeks)
0 < 1,000	2
1,000 < 2,000	5
2,000 < 3,000	22
3,000 < 4,000	50
4,000 < 5,000	62
5,000 < 6,000	40
6,000 < 7,000	15
7,000 < 8,000	3
8,000+	1
	200

Because we haven't provided the actual data, we are also going to assume that the sample mean ($\bar{x}$) equals 4,200 yards and the sample standard deviation (s) equals 1,232 yards, although normally these values would be computed directly from the data as in the previous section.

The first step in performing the chi-square test is to determine the number of observations that *should be* in each frequency range, if the distribution is normal. We start by determining the area (or probability) that should be in each class, using the sample mean and standard deviation. Figure 11.14 shows the theoretical normal distribution, with the area in each range. The area of probability for each range is computed using the normal probability table (Table A.1 in Appendix A), as demonstrated earlier in this chapter. For example, the area less than 1,000 yards is determined by computing the Z statistic for $x = 1{,}000$:

$$Z = \frac{x - \bar{x}}{s}$$

$$= \frac{1{,}000 - 4{,}200}{1{,}232}$$

$$= -2.60$$

FIGURE 11.14
The theoretical normal distribution

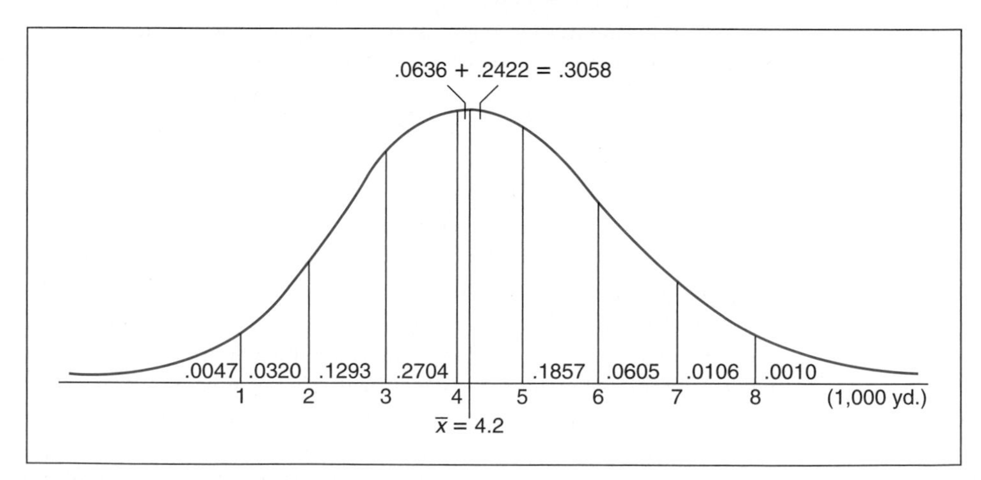

This corresponds to a normal table value of .4953. This is the area from 1,000 to the sample mean (4,200). Subtracting this value from .5000 results in the area less than 1,000, or .5000 − .4953 = .0047. The area in the range of 1,000 to 2,000 yards is computed by subtracting the area from 2,000 to the mean (.4633) from the area from 1,000 to the mean (.4953), or .4953 − .4633 = .0320. The Z values for these ranges and all the range areas in Figure 11.14 are shown in Table 11.2.

TABLE 11.2
The determination of the theoretical range frequencies

Range	Z	Area: $x \rightarrow \bar{x}$	Range Area	Normal Frequency (n = 200)
0 < 1,000	—	.5000	.0047	0.94
1,000 < 2,000	−2.60	.4953	.0320	6.40
2,000 < 3,000	−1.79	.4633	.1293	25.86
3,000 < 4,000	−.97	.3340	.2704	54.08
4,000 < 5,000	{−.16, .65}	{.0636, .2422}	.3058	61.16
5,000 < 6,000	1.46	.4279	.1857	37.14
6,000 < 7,000	2.27	.4884	.0605	12.10
7,000 < 8,000	3.08	.4990	.0106	2.12
8,000+	—	.5000	.0010	0.20

Notice that the area for the range that includes the mean between 4,000 and 5,000 yards is determined by adding the two areas to the immediate right and left of the mean, .0636 + .2422 = .3058.

The next step is to compute the theoretical frequency for each range by multiplying the area in each range by $n = 200$. For example, the frequency in the range from 0 to 1,000 is $(.0047)(200) = 0.94$, and the frequency for the range from 1,000 to 2,000 is $(.0320)(200) = 6.40$. These and the remaining theoretical frequencies are shown in the last column in Table 11.2.

Next, we must compare these theoretical frequencies with the actual frequencies in each range, using the following chi-square test statistic:

$$\chi^2_{k-p-1} = \sum_k \frac{(f_o - f_t)^2}{f_t}$$

where

f_o = observed frequency
f_t = theoretical frequency
k = the number of classes or ranges
p = the number of estimated parameters
$k - p - 1$ = degrees of freedom

However, before we can apply this formula, we must make an important adjustment. Before we can apply the chi-square test, each range must include at least five theoretical observations. Thus, we need to combine some of the ranges so that they will contain at least five theoretical observations. For our distribution we can accomplish this by combining the two lower class ranges (0–1,000 and 1,000–2,000) and the three higher ranges (6,000–7,000; 7,000–8,000; and 8,000+). This results in a revised frequency distribution with six frequency classes, as shown in Table 11.3.

TABLE 11.3
Computation of χ^2 test statistic

Range, Weekly Demand	Observed Frequency f_o	Theoretical Frequency f_t	$(f_o - f_t)^2$	$(f_o - f_t)^2/f_t$
0 < 2,000	7	7.34	.12	.016
2,000 < 3,000	22	25.86	14.90	.576
3,000 < 4,000	50	54.08	16.64	.308
4,000 < 5,000	62	61.16	.71	.012
5,000 < 6,000	40	37.14	8.18	.220
6,000+	19	14.42	21.00	1.456
				2.588

The completed chi-square test statistic is shown in the last column in Table 11.3 and is computed as follows:

$$\chi^2 = \sum_{6} \frac{\left(f_o - f_t^2\right)}{f_t}$$
$$= 2.588$$

Next, we must compare this test statistic with a critical value obtained from the chi-square table (Table A.2) in Appendix A. The degrees of freedom for the critical value are $= k - p - 1$, where k is the number of frequency classes, or 6; and p is the number of parameters that were estimated for the distribution, which in this case is 2, the sample mean and the sample standard deviation. Thus,

$$k - p - 1 = 6 - 2 - 1$$
$$= 3 \text{ degrees of freedom}$$

Using a level of significance (degree of confidence) of .05 (i.e., $\alpha = .05$), from Table A.2:

$$\chi^2_{.05,3} = 7.815$$

Because $7.815 > 2.588$, we accept the hypothesis that the distribution is normal. If the χ^2 value of 7.815 were less than the computed χ^2 test statistic, the distribution would not be considered normal.

Statistical Analysis with Excel

QM for Windows does not have statistical program modules. Therefore, to perform statistical analysis, and specifically to compute the mean and standard deviation from sample data, we must rely on Excel. Exhibit 11.1 shows the Excel spreadsheet for our Armor Carpet Store example. The average demand for the sample data (4,200) is computed in cell C16, using the formula **=AVERAGE(C4:C13)**, which is also shown on the formula bar at the top of the spreadsheet. Cell C17 contains the sample standard deviation (1,231.98), computed by using the formula **=STDEV(C4:C13)**.

A statistical analysis of the sample data can also be obtained by using the "Data Analysis" option from the "Data" tab at the top of the spreadsheet. (If this option is not available on your

EXHIBIT 11.1

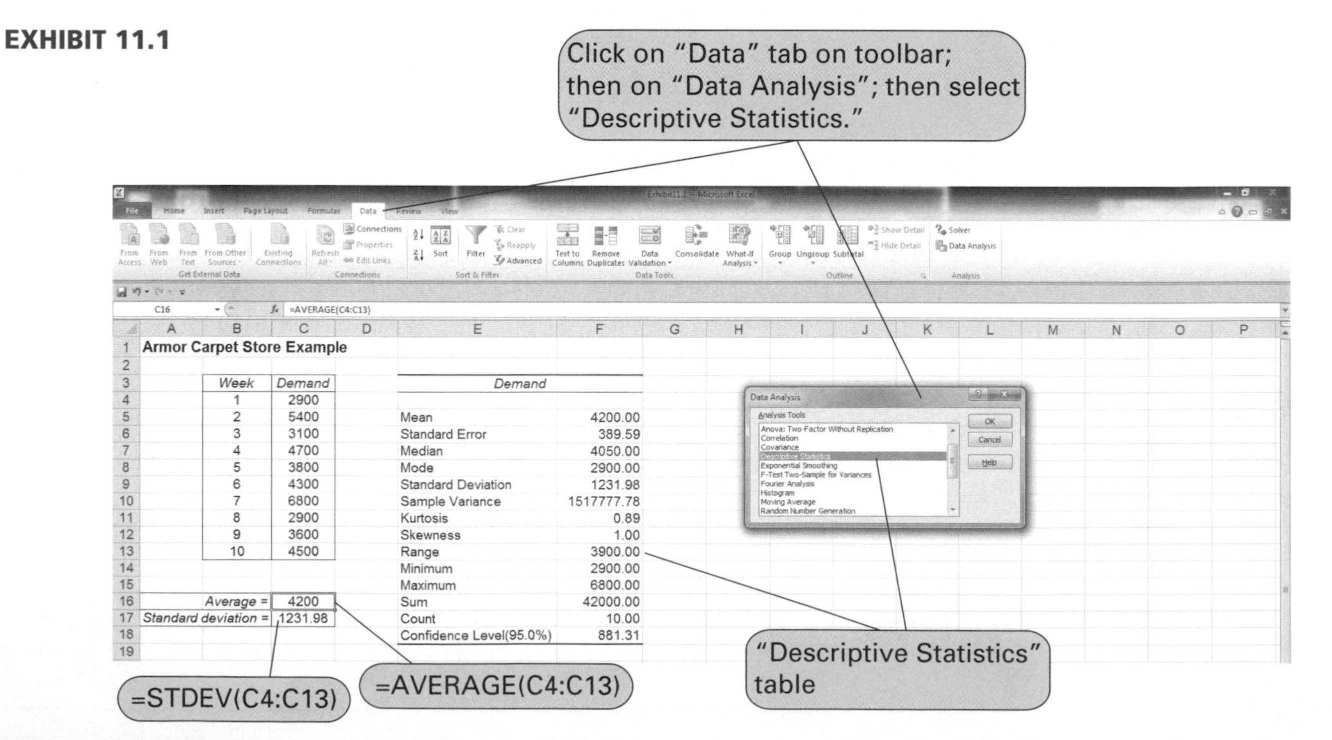

"Data" menu, select the "Add-Ins" option from the "Excel options" menu and then select the "Analysis ToolPak" option. This will insert the "Data Analysis" option on your "Data" toolbar when you return to it.) Selecting the "Data Analysis" option from the menu will result in the "Data Analysis" window shown in Exhibit 11.1.

Select "Descriptive Statistics" from this window. This will result in the dialog window titled Descriptive Statistics shown in Exhibit 11.2. This window, as shown, results in summary statistics for the demand data in our carpet store example. Notice that the input range we entered, **C3:C13**, includes the "Demand" heading on the spreadsheet in cell C3, which we acknowledge by checking the "Labels in first row" box. This will result in our statistical summary being labeled "Demand" on the spreadsheet, as shown in cell E3 in Exhibit 11.1. Notice that we indicated where we wanted to locate the summary statistics on our spreadsheet by typing **E3** in the "Output Range" window. We obtained the summary statistics by checking the "Summary statistics" box at the bottom of the screen.

EXHIBIT 11.2

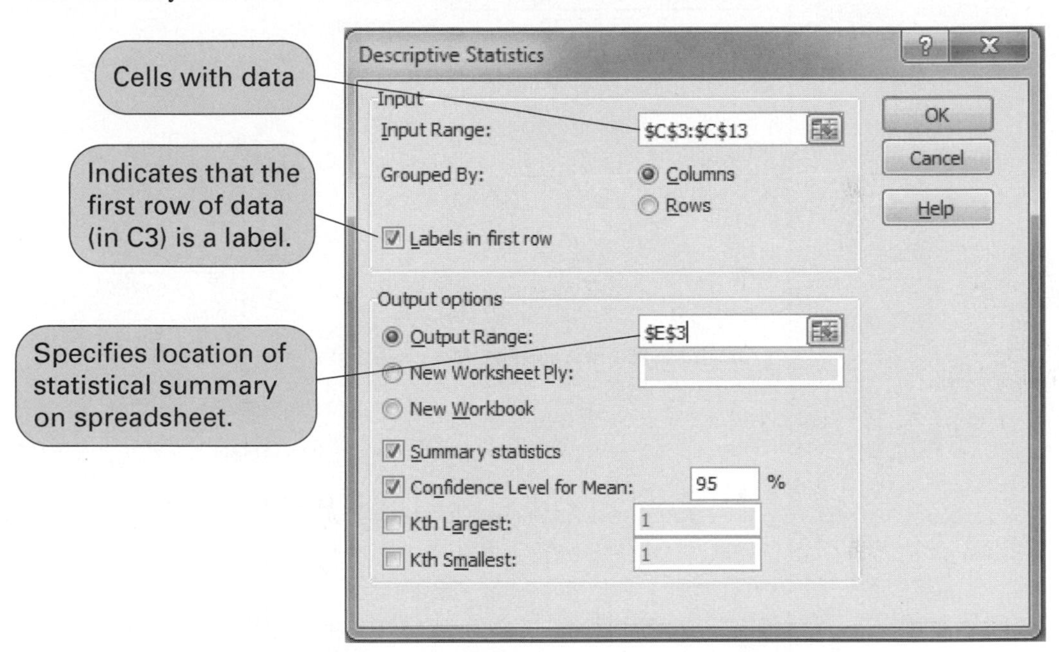

Summary

The field of probability and statistics is quite large and complex, and it contains much more than has been presented in this chapter. This chapter presented the basic principles and fundamentals of probability. The primary purpose of this brief overview was to prepare the reader for other material in the book. The topics of decision analysis (Chapter 12), simulation (Chapter 14), probabilistic inventory models (Chapter 16), and, to a certain extent, project management (Chapter 8) are probabilistic in nature and require an understanding of the fundamentals of probability.

Example Problem Solution

The following example will illustrate the solution of a problem involving a normal probability.

Problem Statement

The Radcliffe Chemical Company and Arsenal produces explosives for the U.S. Army. Because of the nature of its products, the company devotes strict attention to safety, which is also scrutinized by the federal government. Historical records show that the annual number of property damage and personal injury accidents is normally distributed, with a mean of 8.3 accidents and a standard deviation of 1.8 accidents.

A. What is the probability that the company will have fewer than 5 accidents next year? more than 10?
B. The government will fine the company $200,000 if the number of accidents exceeds 12 in a 1-year period. What average annual fine can the company expect?

Solution

Step 1: Set Up the Normal Distribution

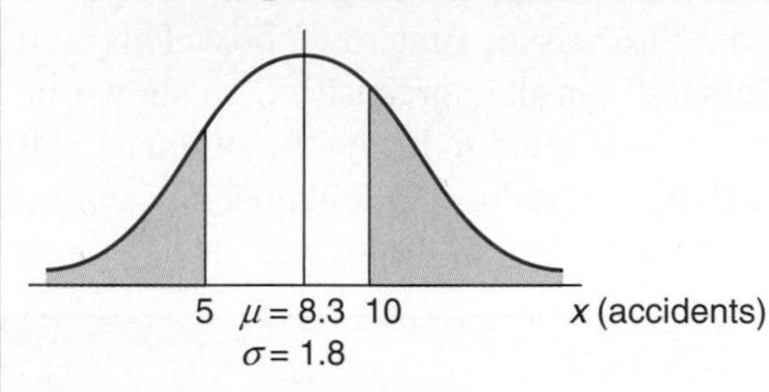

Step 2: Solve Part A

$$P\ (x \leq 5 \text{ accidents})$$

$$Z = \frac{x - \mu}{\sigma} = \frac{5 - 8.3}{1.8} = -1.83$$

From Table A.1 in Appendix A, we see that $Z = -1.83$ corresponds to a probability of .4664; thus,

$$P(x \leq 5) = .5000 - .4664 = .0336$$

$$P\ (x \geq 10 \text{ accidents})$$

$$Z = \frac{x - \mu}{\sigma} = \frac{10 - 8.3}{1.8} = .94$$

From Table A.1 in Appendix A, we see that $Z = .94$ corresponds to a probability of .3264; thus,

$$P(x \geq 10) = .5000 - .3264 = .1736$$

Step 3: Solve Part B

$$P(x \geq 12 \text{ accidents})$$

$$Z = \frac{x - \mu}{\sigma} = \frac{12 - 8.3}{1.8} = 2.06$$

From Table A.1 in Appendix A, we see that $Z = 2.06$ corresponds to a probability of .4803; thus,

$$P(x \geq 12) = .5000 - .4803 = .0197$$

Therefore, the company's expected annual fine is

$$\$200{,}000 \cdot P(x \geq 12) = (\$200{,}000)(.0197) = \$3{,}940$$

Problems

1. Indicate which of the following probabilities are objective and which are subjective. (Note that in some cases, the probabilities may not be entirely one or the other.)
   a. The probability of snow tomorrow
   b. The probability of catching a fish
   c. The probability of the prime interest rate rising in the coming year
   d. The probability that the Cincinnati Reds will win the World Series
   e. The probability that demand for a product will be a specific amount next month
   f. The probability that a political candidate will win an election
   g. The probability that a machine will break down
   h. The probability of being dealt four aces in a poker hand
2. A gambler in Las Vegas is cutting a deck of cards for $1,000. What is the probability that the card for the gambler will be the following?
   a. A face card
   b. A queen
   c. A spade
   d. A jack of spades
3. Downhill Ski Resort in Colorado has accumulated information from records of the past 30 winters regarding the measurable snowfall. This information is as follows:

Snowfall (in.)	*Frequency*
0–19	2
20–29	7
30–39	8
40–49	8
50+	5
	30

   a. Determine the probability of each event in this frequency distribution.
   b. Are all the events in this distribution mutually exclusive? Explain.
4. Employees in the textile industry can be segmented as follows:

Employees	*Number*
Female and union	12,000
Female and nonunion	25,000
Male and union	21,000
Male and nonunion	42,000

   a. Determine the probability of each event in this distribution.
   b. Are the events in this distribution mutually exclusive? Explain.
   c. What is the probability that an employee is male?
   d. Is this experiment collectively exhaustive? Explain.
5. The quality control process at a manufacturing plant requires that each lot of finished units be sampled for defective items. Twenty units from each lot are inspected. If five or more defective units are found, the lot is rejected. If a lot is known to contain 10% defective items, what is the probability that the lot will be rejected? accepted?

6. A manufacturing company has 10 machines in continuous operation during a workday. The probability that an individual machine will break down during the day is .10. Determine the probability that during any given day 3 machines will break down.

7. A polling firm is taking a survey regarding a proposed new law. Of the voters polled, 30% are in favor of the law. If 10 people are surveyed, what is the probability that 4 will indicate that they are opposed to the passage of the new law?

8. An automobile manufacturer has discovered that 20% of all the transmissions it installed in a particular style of truck one year are defective. It has contacted the owners of these vehicles and asked them to return their trucks to the dealer to check the transmission. The Friendly Auto Mart sold seven of these trucks and has two of the new transmissions in stock. What is the probability that the auto dealer will need to order more new transmissions?

9. A new county hospital is attempting to determine whether it needs to add a particular specialist to its staff. Five percent of the general hospital population in the county contracts the illness the specialist would treat. If 12 patients check into the hospital in a day, what is the probability that 4 or more will have the illness?

10. A large research hospital has accumulated statistical data on its patients for an extended period. Researchers have determined that patients who are smokers have an 18% chance of contracting a serious illness such as heart disease, cancer, or emphysema, whereas there is only a .06 probability that a nonsmoker will contract a serious illness. From hospital records, the researchers know that 23% of all hospital patients are smokers, while 77% are nonsmokers. For planning purposes, the hospital physician staff would like to know the probability that a given patient is a smoker if the patient has a serious illness.

11. Two law firms in a community handle all the cases dealing with consumer suits against companies in the area. The Abercrombie firm takes 40% of all suits, and the Olson firm handles the other 60%. The Abercrombie firm wins 70% of its cases, and the Olson firm wins 60% of its cases.
    a. Develop a probability tree showing all marginal, conditional, and joint probabilities.
    b. Develop a joint probability table.
    c. Using Bayes' rule, determine the probability that the Olson firm handled a particular case, given that the case was won.

12. The Senate consists of 100 senators, of whom 34 are Republicans and 66 are Democrats. A bill to increase defense appropriations is before the Senate. Thirty-five percent of the Democrats and 70% of the Republicans favor the bill. The bill needs a simple majority to pass. Using a probability tree, determine the probability that the bill will pass.

13. A retail outlet receives radios from three electrical appliance companies. The outlet receives 20% of its radios from A, 40% from B, and 40% from C. The probability of receiving a defective radio from A is .01; from B, .02; and from C, .08.
    a. Develop a probability tree showing all marginal, conditional, and joint probabilities.
    b. Develop a joint probability table.
    c. What is the probability that a defective radio returned to the retail store came from company B?

14. A metropolitan school system consists of three districts—north, south, and central. The north district contains 25% of all students, the south district contains 40%, and the central district contains 35%. A minimum-competency test was given to all students; 10% of the north district students failed, 15% of the south district students failed, and 5% of the central district students failed.
    a. Develop a probability tree showing all marginal, conditional, and joint probabilities.
    b. Develop a joint probability table.
    c. What is the probability that a student selected at random failed the test?

15. A service station owner sells Goodroad tires, which are ordered from a local tire distributor. The distributor receives tires from two plants, A and B. When the owner of the service station receives

an order from the distributor, there is a .50 probability that the order consists of tires from plant A or plant B. However, the distributor will not tell the owner which plant the tires come from. The owner knows that 20% of all tires produced at plant A are defective, whereas only 10% of the tires produced at plant B are defective. When an order arrives at the station, the owner is allowed to inspect it briefly. The owner takes this opportunity to inspect one tire to see if it is defective. If the owner believes the tire came from plant A, the order will be sent back. Using Bayes' rule, determine the posterior probability that a tire is from plant A, given that the owner finds that it is defective.

16. A metropolitan school system consists of two districts, east and west. The east district contains 35% of all students, and the west district contains the other 65%. A vocational aptitude test was given to all students; 10% of the east district students failed, and 25% of the west district students failed. Given that a student failed the test, what is the posterior probability that the student came from the east district?

17. The Ramshead Pub sells a large quantity of beer every Saturday. From past sales records, the pub has determined the following probabilities for sales:

Barrels	*Probability*
6	.10
7	.20
8	.40
9	.25
10	.05
	1.00

Compute the expected number of barrels that will be sold on Saturday.

18. The following probabilities for grades in management science have been determined based on past records:

Grade	*Probability*
A	.10
B	.30
C	.40
D	.10
F	.10
	1.00

The grades are assigned on a 4.0 scale, where an A is a 4.0, a B a 3.0, and so on. Determine the expected grade and variance for the course.

19. A market in Boston orders oranges from Florida. The oranges are shipped to Boston from Florida by either railroad, truck, or airplane; an order can take 1, 2, 3, or 4 days to arrive in Boston once it is placed. The following probabilities have been assigned to the number of days it takes to receive an order once it is placed (referred to as *lead time*):

Lead Time	*Probability*
1	.20
2	.50
3	.20
4	.10
	1.00

Compute the expected number of days it takes to receive an order and the standard deviation.

20. An investment firm is considering two alternative investments, A and B, under two possible future sets of economic conditions, good and poor. There is a .60 probability of good economic conditions occurring and a .40 probability of poor economic conditions occurring. The

expected gains and losses under each economic type of conditions are shown in the following table:

	Economic Conditions	
Investment	Good	Poor
A	$900,000	–$800,000
B	120,000	70,000

Using the expected value of each investment alternative, determine which should be selected.

21. An investor is considering two investments, an office building and bonds. The possible returns from each investment and their probabilities are as follows:

Office Building		Bonds	
Return	Probability	Return	Probability
$50,000	.30	$30,000	.60
60,000	.20	40,000	.40
80,000	.10		1.00
10,000	.30		
0	.10		
	1.00		

Using expected value and standard deviation as a basis for comparison, discuss which of the two investments should be selected.

22. The Jefferson High School Band Booster Club has organized a raffle. The prize is a $6,000 car. Two thousand tickets to the raffle are to be sold at $1 apiece. If a person purchases four tickets, what will be the expected value of the tickets?

23. The time interval between machine breakdowns in a manufacturing firm is defined according to the following probability distribution:

Time Interval (hr.)	Probability
1	.15
2	.20
3	.40
4	.25
	1.00

Determine the cumulative probability distribution and compute the expected time between machine breakdowns.

24. The life of an electronic transistor is normally distributed, with a mean of 500 hours and a standard deviation of 80 hours. Determine the probability that a transistor will last for more than 400 hours.

25. The grade point average of students at a university is normally distributed, with a mean of 2.6 and a standard deviation of 0.6. A recruiter for a company is interviewing students for summer employment. What percentage of the students will have a grade point average of 3.5 or greater?

26. The weight of bags of fertilizer is normally distributed, with a mean of 50 pounds and a standard deviation of 6 pounds. What is the probability that a bag of fertilizer will weigh between 45 and 55 pounds?

27. The monthly demand for a product is normally distributed, with a mean of 700 units and a standard deviation of 200 units. What is the probability that demand will be greater than 900 units in a given month?

28. The Polo Development Firm is building a shopping center. It has informed renters that their rental spaces will be ready for occupancy in 19 months. If the expected time until the shopping center is completed is estimated to be 14 months, with a standard deviation of 4 months, what is the probability that the renters will not be able to occupy in 19 months?

29. A warehouse distributor of carpet keeps 6,000 yards of deluxe shag carpet in stock during a month. The average demand for carpet from the stores that purchase from the distributor is 4,500 yards per month, with a standard deviation of 900 yards. What is the probability that a customer's order will not be met during a month? (This situation is referred to as a *stockout*.)

30. The manager of the local National Video Store sells videocassette recorders at discount prices. If the store does not have a video recorder in stock when a customer wants to buy one, it will lose the sale because the customer will purchase a recorder from one of the many local competitors. The problem is that the cost of renting warehouse space to keep enough recorders in inventory to meet all demand is excessively high. The manager has determined that if 90% of customer demand for recorders can be met, then the combined cost of lost sales and inventory will be minimized. The manager has estimated that monthly demand for recorders is normally distributed, with a mean of 180 recorders and a standard deviation of 60. Determine the number of recorders the manager should order each month to meet 90% of customer demand.

31. The owner of Western Clothing Company has determined that the company must sell 670 pairs of denim jeans each month to break even (i.e., to reach the point where total revenue equals total cost). The company's marketing department has estimated that monthly demand is normally distributed, with a mean of 805 pairs of jeans and a standard deviation of 207 pairs. What is the probability that the company will make a profit each month?

32. Lauren Moore, a professor in management science, is computing her final grades for her introductory management science class. The average final grade is a 63, with a standard deviation of 10. Professor Moore wants to curve the final grades according to a normal distribution so that 10% of the grades are Fs, 20% are Ds, 40% are Cs, 20% are Bs, and 10% are As. Determine the numeric grades that conform to the curve Professor Moore wants to establish.

33. The SAT scores of all freshmen accepted at State University are normally distributed, with a mean of 1,050 and a standard deviation of 120. The College of Business at State University has accepted 620 of these freshmen into the college. All students in the college who score over 1,200 are eligible for merit scholarships. How many students can the college administration expect to be eligible for merit scholarships?

34. Erin Richards is a junior at Central High School, and she has talked to her guidance counselor about her chances of being admitted to Tech after her graduation. The guidance counselor has told her that Tech generally accepts only those applicants who graduate in the top 10% of their high school class. The average grade point average of the last four senior classes has been 2.67, with a standard deviation of 0.58. What GPA will Erin have to achieve to be in the top 10% of her class?

35. The associate dean in the college of business at Tech is going to purchase a new copying machine for the college. One model he is considering is the Zerox X10. The sales representative has told him that this model will make an average of 125,000 copies, with a standard deviation of 40,000 copies, before breaking down. What is the probability that the copier will make 200,000 copies before breaking down?

36. The Palace Hotel believes its customers may be waiting too long for room service. The hotel operations manager knows that the time for room service orders is normally distributed, and he sampled 10 room service orders during a 3-day period and timed each (in minutes), as follows:

23	23
15	12
26	16
19	18
30	25

The operations manager believes that only 10% of the room service orders should take longer than 25 minutes if the hotel has good customer service. Does the hotel room service meet this goal?

37. Agnes Hammer is a senior majoring in management science. She has been interviewing with several companies for a job when she graduates, and she is curious about what starting salary offers she might receive. There are 140 seniors in the graduating class for her major, and more than half have received job offers. She asked 12 of her classmates at random what their annual starting salary offers were, and she received the following responses:

$28,500	$35,500
32,600	36,000
34,000	25,700
27,500	29,000
24,600	31,500
34,500	26,800

Assume that starting salaries are normally distributed. Compute the mean and standard deviation for these data and determine the probability that Agnes will receive a salary offer of less than $27,000.

38. The owner of Gilley's Ice Cream Parlor has noticed that she sells more ice cream on hotter days during the summer, especially on days when the temperature is 85° or higher. To plan how much ice cream to stock, she would like to know the average daily high temperature for the summer months of July and August. Assuming that daily temperatures are normally distributed, she has gathered the following data for the high temperature for 20 days from a local almanac:

86°	92°
85	94
78	83
91	81
90	84
92	76
83	78
80	78
69	85
74	90

Compute the mean and standard deviation for these data and determine the expected number of days in July and August that the high temperature will be 85° or greater.

39. The state of Virginia has implemented a Standard of Learning (SOL) test that all public school students must pass before they can graduate from high school. A passing grade is 75. Montgomery County High School administrators want to gauge how well their students might do on the SOL test, but they don't want to take the time to test the whole student population. Instead, they selected 20 students at random and gave them the test. The results are as follows:

83	79	56	93
48	92	37	45
72	71	92	71
66	83	81	80
58	95	67	78

Assume that SOL test scores are normally distributed. Compute the mean and standard deviation for these data and determine the probability that a student at the high school will pass the test.

40. The department of management science at Tech has sampled 250 of its majors and compiled the following frequency distribution of grade point averages (on a 4.0 scale) for the previous semester:

GPA	Frequency
0 < 0.5	1
0.5 < 1.0	4
1.0 < 1.5	20
1.5 < 2.0	35
2.0 < 2.5	67
2.5 < 3.0	58
3.0 < 3.5	47
3.5 < 4.0	18
	250

The sample mean ($\bar{x}$) for this distribution is 2.5, and the sample standard deviation (s) is 0.72. Determine whether the student GPAs are normally distributed, using a .05 level of significance (i.e., $\alpha = .05$).

41. Geo-net, a cellular phone company, has collected the following frequency distribution for the length of calls outside its normal customer roaming area:

Length (min.)	Frequency
0 < 5	26
5 < 10	75
10 < 15	139
15 < 20	105
20 < 25	37
25+	18
	400

The sample mean ($\bar{x}$) for this distribution is 14.3 minutes, and the sample standard deviation is 3.7 minutes. Determine whether these data are normally distributed $(\alpha = .05)$.

Case Problem

Valley Swim Club

The Valley Swim Club has 300 stockholders, each holding one share of stock in the club. A share of club stock allows the shareholder's family to use the club's heated outdoor pool during the summer, upon payment of annual membership dues of $175. The club has not issued any new stock in years, and only a few of the existing shares come up for sale each year. The board of directors administers the sale of all stock. When a shareholder wants to sell, he or she turns the stock in to the board, which sells it to the person at the top of the waiting list. For the past few years, the length of the waiting list has remained relatively steady, at approximately 20 names.

However, during the past winter, two events occurred that have increased the demand for shares in the club. The winter was especially severe, and subzero weather and heavy ice storms caused both the town and the county pools to buckle and crack. The problems were not discovered until maintenance crews began to ready the pools for the summer, and repairs cannot be completed until the fall. Also during the winter, the manager of the local country club had an argument with her board of directors and one night burned down the clubhouse. Although the pool itself was not damaged, the dressing room facilities, showers, and snack bar were destroyed. As a result of these two events, the Valley Swim Club was inundated with applications to purchase shares. The waiting list suddenly grew to 250 people as the summer approached.

The board of directors of the swim club had refrained from issuing new shares in the past because there was never a very great demand, and the demand that did exist was usually absorbed within a year by stock turnover. In addition, the board has a real concern about overcrowding. It seemed like the present membership was about right, and there were very few complaints about overcrowding, except on holidays like Memorial Day and the Fourth of July. However, at a recent board meeting, a number of new applicants had attended and asked the board to issue new shares. In addition, a number of current shareholders suggested that this might be an opportunity for the club to raise some capital for needed repairs and to improve some of the existing facilities. This was tempting to the board. Although it had set the share price at $500 in the past, the board could set it at a much higher level now. In addition, any new shares sold would result in almost total profit because the manager, lifeguard, and maintenance costs had already been budgeted for the summer and would not increase with additional members.

Before the board of directors could make a decision on whether to sell more shares and, if so, how many, the board members felt they needed more information. Specifically, they would like to know the average number of people (family members, guests, etc.) that might use the pool each day during the summer. They would also like to know the number of days they could expect more than 500 people to use the pool from June through August, given the current number of shares.

The board of directors has the following daily attendance records for June through August from the previous summer; it thinks the figures would provide accurate estimates for the upcoming summer:

139	380	193	399	177	238
273	367	378	197	161	224
172	359	461	273	308	368
275	463	242	213	256	541
337	578	177	303	391	235
402	287	245	262	400	218
487	247	390	447	224	271
198	356	284	399	239	259
310	322	417	275	274	232
347	419	474	241	205	317
393	516	194	190	361	369
421	478	207	243	411	361
595	303	215	277	419	
497	223	304	241	258	
341	315	331	384	130	
291	258	407	246	195	

The board has developed the following criteria for making a decision on whether to issue new shares:

1. The expected number of days on which attendance would exceed 500 should be no more than 5 with the current membership.
2. The current average daily attendance should be no more than 320.
3. The average daily weekend (Saturday and Sunday) attendance should be no more than 500. (Weekend attendance is every sixth and seventh column entry in each progression of seven entries in the preceding data.)

If these criteria are met, the club will issue one new share, at a price of $1,000, for every two average attendees between the current daily average and an upper limit of 400.

Should the club issue new shares? If so, how many will it issue, and how much additional revenue will it realize?

CHAPTER

12

Decision Analysis

The two categories of decision situations are probabilities that can be assigned to future occurrences and probabilities that cannot be assigned.

See Web site Module E for a chapter on "Game Theory" and Module F for a chapter on "Markov Analysis."

In the previous chapters dealing with linear programming, models were formulated and solved in order to aid the manager in making a decision. The solutions to the models were represented by values for *decision* variables. However, these linear programming models were all formulated under the assumption that certainty existed. In other words, it was assumed that all the model coefficients, constraint values, and solution values were known with certainty and did not vary.

In actual practice, however, many decision-making situations occur under conditions of *uncertainty*. For example, the demand for a product may be not 100 units next week, but 50 or 200 units, depending on the state of the market (which is uncertain). Several decision-making techniques are available to aid the decision maker in dealing with this type of decision situation in which there is uncertainty.

Decision situations can be categorized into two classes: situations in which probabilities *cannot* be assigned to future occurrences and situations in which probabilities *can* be assigned. In this chapter we will discuss each of these classes of decision situations separately and demonstrate the decision-making criterion most commonly associated with each. Decision situations in which there are two or more decision makers who are in competition with each other are the subject of *game theory*, a topic included on the companion Web site that accompanies this text.

Components of Decision Making

*A **state of nature** is an actual event that may occur in the future.*

A decision-making situation includes several components—the decisions themselves *and* the actual events that may occur in the future, known as **states of nature**. At the time a decision is made, the decision maker is uncertain which states of nature will occur in the future and has no control over them.

Suppose a distribution company is considering purchasing a computer to increase the number of orders it can process and thus increase its business. If economic conditions remain good, the company will realize a large increase in profit; however, if the economy takes a downturn, the company will lose money. In this decision situation, the possible decisions are to purchase the computer and to not purchase the computer. The states of nature are *good* economic conditions and *bad* economic conditions. The state of nature that occurs will determine the outcome of the decision, and it is obvious that the decision maker has no control over which state will occur.

*Using a **payoff table** is a means of organizing a decision situation, including the payoffs from different decisions, given the various states of nature.*

As another example, consider a concessions vendor who must decide whether to stock coffee for the concession stands at a football game in November. If the weather is cold, most of the coffee will be sold, but if the weather is warm, very little coffee will be sold. The decision is to order or not to order coffee, and the states of nature are warm and cold weather.

To facilitate the analysis of these types of decision situations so that the best decisions result, they are organized into **payoff tables**. In general, a payoff table is a means of organizing and illustrating the payoffs from the different decisions, given the various states of nature in a decision problem. A payoff table is constructed as shown in Table 12.1.

TABLE 12.1
Payoff table

	State of Nature	
Decision	*a*	*b*
1	Payoff 1*a*	Payoff 1*b*
2	Payoff 2*a*	Payoff 2*b*

Each decision, 1 or 2, in Table 12.1 will result in an outcome, or *payoff*, for the particular state of nature that will occur in the future. Payoffs are typically expressed in terms of profit revenues, or cost (although they can be expressed in terms of a variety of values). For example, if decision 1 is to purchase a computer and state of nature *a* is good economic conditions, payoff 1*a* could be \$100,000 in profit.

It is often possible to assign probabilities to the states of nature to aid the decision maker in selecting the decision that has the best outcome. However, in some cases the decision maker is not able to assign probabilities, and it is this type of decision-making situation that we will address first.

Decision Making Without Probabilities

The following example will illustrate the development of a payoff table without probabilities. An investor is to purchase one of three types of real estate, as illustrated in Figure 12.1. The investor must decide among an apartment building, an office building, and a warehouse. The future states of nature that will determine how much profit the investor will make are good

FIGURE 12.1
Decision situation with real estate investment alternatives

economic conditions and poor economic conditions. The profits that will result from each decision in the event of each state of nature are shown in Table 12.2.

TABLE 12.2
Payoff table for the real estate investments

Decision (purchase)	State of Nature: Good Economic Conditions	State of Nature: Poor Economic Conditions
Apartment building	$ 50,000	$ 30,000
Office building	100,000	−40,000
Warehouse	30,000	10,000

Decision-Making Criteria

Once the decision situation has been organized into a payoff table, several criteria are available for making the actual decision. These decision criteria, which will be presented in this section, include maximax, maximin, minimax regret, Hurwicz, and equal likelihood. On occasion these criteria will result in the same decision; however, often they will yield different decisions. The decision maker must select the criterion or combination of criteria that best suits his or her needs.

The Maximax Criterion

*The **maximax criterion** results in the maximum of the maximum payoffs.*

With the **maximax criterion**, the decision maker selects the decision that will result in the maximum of the maximum payoffs. (In fact, this is how this criterion derives its name—a maximum of a maximum.) The maximax criterion is very optimistic. The decision maker assumes that the most favorable state of nature for each decision alternative will occur. Thus, for example, using this criterion, the investor would optimistically assume that good economic conditions will prevail in the future.

The maximax criterion is applied in Table 12.3. The decision maker first selects the maximum payoff for each decision. Notice that all three maximum payoffs occur under good economic conditions. Of the three maximum payoffs—$50,000, $100,000, and $30,000—the maximum is $100,000; thus, the corresponding decision is to purchase the office building.

TABLE 12.3
Payoff table illustrating a maximax decision

Decision (purchase)	State of Nature: Good Economic Conditions	State of Nature: Poor Economic Conditions
Apartment building	$ 50,000	$ 30,000
Office building	100,000	−40,000
Warehouse	30,000	10,000

Maximum payoff

Although the decision to purchase an office building will result in the largest payoff ($100,000), such a decision completely ignores the possibility of a potential loss of $40,000. The decision maker who uses the maximax criterion assumes a very optimistic future with respect to the state of nature.

Before the next criterion is presented, it should be pointed out that the maximax decision rule as presented here deals with *profit*. However, if the payoff table consisted of costs, the opposite selection would be indicated: the minimum of the minimum costs, or a *minimin* criterion. For the subsequent decision criteria we encounter, the same logic in the case of costs can be used.

The Maximin Criterion

The ***maximin criterion*** *results in the maximum of the minimum payoff.*

In contrast with the maximax criterion, which is very optimistic, the **maximin criterion** is pessimistic. With the maximin criterion, the decision maker selects the decision that will reflect the *maximum* of the *minimum* payoffs. For each decision alternative, the decision maker assumes that the minimum payoff will occur. Of these minimum payoffs, the maximum is selected. The maximin criterion for our investment example is demonstrated in Table 12.4.

TABLE 12.4
Payoff table illustrating a maximin decision

	State of Nature	
Decision (purchase)	**Good Economic Conditions**	**Poor Economic Conditions**
Apartment building	$ 50,000	$ 30,000 ← Maximum payoff
Office building	100,000	−40,000
Warehouse	30,000	10,000

The minimum payoffs for our example are $30,000, −$40,000, and $10,000. The maximum of these three payoffs is $30,000; thus, the decision arrived at by using the maximin criterion would be to purchase the apartment building. This decision is relatively conservative because the alternatives considered include only the worst outcomes that could occur. The decision to purchase the office building as determined by the maximax criterion includes the possibility of a large loss (−$40,000). The worst that can occur from the decision to purchase the apartment building, however, is *a gain of $30,000.* On the other hand, the largest possible gain from purchasing the apartment building is much less than that of purchasing the office building (i.e., $50,000 vs. $100,000).

If Table 12.4 contained costs instead of profits as the payoffs, the conservative approach would be to select the maximum cost for each decision. Then the decision that resulted in the minimum, the minimax, of these costs would be selected.

The Minimax Regret Criterion

Regret *is the difference between the payoff from the best decision and all other decision payoffs.*

In our example, suppose the investor decided to purchase the warehouse, only to discover that economic conditions in the future were better than expected. Naturally, the investor would be disappointed that she had not purchased the office building because it would have resulted in the largest payoff ($100,000) under good economic conditions. In fact, the investor would *regret* the decision to purchase the warehouse, and the *degree of* **regret** *would be $70,000*, the difference between the payoff for the investor's choice and the best choice.

The ***minimax regret criterion*** *minimizes the maximum regret.*

This brief example demonstrates the principle underlying the decision criterion known as **minimax regret criterion**. With this decision criterion, the decision maker attempts to avoid regret by selecting the decision alternative that minimizes the maximum regret.

To use the minimax regret criterion, a decision maker first selects the maximum payoff under each state of nature. For our example, the maximum payoff under good economic conditions is $100,000, and the maximum payoff under poor economic conditions is $30,000. All other payoffs under the respective states of nature are subtracted from these amounts, as follows:

Good Economic Conditions	*Poor Economic Conditions*
$100,000 − 50,000 = $50,000	$30,000 − 30,000 = $0
$100,000 − 100,000 = $0	$30,000 − (−40,000) = $70,000
$100,000 − 30,000 = $70,000	$30,000 − 10,000 = $20,000

These values represent the regret that the decision maker would experience if a decision were made that resulted in less than the maximum payoff. The values are summarized in a modified version of the payoff table known as a *regret table*, shown in Table 12.5. (Such a table is sometimes referred to as an *opportunity loss table*, in which case the term *opportunity loss* is synonymous with *regret*.)

TABLE 12.5
Regret table

Decision (purchase)	State of Nature	
	Good Economic Conditions	Poor Economic Conditions
Apartment building	$50,000	$ 0
Office building	0	70,000
Warehouse	70,000	20,000

To make the decision according to the minimax regret criterion, the maximum regret for *each decision* must be determined. The decision corresponding to the minimum of these regret values is then selected. This process is illustrated in Table 12.6.

TABLE 12.6
Regret table illustrating the minimax regret decision

Decision (purchase)	State of Nature	
	Good Economic Conditions	Poor Economic Conditions
Apartment building	$50,000	$ 0
Office building	0	70,000
Warehouse	70,000	20,000

The minimax regret value

According to the minimax regret criterion, the decision should be to purchase the apartment building rather than the office building or the warehouse. This particular decision is based on the philosophy that the investor will experience the least amount of regret by purchasing the apartment building. In other words, if the investor purchased either the office building or the warehouse, $70,000 worth of regret could result; however, the purchase of the apartment building will result in, at most, $50,000 in regret.

The Hurwicz Criterion

The Hurwicz criterion is a compromise between the maximax and maximin criteria.

*The **coefficient of optimism**, α, is a measure of the decision maker's optimism.*

*The **Hurwicz criterion** multiplies the best payoff by α, the coefficient of optimism, and the worst payoff by $1 - \alpha$, for each decision, and the best result is selected.*

The **Hurwicz criterion** strikes a compromise between the maximax and maximin criteria. The principle underlying this decision criterion is that the decision maker is neither totally optimistic (as the maximax criterion assumes) nor totally pessimistic (as the maximin criterion assumes). With the Hurwicz criterion, the decision payoffs are weighted by a **coefficient of optimism**, a measure of the decision maker's optimism. The coefficient of optimism, which we will define as α, is between zero and one (i.e., $0 \leq \alpha \leq 1.0$). If $\alpha = 1.0$, then the decision maker is said to be completely optimistic; if $\alpha = 0$, then the decision maker is completely pessimistic. (Given this definition, if α is the coefficient of optimism, $1 - \alpha$ is the *coefficient of pessimism*.)

The Hurwicz criterion requires that for each decision alternative, the maximum payoff be multiplied by α and the minimum payoff be multiplied by $1 - \alpha$. For our investment example, if α equals .4 (i.e., the investor is slightly pessimistic), $1 - \alpha = .6$, and the following values will result:

Decision	*Values*
Apartment building	$ 50,000(.4) + 30,000(.6) = $38,000
Office building	$100,000(.4) − 40,000(.6) = $16,000
Warehouse	$ 30,000(.4) + 10,000(.6) = $18,000

The Hurwicz criterion specifies selection of the decision alternative corresponding to the maximum weighted value, which is $38,000 for this example. Thus, the decision would be to purchase the apartment building.

It should be pointed out that when $\alpha = 0$, the Hurwicz criterion is actually the maximin criterion; when $\alpha = 1.0$, it is the maximax criterion. A limitation of the Hurwicz criterion is the fact that α must be determined by the decision maker. It can be quite difficult for a decision maker to accurately determine his or her degree of optimism. Regardless of how the decision maker determines α, it is still a completely *subjective* measure of the decision maker's degree of optimism. Therefore, the Hurwicz criterion is a completely subjective decision-making criterion.

The Equal Likelihood Criterion

*The **equal likelihood**, or **LaPlace**, **criterion** multiplies the decision payoff for each state of nature by an equal weight.*

When the maximax criterion is applied to a decision situation, the decision maker implicitly assumes that the most favorable state of nature for each decision will occur. Alternatively, when the maximin criterion is applied, the least favorable states of nature are assumed. The **equal likelihood**, or **LaPlace, criterion** weights each state of nature equally, thus assuming that the states of nature are equally likely to occur.

Because there are two states of nature in our example, we assign a weight of .50 to each one. Next, we multiply these weights by each payoff for each decision:

Decision	*Values*
Apartment building	\$ 50,000(.50) + 30,000(.50) = \$40,000
Office building	\$100,000(.50) − 40,000(.50) = \$30,000
Warehouse	\$ 30,000(.50) + 10,000(.50) = \$20,000

As with the Hurwicz criterion, we select the decision that has the maximum of these weighted values. Because \$40,000 is the highest weighted value, the investor's decision would be to purchase the apartment building.

In applying the equal likelihood criterion, we are assuming a 50% chance, or .50 probability, that either state of nature will occur. Using this same basic logic, it is possible to weight the states of nature differently (i.e., unequally) in many decision problems. In other words, different probabilities can be assigned to each state of nature, indicating that one state is more likely to occur than another. The application of different probabilities to the states of nature is the principle behind the decision criteria to be presented in the section on expected value.

Summary of Criteria Results

The decisions indicated by the decision criteria examined so far can be summarized as follows:

Criterion	*Decision (Purchase)*
Maximax	Office building
Maximin	Apartment building
Minimax regret	Apartment building
Hurwicz	Apartment building
Equal likelihood	Apartment building

The decision to purchase the apartment building was designated most often by the various decision criteria. Notice that the decision to purchase the warehouse was never indicated by any criterion. This is because the payoffs for an apartment building, under either set of future economic conditions, are always better than the payoffs for a warehouse. Thus, given any situation with these two alternatives (and any other choice, such as purchasing the office building), the decision to purchase an apartment building will always be made over the decision to purchase a warehouse. In fact, the warehouse decision alternative could have been eliminated from consideration under each of our criteria. The alternative of purchasing a warehouse is said to be **dominated** by the alternative of purchasing an apartment building. In general, dominated decision alternatives can be removed from the payoff table and not considered when the various decision-making criteria are applied. This reduces the complexity of the decision analysis somewhat. However, in our discussions throughout this chapter of the application of decision criteria, we will leave the dominated alternative in the payoff table for demonstration purposes.

*A **dominant** decision is one that has a better payoff than another decision under each state of nature.*

The appropriate criterion is dependent on the risk personality and philosophy of the decision maker.

The use of several decision criteria often results in a mix of decisions, with no one decision being selected more than the others. The criterion or collection of criteria used and the resulting decision depend on the characteristics and philosophy of the decision maker. For example, the extremely optimistic decision maker might eschew the majority of the foregoing results and make the decision to purchase the office building because the maximax criterion most closely reflects his or her personal decision-making philosophy.

Solution of Decision-Making Problems Without Probabilities with QM for Windows

QM for Windows includes a module to solve decision analysis problems. QM for Windows will be used to illustrate the use of the maximax, maximin, minimax regret, equal likelihood, and Hurwicz criteria for the real estate problem considered in this section. The problem data are input very easily. A summary of the input and solution output for the maximax, maximin, and Hurwicz criteria is shown in Exhibit 12.1. The decision with the equal likelihood criterion can be determined by using an alpha value for the Hurwicz criterion equal to the equal likelihood weight, which is .5 for our real estate investment example. The solution output with alpha equal to .5 is shown in Exhibit 12.2. The decision with the minimax regret criterion is shown in Exhibit 12.3.

EXHIBIT 12.1

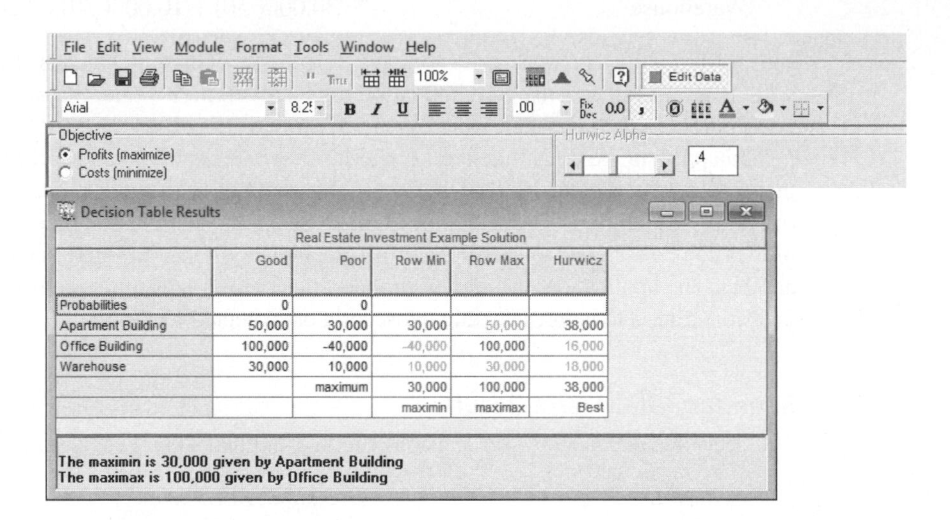

Real Estate Investment Example Solution

	Good	Poor	Row Min	Row Max	Hurwicz
Probabilities	0	0			
Apartment Building	50,000	30,000	30,000	50,000	38,000
Office Building	100,000	-40,000	-40,000	100,000	16,000
Warehouse	30,000	10,000	10,000	30,000	18,000
		maximum	30,000	100,000	38,000
			maximin	maximax	Best

The maximin is 30,000 given by Apartment Building
The maximax is 100,000 given by Office Building

EXHIBIT 12.2

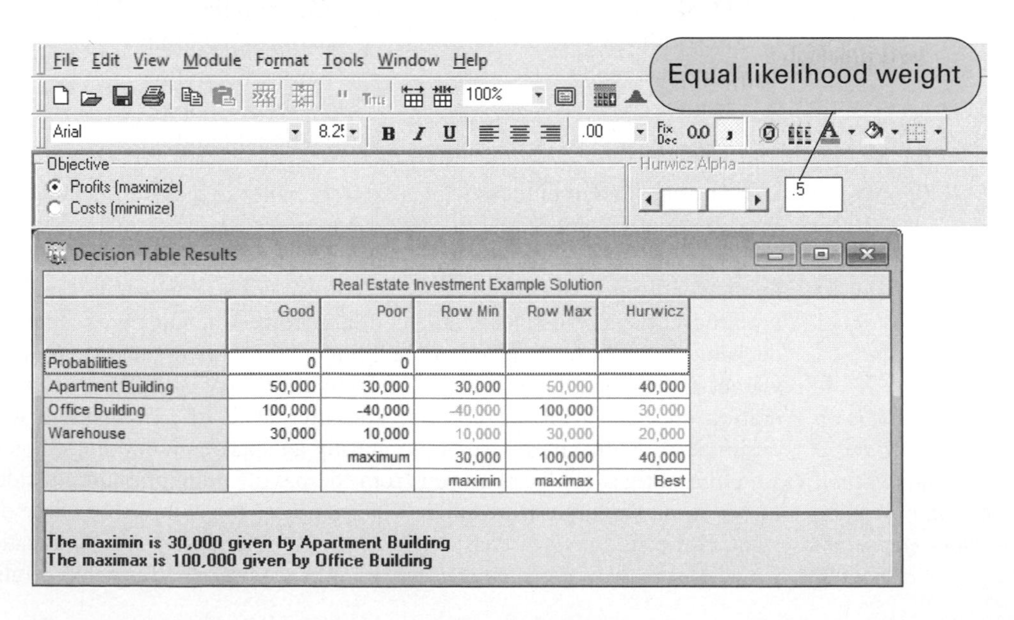

Real Estate Investment Example Solution

	Good	Poor	Row Min	Row Max	Hurwicz
Probabilities	0	0			
Apartment Building	50,000	30,000	30,000	50,000	40,000
Office Building	100,000	-40,000	-40,000	100,000	30,000
Warehouse	30,000	10,000	10,000	30,000	20,000
		maximum	30,000	100,000	40,000
			maximin	maximax	Best

The maximin is 30,000 given by Apartment Building
The maximax is 100,000 given by Office Building

EXHIBIT 12.3

Regret or Opportunity Loss

Real Estate Investment Example Solution				
	Good Regret	Poor Regret	Maximum Regret	Expected Regret
Probabilities	0	0		
Apartment Building	50,000	0	50,000	0
Office Building	0	70,000	70,000	0
Warehouse	70,000	20,000	70,000	0
Minimax regret			50,000	

Solution of Decision-Making Problems Without Probabilities with Excel

Excel can also be used to solve decision analysis problems using the decision-making criteria presented in this section. Exhibit 12.4 illustrates the application of the maximax, minimax, minimax regret, Hurwicz, and equal likelihood criteria for our real estate investment example.

EXHIBIT 12.4

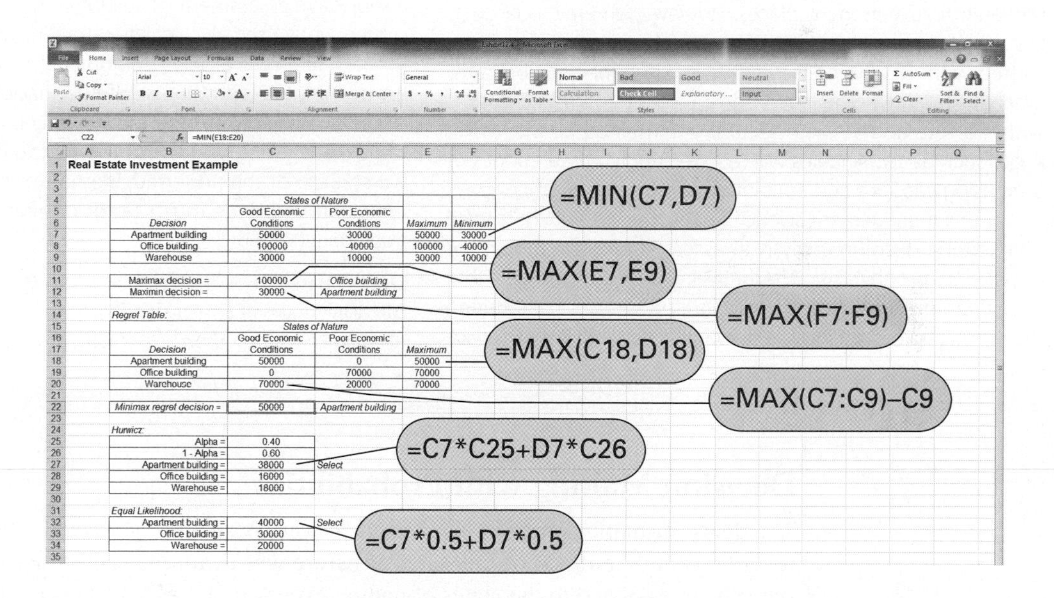

In cell E7, the formula **=MAX(C7,D7)** selects the maximum payoff outcome for the decision to purchase the apartment building. Next, in cell C11 the maximum of the maximum payoffs is determined with the formula **=MAX(E7:E9)**. The maximin decision is determined similarly.

In the regret table in Exhibit 12.4, in cell C18, the formula **=MAX(C7:C9) – C7** computes the regret for the apartment building decision under good economic conditions, and then the maximum regret for the apartment building is determined in cell E18, using the formula **=MAX(C18,D18)**. The minimax regret value is determined in cell C22 with the formula **=MIN(E18:E20)**.

The Hurwicz and equal likelihood decisions are determined using their respective formulas in cells **C27:C29** and **C32:C34**.

Management Science Application

Planning for Terrorist Attacks and Epidemics in Los Angeles County with Decision Analysis

The Centers for Disease Control and Prevention (CDC) has asked all state and local health departments in the United States to develop plans to provide vaccines and antibiotics (referred to as *prophylaxis*) to the general population in the event of a bioterrorism attack or natural epidemic. Outbreaks of diseases such as smallpox and influenza are very contagious, while diseases such as anthrax have a very short incubation period, resulting in deaths within 48 hours. Therefore, any preventive agents need to be distributed within 48 hours. Distributing prophylaxis to a locality such as Los Angeles County, with almost 10 million legal residents, 1 million tourists on any given day, and several million illegal residents, is a complex task.

The CDC requires that points of dispensing (PODs) be used as the primary means for distributing prophylaxis drugs. POD plans are funded by the CDC and have been tested and proven effective in many places. However, an area as large as Los Angeles would require 167 geographically dispersed PODs staffed with 48,000 individuals, who would need to be gathered at staging points, trained, and sent to the POD sites within a few hours—with security maintained all the while. A city such as Los Angeles will require supplemental modes of dispensing vaccines and antibiotics beyond just PODs.

In this management science application, researchers used multicriteria decision analysis to assess alternative modes for dispensing prophylaxis. Alternative distribution modes that have been previously tested on a limited basis or used during the normal flu season include delivery by the U.S. Postal Service (USPS), dispensing through local pharmacies, and use of drive-through PODs. Value measures for each alternative dispensing mode include speed of distribution, staffing requirements, and a subjective assessment of security requirements, which were subsequently used to calculate a total measure of effectiveness for each alternative. The decision analysis indicated that delivery by the USPS and the use of pharmacies were the two best alternatives to supplement the PODs, and the drive-through alternative, although popularly thought to be the best, was the worst. Both of the preferred alternatives had a 100% staffing reduction and were much faster dispensing modes than the traditional POD.

AP Photo/Marcio Jose Sanchez

Source: Based on A. Richter and S. Khan, "Pilot Model: Judging Alternate Modes of Dispensing Prophylaxis in Los Angeles County," *Interfaces* 39, no. 3 (May–June 2009): 238–40.

Decision Making with Probabilities

The decision-making criteria just presented were based on the assumption that no information regarding the likelihood of the states of nature was available. Thus, no *probabilities of occurrence* were assigned to the states of nature, except in the case of the equal likelihood criterion. In that case, by assuming that each state of nature was equally likely and assigning a weight of .50 to each state of nature in our example, we were implicitly assigning a probability of .50 to the occurrence of each state of nature.

It is often possible for the decision maker to know enough about the future states of nature to assign probabilities to their occurrence. Given that probabilities can be assigned, several decision criteria are available to aid the decision maker. We will consider two of these criteria: *expected value* and *expected opportunity loss* (although several others, including the *maximum likelihood criterion*, are available).

***Expected value** is computed by multiplying each decision outcome under each state of nature by the probability of its occurrence.*

Expected Value

To apply the concept of **expected value** as a decision-making criterion, the decision maker must first estimate the probability of occurrence of each state of nature. Once these estimates have been made, the expected value for each decision alternative can be computed. The expected value is computed by multiplying each outcome (of a decision) by the probability of its occurrence and then summing these products. The expected value of a random variable x, written symbolically as $EV(x)$, is computed as follows:

$$EV(x) = \sum_{i=1}^{n} x_i P(x_i)$$

where

n = number of values of the random variable x

Using our real estate investment example, let us suppose that, based on several economic forecasts, the investor is able to estimate a .60 probability that good economic conditions will prevail and a .40 probability that poor economic conditions will prevail. This new information is shown in Table 12.7.

TABLE 12.7
Payoff table with probabilities for states of nature

	State of Nature	
Decision (purchase)	Good Economic Conditions .60	Poor Economic Conditions .40
Apartment building	$ 50,000	$30,000
Office building	100,000	−40,000
Warehouse	30,000	10,000

The expected value (EV) for each decision is computed as follows:

$$EV(\text{apartment}) = \$50{,}000(.60) + 30{,}000(.40) = \$42{,}000$$
$$EV(\text{office}) = \$100{,}000(.60) - 40{,}000(.40) = \$44{,}000$$
$$EV(\text{warehouse}) = \$30{,}000(.60) + 10{,}000(.40) = \$22{,}000$$

The best decision is the one with the greatest expected value. Because the greatest expected value is $44,000, the best decision is to purchase the office building. This does not mean that $44,000 will result if the investor purchases the office building; rather, it is assumed that one of the payoff values will result (either $100,000 or −$40,000). The expected value means that if this decision situation occurred a large number of times, an *average* payoff of $44,000 would result. Alternatively, if the payoffs were in terms of costs, the best decision would be the one with the lowest expected value.

Expected Opportunity Loss

Expected opportunity loss is the expected value of the regret for each decision.

A decision criterion closely related to expected value is **expected opportunity loss.** To use this criterion, we multiply the probabilities by the regret (i.e., opportunity loss) for each decision outcome rather than multiplying the decision outcomes by the probabilities of their occurrence, as we did for expected monetary value.

The concept of regret was introduced in our discussion of the minimax regret criterion. The regret values for each decision outcome in our example were shown in Table 12.6. These values are repeated in Table 12.8, with the addition of the probabilities of occurrence for each state of nature.

TABLE 12.8
Regret (opportunity loss) table with probabilities for states of nature

	State of Nature	
Decision (purchase)	Good Economic Conditions .60	Poor Economic Conditions .40
Apartment building	$50,000	$ 0
Office building	0	70,000
Warehouse	70,000	20,000

The expected opportunity loss (*EOL*) for each decision is computed as follows:

$$EOL(\text{apartment}) = \$50{,}000(.60) + 0(.40) = \$30{,}000$$
$$EOL(\text{office}) = \$0(.60) + 70{,}000(.40) = \$28{,}000$$
$$EOL(\text{warehouse}) = \$70{,}000(.60) + 20{,}000(.40) = \$50{,}000$$

As with the minimax regret criterion, the best decision results from minimizing the regret, or, in this case, minimizing the *expected* regret or opportunity loss. Because $28,000 is the minimum expected regret, the decision is to purchase the office building.

The expected value and expected opportunity loss criteria result in the same decision.

Notice that the decisions recommended by the expected value and expected opportunity loss criteria were the same—to purchase the office building. This is not a coincidence because these two methods always result in the same decision. Thus, it is repetitious to apply both methods to a decision situation when one of the two will suffice.

In addition, note that the decisions from the expected value and expected opportunity loss criteria are totally dependent on the probability estimates determined by the decision maker. Thus, if inaccurate probabilities are used, erroneous decisions will result. It is therefore important that the decision maker be as accurate as possible in determining the probability of each state of nature.

Solution of Expected Value Problems with QM for Windows

QM for Windows not only solves decision analysis problems without probabilities but also has the capability to solve problems using the expected value criterion. A summary of the input data and the solution output for our real estate example is shown in Exhibit 12.5. Notice that the expected value results are included in the third column of this solution screen.

EXHIBIT 12.5

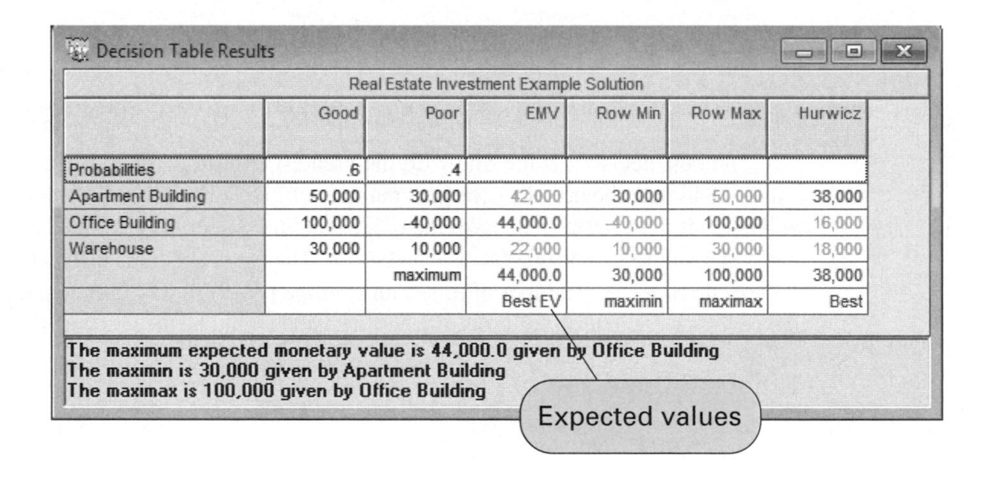

Decision Table Results

Real Estate Investment Example Solution

	Good	Poor	EMV	Row Min	Row Max	Hurwicz
Probabilities	.6	.4				
Apartment Building	50,000	30,000	42,000	30,000	50,000	38,000
Office Building	100,000	-40,000	44,000.0	-40,000	100,000	16,000
Warehouse	30,000	10,000	22,000	10,000	30,000	18,000
		maximum	44,000.0	30,000	100,000	38,000
			Best EV	maximin	maximax	Best

The maximum expected monetary value is 44,000.0 given by Office Building
The maximin is 30,000 given by Apartment Building
The maximax is 100,000 given by Office Building

Solution of Expected Value Problems with Excel and Excel QM

This type of expected value problem can also be solved by using an Excel spreadsheet. Exhibit 12.6 shows our real estate investment example set up in a spreadsheet format. Cells E7, E8, and E9

EXHIBIT 12.6

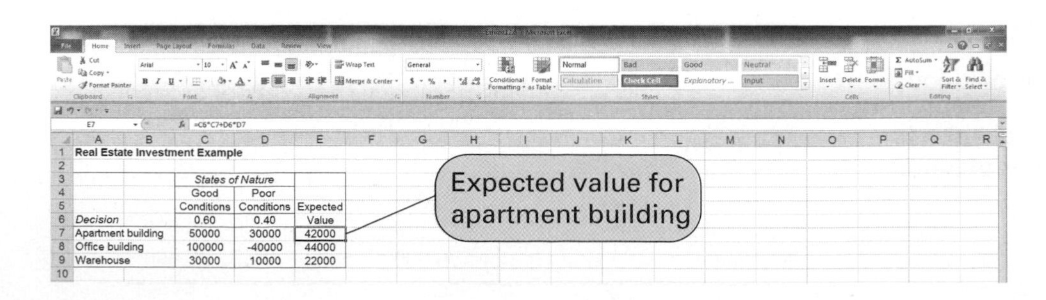

E7 =C6*C7+D6*D7

Real Estate Investment Example

	States of Nature		
	Good Conditions	Poor Conditions	Expected
Decision	0.60	0.40	Value
Apartment building	50000	30000	42000
Office building	100000	-40000	44000
Warehouse	30000	10000	22000

include the expected value formulas for this example. The expected value formula for the first decision, purchasing the apartment building, is embedded in cell E7 and is shown on the formula bar at the top of the spreadsheet.

Excel QM is a set of spreadsheet macros that is included on the companion Web site that accompanies this text, and it has a macro to solve decision analysis problems. Once activated, clicking on "Decision Analysis" will result in a Spreadsheet Initialization window. After entering several problem parameters, including the number of decisions and states of nature, and then clicking on "OK," the spreadsheet shown in Exhibit 12.7 will result. Initially, this spreadsheet contains example values in cells **B8:C11**. Exhibit 12.7 shows the spreadsheet with our problem data already typed in. The results are computed automatically as the data are entered, using the cell formulas already embedded in the macro.

EXHIBIT 12.7

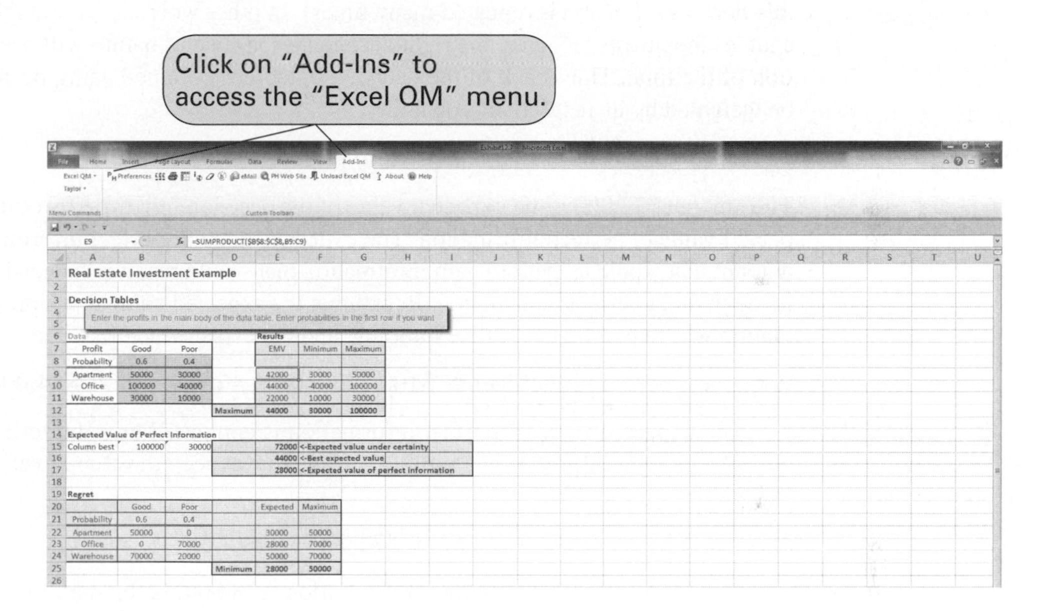

Expected Value of Perfect Information

It is often possible to purchase additional information regarding future events and thus make a better decision. For example, a real estate investor could hire an economic forecaster to perform an analysis of the economy to more accurately determine which economic condition will occur in the future. However, the investor (or any decision maker) would be foolish to pay more for this information than he or she stands to gain in extra profit from having the information. That is, the information has some maximum value that represents the limit of what the decision maker would be willing to spend. This value of information can be computed as an expected value—hence its name, the **expected value of perfect information** (also referred to as EVPI).

The expected value of perfect information (EVPI) is the maximum amount a decision maker would pay for additional information.

To compute the expected value of perfect information, we first look at the decisions under each state of nature. If we could obtain information that assured us which state of nature was going to occur (i.e., perfect information), we could select the best decision for that state of nature. For example, in our real estate investment example, if we know for sure that good economic conditions will prevail, then we will decide to purchase the office building. Similarly, if we know for sure that poor economic conditions will occur, then we will decide to purchase the apartment building. These hypothetical "perfect" decisions are summarized in Table 12.9.

TABLE 12.9
Payoff table with decisions, given perfect information

Decision (purchase)	State of Nature: Good Economic Conditions .60	State of Nature: Poor Economic Conditions .40
Apartment building	$ 50,000	$30,000
Office building	100,000	−40,000
Warehouse	30,000	10,000

The probabilities of each state of nature (i.e., .60 and .40) tell us that good economic conditions will prevail 60% of the time and poor economic conditions will prevail 40% of the time (if this decision situation is repeated many times). In other words, even though perfect information enables the investor to make the right decision, each state of nature will occur only a certain portion of the time. Thus, each of the decision outcomes obtained using perfect information must be weighted by its respective probability:

$$\$100{,}000(.60) + 30{,}000(.40) = \$72{,}000$$

The amount $72,000 is the expected value of the decision, *given* perfect information, not the expected value *of* perfect information. The expected value of perfect information is the maximum amount that would be paid to gain information that would result in a decision better than the one made without perfect information. Recall that the expected value decision without perfect information was to purchase an office building, and the expected value was computed as

$$EV(\text{office}) = \$100{,}000(.60) - 40{,}000(.40) = \$44{,}000$$

EVPI equals the expected value, given perfect information, minus the expected value without perfect information.

The expected value of perfect information is computed by subtracting the expected value without perfect information ($44,000) from the expected value given perfect information ($72,000):

$$EVPI = \$72{,}000 - 44{,}000 = \$28{,}000$$

The expected value of perfect information, $28,000, is the maximum amount that the investor would pay to purchase perfect information from some other source, such as an economic forecaster. Of course, perfect information is rare and usually unobtainable. Typically, the decision maker would be willing to pay some amount less than $28,000, depending on how accurate (i.e., close to perfection) the decision maker believed the information was.

The expected value of perfect information equals the expected opportunity loss for the best decision.

It is interesting to note that the expected value of perfect information, $28,000 for our example, is the same as the *expected opportunity loss (EOL)* for the decision selected, using this later criterion:

$$EOL(\text{office}) = \$0(.60) + 70{,}000(.40) = \$28{,}000$$

This will always be the case, and logically so, because regret reflects the *difference between the best decision under a state of nature and the decision actually made.* This is actually the same thing determined by the expected value of perfect information.

Excel QM for decision analysis computes the expected value of perfect information, as shown in cell E17 at the bottom of the spreadsheet in Exhibit 12.7. The expected value of perfect information can also be determined by using Excel. Exhibit 12.8 shows the EVPI for our real estate investment example.

*A **decision tree** is a diagram consisting of square decision nodes, circle probability nodes, and branches representing decision alternatives.*

Decision Trees

Another useful technique for analyzing a decision situation is using a **decision tree**. A decision tree is a graphical diagram consisting of nodes and branches. In a decision tree, the user computes the expected value of each outcome and makes a decision based on these expected values.

EXHIBIT 12.8

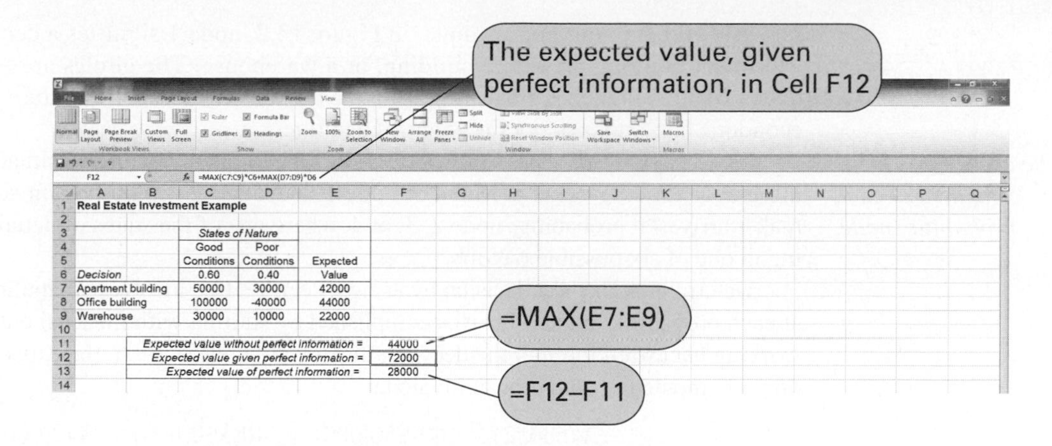

	States of Nature		
	Good Conditions	Poor Conditions	Expected
Decision	0.60	0.40	Value
Apartment building	50000	30000	42000
Office building	100000	-40000	44000
Warehouse	30000	10000	22000

Expected value without perfect information =	44000
Expected value given perfect information =	72000
Expected value of perfect information =	28000

The primary benefit of a decision tree is that it provides an illustration (or picture) of the decision-making process. This makes it easier to correctly compute the necessary expected values and to understand the process of making the decision.

We will use our example of the real estate investor to demonstrate the fundamentals of decision tree analysis. The various decisions, probabilities, and outcomes of this example, initially presented in Table 12.7, are repeated in Table 12.10. The decision tree for this example is shown in Figure 12.2.

The circles (●) and the square (■) in Figure 12.2 are referred to as *nodes*. The square is a decision node, and the *branches* emanating from a decision node reflect the alternative decisions

TABLE 12.10
Payoff table for real estate investment example

	State of Nature	
Decision (purchase)	Good Economic Conditions .60	Poor Economic Conditions .40
Apartment building	$ 50,000	$30,000
Office building	100,000	−40,000
Warehouse	30,000	10,000

FIGURE 12.2
Decision tree for real estate investment example

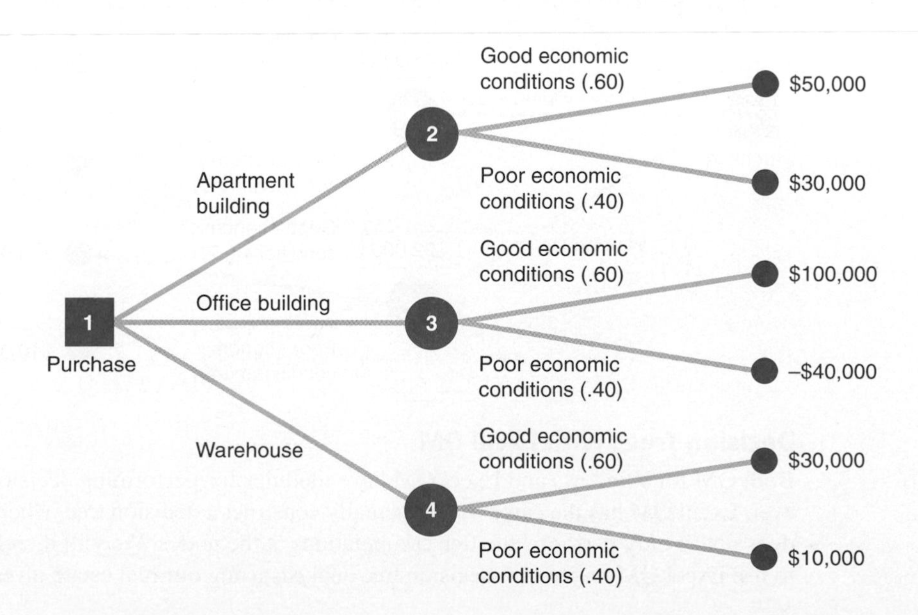

possible at that point. For example, in Figure 12.2, node 1 signifies a decision to purchase an apartment building, an office building, or a warehouse. The circles are probability, or event, nodes, and the branches emanating from them indicate the states of nature that can occur: good economic conditions or poor economic conditions.

The expected value is computed at each probability node.

The decision tree represents the sequence of events in a decision situation. First, one of the three decision choices is selected at node 1. Depending on the branch selected, the decision maker arrives at probability node 2, 3, or 4, where one of the states of nature will prevail, resulting in one of six possible payoffs.

Determining the best decision by using a decision tree involves computing the expected value at each probability node. This is accomplished by starting with the final outcomes (payoffs) and working backward through the decision tree toward node 1. First, the expected value of the payoffs is computed at each probability node:

$$EV(\text{node } 2) = .60(\$50{,}000) + .40(\$30{,}000) = \$42{,}000$$
$$EV(\text{node } 3) = .60(\$100{,}000) + .40(-\$40{,}000) = \$44{,}000$$
$$EV(\text{node } 4) = .60(\$30{,}000) + .40(\$10{,}000) = \$22{,}000$$

Branches with the greatest expected value are selected.

These values are now shown as the *expected* payoffs from each of the three branches emanating from node 1 in Figure 12.3. Each of these three expected values at nodes 2, 3, and 4 is the outcome of a possible decision that can occur at node 1. Moving toward node 1, we select the branch that comes from the probability node with the highest expected payoff. In Figure 12.3, the branch corresponding to the highest payoff, $44,000, is from node 1 to node 3. This branch represents the decision to purchase the office building. The decision to purchase the office building, with an expected payoff of $44,000, is the same result we achieved earlier by using the expected value criterion. In fact, when only one decision is to be made (i.e., there is not a series of decisions), the decision tree will always yield the same decision and expected payoff as the expected value criterion. As a result, in these decision situations a decision tree is not very useful. However, when a sequence or series of decisions is required, a decision tree can be very useful.

FIGURE 12.3
Decision tree with expected value at probability nodes

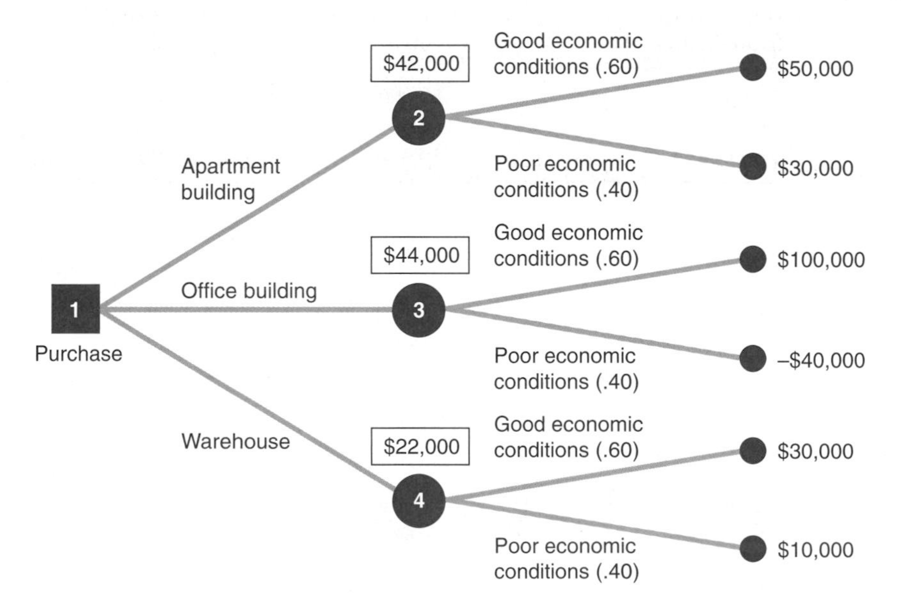

Decision Trees with Excel QM

Both QM for Windows and Excel QM have modules for performing decision tree analysis. However, Excel QM has the capability to actually construct a decision tree, whereas QM for Windows does not; it only performs the tree computations at the nodes. We will therefore demonstrate how to use Excel QM to perform decision tree analysis using our real estate investment example.

After opening Excel QM, click on "Add-Ins" then on "Taylor," and from the drop down menu select "Decision Analysis," then click on "Decision Trees." A window like the one in Exhibit 12.9 will appear, with only node "1" in cell A10. A window titled Decision Tree Creation will also be on your screen; this is the primary tool for developing the decision tree. The "Decision Tree Creation" window automatically will show "1" as the "Selected node" and the "Number of branches to add" as "2," so the first step is to increase this to "3" and then click on "Add 3 DECISIONS from node 1." This will result in the three new branches connected to nodes 2, 3 and 4, as shown in Exhibit 12.9.

EXHIBIT 12.9

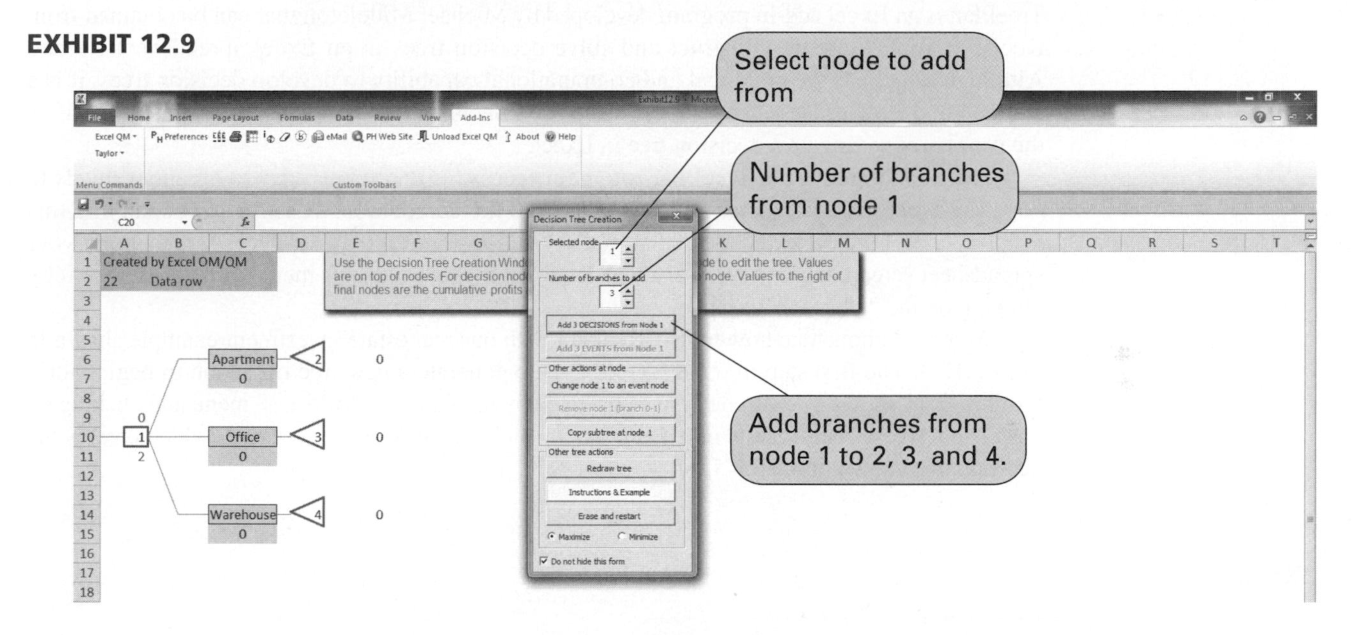

Next, we use the Decision Tree Creation window to add 2 new "Event" branches from nodes 2, 3, and 4. The resulting new window is shown in Exhibit 12.10. Notice that six new

EXHIBIT 12.10

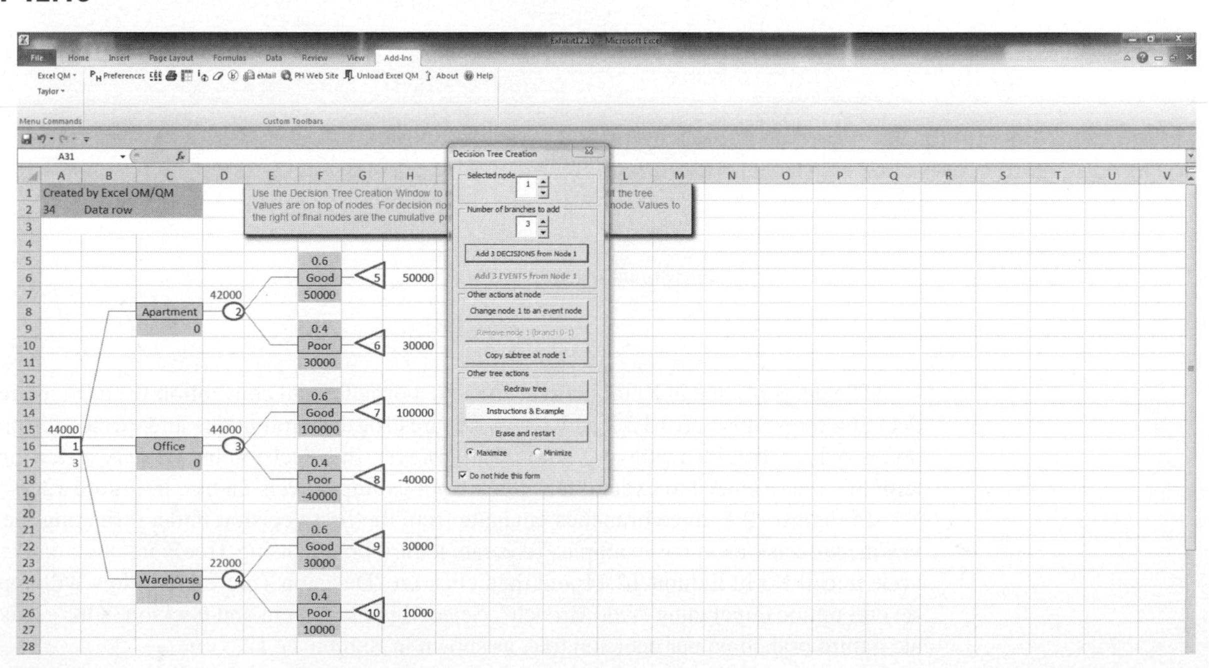

branches have been created. The cells next to node 5 are initially empty, and we enter "0.6" into cell F5 and "50,000" in cell F7. When we enter "0.6" in cell F5, "0.4" automatically appears in cell F9. We repeat this process for the next 4 new event branches to re-create the decision tree in Figure 12.3. Notice that the maximum expected value of $44,000 is shown in cell A15, just above node 1, and in cell D15, above node 3, indicating that the office building is the best decision.

Decision Trees with Excel and TreePlan (www.treeplan.com)

TreePlan is an Excel add-in program developed by Michael Middleton that can be obtained from Decision Tool Works to construct and solve decision trees in an Excel spreadsheet format. Although Excel has the graphical and computational capability to develop decision trees, it is a difficult and slow process. TreePlan is basically a decision tree template that greatly simplifies the process of setting up a decision tree in Excel.

The first step in using TreePlan is to gain access to it. The best way to go about this is to copy the TreePlan add-in file, TreePlan.xla, from the companion Web site accompanying this text onto your hard drive and then add it to the "Add-Ins" menu that you access at the top of your spreadsheet screen. Once you have added TreePlan to the "Add-Ins" menu, you can invoke it by clicking on the "Decision Trees" menu item.

We will demonstrate how to use TreePlan with our real estate investment example shown in Figure 12.3. The first step in using TreePlan is to generate a new tree on which to begin work. Exhibit 12.11 shows a new tree that we generated by invoking the "Add-ins" menu and clicking on "Decision Trees." This results in a menu from which we click on "New Tree," which creates the decision tree shown in Exhibit 12.11.

EXHIBIT 12.11

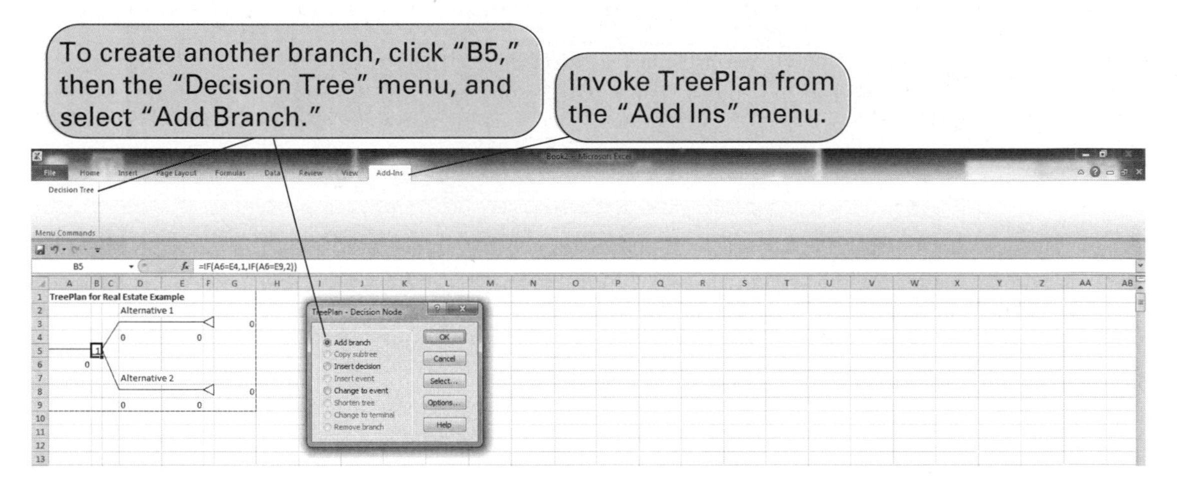

The decision tree in Exhibit 12.11 uses the normal nodal convention we used in creating the decision trees in Figures 12.2 and 12.3—squares for decision nodes and circles for probability nodes (which TreePlan calls *event nodes*). However, this decision tree is only a starting point or template that we need to expand to replicate our example decision tree in Figure 12.3.

In Figure 12.3, three branches emanate from the first decision node, reflecting the three investment decisions in our example. To create a third branch using TreePlan, click on the decision node in cell B5 in Exhibit 12.11 and then click on "Decision Tree". A window will appear, with several options, including "Add Branch". Select this menu item and click on "OK." This will create a third branch on our decision tree, as shown in Exhibit 12.12.

EXHIBIT 12.12

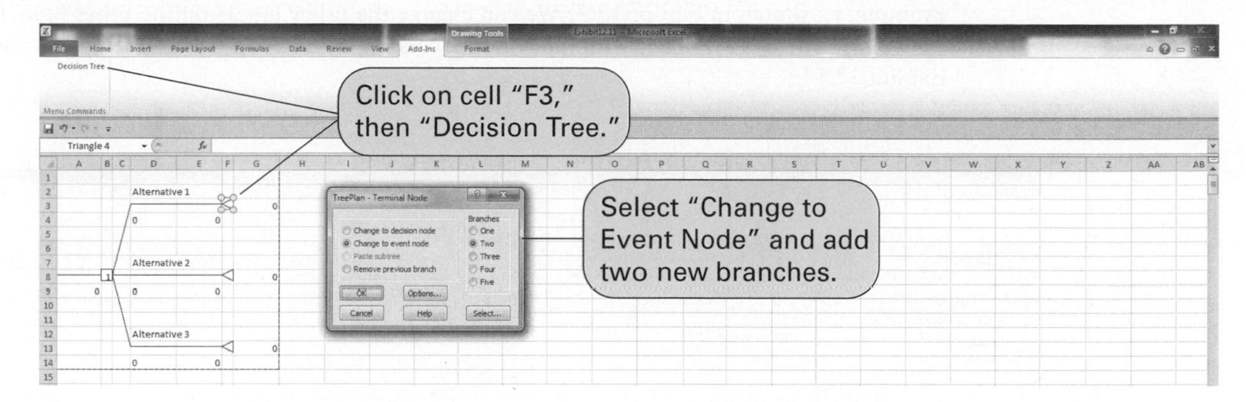

Next, we need to expand our decision tree in Exhibit 12.12 by adding probability (event) nodes (2, 3, and 4 in Figure 12.3) and branches from these nodes for our example. To add a new node, click on the end node in cell F3 in Exhibit 12.12 and then "Decision Tree." From the menu window that appears, select "Change to Event Node" and then select "Two" Branches from the same window and click on "OK." This process must be repeated two more times for the other two end nodes to create our three probability (event) nodes. The resulting decision tree is shown in Exhibit 12.13, with the new probability nodes at cells F5, F15, and F25 and with accompanying branches.

EXHIBIT 12.13

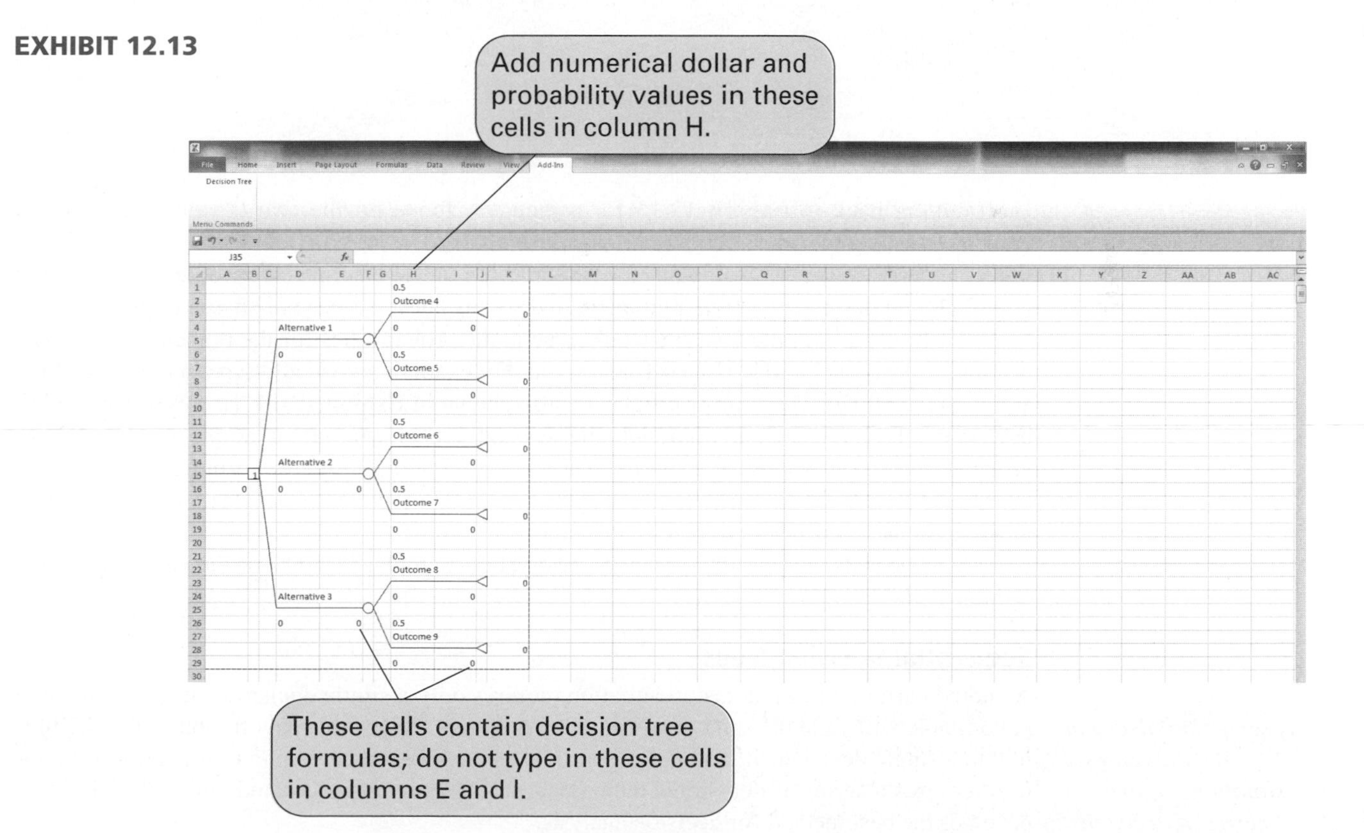

The next step is to edit the decision tree labels and add the numeric data from our example. Generic descriptive labels are shown above each branch in Exhibit 12.13—for example, "Alternative 1" in cell D4 and "Outcome 4" in cell H2. We edit the labels the same way we would edit any spreadsheet. For example, if we click on cell D4, we can type in "Apartment

Building" in place of "Decision 1," reflecting the decision corresponding to this branch in our example, as shown in Figure 12.3. We can change the other labels on the other branches the same way. The decision tree with the edited labels corresponding to our example is shown in Exhibit 12.14.

EXHIBIT 12.14

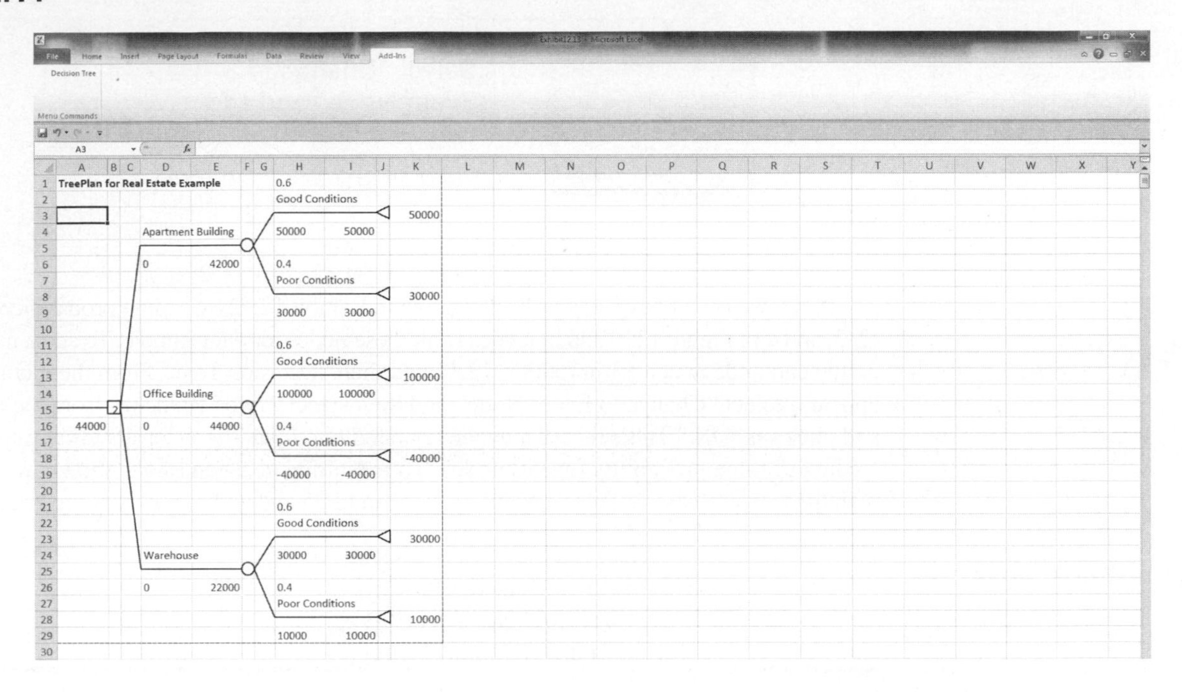

Looking back to Exhibit 12.13 for a moment, focus on the two 0 values below each branch—for example, in cells D6 and E6 and in cells H9 and I4. The first 0 cell is where we type in the numeric value (i.e., $ amount) for that branch. For our example, we would type in 50,000 in cell H4, 30,000 in H9, 100,000 in H14, and so on. These values are shown on the decision tree in Exhibit 12.14. Likewise, we would type in the probabilities for the branches in the cells just above the branch—H1, H6, H11, and so on. For example, we would type in 0.60 in cell H1 and 0.40 in cell H6. These probabilities are also shown in Exhibit 12.14. However, we need to be very careful not to type anything into the second 0 branch cell—for example, E6, I4, I9, E16, I14, I19, and so on. These cells automatically contain the decision tree formulas that compute the expected values at each node and select the best decision branches, so we do not want to type anything in these cells that would eliminate these formulas.

The expected value for this decision tree and our example, $44,000, is shown in cell A16 in Exhibit 12.14.

Sequential Decision Trees

A sequential decision tree illustrates a situation requiring a series of decisions.

As noted earlier, when a decision situation requires only a single decision, an expected value payoff table will yield the same result as a decision tree. However, a payoff table is usually limited to a single decision situation, as in our real estate investment example. If a decision situation requires a series of decisions, then a payoff table cannot be created, and a decision tree becomes the best method for decision analysis.

To demonstrate the use of a decision tree for a sequence of decisions, we will alter our real estate investment example to encompass a 10-year period during which several decisions must be made. In this new example, the *first decision* facing the investor is whether to purchase an apartment building or land. If the investor purchases the apartment building, two states of nature are possible: Either the population of the town will grow (with a probability of .60) or the population will

Management Science Application

Evaluating Electric Generator Maintenance Schedules Using Decision Tree Analysis

Electric utility companies plan annual outage periods for preventive maintenance on generators. The outages are typically part of 5- to-20-year master schedules. However, at Entergy Electric Systems these scheduled outages were traditionally based on averages that did not reflect short-term fluctuations in demand due to breakdowns and bad weather conditions. The master schedule had to be reviewed each week by outage planners who relied on their experience to determine whether the schedule needed to be changed. A user-friendly software system was developed to assist planners at Entergy Electric Systems in making changes in their schedule. The system is based on decision tree analysis.

Each week in the master schedule is represented by a decision tree that is based on changes in customer demand, unexpected generator breakdowns, and delays in returning generators from planned outages. The numeric outcome of the decision tree is the average reserve margin of megawatts (MW) for a specific week. This value enables planners to determine whether customer demand will be met and whether the maintenance schedule planned for the week is acceptable. The planner's objective is to avoid negative power reserves by making changes in the generators' maintenance schedule. The branches of the decision tree, their probabilities of occurrence, and the branch MW values are based on historical data. The system has enabled Entergy Electric Systems to isolate high-risk weeks and to develop timely maintenance schedules on short notice. The new computerized system has reduced the maintenance schedules review time for as many as 260 weeks from several days to less than an hour.

© Corbis Flirt/Alamy

Source: Based on H. A. Taha and H. M. Wolf, "Evaluation of Generator Maintenance Schedules at Entergy Electric Systems," *Interfaces* 26, no. 4 (July–August 1996): 56–65.

not grow (with a probability of .40). Either state of nature will result in a payoff. On the other hand, if the investor chooses to purchase land, 3 years in the future another decision will have to be made regarding the development of the land. The decision tree for this example, shown in Figure 12.4, contains all the pertinent data, including decisions, states of nature, probabilities, and payoffs.

At decision node 1 in Figure 12.4, the decision choices are to purchase an apartment building or to purchase land. Notice that the cost of each venture ($800,000 and $200,000, respectively) is shown in parentheses. If the apartment building is purchased, two states of nature are possible at probability node 2: The town may exhibit population growth, with a probability of .60, or there may be no population growth or a decline, with a probability of .40. If the population grows, the investor will achieve a payoff of $2,000,000 over a 10-year period. (Note that this whole decision situation encompasses a 10-year time span.) However, if no population growth occurs, a payoff of only $225,000 will result.

If the decision is to purchase land, two states of nature are possible at probability node 3. These two states of nature and their probabilities are identical to those at node 2; however, the payoffs are different. If population growth occurs *for a 3-year period*, no payoff will occur, but the investor will make another decision at node 4 regarding development of the land. At that point, either apartments will be built, at a cost of $800,000, or the land will be sold, with a payoff of $450,000. Notice that the decision situation at node 4 can occur only if population growth occurs first. If no population growth occurs at node 3, there is no payoff, and another decision situation becomes necessary at node 5: The land can be developed commercially at a cost of $600,000, or the land can be sold for

FIGURE 12.4
Sequential decision tree

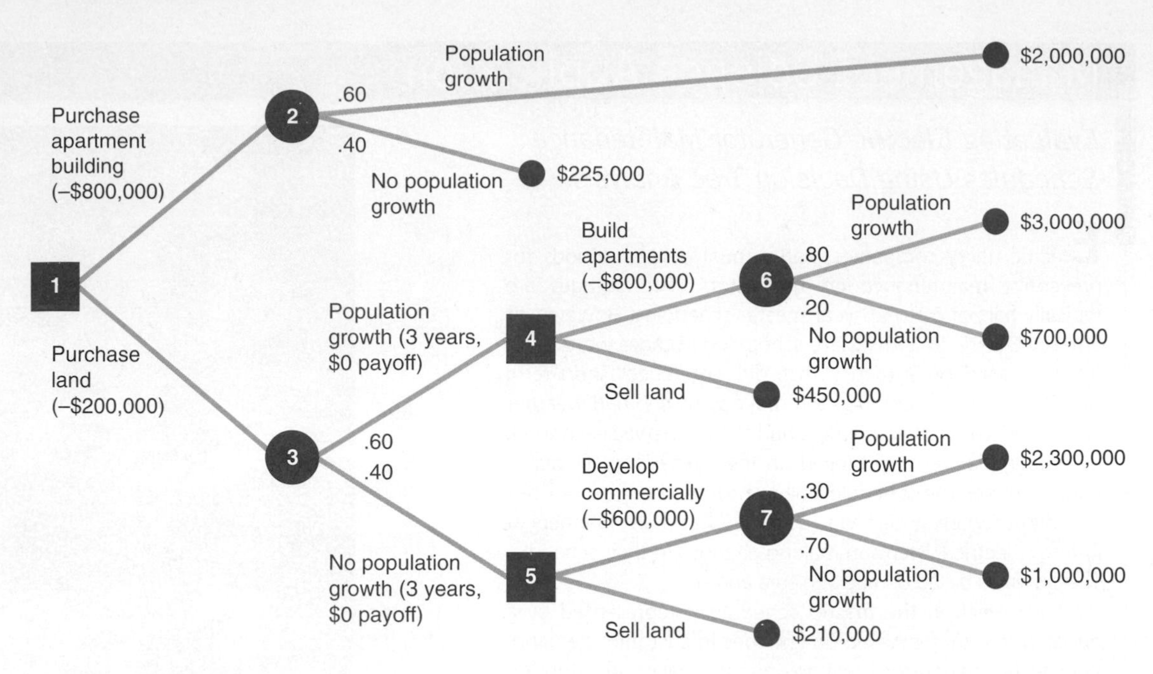

\$210,000. (Notice that the sale of the land results in less profit if there is no population growth than if there is population growth.)

If the decision at decision node 4 is to build apartments, two states of nature are possible: The population may grow, with a conditional probability of .80, or there may be no population growth, with a conditional probability of .20. The probability of population growth is higher (and the probability of no growth is lower) than before because there has already been population growth for the first 3 years, as shown by the branch from node 3 to node 4. The payoffs for these two states of nature at the end of the 10-year period are \$3,000,000 and \$700,000, respectively, as shown in Figure 12.4.

If the investor decides to develop the land commercially at node 5, then two states of nature can occur: Population growth can occur, with a probability of .30 and an eventual payoff of \$2,300,000, or no population growth can occur, with a probability of .70 and a payoff of \$1,000,000. The probability of population growth is low (i.e., .30) because there has already been no population growth, as shown by the branch from node 3 to node 5.

This decision situation encompasses several sequential decisions that can be analyzed by using the decision tree approach outlined in our earlier (simpler) example. As before, we start at the end of the decision tree and work backward toward a decision at node 1.

First, we must compute the expected values at nodes 6 and 7:

$$EV(\text{node } 6) = .80(\$3{,}000{,}000) + .20(\$700{,}000) = \$2{,}540{,}000$$
$$EV(\text{node } 7) = .30(\$2{,}300{,}000) + .70(\$1{,}000{,}000) = \$1{,}390{,}000$$

These expected values (and all other nodal values) are shown in boxes in Figure 12.5.

At decision nodes 4 and 5, we must make a decision. As with a normal payoff table, we make the decision that results in the greatest expected value. At node 4 we have a choice between two values: \$1,740,000, the value derived by subtracting the cost of building an apartment building (\$800,000) from the expected payoff of \$2,540,000, *or* \$450,000, the expected value of selling the land computed with a probability of 1.0. The decision is to build the apartment building, and the value at node 4 is \$1,740,000 (\$2,540,000 – 800,000).

This same process is repeated at node 5. The decisions at node 5 result in payoffs of \$790,000 (i.e., \$1,390,000 − 600,000 = \$790,000) and \$210,000. Because the value \$790,000 is higher, the decision is to develop the land commercially.

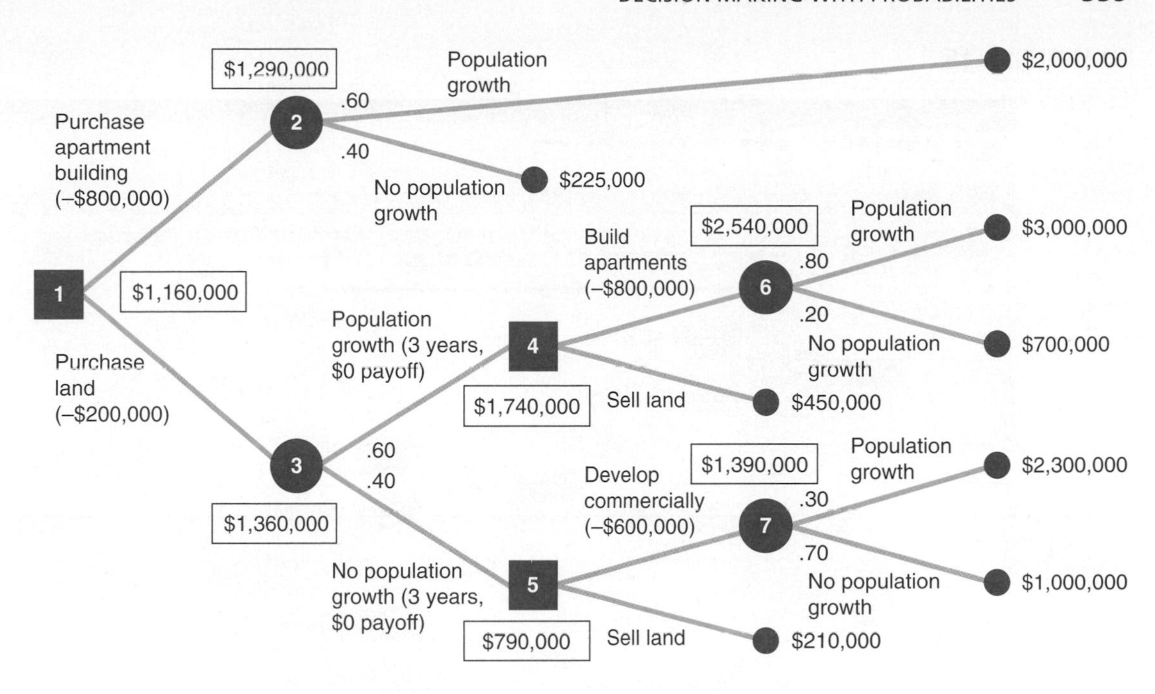

FIGURE 12.5
Sequential decision tree with nodal expected values

Next, we must compute the expected values at nodes 2 and 3:

$$EV(\text{node 2}) = .60(\$2{,}000{,}000) + .40(\$225{,}000) = \$1{,}290{,}000$$
$$EV(\text{node 3}) = .60(\$1{,}740{,}000) + .40(\$790{,}000) = \$1{,}360{,}000$$

(Note that the expected value for node 3 is computed from the decision values previously determined at nodes 4 and 5.)

Now we must make the final decision for node 1. As before, we select the decision with the greatest expected value *after the cost of each decision is subtracted out*:

$$\text{apartment building: } \$1{,}290{,}000 - 800{,}000 = \$490{,}000$$
$$\text{land: } \$1{,}360{,}000 - 200{,}000 = \$1{,}160{,}000$$

Because the highest *net* expected value is $1,160,000, the decision is to purchase land, and the payoff of the decision is $1,160,000.

This example demonstrates the usefulness of decision trees for decision analysis. A decision tree allows the decision maker to see the logic of decision making because it provides a picture of the decision process. Decision trees can be used for decision problems more complex than the preceding example without too much difficulty.

Sequential Decision Tree Analysis with Excel QM

We have already demonstrated the capability of Excel QM to build decision trees (Exhibits 12.9 and 12.10). For the sequential decision tree example described in the preceding section and illustrated in Figures 12.4 and 12.5, the Excel QM decision tree is shown in Exhibit 12.15. Notice that the expected value for the decision tree (i.e., the investment decision), $1,160,000, is shown in cell A17. You might notice that the node numbers are a little different in Exhibit 12.15 from Figure 12.3; the reason is that Excel QM assigns node numbers sequentially for every node, while the decision tree drawn in Figure 12.3 did not assign node numbers to end nodes.

Sequential Decision Tree Analysis with Excel and TreePlan

The sequential decision tree example shown in Figure 12.5, developed and solved by using TreePlan, is shown in Exhibit 12.16. Although this TreePlan decision tree is larger than the one we previously developed in Exhibit 12.14, it was accomplished in exactly the same way.

EXHIBIT 12.15

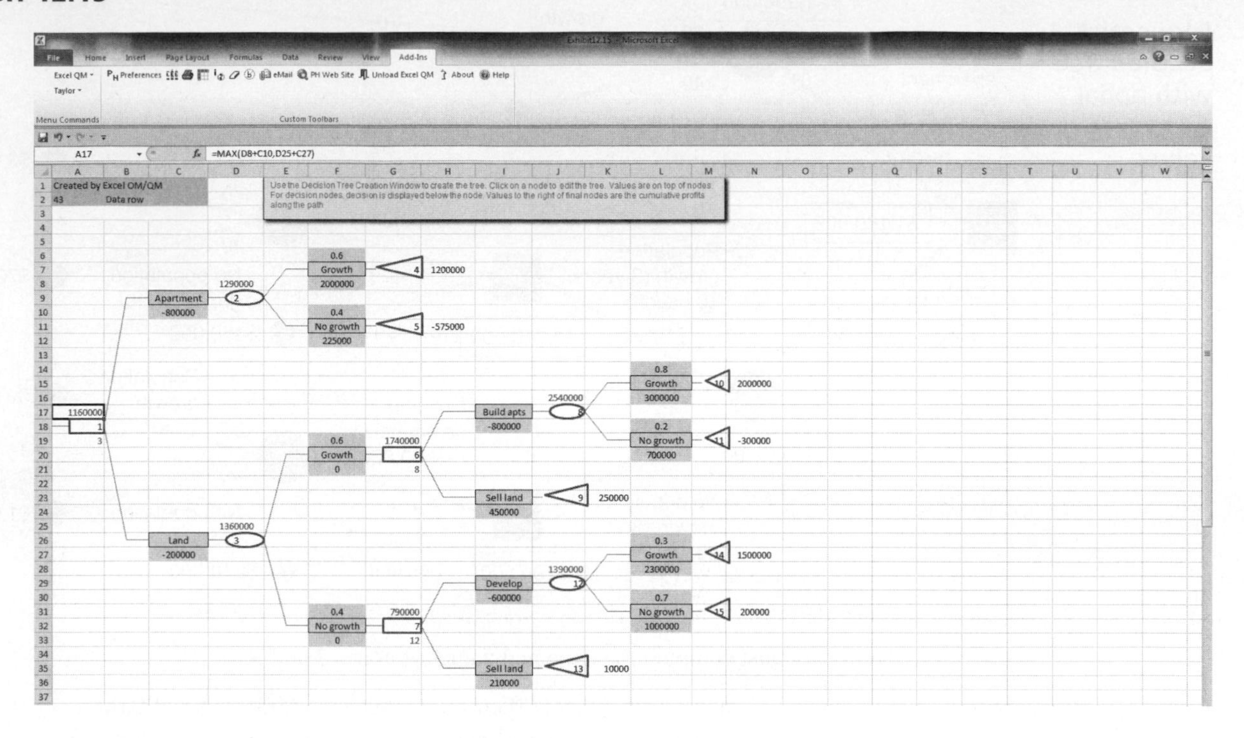

EXHIBIT 12.16

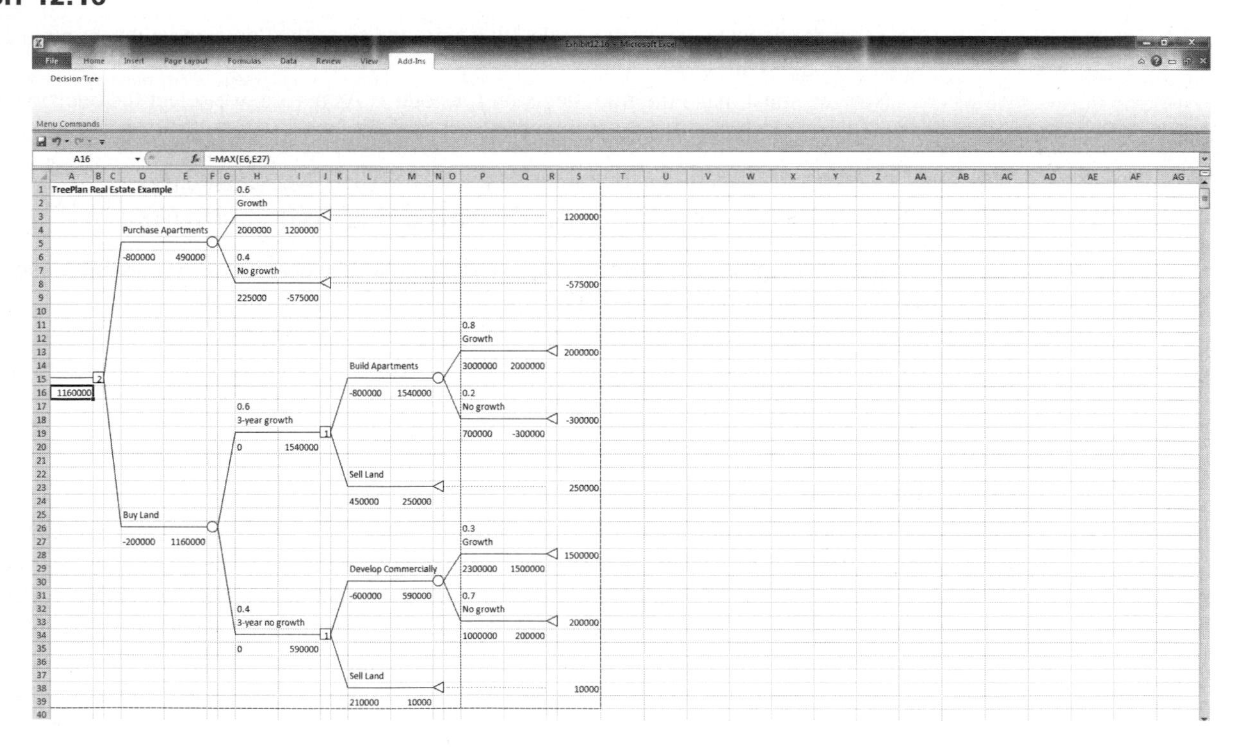

Decision Analysis with Additional Information

Earlier in this chapter, we discussed the concept of the expected value of perfect information. We noted that if perfect information could be obtained regarding which state of nature would occur in the future, the decision maker could obviously make better decisions. Although perfect information about the future is rare, it is often possible to gain some amount of additional (imperfect) information that will improve decisions.

In ***Bayesian analysis,*** *additional information is used to alter the marginal probability of the occurrence of an event.*

In this section, we will present a process for using additional information in the decision-making process by applying **Bayesian analysis**. We will demonstrate this process using the real estate investment example employed throughout this chapter. Let's review this example briefly: A real estate investor is considering three alternative investments, which will occur under one of the two possible economic conditions (states of nature) shown in Table 12.11.

TABLE 12.11
Payoff table for the real estate investment example

	State of Nature	
Decision (purchase)	Good Economic Conditions .60	Poor Economic Conditions .40
Apartment building	$ 50,000	$30,000
Office building	100,000	−40,000
Warehouse	30,000	10,000

Recall that, using the expected value criterion, we found the best decision to be the purchase of the office building, with an expected value of $44,000. We also computed the expected value of perfect information to be $28,000. Therefore, the investor would be willing to pay up to $28,000 for information about the states of nature, depending on how close to perfect the information was.

Now suppose that the investor has decided to hire a professional economic analyst who will provide additional information about future economic conditions. The analyst is constantly researching the economy, and the results of this research are what the investor will be purchasing.

A ***conditional probability*** *is the probability that an event will occur, given that another event has already occurred.*

The economic analyst will provide the investor with a report predicting one of two outcomes. The report will be either positive, indicating that good economic conditions are most likely to prevail in the future, or negative, indicating that poor economic conditions will probably occur. Based on the analyst's past record in forecasting future economic conditions, the investor has determined **conditional probabilities** of the different report outcomes, given the occurrence of each state of nature in the future. We will use the following notations to express these conditional probabilities:

$$\begin{aligned} \text{g} &= \text{good economic conditions} \\ \text{p} &= \text{poor economic conditions} \\ \text{P} &= \text{positive economic report} \\ \text{N} &= \text{negative economic report} \end{aligned}$$

The conditional probability of each report outcome, given the occurrence of each state of nature, follows:

$$\begin{aligned} P(\text{P}|\text{g}) &= .80 \\ P(\text{N}|\text{g}) &= .20 \\ P(\text{P}|\text{p}) &= .10 \\ P(\text{N}|\text{p}) &= .90 \end{aligned}$$

For example, if the future economic conditions are, in fact, good (g), the probability that a positive report (P) will have been given by the analyst, $P(\text{P}|\text{g})$, is .80. The other three conditional probabilities can be interpreted similarly. Notice that these probabilities indicate that the analyst is a relatively accurate forecaster of future economic conditions.

The investor now has quite a bit of probabilistic information available—not only the conditional probabilities of the report but also the *prior probabilities* that each state of nature will occur. These prior probabilities that good or poor economic conditions will occur in the future are

$$P(\text{g}) = .60$$
$$P(\text{p}) = .40$$

*A **posterior probability** is the altered marginal probability of an event, based on additional information.*

Given the conditional probabilities, the prior probabilities can be revised to form **posterior probabilities** by means of Bayes' rule. If we know the conditional probability that a positive report was presented, given that good economic conditions prevail, $P(\text{P} \mid \text{g})$, the posterior probability of good economic conditions, given a positive report, $P(g \mid \text{P})$, can be determined using Bayes' rule, as follows:

$$P(\text{g}|\text{P}) = \frac{P(\text{P}|\text{g})P(\text{g})}{P(\text{P}|\text{g})P(\text{g}) + P(\text{P}|\text{p})P(\text{p})}$$
$$= \frac{(.80)(.60)}{(.80)(.60) + (.10)(.40)}$$
$$= .923$$

The prior probability that good economic conditions will occur in the future is .60. However, by obtaining the additional information of a positive report from the analyst, the investor can revise the prior probability of good conditions to a .923 probability that good economic conditions will occur. The remaining posterior (revised) probabilities are

$$P(\text{g}|\text{N}) = .250$$
$$P(\text{p}|\text{P}) = .077$$
$$P(\text{p}|\text{N}) = .750$$

Decision Trees with Posterior Probabilities

The original decision tree analysis of the real estate investment example is shown in Figures 12.2 and 12.3. Using these decision trees, we determined that the appropriate decision was the purchase of an office building, with an expected value of $44,000. However, if the investor hires an economic analyst, the decision regarding which piece of real estate to invest in will not be made until after the analyst presents the report. This creates an additional stage in the decision-making process, which is shown in the decision tree in Figure 12.6. This decision tree differs in two respects from the decision trees in Figures 12.2 and 12.3. The first difference is that there are two new branches at the beginning of the decision tree that represent the two report outcomes. Notice, however, that given either report outcome, the decision alternatives, the possible states of nature, and the payoffs are the same as those in Figures 12.2 and 12.3.

The second difference is that the probabilities of each state of nature are no longer the prior probabilities given in Figure 12.2; instead, they are the revised posterior probabilities computed in the previous section, using Bayes' rule. If the economic analyst issues a positive report, then the upper branch in Figure 12.6 (from node 1 to node 2) will be taken. If an apartment building is purchased (the branch from node 2 to node 4), the probability of good economic conditions is .923, whereas the probability of poor conditions is .077. These are the revised posterior probabilities of the economic conditions, given a positive report. However, before we can perform an expected value analysis using this decision tree, one more piece of probabilistic information must be determined—the initial branch probabilities of positive and negative economic reports.

The probability of a positive report, $P(\text{P})$, and of a negative report, $P(\text{N})$, can be determined according to the following logic. The probability that two dependent events, A and B, will both occur is

$$P(\text{AB}) = P(\text{A} \mid \text{B})P(\text{B})$$

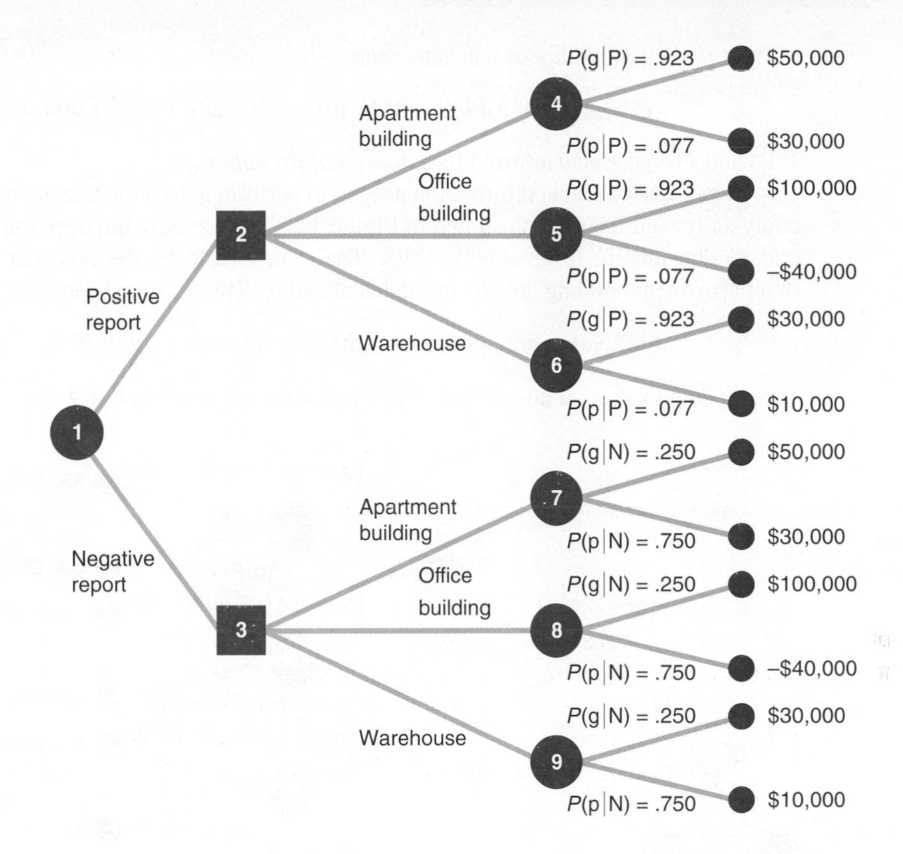

FIGURE 12.6
Decision tree with posterior probabilities

If event A is a positive report and event B is good economic conditions, then according to the preceding formula,

$$P(\mathrm{Pg}) = P(\mathrm{P}\,|\,\mathrm{g})P(\mathrm{g})$$

We can also determine the probability of a positive report and poor economic conditions the same way:

$$P(\mathrm{Pp}) = P(\mathrm{P}\,|\,\mathrm{p})P(\mathrm{p})$$

Next, we consider the two probabilities $P(\mathrm{Pg})$ and $P(\mathrm{Pp})$, also called *joint probabilities*. These are, respectively, the probability of a positive report and good economic conditions and the probability of a positive report and poor economic conditions. These two sets of occurrences are **mutually exclusive** because both good and poor economic conditions cannot occur simultaneously in the immediate future. Conditions will be either good or poor, but not both. To determine the probability of a positive report, we add the mutually exclusive probabilities of a positive report with good economic conditions and a positive report with poor economic conditions, as follows:

*Events are **mutually exclusive** if only one can occur at a time.*

$$P(\mathrm{P}) = P(\mathrm{Pg}) + P(\mathrm{Pp})$$

Now, if we substitute into this formula the relationships for $P(\mathrm{Pg})$ and $P(\mathrm{Pp})$ determined earlier, we have

$$P(\mathrm{P}) = P(\mathrm{P}\,|\,\mathrm{g})P(\mathrm{g}) + P(\mathrm{P}\,|\,\mathrm{p})P(\mathrm{p})$$

You might notice that the right-hand side of this equation is the denominator of the Bayesian formula we used to compute $P(\mathrm{g}\,|\,\mathrm{P})$ in the previous section. Using the conditional and prior probabilities that have already been established, we can determine that the probability of a positive report is

$$P(\mathrm{P}) = P(\mathrm{P}\,|\,\mathrm{g})P(\mathrm{g}) + P(\mathrm{P}\,|\,\mathrm{p})P(\mathrm{p}) = (.80)(.60) + (.10)(.40) = .52$$

Similarly, the probability of a negative report is

$$P(\text{N}) = P(\text{N}|\text{g})P(\text{g}) + P(\text{N}|\text{p})P(\text{p}) = (.20)(.60) + (.90)(.40) = .48$$

$P(\text{P})$ and $P(\text{N})$ are also referred to as *marginal probabilities*.

Now we have all the information needed to perform a decision tree analysis. The decision tree analysis for our example is shown in Figure 12.7. To see how the decision tree analysis is conducted, consider the result at node 4 first. The value \$48,460 is the expected value of the purchase of an apartment building, given both states of nature. This expected value is computed as follows:

$$EV(\text{apartment building}) = \$50{,}000(.923) + 30{,}000(.077) = \$48{,}460$$

The expected values at nodes 5, 6, 7, 8, and 9 are computed similarly.

FIGURE 12.7
Decision tree analysis for real estate investment example

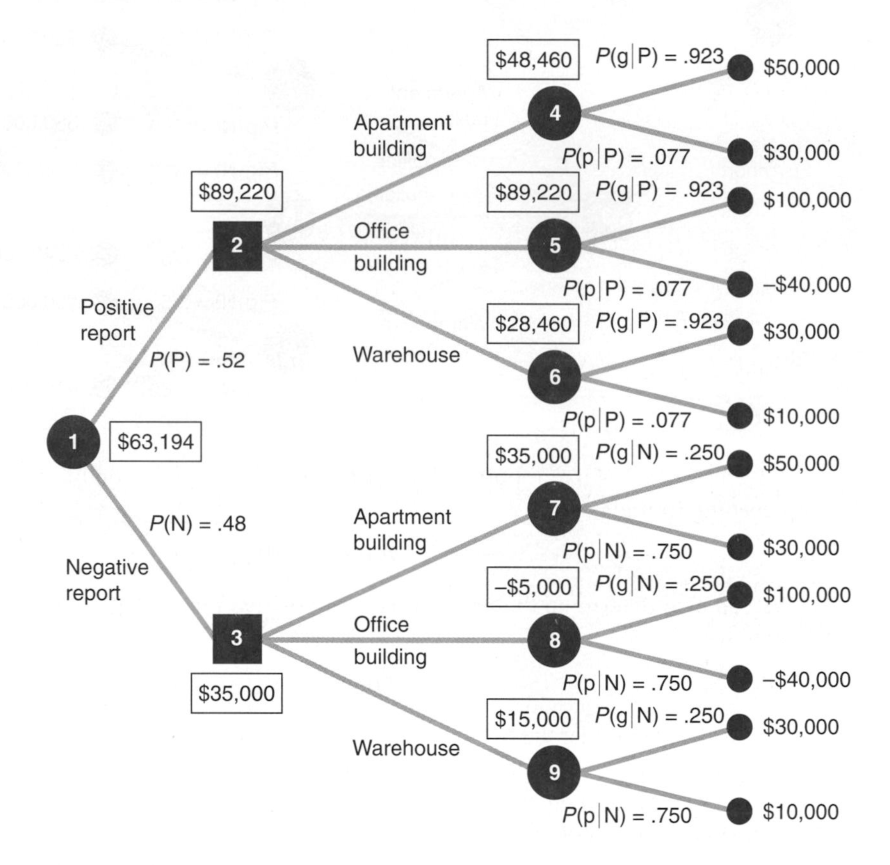

The investor will actually make the decision about the investment at nodes 2 and 3. It is assumed that the investor will make the best decision in each case. Thus, the decision at node 2 will be to purchase an office building, with an expected value of \$89,220; the decision at node 3 will be to purchase an apartment building, with an expected value of \$35,000. These two results at nodes 2 and 3 are referred to as *decision strategies*. They represent a plan of decisions to be made, given either a positive or a negative report from the economic analyst.

The final step in the decision tree analysis is to compute the expected value of the decision strategy, given that an economic analysis is performed. This expected value, shown as \$63,194 at node 1 in Figure 12.7, is computed as follows:

$$\begin{aligned} EV(\text{strategy}) &= \$89{,}220(.52) + 35{,}000(.48) \\ &= \$63{,}194 \end{aligned}$$

This amount, \$63,194, is the expected value of the investor's decision strategy, given that a report forecasting future economic condition is generated by the economic analyst.

Computing Posterior Probabilities with Tables

One of the difficulties that can occur with decision analysis with additional information is that as the size of the problem increases (i.e., as we add more decision alternatives and states of nature), the application of Bayes' rule to compute the posterior probabilities becomes more complex. In such cases, the posterior probabilities can be computed by using tables. This tabular approach will be demonstrated with our real estate investment example. The table for computing posterior probabilities for a positive report and $P(\mathrm{P})$ is initially set up as shown in Table 12.12.

TABLE 12.12
Computation of posterior probabilities

(1) State of Nature	(2) Prior Probability	(3) Conditional Probability	(4) Prior Probability × Conditional Probability: (2) × (3)	(5) Posterior Probability: (4) ÷ Σ(4)
Good conditions	$P(\mathrm{g}) = .60$	$P(\mathrm{P}\vert\mathrm{g}) = .80$	$P(\mathrm{Pg}) = .48$	$P(\mathrm{g}\vert\mathrm{P}) = \frac{.48}{.52} = .923$
Poor conditions	$P(\mathrm{p}) = .40$	$P(\mathrm{P}\vert\mathrm{p}) = .10$	$P(\mathrm{Pp}) = .04$	$P(\mathrm{p}\vert\mathrm{P}) = \frac{.04}{.52} = .077$
			$\Sigma = P(\mathrm{P}) = .52$	

The posterior probabilities for either state of nature (good or poor economic conditions), given a negative report, are computed similarly.

No matter how large the decision analysis, the steps of this tabular approach can be followed the same way as in this relatively small problem. This approach is more systematic than the direct application of Bayes' "rule," making it easier to compute the posterior probabilities for larger problems.

Computing Posterior Probabilities with Excel

The posterior probabilities computed in Table 12.12 can also be computed by using Excel. Exhibit 12.17 shows Table 12.12 set up in an Excel spreadsheet format, as well as the table for computing $P(\mathrm{N})$.

EXHIBIT 12.17

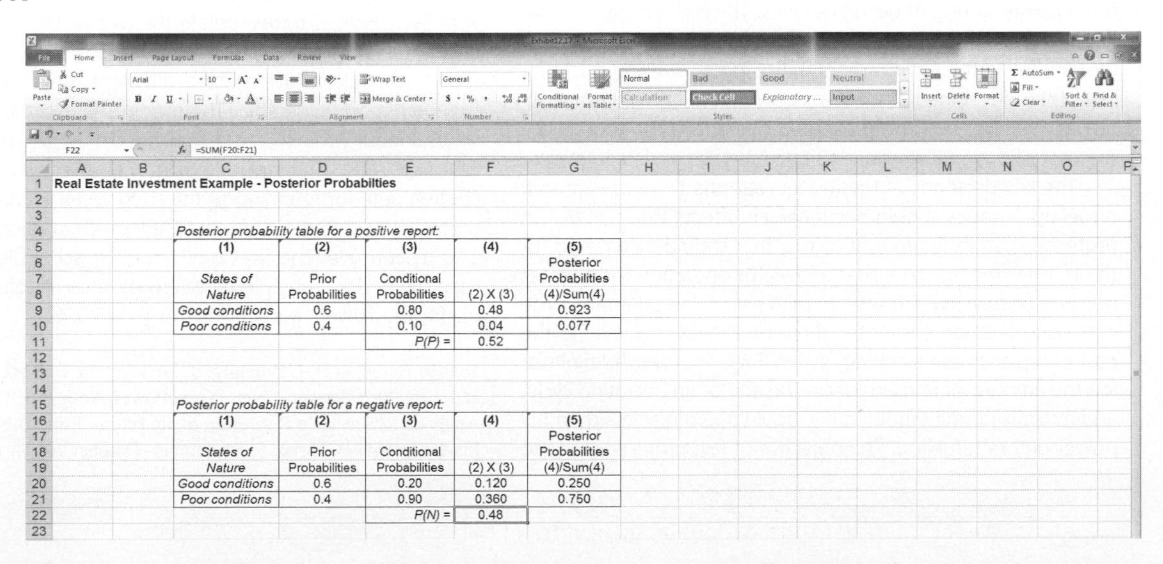

Real Estate Investment Example - Posterior Probabilties

Posterior probability table for a positive report:

(1) States of Nature	(2) Prior Probabilities	(3) Conditional Probabilities	(4) (2) X (3)	(5) Posterior Probabilities (4)/Sum(4)
Good conditions	0.6	0.80	0.48	0.923
Poor conditions	0.4	0.10	0.04	0.077
		P(P) =	0.52	

Posterior probability table for a negative report:

(1) States of Nature	(2) Prior Probabilities	(3) Conditional Probabilities	(4) (2) X (3)	(5) Posterior Probabilities (4)/Sum(4)
Good conditions	0.6	0.20	0.120	0.250
Poor conditions	0.4	0.90	0.360	0.750
		P(N) =	0.48	

The Expected Value of Sample Information

Recall that we computed the expected value of our real estate investment example to be \$44,000 when we did not have any additional information. After obtaining the additional information provided by the economic analyst, we computed an expected value of \$63,194, using the decision tree in Figure 12.7. The difference between these two expected values is called the **expected value of sample information (EVSI)**, and it is computed as follows:

The expected value of sample information (EVSI) is the difference between the expected value with and without additional information.

$$EVSI = EV_{\text{with information}} - EV_{\text{without information}}$$

For our example, the expected value of sample information is

$$EVSI = \$63{,}194 - 44{,}000 = \$19{,}194$$

Management Science Application

Scheduling Refueling at the Indian Point 3 Nuclear Power Plant

AP Photo/Ed Bailey

The New York Power Authority (NYPA) owns and operates 12 power projects that provide roughly one-fourth of all the electricity used in New York State. Approximately 20% of NYPA's electrical power (supplying Westchester County and New York City) is generated by the Indian Point 3 (IP3) nuclear power plant, located on the Hudson River. IP3 withdraws 840,000 gallons of water per minute from the river for cooling steam and then returns it to the river. When water is withdrawn from the river, fish, especially small fish and eggs, do not always survive as they pass through the cooling system. The NYPA can reduce the negative effects by scheduling plant shutdowns to refuel IP3 when fish eggs and small fish are most abundant in the Hudson River. In the past, NYPA developed a refueling schedule according to a 10-year planning horizon, although unforeseen events often altered this schedule. There is uncertainty in the future about possible deregulation of the electric utility industry and its effect on the price of replacement power during refueling outages. The NYPA must consider these uncertainties in its attempt to provide low-cost power and minimize the environmental effects of refueling outages. The NYPA developed a decision analysis model to compare alternative strategies for refueling that balanced fish protection, the cost of buying fuel, and the uncertainties of deregulation.

The key decisions of the model were the times of year that refueling outages should occur, which affects the level of fish protection, and the amount of fuel that should be ordered for the nuclear reactor core that allows for operation for a target number of days, which affects the cost of the operation. The cost consideration comprises three objectives: minimizing the cost of replacement electricity when IP3 is shut down, minimizing the amount of unused fuel at the end of an operating cycle, and minimizing the time that IP3 would operate at less than full power before refueling. The objective of fish protection was to minimize the sum of the average percentage reduction in the mortality rate caused by water removal at IP3 for five types of fish over the 10-year planning horizon. The three major uncertainties in the model were the cost of refueling, how long it takes IP3 to refuel and how well it operates, and when New York State is likely to deregulate the electric utility industry. The decision analysis model considered five strategies based on different schedules that reflected different time windows when fish were most vulnerable. The model showed that no strategy simultaneously minimized refueling cost while minimizing fish mortality rates. The strategy that was selected restricted the starting date for refueling to the third week in May, when fish protection would meet accepted standards, at a cost savings of \$10 million over the previous refueling schedule. The NYPA used the decision analysis model to develop its refueling schedule for the 10-year period from 1999 to 2008.

Source: Based on D. J. Dunning, S. Lockfort, Q. E. Ross, P. C. Beccue, and J. S. Stonebraker, "New York Power Authority Uses Decision Analysis to Schedule Refueling of Its Indian Point 3 Nuclear Power Plant," *Interfaces* 31, no. 5 (September–October 2001): 121–35.

This means that the real estate investor would be willing to pay the economic analyst up to $19,194 for an economic report that forecasted future economic conditions.

After we computed the expected value of the investment without additional information, we computed the expected value of perfect information, which equaled $28,000. However, the expected value of the sample information was only $19,194. This is a logical result because it is rare that absolutely perfect information can be determined. Because the additional information that is obtained is less than perfect, it will be worth less to the decision maker. We can determine how close to perfect our sample information is by computing the **efficiency of sample information** as follows:

*The **efficiency of sample information** is the ratio of the expected value of sample information to the expected value of perfect information.*

$$\text{efficiency} - EVSI \div EVPI = \$19{,}194 / 28{,}000 = .69$$

Thus, the analyst's economic report is viewed by the investor to be 69% as efficient as perfect information. In general, a high efficiency rating indicates that the information is very good, or close to being perfect information, and a low rating indicates that the additional information is not very good. For our example, the efficiency of .69 is relatively high; thus, it is doubtful that the investor would seek additional information from an alternative source. (However, this is usually dependent on how much money the decision maker has available to purchase information.) If the efficiency had been lower, however, the investor might seek additional information elsewhere.

Utility

All the decision-making criteria presented so far in this chapter have been based on monetary value. In other words, decisions have been based on the potential dollar payoffs of the alternatives. However, there are certain decision situations in which individuals do not make decisions based on the expected dollar gain or loss.

For example, consider an individual who purchases automobile insurance. The decisions are to purchase and to not purchase, and the states of nature are *an accident* and *no accident*. The payoff table for this decision situation, including probabilities, is shown in Table 12.13.

TABLE 12.13
Payoff table for auto insurance example

	State of Nature	
	No Accident	Accident
Decision	.992	.008
Purchase insurance	$500	$ 500
Do not purchase insurance	0	10,000

The dollar outcomes in Table 12.13 are the *costs* associated with each outcome. The insurance costs $500 whether there is an accident or no accident. If the insurance is not purchased and there is no accident, then there is no cost at all. However, if an accident does occur, the individual will incur a cost of $10,000.

The expected cost (*EC*) for each decision is

$$EC(\text{insurance}) = .992(\$500) + .008(\$500) = \$500$$

$$EC(\text{no insurance}) = .992(\$0) + .008(\$10{,}000) = \$80$$

Because the *lower* expected cost is $80, the decision *should be* not to purchase insurance. However, people almost always purchase insurance (even when they are not legally required to do so). This is true of all types of insurance, such as accident, life, or fire.

People who forgo a high expected value to avoid a disaster with a low probability are ***risk averters****.*

Why do people shun the greater *expected* dollar outcome in this type of situation? The answer is that people want to avoid a ruinous or painful situation. When faced with a relatively small dollar cost versus a disaster, people typically pay the small cost to avert the disaster. People who display this characteristic are referred to as **risk averters** because they avoid risky situations.

People who take a chance on a bonanza with a very low probability of occurrence in lieu of a sure thing are ***risk takers****.*

Alternatively, people who go to the track to wager on horse races, travel to Atlantic City to play roulette, or speculate in the commodities market decide to take risks even though the greatest *expected value* would occur if they simply held on to the money. These people shun the greater expected value accruing from a sure thing (keeping their money) in order to take a chance on receiving a "bonanza." Such people are referred to as **risk takers**.

Utility *is a measure of personal satisfaction derived from money.*

For both risk averters and risk takers (as well as those who are indifferent to risk), the decision criterion is something other than the expected dollar outcome. This alternative criterion is known as **utility**. Utility is a measure of the satisfaction derived from money. In our examples of risk averters and risk takers presented earlier, the utility derived from their decisions *exceeded* the expected dollar value. For example, the utility to the average decision maker of having insurance is much greater than the utility of not having insurance.

As another example, consider two people, each of whom is offered $100,000 to perform some particularly difficult and strenuous task. One individual has an annual income of $10,000; the other individual is a multimillionaire. It is reasonable to assume that the average person with an annual income of only $10,000 would leap at the opportunity to earn $100,000, whereas the multimillionaire would reject the offer. Obviously, $100,000 has more *utility* (i.e., value) for one individual than for the other.

In general, the same incremental amount of money does not have the same intrinsic value to every person. For individuals with a great deal of wealth, more money does not usually have as much intrinsic value as it does for individuals who have little money. In other words, although the dollar value is the same, the value as measured by utility is different, depending on how much wealth a person has. Thus, utility in this case is a measure of the pleasure or satisfaction an individual would receive from an incremental increase in wealth.

Utiles *are units of subjective measures of utility.*

In some decision situations, decision makers attempt to assign a subjective value to utility. This value is typically measured in terms of units called **utiles**. For example, the $100,000 offered to the two individuals may have a utility value of 100 utiles to the person with a low income and 0 utiles to the multimillionaire.

In our automobile insurance example, the *expected utility* of purchasing insurance could be 1,000 utiles, and the expected utility of not purchasing insurance only 1 utile. These utility values are completely reversed from the *expected monetary values* computed from Table 12.13, which explains the decision to purchase insurance.

As might be expected, it is usually very difficult to measure utility and the number of utiles derived from a decision outcome. The process is a very subjective one in which the decision maker's psychological preferences must be determined. Thus, although the concept of utility is realistic and often portrays actual decision-making criteria more accurately than does expected monetary value, its application is difficult and, as such, somewhat limited.

Summary

The purpose of this chapter was to demonstrate the concepts and fundamentals of decision making when uncertainty exists. Within this context, several decision-making criteria were presented. The maximax, maximin, minimax regret, equal likelihood, and Hurwicz decision criteria were demonstrated for cases in which probabilities could not be attached to the occurrence of outcomes. The expected value criterion and decision trees were discussed for cases in which probabilities could be assigned to the states of nature of a decision situation.

All the decision criteria presented in this chapter were demonstrated by rather simplified examples; actual decision-making situations are usually more complex. Nevertheless, the process of analyzing decisions presented in this chapter is the logical method that most decision makers follow to make a decision.

Example Problem Solution

The following example will illustrate the solution procedure for a decision analysis problem.

Problem Statement

T. Bone Puckett, a corporate raider, has acquired a textile company and is contemplating the future of one of its major plants, located in South Carolina. Three alternative decisions are being considered: (1) expand the plant and produce lightweight, durable materials for possible sales to the military, a market with little foreign competition; (2) maintain the status quo at the plant, continuing production of textile goods that are subject to heavy foreign competition; or (3) sell the plant now. If one of the first two alternatives is chosen, the plant will still be sold at the end of a year. The amount of profit that could be earned by selling the plant in a year depends on foreign market conditions, including the status of a trade embargo bill in Congress. The following payoff table describes this decision situation:

	State of Nature	
Decision	*Good Foreign Competitive Conditions*	*Poor Foreign Competitive Conditions*
Expand	$ 800,000	$ 500,000
Maintain status quo	1,300,000	−150,000
Sell now	320,000	320,000

A. Determine the best decision by using the following decision criteria:
   1. Maximax
   2. Maximin
   3. Minimax regret
   4. Hurwicz ($\alpha = .3$)
   5. Equal likelihood

B. Assume that it is now possible to estimate a probability of .70 that good foreign competitive conditions will exist and a probability of .30 that poor conditions will exist. Determine the best decision by using expected value and expected opportunity loss.

C. Compute the expected value of perfect information.

D. Develop a decision tree, with expected values at the probability nodes.

E. T. Bone Puckett has hired a consulting firm to provide a report on future political and market situations. The report will be positive (P) or negative (N), indicating either a good (g) or poor (p) future foreign competitive situation. The conditional probability of each report outcome, given each state of nature, is

$$P(\text{P} \mid \text{g}) = .70$$
$$P(\text{N} \mid \text{g}) = .30$$
$$P(\text{P} \mid \text{p}) = .20$$
$$P(\text{N} \mid \text{p}) = .80$$

Determine the posterior probabilities by using Bayes' rule.

F. Perform a decision tree analysis by using the posterior probability obtained in (E).

Solution

Step 1 (part A): Determine Decisions Without Probabilities

Maximax:

Expand	$ 800,000		
Status quo	1,300,000	←	Maximum
Sell	320,000		

Decision: Maintain status quo.

Maximin:

Expand	$ 500,000	←	Maximum
Status quo	−150,000		
Sell	320,000		

Decision: Expand.

Minimax regret:

Expand	$500,000	←	Minimum
Status quo	650,000		
Sell	980,000		

Decision: Expand.

Hurwicz ($\alpha = .3$):

Expand	$800,000(.3) + 500,000(.7) = $590,000
Status quo	$1,300,000(.3) − 150,000(.7) = $285,000
Sell	$320,000(.3) + 320,000(.7) = $320,000

Decision: Expand.

Equal likelihood:

Expand	$800,000(.50) + 500,000(.50) = $650,000
Status quo	$1,300,000(.50) − 150,000(.50) = $575,000
Sell	$320,000(.50) + 320,000(.50) = $320,000

Decision: Expand.

Step 2 (part B): Determine Decisions with *EV* and *EOL*

Expected value:

Expand	$800,000(.70) + 500,000(.30) = $710,000
Status quo	$1,300,000(.70) − 150,000(.30) = $865,000
Sell	$320,000(.70) + 320,000(.30) = $320,000

Decision: Maintain status quo.

Expected opportunity loss:

Expand	$500,000(.70) + 0(.30) = $350,000
Status quo	$0(.70) + 650,000(.30) = $195,000
Sell	$980,000(.70) + 180,000(.30) = $740,000

Decision: Maintain status quo.

Step 3 (part C): Compute *EVPI*

$$\text{expected value given perfect information} = 1{,}300{,}000(.70) + 500{,}000(.30) = \$1{,}060{,}000$$

$$\text{expected value without perfect information} = \$1{,}300{,}000(.70) - 150{,}000(.30) = \$865{,}000$$

$$EVPI = \$1{,}060{,}000 - 865{,}000 = \$195{,}000$$

Step 4 (part D): Develop a Decision Tree

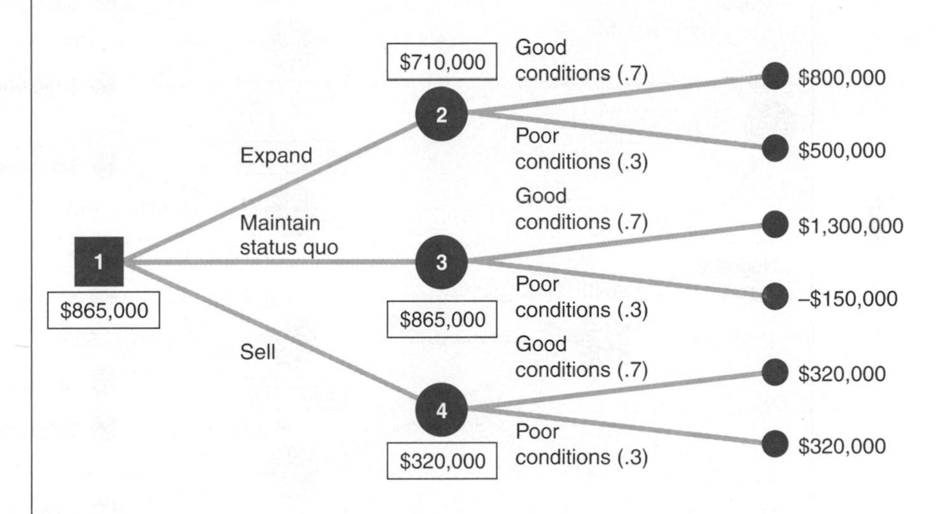

Step 5 (part E): Determine Posterior Probabilities

$$P(\text{g}|\text{P}) = \frac{P(\text{P}|\text{g})P(\text{g})}{P(\text{P}|\text{g})P(\text{g}) + P(\text{P}|\text{p})P(\text{p})} = \frac{(.70)(.70)}{(.70)(.70) + (.20)(.30)} = .891$$

$$P(\text{p}|\text{P}) = .109$$

$$P(\text{g}|\text{N}) = \frac{P(\text{N}|\text{g})P(\text{g})}{P(\text{N}|\text{g})P(\text{g}) + P(\text{N}|\text{p})P(\text{p})} = \frac{(.30)(.70)}{(.30)(.70) + (.80)(.30)} = .467$$

$$P(\text{p}|\text{N}) = .533$$

Step 6 (part F): Perform Decision Tree Analysis with Posterior Probabilities

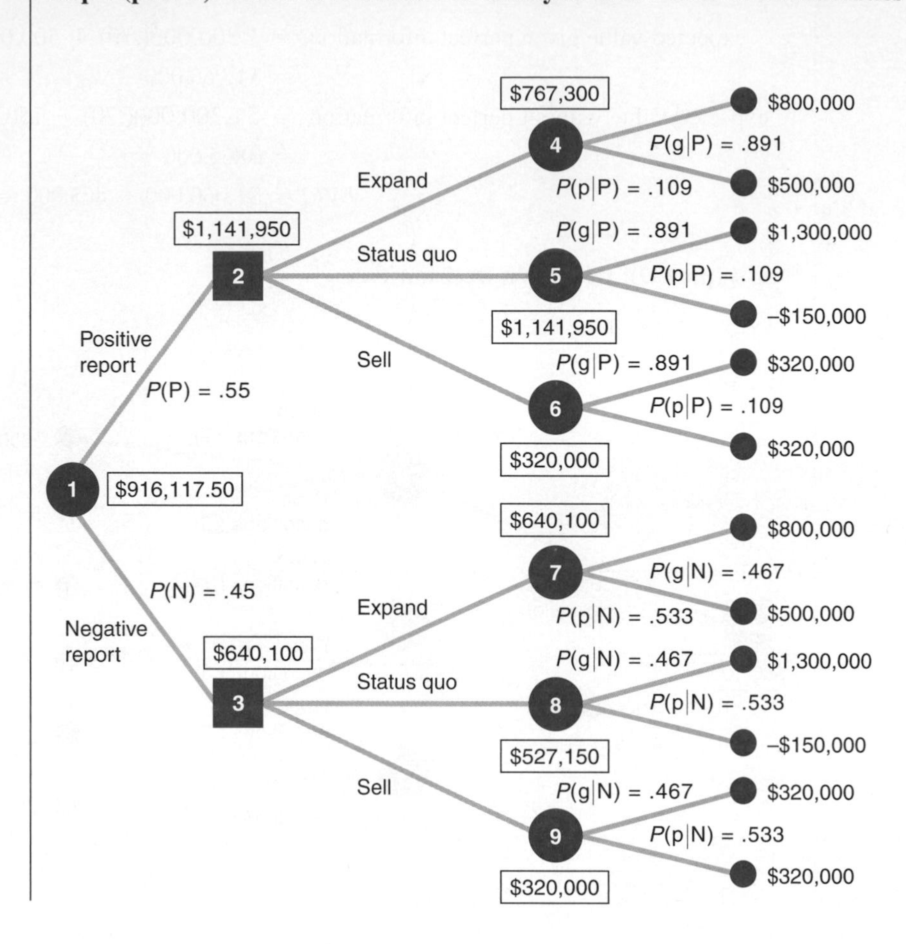

Problems

1. A farmer in Iowa is considering either leasing some extra land or investing in savings certificates at the local bank. If weather conditions are good next year, the extra land will give the farmer an excellent harvest. However, if weather conditions are bad, the farmer will lose money. The savings certificates will result in the same return, regardless of the weather conditions. The return for each investment, given each type of weather condition, is shown in the following payoff table:

	Weather	
Decision	Good	Bad
Lease land	$90,000	$−40,000
Buy savings certificate	10,000	10,000

Select the best decision, using the following decision criteria:

a. Maximax

b. Maximin

2. The owner of the Burger Doodle Restaurant is considering two ways to expand operations: open a drive-up window or serve breakfast. The increase in profits resulting from these proposed expansions depends on whether a competitor opens a franchise down the street. The possible profits

from each expansion in operations, given both future competitive situations, are shown in the following payoff table:

Decision	Competitor	
	Open	Not Open
Drive-up window	$−6,000	$20,000
Breakfast	4,000	8,000

Select the best decision, using the following decision criteria.

a. Maximax
b. Maximin

3. Stevie Stone, a bellhop at the Royal Sundown Hotel in Atlanta, has been offered a management position. Although accepting the offer would assure him a job if there was a recession, if good economic conditions prevailed, he would actually make less money as a manager than as a bellhop (because of the large tips he gets as a bellhop). His salary during the next 5 years for each job, given each future economic condition, is shown in the following payoff table:

Decision	Economic Conditions	
	Good	Recession
Bellhop	$120,000	$60,000
Manager	85,000	85,000

Select the best decision, using the following decision criteria.

a. Minimax regret
b. Hurwicz ($\alpha = .4$)
c. Equal likelihood

4. Brooke Bentley, a student in business administration, is trying to decide which management science course to take next quarter—I, II, or III. "Steamboat" Fulton, "Death" Ray, and "Sadistic" Scott are the three management science professors who teach the courses. Brooke does not know who will teach what course. Brooke can expect a different grade in each of the courses, depending on who teaches it next quarter, as shown in the following payoff table:

Course	Professor		
	Fulton	Ray	Scott
I	B	D	D
II	C	B	F
III	F	A	C

Determine the best course to take next quarter, using the following criteria.

a. Maximax
b. Maximin

5. A farmer in Georgia must decide which crop to plant next year on his land: corn, peanuts, or soybeans. The return from each crop will be determined by whether a new trade bill with Russia

passes the Senate. The profit the farmer will realize from each crop, given the two possible results on the trade bill, is shown in the following payoff table:

	Trade Bill	
Crop	Pass	Fail
Corn	$35,000	$ 8,000
Peanuts	18,000	12,000
Soybeans	22,000	20,000

Determine the best crop to plant, using the following decision criteria.

a. Maximax
b. Maximin
c. Minimax regret
d. Hurwicz ($\alpha = .3$)
e. Equal likelihood

6. A company must decide now which of three products to make next year to plan and order proper materials. The cost per unit of producing each product will be determined by whether a new union labor contract passes or fails. The cost per unit for each product, given each contract result, is shown in the following payoff table:

	Contract Outcome	
Product	Pass	Fail
1	$7.50	$6.00
2	4.00	7.00
3	6.50	3.00

Determine which product should be produced, using the following decision criteria.

a. Minimin
b. Minimax

7. The owner of the Columbia Construction Company must decide between building a housing development, constructing a shopping center, and leasing all the company's equipment to another company. The profit that will result from each alternative will be determined by whether material costs remain stable or increase. The profit from each alternative, given the two possibilities for material costs, is shown in the following payoff table:

	Material Costs	
Decision	Stable	Increase
Houses	$70,000	$30,000
Shopping center	105,000	20,000
Leasing	40,000	40,000

Determine the best decision, using the following decision criteria.

a. Maximax
b. Maximin
c. Minimax regret
d. Hurwicz ($\alpha = .2$)
e. Equal likelihood

8. A local real estate investor in Orlando is considering three alternative investments: a motel, a restaurant, or a theater. Profits from the motel or restaurant will be affected by the availability of gasoline and the number of tourists; profits from the theater will be relatively stable under any conditions. The following payoff table shows the profit or loss that could result from each investment:

	Gasoline Availability		
Investment	Shortage	Stable Supply	Surplus
Motel	$−8,000	$15,000	$20,000
Restaurant	2,000	8,000	6,000
Theater	6,000	6,000	5,000

Determine the best investment, using the following decision criteria.

a. Maximax
b. Maximin
c. Minimax regret
d. Hurwicz ($\alpha = .4$)
e. Equal likelihood

9. A television network is attempting to decide during the summer which of the following three football games to televise on the Saturday following Thanksgiving Day: Alabama versus Auburn, Georgia versus Georgia Tech, or Army versus Navy. The estimated viewer ratings (millions of homes) for the games depend on the win–loss records of the six teams, as shown in the following payoff table:

	Number of Viewers (1,000,000s)		
Game	Both Teams Have Winning Records	One Team Has Winning Record; One Team Has Losing Record	Both Teams Have Losing Records
Alabama vs. Auburn	10.2	7.3	5.4
Georgia vs. Georgia Tech	9.6	8.1	4.8
Army vs. Navy	12.5	6.5	3.2

Determine the best game to televise, using the following decision criteria.

a. Maximax
b. Maximin
c. Equal likelihood

10. Ann Tyler has come into an inheritance from her grandparents. She is attempting to decide among several investment alternatives. The return after 1 year is primarily dependent on the interest rate during the next year. The rate is currently 7%, and Ann anticipates that it will stay the same or go up or down by at most two points. The various investment alternatives plus their returns ($10,000s), given the interest rate changes, are shown in the following table:

	Interest Rate				
Investment	5%	6%	7%	8%	9%
Money market fund	2	3.1	4	4.3	5
Stock growth fund	−3	−2	2.5	4	6
Bond fund	6	5	3	3	2
Government fund	4	3.6	3.2	3	2.8
Risk fund	−9	−4.5	1.2	8.3	14.7
Savings bonds	3	3	3.2	3.4	3.5

Determine the best investment, using the following decision criteria.

a. Maximax
b. Maximin
c. Equal likelihood

11. The Tech football coaching staff has six basic offensive plays it runs every game. Tech has an upcoming game against State on Saturday, and the Tech coaches know that State employs five different defenses. The coaches have estimated the number of yards Tech will gain with each play against each defense, as shown in the following payoff table:

	Defense				
Play	54	63	Wide Tackle	Nickel	Blitz
Off tackle	3	−2	9	7	−1
Option	−1	8	−2	9	12
Toss sweep	6	16	−5	3	14
Draw	−2	4	3	10	−3
Pass	8	20	12	−7	−8
Screen	−5	−2	8	3	16

a. If the coaches employ an offensive game plan, they will use the maximax criterion. What will be their best play?
b. If the coaches employ a defensive plan, they will use the maximin criterion. What will be their best play?
c. What will be their best offensive play if State is equally likely to use any of its five defenses?

12. Microcomp is a U.S.-based manufacturer of personal computers. It is planning to build a new manufacturing and distribution facility in either South Korea, China, Taiwan, the Philippines, or Mexico. It will take approximately 5 years to build the necessary infrastructure (roads, etc.), construct the new facility, and put it into operation. The eventual cost of the facility will differ between countries and will even vary within countries depending on the financial, labor, and political climate, including monetary exchange rates. The company has estimated the facility cost (in $1,000,000s) in each country under three different future economic and political climates, as follows:

	Economic/Political Climate		
Country	Decline	Same	Improve
South Korea	21.7	19.1	15.2
China	19.0	18.5	17.6
Taiwan	19.2	17.1	14.9
Philippines	22.5	16.8	13.8
Mexico	25.0	21.2	12.5

Determine the best decision, using the following decision criteria.

a. Minimin
b. Minimax

c. Hurwicz ($\alpha = .4$)
d. Equal likelihood

13. Place-Plus, a real estate development firm, is considering several alternative development projects. These include building and leasing an office park, purchasing a parcel of land and building an office building to rent, buying and leasing a warehouse, building a strip mall, and building and selling condominiums. The financial success of these projects depends on interest rate movement in the next 5 years. The various development projects and their 5-year financial return (in $1,000,000s) given that interest rates will decline, remain stable, or increase, are shown in the following payoff table:

	Interest Rate		
Project	Decline	Stable	Increase
Office park	$0.5	$1.7	$4.5
Office building	1.5	1.9	2.5
Warehouse	1.7	1.4	1.0
Mall	0.7	2.4	3.6
Condominiums	3.2	1.5	0.6

Determine the best investment, using the following decision criteria.
a. Maximax
b. Maximin
c. Equal likelihood
d. Hurwicz ($\alpha = .3$)

14. The Oakland Bombers professional basketball team just missed making the playoffs last season and believes it needs to sign only one very good free agent to make the playoffs next season. The team is considering four players: Barry Byrd, Rayneal O'Neil, Marvin Johnson, and Michael Gordan. Each player differs according to position, ability, and attractiveness to fans. The payoffs (in $1,000,000s) to the team for each player, based on the contract, profits from attendance, and team product sales for several different season outcomes, are provided in the following table:

	Season Outcome		
Player	Loser	Competitive	Makes Playoffs
Byrd	$−3.2	$ 1.3	4.4
O'Neil	−5.1	1.8	6.3
Johnson	−2.7	0.7	5.8
Gordan	−6.3	−1.6	9.6

Determine the best decision, using the following decision criteria.
a. Maximax
b. Maximin
c. Hurwicz ($\alpha = .60$)
d. Equal likelihood

15. A machine shop owner is attempting to decide whether to purchase a new drill press, a lathe, or a grinder. The return from each will be determined by whether the company succeeds in getting

a government military contract. The profit or loss from each purchase and the probabilities associated with each contract outcome are shown in the following payoff table:

Purchase	Contract .40	No Contract .60
Drill press	$40,000	$−8,000
Lathe	20,000	4,000
Grinder	12,000	10,000

Compute the expected value for each purchase and select the best one.

16. A concessions manager at the Tech versus A&M football game must decide whether to have the vendors sell sun visors or umbrellas. There is a 30% chance of rain, a 15% chance of overcast skies, and a 55% chance of sunshine, according to the weather forecast in College Junction, where the game is to be held. The manager estimates that the following profits will result from each decision, given each set of weather conditions:

	Weather Conditions		
Decision	Rain .30	Overcast .15	Sunshine .55
Sun visors	$−500	$−200	$1,500
Umbrellas	2,000	0	−900

a. Compute the expected value for each decision and select the best one.
b. Develop the opportunity loss table and compute the expected opportunity loss for each decision.

17. Allen Abbott has a wide-curving, uphill driveway leading to his garage. When there is a heavy snow, Allen hires a local carpenter, who shovels snow on the side in the winter, to shovel his driveway. The snow shoveler charges $30 to shovel the driveway. Following is a probability distribution of the number of heavy snows each winter:

Heavy Snows	Probability
1	.13
2	.18
3	.26
4	.23
5	.10
6	.07
7	.03
	1.00

Allen is considering purchasing a new self-propelled snowblower for $625 that would allow him, his wife, or his children to clear the driveway after a snow. Discuss what you think Allen's decision should be and why.

18. The Miramar Company is going to introduce one of three new products: a widget, a hummer, or a nimnot. The market conditions (favorable, stable, or unfavorable) will determine the profit or loss the company realizes, as shown in the following payoff table:

Product	Market Conditions Favorable .2	Stable .7	Unfavorable .1
Widget	$120,000	$70,000	$−30,000
Hummer	60,000	40,000	20,000
Nimnot	35,000	30,000	30,000

a. Compute the expected value for each decision and select the best one.
b. Develop the opportunity loss table and compute the expected opportunity loss for each product.
c. Determine how much the firm would be willing to pay to a market research firm to gain better information about future market conditions.

19. The financial success of the Downhill Ski Resort in the Blue Ridge Mountains is dependent on the amount of snowfall during the winter months. If the snowfall averages more than 40 inches, the resort will be successful; if the snowfall is between 20 and 40 inches, the resort will receive a moderate financial return; and if snowfall averages less than 20 inches, the resort will suffer a financial loss. The financial return and probability, given each level of snowfall, follow:

Snowfall Level (in.)	Financial Return
>40, .4	$ 120,000
20–40, .2	40,000
<20, .4	−40,000

A large hotel chain has offered to lease the resort for the winter for $40,000. Compute the expected value to determine whether the resort should operate or lease. Explain your answer.

20. An investor must decide between two alternative investments—stocks and bonds. The return for each investment, given two future economic conditions, is shown in the following payoff table:

Investment	Economic Conditions Good	Bad
Stocks	$10,000	$−4,000
Bonds	7,000	2,000

What probability for each economic condition would make the investor indifferent to the choice between stocks and bonds?

21. In Problem 10, Ann Tyler, with the help of a financial newsletter and some library research, has been able to assign probabilities to each of the possible interest rates during the next year, as follows:

Interest Rate (%)	Probability
5	.2
6	.3
7	.3
8	.1
9	.1

Using expected value, determine her best investment decision.

22. In Problem 11, the Tech coaches have reviewed game films and have determined the following probabilities that State will use each of its defenses:

Defense	Probability
54	.40
63	.10
Wide tackle	.20
Nickel	.20
Blitz	.10

a. Using expected value, rank Tech's plays from best to worst.
b. During the actual game, a situation arises in which Tech has a third down and 10 yards to go, and the coaches are 60% certain State will blitz, with a 10% chance of any of the other four defenses. What play should Tech run? Is it likely the team will make the first down?

23. A global economist hired by Microcomp, the U.S.-based computer manufacturer in Problem 12, estimates that the probability that the economic and political climate overseas and in Mexico will decline during the next 5 years is .40, the probability that it will remain approximately the same is .50, and the probability that it will improve is .10. Determine the best country to construct the new facility in and the expected value of perfect information.

24. In Problem 13, the Place-Plus real estate development firm has hired an economist to assign a probability to each direction interest rates may take over the next 5 years. The economist has determined that there is a .50 probability that interest rates will decline, a .40 probability that rates will remain stable, and a .10 probability that rates will increase.
a. Using expected value, determine the best project.
b. Determine the expected value of perfect information.

25. Fenton and Farrah Friendly, husband-and-wife car dealers, are soon going to open a new dealership. They have three offers: from a foreign compact car company, from a U.S.-producer of full-sized cars, and from a truck company. The success of each type of dealership will depend on how much gasoline is going to be available during the next few years. The profit from each type of dealership, given the availability of gas, is shown in the following payoff table:

	Gasoline Availability	
Dealership	Shortage .6	Surplus .4
Compact cars	$ 300,000	$150,000
Full-sized cars	−100,000	600,000
Trucks	120,000	170,000

Determine which type of dealership the couple should purchase.

26. The Steak and Chop Butcher Shop purchases steak from a local meatpacking house. The meat is purchased on Monday at $2.00 per pound, and the shop sells the steak for $3.00 per pound. Any steak left over at the end of the week is sold to a local zoo for $.50 per pound. The possible demands for steak and the probability of each are shown in the following table:

Demand (lb.)	Probability
20	.10
21	.20
22	.30
23	.30
24	.10
	1.00

The shop must decide how much steak to order in a week. Construct a payoff table for this decision situation and determine the amount of steak that should be ordered, using expected value.

27. The Loebuck Grocery must decide how many cases of milk to stock each week to meet demand. The probability distribution of demand during a week is shown in the following table:

Demand (cases)	Probability
15	.20
16	.25
17	.40
18	.15
	1.00

Each case costs the grocer $10 and sells for $12. Unsold cases are sold to a local farmer (who mixes the milk with feed for livestock) for $2 per case. If there is a shortage, the grocer considers the cost of customer ill will and lost profit to be $4 per case. The grocer must decide how many cases of milk to order each week.

a. Construct the payoff table for this decision situation.
b. Compute the expected value of each alternative amount of milk that could be stocked and select the best decision.
c. Construct the opportunity loss table and determine the best decision.
d. Compute the expected value of perfect information.

28. The manager of the greeting card section of Mazey's department store is considering her order for a particular line of Christmas cards. The cost of each box of cards is $3; each box will be sold for $5 during the Christmas season. After Christmas, the cards will be sold for $2 a box. The card section manager believes that all leftover cards can be sold at that price. The estimated demand during the Christmas season for the line of Christmas cards, with associated probabilities, is as follows:

Demand (boxes)	Probability
25	.10
26	.15
27	.30
28	.20
29	.15
30	.10

a. Develop the payoff table for this decision situation.
b. Compute the expected value for each alternative and identify the best decision.
c. Compute the expected value of perfect information.

29. The Palm Garden Greenhouse specializes in raising carnations that are sold to florists. Carnations are sold for \$3.00 per dozen; the cost of growing the carnations and distributing them to the florists is \$2.00 per dozen. Any carnations left at the end of the day are sold to local restaurants and hotels for \$0.75 per dozen. The estimated cost of customer ill will if demand is not met is \$1.00 per dozen. The expected daily demand (in dozens) for the carnations is as follows:

Daily Demand	Probability
20	.05
22	.10
24	.25
26	.30
28	.20
30	.10
	1.00

a. Develop the payoff table for this decision situation.
b. Compute the expected value of each alternative number of (dozens of) carnations that could be stocked and select the best decision.
c. Construct the opportunity loss table and determine the best decision.
d. Compute the expected value of perfect information.

30. Assume that the probabilities of demand in Problem 28 are no longer valid; the decision situation is now one without probabilities. Determine the best number of boxes of cards to stock, using the following decision criteria.
a. Maximin
b. Maximax
c. Hurwicz ($\alpha = .4$)
d. Minimax regret

31. Federated Electronics, Ltd., manufactures display screens and monitors for computers and televisions, which it sells to companies around the world. It wants to construct a new warehouse and distribution center in Asia to serve emerging markets there. It has identified potential sites in the port cities Shanghai, Singapore, Pusan, Kaohsiung, and Hong Kong, and it has estimated the possible revenues for each (minus construction costs, which are higher in some cities, such as Hong Kong). At each site the projected revenues are primarily based on two factors—the economic conditions at the port, including the projected traffic, infrastructure, labor rates and availability, and expansion and modernization; and the future government situation, which inclues the political stability, fees, tariffs, duties, and trade regulations. Following is a payoff table that shows the projected revenues (in \$billions) for 6 years, given the four possible combinations for positive and negative port and government conditions:

	Government Conditions			
Port	Port Negative/ Government Negative	Port Negative/ Government Positive	Port Positive/ Government Negative	Port Positive/ Government Positive
Shanghai	–0.271	\$0.437	\$0.523	\$1.08
Singapore	–0.164	0.329	0.441	0.873
Pusan	0.119	0.526	0.337	0.732
Kaoshiung	–0.235	0.522	0.226	1.116
Hong Kong	–0.317	0.256	0.285	1.653

Determine the port city Federated should select for its new distribution center using the following decision criteria.

a. Maximax
b. Maximin
c. Equal likelihood
d. Hurwicz ($\alpha = .55$)

32. In Problem 31, Federated Electronics, Ltd., has hired a Washington, DC–based global trade research firm to assess the probabilities of each combination of port and government conditions for the five ports. The research firm probability estimates for the five ports are as follows:

	Government Conditions			
Port	Port Negative/ Government Negative	Port Negative/ Government Positive	Port Positive/ Government Negative	Port Positive/ Government Positive
Shanghai	0.09	0.27	0.32	0.32
Singapore	0.05	0.22	0.22	0.51
Pusan	0.08	0.36	0.27	0.29
Kaoshiung	0.11	0.12	0.46	0.31
Hong Kong	0.10	0.23	0.30	0.37

a. Using expected value, determine the best port to construct the distribution center.
b. Using any decision criteria, determine the port you think would be the best location for the distribution center and justify your answer.

33. In Problem 14, the Bombers' management has determined the following probabilities of the occurrence of each future season outcome for each player:

	Probability		
Player	Loser	Competitive	Makes Playoffs
Byrd	.15	.55	.30
O'Neil	.18	.26	.56
Johnson	.21	.32	.47
Gordan	.30	.25	.45

Compute the expected value for each player and indicate which player the team should try to sign.

34. The director of career advising at Orange Community College wants to use decision analysis to provide information to help students decide which 2-year degree program they should pursue. The director has set up the following payoff table for six of the most popular and successful degree programs at OCC that shows the estimated 5-year gross income ($) from each degree for four future economic conditions:

	Economic Conditions			
Degree Program	Recession	Average	Good	Robust
Graphic design	145,000	175,000	220,000	260,000
Nursing	150,000	180,000	205,000	215,000
Real estate	115,000	165,000	220,000	320,000
Medical technology	130,000	180,000	210,000	280,000
Culinary technology	115,000	145,000	235,000	305,000
Computer information technology	125,000	150,000	190,000	250,000

Determine the best degree program in terms of projected income, using the following decision criteria:

a. Maximax
b. Maximin
c. Equal likelihood
d. Hurwicz (α = .50)

35. In Problem 34, the director of career advising at Orange Community College has paid a small fee to a local investment firm to indicate a probability for each future economic condition over the next 5 years. The firm estimates that there is a .20 probability of a recession, a .40 probability that the economy will be average, a .30 probability that the economy will be good, and a .10 probability that it will be robust. Using expected value determine the best degree program in terms of projected income. If you were the director of career advising, which degree program would you recommend?

36. The Blue Sox American League baseball team is going to enter the free-agent market over the winter to sign a new starting pitcher. They are considering five prospects who will enter the free-agent market. All five pitchers are in their mid-20s, have been in the major leagues for approximately 5 years, and have been relatively successful. The team's general manager has compiled a lot of information about the pitchers from scouting reports and their playing histories since high school. He has developed a chart projecting how many wins each pitcher will likely have during the next 10 years given three possible future states of nature: the pitchers will be relatively injury free, they will have a normal career with injuries, or they will have excessive injuries, as shown in the following payoff table:

	Physical Condition		
Pitcher	**No injuries**	**Normal**	**Excessive Injuries**
Jose Diaz	153	122	76
Jerry Damon	173	135	46
Frank Thompson	133	115	88
Derek Rodriguez	105	101	95
Ken Griffin	127	98	75

Determine the best pitcher to sign, using the following decision criteria:

a. Maximax
b. Maximin
c. Equal likelihood
d. Hurwicz (α = .35)

37. In Problem 36, the Blue Sox general manager has asked a superscout to assign a probability to each of the three states of nature for the pitchers during the next 10 years. The scout estimates there is a .10 probability that these pitchers at this stage of their careers will have no injuries, a .60 probability that they will have a career with the normal number of injuries, and a .30 probability that they will have excessive injuries.
    a. Using expected value, determine the best pitcher to sign.
    b. Given the following 10-year contract price for each pitcher (in $millions), which would you recommend signing?

Jose Diaz	$ 97.3
Jerry Damon	$121.5
Frank Thompson	$ 73.5
Derek Rodriguez	$103.4
Ken Griffin	$ 85.7

c. Suppose that the general manager asked the superscout to determine the probabilities of each state of nature for each individual pitcher, and the results were as follows:

Pitcher	Physical Condition		
	No injuries	Normal	Excessive Injuries
Jose Diaz	0.30	0.45	0.25
Jerry Damon	0.13	0.47	0.40
Frank Thompson	0.45	0.35	0.20
Derek Rodriguez	0.50	0.40	0.10
Ken Griffin	0.15	0.60	0.25

Determine the expected number of wins for each pitcher, and combined with the contract price in part b, indicate which pitcher you would recommend signing.

38. Construct a decision tree for the decision situation described in Problem 25 and indicate the best decision.

39. Given the following sequential decision tree, determine which is the optimal investment, A or B:

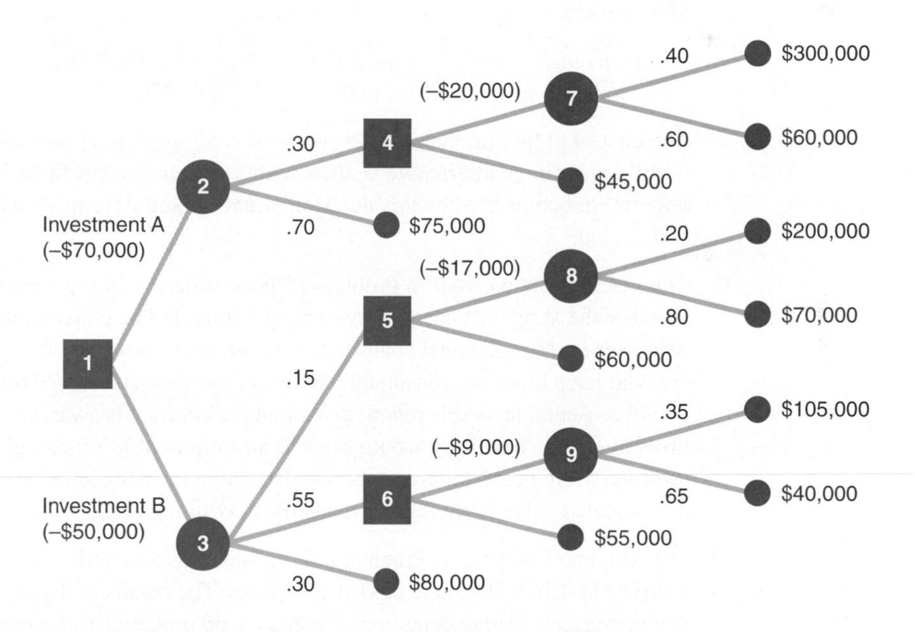

40. The management of First American Bank was concerned about the potential loss that might occur in the event of a physical catastrophe such as a power failure or a fire. The bank estimated that the loss from one of these incidents could be as much as $100 million, including losses due to interrupted service and customer relations. One project the bank is considering is the installation of an emergency power generator at its operations headquarters. The cost of the emergency generator is $800,000, and if it is installed, no losses from this type of incident will be incurred. However, if the generator is not installed, there is a 10% chance that a power outage will occur during the next year. If there is an outage, there is a .05 probability that the resulting losses will be very large, or approximately $80 million in lost earnings. Alternatively, it is estimated that there is a .95 probability of only slight losses of around $1 million. Using decision tree analysis, determine whether the bank should install the new power generator.

41. The Americo Oil Company is considering making a bid for a shale oil development contract to be awarded by the federal government. The company has decided to bid $112 million. The company estimates that it has a 60% chance of winning the contract with this bid. If the firm wins the contract, it can choose one of three methods for getting the oil from the shale. It can develop a new method for oil extraction, use an existing (inefficient) process, or subcontract the processing to a number of smaller companies once the shale has been excavated. The results from these alternatives are as follows:

    Develop new process:

Outcomes	*Probability*	*Profit ($1,000,000s)*
Great success	.30	$ 600
Moderate success	.60	300
Failure	.10	−100

    Use present process:

Outcomes	*Probability*	*Profit ($1,000,000s)*
Great success	.50	$ 300
Moderate success	.30	200
Failure	.20	−40

    Subcontract:

Outcome	*Probability*	*Profit ($1,000,000s)*
Moderate success	1.00	250

    The cost of preparing the contract proposal is $2 million. If the company does not make a bid, it will invest in an alternative venture with a guaranteed profit of $30 million. Construct a sequential decision tree for this decision situation and determine whether the company should make a bid.

42. The machine shop owner in Problem 15 is considering hiring a military consultant to ascertain whether the shop will get the government contract. The consultant is a former military officer who uses various personal contacts to find out such information. By talking to other shop owners who have hired the consultant, the owner has estimated a .70 probability that the consultant would present a favorable report, given that the contract is awarded to the shop, and a .80 probability that the consultant would present an unfavorable report, given that the contract is not awarded. Using decision tree analysis, determine the decision strategy the owner should follow, the expected value of this strategy, and the maximum fee the owner should pay the consultant.

43. The Miramar Company in Problem 18 is considering contracting with a market research firm to do a survey to determine future market conditions. The results of the survey will indicate either positive or negative market conditions. There is a .60 probability of a positive report, given favorable conditions; a .30 probability of a positive report, given stable conditions; and a .10 probability of a positive report, given unfavorable conditions. There is a .90 probability of a negative report, given unfavorable conditions; a .70 probability, given stable conditions; and a .40 probability, given favorable conditions. Using decision tree analysis *and* posterior probability tables, determine the decision strategy the company should follow, the expected value of the strategy, and the maximum amount the company should pay the market research firm for the survey results.

44. The Friendlys in Problem 25 are considering hiring a petroleum analyst to determine the future availability of gasoline. The analyst will report that either a shortage or a surplus will occur. The

probability that the analyst will indicate a shortage, given that a shortage actually occurs is .90; the probability that the analyst will indicate a surplus, given that a surplus actually occurs is .70.

a. Determine the decision strategy the Friendlys should follow, the expected value of this strategy, and the maximum amount the Friendlys should pay for the analyst's services.
b. Compute the efficiency of the sample information for the Friendly car dealership.

45. Jeffrey Mogul is a Hollywood film producer, and he is currently evaluating a script by a new screenwriter and director, Betty Jo Thurston. Jeffrey knows that the probability of a film by a new director being a success is about .10 and that the probability it will flop is .90. The studio accounting department estimates that if this film is a hit, it will make $25 million in profit, whereas if it is a box office failure, it will lose $8 million. Jeffrey would like to hire noted film critic Dick Roper to read the script and assess its chances of success. Roper is generally able to correctly predict a successful film 70% of the time and correctly predict an unsuccessful film 80% of the time. Roper wants a fee of $1 million. Determine whether Roper should be hired, the strategy Mogul should follow if Roper is hired, and the expected value.

46. Tech is playing State in the last conference game of the season. Tech is trailing State 21 to 14, with 7 seconds left in the game, when Tech scores a touchdown. Still trailing 21 to 20, Tech can either go for 2 points and win or go for 1 point to send the game into overtime. The conference championship will be determined by the outcome of this game. If Tech wins, it will go to the Sugar Bowl, with a payoff of $7.2 million; if it loses, it will go to the Gator Bowl, with a payoff of $1.7 million. If Tech goes for 2 points, there is a 33% chance it will be successful and win (and a 67% chance it will fail and lose). If it goes for 1 point, there is a 0.98 probability of success and a tie and a 0.02 probability of failure. If the teams tie, they will play overtime, during which Tech believes it has only a 20% chance of winning because of fatigue.

a. Use decision tree analysis to determine whether Tech should go for 1 or 2 points.
b. What would Tech's probability of winning the game in overtime have to be to make Tech indifferent to going for either 1 or 2 points?

47. Jay Seago is suing the manufacturer of his car for $3.5 million because of a defect that he believes caused him to have an accident. The accident kept him out of work for a year. The company has offered him a settlement of $700,000, of which Jay would receive $600,000 after attorneys' fees. His attorney has advised him that he has a 50% chance of winning his case. If he loses, he will incur attorneys' fees and court costs of $75,000. If he wins, he is not guaranteed his full requested settlement. His attorney believes that there is a 50% chance he could receive the full settlement, in which case Jay would realize $2 million after his attorney takes her cut, and a 50% chance that the jury will award him a lesser amount of $1 million, of which Jay would get $500,000.

Using decision tree analysis, decide whether Jay should proceed with his lawsuit against the manufacturer.

48. Tech has three health care plans for its faculty and staff to choose from, as follows:

Plan 1—monthly cost of $32, with a $500 deductible; the participants pay the first $500 of medical costs for the year; the insurer pays 90% of all remaining expenses.

Plan 2—monthly cost of $5 but a deductible of $1,200, with the insurer paying 90% of medical expenses after the insurer pays the first $1,200 in a year.

Plan 3—monthly cost of $24, with no deductible; the participants pay 30% of all expenses, with the remainder paid by the insurer.

Tracy McCoy, an administrative assistant in the management science department, estimates that her annual medical expenses are defined by the following probability distribution:

Annual Medical Expenses	Probability
$ 100	.15
500	.30
1,500	.35
3,000	.10
5,000	.05
10,000	.05

Determine which medical plan Tracy should select.

49. The Valley Wine Company purchases grapes from one of two nearby growers each season to produce a particular red wine. It purchases enough grapes to produce 3,000 bottles of the wine. Each grower supplies a certain portion of poor-quality grapes, resulting in a percentage of bottles being used as fillers for cheaper table wines, according to the following probability distribution:

	Probability of Percentage Defective	
Defective (%)	Grower A	Grower B
2	.15	.30
4	.20	.30
6	.25	.20
8	.30	.10
10	.10	.10

The two growers charge different prices for their grapes and, because of differences in taste, the company charges different prices for its wine, depending on which grapes it uses. Following is the annual profit from the wine produced from each grower's grapes for each percentage defective:

	Profit	
Defective (%)	Grower A	Grower B
2	$44,200	$42,600
4	40,200	40,300
6	36,200	38,000
8	32,200	35,700
10	28,200	33,400

Use decision tree analysis to determine from which grower the company should purchase grapes.

50. Kroft Food Products is attempting to decide whether it should introduce a new line of salad dressings called Special Choices. The company can test market the salad dressings in selected geographic areas or bypass the test market and introduce the product nationally. The cost of the test market is $150,000. If the company conducts the test market, it must wait to see the results before deciding whether to introduce the salad dressings nationally. The probability of a positive test market result is estimated to be 0.6. Alternatively, the company can decide not to conduct

the test market and go ahead and make the decision to introduce the dressings or not. If the salad dressings are introduced nationally and are a success, the company estimates that it will realize an annual profit of $1.6 million, whereas if the dressings fail, it will incur a loss of $700,000. The company believes the probability of success for the salad dressings is 0.50 if they are introduced without the test market. If the company does conduct the test market and it is positive, then the probability of successfully introducing the salad dressings increases to 0.8. If the test market is negative and the company introduces the salad dressings anyway, the probability of success drops to 0.30.

Using decision tree analysis, determine whether the company should conduct the test market.

51. In Problem 50, determine the expected value of sample information (EVSI) (i.e., the test market value) and the expected value of perfect information (EVPI).

52. Ellie Daniels has $200,000 and is considering three mutual funds for investment—a global fund, an index fund, and an Internet stock fund. During the first year of investment, Ellie estimates that there is a .70 probability that the market will go up and a .30 probability that the market will go down. Following are the returns on her $200,000 investment at the end of the year under each market condition:

	Market Conditions	
Fund	Up	Down
Global	$25,000	$ −8,000
Index	35,000	5,000
Internet	60,000	−35,000

At the end of the first year, Ellie will either reinvest the entire amount plus the return or sell and take the profit or loss. If she reinvests, she estimates that there is a .60 probability the market will go up and a .40 probability the market will go down. If Ellie reinvests in the global fund after it has gone up, her return on her initial $200,000 investment plus her $25,000 return after 1 year will be $45,000. If the market goes down, her loss will be $15,000. If she reinvests after the market has gone down, her return will be $34,000, and her loss will be $17,000. If Ellie reinvests in the index fund after the market has gone up, after 2 years her return will be $65,000 if the market continues upward, but only $5,000 if the market goes down. Her return will be $55,000 if she reinvests and the market reverses itself and goes up after initially going down, and it will be $5,000 if the market continues to go down. If Ellie invests in the Internet fund, she will make $60,000 if the market goes up, but she will lose $35,000 if it goes down. If she reinvests as the market continues upward, she will make an additional $100,000; but if the market reverses and goes down, she will lose $70,000. If she reinvests after the market has initially gone down, she will make $65,000, but if the market continues to go down, she will lose an additional $75,000.

Using decision tree analysis, determine which fund Ellie should invest in and its expected value.

53. Blue Ridge Power and Light is an electric utility company with a large fleet of vehicles, including automobiles, light trucks, and construction equipment. The company is evaluating four alternative strategies for maintaining its vehicles at the lowest cost: (1) do no preventive maintenance at all and repair vehicle components when they fail; (2) take oil samples at regular intervals and perform whatever preventive maintenance is indicated by the oil analysis; (3) change the vehicle oil on a regular basis and perform repairs when needed; (4) change the oil at regular intervals, take oil samples regularly, and perform maintenance repairs as indicated by the sample analysis.
For autos and light trucks, strategy 1 (no preventive maintenance) costs nothing to implement and results in two possible outcomes: There is a .10 probability that a defective component will

occur, requiring emergency maintenance at a cost of $1,200, or there is a .90 probability that no defects will occur and no maintenance will be necessary.

Strategy 2 (take oil samples) costs $20 to implement (i.e., take a sample), and there is a .10 probability that there will be a defective part and .90 probability that there will not be a defect. If there is actually a defective part, there is a .70 probability that the sample will correctly identify it, resulting in preventive maintenance at a cost of $500. However, there is a .30 probability that the sample will not identify the defect and indicate that everything is okay, resulting in emergency maintenance later at a cost of $1,200. On the other hand, if there are actually no defects, there is a .20 probability that the sample will erroneously indicate that there is a defect, resulting in unnecessary maintenance at a cost of $250. There is an .80 probability that the sample will correctly indicate that there are no defects, resulting in no maintenance and no costs.

Strategy 3 (changing the oil regularly) costs $14.80 to implement and has two outcomes: a .04 probability of a defective component, which will require emergency maintenance at a cost of $1,200, and a .96 probability that no defects will occur, resulting in no maintenance and no cost.

Strategy 4 (changing the oil and sampling) costs $34.80 to implement and results in the same probabilities of defects and no defects as strategy 3. If there is a defective component, there is a .70 probability that the sample will detect it and $500 in preventive maintenance costs will be incurred. Alternatively, there is a .30 probability that the sample will not detect the defect, resulting in emergency maintenance at a cost of $1,200. If there is no defect, there is a .20 probability that the sample will indicate that there is a defect, resulting in an unnecessary maintenance cost of $250, and there is an .80 probability that the sample will correctly indicate no defects, resulting in no cost.

Develop a decision strategy for Blue Ridge Power and Light and indicate the expected value of this strategy.[1]

54. In Problem 53, the decision analysis is for automobiles and light trucks. Blue Ridge Power and Light would like to reformulate the problem for its heavy construction equipment. Emergency maintenance is much more expensive for heavy equipment, costing $15,000. Required preventive maintenance costs $2,000, and unnecessary maintenance costs $1,200. The cost of an oil change is $100, and the cost of taking an oil sample and analyzing it is $30. All the probabilities remain the same. Determine the strategy Blue Ridge Power and Light should use for its heavy equipment.

55. In Problem 14, the management of the Oakland Bombers is considering hiring superscout Jerry McGuire to evaluate the team's chances for the coming season. McGuire will evaluate the team, assuming that it will sign one of the four free agents. The team's management has determined the probability that the team will have a losing record with any of the free agents to be .21 by averaging the probabilities of losing for the four free agents in Problem 33. The probability that the team will have a competitive season but not make the playoffs is developed similarly, and it is .35. The probability that the team will make the playoffs is .44. The probability that McGuire will correctly predict that the team will have a losing season is .75, whereas the probability that he will predict a competitive season, given that it has a losing season, is .15, and the probability that he will incorrectly predict a playoff season, given that the team has a losing season, is .10. The probability that he will successfully predict a competitive season is .80, whereas the probability that he will incorrectly predict a losing season, given that the team is competitive, is .10, and the probability that he will incorrectly predict a playoff season, given the team has a competitive season, is .10. The probability that he will correctly predict a playoff season is .85, whereas the probability that he will incorrectly predict a losing season, given that the team

[1]This problem is based on J. Mellichamp, D. Miller, and O.-J. Kwon, "The Southern Company Uses a Probability Model for Cost Justification of Oil Sample Analysis," *Interfaces* 23, no. 3 (May–June 1993): 118–24.

makes the playoffs, is .05, and the probability that he will predict a competitive season, given the team makes the playoffs, is .10. Using decision tree analysis and posterior probabilities, determine the decision strategy the team should follow, the expected value of the strategy, and the maximum amount the team should pay for Jerry McGuire's predictions.

56. The Place-Plus real estate development firm in Problem 24 is dissatisfied with the economist's estimate of the probabilities of future interest rate movement, so it is considering having a financial consulting firm provide a report on future interest rates. The consulting firm is able to cite a track record which shows that 80% of the time when interest rates declined, it had predicted they would, whereas 10% of the time when interest rates declined, the firm had predicted they would remain stable and 10% of the time it had predicted they would increase. The firm has been correct 70% of the time when rates have remained stable, whereas 10% of the time it has incorrectly predicted that rates would decrease, and 20% of the time it has incorrectly predicted that rates would increase. The firm has correctly predicted that interest rates would increase 90% of the time and incorrectly predicted rates would decrease 2% and remain stable 8% of the time. Assuming that the consulting firm could supply an accurate report, determine how much Place-Plus should be willing to pay the consulting firm and how efficient the information will be.

57. A young couple has $5,000 to invest in either savings bonds or a real estate deal. The expected return on each investment, given good and bad economic conditions, is shown in the following payoff table:

	Economic Conditions	
	Good	Bad
Investment	.6	.4
Savings bonds	$ 1,000	$ 1,000
Real estate	10,000	−2,000

The expected value of investing in savings bonds is $1,000, and the expected value of the real estate investment is $5,200. However, the couple decides to invest in savings bonds. Explain the couple's decision in terms of the utility they might associate with each investment.

58. Annie Hays recently sold a condominium she had bought and lived in while she was a college student over 15 years ago. She received $150,000 for the condominium and is considering two investment alternatives. Annie can invest the entire amount in a bank money market for 1 year at 8% interest and thus receive $162,000 at the end of a year, or she can invest in a speculative oil exploration project with a 50–50 chance of doubling her investment at the end of the year or losing everything.
    a. Determine which alternative Annie should invest in, according to the expected value criterion, and indicate whether that is the alternative you would also select.
    b. Suppose Annie determined that the probability of success for the oil exploration investment would have to be .80 before she would be indifferent between the two alternatives. In other words, if the probability of success for the oil exploration investment was less than .80, Annie would invest in the money market, but if the probability of success was greater than .80, she would invest in the oil exploration. In this case, .80 is the utility value of the $162,000 safe return. Determine the preferred alternative, according to the expected utility value.

59. Labran Jones has played for the Cleveland professional basketball team for the past eight seasons and has established himself as one of the top players in the league. He has recently become a free agent, meaning he can sign a new contract with Cleveland or sign a contract with any other team in the league. Cleveland has never seriously contended for a championship, and Labran is strongly considering moving to Miami, where he would have a better chance at a

championship, or to New York, which is a bigger media market and would give him more financial opportunities and endorsements. All three teams are offering Labran a 6-year contract, but because of a salary cap rule, Miami can only offer Labran \$110 million, while New York is offering \$120 million and Cleveland \$125 million. Odds makers give Miami a 70% chance of winning the championship with Labran, while they give Cleveland a 40% chance and New York only a 10% chance. If Labran wins a championship, he will almost certainly finish his career with the team he wins it with. If that team is New York, he will sign a new contract after 6 years that with endorsements and financial deals is estimated to be worth around \$500 million by the end of his career. However, if New York doesn't win a championship during his 6 years, he will either stay with New York for a career total of \$200 million, or sign a new 4-year contract with a new team. If he signs with a new team, he assumes he will sign with a team good enough to give him a 50–50 chance of winning a championship but, because of his likely diminished abilities, he will have eventual earnings of only around \$120 million if he wins a championship with his new team. However, if he doesn't win a championship, his eventual earnings would only be about \$65 million. If he signs with Miami and Miami wins the championship, he will stay with the team and have eventual long-term earnings of \$375 million, but if Miami doesn't win a championship, he will either stay with Miami for earnings of about \$90 million or sign with a new team, with the same expected outcomes if he signs with a new team after playing in New York. If he re-signs with Cleveland and Cleveland wins a championship, his eventual earnings are expected to be \$300 million by the end of his playing career, but if Cleveland doesn't win, he can stay with the team and expect eventual possible earnings of around \$145 million or sign with a new team with the same expected outcomes if he signs with a new team after playing in New York and Miami. Using decision tree analysis, determine which team Labran should sign a new contract with and the expected value of his decision.

Case Problem

Steeley Associates Versus Concord Falls

Steeley Associates, Inc., a property development firm, purchased an old house near the town square in Concord Falls, where State University is located. The old house was built in the mid-1800s, and Steeley Associates restored it. For almost a decade, Steeley has leased it to the university for academic office space. The house is located on a wide lawn and has become a town landmark.

However, in 2008, the lease with the university expired, and Steeley Associates decided to build high-density student apartments on the site, using all the open space. The community was outraged and objected to the town council. The legal counsel for the town spoke with a representative from Steeley and hinted that if Steeley requested a permit, the town would probably reject it. Steeley had reviewed the town building code and felt confident that its plan was within the guidelines, but that did not necessarily mean that it could win a lawsuit against the town to force the town to grant a permit.

The principals at Steeley Associates held a series of meetings to review their alternatives. They decided that they had three options: They could request the permit, they could sell the property, or they could request a permit for a low-density office building, which the town had indicated it would not fight. Regarding the last two options, if Steeley sells the house and property, it thinks it can get \$900,000. If it builds a new office building, its return will depend on town business growth in the future. It feels that there is a 70% chance of future growth, in which case Steeley will see a return of \$1.3 million (over a 10-year planning horizon); if no growth (or erosion) occurs, it will make only \$200,000.

If Steeley requests a permit for the apartments, a host of good and bad outcomes are possible. The immediate good outcome is approval of its permit, which it estimates will result in a return of \$3 million. However, Steeley gives that result only a 10% chance that it will occur. Alternatively, Steeley thinks there is a 90% chance that the town will reject its application, which will result in another set of decisions.

Steeley can sell the property at that point. However, the rejection of the permit will undoubtedly decrease the value to potential buyers, and Steeley estimates that it will get only \$700,000. Alternatively, it can construct the office building and face the same potential outcomes it did earlier, namely, a 30% chance of no town growth and a \$200,000 return or a 70% chance of growth with a return of \$1.3 million. A third option is to sue the town. On the surface, Steeley's case looks good, but the town building code is vague, and a sympathetic judge could throw out its suit. Whether or not it wins, Steeley estimates its possible legal fees to be \$300,000, and it feels it has only a 40% chance of winning. However, if Steeley does win, it estimates that the award will be approximately \$1 million, and it will also get its \$3 million return for building the apartments. Steeley also estimates that there is a 10% chance that the suit could linger on in the courts for such a long time that any future return would be negated during its planning horizon, and it would incur an additional \$200,000 in legal fees.

If Steeley loses the suit, it will then be faced with the same options of selling the property or constructing an office building. However, if the suit is carried this far into the future, it feels that the selling price it can ask will be somewhat dependent on the town's growth prospects at that time, which it feels it can estimate at only 50–50. If the town is in a growth mode that far in the future, Steeley thinks that \$900,000 is a conservative estimate of the potential sale price, whereas if the town is not growing, it thinks \$500,000 is a more likely estimate. Finally, if Steeley constructs the office building, it feels that the chance of town growth is 50%, in which case the return will be only \$1.2 million. If no growth occurs, it conservatively estimates only a \$100,000 return.

A. Perform a decision tree analysis of Steeley Associates's decision situation, using expected value, and indicate the appropriate decision with these criteria.
B. Indicate the decision you would make and explain your reasons.

Case Problem

Transformer Replacement at Mountain States Electric Service

Mountain States Electric Service is an electrical utility company serving several states in the Rocky Mountains region. It is considering replacing some of its equipment at a generating substation and is attempting to decide whether it should replace an older, existing PCB transformer. (PCB is a toxic chemical known formally as polychlorinated biphenyl.) Even though the PCB generator meets all current regulations, if an incident occurred, such as a fire, and PCB contamination caused harm either to neighboring businesses or farms or to the environment, the company would be liable for damages. Recent court cases have shown that simply meeting utility regulations does not relieve a utility of liability if an incident causes harm to others. Also, courts have been awarding large damages to individuals and businesses harmed by hazardous incidents.

If the utility replaces the PCB transformer, no PCB incidents will occur, and the only cost will be the cost of the transformer, \$85,000. Alternatively, if the company decides to keep the existing PCB transformer, then management estimates that there is a 50–50 chance of there being a high likelihood of an incident or a low likelihood of an incident. For the case in which there is a high likelihood that an incident will occur, there is a .004 probability that a fire will occur sometime during the remaining life of the transformer and a .996 probability that no fire will occur. If a fire occurs, there is a .20 probability that it will be bad and the utility will incur a very high cost of approximately \$90 million for the cleanup, whereas there is an .80 probability that the fire will be minor and cleanup can be accomplished at a low cost of approximately \$8 million. If no fire occurs, no cleanup costs will occur. For the case in which there is a low likelihood that an incident will occur, there is a .001 probability that a fire will occur during the life of the existing transformer and a .999 probability that a fire will not occur. If a fire does occur, the same probabilities exist for the incidence of high and low cleanup costs, as well as the same cleanup costs, as indicated for the previous case. Similarly, if no fire occurs, there is no cleanup cost.

Perform a decision tree analysis of this problem for Mountain States Electric Service and indicate the recommended solution. Is this the decision you believe the company should make? Explain your reasons.

This case is based on W. Balson, J. Welsh, and D. Wilson, "Using Decision Analysis and Risk Analysis to Manage Utility Environmental Risk," *Interfaces* 22, no. 6 (November–December 1992): 126–39.

Case Problem

THE CAROLINA COUGARS

The Carolina Cougars is a major league baseball expansion team beginning its third year of operation. The team had losing records in each of its first 2 years and finished near the bottom of its division. However, the team was young and generally competitive. The team's general manager, Frank Lane, and manager, Biff Diamond, believe that with a few additional good players, the Cougars can become a contender for the division title and perhaps even for the pennant. They have prepared several proposals for free-agent acquisitions to present to the team's owner, Bruce Wayne.

Under one proposal the team would sign several good available free agents, including two pitchers, a good fielding shortstop, and two power-hitting outfielders for $52 million in bonuses and annual salary. The second proposal is less ambitious, costing $20 million to sign a relief pitcher, a solid, good-hitting infielder, and one power-hitting outfielder. The final proposal would be to stand pat with the current team and continue to develop.

General Manager Lane wants to lay out a possible season scenario for the owner so he can assess the long-run ramifications of each decision strategy. Because the only thing the owner understands is money, Frank wants this analysis to be quantitative, indicating the money to be made or lost from each strategy. To help develop this analysis, Frank has hired his kids, Penny and Nathan, both management science graduates from Tech.

Penny and Nathan analyzed league data for the previous five seasons for attendance trends, logo sales (i.e., clothing, souvenirs, hats, etc.), player sales and trades, and revenues. In addition, they interviewed several other owners, general managers, and league officials. They also analyzed the free agents that the team was considering signing.

Based on their analysis, Penny and Nathan feel that if the Cougars do not invest in any free agents, the team will have a 25% chance of contending for the division title and a 75% chance of being out of contention most of the season. If the team is a contender, there is a .70 probability that attendance will increase as the season progresses and the team will have high attendance levels (between 1.5 million and 2.0 million) with profits of $170 million from ticket sales, concessions, advertising sales, TV and radio sales, and logo sales. They estimate a .25 probability that the team's attendance will be mediocre (between 1.0 million and 1.5 million) with profits of $115 million and a .05 probability that the team will suffer low attendance (less than 1.0 million) with profit of $90 million. If the team is not a contender, Penny and Nathan estimate that there is .05 probability of high attendance with profits of $95 million, a .20 probability of medium attendance with profits of $55 million, and a .75 probability of low attendance with profits of $30 million.

If the team marginally invests in free agents at a cost of $20 million, there is a 50–50 chance it will be a contender. If it is a contender, then later in the season it can either stand pat with its existing roster or buy or trade for players that could improve the team's chances of winning the division. If the team stands pat, there is a .75 probability that attendance will be high and profits will be $195 million. There is a .20 probability that attendance will be mediocre with profits of $160 million and a .05 probability of low attendance and profits of $120 million. Alternatively, if the team decides to buy or trade for players, it will cost $8 million, and the probability of high attendance with profits of $200 million will be .80. The probability of mediocre attendance with $170 million in profits will be .15, and there will be a .05 probability of low attendance, with profits of $125 million.

If the team is not in contention, then it will either stand pat or sell some of its players, earning approximately $8 million in profit. If the team stands pat, there is a .12 probability of high attendance, with profits of $110 million; a .28 probability of mediocre attendance, with profits of $65 million; and a .60 probability of low attendance, with profits of $40 million. If the team sells players, the fans will likely lose interest at an even faster rate, and the probability of high attendance with profits of $100 million will drop to .08, the probability of mediocre attendance with profits of $60 million will be .22, and the probability of low attendance with profits of $35 million will be .70.

The most ambitious free-agent strategy will increase the team's chances of being a contender to 65%. This strategy will also excite the fans most during the off-season and boost ticket sales and advertising and logo sales early in the year. If the team does contend for the division title, then later in the season it will have to decide whether to invest in more players. If the Cougars stand pat, the probability of high attendance with profits of $210 million will be .80, the probability of mediocre attendance with profits of $170 million will be .15, and the probability of low attendance with profits of $125 million will be .05. If the team buys players at a cost of $10 million, then the probability of having high attendance with profits of $220 million will increase to .83, the probability of mediocre attendance with profits of $175 million will be .12, and the probability of low attendance with profits of $130 million will be .05.

If the team is not in contention, it will either sell some players' contracts later in the season for profits of around $12 million or stand pat. If it stays with its roster, the probability of high attendance with profits of $110 million will be .15, the probability of mediocre attendance with profits of $70 million will be .30, and the probability of low attendance with profits of $50 million will be .55. If the team sells players late in the season, there will be a .10 probability of high attendance with profits of $105 million, a .30 probability of mediocre attendance with profits of $65 million, and a .60 probability of low attendance with profits of $45 million.

Assist Penny and Nathan in determining the best strategy to follow and its expected value.

Case Problem

Evaluating R&D Projects at WestCom Systems Products Company

WestCom Systems Products Company develops computer systems and software products for commercial sale. Each year it considers and evaluates a number of different R&D projects to undertake. It develops a road map for each project, in the form of a standardized decision tree that identifies the different decision points in the R&D process from the initial decision to invest in a project's development through the actual commercialization of the final product.

The first decision point in the R&D process is whether to fund a proposed project for 1 year. If the decision is no, then there is no resulting cost; if the decision is yes, then the project proceeds at an incremental cost to the company. The company establishes specific short-term, early technical milestones for its projects after 1 year. If the early milestones are achieved, the project proceeds to the next phase of project development; if the milestones are not achieved, the project is abandoned. In its planning process, the company develops probability estimates of achieving and not achieving the early milestones. If the early milestones are achieved, the project is funded for further development during an extended time frame specific to a project. At the end of this time frame, a project is evaluated according to a second set of (later) technical milestones. Again, the company attaches probability estimates for achieving and not achieving these later milestones. If the later milestones are not achieved, the project is abandoned.

If the later milestones are achieved, technical uncertainties and problems have been overcome, and the company next assesses the project's ability to meet its strategic business objectives. At this stage, the company wants to know if the eventual product coincides with the company's competencies and whether there appears to be an eventual, clear market for the product. It invests in a product "prelaunch" to ascertain the answers to these questions. The outcomes of the prelaunch are that either there is a strategic fit or there is not, and the company assigns probability estimates to each of these two possible outcomes. If there is not a strategic fit at this point, the project is abandoned and the company loses its investment in the prelaunch process. If it is determined that there is a strategic fit, then three possible decisions result: (1) The company can invest in the product's launch, and a successful or unsuccessful outcome will result, each with an estimated probability of occurrence; (2) the company can delay the product's launch and at a later date decide whether to launch or abandon; and (3) if it launches later, the outcomes are success or failure, each with an estimated probability of occurrence.

Also, if the product launch is delayed, there is always a likelihood that the technology will become obsolete or dated in the near future, which tends to reduce the expected return.

The following table provides the various costs, event probabilities, and investment outcomes for five projects the company is considering:

	Project				
Decision Outcomes/Event	1	2	3	4	5
Fund—1 year	$ 200,000	$ 350,000	$ 170,000	$ 230,000	$ 400,000
P(Early milestones—yes)	.70	.67	.82	.60	.75
P(Early milestones—no)	.30	.33	.18	.40	.25
Long-term funding	$ 650,000	780,000	450,000	300,000	450,000
P(Late milestones—yes)	.60	.56	.65	.70	.72
P(Late milestones—no)	.40	.44	.35	.30	.28
Prelaunch funding	$ 300,000	450,000	400,000	500,000	270,000
P(Strategic fit—yes)	.80	.75	.83	.67	.65
P(Strategic fit—no)	.20	.25	.17	.33	.35
P(Invest—success)	.60	.65	.70	.75	.80
P(Invest—failure)	.40	.35	.30	.25	.20
P(Delay—success)	.80	.70	.65	.80	.85
P(Delay—failure)	.20	.30	.35	.20	.15
Invest—success	$ 7,300,000	8,000,000	4,500,000	5,200,000	3,800,000
Invest—failure	−2,000,000	−3,500,000	−1,500,000	−2,100,000	−900,000
Delay—success	4,500,000	6,000,000	3,300,000	2,500,000	2,700,000
Delay—failure	−1,300,000	−4,000,000	−800,000	−1,100,000	−900,000

Determine the expected value for each project and then rank the projects accordingly for the company to consider.

This case is based on R. K. Perdue, W. J. McAllister, P. V. King, and B. G. Berkey, "Valuation of R and D Projects Using Options Pricing and Decision Analysis Models," *Interfaces* 29, no. 6 (November–December 1999): 57–74.

CHAPTER

13

Queuing Analysis

Waiting in queues—waiting lines—is one of the most common occurrences in everyone's life. Anyone who has gone shopping or to a movie has experienced the inconvenience of waiting in line to make purchases or buy a ticket. Not only do people spend a significant portion of their time waiting in lines, but products queue up in production plants, machinery waits in line to be serviced, planes wait to take off and land, and so on. Because time is a valuable resource, the reduction of waiting time is an important topic of analysis.

Providing quick service is an important aspect of quality customer service.

The improvement of service with respect to waiting time has also become more important in recent years because of the increased emphasis on quality, especially in service-related operations. When customers go to a bank to take out a loan, cash a check, or make a deposit; take their car to a dealer for service or repair; or shop at the grocery store; they increasingly equate quality service with rapid service. Aware of this, more and more companies are focusing on reducing waiting time as an important component of quality improvement. In general, companies are able to reduce waiting time and provide faster service by increasing their service capacity, which usually means adding more servers; such as more tellers at a bank, more mechanics at a car dealership, or more checkout clerks at a grocery store. However, increasing service capacity in this manner has a monetary cost, and therein lies the basis of waiting line analysis: the trade-off between the cost of improved service and the cost of making customers wait.

Queuing analysis is the probabilistic analysis of waiting lines.

Like decision analysis, **queuing analysis** is a probabilistic form of analysis, not a deterministic technique. Thus, the results of queuing analysis, referred to as operating characteristics, are probabilistic. These operating statistics (such as the average time a person must wait in line to be served) are used by the manager of the operation containing the queue to make decisions.

A number of different queuing models exist to deal with different queuing systems. We will eventually discuss many of these queuing variations, but we will concentrate on two of the most common types of systems—the single-server system and the multiple-server system.

Elements of Waiting Line Analysis

Waiting lines form because people or things arrive at the servicing function, or server, faster than they can be served. However, this does not mean that the service operation is understaffed or does not have the overall capacity to handle the influx of customers. In fact, most businesses and organizations have sufficient serving capacity available to handle their customers *in the long run*. Waiting lines result because customers do not arrive at a constant, evenly paced rate, nor are they all served in an equal amount of time. Customers arrive at random times, and the time required to serve them individually is not the same. Thus, a waiting line is continually increasing and decreasing in length (and is sometimes empty), and it approaches an average rate of customer arrivals and an average time to serve the customer in the long run. For example, the checkout counters at a grocery store may have enough clerks to serve an average of 100 customers in an hour, and in any particular hour only 60 customers might arrive. However, at specific points in time during the hour, waiting lines may form because more than an average number of customers arrive, and they make more than an average number of purchases.

Operating characteristics are average values for characteristics that describe the performance of a waiting line system.

Decisions about waiting lines and the management of waiting lines are based on these averages for customer arrivals and service times. They are used in queuing formulas to compute **operating characteristics**, such as the average number of customers waiting in line and the average time a customer must wait in line. Different sets of formulas are used, depending on the type of waiting line system being investigated. For example, a bank drive-up teller window that has one bank clerk serving a single line of customers in cars is different from a single line of passengers at an airport ticket counter that is served by three or four airline agents. In the next section we present the different elements and components that make up waiting lines before we look at queuing formulas in the following sections.

The Single-Server Waiting Line System

A single server with a single waiting line is the simplest form of queuing system. As such, it will be used to demonstrate the fundamentals of a queuing system. As an example of this kind of system, consider Fast Shop Market.

Components of a waiting line system include arrivals, servers, and the waiting line structure.

Fast Shop Market has one checkout counter and one employee who operates the cash register at the checkout counter. The combination of the cash register and the operator is the *server* (or service facility) in this queuing system; the customers who line up at the counter to pay for their selections form the *waiting line,* or **queue**. The configuration of this example queuing system is shown in Figure 13.1.

*A **queue** is a waiting line.*

The most important factors to consider in analyzing a queuing system such as the one in Figure 13.1 are the following:

1. The queue discipline (in what order customers are served)
2. The nature of the calling population (where customers come from)
3. The arrival rate (how often customers arrive at the queue)
4. The service rate (how fast customers are served)

We will discuss each of these items as it relates to our example.

FIGURE 13.1
The Fast Shop Market waiting line system

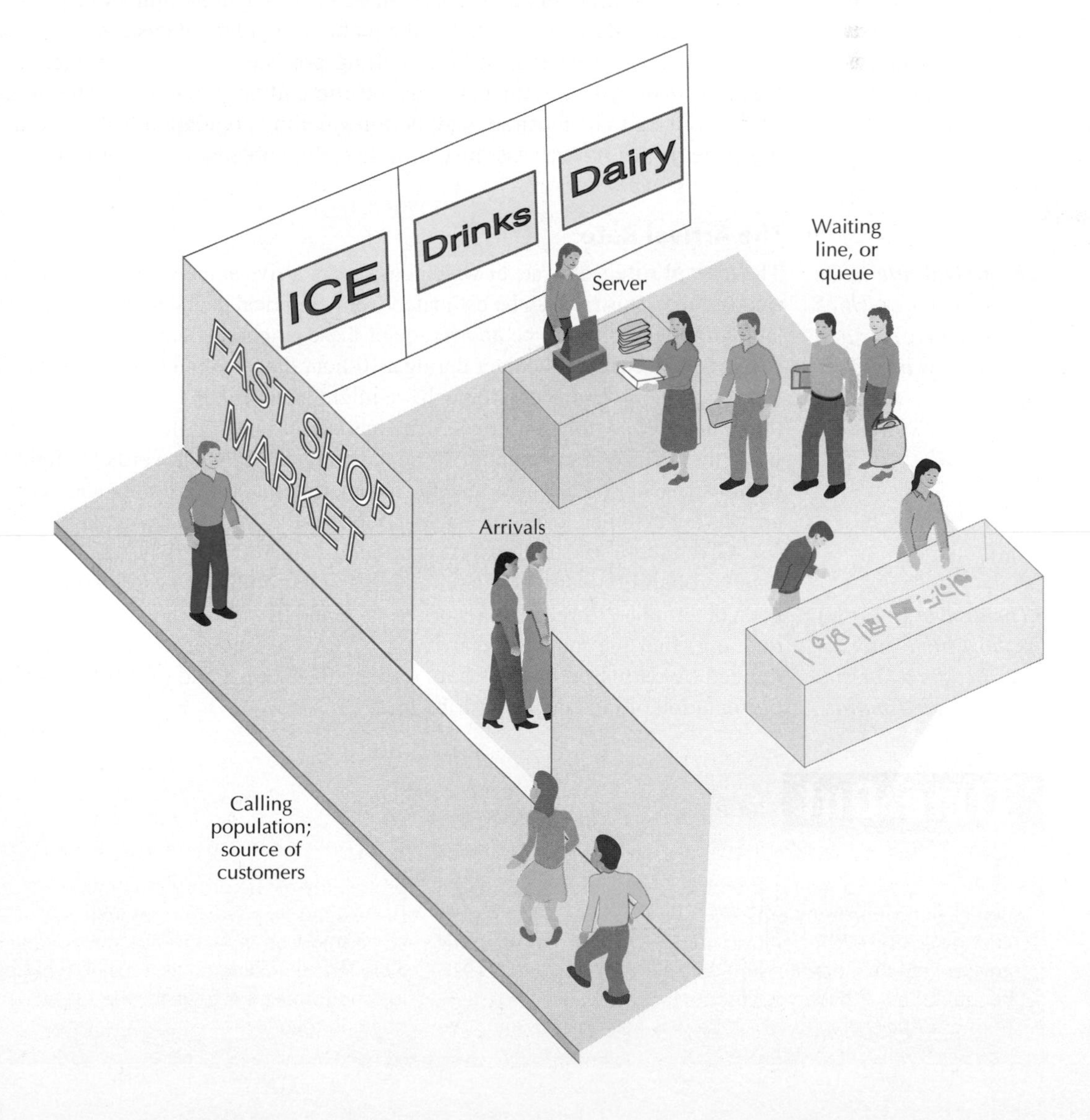

The Queue Discipline

*The **queue discipline** is the order in which waiting customers are served.*

The **queue discipline** is the order in which waiting customers are served. Customers at Fast Shop Market are served on a "first-come, first-served" basis. That is, the first person in line at the checkout counter is served first. This is the most common type of queue discipline. However, other disciplines are possible. For example, a machine operator might stack in-process parts beside a machine so that the last part is on top of the stack and will be selected first. This queue discipline is referred to as "last-in, first-out." Or the machine operator might simply reach into a box full of parts and select one at random. In this case, the queue discipline is random. Often customers are scheduled for service according to a predetermined appointment, such as patients at a doctor's or dentist's office or diners at a restaurant where reservations are required. In this case, the customers are taken according to a prearranged schedule, regardless of when they arrive at the facility. One final example of the many types of queue disciplines is when customers are processed alphabetically according to their last names, such as at school registration or at job interviews.

The Calling Population

*The **calling population** is the source of customers; it may be infinite or finite.*

The **calling population** is the source of the customers to the market, which in this case is assumed to be infinite. In other words, there is such a large number of possible customers in the area where the store is located that the number of potential customers is assumed to be infinite. Some queuing systems have finite calling populations. For example, the repair garage of a trucking firm that has 20 trucks has a finite calling population. The queue is the number of trucks waiting to be repaired, and the finite calling population is the 20 trucks. However, queuing systems that have an assumed infinite calling population are more common.

The Arrival Rate

*The **arrival rate** is the frequency at which customers arrive at a waiting line according to a probability distribution.*

The **arrival rate** is the rate at which customers arrive at the service facility during a specified period of time. This rate can be estimated from empirical data derived from studying the system or a similar system, or it can be an average of these empirical data. For example, if 100 customers arrive at a store checkout counter during a 10-hour day, we could say the arrival rate averages 10 customers per hour. However, although we might be able to determine a rate for arrivals by counting the number of paying customers at the market during a 10-hour day, we would not know exactly when these customers would arrive on the premises. In other words, it might be that no customers would arrive during one hour and 20 customers would arrive during another hour. In general, these arrivals are assumed to be independent of each other and to vary randomly over time.

*The arrival rate (λ) is most frequently described by a **Poisson distribution**.*

Given these assumptions, it is further assumed that arrivals at a service facility conform to some probability distribution. Although arrivals could be described by any distribution, it has been determined (through years of research and the practical experience of people in the field of queuing) that the number of arrivals per unit of time at a service facility can frequently be defined by a **Poisson distribution**. (Appendix C at the end of this text contains a more detailed presentation of the Poisson distribution.)

Time Out for Agner Krarup Erlang

The origin of queuing theory is found in telephone network congestion problems and the work of A. K. Erlang. Erlang (1878–1929), a Danish mathematician, was scientific adviser for the Copenhagen Telephone Company. In 1917 he published a paper outlining the development of telephone traffic theory in which he was able to determine the probability of the different numbers of calls waiting and of the waiting time when the system was in equilibrium. He assumed Poisson inputs (arrivals) from unlimited sources and exponential holding times. Erlang's work provided the stimulus and formed the basis for the subsequent development of queuing theory.

The Service Rate

The service rate is the average number of customers who can be served during a time period.

The service rate is the average number of customers who can be served during a specified period of time. For our Fast Shop Market example, 30 customers can be checked out (served) in 1 hour. A service rate is similar to an arrival rate in that it is a random variable. In other words, such factors as different sizes of customer purchases, the amount of change the cashier must count out, and different forms of payment alter the number of persons that can be served over time. Again, it is possible that only 10 customers might be checked out during one hour and 40 customers might be checked out during the following hour.

The service time can often be described by the negative exponential distribution.

The description of arrivals in terms of a *rate* and of service, in terms of *time* is a convention that has developed in queuing theory. Like arrival rate, service time is assumed to be defined by a probability distribution. It has been determined by researchers in the field of queuing that service times can frequently be defined by a *negative exponential probability* distribution. (Appendix C contains a more detailed presentation of the exponential distribution.) However, to analyze a queuing system, both arrivals and service must be in compatible units of measure. Thus, service time must be expressed as a service rate to correspond with an arrival rate.

The Single-Server Model

The Fast Shop Market checkout counter is an example of a single-server queuing system with the following characteristics:

Assumptions of the basic single-server model.

1. An infinite calling population
2. A first-come, first-served queue discipline
3. Poisson arrival rate
4. Exponential service times

These assumptions have been used to develop a model of a single-server queuing system. However, the analytical derivation of even this simplest queuing model is relatively complex and lengthy. Thus, we will refrain from deriving the model in detail and will consider only the resulting queuing formulas. The reader must keep in mind, however, that these formulas are applicable only to queuing systems having the foregoing conditions.

Given that

λ = mean arrival rate; μ = mean service rate.

$$\lambda = \text{the arrival rate (average number of arrivals per time period)}$$
$$\mu = \text{the service rate (average number served per time period)}$$

and that $\lambda < \mu$ (customers are served at a faster rate than they arrive), we can state the following formulas for the operating characteristics of a single-server model.

Customers must be served faster than they arrive, or an infinitely large queue will build up.

The probability that no customers are in the queuing system (either in the queue or being served) is

$$P_0 = \left(1 - \frac{\lambda}{\mu}\right)$$

Basic single-server queuing formulas.

The probability that n customers are in the queuing system is

$$P_n = \left(\frac{\lambda}{\mu}\right)^n \cdot P_0 = \left(\frac{\lambda}{\mu}\right)^n \left(1 - \frac{\lambda}{\mu}\right)$$

The average number of customers in the queuing system (i.e., the customers being serviced and in the waiting line) is

$$L = \frac{\lambda}{\mu - \lambda}$$

The average number of customers in the waiting line is

$$L_q = \frac{\lambda^2}{\mu(\mu - \lambda)}$$

The average time a customer spends in the total queuing system (i.e., waiting and being served) is

$$W = \frac{1}{\mu - \lambda} = \frac{L}{\lambda}$$

The average time a customer spends waiting in the queue to be served is

$$W_q = \frac{\lambda}{\mu(\mu - \lambda)}$$

*A **utilization factor** is the probability that a server is busy.*

The probability that the server is busy (i.e., the probability that a customer has to wait), known as the **utilization factor**, is

$$U = \frac{\lambda}{\mu}$$

The probability that the server is idle (i.e., the probability that a customer can be served) is

$$I = 1 - U = 1 - \frac{\lambda}{\mu}$$

This last term, $1 - (\lambda/\mu)$, is also equal to P_0. That is, the probability of no customers in the queuing system is the same as the probability that the server is idle.

We can compute these various operating characteristics for Fast Shop Market by simply substituting the average arrival and service rates into the foregoing formulas. For example, if

$$\lambda = 24 \text{ customers per hour arrive at checkout counter}$$
$$\mu = 30 \text{ customers per hour can be checked out}$$

then

$$\begin{aligned} P_0 &= \left(1 - \frac{\lambda}{\mu}\right) \\ &= (1 - 24/30) \\ &= .20 \text{ probability of no customers in the system} \end{aligned}$$

$$\begin{aligned} L &= \frac{\lambda}{\mu - \lambda} \\ &= \frac{24}{30 - 24} \\ &= 4 \text{ customers, on average, in the queuing system} \end{aligned}$$

$$\begin{aligned} L_q &= \frac{\lambda^2}{\mu(\mu - \lambda)} \\ &= \frac{(24)^2}{30(30 - 24)} \\ &= 3.2 \text{ customers, on average, in the waiting line} \end{aligned}$$

$$\begin{aligned} W &= \frac{1}{\mu - \lambda} \\ &= \frac{1}{30 - 24} \\ &= 0.167 \text{ hr. } (10 \text{ min.}) \text{ average time in the system per customer} \end{aligned}$$

$$W_q = \frac{\lambda}{\mu(\mu - \lambda)}$$

$$= \frac{24}{30(30 - 24)}$$

$$= 0.133 \text{ hr. } (8 \text{ min.}) \text{ average time in the waiting line per customer}$$

$$U = \frac{\lambda}{\mu}$$

$$= \frac{24}{30}$$

$$= .80 \text{ probability that the server will be busy and the customer must wait}$$

$$I = 1 - U$$

$$= 1 - .80$$

$$= .20 \text{ probability that the server will be idle and a customer can be served}$$

Several important aspects of both the general model and this particular example will now be discussed in greater detail.

Queuing system operating statistics are steady state, or constant, over time.

First, the operating characteristics are averages. Also, they are assumed to be **steady-state** averages. Steady state is a constant average level that a system realizes after a period of time. For a queuing system, the steady state is represented by the average operating statistics, also determined over a period of time.

Related to this condition is the fact that the utilization factor, U, must be less than one:

$$U < 1$$

or

$$\frac{\lambda}{\mu} < 1.0$$

and

$$\lambda < \mu$$

In other words, the ratio of the arrival rate to the service rate must be less than one, which also means *the service rate must be greater than the arrival rate* if this model is to be used. The server must be able to serve customers faster than they come into the store in the long run, or the waiting line will grow to an infinite size, and the system will never reach a steady state.

The Effect of Operating Characteristics on Managerial Decisions

Now let us consider the operating characteristics of our example as they relate to management decisions. The arrival rate of 24 customers per hour means that, on average, a customer arrives every 2.5 minutes (i.e., 1/24 × 60 minutes). This indicates that the store is very busy. Because of the nature of the store, customers purchase few items and expect quick service. Customers expect to spend a relatively large amount of time in a supermarket because typically they make larger purchases. But customers who shop at a convenience market do so, at least in part, because it is quicker than a supermarket.

Given customers' expectations, the store's manager believes that it is unacceptable for a customer to wait 8 minutes and spend a total of 10 minutes in the queuing system (not including the actual shopping time). The manager wants to test several alternatives for reducing customer waiting time: (1) the addition of another employee to pack up the purchases and (2) the addition of a new checkout counter.

ALTERNATIVE I: THE ADDITION OF AN EMPLOYEE The addition of an extra employee will cost the store manager $150 per week. With the help of the national office's marketing research group, the manager has determined that for each minute that average customer waiting time is reduced, the store avoids a loss in sales of $75 per week. (That is, the store loses money when customers leave prior to shopping because of the long line or when customers do not return.)

If a new employee is hired, customers can be served in less time. In other words, the service rate, which is the number of customers served per time period, will *increase*. The previous service rate was

$$\mu = 30 \text{ customers served per hour}$$

The addition of a new employee will increase the service rate to

$$\mu = 40 \text{ customers served per hour}$$

It will be assumed that the arrival rate will remain the same ($\lambda = 24$ per hour) because the increased service rate will not increase arrivals but instead will minimize the loss of customers. (However, it is not illogical to assume that an increase in service might increase arrivals in the long run.)

Given the new λ and μ values, the operating characteristics can be recomputed as follows:

$$P_0 = \left(1 - \frac{\lambda}{\mu}\right) = \left(1 - \frac{24}{40}\right)$$
$$= .40 \text{ probability of no customers in the system}$$

$$L = \frac{\lambda}{\mu - \lambda} = \frac{24}{40 - 24}$$
$$= 1.5 \text{ customers, or average, in the queuing system}$$

$$L_q = \frac{\lambda^2}{\mu(\mu - \lambda)} = \frac{(24)^2}{40(16)}$$
$$= 0.90 \text{ customer, on average, in the waiting line}$$

$$W = \frac{1}{\mu - \lambda} = \frac{1}{40 - 24}$$
$$= 0.063 \text{ hr. } (3.75 \text{ min.}) \text{ average time in the system per customer}$$

$$W_q = \frac{\lambda}{\mu(\mu - \lambda)} = \frac{24}{40(16)}$$
$$= 0.038 \text{ hr. } (2.25 \text{ min.}) \text{ average time in the waiting line per customer}$$

$$U = \frac{\lambda}{\mu} = \frac{24}{40}$$
$$= .60 \text{ probability that the customer must wait}$$

$$I = 1 - U = 1 - .60$$
$$= .40 \text{ probability that the server will be idle and a customer can be served}$$

Remember that these operating characteristics are *averages* that result over a period of time; they are not absolutes. In other words, customers who arrive at the Fast Shop Market checkout counter will not find 0.90 customer in line. There could be no customers, or one, two, or three customers, for example. The value 0.90 is simply an average that occurs over time, as do the other operating characteristics.

The average waiting time per customer has been reduced from 8 minutes to 2.25 minutes, a significant amount. The savings (that is, the decrease in lost sales) is computed as follows:

$$8.00 \text{ min.} - 2.25 \text{ min.} = 5.75 \text{ min.}$$
$$5.75 \text{ min.} \times \$75/\text{min.} = \$431.25$$

Because the extra employee costs management $150 per week, the total savings will be

$$\$431.25 - \$150 = \$281.25 \text{ per week}$$

The store manager would probably welcome this savings and consider the preceding operating statistics preferable to the previous ones for the condition where the store had only one employee.

ALTERNATIVE II: THE ADDITION OF A NEW CHECKOUT COUNTER Next we will consider the manager's alternative of constructing a new checkout counter. The total cost of this project would be $6,000, plus an extra $200 per week for an additional cashier.

The new checkout counter would be opposite the present counter (so that the servers would have their backs to each other in an enclosed counter area). There would be several display cases and racks between the two lines so that customers waiting in line would not move back and forth between the lines. (Such movement, called *jockeying*, would invalidate the queuing formulas we already developed.) We will assume that the customers would divide themselves equally between the two lines, so the arrival rate for each line would be half of the prior arrival rate for a single checkout counter. Thus, the new arrival rate for each checkout counter is

$$\lambda = 12 \text{ customers per hour}$$

and the service rate remains the same for each of the counters:

$$\mu = 30 \text{ customers served per hour}$$

Substituting this new arrival rate and the service rate into our queuing formulas results in the following operating characteristics:

$$P_0 = .60 \text{ probability of no customers in the system}$$
$$L = 0.67 \text{ customer in the queuing system}$$
$$L_q = 0.27 \text{ customer in the waiting line}$$
$$W = 0.055 \text{ hr. } (3.33 \text{ min.}) \text{ per customer in the system}$$
$$W_q = 0.022 \text{ hr. } (1.33 \text{ min.}) \text{ per customer in the waiting line}$$
$$U = .40 \text{ probability that a customer must wait}$$
$$I = .60 \text{ probability that a server will be idle and a customer can be served}$$

Using the same sales savings of $75 per week for each minute's reduction in waiting time, we find that the store would save

$$8.00 \text{ min.} - 1.33 \text{ min.} = 6.67 \text{ min.}$$
$$6.67 \text{ min.} \times \$75/\text{min.} = \$500.00 \text{ per week}$$

Next we subtract the $200 per week cost for the new cashier from this amount saved:

$$\$500 - 200 = \$300$$

Because the capital outlay of this project is $6,000, it would take 20 weeks ($6,000/$300 = 20 weeks) to recoup the initial cost (ignoring the possibility of interest on the $6,000). Once the cost has been recovered, the store would save $18.75 ($300.00 − 281.25) more per week by adding a new checkout counter rather than simply hiring an extra employee. However, we must not disregard the fact that during the 20-week cost recovery period, the $281.25 savings incurred by simply hiring a new employee would be lost.

Table 13.1 presents a summary of the operating characteristics for each alternative. For the store manager, both of these alternatives seem preferable to the original conditions, which resulted in a lengthy waiting time of 8 minutes per customer. However, the manager might have a difficult time selecting between the two alternatives. It might be appropriate to consider other factors besides waiting time. For example, employee idle time is .40 with the first alternative and .60 with the second, which seems to be a significant difference. An additional factor is the loss of space resulting from a new checkout counter.

TABLE 13.1
Operating characteristics for each alternative system

Operating Characteristics	Present System	Alternative I	Alternative II
L	4.00 customers	1.50 customers	0.67 customer
L_q	3.20 customers	0.90 customer	0.27 customer
W	10.00 min.	3.75 min.	3.33 min.
W_q	8.00 min.	2.25 min.	1.33 min.
U	.80	.60	.40

However, the final decision must be based on the manager's own experience and perceived needs. As we have noted previously, the results of queuing analysis provide information for decision making but do not result in an actual recommended decision, as an optimization model would.

As the level of service improves, the cost of service increases.

Our two example alternatives illustrate the cost trade-offs associated with improved service. As the level of service increases, the corresponding cost of this service also increases. For example, when we added an extra employee in alternative I, the service was improved, and the cost of providing service also increased. But when the level of service was increased, the costs associated with customer waiting decreased. Maintaining an appropriate level of service should minimize the sum of these two costs as much as possible. This cost trade-off relationship is summarized in Figure 13.2. As the level of service increases, the cost of service goes up and the waiting cost goes down. The sum of these costs results in a total cost curve, and the level of service that should be maintained is the level where this total cost curve is at a minimum. (However, this does not mean we can determine an exact optimal minimum cost solution because the service and waiting characteristics we can determine are averages and thus uncertain.)

FIGURE 13.2
Cost trade-off for service levels

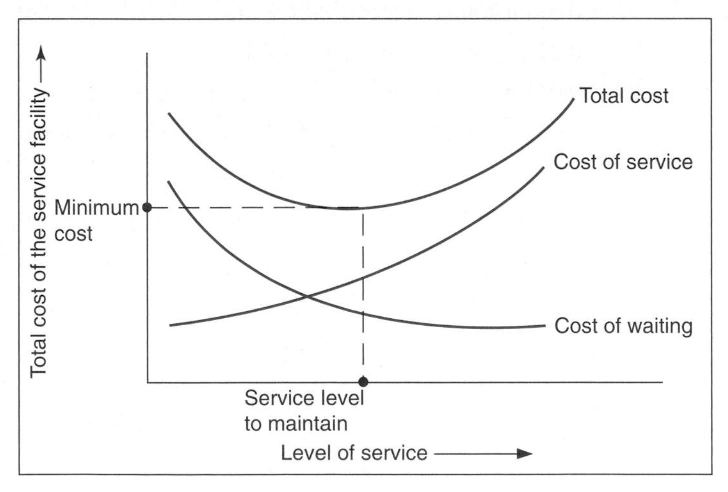

Computer Solution of the Single-Server Model with Excel and Excel QM

Excel spreadsheets can be used to solve queuing problems, although someone must enter all the queuing formulas into spreadsheet cells. Exhibit 13.1 shows a spreadsheet set up to solve the single-server model for the original Fast Shop Market example. Notice that the arrival rate is entered in cell D3, the service rate is entered in cell D4, and our single-server model queuing formulas are entered in cells D6 to D10. For example, cell D7 includes the formula for L_q, the average number of customers in the queue, shown on the formula bar at the top of the spreadsheet.

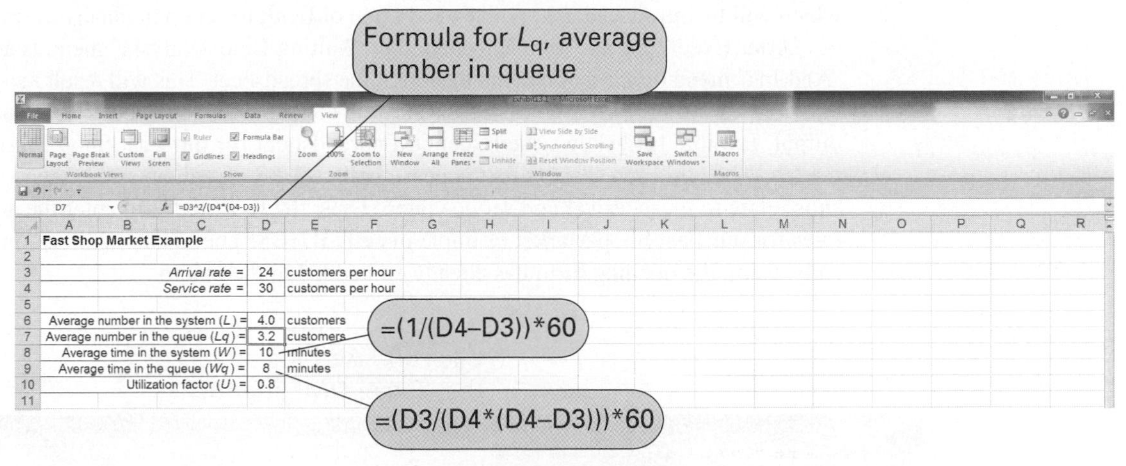

Management Science Application

Reducing Perceived Waiting Time at Bank of America

In a market research study, Bank of America discovered that when a customer stands in line, waiting for teller service for more than about 3 minutes, a gap develops between the actual waiting time and the customer's perceived waiting time. For example, if customers wait for 2 minutes, they feel like they've been waiting for 2 minutes; however, if they have been waiting for 5 minutes, it may seem more like 10 minutes to them. Bank of America also learned from prior studies that long waits have a direct relationship to customer satisfaction. Furthermore, the bank knew from previous psychological studies that if people were distracted with non-boring activities, time would seem to pass more quickly. As a result, Bank of America undertook an experiment to see if customers' perceived waiting time could be reduced if it placed televisions above the tellers in a bank branch lobby to entertain customers waiting in line. The bank installed televisions tuned to CNN in a typical bank branch and meaured actual versus perceived waiting times. The results showed that the amount of waiting time customers overestimated dropped from 32% to 15% when compared to a bank without television. To measure the financial impact of these results, the bank employed a customer satisfaction index (based on a customer survey); every one-point improvement in the index results in $1.40 in increased annual revenue per household customer. The experimental results indicated that the projected reductions in perceived waiting times would result in a 5.9-point increase in the customer satisfaction index. As a result, a bank branch with 10,000 household customers could expect an increase in annual revenue of $82,600, while a bank with only a few thousand household customers could expect to recoup the estimated $10,000 one-time cost of installing televisions in less than a year.

AP Photo/Steven Senne

Source: Based on S. Thomke, "R&D Comes to Services: Bank of America's Pathbreaking Experiments," *Harvard Business Review* 81 (April 4, 2003): 70–79.

Excel QM includes a spreadsheet macro for waiting line (i.e., queuing) analysis. Excel QM includes all the different queuing models that will be presented in this chapter. The queuing macro in Excel QM is one of the more useful ones in this text because some of the queuing formulas, which will be introduced are complex and often difficult to set up manually in spreadsheet cells.

After Excel QM has been activated, the "Waiting Line Analysis" menu is accessed from the "Add-Ins" menu on the menu bar at the top of the spreadsheet. This will result in a Spreadsheet Initialization window, which will enable us to title the problem and specify some of the spreadsheet output. Exhibit 13.2 shows the Excel QM spreadsheet for our Fast Shop Market single-server model. When this spreadsheet first appears, cells **B7:B8** contain example data values, and the results relate to these arrival and service rates. Thus, the first step is to enter the arrival and service rates for our Fast Shop Market example in cells **B7:B8**. The results will be computed automatically, using the queuing formulas already embedded in the macro.

EXHIBIT 13.2

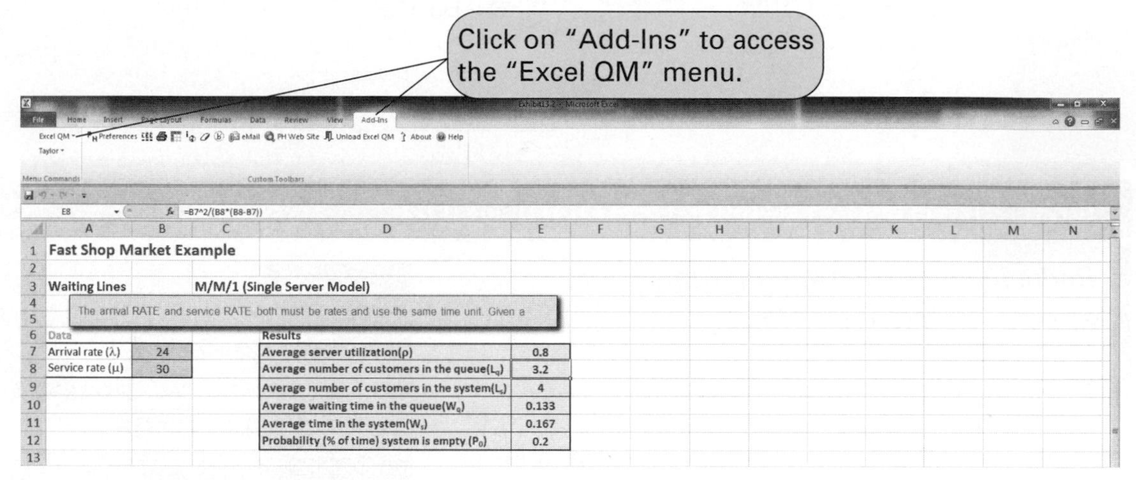

Computer Solution of the Single-Server Model with QM for Windows

QM for Windows has a module that is capable of performing queuing analysis. As an illustration of the computerized analysis of the single-server queuing system, we will use QM for Windows to analyze the Fast Shop Market example. The model input and solution output are shown in Exhibit 13.3.

EXHIBIT 13.3

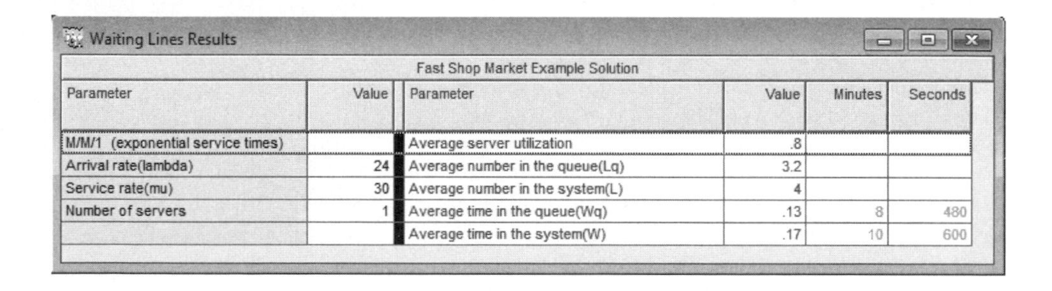

Waiting Lines Results

Fast Shop Market Example Solution

Parameter	Value	Parameter	Value	Minutes	Seconds
M/M/1 (exponential service times)		Average server utilization	.8		
Arrival rate(lambda)	24	Average number in the queue(Lq)	3.2		
Service rate(mu)	30	Average number in the system(L)	4		
Number of servers	1	Average time in the queue(Wq)	.13	8	480
		Average time in the system(W)	.17	10	600

Undefined and Constant Service Times

Constant service times occur with machinery and automated equipment.

Sometimes it cannot be assumed that a waiting line system has an arrival rate that is Poisson distributed or service times that are exponentially distributed. For example, many manufacturing operations use automated equipment or robots that have constant service times. Thus, the single-server model with Poisson arrivals and constant service times is a queuing variation that is of particular interest to manufacturing operations.

Constant service times are a special case of the single-server model with undefined service times.

The constant service time model is actually a special case of a more general variation of the single-server model in which service times cannot be assumed to be exponentially distributed. As such, service times are said to be *general*, or *undefined*. The basic queuing formulas for the operating characteristics of the undefined service time model are as follows:

$$P_0 = 1 - \frac{\lambda}{\mu}$$

$$L_q = \frac{\lambda^2\sigma^2 + (\lambda/\mu)^2}{2(1 - \lambda/\mu)}$$

$$L = L_q + \frac{\lambda}{\mu}$$

$$W_q = \frac{L_q}{\lambda}$$

$$W = W_q + \frac{1}{\mu}$$

$$U = \frac{\lambda}{\mu}$$

The key formula for undefined service times is for L_q, the number of customers in the waiting line. In this formula, μ and σ are the mean and standard deviation, respectively, for any general probability distribution with independent service times.

For the Poisson distribution, the mean is equal to the variance. This same relationship is true for Poisson service rates if service times are exponentially distributed. Thus, if we let $\sigma = \mu$ in the preceding formula for L_q for undefined service times, it becomes the same as our basic formula with exponential service times. In fact, all the queuing formulas become the same as in the basic single-server model.

As an example of the single-server model with undefined service times, consider a business firm with a single fax machine. Employees arrive randomly to use the fax machine, at an average rate of 20 per hour, according to a Poisson distribution. The time an employee spends using the machine is not defined by any probability distribution but has a mean of 2 minutes and a standard deviation of 4 minutes. The operating characteristics for this system are computed as follows:

$$P_0 = 1 - \frac{\lambda}{\mu}$$

$$P_0 = 1 - \frac{20}{30} = .33 \text{ probability that no one is using the machine}$$

$$L_q = \frac{\lambda^2\sigma^2 + (\lambda/\mu)^2}{2(1 - \lambda/\mu)} = \frac{(20)^2(1/15)^2 + (20/30)^2}{2(1 - (20/30))}$$

$$= 3.33 \text{ employees waiting in line}$$

$$L = L_q + \frac{\lambda}{\mu} = 3.33 + (20/30)$$

$$= 4.0 \text{ employees in line and using the machine}$$

$$W_q = \frac{L_q}{\lambda} = \frac{3.33}{20}$$

$$= 0.1665 \text{ hr. (10 min.) waiting in line}$$

$$W = W_q + \frac{1}{\mu} = 0.1665 + \left(\frac{1}{30}\right)$$
$$= 0.1998 \text{ hr. } (12 \text{ min.}) \text{ in the system}$$
$$U = \lambda/\mu = 20/30 = 67\% \text{ machine utilization}$$

In the constant service time model, there is no variability in service times; $\sigma = 0$.

In the case of constant service times, there is no variability in service times (i.e., service time is the same constant value for each customer); thus, $\sigma = 0$. Substituting $\sigma = 0$ into the undefined service time formula for L_q results in the following formula for L_q with constant service times:

$$L_q = \frac{\lambda^2\sigma^2 + (\lambda/\mu)^2}{2(1 - (\lambda/\mu))}$$
$$= \frac{\lambda^2(0)^2 + (\lambda/\mu)^2}{2(1 - (\lambda/\mu))}$$
$$= \frac{(\lambda/\mu)^2}{2(1 - (\lambda/\mu))}$$
$$L_q = \frac{\lambda^2}{2\mu(\mu - \lambda)}$$

Notice that this new formula for L_q for constant service times is simply our basic single-server formula for L_q divided by two. All the remaining formulas for L, W_q, and W are the same as the single-server formulas using this formula for L_q.

The following example illustrates the single-server model with constant service times. The Petroco Service Station has an automatic car wash, and motorists purchasing gas at the station receive a discounted car wash, depending on the number of gallons of gas they buy. The car wash can accommodate one car at a time, and it requires a constant time of 4.5 minutes for a wash. Cars arrive at the car wash at an average rate of 10 per hour (Poisson distributed). The service station manager wants to determine the average length of the waiting line and the average waiting time at the car wash.

First, determine λ and μ such that they are expressed as rates:

$$\lambda = 10 \text{ cars per hr.}$$
$$\mu = 60/4.5 = 13.3 \text{ cars per hr.}$$

Substituting λ and μ into the queuing formulas for constant service time,

$$L_q = \frac{\lambda^2}{2\mu(\mu - \lambda)} = \frac{(10)^2}{2(13.3)(13.3 - 10)}$$
$$= 1.14 \text{ cars waiting}$$
$$W_q = \frac{L_q}{\lambda} = \frac{1.14}{10}$$
$$= 0.114 \text{ hr. } (6.84 \text{ min.}) \text{ waiting in line}$$

Computer Solution of the Constant Service Time Model with Excel

Exhibit 13.4 shows an Excel spreadsheet set up to solve our Petroco Service Station example with constant service times. As in Exhibit 13.1 for the Fast Shop Market example, the queuing formulas are embedded in individual cells. For example, the formula for L_q, the average number in the queue, is entered in cell D6 and is shown on the formula bar at the top of the spreadsheet.

EXHIBIT 13.4

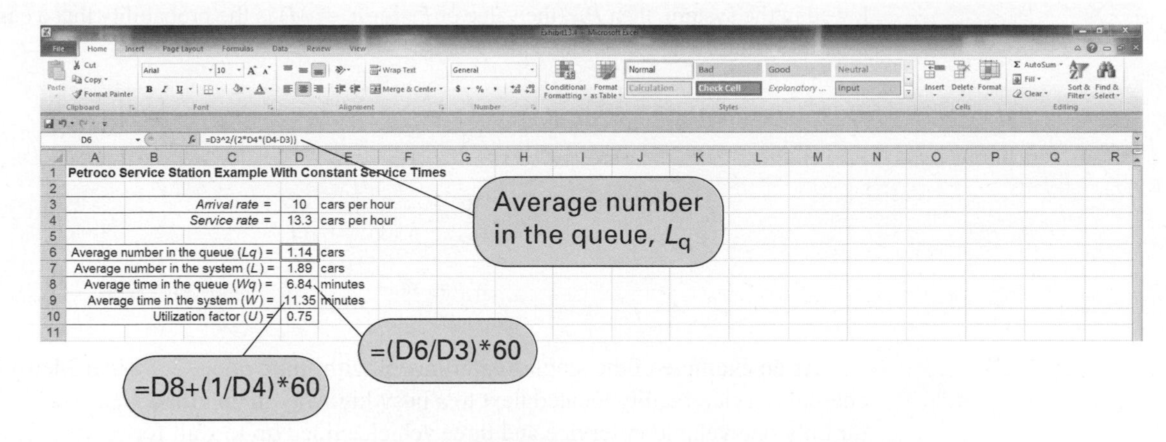

Computer Solution of the Undefined and Constant Service Time Models with QM for Windows

QM for Windows can be used to solve the single-server model with undefined (referred to as "general") and constant service times similarly to the single-server model in the previous section. Exhibit 13.5 shows the QM for Windows solution output screen for our fax machine example with undefined service times (on page 609).

EXHIBIT 13.5

Waiting Lines Results

Fax Machine Example Solution

Parameter	Value	Parameter	Value	Minutes	Seconds
M/G/1 (general service times)		Average server utilization	.6667		
Arrival rate(lambda)	20	Average number in the queue(Lq)	3.3601		
Service rate(mu)	30	Average number in the system(L)	4.0267		
Number of servers	1	Average time in the queue(Wq)	.168	10.0802	604.8121
Standard deviation	.067	Average time in the system(W)	.2013	12.0802	724.8121

Finite Queue Length

In a finite queue, the length of the queue is limited.

For some waiting line systems, the length of the queue may be limited by the physical area in which the queue forms; space may permit only a limited number of customers to enter the queue. Such a waiting line is referred to as a **finite queue** and results in another variation of the single-phase, single-channel queuing model.

The basic single-server model must be modified to consider the finite queuing system. It should be noted that for this case, the service rate does not have to exceed the arrival rate ($\mu > \lambda$) in order to obtain steady-state conditions. The resulting operating characteristics, where M is the maximum number in the system, are as follows:

$$P_0 = \frac{1 - (\lambda/\mu)}{1 - (\lambda/\mu)^{M+1}}$$

$$P_n = (P_0)\left(\frac{\lambda}{\mu}\right)^n \text{ for } n \leq M$$

$$L = \frac{\lambda/\mu}{1 - (\lambda/\mu)} - \frac{(M+1)(\lambda/\mu)^{M+1}}{1 - (\lambda/\mu)^{M+1}}$$

Because P_n is the probability of n units in the system, if we define M as the maximum number allowed in the system, then P_M (the value of P_n for $n = M$) is the probability that a customer will not join the system. The remaining equations are

$$L_q = L - \frac{\lambda(1 - P_M)}{\mu}$$

$$W = \frac{L}{\lambda(1 - P_M)}$$

$$W_q = W - \frac{1}{\mu}$$

As an example of the single-server model with finite queue, consider Metro Quick Lube, a one-bay service facility located next to a busy highway in an urban area. The facility has space for only one vehicle in service and three vehicles lined up to wait for service. There is no space for cars to line up on the busy adjacent highway, so if the waiting line is full (three cars), prospective customers must drive on.

The mean time between arrivals for customers seeking lubrication service is 3 minutes. The mean time required to perform the lube operation is 2 minutes. Both the interarrival times and the service times are exponentially distributed. As stated previously, the maximum number of vehicles in the system is four. The operating characteristics are computed as follows:

$$\lambda = 20$$
$$\mu = 30$$
$$M = 4$$

First, we will compute the probability that the system is full and the customer must drive on, P_M. However, this first requires the determination of P_0, as follows:

$$P_0 = \frac{1 - (\lambda/\mu)}{1 - (\lambda/\mu)^{M+1}} = \frac{1 - (20/30)}{1 - (20/30)^5}$$

$$= .38 \text{ probability of no cars in the system}$$

$$P_M = (P_0)\left(\frac{\lambda}{\mu}\right)^{n=M} = (.38)\left(\frac{20}{30}\right)^4$$

$$= .076 \text{ probability that four cars are in the system, it is full, and a customer must drive on}$$

Next, to compute the average queue length, L_q, the average number of cars in the system, L, must be computed, as follows:

$$L = \frac{\lambda/\mu}{1 - (\lambda/\mu)} - \frac{(M + 1)(\lambda/\mu)^{M+1}}{1 - (\lambda/\mu)^{M+1}}$$

$$L = \frac{20/30}{1 - (20/30)} - \frac{(5)(20/30)^5}{1 - (20/30)^5}$$

$$= 1.24 \text{ cars in the system}$$

$$L_q = L - \frac{\lambda(1 - P_M)}{\mu} = 1.24 - \frac{20(1 - .076)}{30}$$

$$= 0.62 \text{ car waiting}$$

Management Science Application

Providing Telephone Order Service in the Retail Catalog Business

© Image Source/Corbis

In the TIME OUT box for A. K. Erlang on page 600, it was noted that the origin of queuing analysis was in telephone congestion problems in the early 1900s. Today, queuing analysis is still an important tool in analyzing telephone service in catalog phone sales, one of the largest retail businesses in the United States. In a recent year, one of the largest catalog sales companies, L.L.Bean, with over $1 billion in annual sales, received more than 12 million customer calls. During the peak holiday season, its 3,000 customer service representatives took more than 140,000 customer calls on its busiest days. Lands' End, the 15th largest mail-order company in the United States, with sales of over $1.3 billion, handles more than 15 million phone calls each year. On an average day, its 300-plus phone lines handle between 40,000 and 50,000 calls, and during the weeks prior to Christmas, it expands to more than 1,100 phone lines, to handle more than 100,000 phone calls daily. One of the key factors in maintaining a successful catalog phone order system is to provide prompt service; if customers have to wait too long before talking to a customer service representative, they may hang up and not call back. Catalog companies such as L.L.Bean and Lands' End often use queuing analysis to make a number of decisions related to order processing, including the number of telephone trunk lines and customer service representatives they need during various hours of the day and days of the year, the amount of computing capacity they need to handle call volume, the number of workstations and the amount of equipment needed, and the number of full-time and part-time customer service representatives to hire and train.

To compute the average waiting time, W_q, the average time in the system, W, must be computed first:

$$W = \frac{L}{\lambda(1 - P_M)} = \frac{1.24}{20(1 - .076)}$$
$$= 0.067 \text{ hr. } (4.03 \text{ min.}) \text{ waiting in the system}$$
$$W_q = W - \frac{1}{\mu}$$
$$= 0.067 - \frac{1}{30}$$
$$= 0.033 \text{ hr. } (2.03 \text{ min.}) \text{ waiting in line}$$

Computer Solution of the Finite Queue Model with Excel

The Excel spreadsheet solution to the Metro Quick Lube example problem with a finite queue is shown in Exhibit 13.6. Notice that the formula for P_0 in cell D7 is shown on the formula bar at the top of the spreadsheet. The formula for L in cell D9 is shown in the callout box.

Note that Excel QM also has a spreadsheet macro for the finite queue model that can be accessed similarly to the single-server model in Exhibit 13.2.

Computer Solution of the Finite Queue Model with QM for Windows

The single-server model with finite queue can be solved with QM for Windows. Exhibit 13.7 shows the model solution screen for our Metro Quick Lube example.

EXHIBIT 13.6

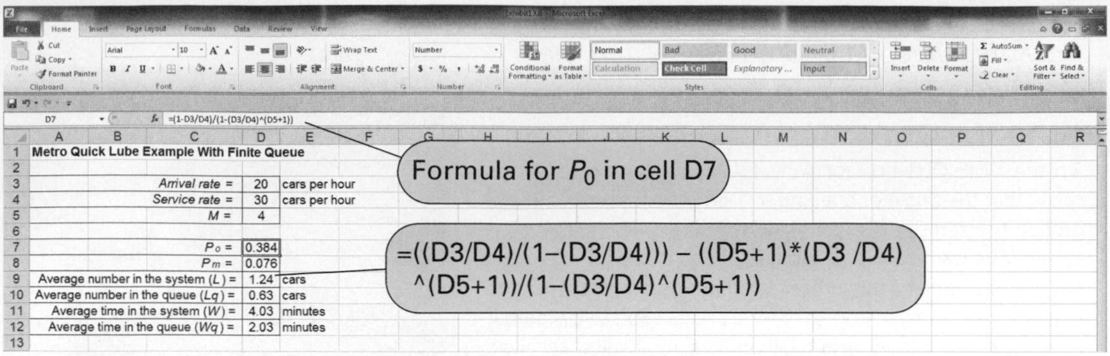

D7 =(1-D3/D4)/(1-(D3/D4)^(D5+1))

	A	B	C	D	E
1	**Metro Quick Lube Example With Finite Queue**				
2					
3			*Arrival rate* =	20	cars per hour
4			*Service rate* =	30	cars per hour
5			*M* =	4	
6					
7			*Po* =	0.384	
8			*Pm* =	0.076	
9	Average number in the system (*L*) =			1.24	cars
10	Average number in the queue (*Lq*) =			0.63	cars
11	Average time in the system (*W*) =			4.03	minutes
12	Average time in the queue (*Wq*) =			2.03	minutes
13					

Formula for P_0 in cell D7

=((D3/D4)/(1–(D3/D4))) – ((D5+1)*(D3 /D4)^(D5+1))/(1–(D3/D4)^(D5+1))

EXHIBIT 13.7

Waiting Lines Results

Metro Quick Lube Example Solution

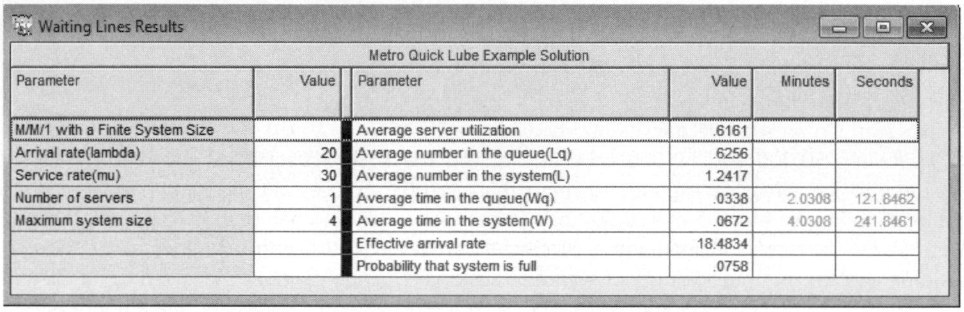

Parameter	Value	Parameter	Value	Minutes	Seconds
M/M/1 with a Finite System Size		Average server utilization	.6161		
Arrival rate(lambda)	20	Average number in the queue(Lq)	.6256		
Service rate(mu)	30	Average number in the system(L)	1.2417		
Number of servers	1	Average time in the queue(Wq)	.0338	2.0308	121.8462
Maximum system size	4	Average time in the system(W)	.0672	4.0308	241.8461
		Effective arrival rate	18.4834		
		Probability that system is full	.0758		

Finite Calling Population

*With a **finite calling population** the customers from which arrivals originate are limited.*

For some waiting line systems there is a specific, limited number of potential customers that can arrive at the service facility. This is referred to as a **finite calling population**. As an example of this type of system, consider the Wheelco Manufacturing Company.

Wheelco Manufacturing operates a shop that includes 20 machines. Due to the type of work performed in the shop, there is a lot of wear and tear on the machines, and they require frequent repair. When a machine breaks down, it is tagged for repair, with the date of the breakdown noted, and a repair person is called. The company has one senior repair person and an assistant. They repair the machines in the same order in which they break down (a first-in, first-out queue discipline). Machines break down according to a Poisson distribution, and the service times are exponentially distributed.

The finite calling population for this example is the 20 machines in the shop, which we will designate as N.

The single-server model with a Poisson arrival and exponential service times and a finite calling population has the following set of formulas for determining operating characteristics. λ in this model is the arrival rate for each member of the population:

$$P_0 = \frac{1}{\displaystyle\sum_{n=0}^{N} \frac{N!}{(N-n)!}\left(\frac{\lambda}{\mu}\right)^n}, \quad \text{where } N = \text{population size}$$

$$P_n = \frac{N!}{(N-n)!}\left(\frac{\lambda}{\mu}\right)^n P_0, \quad \text{where } n = 1, 2, \ldots, N$$

$$L_q = N - \left(\frac{\lambda + \mu}{\lambda}\right)(1 - P_0)$$

$$L = L_q + (1 - P_0)$$

$$W_q = \frac{L_q}{(N - L)\lambda}$$

$$W = W_q + \frac{1}{\mu}$$

The formulas for P_0 and P_n are both relatively complex and can be cumbersome to compute manually. As a result, tables are often used to compute these values, given λ and μ. Alternatively, the queuing module in QM for Windows has the capability to solve the finite calling population model, which we will demonstrate.

Returning to our example of the Wheelco Manufacturing Company, each machine operates an average of 200 hours before breaking down and a repair person is called. The average time to repair a machine is 3.6 hours. The breakdown rate is Poisson distributed, and the service time is exponentially distributed. The company would like an analysis of machine idle time due to breakdowns to determine whether the present repair staff is sufficient.

Using the formulas we developed for the single-server model with finite calling population, we can determine the operating characteristics for the machine repair system, as follows:

$$\lambda = 1/200 \text{ hr.} = .005 \text{ per hr.}$$

$$\mu = 1/3.6 \text{ hr.} = .2778 \text{ per hr.}$$

$$N = 20 \text{ machines}$$

$$P_0 = \frac{1}{\sum_{n=0}^{N} \frac{N!}{(N-n)!}\left(\frac{\lambda}{\mu}\right)^n}$$

$$= \frac{1}{\sum_{n=0}^{20} \frac{20!}{(20-n)!}\left(\frac{.005}{.2778}\right)^n}$$

$$P_0 = .649$$

$$L_q = N - \left(\frac{\lambda + \mu}{\lambda}\right)(1 - P_0) = 20 - \left(\frac{.005 + .2778}{.005}\right)(1 - .652)$$

$$= .169 \text{ machine waiting}$$

$$L = L_q + (1 - P_0) = .169 + (1 - .652)$$

$$= .520 \text{ machine in the system}$$

$$W_q = \frac{L_q}{(N - L)\lambda} = \frac{.169}{(20 - .520)(.005)}$$

$$= 1.74 \text{ hr. waiting for repair}$$

$$W = W_q = \frac{1}{\mu} = 1.74 + \frac{1}{.2778}$$

$$= 5.33 \text{ hr. in the system}$$

These results show that the repair person and assistant are busy 35% of the time repairing machines. Of the 20 machines, an average of .52, or 2.6%, are broken down, waiting for repair, or under repair. Each broken-down machine is idle (broken down, waiting for repair, or under repair) an average of 5.33 hours. Thus the system seems adequate.

Computer Solution of the Finite Calling Population Model with Excel and Excel QM

The finite population queuing model can be tedious to solve by using an Excel spreadsheet because of the complexity of entering the formula for P_0 in the spreadsheet. An array must be created for the N components required by the summation in the denominator of the formula for P_0. Exhibit 13.8 shows the Excel spreadsheet for the Wheelco Manufacturing Company example, with this summation array in cells **F4:G26**.

EXHIBIT 13.8

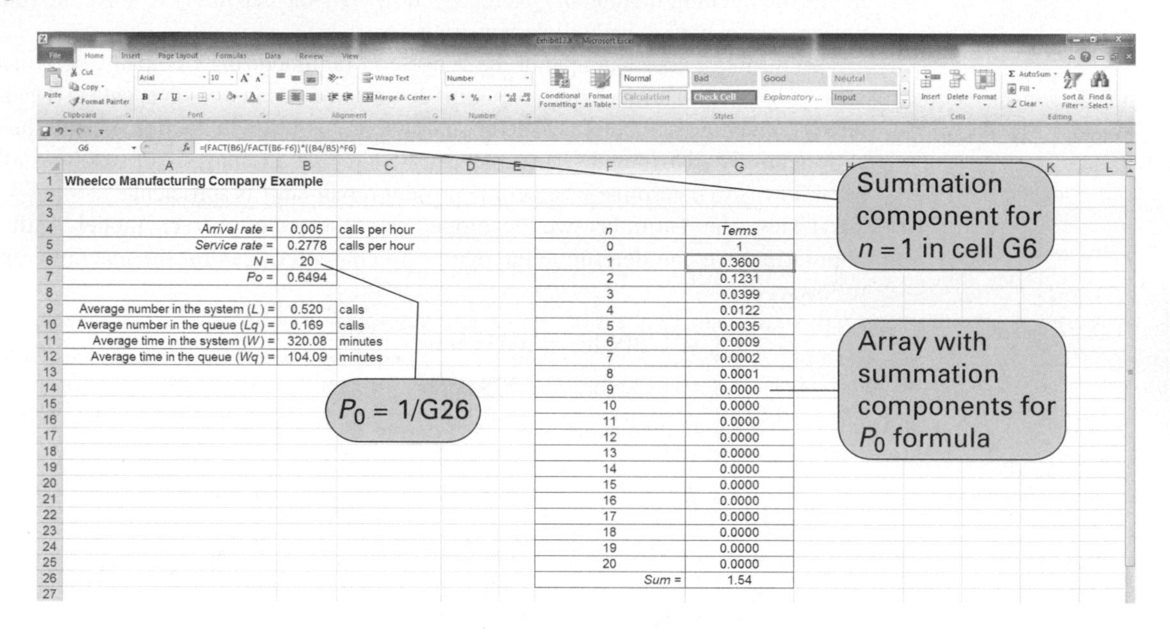

This is a model in which the spreadsheet macro in Excel QM is especially useful. The finite population model is accessed from the "Add-Ins" menu on the menu bar at the top of the spreadsheet. After selecting "Spreadsheet Initialization," the resulting spreadsheet for our Wheelco Manufacturing Company example is shown in Exhibit 13.9. As in our other Excel QM examples, the spreadsheet initially includes example data values. The first step is to enter the parameters for our own example in cells **B7:B9** as shown in Exhibit 13.9.

EXHIBIT 13.9

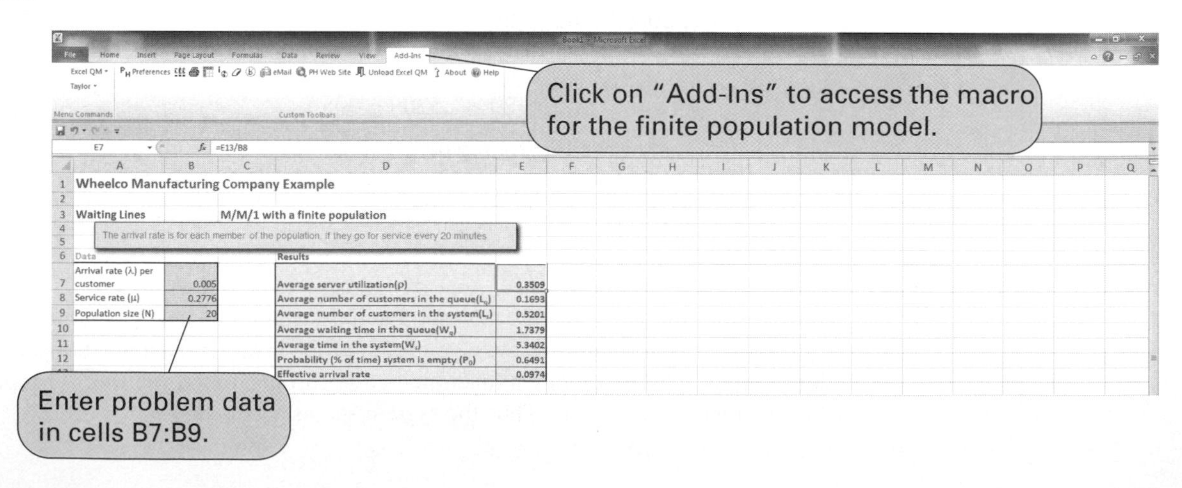

Computer Solution of the Finite Calling Population Model with QM for Windows

QM for Windows is especially useful for solving the finite calling population model because the formulas are complex and their manual solution can be tedious and time-consuming. Exhibit 13.10 includes the QM for Windows solution screen for the Wheelco Manufacturing Company example.

EXHIBIT 13.10

Waiting Lines Results

Wheelco Manufacturing Company Example Solution

Parameter	Value	Parameter	Value	Minutes	Seconds
M/M/1 with a Finite Population		Average server utilization	.3506		
Arvl rt PER CUSTOMER	.005	Average number in the queue(Lq)	.169		
Service rate(mu)	.2778	Average number in the system(L)	.5196		
Number of servers	1	Average time in the queue(Wq)	1.7349	104.0937	6,245.62
Population size	20	Average time in the system(W)	5.3346	320.0764	19,204.58
		Probability (% of time) system is empty (P0)	.6494		
		Effective arrival rate	.0974		

The Multiple-Server Waiting Line

In multiple-server models, two or more independent servers in parallel serve a single waiting line.

Slightly more complex than the single-server queuing system is the single waiting line being serviced by more than one server (i.e., multiple servers). Examples of this type of waiting line include an airline ticket and check-in counter where passengers line up in a single line, waiting for one of several agents for service, and a post office line, where customers in a single line wait for service from several postal clerks. Figure 13.3 illustrates this type of **multiple-server queuing system**.

FIGURE 13.3
A multiple-server waiting line

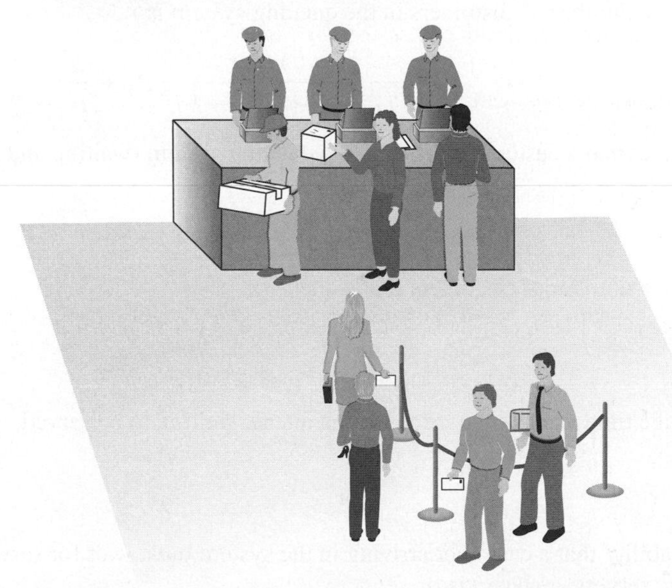

As an example of this type of system, consider the customer service department of the Biggs Department Store. The customer service department of the store has a waiting room in which chairs are placed along the wall, in effect forming a single waiting line. Customers come to this area with questions or complaints or to clarify matters regarding credit card bills. The customers are served by three store representatives, each located in a partitioned stall. Customers are served on a first-come, first-served basis.

The store management wants to analyze this queuing system because excessive waiting times can make customers angry enough to shop at other stores. Typically, customers who come to this area have some problem and thus are impatient anyway. Waiting a long time serves only to increase their impatience.

First, the queuing formulas for a multiple-server queuing system will be presented. These formulas, like single-server model formulas, have been developed on the assumption of a *first-come, first-served queue discipline, Poisson arrivals, exponential service times*, and *an infinite calling population*. The parameters of the multiple-server model are as follows:

$c\mu > \lambda$: The total number of servers must be able to serve customers faster than they arrive.

λ = the arrival rate (average number of arrivals per time period)

μ = the service rate (average number served per time period) *per server* (channel)

c = the number of servers

$c\mu$ = the mean effective service rate for the system, which must exceed the arrival rate

The formulas for the operating characteristics of the multiple-server model are as follows. The probability that there are no customers in the system (all servers are idle) is

$$P_0 = \frac{1}{\left[\sum_{n=0}^{n=c-1} \frac{1}{n!}\left(\frac{\lambda}{\mu}\right)^n\right] + \frac{1}{c!}\left(\frac{\lambda}{\mu}\right)^c\left(\frac{c\mu}{c\mu - \lambda}\right)}$$

The probability of n customers in the queuing system is

$$P_n = \frac{1}{c!c^{n-c}}\left(\frac{\lambda}{\mu}\right)^n P_0, \quad \text{for } n > c; \; P_n = \frac{1}{n}\left(\frac{\lambda}{\mu}\right)^n P_0, \quad \text{for } n \leq c$$

The average number of customers in the queuing system is

$$L = \frac{\lambda\mu(\lambda/\mu)^c}{(c-1)!(c\mu - \lambda)^2} P_0 + \frac{\lambda}{\mu}$$

The average time a customer spends in the queuing system (waiting and being served) is

$$W = \frac{L}{\lambda}$$

The average number of customers in the queue is

$$L_q = L - \frac{\lambda}{\mu}$$

The average time a customer spends in the queue, waiting to be served, is

$$W_q = W - \frac{1}{\mu} = \frac{L_q}{\lambda}$$

The probability that a customer arriving in the system must wait for service (i.e., the probability that all the servers are busy) is

$$P_w = \frac{1}{c!}\left(\frac{\lambda}{\mu}\right)^c \frac{c\mu}{c\mu - \lambda} P_0$$

Notice in the foregoing formulas that if $c = 1$ (i.e., if there is one server), then these formulas become the single-server formulas presented previously in this chapter.

Returning to our example, let us assume that a survey of the customer service department for a 12-month period shows that the arrival rate and service rate are as follows:

$$\lambda = 10 \text{ customers per hr. arrive at the service department}$$

$$\mu = 4 \text{ customers per hr. can be served by each store representative}$$

In addition, recall that this is a three-server queuing system; therefore,

$$c = 3 \text{ store representatives}$$

Using the multiple-server model formulas, we can compute the following operating characteristics for the service department:

$$P_0 = \frac{1}{\left[\sum_{n=0}^{n=c-1} \frac{1}{n!}\left(\frac{\lambda}{\mu}\right)^n\right] + \frac{1}{c!}\left(\frac{\lambda}{\mu}\right)^c\left(\frac{c\mu}{c\mu - \lambda}\right)}$$

$$= \frac{1}{\left[\frac{1}{0!}\left(\frac{10}{4}\right)^0 + \frac{1}{1!}\left(\frac{10}{4}\right)^1 + \frac{1}{2!}\left(\frac{10}{4}\right)^2\right] + \frac{1}{3!}\left(\frac{10}{4}\right)^3 \frac{3(3)}{3(4) - 10}}$$

$$= 0.045 \text{ probability that no customers are in the service department}$$

$$L = \frac{\lambda\mu(\lambda/\mu)^c}{(c-1)!(c\mu - \lambda)^2} P_0 + \frac{\lambda}{\mu}$$

$$= \frac{(10)(4)(10/4)^3}{(3-1)![3(4) - 10]^2}(.045) + \frac{10}{4}$$

$$= 6 \text{ customers, on average, in the service department}$$

$$W = \frac{L}{\lambda}$$

$$= \frac{6}{10}$$

$$= 0.60 \text{ hr. (36 min.) average time in the service department per customer}$$

$$L_q = L - \frac{\lambda}{\mu}$$

$$= 6 - \frac{10}{4}$$

$$= 3.5 \text{ customers, on average, waiting to be served}$$

$$W_q = \frac{L_q}{\lambda}$$

$$= \frac{3.5}{10}$$

$$= 0.35 \text{ hr. (21 min.) average time waiting in line per customer}$$

$$P_w = \frac{1}{c!}\left(\frac{\lambda}{\mu}\right)^c \frac{c\mu}{c\mu - \lambda} P_0$$

$$= \frac{1}{3!}\left(\frac{10}{4}\right)^3 \frac{3(4)}{3(4) - 10}(.045)$$

$$= .703 \text{ probability that a customer must wait for service (i.e., that there are three or more customers in the system)}$$

Management Science Application

Making Sure 911 Calls Get Through at AT&T

In 2001 AT&T received complaints from customers in one of its service areas that they were not getting a dial tone at certain times of the day. AT&T considered this to be a very serious problem, especially for customers making emergency 911 calls. Based on national statistics, the number of 911 calls related to life-threatening emergencies in this service area alone would be around 90 per day. Initially it was suspected that the problem was the result of maintenance activity related to the addition of new customer lines; however, this proved not to be the case, since maintenance was performed during the day and the dial-tone problems were occurring mainly in the evening. Using queuing analysis it was subsequently discovered that the problem was likely due to long Internet dial-up calls, which typically last much longer than voice calls. An analysis of data for 4.5 million residential calls in a service area showed a distribution with a mean call of 297 seconds, which is significantly longer than 3 minutes, the generally accepted standard for average voice-call duration. It was also determined that only 6% of the calls were Internet calls. Thus, it needed to be determined whether such a small fraction of Internet calls could create the kind of congestion that would block dial tones. Examining Internet calls in detail, analysts found that calls to Internet service providers (ISPs) had a mean of 1,956 seconds (i.e., over 30 minutes). It was ultimately concluded that when congestion occurred and blocked dial tones, many circuits were likely being held by ISP dial-up connections that lasted for a long time. To ensure that customers calling 911 calls would get a dial tone AT&T developed a solution approach that encompassed automated controls so that under congested conditions new ISP calls would not be admitted on the circuit, and under extremely congested conditions, ongoing ISP calls would be terminated to make room for voice calls. The controls were developed such that the chances of terminating an ongoing ISP call and repeatedly terminating the same caller were small. The analysis also revealed that many Internet calls were being routed to ISPs through third-party switches, and this discovery enabled AT&T to more efficiently route calls to other carriers, saving approximately $15 million annually.

SIPA USA/SIPA/Newscom

Source: Based on V. Ramaswami, D. Poole, S. Ahn, S. Byer, and A. Kaplan, "Ensuring Access to Emergency Services in the Presence of Long Internet Dial-Up Calls," *Interfaces* 35, no. 5 (September–October 2005): 411–22.

The department store's management has observed that customers are frustrated by the relatively long waiting time of 21 minutes and the .703 probability of waiting. To try to improve matters, management has decided to consider the addition of an extra service representative. The operating characteristics for this system must be recomputed with $c = 4$ service representatives.

Substituting this value along with λ and μ into our queuing formulas results in the following operating characteristics:

$P_0 = .073$ probability that no customers are in the service department

$L = 3.0$ customers, on average, in the service department

$W = 0.30$ hr. (18 min.) average time in the service department per customer

$L_q = 0.5$ customer, on average, waiting to be served

$W_q = 0.05$ hr. (3 min.) average time waiting in line per customer

$P_w = .31$ probability that a customer must wait for service

As in our previous example of the single-server system, the queuing operating characteristics provide input into the decision-making process, and the decision criteria are the waiting costs and service costs. The department store management would have to consider the cost of the extra service representative, as compared to the dramatic decrease in customer waiting time from 21 minutes to 3 minutes, in making a decision.

Computer Solution of the Multiple-Server Model with Excel and Excel QM

The multiple-server model is somewhat cumbersome to set up in a spreadsheet format because of the necessity to enter some of the complex queuing formulas for this model into spreadsheet cells. Exhibit 13.11 shows a spreadsheet set up for our Biggs Department Store multiple-server example.

EXHIBIT 13.11

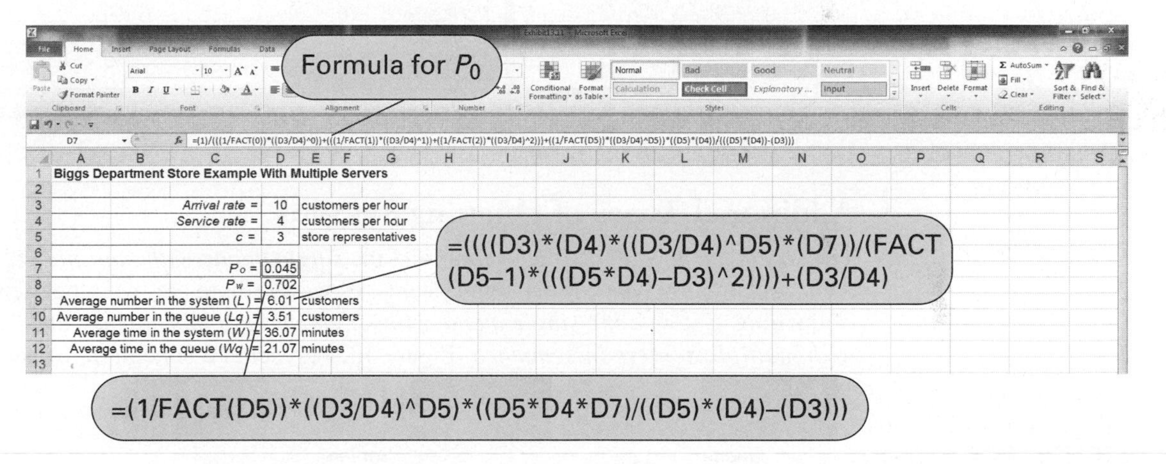

The formula for P_0 in cell D7 is shown on the formula bar at the top of the spreadsheet. As you can see, it is long and complex. The specific summation terms in P_0 for our example are entered directly into the formula. Thus, for a larger problem with a larger number of servers, additional summation terms would have to be entered. The term FACT() takes the factorial of a number or numbers in a cell. For example, FACT(1) is 1!. Notice also that two of the other more complex formulas in this model for P_w and L are shown in callout boxes attached to Exhibit 13.11.

Solution of the multiple-server model, as with some of the other more complex queuing models, can be tedious using Excel spreadsheets. Excel QM is particularly useful for solving these more complex queuing models. Exhibit 13.12 shows the Excel QM spreadsheet for our Biggs Department Store example. As with other queuing models we have solved with an Excel QM queuing macro, all the formulas are already embedded in this spreadsheet; all that is required in this case is that the arrival and service rates and number of servers be entered in cells **B7:B9**; the operating characteristics in cells **E7:E12** are computed automatically.

EXHIBIT 13.12

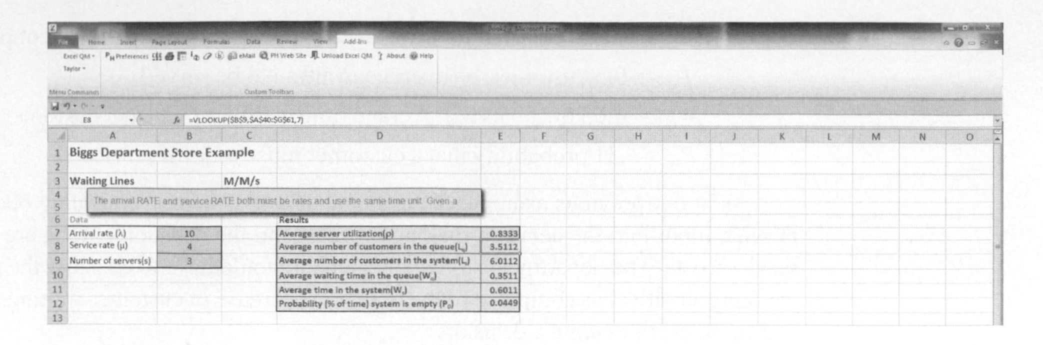

Computer Solution of the Multiple-Server Model with QM for Windows

Exhibit 13.13 shows the QM for Windows solution screen for the Biggs Department Store multiple-server model example.

EXHIBIT 13.13

Waiting Lines Results

Biggs Department Store Example Solution

Parameter	Value	Parameter	Value	Minutes	Seconds
M/M/s		Average server utilization	.8333		
Arrival rate(lambda)	10	Average number in the queue(Lq)	3.5112		
Service rate(mu)	4	Average number in the system(L)	6.0112		
Number of servers	3	Average time in the queue(Wq)	.3511	21.0674	1,264.045
		Average time in the system(W)	.6011	36.0674	2,164.045

Additional Types of Queuing Systems

The *single queue with a single server* and the *single queue with multiple servers* are two of the most common types of queuing systems. However, there are two other general categories of queuing systems: *the single queue with single servers in sequence* and the *single queue with multiple servers in sequence*. Figure 13.4 presents a schematic of each of these two systems.

FIGURE 13.4
Single queues with single and multiple servers in sequence

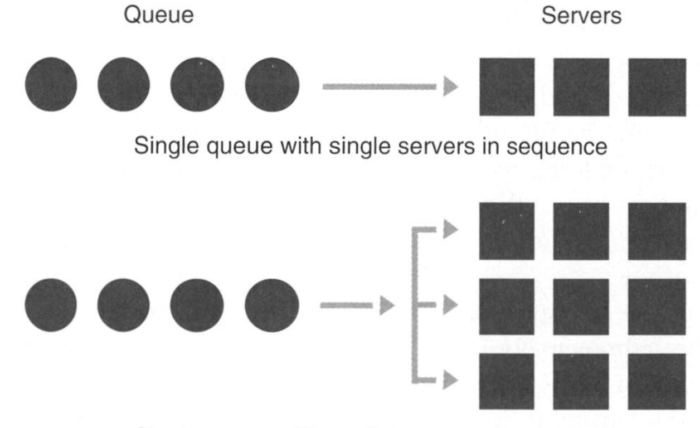

An example of a queuing system that has a single queue leading into a sequence of single servers is the personnel office of a company where job applicants line up to apply for a specific job. All the applicants wait in one area and are called alphabetically. The application process consists of moving from one interview to the next in a single sequence to take tests, answer questions, fill out forms, and so on. Another example of this type of system is an assembly line, in which products are queued up prior to being worked on by a sequenced line of machines.

If in the personnel office example an extra sequence of interviews were added, the result would be a queuing system with a single queue and multiple servers in sequence. Likewise, if products were lined up in a single queue prior to being worked on by machines in any one of three assembly lines, the result would be a sequence of multiple servers.

These are four general categories of queuing systems. Other items that can contribute to the variety of possible queuing systems include the following:

- Queuing systems in which customers *balk* from entering the system or leave the line if it is too long (i.e., *renege*)
- Servers who provide service on some basis other than first-come, first-served (such as alphabetical order or appointment, as in a doctor's office)
- Service times that are not exponentially distributed or are undefined or constant
- Arrival rates that are not Poisson distributed
- Jockeying (i.e., movement between queues), which can often occur when there are multiple servers, each preceded by a separate queue (such as in a bank with several tellers or in a grocery store with several cash registers)

Summary

The various forms of queuing systems can make queuing a potentially complex field of analysis. However, because queues are encountered often in our everyday life, the analysis of queues is an important and widely explored area of management science. We have considered only the fundamentals of a few basic types of queuing systems; a number of other analysis models have been developed to analyze the more complex queuing systems.

Some queuing situations, however, are so complex that it is impossible to develop an analytical model. When these situations occur, an alternative form of analysis is *simulation*, in which the real-life queuing system is simulated by a computerized mathematical model. The operating characteristics are determined by observing the simulated queuing system. This alternative technique, simulation, is the subject of the next chapter.

Example Problem Solution

The following example illustrates the analysis of both a single-server and a multiple-server queuing system, including the determination of the operating characteristics for each system.

Problem Statement

The new accounts loan officer of the Citizens Northern Savings Bank interviews all customers for new accounts. The customers desiring to open new accounts arrive at the rate of 4 per hour, according to a Poisson distribution, and the accounts officer spends an average of 12 minutes with each customer, setting up a new account.

A. Determine the operating characteristics $(P_0, L, L_q, W, W_q, \text{and } P_w)$ for this system.
B. Add an additional accounts officer to the system described in this problem so that it is now a multiple-server queuing system with two channels and determine the operating characteristics required in part A.

Solution

Step 1: Determine Operating Characteristics for the Single-Server System

$$\lambda = 4 \text{ customers per hr. arrive}$$

$$\mu = 5 \text{ customers per hr. are served}$$

$$P_0 = \left(1 - \frac{\lambda}{\mu}\right) = \left(1 - \frac{4}{5}\right)$$
$$= .20 \text{ probability of no customers in the system}$$

$$L = \frac{\lambda}{\mu - \lambda} = \frac{4}{5 - 4}$$
$$= 4 \text{ customers, on average, in the queuing system}$$

$$L_q = \frac{\lambda^2}{\mu(\mu - \lambda)} = \frac{(4)^2}{5(5 - 4)}$$
$$= 3.2 \text{ customers, on average, in the waiting line}$$

$$W = \frac{1}{\mu - \lambda} = \frac{1}{5 - 4}$$
$$= 1 \text{ hr. on average, in the system}$$

$$W_q = \frac{\lambda}{\mu(\mu - \lambda)} = \frac{(4)}{5(5 - 4)}$$
$$= 0.80 \text{ hr. (48 min.) average time in the waiting line}$$

$$P_w = \frac{\lambda}{\mu} = \frac{4}{5}$$
$$= .80 \text{ probability that the new accounts officer is busy and a customer must wait}$$

Step 2: Determine the Operating Characteristics for the Multiple-Server System

$$\lambda = 4 \text{ customers per hr. arrive}$$

$$\mu = 5 \text{ customers per hr. are served}$$

$$c = 2 \text{ servers}$$

$$P_0 = \frac{1}{\left[\sum_{n=0}^{n=c-1} \frac{1}{n!}\left(\frac{\lambda}{\mu}\right)^n\right] + \frac{1}{c!}\left(\frac{\lambda}{\mu}\right)^c\left(\frac{c\mu}{c\mu - \lambda}\right)}$$

$$= \frac{1}{\left[\frac{1}{0!}\left(\frac{4}{5}\right)^0 + \frac{1}{1!}\left(\frac{4}{5}\right)^1\right] + \frac{1}{2!}\left(\frac{4}{5}\right)^2 \frac{(2)(5)}{(2)(5) - 4}}$$
$$= .429 \text{ probability that no customers are in the system}$$

$$L = \frac{\lambda\mu\left(\frac{\lambda}{\mu}\right)^c}{(c-1)!(c\mu-\lambda)^2}P_0 + \frac{\lambda}{\mu}$$

$$= \frac{(4)(5)\left(\frac{4}{5}\right)^2}{1![(2)(5)-(4)]^2}(.429) + \frac{4}{5}$$

$= 0.952$ customer, on average, in the system

$$L_q = L - \frac{\lambda}{\mu} = .952 - \frac{4}{5}$$

$= 0.152$ customer, on average, waiting to be served

$$W = \frac{L}{\lambda} = \frac{.952}{4}$$

$= 0.238$ hr. (14.3 min.) average time in the system

$$W_q = \frac{L_q}{\lambda} = \frac{.152}{4}$$

$= 0.038$ hr. (2.3 min.) average time spent waiting in line

$$P_w = \frac{1}{c!}\left(\frac{\lambda}{\mu}\right)^c \frac{c\mu}{c\mu-\lambda}P_0 = \frac{1}{2!}\left(\frac{4}{5}\right)^2 \frac{(2)(5)}{(2)(5)-4}(.429)$$

$= .229$ probability that a customer must wait for service

Problems

1. For each of the following queuing systems, indicate whether it is a single- or multiple-server model, the queue discipline, and whether its calling population is infinite or finite.
   a. Hair salon
   b. Bank
   c. Laundromat
   d. Doctor's office
   e. Adviser's office
   f. Airport runway
   g. Service station
   h. Copy center
   i. Team trainer
   j. Web site
2. In the Fast Shop Market example in this chapter, Alternative II was to add a new checkout counter at the market. This alternative was analyzed using the single-server model. Why was the multiple-server model not used?
3. a. Is the following statement true or false? "The single-phase, single-channel model with Poisson arrivals and undefined service times will always have larger (i.e., greater) operating characteristic values (i.e., W, W_q, L, L_q) than the same model with exponentially distributed service times." Explain your answer.

b. Is the following statement true or false? "The single-phase, single-channel model with Poisson arrivals and constant service times will always have smaller (i.e., lower) operating characteristic values (i.e., W, W_q, L, L_q) than the same model with exponentially distributed service times." Explain your answer.

4. Why do waiting lines form at a service facility even though there may be more than enough service capacity to meet normal demand in the long run?

5. Provide an example of when a first-in, first-out (FIFO) rule for queue discipline would not be appropriate.

6. Under what conditions will the single-server queuing model with Poisson arrivals and undefined service times provide the same operating characteristics as the basic model with exponentially distributed service times?

7. A single-server queuing system with an infinite calling population and a first-come, first-served queue discipline has the following arrival and service rates (poisson distributed):

$$\lambda = 16 \text{ customers per hour}$$
$$\mu = 24 \text{ customers per hour}$$

Determine P_0, P_3, L, L_q, W, W_q, and U.

8. The ticket booth on the Tech campus is operated by one person, who is selling tickets for the annual Tech versus State football game on Saturday. The ticket seller can serve an average of 12 customers per hour; on average, 10 customers arrive to purchase tickets each hour (poisson distributed). Determine the average time a ticket buyer must wait and the portion of time the ticket seller is busy.

9. The Petroco Service Station has one pump for regular unleaded gas, which (with an attendant) can service 10 customers per hour. Cars arrive at the regular unleaded pump at a rate of 6 per hour. Determine the average queue length, the average time a car is in the system, and the average time a car must wait. If the arrival rate increases to 12 cars per hour, what will be the effect on the average queue length?

10. The Dynaco Manufacturing Company produces a particular product in an assembly line operation. One of the machines on the line is a drill press that has a single assembly line feeding into it. A partially completed unit arrives at the press to be worked on every 7.5 minutes, on average. The machine operator can process an average of 10 parts per hour. Determine the average number of parts waiting to be worked on, the percentage of time the operator is working, and the percentage of time the machine is idle.

11. The management of Dynaco Manufacturing Company (see Problem 10) likes to have its operators working 90% of the time. What must the assembly line arrival rate be for the operators to be as busy as management would like?

12. The Peachtree Airport in Atlanta serves light aircraft. It has a single runway and one air traffic controller to land planes. It takes an airplane 12 minutes to land and clear the runway. Planes arrive at the airport at the rate of four per hour.
    a. Determine the average number of planes that will stack up, waiting to land.
    b. Find the average time a plane must wait in line before it can land.
    c. Calculate the average time it takes a plane to clear the runway once it has notified the airport that it is in the vicinity and wants to land.
    d. The FAA has a rule that an air traffic controller can, on average, land planes a maximum of 45 minutes out of every hour. There must be 15 minutes of idle time available to relieve the tension. Will this airport have to hire an extra air traffic controller?

13. The First American Bank of Rapid City has one outside drive-up teller. It takes the teller an average of 4 minutes to serve a bank customer. Customers arrive at the drive-up window at a rate

of 12 per hour. The bank operations officer is currently analyzing the possibility of adding a second drive-up window, at an annual cost of $20,000. It is assumed that arriving cars would be equally divided between both windows. The operations officer estimates that each minute's reduction in customer waiting time would increase the bank's revenue by $2,000 annually. Should the second drive-up window be installed?

14. During registration at State University every semester, students in the college of business must have their courses approved by the college adviser. It takes the adviser an average of 2 minutes to approve each schedule, and students arrive at the adviser's office at the rate of 28 per hour.
    a. Compute L, L_q, W, W_q, and U.
    b. The dean of the college has received a number of complaints from students about the length of time they must wait to have their schedules approved. The dean feels that waiting 10.00 minutes to get a schedule approved is not unreasonable. Each assistant the dean assigns to the adviser's office will reduce the average time required to approve a schedule by 0.25 minute, down to a minimum time of 1.00 minute to approve a schedule. How many assistants should the dean assign to the adviser?

15. All trucks traveling on Interstate 40 between Albuquerque and Amarillo are required to stop at a weigh station. Trucks arrive at the weigh station at a rate of 200 per 8-hour day, and the station can weigh, on the average, 220 trucks per day.
    a. Determine the average number of trucks waiting, the average time spent waiting and being weighed at the weigh station by each truck, and the average waiting time before being weighed for each truck.
    b. If the truck drivers find out they must remain at the weigh station longer than 15 minutes, on average, they will start taking a different route or traveling at night, thus depriving the state of taxes. The state of New Mexico estimates that it loses $10,000 in taxes per year for each extra minute trucks must remain at the weigh station. A new set of scales would have the same service capacity as the present set of scales, and it is assumed that arriving trucks would line up equally behind the two sets of scales. It would cost $50,000 per year to operate the new scales. Should the state install the new set of scales?

16. In Problem 15, suppose passing truck drivers look to see how many trucks are waiting to be weighed at the weigh station. If they see four or more trucks in line, they will pass by the station and risk being caught and ticketed. What is the probability that a truck will pass by the station?

17. In Problem 14, the dean of the college of business at State University is considering the addition of a second adviser in the college advising office to serve students waiting to have their schedules approved. This new adviser could serve the same number of students per hour as the present adviser. Determine L, L_q, W, and W_q for this altered advising system. As a student, would you recommend adding the adviser?

18. The Dynaco Manufacturing Company has an assembly line that feeds two drill presses. As partially completed products come off the line, they are lined up to be worked on as drill presses become available. The units arrive at the workstation (containing both presses) at the rate of 100 per hour. Each press operator can process an average of 60 units per hour. Compute L, L_q, W, and W_q.

19. The Acme Machine Shop has five machines that periodically break down and require service. The average time between breakdowns is 4 days, distributed according to an exponential distribution. The average time to repair a machine is 1 day, distributed according to an exponential distribution. One mechanic repairs the machines in the order in which they break down.
    a. Determine the probability that the mechanic is idle.
    b. Determine the mean number of machines waiting to be repaired.
    c. Determine the mean time machines wait to be repaired.
    d. Determine the probability that three machines are not operating (are being repaired or waiting to be repaired).

20. McBurger's fast-food restaurant has a drive-through window with a single server who takes orders from an intercom and also is the cashier. The window operator is assisted by other employees who prepare the orders. Customers arrive at the ordering station prior to the drive-through window every 4.5 minutes (Poisson distributed), and the service time is 2.8 minutes (exponentially distributed). Determine the average length of the waiting line and the waiting time. Discuss the quality of the service.

21. Game World, a video game arcade at Tanglewood Mall, has just installed a new virtual reality battle game. It requires exactly 2.7 minutes to play. Customers arrive, on average, every 2.9 minutes (Poisson distributed) to play the game. How long will the line of customers waiting to play the game be, and how long, on average, must a customer wait?

22. Drivers who come to get their licenses at the Department of Motor Vehicles have their photograph taken by an automated machine that develops the photograph onto the license card and laminates the complete license. The machine requires a constant time of 4.5 minutes to prepare a completed license. If drivers arrive at the machine at the mean rate of 10 per hour (Poisson distributed), determine the average length of the waiting line and the average waiting time.

23. A vending machine at City Airport dispenses hot coffee, hot chocolate, or hot tea, in a constant service time of 20 seconds. Customers arrive at the vending machine at a mean rate of 60 per hour (Poisson distributed). Determine the average length of the waiting line and the average time a customer must wait.

24. Norfolk, Virginia, a major seaport on the East Coast, has a ship coal-loading facility. Coal trucks filled with coal arrive at the port facility at the mean rate of 149 per day (Poisson distributed). The facility operates 24 hours a day. The coal trucks are unloaded one at a time, on a first-come, first-served basis, by automated mechanical equipment that empties the trucks in a constant time of 8 minutes per truck, regardless of truck size. The port authority is negotiating with a coal company for an additional 30 trucks per day. However, the coal company will not use this port facility unless the port authority can assure it that its coal trucks will not have to wait to be unloaded at the port facility for more than 12 hours per truck, on average. Can the port authority provide this assurance?

25. The Bay City Police Department has eight patrol cars that are on constant call 24 hours per day. A patrol car requires repairs every 20 days, on average, according to an exponential distribution. When a patrol car is in need of repair, it is driven into the motor pool, which has a repair person on duty at all times. The average time required to repair a patrol car is 18 hours (exponentially distributed). Determine the average time a patrol car is not available for use and the average number of patrol cars out of service at any one time. Indicate whether the repair service seems adequate.

26. The Rowntown Cab Company has four cabs that are on duty during normal business hours. The cab company dispatcher receives requests for service every 8 minutes, on average, according to an exponential distribution. The average time to complete a trip is 20 minutes (exponentially distributed). Determine the average number of customers waiting for service and the average time a customer must wait for a cab.

27. In Problem 14, in addition to course approvals, suppose that the advisor also answers questions and provides course counseling, such that the service time can no longer be assumed to be exponentially distributed. Instead, the service time distribution is undefined (i.e., general), with a mean of 2 minutes and a standard deviation of 5 minutes. Compute L, L_q, W, and W_q and compare these results with those in part (a) in Problem 14.

28. The College Avenue Sub Shop is located in a small college town. Virtually all of its phone delivery orders are in the evening, between 5:00 P.M. and 1:00 A.M., so the shop has a delivery person for that time period. It receives an average of about 3 orders per hour (Poisson distributed). However, because of the different distances the delivery person must travel, the service time is undefined, with a mean of 15 minutes and a standard deviation of 6 minutes. Compute L, L_q, W, and W_q.

29. The Riverton Post Office has four stations for service. Customers line up in single file for service on a FIFO basis. The mean arrival rate is 40 per hour, Poisson distributed, and the mean service time per server is 4 minutes, exponentially distributed. Compute the operating characteristics for this operation. Indicate whether the operation appears to be satisfactory in terms of the following: (a) postal workers' (servers') idle time; (b) customer waiting time and/or the number waiting for service; and (c) the percentage of the time a customer can walk in and get served without waiting at all.

30. In Problem 18, the Dynaco Company has found that if more than three units (average) are waiting to be processed at any one workstation, then too much money is being tied up in work-in-process inventory (i.e., units waiting to be processed). The company estimates that (on average) each unit waiting to be processed costs $50 per day. Alternatively, operating a third press would cost $150 per day. Should the company operate a third press at this workstation?

31. Cakes baked by the Freshfood Bakery are transported from the ovens to be packaged by one of three wrappers. Each wrapper can wrap an average of 200 cakes per hour. The cakes are brought to the wrappers at the rate of 500 per hour. If a cake sits longer than 5 minutes before being wrapped, it will not be fresh enough to meet the bakery's quality control standards. Does the bakery need to hire another wrapper?

32. The associate dean of the college of business at Tech is determining which of two copiers he should lease for the college's administrative suite. A regular copier leases for $8 per hour, and it takes an employee an average of 6 minutes (exponentially distributed) to complete a copying job. A deluxe, high-speed copier leases for $16 per hour, and it requires an average of 3 minutes to complete a copying job. Employees arrive at the copying machine at a rate of 7 per hour (Poisson distributed), and an employee's time is valued at $10 per hour. Determine which copier the college should lease.

33. The Quick Wash 24-hour Laundromat has 16 washing machines. A machine breaks down every 20 days (exponentially distributed). The repair service with which the Laundromat contracts takes an average of 1 day to repair a machine (exponentially distributed). A washing machine averages $5 per hour in revenue. The Laundromat is considering a new repair service that guarantees repairs in 0.50 day, but it charges $10 more per hour than the current repair service. Should the Laundromat switch to the new repair service?

34. The Regency Hotel has enough space at its entrance for six taxicabs to line up, wait for guests, and then load passengers. Cabs arrive at the hotel every 8 minutes; if a taxi drives by the hotel and the line is full, it must drive on. Hotel guests require a taxi every 5 minutes, on average. It takes a cab driver an average of 3.5 minutes to load passengers and luggage and leave the hotel (exponentially distributed).
    a. What is the average time a cab must wait for a fare?
    b. What is the probability that the line will be full when a cab drives by, causing it to drive on?

35. The local Burger Doodle fast-food restaurant has a drive-through window. Customers in cars arrive at the window at the rate of 10 per hour (Poisson distributed). It requires an average of 4 minutes (exponentially distributed) to take and fill an order. The restaurant chain has a service goal of an average waiting time of 3 minutes.
    a. Will the current system meet the restaurant's service goal?
    b. If the restaurant is not meeting its service goal, it can add a second drive-through window that will reduce the service time per customer to 2.5 minutes. Will the additional window enable the restaurant to meet its service goal?
    c. During the 2-hour lunch period, the arrival rate of drive-in customers increases to 20 per hour. Will the two-window system be able to achieve the restaurant's service goal during this rush period?

36. From 3:00 P.M. to 8:00 P.M. the local K-Star Supermarket has a steady stream of customers. Customers finish shopping and arrive at the checkout area at a rate of 70 per hour (Poisson distributed).

It is assumed that when customers arrive at the cash registers, they will divide themselves relatively evenly so that all the checkout lines contain an equal number. The average checkout time at a register is 7 minutes (exponentially distributed). The store manager's service goal is for customers to be out of the store within 12 minutes (on average) after they complete their shopping and arrive at a cash register. How many cash registers must the store have open to achieve the manager's service goal?

37. Customers arrive to check in at the exclusive and expensive Regency Hotel's lobby at a rate of 40 per hour (Poisson distributed). The hotel normally has three clerks available at the desk to check in guests. The average time for a clerk to check in a guest is 4 minutes (exponentially distributed). Clerks at the Regency are paid $24 per hour, and the hotel assigns a goodwill cost of $2 per minute for the time a guest must wait in line. Determine whether the present check-in system is cost-effective. If it is not, recommend what hotel management should do.

38. The Riverview Clinic has two general practitioners who see patients daily. An average of six patients arrive at the clinic per hour. Each doctor spends an average of 15 minutes with a patient. The patients wait in a waiting area until one of the two doctors is able to see them. However, because patients typically do not feel well when they come to the clinic, the doctors do not believe it is good practice to have a patient wait longer than an average of 15 minutes. Should this clinic add a third doctor, and, if so, will this alleviate the waiting problem?

39. The Footrite Shoe Company is going to open a new branch at a mall, and company managers are attempting to determine how many salespeople to hire. Based on an analysis of mall traffic, the company estimates that customers will arrive at the store at a rate of 10 per hour, and from past experience at its other branches, the company knows that salespeople can serve an average of 6 customers per hour. How many salespeople should the company hire to uphold a company policy that on average the probability of a customer having to wait for service be no more than .30?

40. When customers arrive at Gilley's Ice Cream Shop, they take a number and wait to be called to purchase ice cream from one of the counter servers. From experience in past summers, the store's staff knows that customers arrive at a rate of 40 per hour on summer days between 3:00 P.M. and 10:00 P.M., and a server can serve 15 customers per hour on average. Gilley's wants to make sure that customers wait no longer than 10 minutes for service. Gilley's is contemplating keeping three servers behind the ice cream counter during the peak summer hours. Will this number be adequate to meet the waiting time policy?

41. Moore's television repair service receives an average of six TV sets per 8-hour day to be repaired. The service manager would like to be able to tell customers that they can expect 1-day service. What average repair time per set will the repair shop have to achieve to provide 1-day service, on average? (Assume that the arrival rate is Poisson distributed and repair times are exponentially distributed.)

42. In Problem 41, suppose that Moore's television repair service cannot accommodate more than 30 TV sets at a time. What is the probability that the number of TV sets on hand (under repair and waiting for service) will exceed the shop capacity?

43. Maggie Attaberry is a nurse on the evening shift from 10:00 P.M. to 6:00 A.M. at Community Hospital. She has 15 patients for whom she is responsible in her area. She averages two calls from each of her patients every evening, on average (Poisson distributed), and she must spend an average of 10 minutes (negative exponential distribution) with each patient who calls. Nurse Attaberry has indicated to her shift supervisor that, although she has not kept records, she believes her patients must wait about 10 minutes, on average, for her to respond, and she has requested that her supervisor assign a second nurse to her area. The supervisor believes 10 minutes is too long for a patient to wait, but she does not want her nurses to be idle more than 40% of the time. Determine what the supervisor should do.

44. The Escargot is a small French restaurant with six waiters and waitresses. The average service time at the restaurant for a table (of any size) is 85 minutes (Poisson distributed). The restaurant does not take reservations, and parties arrive for dinner (and stay and wait) every 18 minutes (negative exponential distribution). The restaurant owner is concerned that a lengthy waiting time might hurt its business in the long run. What are the current waiting time and queue length for the restaurant? Discuss the business implications of the current waiting time and any actions the restaurant owner might take.

45. Hudson Valley Books is a small, independent publisher of fiction and nonfiction books. Each week the publisher receives an average of seven unsolicited manuscripts to review (Poisson distributed). The publisher has 12 freelance reviewers in the area who read and evaluate manuscripts. It takes a reviewer an average of 10 days (exponentially distributed) to read a manuscript and write a brief synopsis. (Reviewers work on their own, 7 days a week.) Determine how long the publisher must wait, on average, to receive a reviewer's manuscript evaluation, how many manuscripts are waiting to be reviewed, and how busy the reviewers are.

46. Amanda Fall is starting up a new house painting business, Fall Colors. She has been advertising in the local newspaper for several months. Based on inquiries and informal surveys of the local housing market, she anticipates that she will get painting jobs at the rate of four per week (Poisson distributed). Amanda has also determined that it will take a four-person team of painters an average of 0.7 week (exponentially distributed) for a typical painting job.
    a. Determine the number of teams of painters Amanda needs to hire so that customers will have to wait no longer than 2 weeks to get their houses painted.
    b. If the average price for a painting job is $1,700 and Amanda pays a team of painters $500 per week, will she make any money?

47. Partially completed products arrive at a workstation in a manufacturing operation at a mean rate of 40 per hour (Poisson distributed). The processing time at the workstation averages 1.2 minutes per unit (exponentially distributed). The manufacturing company estimates that each unit of work-in-process inventory at the workstation costs $31 per day (on average). However, the company can add extra employees and reduce the processing time to 0.90 minute per unit, at a cost of $52 per day. Determine whether the company should continue the present operation or add extra employees.

48. The Atlantic Coast Shipping Company has a warehouse terminal in Spartanburg, South Carolina. The capacity of each terminal dock is three trucks. As trucks enter the terminal, the drivers receive numbers; and when one of the three dock spaces becomes available, the truck with the lowest number enters the vacant dock. Truck arrivals are Poisson distributed, and the unloading and loading times (service times) are exponentially distributed. The average arrival rate at the terminal is five trucks per hour, and the average service rate per dock is two trucks per hour (30 minutes per truck).
    a. Compute L, L_q, W, and W_q.
    b. The management of the shipping company is considering adding extra employees and equipment to improve the average service time per terminal dock to 25 minutes per truck. It would cost the company $18,000 per year to achieve this improved service. Management estimates that it will increase its profit by $750 per year for each minute it is able to reduce a truck's waiting time. Determine whether management should make the investment.
    c. Now suppose that the managers of the shipping company have decided that truck waiting time from part (a) is excessive and they want to reduce the waiting time. They have determined that there are two alternatives available for reducing the waiting time. They can add a fourth dock, or they can add extra employees and equipment at the existing docks, which will reduce the average service time per location from the original 30 minutes per truck to 23 minutes per truck. The costs of these alternatives are approximately equal. Management desires to implement the alternative that reduces waiting time by the greatest amount. Which alternative should be selected? (Computer solution is suggested.)

49. The Waterfall Buffet in the lower level of the National Art Gallery serves food cafeteria-style daily to visitors and employees. The buffet is self-service. From 7:00 A.M. to 9:00 A.M. customers arrive at the buffet at a rate of 10 per minute; from 9:00 A.M. to noon, at 4 per minute; from noon to 2:00 P.M., at 14 per minute; and from 2:00 P.M. to closing at 5:00 P.M., at 8 per minute. All the customers take about the same amount of time to serve themselves from the buffet. Once a customer goes through the buffet, it takes 0.4 minute to pay the cashier. The gallery does not want a customer to have to wait longer than 4 minutes to pay. How many cashiers should be working at each of the four times during the day?

50. The Clip Joint is a hairstyling salon at University Mall. Four stylists are always available to serve customers on a first-come, first-served basis. Customers arrive at an average rate of five per hour, and the stylists spend an average of 35 minutes on each customer.
   a. Determine the average number of customers in the salon, the average time a customer must wait, and the average number waiting to be served.
   b. The salon manager is considering adding a fifth stylist. Would this have a significant impact on waiting time?

51. The Delacroix Inn in Alexandria is a small, exclusive hotel with 20 rooms. Guests can call housekeeping from 8:00 A.M. to midnight for any of their service needs. Housekeeping keeps one person on duty during this time to respond to guest calls. Each room averages 0.7 call per day to housekeeping (Poisson distributed), and a guest request requires an average of 30 minutes (exponentially distributed) for the staff person to respond to and take care of. Determine the portion of time the staff person is busy and how long a guest must wait for his or her request to be addressed. Does the housekeeping system seem adequate?

52. Jim Carter builds custom furniture, primarily cabinets, bookcases, small tables, and chairs. He works on only one piece of furniture for a customer at a time. It takes him an average of 5 weeks (exponentially distributed) to build a piece of furniture. An average of 14 customers approach Jim to order pieces of furniture each year (Poisson distributed); however, Jim will take only a maximum of 8 advance orders. Determine the average time a customer must wait to receive a furniture order once it is placed and how busy Jim is. What is the probability that a customer will be able to place an order with Jim?

53. Judith Lewis is a doctoral student at State University, and she also works full-time as an academic tutor for 10 scholarshiped student athletes. She took the job, hoping it would leave her free time between tutoring to devote to her own studies. An athlete visits her for tutoring an average of every 16 hours, and she spends an average of 1.5 hours (exponentially distributed) with him or her. She is able to tutor only one athlete at a time, and athletes study while they are waiting.
   a. Determine how long a player must wait to see her and the percentage of time Judith is busy. Does the job seem to meet Judith's expectations, and does the system seem adequate to meet the athletes' needs?
   b. If the results in (a) indicate that the tutoring arrangement is ineffective, suggest an adjustment that could improve it for both the athletes and Judith.

54. TSA security agents at the security gate at the Tri-Cities Regional Airport are able to check passengers and process them through the security gate at a rate of 50 per hour (Poisson distributed). Passengers arrive at the gate throughout the day at a rate of 40 per hour (Poisson distributed). Determine the average waiting time and waiting line.

55. Tech has a student population of 30,000 and is located in a small college town in Virginia. DirectCast Cable TV has a small service staff or three installation trucks and technicians that is sufficient to handle service calls and installations for almost the entire year. However, for the month-long period right before and during the beginning of fall semester in August, when all the students return, the cable TV service is overwhelmed. During the normal year DirectCast averages about 15 service requests per day (Poisson distributed), and service calls average about

50 minutes (exponentially distributed) during their 8-hour workday. However, during the month before school starts, the demand for service calls triples, and almost all of these are requests for installations, so the average service time increases to 80 minutes. During normal demand periods, DirectCast guarantees service within a 24-hour period.

a. What will the cable service and customer waiting time be like during the month before the semester starts?
b. It is possible for DirectCast to "borrow" trucks and crews from other company offices. How many extra trucks would it need to provide its normal service guarantee?

56. In Problem 55, as an alternative to borrowing trucks and crews from other offices (which can be costly), DirectCast is considering extending the normal workday to 16 hours and paying its own technicians overtime. Would this satisfy demand to the extent that additional trucks and crews would not have to be borrowed from other offices? Discuss the financial information DirectCast would likely have to consider in making these decisions.

57. In Problem 28, when the students come back for fall semester, the Sub Shop experiences a large increase in demand for about a month before it begins to level off again, such that the number of phone orders increases to an average of 8 per hour (Poisson distributed). If the shop now assumes that service times are exponentially distributed, with a mean of 20 minutes, how many delivery people will it need to provide a reasonable service level?

58. In Problem 54 the passenger traffic arriving at the airport security gate varies significantly during the day. At times close to flight takeoffs, the traffic is very heavy, while at other times there is little or no passenger traffic through the security gate. At the times leading up to flight takeoffs, passengers arrive at a rate of 110 per hour (Poisson distributed). Suggest a waiting line system that can accommodate this level of passenger traffic.

59. The inland port at Front Royal, Virginia, is a transportation hub that primarily transfers shipping containers from trucks coming from eastern seaports to rail cars for shipment to inland destinations. Containers arrive at the inland ports at a rate of eight per hour. It takes 28 minutes (exponentially distributed) for a crane to unload a container from a truck, place it on a flatbed railcar, and secure it. Suggest a waiting line system that will effectively handle this level of container traffic at the inland port.

60. The Old Colony theme park has a new ride, the Double Cyclone. The ride holds 30 people in double roller coaster cars, and it takes 3.8 minutes to complete the ride circuit. It also takes the ride attendants another 3 minutes (with virtually no variation) to load and unload passengers. Passengers arrive at the ride during peak park hours at a rate of 4 per minute (Poisson distributed). Determine the length of the waiting line for the ride.

Case Problem

The College of Business Copy Center

The copy center at the college of business at State University has become an increasingly contentious item among the college administrators. The department heads have complained to the associate dean about the long lines and waiting times for their secretaries at the copy center. They claim that it is a waste of scarce resources for the secretaries to stand in line talking when they could be doing more productive work in the office. Alternatively, Handford Burris, the associate dean, says the limited operating budget will not allow the college to purchase a new copier, or several copiers, to relieve the problem. This standoff has been going on for several years.

To make her case for improved copying facilities, Lauren Moore, the department head for management science, assigned students a class project to gather some information about the copy center. The students were to record the arrivals at the center and the length of time it took to do a copy job once the secretary actually reached a copy machine. In addition, the students were to describe how the copy center system worked.

When the students completed the project, they turned in a report to Professor Moore. The report described the copy center as containing two machines. When secretaries arrived for a copy job, they joined a queue, which looked more like milling around to the students. But they acknowledged that the secretaries knew when it was their turn, and, in effect, they formed a single queue for the first available copy machine. Also, because copy jobs are assigned tasks, secretaries always stayed to do the job, no matter how long the line was or how long they had to wait. They never left the queue.

From the data the students gathered, Professor Moore is able to determine that secretaries arrive every 8 minutes for a copy job and that the arrival rate is Poisson distributed. Furthermore, she was able to determine that the average time it takes to complete a job is 12 minutes and that this is exponentially distributed.

Using her own personnel records and some data from the university's personnel office, Dr. Moore determines that a secretary's average salary is $8.50 per hour. From her academic calendar she adds up the actual days in the year when the college and departmental offices are open and finds there are 247. However, as she adds up working days, it occurs to her that during the summer months, the workload is much lighter, and the copy center also probably gets less traffic. The summer includes about 70 days, during which she expects the copy center traffic to be about half of what it is during the normal year, but she speculates that the average time of a copying job remains about the same.

Professor Moore next calls a local office supply firm to check the prices on copiers. A new copier of the type in the copy center now would cost $36,000. It would also require $8,000 per year for maintenance and would have a normal useful life of 6 years.

Do you think Dr. Moore will be able to convince the associate dean that a new copier machine will be cost-effective?

Case Problem

NORTHWOODS BACKPACKERS

Bob and Carol Packer operate a successful outdoor-wear store in Vermont called Northwoods Backpackers. They stock mostly cold weather outdoor items such as hiking and backpacking clothes, gear, and accessories. They established an excellent reputation throughout New England for quality products and service. Eventually, Bob and Carol noticed that more and more of their sales were to customers who did not live in the immediate vicinity but were calling in orders on the telephone. As a result, the Packers decided to distribute a catalog and establish a phone-order service. The order department consisted of five operators working 8 hours per day from 10:00 A.M. to 6:00 P.M., Monday through Friday. For a few years the mail-order service was only moderately successful; the Packers just about broke even on their investment. However, during the holiday season of the third year of the catalog-order service, they were overwhelmed with phone orders. Although they made a substantial profit, they were concerned about the large number of lost sales they estimated they had incurred. Based on information provided by the telephone company regarding call volume and complaints from customers, the Packers estimated that they lost sales of approximately $100,000. Also, they felt they had lost a substantial number of old and potentially new customers because of the poor service of the catalog order department.

Prior to the next holiday season, the Packers explored several alternatives for improving the catalog-order service. The current system includes the five original operators with computer terminals who work 8-hour days, 5 days per week. The Packers hired a consultant to study this system, and she reported that the time for an operator to take a customer order is exponentially distributed, with a mean of 3.6 minutes. Calls are expected to arrive at the telephone center during the 6-week holiday season, according to a Poisson distribution, with a mean rate of 175 calls per hour. When all operators are busy, callers are put on hold, listening to music until an operator can answer. Waiting calls are answered on a FIFO basis. Based on her experience with other catalog telephone-order operations and data from Northwoods Backpackers, the consultant has determined that if Northwoods Backpackers can reduce customer call-waiting time to approximately 0.5 minute or less, the company will save $135,000 in lost sales during the coming holiday season.

Therefore, the Packers have adopted this level of call service as their goal. However, in addition to simply avoiding lost sales, the Packers believe it is important to reduce waiting time in order to maintain their reputation for good customer service. Thus, they would like for about 70% of their callers to receive immediate service.

The Packers can maintain the same number of workstations and computer terminals they currently have and increase their service to 16 hours per day with two operator shifts running from 8:00 A.M. to midnight. The Packers believe when customers become aware of their extended hours, the calls will spread out uniformly, resulting in a new call average arrival rate of 87.5 calls per hour (still Poisson distributed). This schedule change would cost Northwoods Backpackers approximately $11,500 for the 6-week holiday season.

Another alternative for reducing customer waiting times is to offer weekend service. However, the Packers believe that if they offer weekend service, it must coincide with whatever service they offer during the week. In other words, if they have phone-order service 8 hours per day during the week, they must have the same service during the weekend; the same is true with 16-hours-per-day service. They feel that if weekend hours differ from week-day hours, it will confuse customers. If 8-hour service is offered 7 days per week, the new call arrival rate will be reduced to 125 calls per hour, at a cost of $3,600. If they offer 16-hour service, the mean call arrival rate will be reduced to 62.5 calls per hour, at a cost of $7,200.

Still another possibility is to add more operator stations. Each station includes a desk, an operator, a phone, and a computer terminal. An additional station that is in operation 5 days per week, 8 hours per day, will cost $2,900 for the holiday season. For a 16-hour day the cost per new station is $4,700. For 7-day service the cost of an additional station for 8-hours-per-day service is $3,800; for 16-hours-per-day service, the cost is $6,300.

The facility Northwoods Backpackers uses to house its operators can accommodate a maximum of 10 stations. Additional operators in excess of 10 would require the Packers to lease, remodel, and wire a new facility, which is a capital expenditure they do not want to undertake this holiday season. Alternatively, the Packers do not want to reduce their current number of operator stations.

Determine what order service configuration the Packers should use to achieve their goals and explain your recommendation.

Case Problem

Analyzing Disaster Situations at Tech

Two area hospitals have jointly initiated several planning projects to determine how effectively their emergency facilities can handle disaster-related situations at nearby Tech University. These disasters could be weather related, such as a tornado, a fire, accidents (such as a gas main explosion or a building collapse), or acts of terrorism. One of these projects has focused on the transport of disaster victims from the Tech campus to the two hospitals in the area, Montgomery Regional and Radford Memorial. When a disaster occurs at Tech, emergency vehicles are dispatched from Tech police, local EMT units, hospitals, and local county and city police departments. Victims are brought to a staging area near the disaster scene and wait for transport to one of the two area hospitals. Aspects of the project analysis include the waiting times victims might experience at the disaster scene for emergency vehicles to transport them to the hospital, and waiting times for treatment once victims arrive at the hospital. The project team is analyzing various waiting line models, as follows. (Unless stated otherwise, arrivals are Poisson distributed, and service times are exponentially distributed.)

a. First, consider a single-server waiting line model in which the available emergency vehicles are considered to be the server. Assume that victims arrive at the staging area ready to be transported to a hospital on average every 7 minutes and that emergency vehicles are plentiful and available to pick up and transport victims every 4.5 minutes. Compute the average waiting time for victims. Next, assume that the distribution of service times is undefined, with a mean of 4.5 minutes and a standard deviation of 5 minutes. Compute the average waiting time for the victims.
b. Next, consider a multiple-server model in which there are eight emergency vehicles available for transporting victims to the hospitals, and the mean time required for a vehicle to pick up and transport a victim to a hospital is 20 minutes. (Assume the same arrival rate as in part a.) Compute the average waiting line, the average waiting time for a victim, and the average time in the system for a victim (waiting and being transported).
c. For the multiple-server model in part b now assume that there are a finite number of victims, 18. Determine the average waiting line, the average waiting time, and the average time in the system. (Note that a finite calling population model with multiple servers will require the use of the QM for Windows software.)
d. From the two hospitals' perspectives, consider a multiple-server model in which the two hospitals are the servers. The emergency vehicles at the disaster scene constitute a single waiting line, and each driver calls ahead to see which hospital is most likely to admit the victim first, and travels to that hospital. Vehicles arrive at a hospital every 8.5 minutes, on average, and

the average service time for the emergency staff to admit and treat a victim is 12 minutes. Determine the average waiting line for victims, the average waiting time, and the average time in the system.

e. Next, consider a single hospital, Montgomery Regional, which in an emergency disaster situation has five physicians with supporting staff available. Victims arrive at the hospital on average every 8.5 minutes. It takes an emergency room team, on average, 21 minutes to treat a victim. Determine the average waiting line, the average waiting time, and the average time in the system.

f. For the multiple-server model in part e now assume that there are a finite number of victims, 23. Determine the average waiting line, the average waiting time, and the average time in the system. (Note that a finite calling population model with multiple servers will require the use of the QM for Windows software.)

g. Which of these waiting line models do you think would be the most useful in analyzing a disaster situation? How do you think some, or all, of the models might be used together to analyze a disaster situation? What other type(s) of waiting line model(s) do you think might be useful in analyzing a disaster situation?

Case Problem

Forecasting Airport Passenger Arrivals—Continued

In the "Forecasting Airport Passenger Arrivals" case problem in Chapter 15 (Forecasting), the objective is to develop a forecasting model to predict daily airline passenger arrivals for 2-hour time segments from 4:00 A.M. to 10:00 P.M. for July at Berry International Airport (BEI). Such a forecasting model is necessary in order to determine how many security gates will be needed at the South concourse during each of the daily time segments for any day in July, the airport's busiest travel month. Use the forecast developed in the Chapter 15 case to perform this type of waiting line analysis to determine how many security checkpoints are needed during each time segment. Assume that as passengers arrive at the South concourse security gate they join a single line to have their boarding pass and identification checked at one of several stations. When passengers leave these stations they again form a single line and are approximately equally distributed among the security checkpoints by security personnel before going through the various detection machines. For July the airport plans to staff six security checkpoints for each 2-hour time segment from 4:00 A.M. to 6:00 P.M., and then three checkpoints from 6:00 P.M. to 8:00 P.M., and two checkpoints from 8:00 P.M. to 10:00 P.M. Assume that the arrival rate at this point in the security system is Poisson distributed, with the forecasted passenger arrivals developed in the Chapter 15 case as the mean arrival rate. Further, assume that service times are exponentially distributed with a mean of 11.6 seconds. Determine whether the number of security checkpoints the airport plans to use for each 2-hour time segment is sufficient to keep passengers moving freely through the security system without excessive delays. If the current number of checkpoints is not sufficient, what is the likely result? If the planned system is not likely to be sufficient, determine the number of checkpoints that would be needed for each 2-hour segment in order for passengers to move quickly through the security checkpoints without excessive waiting times.

CHAPTER

14

Simulation

Simulation represents a major divergence from the topics presented in the previous chapters of this text. Previous topics usually dealt with mathematical models and formulas that could be applied to certain types of problems. The solution approaches to these problems were, for the most part, analytical. However, not all real-world problems can be solved by applying a specific type of technique and then performing the calculations. Some problem situations are too complex to be represented by the concise techniques presented so far in this text. In such cases, *simulation* is an alternative form of analysis.

***Analogue simulation** replaces a physical system with an analogous physical system that is easier to manipulate.*

Analogue simulation is a form of simulation that is familiar to most people. In analogue simulation, an original physical system is replaced by an analogous physical system that is easier to manipulate. Much of the experimentation in staffed spaceflight was conducted using physical simulation that re-created the conditions of space. For example, conditions of weightlessness were simulated using rooms filled with water. Other examples include wind tunnels that simulate the conditions of flight and treadmills that simulate automobile tire wear in a laboratory instead of on the road.

*In **computer mathematical simulation**, a system is replicated with a mathematical model that is analyzed by using the computer.*

This chapter is concerned with an alternative type of simulation, **computer mathematical simulation**. In this form of simulation, systems are replicated with mathematical models, which are analyzed using a computer. This form of simulation has become very popular and has been applied to a wide variety of business problems. One reason for its popularity is that it offers a means of analyzing very complex systems that cannot be analyzed by using the other management science techniques in this text. However, because such complex systems are beyond the scope of this text, we will not present actual simulation models; instead, we will present simplified simulation models of systems that can also be analyzed analytically. We will begin with one of the simplest forms of simulation models, which encompasses the Monte Carlo process for simulating random variables.

The Monte Carlo Process

One characteristic of some systems that makes them difficult to solve analytically is that they consist of random variables represented by probability distributions. Thus, a large proportion of the applications of simulations are for probabilistic models.

***Monte Carlo** is a technique for selecting numbers randomly from a probability distribution.*

The Monte Carlo process is analogous to gambling devices.

The term **Monte Carlo** has become synonymous with probabilistic simulation in recent years. However, the Monte Carlo technique can be more narrowly defined as a technique for selecting numbers *randomly* from a probability distribution (i.e., "sampling") for use in a *trial* (computer) run of a simulation. The Monte Carlo technique is not a type of simulation model but rather a mathematical process used within a simulation.

The name *Monte Carlo* is appropriate because the basic principle behind the process is the same as in the operation of a gambling casino in Monaco. In Monaco such devices as roulette wheels, dice, and playing cards are used. These devices produce numbered results at random from well-defined populations. For example, a 7 resulting from a pair of thrown dice is a random value from a population of 11 possible numbers (i.e., 2 through 12). This same process is employed, in principle, in the Monte Carlo process used in simulation models.

The Use of Random Numbers

The Monte Carlo process of selecting random numbers according to a probability distribution will be demonstrated using the following example. The manager of ComputerWorld, a store that sells computers and related equipment, is attempting to determine how many laptop PCs the store should order each week. A primary consideration in this decision is the average number of laptop computers that the store will sell each week and the average weekly revenue generated from the sale of laptop PCs. A laptop sells for $4,300. The number of laptops demanded each week is a random variable (which we will define as x) that ranges from 0 to 4. From past sales records, the manager has determined the frequency of demand for laptop PCs for the past 100 weeks. From this frequency distribution, a probability distribution of demand can be developed, as shown in Table 14.1.

TABLE 14.1
Probability distribution of demand for laptop PCs

PCs Demanded per Week	Frequency of Demand	Probability of Demand, $P(x)$
0	20	.20
1	40	.40
2	20	.20
3	10	.10
4	10	.10
	100	1.00

In the Monte Carlo process, values for a random variable are generated by sampling from a probability distribution.

The purpose of the Monte Carlo process is to generate the random variable, demand, by sampling from the probability distribution, $P(x)$. The demand per week can be randomly generated according to the probability distribution by spinning a wheel that is partitioned into segments corresponding to the probabilities, as shown in Figure 14.1.

FIGURE 14.1
A roulette wheel for demand

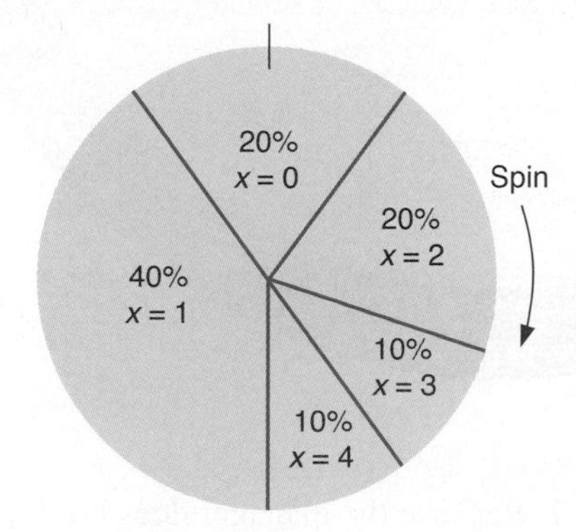

Because the surface area on the roulette wheel is partitioned according to the probability of each weekly demand value, the wheel replicates the probability distribution for demand if the values of demand occur in a random manner. To simulate demand for 1 week, the manager spins the wheel; the segment at which the wheel stops indicates demand for 1 week. Over a period of weeks (i.e., many spins of the wheel), the frequency with which demand values occur will approximate the probability distribution, $P(x)$. This method of generating values of a variable, x, by randomly selecting from the probability distribution—the wheel—is the Monte Carlo process.

*A long period of real time is represented by a short period of **simulated time**.*

By spinning the wheel, the manager artificially reconstructs the purchase of PCs during a week. In this reconstruction, a long period of *real time* (i.e., a number of weeks) is represented by a short period of **simulated time** (i.e., several spins of the wheel).

Now let us slightly reconstruct the roulette wheel. In addition to partitioning the wheel into segments corresponding to the probability of demand, we will put numbers along the outer rim, as on a real roulette wheel. This reconstructed roulette wheel is shown in Figure 14.2.

There are 100 numbers from 0 to 99 on the outer rim of the wheel, and they have been partitioned according to the probability of each demand value. For example, 20 numbers from 0 to 19 (i.e., 20% of the total 100 numbers) correspond to a demand of no (0) PCs. Now we can determine the value of demand by seeing which number the wheel stops at as well as by looking at the segment of the wheel.

When the manager spins this new wheel, the actual demand for PCs will be determined by a number. For example, if the number 71 comes up on a spin, the demand is 2 PCs per week; the

FIGURE 14.2
Numbered roulette wheel

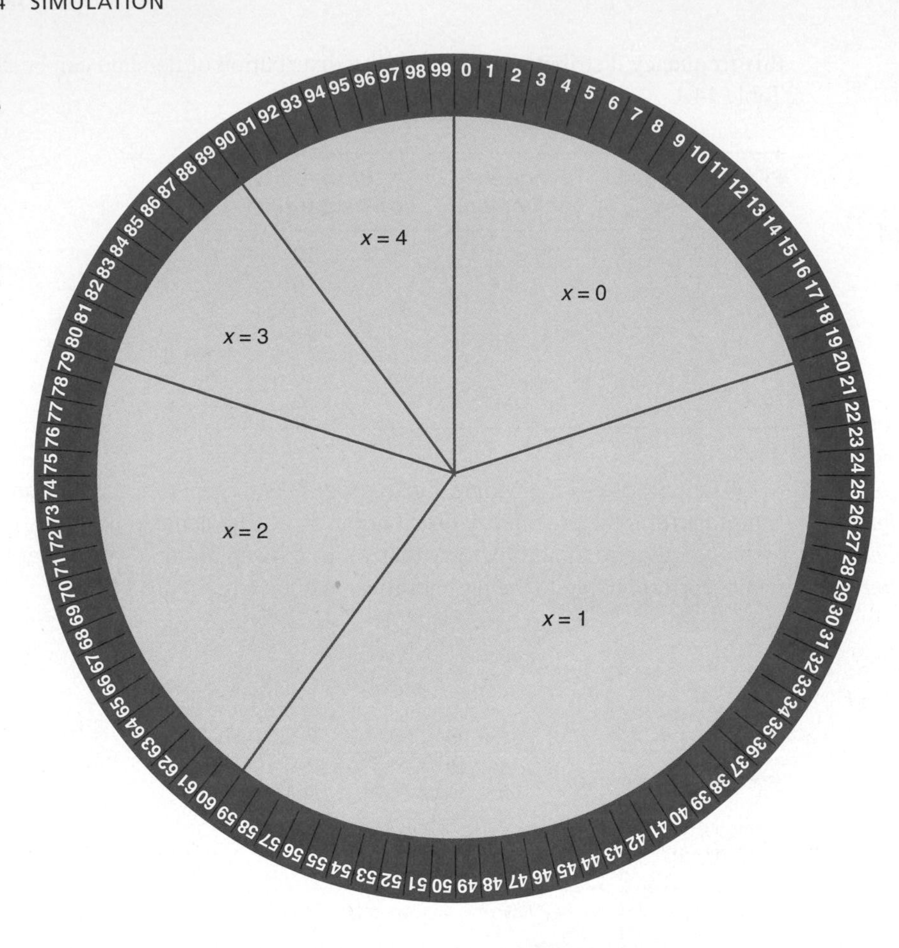

Random numbers are numbers equally likely to be chosen from a large population of numbers.

In a random number table, the random numbers are derived from some artificial process, like a computer program.

Random numbers are equally likely to occur.

number 30 indicates a demand of 1. Because the manager does not know which number will come up prior to the spin and there is an equal chance that any of the 100 numbers will occur, the numbers occur at random; that is, they are **random numbers**.

Obviously, it is not generally practical to generate weekly demand for PCs by spinning a wheel. Alternatively, the process of spinning a wheel can be replicated by using random numbers alone.

First, we will transfer the ranges of random numbers for each demand value from the roulette wheel to a table, as in Table 14.2. Next, instead of spinning the wheel to get a random number, we will select a random number from Table 14.3, which is referred to as a **random number table**. (These random numbers have been generated by computer so that they are all *equally likely to occur*, just as if we had spun a wheel. The development of random numbers is discussed in more detail later in this chapter.) As an example, let us select the number 39, the first entry in Table 14.3. Looking again at Table 14.2, we can see that the random number 39 falls in the range 20–59, which corresponds to a weekly demand of 1 laptop PC.

TABLE 14.2
Generating demand from random numbers

Demand, x	Ranges of Random Numbers, r	
0	0–19	
1 ←	20–59 ←	$r = 39$
2	60–79	
3	80–89	
4	90–99	

TABLE 14.3
Random number table

39 65 76 45 45	19 90 69 64 61	20 26 36 31 62	58 24 97 14 97	95 06 70 99 00
73 71 23 70 90	65 97 60 12 11	31 56 34 19 19	47 83 75 51 33	30 62 38 20 46
72 18 47 33 84	51 67 47 97 19	98 40 07 17 66	23 05 09 51 80	59 78 11 52 49
75 12 25 69 17	17 95 21 78 58	24 33 45 77 48	69 81 84 09 29	93 22 70 45 80
37 17 79 88 74	63 52 06 34 30	01 31 60 10 27	35 07 79 71 53	28 99 52 01 41
02 48 08 16 94	85 53 83 29 95	56 27 09 24 43	21 78 55 09 82	72 61 88 73 61
87 89 15 70 07	37 79 49 12 38	48 13 93 55 96	41 92 45 71 51	09 18 25 58 94
98 18 71 70 15	89 09 39 59 24	00 06 41 41 20	14 36 59 25 47	54 45 17 24 89
10 83 58 07 04	76 62 16 48 68	58 76 17 14 86	59 53 11 52 21	66 04 18 72 87
47 08 56 37 31	71 82 13 50 41	27 55 10 24 92	28 04 67 53 44	95 23 00 84 47
93 90 31 03 07	34 18 04 52 35	74 13 39 35 22	68 95 23 92 35	36 63 70 35 33
21 05 11 47 99	11 20 99 45 18	76 51 94 84 86	13 79 93 37 55	98 16 04 41 67
95 89 94 06 97	27 37 83 28 71	79 57 95 13 91	09 61 87 25 21	56 20 11 32 44
97 18 31 55 73	10 65 81 92 59	77 31 61 95 46	20 44 90 32 64	26 99 76 75 63
69 08 88 86 13	59 71 74 17 32	48 38 75 93 29	73 37 32 04 05	60 82 29 20 25
41 26 10 25 03	87 63 93 95 17	81 83 83 04 49	77 45 85 50 51	79 88 01 97 30
91 47 14 63 62	08 61 74 51 69	92 79 43 89 79	29 18 94 51 23	14 85 11 47 23
80 94 54 18 47	08 52 85 08 40	48 40 35 94 22	72 65 71 08 86	50 03 42 99 36
67 06 77 63 99	89 85 84 46 06	64 71 06 21 66	89 37 20 70 01	61 65 70 22 12
59 72 24 13 75	42 29 72 23 19	06 94 76 10 08	81 30 15 39 14	81 33 17 16 33
63 62 06 34 41	79 53 36 02 95	94 61 09 43 62	20 21 14 68 86	84 95 48 46 45
78 47 23 53 90	79 93 96 38 63	34 85 52 05 09	85 43 01 72 73	14 93 87 81 40
87 68 62 15 43	97 48 72 66 48	53 16 71 13 81	59 97 50 99 52	24 62 20 42 31
47 60 92 10 77	26 97 05 73 51	88 46 38 03 58	72 68 49 29 31	75 70 16 08 24
56 88 87 59 41	06 87 37 78 48	65 88 69 58 39	88 02 84 27 83	85 81 56 39 38
22 17 68 65 84	87 02 22 57 51	68 69 80 95 44	11 29 01 95 80	49 34 35 36 47
19 36 27 59 46	39 77 32 77 09	79 57 92 36 59	89 74 39 82 15	08 58 94 34 74
16 77 23 02 77	28 06 24 25 93	22 45 44 84 11	87 80 61 65 31	09 71 91 74 25
78 43 76 71 61	97 67 63 99 61	30 45 67 93 82	59 73 19 85 23	53 33 65 97 21
03 28 28 26 08	69 30 16 09 05	53 58 47 70 93	66 56 45 65 79	45 56 20 19 47
04 31 17 21 56	33 73 99 19 87	26 72 39 27 67	53 77 57 68 93	60 61 97 22 61
61 06 98 03 91	87 14 77 43 96	43 00 65 98 50	45 60 33 01 07	98 99 46 50 47
23 68 35 26 00	99 53 93 61 28	52 70 05 48 34	56 65 05 61 86	90 92 10 70 80
15 39 25 70 99	93 86 52 77 65	15 33 59 05 28	22 87 26 07 47	86 96 98 29 06
58 71 96 30 24	18 46 23 34 27	85 13 99 24 44	49 18 09 79 49	74 16 32 23 02
93 22 53 64 39	07 10 63 76 35	87 03 04 79 88	08 13 13 85 51	55 34 57 72 69
78 76 58 54 74	92 38 70 96 92	52 06 79 79 45	82 63 18 27 44	69 66 92 19 09
61 81 31 96 82	00 57 25 60 59	46 72 60 18 77	55 66 12 62 11	08 99 55 64 57
42 88 07 10 05	24 98 65 63 21	47 21 61 88 32	27 80 30 21 60	10 92 35 36 12
77 94 30 05 39	28 10 99 00 27	12 73 73 99 12	49 99 57 94 82	96 88 57 17 91

By repeating this process of selecting random numbers from Table 14.3 (starting anywhere in the table and moving in any direction but not repeating the same sequence) and then determining weekly demand from the random number, we can simulate demand for a period of time. For example, Table 14.4 shows demand for a period of 15 consecutive weeks.

From Table 14.4 the manager can compute the estimated average weekly demand and revenue:

$$\text{estimated average demand} = \frac{31}{15} = 2.07 \text{ laptop PCs per week}$$

$$\text{estimated average revenue} = \frac{\$133{,}300}{15} = \$8{,}886.67$$

TABLE 14.4
Randomly generated demand for 15 weeks

Week	r	Demand, x	Revenue
1	39	1	$ 4,300
2	73	2	8,600
3	72	2	8,600
4	75	2	8,600
5	37	1	4,300
6	02	0	0
7	87	3	12,900
8	98	4	17,200
9	10	0	0
10	47	1	4,300
11	93	4	17,200
12	21	1	4,300
13	95	4	17,200
14	97	4	17,200
15	69	2	8,600
		Σ = 31	$133,300

The manager can then use this information to help determine the number of PCs to order each week.

Although this example is convenient for illustrating how simulation works, the average demand could have more appropriately been calculated *analytically* using the formula for expected value. The *expected value* or average for weekly demand can be computed analytically from the probability distribution, $P(x)$:

$$EV(x) = \sum_{i=1}^{n} P(x_i)x$$

where

x_i = demand value i
$P(x_i)$ = probability of demand
n = the number of different demand values

Therefore,

$$\begin{aligned} EV(x) &= (.20)(0) + (.40)(1) + (.20)(2) + (.10)(3) + (.10)(4) \\ &= 1.5 \text{ PCs per week} \end{aligned}$$

Simulation results will not equal analytical results unless enough trials of the simulation have been conducted to reach steady state.

The analytical result of 1.5 PCs is close to the simulated result of 2.07 PCs, but clearly there is some difference. The margin of difference (0.57 PC) between the simulated value and the analytical value is a result of the number of periods over which the simulation was conducted. The results of any simulation study are subject to the number of times the simulation occurred (i.e., the number of *trials*). Thus, the more periods for which the simulation is conducted, the more accurate the result. For example, if demand was simulated for 1,000 weeks, in all likelihood an average value exactly equal to the analytical value (1.5 laptop PCs per week) would result.

Once a simulation has been repeated enough times that it reaches an average result that remains constant, this result is analogous to the *steady-state* result, a concept we discussed previously in our presentation of queuing. For this example, 1.5 PCs is the long-run average or steady-state result, but we have seen that the simulation might have to be repeated more than 15 times (i.e., weeks) before this result is reached.

It is often difficult to validate that the results of a simulation truly replicate reality.

Comparing our simulated result with the analytical (expected value) result for this example points out one of the problems that can occur with simulation. It is often difficult to *validate* the results of a simulation model—that is, to make sure that the true steady-state average result has

Time Out for John Von Neumann

The mathematics of the Monte Carlo method have been known for years; the British mathematician Lord Kelvin used the technique in a paper in 1901. However, it was formally identified and given this name by the Hungarian mathematician John Von Neumann while working on the Los Alamos atomic bomb project during World War II. During this project, physicists confronted a problem in determining how far neutrons would travel through various materials (i.e., neutron diffusion in fissile material). The Monte Carlo process was suggested to Von Neumann by a colleague at Los Alamos, Stanislas Ulam, as a means to solve this problem—that is, by selecting random numbers to represent the random actions of neutrons. However, the Monte Carlo method as used in simulation did not gain widespread popularity until the development of the modern electronic computer after the war. Interestingly, this remarkable man, John Von Neumann, is credited with being the key figure in the development of the computer.

been reached. In this case we were able to compare the simulated result with the expected value (which is the true steady-state result), and we found there was a slight difference. We logically deduced that the 15 trials of the simulation were not sufficient to determine the steady-state average. However, simulation most often is employed whenever analytical analysis is not possible (this is one of the reasons that simulation is generally useful). In these cases, there is no analytical standard of comparison, and validation of the results becomes more difficult. We will discuss this problem of validation in more detail later in the chapter.

Computer Simulation with Excel Spreadsheets

Simulations are normally done on the computer.

The simulation we performed manually for the ComputerWorld example was not too difficult. However, if we had performed the simulation for 1,000 weeks, it would have taken several hours. On the other hand, this simulation could be done on a computer in several seconds. Also, our simulation example was not very complex. As simulation models get progressively more complex, it becomes virtually impossible to perform them manually, thus making the computer a necessity.

Although we will not develop a simulation model in a computer language for this example, we will demonstrate how a computerized simulation model is developed by using Excel spreadsheets.

Random numbers generated by a mathematical process instead of a physical process are ***pseudorandom numbers****.*

The first step in developing a simulation model is to generate a random number, r. Numerous subroutines that are available on practically every computer system generate random numbers. Most are quite easy to use and require the insertion of only a few statements in a program. These random numbers are generated by mathematical processes as opposed to a physical process, such as spinning a roulette wheel. For this reason, they are referred to as **pseudorandom numbers**. It should be apparent from the previous discussion that random numbers play a very important part in a probabilistic simulation. Some of the random numbers we used came from Table 14.3, a table of random numbers. However, random numbers do not come just from tables, and their generation is not as simple as one might initially think. If random numbers are not truly random, the validity of simulation results can be significantly affected.

Random numbers are typically generated on the computer by using a numerical technique.

The random numbers in Table 14.3 were generated by using a *numerical technique*. Thus, they are not true random numbers but *pseudorandom numbers*. True random numbers can be produced only by a physical process, such as spinning a roulette wheel over and over. However, a physical process, such as spinning a roulette wheel, cannot be conveniently employed in a computerized simulation model. Thus, there is a need for a numerical method that artificially creates random numbers.

A table of random numbers must be uniform, efficiently generated, and absent of patterns.

To truly reflect the system being simulated, the artificially created random numbers must have the following characteristics:

1. The random numbers must be uniformly distributed. This means that each random number in the interval of random numbers (i.e., 0 to 1 or 0 to 100) has an equal chance of being

selected. If this condition is not met, then the simulation results will be biased by the random numbers that have a *more likely* chance of being selected.

2. The numerical technique for generating random numbers should be efficient. This means that the random numbers should not degenerate into constant values or recycle too frequently.
3. The sequence of random numbers should not reflect any pattern. For example, the sequence of numbers 0, 1, 2, 3, 4, 5, 6, 7, 8, 9, 0, 1, 2, 3, 4, 5, 6, 7, 8, 9, 0, 1, 2, 3, 4, 5, 6, 7, 8, 9, 0, and so on, although uniform, is not random.

Random numbers between 0 and 1 can be generated in Excel by entering the formula **=RAND()** in a cell. The random numbers generated by this formula include all the necessary characteristics for randomness and uniformity that we discussed earlier. Exhibit 14.1 is an Excel spreadsheet with 100 random numbers generated by entering the formula **=RAND()** in cell A3 and copying to the cells in the range **A3:J12**. Recall that you can copy things in a range of cells in two ways. You can first cover cells **A3:J12** with the cursor and then type the formula **=RAND()** into cell A3. Then you press the "Ctrl" and "Enter" keys simultaneously. Alternatively, you can type **=RAND()** in cell A3, copy this cell (using the right mouse button), then cover cells **A3:J12** with the cursor, and (again with the right mouse button) paste this formula in these cells.

EXHIBIT 14.1

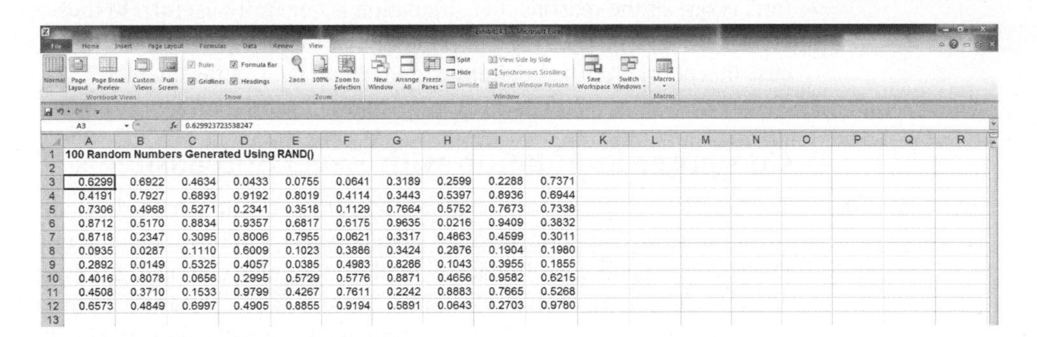

A3 = 0.629923723538247

	A	B	C	D	E	F	G	H	I	J
1	100 Random Numbers Generated Using RAND()									
2										
3	0.6299	0.6922	0.4634	0.0433	0.0755	0.0641	0.3189	0.2599	0.2288	0.7371
4	0.4191	0.7927	0.6893	0.9192	0.8019	0.4114	0.3443	0.5397	0.8936	0.6944
5	0.7306	0.4968	0.5271	0.2341	0.3518	0.1129	0.7664	0.5752	0.7673	0.7338
6	0.8712	0.5170	0.8834	0.9357	0.6817	0.6175	0.9635	0.0216	0.9409	0.3832
7	0.8718	0.2347	0.3095	0.8006	0.7955	0.0621	0.3317	0.4863	0.4599	0.3011
8	0.0935	0.0287	0.1110	0.6009	0.1023	0.3886	0.3424	0.2876	0.1904	0.1980
9	0.2892	0.0149	0.5325	0.4057	0.0385	0.4983	0.8286	0.1043	0.3955	0.1855
10	0.4016	0.8078	0.0656	0.2995	0.5729	0.5776	0.8871	0.4656	0.9582	0.6215
11	0.4508	0.3710	0.1533	0.9799	0.4267	0.7611	0.2242	0.8883	0.7665	0.5268
12	0.6573	0.4849	0.6997	0.4905	0.8855	0.9194	0.5891	0.0643	0.2703	0.9780
13										

If you attempt to replicate this spreadsheet, you will generate different random numbers from those shown in Exhibit 14.1. Every time you generate random numbers, they will be different. In fact, any time you recalculate anything on your spreadsheet, the random numbers will change. You can see this by pressing the "F9" key and observing that all the random numbers change. However, sometimes it is useful in a simulation model to be able to use the same set (or stream) of random numbers over and over. You can freeze the random numbers you are using on your spreadsheet by first covering the cells with random numbers in them with the cursor—for example, cells **A3:J12** in Exhibit 14.1. Next, you copy these cells (using the right mouse button); then you click on the "Edit" menu at the top of your spreadsheet and select "Paste Special" from this menu. Next, you select the "Values" option and click on "OK." This procedure pastes a copy of the numbers in these cells over the same cells with **=RAND()** formulas in them, thus freezing the numbers in place.

Notice one more thing from Exhibit 14.1: The random numbers are all between 0 and 1, whereas the random numbers in Table 14.3 are whole numbers between 0 and 100. We used whole random numbers previously for illustrative purposes; however, computer programs such as Excel generally provide random numbers between 0 and 1.

Now we are ready to duplicate our example simulation model for the ComputerWorld store by using Excel. The spreadsheet in Exhibit 14.2 includes the simulation model originally developed in Table 14.4.

Note that the probability distribution for the weekly demand for laptops has been entered in cells **A6:C10**. Also notice that we have entered a new set of cumulative probability values in column B. We generated these cumulative probabilities by first entering 0 in cell B6, then

EXHIBIT 14.2

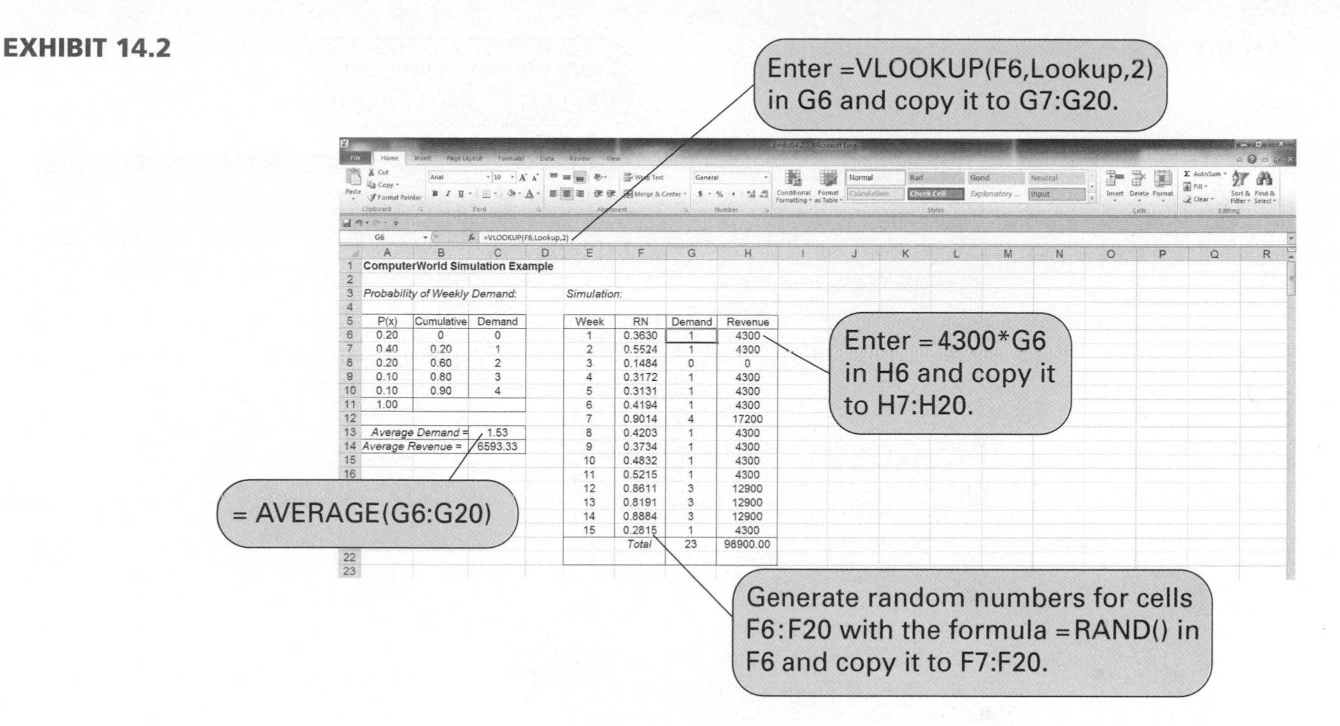

entering the formula **=A6+B6** in cell B7, and copying this formula to cells **B8:B10**. This cumulative probability creates a range of random numbers for each demand value. For example, any random number less than 0.20 will result in a demand value of 0, and any random number greater than 0.20 but less than 0.60 will result in a demand value of 1, and so on. (Notice that there is no value of 1.00 in cell B11; the last demand value, 4, will be selected for any random number equal to or greater than .90.)

Random numbers are generated in cells **F6:F20** by entering the formula **=RAND()** in cell F6 and copying it to the range of cells in **F7:F20**.

Now we need to be able to generate demand values for each of these random numbers in column F. We accomplish this by first covering the cumulative probabilities and the demand values in cells **B6:C10** with the cursor. Then we give this range of cells the name "Lookup." This can be done by typing "Lookup" directly on the formula bar in place of B6 or by clicking on the "Insert" button at the top of the spreadsheet and selecting "Name" and "Define" and then entering the name "Lookup." This has the effect of creating a table called "Lookup" with the ranges of random numbers and associated demand values in it. Next, we enter the formula **=VLOOKUP(F6,Lookup,2)** in cell G6 and copy it to the cells in the range **G7:G20**. This formula will compare the random numbers in column F with the cumulative probabilities in **B6:B10** and generate the correct demand value from cells **C6:C10**.

Once the demand values have been generated in column G, we can determine the weekly revenue values by entering the formula **=4300*G6** in H6 and copying it to cells **H7:H20**.

Average weekly demand is computed in cell C13 by using the formula **=AVERAGE(G6:G20)**, and the average weekly revenue is computed by entering a similar formula in cell C14.

Notice that the average weekly demand value of 1.53 in Exhibit 14.2 is different from the simulation result (2.07) we obtained from Table 14.4. This is because we used a different stream of random numbers. As mentioned previously, to acquire an average closer to the true steady-state value, the simulation probably needs to include more repetitions than 15 weeks. As an example, Exhibit 14.3 simulates demand for 100 weeks. The window has been "frozen" at row 16 and scrolled up to show the first 10 weeks and the last 6 weeks on the screen in Exhibit 14.3.

EXHIBIT 14.3

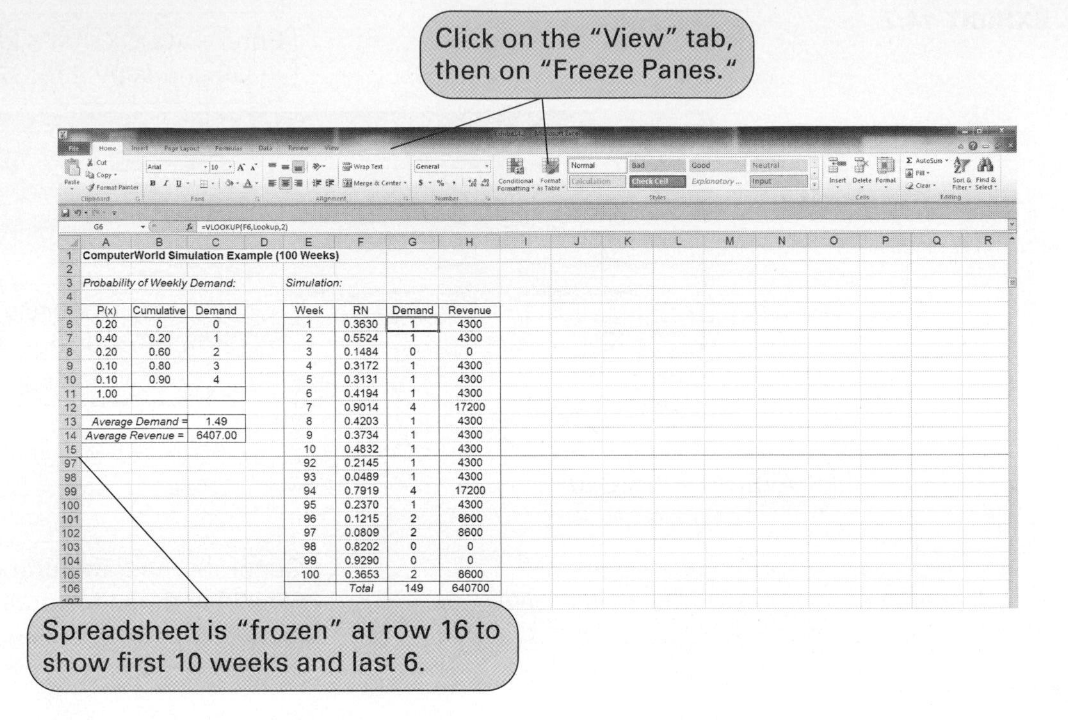

	A	B	C	D	E	F	G	H
1	ComputerWorld Simulation Example (100 Weeks)							
2								
3	*Probability of Weekly Demand:*				*Simulation:*			
4								
5	P(x)	Cumulative	Demand		Week	RN	Demand	Revenue
6	0.20	0	0		1	0.3630	1	4300
7	0.40	0.20	1		2	0.5524	1	4300
8	0.20	0.60	2		3	0.1484	0	0
9	0.10	0.80	3		4	0.3172	1	4300
10	0.10	0.90	4		5	0.3131	1	4300
11	1.00				6	0.4194	1	4300
12					7	0.9014	4	17200
13	*Average Demand =*		1.49		8	0.4203	1	4300
14	*Average Revenue =*		6407.00		9	0.3734	1	4300
15					10	0.4832	1	4300
97					92	0.2145	1	4300
98					93	0.0489	1	4300
99					94	0.7919	4	17200
100					95	0.2370	1	4300
101					96	0.1215	2	8600
102					97	0.0809	2	8600
103					98	0.8202	0	0
104					99	0.9290	0	0
105					100	0.3653	2	8600
106						*Total*	149	640700

Decision Making with Simulation

In our previous example, the manager of ComputerWorld acquired some useful information about the weekly demand and revenue for laptops that would be helpful in making a decision about how many laptops would be needed each week to meet demand. However, this example did not lead directly to a decision. Next, we will expand our ComputerWorld store example so that a possible decision will result.

From the simulation in Exhibit 14.3 the manager of the store knows that the average weekly demand for laptop PCs will be approximately 1.49; however, the manager cannot order 1.49 laptops each week. Because fractional laptops are not possible, either one or two must be ordered. Thus, the manager wants to repeat the earlier simulation with two possible order sizes, 1 and 2. The manager also wants to include some additional information in the model that will affect the decision.

If too few laptops are on hand to meet demand during the week, then not only will there be a loss of revenue, but there will also be a shortage cost of $500 per unit incurred because the customer will be unhappy. However, each laptop still in stock at the end of each week that has not been sold will incur an inventory or storage cost of $50. Thus, it costs the store money to have either too few or too many laptops on hand each week. Given this scenario, the manager wants to order either one or two laptops, depending on which order size will result in the greatest average weekly revenue.

Exhibit 14.4 shows the Excel spreadsheet for this revised example. The simulation is for 100 weeks. The columns labeled "1," "2," and "4" for "Week," "RN," and "Demand" were constructed similarly to the model in Exhibit 14.3. The array of cells **B6:C10** was given the name "Lookup," and the formula **=VLOOKUP(F6,Lookup,2)** was entered in cell H6 and copied to cells **H7:H105**.

The simulation in Exhibit 14.4 is for an order size of one laptop each week. The "Inventory" column (3) keeps track of the amount of inventory available each week—the one laptop that comes in on order plus any laptops carried over from the previous week. The cumulative inventory is computed each week by entering the formula **=1+MAX(G6−H6,0)** in cell G7 and copying it to cells **G8:G105**. This formula adds the one laptop on order to either the number left over from the

EXHIBIT 14.4

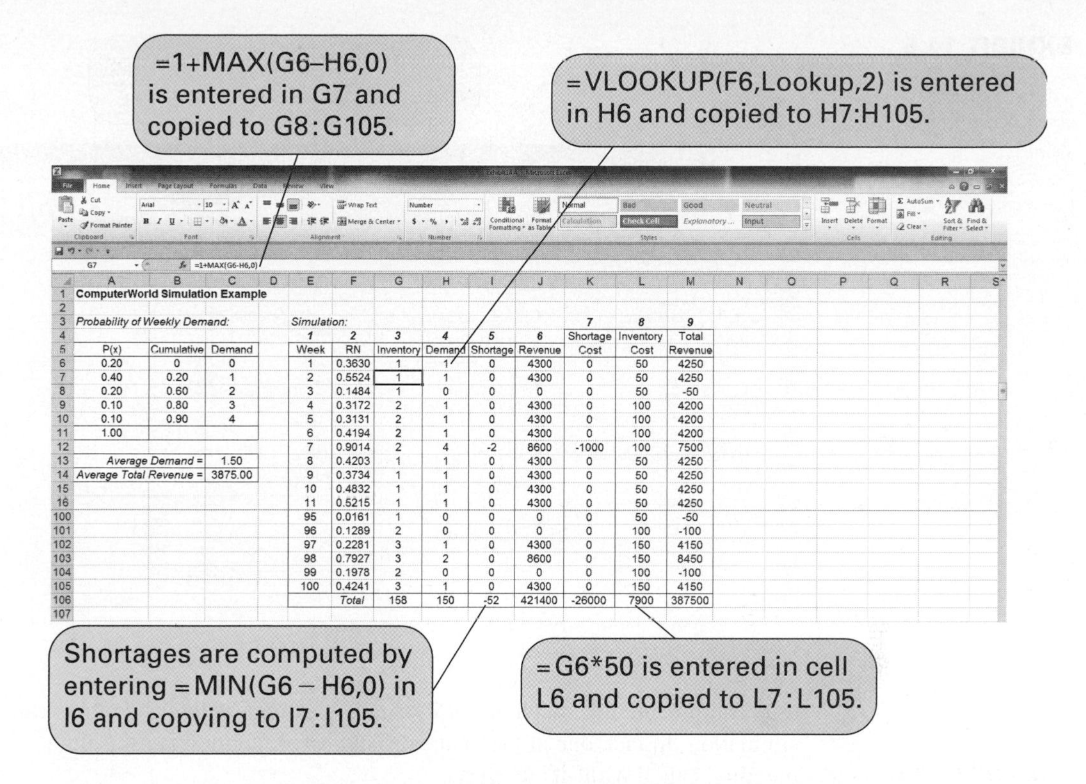

ComputerWorld Simulation Example

Probability of Weekly Demand:

P(x)	Cumulative	Demand
0.20	0	0
0.40	0.20	1
0.20	0.60	2
0.10	0.80	3
0.10	0.90	4
1.00		
	Average Demand =	1.50
	Average Total Revenue =	3875.00

Simulation:

1	2	3	4	5	6	7	8	9
Week	RN	Inventory	Demand	Shortage	Revenue	Shortage Cost	Inventory Cost	Total Revenue
1	0.3630	1	1	0	4300	0	50	4250
2	0.5524	1	1	0	4300	0	50	4250
3	0.1484	1	0	0	0	0	50	-50
4	0.3172	2	1	0	4300	0	100	4200
5	0.3131	2	1	0	4300	0	100	4200
6	0.4194	2	1	0	4300	0	100	4200
7	0.9014	2	4	-2	8600	-1000	100	7500
8	0.4203	1	1	0	4300	0	50	4250
9	0.3734	1	1	0	4300	0	50	4250
10	0.4832	1	1	0	4300	0	50	4250
11	0.5215	1	1	0	4300	0	50	4250
95	0.0161	1	0	0	0	0	50	-50
96	0.1289	2	0	0	0	0	100	-100
97	0.2281	3	1	0	4300	0	150	4150
98	0.7927	3	2	0	8600	0	150	8450
99	0.1978	2	0	0	0	0	100	-100
100	0.4241	3	1	0	4300	0	150	4150
	Total	158	150	-52	421400	-26000	7900	387500

previous week (**G6–H6**) or 0, if there were not enough laptops on hand to meet demand. It does not allow for negative inventory levels, called *back orders*. In other words, if a sale cannot be made owing to a shortage, then it is gone. The inventory values in column 3 are eventually multiplied by the inventory cost of $50 per unit in column 8, using the formula **=G6*50**.

If there is a shortage, it is recorded in column 5, labeled "Shortage." The shortage is computed by entering the formula **=MIN(G6–H6,0)** in cell I6 and copying it to cells **I7:I105**. Shortage costs are computed in column 7 by multiplying the shortage values in column 5 by $500 by entering the formula **=I6*500** in cell K6 and copying it to cells **K7:K105**.

Weekly revenues are computed in column 6 by entering the formula **=4300*MIN(H6,G6)** in cell J6 and copying it to cells **J7:J105**. In other words, the revenue is determined by either the inventory level in column 3 or the demand in column 4, whichever is smaller.

Total weekly revenue is computed in column 9 by subtracting shortage costs and inventory costs from revenue by entering the formula **=J6-K6-L6** in cell M6 and copying it to cells **M7:M105**.

The average weekly demand, 1.50, is shown in cell C13. The average weekly revenue, $3,875, is computed in cell C14.

Next, we must repeat this same simulation for an order size of two laptops each week. The spreadsheet for an order size of two is shown in Exhibit 14.5. Notice that the only actual difference is the use of a new formula to compute the weekly inventory level in column 3. This formula, **=2+MAX(G6–H6,0)**, in cell G7, reflecting two laptops ordered each week, is shown on the formula bar at the top of the spreadsheet.

This second simulation, in Exhibit 14.5, results in an average weekly demand of 1.50 laptops and an average weekly total revenue of $4,927.50. This is higher than the total weekly revenue of $3,875 achieved in the first simulation run in Exhibit 14.4, even though the store would incur significantly higher inventory costs. Thus, the correct decision—based on weekly revenue—would be to order two laptops per week. However, there are probably additional aspects of this problem the manager would want to consider in the decision-making process, such as the increasingly high inventory levels as the simulation progresses. For example, there may not be enough storage space

EXHIBIT 14.5

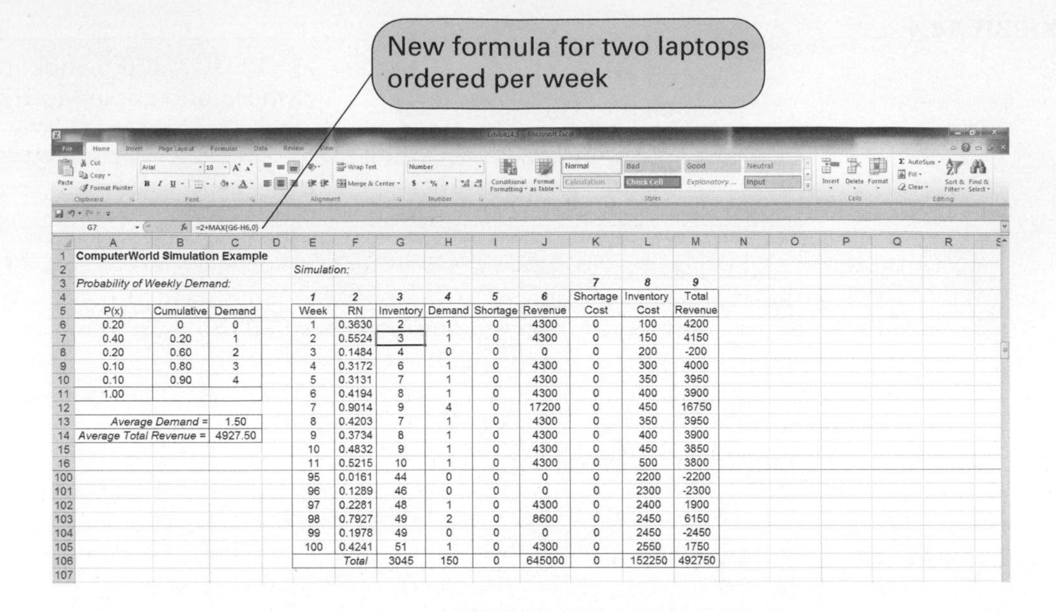

ComputerWorld Simulation Example

Probability of Weekly Demand:

P(x)	Cumulative	Demand
0.20	0	0
0.40	0.20	1
0.20	0.60	2
0.10	0.80	3
0.10	0.90	4
1.00		

Average Demand =	1.50
Average Total Revenue =	4927.50

Simulation:

1 Week	2 RN	3 Inventory	4 Demand	5 Shortage	6 Revenue	7 Shortage Cost	8 Inventory Cost	9 Total Revenue
1	0.3630	2	1	0	4300	0	100	4200
2	0.5524	3	1	0	4300	0	150	4150
3	0.1484	4	0	0	0	0	200	-200
4	0.3172	6	1	0	4300	0	300	4000
5	0.3131	7	1	0	4300	0	350	3950
6	0.4194	8	1	0	4300	0	400	3900
7	0.9014	9	4	0	17200	0	450	16750
8	0.4203	7	1	0	4300	0	350	3950
9	0.3734	8	1	0	4300	0	400	3900
10	0.4832	9	1	0	4300	0	450	3850
11	0.5215	10	1	0	4300	0	500	3800
95	0.0161	44	0	0	0	0	2200	-2200
96	0.1289	46	0	0	0	0	2300	-2300
97	0.2281	48	1	0	4300	0	2400	1900
98	0.7927	49	2	0	8600	0	2450	6150
99	0.1978	49	0	0	0	0	2450	-2450
100	0.4241	51	1	0	4300	0	2550	1750
	Total	3045	150	0	645000	0	152250	492750

to accommodate this much inventory. Such questions as this and others can also be analyzed with simulation. In fact, one of the main attributes of simulation is its usefulness as a model for experimenting, called **what-if? analysis**.

What-if? analysis is a form of model experimentation using a computer to ascertain the results of making changes in a model.

This example briefly demonstrates how simulation can be used to make a decision (i.e., to "optimize"). In this example we experimented with two order sizes and determined the one that resulted in the greater revenue. The same basic modeling principles can be used to solve larger problems with hundreds of possible order sizes and a probability distribution for demand with many more values plus variable lead times (i.e., the time it takes to receive an order), the ability to back-order, and other complicating factors. These factors make the simulation model larger and more complex, but such models are frequently developed and used in business.

Simulation of a Queuing System

To demonstrate the simulation of a queuing system, we will use an example. Burlingham Mills produces denim, and one of the main steps in the production process is dyeing the cotton yarn that is subsequently woven into denim cloth. The yarn is dyed in a large concrete vat like a narrow swimming pool. The yarn is strung over a series of rollers so that it passes through the vat, up over a set of rollers, back down into the vat, back up and over another set of rollers, and so on. Yarn is dyed in batches that arrive at the dyeing vat in 1-, 2-, 3-, or 4-day intervals, according to the probability distribution shown in Table 14.5. Once a batch of yarn arrives at the dyeing facility, it takes either 0.5, 1.0, or 2.0 days to complete the dyeing process, according to the probability distribution shown in Table 14.6.

TABLE 14.5
Distribution of arrival intervals

Arrival Interval (days), x	Probability, $P(x)$	Cumulative Probability	Random Number Range, r_1
1.0	.20	.20	1–20
2.0	.40	.60	21–60
3.0	.30	.90	61–90
4.0	.10	1.00	91–99, 00

TABLE 14.6
Distribution of service times

Service Time (days), y	Probability, $P(y)$	Cumulative Probability	Random Number Range, r_2
0.5	.20	.20	1–20
1.0	.50	.70	21–70
2.0	.30	1.00	71–99, 00

Developing the cumulative probability distribution helps to determine random number ranges.

Table 14.5 defines the interarrival time, or how often batches arrive at the dyeing facility. For example, there is a .20 probability that a batch will arrive *1 day* after the previous batch. Table 14.6 defines the service time for a batch. Notice that cumulative probabilities have been included in Tables 14.5 and 14.6. The cumulative probability provides a means for determining the ranges of random numbers associated with each probability, as we demonstrated with our Excel example in the previous section. For example, in Table 14.5 the first random number range for r_1 is from 1 to 20, which corresponds to the cumulative probability of .20. The second range of random numbers is from 21 to 60, which corresponds to a cumulative probability of .60. Although the cumulative probability goes up to 1.00, Table 14.3 contains only random number values from 0 to 99. Thus, the number 00 is used in place of 100 in the last random number range of each table.

Table 14.7 illustrates the simulation of 10 batch arrivals at the dyeing vat.

TABLE 14.7
Simulation at the Burlingham Mills Dyeing Facility

Batch	r_1	Arrival Interval, x	Arrival Clock	Enter Facility Clock	Waiting Time	r_2	Service Time, y	Departure Clock	Time in System
1	—	—	0.0	0.0	0.0	65	1.0	1.0	1.0
2	71	3.0	3.0	3.0	0.0	18	0.5	3.5	0.5
3	12	1.0	4.0	4.0	0.0	17	0.5	4.5	0.5
4	48	2.0	6.0	6.0	0.0	89	2.0	8.0	2.0
5	18	1.0	7.0	8.0	1.0	83	2.0	10.0	3.0
6	08	1.0	8.0	10.0	2.0	90	2.0	12.0	4.0
7	05	1.0	9.0	12.0	3.0	89	2.0	14.0	5.0
8	18	1.0	10.0	14.0	4.0	08	0.5	14.5	4.5
9	26	2.0	12.0	14.5	2.5	47	1.0	15.5	3.5
10	94	4.0	16.0	16.0	0.0	06	0.5	16.5	0.5
					12.5				24.5

The manual simulation process illustrated in Table 14.7 can be interpreted as follows:

1. Batch 1 arrives at time 0, which is recorded on an *arrival clock*. Because there are no batches in the system, batch 1 approaches the dyeing facility immediately, also at time 0. The waiting time is 0.
2. Next, a random number, $r_2 = 65$, is selected from the second column in Table 14.3. From Table 14.6 we see that the random number 65 results in a service time, y, of 1 day. After leaving the dyeing vat, the batch departs at time 1 day, having been in the queuing system a total of 1 day.
3. The next random number, $r_1 = 71$, is selected from Table 14.3, which specifies that batch 2 arrives 3 days after batch 1, or at time 3.0, as shown on the arrival clock. Because batch 1 departed the service facility at time 1.0, batch 2 can be served immediately and thus incurs no waiting time.
4. The next random number, $r_2 = 18$, is selected from Table 14.3, which indicates that batch 2 will spend 0.5 day being served and will depart at time 3.5.

This process of selecting random numbers and generating arrival intervals and service times continues until 10 batch arrivals have been simulated, as shown in Table 14.7.

Once the simulation is complete, we can compute operating characteristics from the simulation results, as follows:

$$\text{average waiting time} = \frac{12.5 \text{ days}}{10 \text{ batches}} = 1.25 \text{ days per batch}$$

$$\text{average time in the system} = \frac{24.5 \text{ days}}{10 \text{ batches}} = 2.45 \text{ days per batch}$$

However, as in our previous example, these results must be viewed with skepticism. Ten trials of the system do not ensure steady-state results. In general, we can expect a considerable difference between the true average values and the values estimated from only 10 random draws. One reason is that we cannot be sure that the random numbers we selected in this example replicated the actual probability distributions because we used so few random numbers. The stream of random numbers that was used might have had a preponderance of high or low values, thus biasing the final model results. For example, of nine arrivals, five had interarrival times of 1 day. This corresponds to a probability of .55 (i.e., 5/9); however, the actual probability of an arrival interval of 1 day is .20 (from Table 14.5). This excessive number of short interarrival times (caused by the sequence of random numbers) has probably artificially inflated the operating statistics of the system.

As the number of random trials is increased, the probabilities in the simulation will more closely conform to the actual probability distributions. That is, if we simulated the queuing system for 1,000 arrivals, we could more reasonably expect that 20% of the arrivals would have an interarrival time of 1 day.

A factor that can affect simulation results is the starting conditions.

An additional factor that can affect simulation results is the starting conditions. If we start our queuing system with no batches in the system, we must simulate a length of time before the system replicates normal operating conditions. In this example, it is logical to start simulating at the time the vat starts operating in the morning, especially if we simulate an entire working day. Some queuing systems, however, start with items already in the system. For example, the dyeing facility might logically start each day with partially completed batches from the previous day. In this case, it is necessary to begin the simulation with batches already in the system.

By adding a second random variable to a simulation model, such as the one just shown, we increase the complexity and therefore the manual operations. To simulate the example in Table 14.7 manually for 1,000 trials would require several hours. It would be far preferable to perform this type of simulation on a computer. A number of mathematical computations would be required to determine the various column values of Table 14.7, as we will demonstrate in the Excel spreadsheet simulation of this example in the next section.

Computer Simulation of the Queuing Example with Excel

The simulation of the dyeing process at Burlingham Mills shown in Table 14.7 can also be done in Excel. Exhibit 14.6 shows the spreadsheet simulation model for this example.

The stream of random numbers in column C is generated by the formula **=RAND()**, used in the ComputerWorld examples in Exhibits 14.2 through 14.5. (Notice that for this computer simulation, we have changed our random numbers from whole numbers to numbers between 0.0 and 1.0.) The arrival times are generated from the cumulative probability distribution of arrival intervals in cells **B6:C9**. This array of cells is named "Lookup1" because there are two probability distributions in the model. The formula **=VLOOKUP(C15, Lookup1,2)** is entered in cell D15 and copied to cells **D16:D23** to generate the arrival times in column D. The arrival clock times are computed by entering the formula **=E14+D15** in cell E15 and copying it to cells **E16:E23**.

A batch of yarn can enter the dyeing facility as soon as it arrives (in column E) if there is not another batch being dyed or as soon as any batches being dyed or waiting to be dyed have

EXHIBIT 14.6

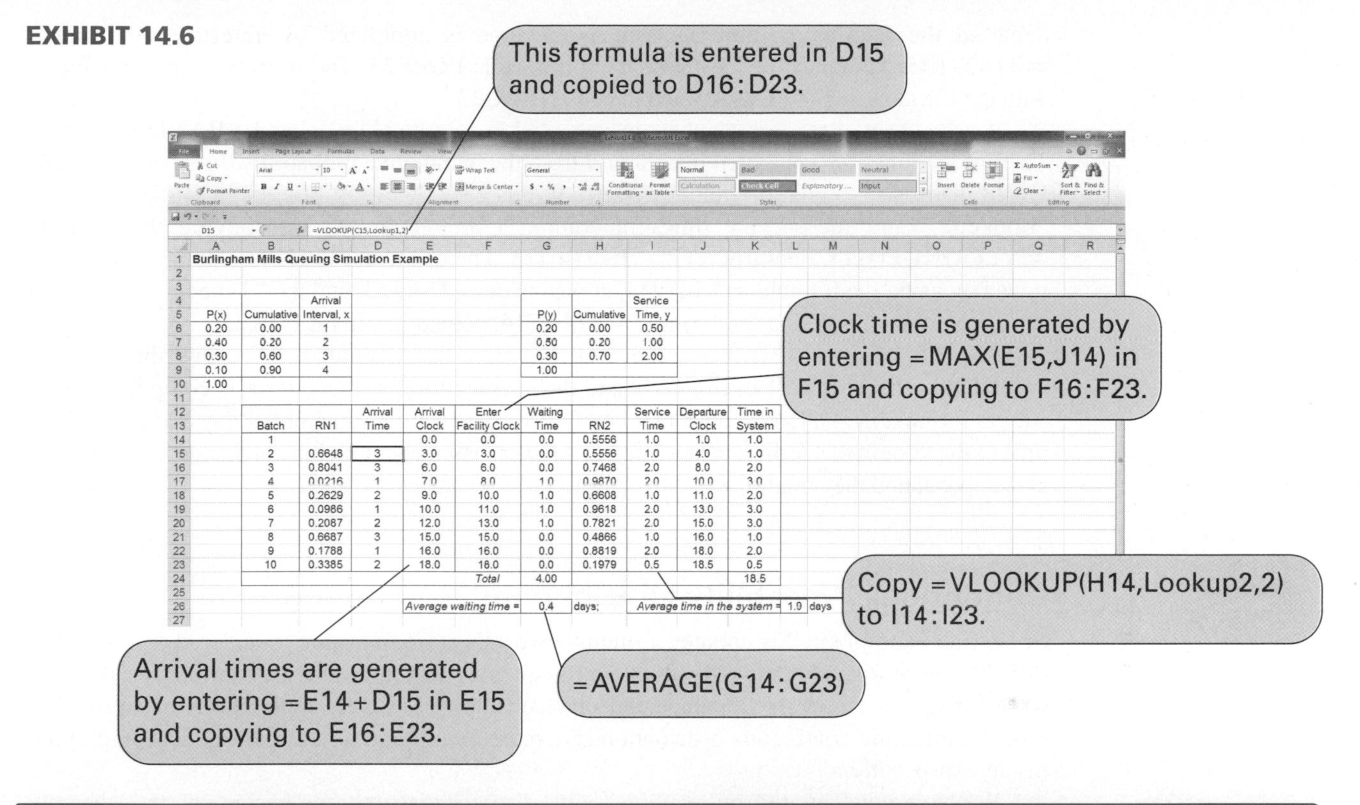

P(x)	Cumulative	Arrival Interval, x
0.20	0.00	1
0.40	0.20	2
0.30	0.60	3
0.10	0.90	4
1.00		

P(y)	Cumulative	Service Time, y
0.20	0.00	0.50
0.50	0.20	1.00
0.30	0.70	2.00
1.00		

Batch	RN1	Arrival Time	Arrival Clock	Enter Facility Clock	Waiting Time	RN2	Service Time	Departure Clock	Time in System
1			0.0	0.0	0.0	0.5556	1.0	1.0	1.0
2	0.6648	3	3.0	3.0	0.0	0.5556	1.0	4.0	1.0
3	0.8041	3	6.0	6.0	0.0	0.7468	2.0	8.0	2.0
4	0.0216	1	7.0	8.0	1.0	0.9870	2.0	10.0	3.0
5	0.2629	2	9.0	10.0	1.0	0.6608	1.0	11.0	2.0
6	0.0986	1	10.0	11.0	1.0	0.9618	2.0	13.0	3.0
7	0.2087	2	12.0	13.0	1.0	0.7821	2.0	15.0	3.0
8	0.6687	3	15.0	15.0	0.0	0.4866	1.0	16.0	1.0
9	0.1788	1	16.0	16.0	0.0	0.8819	2.0	18.0	2.0
10	0.3385	2	18.0	18.0	0.0	0.1979	0.5	18.5	0.5
				Total	4.00				18.5

Management Science Application

Planning for Health Emergencies Created by Terrorist Attacks

In today's world the threat of terrorist attacks cannot be underestimated or discounted, especially in densely populated or strategically important areas of the country, like Washington, DC. Because such attacks could take the form of outbreaks of contagious diseases, public health departments have been prompted to plan how to respond to such health disasters.

If terrorists released a lethal virus like smallpox into the general population, everyone would have to be vaccinated within a few days. This would require a massive response including the mass-dispensing of vaccinations through clinics staffed with the necessary people to perform the tasks. For example, Montgomery County, Maryland, which borders the District of Columbia, would need to vaccinate nearly a million people in just a few days. To vaccinate such a large number of people would require mass-vaccination clinics throughout the county. Any type of planning model would need to include the number of people to train beforehand, the clinics' capacities, and the time residents would spend at a clinic (i.e., time in the system).

A simulation study was used to evaluate alternative mass-vaccination clinic designs—specifically, clinic operations, clinic capacity, and resident time in the system. Model data for arrivals, the time residents stayed at each station, and the total time at the clinic was derived from a full-scale test exercise at a mock-up clinic with actual workers and volunteers. The simulation model included animation so that the user could visualize residents moving through the clinic. The simulation study provided guidelines on how many staff members would be required (fewer than estimated) and how to manage bus arrivals. The simulation study provided input into additional quantitative models for capacity planning, clinic layout, overall clinic operations, and the flow of residents through the clinic.

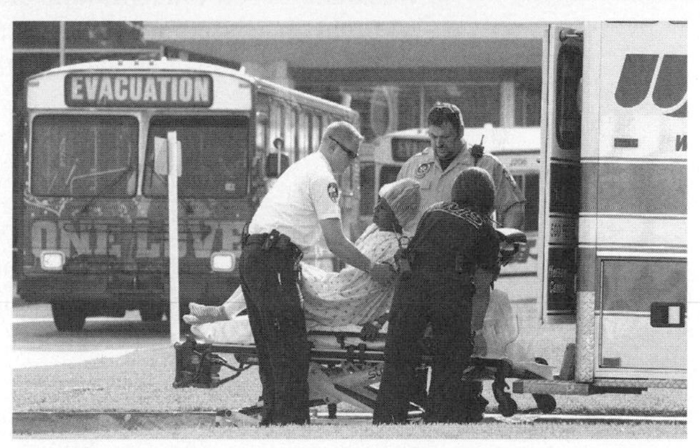

AP Photo/Brian Lawdermilk

Source: Based on K. Aaby, J. Herrmann, C. Jordan, M. Treadwell, and K. Wood, "Montgomery County's Public Health Service Uses Operations Research to Plan Emergency Mass Dispensing and Vaccination Clinics," *Interfaces* 36, no. 6 (November–December 2006): 569–79.

departed the facility (column J). This clock time is computed by entering the formula **=MAX(E15, J14)** in cell F15 and copying it to cells **F16:F23**. The waiting time is computed with the formula **=F14−E14**, copied in cells **G14:G23**.

A second set of random numbers is generated in column H by using the **RAND()** function. The service times are generated in column I from the cumulative probability distribution in cells **H6:I8**, using the "Lookup" function again. In this case the array of cells in **H6:I8** is named "Lookup2," and the service times in column I are generated by copying the formula **=VLOOKUP(H14, Lookup2,2)** in cells **I14:I23**. The departure times in column J are determined by using the formula **=F14+I14**, copied in cells **J14:J23**, and the "Time in System" values are computed by using the formula **=J14−E14**, copied in cells **K14:K23**.

The operating statistic, average waiting time, is computed by using the formula **=AVERAGE(G14:G23)** in cell G26, and the average time in the system is computed with a similar formula in cell L26. Notice that both the average waiting time of 0.4 day and the average time in the system of 1.9 days are significantly lower than the simulation conducted in Table 14.7, as we speculated they might be.

Continuous Probability Distributions

In the first example in this chapter, ComputerWorld's store manager considered a probability distribution of discrete demand values. In the queuing example, the probability distributions were for discrete interarrival times and service times. However, applications of simulation models reflecting continuous distributions are more common than those of models employing discrete distributions.

We have concentrated on examples with discrete distributions because with a discrete distribution, the ranges of random numbers can be explicitly determined and are thus easier to illustrate. When random numbers are being selected according to a continuous probability distribution, a continuous function must be used. For example, consider the following continuous probability function, $f(x)$, for time (minutes), x:

$$f(x) = \frac{x}{8}, 0 \leq x \leq 4$$

The area under the curve, $f(x)$, represents the probability of the occurrence of the random variable x. Therefore, the area under the curve must equal 1.0 because the sum of all probabilities of the occurrence of a random variable must equal 1.0. By computing the area under the curve from 0 to any value of the random variable x, we can determine the cumulative probability of that value of x, as follows:

$$F(x) = \int_0^x \frac{x}{8} dx = \frac{1}{8}\int_0^x x dx = \frac{1}{8}\left(\frac{1}{2}x^2\right)\Big]_0^x$$

$$F(x) = \frac{x^2}{16}$$

Cumulative probabilities are analogous to the discrete ranges of random numbers we used in previous examples. Thus, we let this function, $F(x)$, equal the random number r,

$$r = \frac{x^2}{16}$$

and solve for x,

$$x = 4\sqrt{r}$$

By generating a random number, r, and substituting it into this function, we determine a value for x, "time." (However, for a continuous function, the range of random numbers must be between zero and one to correspond to probabilities between 0.0 and 1.00.) For example, if $r = .25$, then

$$x = 4\sqrt{.25} = 2 \text{ min.}$$

The purpose of briefly presenting this example is to demonstrate the difference between discrete and continuous functions. This continuous function is relatively simple; as functions become more complex, it becomes more difficult to develop the equation for determining the random variable x from r. Even this simple example required some calculus, and developing more complex models would require a higher level of mathematics.

Simulation of a Machine Breakdown and Maintenance System

A continuous probability distribution of the time between machine breakdowns.

In this example we will demonstrate the use of a continuous probability distribution. The Bigelow Manufacturing Company produces a product on a number of machines. The elapsed time between breakdowns of the machines is defined by the following continuous probability distribution:

$$f(x) = \frac{x}{8}, \; 0 \leq x \leq 4 \text{ weeks}$$

where

x = weeks between machine breakdowns

As indicated in the previous section on continuous probability distributions, the equation for generating x, given the random number r_1, is

$$x = 4\sqrt{r_1}$$

When a machine breaks down, it must be repaired, and it takes either 1, 2, or 3 days for the repair to be completed, according to the discrete probability distribution shown in Table 14.8. Every time a machine breaks down, the cost to the company is an estimated $2,000 per day in lost production until the machine is repaired.

TABLE 14.8 Probability distribution of machine repair time

Machine Repair Time, y (days)	Probability of Repair Time, $P(y)$	Cumulative Probability	Random Number Range, r_2
1	.15	.15	0.00–.15
2	.55	.70	.16–.70
3	.30	1.00	.71–1.00

The company would like to know whether it should implement a machine maintenance program at a cost of $20,000 per year that would reduce the frequency of breakdowns and thus the time for repair. The maintenance program would result in the following continuous probability function for time between breakdowns:

$$f(x) = x/18, \quad 0 \leq x \leq 6 \text{ weeks}$$

where

x = weeks between machine breakdowns

The equation for generating x, given the random number r_1, for this probability distribution is

$$x = 6\sqrt{r_1}$$

The reduced repair time resulting from the maintenance program is defined by the discrete probability distribution shown in Table 14.9.

TABLE 14.9
Revised probability distribution of machine repair time with the maintenance program

Machine Repair Time, y (days)	Probability of Repair Time, $P(y)$	Cumulative Probability	Random Number Range, r_2
1	.40	.40	0.00–.40
2	.50	.90	.41–.90
3	.10	1.00	.91–1.00

To solve this problem, we must first simulate the existing system to determine an estimate of the average annual repair costs. Then we must simulate the system with the maintenance program installed to see what the average annual repair costs will be with the maintenance program. We will then compare the average annual repair cost with and without the maintenance program and compute the difference, which will be the average annual savings in repair costs with the maintenance program. If this savings is more than the annual cost of the maintenance program ($20,000), we will recommend that it be implemented; if it is less, we will recommend that it not be implemented.

First, we will manually simulate the existing breakdown and repair system without the maintenance program, to see how the simulation model is developed. Table 14.10 illustrates the simulation of machine breakdowns and repair for 20 breakdowns that occur over a period of approximately 1 year (i.e., 52 weeks).

TABLE 14.10
Simulation of machine breakdowns and repair times

Breakdowns	r_1	Time Between Breakdowns, x (weeks)	r_2	Repair Time, y (days)	Cost, $2,000$y$	Cumulative Time, Σx (weeks)
1	.45	2.68	.19	2	$4,000	2.68
2	.90	3.80	.65	2	4,000	6.48
3	.84	3.67	.51	2	4,000	10.15
4	.17	1.65	.17	2	4,000	11.80
5	.74	3.44	.63	2	4,000	15.24
6	.94	3.88	.85	3	6,000	19.12
7	.07	1.06	.37	2	4,000	20.18
8	.15	1.55	.89	3	6,000	21.73
9	.04	0.80	.76	3	6,000	22.53
10	.31	2.23	.71	3	6,000	24.76
11	.07	1.06	.34	2	4,000	25.82
12	.99	3.98	.11	1	2,000	29.80
13	.97	3.94	.27	2	4,000	33.74
14	.73	3.42	.10	1	2,000	37.16
15	.13	1.44	.59	2	4,000	38.60
16	.03	0.70	.87	3	6,000	39.30
17	.62	3.15	.08	1	2,000	42.45
18	.47	2.74	.08	1	2,000	45.19
19	.99	3.98	.89	3	6,000	49.17
20	.75	3.46	.42	2	4,000	52.63
					$84,000	

The simulation in Table 14.10 results in a total annual repair cost of $84,000. However, this is for only 1 year, and thus it is probably not very accurate.

The next step in our simulation analysis is to simulate the machine breakdown and repair system with the maintenance program installed. We will use the revised continuous probability distribution for time between breakdowns and the revised discrete probability distribution for repair time shown in Table 14.9. Table 14.11 illustrates the manual simulation of machine breakdowns and repair for 1 year.

TABLE 14.11
Simulation of machine breakdowns and repair with the maintenance program

Breakdowns	r_1	Time Between Breakdowns, x (weeks)	r_2	Repair Time, y (days)	Cost, \$2,000y	Cumulative Time, Σx (weeks)
1	.45	4.03	.19	1	$2,000	4.03
2	.90	5.69	.65	2	4,000	9.72
3	.84	5.50	.51	2	4,000	15.22
4	.17	2.47	.17	1	2,000	17.69
5	.74	5.16	.63	2	4,000	22.85
6	.94	5.82	.85	2	4,000	28.67
7	.07	1.59	.37	1	2,000	30.29
8	.15	2.32	.89	2	4,000	32.58
9	.04	1.20	.76	2	4,000	33.78
10	.31	3.34	.71	2	4,000	37.12
11	.07	1.59	.34	1	2,000	38.71
12	.99	5.97	.11	1	2,000	44.68
13	.97	5.91	.27	1	2,000	50.59
14	.73	5.12	.10	1	2,000	55.71
					$42,000	

Table 14.11 shows that the annual repair cost with the maintenance program totals $42,000. Recall that in the manual simulation shown in Table 14.10, the annual repair cost was $84,000 for the system without the maintenance program. The difference between the two annual repair costs is $84,000 − 42,000 = $42,000. This figure represents the savings in average annual repair cost with the maintenance program. Because the maintenance program will cost $20,000 per year, it would seem that the recommended decision would be to implement the maintenance program and generate an expected annual savings of $22,000 per year (i.e., $42,000 − 20,000 = $22,000).

Manual simulation is limited because of the amount of real time required to simulate even one trial.

However, let us now concern ourselves with the potential difficulties caused by the fact that we simulated each system (the existing one and the system with the maintenance program) only *once*. Because the time between breakdowns and the repair times are probabilistic, the simulation results could exhibit significant variation. The only way to be sure of the accuracy of our results is to simulate each system many times and compute an average result. Performing these many simulations manually would obviously require a great deal of time and effort. However, Excel can be used to accomplish the required simulation analysis.

Computer Simulation of the Machine Breakdown Example Using Excel

Exhibit 14.7 shows the Excel spreadsheet model of the simulation of our original machine breakdown example simulated manually in Table 14.10. The Excel simulation is for 100 breakdowns. The random numbers in **C14:C113** are generated using the **RAND()** function, which was used in our previous Excel examples. The "Time Between Breakdowns" values in column D are developed using the formula for the continuous cumulative probability function, **=4*SQRT(C14)**, typed in cell D14 and copied in cells **D15:D113**.

The "Cumulative Time" in column E is computed by copying the formula **=E14+D15** in cells **E15:E113.** The second stream of random numbers in column F is generated using the **RAND()** function. The "Repair Time" values in column G are generated from the cumulative probability distribution in the array of cells **B6:C8**. As in our previous examples, we name this array "Lookup" and copy the formula **=VLOOKUP(F14,Lookup,2)** in cells **G14:G113**. The cost values in column H are computed by entering the formula **=2000*G14** in cell H14 and copying it to cells **H15:H113**.

The "Average Annual Cost" in cell H7 is computed with the formula **=SUM(H14:H113)/(E113/52)**. For this, the original problem, the annual cost is $82,397.35, which is not too different from the manual simulation in Table 14.10.

EXHIBIT 14.7

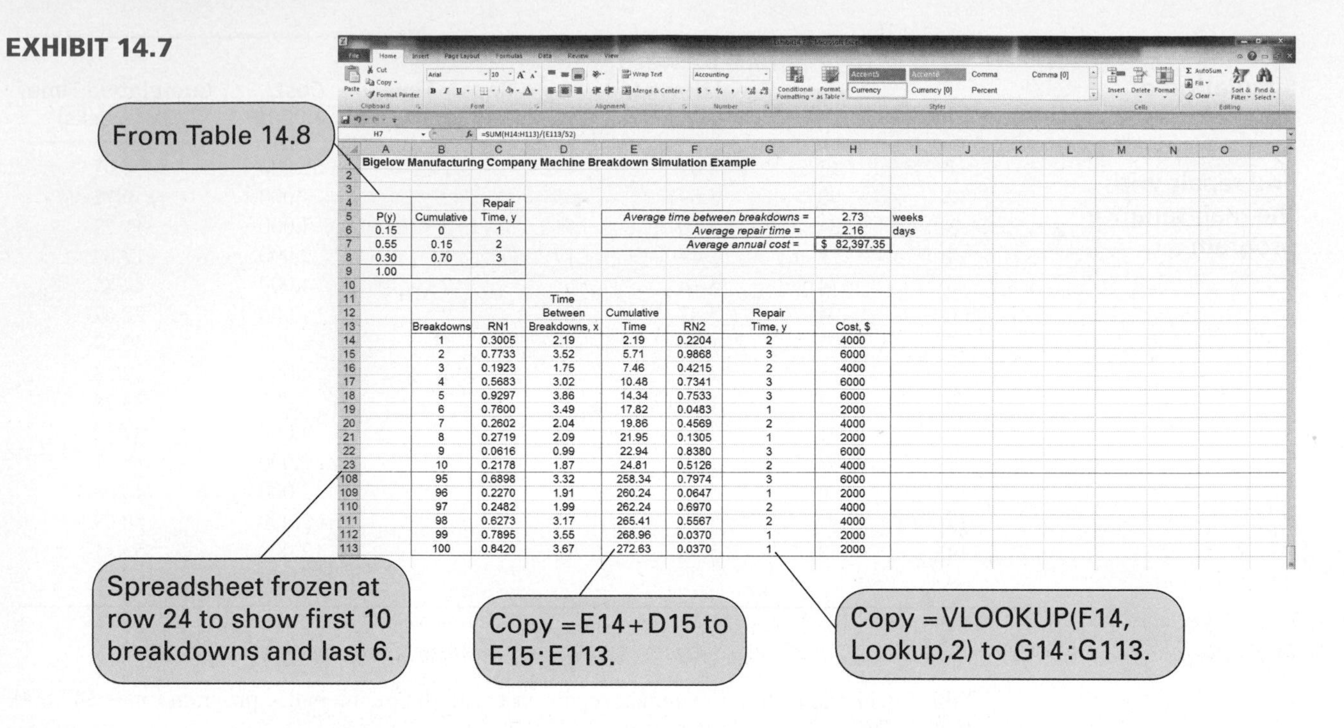

Bigelow Manufacturing Company Machine Breakdown Simulation Example

P(y)	Cumulative	Repair Time, y
0.15	0	1
0.55	0.15	2
0.30	0.70	3
1.00		

Average time between breakdowns =	2.73	weeks
Average repair time =	2.16	days
Average annual cost =	$ 82,397.35	

Breakdowns	RN1	Time Between Breakdowns, x	Cumulative Time	RN2	Repair Time, y	Cost, $
1	0.3005	2.19	2.19	0.2204	2	4000
2	0.7733	3.52	5.71	0.9868	3	6000
3	0.1923	1.75	7.46	0.4215	2	4000
4	0.5683	3.02	10.48	0.7341	3	6000
5	0.9297	3.86	14.34	0.7533	3	6000
6	0.7600	3.49	17.82	0.0483	1	2000
7	0.2602	2.04	19.86	0.4569	2	4000
8	0.2719	2.09	21.95	0.1305	1	2000
9	0.0616	0.99	22.94	0.8380	3	6000
10	0.2178	1.87	24.81	0.5126	2	4000
95	0.6898	3.32	258.34	0.7974	3	6000
96	0.2270	1.91	260.24	0.0647	1	2000
97	0.2482	1.99	262.24	0.6970	2	4000
98	0.6273	3.17	265.41	0.5567	2	4000
99	0.7895	3.55	268.96	0.0370	1	2000
100	0.8420	3.67	272.63	0.0370	1	2000

Exhibit 14.8 shows the Excel spreadsheet simulation for the modified breakdown system with the new maintenance program, which was simulated manually in Table 14.11. The two differences in this simulation model are the cumulative probability distribution formulas for the time between breakdowns and the reduced repair time distributions from Table 14.9 in cells **A6:C8**.

The average annual cost for this model, shown in cell H8, is $44,504.74. This annual cost is only slightly higher than the $42,000 obtained from the manual simulation in Table 14.11. Thus, as before, the decision should be to implement the new maintenance system.

EXHIBIT 14.8

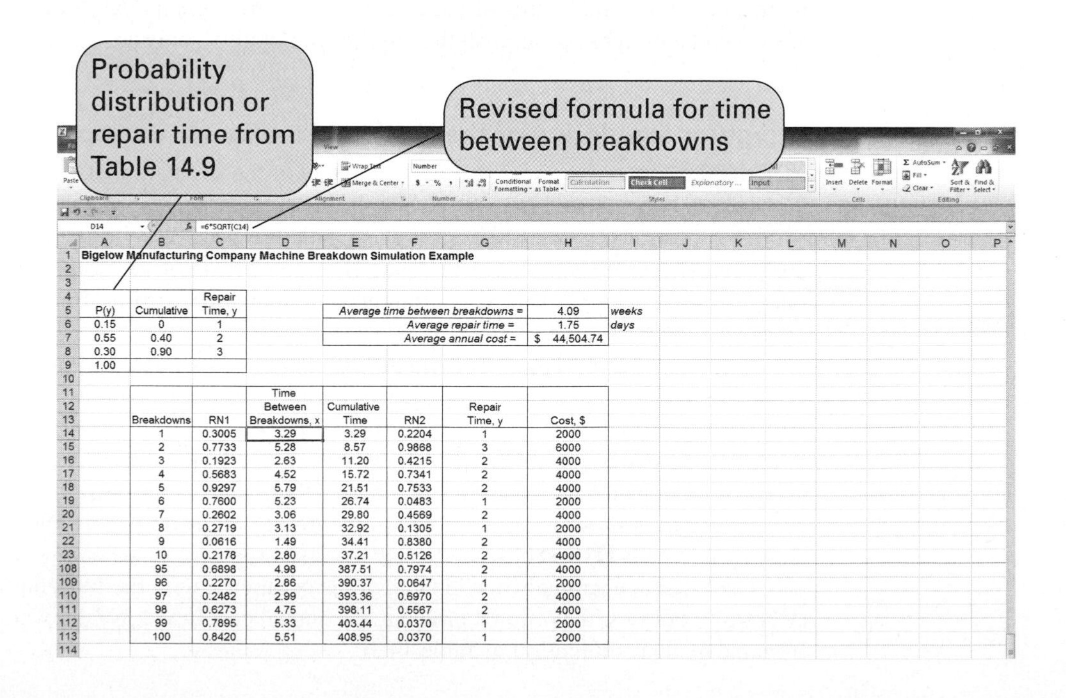

Bigelow Manufacturing Company Machine Breakdown Simulation Example

P(y)	Cumulative	Repair Time, y
0.15	0	1
0.55	0.40	2
0.30	0.90	3
1.00		

Average time between breakdowns =	4.09	weeks
Average repair time =	1.75	days
Average annual cost =	$ 44,504.74	

Breakdowns	RN1	Time Between Breakdowns, x	Cumulative Time	RN2	Repair Time, y	Cost, $
1	0.3005	3.29	3.29	0.2204	1	2000
2	0.7733	5.28	8.57	0.9868	3	6000
3	0.1923	2.63	11.20	0.4215	2	4000
4	0.5683	4.52	15.72	0.7341	2	4000
5	0.9297	5.79	21.51	0.7533	2	4000
6	0.7600	5.23	26.74	0.0483	1	2000
7	0.2602	3.06	29.80	0.4569	2	4000
8	0.2719	3.13	32.92	0.1305	1	2000
9	0.0616	1.49	34.41	0.8380	2	4000
10	0.2178	2.80	37.21	0.5126	2	4000
95	0.6898	4.98	387.51	0.7974	2	4000
96	0.2270	2.86	390.37	0.0647	1	2000
97	0.2482	2.99	393.36	0.6970	2	4000
98	0.6273	4.75	398.11	0.5567	2	4000
99	0.7895	5.33	403.44	0.0370	1	2000
100	0.8420	5.51	408.95	0.0370	1	2000

Statistical Analysis of Simulation Results

In general, the outcomes of a simulation model are statistical measures such as averages, as in the examples presented in this chapter. In our ComputerWorld example, we generated average revenue as a measure of the system we simulated; in the Burlingham Mills queuing example, we generated the average waiting time for batches to be dyed; and in the Bigelow Manufacturing Company machine breakdown example, the system was measured in terms of average repair costs. However, we also discussed the care that must be taken in accepting the accuracy of these statistical results because they were frequently based on relatively few observations (i.e., simulation replications). Thus, as part of the simulation process, these statistical results are typically subjected to additional statistical analysis to determine their degree of accuracy.

One of the most frequently used tools for the analysis of the statistical validity of simulations results is confidence limits. Confidence limits can be developed within Excel for the averages resulting from simulation models in several different ways. Recall that the statistical formulas for 95% confidence limits are:

$$\text{upper confidence limit} = \bar{x} + (1.96)\,(s/\sqrt{n})$$
$$\text{lower confidence limit} = \bar{x} - (1.96)\,(s/\sqrt{n})$$

where $\bar{x}$ is the mean and s is the sample standard deviation from a sample of size n from any population. Although we cannot be sure that the sample mean will exactly equal the population mean, we can be 95% confident that the true population mean will be between the upper confidence limit (UCL) and lower confidence limit (LCL) computed using these formulas.

Exhibit 14.9 shows the Excel spreadsheet for our machine breakdown example (from Exhibit 14.8), with the upper and lower confidence limits for average repair cost in cells L13 and L14. Cell L11 contains the average repair cost (for each incidence of a breakdown), computed by using the formula **=AVERAGE(H14:H113)**. Cell L12 contains the sample standard deviation, computed by using the formula **=STDEV(H14:H113)**. The upper confidence limit is computed in cell L13 by using the formula **=L11+1.96*L12/SQRT(100)**, shown on the formula bar at the top of the spreadsheet, and the lower control limit is computed similarly. Thus, we can be 95% confident that the true average repair cost for the population is between $3,248.50 and $3,751.50.

EXHIBIT 14.9

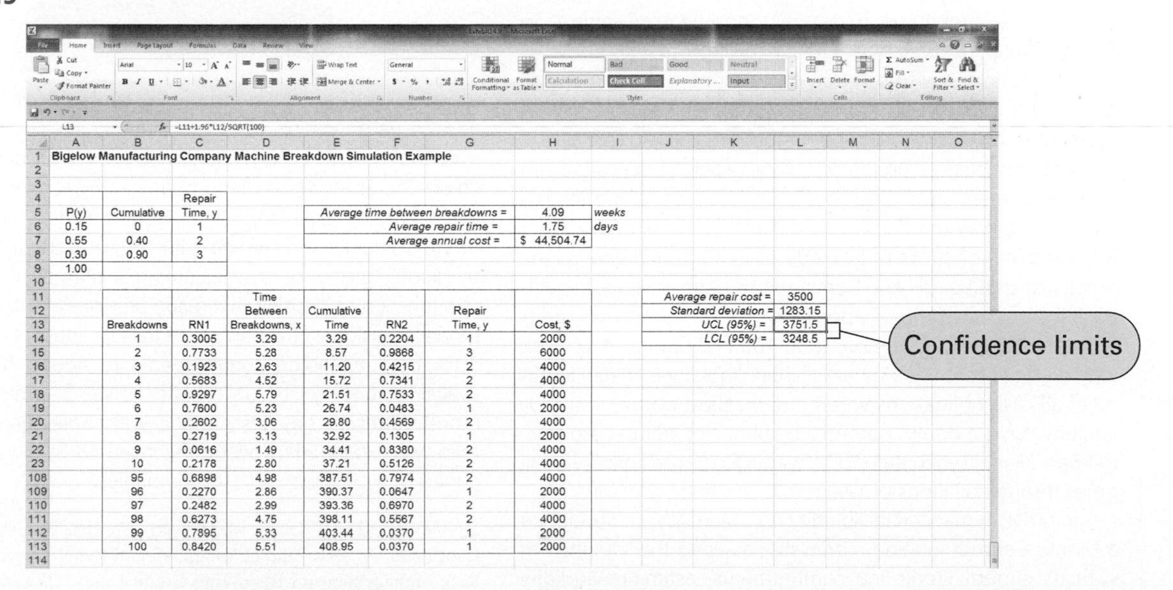

Bigelow Manufacturing Company Machine Breakdown Simulation Example

P(y)	Cumulative	Repair Time, y
0.15	0	1
0.55	0.40	2
0.30	0.90	3
1.00		

Average time between breakdowns =	4.09	weeks
Average repair time =	1.75	days
Average annual cost =	$ 44,504.74	

Average repair cost =	3500
Standard deviation =	1283.15
UCL (95%) =	3751.5
LCL (95%) =	3248.5

Breakdowns	RN1	Time Between Breakdowns, x	Cumulative Time	RN2	Repair Time, y	Cost, $
1	0.3005	3.29	3.29	0.2204	1	2000
2	0.7733	5.28	8.57	0.9868	3	6000
3	0.1923	2.63	11.20	0.4215	2	4000
4	0.5683	4.52	15.72	0.7341	2	4000
5	0.9297	5.79	21.51	0.7533	2	4000
6	0.7600	5.23	26.74	0.0483	1	2000
7	0.2602	3.06	29.80	0.4569	2	4000
8	0.2719	3.13	32.92	0.1305	1	2000
9	0.0616	1.49	34.41	0.8380	2	4000
10	0.2178	2.80	37.21	0.5126	2	4000
95	0.6898	4.98	387.51	0.7974	2	4000
96	0.2270	2.86	390.37	0.0647	1	2000
97	0.2482	2.99	393.36	0.6970	2	4000
98	0.6273	4.75	398.11	0.5567	2	4000
99	0.7895	5.33	403.44	0.0370	1	2000
100	0.8420	5.51	408.95	0.0370	1	2000

Confidence limits plus several additional statistics can also be obtained by using the "Data Analysis" option from the "Data" menu. Select the "Data Analysis" option from the "Data" menu at the top of the spreadsheet, and then from the resulting menu select "Descriptive Statistics." This will result in a Descriptive Statistics dialog box like the one shown on the right in Exhibit 14.10.

EXHIBIT 14.10

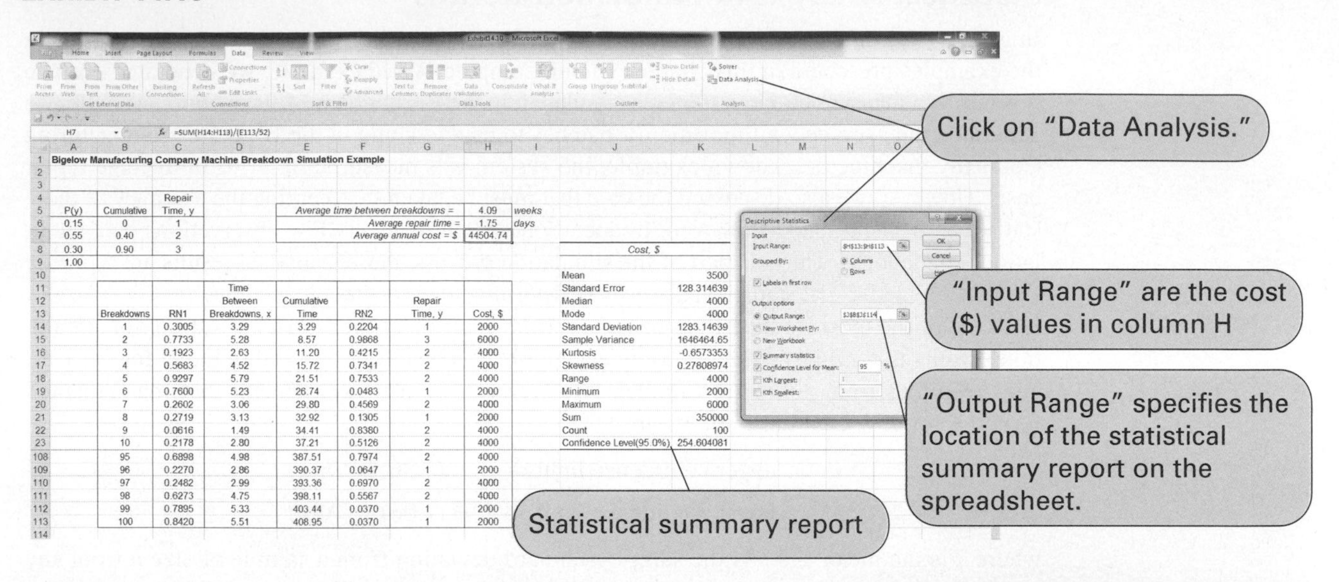

H7 =SUM(H14:H113)/(E113/52)

Bigelow Manufacturing Company Machine Breakdown Simulation Example

P(y)	Cumulative	Repair Time, y
0.15	0	1
0.55	0.40	2
0.30	0.90	3
1.00		

Average time between breakdowns =	4.09	*weeks*
Average repair time =	1.75	*days*
Average annual cost = $	44504.74	

Breakdowns	RN1	Time Between Breakdowns, x	Cumulative Time	RN2	Repair Time, y	Cost, $
1	0.3005	3.29	3.29	0.2204	1	2000
2	0.7733	5.28	8.57	0.9868	3	6000
3	0.1923	2.63	11.20	0.4215	2	4000
4	0.5683	4.52	15.72	0.7341	2	4000
5	0.9297	5.79	21.51	0.7533	2	4000
6	0.7600	5.23	26.74	0.0483	1	2000
7	0.2602	3.06	29.80	0.4569	2	4000
8	0.2719	3.13	32.92	0.1305	1	2000
9	0.0616	1.49	34.41	0.8380	2	4000
10	0.2178	2.80	37.21	0.5126	2	4000
95	0.6898	4.98	387.51	0.7974	2	4000
96	0.2270	2.86	390.37	0.0647	1	2000
97	0.2482	2.99	393.36	0.6970	2	4000
98	0.6273	4.75	398.11	0.5567	2	4000
99	0.7895	5.33	403.44	0.0370	1	2000
100	0.8420	5.51	408.95	0.0370	1	2000

	Cost, $
Mean	3500
Standard Error	128.314639
Median	4000
Mode	4000
Standard Deviation	1283.14639
Sample Variance	1646464.65
Kurtosis	-0.6573353
Skewness	0.27808974
Range	4000
Minimum	2000
Maximum	6000
Sum	350000
Count	100
Confidence Level(95.0%)	254.604081

Management Science Application

A Monte Carlo Simulation Model for Managing Liquidity Risk at Merrill Lynch Bank USA

Merrill Lynch, founded in 1914 by Charles E. Merrill, has two major divisions: Global Private Client Group, the retail part of the company, and the Global Markets and Investment Banking division, which serves major corporations by helping them raise capital through new issues of equity and debt. The Global Private Client Group that the public is most familiar with provides brokerage, investment, and banking services to about 3 million household clients through a sales force of 14,000 financial advisors in 500 offices around the world, and it manages assets of $3 trillion. Merrill Lynch Bank USA, with assets of over $60 billion, supports both of these divisions.

The bank acts as an intermediary, receiving deposits from retail customers and using these funds for loans and investments. One way the bank uses these assets is by providing a revolving line of short-term credit to institutional borrowers, much like credit cards and home equity loans work for individuals; it has a portfolio of approximately $13 billion in credit-line commitments with more than 100 institutions. A key risk with credit lines is liquidity risk, where liquidity is the ability to meet all cash obligations when due on short notice, and thus liquidity risk is a bank's potential inability to meet its cash obligations. Merrill Lynch Bank USA wants to ensure liquidity at all times through all kinds of financial cycles and conditions.

In order to manage its liquidity risk, Merrill Lynch developed a Monte Carlo simulation model that assesses the liquidity risk with its current credit-line commitments, estimates liquidity needs in extreme financial situations, identifies potential risk exposure, such as concentrations of credit in specific industries, and evaluates possible portfolios of credit line commitments for potential growth in the future. The model simulates monthly credit-line usage for each company and each type of security over a 5-year time horizon; it also provides the bank with monthly liquidity requirements for the total credit-line portfolio by modeling each borrower's timing and amount of credit use as a function of its credit rating and previous month's usage. The model has allowed the bank to reduce its liquidity reserves from 50% of outstanding commitments to about 20%, making available about $4 billion for other uses. During the first 21 months after the simulation model was implemented, the portfolio was expanded by over 60%, from $8 billion and 80 companies to $13 billion and 100 companies.

© Steven Widoff / Alamy

Source: Based on T. Duffy, M. Hatzakis, W. Hsu, R. Labe, B. Liao, X. Luo, J. Oh, A. Setya, and L. Yang, "Merrill Lynch Improves Liquidity Risk Management for Revolving Credit Lines," *Interfaces* 35, no. 5 (September–October 2005): 353–69.

This box, completed as shown, results in the summary statistics for repair costs shown in cells **J8:K23** in Exhibit 14.10. These summary statistics include the mean, standard deviation, and confidence limits.

Crystal Ball

So far in this chapter we have used simulation examples that included mostly discrete probability distributions that we set up on an Excel spreadsheet. These are the easiest types of probability distributions to work with in spreadsheets. However, many realistic problems contain more complex probability distributions, like the normal distribution, which are not discrete but are continuous, or they include discrete probability distributions that are more difficult to work with than the simple ones we have used. However, there are several simulation add-ins for Excel that provide the user with the capability to perform simulation analysis, using a variety of different probability distributions in a spreadsheet environment. One of these add-ins is Crystal Ball, published by Oracle; it is available for download from the Crystal Ball Web site, using the licensing instructions that accompany this text. Crystal Ball is a risk analysis and forecasting program that uses Monte Carlo simulation to forecast a statistical range of results possible for a given situation. In this section we will provide an overview of how to apply Crystal Ball to a simple example for profit (break-even) analysis that we first introduced in Chapter 1.

Simulation of a Profit Analysis Model

In Chapter 1 we used a simple example for the Western Clothing Company to demonstrate break-even and profit analysis. In that example, Western Clothing Company produced denim jeans. The price (p) for jeans was $23, the variable cost ($c_v$) was $8 per pair of jeans, and the fixed cost (c_f) was $10,000. Given these parameters, we formulated a profit (Z) function as follows:

$$Z = vp - c_f - vc_v$$

Our objective in that analysis was to determine the break-even volume, v, that would result in no profit or loss. This was accomplished by setting $Z = 0$ and solving the profit function for v, as follows:

$$v = \frac{c_f}{p - c_v}$$

Substituting the values for p, c_f, and c_v into this formula resulted in the break-even volume:

$$v = \frac{10{,}000}{23 - 8}$$
$$= 666.7 \text{ pairs of jeans}$$

To demonstrate the use of Crystal Ball, we will modify that example. First, we will assume that volume is actually volume *demanded* and that it is a random variable defined by a normal probability distribution, with a mean value of 1,050 pairs of jeans and a standard deviation of 410.

Furthermore, we will assume that the price is not fixed but is also uncertain and defined by a uniform probability distribution (from $20 to $26) and that variable cost is not a constant value but defined by a triangular probability distribution. Instead of seeking to determine the break-even volume, we will simulate the profit model, given probabilistic demand, price, and variable costs to determine average profit and the probability that Western Clothing will break even.

The first thing we need to do is access Crystal Ball, which you can download according to the instructions provided with this text.

Exhibit 14.11 shows the Excel spreadsheet for our example.[1] We have described the parameters of each probability distribution in our profit model next to its corresponding cell. For example, cell C4 contains the probability distribution for demand. We want to generate demand values in this cell according to the probability distribution for demand (i.e., Monte Carlo simulation). We also want to do this in cell C5 for price and in cell C7 for variable cost. This is the same process we used in our earlier ComputerWorld example to generate demand values from a discrete probability distribution, using random numbers. Notice that cell C9 contains our formula for profit, **=C4*C5−C6−(C4*C7).** This is the only cell formula in our spreadsheet.

EXHIBIT 14.11

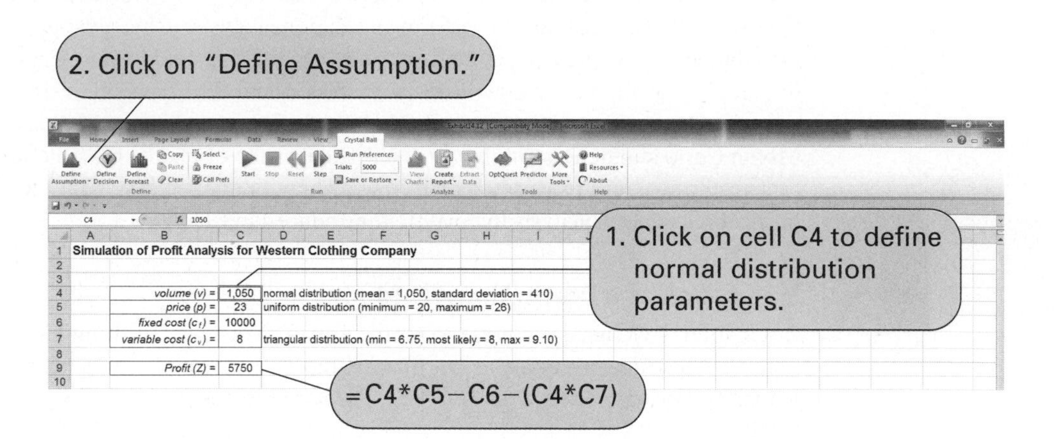

To set up the normal probability distribution for demand, we first enter the mean value 1,050 in cell C4. Cells require some initial value to start with. Next we click on "Define Assumption" from the top of the spreadsheet, as shown in Exhibit 14.11, which will result in a drop-down menu of distributions.

The distribution drop-down menu window includes several different probability distributions we can use. Because we have indicated that demand is defined by a normal distribution, click on it. This will result in the normal distribution window shown in Exhibit 14.12.

The "Name" value in the box at the top of the window in Exhibit 14.12 was automatically pulled from the spreadsheet, where it is the heading "volume (v) ="; however, a new or different name could be typed in. Next, we click on "Mean" or use the Tab key to toggle down to the "Mean" display in the lower-left-hand corner of this window. Because we entered the mean value of 1,050 in cell C4 on our spreadsheet, this value will already be shown in this window. Next, we click on "Std. Dev." or use the Tab key to move to the Std. Dev. window and enter the standard deviation of 410. Then we click on the "Enter" button, which will configure the normal distribution figure in the window, and then we click on "OK."

We will repeat this same process to enter the parameters for the uniform distribution for price in cell C5. First, we enter the value for price, 23, in cell C5. Next (with cell C5 activated), we click on "Define Assumption" at the top of the spreadsheet, as shown earlier in Exhibit 14.11.

[1] This example in Exhibit 14.11 is located on the Web site accompanying this text, in the "Crystal Ball" folder.

EXHIBIT 14.12

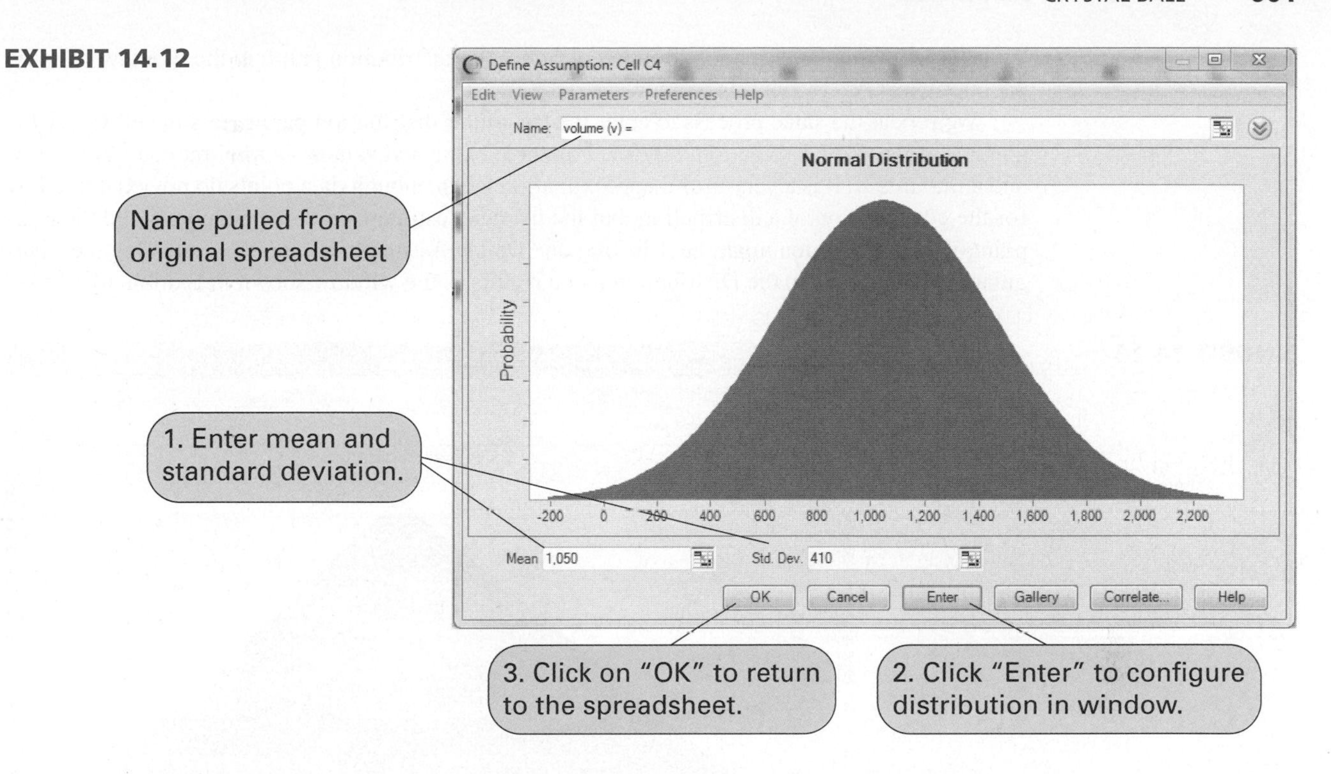

The Distribution menu window will again appear, and this time we click on "Uniform Distribution." This results in the Uniform Distribution window shown in Exhibit 14.13.

EXHIBIT 14.13

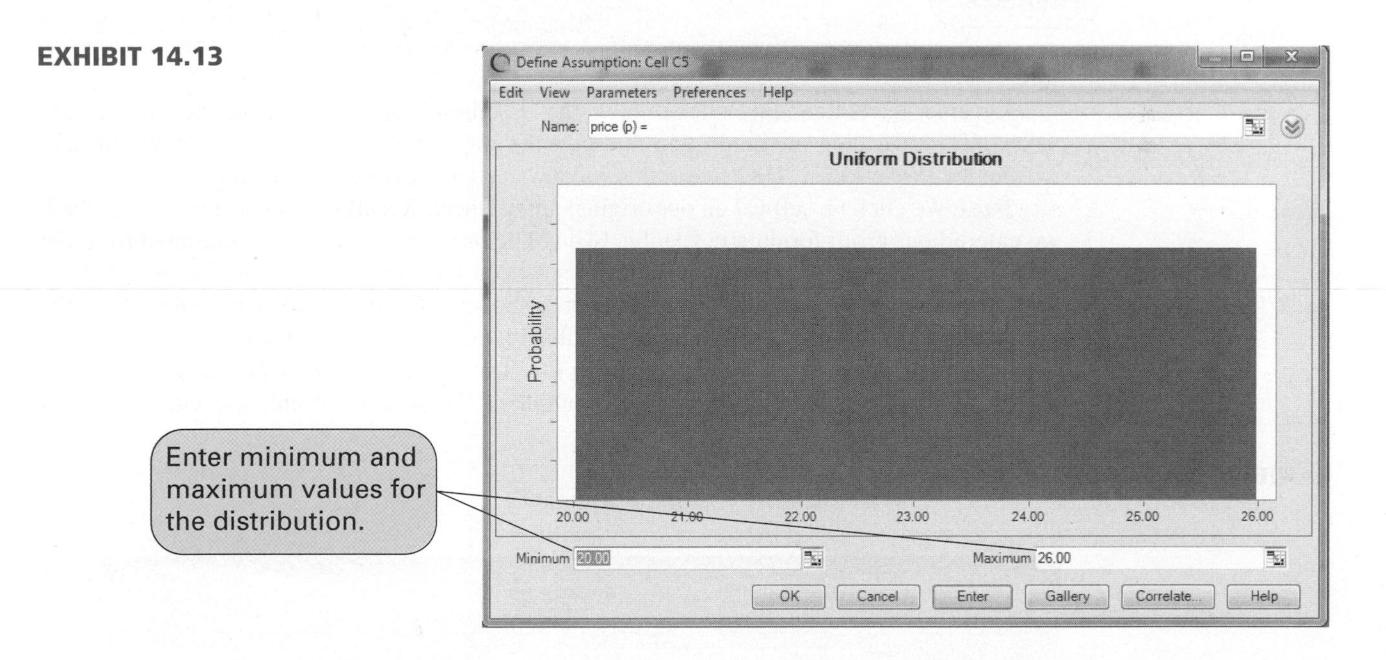

As before, the "Name" value, "price (p)," was pulled from the original spreadsheet in Exhibit 14.11. Next, we click on "Minimum" or use the Tab key to move to the "Minimum" display at the bottom of the window and enter 20, the lower limit of the uniform distribution specified in the problem statement. Next, we activate the "Maximum" display window and enter 26.

Then we click on the "Enter" button to configure the distribution graph in the window. Finally, we click on "OK" to exit this window.

We repeat the same process to enter the triangular distribution parameters in cell C7. A triangular probability distribution is defined by three estimated values—a minimum, a most likely, and a maximum. It is a very useful approximation when enough data points do not exist to allow for the construction of a distribution, but the user can estimate what the endpoints and the midpoint of the distribution might be. Clicking on "Define Assumption" and then selecting the triangular distribution from the Distribution menu results in the window shown in Exhibit 14.14.

EXHIBIT 14.14

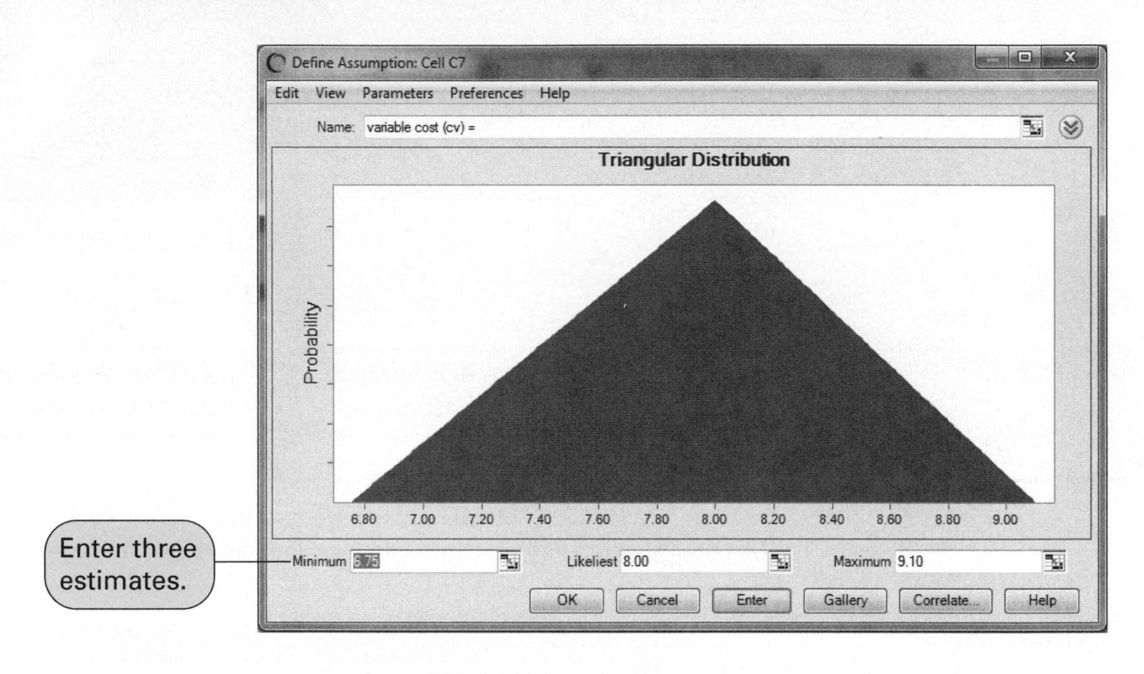

We enter the "Minimum" value of 6.75, the "Likeliest" value of 8.00, and the "Maximum" value of 9.10. Clicking on "Enter" will configure the graph of the triangular distribution shown in the window. We click on "OK" to exit this window and return to the spreadsheet.

Next, we click on cell C9 on our original spreadsheet. Recall that this is the cell in which we entered our profit formula in Exhibit 14.11. The profit value of 5,750, computed from the other cell values entered on the original spreadsheet, will be shown in cell C9. We click on "Define Forecast" at the top of the spreadsheet as shown in Exhibit 14.15. This will result in the "Define Forecast" window also shown in Exhibit 14.15. The heading "Profit(Z) =" will already be entered from the spreadsheet. We click on the "Units" display and enter "dollars." We then click on "OK" to exit this window. This completes the process of entering our simulation

EXHIBIT 14.15

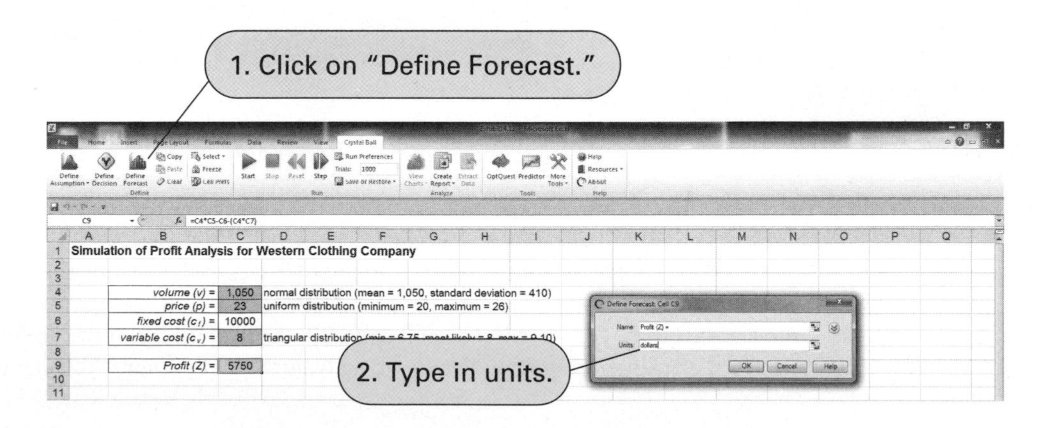

parameters and data. Exhibit 14.16 shows the spreadsheet with changes resulting from the parameter inputs. The next step is to run the simulation.

EXHIBIT 14.16

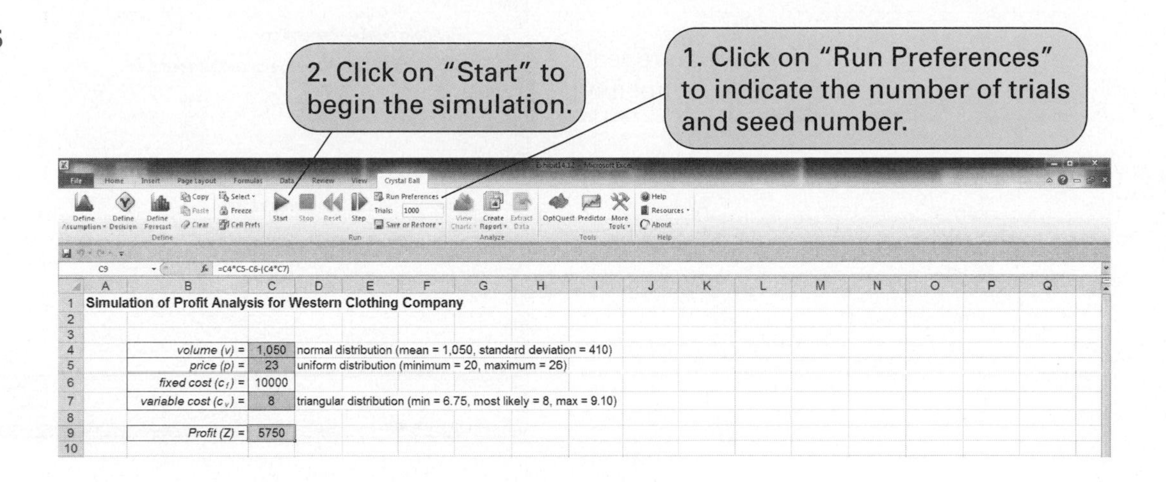

The mechanics of the simulation are similar to those of our previous Excel spreadsheet models. Using random numbers, we want to generate a value for demand in cell C4, then a value for price in C5, and then a value for variable cost in C7. These three values are then substituted into the profit formula in cell C9 to compute a profit value. This represents one repetition, or *trial*, of the simulation. The simulation is run for many trials in order to develop a distribution for profit.

To run the simulation, we click on "Run Preferences" at the top of the spreadsheet in Exhibit 14.16. This activates the window shown in Exhibit 14.17. We then enter the number of simulations for the simulation run. For this example we will run the simulation for 5,000 trials. Next, we click on "Sampling" at the top of this window to activate the window shown in Exhibit 14.18. In this window we must enter the seed value for a sequence of random numbers for the simulation, which is always 999. We click on "OK" and then go back to the spreadsheet. From the spreadsheet menu (Exhibit 14.19), we click on "Start," which will run the simulation. Exhibit 14.19 shows the simulation window with the simulation completed for 5,000 trials and the frequency distribution for this simulation.

EXHIBIT 14.17

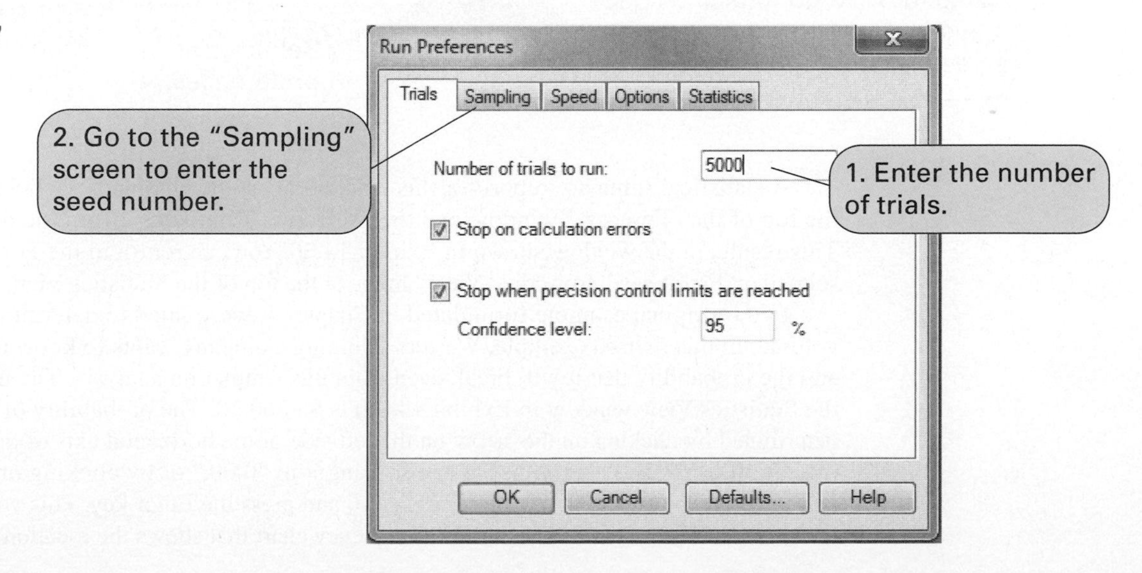

EXHIBIT 14.18

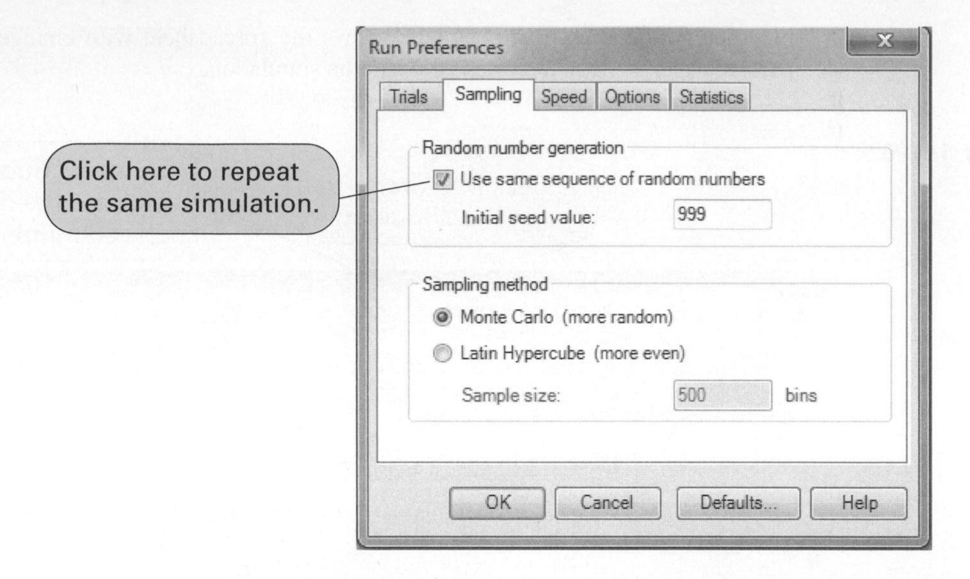

EXHIBIT 14.19

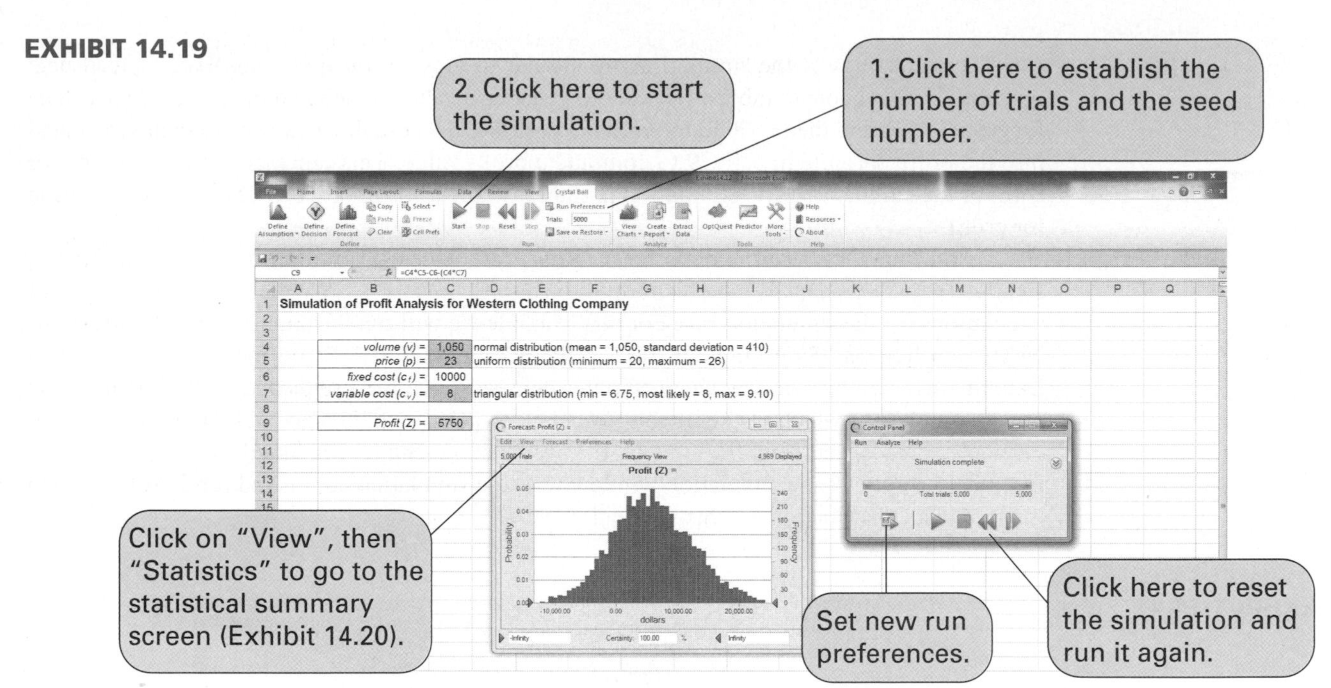

A statistical summary report for this simulation can be obtained by clicking on "View" at the top of the "Forecast" window and then selecting "Statistics" from the drop-down menu. This results in the window shown in Exhibit 14.20. You can return to the Forecast window by selecting "Frequency" from the "View" menu at the top of the Statistics View window.

In our original example formulated in Chapter 1, we wanted to determine the break-even volume. In this revised example, Western Clothing Company wants to know the average profit and the probability that it will break even from this simulation analysis. The mean profit (from the Statistics View window in Exhibit 14.20) is $5,860.50. The probability of breaking even is determined by clicking on the arrow on the left side of the horizontal axis of the window shown in Exhibit 14.21 and "grabbing" it and moving it to "0.00," or by clicking on the lower limit, currently set at "Infinity"; we change this to 0 and press the Enter key. This will shift the lower limit to zero, the break-even point. The frequency chart that shows the location of the new lower

EXHIBIT 14.20

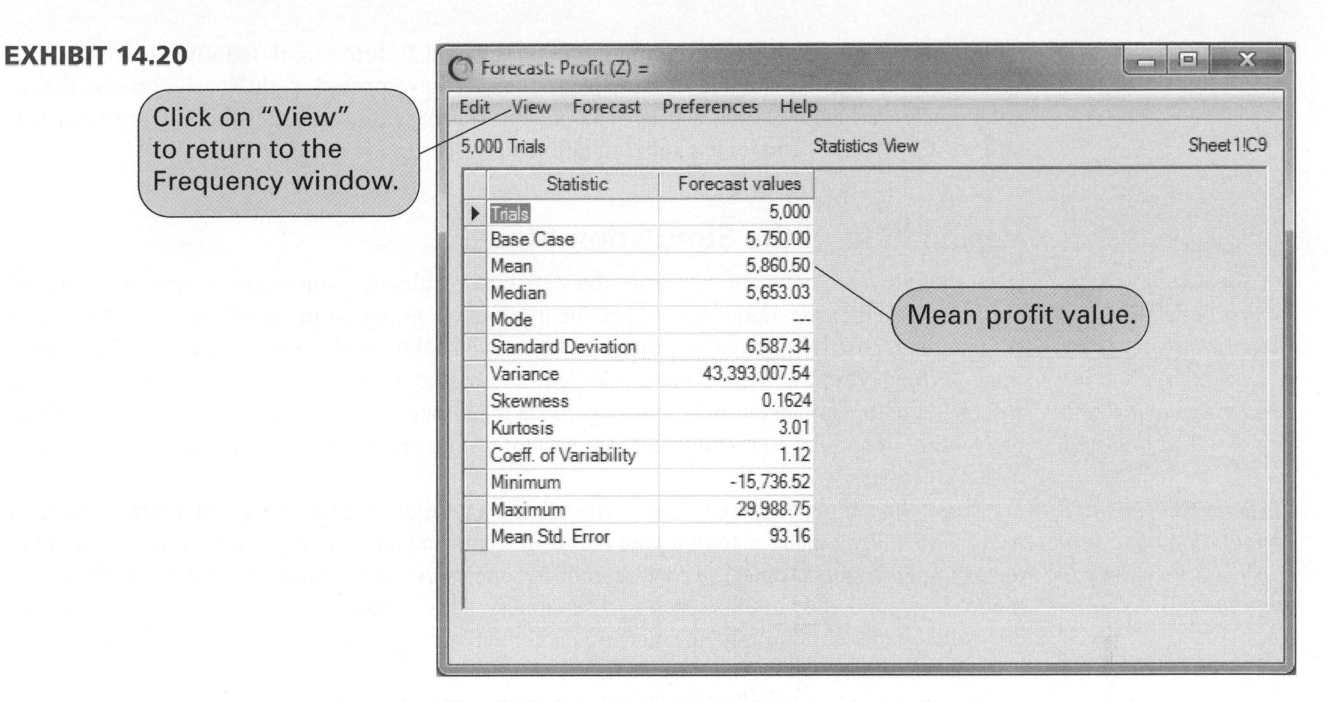

limit and the "Certainty" of zero profit is shown as 81.61% at the bottom of the window as shown in Exhibit 14.21. Thus, there is a .8161 probability that the company will break even.

EXHIBIT 14.21

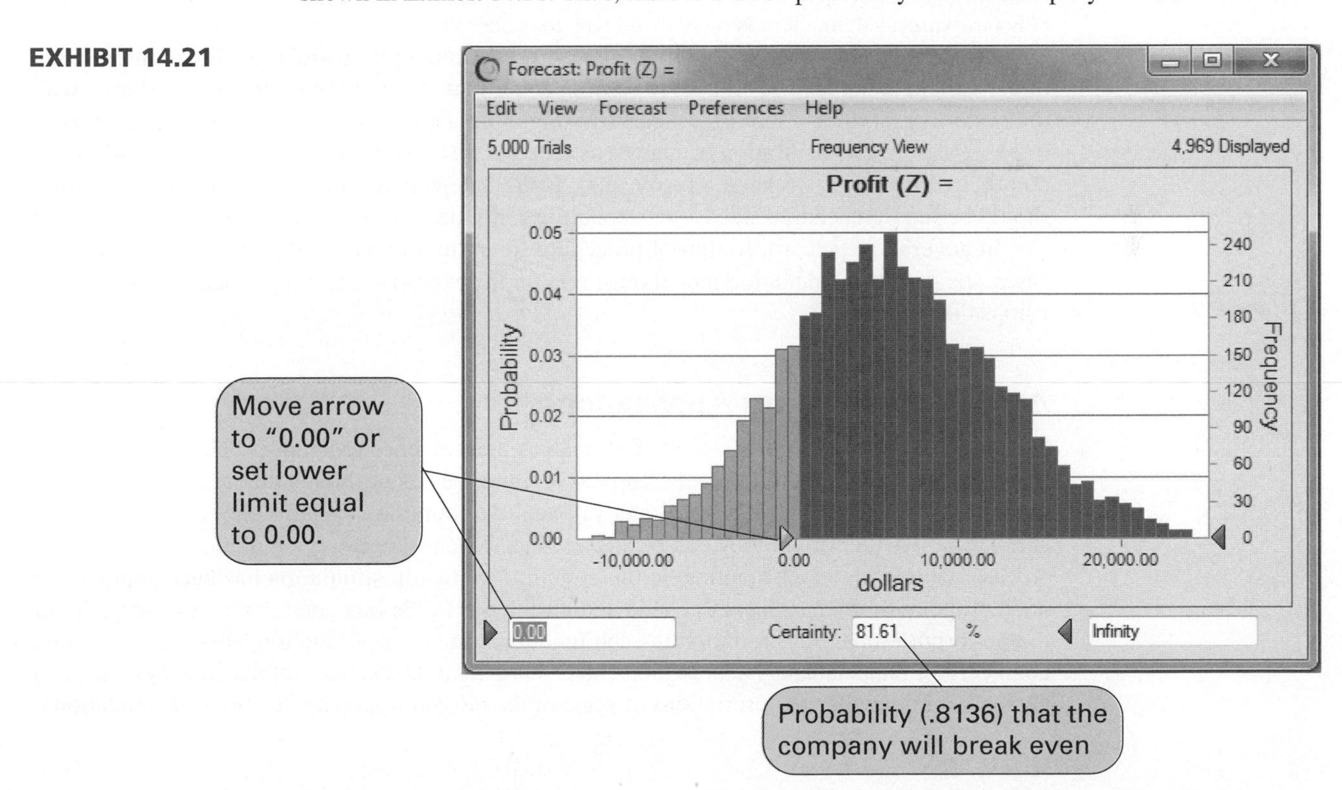

We have demonstrated using Crystal Ball with a straightforward example that was not very complex or detailed. Crystal Ball has the capability to perform much more sophisticated simulation analyses than what we have shown in this section. However, the demonstration of these capabilities and other features of Crystal Ball would require more space and in-depth coverage than is possible here. However, although using Crystal Ball to simulate more complex situations requires a greater degree of knowledge than we have provided, this basic introduction to and demonstration of Crystal Ball provide a good starting point to understanding the basic features of Crystal Ball and its use for simulation analysis.

Verification of the Simulation Model

Simulation models must be validated to make sure they are accurately replicating the system being simulated.

Even though we may be able to verify the statistical results of a simulation model, we still may not know whether the model actually replicates what is going on in the real world. The user of simulation generally wants to be certain that the model is internally correct and that all the operations performed in the simulation are logical and mathematically correct. An old adage often associated with computer simulation is "garbage in, garbage out." To gain some assurances about the validity of simulation results, there are several testing procedures that the user of a simulation model can apply.

Sometimes manual simulation of several trials is a good way to validate a simulation.

First, the simulation model can be run for short periods of time or for only a few simulation trials. This allows the user to compare the results with manually derived solutions (as we did in the examples in this chapter) to check for discrepancies. Another means of testing is to divide the model into parts and simulate each part separately. This reduces the complexity of seeking out errors in the model. Similarly, the mathematical relationships in the simulation model can be simplified so that they can more easily be tested to see if the model is operating correctly.

To determine whether the model reliably represents the system being simulated, the simulation results can sometimes be compared with actual real-world data. Several statistical tests are available for performing this type of analysis. However, when a model is developed to simulate a *new* or *unique* system, there is no realistic way to ensure that the results are valid.

An additional problem in determining whether a simulation model is a valid representation of the system under analysis relates to starting conditions. Should the simulation be started with the system empty (e.g., should we start by simulating a queuing system with no customers in line), or should the simulation be started as close as possible to normal operating conditions? Another problem, as we have already seen, is the determination of how long the simulation should be run to reach true steady-state conditions, if indeed a steady state exists.

In general, a standard, foolproof procedure for validation is simply not possible. In many cases, the user of a simulation model must rely on the expertise and experience of whoever develops the model.

Areas of Simulation Application

Simulation is one of the most useful of all management science techniques. The reason for this popularity is that simulation can be applied to a number of problems that are too difficult to model and solve analytically. Some analysts feel that complex systems should be studied by simulation whether or not they can be analyzed analytically because simulation provides such an easy vehicle for experimenting on the system. As a result, simulation has been applied to a wide range of problems. Surveys conducted during the 1990s indicated that a large majority of major corporations use simulation in such functional areas as production, corporate planning, engineering, financial analysis, research and development, marketing, information systems, and personnel. Following are descriptions of some of the most common applications of simulation.

Queuing

A major application of simulation has been in the analysis of queuing systems. As indicated in Chapter 13, the assumptions required to solve the operating characteristic formulas are relatively restrictive. For the more complex queuing systems (which result from a relaxation of these assumptions), it is not possible to develop analytical formulas, and simulation is often the only available means of analysis.

Inventory Control

Most people are aware that product demand is an essential component in determining the amount of inventory a commercial enterprise should keep. Most of the mathematical formulas used to analyze inventory systems make the assumption that this demand is certain (i.e., not a random variable). In practice, however, demand is rarely known with certainty. Simulation is one of the few means for analyzing inventory systems in which demand is a random variable, reflecting demand uncertainty. Inventory control is discussed in Chapter 16.

Production and Manufacturing

Simulation is often applied to production problems, such as production scheduling, production sequencing, assembly line balancing (of work-in-process inventory), plant layout, and plant location analysis. It is surprising how often various production processes can be viewed as queuing systems that can be analyzed only by using simulation. Because machine breakdowns typically occur according to some probability distributions, maintenance problems are also frequently analyzed using simulation.

Finance

Capital budgeting problems require estimates of cash flows, which are often a result of many random variables. Simulation has been used to generate values of the various contributing factors to derive estimates of cash flows. Simulation has also been used to determine the inputs into rate of return calculations in which the inputs are random variables, such as market size, selling price, growth rate, and market share.

Marketing

Marketing problems typically include numerous random variables, such as market size and type, and consumer preferences. Simulation can be used to ascertain how a particular market might react to the introduction of a product or to an advertising campaign for an existing product. Another area in marketing where simulation is applied is the analysis of distribution channels to determine the most efficient distribution system.

Public Service Operations

The operations of police departments, fire departments, post offices, hospitals, court systems, airports, and other public systems have all been analyzed by using simulation. Typically, such operations are so complex and contain so many random variables that no technique except simulation can be employed for analysis.

Environmental and Resource Analysis

Some of the more recent innovative applications of simulation have been directed at problems in the environment. Simulation models have been developed to ascertain the impact on the environment of projects such as nuclear power plants, reservoirs, highways, and dams. In many cases, these models include measures to analyze the financial feasibility of such projects. Other models have been developed to simulate pollution conditions. In the area of resource analysis, numerous models have been developed in recent years to simulate energy systems and the feasibility of alternative energy sources.

Summary

Simulation has become an increasingly important management science technique in recent years. Various surveys have shown simulation to be one of the techniques most widely applied to real-world problems. Evidence of this popularity is the number of specialized simulation languages that have been developed by the computer industry and academia to deal with complex problem areas.

Simulation is one of the most important and widely used management science techniques.

Simulation provides a laboratory for experimentation on a real system.

Simulation does not usually provide a recommended decision as does an optimization model; it provides operating statistics.

Simulation has certain limitations.

The popularity of simulation is due in large part to the flexibility it allows in analyzing systems, compared to more confining analytical techniques. In other words, the problem does not have to fit the model (or technique)—the simulation model can be constructed to fit the problem. A primary benefit of simulation analysis is that it enables us to experiment with the model. For example, in our queuing example we could expand the model to represent more service facilities, more queues, and different arrival and service times; and we could observe their effects on the results. In many analytical cases, such experimentation is limited by the availability of an applicable formula. That is, by changing various parts of the problem, we may create a problem for which we have no specific analytical formula. Simulation, however, is not subject to such limitations. Simulation is limited only by one's ability to develop a computer program.

Simulation is a management science technique that does not usually result in an optimal solution. Generally, a simulation model reflects the *operation of a system*, and the results of the model are in the form of operating statistics, such as averages. However, optimal solutions can sometimes be obtained for simulation models by employing *search techniques*.

However, in spite of its versatility, simulation has limitations and must be used with caution. One limitation is that simulation models are typically unstructured and must be developed for a system or problem that is also unstructured. Unlike some of the structured techniques presented in this text, they cannot simply be applied to a specific type of problem. As a result, developing simulation models often requires imagination and intuitiveness that are not required by some of the more straightforward solution techniques we have presented. In addition, the validation of simulation models is an area of serious concern. It is often impossible realistically to validate simulation results, to know if they accurately reflect the system under analysis. This problem has become an area of such concern that "output analysis" of simulation results is developing into a new field of study. Another limiting factor in simulation is the cost in money and time of model building. Because simulation models are developed for unstructured systems, they often take large amounts of staff time, computer time, and money to develop and run. For many business companies, these costs can be prohibitive.

Example Problem Solution

The following example problem demonstrates a manual simulation using discrete probability distributions.

Problem Statement

Members of the Willow Creek Emergency Rescue Squad know from past experience that they will receive between zero and six emergency calls each night, according to the following discrete probability distribution:

Calls	Probability
0	.05
1	.12
2	.15
3	.25
4	.22
5	.15
6	.06
	1.00

The rescue squad classifies each emergency call into one of three categories: minor, regular, or major emergency. The probability that a particular call will be each type of emergency is as follows:

Emergency Type	Probability
Minor	.30
Regular	.56
Major	.14
	1.00

The type of emergency call determines the size of the crew sent in response. A minor emergency requires a two-person crew, a regular call requires a three-person crew, and a major emergency requires a five-person crew.

Simulate the emergency calls received by the rescue squad for 10 nights, compute the average number of each type of emergency call each night, and determine the maximum number of crew members that might be needed on any given night.

Solution

Step 1: Develop Random Number Ranges for the Probability Distributions

Calls	Probability	Cumulative Probability	Random Number Range, r_1
0	.05	.05	1–5
1	.12	.17	6–17
2	.15	.32	18–32
3	.25	.57	33–57
4	.22	.79	58–79
5	.15	.94	80–94
6	.06	1.00	95–99, 00
	1.00		

Emergency Type	Probability	Cumulative Probability	Random Number Range, r_2
Minor	.30	.30	1–30
Regular	.56	.86	31–86
Major	.14	1.00	87–99, 00
	1.00		

Step 2: Set Up a Tabular Simulation

Use the second column of random numbers in Table 14.3:

Night	r_1	Number of Calls	r_2	Emergency Type	Crew Size	Total per Night
1	65	4	71	Regular	3	
			18	Minor	2	
			12	Minor	2	
			17	Minor	2	9
2	48	3	89	Major	5	
			18	Minor	2	
			83	Regular	3	10
3	08	1	90	Major	5	5
4	05	0	—	—	—	—
5	89	5	18	Minor	2	
			08	Minor	2	
			26	Minor	2	
			47	Regular	3	
			94	Major	5	14
6	06	1	72	Regular	3	3
7	62	4	47	Regular	3	
			68	Regular	3	
			60	Regular	3	
			88	Major	5	14
8	17	1	36	Regular	3	3
9	77	4	43	Regular	3	
			28	Minor	2	
			31	Regular	3	
			06	Minor	2	10
10	68	4	39	Regular	3	
			71	Regular	3	
			22	Minor	2	
			76	Regular	3	11

Step 3: Compute Results

$$\text{average number of minor emergency calls per night} = \frac{10}{10} = 1.0$$

$$\text{average number of regular emergency calls per night} = \frac{13}{10} = 1.3$$

$$\text{average number of major emergency calls per night} = \frac{4}{10} = 0.40$$

If all the calls came in at the same time, the maximum number of squad members required during any 1 night would be 14.

Problems

1. The Hoylake Rescue Squad receives an emergency call every 1, 2, 3, 4, 5, or 6 hours, according to the following probability distribution. The squad is on duty 24 hours per day, 7 days per week:

Time Between Emergency Calls (hr.)	Probability
1	.05
2	.10
3	.30
4	.30
5	.20
6	.05
	1.00

   a. Simulate the emergency calls for 3 days (note that this will require a "running," or cumulative, hourly clock), using the random number table.
   b. Compute the average time between calls and compare this value with the expected value of the time between calls from the probability distribution. Why are the results different?
   c. How many calls were made during the 3-day period? Can you logically assume that this is an average number of calls per 3-day period? If not, how could you simulate to determine such an average?

2. The time between arrivals of cars at the Petroco Service Station is defined by the following probability distribution:

Time Between Arrivals (min.)	Probability
1	.15
2	.30
3	.40
4	.15
	1.00

   a. Simulate the arrival of cars at the service station for 20 arrivals and compute the average time between arrivals.
   b. Simulate the arrival of cars at the service station for 1 hour, using a different stream of random numbers from those used in (a) and compute the average time between arrivals.
   c. Compare the results obtained in (a) and (b).

3. The Dynaco Manufacturing Company produces a product in a process consisting of operations of five machines. The probability distribution of the number of machines that will break down in a week follows:

Machine Breakdowns per Week	Probability
0	.10
1	.10
2	.20
3	.25
4	.30
5	.05
	1.00

a. Simulate the machine breakdowns per week for 20 weeks.
b. Compute the average number of machines that will break down per week.

4. Solve Problem 19 at the end of Chapter 12 by using simulation.

5. Simulate the decision situation described in Problem 16(a) at the end of Chapter 12 for 20 weeks, and recommend the best decision.

6. Every time a machine breaks down at the Dynaco Manufacturing Company (Problem 3), either 1, 2, or 3 hours are required to fix it, according to the following probability distribution:

Repair Time (hr.)	Probability
1	.30
2	.50
3	.20
	1.00

a. Simulate the repair time for 20 weeks and then compute the average weekly repair time.
b. If the random numbers that are used to simulate breakdowns per week are also used to simulate repair time per breakdown, will the results be affected in any way? Explain.
c. If it costs $50 per hour to repair a machine when it breaks down (including lost productivity), determine the average weekly breakdown cost.
d. The Dynaco Company is considering a preventive maintenance program that would alter the probabilities of machine breakdowns per week as shown in the following table:

Machine Breakdowns per Week	Probability
0	.20
1	.30
2	.20
3	.15
4	.10
5	.05
	1.00

The weekly cost of the preventive maintenance program is $150. Using simulation, determine whether the company should institute the preventive maintenance program.

7. Sound Warehouse in Georgetown sells CD players (with speakers), which it orders from Fuji Electronics in Japan. Because of shipping and handling costs, each order must be for five CD players.

Because of the time it takes to receive an order, the warehouse outlet places an order every time the present stock drops to five CD players. It costs $100 to place an order. It costs the warehouse $400 in lost sales when a customer asks for a CD player and the warehouse is out of stock. It costs $40 to keep each CD player stored in the warehouse. If a customer cannot purchase a CD player when it is requested, the customer will not wait until one comes in but will go to a competitor. The following probability distribution for demand for CD players has been determined:

Demand per Month	Probability
0	.04
1	.08
2	.28
3	.40
4	.16
5	.02
6	.02
	1.00

The time required to receive an order once it is placed has the following probability distribution:

Time to Receive an Order (mo.)	Probability
1	.60
2	.30
3	.10
	1.00

The warehouse has five CD players in stock. Orders are always received at the beginning of the week. Simulate Sound Warehouse's ordering and sales policy for 20 months, using the first column of random numbers in Table 14.3. Compute the average monthly cost.

8. First American Bank is trying to determine whether it should install one or two drive-through teller windows. The following probability distributions for arrival intervals and service times have been developed from historical data:

Time Between Automobile Arrivals (min.)	Probability
1	.20
2	.60
3	.10
4	.10
	1.00

Service Time (min.)	Probability
2	.10
3	.40
4	.20
5	.20
6	.10
	1.00

Assume that in the two-server system, an arriving car will join the shorter queue. When the queues are of equal length, there is a 50–50 chance the driver will enter the queue for either window.

a. Simulate both the one- and two-teller systems. Compute the average queue length, waiting time, and percentage utilization for each system.
b. Discuss your results in (a) and suggest the degree to which they could be used to make a decision about which system to employ.

9. The time between arrivals of oil tankers at a loading dock at Prudhoe Bay is given by the following probability distribution:

Time Between Ship Arrivals (days)	Probability
1	.05
2	.10
3	.20
4	.30
5	.20
6	.10
7	.05
	1.00

The time required to fill a tanker with oil and prepare it for sea is given by the following probability distribution:

Time to Fill and Prepare (days)	Probability
3	.10
4	.20
5	.40
6	.30
	1.00

a. Simulate the movement of tankers to and from the single loading dock for the first 20 arrivals. Compute the average time between arrivals, average waiting time to load, and average number of tankers waiting to be loaded.
b. Discuss any hesitation you might have about using your results for decision making.

10. The Saki automobile dealer in the Minneapolis–St. Paul area orders the Saki sport compact, which gets 50 miles per gallon of gasoline, from the manufacturer in Japan. However, the dealer never knows for sure how many months it will take to receive an order once it is placed. It can take 1, 2, or 3 months, with the following probabilities:

Months to Receive an Order	Probability
1	.50
2	.30
3	.20
	1.00

The demand per month is given by the following distribution:

Demand per Month (cars)	Probability
1	.10
2	.30
3	.40
4	.20
	1.00

The dealer orders when the number of cars on the lot gets down to a certain level. To determine the appropriate level of cars to use as an indicator of when to order, the dealer needs to know how many cars will be demanded during the time required to receive an order. Simulate the demand for 30 orders and compute the average number of cars demanded during the time required to receive an order. At what level of cars in stock should the dealer place an order?

11. State University is playing Tech in their annual football game on Saturday. A sportswriter has scouted each team all season and accumulated the following data: State runs four basic plays—a sweep, a pass, a draw, and an off tackle; Tech uses three basic defenses—a wide tackle, an Oklahoma, and a blitz. The number of yards State will gain for each play against each defense is shown in the following table:

	Tech Defense		
State Play	Wide Tackle	Oklahoma	Blitz
Sweep	−3	5	12
Pass	12	4	−10
Draw	2	1	20
Off tackle	7	3	−3

The probability that State will run each of its four plays is shown in the following table:

Play	Probability
Sweep	.10
Pass	.20
Draw	.20
Off tackle	.50

The probability that Tech will use each of its defenses follows:

Defense	Probability
Wide tackle	.30
Oklahoma	.50
Blitz	.20

The sportswriter estimates that State will run 40 plays during the game. The sportswriter believes that if State gains 300 or more yards, it will win; however, if Tech holds State to fewer than 300 yards, it will win. Use simulation to determine which team the sportswriter will predict to win the game.

12. Each semester, the students in the college of business at State University must have their course schedules approved by the college adviser. The students line up in the hallway outside the adviser's office. The students arrive at the office according to the following probability distribution:

Time Between Arrivals (min.)	Probability
4	.20
5	.30
6	.40
7	.10
	1.00

The time required by the adviser to examine and approve a schedule corresponds to the following probability distribution:

Schedule Approval (min.)	Probability
6	.30
7	.50
8	.20
	1.00

Simulate this course approval system for 90 minutes. Compute the average queue length and the average time a student must wait. Discuss these results.

13. A city is served by two newspapers—the *Tribune* and the *Daily News*. Each Sunday readers purchase one of the newspapers at a stand. The following matrix contains the probabilities of a customer's buying a particular newspaper in a week, given the newspaper purchased the previous Sunday:

This Sunday	***Next Sunday***	
	Tribune	*Daily News*
Tribune	.65	.35
Daily News	.45	.55

Simulate a customer's purchase of newspapers for 20 weeks to determine the steady-state probabilities that a customer will buy each newspaper in the long run.

14. Loebuck Grocery orders milk from a dairy on a weekly basis. The manager of the store has developed the following probability distribution for demand per week (in cases):

Demand (cases)	Probability
15	.20
16	.25
17	.40
18	.15
	1.00

The milk costs the grocery \$10 per case and sells for \$16 per case. The carrying cost is \$0.50 per case per week, and the shortage cost is \$1 per case per week. Simulate the ordering system for

Loebuck Grocery for 20 weeks. Use a weekly order size of 16 cases of milk and compute the average weekly profit for this order size. Explain how the complete simulation for determining order size would be developed for this problem.

15. The Paymore Rental Car Agency rents cars in a small town. It wants to determine how many rental cars it should maintain. Based on market projections and historical data, the manager has determined probability distributions for the number of rentals per day and rental duration (in days only) as shown in the following tables:

Number of Customers/Day	Probability
0	.20
1	.20
2	.50
3	.10
	1.00

Rental Duration (days)	Probability
1	.10
2	.30
3	.40
4	.10
5	.10
	1.00

Design a simulation experiment for the car agency and simulate using a fleet of four rental cars for 10 days. Compute the probability that the agency will not have a car available upon demand. Should the agency expand its fleet? Explain how a simulation experiment could be designed to determine the optimal fleet size for the Paymore Agency.

16. A CPM/PERT project network has probabilistic activity times (x) as shown on each branch of the network; for example, activity 1–3 has a .40 probability that it will be completed in 6 weeks and a .60 probability it will be completed in 10 weeks:

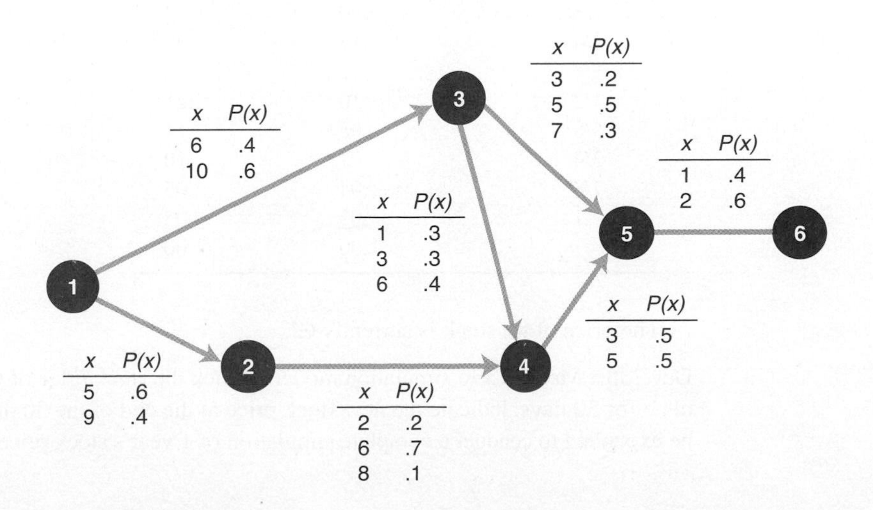

Simulate the project network 10 times and determine the critical path each time. Compute the average critical path time and the frequency at which each path is critical. How does this simulation analysis of the critical path method compare with regular CPM/PERT analysis?

17. A robbery has just been committed at the Corner Market in the downtown area of the city. The market owner was able to activate the alarm, and the robber fled on foot. Police officers arrived a few minutes later and asked the owner, "How long ago did the robber leave?" "He left only a few minutes ago," the store owner responded. "He's probably 10 blocks away by now," one of the officers said to the other. "Not likely," said the store owner. "He was so stoned on drugs that I bet even if he has run 10 blocks, he's still only within a few blocks of here! He's probably just running in circles!"

    Perform a simulation experiment that will test the store owner's hypothesis. Assume that at each corner of a city block there is an equal chance that the robber will go in any one of the four possible directions: north, south, east, or west. Simulate for five trials and then indicate in how many of the trials the robber is within 2 blocks of the store.

18. Compcomm, Inc., is an international communications and information technology company that has seen the value of its common stock appreciate substantially in recent years. A stock analyst would like to use simulation to predict the stock prices of Compcomm for an extended period. Based on historical data, the analyst has developed the following probability distribution for the movement of Compcomm stock prices per day:

Stock Price Movement	Probability
Increase	.45
Same	.30
Decrease	.25
	1.00

    The analyst has also developed the following probability distributions for the amount of the increases or decreases in the stock price per day:

	Probability	
Stock Price Change	Increase	Decrease
1/8	.40	.12
1/4	.17	.15
3/8	.12	.18
1/2	.10	.21
5/8	.08	.14
3/4	.07	.10
7/8	.04	.05
1	.02	.05
	1.00	1.00

    The price of the stock is currently 62.

    Develop a Monte Carlo simulation model to track the stock price of Compcomm stock and simulate for 30 days. Indicate the new stock price at the end of the 30 days. How would this model be expanded to conduct a complete simulation of 1 year's stock price movement?

19. The emergency room of the community hospital in Farmburg has one receptionist, one doctor, and one nurse. The emergency room opens at time zero, and patients begin to arrive some time later. Patients arrive at the emergency room according to the following probability distribution:

Time Between Arrivals (min.)	Probability
5	.06
10	.10
15	.23
20	.29
25	.18
30	.14
	1.00

The attention needed by a patient who comes to the emergency room is defined by the following probability distribution:

Patient Needs to See	Probability
Doctor alone	.50
Nurse alone	.20
Both	.30
	1.00

If a patient needs to see both the doctor and the nurse, he or she cannot see one before the other—that is, the patient must wait to see both together.

The length of the patient's visit (in minutes) is defined by the following probability distributions:

Doctor	Probability	Nurse	Probability	Both	Probability
10	.22	5	.08	15	.07
15	.31	10	.24	20	.16
20	.25	15	.51	25	.21
25	.12	20	.17	30	.28
30	.10		1.00	35	.17
	1.00			40	.11
					1.00

Simulate the arrival of 20 patients to the emergency room and compute the probability that a patient must wait and the average waiting time. Based on this one simulation, does it appear that this system provides adequate patient care?

20. The Western Outfitters Store specializes in denim jeans. The variable cost of the jeans varies according to several factors, including the cost of the jeans from the distributor, labor costs, handling, packaging, and so on. Price also is a random variable that varies according to competitors' prices.

Sales volume also varies each month. The probability distributions for volume, price, and variable costs each month are as follows:

Sales Volume	Probability
300	.12
400	.18
500	.20
600	.23
700	.17
800	.10
	1.00

Price	Probability
$22	.07
23	.16
24	.24
25	.25
26	.18
27	.10
	1.00

Variable Cost	Probability
$ 8	.17
9	.32
10	.29
11	.14
12	.08
	1.00

Fixed costs are $9,000 per month for the store.

Simulate 20 months of store sales and compute the probability that the store will at least break even and the average profit (or loss).

21. Randolph College and Salem College are within 20 miles of each other, and the students at each college frequently date each other. The students at Randolph College are debating how good their dates are at Salem College. The Randolph students have sampled several hundred of their fellow students and asked them to rate their dates from 1 to 5 (in which 1 is excellent and 5 is poor) according to physical attractiveness, intelligence, and personality. Following are the resulting probability distributions for these three traits for students at Salem College:

Physical Attractiveness	Probability
1	.27
2	.35
3	.14
4	.09
5	.15
	1.00

Intelligence	Probability
1	.10
2	.16
3	.45
4	.17
5	.12
	1.00

Personality	Probability
1	.15
2	.30
3	.33
4	.07
5	.15
	1.00

Simulate 20 dates and compute an average overall rating of the Salem students.

22. In Problem 21, discuss how you might assess the accuracy of the average rating for Salem College students based on only 20 simulated dates.

23. Burlingham Mills produces denim cloth that it sells to jeans manufacturers. It is negotiating a contract with Troy Clothing Company to provide denim cloth on a weekly basis. Burlingham has established its monthly available production capacity for this contract to be between 0 and 600 yards, according to the following probability distribution:

$$f(x) = \frac{x}{180{,}000}, \quad 0 \leq x \leq 600 \text{ yd.}$$

Troy Clothing's weekly demand for denim cloth varies according to the following probability distribution:

Demand (yd.)	Probability
0	.03
100	.12
200	.20
300	.35
400	.20
500	.10
	1.00

Simulate Troy Clothing's cloth orders for 20 weeks and determine the average weekly capacity and demand. Also determine the probability that Burlingham will have sufficient capacity to meet demand.

24. A baseball game consists of plays that can be described as follows:

Play	Description
No advance	An out where no runners can advance. This includes strikeouts, pop-ups, short flies, and the like.
Groundout	Each runner can advance one base.
Possible double play	Double play if there is a runner on first base and fewer than two outs. The lead runner who can be forced is out; runners not out advance one base. If there is no runner on first or there are two outs, this play is treated as a "no advance."
Long fly	A runner on third base can score.
Very long fly	Runners on second and third base advance one base.
Walk	Includes a hit batter.
Infield single	All runners advance one base.
Outfield single	A runner on first base advances one base, but a runner on second or third base scores.
Long single	All runners can advance a maximum of two bases.
Double	Runners can advance a maximum of two bases.
Long double	All runners score.
Triple	
Home run	

Note: Singles also include a factor for errors, allowing the batter to reach first base.

Distributions for these plays for two teams, the White Sox (visitors) and the Yankees (home), are as follows:

Team: White Sox

Play	Probability
No advance	.03
Groundout	.39
Possible double play	.06
Long fly	.09
Very long fly	.08
Walk	.06
Infield single	.02
Outfield single	.10
Long single	.03
Double	.04
Long double	.05
Triple	.02
Home run	.03
	1.00

Team: Yankees

Play	Probability
No advance	.04
Groundout	.38
Possible double play	.04
Long fly	.10
Very long fly	.06
Walk	.07
Infield single	.04
Outfield single	.10
Long single	.04
Double	.05
Long double	.03
Triple	.01
Home run	.04
	1.00

Simulate a nine-inning baseball game using the preceding information.[2]

[2]This problem is based on R. E. Trueman, "A Computer Simulation Model of Baseball: With Particular Application to Strategy Analysis," in R. E. Machol, S. P. Ladany, and D. G. Morrison, eds., *Management Science in Sports* (New York: North Holland Publishing Co., 1976), 1–14.

25. For the ComputerWorld example in this chapter, recreate the simulation shown in Exhibit 14.2 using Crystal Ball. Assume that demand is normally distributed with a mean of 1.5 laptops and a standard deviation of 0.8. Using Crystal Ball determine average weekly demand and average weekly revenue.

26. Perform the Crystal Ball simulation in Problem 25, assuming that demand is normally distributed with a mean of 2.5 and a standard deviation of 1.2 laptops.

27. For the Bigelow Manufacturing example in this chapter, re-create the simulation shown in Table 14.10 and Exhibit 14.7 using Crystal Ball. Assume that the repair time is normally distributed with a mean of 2.15 days and a standard deviation of 0.8 day. Assume that the time between breakdowns is defined by the triangular distribution used in the example. Using Crystal Ball, determine the average annual number of breakdowns, average annual repair time, and average annual repair cost.

28. For the Bigelow Manufacturing example in this chapter, re-create the simulation for the improved maintenance program shown in Table 14.11 and Exhibit 14.8 using Crystal Ball. For the improved program assume that the repair time is normally distributed with a mean of 1.70 days and a standard deviation of 0.6 day. Assume that the time between breakdowns is defined by the triangular distribution for the improved maintenance program used in the example. Using Crystal Ball, determine the average annual number of breakdowns, average annual repair time, and average annual repair cost. Compare this improved maintenance system with the current one (Problem 27), and indicate whether it should be adopted given the cost of improving the system (i.e., $20,000).

29. Tracy McCoy is shopping for a new car. She has identified a particular sports utility vehicle she likes but has heard that it has high maintenance costs. Tracy has decided to develop a simulation model to help her estimate maintenance costs for the life of the car. Tracy estimates that the projected life of the car with the first owner (before it is sold) is uniformly distributed with a minimum of 2.0 years and a maximum of 8.0 years. Furthermore, she believes that the miles she will drive the car each year can be defined by a triangular distribution with a minimum value of 3,700 miles, a maximum value of 14,500 miles, and a most likely value of 9,000 miles. She has determined from automobile association data that the maintenance cost per mile driven for the vehicle she is interested in is normally distributed, with a mean of $0.08 per mile and a standard deviation of $0.02 per mile. Using Crystal Ball, develop a simulation model (using 1,000 trials) and determine the average maintenance cost for the life of the car with Tracy and the probability that the cost will be less than $3,000.

30. In Problem 20, assume that the sales volume for Western Outfitters Store is normally distributed, with a mean of 600 pairs of jeans and a standard deviation of 200; the price is uniformly distributed, with a minimum of $22 and a maximum of $28; and the variable cost is defined by a triangular distribution with a minimum value of $6, a maximum of $11, and a most likely value of $9. Develop a simulation model by using Crystal Ball (with 1,000 trials) and determine the average profit and the probability that Western Outfitters will break even.

31. In Problem 21, assume that the students at Randolph College have redefined the probability distributions of their dates at Salem College as follows: Physical attractiveness is uniformly distributed from 1 to 5; intelligence is defined by a triangular distribution with a minimum rating of 1, a maximum of 5, and a most likely of 2; and personality is defined by a triangular distribution with a minimum of 1, a maximum of 5, and a most likely rating of 3. Develop a simulation model by using Crystal Ball and determine the average date rating (for 1,000 trials). Also compute the probability that the rating will be "better" than 3.0.

32. In Problem 23, assume that production capacity at Burlingham Mills for the Troy Clothing Company contract is normally distributed, with a mean of 320 yards per month and a standard deviation of 120 yards, and that Troy Clothing's demand is uniformly distributed between 0 and 500 yards. Develop a simulation model by using Crystal Ball and determine the average monthly shortage or surplus for denim cloth (for 1,000 trials). Also determine the probability that Burlingham will always have sufficient production capacity.

33. Erin Jones has $100,000 and, to diversify, she wants to invest equal amounts of $50,000 each in two mutual funds selected from a list of four possible mutual funds. She wants to invest for a 3-year period. She has used historical data from the four funds plus data from the market to determine the mean and standard deviation (normally distributed) of the annual return for each fund, as follows:

	Return (r)	
Fund	μ	σ
1. Internet	.20	.09
2. Index	.12	.04
3. Entertainment	.16	.10
4. Growth	.14	.06

The possible combinations of two investment funds are (1,2), (1,3), (1,4), (2,3), (2,4), and (3,4).

a. Use Crystal Ball to simulate each of the investment combinations to determine the expected return in 3 years. (Note that the formula for the future value, FV, of a current investment, P, with return r for n years in the future is $FV_n = P_r(1 + r)^n$.) Indicate which investment combination has the highest expected return.

b. Erin wants to reduce her risk as much as possible. She knows that if she invests her $100,000 in a CD at the bank, she is guaranteed a return of $20,000 after 3 years. Using the frequency charts for the simulation runs in Crystal Ball, determine which combination of investments will result in the greatest probability of receiving a return of $120,000 or greater.

34. In Chapter 16, the formula for the optimal order quantity of an item, Q, given its demand, D, ordering cost, C_o, and the cost of holding, or carrying, an item in inventory, C_c, is as follows:

$$Q = \sqrt{\frac{2C_0D}{C_c}}$$

The total inventory cost formula is

$$TC = \frac{C_oD}{Q} + C_c\frac{Q}{2}$$

Ordering cost, C_o, and carrying cost, C_c, are generally values that the company is often able to determine with certainty because they are internal costs, whereas demand, D, is usually not known with certainty because it is external to the company. However, in the order quantity formula given here, demand is treated as if it were certain. To consider the uncertainty of demand, it must be simulated.

Using Crystal Ball, simulate the preceding formulas for Q and TC to determine their average values for an item, with $C_o = \$150$, $C_c = \$0.75$, and demand, D, that is normally distributed with a mean of 10,000 and a standard deviation of 4,000.

35. The Management Science Association (MSA) has arranged to hold its annual conference at the Riverside Hotel in Orlando next year. Based on historical data, the MSA believes the number of rooms it will need for its members attending the conference is normally distributed, with a mean of 800 and a standard deviation of 270. The MSA can reserve rooms now (1 year prior to the conference) for $80; however, for any rooms not reserved now, the cost will be at the hotel's regular room rate of $120. The MSA guarantees the room rate of $80 to its members. If its members reserve fewer than the number of rooms it reserves, MSA must pay the hotel for the difference, at the $80 room rate. If MSA does not reserve enough rooms, it must pay the extra cost—that is, $40 per room.

a. Using Crystal Ball, determine whether the MSA should reserve 600, 700, 800, 900, or 1,000 rooms in advance to realize the lowest total cost.

b. Can you determine a more exact value for the number of rooms to reserve to minimize cost?

36. In Chapter 8, Figure 8.6 shows a simplified project network for building a house, as follows:

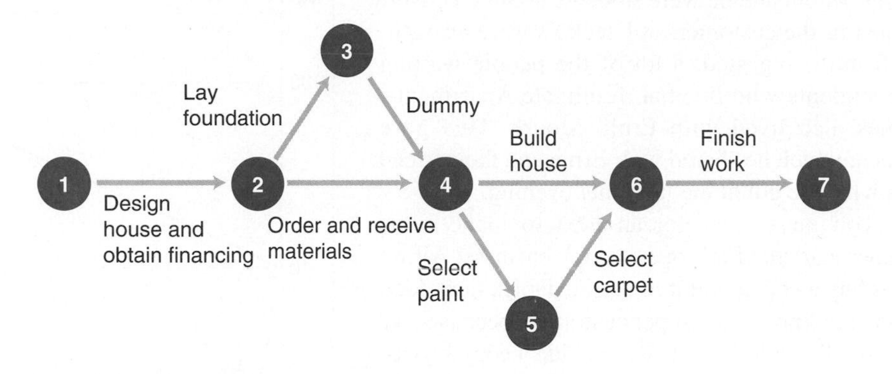

There are four paths through this network:

Path A: 1–2–3–4–6–7
Path B: 1–2–3–4–5–6–7
Path C: 1–2–4–6–7
Path D: 1–2–4–5–6–7

The time parameters (in weeks) defining a triangular probability distribution for each activity are provided as follows:

	Time Parameters		
Activity	**Minimum**	**Likeliest**	**Maximum**
1–2	1	3	5
2–3	1	2	4
2–4	0.5	1	2
3–4	0	0	0
4–5	1	2	3
4–6	1	3	6
5–6	1	2	4
6–7	1	2	4

a. Using Crystal Ball, simulate each path in the network and identify the longest path (i.e., the critical path).
b. Observing the simulation run frequency chart for path A, determine the probability that this path will exceed the critical path time. What does this tell you about the simulation results for a project network versus an analytical result?

Case Problem

JET COPIES

James Banks was standing in line next to Robin Cole at Klecko's Copy Center, waiting to use one of the copy machines. "Gee, Robin, I hate this," he said. "We have to drive all the way over here from Southgate and then wait in line to use these copy machines. I hate wasting time like this."

"I know what you mean," said Robin. "And look who's here. A lot of these students are from Southgate Apartments or one of the other apartments near us. It seems as though it would be more logical if Klecko's would move its operation over to us, instead of all of us coming over here."

James looked around and noticed what Robin was talking about. Robin and he were students at State University, and most of the customers at Klecko's were also students. As Robin suggested, a lot of the people waiting were State students who lived at Southgate Apartments, where James also lived with Ernie Moore. This gave James an idea, which he shared with Ernie and their friend Terri Jones when he got home later that evening.

"Look, you guys, I've got an idea to make some money," James started. "Let's open a copy business! All we have to do is buy a copier, put it in Terri's duplex next door, and sell copies. I know we can get customers because I've just seen them all at Klecko's. If we provide a copy service right here in the Southgate complex, we'll make a killing."

Terri and Ernie liked the idea, so the three decided to go into the copying business. They would call it JET Copies, named for James, Ernie, and Terri. Their first step was to purchase a copier. They bought one like the one used in the college of business office at State for $18,000. (Terri's parents provided a loan.) The company that sold them the copier touted the copier's reliability, but after they bought it, Ernie talked with someone in the dean's office at State, who told him that the University's copier broke down frequently and when it did, it often took between 1 and 4 days to get it repaired. When Ernie told this to Terri and James, they became worried. If the copier broke down frequently and was not in use for long periods while they waited for a repair person to come fix it, they could lose a lot of revenue. As a result, James, Ernie, and Terri thought they might need to purchase a smaller backup copier for $8,000 to use when the main copier broke down. However, before they approached Terri's parents for another loan, they wanted to have an estimate of just how much money they might lose if they did not have a backup copier. To get this estimate, they decided to develop a simulation model because they were studying simulation in one of their classes at State.

To develop a simulation model, they first needed to know how frequently the copier might break down—specifically, the time between breakdowns. No one could provide them with an exact probability distribution, but from talking to staff members in the college of business, James estimated that the time between breakdowns was probably between 0 and 6 weeks, with the probability increasing the longer the copier went without breaking down. Thus, the probability distribution of breakdowns generally looked like the following:

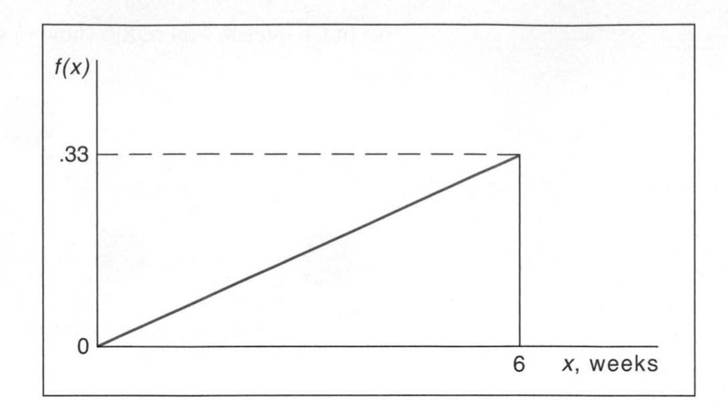

Next, they needed to know how long it would take to get the copier repaired when it broke down. They had a service contract with the dealer that "guaranteed" prompt repair service. However, Terri gathered some data from the college of business from which she developed the following probability distribution of repair times:

Repair Time (days)	Probability
1	.20
2	.45
3	.25
4	.10
	1.00

Finally, they needed to estimate how much business they would lose while the copier was waiting for repair. The three of them had only a vague idea of how much business they would do but finally estimated that they would sell between 2,000 and 8,000 copies per day at $0.10 per copy. However, they had no idea about what kind of probability distribution to use for this range of values. Therefore, they decided to use a uniform probability distribution between 2,000 and 8,000 copies to estimate the number of copies they would sell per day.

James, Ernie, and Terri decided that if their loss of revenue due to machine downtime during 1 year was $12,000 or more, they should purchase a backup copier. Thus, they needed to simulate the breakdown and repair process for a number of years to obtain an average annual loss of revenue. However, before programming the simulation model, they decided to conduct a manual simulation of this process for 1 year to see if the model was working correctly. Perform this manual simulation for JET Copies and determine the loss of revenue for 1 year.

Case Problem

BENEFIT–COST ANALYSIS OF THE SPRADLIN BLUFF RIVER PROJECT

The U.S. Army Corps of Engineers has historically constructed dams on various rivers in the southeastern United States. Its primary instrument for evaluating and selecting among many projects under consideration is benefit–cost analysis. The Corps estimates both the annual benefits deriving from a project in several different categories and the annual costs and then divides the total benefits by the total costs to develop a benefit–cost ratio. This ratio is then used by the Corps and Congress to compare numerous projects under consideration and select those for funding. A benefit–cost ratio greater than 1.0 indicates that the benefits are greater than the costs; and the higher a project's benefit–cost ratio, the more likely it is to be selected over projects with lower ratios.

The Corps is evaluating a project to construct a dam over the Spradlin Bluff River in southwest Georgia. The Corps has identified six traditional areas in which benefits will accrue: flood control, hydroelectric power, improved navigation, recreation, fish and wildlife, and area commercial redevelopment. The Corps has made three estimates (in dollars) for each benefit—a minimum possible value, a most likely value, and a maximum benefit value. These benefit estimates are as follows:

	Estimate		
Benefit	*Minimum*	*Most Likely*	*Maximum*
Flood control	$ 1,695,200	$ 2,347,800	$ 3,570,600
Hydroelectric power	8,068,250	11,845,000	14,845,000
Navigation	50,400	64,000	109,500
Recreation	6,404,000	9,774,000	14,566,000
Fish and wildlife	104,300	255,000	455,300
Area redevelopment	0	1,630,000	2,385,000

There are two categories of costs associated with a construction project of this type—the total capital cost, annualized over 100 years (at a rate of interest specified by the government), and annual operation and maintenance costs. The cost estimates for this project are as follows:

	Estimate		
Cost	*Minimum*	*Most Likely*	*Maximum*
Annualized capital cost	$12,890,750	$14,150,500	$19,075,900
Operation and maintenance	3,483,500	4,890,000	7,350,800

Using Crystal Ball, determine a simulated mean benefit–cost ratio and standard deviation.

What is the probability that this project will have a benefit–cost ratio greater than 1.0?

Case Problem

DISASTER PLANNING AT TECH

Concerned about recent weather-related disasters, fires, and other calamities at universities around the country, university administrators at Tech have initiated several planning projects to determine how effectively local emergency facilities can handle such situations. One of these projects has focused on the transport of disaster victims from campus to the five major hospitals in the area: Montgomery Regional, Raeford Memorial, County General, Lewis Galt, and HGA Healthcare. The project team would like to determine how many victims each hospital might expect in a disaster and how long it would take to transport victims to the hospitals. However, one of the problems the project team faces is the lack of data on disasters, since they occur so infrequently. The project team has looked at disasters at other schools and has estimated that the minimum number of victims that would qualify an event as a disaster for the purpose of initiating a disaster plan is 10. The team has further estimated that the largest number of victims in any disaster would be 200, and based on limited data from other schools, they believe the most likely number of disaster victims is approximately 50. Because of the lack of data, it is assumed that these parameters best define a triangular distribution. The emergency facilities and capabilities at the five area hospitals vary. It has been estimated that in the event of a disaster situation, the victims should be dispersed to the hospitals on a percentage basis based on the hospitals' relative emergency capabilities, as follows: 25% should be sent to Montgomery Regional, 30% to Raeford Memorial, 15% to County General, 10% to Lewis Galt, and 20% to HGA Healthcare. The proximity of the hospitals to Tech also varies. It is estimated that transport times to each of the hospitals is exponentially distributed with an average time of 5 minutes to Montgomery Regional, 10 minutes to Raeford Memorial, 20 minutes to County General, 20 minutes to Lewis Galt, and 15 minutes to HGA Healthcare. (It is assumed that each hospital has two emergency vehicles, so that one leaves Tech when the other leaves the hospital, and consequently, one arrives at Tech when the other arrives at the hospital. Thus, the total transport time will be the sum of transporting each victim to a specific hospital.)

a. Perform a simulation analysis using Crystal Ball to determine the average number of victims that can be expected at each hospital and the average total time required to transport victims to each hospital.
b. Suppose that the project team believes they cannot confidently assume that the number of victims will follow a triangular distribution using the parameters they have estimated. Instead, they believe that the number of victims is best estimated using a normal distribution with the following parameters for each hospital: a mean of 6 minutes and a standard deviation of 4 minutes for Montgomery Regional; a mean of 11 minutes and a standard deviation of 4 minutes for Raeford Memorial; a mean of 22 minutes and a standard deviation of 8 minutes for County General; a mean of 22 minutes and a standard deviation of 9 minutes for Lewis Galt; and a mean of 15 minutes and a standard deviation of 5 minutes for HGA Healthcare. Perform a simulation analysis using this revised information.
c. Discuss how this information might be used for planning purposes. How might the simulation be altered or changed to provide additional useful information?

CHAPTER

15

Forecasting

A *forecast* is a prediction of what will occur in the future. Meteorologists forecast the weather, sportscasters predict the winners of football games, and managers of business firms attempt to predict how much of their product will be demanded in the future. In fact, managers are constantly trying to predict the future, making decisions in the present that will ensure the continued success of their firms. Often a manager will use judgment, opinion, or past experiences to forecast what will occur in the future. However, a number of mathematical methods are also available to aid managers in making decisions. In this chapter, we present two of the traditional forecasting methods: time series analysis and regression. Although no technique will result in a totally accurate forecast (i.e., it is impossible to predict the future exactly), these forecasting methods can provide reliable guidelines for decision making.

Forecasting Components

A variety of forecasting methods exist, and their applicability is dependent on the *time frame* of the forecast (i.e., how far in the future we are forecasting), the *existence of patterns* in the forecast (i.e., seasonal trends, peak periods), and the *number of variables* to which the forecast is related. We will discuss each of these factors separately.

__Short-range forecasts__ are daily operations.

__Medium-range forecasts__ are usually from a month up to a year.

__Long-range forecasts__ are more strategic and for over a year.

In general, forecasts can be classified according to three time frames: short range, medium range, and long range. **Short-range forecasts** typically encompass the immediate future and are concerned with the daily operations of a business firm, such as daily demand or resource requirements. A short-range forecast rarely goes beyond a couple of months into the future. A **medium-range forecast** typically encompasses anywhere from 1 or 2 months to 1 year. A forecast of this length is generally more closely related to a yearly production plan and will reflect such items as peaks and valleys in demand and the necessity to secure additional resources for the upcoming year. A **long-range forecast** typically encompasses a period longer than 1 or 2 years. Long-range forecasts are related to management's attempt to plan new products for changing markets, build new facilities, or secure long-term financing. In general, the farther into the future one seeks to predict, the more difficult forecasting becomes.

These classifications should be viewed as generalizations. The line of demarcation between medium- and long-range forecasts is often quite arbitrary and not always distinct. For some firms a medium-range forecast could be several years, and for other firms a long-range forecast could be in terms of months.

A __trend__ is a gradual, long-term, up-or-down movement of demand.

__Random variations__ are unpredictable movements in demand that follow no pattern.

Forecasts often exhibit patterns, or trends. A **trend** is a long-term movement of the item being forecast. For example, the demand for personal computers has shown an upward trend during the past decade, without any long downward movement in the market. Trends are the easiest patterns of demand behavior to detect and are often the starting point for developing a forecast. Figure 15.1(a) illustrates a demand trend in which there is a general upward movement or increase. Notice that Figure 15.1(a) also includes several random movements up and down. **Random variations** are movements that are not predictable and follow no pattern (and thus are virtually unpredictable).

A __cycle__ is an up-and-down repetitive movement in demand.

A **cycle** is an undulating movement in demand, up and down, that repeats itself over a lengthy time span (i.e., more than 1 year). For example, new housing starts and thus construction-related products tend to follow cycles in the economy. Automobile sales tend to follow cycles in the same fashion. The demand for winter sports equipment increases every 4 years, before and after the Winter Olympics. Figure 15.1(b) shows the general behavior of a demand cycle.

A __seasonal pattern__ is an up-and-down, repetitive movement within a trend occurring periodically.

A **seasonal pattern** is an oscillating movement in demand that occurs periodically (in the short run) and is repetitive. Seasonality is often weather related. For example, every winter the demand for snowblowers and skis increases dramatically, and retail sales in general increase during the Christmas season. However, a seasonal pattern can occur on a daily or weekly basis. For example, some restaurants are busier at lunch than at dinner, and shopping mall stores and

FIGURE 15.1
Forms of forecast movement: (a) trend, (b) cycle, (c) seasonal pattern, and (d) trend with seasonal pattern

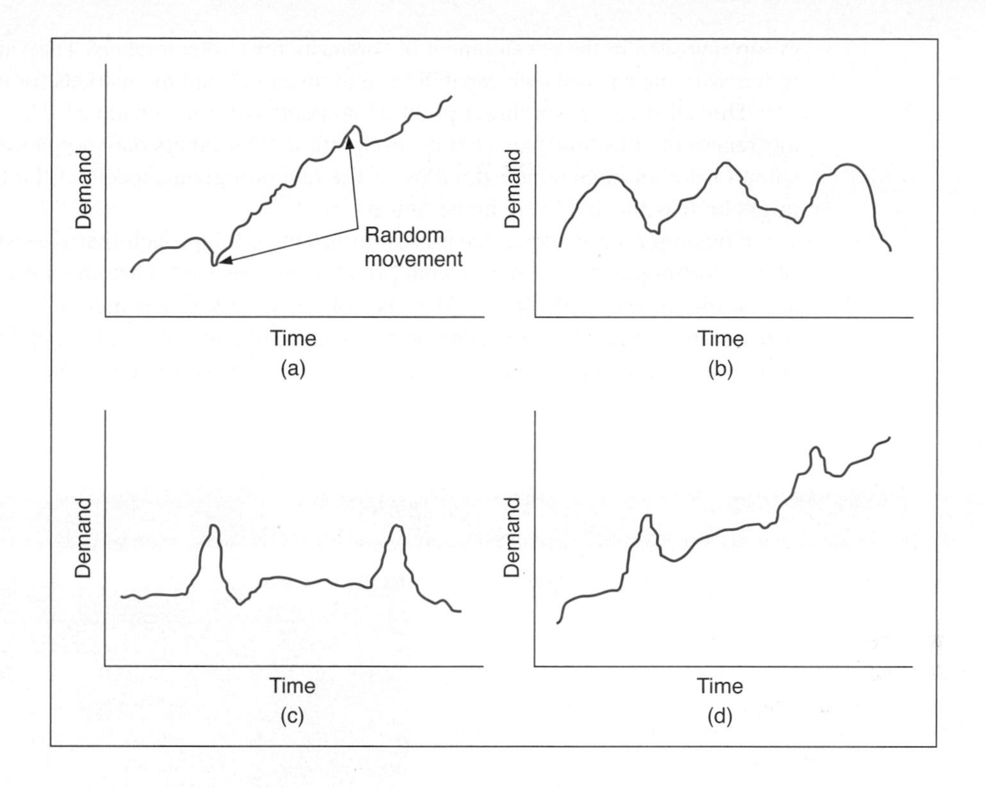

theaters tend to have higher demand on weekends. Figure 15.1(c) illustrates a seasonal pattern in which the same demand behavior is repeated each period at the same time.

Of course, demand behavior will frequently display several of these characteristics simultaneously. Although housing starts display cyclical behavior, there has been an upward trend in new house construction over the years. As we noted, demand for skis is seasonal; however, there has been a general upward trend in the demand for winter sports equipment during the past 2 decades. Figure 15.1(d) displays the combination of two demand patterns, a trend with a seasonal pattern.

Types of forecasting methods are time series, regression, and qualitative.

There are instances in which demand behavior exhibits no pattern. These are referred to as *irregular movements*, or variations. For example, a local flood might cause a momentary increase in carpet demand, or negative publicity resulting from a lawsuit might cause product demand to drop for a period of time. Although this behavior is causal, and thus not totally random, it still does not follow a pattern that can be reflected in a forecast.

Time series forecasts are statistical techniques that use historical data.

Forecasting Methods

Regression develops a mathematical relationship between the forecasted item and factors that cause it to behave the way it does.

The factors discussed previously determine to a certain extent the type of forecasting method that can or should be used. In this chapter we discuss the basic types of forecasting: *time series*, *regression methods*, and *qualitative methods*. **Time series** is a category of statistical techniques that uses historical data to predict future behavior. **Regression** (or causal) methods attempt to develop a mathematical relationship (in the form of a regression model) between the item being forecast and factors that cause it to behave the way it does. Most of the remainder of this chapter is about time series and regression forecasting methods. In this section we focus our discussion on qualitative forecasting.

Qualitative methods use judgment, expertise, and opinion to make forecasts.

Qualitative methods use management judgment, expertise, and opinion to make forecasts. Often called "the jury of executive opinion," they are the most common type of forecasting method for the long-term strategic planning process. There are normally individuals or groups within an organization whose judgments and opinions regarding the future are as valid or more valid than those of outside experts or other structured approaches. Top managers are the key group involved in the development of forecasts for strategic plans. They are generally most familiar with their firms' own capabilities and resources and the markets for their products.

The sales force is a direct point of contact with the consumer. This contact provides an awareness of consumer expectations in the future that others may not possess. Engineering personnel have an innate understanding of the technological aspects of the type of products that might be feasible and likely in the future.

Consumer, or market, research is an organized approach that uses surveys and other research techniques to determine what products and services customers want and will purchase, and to identify new markets and sources of customers. Consumer and market research is normally conducted by the marketing department within an organization, by industry organizations and groups, and by private marketing or consulting firms. Although market research can provide

Management Science Application

Forecasting Advertising Demand at NBC

NBC Universal, a subsidiary of General Electric Company, owns and operates the most profitable television network in the United States, with revenues of over $14 billion. More than 60% of these revenues are generated by on-air advertising time on its television networks and stations. The major television networks announce their new programming schedules for the upcoming season (which starts in late September) in mid-May. NBC begins selling advertising time very soon after the new schedule is announced in May, and 60% to 80% of its airtime inventory is sold in the following 2- to 3-week period (to approximately 400 advertisers), known as the *upfront market*.

Immediately following the announcement of its new season schedule in May, NBC forecasts ratings and estimates market demand for its shows. The ratings forecasts are estimates of the number of people in several demographic groups who are expected to watch each airing of the shows in the schedule for the entire year, which are, in turn, based on such factors as a show's strength, historical time-slot ratings, and the ratings performance of adjacent shows. Total market demand depends primarily on the strength of the economy and the expected performance of the network's schedule. Based on this ratings forecast and market demand, NBC develops pricing strategies and sets the prices for commercials on its shows.

Forecasting the upfront market has always been a challenging process for NBC. The network used to use historical patterns, expert knowledge, and intuition to forecast demand, then later it switched to time series forecasting models based on historical demand. However, these models were unsatisfactory because of the unique nature of NBC's demand population of advertisers.

NBC ultimately developed a unique approach to forecasting its upfront market for demand that includes a combination of the Delphi technique and a "grassroots" forecasting approach.

© Richard Levine/Alamy

The Delphi technique seeks to develop a consensus (or at least a compromise) forecast from among a group of experts, whereas a grassroots approach to forecasting asks the individuals closest to the final customer, such as salespeople, about the customers' purchasing plans. In this forecasting approach, NBC used its (more than 100) account executives, who interact closely with the network's (more than 400) advertisers, to build a knowledge base in order to estimate individual advertiser demand, aggregate this demand into a total demand forecast, and then continuously and iteratively update demand estimates based on the account executives' expertise. Previous forecasting methods had resulted in forecast errors of 5% to 12%, whereas this new forecasting process resulted in a forecast error of only 2.8%, the most accurate forecast NBC had ever achieved for its upfront market.

Source: Based on S. Bollapragada, S. Gupta, B. Hurwitz, P. Miles, and R. Tyagi, "NBC-Universal Uses a Novel Qualitative Forecasting Technique to Predict Advertising Demand," *Interfaces* 38, no. 2 (March–April 2008): 103–11.

accurate and useful forecasts of product demand, it must be skillfully and correctly conducted, and it can be expensive.

*The **Delphi method** forecasts future using informed judgment and opinions from knowledgeable individuals.*

The **Delphi method** is a procedure for acquiring informed judgments and opinions from knowledgeable individuals, using a series of questionnaires to develop a consensus forecast about what will occur in the future. It was developed at the RAND Corporation shortly after World War II to forecast the impact of a hypothetical nuclear attack on the United States. Although the Delphi method has been used for a variety of applications, forecasting has been one of its primary uses. It has been especially useful for forecasting technological change and advances.

Technological forecasting has become increasingly crucial for successful competition in today's global business environment. New enhanced computer technology, new production methods, and advanced machinery and equipment are constantly being made available to companies. These advances enable them to introduce more new products into the marketplace faster than ever before. The companies that succeed do so by getting a technological jump on their competitors through accurate prediction of future technology and its capabilities. What new products and services will be technologically feasible, when they can be introduced, and what their demand will be are questions about the future for which answers cannot be predicted from historical data. Instead, the informed opinion and judgment of experts are necessary to make these types of single, long-term forecasts.

Time Series Methods

Time series methods are statistical techniques that make use of historical data accumulated over a period of time. Time series methods assume that what has occurred in the past will continue to occur in the future. As the name *time series* suggests, these methods relate the forecast to only one factor—*time*. Time series methods tend to be most useful for short-range forecasting, although they can be used for longer-range forecasting. We will discuss two types of time series methods: the *moving average* and *exponential smoothing*.

Moving Average

A time series forecast can be as simple as using demand in the current period to predict demand in the next period. For example, if demand is 100 units this week, the forecast for next week's demand would be 100 units; if demand turned out to be 90 units instead, then the following week's demand would be 90 units, and so forth. This is sometimes referred to as *naïve* forecasting. However, this type of forecasting method does not take into account any type of historical demand behavior; it relies only on demand in the current period. As such, it reacts directly to the normal, random up-and-down movements in demand.

*The **moving average** method is good for stable demand with no pronounced behavioral patterns.*

Alternatively, the **moving average** method uses several values during the recent past to develop a forecast. This tends to *dampen*, or *smooth out*, the random increases and decreases of a forecast that uses only one period. As such, the simple moving average is particularly useful for forecasting items that are relatively stable and do not display any pronounced behavior, such as a trend or seasonal pattern.

Moving averages are computed for specific periods, such as 3 months or 5 months, depending on how much the forecaster desires to smooth the data. The longer the moving average period, the smoother the data will be. The formula for computing the simple moving average is as follows:

$$MA_n = \frac{\sum_{i=1}^{n} D_i}{n}$$

where

n = number of periods in the moving average
D_i = data in period i

To demonstrate the *moving average* forecasting method, we will use an example. The Instant Paper Clip Supply Company sells and delivers office supplies to various companies, schools, and agencies within a 30-mile radius of its warehouse. The office supply business is extremely competitive, and the ability to deliver orders promptly is an important factor in getting new customers and keeping old ones. (Offices typically order not when their inventory of supplies is getting low but when they completely run out. As a result, they need their orders immediately.) The manager of the company wants to be certain that enough drivers and delivery vehicles are available so that orders can be delivered promptly. Therefore, the manager wants to be able to forecast the number of orders that will occur during the next month (i.e., to forecast the demand for deliveries).

From records of delivery orders, the manager has accumulated data for the past 10 months. These data are shown in Table 15.1.

TABLE 15.1
Orders for 10-month period

Month	Orders Delivered per Month
January	120
February	90
March	100
April	75
May	110
June	50
July	75
August	130
September	110
October	90

The moving average forecast is computed by dividing the sum of the values of the forecast variable, orders per month for a sequence of months, by the number of months in the sequence. Frequently, a moving average is calculated for three or five time periods. The forecast resulting from either the 3- or the 5-month moving average is typically for the next month in the sequence, which in this case is November. The moving average is computed from the demand for orders for the last 3 months in the sequence, according to the following formula:

$$MA_3 = \frac{\sum_{i=1}^{3} D_i}{3} = \frac{90 + 110 + 130}{3} = 110 \text{ orders}$$

The 5-month moving average is computed from the last 5 months of demand data, as follows:

$$MA_5 = \frac{\sum_{i=1}^{5} D_i}{5} = \frac{90 + 110 + 130 + 75 + 50}{5} = 91 \text{ orders}$$

The 3- and 5-month moving averages for all the months of demand data are shown in Table 15.2. Notice that we have computed forecasts for all the months. Actually, only the forecast for November, based on the most recent monthly demand, would be used by the manager. However, the earlier forecasts for prior months allow us to compare the forecast with actual demand to see how accurate the forecasting method is (i.e., how well it does).

Both moving average forecasts in Table 15.2 tend to smooth out the variability occurring in the actual data. This smoothing effect can be observed in Figure 15.2, in which the 3-month and 5-month averages have been superimposed on a graph of the original data. The extremes in the actual orders per month have been reduced. This is beneficial if these extremes simply reflect random fluctuations in orders per month because our moving average forecast will not be strongly influenced by them.

TABLE 15.2
Three- and 5-month averages

Month	Orders per Month	3-Month Moving Average	5-Month Moving Average
January	120	—	—
February	90	—	—
March	100	—	—
April	75	103.3	—
May	110	88.3	—
June	50	95.0	99.0
July	75	78.3	85.0
August	130	78.3	82.0
September	110	85.0	88.0
October	90	105.0	95.0
November	—	110.0	91.0

FIGURE 15.2
Three- and 5-month moving averages

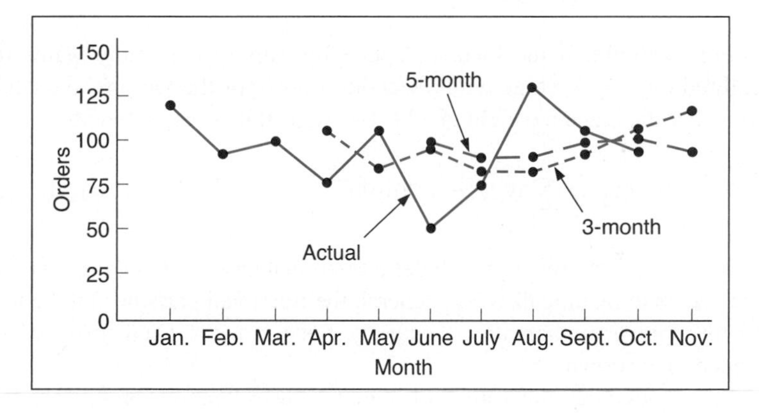

Longer-period moving averages react more slowly to recent demand changes than do shorter-period moving averages.

Notice that the 5-month moving average in Figure 15.2 smoothes out fluctuations to a greater extent than the 3-month moving average. However, the 3-month average more closely reflects the most recent data available to the office supply manager. (The 5-month average forecast considers data all the way back to June; the 3-month average does so only to August.) In general, forecasts computed using the longer-period moving average are slower to react to recent changes in demand than those made using shorter-period moving averages. The extra periods of data dampen the speed with which the forecast responds. Establishing the appropriate number of periods to use in a moving average forecast often requires some amount of trial-and-error experimentation.

One additional point to mention relates to the period of the forecast. Sometimes a forecaster will need to forecast for a short planning horizon rather than just a single period. For both the 3- and 5-month moving average forecasts computed in Table 15.2, the final forecast is for one period only in the future. Because the moving average includes multiple periods of data for each forecast, and it is most often used to forecast in a stable demand environment, it would be appropriate to use the forecast for multiple periods in the future. It could then be updated as actual demand data became available.

The major disadvantage of the moving average method is that it does not react well to variations that occur for a reason, such as trends and seasonal effects (although this method does reflect trends to a moderate extent). Those factors that cause changes are generally ignored. It is basically a "mechanical" method that reflects historical data in a consistent fashion. However, the moving average method does have the advantage of being easy to use, quick, and relatively inexpensive, although moving averages for a substantial number of periods for many different items can result in the accumulation and storage of a large amount of data. In general, this method can provide a good forecast for the short run, but an attempt should not be made to push the forecast too far into the distant future.

Weighted Moving Average

*In a **weighted moving average**, weights are assigned to the most recent data.*

The moving average method can be adjusted to reflect more closely more recent fluctuations in the data and seasonal effects. This adjusted method is referred to as a **weighted moving average** method. In this method, weights are assigned to the most recent data according to the following formula:

$$WMA_n = \sum_{i=1}^{n} W_i D_i$$

where

W_i = the weight for period i, between 0% and 100% (i.e., between 0 and 1.0)

$\Sigma W_i = 1.00$

For example, if the Instant Paper Clip Supply Company wants to compute a 3-month weighted moving average with a weight of 50% for the October data, a weight of 33% for the September data, and a weight of 17% for August, it is computed as

$$WMA_3 = \sum_{i=1}^{3} W_i D_i = (.50)(90) + (.33)(110) + (.17)(130) = 103.4 \text{ orders}$$

Notice that the forecast includes a fractional part, .4. Because .4 order would be impossible, this appears to be unrealistic. In general, the fractional parts need to be included in the computation to achieve mathematical accuracy, but when the final forecast is achieved, it must be rounded up or down.

Also notice that this forecast is slightly lower than our previously computed 3-month average forecast of 110 orders, reflecting the lower number of orders in October (the most recent month in the sequence).

Determining the precise weights to use for each period of data frequently requires some trial-and-error experimentation, as does determining the exact number of periods to include in the moving average. If the most recent months are weighted too heavily, the forecast might overreact to a random fluctuation in orders; if they are weighted too lightly, the forecast might under react to an actual change in the pattern of orders.

Exponential Smoothing

***Exponential smoothing** is an averaging method that reacts more strongly to recent changes in demand than to more distant past data.*

The **exponential smoothing** forecast method is an averaging method that weights the most recent past data more strongly than more distant past data. Thus, the forecast will react more strongly to immediate changes in the data. This is very useful if the recent changes in the data are the results of an *actual* change (e.g., a seasonal pattern) instead of just random fluctuations (for which a simple moving average forecast would suffice).

We will consider two forms of exponential smoothing: *simple exponential smoothing* and *adjusted exponential smoothing* (adjusted for trends, seasonal patterns, etc.). We will discuss the simple exponential smoothing case first, followed by the adjusted form.

Management Science Application

Forecasting at Goodyear Tire and Rubber Company

Landov Photo

Goodyear, with net sales of almost $20 billion, is one of the world's leading tire manufacturing companies and the number one tire maker in North America. Forecasting customer demand has always been recognized as an important part of its production and distribution planning processes, and it has historically used such forecasting techniques as time series analysis and exponential smoothing. For example, in the early 1990s in the United States, Goodyear was experiencing large errors in its 60-day advance planning forecasts, so it replaced its linear regression models with a single exponential smoothing model, which increased forecast accuracy by 20%.

In the mid-1990s the company developed a forecast system that analyzed the performance of more than 20 mathematical forecasting models for different products using historical demand data. It then picked the forecasting model that resulted in the lowest mean absolute percent error (*MAPE*).

In the late 1990s Goodyear upgraded its planning systems and simultaneously consolidated all its forecasting systems in North America (i.e., the United States and Canada) into a single integrated system. However, as the complexity of its business increased, so did its forecasting errors. To resolve this problem Goodyear created forecasts based on products, customers, and location to enhance its customer-specific forecasting capabilities. This required a collaborative relationship between Goodyear and its customers that included shared demand data and forecasts. For example, Goodyear used EDI technology to receive a customer's forecast and incorporate it into its own production planning process. By the mid-2000s Goodyear had experienced a 10% improvement in the accuracy of its 60-day forecast.

Source: Based on S. D. Miller and K. M. Liem, "Collaborative Forecasting: Goodyear Tire and Rubber Company's Journey," *Journal of Business Forecasting* 22, no. 1 (Fall 2004): 23–27.

To demonstrate simple exponential smoothing, we will return to the Instant Paper Clip Supply Company example. The simple exponential smoothing forecast is computed by using the formula

$$F_{t+1} = \alpha D_t + (1 - \alpha)F_t$$

where

F_{t+1} = the forecast for the next period
D_t = actual demand in the present period
F_t = the previously determined forecast for the present period
α = a weighting factor referred to as the *smoothing constant*

The smoothing constant, α, is between zero and one. It reflects the weight given to the most recent demand data. For example, if $\alpha = .20$,

$$F_{t+1} = .20D_t + .80F_t$$

which means that our forecast for the next period is based on 20% of recent demand (D_t) and 80% of past demand (in the form of the forecast F_t because F_t is derived from previous demands and forecasts). If we go to one extreme and let $\alpha = 0.0$, then

$$\begin{aligned} F_{t+1} &= 0D_t + 1F_t \\ &= F_t \end{aligned}$$

The closer α is to one, the greater the reaction to the most recent demand.

and the forecast for the next period is the same as for this period. In other words, *the forecast does not reflect the most recent demand at all.*

On the other hand, if $\alpha = 1.0$, then

$$\begin{aligned} F_{t+1} &= 1D_t + 0F_t \\ &= D_t \end{aligned}$$

and we have considered only the most recent occurrence in our data (demand in the present period) and nothing else. Thus, we can conclude that the higher α is, the more sensitive the forecast will be to changes in recent demand. Alternatively, the closer α is to zero, the greater will be the dampening or smoothing effect. As α approaches zero, the forecast will react and adjust more slowly to differences between the actual demand and the forecasted demand. The most commonly used values of α are in the range from .01 to .50. However, the determination of α is usually judgmental and subjective and will often be based on trial-and-error experimentation. An inaccurate estimate of α can limit the usefulness of this forecasting technique.

To demonstrate the computation of an exponentially smoothed forecast, we will use an example. PM Computer Services assembles customized personal computers from generic parts. The company was formed and is operated by two part-time State University students, Paulette Tyler and Maureen Becker, and has had steady growth since it started. The company assembles computers mostly at night, using other part-time students as labor. Paulette and Maureen purchase generic computer parts in volume at a discount from a variety of sources whenever they see a good deal. It is therefore important that they develop a good forecast of demand for their computers so that they will know how many computer component parts to purchase and stock.

The company has accumulated the demand data in Table 15.3 for its computers for the past 12 months, from which it wants to compute exponential smoothing forecasts, using smoothing constants (α) equal to .30 and .50.

TABLE 15.3
Demand for personal computers

Period	Month	Demand
1	January	37
2	February	40
3	March	41
4	April	37
5	May	45
6	June	50
7	July	43
8	August	47
9	September	56
10	October	52
11	November	55
12	December	54

To develop the series of forecasts for the data in Table 15.3, we will start with period 1 (January) and compute the forecast for period 2 (February) by using $\alpha = .30$. The formula for exponential smoothing also requires a forecast for period 1, which we do not have, so we will use the demand for period 1 as both demand and *the forecast for period 1.* Other ways to determine a starting forecast include averaging the first three or four periods and making a subjective estimate. Thus, the forecast for February is

$$F_2 = \alpha D_1 + (1 - \alpha)F_1 = (.30)(37) + (.70)(37) = 37 \text{ units}$$

The forecast for period 3 is computed similarly:

$$F_3 = \alpha D_2 + (1 - \alpha)F_2 = (.30)(40) + (.70)(37) = 37.9 \text{ units}$$

The remaining monthly forecasts are shown in Table 15.4. The final forecast is for period 13, January, and is the forecast of interest to PM Computer Services:

$$F_{13} = \alpha D_{12} + (1 - \alpha)F_{12} = (.30)(54) + (.70)(50.84) = 51.79 \text{ units}$$

Table 15.4 also includes the forecast values by using $\alpha = .50$. Both exponential smoothing forecasts are shown in Figure 15.3, together with the actual data.

In Figure 15.3, the forecast using the higher smoothing constant, $\alpha = .50$, reacts more strongly to changes in demand than does the forecast with $\alpha = .30$, although both smooth out the random fluctuations in the forecast. Notice that both forecasts lag the actual demand. For example, a pronounced downward change in demand in July is not reflected in the forecast until August. If these changes mark a change in trend (i.e., a long-term upward or downward movement) rather than just a random fluctuation, then the forecast will always lag this trend. We can see a general upward trend in delivered orders throughout the year. Both forecasts tend to be consistently lower than the actual demand; that is, the forecasts lag behind the trend.

TABLE 15.4
Exponential smoothing forecasts, $\alpha = .30$ and $\alpha = .50$

			Forecast, F_{t+1}	
Period	**Month**	**Demand**	$\alpha = .30$	$\alpha = .50$
1	January	37	—	—
2	February	40	37.00	37.00
3	March	41	37.90	38.50
4	April	37	38.83	39.75
5	May	45	38.28	38.37
6	June	50	40.29	41.68
7	July	43	43.20	45.84
8	August	47	43.14	44.42
9	September	56	44.30	45.71
10	October	52	47.81	50.85
11	November	55	49.06	51.42
12	December	54	50.84	53.21
13	January	—	51.79	53.61

FIGURE 15.3
Exponential smoothing forecasts

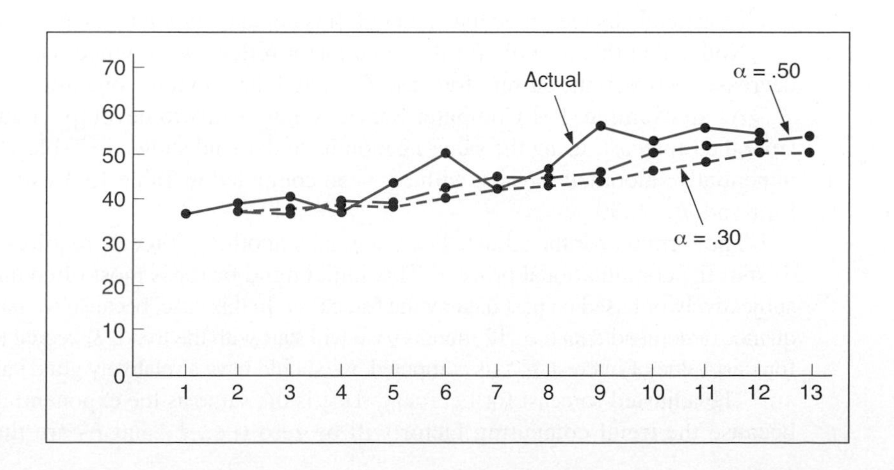

Based on simple observation of the two forecasts in Figure 15.3, $\alpha = .50$ seems to be the more accurate of the two, in the sense that it seems to follow the actual data more closely. (Later in this chapter we will discuss several quantitative methods for determining forecast accuracy.) In general, when demand is relatively stable, without any trend, using a small value for α is more appropriate to simply smooth out the forecast. Alternatively, when actual demand displays an increasing (or decreasing) trend, as is the case in Figure 15.3, a larger value of α is generally better. It will react more quickly to the more recent upward or downward movements in the actual data. In some approaches to exponential smoothing, the accuracy of the forecast is monitored in terms of the difference between the actual values and the forecasted values. If these differences become larger, then α is changed (higher or lower) in an attempt to adapt the forecast to the actual data. However, the exponential smoothing forecast can also be adjusted for the effects of a trend.

As we noted with the moving average forecast, the forecaster sometimes needs a forecast for more than one period into the future. In our PM Computer Services example, the final forecast computed was for 1 month, January. A forecast for 2 or 3 months could have been computed by grouping the demand data into the required number of periods and then using these values in the exponential smoothing computations. For example, if a 3-month forecast was needed, demand for January, February, and March could be summed and used to compute the forecast for the next 3-month period, and so on, until a final 3-month forecast resulted. Alternatively, if a trend was present, the final period forecast could be used for an extended forecast by adjusting it by a trend factor.

Adjusted Exponential Smoothing

Adjusted exponential ***smoothing*** *is the exponential smoothing forecast with an adjustment for a trend added to it.*

The **adjusted exponential smoothing** forecast consists of the exponential smoothing forecast with a trend adjustment factor added to it. The formula for the adjusted forecast is

$$AF_{t+1} = F_{t+1} + T_{t+1}$$

where
T = an exponentially smoothed trend factor

The trend factor is computed much the same as the exponentially smoothed forecast. It is, in effect, a forecast model for trend:

$$T_{t+1} = \beta(F_{t+1} - F_t) + (1 - \beta)T_t$$

where
T_t = the last period trend factor
β = a smoothing constant for trend

The closer β is to one, the stronger a trend is reflected.

Like α, β is a value between zero and one. It reflects the weight given to the most recent trend data. Also like α, β is often determined subjectively, based on the judgment of the forecaster. A high β reflects trend changes more than a low β. It is not uncommon for β to equal α in this method.

Notice that this formula for the trend factor reflects a weighted measure of the increase (or decrease) between the current forecast, F_{t+1}, and the previous forecast, F_t.

As an example, PM Computer Services now wants to develop an adjusted exponentially smoothed forecast, using the same 12 months of demand shown in Table 15.3. It will use the exponentially smoothed forecast with $\alpha = .50$ computed in Table 15.4 with a smoothing constant for trend, β, of .30.

The formula for the adjusted exponential smoothing forecast requires an initial value for T_t to start the computational process. This initial trend factor is most often an estimate determined subjectively or based on past data by the forecaster. In this case, because we have a relatively long sequence of demand data (i.e., 12 months), we will start with the trend, T_t, equal to zero. By the time the forecast value of interest, F_{13}, is computed, we should have a relatively good value for the trend factor.

The adjusted forecast for February, AF_2, is the same as the exponentially smoothed forecast because the trend computing factor will be zero (i.e., F_1 and F_2 are the same and $T_2 = 0$).

Thus, we will compute the adjusted forecast for March, AF_3, as follows, starting with the determination of the trend factor, T_3:

$$T_3 = \beta(F_3 - F_2) + (1 - \beta)T_2 = (.30)(38.5 - 37.0) + (.70)(0) = 0.45$$

and

$$AF_3 = F_3 + T_3 = 38.5 + 0.45 = 38.95$$

This adjusted forecast value for period 3 is shown in Table 15.5, with all other adjusted forecast values for the 12-month period plus the forecast for period 13, computed as follows:

$$T_{13} = \beta(F_{13} - F_{12}) + (1 - \beta)T_{12} = (.30)(53.61 - 53.21) + (.70)(1.77) = 1.36$$

TABLE 15.5 Adjusted exponentially smoothed forecast values

Period	Month	Demand	Forecast (F_{t+1})	Trend (T_{t+1})	Adjusted Forecast (AF_{t+1})
1	January	37	37.00	—	—
2	February	40	37.00	0.00	37.00
3	March	41	38.50	0.45	38.95
4	April	37	39.75	0.69	40.44
5	May	45	38.37	0.07	38.44
6	June	50	41.68	1.04	42.73
7	July	43	45.84	1.97	47.82
8	August	47	44.42	0.95	45.37
9	September	56	45.71	1.05	46.76
10	October	52	50.85	2.28	53.13
11	November	55	51.42	1.76	53.19
12	December	54	53.21	1.77	54.98
13	January	—	53.61	1.36	54.96

and

$$AF_{13} = F_{13} + T_{13} = 53.61 + 1.36 = 54.96 \text{ units}$$

The adjusted exponentially smoothed forecast values shown in Table 15.5 are compared with the exponentially smoothed forecast values and the actual data in Figure 15.4.

FIGURE 15.4 Adjusted, exponentially smoothed forecast

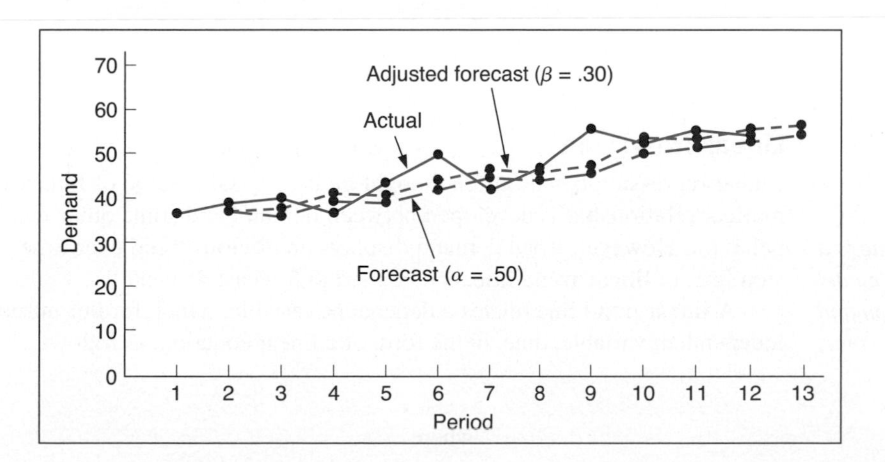

Notice that the adjusted forecast is consistently higher than the exponentially smoothed forecast and is thus more reflective of the generally increasing trend of the actual data. However, in general, the pattern, or degree of smoothing, is very similar for both forecasts.

MANAGEMENT SCIENCE APPLICATION

Forecasting at Heineken USA

Heineken is an international brewer based in Amsterdam with more than 200 brands of beer sold in more than 100 countries. Its annual revenues are over €14,000 million. Heineken brews, packages, and ships its Dutch brands, including Heineken, Heineken Premium Light, and Amstel Light, to Heineken USA on ocean carriers to seven U.S. ports (Miami, Charleston, Norfolk, New York, Houston, Long Beach, and Oakland), where it maintains warehouses and demand centers. Heineken thus operates a three-tier supply chain with a brewer, distributors, and retailers. Heineken ships full container loads directly from the ports to the warehouses of its largest distributors; for smaller distributors, loads are shipped from Heineken demand centers near the ports.

A "depletion" forecast showing how much beer distributors are currently and historically selling to retailers is converted to a sales demand forecast, which indicates the amount of inventory the distributors will be buying from Heineken. Promotions and seasonal sales are factored into this demand forecast. The sales demand forecast is then used in Holland to create a production/replenishment plan for product supply. Distributors order beer over the Internet on a weekly basis, at which time they also provide their 3-month depletion forecast, which is a major input to the demand forecast that drives production in Holland. One week later, they receive delivery from a port or demand center. Retailers order directly from the distributors. Since lead times are long (4 to 6 weeks) and many consumers will switch beers if a particular one is out of stock, forecasting at Heineken is critical. Under-forecasting results in lost sales; over-forecasting results in costly excess inventory, which becomes obsolete and is destroyed if it sits in inventory

too long and is not absolutely fresh. However, there are no negative effects for the distributor if it provides a bad forecast; if it over-forecasts, Heineken absorbs the excess inventory; if it under-forecasts, Heineken transfers products from other demand centers to avoid stockouts.

In the mid-2000s, because distributors suffered no ill effects from bad forecasts, the forecasting process had deteriorated badly in the southeastern region, so an incentive plan was developed to award American Express gift cards to individual distributor forecasters with the best monthly, quarterly, and yearly forecast accuracy. The incentive program was a success; average monthly forecasts improved 10% in 1 year, which resulted in dependable forecasts to drive Heineken's supply chain.

Source: Based on B. Dershem, "Financial Incentives Improve the Supply Chain: Heineken's Journey," *Journal of Business Forecasting* 26, no. 1 (Spring 2007): 32–37.

Linear Trend Line

*A **linear trend line** is a linear regression model that relates demand to time.*

Linear regression is most often thought of as a causal method of forecasting in which a mathematical relationship is developed between demand and some other factor that affects demand behavior. However, when demand displays an obvious trend over time, a least squares regression line, or **linear trend line**, can be used to forecast demand.

A linear trend line relates a dependent variable, which for our purposes is demand, to one independent variable, time, in the form of a linear equation, as follows:

$$y = a + bx$$

where

a = intercept (at period 0)
b = slope of the line
x = the time period
y = forecast for demand for period x

These parameters of the linear trend line can be calculated by using the least squares formulas for linear regression:

$$b = \frac{\Sigma xy - n\bar{x}\bar{y}}{\Sigma x^2 - n\bar{x}^2}$$

$$a = \bar{y} - b\bar{x}$$

where

n = number of periods

$$\bar{x} = \frac{\Sigma x}{n}$$

$$\bar{y} = \frac{\Sigma y}{n}$$

As an example, consider the demand data for PM Computer Services shown in Table 15.3. They appear to follow an increasing linear trend. As such, the company wants to compute a linear trend line as an alternative to the exponential smoothing and adjusted exponential smoothing forecasts shown in Tables 15.4 and 15.5. The values that are required for the least squares calculations are shown in Table 15.6.

TABLE 15.6
Least squares calculations

x (period)	y (demand)	xy	x^2
1	37	37	1
2	40	80	4
3	41	123	9
4	37	148	16
5	45	225	25
6	50	300	36
7	43	301	49
8	47	376	64
9	56	504	81
10	52	520	100
11	55	605	121
12	54	648	144
78	557	3,867	650

Using these values for $\bar{x}$ and $\bar{y}$ and the values from Table 15.6, the parameters for the linear trend line are computed as follows:

$$\bar{x} = \frac{78}{12} = 6.5$$

$$\bar{y} = \frac{557}{12} = 46.42$$

$$b = \frac{\Sigma xy - n\bar{x}\bar{y}}{\Sigma x^2 - n\bar{x}^2} = \frac{3{,}867 - (12)(6.5)(46.42)}{650 - 12(6.5)^2} = 1.72$$

$$a = \bar{y} - b\bar{x} = 46.42 - (1.72)(6.5) = 35.2$$

Therefore, the linear trend line is

$$y = 35.2 + 1.72x$$

To calculate a forecast for period 13, $x = 13$ would be substituted in the linear trend line:

$$y = 35.2 + 1.72(13) = 57.56$$

A linear trend line will not adjust to a change in trend as will exponential smoothing.

Figure 15.5 shows the linear trend line in comparison to the actual data. The trend line visibly appears to closely reflect the actual data (i.e., to be a "good fit") and would thus be a good forecast model for this problem. However, a disadvantage of the linear trend line is that it will not adjust to a change in the trend as the exponential smoothing forecast methods will (i.e., it is assumed that all future forecasts will follow a straight line). This limits the use of this method to a shorter time frame in which the forecaster is relatively certain that the trend will not change.

FIGURE 15.5
Linear trend line

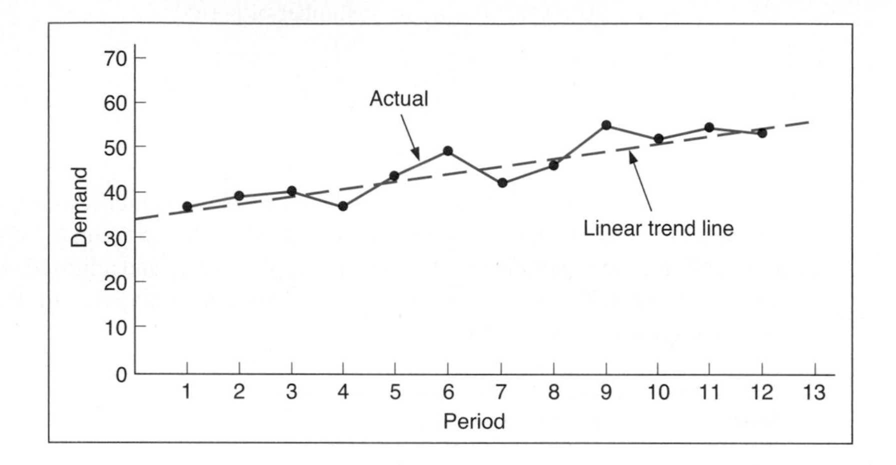

Seasonal Adjustments

As we mentioned at the beginning of this chapter, a seasonal pattern is a repetitive up-and-down movement in demand. Many demand items exhibit seasonal behavior. Clothing sales follow annual seasonal patterns, with demand for warm clothes increasing in the fall and winter and declining in the spring and summer, as the demand for cooler clothing increases. Demand for many retail items—including toys, sports equipment, clothing, electronic appliances, hams, turkeys, wine, and fruit—increases during the Christmas season. Greeting card demand increases in conjunction with various special days, such as Valentine's Day and Mother's Day. Seasonal patterns can also occur on a monthly, weekly, or even daily basis. Some restaurants have higher demand in the evening than at lunch, or on weekends as opposed to weekdays. Traffic and sales picks up at shopping malls on Friday and Saturday.

It is possible to adjust for seasonality by multiplying the normal forecast by a seasonal factor.

There are several methods available for reflecting seasonal patterns in a time series forecast. We will describe one of the simpler methods—using a **seasonal factor**, which is a numerical value that is multiplied by the normal forecast to get a seasonally adjusted forecast.

One method for developing a demand for seasonal factors is dividing the actual demand for each seasonal period by the total annual demand, according to the following formula:

$$S_i = \frac{D_i}{\Sigma D}$$

The resulting seasonal factors between zero and one are, in effect, the portion of total annual demand assigned to each season. These seasonal factors are thus multiplied by the annual forecasted demand to yield seasonally adjusted forecasts for each period.

As an example, Wishbone Farms is a company that raises turkeys, which it sells to a meat-processing company throughout the year. However, the peak season obviously occurs during the fourth quarter of the year, October to December. Wishbone Farms has experienced a demand for turkeys for the past 3 years as shown in Table 15.7.

TABLE 15.7
Demand for turkeys at Wishbone Farms

Year	Demand (1,000s)				
	Quarter 1	Quarter 2	Quarter 3	Quarter 4	Total
1	12.6	8.6	6.3	17.5	45.0
2	14.1	10.3	7.5	18.2	50.1
3	15.3	10.6	8.1	19.6	53.6
Total	42.0	29.5	21.9	55.3	148.7

Because we have 3 years of demand data, we can compute the seasonal factors by dividing total quarterly demand for the 3 years by total demand across all 3 years:

$$S_1 = \frac{D_1}{\Sigma D} = \frac{42.0}{148.7} = 0.28$$

$$S_2 = \frac{D_2}{\Sigma D} = \frac{29.5}{148.7} = 0.20$$

$$S_3 = \frac{D_3}{\Sigma D} = \frac{21.9}{148.7} = 0.15$$

$$S_4 = \frac{D_4}{\Sigma D} = \frac{55.3}{148.7} = 0.37$$

Next, we want to multiply the forecasted demand for year 4 by each of the seasonal factors to get the forecasted demand for each quarter. However, to accomplish this, we need a demand forecast for year 4. In this case, because the demand data in Table 15.7 seem to exhibit a generally increasing trend, we compute a linear trend line for the 3 years of data in Table 15.7 to use as a rough forecast estimate:

$$y = 40.97 + 4.30x = 40.97 + 4.30(4) = 58.17$$

Thus, the forecast for year 4 is 58.17, or 58,170 turkeys.

Using this annual forecast of demand, the seasonally adjusted forecasts, SF_i, for year 4 are as follows:

$$SF_1 = (S_1)(F_5) = (.28)(58.17) = 16.28$$
$$SF_2 = (S_2)(F_5) = (.20)(58.17) = 11.63$$
$$SF_3 = (S_3)(F_5) = (.15)(58.17) = 8.73$$
$$SF_4 = (S_4)(F_5) = (.37)(58.17) = 21.53$$

Comparing these quarterly forecasts with the actual demand values in Table 15.7 shows them to be relatively good forecast estimates, reflecting both the seasonal variations in the data and the general upward trend.

Forecast Accuracy

*The **forecast error** is the difference between the forecast and actual demand.*

It is not probable that a forecast will be completely accurate; forecasts will always deviate from the actual demand. This difference between the forecast and the actual demand is referred to as the **forecast error**. Although some amount of forecast error is inevitable, the objective of forecasting is for the error to be as slight as possible. Of course, if the degree of error is not small, this may indicate either that the forecasting technique being used is the wrong one or that the technique needs to be adjusted by changing its parameters (for example, α in the exponential smoothing forecast).

There are a variety of different measures of forecast error, and in this section we discuss several of the most popular ones: mean absolute deviation (*MAD*), mean absolute percent deviation (*MAPD*), cumulative error (E), average error or bias ($\bar{E}$), and mean squared error (*MSE*).

Mean Absolute Deviation

MAD is the average, absolute difference between the forecast and the demand.

The **mean absolute deviation (MAD)** is one of the most popular and simplest-to-use measures of forecast error. *MAD* is an average of the difference between the forecast and actual demand, as computed by the following formula:

$$MAD = \frac{\Sigma|D_t - F_t|}{n}$$

where

t = the period number
D_t = demand in period t
F_t = the forecast for period t
n = the total number of periods
$||$ = the absolute value

In our examples for PM Computer Services, forecasts were developed using exponential smoothing (with $\alpha = .30$ and with $\alpha = .50$), adjusted exponential smoothing $(\alpha = .50, \beta = .30)$, and a linear trend line for the demand data. The company wants to compare the accuracy of these different forecasts by using *MAD*.

We will compute *MAD* for all four forecasts; however, we will only present the computational detail for the exponential smoothing forecast with $\alpha = .30$. Table 15.8 shows the values necessary to compute *MAD* for the exponential smoothing forecast.

TABLE 15.8 Computational values for *MAD* and error

Period	Demand, D_t	Forecast, F_t ($\alpha = .30$)	Error ($D_t - F_t$)	$\lvert D_t - F_t \rvert$	Error^2 $(D_t - F_t)^2$
1	37	37.00	—	—	—
2	40	37.00	3.00	3.00	9.00
3	41	37.90	3.10	3.10	9.61
4	37	38.83	−1.83	1.83	3.35
5	45	38.28	6.72	6.72	45.15
6	50	40.29	9.71	9.71	94.28
7	43	43.20	−0.20	0.20	0.04
8	47	43.14	3.86	3.86	14.90
9	56	44.30	11.70	11.70	136.89
10	52	47.81	4.19	4.19	17.56
11	55	49.06	5.94	5.94	35.28
12	54	50.84	3.16	3.16	9.98
	520*		49.31	53.41	376.04

*Because the computation of *MAD* will be based on the 11 periods 2 through 12, $\Sigma D_t = 520$, excluding the demand value of 37.

Using the data in Table 15.8, we compute *MAD* as

$$MAD = \frac{\Sigma|D_t - F_t|}{n} = \frac{53.41}{11} = 4.85$$

The lower the value of MAD relative to the magnitude of the data, the more accurate the forecast.

In general, the smaller the value of *MAD*, the more accurate the forecast, although, viewed alone, *MAD* is difficult to assess. In this example, the data values were relatively small, and the *MAD* value of 4.85 should be judged accordingly. Overall, it would seem to be a "low" value (i.e., the forecast appears to be relatively accurate). However, if the magnitude of the data values were in the thousands or millions, then a *MAD* value of a similar magnitude might not be bad either. The point is that you cannot compare a *MAD* value of 4.85 with a *MAD* value of 485 and say the former is good and the latter is bad; they depend to a certain extent on the relative magnitude of the data.

One benefit of *MAD* is being able to compare the accuracy of several different forecasting techniques, as we are doing in this example. The *MAD* values for the remaining forecasts are:

- Exponential smoothing ($\alpha = .50$): $MAD = 4.04$
- Adjusted exponential smoothing ($\alpha = .50, \beta = .30$): $MAD = 3.81$
- Linear trend line: $MAD = 2.29$

When we compare all four forecasts, the linear trend line has the lowest *MAD* value, of 2.29. It would seem to be the most accurate, although it does not appear to be significantly better than the adjusted exponential smoothing forecast. Furthermore, we can deduce from these *MAD* values that increasing α from .30 to .50 enhanced the accuracy of the exponentially smoothed forecast. The adjusted forecast is even more accurate.

MAPD is absolute error as a percentage of demand.

A variation of *MAD* is the **mean absolute percent deviation (*MAPD*)**. It measures the absolute error as a percentage of demand rather than per period. As a result, it eliminates the problem of interpreting the measure of accuracy relative to the magnitude of the demand and forecast values, as *MAD* does. *MAPD* is computed according to the following formula:

$$MAPD = \frac{\Sigma|D_t - F_t|}{\Sigma D_t}$$

Using the data from Table 15.8 for the exponential smoothing forecast ($\alpha = .30$) for PM Computer Services, *MAPD* is computed as

$$MAPD = \frac{53.41}{520} = .103, \text{ or } 10.3\%$$

A lower *MAPD* implies a more accurate forecast. The *MAPD* values for our other three forecasts are:

- Exponential smoothing ($\alpha = .50$): $MAPD = 8.5\%$
- Adjusted exponential smoothing ($\alpha = .50, \beta = .30$): $MAPD = 8.1\%$
- Linear trend line: $MAPD = 4.9\%$

Cumulative Error

Cumulative error is the sum of the forecast errors.

Cumulative error is computed simply by summing the forecast errors, as shown in the following formula:

$$E = \Sigma e_t$$

where

$$e_i = D_t - F_t$$

A relatively large positive value indicates that the forecast is probably consistently lower than the actual demand or is biased low. A large negative value implies that the forecast is consistently higher than actual demand or is biased high. Also, when the errors for each period are scrutinized and there appears to be a preponderance of positive values, this shows that the forecast is consistently less than actual, and vice versa.

The cumulative error for the exponential smoothing forecast for PM Computer Services can be read directly from Table 15.8; it is simply the sum of the values in the "Error" column:

$$E = \Sigma e_t = 49.31$$

This relatively large-value positive error for cumulative error and the fact that the individual errors for each period in Table 15.8 are positive indicate that this forecast is frequently below the actual demand. A quick glance back at the plot of the exponential smoothing ($\alpha = .30$) forecast in Figure 15.3 visually verifies this result.

The cumulative errors for the other forecasts are:

- Exponential smoothing ($\alpha = .50$): $E = 33.21$
- Adjusted exponential smoothing ($\alpha = .50, \beta = .30$): $E = 21.14$

We did not show the cumulative error for the linear trend line. E will always be near zero for the linear trend line; thus, it is not a good measure on which to base comparisons with other forecast methods.

***Average error** is the per-period average of cumulative error.*

A measure closely related to cumulative error is the **average error**. It is computed by averaging the cumulative error over the number of time periods:

$$\overline{E} = \frac{\Sigma e_t}{n}$$

For example, the average error for the exponential smoothing forecast ($\alpha = .30$) is computed as follows (notice that the value of 11 was used for n because we used actual demand for the first-period forecast, resulting in no error, i.e., $D_1 = F_1 = 37$):

$$\overline{E} = \frac{49.31}{11} = 4.48$$

Large $+\overline{E}$ indicates that a forecast is biased low; large $-\overline{E}$ indicates that a forecast is biased high.

The average error is interpreted similarly to the cumulative error. A positive value indicates low bias, and a negative value indicates high bias. A value close to zero implies a lack of bias.

***Mean squared error** individual error values are squared and then summed and averaged.*

Another measure of forecast accuracy related to error is **mean squared error** (*MSE*). With *MSE*, each individual error value is squared, and then these values are summed and averaged. The last column in Table 15.8 shows the sum of the squared forecast errors (i.e., 376.04) for the PM Computer example forecast ($\alpha = 0.30$). The *MSE* is computed as

$$MSE = \frac{376.04}{11} = 34.18$$

As with other measures of forecast accuracy, the smaller the *MSE*, the better.

Table 15.9 summarizes the measures of forecast accuracy we have discussed in this section for the four example forecasts we developed in the previous section for PM Computer Services. The results are consistent for all four forecasts, indicating that for the PM Computer Services example data, a larger value of α is preferable for the exponential smoothing forecast. The adjusted forecast is more accurate than the exponential smoothing forecasts, and the linear trend is more accurate than all the others. Although these results are example specific, they do indicate how the different forecast measures for accuracy can be used to adjust a forecasting method or select the best method.

Management Science Application

Forecasting at Bayer Consumer Care

Andy Kropa/Redux

Since its development of aspirin in 1899, The Bayer Group has grown into a company with a global workforce of 110,000, more than 5,000 products, and annual sales exceeding 27,380 million Euros. Bayer HealthCare, one of the three operational subgroups of Bayer AG, researches, develops, manufactures, and markets products for the prevention, diagnosis, and treatment of diseases. The Consumer Care Division of Bayer HealthCare markets over-the-counter medications such as aspirin, Aleve, Midol, Alka-Seltzer, milk of magnesia, and One-A-Day and Flintstones vitamins.

Forecasting is an important function at Bayer Consumer Care because of the competitiveness in the over-the-counter pharmaceutical market and the seasonality of its products. The supply chain planning process at Bayer's Consumer Care division in the EU is based on each country customer's forecasts of its future sales for 18 months into the future, which they share (along with inventory information) with Bayer. Bayer uses this information to develop its production plans, generate customer orders, and establish shipment schedules. This forecasting approach is necessary because the pharmaceutical industry has long lead times for raw materials and ingredients, plus uncertainty created by irregular product trends, seasonality, and promotions. To remain competitive, Bayer guarantees its customers close to a 100% service level, and thus it seeks an inventory level that will meet this goal, which decreases as its forecast accuracy increases.

Initially, the company discovered that forecast errors resulted from the number of different forecasts that were being used by different supply chain members in each country. To resolve this problem, monthly round-table meetings were established to develop a consensus forecast in each EU country. To determine forecast accuracy three performance measures were used: absolute percent forecast error (for a 3-month horizon), percent forecast error, and standard deviation of percent forecast error. Forecast accuracy in most countries is around 80% and forecasting bias approaches zero.

Source: Based on H. Petersen, "Integrating the Forecasting Process with the Supply Chain: Bayer HealthCare's Journey," *Journal of Business Forecasting* 22, no. 4 (Winter 2003–2004): 11–15.

TABLE 15.9
Comparison of forecasts for PM Computer Services

Forecast	*MAD*	*MAPD* (%)	*E*	$\bar{E}$	*MSE*
Exponential smoothing ($\alpha = .30$)	4.85	10.3	49.31	4.48	34.18
Exponential smoothing ($\alpha = .50$)	4.04	8.5	33.21	3.02	24.64
Adjusted exponential smoothing ($\alpha = .50, \beta = .30$)	3.81	8.1	21.14	1.92	21.58
Linear trend line	2.29	4.9	—	—	8.67

Time Series Forecasting Using Excel

All the time series forecasts we have presented can also be developed with Excel spreadsheets. Exhibit 15.1 shows an Excel spreadsheet set up to compute the exponentially smoothed forecast and adjusted exponentially smoothed forecast for the PM Computer Services example summarized in Table 15.5. Notice that the formula for computing the trend factor in cell D10 is shown on the formula bar on the top of the spreadsheet. The adjusted forecast in column E is computed by typing the formula **=C9+D9** in cell E9 and copying it to cells **E10:E20** (using the "Copy" and "Paste" options that appear after clicking the right mouse button).

EXHIBIT 15.1

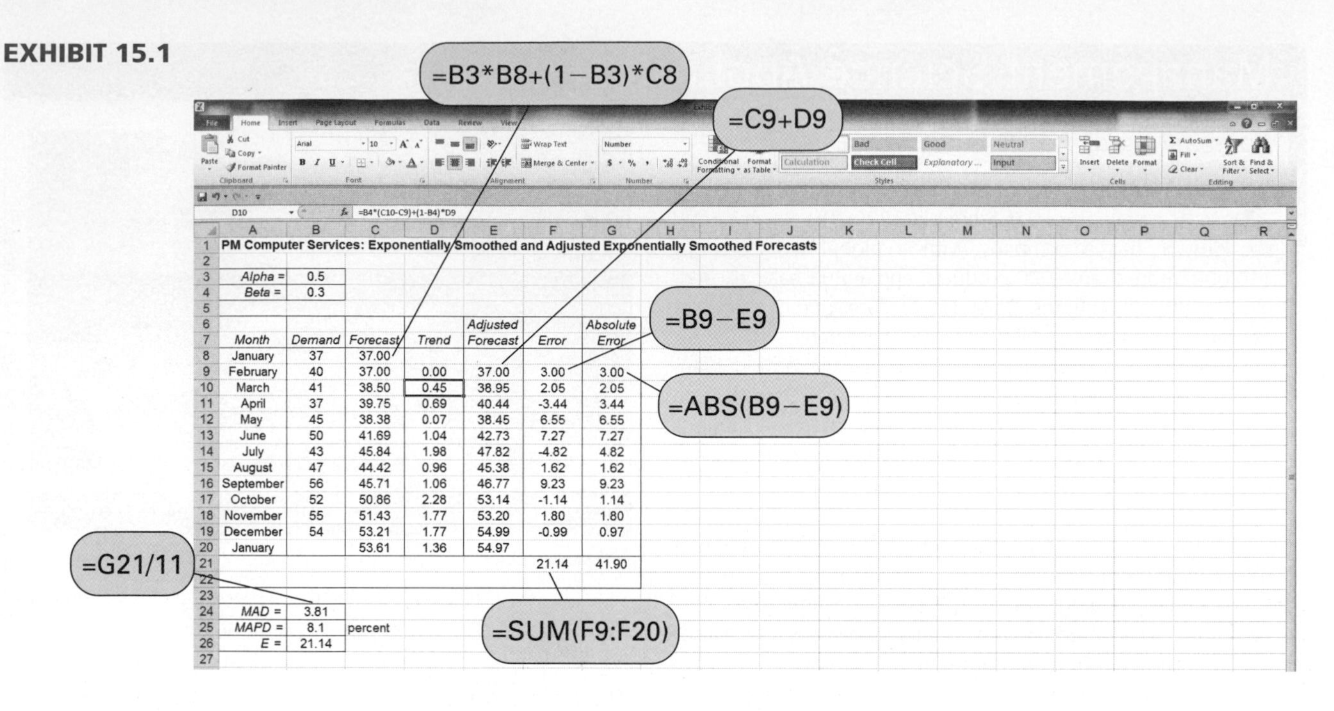

PM Computer Services: Exponentially Smoothed and Adjusted Exponentially Smoothed Forecasts

Alpha =	0.5
Beta =	0.3

Month	Demand	Forecast	Trend	Adjusted Forecast	Error	Absolute Error
January	37	37.00				
February	40	37.00	0.00	37.00	3.00	3.00
March	41	38.50	0.45	38.95	2.05	2.05
April	37	39.75	0.69	40.44	-3.44	3.44
May	45	38.38	0.07	38.45	6.55	6.55
June	50	41.69	1.04	42.73	7.27	7.27
July	43	45.84	1.98	47.82	-4.82	4.82
August	47	44.42	0.96	45.38	1.62	1.62
September	56	45.71	1.06	46.77	9.23	9.23
October	52	50.86	2.28	53.14	-1.14	1.14
November	55	51.43	1.77	53.20	1.80	1.80
December	54	53.21	1.77	54.99	-0.99	0.97
January		53.61	1.36	54.97		
					21.14	41.90

MAD =	3.81	
MAPD =	8.1	percent
E =	21.14	

The exponential smoothing forecast can also be developed directly from Excel without "customizing" a spreadsheet and entering our own formulas, as we did in Exhibit 15.1. From the "Data" menu at the top of the spreadsheet, select the "Data Analysis" option. (If your "Data" menu does not include this menu item, you should add it by accessing the "Add-Ins" option from the "Data" menu or by loading from the original Excel or Office software.) Exhibit 15.2 shows the Data Analysis window and the "Exponential Smoothing" menu item you should select and then click on "OK." The resulting Exponential Smoothing window is shown in Exhibit 15.3. The input range includes the demand values in column B in Exhibit 15.1, the damping factor is alpha (α), which in this case is 0.5, and the output should be placed in column C in Exhibit 15.1. Clicking on "OK" will result in the same forecast values in column C of Exhibit 15.1 that we computed using our own exponential smoothing formula. Note that the "Data Analysis" group of analysis tools does not have an adjusted exponential smoothing selection; that is the reason we developed our own customized spreadsheet in Exhibit 15.1. The "Data Analysis" tools also have a "Moving Average" menu item from which a moving average forecast can be computed.

EXHIBIT 15.2

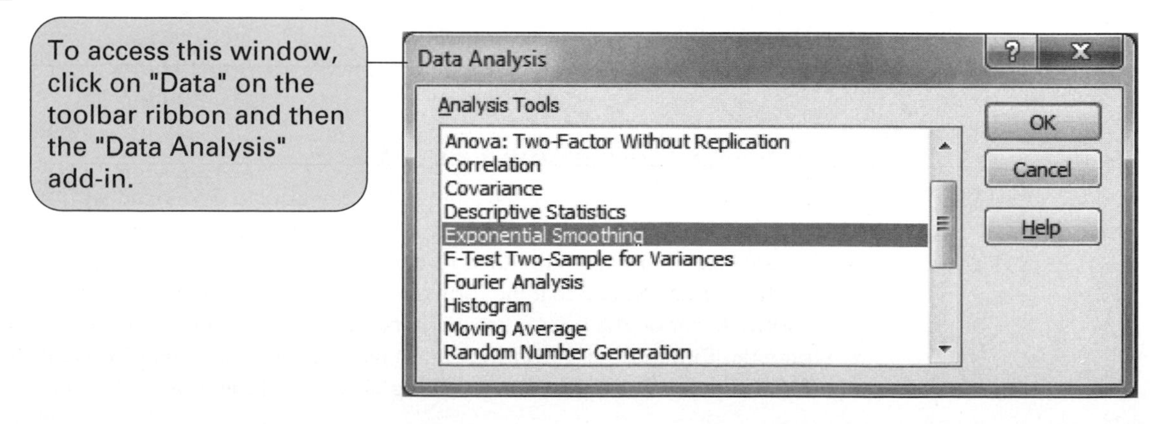

EXHIBIT 15.3

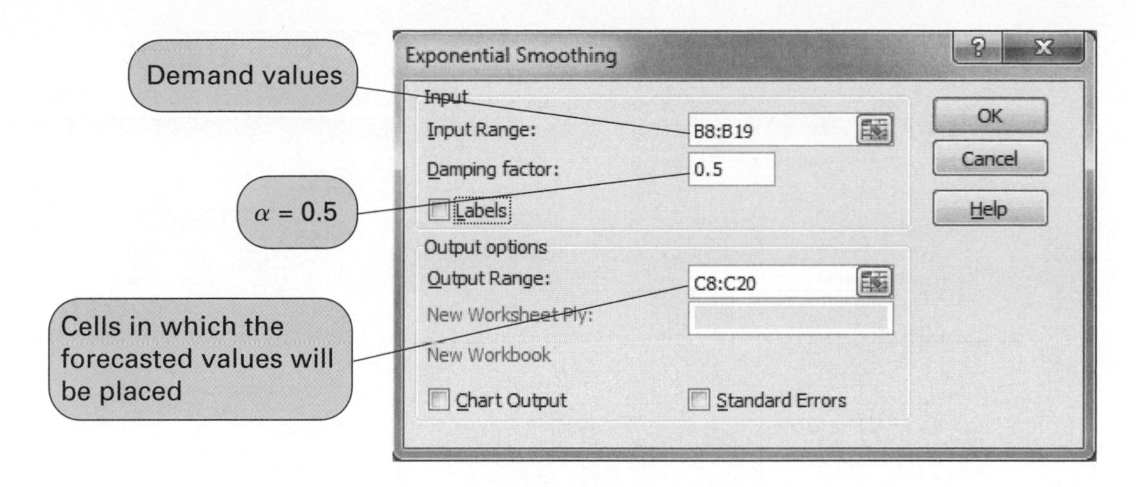

EXHIBIT 15.4

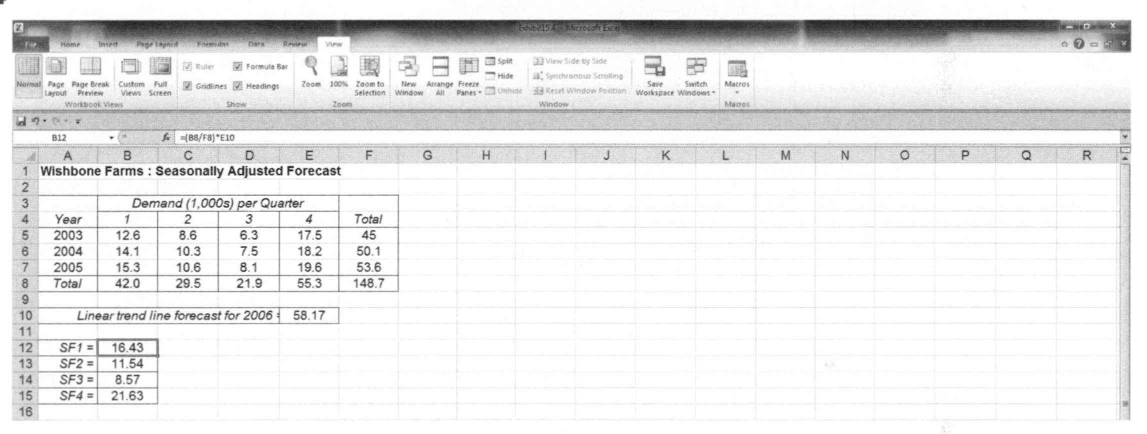

B12 =(B8/F8)*E10

Wishbone Farms : Seasonally Adjusted Forecast

	Demand (1,000s) per Quarter				
Year	1	2	3	4	Total
2003	12.6	8.6	6.3	17.5	45
2004	14.1	10.3	7.5	18.2	50.1
2005	15.3	10.6	8.1	19.6	53.6
Total	42.0	29.5	21.9	55.3	148.7

Linear trend line forecast for 2006 =	58.17
SF1 =	16.43
SF2 =	11.54
SF3 =	8.57
SF4 =	21.63

Excel can also be used to develop seasonally adjusted forecasts. Exhibit 15.4 shows an Excel spreadsheet set up to develop the seasonally adjusted forecast for the demand for turkeys at Wishbone Farms. You will notice that the seasonally adjusted forecasts for each quarter are slightly different from the forecasts computed manually (e.g., SF_1 with Excel equals 16.43, whereas SF_1 computed manually equals 16.28). This is due to the rounding done in the manual computations.

Computing the Exponential Smoothing Forecast with Excel QM

In Chapter 1 we introduced Excel QM, a set of spreadsheet macros that we also used in several other chapters. Excel QM includes a spreadsheet macro for exponential smoothing. After it is activated, the Excel QM menu is accessed by clicking on "Add-Ins" on the menu bar at the top of the spreadsheet. Clicking on "Forecasting" from this menu results in a Spreadsheet Initialization window, in which you enter the problem title and the number of periods of past demand. Clicking on "OK" will result in the spreadsheet shown in Exhibit 15.5. Initially, this spreadsheet will have example values in the (shaded) data cells, **B7** and **B10:B21**. Thus, the first step in using this macro is to type in the data for our PM Computer Services problem: alpha (α) = 0.50 in cell B7 and our demand values in cells **B10:B21**. The forecast results are computed automatically from formulas already embedded in the spreadsheet. The resulting next-period forecast for January is shown in cell B26, and the monthly forecasts are shown in cells **D10:D21**. (Do not confuse this exponentially smoothed forecast and the values for *MAD* and the like with the *adjusted* exponentially smoothed forecast and *MAD* value in Exhibit 15.1.)

EXHIBIT 15.5

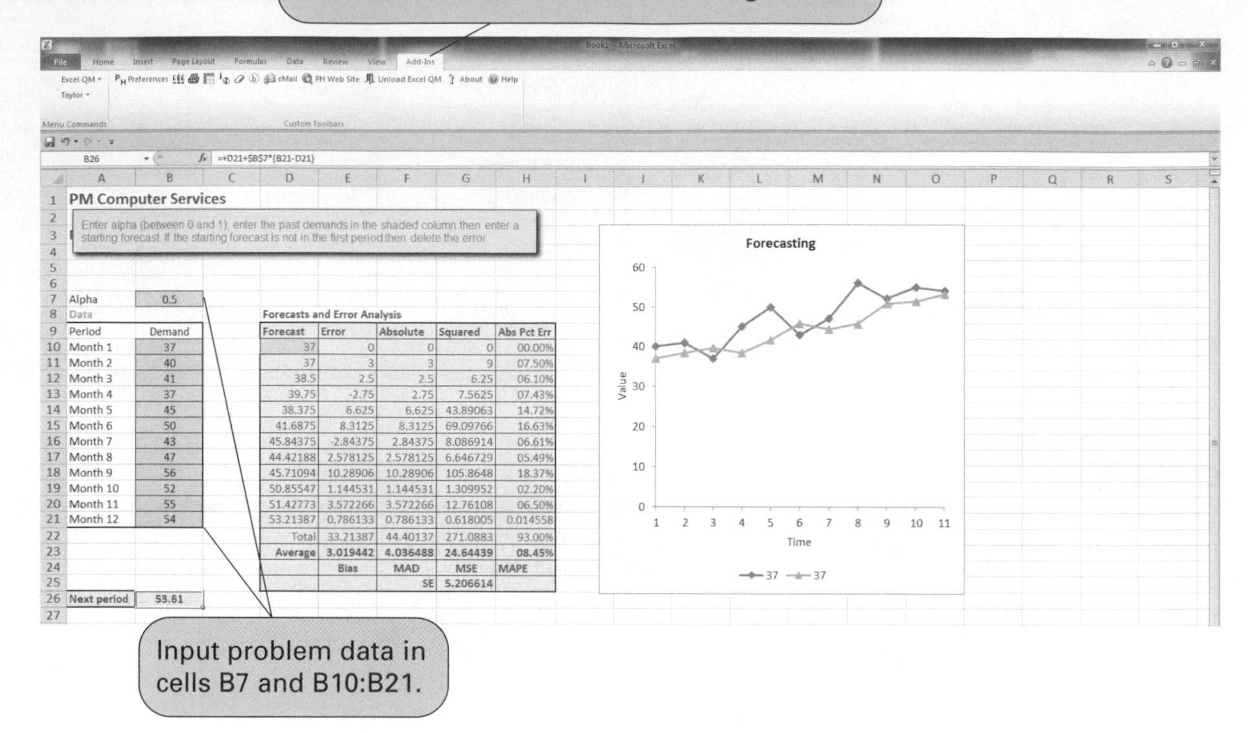

Alpha 0.5

Period	Demand	Forecast	Error	Absolute	Squared	Abs Pct Err
Month 1	37	37	0	0	0	00.00%
Month 2	40	37	3	3	9	07.50%
Month 3	41	38.5	2.5	2.5	6.25	06.10%
Month 4	37	39.75	-2.75	2.75	7.5625	07.43%
Month 5	45	38.375	6.625	6.625	43.89063	14.72%
Month 6	50	41.6875	8.3125	8.3125	69.09766	16.63%
Month 7	43	45.84375	-2.84375	2.84375	8.086914	06.61%
Month 8	47	44.42188	2.578125	2.578125	6.646729	05.49%
Month 9	56	45.71094	10.28906	10.28906	105.8648	18.37%
Month 10	52	50.85547	1.144531	1.144531	1.309952	02.20%
Month 11	55	51.42773	3.572266	3.572266	12.76108	06.50%
Month 12	54	53.21387	0.786133	0.786133	0.618005	0.014558
		Total	33.21387	44.40137	271.0883	93.00%
		Average	3.019442	4.036488	24.64439	08.45%
			Bias	MAD	MSE	MAPE
				SE	5.206614	
Next period	53.61					

Time Series Forecasting Using QM for Windows

QM for Windows has the capability to perform forecasting for all the time series methods we have described so far. QM for Windows has modules for moving averages, exponential smoothing and adjusted exponential smoothing, and linear regression.

To demonstrate the forecasting capability of QM for Windows, we will generate the exponential smoothing ($\alpha = .30$) forecast computed manually for PM Computer Services (Table 15.4). The solution output is shown in Exhibit 15.6.

EXHIBIT 15.6

Details and Error Analysis

PM Computer Services Example Solution

	Demand(y)	Forecast	Error	|Error|	Error^2	|Pct Error|
1	37					
2	40	37	3	3	9	.08
3	41	37.9	3.1	3.1	9.61	.08
4	37	38.83	-1.83	1.83	3.35	.05
5	45	38.28	6.72	6.72	45.14	.15
6	50	40.3	9.7	9.7	94.15	.19
7	43	43.21	-.21	.21	.04	.0
8	47	43.15	3.85	3.85	14.86	.08
9	56	44.3	11.7	11.7	136.85	.21
10	52	47.81	4.19	4.19	17.55	.08
11	55	49.07	5.93	5.93	35.19	.11
12	54	50.85	3.15	3.15	9.94	.06
TOTALS	557		49.31	53.39	375.68	1.09
AVERAGE	46.42		4.48	4.85	34.15	.1
Next period forecast		51.79	(Bias)	(MAD)	(MSE)	(MAPE)
				Std err	6.46	

Notice that the solution summary includes the forecast per period and the forecast for the next period (13), as well as four measures of forecast accuracy: average error (bias), mean absolute deviation (*MAD*), mean squared error (*MSE*), and mean absolute percent error (*MAPE*), which we have also referred to as *MAPD*.

The least squares module or the simple linear regression module in QM for Windows can be used to develop a linear trend line forecast. Using the least squares module, the solution summary for the linear trend line forecast we developed for PM Computer Services is shown in Exhibit 15.7.

EXHIBIT 15.7

Forecasting Results

PM Computer Services Example Summary

Measure	Value	Future Period	Forecast
Error Measures		13	57.6212
Bias (Mean Error)	0	14	59.345
MAD (Mean Absolute Deviation)	2.2892	15	61.0688
MSE (Mean Squared Error)	8.6672	16	62.7925
Standard Error (denom=n-2=10)	3.225	17	64.5163
MAPE (Mean Absolute Percent Error)	.0499	18	66.2401
Regression line		19	67.9639
Demand(y) = 35.2121		20	69.6876
+ 1.7238 * Time(x)		21	71.4114
Statistics		22	73.1352
Correlation coefficient	.8963	23	74.859
Coefficient of determination (r^2)	.8034	24	76.5827
		25	78.3065
		26	80.0303

Regression Methods

The time series techniques of exponential smoothing and moving average relate a single variable being forecast (such as demand) to *time*. In contrast, *regression* is a forecasting technique that measures the relationship of one variable to one or more other variables. For example, if we know that something has caused product demand to behave in a certain way in the past, we might like to identify that relationship. If the same thing happens again in the future, we can then predict what demand will be. For example, there is a well-known relationship between increased demand in new housing and lower interest rates. Correspondingly, a myriad of building products and services display increased demand if new housing starts increase. Similarly, an increase in sales of DVD players results in an increase in demand for DVDs.

The simplest form of regression is linear regression, which you will recall we used previously to develop a linear trend line for forecasting. In the following section we will show how to develop a regression model for variables related to items other than time.

Linear Regression

Linear regression relates demand (dependent variable) to an independent variable.

Simple **linear regression** relates one dependent variable to one independent variable in the form of a linear equation:

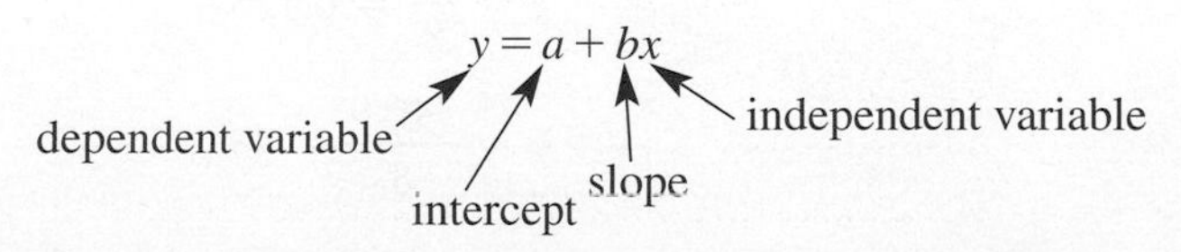

To develop the linear equation, the slope, b, and the intercept, a, must first be computed by using the following least squares formulas:

$$a = \bar{y} - b\bar{x}$$

$$b = \frac{\Sigma xy - n\bar{x}\bar{y}}{\Sigma x^2 - n\bar{x}^2}$$

where

$$\bar{x} = \frac{\Sigma x}{n} = \text{mean of the } x \text{ data}$$

$$\bar{y} = \frac{\Sigma y}{n} = \text{mean of the } y \text{ data}$$

We will consider regression within the context of an example. The State University athletic department wants to develop its budget for the coming year, using a forecast for football attendance. Football attendance accounts for the largest portion of its revenues, and the athletic director believes attendance is directly related to the number of wins by the team. The business manager has accumulated total annual attendance figures for the past 8 years:

Wins	Attendance
4	36,300
6	40,100
6	41,200
8	53,000
6	44,000
7	45,600
5	39,000
7	47,500

Given the number of returning starters and the strength of the schedule, the athletic director believes the team will win at least seven games next year. He wants to develop a simple regression equation for these data to forecast attendance for this level of success.

The computations necessary to compute a and b, using the least squares formulas, are summarized in Table 15.10. (Note that the magnitude of y has been reduced to make manual computation easier.)

TABLE 15.10
Least squares computations

x (wins)	y (attendance, 1,000s)	xy	x^2
4	36.3	145.2	16
6	40.1	240.6	36
6	41.2	247.2	36
8	53.0	424.0	64
6	44.0	264.0	36
7	45.6	319.2	49
5	39.0	195.0	25
7	47.5	332.5	49
49	346.7	2,167.7	311

$$\bar{x} = \frac{49}{8} = 6.125$$

$$\bar{y} = \frac{346.9}{8} = 43.34$$

$$b = \frac{\Sigma xy - n\bar{x}\,\bar{y}}{\Sigma x^2 - n\bar{x}^2}$$

$$= \frac{(2{,}167.7) - (8)(6.125)(43.34)}{(311) - (8)(6.125)^2} = 4.06$$

$$a = \bar{y} - b\bar{x} = 43.34 - (4.06)(6.125) = 18.46$$

Substituting these values for a and b into the linear equation line, we have

$$y = 18.46 + 4.06x$$

Thus, for $x = 7$ (wins), the forecast for attendance is

$$y = 18.46 + 4.06(7) = 46.88 \text{ or } 46{,}880$$

The data points with the regression line are shown in Figure 15.6. Observing the regression line relative to the data points, it would appear that the data follow a distinct upward linear trend, which would indicate that the forecast should be relatively accurate. In fact, the *MAD* value for this forecasting model is 1.41, which suggests an accurate forecast.

FIGURE 15.6
Linear regression line

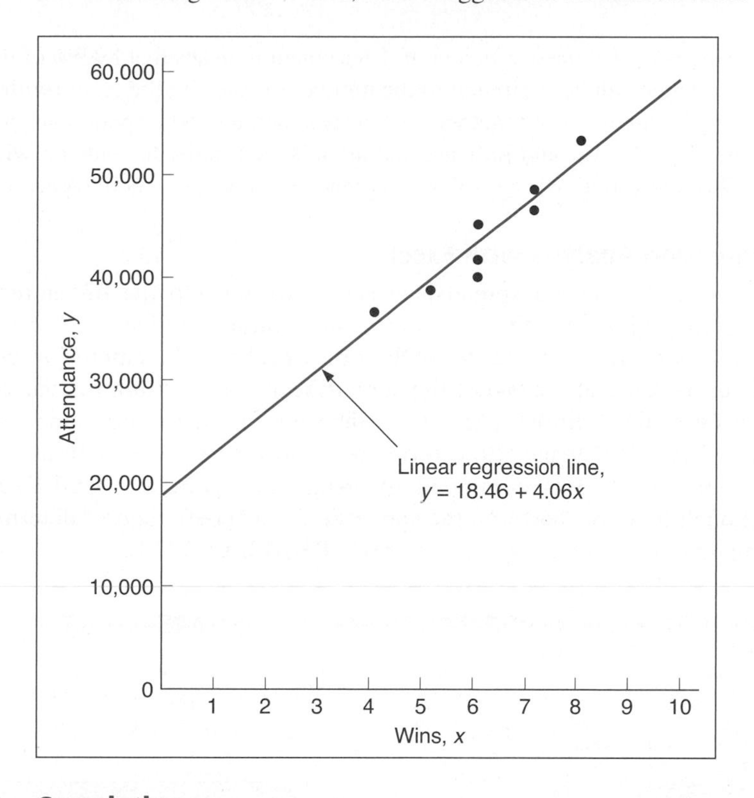

Correlation

Correlation is a measure of the strength of the relationship between independent and dependent variables.

Correlation in a linear regression equation is a measure of the strength of the relationship between the independent and dependent variables. The formula for the correlation coefficient is

$$r = \frac{n\Sigma xy - \Sigma x \Sigma y}{\sqrt{[n\Sigma x^2 - (\Sigma x)^2][n\Sigma y^2 - (\Sigma y)^2]}}$$

The value of r varies between -1.00 and $+1.00$, with a value of ± 1.00 indicating a strong linear relationship between the variables. If $r = 1.00$, then an increase in the independent variable will result in a corresponding linear increase in the dependent variable. If $r = -1.00$, an increase in the dependent variable will result in a linear decrease in the dependent variable. A value of r near zero implies that there is little or no linear relationship between variables.

We can determine the correlation coefficient for the linear regression equation determined in our State University example by substituting most of the terms calculated for the least squares formula (except for Σy^2) into the formula for r:

$$r = \frac{(8)(2{,}167.7) - (49)(346.7)}{\sqrt{[(8)(311) - (49)^2][(8)(15{,}224.7) - (346.7)^2]}}$$
$$= .948$$

This value for the correlation coefficient is very close to one, indicating a strong linear relationship between the number of wins and home attendance.

The coefficient of determination is the percentage of the variation in the dependent variable that results from the independent variable.

Another measure of the strength of the relationship between the variables in a linear regression equation is the **coefficient of determination**. It is computed by simply squaring the value of r. It indicates the percentage of the variation in the dependent variable that is a result of the behavior of the independent variable. For our example, $r = .948$; thus, the coefficient of determination is

$$r^2 = (.948)^2$$
$$= .899$$

This value for the coefficient of determination means that 89.9% of the amount of variation in attendance can be attributed to the number of wins by the team (with the remaining 10.1% due to other unexplained factors, such as weather, a good or poor start, publicity, etc.). A value of one (or 100%) would indicate that attendance totally depends on wins. However, because 10.1% of the variation is a result of other factors, some amount of forecast error can be expected.

Regression Analysis with Excel

Exhibit 15.8 shows a spreadsheet set up to develop the linear regression forecast for our State University athletic department example. Notice that Excel computes the slope directly with the formula **=SLOPE(B5:B12,A5:A12)** entered in cell E7 and shown on the formula bar at the top of the spreadsheet. The formula for the intercept in cell E6 is **=INTERCEPT(B5:B12,A5:A12)**. The values for the slope and intercept are subsequently entered in cells E9 and G9 to form the linear regression equation. The correlation coefficient in cell E13 is computed by using the formula **=CORREL(B5:B12,A5:A12)**. Although it is not shown on the spreadsheet, the coefficient of determination (r^2) could be computed by using the formula **=RSQ(B5:B12,A5:A12)**.

EXHIBIT 15.8

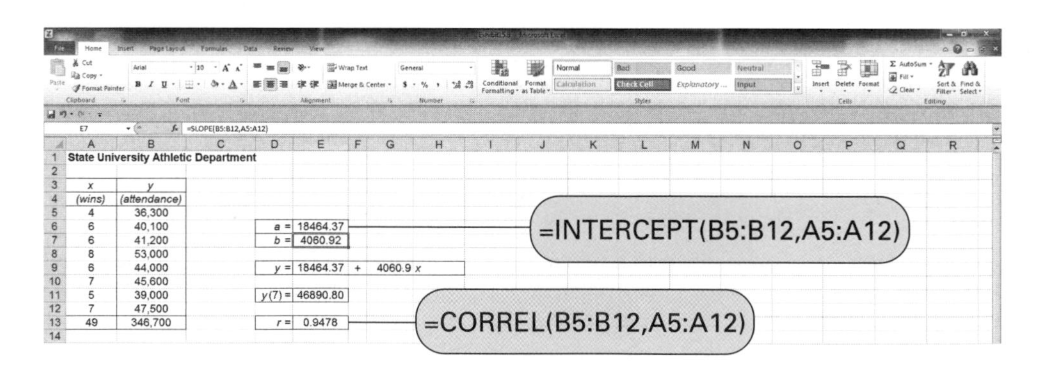

The same linear regression equation could be computed in Excel if we had developed and entered the mathematical formulas for computing the slope and intercept we developed in the previous section, although that would have been more time-consuming and tedious.

It is also possible to develop a scatter diagram of our example data similar to the chart shown in Figure 15.6 by using Excel. First, cover the example data in cells **A5:B12** on the

Management Science Application

An Airline Passenger Forecasting Model

Following the terrorist attacks of September 11, 2001, Congress created the Transportation Security Administration (TSA), which became part of the Department of Homeland Security in March 2003. The TSA has responsibility for national transportation security and, specifically, U.S. commercial aviation system security. The TSA put into effect a number of procedures, processes, and regulations to safeguard against terrorism, the unintended consequences of which were long lines and waiting times at security checkpoints in airports.

To address the problems created by the new security measures, the TSA joined with the commercial airlines to form the U.S. Commercial Aviation Partnership (USCAP). The goal of this partnership was to maintain the health of U.S. commercial aviation while ensuring that security requirements were implemented efficiently and effectively. To achieve this goal, USCAP used its technical professionals to develop mathematical models to analyze, define, and predict the operational and economic effects of air travel security measures. The key component of this modeling effort was a model to forecast air travel passenger demand.

Previously most air travel forecasting models focused on long-term trends with time horizons of 5 to 20 years, which weren't effective for making near-term tactical decisions and reflecting the immediate impact of security policies. The model developed by USCAP used multiple regression to forecast passenger demand based on six causal factors including the economy (reflected by U.S. gross domestic product); ticket prices (represented by the average fare for a 1,000-mile trip); travel time (from door to door); fear (represented by the reduction in plane boardings owing to an event causing fear); a hassle factor (created by the reduction in plane boardings owing to an event causing hassle); and an operations factor (resulting from a reduction in plane boardings owing to a reduction in operations caused by an event).

A detailed regression analysis Solver tool with special capabilities was developed for this model. The forecasting model was subsequently used as the basis for other analytical models to explore numerous questions concerning airport security procedures, such as staffing for security screening, the number of screening lanes at airports, security processing times, security credentials for employees, and what-if? analyses of proposed security measures such as passenger security fees. The immediate financial savings of the models developed by USCAP, in addition to improved security procedures, were in the tens of billions of dollars with an estimated 20-year impact of more than $100 billion.

Source: Based on R. Peterson, R. Bittel, C. Forgie, W. Lee, and J. Nestor, "Using USCAP's Analytical Models, the Transportation Security Administration Balances the Impacts of Aviation Security Policies on Passengers and Airlines," *Interfaces* 37, no. 1 (January–February 2007): 52–67.

spreadsheet in Exhibit 15.8. Next click on the "Insert" tab on the toolbar at the top of the spreadsheet. This will result in the toolbar menu shown in Exhibit 15.9.

On the "Charts" segment from this menu select the "Scatter" chart, as shown in Exhibit 15.9. Exhibit 15.9 shows the spreadsheet with the scatter diagram chart for our example data.

A linear regression forecast can also be developed directly with Excel by using the "Data Analysis" add-in option from the "Data" toolbar we accessed previously to develop an exponentially smoothed forecast. Exhibit 15.10 shows the "Regression" selection from the Data Analysis window, and Exhibit 15.11 shows the Regression window. We first enter the cells from Exhibit 15.8 that include the y values (for attendance), **B5:B12**. Next, we enter the x value cells, **A5:A12**. The output range is the location on the spreadsheet where we want to put the output results. This range needs to be large (18 cells by 9 cells) and must not overlap with anything else on the spreadsheet. Clicking on "OK" will result in the

EXHIBIT 15.9

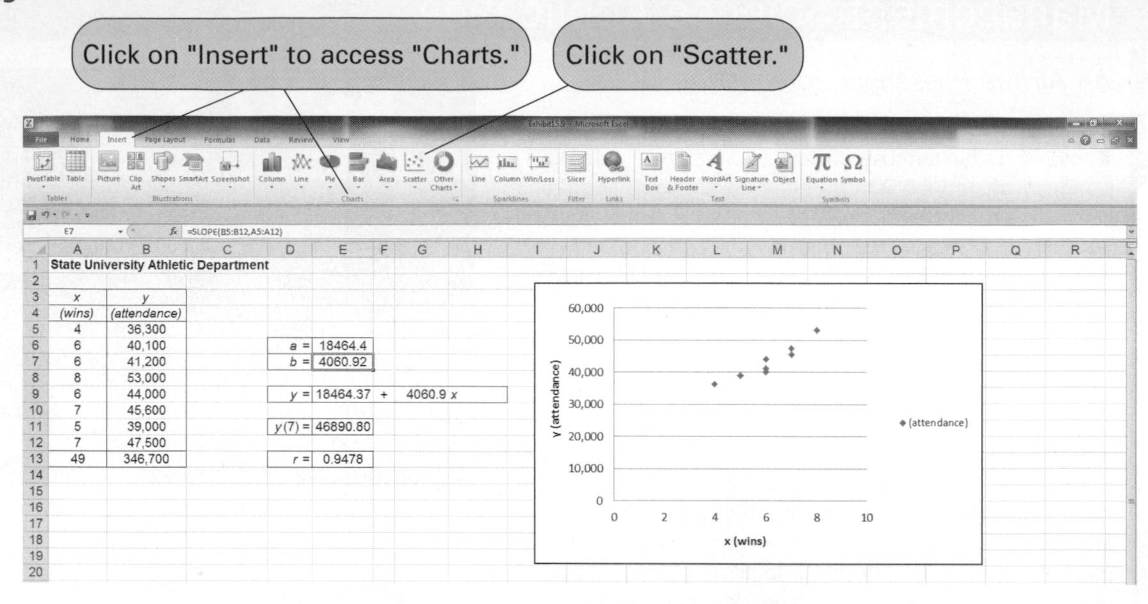

EXHIBIT 15.10

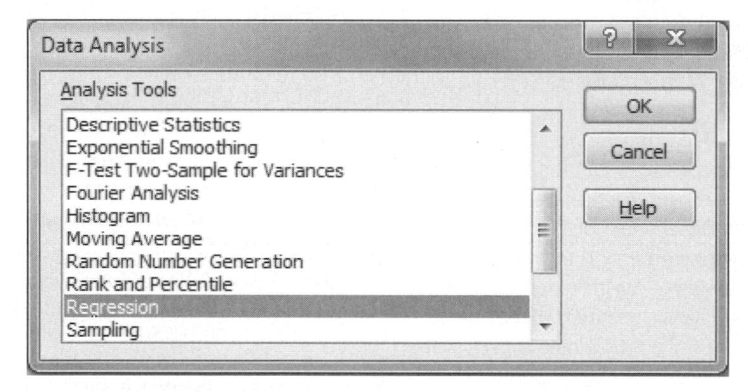

EXHIBIT 15.11

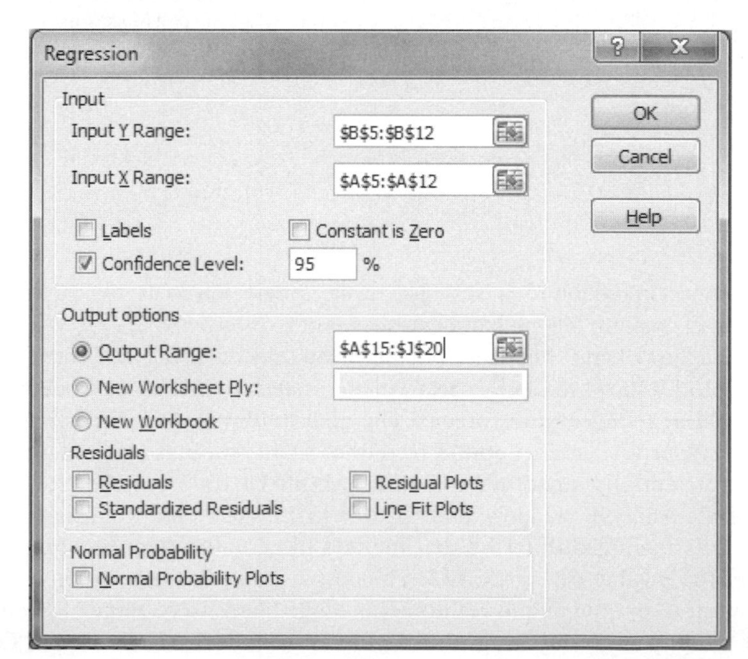

spreadsheet shown in Exhibit 15.12. (Note that the "Summary Output" section has been slightly edited—i.e., moved around—so that all the results could be included on the screen in Exhibit 15.12.)

EXHIBIT 15.12

State University Athletic Department

x (wins)	y (attendance)
4	36,300
6	40,100
6	41,200
8	53,000
6	44,000
7	45,600
5	39,000
7	47,500
49	346,700

SUMMARY OUTPUT

Regression Statistics	
Multiple R	0.9478
R Square	0.8983
Adjusted R Square	0.8814
Standard Error	1839.3110
Observations	8

ANOVA

	df	*SS*	*MS*	*F*	*Significance F*
Regression	1	179340359.20	179340359.20	53.01	0.00
Residual	6	20298390.80	3383065.13		
Total	7	199638750			

	Coefficients	*Standard Error*	*t Stat*	*P-value*	*Lower 95%*	*Upper 95%*	*Lower 95.0%*	*Upper 95.0%*
Intercept	18464.368	3477.569	5.310	0.002	9955.057	26973.679	9955.057	26973.679
X Variable 1	4060.920	557.751	7.281	0.000	2696.150	5425.689	2696.150	5425.689

The "Summary Output" section in Exhibit 15.12 provides a large amount of statistical information, the explanation and use of which are beyond the scope of this text. The essential items that interest us are the intercept and slope (labeled "X Variable 1") in the "Coefficients" column at the bottom of the spreadsheet and the "Multiple R" (or correlation coefficient) value shown under "Regression Statistics."

Note that Excel QM also has a spreadsheet macro for regression analysis that can be accessed in much the same way as was the exponentially smoothed forecast in Exhibit 15.5.

Regression Analysis with QM for Windows

QM for Windows has the capability to perform linear regression, as demonstrated earlier. To demonstrate this program module, we will use our State University athletic department example. The program output, including the linear equation and correlation coefficient, is shown in Exhibit 15.13.

Multiple Regression with Excel

Multiple regression relates demand to two or more independent variables.

Another causal method of forecasting is **multiple regression**, a more powerful extension of linear regression. Linear regression relates a dependent variable such as demand to one other independent variable, whereas multiple regression reflects the relationship between a dependent variable and *two or more* independent variables. A multiple regression model has the following general form:

$$y = \beta_0 + \beta_1 x_1 + \beta_2 x_2 + \cdots + \beta_k x_k$$

where

β_0 = the intercept

$\beta_1 \ldots \beta_k$ = parameters representing the contribution of the independent variables

$x_1 \ldots x_k$ = independent variables

EXHIBIT 15.13

Forecasting Results

State University Athletic Department Summary	
Measure	Value
Error Measures	
Bias (Mean Error)	-.0005
MAD (Mean Absolute Deviation)	1,412.644
MSE (Mean Squared Error)	2,537,301.0
Standard Error (denom=n-2=6)	1,839.312
MAPE (Mean Absolute Percent Error)	.033
Regression line	
Dpndnt var, Y = 18,464.37	
+ 4,060.919 * X1	
Statistics	
Correlation coefficient	.9478
Coefficient of determination (r^2)	.8983

For example, the demand for new housing (y) in a region or an urban area might be a function of several independent variables, including interest rates, population, housing prices, and personal income. Development and computation of the multiple regression equation, including the compilation of data, are quite a bit more complex than linear regression. Therefore, the only viable means for forecasting using multiple regression problems is by using a computer.

To demonstrate the capability to solve multiple regression problems with Excel spreadsheets, we will expand our State University athletic department example for forecasting attendance at football games that we used to demonstrate linear regression. Instead of attempting to predict attendance based on only one variable (wins), we will include a second variable for advertising and promotional expenditures, as follows:

Wins	Promotion	Attendance
4	$29,500	36,300
6	55,700	40,100
6	71,300	41,200
8	87,000	53,000
6	75,000	44,000
7	72,000	45,600
5	55,300	39,000
7	81,600	47,500

We will use the "Data Analysis" option (add-in) from the "Data" menu at the top of the spreadsheet that we used in the previous section to develop our linear regression equation, and then we will use the "Regression" option from the "Data Analysis" menu. The resulting spreadsheet, with the multiple regression statistics, is shown in Exhibit 15.14.

Note that the data need to be set up on the spreadsheet so that the x variables are in adjacent columns (in this case, columns A and B). Then we enter the "Input X Range" as **A4:B12**, as shown in Exhibit 15.15. Notice that we have also included cells A4, B4, and C4, which include our variable headings (i.e., "wins," "$ promotion," and "attendance"), in the input ranges. By clicking on "Labels," headings can be placed on our spreadsheet in cells A27 and A28.

EXHIBIT 15.14

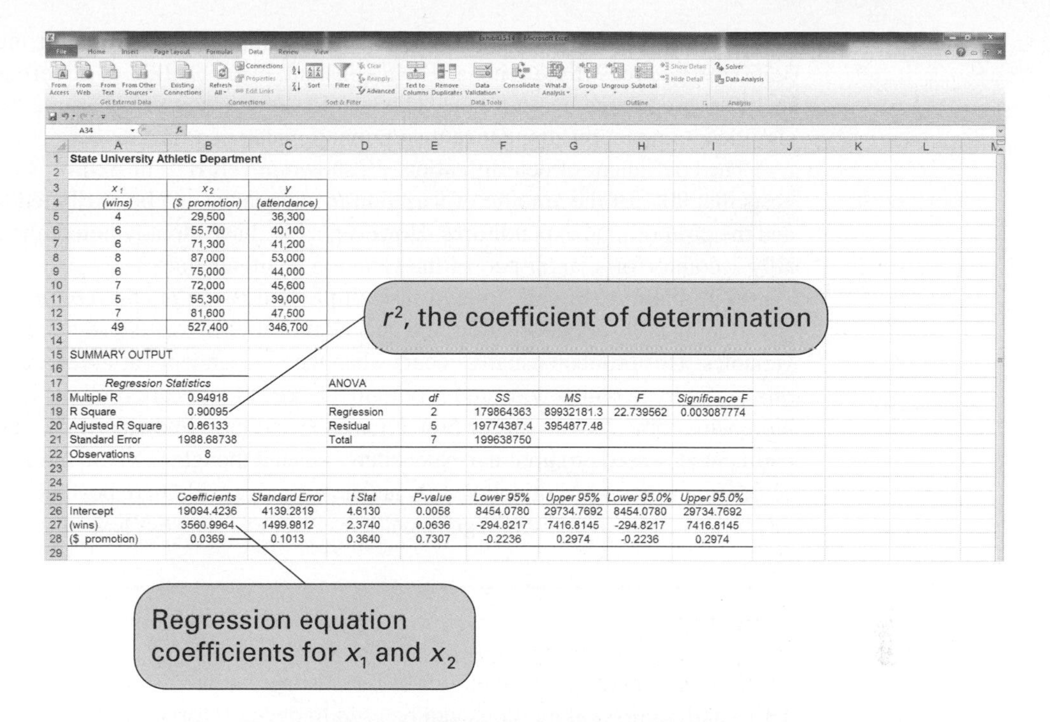

State University Athletic Department

x_1 (wins)	x_2 ($ promotion)	y (attendance)
4	29,500	36,300
6	55,700	40,100
6	71,300	41,200
8	87,000	53,000
6	75,000	44,000
7	72,000	45,600
5	55,300	39,000
7	81,600	47,500
49	527,400	346,700

SUMMARY OUTPUT

Regression Statistics	
Multiple R	0.94918
R Square	0.90095
Adjusted R Square	0.86133
Standard Error	1988.68738
Observations	8

ANOVA

	df	SS	MS	F	Significance F
Regression	2	179864363	89932181.3	22.739562	0.003087774
Residual	5	19774387.4	3954877.48		
Total	7	199638750			

	Coefficients	Standard Error	t Stat	P-value	Lower 95%	Upper 95%	Lower 95.0%	Upper 95.0%
Intercept	19094.4236	4139.2819	4.6130	0.0058	8454.0780	29734.7692	8454.0780	29734.7692
(wins)	3560.9964	1499.9812	2.3740	0.0636	-294.8217	7416.8145	-294.8217	7416.8145
($ promotion)	0.0369	0.1013	0.3640	0.7307	-0.2236	0.2974	-0.2236	0.2974

EXHIBIT 15.15

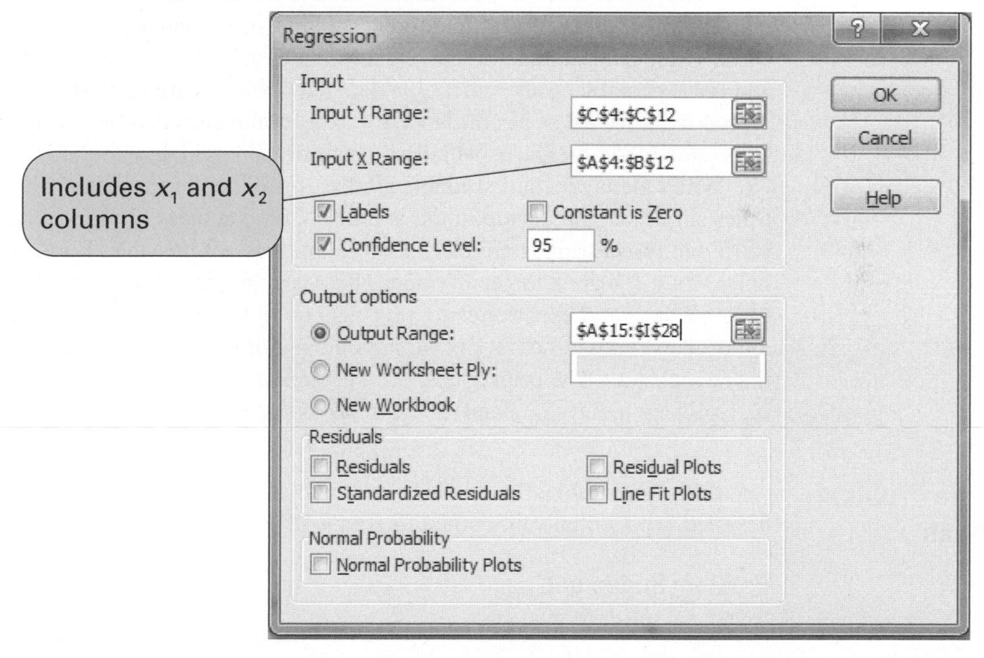

The regression coefficients for our x variables, wins and promotion, are shown in cells B27 and B28 in Exhibit 15.14. Thus, the multiple regression equation is formulated as

$$y = 19{,}094.42 + 3{,}560.99x_1 + .0368x_2$$

This equation can now be used to forecast attendance based on both projected football wins and promotional expenditure. For example, if the athletic department expects the team to win seven games and plans to spend $60,000 on promotion and advertising, the forecasted attendance is

$$\begin{aligned} y &= 19{,}094.42 + 3{,}560.99(7) + .0368(60{,}000) \\ &= 46{,}229.35 \end{aligned}$$

If the promotional expenditure is held constant, every win will increase attendance by 3,560.99, whereas if the wins are held constant, every $1,000 of advertising money spent will increase attendance by 36.8 fans. This would seem to suggest that number of wins has a more significant impact on attendance than promotional expenditures.

The coefficient of determination, r^2, shown in cell B19 in Exhibit 15.14, is .90, which suggests that 90% of the amount of variation in attendance can be attributed to the number of wins and the promotional expenditures. However, as we have already noted, the number of wins probably accounts for a larger part of the variation in attendance.

A problem often encountered in multiple regression is *multicollinearity*, or the amount of "overlapping" information about the dependent variable that is provided by several independent variables. This problem usually occurs when the independent variables are highly correlated, as in this example, in which wins and promotional expenditures are both positively correlated; that is, more wins coincide with higher promotional expenditures and vice versa. (Possibly the athletic department increased promotional expenditures when it thought it would have a better team that would achieve more wins.) Multicollinearity and how to cope with it are beyond the scope of this text and this brief section on multiple regression; however, most statistics texts discuss this topic in detail.

Summary

We have presented several methods of forecasting that are useful for different time frames. Time series and regression methods can be used to develop forecasts encompassing horizons of any length time, although they tend to be used most frequently for short- and medium-range forecasts. These quantitative forecasting techniques are generally easy to understand, simple to use, and not especially costly, unless the data requirements are substantial. They also have exhibited a good track record of performance for many companies that have used them. For these reasons, regression methods, and especially times series, are widely popular.

When managers and students are first introduced to forecasting methods, they are sometimes surprised and disappointed at the lack of exactness of the forecasts. However, they soon learn that forecasting is not easy and exactness is not possible. Forecasting is a lot like playing horseshoes: It's great to get a "ringer," but you can win by just getting close, although those who have the skill and experience to get ringers will beat those who just get close. Often forecasts are used as inputs to other decision models—for example, in inventory models, the subject of the next chapter. The primary factor in determining the amount of inventory a firm should order (i.e., prepare to have on hand) is the demand that will occur in the future—forecasted demand.

Example Problem Solutions

The following problem provides an example of the computation of exponentially smoothed and adjusted exponentially smoothed forecasts.

Problem Statement

A computer software firm has experienced the following demand for its Personal Finance software package:

Period	Units
1	56
2	61
3	55
4	70
5	66
6	65
7	72
8	75

Develop an exponential smoothing forecast, using $\alpha = .40$, and an adjusted exponential smoothing forecast, using $\alpha = .40$ and $\beta = .20$. Compare the accuracy of the two forecasts, using *MAD* and cumulative error.

Solution

Step 1: Compute the Exponential Smoothing Forecast

$$F_{t+1} = \alpha D_t + (1 - \alpha)F_t$$

For period 2 the forecast (assuming that $F_1 = 56$) is

$$F_2 = \alpha D_1 + (1 - \alpha)F_1 = (.40)(56) + (.60)(56) = 56$$

For period 3 the forecast is

$$F_3 = (.40)(61) + (.60)(56) = 58$$

The remaining forecasts are computed similarly and are shown in the table below.

Step 2: Compute the Adjusted Exponential Smoothing Forecast

$$AF_{t+1} = F_{t+1} + T_{t+1}$$
$$T_{t+1} = \beta(F_{t+1} - F_t) + (1 - \beta)T_t$$

Starting with the forecast for period 3 (because $F_1 = F_2$ and we will assume that $T_2 = 0$),

$$T_3 = \beta(F_3 - F_2) + (1 - \beta)T_2 = (.20)(58 - 56) + (.80)(0) = .40$$
$$AF_3 = F_3 + T_3 = 58 + .40 = 58.40$$

The remaining adjusted forecasts are computed similarly and are shown in the following table:

Period	D_t	F_t	AF_t	$D_t - F_t$	$D_t - AF_t$
1	56	—	—	—	—
2	61	56.00	56.00	5.00	5.00
3	55	58.00	58.40	−3.00	−3.40
4	70	56.80	56.88	13.20	13.12
5	66	62.08	63.20	3.92	2.80
6	65	63.65	64.86	1.35	0.14
7	72	64.18	65.26	7.81	6.73
8	75	67.31	68.80	7.68	6.20
9	—	70.39	72.19	—	—
				35.97	30.60

Step 3: Compute the *MAD* Values

$$MAD(F_t) = \frac{\Sigma|D_t - F_t|}{n} = \frac{41.97}{7} = 5.99$$

$$MAD(AF_t) = \frac{\Sigma|D_t - AF_t|}{n} = \frac{37.39}{7} = 5.34$$

Step 4: Compute the Cumulative Error

$$E(F_t) = 35.97$$
$$E(AF_t) = 30.60$$

Because both *MAD* and the cumulative error are less for the adjusted forecast, it would appear to be the most accurate.

The following problem provides an example of the computation of a linear regression forecast.

Problem Statement

A local building products store has accumulated sales data for two-by-four lumber (in board feet) and the number of building permits in its area for the past 10 quarters:

Quarter	Building Permits, x	Lumber Sales (1,000s of bd. ft.), y
1	8	12.6
2	12	16.3
3	7	9.3
4	9	11.5
5	15	18.1
6	6	7.6
7	5	6.2
8	8	14.2
9	10	15.0
10	12	17.8

Develop a linear regression model for these data and determine the strength of the linear relationship by using correlation. If the model appears to be relatively strong, determine the forecast for lumber, given 10 building permits in the next quarter.

Solution

Step 1: Compute the Components of the Linear Regression Equation

$$\bar{x} = \frac{92}{10} = 9.2$$

$$\bar{y} = \frac{128.6}{10} = 12.86$$

$$b = \frac{\Sigma xy - n\bar{x}\bar{y}}{\Sigma x^2 - n\bar{x}^2} = \frac{(1{,}290.3) - (10)(9.2)(12.86)}{(932) - (10)(9.2)^2}$$

$$b = 1.25$$

$$a = \bar{y} - b\bar{x} = 12.86 - (1.25)(9.2)$$

$$a = 1.36$$

Step 2: Develop the Linear Regression Equation

$$y = a + bx$$

$$y = 1.36 + 1.25x$$

Step 3: Compute the Correlation Coefficient

$$r = \frac{n\Sigma xy - \Sigma x\Sigma y}{\sqrt{[n\Sigma x^2 - (\Sigma x)^2][n\Sigma y^2 - (\Sigma y)^2]}}$$

$$= \frac{(10)(1{,}290.3) - (92)(128.6)}{\sqrt{[(10)(932) - (92)^2][(10)(1{,}810.48) - (128.6)^2]}}$$

$$= .925$$

Thus, there appears to be a strong linear relationship.

Step 4: Calculate the Forecast for $x = 10$ Permits

$$y = a + bx$$
$$= 1.36 + 1.25(10)$$
$$= 13.86 \text{ or } 1{,}386 \text{ bd. ft.}$$

Problems

1. The Saki motorcycle dealer in the Minneapolis–St. Paul area wants to make an accurate forecast of demand for the Saki Super TXII motorcycle during the next month. Because the manufacturer is in Japan, it is difficult to send motorcycles back or reorder if the proper number is not ordered a month ahead. From sales records, the dealer has accumulated the following data for the past year:

Month	Motorcycle Sales
January	9
February	7
March	10
April	8
May	7
June	12
July	10
August	11
September	12
October	10
November	14
December	16

   a. Compute a 3-month moving average forecast of demand for April through January (of the next year).
   b. Compute a 5-month moving average forecast for June through January.
   c. Compare the two forecasts computed in (a) and (b), using *MAD*. Which one should the dealer use for January of the next year?

2. The manager of the Carpet City outlet needs to make an accurate forecast of the demand for Soft Shag carpet (its biggest seller). If the manager does not order enough carpet from the carpet mill, customers will buy their carpet from one of Carpet City's many competitors. The manager has collected the following demand data for the past 8 months:

Month	Demand for Soft Shag Carpet (1,000 yd.)
1	8
2	12
3	7
4	9
5	15
6	11
7	10
8	12

   a. Compute a 3-month moving average forecast for months 4 through 9.
   b. Compute a weighted 3-month moving average forecast for months 4 through 9. Assign weights of .55, .33, and .12 to the months in sequence, starting with the most recent month.
   c. Compare the two forecasts by using *MAD*. Which forecast appears to be more accurate?

3. The Fastgro Fertilizer Company distributes fertilizer to various lawn and garden shops. The company must base its quarterly production schedule on a forecast of how many tons of fertilizer will be demanded from it. The company has gathered the following data for the past 3 years from its sales records:

Year	Quarter	Demand for Fertilizer (tons)
1	1	105
	2	150
	3	93
	4	121
2	5	140
	6	170
	7	105
	8	150
3	9	150
	10	170
	11	110
	12	130

a. Compute a three-quarter moving average forecast for quarters 4 through 13 and compute the forecast error for each quarter.
b. Compute a five-quarter moving average forecast for quarters 6 through 13 and compute the forecast error for each quarter.
c. Compute a weighted three-quarter moving average forecast, using weights of .50, .33, and .17 for the most recent, next most recent, and most distant data, respectively, and compute the forecast error for each quarter.
d. Compare the forecasts developed in (a), (b), and (c), using cumulative error. Which forecast appears to be most accurate? Do any of them exhibit any bias?

4. Graph the demand data in Problem 3. Can you identify any trends, cycles, or seasonal patterns?

5. The chairperson of the department of management at State University wants to forecast the number of students who will enroll in production and operations management (POM) next semester, in order to determine how many sections to schedule. The chair has accumulated the following enrollment data for the past eight semesters:

Semester	Students Enrolled in POM
1	400
2	450
3	350
4	420
5	500
6	575
7	490
8	650

a. Compute a three-semester moving average forecast for semesters 4 through 9.
b. Compute the exponentially smoothed forecast ($\alpha = .20$) for the enrollment data.
c. Compare the two forecasts by using *MAD* and indicate the more accurate of the two.

6. The manager of the Petroco Service Station wants to forecast the demand for unleaded gasoline next month so that the proper number of gallons can be ordered from the distributor. The owner has accumulated the following data on demand for unleaded gasoline from sales during the past 10 months:

Month	Gasoline Demanded (gal.)
October	800
November	725
December	630
January	500
February	645
March	690
April	730
May	810
June	1,200
July	980

a. Compute an exponentially smoothed forecast, using an α value of .30.
b. Compute an adjusted exponentially smoothed forecast (with $\alpha = .30$ and $\beta = .20$).
c. Compare the two forecasts by using *MAPD* and indicate which seems to be more accurate.

7. The Victory Plus Mutual Fund of growth stocks has had the following average monthly price for the past 10 months:

Month	Fund Price
1	62.7
2	63.9
3	68.0
4	66.4
5	67.2
6	65.8
7	68.2
8	69.3
9	67.2
10	70.1

Compute the exponentially smoothed forecast with $\alpha = .40$ the adjusted exponential smoothing forecast with $\alpha = .40$ and $\beta = .30$, and the linear trend line forecast. Compare the accuracy of the three forecasts, using cumulative error and *MAD*, and indicate which forecast appears to be most accurate.

8. The Bayside Fountain Hotel is adjacent to County Coliseum, a 24,000-seat arena that is home to the city's professional basketball and ice hockey teams and that hosts a variety of concerts, trade shows, and conventions throughout the year. The hotel has experienced the following occupancy rates for the past 9 years, since the coliseum opened:

Year	Occupancy Rate (%)
1	83
2	78
3	75
4	81
5	86
6	85
7	89
8	90
9	86

Compute an exponential smoothing forecast with $\alpha = .20$, an adjusted exponential smoothing forecast with $\alpha = .20$ and $\beta = .20$, and a linear trend line forecast. Compare the three forecasts, using *MAD* and average error ($\overline{E}$), and indicate which seems to be most accurate.

9. Emily Andrews has invested in a science and technology mutual fund. Now she is considering liquidating and investing in another fund. She would like to forecast the price of the science and technology fund for the next month before making a decision. She has collected the following data on the average price of the fund during the past 20 months:

Month	Fund Price
1	$ 63 1/4
2	60 1/8
3	61 3/4
4	64 1/4
5	59 3/8
6	57 7/8
7	62 1/4
8	65 1/8
9	68 1/4
10	65 1/2
11	68 1/8
12	63 1/4
13	64 3/8
14	68 5/8
15	70 1/8
16	72 3/4
17	74 1/8
18	71 3/4
19	75 1/2
20	76 3/4

a. Using a 3-month average, forecast the fund price for month 21.
b. Using a 3-month weighted average with the most recent month weighted 0.60, the next most recent month weighted 0.30, and the third month weighted 0.10, forecast the fund price for month 21.
c. Compute an exponentially smoothed forecast, using $\alpha = .40$, and forecast the fund price for month 21.
d. Compare the forecasts in (a), (b), and (c), using *MAD*, and indicate the most accurate.

10. Eurotronics manufactures components for use in small electronic products such as computers, CD players, and radios at plants in Belgium, Germany, and France. The parts are transported by truck to Hamburg, where they are shipped overseas to customers in Mexico, South America, the United States, and the Pacific Rim. The company has to reserve space on ships months and sometimes years in advance. This requires an accurate forecasting model. Following are the number of cubic feet of container space the company has used in each of the past 18 months:

Month	Space (1,000s ft.^{3})
1	10.6
2	12.7
3	9.8
4	11.3
5	13.6
6	14.4
7	12.2
8	16.7
9	18.1
10	19.2
11	16.3
12	14.7
13	18.2
14	19.6
15	21.4
16	22.8
17	20.6
18	18.7

Develop a forecasting model that you believe would provide the company with relatively accurate forecasts for the next year and indicate the forecasted shipping space required for the next 3 months.

11. The Whistle Stop Cafe in Weems, Georgia, is well known for its popular homemade ice cream, made in a small plant in back of the cafe. People drive all the way from Atlanta and Macon to buy the ice cream. The two women who own the cafe want to develop a forecasting model so they can plan their ice cream production operation and determine the number of employees they need to sell ice cream in the cafe. They have accumulated the following sales records for their ice cream for the past 12 quarters:

Year	Quarter	Ice Cream Sales (gal.)
1	1	350
	2	510
	3	750
	4	420
2	5	370
	6	480
	7	860
	8	500
3	9	450
	10	550
	11	820
	12	570

Develop an adjusted exponential smoothing model with $\alpha = .50$ and $\beta = .50$ to forecast demand and assess its accuracy using cumulative error (E) and average error ($\bar{E}$). Does there appear to be any bias in the forecast?

12. For the demand data in Problem 11, develop a seasonally adjusted forecast for year 4. (Use a linear trend line model to develop a forecast estimate for year 4.) Which forecast model do you perceive to be more accurate: the exponential smoothing model from Problem 11 or the seasonally adjusted forecast?

13. Develop a seasonally adjusted forecast for the demand data for fertilizer found in Problem 3. Then use a linear trend line model to compute a forecast estimate for demand in year 4.

14. Monaghan's Pizza delivery service has randomly selected 8 weekdays during the past month and recorded orders for pizza at four different time periods per day:

	Day							
Time Period	1	2	3	4	5	6	7	8
10:00 A.M.–3:00 P.M.	62	49	53	35	43	48	56	43
3:00 P.M.–7:00 P.M.	73	55	81	77	60	66	85	70
7:00 P.M.–11:00 P.M.	42	38	45	50	29	37	35	44
11:00 P.M.–12:00 A.M.	35	40	36	39	26	25	36	31

Develop a seasonally adjusted forecasting model for daily pizza demand and forecast demand for each of the time periods for a single upcoming day.

15. The Cat Creek Mining Company mines and ships coal. It has experienced the following demand for coal during the past 8 years:

Year	Coal Sales (tons)
1	4,260
2	4,510
3	4,050
4	3,720
5	3,900
6	3,470
7	2,890
8	3,100

Develop an adjusted exponential smoothing model ($\alpha = .30$, $\beta = .20$) and a linear trend line model and compare the forecast accuracy of the two by using *MAD*. Indicate which forecast seems to be more accurate.

16. *The River* is a literary magazine published by the English Department at Tech. It has a 70-year history and enjoys an excellent reputation for providing a literary outlet for aspiring Appalachian writers in particular. However, its sales, mostly through independent bookstores and college libraries, have never been very large. Several years ago, the magazine's advisory board decided to create a Web site and post the bi-monthly issues online for free access. Due to poor economic conditions and state budget cuts, Tech recently withdrew funding for the magazine, and in order to make up for this financial loss, the magazine's staff has decided to try to sell advertising (especially to local and regional businesses) on the magazine's Web site. In order to

sell advertising, the magazine's staff would like to be able to provide potential advertisers with a forecast of the number of "visits" the Web site might receive in future months. Following is the number of monthly visits the Web site has received for the 24 months that it has existed:

Month	Web Site Visits	Month	Web Site Visits
1	537	13	822
2	375	14	677
3	419	15	1,031
4	276	16	657
5	445	17	983
6	512	18	774
7	670	19	1,210
8	561	20	811
9	705	21	1,137
10	619	22	763
11	768	23	1,225
12	645	24	941

Develop a linear trend line forecast, an exponential smoothing forecast ($\alpha = 0.60$), and a 3-month weighted moving average (with the most recent month weighted by 0.50, the next closest month by 0.30, and the final month by 0.20). Indicate which one you believe is the most accurate forecast model and the forecast for month 25.

17. The Beaver Creek Pottery Company sells bowls and mugs, hand-made by Native American artisans, at a craft store and through a Web site. Making these items requires a special type of clay and a large amount of individual person-hours, so for planning purposes, the company would like to forecast future demand, specifically through its Web site, which has increasingly become the primary source of sales. Following is the company's Web site demand (in items sold) for the past 36 months:

Month	Sales	Month	Sales	Month	Sales
1	345	13	415	25	344
2	411	14	395	26	286
3	266	15	298	27	455
4	347	16	377	28	634
5	506	17	418	29	502
6	278	18	522	30	388
7	411	19	421	31	427
8	510	20	384	32	561
9	198	21	455	33	447
10	387	22	506	34	395
11	344	23	478	35	414
12	412	24	613	36	522

Develop a linear trend forecast model, an exponentially smoothed model ($\alpha = 0.20$), and a 5-month moving average forecast model and indicate which one you think should be used to forecast Web site demand.

18. The Northwoods Outdoor Company is a catalog sales operation that specializes in outdoor recreational clothing. Demand for its items is very seasonal, peaking during the Christmas season and during the spring. It has accumulated the following data for orders per season (quarter) during the past 5 years:

	Orders (1,000s)/Year				
Quarter	1	2	3	4	5
January–March	18.6	18.1	22.4	23.2	24.5
April–June	23.5	24.7	28.8	27.6	31.0
July–September	20.4	19.5	21.0	24.4	23.7
October–December	41.9	46.3	45.5	47.1	52.8

a. Develop a seasonally adjusted forecast model for these order data. Forecast demand for each quarter for year 6 (using a linear trend line forecast estimate for orders in year 6).
b. Develop a separate linear trend line forecast for each of the four seasons and forecast each season for year 6.
c. Which of the two approaches used in (a) and (b) appears to be the more accurate? Use *MAD* to verify your selection.

19. Metro Food Vending operates vending machines in office buildings, the airport, bus stations, colleges, and other businesses and agencies around town, and it operates vending trucks for building and construction sites. The company believes its sandwich sales follow a seasonal pattern. It has accumulated the following data for sandwich sales per season during the past 4 years:

	Sandwich Sales (1,000s)/Year			
Season	1	2	3	4
Fall	42.7	44.3	45.7	40.6
Winter	36.9	42.7	34.8	41.5
Spring	51.3	55.6	49.3	47.3
Summer	62.9	64.8	71.2	74.5

Develop a seasonally adjusted forecast model for these sandwich sales data. Forecast demand for each season for year 5 by using a linear trend line estimate for sales in year 5. Do the data appear to have a seasonal pattern?

20. The Hillsboro Aquatic Center has an indoor pool with lanes for lap swimming and an open area for recreational swimming and various exercise and water aerobics programs. From June to mid-August it operates on a summer schedule, and from mid-August to the end of May it operates according to normal weekday and weekend schedules. The center's policy for the pool is to have a lifeguard-to-patron ratio of 1:40. The center director wants to develop a forecast of pool attendance for the weekday schedule in order to determine the number of lifeguards to hire. The director believes pool attendance follows a seasonal pattern during the day and has accumulated the following data for average daily attendance for each hour of the day that the pool is open to the public (i.e., there are no swim team practices):

Time	Year					
	1	2	3	4	5	6
7:00 A.M.	56	64	66	60	72	65
8:00	31	41	37	44	52	46
9:00	15	22	24	30	19	26
10:00	34	35	38	31	28	33
11:00	45	52	55	49	57	50
Noon	63	71	57	65	75	70
1:00 P.M.	35	30	41	42	33	45
2:00	24	28	32	30	35	33
3:00	27	19	24	23	25	27
6:00	31	47	36	45	40	46
7:00	25	35	41	43	39	45
8:00	14	20	18	17	23	27
9:00	10	8	16	14	15	18

Develop a seasonally adjusted forecast model for these data for hourly pool attendance. Forecast attendance for each hour for year 7 by using a linear trend line estimate for pool attendance in year 7. Do the data appear to have a seasonal pattern?

21. Develop an adjusted exponential smoothing forecast ($\alpha = .30$, $\beta = .20$) for the annual pool attendance data in Problem 20. Does this forecast appear to be more or less accurate than the linear trend line model for forecasting annual pool attendance developed in Problem 20?

22. The emergency room at the new Community Hospital selected every other week during the past 5 months to observe the number of patients during two parts of each week—the weekend (Friday through Sunday) and weekdays (Monday through Thursday). They typically experience greater patient traffic on weekends than during the week:

Week	Number of Patients	
	Weekend	Weekdays
1	116	83
2	126	92
3	125	97
4	132	91
5	128	103
6	139	88
7	145	96
8	137	106
9	151	95
10	148	102

Develop a seasonally adjusted forecasting model for the number of patients during each part of the week for week 11.

23. Aztec Industries has developed a forecasting model that was used to forecast during a 10-month period. The forecasts and actual demand were as follows:

Month	Actual Demand	Forecast Demand
1	160	170
2	150	165
3	175	157
4	200	166
5	190	183
6	220	186
7	205	203
8	210	204
9	200	207
10	220	203

Measure the accuracy of the forecast by using *MAD*, *MAPD*, and cumulative error. Does the forecast method appear to be accurate?

24. RAP Computers assembles personal computers from generic parts it purchases at a discount, and it sells the units by phone orders it receives from customers responding to the company's ads in trade journals. The business has developed an exponential smoothing forecast model to forecast future computer demand. Actual demand for the company's computers for the past 8 months as well as a forecast are shown in the following table:

Month	Demand	Forecast
March	120	—
April	110	120.0
May	150	116.0
June	130	129.6
July	160	129.7
August	165	141.8
September	140	151.1
October	155	146.7
November	—	150.0

a. Using a measure of forecast accuracy of your choice, ascertain whether the forecast appears to be accurate.
b. Determine whether a 3-month moving average would provide a better forecast.

25. Develop an exponential smoothing forecast with $\alpha = .20$ for the demand data in Problem 1. Compare this forecast with the 3-month moving average computed in part (a) of Problem 1, using *MAD*, and indicate which forecast seems to be more accurate.

26. The Jersey Dairy Products Company produces cheese, which it sells to supermarkets and food-processing companies. Because of concerns about cholesterol and fat in cheese, the company has seen demand for its products decline during the past decade. It is now considering introducing some alternative low-fat dairy products and wants to determine how much available plant capacity it will have next year. The company has developed an exponential smoothing forecast with $\alpha = .40$ to forecast cheese demand. The actual demand and the forecasts from the model are as follows:

Year	Demand (1,000 lb.)	Forecast
1	16.8	—
2	14.1	16.8
3	15.3	15.7
4	12.7	15.5
5	11.9	14.4
6	12.3	13.4
7	11.5	12.9
8	10.8	12.4

Assess the accuracy of the forecast model by using *MAD* and cumulative error. If the exponential smoothing forecast model does not appear to be accurate, determine whether a linear trend model would provide a more accurate forecast.

27. The manager of the Ramona Inn Hotel near Cloverleaf Stadium believes that how well the local Blue Sox professional baseball team is playing has an impact on the occupancy rate at the hotel during the summer months. Following are the number of victories for the Blue Sox (in a 162-game schedule) for the past 8 years and the hotel occupancy rates:

Year	Blue Sox Wins	Occupancy Rate (%)
1	75	83
2	70	78
3	85	86
4	91	85
5	87	89
6	90	93
7	87	92
8	67	91

Develop a linear regression model for these data and forecast the occupancy rate for next year if the Blue Sox win 88 games.

28. Carpet City wants to develop a means to forecast its carpet sales. The store manager believes that the store's sales are directly related to the number of new housing starts in town. The manager has gathered data from county records on monthly house construction permits and from store records on monthly sales. These data are as follows:

Monthly Carpet Sales (1,000 yd.)	Monthly Construction Permits
5	21
10	35
4	10
3	12
8	16
2	9
12	41
11	15
9	18
14	26

a. Develop a linear regression model for these data and forecast carpet sales if 30 construction permits for new homes are filed.
b. Determine the strength of the causal relationship between monthly sales and new home construction by using correlation.

29. The DirectCast cable TV company is a national chain that services the small college town that is home to Tech, with a student population of almost 30,000. The company has generally been able to handle service calls and installations in the past, but a growth trend in the town and occasional significant jumps in service requests, especially for new high-definition installations, have made it more difficult for the company to determine the number of technicians and trucks it needs to maintain good service to its customers. Following is the demand for service calls during the past 36 months:

	Year 1	Year 2	Year 3
Month	**Service Calls**	**Service Calls**	**Service Calls**
January	1,048	1155	1135
February	326	319	365
March	303	324	341
April	351	344	370
May	673	712	694
June	274	306	310
July	219	245	266
August	1347	1455	1505
September	973	1056	981
October	536	545	555
November	312	298	317
December	577	481	562

Develop a monthly forecast for the upcoming year for DirectCast. Explain why you think there are occasional increases and decreases in the monthly demand for service calls.

30. Bell Inc. is a computer company that gets most of its component parts from suppliers in Southeast Asia, who ship to small warehouses called "revolvers" (for revolving inventory) near Bell's main assembly plants near Fort Worth, Texas. Bell withdraws inventory for these warehouses every few hours, and most of Bell's suppliers deliver to their revolvers every 3 or 4 days. One particular supplier provides most of the workings for the Bell Intrepid laptop, including the screen and keyboard, and the hard drive and mother board plus a few other parts are added at the assembly plant. These revolvers allow Bell to carry little of its own inventory by pushing it down to its suppliers. However, the supplier also wants to minimize its inventory, and one way it can do that is to develop an accurate demand forecast. Following is the supplier's demand data (units) from Bell for the past 24 months:

Month	Year 1 Demand	Year 2 Demand
January	2,447	2,561
February	1,826	1,733
March	1,755	1,693
April	1,456	1,484
May	1,529	1,501
June	1,633	1,655
July	2,346	2,412
August	3,784	4,017
September	4,106	3,886
October	3,006	2,844
November	2,257	2,107
December	3,212	3,410

Closely observe the demand data and develop a forecast model for year 3 that you believe will be accurate.

31. The manager of Gilley's Ice Cream Parlor needs an accurate forecast of the demand for ice cream. The store orders ice cream from a distributor a week ahead; if the store orders too little, it loses business, and if it orders too much, the extra must be thrown away. The manager believes that a major determinant of ice cream sales is temperature (i.e., the hotter the weather, the more ice cream people buy). Using an almanac, the manager has determined the average daytime temperature for 10 weeks, selected at random, and from store records he has determined the ice cream consumption for the same 10 weeks. These data are summarized as follows:

Week	Average Temperature (degrees)	Ice Cream Sold (gal.)
1	73	110
2	65	95
3	81	135
4	90	160
5	75	97
6	77	105
7	82	120
8	93	175
9	86	140
10	79	121

a. Develop a linear regression model for these data and forecast the ice cream consumption if the average weekly daytime temperature is expected to be 85 degrees.
b. Determine the strength of the linear relationship between temperature and ice cream consumption by using correlation.

32. Compute the coefficient of determination for the data in Problem 31 and explain its meaning.

33. Administrators at State University believe that decreases in the number of freshman applications that they have experienced are directly related to tuition increases. They have collected the following enrollment and tuition data for the past decade:

Year	Freshman Applications	Annual Tuition
1	6,050	$3,600
2	4,060	3,600
3	5,200	4,000
4	4,410	4,400
5	4,380	4,500
6	4,160	5,700
7	3,560	6,000
8	2,970	6,000
9	3,280	7,500
10	3,430	8,000

a. Develop a linear regression model for these data and forecast the number of applications for State University if tuition increases to $9,000 per year and if tuition is lowered to $7,000 per year.
b. Determine the strength of the linear relationship between freshman applications and tuition by using correlation.
c. Describe the various planning decisions for State University that would be affected by the forecast for freshman applications.

34. Develop a linear trend line model for the freshman applications data at State University in Problem 33.
    a. Does this forecast appear to be more or less accurate than the linear regression forecast developed in Problem 33 Justify your answer.
    b. Compute the correlation coefficient for the linear trend line forecast and explain its meaning.

35. Explain the numerical value of the slope of the linear regression equation in Problem 31.

36. Some members of management of the Fairface Cosmetics Firm believe that demand for its products is related to the promotional activities of local department stores where the cosmetics are sold. However, others in management believe that other factors, such as local demographics, are stronger determinants of demand behavior. The following data for local annual promotional expenditures for Fairface products and local annual unit sales for Fairface lip gloss have been collected from 20 stores selected at random from different localities:

Store	Annual Unit Sales (1,000s)	Annual Promotional Expenditures ($1,000s)
1	3.5	$12.6
2	7.2	15.5
3	3.1	10.8
4	1.6	8.7
5	8.9	20.3
6	5.7	21.9
7	6.3	25.6
8	9.1	14.3
9	10.2	15.1
10	7.3	18.7
11	2.5	9.6
12	4.6	12.7
13	8.1	16.3
14	2.5	8.1
15	3.0	7.5
16	4.8	12.4
17	10.2	17.3
18	5.1	11.2
19	11.3	18.5
20	10.4	16.7

Based on these data, does it appear that the strength of the relationship between sales and promotional expenditures is sufficient to warrant using a linear regression forecasting model? Explain your response.

37. Employees at Precision Engine Parts Company produce parts according to exact design specifications. The employees are paid according to a piece-rate system, wherein the faster they work and the more parts they produce, the greater their chances for monthly bonuses. Management suspects that this method of pay may contribute to an increased number of defective parts. A specific part requires a normal, standard time of 23 minutes to produce. The quality control manager has checked the actual average times to produce this part for 10 different employees during 20 days selected at random during the past month and determined the corresponding percentage of defective parts, as follows:

Average Time (min.)	Defective (%)	Average Time (min.)	Defective (%)
21.6	3.1	20.8	2.7
22.5	4.6	18.9	4.5
23.1	2.7	21.4	2.8
24.6	1.8	23.7	1.9
22.8	3.5	23.8	1.1
23.7	3.2	24.9	1.2
20.9	3.7	19.8	2.3
19.7	4.5	19.7	5.1
24.5	0.8	21.2	3.6
26.7	1.2	20.8	4.2

Develop a linear regression model relating average production time to percentage defects to determine whether a relationship exists and the percentage of defective items that would be expected with a normal production time of 23 minutes.

38. Apperson and Fitz is a chain of clothing stores that caters to high school and college students. It publishes a quarterly catalog and operates a Web site that features provocatively attired males and females. The Web site is very expensive to maintain, and company executives are not sure whether the number of hits at the site relate to sales (i.e., people may be looking at the site's pictures only). The Web master has accumulated the following data for hits per month and orders placed at the Web site for the past 20 months:

Month	Hits (1,000s)	Orders (1,000s)
1	34.2	7.6
2	28.5	6.3
3	36.7	8.9
4	42.3	5.7
5	25.8	5.9
6	52.3	6.3
7	35.2	7.2
8	27.9	4.1
9	31.4	3.7
10	29.4	5.9
11	46.7	10.8
12	43.5	8.7
13	52.6	9.3
14	61.8	6.5
15	37.3	4.8
16	28.9	3.1
17	26.4	6.2
18	39.4	5.9
19	44.7	7.2
20	46.3	5.5

Develop a linear regression model for these data and indicate whether there appears to be a strong relationship between Web site hits and orders. What would be the forecast for orders with 50,000 hits per month?

39. The Gametime Hat Company manufactures baseball caps that have various team logos in an assortment of designs and colors. The company has had monthly sales for the past 24 months as follows:

Month	Demand (1,000s)
1	8.2
2	7.5
3	8.1
4	9.3
5	9.1
6	9.5
7	10.4
8	9.7
9	10.2
10	10.6
11	8.2
12	9.9
13	10.3
14	10.5
15	11.7
16	9.8
17	10.8
18	11.3
19	12.6
20	11.5
21	10.8
22	11.7
23	12.5
24	12.8

Develop a forecast model using the method you believe best and justify your selection by using a measure (or measures) of forecast accuracy.

40. Infoworks is a large computer discount store that sells computers and ancillary equipment and software in the town where State University is located. Infoworks has collected historical data on computer sales and printer sales for the past 10 years, as follows:

Year	Personal Computers Sold	Printers Sold
1	1,045	326
2	1,610	510
3	860	296
4	1,211	478
5	975	305
6	1,117	506
7	1,066	612
8	1,310	560
9	1,517	590
10	1,246	676

a. Develop a linear trend line forecast to forecast printer demand in year 11.
b. Develop a linear regression model relating printer sales to computer sales in order to forecast printer demand in year 11 if 1,300 computers are sold.
c. Compare the forecasts developed in (a) and (b) and indicate which one appears to be more accurate.

41. Develop an exponential smoothing model with $\alpha = .30$ for the data in Problem 40 to forecast printer demand in year 11 and compare its accuracy to the linear trend line forecast developed in (a).

42. Arrow Air is a regional East Coast airline that has collected data for the percentage available seats occupied on its flights for four quarters—(1) January–March, (2) April–June, (3) July–September, and (4) October–December—for the past 5 years. Arrow Air also has collected data for the average percentage fare discount for each of these quarters, as follows:

Year	Quarter	Average Fare Discount (%)	Seat Occupancy (%)
1	1	63	21
	2	75	34
	3	76	18
	4	58	26
2	1	59	18
	2	62	40
	3	81	25
	4	76	30
3	1	65	23
	2	70	28
	3	78	30
	4	69	35
4	1	59	20
	2	61	35
	3	83	26
	4	71	30
5	1	60	25
	2	66	37
	3	86	25
	4	74	30

a. Develop a seasonally adjusted forecast model for seat occupancy. Forecast seat occupancy for year 6 by using a linear trend line forecast estimate for seat occupancy in year 6.
b. Develop linear regression models relating seat occupancy to discount fares in order to forecast seat occupancy for each quarter in year 6. Assume a fare discount of 20% for quarter 1, 36% for quarter 2, 25% for quarter 3, and 30% for quarter 4.
c. Compare the forecasts developed in (a) and (b) and indicate which one appears to be the best.

43. Develop an adjusted exponential smoothing forecast model ($\alpha = .40$ and $\beta = .40$) for the data in Problem 42 to forecast seat occupancy, and compare its accuracy with the seasonally adjusted model developed in (a).

44. The consumer loan department at Central Union Bank and Trust wants to develop a forecasting model to help determine its potential loan application volume for the coming year. Because adjustable-rate home mortgages are based on government long-term treasury note rates, the department collected the following data for 3- to 5-year treasury note interest rates for the past 24 years:

Year	Rate	Year	Rate	Year	Rate
1	5.77	9	9.71	17	7.68
2	5.85	10	11.55	18	8.26
3	6.92	11	14.44	19	8.55
4	7.82	12	12.92	20	8.26
5	7.49	13	10.45	21	6.80
6	6.67	14	11.89	22	6.12
7	6.69	15	9.64	23	5.48
8	8.29	16	7.06	24	6.09

Develop an appropriate forecast model for the bank to use to forecast treasury note rates in the future and indicate how accurate it appears to be compared to historical data.

45. The busiest time of the day at the Taco Town fast-food restaurant is between 11:00 A.M. and 2:00 P.M. Taco Town's service is very labor dependent, and a critical factor for providing quick service is the number of employees on hand during this 3-hour period. To determine the number of employees it needs during each hour of the 3-hour lunch period, Taco Town requires an accurate forecasting model. Following are the number of customers served at Taco Town during each hour of the lunch period for the past 20 weekdays:

	Hour		
Day	11–12	12–1	1–2
1	90	125	87
2	76	131	93
3	87	112	99
4	83	149	78
5	71	156	83
6	94	178	89
7	56	101	124
8	63	91	66
9	73	146	119
10	101	104	96
11	57	114	106
12	68	125	95
13	75	206	102
14	94	117	118
15	103	145	122
16	67	121	93
17	94	113	76
18	83	166	94
19	79	124	87
20	81	118	115

Develop a forecast model that you believe will best forecast Taco Town's customer demand for the next day and explain why you selected this model.

46. In order to take advantage of the burgeoning public interest in sustainability, Klorax has developed a nonsynthetic line of natural cleaning products called GreenClean. One of the first sustainable products Klorax introduced was a natural cold-water detergent. The company has now gathered several years' worth of sales data for this product in order to develop various forecast models. The company's marketing department believes that sales increase when there are several major environmental (i.e., "green") episodes in the national media, such as a government energy report, an increase in gasoline or energy prices, an oil spill, a pollution episode, a company getting positive press for undertaking a major green project, etc. Therefore, the marketing department wants to develop a linear regression forecasting model that relates the number of green episodes reported monthly in the national media to monthly sales of the detergent. The marketing department had several staff members go through various media sources for the past 2 years and count the number of major national green episodes reported on, which with monthly sales, is shown in the following table:

Month	Sales	Green Episodes	Month	Sales	Green Episodes
1	34,175	3	13	55,732	7
2	28,366	2	14	26,004	4
3	41,819	4	15	49,188	5
4	27,666	1	16	40,005	2
5	31,299	1	17	38,912	2
6	37,456	4	18	31,777	1
7	52,444	5	19	30,367	0
8	46,712	3	20	34,566	1
9	37,222	2	21	29,078	1
10	44,981	2	22	45,876	3
11	40,006	2	23	48,556	4
12	47,321	4	24	51,022	6

Develop a linear regression model and determine the sales forecast if three green episodes are reported in the media next month. Discuss the value and usefulness of this type of relational forecast for Klorax.

47. In Problem 46, one of the new products in the GreenClean line that Klorax is ready to introduce is an all-purpose cleaner, and Klorax is attempting to forecast what demand might be like 5 years into the future. However, it has no historical demand data to use to develop a forecast model. The marketing department has made its own estimate of first-year sales and has acquired estimates for the new cleaner from four other sources—top management, the sales force, a cleaning products trade association, and an independent marketing firm. Top management estimates first-year sales to be 34,000 units; the marketing department's estimate is 47,000 units; the sales force suggests 41,000 units; the trade association estimates that first-year sales will be 28,000 units; and the independent marketing firm estimates sales will be 51,000 units. Using these estimates and the demand data and forecast model for the detergent in Problem 46, develop a forecast for planning purposes for year 5 for the new cleaner.

48. The Wellton Fund is a balanced mutual fund that includes a mix of stocks and bonds. Following are the year-end share prices of the fund and Dow Jones Industrial Average (DJIA) for a 20-year period:

Year	Share Price	DJIA
1	$14.75	1,046
2	15.06	1,258
3	14.98	1,211
4	15.73	1,546
5	16.11	1,895
6	16.07	1,938
7	16.78	2,168
8	17.69	2,753
9	16.90	2,633
10	17.81	3,168
11	19.08	3,301
12	20.40	3,754
13	19.39	3,834
14	24.43	5,117
15	26.46	6,448
16	29.45	7,908
17	29.35	9,181
18	27.96	11,497
19	28.21	10,786
20	27.26	10,150

Develop a linear regression model for these data and forecast the fund share price for a DJIA of 12,000. Does there appear to be a strong relationship between the fund's share price and the DJIA?

49. The Valley United Soccer Club has boys' and girls' travel soccer teams at all age levels up to 18 years. The club has been successful and grown in popularity over the years; however, an obstacle to its continued growth is a shortage of practice and game soccer fields in the area. The club has tried to make a case to the town council and the parks and recreation committee that it needs more soccer fields to accommodate the increasing number of kids who want to play on club teams. The number of kids who have played soccer on club teams and the town's population for the past 15 years are as follows:

Year	Club Soccer Players	Town Population
1	146	18,060
2	135	18,021
3	159	18,110
4	161	18,125
5	176	18,240
6	190	18,231
7	227	18,306
8	218	18,477
9	235	18,506
10	231	18,583
11	239	18,609
12	251	18,745
13	266	19,003
14	301	19,062
15	327	19,114

The soccer club wants to develop a forecasting model to demonstrate to the town council its expected growth in the future.

a. Develop a linear trend line forecast to predict the number of soccer players the club can expect next year.
b. The town planning department has told the soccer club that the town expects to grow to a population of 19,300 by next year and to 20,000 in 5 years. Develop a linear regression model, using the town's population as a predictor of the number of club soccer players, and compare this forecasting model to the one developed in part (a). Which forecasting model should the club use to support its request for new fields?

50. The Port of Savannah is considering an expansion of its container terminal. The port has experienced the following container throughput during the past 12 years, expressed as TEUs (i.e., 20-foot equivalent units, a standard unit of measure for containers):

Year	TEUs (1,000s)
1	526.1
2	549.4
3	606.0
4	627.0
5	695.7
6	734.9
7	761.1
8	845.4
9	1,021.1
10	1,137.1
11	1,173.6
12	1,233.4

a. Develop a linear trend line forecast for these data and forecast the number of TEUs for year 13.
b. How strong is the linear relationship for these data?

51. The admission data for freshmen at Tech during the past 10 years are as follows:

Year	Applicants	Offers	% Offers	Acceptances	% Acceptances
1	13,876	11,200	80.7	4,112	36.7
2	14,993	11,622	77.8	4,354	37.3
3	14,842	11,579	78.0	4,755	41.1
4	16,285	13,207	81.1	5,068	38.0
5	16,922	11,382	73.2	4,532	39.8
6	16,109	11,937	74.1	4,655	39.0
7	15,883	11,616	73.1	4,659	40.1
8	18,407	11,539	62.7	4,620	40.0
9	18,838	13,138	69.7	5,054	38.5
10	17,756	11,952	67.3	4,822	40.3

Tech's admission objective is a class of 5,000 entering freshmen, and Tech wants to forecast the percentage of offers it will likely have to make in order to achieve this objective.

a. Develop a linear trend line to forecast next year's applicants and percentage of acceptances and use these results to estimate the percentage of offers that Tech should expect to make.
b. Develop a linear trend line to forecast the percentage of offers that Tech should expect to make and compare this result with the result in (a). Which forecast do you think is more accurate?
c. Assume that Tech receives 18,300 applicants in year 11. How many offers do you think it should make to get 5,000 acceptances?

52. The State of Virginia has instituted a series of standards of learning (SOL) tests in math, history, English, and science that all high school students must pass with a grade of 70 before they are allowed to graduate and receive their diplomas. The school superintendent of Montgomery County believes the tests are unfair because the test scores are closely related to teacher salary and tenure (i.e., the years a teacher has been at a school). The superintendent has sampled 12 other county school systems in the state and accumulated the following data for average teacher salary and average teacher tenure:

School	Average SOL Score	Average Teacher Salary	Average Teacher Tenure (yr.)
1	81	$34,300	9.3
2	78	28,700	10.1
3	76	26,500	7.6
4	77	36,200	8.2
5	84	35,900	8.8
6	86	32,500	12.7
7	79	31,800	8.4
8	91	38,200	11.5
9	68	27,100	8.3
10	73	31,500	7.3
11	90	37,600	12.3
12	85	40,400	14.2

a. Using Excel or QM for Windows, develop the multiple regression equation for these data.
b. What is the coefficient of determination for this regression equation? Do you think the superintendent is correct in his beliefs?
c. Montgomery County has an average SOL score of 74, with an average teacher salary of $27,500 and an average teacher tenure of 7.8 years. The superintendent has proposed to the school board a salary increase that would raise the average salary to $30,000 as well as a benefits program, with the goal of increasing the average tenure to 9 years. He has suggested that if the board passes his proposals, then the average SOL score will increase to 80. Is he correct, according to the forecasting model?

53. Tech administrators believe their freshman applications are influenced by two variables: tuition and the size of the applicant pool of eligible high school seniors in the state. The following data for an 8-year period show the tuition rates (per semester) and the sizes of the applicant pool for each year:

Tuition	Applicant Pool	Applicants
$ 900	76,200	11,060
1,250	78,050	10,900
1,375	67,420	8,670
1,400	70,390	9,050
1,550	62,550	7,400
1,625	59,230	7,100
1,750	57,900	6,300
1,930	60,080	6,100

a. Using Excel, develop the multiple regression equation for these data.
b. What is the coefficient of determination for this regression equation?
c. Determine the forecast for freshman applicants for a tuition rate of $1,500 per semester, with a pool of applicants of 60,000.

54. In Problem 40, Infoworks believes its printer sales are also related to the average price of its printers. It has collected historical data on average printer prices for the past 10 years, as follows:

Year	Average Printer Price
1	$475
2	490
3	520
4	420
5	410
6	370
7	350
8	300
9	280
10	250

a. Using Excel, develop the multiple regression equation for these data.
b. What is the coefficient of determination for this regression equation?
c. Determine the forecast for printer sales, based on personal computer sales of 1,500 units and an average printer price of $300.

55. The manager of the Bayville police department motor pool wants to develop a forecast model for annual maintenance on police cars, based on mileage in the past year and age of the cars. The following data have been collected for eight different cars:

Miles Driven	Car Age (yr.)	Maintenance Cost
16,320	7	$1,200
15,100	8	1,400
18,500	8	1,820
10,200	3	900
9,175	3	650
12,770	7	1,150
8,600	2	875
7,900	3	900

a. Using Excel, develop a multiple regression equation for these data.
b. What is the coefficient of determination for this regression equation?
c. Forecast the annual maintenance cost for a police car that is 5 years old and will be driven 10,000 miles in 1 year.

56. The dean of the college of business at Tech has initiated a fund-raising campaign. One of the selling points he plans to use with potential donors is that increasing the college's private endowment will improve its ranking among all business schools, as published each year by the magazine *The Global News and Business Report*. He would like to demonstrate that there is a

relationship between funding and the rankings. He has collected the following data, showing the private endowments ($1,000,000s) and annual budgets ($1,000,000s) from state and private sources for eight of Tech's peer institutions plus Tech, and the ranking of each school:

Private Endowment ($1,000,000s)	Annual Budget ($1,000,000s)	Ranking
$ 2.5	$ 8.1	87
52.0	26.0	20
12.7	7.5	122
63.0	33.0	32
46.0	12.0	54
27.1	16.1	76
23.3	17.0	103
46.4	14.9	40
48.9	21.8	98

a. Using Excel, develop a linear regression model for the amount of the private endowment and the ranking and forecast a ranking for a private endowment of $70 million. Does there appear to be a strong relationship between the endowment and the ranking?
b. Using Excel, develop a multiple regression equation for all these data, including private endowment and annual budget, and forecast a ranking for a private endowment of $70 million and an annual budget of $40 million. How does this forecast compare with the forecast in part (a)?

Case Problem

Forecasting at State University

During the past few years, the legislature has severely reduced funding for State University. In reaction, the administration at State has significantly raised tuition each year for the past 5 years. Perceived as a bargain 5 years ago, State is now considered one of the more expensive state-supported universities. This has led some parents and students to question the value of a State education, and applications for admission have declined. Because a portion of state educational funding is based on a formula tied to enrollments, State has maintained its enrollment levels by going deeper into its applicant pool and accepting less-qualified students.

On top of these problems, a substantial increase in the college-age population is expected in the next decade. Key members of the state legislature have told the university administration that State will be expected to absorb additional students during the next decade. However, because of the economic outlook and the budget situation, the university should not expect any funding increases for additional facilities, classrooms, dormitory rooms, or faculty. The university already has a classroom deficit in excess of 25%, and class sizes are above the averages of the peer institutions.

The president of the university, Alva McMahon, established several task forces consisting of faculty and administrators to address these problems. These groups made several wide-ranging general recommendations, including the implementation of appropriate management practices and more in-depth, focused planning.

Discuss in general terms how forecasting might be used for university planning to address these specific problem areas. Include in your discussion the types of forecasting methods that might be used.

Case Problem

THE UNIVERSITY BOOKSTORE STUDENT COMPUTER PURCHASE PROGRAM

The University Bookstore is owned and operated by State University through an independent corporation with its own board of directors. The bookstore has three locations on or near the State University campus. It stocks a range of items, including textbooks, trade books, logo apparel, drawing and educational supplies, and computers and related products, including printers, modems, and software. The bookstore has a program to sell personal computers to incoming freshmen and other students at a substantial educational discount, partly passed on from computer manufacturers. This means that the bookstore just covers computer costs, with a very small profit margin remaining.

Each summer all incoming freshmen and their parents come to the State campus for a 3-day orientation program. The students come in groups of 100 throughout the summer. During their visit the students and their parents are given details about the bookstore's computer purchase program. Some students place their computer orders for the fall semester at this time, whereas others wait until later in the summer. The bookstore also receives orders from returning students throughout the summer. This program presents a challenging management problem for the bookstore.

Orders come in throughout the summer, many only a few weeks before school starts in the fall, and the computer suppliers require at least 6 weeks for delivery. Thus, the bookstore must forecast computer demand to build up inventory to meet student demand in the fall. The student computer program and the forecast of computer demand have repercussions all along the bookstore supply chain. The bookstore has a warehouse near campus where it must store all computers because it has no storage space at its retail locations. Ordering too many computers not only ties up the bookstore's cash reserves, it also takes up limited storage space and limits inventories for other bookstore products during the bookstore's busiest sales period. Because the bookstore has such a low profit margin on computers, its bottom line depends on these other products. Because competition for good students has increased, the university has become very quality conscious and insists that all university facilities provide exemplary student service, which for the bookstore means meeting all student demands for computers when fall semester starts. The number of computers ordered also affects the number of temporary warehouse and bookstore workers who must be hired for handling and assisting with PC installations. The number of truck trips from the warehouse to the bookstore each day of fall registration is also affected by computer sales.

The bookstore student computer purchase program has been in place for 14 years. Although the student population has remained stable during this period, computer sales have been somewhat volatile. Following are the historical sales data for computers during the first month of fall registration:

Year	Computers Sold	Year	Computers Sold
1	518	8	792
2	651	9	877
3	708	10	693
4	921	11	841
5	775	12	1,009
6	810	13	902
7	856	14	1,103

Develop an appropriate forecast model for bookstore management to use to forecast computer demand for next fall semester and indicate how accurate it appears to be. What other forecasts might be useful to the bookstore?

Case Problem

VALLEY SWIM CLUB

The Valley Swim Club has 300 stockholders, each holding one share of stock in the club. A share of club stock allows the shareholder's family to use the club's heated outdoor pool during the summer, upon payment of annual membership dues of $175. The club has not issued any stock in years, and only a few of the existing shares come up for sale each year. The board of directors administers the sale of all stock. When a shareholder wants to sell, he or she turns the stock in to the board, which sells it to the person at the top of the waiting list. For the past few years, the length of the waiting list has remained relatively steady, at approximately 20 names.

However, during the past winter two events occurred that have suddenly increased the demand for shares in the club. The winter was especially severe, and subzero weather and heavy ice storms caused both the town and the county pools to buckle and crack. The problems were not discovered until maintenance crews began to prepare the pools for the summer, and repairs cannot be completed until the fall. Also during the winter, the manager of the local country club had an argument with her board of directors and one night burned down the clubhouse. Although the pool itself was not damaged, the dressing room facilities, showers, and snack bar were destroyed. As a result of these two events, the Valley Swim Club was inundated with applications to purchase shares. The waiting list suddenly grew to 250 people as the summer approached.

The board of directors of the swim club had refrained from issuing new shares in the past because there never was a very great demand, and the demand that did exist was usually absorbed within a year by stock turnover. In addition, the board has a real concern about overcrowding. It seemed like the present membership was about right, and there were very few complaints about overcrowding, except on holidays such as Memorial Day and the Fourth of July. However, at a recent board meeting, a number of new applicants had attended and asked the board to issue new shares. In addition, a number of current shareholders suggested that this might be an opportunity for the club to raise some capital for needed repairs and to improve some of the existing facilities. This was tempting to the board. Although it had set the share price at $500 in the past, the board could set it at a much higher level now. In addition, an increase in attendance could create a need for more lifeguards.

Before the board of directors could make a decision on whether to sell more shares and, if so, how many, the board members felt they needed more information. Specifically, they would like a forecast of the average number of people (family members, guests, etc.) who might attend the pool each day during the summer, with the current number of shares.

The board of directors has the following daily attendance records for June through August from the previous summer; it thinks the figures would provide accurate estimates for the upcoming summer:

M-139	W-380	F-193	Su-399	T-177	Th-238
T-273	Th-367	Sa-378	M-197	W-161	F-224
W-172	F-359	Su-461	T-273	Th-308	Sa-368
Th-275	Sa-463	M-242	W-213	F-256	Su-541
F-337	Su-578	T-177	Th-303	Sa-391	M-235
Sa-402	M-287	W-245	F-262	Su-400	T-218
Su-487	T-247	Th-390	Sa-447	M-224	W-271
M-198	W-356	F-284	Su-399	T-239	Th-259
T-310	Th-322	Sa-417	M-275	W-274	F-232
W-347	F-419	Su-474	T-241	Th-205	Sa-317
Th-393	Sa-516	M-194	W-190	F-361	Su-369
F-421	Su-478	T-207	Th-243	Sa-411	M-361
Sa-595	M-303	W-215	F-277	Su-419	
Su-497	T-223	Th-304	Sa-241	M-258	
M-341	W-315	F-331	Su-384	T-130	
T-291	Th-258	Sa-407	M-246	W-195	

Develop a forecasting model to forecast daily demand during the summer.

Case Problem

Forecasting Airport Passenger Arrivals

Since the terrorist attacks of 9/11 and because of the ensuing measures to increase airline security, airports have faced the problem of long waiting lines and waiting times at security gates. Waiting lines can be as long as hundreds of yards, and waiting times can sometimes be hours. In their efforts to reduce waiting lines and times, or at least to not have them become longer as airline demand increases, airports have analyzed their existing security systems and sought quantitative solutions. One of the key components of any effort to operationally improve airport security procedures is forecasting passenger arrivals at security checkpoints in order to determine how many security checkpoints and staff are needed. At Berry International Airport (BEI), security analysts would like to forecast passenger arrivals for next July, the airport's busiest travel month of the year, for the purpose of determining how many security checkpoints they should staff during the month so that waiting lines and times will not be excessively long. Demand for airline travel has generally been increasing during the past 3 years. There are two main concourses at BEI, North and South, each serving different airlines. The following table shows passenger arrivals at the South concourse for 10 days (selected randomly) in 2-hour segments from 4:00 A.M. to 10:00 P.M. for the month of July for the past 3 years.

	Day	*4–6 A.M.*	*6–8 A.M.*	*8–10 A.M.*	*10–Noon*	*Noon–2 P.M.*	*2–4 P.M.*	*4–6 P.M.*	*6–8 P.M.*	*8–10 P.M.*
Year One	1	2400	2700	3200	1400	1700	1800	1600	800	200
	2	1900	2500	3100	1600	1800	2000	1800	900	300
	3	2300	3100	2500	1500	1500	1800	1900	1100	200
	4	2200	3200	3100	2200	1900	2400	2100	1200	400
	5	2400	3300	3400	1700	2200	2100	2000	1000	600
	6	2600	2800	3500	1500	1700	1900	1500	1100	300
	7	1900	2800	3100	1200	1500	2000	1400	900	400
	8	2000	2700	2500	1500	2000	2300	1900	1000	200
	9	2400	3200	3600	1600	2100	2500	1800	1400	200
	10	2600	3300	3100	200	2500	2600	2400	1100	400
Year Two	11	3100	3900	4100	2200	2600	2300	2500	1100	300
	12	2800	3400	3900	1900	2100	2500	2000	1200	300
	13	2700	3800	4300	2100	2400	2400	2400	1200	400
	14	2400	3500	4100	2400	3000	3200	2600	1200	700
	15	3300	3700	4000	2600	2600	2700	2900	1000	300
	16	3500	4000	3800	2300	2700	3100	3000	900	200
	17	2900	4100	3900	2400	3000	3200	2500	1100	500
	18	3400	3800	4200	2000	2500	3000	2200	1000	300
	19	3600	3600	4000	2300	2600	2800	2600	1200	200
	20	3700	3700	4000	2200	2600	2700	2400	1200	200
Year Three	21	4400	4400	4500	2600	3300	3400	3000	1200	400
	22	4200	4500	4300	2500	3400	3600	3100	1400	300
	23	4500	4500	4700	2700	3400	3500	2900	1200	300
	24	4600	4600	4600	2500	3200	3500	2800	1300	300
	25	4500	4300	4400	2900	3300	3300	3300	1500	400
	26	4200	4300	4500	3000	4000	3400	3000	1500	600
	27	4500	4500	5100	3300	4000	3700	3100	1200	300
	28	4300	4200	4300	2800	3500	4000	3300	1100	400
	29	4900	4100	4200	3100	3600	3900	3400	1400	500
	30	4700	4500	4100	3000	4000	3700	3400	1200	500

Develop a forecast for daily passenger arrivals at the South concourse at BEI for each time period for July of year 4. Discuss the various forecast model variations that might be used to develop this forecast.

CHAPTER

16

Inventory Management

Inventory analysis is one of the most popular topics in management science. One reason is that almost all types of business organizations have inventory. Although we tend to think of inventory only in terms of stock on a store shelf, it can take on a variety of forms, such as partially finished products at different stages of a manufacturing process, raw materials, resources, labor, or cash. In addition, the purpose of inventory is not always simply to meet customer demand. For example, companies frequently stock large inventories of raw materials as a hedge against strikes. Whatever form inventory takes or whatever its purpose, it often represents a significant cost to a business firm. It is estimated that the average annual cost of manufactured goods inventory in the United States is approximately 30% of the total value of the inventory. Thus, if a company has $10.0 million worth of products in inventory, the cost of holding the inventory (including insurance, obsolescence, depreciation, interest, opportunity costs, storage costs, etc.) would be approximately $3.0 million. If the amount of inventory could be reduced by half to $5.0 million, then $1.5 million would be saved in inventory costs, a significant cost reduction.

In this chapter we describe the classic economic order quantity models, which represent the most basic and fundamental form of inventory analysis. These models provide a means for determining how much to order (the order quantity) and when to place an order so that inventory-related costs are minimized. The underlying assumption of these models is that demand is known with certainty and is constant. In addition, we will describe models for determining the order size and reorder points (when to place an order) when demand is uncertain.

Elements of Inventory Management

__Inventory__ is a stock of items kept on hand to meet demand.

Inventory is defined as a stock of items kept on hand by an organization to use to meet customer demand. Virtually every type of organization maintains some form of inventory. A department store carries inventories of all the retail items it sells; a nursery has inventories of different plants, trees, and flowers; a rental car agency has inventories of cars; and a major league baseball team maintains an inventory of players on its minor league teams. Even a family household will maintain inventories of food, clothing, medical supplies, personal hygiene products, and so on.

The Role of Inventory

A company or an organization keeps stocks of inventory for a variety of important reasons. The most prominent is holding finished goods inventories to meet customer demand for a product, especially in a retail operation. However, customer demand can also be in the form of a secretary going to a storage closet to get a printer cartridge or paper, or a carpenter getting a board or nail from a storage shed. A level of inventory is normally maintained that will meet anticipated or expected customer demand. However, because demand is usually not known with certainty, additional amounts of inventory, called **safety**, or *buffer*, **stocks**, are often kept on hand to meet unexpected variations in excess of expected demand.

__Safety stocks__ are additional inventory to compensate for demand uncertainty.

Additional stocks of inventories are sometimes built up to meet seasonal or cyclical demand. Companies will produce items when demand is low to meet high seasonal demand for which their production capacity is insufficient. For example, toy manufacturers produce large inventories during the summer and fall to meet anticipated demand during the Christmas season. Doing so enables them to maintain a relatively smooth production flow throughout the year. They would not normally have the production capacity or logistical support to produce enough to meet all of the Christmas demand during that season. Correspondingly, retailers might find it necessary to keep large stocks of inventory on their shelves to meet peak seasonal demand, such as at Christmas, or for display purposes to attract buyers.

A company will often purchase large amounts of inventory to take advantage of price discounts, as a hedge against anticipated future price increases, or because it can get a lower price by purchasing in volume. For example, Walmart has long been known to purchase an entire manufacturer's stock of soap powder or other retail items because it can get a very low price, which it subsequently passes on to its customers. Companies will often purchase large stocks of items when a supplier liquidates to get a low price. In some cases, large orders will be made simply because the cost of an order may be very high, and it is more cost-effective to have higher inventories than to make a lot of orders.

Many companies find it necessary to maintain in-process inventories at different stages in a manufacturing process to provide independence between operations and to avoid work stoppages or delays. Inventories of raw materials and purchased parts are kept on hand so that the production process will not be delayed as a result of missed or late deliveries or shortages from a supplier. Work-in-process inventories are kept between stages in the manufacturing process so that production can continue smoothly if there are temporary machine breakdowns or other work stoppages. Similarly, a stock of finished parts or products allows customer demand to be met in the event of a work stoppage or problem with the production process.

Demand

A crucial component and the basic starting point for the management of inventory is customer demand. Inventory exists for the purpose of meeting the demand of customers. Customers can be inside the organization, such as a machine operator waiting for a part or a partially completed product to work on, or outside the organization, such as an individual purchasing groceries or a new stereo. As such, an essential determinant of effective inventory management is an accurate forecast of demand. For this reason the topics of forecasting (Chapter 15) and inventory management are directly interrelated.

__Dependent demand__ items are used internally to produce a final product.

__Independent demand__ items are final products demanded by an external customer.

In general, the demand for items in inventory is classified as dependent or independent. **Dependent demand** items are typically component parts, or materials, used in the process of producing a final product. For example, if an automobile company plans to produce 1,000 new cars, it will need 5,000 wheels and tires (including spares). In this case the demand for wheels is dependent on the production of cars; that is, the demand for one item is a function of demand for another item.

Alternatively, cars are an example of an **independent demand** item. In general, independent demand items are final or finished products that are not a function of, or dependent upon, internal production activity. Independent demand is usually external, and, thus, beyond the direct control of the organization. In this chapter we will focus on the management of inventory for independent demand items.

Inventory Costs

Inventory costs include __carrying__, __ordering__, and __shortage costs__.

__Carrying costs__ are the costs of holding inventory in storage.

There are three basic costs associated with inventory: carrying (or holding) costs, ordering costs, and shortage costs. **Carrying costs** are the costs of holding items in storage. These vary with the level of inventory and occasionally with the length of time an item is held; that is, the greater the level of inventory over time, the higher the carrying cost(s). Carrying costs can include the cost of losing the use of funds tied up in inventory; direct storage costs, such as rent, heating, cooling, lighting, security, refrigeration, record keeping, and logistics; interest on loans used to purchase inventory; depreciation; obsolescence as markets for products in inventory diminish; product deterioration and spoilage; breakage; taxes; and pilferage.

Carrying costs are normally specified in one of two ways. The most general form is to assign total carrying costs, determined by summing all the individual costs mentioned previously, on a per-unit basis per time period, such as a month or a year. In this form, carrying costs would commonly be expressed as a per-unit dollar amount on an annual basis (for example, $10 per year). Alternatively, carrying costs are sometimes expressed as a percentage of the value of an item or as a percentage of average inventory value. It is generally estimated that carrying costs range from 10% to 40% of the value of a manufactured item.

__Ordering costs__ are the cost of replenishing inventory.

Ordering costs are the costs associated with replenishing the stock of inventory being held. These are normally expressed as a dollar amount per order and are independent of the order size. Thus, ordering costs vary with the number of orders made (i.e., as the number of

Management Science Application

Evaluating Inventory Costs at Hewlett-Packard

Hewlett-Packard, with annual revenues exceeding $90 billion and 150,000 employees worldwide, is the *Fortune 500* 11th-ranked company. Although demand for PCs increased by five-fold in the 1990s, becoming a veritable household product, many PC companies struggled to remain profitable. By the end of the 1990s HP was struggling to make a profit in the increasingly competitive global PC market because of price cuts throughout the decade. Because prices were not really controllable, inventory costs became especially critical in the PC profit equation. Rapid technological advancements render new PC products obsolete in a few months, and, in general, it's believed that the value of a PC decreases at the rate of 1% per week. Consequently, holding any excess inventory is very costly.

HANDOUT/Newscom

In the late 1990s, to return its PC business to more sustainable profitability, HP undertook an extensive evaluation of its inventory costs. It discovered that inventory-related costs were the main determinants of overall PC cost, and, in fact, in 1 year alone inventory-related costs equaled the PC business's total operating margin. Further, HP determined that the traditional inventory carrying (or holding) costs (which encompass capital costs plus storage, taxes, insurance, breakage, etc.) accounted for less than 10% of the total inventory-related costs. HP identified four additional inventory costs in their PC business that were a major factor in overall supply chain costs. The single biggest inventory cost was determined to be the "component devaluation cost." This is the penalty cost HP incurred when the price dropped for excess components and parts (for example, CPUs, memory, and chips) being held in inventory. HP held inventories of parts and components in factories, in distribution centers, and in transit and incurred a devaluation cost at all these points in its supply chain whenever a price reduction occurred. Another inventory cost is the "price protection cost," which occurs when the retail price of a product drops after it has already been shipped to the sales outlet. HP had to reimburse its sales partners for the difference in price for any unsold units so its partners wouldn't incur a loss. Given how fast PC products lose their value, excess inventory can result in large protection costs. A third inventory-related cost in the PC business is the "product return cost." This is the cost of a full refund HP paid its distributors when unsold products were returned; essentially it is a 100% price protection cost. In some cases sales partners and distributors returned excess unsold inventory valued at more than 10% of a product's revenue. The fourth inventory cost is "obsolescence cost," which is the cost of writing-off unsold products in inventory after the life of the product ends. Because PC products' life cycles are so short, there is the potential for large costs if excessive inventories are held. Related costs include price discounts for products that are about to be discontinued and the marketing costs to quickly reduce inventory. The Mobile Computing Division (which manufactures notebooks) was the first HP PC business unit to focus on all these inventory-related costs in redesigning its supply chain. The original supply chain consisted of a central manufacturing facility with local product configuration occurring at regional sites. (In this configuration about 40% of the total supply chain cost was related to inventory.) In the redesigned supply chain there is a central manufacturing facility, and products are air freighted directly to customers around the world. Positive results were immediate; in a 2-year period inventory-related costs dropped from almost 19% of total revenue to less than 4%, and the notebook division became profitable. As a result, all other HP PC operations began using these inventory costs to evaluate and redesign their supply chains.

Source: Based on "Inventory Driven Costs," by G. Callioni, X. de Montgros, R. Slagmulder, L. N. Van Wassenove, and L. Wright. *Harvard Business Review*, March 2005.

orders increases, the ordering cost increases). Costs incurred each time an order is made can include requisition costs, purchase orders, transportation and shipping, receiving, inspection, handling and placing in storage, and accounting and auditing.

Ordering costs generally react inversely to carrying costs. As the size of orders increases, fewer orders are required, thus reducing annual ordering costs. However, ordering larger amounts results in higher inventory levels and higher carrying costs. In general, as the order size increases, annual ordering costs decrease and annual carrying costs increase.

Shortage costs are incurred when customer demand cannot be met.

Shortage costs, also referred to as *stockout costs*, occur when customer demand cannot be met because of insufficient inventory on hand. If these shortages result in a permanent loss of sales for items demanded but not provided, shortage costs include the loss of profits. Shortages can also cause customer dissatisfaction and a loss of goodwill that can result in a permanent loss of customers and future sales. In some instances the inability to meet customer demand or lateness in meeting demand results in specified penalties in the form of price discounts or rebates. When demand is internal, a shortage can cause work stoppages in the production process and create delays, resulting in downtime costs and the cost of lost production (including indirect and direct production costs).

Costs resulting from immediate or future lost sales because demand could not be met are more difficult to determine than carrying or ordering costs. As a result, shortage costs are frequently subjective estimates and many times no more than educated guesses.

Shortages occur because it is costly to carry inventory in stock. As a result, shortage costs have an inverse relationship to carrying costs; as the amount of inventory on hand increases, the carrying cost increases, while shortage costs decrease.

*The purpose of **inventory management** is to determine how much and when to order.*

The objective of **inventory management** is to employ an inventory control system that will indicate how much should be ordered and when orders should take place to minimize the sum of the three inventory costs described here.

Inventory Control Systems

An inventory system is a structure for controlling the level of inventory by determining how much to order (the level of replenishment) and when to order. There are two basic types of inventory systems: a *continuous* (or *fixed–order quantity*) *system* and a *periodic* (or *fixed–time period*) *system*. The primary difference between the two systems is that in a continuous system, an order is placed for the same constant amount whenever the inventory on hand decreases to a certain level, whereas in a periodic system, an order is placed for a variable amount after an established passage of time.

Continuous Inventory Systems

In a continuous inventory system, a constant amount is ordered when inventory declines to a predetermined level.

In a *continuous inventory system*, alternatively referred to as a *perpetual system* or a *fixed–order quantity system*, a continual record of the inventory level for every item is maintained. Whenever the inventory on hand decreases to a predetermined level, referred to as the *reorder point*, a new order is placed to replenish the stock of inventory. The order that is placed is for a "fixed" amount that minimizes the total inventory carrying, ordering, and shortage costs. This fixed order quantity is called the *economic order quantity*; its determination will be discussed in greater detail in a later section.

A positive feature of a continuous system is that the inventory level is closely and continuously monitored so that management always knows the inventory status. This is especially advantageous for critical inventory items such as replacement parts or raw materials and supplies. However, the cost of maintaining a continual record of the amount of inventory on hand can also be a disadvantage of this type of system.

A simple example of a continuous inventory system is a ledger-style checkbook that many of us use on a daily basis. Our checkbook comes with 300 checks; after the 200th check has been used (and there are 100 left), there is an order form for a new batch of checks that has been inserted by the printer. This form, when turned in at the bank, initiates an order for a new batch of 300 checks from the printer. Many office inventory systems use "reorder" cards that are placed within stacks of stationery or at the bottom of a case of pens or paper clips to signal when a new order should be placed. If you look behind the items on a hanging rack in a Kmart store, you will see a card indicating that it is time to place an order for this item, for an amount indicated on the card.

Time Out for Ford Harris

The earliest published derivation of the classic economic lot size model is credited to Ford Harris of Westinghouse Corporation in 1915. His equation determined a minimum sum of inventory costs and setup costs, given demand that was known and constant and a rate of production that was assumed to be higher than demand.

A more sophisticated example of a continuous inventory system is a computerized checkout system with a laser scanner, used by many supermarkets and retail stores. In this system a laser scanner reads the Universal Product Code (UPC), or bar code, off the product package, and the transaction is instantly recorded and the inventory level updated. Such a system is not only quick and accurate, but it also provides management with continuously updated information on the status of inventory levels. Although not as publicly visible as supermarket systems, many manufacturing companies, suppliers, and distributors also use bar code systems and handheld laser scanners to inventory materials, supplies, equipment, in-process parts, and finished goods.

Because continuous inventory systems are much more common than periodic systems, models that determine fixed order quantities and the time to order will receive most of our attention in this chapter.

Periodic Inventory Systems

*In a **periodic inventory system**, an order is placed for a variable amount after a fixed passage of time.*

In a **periodic inventory system**, also referred to as a *fixed–time period system* or *periodic review system*, the inventory on hand is counted at specific time intervals—for example, every week or at the end of each month. After the amount of inventory in stock is determined, an order is placed for an amount that will bring inventory back up to a desired level. In this system the inventory level is not monitored at all during the time interval between orders, so it has the advantage of requiring little or no record keeping. However, it has the disadvantage of less direct control. This typically results in larger inventory levels for a periodic inventory system than in a continuous system, to guard against unexpected stockouts early in the fixed period. Such a system also requires that a new order quantity be determined each time a periodic order is made.

Periodic inventory systems are often found at a college or university bookstore. Textbooks are normally ordered according to a periodic system, wherein a count of textbooks in stock (for every course) is made after the first few weeks or month during the semester or quarter. An order for new textbooks for the next semester is then made according to estimated course enrollments for the next term (i.e., demand) and the number remaining in stock. Smaller retail stores, drugstores, grocery stores, and offices often use periodic systems; the stock level is checked every week or month, often by a vendor, to see how much (if anything) should be ordered.

Economic Order Quantity Models

You will recall that in a continuous or fixed–order quantity system, when inventory reaches a specific level, referred to as the *reorder point*, a fixed amount is ordered. The most widely used and traditional means for determining how much to order in a continuous system is the **economic order quantity (EOQ)** model, also referred to as the economic lot size model.

EOQ is a continuous inventory system.

The function of the EOQ model is to determine the optimal order size that minimizes total inventory costs. There are several variations of the EOQ model, depending on the assumptions made about the inventory system. In this and following sections we will describe three model versions: the basic EOQ model, the EOQ model with noninstantaneous receipt, and the EOQ model with shortages.

The Basic EOQ Model

EOQ is the optimal order quantity that will minimize total inventory costs.

The simplest form of the economic order quantity model on which all other model versions are based is called the basic EOQ model. It is essentially a single formula for determining the optimal order size that minimizes the sum of carrying costs and ordering costs. The model formula is derived under a set of simplifying and restrictive assumptions, as follows:

Assumptions of the EOQ model include constant demand, no shortages, constant lead time, and instantaneous order receipt.

- Demand is known with certainty and is relatively constant over time.
- No shortages are allowed.
- Lead time for the receipt of orders is constant.
- The order quantity is received all at once.

The graph in Figure 16.1 reflects these basic model assumptions.

FIGURE 16.1
The inventory order cycle

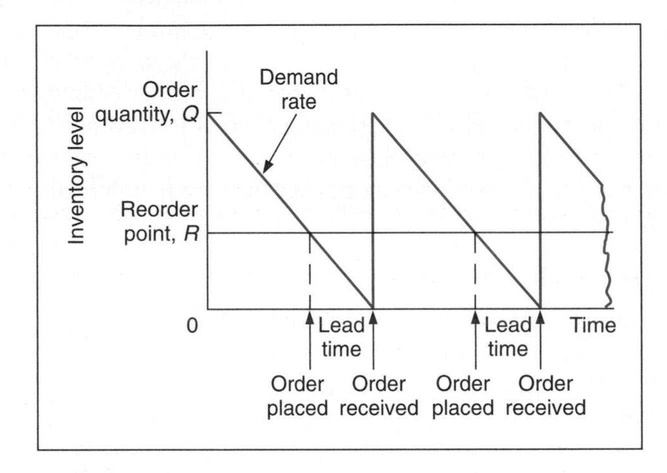

Figure 16.1 describes the continuous inventory order cycle system inherent in the EOQ model. An order quantity, Q, is received and is used up over time at a constant rate. When the inventory level decreases to the reorder point, R, a new order is placed, and a period of time, referred to as the *lead time*, is required for delivery. The order is received all at once, just at the moment when demand depletes the entire stock of inventory (and the inventory level reaches zero), thus allowing no shortages. This cycle is continuously repeated for the same order quantity, reorder point, and lead time.

As we mentioned earlier, Q is the order size that minimizes the sum of carrying costs and holding costs. These two costs react inversely to each other in response to an increase in the order size. As the order size increases, fewer orders are required, causing the ordering cost to decline, whereas the average amount of inventory on hand increases, resulting in an increase in carrying costs. Thus, in effect, the optimal order quantity represents a compromise between these two conflicting costs.

Carrying Cost

Carrying cost is usually expressed on a per-unit basis for some period of time (although it is sometimes given as a percentage of average inventory). Traditionally, the carrying cost is referred to on an annual basis (i.e., per year).

The total carrying cost is determined by the amount of inventory on hand during the year. The amount of inventory available during the year is illustrated in Figure 16.2.

In Figure 16.2, Q represents the size of the order needed to replenish inventory, which is what a manager wants to determine. The line connecting Q to time, t, in our graph represents the rate at which inventory is depleted, or *demand*, during the time period, t. Demand is assumed to be *known with certainty* and is thus constant, which explains why the line representing demand

FIGURE 16.2
Inventory usage

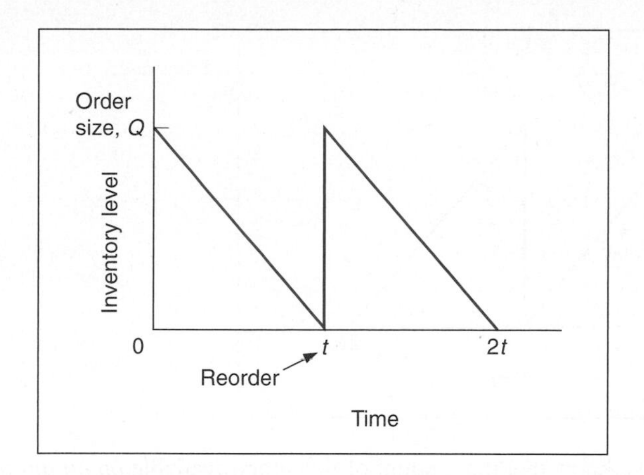

is straight. Also, notice that inventory never goes below zero; shortages do not exist. In addition, when the inventory level does reach zero, it is assumed that an order arrives immediately after an infinitely small passage of time, a condition referred to as **instantaneous receipt**. This is a simplifying assumption that we will maintain for the moment.

Instantaneous receipt *means inventory is received almost immediately after it is ordered.*

Referring to Figure 16.2, we can see that the amount of inventory is Q, the size of the order, for an infinitely small period of time because Q is always being depleted by demand. Similarly, the amount of inventory is zero for an infinitely small period of time because the only time there is no inventory is at the specific time t. Thus, the amount of inventory available is somewhere between these two extremes. A logical deduction is that the amount of inventory available is the *average inventory* level, defined as

$$\text{average inventory} = \frac{Q}{2}$$

To verify this relationship, we can specify any number of points—values of Q—over the entire time period, t, and divide by the number of points. For example, if $Q = 5{,}000$, the six points designated from 5,000 to 0, as shown in Figure 16.3, are summed and divided by 6:

$$\begin{aligned}\text{average inventory} &= \frac{5{,}000 + 4{,}000 + 3{,}000 + 2{,}000 + 1{,}000 + 0}{6} \\ &= 2{,}500\end{aligned}$$

FIGURE 16.3
Levels of *Q*

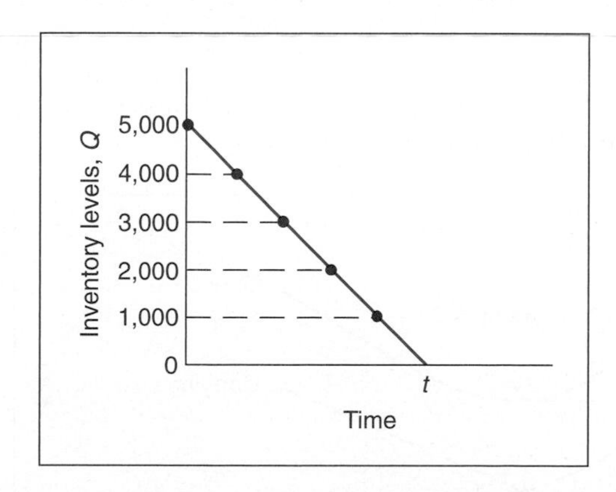

Alternatively, we can sum just the two extreme points (which also encompass the range of time, t) and divide by 2. This also equals 2,500. This computation is the same, in principle, as adding Q and 0 and dividing by 2, which equals $Q/2$. This relationship for average inventory is maintained, regardless of the size of the order, Q, or the frequency of orders (i.e., the time period, t). Thus, the average inventory on an *annual basis* is also $Q/2$, as shown in Figure 16.4.

FIGURE 16.4
Annual average inventory

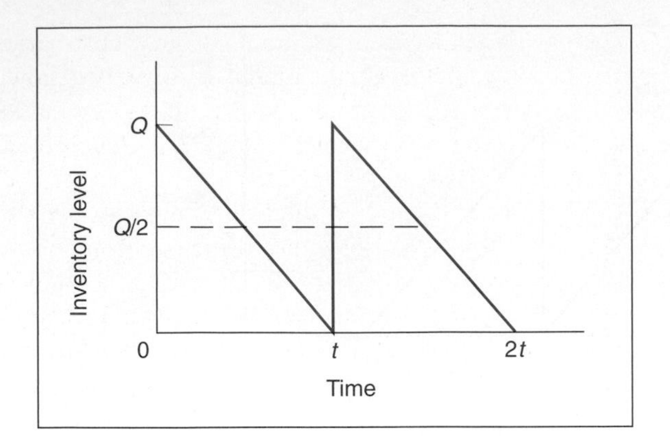

Now that we know that the amount of inventory available *on an annual basis* is the average inventory, $Q/2$, we can determine the total annual carrying cost by multiplying the average number of units in inventory by the carrying cost per unit per year, C_c:

$$\text{annual carrying cost} = C_c \frac{Q}{2}$$

Ordering Cost

The total annual ordering cost is computed by multiplying the cost per order, designated as C_o, by the number of orders per year. Because annual demand is assumed to be known and constant, the number of orders will be D/Q, where Q is the order size:

$$\text{annual ordering cost} = C_o \frac{D}{Q}$$

The only variable in this equation is Q; both C_o and D are constant parameters. In other words, demand is known with certainty. Thus, the relative magnitude of the ordering cost is dependent upon the order size.

Total Inventory Cost

The total annual inventory cost is simply the sum of the ordering and carrying costs:

$$TC = C_o \frac{D}{Q} + C_c \frac{Q}{2}$$

These cost functions are shown in Figure 16.5. Notice the inverse relationship between ordering cost and carrying cost, resulting in a convex total cost curve.

FIGURE 16.5
The EOQ cost model

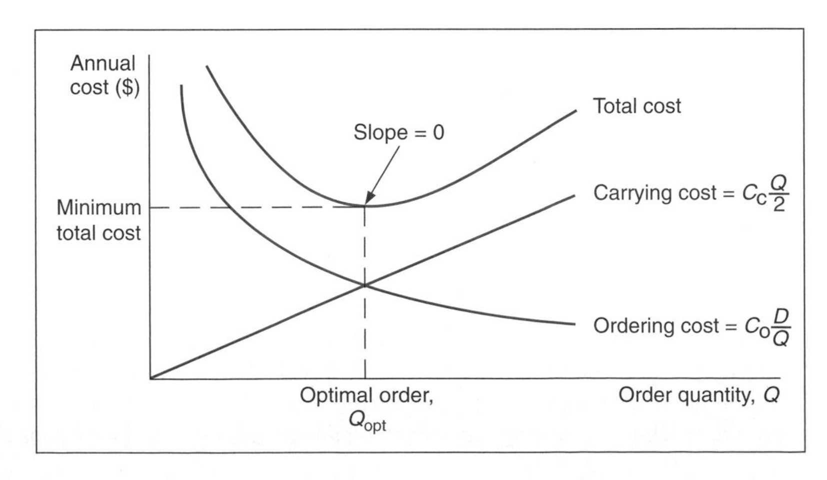

Observe the general upward trend of the total carrying cost curve. As the order size Q (shown on the horizontal axis) increases, the total carrying cost (shown on the vertical axis) increases. This is logical because larger orders will result in more units carried in inventory. Next, observe the ordering cost curve in Figure 16.5. As the order size, Q, increases, the ordering cost *decreases* (just the opposite of what occurred with the carrying cost). This is logical, because an increase in the size of the orders will result in fewer orders being placed each year. Because one cost increases as the other decreases, the result of summing the two costs is a convex total cost curve.

The optimal value of Q corresponds to the lowest point on the total cost curve.

The optimal order quantity occurs at the point in Figure 16.5 where the total cost curve is at a minimum, which also coincides exactly with the point where the ordering cost curve intersects with the carrying cost curve. This enables us to determine the optimal value of Q by equating the two cost functions and solving for Q, as follows:

$$C_o \frac{D}{Q} = C_c \frac{Q}{2}$$

$$Q^2 = \frac{2C_oD}{C_c}$$

$$Q_{opt} = \sqrt{\frac{2C_oD}{C_c}}$$

Alternatively, the optimal value of Q can be determined by differentiating the total cost curve with respect to Q, setting the resulting function equal to zero (the slope at the minimum point on the total cost curve), and solving for Q, as follows:

$$TC = C_o \frac{D}{Q} + C_c \frac{Q}{2}$$

$$\frac{\delta TC}{\delta Q} = -\frac{C_oD}{Q^2} + \frac{C_c}{2}$$

$$0 = -\frac{C_oD}{Q^2} + \frac{C_c}{2}$$

$$Q_{opt} = \sqrt{\frac{2C_oD}{C_c}}$$

The total minimum cost is determined by substituting the value for the optimal order size, Q_{opt}, into the total cost equation:

$$TC_{min} = C_o \frac{D}{Q_{opt}} + C_o \frac{Q_{opt}}{2}$$

We will use the following example to demonstrate how the optimal value of Q is computed. The I-75 Carpet Discount Store in north Georgia stocks carpet in its warehouse and sells it through an adjoining showroom. The store keeps several brands and styles of carpet in stock; however, its biggest seller is Super Shag carpet. The store wants to determine the optimal order size and total inventory cost for this brand of carpet, given an estimated annual demand of 10,000 yards of carpet, an annual carrying cost of \$0.75 per yard, and an ordering cost of \$150. The store would also like to know the number of orders that will be made annually and the time between orders (i.e., the order cycle), given that the store is open every day except Sunday, Thanksgiving Day, and Christmas Day (that is not on a Sunday).

We can summarize the model parameters as follows:

$$C_c = \$0.75$$

$$C_o = \$150$$

$$D = 10{,}000 \text{ yd.}$$

The optimal order size is computed as follows:

$$Q_{opt} = \sqrt{\frac{2C_oD}{C_c}}$$

$$= \sqrt{\frac{2(150)(10{,}000)}{(0.75)}}$$

$$Q_{opt} = 2{,}000 \text{ yd.}$$

The total annual inventory cost is determined by substituting Q_{opt} into the total cost formula, as follows:

$$TC_{min} = C_o\frac{D}{Q_{opt}} + C_c\frac{Q_{opt}}{2}$$

$$= (150)\frac{10{,}000}{2{,}000} + (0.75)\frac{(2{,}000)}{2}$$

$$= \$750 + 750$$

$$TC_{min} = \$1{,}500$$

The number of orders per year is computed as follows:

$$\text{number of orders per year} = \frac{D}{Q_{opt}} = \frac{10{,}000}{2{,}000} = 5$$

Given that the store is open 311 days annually (365 days minus 52 Sundays, plus Thanksgiving and Christmas), the order cycle is determined as follows:

$$\text{order cycle time} = \frac{311 \text{ days}}{D/Q_{opt}} = \frac{311}{5} = 62.2 \text{ store days}$$

The EOQ model is robust; because Q is a square root, errors in the estimation of D, C_c, *and* C_o *are dampened.*

It should be noted that the optimal order quantity determined in this example, and in general, is an approximate value because it is based on estimates of carrying and ordering costs as well as uncertain demand (although all these parameters are treated as known, certain values in the EOQ model). Thus, in practice it is acceptable to round off the Q values to the nearest whole number. The precision of a decimal place generally is neither necessary nor appropriate. In addition, because the optimal order quantity is computed from a square root, errors or variations in the cost parameters and demand tend to be dampened. For instance, if the order cost had actually been a third higher, or \$200, the resulting optimal order size would have varied by about 15% (i.e., 2,390 yards instead of 2,000 yards). In addition, variations in both inventory costs will tend to offset each other because they have an inverse relationship. As a result, the EOQ model is relatively robust, or resilient to errors in the cost estimates and demand, which has tended to enhance its popularity.

EOQ Analysis Over Time

One aspect of inventory analysis that can be confusing is the time frame encompassed by the analysis. Therefore, we will digress for just a moment to discuss this aspect of EOQ analysis.

Recall that previously we developed the EOQ model "regardless of order size, Q, and time, t." Now we will verify this condition. We will do so by developing our EOQ model on a *monthly basis*. First, demand is equal to 833.3 yards per month (which we determined by dividing the annual demand of 10,000 yards by 12 months). Next, by dividing the annual carrying cost, C_c, of \$0.75 by 12, we get the monthly (per-unit) carrying cost: $C_c = \$0.0625$. (The ordering cost of \$150 is not related to time.) We thus have the values

$$D = 833.3 \text{ yd. per month}$$

$$C_c = \$0.0625 \text{ per yd. per month}$$

$$C_o = \$150 \text{ per order}$$

which we can substitute into our EOQ formula:

$$Q_{\text{opt}} = \sqrt{\frac{2C_oD}{C_c}}$$

$$= \sqrt{\frac{2(150)(833.3)}{(0.0625)}} = 2{,}000 \text{ yd.}$$

This is the same optimal order size that we determined on an annual basis. Now we will compute total monthly inventory cost:

$$\text{total monthly inventory cost} = C_c\frac{Q_{\text{opt}}}{2} + C_o\frac{D}{Q_{\text{opt}}}$$

$$= (\$0.0625)\frac{(2{,}000)}{2} + (\$150)\frac{(833.3)}{(2{,}000)}$$

$$= \$125 \text{ per month}$$

To convert this monthly total cost to an annual cost, we multiply it by 12 (months):

$$\text{total annual inventory cost} = (\$125)(12) = \$1{,}500$$

This brief example demonstrates that regardless of the time period encompassed by EOQ analysis, the economic order quantity (Q_{opt}) is the same.

The EOQ Model with Noninstantaneous Receipt

*The **noninstantaneous receipt model** relaxes the assumption that Q is received all at once.*

A variation of the basic EOQ model is achieved when the assumption that orders are received all at once is relaxed. This version of the EOQ model is known as the **noninstantaneous receipt model**, also referred to as the *gradual usage*, or *production lot size, model.* In this EOQ variation, the order quantity is received gradually over time and the inventory level is depleted at the same time it is being replenished. This is a situation most commonly found when the inventory user is also the producer, as, for example, in a manufacturing operation where a part is produced to use in a larger assembly. This situation can also occur when orders are delivered gradually over time or the retailer and producer of a product are one and the same. The noninstantaneous receipt model is illustrated graphically in Figure 16.6, which highlights the difference between this variation and the basic EOQ model.

FIGURE 16.6
The EOQ model with noninstantaneous order receipt

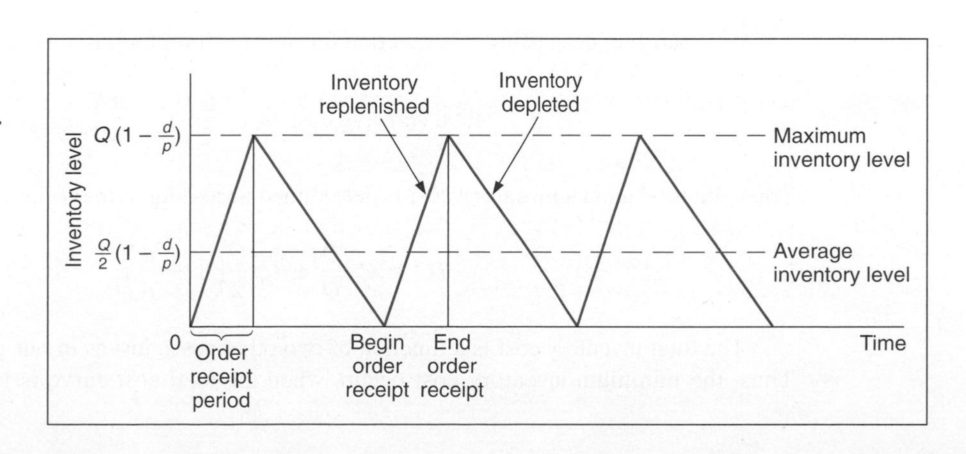

The ordering cost component of the basic EOQ model does not change as a result of the gradual replenishment of the inventory level because it is dependent only on the number of annual orders. However, the carrying cost component is not the same for this model variation because average inventory is different. In the basic EOQ model, average inventory was half the maximum inventory level, or $Q/2$, but in this variation, the maximum inventory level is not simply Q; it is an amount somewhat lower than Q, adjusted for the fact that the order quantity is depleted during the order receipt period.

To determine the average inventory level, we define the following parameters that are unique to this model:

p = daily rate at which the order is received over time, also known as the *production rate*

d = the daily rate at which inventory is demanded

The demand rate cannot exceed the production rate because we are still assuming that no shortages are possible, and if $d = p$, then there is no order size because items are used as fast as they are produced. Thus, for this model, the production rate must exceed the demand rate, or $p > d$.

Figure 16.6 shows that the time required to receive an order is the order quantity divided by the rate at which the order is received, or Q/p. For example, if the order size is 100 units and the production rate, p, is 20 units per day, the order will be received in 5 days. The amount of inventory that will be depleted or used up during this time period is determined by multiplying by the demand rate, or $(Q/p)d$. For example, if it takes 5 days to receive the order and during this time inventory is depleted at the rate of 2 units per day, then a total of 10 units is used. As a result, the maximum amount of inventory that is on hand is the order size minus the amount depleted during the receipt period, computed as follows and shown earlier in Figure 16.6:

$$\text{maximum inventory level} = Q - \frac{Q}{p}d = Q\left(1 - \frac{d}{p}\right)$$

Because this is the maximum inventory level, the average inventory level is determined by dividing this amount by 2, as follows:

$$\text{average inventory level} = \frac{1}{2}\left[Q\left(1 - \frac{d}{p}\right)\right] = \frac{Q}{2}\left(1 - \frac{d}{p}\right)$$

The total carrying cost, using this function for average inventory, is

$$\text{total carrying cost} = C_c\frac{Q}{2}\left(1 - \frac{d}{p}\right)$$

Thus, the total annual inventory cost is determined according to the following formula:

$$TC = C_o\frac{D}{Q} + C_c\frac{Q}{2}\left(1 - \frac{d}{p}\right)$$

The total inventory cost is a function of two other costs, just as in our previous EOQ model. Thus, the minimum inventory cost occurs when the total cost curve is lowest and where the carrying cost curve and ordering cost curve intersect (see Figure 16.5). Therefore, to find optimal Q_{opt}, we equate total carrying cost with total ordering cost:

$$C_c \frac{Q}{2}\left(1 - \frac{d}{p}\right) = C_o \frac{D}{Q}$$

$$C_c \frac{Q^2}{2}\left(1 - \frac{d}{p}\right) = C_o D$$

$$Q_{opt} = \sqrt{\frac{2C_o D}{C_c(1 - d/p)}}$$

For our previous example we will now assume that the I-75 Carpet Discount Store has its own manufacturing facility, in which it produces Super Shag carpet. We will further assume that the ordering cost, C_o, is the cost of setting up the production process to make Super Shag carpet. Recall that $C_c = \$0.75$ per yard and $D = 10{,}000$ yards per year. The manufacturing facility operates the same days the store is open (i.e., 311 days) and produces 150 yards of the carpet per day. The optimal order size, the total inventory cost, the length of time to receive an order, the number of orders per year, and the maximum inventory level are computed as follows:

$$\begin{aligned} C_o &= \$150 \\ C_c &= \$0.75 \text{ per unit} \\ D &= 10{,}000 \text{ yd.} \\ d &= \frac{10{,}000}{311} = 32.2 \text{ yd. per day} \\ p &= 150 \text{ yd. per day} \end{aligned}$$

The optimal order size is determined as follows:

$$\begin{aligned} Q_{opt} &= \sqrt{\frac{2C_o D}{C_c\left(1 - \frac{d}{p}\right)}} \\ &= \sqrt{\frac{2(150)(10{,}000)}{0.75\left(1 - \frac{32.2}{150}\right)}} \\ Q_{opt} &= 2{,}256.8 \text{ yd.} \end{aligned}$$

This value is substituted into the following formula to determine total minimum annual inventory cost:

$$\begin{aligned} TC_{min} &= C_o \frac{D}{Q} + C_c \frac{Q}{2}\left(1 - \frac{d}{p}\right) \\ &= (150)\frac{(10{,}000)}{(2{,}256.8)} + (0.75)\frac{(2{,}256.8)}{2}\left(1 - \frac{32.2}{150}\right) \\ &= \$1{,}329 \end{aligned}$$

The length of time to receive an order for this type of manufacturing operation is commonly called the length of the *production run*. It is computed as follows:

$$\text{production run length} = \frac{Q}{p} = \frac{2{,}256.8}{150} = 15.05 \text{ days}$$

The number of orders per year is actually the number of production runs that will be made, computed as follows:

$$\text{number of production runs} = \frac{D}{Q} = \frac{10{,}000}{2{,}256.8} = 4.43 \text{ runs}$$

Finally, the maximum inventory level is computed as follows:

$$\text{maximum inventory level} = Q\left(1 - \frac{d}{p}\right) = 2{,}256.8\left(1 - \frac{32.2}{150}\right) = 1{,}755 \text{ yd.}$$

The EOQ Model with Shortages

The EOQ model with shortages relaxes the assumption that shortages cannot exist.

One of the assumptions of our basic EOQ model is that shortages and back ordering are not allowed. The third model variation that we will describe, the EOQ model with shortages, relaxes this assumption. However, it will be assumed that all demand not met because of inventory shortage can be back-ordered and delivered to the customer later. Thus, all demand is eventually met. The EOQ model with shortages is illustrated in Figure 16.7.

FIGURE 16.7
The EOQ model with shortages

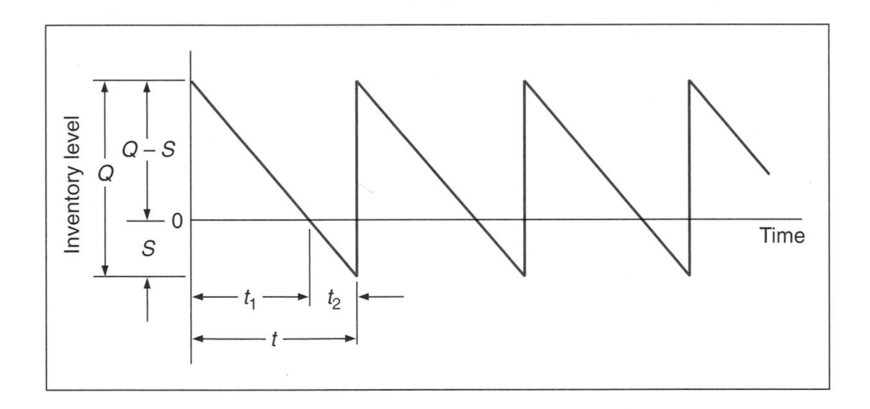

Because back-ordered demand, or shortages, S, are filled when inventory is replenished, the maximum inventory level does not reach Q, but instead a level equal to $Q - S$. It can be seen from Figure 16.7 that the amount of inventory on hand ($Q - S$) decreases as the amount of the shortage increases, and vice versa. Therefore, the cost associated with shortages, which we described earlier in this chapter as primarily the cost of lost sales and lost customer goodwill, has an inverse relationship to carrying costs. As the order size, Q, increases, the carrying cost increases and the shortage cost declines. This relationship between carrying and shortage cost as well as ordering cost is shown in Figure 16.8.

We will forgo the lengthy derivation of the individual cost components of the EOQ model with shortages, which requires the application of plane geometry to the graph in Figure 16.8.

FIGURE 16.8
Cost model with shortages

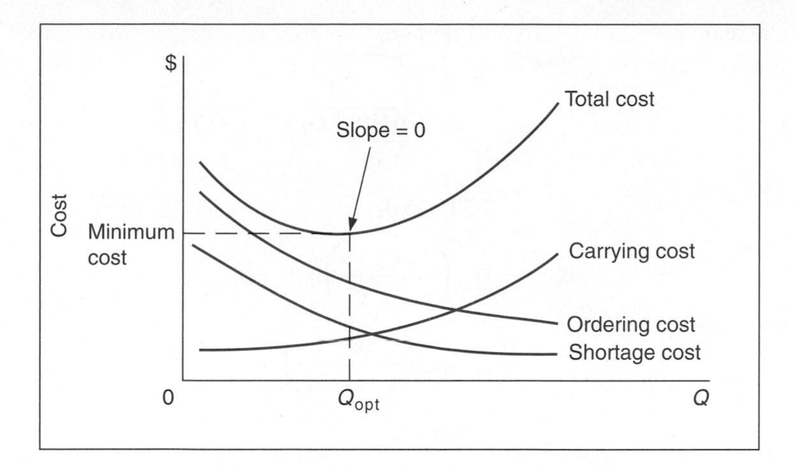

The individual cost functions are provided as follows, where S equals the shortage level and C_s equals the annual per-unit cost of shortages:

$$\text{total shortage cost} = C_s \frac{S^2}{2Q}$$

$$\text{total carrying cost} = C_c \frac{(Q - S)^2}{2Q}$$

$$\text{total ordering cost} = C_o \frac{D}{Q}$$

Combining these individual cost components results in the total inventory cost formula:

$$TC = C_s \frac{S^2}{2Q} + C_c \frac{(Q - S)^2}{2Q} + C_o \frac{D}{Q}$$

You will notice in Figure 16.8 that the three cost component curves do not intersect at a common point, as was the case in the basic EOQ model. As a result, the only way to determine the optimal order size *and the optimal shortage level, S,* is to differentiate the total cost function with respect to Q and S, set the two resulting equations equal to zero, and solve them simultaneously. Doing so results in the following formulas for the optimal order quantity and shortage level:

$$Q_{opt} = \sqrt{\frac{2C_o D}{C_c}\left(\frac{C_s + C_c}{C_s}\right)}$$

$$S_{opt} = Q_{opt}\left(\frac{C_c}{C_c + C_s}\right)$$

For example, we will now assume that the I-75 Carpet Discount Store allows shortages and the shortage cost, C_s, is \$2 per yard per year. All other costs and demand remain the same (C_c = \$0.75, C_o = \$150, and D = 10,000 yd.). The optimal order size and shortage level and total minimum annual inventory cost are computed as follows:

$$\begin{aligned} C_o &= \$150 \\ C_c &= \$0.75 \text{ per yd.} \\ C_s &= \$2 \text{ per yd.} \\ D &= 10{,}000 \text{ yd.} \end{aligned}$$

$$
\begin{aligned}
Q_{opt} &= \sqrt{\frac{2C_oD}{C_c}\left(\frac{C_s + C_c}{C_s}\right)} \\
&= \sqrt{\frac{2(150)(10{,}000)}{0.75}\left(\frac{2 + 0.75}{2}\right)} \\
&= 2{,}345.2 \text{ yd.} \\
S_{opt} &= Q_{opt}\left(\frac{C_c}{C_c + C_s}\right) \\
&= 2{,}345.2\left(\frac{0.75}{2 + 0.75}\right) \\
&= 639.6 \text{ yd.} \\
TC_{min} &= \frac{C_sS^2}{2Q} + \frac{C_c(Q - S)^2}{2Q} + C_o\frac{D}{Q} \\
&= \frac{(2)(639.6)^2}{2(2{,}345.2)} + \frac{(0.75)(1{,}705.6)^2}{2(2{,}345.2)} + \frac{(150)(10{,}000)}{2{,}345.2} \\
&= \$174.44 + 465.16 + 639.60 \\
&= \$1{,}279.20
\end{aligned}
$$

Several additional parameters of the EOQ model with shortages can be computed for this example, as follows:

$$
\begin{aligned}
\text{number of orders} &= \frac{D}{Q} = \frac{10{,}000}{2{,}345.2} = 4.26 \text{ orders per year} \\
\text{maximum inventory level} &= Q - S = 2{,}345.2 - 639.6 = 1{,}705.6 \text{ yd.}
\end{aligned}
$$

The time between orders, identified as t in Figure 16.7, is computed as follows:

$$
t = \frac{\text{days per year}}{\text{number of orders}} = \frac{311}{4.26} = 73.0 \text{ days between orders}
$$

The time during which inventory is on hand, t_1 in Figure 16.7, and the time during which there is a shortage, t_2 in Figure 16.7, during each order cycle can be computed using the following formulas:

$$
\begin{aligned}
t_1 &= \frac{Q - S}{D} \\
&= \frac{2{,}345.2 - 639.6}{10{,}000} \\
&= 0.171 \text{ year, or } 53.2 \text{ days} \\
t_2 &= \frac{S}{D} \\
&= \frac{639.6}{10{,}000} \\
&= 0.064 \text{ year, or } 19.9 \text{ days}
\end{aligned}
$$

Management Science Application

Determining Inventory Ordering Policy at Dell

Dell Inc., the world's largest computer-systems company, bypasses retailers and sells directly to customers via phone or the Internet. After an order is processed, it is sent to one of its assembly plants in Austin, Texas, where the product is built, tested, and packaged within 8 hours.

Dell carries very little components inventory itself. Technology changes occur so fast that holding inventory can be a huge liability; some components lose 0.5% to 2.0% of their value per week. In addition, many of Dell's suppliers are located in Southeast Asia, and their shipping times to Austin range from 7 days for air transport to 30 days for water and ground transport. To compensate for these factors, Dell's suppliers keep inventory in small warehouses called "revolvers" (for revolving inventory), which are a few miles from Dell's assembly plants. Dell keeps very little inventory at its own plants, so it withdraws inventory from the revolvers every few hours, and most of Dell's suppliers deliver to their revolvers three times per week.

The cost of carrying inventory by Dell's suppliers is ultimately reflected in the final price of a computer. Thus, in order to maintain a competitive price advantage in the market, Dell strives to help its suppliers reduce inventory costs. Dell has a vendor-managed inventory (VMI) arrangement with its suppliers, who decide how much to order and when to send their orders to the revolvers. Dell's suppliers order in batches (to offset ordering costs), using a continuous ordering system with a batch order size, Q, and a reorder point, R, where R is the sum of the inventory on order and a safety stock. The order size estimate, based on long-term data and forecasts, is held constant. Dell sets target inventory levels for its suppliers—typically 10 days of inventory—and keeps track of how much suppliers deviate from these targets and reports this information back to suppliers so that they can make adjustments accordingly.

Source: Based on R. Kapuscinski, R. Zhang, P. Carbonneau, R. Moore, and B. Reeves, "Inventory Decisions in Dell's Supply Chain," *Interfaces* 34, no. 3 (May–June 2004): 191–205.

EOQ Analysis with QM for Windows

QM for Windows has modules for all the EOQ models we have presented, including the basic model, the noninstantaneous receipt model, and the model with shortages. To demonstrate the capabilities of this program, we will use our basic EOQ example, for which the solution output summary is shown in Exhibit 16.1.

EXHIBIT 16.1

Inventory Results

I-75 Carpet Discount Store Solution

Parameter	Value		Parameter	Value
Demand rate(D)	10,000		Optimal order quantity (Q*)	2,000
Setup/Ordering cost(S)	150		Maximum Inventory Level (Imax)	2,000
Holding cost(H)	.75		Average inventory	1,000
Unit cost	0		Orders per period(year)	5
			Annual Setup cost	750
			Annual Holding cost	750
			Unit costs (PD)	0
			Total Cost	1,500

EOQ Analysis with Excel and Excel QM

Exhibit 16.2 shows an Excel spreadsheet set up to perform EOQ analysis for our noninstantaneous receipt model example. The parameters of the model have been input in cells **D3:D8**, and all the formulas for optimal Q, total cost, and so on have been embedded in cells **D10:D14**. Notice that the formula for computing optimal Q in cell D10 is shown on the formula bar at the top of the screen.

EXHIBIT 16.2

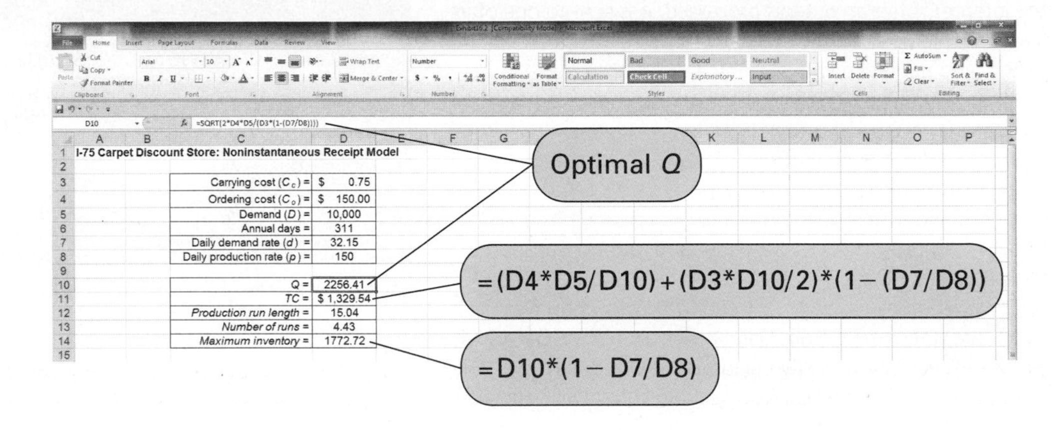

In Chapter 1, we introduced Excel QM, a set of spreadsheet macros that we also used in several other chapters. Excel QM includes a set of spreadsheet macros for "Inventory" that includes EOQ analysis. After Excel QM is activated, the "Excel QM" menu is accessed by clicking on "Add-Ins" on the menu bar at the top of the spreadsheet. Clicking on "Inventory" from this menu results in a Spreadsheet Initialization window, in which you enter the problem title and the form of holding (or carrying) cost. Clicking on "OK" will result in the spreadsheet shown in Exhibit 16.3. Initially, this spreadsheet will have example values in the data cells **B8:B13**. Thus, the first step in using this macro is to type in cells **B8:B13** the data for the noninstantaneous receipt model for our I-75 Carpet Discount Store problem. The model results are computed automatically in cells **B16:B26** from formulas already embedded in the spreadsheet.

EXHIBIT 16.3

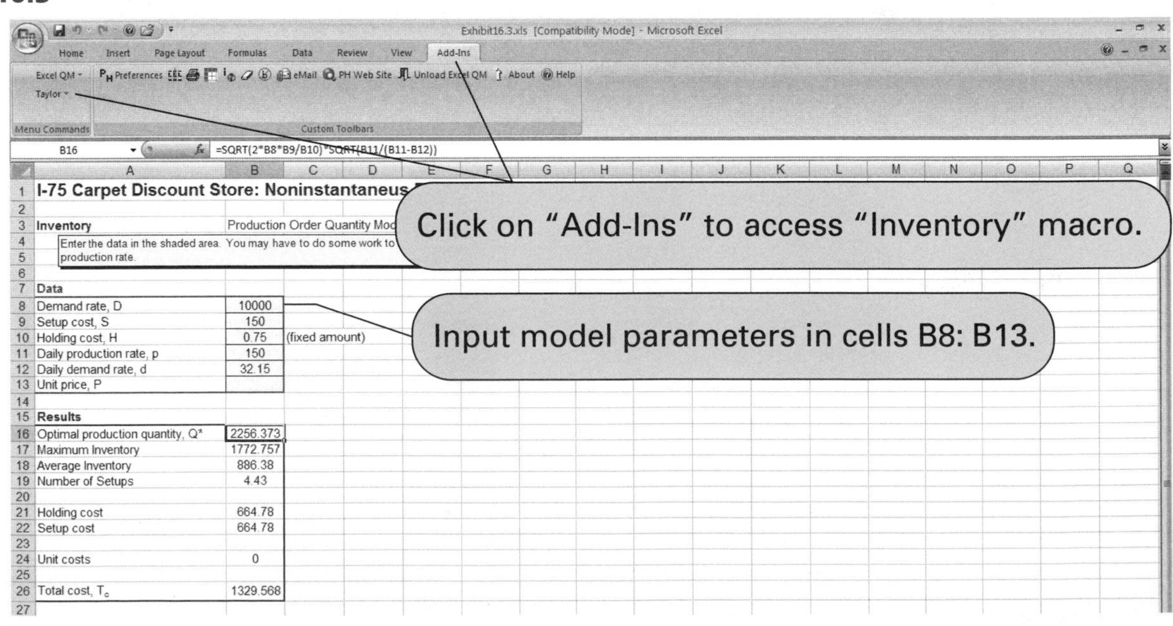

Quantity Discounts

It is often possible for a customer to receive a price discount on an item if a predetermined number of units are ordered. For example, occasionally in the back of a magazine you might see an advertisement for a firm that will produce a coffee mug (or hat) with a company or organizational logo on it, and the price will be $5 per mug if you purchase 100, $4 per mug if you purchase 200, or $3 per mug if you purchase 500 or more. Many manufacturing companies receive price discounts for ordering materials and supplies in high volume, and retail stores receive price discounts for ordering merchandise in large quantities.

Determining whether an order size with a discount is more cost-effective than optimal Q.

The basic EOQ model can be used to determine the optimal order size with quantity discounts; however, the application of the model is slightly altered. The total inventory cost function must now include the purchase price for the order, as follows:

$$TC = C_o \frac{D}{Q} + C_c \frac{Q}{2} + PD$$

where
P = per unit price of the item
D = annual demand

Purchase price was not considered as part of our basic EOQ formulation earlier because it had no real impact on the optimal order size. PD in the foregoing formula is a constant value that would not alter the basic shape of the total cost curve (i.e., the minimum point on the cost curve would still be at the same location, corresponding to the same value of Q). Thus, the optimal order size will be the same, no matter what the purchase price. However, when a discount price is available, it is associated with a specific order size that may be different from the optimal order size, and the customer must evaluate the trade-off between possibly higher carrying costs with the discount quantity versus EOQ cost. As a result, the purchase price does influence the order size decision when a discount is available.

Quantity discounts are evaluated with constant C_c and as a percentage of price.

Quantity discounts can be evaluated using the basic EOQ model under two different scenarios—with constant carrying costs and with carrying costs as a percentage of the purchase price. It is not uncommon for carrying costs to be determined as a percentage of purchase price, although it was not considered as such in our previous basic EOQ model. Carrying cost very well could have been a percentage of purchase price, but it was reflected as a constant value, C_c, in the basic EOQ model because the purchase price was not part of the EOQ formula. However, in the case of a quantity discount, carrying cost will vary with the change in price if it is computed as a percentage of purchase price.

Quantity Discounts with Constant Carrying Costs

In the EOQ cost model with constant carrying costs, the optimal order size, Q_{opt}, is the same, regardless of the discount price. Although total cost decreases with each discount in price because ordering and carrying cost are constant, the optimal order size, Q_{opt}, does not change. The total cost with Q_{opt} must be compared with any lower total cost with a discount price to see which is the minimum.

The following example will illustrate the evaluation of an EOQ model with a quantity discount when the carrying cost is a constant value. Comptek Computers wants to reduce a large stock of personal computers it is discontinuing. It has offered the University Bookstore at Tech a quantity discount pricing schedule if the store will purchase the personal computers in volume, as follows:

Quantity	*Price*
1–49	$1,400
50–89	1,100
90+	900

The annual carrying cost for the bookstore for a computer is $190, the ordering cost is $2,500, and annual demand for this particular model is estimated to be 200 units. The bookstore wants to determine whether it should take advantage of this discount or order the basic EOQ order size.

First, determine both the optimal order size and the total cost by using the basic EOQ model:

$$C_o = \$2{,}500$$
$$C_c = \$190 \text{ per unit}$$
$$D = 200$$
$$Q_{opt} = \sqrt{\frac{2C_oD}{C_c}} = \sqrt{\frac{2(2{,}500)(200)}{190}}$$
$$Q_{opt} = 72.5$$

This order size is eligible for the first discount of $1,100; therefore, this price is used to compute total cost, as follows:

$$TC = \frac{C_oD}{Q_{opt}} + C_c\frac{Q_{opt}}{2} + PD = \frac{(2{,}500)(200)}{(72.5)} + (190)\frac{(72.5)}{2} + (1{,}100)\,(200)$$
$$TC_{min} = \$233{,}784$$

Because there is a discount for an order size larger than 72.5, this total cost of $233,784 must be compared with total cost with an order size of 90 and a price of $900, as follows:

$$TC = \frac{C_oD}{Q} + C_c\frac{Q}{2} + PD = \frac{(2{,}500)(200)}{(90)} + \frac{(190)(90)}{2} + (900)(200) = \$194{,}105$$

Because this total cost is lower ($\$194{,}105 < \$233{,}784$), the maximum discount price should be taken and 90 units ordered.

Quantity Discounts with Constant Carrying Costs as a Percentage of Price

The difference between the model in the previous section and the quantity discount model with carrying cost as a percentage of price is that there is a different optimal order size, Q_{opt}, for each price discount. This requires that the optimal order size with a discount be determined a little differently from the case for a constant carrying cost.

The optimal order size and total cost are determined by using the basic EOQ model for the case with no quantity discount. This total cost value is then compared with the various discount quantity order sizes to determine the minimum cost order size. However, once this minimum cost order size is determined, it must be compared with the EOQ-determined order size for the specific discount price because the EOQ order size, Q_{opt}, will change for every discount level.

Reconsider our previous example, except now assume that the annual carrying cost for a computer at the University Bookstore is 15% of the purchase price. Using the same discount pricing schedule, determine the optimal order size.

The annual carrying cost is determined as follows:

Quantity	*Price*	*Carrying Cost*
1–49	$1,400	1,400(.15) = $210
50–89	1,100	1,100(.15) = 165
90+	900	900(.15) = 135

$$C_o = \$2{,}500$$
$$D = 200 \text{ computers per year}$$

First, compute the optimal order size for the purchase price without a discount and with $C_c = \$210$, as follows:

$$Q_{opt} = \sqrt{\frac{2C_oD}{C_c}}$$
$$= \sqrt{\frac{2(2{,}500)(200)}{210}}$$
$$Q_{opt} = 69$$

Because this order size exceeds 49 units, it is not feasible for this price, and a lower total cost will automatically be achieved with the first price discount of $1,100. However, the optimal order size will be different for this price discount because carrying cost is no longer constant. Thus, the new order size is computed as follows:

$$Q_{opt} = \sqrt{\frac{2(2{,}500)(200)}{165}} = 77.8$$

This order size is the true optimum for this price discount instead of the 50 units required to receive the discount price; thus, it will result in the minimum total cost, computed as follows:

$$TC = \frac{C_oD}{Q} + C_c\frac{Q}{2} + PD$$
$$= \frac{(2{,}500)(200)}{77.8} + 165\frac{(77.8)}{2} + (1{,}100)(200)$$
$$= \$232{,}845$$

This cost must still be compared with the total cost for lowest discount price ($900) and order quantity (90 units), computed as follows:

$$TC = \frac{(2{,}500)(200)}{90} + \frac{(135)(90)}{2} + (900)(200)$$
$$= \$191{,}630$$

Because this total cost is lower ($\$191{,}630 < \$232{,}845$), the maximum discount price should be taken and 90 units ordered. However, as before, we still must check to see whether there is a new optimal order size for this discount that will result in an even lower cost. The optimal order size with $C_c = \$135$ is computed as follows:

$$Q_{opt} = \sqrt{\frac{2(2{,}500)(200)}{135}} = 86.1$$

Because this order size is less than the 90 units required to receive the discount, it is not feasible; thus, the optimal order size is 90 units.

Management Science Application

Quantity Discount Orders at Mars

Mars is one of the world's largest privately owned companies, with over $14 billion in annual sales. It has grown from making and selling buttercream candies door-to-door to a global business spanning 100 countries that includes food, pet care, beverage vending, and electronic payment systems. It produces such well-known products as Mars candies, M&M's, Snickers, and Uncle Ben's rice.

Mars relies on a small number of suppliers for each of the huge number of materials it uses in its products. One way that Mars purchases materials from its suppliers is through online electronic auctions, in which Mars buyers negotiate bids for orders from suppliers. The most important purchases are those of high value and large volume, for which the suppliers provide quantity discounts. The suppliers provide a pricing schedule that includes quantity ranges associated with each price level. Such quantity-discount auctions are tailored (by online brokers) for industries in which volume discounts are common, such as bulk chemicals and agricultural commodities.

A Mars buyer selects the bids that minimize total purchasing costs, subject to several rules: There must be a minimum and maximum number of suppliers so that Mars is not dependent on too few suppliers nor loses quality control by having too many; there must be a maximum amount purchased from each supplier to limit the influence of any one supplier; and a minimum amount must be ordered to avoid economically inefficient orders (i.e., less than a full truckload).

© Eric Carr/Alamy

Source: Based on G. Hohner, J. Rich, E. Ng, A. Davenport, J. Kalagnanam, H. Lee, and C. An, "Combinatorial and Quantity-Discount Procurement Auctions Benefit Mars, Incorporated and Its Suppliers," *Interfaces* 33, no. 1 (January–February 2003): 23–35.

Quantity Discount Model Solution with QM for Windows

QM for Windows has the capability to perform EOQ analysis with quantity discounts when carrying costs are constant. Exhibit 16.4 shows the solution summary for our University Bookstore example.

EXHIBIT 16.4

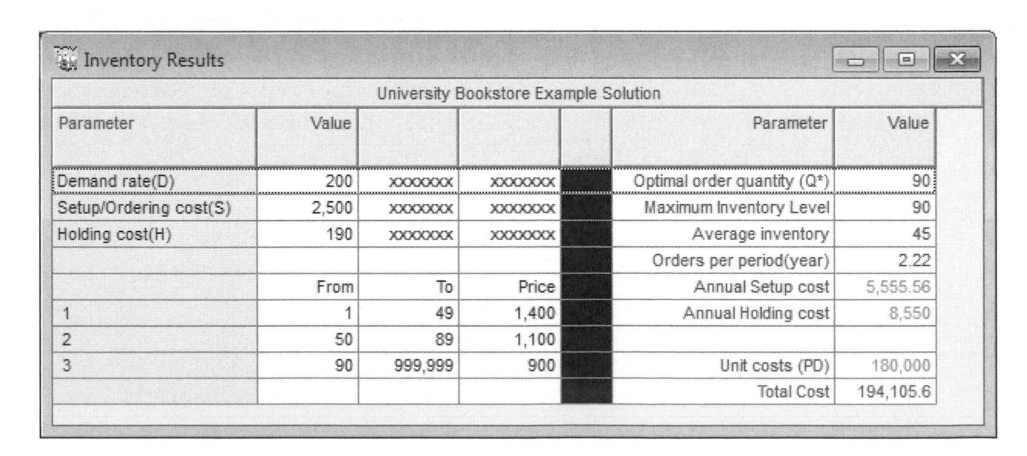

Inventory Results

University Bookstore Example Solution

Parameter	Value				Parameter	Value
Demand rate(D)	200	xxxxxxx	xxxxxxx		Optimal order quantity (Q*)	90
Setup/Ordering cost(S)	2,500	xxxxxxx	xxxxxxx		Maximum Inventory Level	90
Holding cost(H)	190	xxxxxxx	xxxxxxx		Average inventory	45
					Orders per period(year)	2.22
	From	To	Price		Annual Setup cost	5,555.56
1	1	49	1,400		Annual Holding cost	8,550
2	50	89	1,100			
3	90	999,999	900		Unit costs (PD)	180,000
					Total Cost	194,105.6

Quantity Discount Model Solution with Excel

It is also possible to use Excel to solve the quantity discount model with constant carrying costs. Exhibit 16.5 shows the Excel solution screen for the University Bookstore example. Notice that the selection of the appropriate order size, Q, that results in the minimum total cost for each discount range is determined by the formulas embedded in cells E8, E9, and E10. For example, the formula for the first quantity discount range, 1–49, is embedded in cell E8 and shown on the formula bar at

EXHIBIT 16.5

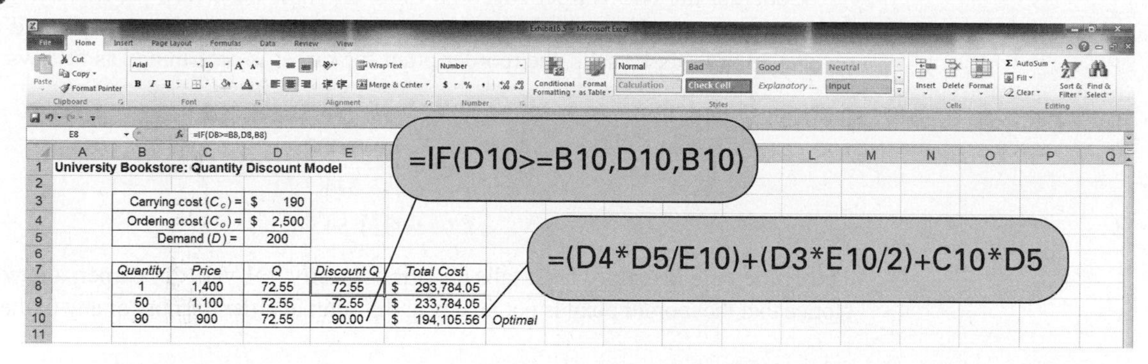

University Bookstore: Quantity Discount Model

Carrying cost (C_c) =	$	190
Ordering cost (C_o) =	$	2,500
Demand (D) =		200

Quantity	Price	Q	Discount Q	Total Cost	
1	1,400	72.55	72.55	$ 293,784.05	
50	1,100	72.55	72.55	$ 233,784.05	
90	900	72.55	90.00	$ 194,105.56	Optimal

the top of the screen, **=IF(D8>=B8,D8,B8)**. This means that if the discount order size in cell D8 (i.e., $Q = 72.55$) is greater than or equal to the quantity in cell B8 (i.e., 1), the quantity in cell D8 (72.55) is selected; otherwise, the amount in cell B8 is selected. The formulas in cells E9 and E10 are constructed similarly. The result is that the order quantity for the final discount range, $Q = 90$, is selected.

Reorder Point

In our presentation of the basic EOQ model in the previous section, we addressed one of the two primary questions related to inventory management: *How much should be ordered?* In this section we will discuss the other aspect of inventory management: *When to order?* The determinant of when to order in a continuous inventory system is the **reorder point**, the inventory level at which a new order is placed.

The ***reorder point*** *is the level of inventory at which a new order is placed.*

The concept of lead time is illustrated graphically in Figure 16.9. Notice that the order must be made prior to the time when the level of inventory falls to zero. Because demand is consuming the inventory while the order is being shipped, the order must be made while there is enough inventory in stock to meet demand during the lead-time period. This level of inventory is referred to as the *reorder point* and is so designated in Figure 16.9.

FIGURE 16.9
Reorder point and lead time

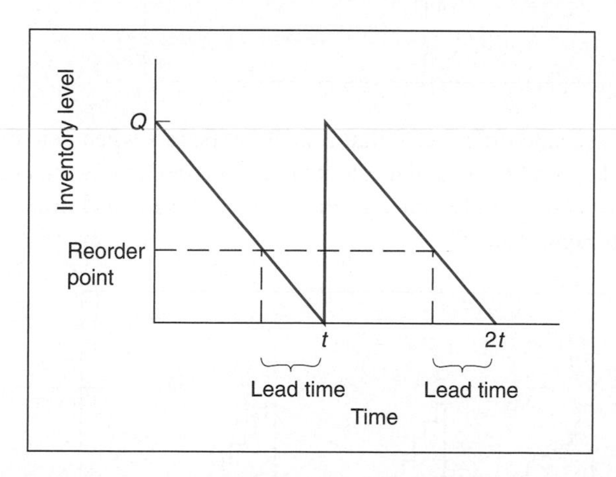

The reorder point for our basic EOQ model with constant demand and a constant lead time to receive an order is relatively straightforward. It is simply equal to the amount demanded during the lead-time period, computed using the following formula:

$$R = dL$$

where

d = demand rate per time period (e.g., daily)
L = lead time

Consider the I-75 Carpet Discount Store example described previously. The store is open 311 days per year. If annual demand is 10,000 yards of Super Shag carpet and the lead time to receive an order is 10 days, the reorder point for carpet is determined as follows:

$$R = dL$$
$$= \left(\frac{10{,}000}{311}\right)(10)$$
$$= 321.54 \text{ yd.}$$

Thus, when the inventory level falls to approximately 321 yards of carpet, a new order is placed. Notice that the reorder point is not related to the optimal order quantity or any of the inventory costs.

Safety Stocks

In our previous example for determining the reorder point, an order is made when the inventory level reaches the reorder point. During the lead time, the remaining inventory in stock is depleted as a constant demand rate, such that the new order quantity arrives at exactly the same moment as the inventory level reaches zero in Figure 16.9. Realistically, however, demand—and to a lesser extent lead time—is uncertain. The inventory level might be depleted at a slower or faster rate during lead time. This is depicted in Figure 16.10 for uncertain demand and a constant lead time.

FIGURE 16.10
Inventory model with uncertain demand

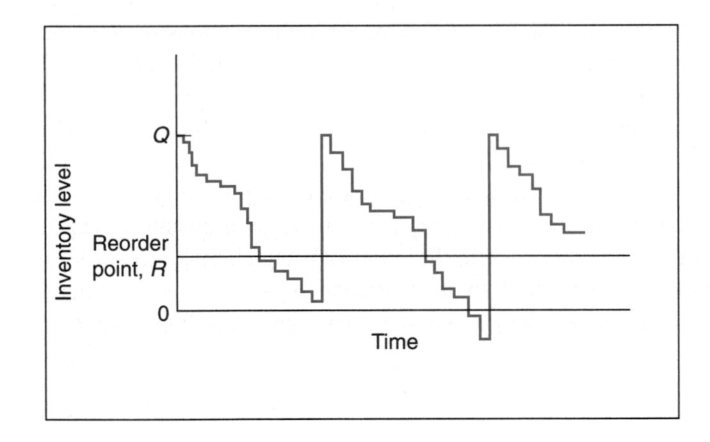

Notice in the second order cycle that a stockout occurs when demand exceeds the available inventory in stock. As a hedge against stockouts when demand is uncertain, a safety (or buffer) stock of inventory is frequently added to the demand during lead time. The addition of a safety stock is shown in Figure 16.11.

FIGURE 16.11
Inventory model with safety stock

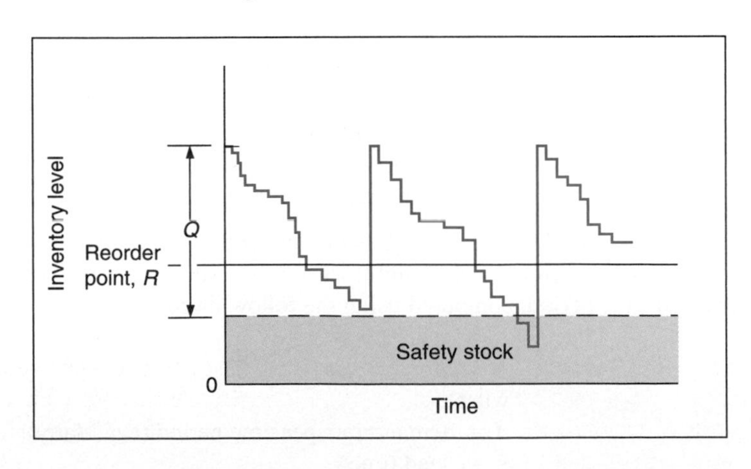

Determining Safety Stock by Using Service Levels

The service level is the probability that the inventory available during the lead time will meet demand.

There are several approaches to determining the amount of the safety stock needed. One of the most popular methods is to establish a safety stock that will meet a specified service level. The service level is the probability that the amount of inventory on hand during the lead time is sufficient to meet expected demand (i.e., the probability that a stockout will not occur). The word *service* is used because the higher the probability that inventory will be on hand, the more likely that customer demand will be met (i.e., the customer can be served). For example, a service level of 90% means that there is a .90 probability that demand will be met during the lead time period and a .10 probability that a stockout will occur. The specification of the service level is typically a policy decision based on a number of factors, including costs for the "extra" safety stock and present and future lost sales if customer demand cannot be met.

Reorder Point with Variable Demand

To compute the reorder point with a safety stock that will meet a specific service level, we will assume that the individual demands during each day of lead time are uncertain and independent and can be described by a normal probability distribution. The average demand for the lead-time period is the sum of the average daily demands for the days of the lead time, which is also the product of the average daily demand multiplied by the lead time. Likewise, the variance of the distribution is the sum of the daily variances for the number of days in the lead-time period. Using these parameters, the reorder point to meet a specific service level can be computed as follows:

$$R = \bar{d}L + Z\sigma_d\sqrt{L}$$

where

$\bar{d}$ = average daily demand
L = lead time
σ_d = the standard deviation of daily demand
Z = number of standard deviations corresponding to the service level probability
$Z\sigma_d\sqrt{L}$ = safety stock

The term $\sigma_d\sqrt{L}$ in this formula for reorder point is the square root of the sum of the daily variances during lead time:

$$\text{variance} = (\text{daily variances}) \times (\text{number of days of lead time}) = \sigma_d^2 L$$
$$\text{standard deviation} = \sqrt{\sigma_d^2 L} = \sigma_d\sqrt{L}$$

The reorder point relative to the service level is shown in Figure 16.12. The service level is the shaded area, or probability, to the left of the reorder point, R.

FIGURE 16.12
Reorder point for a service level

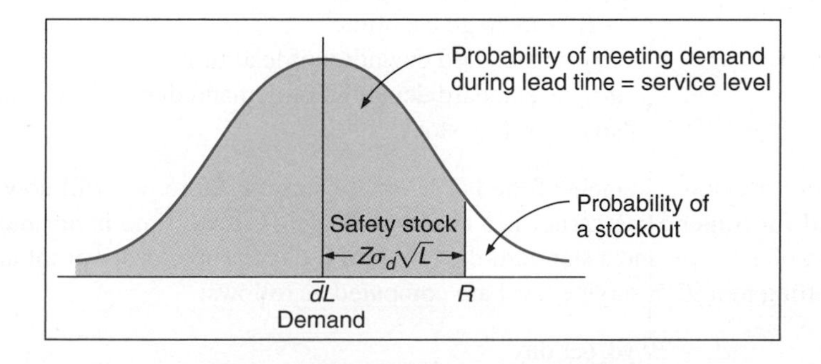

The I-75 Carpet Discount Store sells Super Shag carpet. The average daily customer demand for the carpet stocked by the store is normally distributed, with a mean daily demand of 30 yards and a standard deviation of 5 yards of carpet per day. The lead time for receiving a new order of carpet is 10 days. The store wants a reorder point and safety stock for a service level of 95%, with the probability of a stockout equal to 5%:

$$\bar{d} = 30 \text{ yd. per day}, \ L = 10 \text{ days}, \ \sigma_d = 5 \text{ yd. per day}$$

For a 95% service level, the value of Z (from Table A.1 in Appendix A) is 1.65. The reorder point is computed as follows:

$$R = \bar{d}L + Z\sigma_d\sqrt{L} = 30(10) + (1.65)(5)(\sqrt{10}) = 300 + 26.1 = 326.1 \text{ yd.}$$

The safety stock is the second term in the reorder point formula:

$$\text{safety stock} = Z\sigma_d\sqrt{L} = (1.65)(5)(\sqrt{10}) = 26.1 \text{ yd.}$$

Determining the Reorder Point Using Excel

Excel can be used to determine the reorder point with variable demand. Exhibit 16.6 shows an Excel spreadsheet set up to compute the reorder point for our I-75 Carpet Discount Store example. Notice that the formula for computing the reorder point in cell E7 is shown on the formula bar at the top of the spreadsheet.

EXHIBIT 16.6

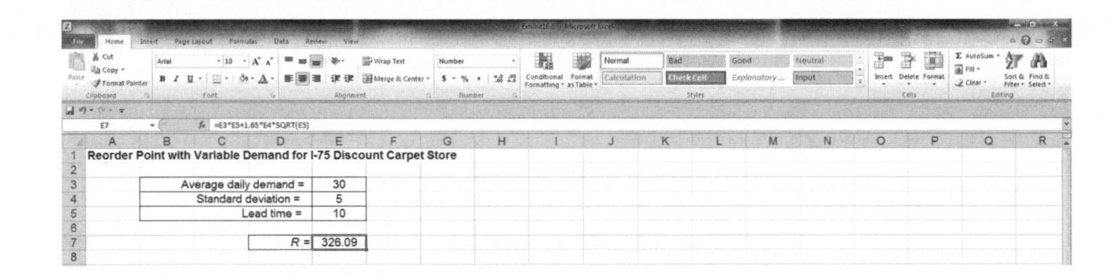

Reorder Point with Variable Lead Time

In the model in the previous section for determining the reorder point, we assumed a variable demand rate and a constant lead time. In the case where demand is constant and the lead time varies, we can use a similar formula, as follows:

$$R = d\bar{L} + Zd\sigma_L$$

where

d = constant daily demand
$\bar{L}$ = average lead time
σ_L = standard deviation of lead time
$d\sigma_L$ = standard deviation of demand during lead time
$Zd\sigma_L$ = safety stock

For our previous example of the I-75 Carpet Discount Store, we will now assume that daily demand for Super Shag carpet is a constant 30 yards. Lead time is normally distributed, with a mean of 10 days and a standard deviation of 3 days. The reorder point and safety stock corresponding to a 95% service level are computed as follows:

$$d = 30 \text{ yd. per day}$$
$$\bar{L} = 10 \text{ days}$$
$$\sigma_L = 3 \text{ days}$$
$$Z = 1.65 \text{ for a } 95\% \text{ service level}$$
$$R = d\bar{L} + Zd\sigma_L = (30)(10) + (1.65)(30)(3) = 300 + 148.5 = 448.5 \text{ yd.}$$

Management Science Application

Establishing Inventory Safety Stocks at Kellogg's

© Richard Levine/Alamy

Kellogg's is the world's largest cereal producer and a leading maker of convenience foods. The company started with a single product, Kellogg's Corn Flakes, in 1906, and over the years has developed a product line of other cereals, including Rice Krispies and Corn Pops, as well as convenience foods, such as Pop-Tarts and Nutri-Grain cereal bars. Kellogg's operates five plants in the United States and Canada and seven distribution centers, and it contracts with 15 co-packers to produce or pack some Kellogg's products. Kellogg's must coordinate the production, packaging, inventory, and distribution of roughly 80 cereal products alone at these various facilities.

Kellogg's uses a model called the Kellogg Planning System (KPS) to plan its weekly production, inventory, and distribution decisions. The data used in the model are subject to much uncertainty, and the greatest uncertainty is in product demand. Demand in the first few weeks of a planning horizon is based on customer orders and is fairly accurate; however, demand in the third and fourth weeks may be significantly different from marketing forecasts. However, Kellogg's primary goal is to meet customer demand, and in order to achieve this goal, Kellogg's employs safety stocks as a buffer against uncertain demand. The safety stock for a product at a specific production facility in week t is the sum of demands for weeks t and $t + 1$. However, for a product that is being promoted in an advertising campaign, the safety stock is the sum of forecasted demand for a 4-week horizon or longer. KPS has saved Kellogg's many millions of dollars since the mid-1990s. The tactical version of KPS recently helped the company consolidate production capacity, with estimated projected savings of almost $40 million.

Source: Based on G. Brown, J. Keegan, B. Vigus, and K. Wood, "The Kellogg Company Optimizes Production, Inventory, and Distribution," *Interfaces* 31, no. 6 (November–December 2001): 1–15.

Reorder Point with Variable Demand and Lead Time

The final reorder point case we will consider is the case in which both demand and lead time are variables. The reorder point formula for this model is as follows:

$$R = \bar{d}\,\bar{L} + Z\sqrt{\sigma_d^2\bar{L} + \sigma_L^2\bar{d}^2}$$

where

$$\bar{d} = \text{average daily demand}$$
$$\bar{L} = \text{average lead time}$$
$$\sqrt{\sigma_d^2\bar{L} + \sigma_L^2\bar{d}^2} = \text{standard deviation of demand during lead time}$$
$$Z\sqrt{\sigma_d^2\bar{L} + \sigma_L^2\bar{d}^2} = \text{safety stock}$$

Again we will consider the I-75 Carpet Discount Store example, used previously. In this case, daily demand is normally distributed, with a mean of 30 yards and a standard deviation of 5 yards. Lead time is also assumed to be normally distributed, with a mean of 10 days and a standard deviation of 3 days. The reorder point and safety stock for a 95% service level are computed as follows:

$$\bar{d} = 30 \text{ yd./day}$$
$$\sigma_d = 5 \text{ yd./day}$$
$$\bar{L} = 10 \text{ days}$$
$$\sigma_L = 3 \text{ days}$$
$$Z = 1.65 \text{ for a } 95\% \text{ service level}$$
$$R = \bar{d}\,\bar{L} + Z\sqrt{\sigma_d^2\bar{L} + \sigma_L^2\bar{d}^2}$$
$$= (30)(10) + (1.65)\sqrt{(5)^2(10) + (3)^2(30)^2}$$
$$= 300 + 150.8$$
$$R = 450.8 \text{ yd.}$$

Thus, the reorder point is 450.8 yards, with a safety stock of 150.8 yards. Notice that this reorder point encompasses the largest safety stock of our three reorder point examples, which would be anticipated, given the increased variability resulting from variable demand and lead time.

Order Quantity for a Periodic Inventory System

*A **periodic inventory system** uses variable order sizes at fixed time intervals.*

Previously we defined a continuous, or fixed–order quantity, inventory system as one in which the order quantity was constant and the time between orders varied. So far, this type of inventory system has been the primary focus of our discussion. The less common **periodic**, or *fixed–time period*, **inventory system** is one in which the time between orders is constant and the order size varies. A drugstore is one example of a business that sometimes uses a fixed-period inventory system. Drugstores stock a number of personal hygiene and health-related products, such as shampoo, toothpaste, soap, bandages, cough medicine, and aspirin. Normally, the vendors that provide these items to the store will make periodic visits—for example, every few weeks or every month—and count the stock of inventory on hand for their products. If the inventory is exhausted or at some predetermined reorder point, a new order will be placed for an amount that will bring the inventory level back up to the desired level. The drugstore managers will generally not monitor the inventory level between vendor visits but instead rely on the vendor to take inventory at the time of the scheduled visit.

A periodic inventory system normally requires a larger safety stock.

A limitation of this type of inventory system is that inventory can be exhausted early in the time period between visits, resulting in a stockout that will not be remedied until the next scheduled order. Alternatively, in a fixed–order quantity system, when inventory reaches a reorder point, an order is made that minimizes the time during which a stockout might exist. As a result of this drawback, a larger safety stock is normally required for the fixed-interval system.

Order Quantity with Variable Demand

If the demand rate and lead time are constant, then the fixed-period model will have a fixed order quantity that will be made at specified time intervals, which is the same as the fixed quantity (EOQ) model under similar conditions. However, as we have already explained, the fixed-period model reacts significantly differently from the fixed–order quantity model when demand is a variable.

The order size for a fixed-period model, given variable daily demand that is normally distributed, is determined by the following formula:

$$Q = \bar{d}(t_b + L) + Z\sigma_d\sqrt{t_b + L} - I$$

where

$\bar{d}$ = average demand rate
t_b = the fixed time between orders
L = lead time
σ_d = standard deviation of demand
$Z\sigma_d\sqrt{t_b + L}$ = safety stock
I = inventory in stock

The first term in the preceding formula, $d(t_b + L)$, is the average demand during the order cycle time plus the lead time. It reflects the amount of inventory that will be needed to protect against shortages during the entire time from this order to the next and the lead time, until the order is received. The second term, $Z\sigma_d\sqrt{t_b + L}$, is the safety stock for a specific service level, determined in much the same way as previously described for a reorder point. The final term, I, is the amount of inventory on hand when the inventory level is checked and an order is made. We will demonstrate the computation of Q with an example.

The Corner Drug Store stocks a popular brand of sunscreen. The average demand for the sunscreen is 6 bottles per day, with a standard deviation of 1.2 bottles. A vendor for the sunscreen producer checks the drugstore stock every 60 days, and during a particular visit the drugstore had 8 bottles in stock. The lead time to receive an order is 5 days. The order size for this order period that will enable the drugstore to maintain a 95% service level is computed as follows:

$$
\begin{aligned}
\bar{d} &= 6 \text{ bottles per day} \\
\sigma_d &= 1.2 \text{ bottles} \\
t_b &= 60 \text{ days} \\
L &= 5 \text{ days} \\
I &= 8 \text{ bottles} \\
Z &= 1.65 \text{ for } 95\% \text{ service level} \\
Q &= \bar{d}(t_b + L) + Z\sigma_d\sqrt{t_b + L} - I \\
&= (6)(60 + 5) + (1.65)(1.2)\sqrt{60 + 5} - 8 \\
&= 398 \text{ bottles}
\end{aligned}
$$

Determining the Order Quantity for the Fixed-Period Model with Excel

Exhibit 16.7 shows an Excel spreadsheet set up to compute the order quantity for the fixed-period model with variable demand for our Corner Drug Store example. Notice that the formula for the order quantity in cell D10 is shown on the formula bar at the top of the spreadsheet.

EXHIBIT 16.7

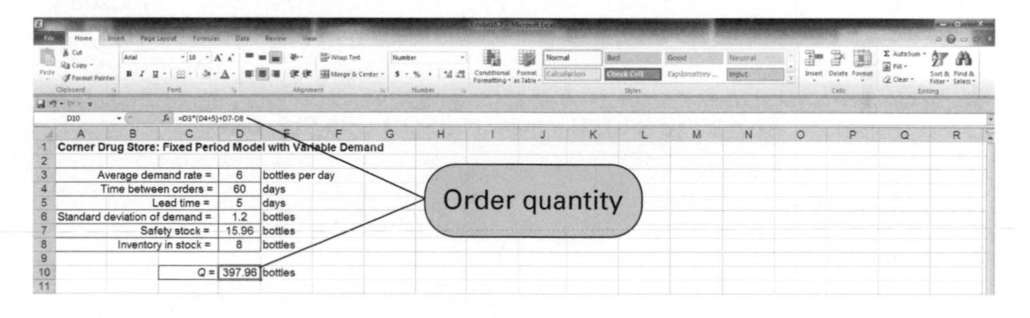

Summary

In this chapter the classical economic order quantity model was presented. The basic form of the EOQ model we discussed included simplifying assumptions regarding order receipt, no shortages, and constant demand known with certainty. By relaxing some of these assumptions, we were able to create increasingly complex but realistic models. These EOQ variations included the reorder point model, the noninstantaneous receipt model, the model with shortages, and models with safety stocks. The techniques for inventory analysis presented in this chapter are not widely used to analyze other types of problems. Conversely, however, many of the techniques presented in this text are used for inventory analysis (in addition to the methods presented in these chapters). The wide use of management science techniques for inventory analysis attests to the importance of inventory to all types of organizations.

Example Problem Solutions

The following example will demonstrate EOQ analysis for the classical model and the model with shortages and back ordering.

Problem Statement

Electronic Village stocks and sells a particular brand of personal computer. It costs the store $450 each time it places an order with the manufacturer for the personal computers. The annual cost of carrying the PCs in inventory is $170. The store manager estimates the annual demand for the PCs will be 1,200 units.

a. Determine the optimal order quantity and the total minimum inventory cost.
b. Assume that shortages are allowed and that the shortage cost is $600 per unit per year. Compute the optimal order quantity and the total minimum inventory cost.

Solution

Step 1: (part a): Determine the Optimal Order Quantity

$$\begin{aligned} D &= 1{,}200 \text{ personal computers} \\ C_c &= \$170 \\ C_o &= \$450 \\ Q &= \sqrt{\frac{2C_oD}{C_c}} \\ &= \sqrt{\frac{2(450)(1{,}200)}{170}} \\ &= 79.7 \text{ personal computers} \\ \text{total cost} &= C_c\frac{Q}{2} + C_o\frac{D}{Q} \\ &= 170\left(\frac{79.7}{2}\right) + 450\left(\frac{1{,}200}{79.7}\right) \\ &= \$13{,}549.91 \end{aligned}$$

Step 2: (part b): Compute the EOQ with Shortages

$$\begin{aligned} C_s &= \$600 \\ Q &= \sqrt{\frac{2C_oD}{C_c}\left(\frac{C_s + C_c}{C_s}\right)} \\ &= \sqrt{\frac{2(450)(1{,}200)}{170}\left(\frac{600 + 170}{600}\right)} \\ &= 90.3 \text{ personal computers} \\ S &= Q\left(\frac{C_c}{C_c + C_s}\right) \\ &= 90.3\left(\frac{170}{170 + 600}\right) \\ &= 19.9 \text{ personal computers} \\ \text{total cost} &= \frac{C_sS^2}{2Q} + C_c\frac{(Q - S)^2}{2Q} + \frac{C_oD}{Q} \\ &= \frac{(600)(19.9)^2}{2(90.3)} + 170\frac{(90.3 - 19.9)^2}{2(90.3)} + 450\left(\frac{1{,}200}{90.3}\right) \\ &= \$1{,}315.65 + 4{,}665.27 + 5{,}980.07 \\ &= \$11{,}960.98 \end{aligned}$$

Problem Statement

A computer products store stocks color graphics monitors, and the daily demand is normally distributed, with a mean of 1.6 monitors and a standard deviation of 0.4 monitor. The lead time to receive an order from the manufacturer is 15 days. Determine the reorder point that will achieve a 98% service level.

Solution

Step 1: Identify Parameters

$$\bar{d} = 1.6 \text{ monitors per day}$$
$$L = 15 \text{ days}$$
$$\sigma_d = 0.4 \text{ monitor per day}$$
$$Z = 2.05 \text{ (for a 98\% service level)}$$

Step 2: Solve for R

$$R = \bar{d}L + Z\sigma_d\sqrt{L} = (1.6)(15) + (2.05)(0.4)\sqrt{15} = 24 + 3.18 = 27.18 \text{ monitors}$$

Problems

1. Hayes Electronics stocks and sells a particular brand of personal computer. It costs the firm $450 each time it places an order with the manufacturer for the personal computers. The cost of carrying one PC in inventory for a year is $170. The store manager estimates that total annual demand for the computers will be 1,200 units, with a constant demand rate throughout the year. Orders are received within minutes after placement from a local warehouse maintained by the manufacturer. The store policy is never to have stockouts of the PCs. The store is open for business every day of the year except Christmas Day. Determine the following:
   a. The optimal order quantity per order
   b. The minimum total annual inventory costs
   c. The optimal number of orders per year
   d. The optimal time between orders (in working days)

2. Hayes Electronics in Problem 1 assumed with certainty that the ordering cost is $450 per order and the inventory carrying cost is $170 per unit per year. However, the inventory model parameters are frequently only estimates that are subject to some degree of uncertainty. Consider four cases of variation in the model parameters: (a) Both ordering cost and carrying cost are 10% less than originally estimated, (b) both ordering cost and carrying cost are 10% higher than originally estimated, (c) ordering cost is 10% higher and carrying cost is 10% lower than originally estimated, and (d) ordering cost is 10% lower and carrying cost is 10% higher than originally estimated. Determine the optimal order quantity and total inventory cost for each of the four cases. Prepare a table with values from all four cases and compare the sensitivity of the model solution to changes in parameter values.

3. A firm is faced with the attractive situation in which it can obtain immediate delivery of an item it stocks for retail sale. The firm has therefore not bothered to order the item in any systematic way. Recently, however, profits have been squeezed due to increasing competitive pressures, and the firm has retained a management consultant to study its inventory management. The consultant has determined that the various costs associated with making an order for the item stocked are approximately $30 per order. She has also determined that the costs of carrying the item in inventory amount to approximately $20 per unit per year (primarily direct storage costs and foregone profit on investment in inventory). Demand for the item is reasonably constant over time,

and the forecast is for 19,200 units per year. When an order is placed for the item, the entire order is immediately delivered to the firm by the supplier. The firm operates 6 days a week plus a few Sundays, or approximately 320 days per year. Determine the following:
a. The optimal order quantity per order
b. The total annual inventory costs
c. The optimal number of orders to place per year
d. The number of operating days between orders, based on the optimal number of orders

4. The Western Jeans Company purchases denim from Cumberland Textile Mills. The Western Jeans Company uses 35,000 yards of denim per year to make jeans. The cost of ordering denim from the textile company is $500 per order. It costs Western $0.35 per yard annually to hold a yard of denim in inventory. Determine the optimal number of yards of denim the Western Jeans Company should order, the minimum total annual inventory cost, the optimal number of orders per year, and the optimal time between orders.

5. The Metropolitan Book Company purchases paper from the Atlantic Paper Company. Metropolitan produces magazines and paperbacks that require 1,215,000 pounds of paper per year. The cost per order for the company is $1,200; the cost of holding 1 pound of paper in inventory is $0.08 per year. Determine the following:
a. The economic order quantity
b. The minimum total annual cost
c. The optimal number of orders per year
d. The optimal time between orders

6. The Simple Simon Bakery produces fruit pies for freezing and subsequent sale. The bakery, which operates 5 days per week, 52 weeks per year, can produce pies at the rate of 64 pies per day. The bakery sets up the pie production operation and produces until a predetermined number (Q) of pies has been produced. When not producing pies, the bakery uses its personnel and facilities for producing other bakery items. The setup cost for a production run of fruit pies is $500. The cost of holding frozen pies in storage is $5 per pie per year. The annual demand for frozen fruit pies, which is constant over time, is 5,000 pies. Determine the following:
a. The optimal production run quantity (Q)
b. The total annual inventory costs
c. The optimal number of production runs per year
d. The optimal cycle time (time between run starts)
e. The run length, in working days

7. The Pedal Pusher Bicycle Shop operates 7 days per week, closing only on Christmas Day. The shop pays $300 for a particular bicycle purchased from the manufacturer. The annual holding cost per bicycle is estimated to be 25% of the dollar value of inventory. The shop sells an average of 25 bikes per week. Frequently, the dealer does not have a bike in stock when a customer purchases it, and the bike is back-ordered. The dealer estimates his shortage cost per unit back-ordered, on an annual basis, to be $250 due to lost future sales (and profits). The ordering cost for each order is $100. Determine the optimal order quantity and shortage level and the total minimum cost.

8. The Petroco Company uses a highly toxic chemical in one of its manufacturing processes. It must have the product delivered by special cargo trucks designed for safe shipment of chemicals. As such, ordering (and delivery) costs are relatively high, at $2,600 per order. The chemical product is packaged in 1-gallon plastic containers. The cost of holding the chemical in storage is $50 per gallon per year. The annual demand for the chemical, which is constant over time, is 2,000 gallons per year. The lead time from time of order placement until receipt is 10 days. The company operates 310 working days per year. Compute the optimal order quantity, the total minimum inventory cost, and the reorder point.

9. The Big Buy Supermarket stocks Munchies Cereal. Demand for Munchies is 4,000 boxes per year (365 days). It costs the store $60 per order of Munchies, and it costs $0.80 per box per year to keep the cereal in stock. Once an order for Munchies is placed, it takes 4 days to receive the order from a food distributor. Determine the following:
   a. The optimal order size
   b. The minimum total annual inventory cost
   c. The reorder point

10. The Wood Valley Dairy makes cheese to supply to stores in its area. The dairy can make 250 pounds of cheese per day (365 days per year), and the demand at area stores is 180 pounds per day. Each time the dairy makes cheese, it costs $125 to set up the production process. The annual cost of carrying a pound of cheese in a refrigerated storage area is $12. Determine the optimal order size and the minimum total annual inventory cost.

11. The Rainwater Brewery produces Rainwater Light Beer, which it stores in barrels in its warehouse and supplies to its distributors on demand. The demand for Rainwater is 1,500 barrels of beer per day (365 days per year). The brewery can produce 2,000 barrels of Rainwater per day. It costs $6,500 to set up a production run for Rainwater. Once it is brewed, the beer is stored in a refrigerated warehouse at an annual cost of $50 per barrel. Determine the economic order quantity and the minimum total annual inventory cost.

12. The purchasing manager for the Atlantic Steel Company must determine a policy for ordering coal to operate 12 converters. Each converter requires exactly 5 tons of coal per day to operate, and the firm operates 360 days per year. The purchasing manager has determined that the ordering cost is $80 per order and the cost of holding coal is 20% of the average dollar value of inventory held. The purchasing manager has negotiated a contract to obtain the coal for $12 per ton for the coming year.
   a. Determine the optimal quantity of coal to receive in each order.
   b. Determine the total inventory-related costs associated with the optimal ordering policy (do not include the cost of the coal).
   c. If 5 days of lead time are required to receive an order of coal, how much coal should be on hand when an order is placed?

13. In Problem 1 in Chapter 15, the Saki motorcycle dealer in Minneapolis–St. Paul orders the Saki Super TXII motorcycle it sells from the manufacturer in Japan. Using the 3-month moving average forecast of demand for January as the monthly forecast for the next year, an annual carrying cost of $375, an ordering cost of $3,200, and a lead time for receiving an order of 1 month, determine the optimal order size, the minimum total annual inventory cost, the optimal time between orders, the number of orders, and the reorder point.

14. In Problem 2 in Chapter 15, Carpet City orders Soft Shag carpet from its own mill. Using the 3-month moving average forecast of demand for month 9 as the monthly forecasts for all of next year, a production rate at the mill of 1,200 yards per day (with the mill operating 260 days per year), an annual carrying cost of $0.63, a $425 cost for setting up a production run and delivering the carpet to the store, and a lead time for receiving an order of 7 days, determine the optimal order size, the minimum total annual inventory cost, and the reorder point (given that Carpet City is open 360 days per year).

15. In Problem 30 in Chapter 15, the supplier receives shipments of partially completed laptops from its manufacturing facility in Southeast Asia, which has maximum production rate of 200 units per day. Using the forecast of annual demand developed in that problem, an annual carrying cost of $115.75 (which includes an average obsolescence cost), a shipping cost from Asia of $6,500 per shipment, and a lead time for receiving an order of 25 days, determine the optimal order size, the minimum total annual inventory cost, the maximum inventory level, and the reorder point (given that the Bell assembly operation operates 365 days per year).

16. Craftwood Furniture Company is a U.S.–based furniture manufacturer that offshored all of its actual manufacturing operations to China about a decade ago. It set up a distribution center in Hong Kong, from which the company ships its items to the United States on container ships. The company learned early on that it could not rely on local Chinese freight forwarders to arrange for sufficient containers for the company's shipments, so it contracted to purchase containers from a Taiwanese manufacturer and then sell them to shipping companies at the U.S. ports the containers are shipped to. Craftwood needs 715 containers each year. It costs $265 to hold a container at its distribution center, and it costs $6,000 to process and receive an order for the containers. The cost of not having sufficient containers and delaying a shipment is $14,000 per container. Determine the optimal order size, minimum total annual inventory cost, and maximum shortage level.

17. The Pacific Lumber Company and Mill processes 10,000 logs annually, operating 250 days per year. Immediately upon receiving an order, the logging company's supplier begins delivery to the lumber mill, at a rate of 60 logs per day. The lumber mill has determined that the ordering cost is $1,600 per order and the cost of carrying logs in inventory before they are processed is $15 per log on an annual basis. Determine the following:
    a. The optimal order size
    b. The total inventory cost associated with the optimal order quantity
    c. The number of operating days between orders
    d. The number of operating days required to receive an order

18. The Roadking Tire Store sells a brand of tires called the Roadrunner. The annual demand from the store's customers for Roadrunner tires is 3,700. The cost to order tires from the tire manufacturer is $420 per order. The annual carrying cost is $1.75 per tire. The store allows shortages, and the annual shortage cost per tire is $4. Determine the optimal order size, maximum shortage level, and minimum total annual inventory cost.

19. The Laurel Creek Lawn Shop sells Fastgro Fertilizer. The annual demand for the fertilizer is 270,000 pounds. The cost to order the fertilizer from the Fastgro Company is $105 per order. The annual carrying cost is $0.25 per pound. The store operates with shortages, and the annual shortage cost is $0.70 per pound. Compute the optimal order size, minimum total annual inventory cost, and maximum shortage level.

20. Videoworld is a discount store that sells color televisions. The annual demand for color television sets is 400. The cost per order from the manufacturer is $650. The carrying cost is $45 per set each year. The store has an inventory policy that allows shortages. The shortage cost per set is estimated at $60. Determine the following:
    a. The optimal order size
    b. The maximum shortage level
    c. The minimum total annual inventory cost

21. The University Bookstore at Tech stocks the required textbook for Management Science 2405. The demand for this text is 1,200 copies per year. The cost of placing an order is $350, and the annual carrying cost is $2.75 per book. If a student requests the book and it is not in stock, the student will likely go to the privately owned Tech Bookstore. It is likely that the student will not buy books at the University Bookstore in the future; thus the shortage cost to the University Bookstore is estimated to be $45 per book. Determine the optimal order size, the maximum shortage level, and the total inventory cost.

22. The A-to-Z Office Supply Company is open from 8:00 A.M. to 6:00 P.M., and it receives 200 calls per day for delivery orders. It costs A-to-Z $20 to send out its trucks to make deliveries. The company estimates that each minute a customer spends waiting for an order costs A-to-Z $0.20 in lost sales.

a. How frequently should A-to-Z send out its delivery trucks each day? Indicate the total daily cost of deliveries.
b. If a truck could carry only six orders, how often would deliveries be made, and what would be the cost?

23. The Union Street Microbrewery makes 1220 Union beer, which it bottles and sells in its adjoining restaurant and by the case. It costs $1,700 to set up, brew, and bottle a batch of the beer. The annual cost to store the beer in inventory is $1.25 per bottle. The annual demand for the beer is 18,000 bottles, and the brewery has the capacity to produce 30,000 bottles annually.
    a. Determine the optimal order quantity, the total annual inventory cost, the number of production runs per year, and the maximum inventory level.
    b. If the microbrewery has only enough storage space to hold a maximum of 2,500 bottles of beer in inventory, how will that affect total inventory costs?

24. Eurotronics is a European manufacturer of electronic components. During the course of a year, it requires container cargo space on ships leaving Hamburg bound for the United States, Mexico, South America, and Canada. Annually, the company needs 160,000 cubic feet of cargo space. The cost of reserving cargo space is $7,000, and the cost of holding cargo space is $0.80 per cubic foot. Determine how much storage space Eurotronics should optimally order, the total cost, and how many times per year it should place orders to reserve space.

25. The Summer Outdoor Furniture Company produces wooden lawn chairs. The annual demand from its store customers is 17,400 chairs per year. The transport and handling costs are $2,600 each time a shipment of chairs is delivered to stores from its warehouse. The annual carrying cost is $3.75 per chair.
    a. Determine the optimal order quantity and minimum total annual cost.
    b. The company is thinking about relocating its warehouse closer to its customers, which would reduce transport and handling costs to $1,900 per order but increase carrying costs to $4.50 per chair per year. Should the company relocate based on inventory costs?

26. The Spruce Creek Vegetable Farm produces organically grown greenhouse tomatoes that are sold to area grocery stores. The annual demand for Spruce Creek's tomatoes is 270,000 pounds. The farm is able to produce 305,000 pounds annually. The cost to transport the tomatoes from the farm to the stores is $620 per load. The annual carrying cost is $0.12 per pound.
    a. Compute the optimal order size, the maximum inventory level, and the total minimum cost.
    b. If Spruce Creek can increase production capacity to 360,000 tomatoes per year, will it reduce total inventory cost?

27. The Uptown Kiln is an importer of ceramics from overseas. It has arranged to purchase a particular type of ceramic pottery from a Korean artisan. The artisan makes the pottery in 120-unit batches and will ship only that exact number of units. The transportation and handling cost of a shipment is $7,600 (not including the unit cost). The Uptown Kiln estimates its annual demand to be 900 units. What storage and handling cost per unit does it need to achieve in order to minimize its inventory cost?

28. The I-75 Carpet Discount Store has an annual demand of 10,000 yards of Super Shag carpet. The annual carrying cost for a yard of this carpet is $0.75, and the ordering cost is $150. The carpet manufacturer normally charges the store $8 per yard for the carpet; however, the manufacturer has offered a discount price of $6.50 per yard if the store will order 5,000 yards. How much should the store order, and what will be the total annual inventory cost for that order quantity?

29. The Fifth Quarter Bar buys Old World draft beer by the barrel from a local distributor. The bar has an annual demand of 900 barrels, which it purchases at a price of $205 per barrel. The annual carrying cost is 12% of the price, and the cost per order is $160. The distributor has offered

the bar a reduced price of $190 per barrel if it will order a minimum of 300 barrels. Should the bar take the discount?

30. The bookstore at State University purchases from a vendor sweatshirts emblazoned with the school name and logo. The vendor sells the sweatshirts to the store for $38 apiece. The cost to the bookstore for placing an order is $120, and the carrying cost is 25% of the average annual inventory value. The bookstore manager estimates that 1,700 sweatshirts will be sold during the year. The vendor has offered the bookstore the following volume discount schedule:

Order Size	Discount %
1–299	0
300–499	2
500–799	4
800+	5

The bookstore manager wants to determine the bookstore's optimal order quantity, given the foregoing quantity discount information.

31. Determine the optimal order quantity of sweatshirts and total annual cost in Problem 30 if the carrying cost is a constant $8 per sweatshirt per year.

32. The office manager for the Gotham Life Insurance Company orders letterhead stationery from an office products firm in boxes of 500 sheets. The company uses 6,500 boxes per year. Annual carrying costs are $3 per box, and ordering costs are $28. The following discount price schedule is provided by the office supply company:

Order Quantity (boxes)	Price per Box
200–999	$16
1,000–2,999	14
3,000–5,999	13
6,000+	12

Determine the optimal order quantity and the total annual inventory cost.

33. Determine the optimal order quantity and total annual inventory cost for boxes of stationery in Problem 32 if the carrying cost is 20% of the price of a box of stationery.

34. The 23,000-seat City Coliseum houses the local professional ice hockey, basketball, indoor soccer, and arena football teams, as well as various trade shows, wrestling and boxing matches, tractor pulls, and circuses. Coliseum vending annually sells large quantities of soft drinks and beer in plastic cups, with the name of the coliseum and the various team logos on them. The local container cup manufacturer that supplies the cups in boxes of 100 has offered coliseum management the following discount price schedule for cups:

Order Quantity (boxes)	Price per Box
2,000–6,999	$47
7,000–11,999	43
12,000–19,999	41
20,000+	38

The annual demand for cups is 2.3 million, thc annual carrying cost per box of cups is $1.90, and the ordering cost is $320. Determine the optimal order quantity and total annual inventory cost.

35. Community Hospital orders latex sanitary gloves from a hospital supply firm. The hospital expects to use 40,000 pairs of gloves per year. The cost to order and to have the gloves delivered is $180. The annual carrying cost is $0.18 per pair of gloves. The hospital supply firm offers the following quantity discount pricing schedule:

Quantity	Price
0–9,999	$0.34
10,000–19,999	0.32
20,000–29,999	0.30
30,000–39,999	0.28
40,000–49,999	0.26
50,000+	0.24

Determine the optimal order size for the hospital.

36. Tracy McCoy is the office administrator for the department of management science at Tech. The faculty uses a lot of printer paper, and although Tracy is constantly reordering, paper frequently runs out. She orders the paper from the university central stores. Several faculty members have determined that the lead time to receive an order is normally distributed, with a mean of 2 days and a standard deviation of 0.5 day. The faculty has also determined that daily demand for the paper is normally distributed, with a mean of 2 packages and a standard deviation of 0.8 package. What reorder point should Tracy use in order not to run out 99% of the time?

37. Determine the optimal order quantity and total annual inventory cost for cups in Problem 34 if the carrying cost is 5% of the price of a box of cups.

38. The amount of denim used daily by the Western Jeans Company in its manufacturing process to make jeans is normally distributed, with an average of 3,000 yards of denim and a standard deviation of 600 yards. The lead time required to receive an order of denim from the textile mill is a constant 6 days. Determine the safety stock and reorder point if the Western Jeans Company wants to limit the probability of a stockout and work stoppage to 5%.

39. In Problem 38, what level of service would a safety stock of 2,000 yards provide?

40. The Atlantic Paper Company produces paper from wood pulp ordered from a lumber products firm. The paper company's daily demand for wood pulp is a constant 8,000 pounds. Lead time is normally distributed, with an average of 7 days and a standard deviation of 1.6 days. Determine the reorder point if the paper company wants to limit the probability of a stockout and work stoppage to 2%.

41. The Uptown Bar and Grill serves Rainwater draft beer to its customers. The daily demand for beer is normally distributed, with an average of 18 gallons and a standard deviation of 4 gallons. The lead time required to receive an order of beer from the local distributor is normally distributed, with a mean of 3 days and a standard deviation of 0.8 day. Determine the safety stock and reorder point if the restaurant wants to maintain a 90% service level. What would be the increase in the safety stock if a 95% service level were desired?

42. In Problem 41, the manager of the Uptown Bar and Grill has negotiated with the beer distributor for the lead time to receive orders to be a constant 3 days. What effect does this have on the reorder point developed in Problem 41 for a 90% service level?

43. The daily demand for Sunlight paint at the Rainbow Paint Store in East Ridge is normally distributed, with a mean of 26 gallons and a standard deviation of 10 gallons. The lead time for receiving an order of paint from the Sunlight distributor is 9 days. Because this is the only paint store in East Ridge, the manager is interested in maintaining only a 75% service level. What reorder point should be used to meet this service level? The manager subsequently has learned that a new paint store will open soon in East Ridge, which has prompted her to increase the service level to 95%. What reorder point will maintain this service level?

44. PM Computers assembles personal computers from generic components. It purchases its color monitors from a manufacturer in Taiwan; thus, there is a long and uncertain lead time for receiving orders. Lead time is normally distributed, with a mean of 25 days and a standard deviation of 10 days. Daily demand is also normally distributed, with a mean of 2.5 monitors and a standard deviation of 1.2 monitors. Determine the safety stock and reorder point corresponding to a 90% service level.

45. PM Computers in Problem 44 is considering purchasing monitors from an American manufacturer that would guarantee a lead time of 8 days, instead of the Taiwanese company. Determine the new reorder point, given this lead time, and identify the factors that would enter into the decision to change manufacturers.

46. The Corner Drug Store fills prescriptions for a popular children's antibiotic, amoxicillin. The daily demand for amoxicillin is normally distributed, with a mean of 200 ounces and a standard deviation of 80 ounces. The vendor for the pharmaceutical firm that supplies the drug calls the drugstore pharmacist every 30 days to check the inventory of amoxicillin. During a call, the druggist indicated that the store had 60 ounces of the antibiotic in stock. The lead time to receive an order is 4 days. Determine the order size that will enable the drugstore to maintain a 95% service level.

47. The Fast Service Food Mart stocks frozen pizzas in a refrigerated display case. The average daily demand for the pizzas is normally distributed, with a mean of 8 pizzas and a standard deviation of 2.5 pizzas. A vendor for a packaged food distributor checks the market's inventory of frozen foods every 10 days, and during a particular visit, there were no pizzas in stock. The lead time to receive an order is 3 days. Determine the order size for this order period that will result in a 99% service level. During the vendor's following visit, there were 5 frozen pizzas in stock. What is the order size for the next order period?

48. The Impanema Restaurant stocks a red Brazilian table wine it purchases from a wine merchant in a nearby city. The daily demand for the wine at the restaurant is normally distributed, with a mean of 18 bottles and a standard deviation of 4 bottles. The wine merchant sends a representative to check the restaurant's wine cellar every 30 days, and during a recent visit, there were 25 bottles in stock. The lead time to receive an order is 2 days. The restaurant manager has requested an order size that will enable him to limit the probability of a stockout to 2%. Determine the order size.

49. The concession stand at the Blacksburg High School stadium sells slices of pizza during soccer games. Concession stand sales are a primary source of revenue for high school athletic programs, so the athletic director wants to sell as much food as possible. However, any pizza not sold is given away to the players, coaches, and referees, or it is thrown away. The athletic director wants to determine a reorder point that will meet, not exceed, the demand for pizza. Pizza sales are normally distributed, with a mean of 6 pizzas per hour and a standard deviation of 2.5 pizzas. The pizzas are ordered from Pizza Town restaurant, and the mean delivery time is 30 minutes, with a standard deviation of 8 minutes.
    a. Currently, the concession stand places an order when it has 1 pizza left. What level of service does this result in?
    b. What should the reorder point be to have a 98% service level?

Case Problem

The Northwoods General Store

The Northwoods General Store in Vermont sells a variety of outdoor clothing items and equipment and several food products at its modern but rustic-looking retail store. Its food products include salmon and maple syrup. The store also runs a lucrative catalog operation. One of its most popular products is maple syrup, which is sold in metal half-gallon cans with a picture of the store on the front.

Maple syrup was one of the first products the store produced and sold, and it continues to do so. Setting up the syrup-making equipment to produce a batch of syrup costs $450. Storing the syrup for sales throughout the year is a tricky process because the syrup must be kept in a temperature-controlled facility. The annual cost of carrying a gallon of the syrup is $15. Based on past sales data, the store has forecasted a demand of 7,500 gallons of maple syrup for the coming year. The store can produce approximately 100 gallons of syrup per day during the maple syrup season, which runs from February through May.

Because of the short season when the store can actually get sap out of trees, it obviously must produce enough during this 4-month season to meet demand for the whole year. Specifically, store management would like a production and inventory schedule that minimizes costs and indicates when during the year they need to start operating the syrup-making facility full time on a daily basis to meet demand for the remaining 8 months.

Develop a syrup production and inventory schedule for the Northwoods General Store.

Case Problem

The Texans Stadium Store

The Fort Worth Texans have won three Super Bowls in the past 5 years, including two in a row the past 2 years. As a result, sportswear such as hats, sweatshirts, sweatpants, and jackets with the Texans logo are particularly popular in Texas. The Texans operate a stadium store outside the football stadium where they play. It is near a busy highway, so the store has heavy customer traffic throughout the year, not just on game days. In addition, the stadium holds high school or college football and soccer games almost every week in the fall, and it holds baseball games in the spring and summer. The most popular single item the stadium store sells is a blue and silver baseball cap with the Texans logo embroidered on it in a special and very attractive manner. The cap has an elastic headband inside it, which automatically conforms to different head sizes. However, the store has had a difficult time keeping the cap in stock, especially during the time between the placement and receipt of an order. Often customers come to the store just for the hat; when it is not in stock, customers are visibly upset, and the store management believes they tend to go to competing stores to purchase their Texans clothing. To rectify this problem, the store manager, Jenny Jones, would like to develop an inventory control policy that would ensure that customers would be able to purchase the cap 99% of the time they asked for it. Jenny has accumulated the following demand data for the cap for a 30-week period:

Week	*Demand*	*Week*	*Demand*	*Week*	*Demand*
1	38	11	28	21	52
2	51	12	41	22	38
3	25	13	37	23	49
4	60	14	44	24	46
5	35	15	45	25	47

Week	*Demand*	*Week*	*Demand*	*Week*	*Demand*
6	42	16	56	26	41
7	29	17	62	27	39
8	46	18	53	28	50
9	55	19	46	29	28
10	19	20	41	30	34

(Demand includes actual sales plus a record of the times a cap has been requested but not available and an estimate of the number of times a customer wanted a cap when it was not available but did not ask for it.)

The store purchases the hats from a small manufacturing company in Jamaica. The shipments from Jamaica are somewhat erratic, with a lead time anywhere between

10 days and 1 month. The following lead time data (in days) were accumulated during approximately a 1-year period:

Order	*Lead Time*	*Order*	*Lead Time*
1	12	11	14
2	16	12	16
3	25	13	23
4	18	14	18
5	10	15	21
6	30	16	19
7	24	17	17
8	19	18	16
9	17	19	22
10	15	20	18

In the past, Jenny placed an order whenever the stock got down to 150 caps. To what level of service does this reorder point correspond? What would the reorder point and safety stock need to be to attain the desired service level? Discuss how Jenny might determine the order size for caps and what additional, if any, information would be needed to determine the order size.

Case Problem

The A-to-Z Office Supply Company

Christine Yamaguchi is the manager of the A-to-Z Office Supply Company in Charlotte. The company attempts to gain an advantage over its competitors by providing quality customer service, which includes prompt delivery of orders by truck or van and always being able to meet customer demand from its stock. In order to achieve this degree of customer service, A-to-Z must stock a large volume of items on a daily basis at a central warehouse and at three retail stores in the city and suburbs. Christine maintains these inventory levels by borrowing cash on a daily basis from the First Piedmont Bank. She estimates that for the coming fiscal year, the company's demand for cash to pay for inventory will be $17,000 per day for 305 working days. Any money she borrows during the year must be repaid with interest by the end of the year. The annual interest rate currently charged by the bank is 9%. Any time Christine takes out a loan to purchase inventory, the bank charges the company a loan origination fee of $1,200 plus 2¼ points (2.25% of the amount borrowed).

Christine often uses EOQ analysis to determine optimal amounts of inventory to order for different office supplies. Now she is wondering if she can use the same type of analysis to determine an optimal borrowing policy. Determine the amount of the loan Christine should secure from the bank, the total annual cost of the company's borrowing policy, and the number of loans the company should obtain during the year. Also determine the level of cash on hand at which the company should apply for a new loan, given that it takes 15 days for a loan to be processed by the bank.

Suppose the bank offers Christine a discount, as follows: On any loan amount equal to or greater than $500,000, the bank will lower the number of points charged on the loan origination fee from 2.25% to 2.00%. What would the company's optimal loan amount be?

Case Problem

Diamant Foods Company

Diamant Foods Company produces a variety of food products, including a line of candies. One of its most popular candy items is Divine Diamonds, a bag of a dozen individually wrapped diamond-shaped candies made primarily from a blend of dark and milk chocolates, macadamia nuts, and a blend of heavy cream fillings. The item is relatively expensive, so Diamant Foods produces it only for its eastern market, encompassing urban areas such as New York, Atlanta, Philadelphia, and Boston. The item is not sold in grocery or discount stores but mainly in specialty shops and specialty groceries, candy stores, and department stores. Diamant Foods supplies the candy to a single food distributor, which has several warehouses on the East Coast. The candy is shipped in cases of 60 bags of the candy per case. Diamonds sell well, despite the fact that they are expensive, at $9.85 per bag (wholesale). Diamant uses high-quality, fresh ingredients and does not store large stocks of the candy in inventory for very long periods of time.

Diamant's distributor believes that demand for the candy follows a seasonal pattern. It has collected demand data (i.e., cases sold) for Diamonds from its warehouses and the stores it supplies for the past 3 years, as follows:

Month	Demand (cases)		
	Year 1	*Year 2*	*Year 3*
January	192	212	228
February	210	223	231
March	205	216	226
April	260	252	293
May	228	235	246
June	172	220	229
July	160	209	217
August	147	231	226
September	256	263	302
October	342	370	411
November	251	260	279
December	273	277	293

The distributor must hold the candy inventory in climate-controlled warehouses and be careful in handling it. The annual carrying cost is $116 per case. Diamonds must be shipped a long distance from the manufacturer to the distributor, and in order to keep the candy as fresh as possible, trucks must be air-conditioned, shipments must be direct, and shipments are often less than a truckload. As a result, the ordering cost is $4,700.

Diamant Foods makes Diamonds from three primary ingredients it orders from different suppliers: dark and milk chocolate, macadamia nuts, and a special heavy cream filling. Except for its unique shape, a Diamond is almost like a chocolate truffle. Each Diamond weighs 1.2 ounces and requires 0.70 ounce of blended chocolates, 0.50 ounce of macadamia nuts, and 0.40 ounce of filling to produce (including spillage and waste). Diamant Foods orders chocolate, nuts, and filling from its suppliers by the pound. The annual ordering cost is $5,700 for chocolate, and the annual carrying cost is $0.45 per pound. The ordering cost for macadamia nuts is $6,300, and the annual carrying cost is $0.63 per pound. The ordering cost for filling is $4,500, and the annual carrying cost is $0.55 per pound.

Each of the suppliers offers the candy manufacturer a quantity discount price schedule for the ingredients, as follows:

Chocolate		Macadamia Nuts		Filling	
Price	*Quantity (lb.)*	*Price*	*Quantity (lb.)*	*Price*	*Quantity (lb.)*
$3.05	0–50,000	$6.50	0–30,000	$1.50	0–40,000
2.90	50,001–100,000	6.25	30,001–70,000	1.35	40,001–80,000
2.75	100,001–150,000	5.95	70,001+	1.25	80,001+
2.60	150,001+				

Determine the inventory order quantity for Diamant's distributor. Compare the optimal order quantity with a seasonally adjusted forecast for demand. Does the order quantity seem adequate to meet the seasonal demand pattern for Diamonds (i.e., is it likely that shortages or excessive inventories will occur)? Can you identify the causes of the seasonal demand pattern for Diamonds? Determine the inventory order quantity for each of the three primary ingredients that Diamant Foods orders from its suppliers.

Appendix A
Normal and Chi-Square Tables

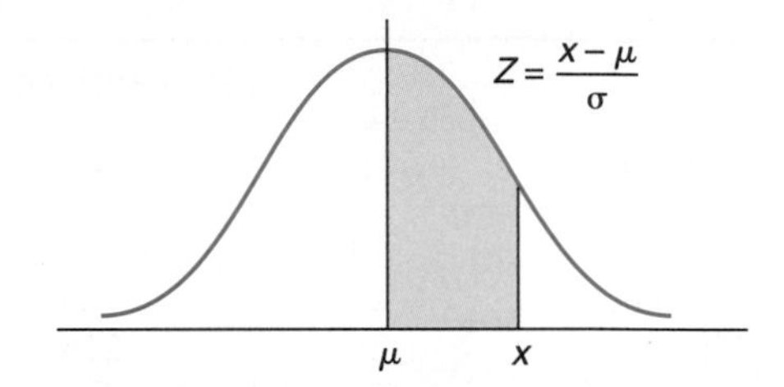

TABLE A.1
The normal table
Normal curve areas

Z	.00	.01	.02	.03	.04	.05	.06	.07	.08	.09
0.0	.0000	.0040	.0080	.0120	.0160	.0199	.0239	.0279	.0319	.0359
0.1	.0398	.0438	.0478	.0517	.0557	.0596	.0636	.0675	.0714	.0753
0.2	.0793	.0832	.0871	.0910	.0948	.0987	.1026	.1064	.1103	.1141
0.3	.1179	.1217	.1255	.1293	.1331	.1368	.1406	.1443	.1480	.1517
0.4	.1554	.1591	.1628	.1664	.1700	.1736	.1772	.1808	.1844	.1879
0.5	.1915	.1950	.1985	.2019	.2054	.2088	.2123	.2157	.2190	.2224
0.6	.2257	.2291	.2324	.2357	.2389	.2422	.2454	.2486	.2517	.2549
0.7	.2580	.2611	.2642	.2673	.2704	.2734	.2764	.2794	.2823	.2852
0.8	.2881	.2910	.2939	.2967	.2995	.3023	.3051	.3078	.3106	.3133
0.9	.3159	.3186	.3212	.3238	.3264	.3289	.3315	.3340	.3365	.3389
1.0	.3413	.3438	.3461	.3485	.3508	.3531	.3554	.3577	.3599	.3621
1.1	.3643	.3665	.3686	.3708	.3729	.3749	.3770	.3790	.3810	.3830
1.2	.3849	.3869	.3888	.3907	.3925	.3944	.3962	.3980	.3997	.4015
1.3	.4032	.4049	.4066	.4082	.4099	.4115	.4131	.4147	.4162	.4177
1.4	.4192	.4207	.4222	.4236	.4251	.4265	.4279	.4292	.4306	.4319
1.5	.4332	.4345	.4357	.4370	.4382	.4394	.4406	.4418	.4429	.4441
1.6	.4452	.4463	.4474	.4484	.4495	.4505	.4515	.4525	.4535	.4545
1.7	.4554	.4564	.4573	.4582	.4591	.4599	.4608	.4616	.4625	.4633
1.8	.4641	.4649	.4656	.4664	.4671	.4678	.4686	.4693	.4699	.4706
1.9	.4713	.4719	.4726	.4732	.4738	.4744	.4750	.4756	.4761	.4767
2.0	.4772	.4778	.4783	.4788	.4793	.4798	.4803	.4808	.4812	.4817
2.1	.4821	.4826	.4830	.4834	.4838	.4842	.4846	.4850	.4854	.4857
2.2	.4861	.4864	.4868	.4871	.4875	.4878	.4881	.4884	.4887	.4890
2.3	.4893	.4896	.4898	.4901	.4904	.4906	.4909	.4911	.4913	.4916
2.4	.4918	.4920	.4922	.4925	.4927	.4929	.4931	.4932	.4934	.4936
2.5	.4938	.4940	.4941	.4943	.4945	.4946	.4948	.4949	.4951	.4952
2.6	.4953	.4955	.4956	.4957	.4959	.4960	.4961	.4962	.4963	.4964
2.7	.4965	.4966	.4967	.4968	.4969	.4970	.4971	.4972	.4973	.4974
2.8	.4974	.4975	.4976	.4977	.4977	.4978	.4979	.4979	.4980	.4981
2.9	.4981	.4982	.4982	.4983	.4984	.4984	.4985	.4985	.4986	.4986
3.0	.4987	.4987	.4987	.4988	.4988	.4989	.4989	.4989	.4990	.4990

TABLE A.2

Chi-square table

For a particular number of degrees of freedom, an entry represents the critical value of χ^2 corresponding to a specified upper tail area (α)

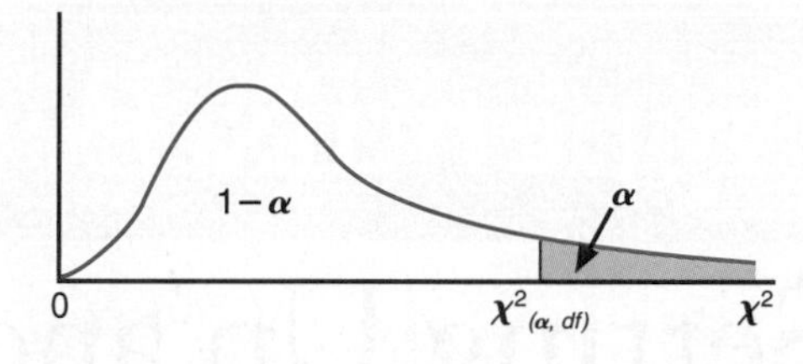

	Upper Tail Areas (α)											
Degrees of Freedom	**.995**	**.99**	**.975**	**.95**	**.90**	**.75**	**.25**	**.10**	**.05**	**.025**	**.01**	**.005**
1			0.001	0.004	0.016	0.102	1.323	2.706	3.841	5.024	6.635	7.879
2	0.010	0.020	0.051	0.103	0.211	0.575	2.773	4.605	5.991	7.378	9.210	10.597
3	0.072	0.115	0.216	0.352	0.584	1.213	4.108	6.251	7.815	9.348	11.345	12.838
4	0.207	0.297	0.484	0.711	1.064	1.923	5.385	7.779	9.488	11.143	13.277	14.860
5	0.412	0.554	0.831	1.145	1.610	2.675	6.626	9.236	11.071	12.833	15.086	16.750
6	0.676	0.872	1.237	1.635	2.204	3.455	7.841	10.645	12.592	14.449	16.812	18.548
7	0.989	1.239	1.690	2.167	2.833	4.255	9.037	12.017	14.067	16.013	18.475	20.278
8	1.344	1.646	2.180	2.733	3.490	5.071	10.219	13.362	15.507	17.535	20.090	21.955
9	1.735	2.088	2.700	3.325	4.168	5.899	11.389	14.684	16.919	19.023	21.666	23.589
10	2.156	2.558	3.247	3.940	4.865	6.737	12.549	15.987	18.307	20.483	23.209	25.188
11	2.603	3.053	3.816	4.575	5.578	7.584	13.701	17.275	19.675	21.920	24.725	26.757
12	3.074	3.571	4.404	5.226	6.304	8.438	14.845	18.549	21.026	23.337	26.217	28.299
13	3.565	4.107	5.009	5.892	7.042	9.299	15.984	19.812	22.362	24.736	27.688	29.819
14	4.075	4.660	5.629	6.571	7.790	10.165	17.117	21.064	23.685	26.119	29.141	31.319
15	4.601	5.229	6.262	7.261	8.547	11.037	18.245	22.307	24.996	27.488	30.578	32.801
16	5.142	5.812	6.908	7.962	9.312	11.912	19.369	23.542	26.296	28.845	32.000	34.267
17	5.697	6.408	7.564	8.672	10.085	12.792	20.489	24.769	27.587	30.191	33.409	35.718
18	6.265	7.015	8.231	9.390	10.865	13.675	21.605	25.989	28.869	31.526	34.805	37.156
19	6.844	7.633	8.907	10.117	11.651	14.562	22.718	27.204	30.144	32.852	36.191	38.582
20	7.434	8.260	9.591	10.851	12.443	15.452	23.828	28.412	31.410	34.170	37.566	39.997
21	8.034	8.897	10.283	11.591	13.240	16.344	24.935	29.615	32.671	35.479	38.932	41.401
22	8.643	9.542	10.982	12.338	14.042	17.240	26.039	30.813	33.924	36.781	40.289	42.796
23	9.260	10.196	11.689	13.091	14.848	18.137	27.141	32.007	35.172	38.076	41.638	44.181
24	9.886	10.856	12.401	13.848	15.659	19.037	28.241	33.196	36.415	39.364	42.980	45.559
25	10.520	11.524	13.120	14.611	16.473	19.939	29.339	34.382	37.652	40.646	44.314	46.928
26	11.160	12.198	13.844	15.379	17.292	20.843	30.435	35.563	38.885	41.923	45.642	48.290
27	11.808	12.879	14.573	16.151	18.114	21.749	31.528	36.741	40.113	43.194	46.963	49.645
28	12.461	13.565	15.308	16.928	18.939	22.657	32.620	37.916	41.337	44.461	48.278	50.993
29	13.121	14.257	16.047	17.708	19.768	23.567	33.711	39.087	42.557	45.722	49.588	52.336
30	13.787	14.954	16.791	18.493	20.599	24.478	34.800	40.256	43.773	46.979	50.892	53.672

Appendix B

Setting Up and Editing a Spreadsheet

One of the benefits of using Excel spreadsheets to solve management science problems is that spreadsheets provide a nice medium for visual presentation. They enable the user to customize the problem presentation in any manner or format desired. However, problems and models do not just "appear" on spreadsheets as they do in this book without some careful editing. The purpose of this brief appendix is to review some of the steps required to set up and edit spreadsheets like the ones shown in this book.

Virtually all the spreadsheet editing functions and tools can be accessed directly from the toolbars at the top of the spreadsheet window. Exhibit B.1 shows our Excel spreadsheet originally

EXHIBIT B.1

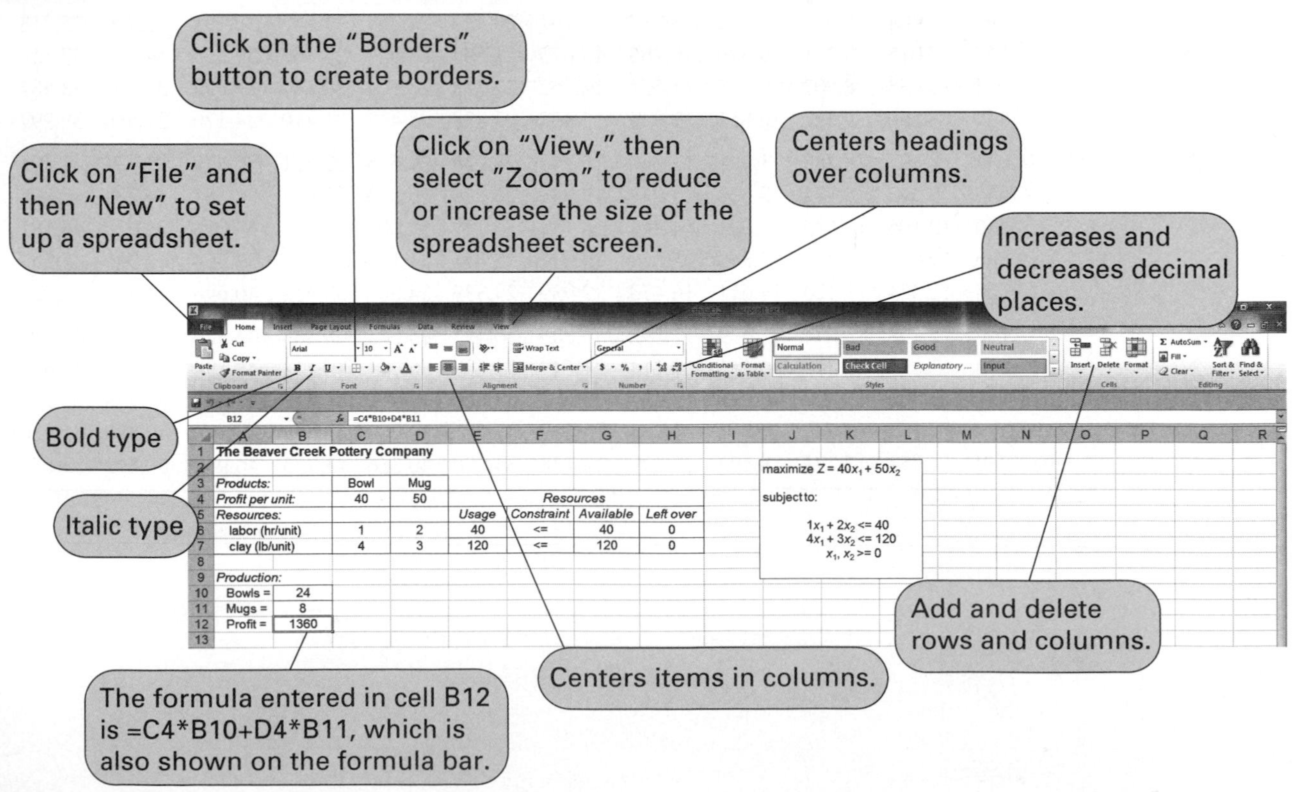

shown in Exhibit 3.4 in Chapter 3 with the solution to the Beaver Creek Pottery Company lincar programming example. We will describe the Excel editing features as they apply to this example spreadsheet. If you have not read Chapter 3 yet, don't worry; it's not necessary to know how to solve a linear programming problem in order to understand the spreadsheet editing tools.

Titles and Headings

The title "The Beaver Creek Pottery Company" was typed in cell A1. The **bold type** for this heading was created by activating the bold type button, "**B**," on the toolbar. Note that we could have centered this heading across the top of the spreadsheet by covering row 1 with the cursor and clicking on the button with an "a" with arrows on both sides of it (Merge & Center). *Italic type* for headings such as the "*Products*" heading in cell A3 is created by using the italic button, "*I*," on the toolbar.

Borders

You will notice that all the data and headings in Exhibit B.1 are enclosed by boxes created by black line borders. These borders are created by clicking on the "Borders" button on the toolbar at the top of the spreadsheet. Clicking on this button will bring down a selection of different line locations and line intensities for borders. Activate the border you want, cover the area on your spreadsheet where you want to include this border, and then click on the activated border that you selected. Most of the borders in Exhibit B.1 were created by using a border that results in a closed box. For example, the border that surrounds the headings and numbers in cells **A3:D7** was created by covering this area with the cursor and clicking on the "Borders" button at the top of the spreadsheet.

Column Centering

Headings and numbers can be centered in the cells (or moved to the right or left side of the cells) by using the "Merge & Center" button on the toolbar. To center a column of values, such as column C in Exhibit B.1, cover cells **C3:C7** with the cursor and then click on the "Center" button.

Deleting and Inserting Rows and Columns

Rows and columns can be inserted in your spreadsheet by clicking the cursor at the row or column you want a new row or column to appear next to; then click on "Insert" at the top of the spreadsheet window and select to insert either rows or columns from the menu. Rows and columns can be deleted by clicking the cursor on the row or column you want to delete and then selecting "Delete" at the top of the spreadsheet window.

Decimal Places

The number of decimal places in a numeral can be reduced, increased, or eliminated by using the "Decimal" buttons on the toolbar. Notice that there is one button to increase the number of decimal places and one button to decrease the number of decimal places.

Increasing or Decreasing the Spreadsheet Area

The spreadsheet in Exhibit B.1 has 14 columns (A through R) and 19 rows, which are less than the numbers of rows and columns that are present when a new spreadsheet is activated. Click on the "View" tab on the toolbar at the top of the spreadsheet, then on the "Zoom" button. Notice that the "Magnification" is 140%, which indicates that the spreadsheet is 140% its normal

(100%) size, thus reducing the number of rows and columns visible on the screen. For this example the spreadsheet was blown up a little, strictly for presentation purposes, so the typed material in the spreadsheet would be easier to read. To increase or reduce the visible spreadsheet, click on "View" at the top of the window, then click on "Zoom," and this will display a window in which you can alter the size of the spreadsheet. Increasing the size above 100% will reduce the number of rows and columns, and decreasing the spreadsheet size below 100% will increase the number of rows and columns that will appear on the spreadsheet screen.

Expanding or Reducing Column and Row Widths

Sometimes it is desirable to expand the width of a column so that you can see a longer heading or a larger number, or to reduce the width to reflect a smaller number. To accomplish this, click and hold the left mouse button at the very top of the column between the two column letters on the line separating the column you want to expand and the column next to it. This should result in a set of "crossed arrows," which you can move left or right to expand or reduce the width of the column by the amount you want. The same procedure can be followed to expand row widths.

Inserting an Equation or a Formula into a Cell

To insert a number or word in a cell, you normally position the cursor on the target cell and type the word or number, then press Enter or click the cursor on another cell. To type a formula or an equation into a cell, you click the cursor on the target cell and then an = sign, then your formula (for example, the formula in cell B12 in Exhibit B.1). The formula is shown on the formula bar at the top of the spreadsheet. Be aware that Excel is very sensitive to the order of mathematical operations, so your formulas must be very clearly stated. In general, you use an asterisk (*) to show multiplication, a slash (/) to show division, and "^2" for squaring.

Printing a Spreadsheet

The spreadsheets shown in the various exhibits in this book were developed by using a "screen capture" program, which allows the entire screen to be printed just as you would see it on the screen. However, unless your computer has a screen capture program, you should not expect your spreadsheets to look exactly like the ones in this book when you print them out. Instead, you will print your spreadsheet out by using the normal Excel and Windows print routines. To set up your spreadsheet document for printing, click on the "File" button located in the top-left-hand corner of the window and then click on "Print," then "Page Setup." This will result in the window shown in Exhibit B.2. This screen enables you to reduce or increase the spreadsheet size as well as position the spreadsheet on the page in portrait or landscape format. Notice the "Header/Footer" tab at the top of this window. By clicking on this tab, you can delete the page numbers and other spreadsheet information, called headers and footers, that normally appears at the top and bottom of a printed spreadsheet, or you can customize your own header and footer. By clicking on the "Sheet" tab at the top of the "Page Setup" window, you can remove the grid lines that separate the cells on every spreadsheet. When you do so, only the borders that you have created will show up. You can preview what your spreadsheet will look like by selecting "Print Preview."

EXHIBIT B.2

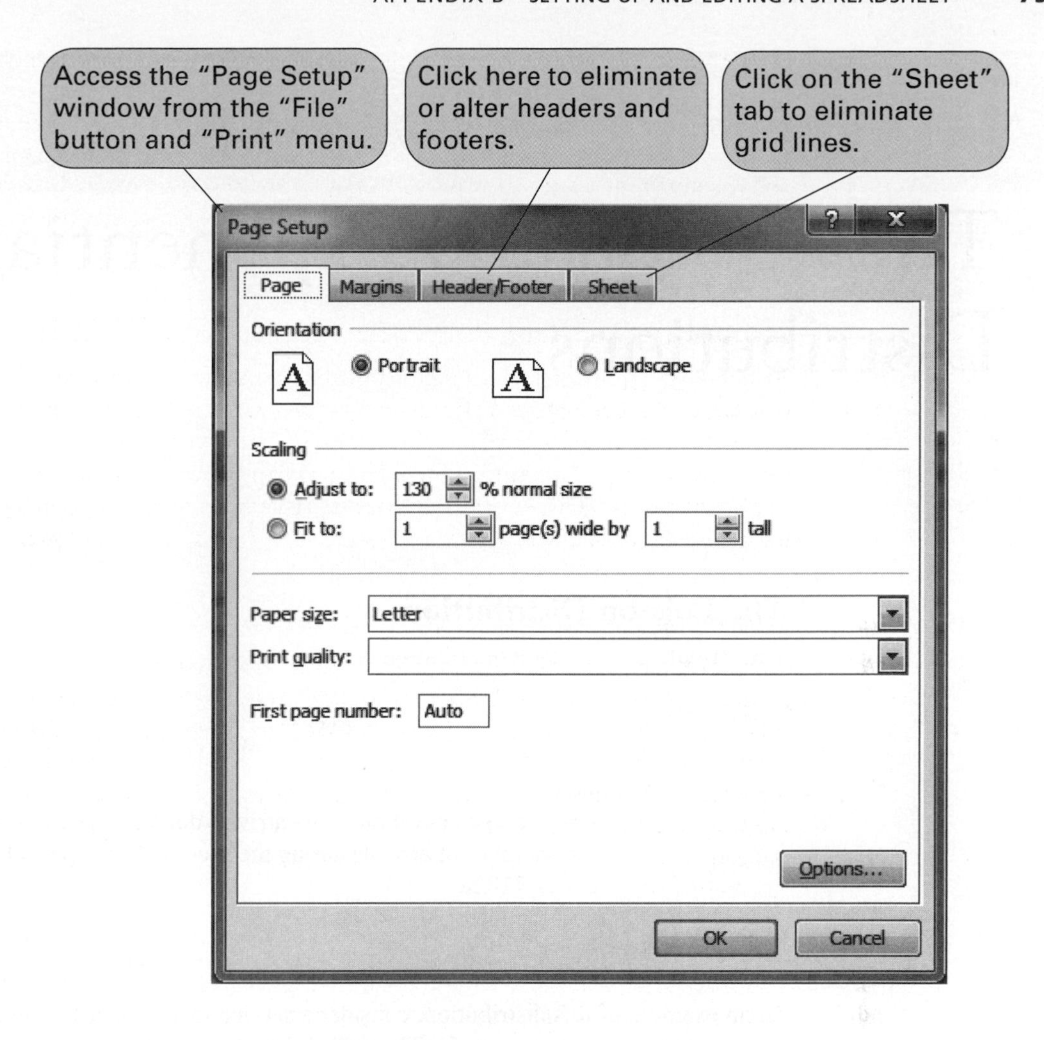

The editing features just described are the most basic and are provided to help you get started in setting up and solving problems by using Excel spreadsheets. More sophisticated Excel functions that can be used to solve problems are located in specific chapters where they are applied and can be located by using the book's index.

Appendix C
The Poisson and Exponential Distributions

The Poisson Distribution

The formula for a Poisson distribution is

$$P(x) = \frac{\lambda^x e^{-\lambda}}{x!}$$

where

λ = average arrival rate (i.e., arrivals during a specified period of time)
x = number of arrivals during the specified time period
e = 2.71828
$x!$ = the factorial of a value x

$$[\text{i.e., } x! = x(x-1)(x-2)\cdots(3)(2)(1)]$$

As an example of this distribution, consider a service facility that has an average arrival rate of five customers per hour ($\lambda = 5$). The probability that exactly two customers will arrive at the service facility is found by letting $x = 2$ in the preceding Poisson formula:

$$P(x = 2) = \frac{5^2 e^{-5}}{2!} = \frac{25(.007)}{(2)(1)} = .084$$

The value .084 is the probability of exactly two customers arriving at the service facility.

By substituting values of x into the Poisson formula, we can develop a distribution of customer arrivals during a 1-hour period, as shown in Figure C.1. However, remember that this distribution is for an arrival rate of five customers per hour. Other values of λ will result in distributions different from the one in Figure C.1.

FIGURE C.1
Poisson distribution for λ = 5

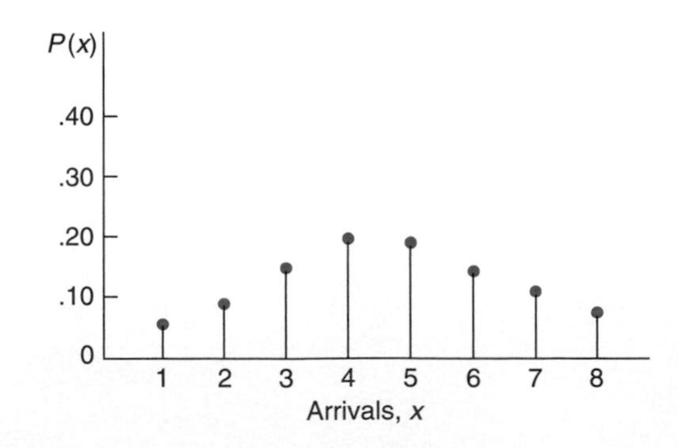

The Exponential Distribution

The formula for the exponential distribution is

$$f(t) = \mu e^{-\mu t}, \; t \geq 0$$

where

μ = average number of customers served during a specified period of time
t = service time
e = 2.71828

The *probability* that a customer will be served within a specified time period can be determined by using the exponential distribution, in the following form:

$$P(T \leq t) = 1 - e^{-\mu t}$$

If, for example, the service rate is six customers per hour, then the probability that a customer will be served within 10 minutes (.17 hour) is determined as follows:

$$\begin{aligned} P(T \leq .17) &= 1 - e^{-6(.17)} \\ &= 1 - e^{-1.0} \\ &= 1 - .368 \\ &= .632 \end{aligned}$$

Thus, the probability of a customer's being served within 10 minutes is .632. Figure C.2 is the exponential probability distribution for this service rate ($\mu = 6$).

FIGURE C.2
Exponential distribution for $\mu = 6$

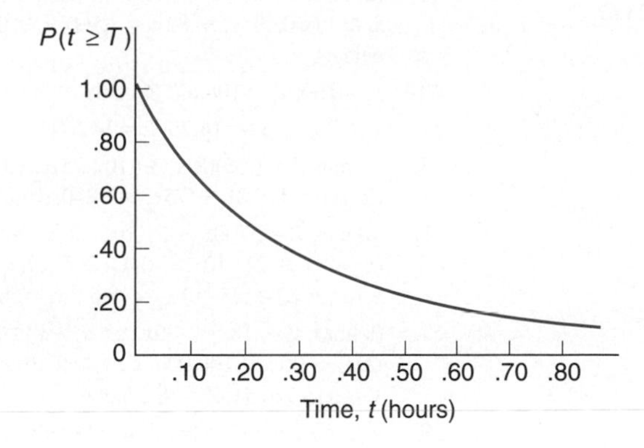

Solutions to Selected Odd-Numbered Problems

Chapter 1

1. (a) $TC = \$27{,}500$; $TR = \$54{,}000$; $Z = \$26{,}500$; (b) $v = 69.56$ tables
3. (a) $TC = \$29{,}100$; $TR = \$23{,}400$; $Z = -\$5{,}700$; (b) $v = 24{,}705.88$ yd./month
7. (a) $v = 1{,}250$ dolls
9. 98.8%
11. reduces v to 2,727.3 tires
13. increases v to 65,789.47 lb.
15. do not raise price
17. yes, v will be reduced
19. (a) executive plan; (b) 937.5 min. per month
21. (a) $v = 26.9$ pupils; (b) $v = 106.3$ pupils; (c) $p = \$123.67$
23. $v = 209.2$ teams
25. $Z = \$138{,}725$; $p = \$6.25$; $v = 47{,}125$
27. (a) $v = 31.25$; (b) $Z = \$1{,}804$; (c) $Z = \$1{,}612$; no
29. $x = 0$, $y = 50$
31. bowls = 2, mugs = 2, $Z = 1{,}100$
33. 34,500 visits
35. 4 registers

Chapter 2

1. $x_1 = 4, x_2 = 0, Z = 40$
3. (a) min. $Z = .05x_1 + .03x_2$; s.t. $8x_1 + 6x_2 \geq 48$, $x_1 + 2x_2 \geq 12$, $x_i \geq 0$; (b) $x_1 = 0, x_2 = 8, Z = 0.24$
5. (a) min. $Z = 3x_1 + 5x_2$; s.t. $10x_1 + 2x_2 \geq 20$, $6x_1 + 6x_2 \geq 36$, $x_2 \geq 2$, $x_i \geq 0$; (b) $x_1 = 4, x_2 = 2, Z = 22$
7. No labor, 4.8 wood
9. (a) max. $Z = x_1 + 5x_2$; s.t. $5x_1 + 5x_2 \leq 25$, $2x_1 + 4x_2 \leq 16$, $x_1 \leq 5$, $x_1, x_2 \geq 0$; (b) $x_1 = 0, x_2 = 4, Z = 20$
11. $x_1 = 0, x_2 = 9, Z = 54$
13. (a) max. $Z = 300x_1 + 400x_2$; s.t. $3x_1 + 2x_2 \leq 18$, $2x_1 + 4x_2 \leq 20$, $x_2 \leq 4$, $x_1, x_2 \geq 0$; (b) $x_1 = 4, x_2 = 3, Z = 2{,}400$
15. (a) maximum demand is not achieved by one bracelet; (b) \$600
17. $x_1 = 15.8, x_2 = 20.5, Z = 1{,}610$
19. (a) no; (b) $x_1 = 55, x_2 = 16.25, Z = 1{,}851$; (c) $Z = 2{,}435$; (d) $Z = 2{,}073$.
21. A: $s_1 = 4, s_2 = 1, s_3 = 0$; B: $s_1 = 0, s_2 = 5, s_3 = 0$; C: $s_1 = 0, s_2 = 6, s_3 = 1$
23. A: $s_1 = 0, s_2 = 0, s_3 = 8, s_4 = 0$; B: $s_1 = 0, s_2 = 3.2, s_3 = 0, s_4 = 4.8$; C: $s_1 = 26, s_2 = 24, s_3 = 0, s_4 = 10$
25. changes the optimal solution
27. $x_1 = 28.125, x_2 = 0, Z = \$1{,}671.95$; no effect
29. infeasible solution
31. $x_1 = 4, x_2 = 1, Z = 18$
33. $x_1 = 4.8, x_2 = 2.4, Z = 26.4$
35. $x_1 = 3.2, x_2 = 6, Z = 37.6$
37. no additional profit
39. (a) max. $Z = 800x_1 + 900x_2$; s.t. $2x_1 + 4x_2 \leq 30$, $4x_1 + 2x_2 \leq 30$, $x_1 + x_2 \geq 9$, $x_i \geq 0$; (b) $x_1 = 5, x_2 = 5, Z = 8{,}500$
41. $x_1 = 5.3, x_2 = 4.7, Z = 806$
43. (a) 12 hr.; (b) new solution: $x_1 = 5.09, x_2 = 5.45, Z = 111.27$
45. $x_1 = 38.4, x_2 = 57.6, Z = 19.78$; profit reduced
47. (a) min. $Z = .09x_1 + .18x_2$, s.t. $.46x_1 + .35x_2 \leq 2{,}000$, $x_1 \geq 1{,}000$, $x_2 \geq 1{,}000$, $.91x_1 - .82x_2 = 3{,}500$, $x_1 \geq 0, x_2 \geq 0$; (b) 32 fewer defects
49. $x_1 = 160, x_2 = 106.67, Z = 568$
51. $x_1 = 25.71, x_2 = 14.29, Z = 14{,}571$
53. (a) min. $Z = (.05)(8)x_1 + (.10)(.75)x_2$; s.t. $5x_1 + 8x_2 \geq 800$, $(5x_1/x_2) = 1.5$, $8x_1 + .75x_2 \leq 1200$; (b) $x_1 = 96, x_2 = 320, Z = 62.40$
55. (a) min. $Z = 3700x_1 + 5100x_2$; s.t. $x_1 + x_2 = 45$, $(32x_1 + 14x_2)/(x_1 + x_2) \leq 21$, $.10x_1 + .04x_2 \leq 6$, $x_1/(x_1 + x_2) \geq .25$, $x_2/(x_2 + x_2) \geq .25$; (b) $x_1 = 18.67$; $x_2 = 26.3$; $Z = 203{,}367$
57. (a) max. $Z = .18x_1 + .06x_2$, s.t. $x_1 + x_2 \leq 720{,}000$, $x_1/(x_1 + x_2) \leq .65$, $.22x_1 + .05x_2 \leq 100{,}000$, $x_1, x_2 \geq 0$; (b) $x_1 = 376{,}470.59$, $x_2 = 343{,}526.41$, $Z = 88{,}376.47$
59. one more hour for Sarah would reduce the regraded exams from 10 to 9.8; another hour for Brad has no effect
61. only more Colombian would affect the solution; 1 lb of Colombian would increase sales to \$463.20; increasing the brewing capacity has no effect; extra advertising increases sales
63. infeasible

Chapter 3

3. Cells: B10:B12; Constraints: B10:B12 ≥ 0, G6 ≤ F6, G7 ≤ F7; Profit: = B10*C4+B11*D4+B12*E4; Target cell = B13.
5. F6 = C6*B12 + D6*B13, F7 = C7*B12 + D7*B13, G6 = E6 − F6 G7 = E7 − F7, B14 = C4*B12 + D4*B13, $x_1 = 8, x_2 = 5.2, Z = 81.6$
7. (a and b) max. $Z = 12x_1 + 16x_2 + 0s_1 + 0s_2$; s.t. $3x_1 + 2x_2 + s_1 = 500$, $4x_1 + 5x_2 + s_2 = 800$, $x_i \geq 0$, $s_i \geq 0$
9. (a and b) $-\infty \leq c_1 \leq 12.8$, $15 \leq c_2 \leq \infty$, $320 \leq q_1 \leq \infty$, $0 \leq q_2 \leq 1{,}250$; (c) \$0, \$3.20

11. (a) $x_1 = 4, x_2 = 3, Z = 57$, A: $s_1 = 40, s_2 = 0$, B: $s_1 = 0, s_2 = 0$, C: $s_1 = 0, s_2 = 20$; (b) $x_1 = 2, x_2 = 4$; (c) solution point same for profit = \$15, new solution $x_1 = 0, x_2 = 5, Z = 100$ for profit = \$20
13. (a and b) max. $Z = 2.25x_1 + 3.10x_2 + 0s_1 + 0s_2 + 0s_3$; s.t. $5.0x_1 + 7.5x_2 + s_1 = 6{,}500$, $3.0x_1 + 3.2x_2 + s_2 = 3{,}000$, $x_2 + s_3 = 510$, $x_i \geq 0$, $s_j \geq 0$
15. (a) Additional processing time, \$0.75/hr.; (b) $0 \leq c_1 \leq 2.906$, $2.4 \leq c_2 \leq \infty$, $6{,}015 \leq q_1 \leq \infty$, $1{,}632 \leq q_2 \leq 3{,}237$, $0 \leq q_3 \leq 692.308$
17. (a) $x_1 = 4, x_2 = 0, s_1 = 12, s_2 = 0, s_3 = 11, Z = 24{,}000$; (b) $x_1 = 1, x_2 = 3, Z = 28{,}500$; (c) C still optimal
19. (a and b) max. $Z = 300x_1 + 520x_2$; s.t. $x_1 + x_2 = 410$, $105x_1 + 210x_2 = 52{,}500$, $x_2 = 100$, $x_i \geq 0$
21. (a) no, max. price = \$80; (b) \$2.095
23. (a) $x_1 = 300, x_2 = 100, s_1 = 0$ lb., $s_2 = 15$ lb., $s_4 = 0.6$ hr., $Z = 230$; (b) $c_1 = \$0.60, x_1 = 257, x_2 = 143, Z = 240$; (c) $x_1 = 300, x_2 = 125, Z = 242.50$
25. (a and b) min. $Z = 50x_1 + 70x_2$; s.t. $80x_1 + 40x_2 = 3{,}000$, $80x_1 = 1{,}000$, $40x_2 = 800$, x_i
27. (a) personal interviews, \$0.625/interview; (b) $25 \leq c_2 \leq \infty$, $1{,}800 \leq q_1 \leq \infty$
29. (a) $x_1 = 333.3, x_2 = 166.7, s_1 = 100$ gal., $s_2 = 133.3$ gal., $s_3 = 83.3$ gal., $s_4 = 100$ gal., $Z = 1{,}666$; (b) any values of c
31. (a) max. $Z = 8.65x_1 + 10.95x_2$; s.t. $x_1 + x_2 \leq 250$, $x_1 + x_2 \geq 120$, $x_1 - 2x_2 \leq 0$, $-x_1 + 1.2x_2 \leq 0$, $x_1 \leq 150$, $x_2 \leq 110$, $x_i \geq 0$; (b) $x_1 = 140, x_2 = 110, Z = 2{,}415.50$
33. min. $Z = 11x_1 + 16x_2$; s.t. $x_1 + x_2 = 500$, $0.7x_1 + .02x_2 \leq 25$, $x_1/(x_1 + x_2) \geq .20$, $x_2/(x_1 + x_2) \geq .20$, $x_i \geq 0$
35. max. $Z = 1.20x_1 + 1.30x_2$; s.t. $x_1 + x_2 \leq 95{,}000$, $.18x_1 + .30x_2 \leq 20{,}000$, $x_i \geq 0$
37. (a) 5%, \$16,111.11; (b) $x_1 = 0, x_2 = 66{,}666.7, Z = 86{,}666.67$
39. (a and b) min. $Z = 400x_1 + 180x_2 + 90x_3$; s.t. $x_1 \geq 200$, $x_2 \geq 300$, $x_3 \geq 100$, $4x_3 - x_1 - x_2 \leq 0$, $x_1 + x_2 + x_3 = 1{,}000$, $x_i \geq 0$; (c) $x_1 = 200, x_2 = 600, x_3 = 200, Z = 206{,}000$
41. max. $Z = 0.50x_1 + 0.75x_2$; s.t. $0.17x_1 + 0.25x_2 \leq 4{,}000$, $x_1 + x_2 \leq 18{,}000$, $x_1 \geq 8{,}000$, $x_2 \geq 8{,}000$, $x_1, x_2 \geq 0$; (b) max. $Z = 0.50x_1 + 0.75x_2$; $0.17x_1 + 0.25x_2 = 4{,}000$, $x_1 + x_2 = 18{,}000$, $x_1 = 8{,}000$, $x_2 = 8{,}000$, $x_i \geq 0$
43. $x_1 = 8{,}000, x_2 = 10{,}000, Z = 11{,}500$; (a) \$375; increase, $x_1 = 8{,}000, x_2 = 10{,}500, Z = \$11{,}875$; $x_1 = 8{,}000, x_2 = 10{,}560, Z = \$11{,}920$
45. (a) purchase land; (b) not purchase land
47. (a) \$0.78, 360 cartons; (b) \$0; (c) $x_1 = 108, x_2 = 54, x_3 = 162, Z = 249.48$, no discount
49. $x_1 = 3, x_3 = 6, Z = 3{,}600$; (a) more assembly hr.; (b) additional profit = \$600; (c) no effect
51. $x_1 = 1{,}000, x_2 = 800, x_3 = 200, Z = 760$; (a) increase by 100, \$38 in additional profit; (b) $x_1 = 1{,}000, x_2 = 1{,}000, Z = 770$; (c) $Z = 810$; $x_1 = 1{,}600, x_2 = 200, x_3 = 200$
53. $x_1 = 20, x_2 = 33.33, x_3 = 26.67, Z = \$703{,}333.4$
55. $x_{13} = 350, x_{21} = 158.33, x_{22} = 296.67, x_{23} = 75, x_{31} = 610, x_{42} = 240, Z = \$77{,}910$
57. max. $Z = 12x_1 + 8x_2 + 10x_3 + 6x_4$; s.t. $2x_1 + 9x_2 + 1.3x_3 + 2.5x_4 \leq 74$; $2x_1 + .25x_2 + x_3 + x_4 \leq 24$; $2x_1 + 5x_3 + 2x_4 \leq 36$; $45x_1 + 35x_2 + 50x_3 + 16x_4 \leq 480$, $x_i \geq 0$
59. max. $Z = \$5x_1 + 3x_2 + 7x_3 + 16x_4$; s.t. $8x_1 + 12x_2 + 32x_3 + 80x_4 \leq 80{,}000$, $\frac{55x_1}{4} + \frac{15x_2}{2} \leq 7{,}200$, $6x_1 + 4x_2 + 10x_3 + 1x_4 \leq 18{,}000$
61. (a) max. $Z = \$23x_1 + 61x_2$; s.t. $1.2x_1 + 5x_2 \leq 96$, $12x_1 + 20x_2 \leq 700$, $x_1 \leq 50$, $x_2 \leq 15$; (b) $x_1 = 43.89, x_2 = 8.67, Z = \$1{,}538.11$

Chapter 4

1. Model must be resolved; $Z = 43{,}310$, do not implement; no; $x_1 = x_2 = x_3 = x_4 = 112.5$
3. No effect; \$740; $x_1 = 22{,}363.636, x_3 = 43{,}636.364, x_4 = 14{,}000$
5. Add slack variables for 3 warehouses ≤ constraints; coefficients in objective function—\$9 for s_1, \$6 for s_2, \$7 for s_3, solution does not change
7. (a) max. $Z = 190x_1 + 170x_2 + 155x_3$; s.t. $3.5x_1 + 5.2x_2 + 2.8x_3 \leq 500$, $1.2x_1 + 0.8x_2 + 1.5x_3 \leq 240$, $40x_1 + 55x_2 + 20x_3 \leq 6{,}500$, $x_i \geq 0$; (b) $x_1 = 41.27, x_2 = 0, x_3 = 126.98, Z = 27{,}523.81$
9. (a) min. $Z = 4x_1 + 3x_2 + 2x_3$; s.t. $2x_1 + 4x_2 + x_3 \geq 16$, $3x_1 + 2x_2 + x_3 \geq 12$, $x_i \geq 0$; (b) $x_1 = 2, x_2 = 3, Z = \$0.17$
11. (a) max. $Z = 8x_1 + 10x_2 + 7x_3$; s.t. $7x_1 + 10x_2 + 5x_3 \leq 2{,}000$, $2x_1 + 3x_2 + 2x_3 \leq 660$, $x_1 \leq 200$, $x_2 \leq 300$, $x_3 \leq 150$, $x_1 \geq 0$, $x_2 \geq 0$, $x_3 \geq 0$; (b) $x_1 = 178.57, x_3 = 150, Z = 2{,}478.57$
13. (a) max. $Z = 7x_1 + 5x_2 + 5x_3 + 4x_4$; s.t. $2x_1 + 4x_2 + 2x_3 + 3x_4 \leq 45{,}000$, $x_1 + x_2 \leq 6{,}000$, $x_3 + x_4 \leq 7{,}000$, $x_1 + x_3 \leq 5{,}000$, $x_2 + x_4 \leq 6{,}000$, $x_1 \geq 0$, $x_2 \geq 0$, $x_3 \geq 0$, $x_4 \geq 0$; (b) $x_1 = 5{,}000, x_2 = 1{,}000, x_4 = 5{,}000, Z = 60{,}000$
15. (a) max. $Z = 1{,}800x_{1a} + 2{,}100x_{1b} + 1{,}600x_{1c} + 1{,}000x_{2a} + 700x_{2b} + 900x_{2c} + 1{,}400x_{3a} + 800x_{3b} + 2{,}200x_{3c}$; s.t. $x_{1a} + x_{1b} + x_{1c} = 30$, $x_{2a} + x_{2b} + x_{2c} = 30$, $x_{3a} + x_{3b} + x_{3c} = 30$, $x_{1a} + x_{2a} + x_{3a} \leq 40$, $x_{1b} + x_{2b} + x_{3b} \leq 60$, $x_{1c} + x_{2c} + x_{3c} \leq 50$, $x_{ij} \geq 0$; (b) $x_{1b} = 30, x_{2a} = 30, x_{3c} = 30, Z = 159{,}000$
17. (a) min. $Z = 1.7(x_{t1} + x_{t2} + x_{t3}) + 2.8(x_{m1} + x_{m2} + x_{m3}) + 3.25(x_{b1} + x_{b2} + x_{b3})$; s.t. $.50x_{t1} - .50x_{m1} - .50x_{b1} \leq 0$, $-.20x_{t1} + .80x_{m1} - .20x_{b1} \geq 0$, $-.30x_{t2} - .30x_{m2} + .70x_{b2} \geq 0$, $-.30x_{t2} + .70x_{m2} - .30x_{b2} \geq 0$, $.80x_{t2} - .20x_{m2} - .20x_{b2} \leq 0$, $.50x_{t3} - .50x_{m3} - .50x_{b3} \geq 0$, $.30x_{t3} - .70x_{m3} - .70x_{b3} \leq 0$, $-.10x_{t3} - .10x_{m3} + .90x_{b3} \geq 0$, $x_{t1} + x_{m1} + x_{b1} \geq 1{,}200$, $x_{t2} + x_{m2} + x_{b2} \geq 900$, $x_{t3} + x_{m3} + x_{b3} \geq 2{,}400$; $x_{ij} \geq 0$; (b) $x_{t1} = 600, x_{t2} = 180, x_{t3} = 1{,}680, x_{m1} = 600, x_{m2} = 450, x_{m3} = 480, x_{b1} = 0, x_{b2} = 270, x_{b3} = 240, Z = 10{,}123.50$
19. (a) max. $Z = .02x_1 + .09x_2 + .06x_3 + .04x_4$; s.t. $x_1 + x_2 + x_3 + x_4 = 4{,}000{,}000$, $x_1 \leq 1{,}600{,}000$, $x_2 \leq 1{,}600{,}000$, $x_3 \leq 1{,}600{,}000$, $x_4 \leq 1{,}600{,}000$, $x_2 - x_3 - x_4 \leq 0$, $x_1 - x_3 \geq 0$, $x_1 \geq 0$, $x_2 \geq 0$, $x_3 \geq 0$, $x_4 \geq 0$; (b) $x_1 = 800{,}000, x_2 = 1{,}600{,}000, x_3 = 800{,}000, x_4 = 800{,}000, Z = 240{,}000$
21. (a) max. $Z = 7.8x_{11} + 7.8x_{12} + 8.2x_{13} + 7.9x_{14} + 6.7x_{21} + 8.9x_{22} + 9.2x_{23} + 6.3x_{24} + 8.4x_{31} + 8.1x_{32} + 9.0x_{33} + 5.8x_{34}$; s.t. $35x_{11} + 40x_{21} + 38x_{31} \leq 9{,}000$, $41x_{12} + 36x_{22} + 37x_{32} \leq 14{,}400$, $34x_{13} + 32x_{23} + 33x_{33} \leq 12{,}000$, $39x_{14} + 43x_{24} + 40x_{34} \leq 15{,}000$, $x_{11} + x_{12} + x_{13} + x_{14} = 400$, $x_{21} + x_{22} + x_{23} + x_{24} = 570$, $x_{31} + x_{32} + x_{33} + x_{34} = 320$, $x_{ij} \geq 0$; (b) $x_{11} = 15.385, x_{14} = 384.615, x_{22} = 400, x_{23} = 170, x_{31} = 121.212, x_{33} = 198.788, Z = 11{,}089.73$
23. (a) min. $Z = 1.7x_{11} + 1.4x_{12} + 1.2x_{13} + 1.1x_{14} + 1.05x_{15} + 1.0x_{16} + 1.7x_{21} + 1.4x_{22} + 1.2x_{23} + 1.1x_{24} + 1.05x_{25} + 1.7x_{31} + 1.4x_{32} + 1.2x_{33} + 1.1x_{34} + 1.7x_{41} + 1.4x_{42} + 1.2x_{43} + 1.7x_{51} + 1.4x_{52} + 1.7x_{61}$; s.t. $x_{11} + x_{12} + x_{13} + x_{14} + x_{15} + x_{16} = 47{,}000$, $x_{12} + x_{13} + x_{14} + x_{15} + x_{16} + x_{21} + x_{22} + x_{23} + x_{24} + x_{25} = 35{,}000$, $x_{13} + x_{14} + x_{15} + x_{16} + x_{22} + x_{23} + x_{24} + x_{25} + x_{31} + x_{32} + x_{33} + x_{34} = 52{,}000$, $x_{14} + x_{15} + x_{16} + x_{23} + x_{24} + x_{25} + x_{32} + x_{33} + x_{34} + x_{41} + x_{42} + x_{43} = 27{,}000$, $x_{15} + x_{16} + x_{24} + x_{25} + x_{33} + x_{34} + x_{42} + x_{43} + x_{51} + x_{52} = 19{,}000$, $x_{16} + x_{25} + x_{34} + x_{43} + x_{52} + x_{61} = 15{,}000$; (b) $x_{11} = 12{,}000, x_{13} = 25{,}000, x_{14} = 8{,}000, x_{15} = 2{,}000, x_{33} = 2{,}000, x_{34} = 15{,}000, Z = \$80{,}200$; (c) $x_{16} = 52{,}000, Z = \$52{,}000$

25. (a) add $x_{ss} \leq 150$, $x_{ww} \leq 300$, $x_{cc} \leq 250$; $x_{nc} = 700$, $x_{nw} = 0$, $x_{sw} = 150$, $x_{ss} = 150$, $x_{es} = 900$, $x_{wc} = 250$, $x_{ww} = 300$, $x_{cc} = 250$, $x_{cw} = 250$, $x_{ws} = 50$, $Z = 20{,}400$; (b) changes demand constraints from $\leq 1{,}200$ to $= 1{,}000$; $x_{nc} = 400$, $x_{nw} = 300$, $x_{sw} = 150$, $x_{ss} = 150$, $x_{ec} = 50$, $x_{es} = 850$, $x_{wc} = 300$, $x_{ww} = 300$, $x_{cc} = 250$, $x_{cw} = 250$, $Z = 21{,}200$
27. (a) max. $Z = 2x_1 + 4x_2 + 3x_3 + 7x_4$; s.t. $x_2 + x_4 \leq 300$, $6x_1 + 15x_2 \leq 1{,}200$, $5x_3 + 12x_4 \leq 2{,}400$, $x_i \geq 0$; (b) $x_1 = 200$, $x_3 = 480$, $Z = 1{,}840$
29. (a) max. $Z = 35x_1 + 20x_2 + 58x_3$; s.t. $14x_1 + 12x_2 + 35x_3 \leq 35{,}000$, $6x_1 + 3x_2 + 12x_3 \leq 20{,}000$, $x_1 \geq 0$, $x_2 \geq 0$, $x_3 \geq 0$; (b) $x_1 = 2{,}500$, $Z = 87{,}500$
31. (a) min. $Z = 15{,}000x_1 + 4{,}000x_2 + 6{,}000x_3$; s.t. $x_3/x_2 \geq 2/1$, $25{,}000x_1 + 10{,}000x_2 + 15{,}000x_3 \geq 100{,}000$, $(15{,}000x_1 + 3{,}000x_2 + 12{,}000x_3)/(10{,}000x_1 + 7{,}000x_2 + 3{,}000x_3) \geq 2/1$, $(15{,}000x_1 + 4{,}000x_2 + 9{,}000x_3)/(25{,}000x_1 + 10{,}000x_2 + 15{,}000x_3) \geq .30$, $x_2 \leq 7$, $x_1 \geq 0$, $x_2 \geq 0$, $x_3 \geq 0$; (b) $x_2 = 2.5$, $x_3 = 5.0$, $Z = 40{,}000$; (c) x_4, no effect
33. (a) min. $Z = .11x_1 + .05\sum_1^6 y_i$; s.t. $x_1 + y_1 + 20{,}000 - c_1 = 60{,}000$, $c_1 + y_2 + 30{,}000 - c_2 = 60{,}000 + y_1$, $c_2 + y_3 + 40{,}000 - c_3 = 80{,}000 + y_2$, $c_3 + y_4 + 50{,}000 - c_4 = 30{,}000 + y_3$, $c_4 + y_5 + 80{,}000 - c_5 = 30{,}000 + y_4$, $c_5 + y_6 + 100{,}000 - c_6 = 20{,}000 + y_5$, $x_1 + y_6 \leq c_6$, $x_1, y_i, c_i \geq 0$; (b) $x_1 = 70{,}000$, $y_3 = 40{,}000$, $y_4 = 20{,}000$, $y_1 = y_2 = y_5 = y_6 = 0$, $c_1 = 30{,}000$, $c_5 = 30{,}000$, $c_6 = 110{,}000$, $Z = \$10{,}700$; (c) $x_1 = 90{,}000$, $y_3 = 20{,}000$, $c_1 = 50{,}000$, $c_2 = 20{,}000$, $c_5 = 50{,}000$, $c_6 = 130{,}000$, $Z = \$9{,}100$
35. (a) max. $Z = .7x_{cr} + .6x_{br} + .4x_{pr} + .85x_{ar} + 1.05x_{cb} + .95x_{bb} + .75x_{pb} + 1.20x_{ab} + 1.55x_{cm} + 1.45x_{bm} + 1.25x_{pm} + 1.70x_{am}$; s.t. $x_{cr} + x_{cb} + x_{cm} \leq 200$, $x_{br} + x_{bb} + x_{bm} \leq 300$, $x_{pr} + x_{pb} + x_{pm} \leq 150$, $x_{ar} + x_{ab} + x_{am} \leq 400$, $.90x_{br} + .90x_{pr} - .10x_{cr} - .10x_{ar} \leq 0$, $.80x_{cr} - .20x_{br} - .20x_{pr} - .20x_{ar} \geq 0$, $.25x_{bb} + .75x_{cb} - .75x_{pb} - .75x_{ab} \geq 0$, $x_{am} = 0$, $.5x_{bm} + .5x_{pm} - .5x_{cm} - .5x_{am} \leq 0$, $x_{ij} \geq 0$; (b) $x_{cm} = 125$, $x_{ar} = 300$, $x_{cr} = 75$, $x_{bb} = 300$, $x_{pm} = 125$, $x_{ab} = 100$, $Z = 1{,}602.50$
37. (a) min. $Z = 40x_1 + 65x_2 + 70x_3 + 30x_4$; s.t. $x_1 + x_2 = 250$, $x_3 + x_4 = 400$, $x_1 + x_3 = 300$, $x_2 + x_4 = 350$, $x_1 \geq 0$, $x_2 \geq 0$, $x_3 \geq 0$, $x_4 \geq 0$; (b) $x_1 = 250$, $x_3 = 50$, $x_4 = 350$, $Z = 24{,}000$
39. (a) max. $Z = 175\ (7x_1)$; s.t. $8x_1 + 5x_2 + 6.5x_3 \leq 3{,}000$, $x_1 + x_2 + x_3 \leq 120$, $90(7x_1) \leq 10{,}000$, $7x_1 - 12x_2 = 0$, $12x_2 - 10x_3 = 0$, $7x_1 - 10x_3 = 0$, $x_1 \geq 0$, $x_2 \geq 0$, $x_3 \geq 0$; (b) $x_1 = 15.9$, $x_2 = 9.3$, $x_3 = 11.1$, $Z = 19{,}444.44$
41. (a) min. $Z = 3x_{13} + 4x_{14} + 5x_{12} + 2x_{34} + 7x_{45} + 8x_{25}$; s.t. $x_{13} + x_{14} + x_{12} = 5$, $x_{45} + x_{25} = 5$, $x_{13} = x_{34}$, $x_{14} + x_{34} = x_{45}$, $x_{12} = x_{25}$, $x_{ij} \geq 0$; (b) $x_{14} = 5$, $x_{45} = 5$, $Z = 55{,}000$
43. (a) max. $Z = 0.30x_1 + 0.20x_2 + 0.05x_3 + 0.10x_4 + 0.15x_5$; s.t. $0.3x_1 + 0.2x_2 + 0.05x_3 + 0.1x_4 + 0.15x_5 \leq 4.0$, $x_1 + x_2 + x_3 + x_4 + x_5 \leq 24$, $x_1 \leq 4$, $x_2 \leq 8$, $x_3 \leq 10$, $x_4 \leq 3$, $x_4 \geq 2$, $x_5 \leq 10$, $x_5 \geq 3$, $x_i \geq 0$
45. (a) min. $Z = x_1 + x_2 + x_3 + x_4 + x_5 + x_6$; s.t. $3x_1 + 2x_2 + 2x_3 + x_4 = 700$, $x_3 + 2x_4 + x_5 = 1{,}200$, $x_2 + x_5 + 2x_6 = 300$, $x_i \geq 0$; (b) $x_2 = 50$, $x_4 = 600$, $x_6 = 125$, $Z = 775$; (c) min. $Z = 4x_1 + x_2 + 2x_3 + 0x_4 + 6x_5 + 5x_6$; s.t. $3x_1 + 2x_2 + 2x_3 + x_4 \geq 700$, $x_3 + 2x_4 + x_5 \geq 1{,}200$, $x_2 + x_5 + 2x_6 \geq 300$, $x_i \geq 0$; $x_2 = 300$, $x_4 = 600$, $Z = 300$
47. (a) max. $Z = 4x_1 + 8x_2 + 6x_3 + 7x_4 - 5y_1 - 6y_2 - 7y_3$; s.t. $s_0 = 2{,}000$, $x_1 \leq s_0$, $s_1 = s_0 - x_1 + y_1$, $s_1 \leq 10{,}000$, $x_2 \leq s_1$, $s_2 = s_1 - x_2 + y_2$, $s_2 \leq 10{,}000$, $x_3 \leq s_2$, $s_3 = s_2 - x_3 + y_3$, $s_3 \leq 10{,}000$, $x_4 \leq s_3$, $x_i \geq 0$, $y_i \geq 0$, $s_i \geq 0$; (b) $x_1 = 0$, $x_2 = 10{,}000$, $x_4 = 10{,}000$, $y_1 = 8{,}000$, $y_2 = 10{,}000$, $s_0 = 2{,}000$, $s_1 = s_2 = s_3 = 10{,}000$, $Z = 50{,}000$
49. (a) max. $Z = 130x_{1a} + 150x_{1b} + 90x_{1c} + 275x_{2a} + 300x_{2b} + 100x_{2c} + 180x_{3a} + 225x_{3b} + 140x_{3c} + 200x_{4a} + 120x_{4b} + 160x_{4c}$; s.t. $x_{1a} + x_{1b} + x_{1c} = 1$, $x_{2a} + x_{2b} + x_{2c} = 1$, $x_{3a} + x_{3b} + x_{3c} = 1$, $x_{4a} + x_{4b} + x_{4c} = 1$, $1 \leq x_{1a} + x_{2a} + x_{3a} + x_{4a} \leq 2$, $1 \leq x_{1b} + x_{2b} + x_{3b} + x_{4b} \leq 2$, $1 \leq x_{1c} + x_{2c} + x_{3c} + x_{4c} \leq 2$, $x_{ij} \geq 0$; (b) $x_{1a} = 1$, $x_{2b} = 1$, $x_{3b} = 1$, $x_{4c} = 1$, $Z = 815$; (c) max. $Z = 130x_{1a} + 150x_{1b} + 90x_{1c} + 275x_{2a} + 300x_{2b} + 100x_{2c} + 180x_{3a} + 225x_{3b} + 140x_{3c} + 200x_{4a} + 120x_{4b} + 160x_{4c}$; s.t. $x_{1a} + x_{1b} + x_{1c} \leq 1$, $x_{2a} + x_{2b} + x_{2c} \leq 1$, $x_{3a} + x_{3b} + x_{3c} \leq 1$, $x_{4a} + x_{4b} + x_{4c} \leq 1$, $x_{1a} + x_{2a} + x_{3a} + x_{4a} = 1$, $x_{1b} + x_{2b} + x_{3b} + x_{4b} = 1$, $x_{1c} + x_{2c} + x_{3c} + x_{4c} = 1$, $x_{ij} \geq 0$; $x_{2a} = 1$, $x_{3b} = 1$, $x_{4c} = 1$, $Z = 660$
51. Z values: A = 1.000, B = 1.000, C = 1.000; all three efficient
53. (a) min. $Z = x$; s.t. $150x = 650 + y_1$, $150x + y_1 = 450 + y_2$, $150x + y_2 = 600 + y_3$, $150x + y_3 = 500 + y_4$, $150x + y_4 = 700 + y_5$, $150x + y_5 = 650 + y_6$, $150x + y_6 = 750 + y_7$, $150x + y_7 = 900 + y_8$, $150x + y_8 = 800 + y_9$, $150x + y_9 = 650 + y_{10}$, $150x + y_{10} = 700 + y_{11}$, $150x + y_{11} \geq 500$; (b) $Z = x = 4.45$, $y_1 = 18.18$, $y_2 = 236.36$, $y_3 = 304.54$, $y_4 = 472.72$, $y_5 = 440.91$, $y_6 = 459.09$, $y_7 = 377.27$, $y_8 = 145.45$, $y_9 = 13.63$, $y_{10} = 31.81$, $y_{11} = 0$
55. (a) max. $Z = y$; s.t. $y - x_1 = 0$, $y - x_2 = 0$, $y - x_3 = 0$, $10x_1 + 8x_2 + 6x_3 \leq 960$, $9x_1 + 21x_2 + 15x_3 \leq 1{,}440$, $2x_1 - 3x_2 - 2x_3 \leq 60$, $-2x_1 + 3x_2 + 2x_3 \leq 60$, $x_i \geq 0$, $y \geq 0$; (b) $x_1 = x_2 = x_3 = y = 20$; (c) remove balancing requirement, $x_1 = x_2 = x_3 = y = 32$
57. max. $Z = 850x_1 + 600x_n + 750x_s + 1{,}000x_w$; s.t. $x_1 + x_n + x_s + x_w = 18$, $x_1 + x_n + x_s + x_w + y_1 + y_n + y_s + y_w = 60$, $400y_1 + 100y_n + 175y_s + 90y_w \leq 9{,}000$, $10 \leq y_1 \leq 25$, $5 \leq y_n \leq 10$, $5 \leq y_s \leq 10$, $5 \leq y_w \leq 10$, $x_1 \leq 6$, $x_n \leq 6$, $x_s \leq 6$, $x_w \leq 6$, $x_i \geq 0$, $y_i \geq 0$; $x_1 = 6$, $x_n = 0$, $x_s = 6$, $x_w = 6$, $y_1 = 14.44$, $y_n = 10$, $y_s = 7.56$, $y_w = 10$, $Z = 15{,}600$ (multiple optimal)
59. (a) max. $Z = .85x_1 + .90x_2 - y_1 - y_2$; s.t. $x_1 \leq 5{,}000 + 3y_1$, $x_2 \leq 4{,}000 + 5y_2$, $.60x_1 + .85x_2 + y_1 + y_2 \leq 16{,}000$, $x_1 \geq .3(x_1 + x_2)$, $x_1 \leq .6(x_1 + x_2)$, $x_1 \geq 0$, $x_2 \geq 0$, $y_1 \geq 0$, $y_2 \geq 0$; (b) $x_1 = 5{,}458.128$, $x_2 = 12{,}735.63$, $y_1 = 152.709$, $y_2 = 1{,}747.126$, $Z = 14{,}201.64$
61. (a) min. $Z = \Sigma\Sigma$ (ranking)• x_{ij}, s.t. $\Sigma x_{ij} \leq$ hr., $\Sigma x_{ij} =$ project hr., Σ (hourly rate)• $x_{ij} \leq$ budget; (b) $x_{A3} = 400$, $x_{A4} = 50$, $x_{B4} = 250$, $x_{B5} = 350$, $x_{C4} = 175$, $x_{C7} = 274.1$, $x_{C8} = 50.93$, $x_{D2} = 131.7$, $x_{D7} = 15.93$, $x_{E1} = 208.33$, $x_{E8} = 149.07$, $x_{F1} = 291.67$, $x_{F2} = 108.3$, $x_{F6} = 460$, $Z = \$12{,}853.33$
63. $x_1 = 0$, $x_2 = 4$, $x_3 = 18.4$, $x_4 = 6.4$, $x_5 = 24.8$, $y_1 = 72.22$, $y_2 = 72.44$, $y_3 = 64.95$, $y_4 = 62.34$, $y_5 = 52.24$, $y_6 = 38.9$, $y_7 = 28.53$, $y_8 = 43.35$, $Z = \$360{,}196$
65. min. $Z = \Sigma x_{ij}$ (priority ij); s.t. $\sum_{j=1}^{12} x_{ij} = 1$, $\sum\sum_{i=1}^{16} x_{ij} \leq$ available slots j, U11B: 3-5M, U11G: 3-5T, U12B: 3-5T, U12G: 3-5M, U13B: 3-5T, U13G: 3-5M, U14B: 5-7M, U14G: 3-5M, U15B: 5-7T, U15G: 3-5M, U16B: 5-7T, U16G: 5-7T, U17B: 5-7M, U17G: 5-7T, U18B: 3-5T, U18G: 5-7M, $Z = 27$
67. 1-D (1 hr.), 2-E (1 hr.), 3-F (2 hr.), 3-H (6 hr.), 4-I (1 hr.), 5-J (1 hr.), A-5 (8 hr.), B-4 (5 hr.), C-6 (8 hr.), G-7 (2 hr.), $Z = 35$ hr.; 6 crews originate in Pittsburgh, 4 in Orlando
69. min. $Z = .41x_{14} + .57x_{15} + .37x_{24} + .48x_{25} + .51x_{34} + .60x_{35} + .22x_{46} + .10x_{47} + .20x_{48} + .15x_{56} + .16x_{57} + .18x_{58}$; s.t. $x_{14} + x_{15} \leq 24{,}000$, $x_{24} + x_{25} \leq 18{,}000$, $x_{34} + x_{35} \leq 32{,}000$, $x_{14} + x_{24} + x_{34} \leq 48{,}000$, $x_{15} + x_{25} + x_{35} \leq 35{,}000$ $(x_{14} + x_{24} + x_{34})/2 = x_{46} + x_{47} + x_{48}$, $(x_{15} + x_{25} + x_{35})/2 = x_{56} + x_{57} + x_{58}$, $x_{46} + x_{56} = 9{,}000$, $x_{47} + x_{57} = 12{,}000$, $x_{47} + x_{58} = 15{,}000$, $x_{ij} \geq 0$; $x_{14} = 24{,}000$, $x_{24} = 18{,}000$, $x_{34} = 6{,}000$, $x_{35} = 24{,}000$, $x_{47} = 12{,}000$, $x_{48} = 12{,}000$, $x_{56} = 9{,}000$, $x_{58} = 3{,}000$, $Z = \$39{,}450$

Chapter 5

1. $x_1 = 3$, $x_2 = 0$, $Z = 15$
3. (a) max. $Z = 50x_1 + 40x_2$; s.t. $3x_1 + 5x_2 \leq 150$, $10x_1 + 4x_2 \leq 200$, $x_i \geq 0$ and integer; (b) $x_1 = 10$, $x_2 = 24$, $Z = 1{,}460$

5. (a) max. $Z = 50x_1 + 10x_2$; s.t. $x_1 + x_2 \leq 15$, $4x_1 + x_2 \leq 25$, $x_i \geq 0$ and integer; (b) $x_1 = 6$, $x_2 = 1$, $Z = 310$
7. (a) max. $Z = 50x_1 + 40x_2$; s.t. $2x_1 + 5x_2 \leq 35$, $3x_1 + 2x_2 \leq 20$, $x_i \geq 0$ and integer; (b) $x_1 = 4$, $x_2 = 4$, $Z = 360$
9. $x_1 = 1$, $x_2 = 0$, $x_3 = 1$, $Z = 1{,}800$
11. min. $Z = 81x_1 + 50x_2$; s.t. $76x_1 + 53x_2 \geq 600$, $x_1 + x_2 \leq 10$, $1.3x_1 + 4.1x_2 \leq 24$, $x_1, x_2 \geq 0$ and integer; $x_1 = 6$, $x_2 = 3$, $Z = 636$
13. (a) max. $Z = 85{,}000x_1 + 60{,}000x_2 - 18{,}000y_1$; s.t. $x_1 + x_2 \leq 10$, $10{,}000x_1 + 7{,}000x_2 \leq 72{,}000$, $x_1 - 10y_1 \leq 0$, x_1 and $x_2 \geq 0$ and integer, $y_1 = 0$ or 1; (b) $x_1 = 0$, $x_2 = 10$, $Z = 600{,}000$
15. min. $Z = x_1 + x_2 + x_3 + x_4 + x_5 + x_6$; s.t. $x_6 + x_1 \geq 90$, $x_1 + x_2 \geq 215$, $x_2 + x_3 \geq 250$, $x_3 + x_4 \geq 65$, $x_4 + x_5 \geq 300$, $x_5 + x_6 \geq 125$, $x_i \geq 0$; $x_1 = 90$, $x_2 = 250$, $x_4 = 175$, $x_5 = 125$, $Z = 640$
17. (a) min. $Z = 25{,}000x_1 + 7{,}000x_2 + 9{,}000x_3$; s.t. $53{,}000x_1 + 30{,}000x_2 + 41{,}000x_3 \geq 200{,}000$, $(32{,}000x_1 + 20{,}000x_2 + 18{,}000x_3)/(21{,}000x_1 + 10{,}000x_2 + 23{,}000x_3) \geq 1.5$, $(34{,}000x_1 + 12{,}000x_2 + 24{,}000\,x_3)/(53{,}000x_1 + 30{,}000x_2 + 41{,}000x_3) \geq 0.60$, $x_1 \geq 0$, $x_2 \geq 0$, $x_3 \geq 0$ and integer; $x_1 = 4$, $Z = \$99{,}999.99$; (b) $x_1 = 2.9275$, $x_2 = .9713$, $x_3 = .383$, $Z = \$83{,}433.65$
19. max. $Z = 25{,}000x_1 + 18{,}000x_2 + 31{,}000x_3$; s.t. $x_1 + x_2 + x_3 = 100$, $5{,}000x_1 + 11{,}000x_2 + 7{,}000x_3 \leq 700{,}000$, $x_1 \geq 10$, $x_2 \geq 10$, $x_3 \geq 10$, $x_1, x_2, x_3 \geq 0$ and integer; $x_1 = 20$, $x_2 = 10$, $x_3 = 70$, $Z = 2{,}850{,}000$
21. max. $Z = 12{,}100x_1 + 8{,}700x_2 + 10{,}500x_3$; s.t. $360x_1 + 375x_2 + 410x_3 \leq 30{,}000$, $x_1 + x_2 + x_3 \leq 67$, $14x_1 + 10x_2 + 18x_3 \leq 2{,}200$, $x_1/x_3 \geq 2$, $x_2/x_1 \geq 1.5$, $x_1 \geq 0$, $x_2 \geq 0$ and integer, $x_3 \geq 0$; $x_1 = 22$, $x_2 = 33$, $x_3 = 11$, $Z = \$668{,}800$
23. (a) max. $Z = 1{,}650x_1 + 850x_2 + 790x_3$; s.t. $6.3x_1 + 3.9x_2 + 3.1x_3 \leq 125$, $17x_1 + 10x_2 + 7x_3 \leq 320$, $x_i \geq 0$ and integer; (b) $x_1 = 10$, $x_3 = 20$, $Z = 32{,}300$; rounded-down solution: $x_1 = 13$, $x_3 = 12$, $Z = 30{,}930$
25. max. $Z = 575x_1 + 120x_2 + 65x_3$; s.t. $40x_1 + 15x_2 + 4x_3 \leq 600$, $30x_1 + 18x_2 + 5x_3 \leq 480$, $4x_1 - x_2 \leq 0$, $x_3 = 20y_1$, $x_1, x_2, x_3 \geq 0$ and integer, $y_1 = 0$ or 1; $x_1 = 3$, $x_2 = 16$, $x_3 = 20$, $y_1 = 1$, $Z = \$4{,}945$
27. (b) $x_{1C} = 1$, $x_{3D} = 1$, $x_{4B} = 1$, $x_{5A} = 1$, $Z = 83$ parts
29. min. $Z = 120x_1 + 75x_2 + 4.5x_3$; s.t. $220x_1 + 140x_2 + 12x_3 \geq 6{,}300$, $8x_1 + 8x_2 + x_3 \leq 256$, $.4x_1 + .9x_2 + .16x_3 \leq 15$, $x_1, x_2 \geq 0$ and integer, $x_3 \geq 0$; $x_1 = 28$, $x_2 = 0$, $x_3 = 11.67$, $Z = \$3{,}412.50$
31. $y_2 = 1$, $y_3 = 1$, $x_2 = 29$, $x_3 = 64$, $Z = 5{,}289{,}000$
33. $Z = 667$ miles
35. $x_{13} = 1$, $x_{22} = 1$, $x_{32} = 1$, $x_{43} = 1$, $x_{53} = 1$, $x_{61} = 1$, $Z = \$125$ million
37. max. $Z = 127x_1 + 83x_2 + 165x_3 + 96x_4 + 112x_5 + 88x_6 + 135x_7 + 141x_8 + 117x_9 + 94x_{10}$; s.t. $x_1 + x_3 \leq 1$, $x_1 + x_2 + x_4 \leq 1$, $x_4 + x_5 + x_6 \leq 1$, $x_6 + x_7 + x_8 \leq 1$, $x_6 + x_9 \leq 1$, $x_8 + x_{10} \leq 1$, $x_9 + x_{10} \leq 1$, $x_i = 0$ or 1; $x_2 = 1$, $x_3 = 1$, $x_5 = 1$, $x_8 = 1$, $x_9 = 1$, $Z = 618{,}000$
39. A(1,2,3), B(1,2,3), C(4,5,6), D(5,6,7), E(4,5,6), F(1,2,3), G(5,6,7), H(4,5,6), $Z = 100$ hr.
41. max. $Z = .9(3600)x_{A1} + .5(7200)x_{A2} + .9(2400)x_{B1} + .7(3600)x_{B2} + .95(3000)x_{C1} + .4(6000)x_{C2} + .95(3300)x_{D1} + .6(5400)x_{D2}$; s.t. $x_{A1} + x_{A2} = 1$, $x_{B1} + x_{B2} = 1$, $x_{C1} + x_{C2} = 1$, $x_{D1} + x_{D2} = 1$, $.9x_{A1} + .5x_{A2} + .9x_{B1} + .7x_{B2} + .95x_{C1} + .4x_{C2} + .95x_{D1} + .6x_{D2} \geq 3$, $x_{ij} \geq 0$, $x_{A2} = 1$, $x_{B2} = 1$, $x_{C1} = 1$, $x_{D1} = 1$, $Z = \$4{,}035$ per month
43. max. $Z = \sum_{ij} (\text{profit})_i \cdot x_{ij}$; s.t. $\sum_i (\text{load, lb.})_i \cdot x_{ij} \leq 80{,}000$ lb, $\sum_i (\text{load, ft.}^3)_i \cdot x_{ij} \leq 5{,}500$ ft.³, $\sum_i (\text{time})_i \cdot x_{ij} \leq 90$, $\sum_j x_{ij} \leq 1$; $x_{1B} = 1$, $x_{2A} = 1$, $x_{3A} = 1$, $x_{4B} = 1$, $x_{7C} = 1$, $x_{11C} = 1$, $Z = \$78{,}000$
45. (a) max. $Z = \sum_i (\text{annual usage})_i \cdot x_i$; s.t. $\sum_i (\text{acreage})_i \cdot x_i \leq 55$, $\sum_i (\text{cost})_i \cdot x_i \leq \$550{,}000$, $\sum_i (\text{priority})_i \cdot x_i - (\sum x_i \cdot 1.75) \leq 0$; solution: football fields, playground, walking/running trails, and softball fields; $Z = 123{,}500$; (b) min. $Z = \sum_i x_i\,(\text{priority})_i$; s.t. $\sum_i (\text{acreage})_i \cdot x_i \leq 55$, $\sum_i (\text{cost})_i \cdot x_i \leq \$550{,}000$, $\sum (\text{expected usage})_i \cdot x_i \geq 120{,}000$; solution: soccer fields, playground, walking/running trails, $Z = 4.0$ or 1.33 average priority; (c) max. $Z = \sum_i (\text{acreage})_i \cdot x_i$; s.t. $\sum_i (\text{acreage})_i \cdot x_i \leq 55$, $\sum_i (\text{cost})_i \cdot x_i \leq \$550{,}000$, $\sum_i (\text{priority})_i \cdot x_i - (\sum x_i \cdot 1.75) \leq 0$; solution: rugby fields, soccer fields, walking/running trails, $Z = 52$ acres, expected annual usage $= 83{,}700$

Chapter 6

1. St. Louis–Chicago = 250, Richmond–Chicago = 50, Richmond–Atlanta = 350, $Z = 24{,}000$
3. A3 = 100, B1 = 135, B2 = 45, C2 = 130, C3 = 70, $Z = 2{,}350$
5. $x_{11} = 70$, $x_{13} = 20$, $x_{22} = 10$, $x_{23} = 20$, $x_{32} = 100$, $Z = 1{,}240$
7. min. $Z = 14x_{A1} + 9x_{A2} + 16x_{A3} + 18x_{A4} + 11x_{B1} + 8x_{B2} + 100x_{B3} + 16x_{B4} + 16x_{C1} + 12x_{C2} + 10x_{C3} + 22x_{C4}$; s.t. $x_{A1} + x_{A2} + x_{A3} + x_{A4} \leq 150$, $x_{B1} + x_{B2} + x_{B3} + x_{B4} \leq 210$, $x_{C1} + x_{C2} + x_{C3} + x_{C4} \leq 320$, $x_{A1} + x_{B1} + x_{C1} = 130$, $x_{A2} + x_{B2} + x_{C2} = 70$, $x_{A3} + x_{B3} + x_{C3} = 180$, $x_{A4} + x_{B4} + x_{C4} = 240$, $x_{ij} \geq 0$; $x_{A2} = 70$, $x_{A4} = 80$, $x_{B1} = 50$, $x_{B4} = 160$, $x_{C1} = 80$, $x_{C3} = 180$, $Z = 8{,}260$
9. A3 = 90, B1 = 30, B3 = 20, C2 = 80, D1 = 40, D2 = 20, $Z = 1{,}590$
11. 1C = 5, 2C = 10, 3B = 20, 4A = 10, $Z = \$195$; min. $Z = 7x_{1A} + 8x_{1B} + 5x_{1C} + 6x_{2A} + 100x_{2B} + 6x_{2C} + 10x_{3A} + 4x_{3B} + 5x_{3C} + 3x_{4A} + 9x_{4B} + 100x_{4C}$; s.t. $x_{1A} + x_{1B} + x_{1C} \leq 5$, $x_{2A} + x_{2B} + x_{2C} \leq 25$, $x_{3A} + x_{3B} + x_{3C} \leq 20$, $x_{4A} + x_{4B} + x_{4C} \leq 25$, $x_{1A} + x_{2A} + x_{3A} + x_{4A} = 10$, $x_{1B} + x_{2B} + x_{3B} + x_{4B} = 20$, $x_{1C} + x_{2C} + x_{3C} + x_{4C} = 15$, $x_{ij} \geq 0$
13. A2 = 1,800, A4 = 950, A6 = 750, B1 = 1,600, B3 = 1,500, B5 = 1,250, B6 = 650, $Z = 3{,}292.50$
15. 1B = 250, 1D = 170, 2A = 520, 2C = 90, 3C = 130, 3D = 210, $Z = 21{,}930$
17. 1B = 250, 1D = 170, 2A = 520, 2C = 90, 3C = 130, 3D = 210, 4C = 180, $Z = \$26{,}430$; transportation cost = \$21,930, shortage cost = \$4,500
19. 1B = 60, 2A = 45, 2B = 25, 2C = 35, 3B = 5, $Z = 1{,}605$
21. NA = 250, SB = 300, SC = 40, EA = 150, EC = 160, WD = 210, CB = 100, CD = 190, $Z = 20{,}700$ (multiple optimal)
23. A3 = 8, A4 = 18, B3 = 13, B5 = 27, D3 = 5, D6 = 35, E1 = 25, E2 = 15, E3 = 4, $Z = 1{,}528$ (multiple optimal)
25. 17 cases from Albany; \$51
27. 1C = 2, 1E = 5, 2C = 10, 3E = 5, 4D = 8, 5A = 9, 6B = 6, $Z = 1{,}275$ hr.
29. R_J – Jan = 300, O_J – Jan = 110, R_F – Feb = 300, O_F – Feb = 20, O_F – March = 120, R_M – March = 180, R_M – April = 120, O_M – March = 200, R_A – April = 300, O_A – April = 200, R_M – May = 300, O_M – May = 130, R_J – June = 300, O_J – June = 80, $Z = 3{,}010{,}040$
31. Increasing supply at Sacramento, Jacksonville, and Ocala has little effect; increasing supply at San Antonio and Montgomery reduces cost to \$242,500
33. Total cost = \$1,198,500
35. Al—Eagles and Bengals, Barbara—Saints and Jets, Carol—Cowboys and Packers, Dave—Redskins and Cardinals, $Z = 24$, multiple optional solutions

37. Mexico–Houston = 18, Puerto Rico–Miami = 11, Haiti–Miami = 23, Miami–NY = 20, Miami–St. Louis = 12, Miami–LA = 2, Houston–LA = 18, Z = 479,000

39. $x_{1C} = 70, x_{2B} = 80, x_{3A} = 50, x_{BA} = 10, x_{CB} = 30, Z = 14{,}900$

41. $x_{16} = 1{,}600, x_{27} = 1{,}100, x_{35} = 1{,}400, x_{46} = 600, x_{59} = 900, x_{510} = 500, x_{68} = 1{,}200, x_{610} = 600, x_{611} = 400, x_{711} = 1{,}100, Z = 25{,}192$

43. (a) $x_{24} = 5, x_{49} = 2, x_{410} = 3, x_{911} = 1, x_{913} = 1, x_{1012} = 1, x_{1014} = 1, x_{1015} = 1, Z = 144$; (b) $x_{14} = 2, x_{24} = 3, x_{49} = 2, x_{410} = 3, x_{911} = 1, x_{913} = 1, x_{1012} = 1, x_{1014} = 1, x_{1015} = 1, Z = 154$

45. Antwerp = 1, Bremen = 1, Valencia = 1, Hong Kong–Antwerp = 235, Shanghai–Bremen = 170, Busan–Valencia = 165, Mumbai–Antwerp = 285, Mumbai–Valencia = 40, Kaoshiung–Bremen = 300, Kaoshiung–Valencia = 105, Antwerp–Miami = 190, Antwerp–New Orleans = 330, Bremen–New York = 165, Bremen–Savannah = 305, Valencia–New York = 275, Valencia–New Orleans = 35; Z = \$47,986,050; shipping cost = \$6,215,050

47. 1–1, 2–4, 3–2, 5–3, Z = 78

49. 1–B, 2–D, 3–C, 4–A, Z = 32

51. 1–B, 2–E, 3–A, 4–C, 5–D, 6–F, Z = 36

53. 1–C, 2–F, 3–E, 4–A, 5–D, 6–B, Z = 85 defects

55. 1, 4, and 7—Columbia; 2, 6, and 8—Atlanta; 3, 5, and 9—Nashville; Z = 985

57. A1's—parents' brunch; Bon Apetit—post game party; Bon Apetit—lettermen's dinner; Divine—booster club; Epicurean—contributors' dinner; University—alumni brunch; Z = \$103,800

59. Annie—backstroke; Debbie—breaststroke; Erin—freestyle; Fay—butterfly; Z = 10.61 min.

61. A–2, C–1, D–1, E–3, F–3, G–2, H–1, Z = \$1,070

63. A–NJ, B–PA, C–NY, D–FL, E–GA, F–FL, G–VA, Z = 498; success rate = 71.1%

65. $x_{A2} = 1, x_{J2} = 1, x_{B1} = 1, x_{K1} = 1, x_{C1} = 1, x_{L3} = 1, x_{D2} = 1, x_{E1} = 1, x_{F1} = 1, x_{G2} = 1, x_{H3} = 1, Z = 104$

67. Mileage difference = 160 additional miles

69. Jackets-1 = 1, Panthers-1 = 1, Big Red-2 = 1, Lions-2 = 1, Bulldogs-3 = 1, Beavers-3 = 1, Tigers-4 = 1, Bears-4 = 1, Blue Devils-5 = 1, Cavaliers-5 = 1, Blue Jays-6 = 1, Rams-6 = 1, Knights-7 = 1, Hawks-7 = 1, Wasps-8 = 1, Eagles-8 = 1, Z = 3,168 miles

Chapter 7

1. 1–2 = 21, 1–2–5 = 46, 1–3 = 17, 1–3–4 = 29, 1–3–4–6 = 39, 1–3–4–7 = 38

3. 1–2 = 85, 1–3 = 53, 1–4 = 88, 1–3–5 = 114, 1–3–7 = 170, 1–3–7–6 = 194

5. 1–2 = 25, 1–2–3 = 60, 1–2–4 = 43, 1–5 = 48, 1–6 = 50, 1–7 = 32, 1–7–8 = 72, 1–7–9 = 61, 1–5–10 = 72

7. 1–2 = 7, 1–3 = 10, 1–4 = 8, 1–3–5 = 13, 1–3–5–8 = 21, 1–4–6 = 13, 1–4–6–9 = 20, 1–4–7 = 16, 1–4–6–9–10 = 29

9. 1–7–6–10–12 = 18

11. 1–4–7–8 = 42 miles

13. 1–3–11–14 = 13 days

15. Kotzebue: 1–2–8–11–15 = 10 hr.; Nome: 1–2–8–12–13 = 9 hr.; Stebbins: 1–2–7–10 = 8.5 hr.

17. 1–4–7; \$64,500

19. 1–3 = 4.1, 1–4 = 4.8, 2–3 = 3.6, 4–8 = 5.5, 5–6 = 2.1, 6–7 = 2.8, 7–8 = 2.7, 7–9 = 2.7, 9–10 = 4.6; 32,900 ft.

21. 1–3–4–6–5–8, 3–2, 6–7; 320 yd.

23. 1–4–2–6–3, 6–7–5, 7–8; 1,160 yd.

25. 1–2–5–6–8–9, 2–4–3, 4–7; 22 miles

27. 1–2, 1–4–6–3, 6–9–8–5, 9–11, 9–10–7, 10–13, 10–12–14; 1,086 ft.

29. 1–2 = 6, 1–4 = 5, 1–3 = 5, 2–5 = 6, 2–4 = 0, 3–4 = 0, 3–6 = 5, 4–5 = 0, 4–6 = 5, 5–6 = 6, maximum flow = 16

31. 1–2 = 15, 1–4 = 20, 1–3 = 12, 2–5 = 11, 2–4 = 4, 3–4 = 3, 3–6 = 9, 4–6 = 27, 5–6 = 11, maximum flow = 47

33. 1–2 = 7, 1–4 = 7, 1–3 = 4, 2–6 = 6, 2–4 = 1, 3–4 = 3, 3–5 = 1, 4–6 = 0, 4–8 = 4, 4–7 = 7, 5–7 = 1, 6–8 = 6, 7–8 = 8, maximum = 18 flights

35. 1–2 = 4, 1–3 = 4, 1–4 = 4, 2–3 = 0, 2–5 = 4, 3–4 = 0, 3–5 = 2, 3–6 = 2, 4–6 = 4, 5–6 = 0, 5–7 = 6, 6–8 = 6, 7–8 = 0, 7–9 = 6, 8–9 = 6; 12,000 cars

37. 1–2 = 6, 1–3 = 2, 1–4 = 8, 2–5 = 0, 2–6 = 6, 3–5 = 0, 3–6 = 2, 4–5 = 8, 5–7 = 3, 5–8 = 4, 5–9 = 1, 6–8 = 2, 6–9 = 6, 7–10 = 3, 8–10 = 6, 9–10 = 7; 16

39. 1–2 = 70, 1–3 = 50, 1–4 = 60, 1–5 = 70, 2–6 = 25, 2–9 = 45, 3–6 = 30, 3–7 = 20, 4–7 = 35, 4–8 = 25, 5–8 = 40, 5–11 = 30, 6–10 = 65, 6–12 = 5, 7–10 = 40, 7–6 = 15, 8–13 = 60, 8–14 = 5, 9–12 = 45, 10–12 = 55, 10–13 = 50, 11–14 = 30, 12–13 = 60, 12–15 = 45, 13–15 = 170, 14–15 = 35; 250

41. Gdansk–Galveston = 125, Hamburg–Jacksonville = 110, Hamburg–New Orleans = 95, Hamburg–Galveston = 5, Lisbon–Norfolk = 85, Norfolk–KC = 75, Norfolk–Dallas = 10, Jacksonville–FR = 70, Jacksonville–KC = 40, NO–FR = 95, Galveston–Dallas = 130, FR–Denver = 105, FR–Pittsburgh = 60, KC–Cleveland = 65, KC–Nashville = 50, Dallas–Tucson = 85, Dallas–Cleveland = 55

Chapter 8

1. 1–3–4 = 10

3. 1–3–6–8 = 23

5. 1: $ES = 0, EF = 7, LS = 2, LF = 9, S = 2$; 2: $ES = 0, EF = 10, LS = 0, LF = 10, S = 0$; 3: $ES = 7, EF = 13, LS = 9, LF = 15, S = 2$;
4: $ES = 10, EF = 15, LS = 10, LF = 15, S = 0$; 5: $ES = 10, EF = 14, LS = 14, LF = 18, S = 4$; 6: $ES = 15, EF = 18, LS = 15, LF = 18, S = 0$; 7: $ES = 18, EF = 20, LS = 18, LF = 20, S = 0$; 2–4–6–7 = 20

7. 1: $ES = 0, EF = 10, LS = 0, LF = 10, S = 0$; 2: $ES = 0, EF = 7, LS = 5, LF = 12, S = 5$; 4: $ES = 10, EF = 14, LS = 14, LF = 18, S = 4$; 3: $ES = 10, EF = 25, LS = 10, LF = 25, S = 0$; 5: $ES = 7, EF = 13, LS = 12, LF = 18, S = 5$; 6: $ES = 7, EF = 19, LS = 13, LF = 25, S = 6$; 7: $ES = 14, EF = 21, LS = 18, LF = 25, S = 4$; 8: $ES = 25, EF = 34, LS = 25, LF = 34, S = 0$; 1–3–8 = 34

9. 2–4–10–13–14–17 = 78 wk.

11. 1–5–9 = 40.33, $\sigma = 5.95$

13. (e) 2–6–9–11–12; (f) $\mu = 46$ mo., $\sigma = 5$ mo.

15. a–b–e–g–i–j–k–l–m–n = 71.167; $P(x \leq 52) = 0$, $P(x \leq 78) = .9744$

17. $P(x \geq 50) = .2119$

19. c–d–f–g–j–k–o–r = 160.83 min.; $P(x \leq 180) = .9875$

21. $P(x \leq 15) = .0643$
23. a–d–j–n–o, $\mu = 42.3$ wk., $\sigma = 4.10$ wk., 47.59 wk.
25. a–c–d–f = 14 days
27. a–e–f–g–j–o–p = 91.67 days, $\sigma = 3.308$, $P(x \leq 101) = .9976$
29. a–h–l–m–n–o–s–w = 126.67 days; $P(x \leq 150) = .9979$
31. (c) min. $Z = x_6$; s.t. $x_2 - x_1 \geq 16, x_3 - x_1 \geq 14, x_4 - x_2 \geq 8, x_5 - x_2 \geq 5, x_5 - x_3 \geq 4, x_6 - x_3 \geq 6, x_6 - x_4 \geq 10, x_6 - x_5 \geq 15, x_i, x_j \geq 0$
33. min. $Z = x_9$; s.t. $x_2 - x_1 \geq 8, x_3 - x_1 \geq 6, x_4 - x_1 \geq 3, x_5 - x_2 \geq 0, x_6 - x_2 \geq 5, x_5 - x_3 \geq 3, x_5 - x_4 \geq 4, x_7 - x_5 \geq 7, x_7 - x_8 \geq 0, x_8 - x_5 \geq 4, x_8 - x_4 \geq 2, x_9 - x_6 \geq 4, x_9 - x_7 \geq 9, x_i, x_j \geq 0; x_1 = 0, x_2 = 9, x_3 = 6, x_4 = 3, x_5 = 9, x_6 = 14, x_7 = 16, x_8 = 16, x_9 = 25$
35. a–d–g–k; crash cost = \$5,100; total network cost = \$33,900

Chapter 9

1. min. $P_1d_1^-, P_2d_2^-, P_3d_1^+, P_4d_3^+$, s.t. $5x_1 + 2x_2 + 4x_3 + d_1^- - d_1^+ = 240, 3x_1 + 5x_2 + 2x_3 + d_2^- - d_2^+ = 500, 4x_1 + 6x_2 + 3x_3 + d_3^- - d_3^+ = 400$
3. min. $P_1d_1^-, P_2d_2^-, P_3d_3^+, 3P_4d_4^- + 6P_4d_5^- + P_4d_6^- + 2P_4d_7^-$, s.t. $80{,}000x_1 + 24{,}000x_2 + 15{,}000x_3 + 40{,}000x_4 + d_1^- = 600{,}000, 1{,}500x_1 + 3{,}000x_2 + 500x_3 + 1{,}000x_4 + d_2^- - d_2^+ = 20{,}000, 4x_1 + 8x_2 + 3x_3 + 5x_4 + d_3^- - d_3^+ = 50, x_1 + d_4^- - d_4^+ = 7, x_2 + d_5^- - d_5^+ = 10, x_3 + d_6^- - d_6^+ = 8, x_4 + d_7^- - d_7^+ = 12$
5. (a) min. $P_1d_2^+, P_2d_7^+, 4P_3d_4^- + 2P_3d_5^- + P_3d_6^-, P_4d_1^-, P_5d_3^-$; s.t. $x_1 + x_2 + x_3 + d_1^- = 2{,}000, 800x_1 + 1{,}500x_2 + 500x_3 + d_2^- - d_2^+ = 20{,}000, 10x_1 + 12x_2 + 18x_3 + d_3^- - d_3^+ = 800, x_1 + d_4^- - d_4^+ = 800, x_2 + d_5^- - d_5^+ = 900, x_3 + d_6^- - d_6^+ = 1{,}100, d_3^+ + d_7^- - d_7^+ = 100$; (b) $x_1 = 25, d_1^- = 1{,}975, d_3^- = 550, d_4^- = 775, d_5^- = 900, d_6^- = 1{,}100, d_7^- = 650$
7. $x_1 = 20, d_2^- = 10, d_3^- = 50$
9. $x_1 = 15, d_1^+ = 12, d_2^+ = 10, d_4^- = 6$; not satisfied
11. (a) min. $Z = P_1d_1^-, P_2d_4^+, 3P_3d_2^- + 2P_3d_3^-, P_4d_1^+$; s.t. $x_1 + x_2 + d_1^- - d_1^+ = 80, x_1 + d_2^- = 60, x_2 + d_3^- = 35, d_1^+ + d_4^- - d_4^+ = 10$; (b) $x_1 = 60, x_2 = 30, d_1^+ = 10, d_3^- = 5$
13. (a) min. $P_1d_1^-, P_2d_2^+, P_3d_3^-, P_4d_4^-$; s.t. $x_1 + d_1^- - d_1^+ = 30, 20x_1 + 40x_2 + 150x_3 + d_2^- - d_2^+ = 1{,}200, x_2 + x_3 + d_3^- - d_3^+ = 20, x_3 + d_4^- - d_4^+ = 6$; (b) $x_1 = 30, x_2 = 15, d_3^- = 5, d_4^- = 6$
15. (b) x_1 (Sunday) = 0, x_2 (Monday) = 7, x_3 (Tuesday) = 0, x_4 (Wednesday) = 27, x_5 (Thursday) = 0, x_6 (Friday) = 25, x_7 (Saturday) = 1
17. (b) $x_1 = 1, x_2 = 24, x_3 = 5, x_4 = 1, x_5 = 8$, p_1, p_2, p_3, p_4 achieved
19. Profitability – CI/RI = 0.07; Growth – CI/RI = 0.003.
21. CI/RI = .003
23. Global = \$17,225.50; BlueChip = \$43,014.50; Bond = \$24,760.
25. Pawlie = .4575, Cooric = .4243, Brokenaw = .1182
27. Arrington = .2906, Barton =.3817, Claiborne = .3277
29. A = .4937, B = .1781, C = .3282
31. Gym = .215, Field = .539, Tennis = .104, Pool = .142
33. Weather Consistent; Cost-CI/RI = .003; Fun-CI/RI = .14/inconsistent Criteria-CI/RI = .13/inconsistent
39. 1 = .3120, 2 = .3398, 3 = .2772, 4 = .0709
41. A = .2838, B = .4317, C = .2844
45. CI/RI = .0558(consistent)
47. Cleveland = .1529, Miami = .4486, NY = .2192, Chicago = .1794
49. 5(A) = 77.5, 5(B) = 75.5, 5(C) = 74.25, 5(D) = 78.75; select D
53. 5(1) = 77.5, 5(2) = 80.8, 5(3) = 82.
55. 5(South) = 73.8, 5(West A) = 74.5, 5(West B) = 67.25, 5(East) = 76.75, East
57. 5(C) = 76, 5(E) = 75, 5(D) = 74, 5(B) = 70, 5(A) = 69
59. 5(Roanoke) = 84.74, 5(Richmond) = 83.79, 5(Greensboro) = 82.61, 5(Charlotte) = 81.10, 5(Bristol) = 79.02, 5(knoxville) = 76.93

Chapter 10

1. $p = \$186.67$, $v = 176$ chairs, $Z = \$18{,}313.33$
3. $p = \$40.50$, $v = 28{,}269.6$, $Z = \$542{,}641.95$
5. $p = \$25.55$, $v = 78.63$, $Z = -\$1{,}198$
7. $p = \$35.16$, $Z = \$10{,}257.57$
9. $x_1 = 10, x_2 = 45, Z = \$212.50$
11. $x_1 = 0.09, x_3 = 0.497, x_4 = 0.413$; total return = 0.136
13. Lagrange multiplier = 0.27
15. max. $Z = (5.8 - .06p_h + .005p_l)\, p_h + (3.0 - .11p_l + .008p_h)\, p_l$; s.t. $5.8 - .06p_h + .005p_l \leq 2.5, 3.0 - .11p_l + .008p_h \leq 2.5, p_l, p_h \geq 0$; $p_h = \$56.40, p_l = \$16.70, Z = \$167.94$
17. $x = 151.2, y = 484.08, Z = 52{,}684.29$
19. $x_A = .250, x_B = .750, Z = .0382$, total return = 0.11
21. $x_1 = 5, x_2 = 3, x_3 = 8, x_4 = 3, x_5 = 2, x_6 = 5, x_7 = 5, x_8 = 4$, $Z = \$4{,}193{,}450.37$

Chapter 11

1. (a) subjective, (b) subjective, (c) both, (d) subjective, (e) objective, (f) subjective, (g) objective (h) objective
3. (a) 0–19, .066; 20–29, .233; 30–39, .266; 40–49, .266; 50 +, .166; (b) yes, only one of the events can take place each winter
5. $P(r \geq 5) = .0433$ rejected, $P(r < 5) = .9567$ accepted
7. $P(r = 4) = .0367$
9. $P(r \geq 4) = .0024$
11. (a and b) P(WA) = .28, P(LA) = .12, P(WO) = .36, P(LO) = .24; (c) P(O|W) = .563
13. (a and b) P(GA) = .198, P(DA) = .002, P(GB) = .392, P(DB) = .008, P(GC) = .368, P(DC) = .032; (c) P(B|D) = .19
15. P(A|D) = .667
17. E(bbl.) = 7.95
19. E(lead time) = 2.2 days, $\sigma = 0.87$
21. E(building) = \$38,000, $\sigma = 26{,}381$, E(bonds) = 34,000, $\sigma = 4{,}899$; probably bonds, less variation
23. E(time between breakdowns) = 2.75 hr.
25. $P(x \geq 3.5) = .0668$
27. $P(x \geq 900) = .1587$
29. $P(x \geq 6{,}000) = .0475$
31. $P(x \geq 670) = .7422$
33. $P(x \geq 1{,}200) = .1056$, 65.5
35. $P(x \geq 200{,}000) = .9699$
37. $P(x \leq 27{,}000) = .1894$
39. $P(x \geq 75) = .4364$
41. not normal

Chapter 12

1. (a) lease land; (b) savings certificate
3. (a) bellhop; (b) management; (c) bellhop
5. (a) corn; (b) soybeans; (c) corn; (d) soybeans; (e) corn
7. (a) build shopping center; (b) lease equipment; (c) build shopping center; (d) lease equipment; (e) build shopping center
9. (a) Army vs. Navy; (b) Alabama vs. Auburn; (c) Alabama vs. Auburn
11. (a) pass; (b) off tackle or option; (c) toss sweep
13. (a) office park; (b) office building; (c) office park or shopping center; (d) office building
15. press
17. *EV* (shoveler) = $99.6
19. *EV* (operate) same as leasing; conservative decision is to lease
21. bond fund
23. Taiwan, *EVPI* = $0.47 million
25. compact car dealership
27. (b) Stock 16 cases; (c) stock 16 cases; (d) $5.80
29. (b) Stock 26 dozen; (c) stock 26 dozen; (d) $3.10
31. (a) Pusan; (b) Pusan; (c) Hong Kong; (d) Shanghai
33. O'Neil
35. *EV* (graphic design) = 191,000
37. (a) *EV*(Damon) = 121.5, (c) *EV*(Diaz) = 119.8
39. A, *EV*(A) = $23,300
41. make bid, *EV* = $142
43. produce widget, *EV* = $69,966
45. hire Roper, *EVSI* = $5.01M
47. Jay should settle; *EV* = 600,000
49. grower B, *EV* = 39,380
51. *EVSI* = $234,000, *EVPI* = $350,000
53. oil change, *EV* = $62.80
55. *EV* = $3.75 million, *EVSI* = $670,000
57. risk averters
59. Miami, *EV* = $400.3

Chapter 13

1. (a) multiple server, FCFS, finite or infinite;
   (b) multiple server, FCFS, infinite;
   (c) multiple server, FCFS, infinite;
   (d) single or multiple, appt., finite;
   (e) single server, FCFS or appt., finite;
   (f) single server, FCFS, finite;
   (g) multiple server, FCFS, infinite;
   (h) single or multiple, FCFS, infinite;
   (i) single server, FCFS or appt., finite
3. (a) false; (b) true
5. customers served according to a schedule or alphabetically, or at random
7. $P_0 = .33$, $P_3 = .099$, $L = 2$, $L_q = 1.33$, $W = .125$, $W_q = .083$, $U = .67$
9. $L_q = .9$, $W = .25$, $W_q = .15$; an infinite queue would form
11. $\lambda = 9$
13. yes, install the second window
15. (a) $L_q = 9.09$, $W = 24$ min., $W_q = 21.6$ min; (b) $W = 4$ min., the state should install scales
17. $W_q = .56$ min., add an advisor
19. (a) $P_0 = .20$; (b) $L_q = 1$; (c) $W_q = 1.25$ days; (d) $P_3 = .1825$
21. $L_q = 6.34$, $W_q = 18.39$ min
23. $L_q = .083$, $W_q = .083$ min
25. $W = 23.67$ hr.
27. $L_q = 47.97$, $L_q = 47.04$, $W = 1.71$ hr., $W_q = 1.68$ hr.,
29. (a) 33%; (b) $W_q = 1.13$ min, $L_q = .756$; (c) $P_0 = .061$
31. $W_q = .007$ hr.; no
33. current service = $312; new service = $212.27; new service is better
35. (a) no, $W_q = 8$ min.; (b) yes, $W_q = 0.11$ min.; (c) yes, $W_q = .53$ min.
37. hire clerks, 5
39. $P_w = .29976$, three salespeople sufficient
41. 1.14 hr.
43. $W_q = 10.27$ min., $P_0 = .423$
45. $L_q = 2.25$, $L = 12.25$, $W_q = .321$ wk., $W = 1.75$ wk., $U = .833$
47. add extra employees
49. 7–9 A.M. = 5, 9–noon = 2, noon–2 P.M. = 6, 2–5 P.M. = 4
51. adequate; $W_q = 19.1$ min.
53. (a) $W_q = 2.33$ hr., $P_0 = .24$; (b) $W_q = .89$ hr., $P_0 = .51$
55. (a) $L = 1.81$, $L_q = .25$, $W = .97$ hr, $W_q = .16$ hr., (b) 8 trucks required
57. $c = 4$, $W = .43$ hr., $W_q = .09$ hr.
59. $c = 4$ cranes, $W = 2$ hr., $W_q = 1.53$ hours

Chapter 14

1. (a and b) $\mu \approx 3.48$, *EV* = 3.65, the results differ because not enough simulations were done in part a; (c) approximately 21 calls; no; repeat simulations to get enough observations
3. (a and b) $\mu \approx 2.95$
5. sun visors
7. (a) $\mu = \$256$
9. (a) average time between arrivals ≈ 4.3 days, average waiting time ≈ 6.25 days, average number of tankers waiting ≈ 1.16; (b) system has not reached steady state
11. total yardage ≈ 155 yd.; the sportswriter will predict Tech will win
13. [*Tribune Daily News*] = [.50 .50]; too few iterations to approach a steady state
15. expansion is probably warranted
17. two of five trials (depended on random number stream)
19. system inadequate
21. avg. Salem dates = 2.92
23. *P*(capacity > demand) = .75
25. mean weekly demand = 1.52 laptops, mean weekly revenue = $6,516.71
27. mean annual breakdowns = 24.84, mean annual repair time = 54.2 days, mean repair cost = $108,389
29. avg. maintenance cost = $3,594.73; *P*(cost ≤ $3,000) = .435
31. avg. rating = 2.91; $P(x \geq 3.0) = .531$

33. (a) $P(1,2) = .974$, $P(1,3) = .959$, $P(1,4) = .981$, $P(2,3) = .911$, $P(2,4) = .980$, $P(3,4) = .653$; (b) (1,4) and (2,4)

35. (a) 700 rooms; (b) 690 rooms

Chapter 15

1. (a) Apr = 8.67, May = 8.33, Jun = 8.33, Jul = 9.00, Aug = 9.67, Sep = 11.00, Oct = 11.00, Nov = 11.00, Dec = 12.00, Jan = 13.33; (b) Jun = 8.20, Jul = 8.80, Aug = 9.40, Sep = 9.60, Oct = 10.40, Nov = 11.00, Dec = 11.40, Jan = 12.60; (c) $MAD(3) = 1.89$, $MAD(5) = 2.43$

3. (a) $F_4 = 116.00$, $F_5 = 121.33$, $F_6 = 118.00$, $F_7 = 143.67$, $F_8 = 138.33$, $F_9 = 141.67$, $F_{10} = 135.00$, $F_{11} = 156.67$, $F_{12} = 143.33$, $F_{13} = 136.67$; (b) $F_6 = 121.80$, $F_7 = 134.80$, $F_8 = 125.80$, $F_9 = 137.20$, $F_{10} = 143.00$, $F_{11} = 149.00$, $F_{12} = 137.00$, $F_{13} = 142.00$; (c) $F_4 = 113.85$, $F_5 = 116.69$, $F_6 = 125.74$, $F_7 = 151.77$, $F_8 = 132.4$, $F_9 = 138.55$, $F_{10} = 142.35$, $F_{11} = 160.0$, $F_{12} = 136.69$, $F_{13} = 130.20$; (d) 3–qtr *MA*: $E = 32.0$, 5–qtr *MA*: $E = 36.4$, weighted *MA*: $E = 28.09$

5. (a) $F_4 = 400.000$, $F_5 = 406.67$, $F_6 = 423.33$, $F_7 = 498.33$, $F_8 = 521.67$, $F_9 = 571.67$; (b) $F_2 = 400.00$, $F_3 = 410.00$, $F_4 = 398.00$, $F_5 = 402.40$, $F_6 = 421.92$, $F_7 = 452.53$, $F_8 = 460.00$, $F_9 = 498.02$; (c) 3-sem. $MAD = 80.33$, exp. smooth. $MAD = 87.16$

7. F_{11} (exp. smooth) = 68.6, F_{11} (adjusted) = 69.17, F_{11} (linear trend) = 70.22; exp. smooth: $E = 14.75$, $MAD = 1.89$; adjusted: $E = 10.73$, $MAD = 1.72$; linear trend: $MAD = 1.09$

9. (a) $F_{21} = 74.67$, $MAD = 3.12$; (b) $F_{21} = 75.875$, $MAD = 2.98$; (c) $F_{21} = 74.60$, $MAD = 2.87$; (d) 3-mo. moving average and exponentially smoothed

11. $F_{13} = 631.22$, $\overline{E} = 26.30$, $E = 289.33$, biased low

13. $F_1 = 155.6$, $F_2 = 192.9$, $F_3 = 118.2$, $F_4 = 155.6$

15. F_{04} (adjusted) = 3,313.19, F_{04} (linear trend) = 2,785.00; adjusted: $MAD = 431.71$; linear trend: $MAD = 166.25$

17. Linear trend line $-F(37) = 347.33$; exponential smoothing ($\alpha = .20$) $-F(37) = 460.56$; 5-mo. moving avg. $-F(37) = 467.80$

19. fall = 44.61, winter = 40.08, spring = 52.29, summer = 70.34; yes

21. F_9 (exp. smooth) = 492.31, F_9 (adjusted) = 503.27

23. $E = 86.00$, $\overline{E} = 8.60$, $MAD = 15.00$, $MAPD = 0.08$

25. *MAD*: *MA* = 1.89, exp. smooth = 2.16; 3-mo. moving average

27. $y = 0.67 + .223x$, 88.58%

31. (a) $y = -113.4 + 2.986x$, 140.43 gal.; (b) .929

33. (a) $y = 6{,}541.66 - .448x$; $y(\$9{,}000) = 2{,}503.2$, $y(\$7{,}000) = 3{,}400.6$; (b) −.754

35. $b = 2.98$, the number of gallons sold per each degree increase

37. $y = 13.476 - 0.4741x$, $r = -0.76$, $r^2 = 0.58$; yes, relationship; $y = 2.57\%$ defects

39. linear trend line, 3-mo. *MA* and adj. exp. smooth all are relatively accurate

41. exponential smoothing appears to be less accurate than linear trend line

43. F_6 (adj.) = 74.19 seat occupancy, $\overline{E} = 1.08$, $MAD = 8.6$

45. student choice of model

49. (a) $y = 119.27 + 12.28x$, y (lb.) = 315.67; (b) $y = -2504.18 + .147x$, $r^2 = .966$

51. (a) $y = 13{,}803.07 + 470.55x$, 65.15%; (b) $y = 83 - 1.68x$; 64.5%

53. (a) $y = 745.91 - 2.226x_1 + .163x_2$; (b) $r^2 = .992$; (c) $y = 7{,}186.91$

55. (a) $y = 117.128 + .072x_1 + 19.148x_2$; (b) $r^2 = .822$; (c) $y = \$940.60$

Chapter 16

1. (a) $Q = 79.7$; (b) \$13,550; (c) 15.05 orders; (d) 24.18 days

3. (a) $Q = 240$; (b) \$4,800; (c) 80 orders; (d) 4 days

5. (a) $Q = 190{,}918.8$ yd.; (b) \$15,273.51; (c) 6.36 orders; (d) 57.4 days

7. $Q = 67.13$, $S = 15.49$, $TC = \$3{,}872.98$

9. (a) $Q = 774.6$ boxes, $TC = \$619.68$, $R = 43.84$ boxes

11. $Q = 23{,}862$, $TC = \$298{,}276$

13. $Q = 52.3$, $TC = \$19.5\ 95.92$, orders = 3.06, $R = 13.33$

15. $Q = 2{,}346.63$, $TC = \$162{,}484.59$, $R = 2008.9$

17. (a) $Q = 2{,}529.8$ logs; (b) $TC = \$12{,}649.11$; (c) $T_b = 63.3$ days; (d) 42.16 days

19. $Q = 17{,}544.2$; $S = 4{,}616.84$; $TC = \$3{,}231.84$

21. $Q = 569.32$, $S = 32.79$, $TC = \$1{,}475.45$

23. (a) $Q = 11{,}062.62$, $TC = \$5{,}532.14$, runs = 1.63, max. level = 4,425.7; (b) $Q = 6{,}250$, $TC = \$6{,}458.50$

25. (a) $Q = 4{,}912.03$, $TC = \$18{,}420.11$; (b) $Q = 3{,}833.19$, $TC = \$17{,}249.36$; select new location

27. $C_c = \$950$

29. take discount, $Q = 300$

31. $Q = 500$, $TC = \$64{,}424$

33. $Q = 6{,}000$, $TC = \$85{,}230.33$

35. $Q = 30{,}000$, $TC = \$14{,}140$

37. $Q = 20{,}000$, $TC = \$893{,}368$

39. 91%

41. $R(90\%) = 74.61$ gal; increase safety stock to 26.37 gal.

43. 254.4 gal.

45. $R = 24.38$

47. 120 pizzas

49. (a) 15.15%; (b) $R = 6.977$ pizzas

Glossary

A

activities in a CPM/PERT project network, the branches reflecting project operations

activity slack in a CPM/PERT network, the amount of time that the start of an activity can be delayed without exceeding the critical path project time

adjusted exponential smoothing the exponential smoothing forecasting technique adjusted for trend changes and seasonal patterns

analogue simulation replacement of an original physical system with an analogous physical system that is easier to test and manipulate

analytical containing or pertaining to mathematical analysis using formulas or equations

a priori probability one of the two types of objective probabilities; given a set of outcomes for an activity, the ratio of the number of desired outcomes to the total number of outcomes

arrival rate the number of arrivals at a service facility (within a queuing system) during a specified period of time

artificial variable a variable that is added to an = or ≥ constraint so that initial solutions in a linear programming problem can be obtained at the origin

assignment model a type of linear programming model similar to a transportation model, except that the supply at each source is limited to one unit and the demand at each destination is limited to one unit

average error the cumulative error, averaged over the number of time periods

B

back order a customer order that cannot be filled from existing inventory but will be filled when inventory is replenished

backward pass a means of determining the latest event times in a CPM/PERT network

balanced transportation model a model in which supply equals demand

basic feasible solution any solution in linear programming that satisfies the model constraints

basic variables in a linear programming problem, variables that have values (other than zero) at a basic feasible solution point

Bayesian analysis a method for altering marginal probabilities, given additional information; the altered probabilities are referred to as *revised* or *posterior probabilities*

Bernoulli process a probability experiment that has the following properties: (1) each trial has two outcomes, (2) the probabilities remain constant, (3) the outcomes are independent, and (4) the number of trials is discrete

beta distribution a probability distribution used in network analysis for determining activity times

binomial distribution a probability distribution for experiments for which the Bernoulli properties hold

branch in a network diagram, a line that represents the flow of items from one point (i.e., node) to another

branch and bound method a solution approach whereby a total set of feasible solutions is partitioned into smaller subsets, which are then evaluated systematically; this technique is used extensively to solve integer programming problems

break-even analysis the determination of the number of units that must be produced and sold to equate total revenue with total cost

break-even point the volume of units that equates total revenue with total cost

C

calculus a branch of mathematics that is concerned with the rate of change of functions; the two basic forms of calculus are differential calculus and integral calculus

calling population the source of customers to a waiting line

carrying cost the cost incurred by a business for holding items in inventory

causal forecasting methods a class of mathematical techniques that relate variables to other factors that cause forecast behavior

classical optimization the use of calculus to find optimal values for variables

classical probability an a priori probability

coefficient of determination a measure of the strength of the relationship between the variables in a regression equation

coefficient of optimism a measure of a decision maker's optimism

collectively exhaustive events all the possible events of an experiment

concave curve a curve shaped like an inverted bowl

conditional probability the probability that one event will occur, given that another event has already occurred

constrained optimization model a model that has a single objective function and one or more constraints

constraint a mathematical relationship that represents limited resources or minimum levels of activity in a mathematical programming model

continuous distribution a probability distribution in which the random variables can equal an infinite number of values within an interval

continuous random variable a variable that can take on an infinite number of values within some interval

convex curve a curve shaped like an upright bowl

correlation a measure of the strength of the causal relationship between the independent and dependent variables in a linear regression equation

critical path the longest path through a CPM/PERT network; it indicates the minimum time in which a project can be completed

critical path method (CPM) a network technique that uses deterministic activity times for project planning and scheduling

cumulative error a sum of the forecasted errors

cumulative probability distribution a probability distribution in which the probability of an event is added to the sum of the probabilities of the previously listed events

cycle movement up or down during a trend in a forecast

D

data pieces of information

database an organized collection of numeric information

decision analysis the analysis of decision situations in which certainty cannot be assumed

decision support system (DSS) a computer-based information system that a manager can use to assist in and support decision making

decision tree a graphical diagram for analyzing a decision situation

decision variable a variable whose value represents a potential decision on the part of the manager

Delphi method a procedure for acquiring informal judgments and opinions from knowledgeable individuals to use as a subjective forecast

dependent demand typically, component parts used internally to produce a final product

dependent events events for which the probability of one event is affected by the probability of occurrence of other events

derivative in calculus, a transformed form of a mathematical function that defines the slope of the function

deterministic characterized by the assumption that there is no uncertainty

deviational variables in a goal programming model constraint, variables that reflect the possible deviation from a goal level

differential the derivative of a function

directed branch a branch in a network in which flow is possible in only one direction

discrete distribution a probability distribution that consists of values for the random variable that are countable and usually integer

dual an alternative form of a linear programming model that contains useful information regarding the value of resources, which form the constraints of the model

dummy activity a branch in a CPM/PERT network that reflects a precedence relationship but does not represent any passage of time

E

earliest activity times in a CPM/PERT network, the earliest start and earliest finish times at which an activity can be started without exceeding the critical path project time

economic forecast a prediction of the state of the economy in the future

economic order quantity (EOQ) the optimal order size that corresponds to total minimum inventory cost

efficiency of sample information an indicator of how close to perfection sample information is; it is computed by dividing the expected value of sample information by the expected value of perfect information

empirical consisting of (or based on) data or information gained from experiment and observation

equal likelihood criterion a decision-making method in which all states of nature are weighted equally (also known as the *LaPlace criterion*)

equilibrium point the outcome of a game that results from a pure strategy

event the possible result of a probability experiment

events in a CPM/PERT project network, the nodes that reflect the beginning and termination of activities

expected monetary value the expected (average) monetary outcome of a decision, computed by multiplying the outcomes by their probabilities of occurrence and summing these products

expected opportunity loss the expected cost of the loss of opportunity resulting from an incorrect decision by the decision maker

expected value an *average* value computed by multiplying each value of a random variable by its probability of occurrence

expected value of perfect information (*EVPI*) the value of information expressed as the amount of money that a decision maker would be willing to pay to make a better decision

expected value of sample information (*EVSI*) the difference between the expected value of a decision situation with and without additional information

experiment in probability, a particular action, such as tossing a coin

exponential distribution a probability distribution often used to define the service times in a queuing system

exponential smoothing a time series forecasting method similar to a moving average in which more recent data are weighted more heavily than past data

extreme points the maximum and minimum points on a curve (also known as *relative extreme points*); in a linear programming problem, a protrusion in the feasible solution space

F

factorial for a value of n, $n! = n(n-1)(n-2) \cdots (2)(1)$

feasible solution a solution that does not violate any of the restrictions or constraints in a model

feedback decision results that are fed back into a management information system to be used as data

finite calling population a source of customers for a queuing system limited to a finite number

finite queue a queue with a limited (maximum) size

fixed costs costs that are independent of the volume of units produced

forecast a prediction of what will occur in the future

forecast error the difference between actual and forecasted demand

forecast reliability a measure of how closely a forecast reflects reality

forecast time frame how far in the future a forecast projects

forward pass a method for determining earliest activity times in a CPM/PERT network

frequency distribution the organization of events into classes, which shows the frequency with which the events occur

functional relationship an equation that relates a dependent variable to one or more independent variables

G

Gantt chart a graph or bar chart with a bar for each activity in a project, showing the passage of time.

goal constraint a constraint in a goal programming model that contains deviational variables

goal programming a linear programming technique that considers more than one objective in the model

goals the alternative objectives in a goal programming model

H

Hurwicz criterion a method for making a decision in a decision analysis problem; the decision is a compromise between total optimism and total pessimism

I

identity matrix a matrix containing ones along the diagonal and zeros elsewhere

implementation the use of model results

implicit enumeration a method for solving integer programming problems in which obviously infeasible solutions are eliminated and the remaining solutions are systematically evaluated to see which one is best

independent demand final product demanded by an external customer

independent events events for which the probability of occurrence of one event does not affect the probability of occurrence of the other events

inequality a mathematical relationship containing a $\geq$ or $\leq$ sign

infeasible problem a linear programming problem with no feasible solution area and thus no solution

instantaneous receipt the assumption that once inventory level reaches zero, an order is received after the passage of an infinitely small amount of time

integer programming a form of linear programming that generates only integer solution values for the model variables

inventory a stock of items kept on hand by an organization to use to meet customer demand

inventory analysis the analysis of the problems of inventory planning and control, with the objective of minimizing inventory-related costs

J

joint probability the probability of several events occurring jointly in an experiment

L

LaPlace criterion a decision-making method in which all states of nature are weighted equally (more commonly known as the *equal likelihood criterion*)

latest activity times in a CPM/PERT network, the latest start and latest finish times at which an activity can be started without exceeding the critical path project time

linear programming a management science technique used to determine the optimal way to achieve an objective, subject to restrictions, in cases in which all the mathematical relationships are linear

linear (simple) regression a form of regression that relates a dependent variable to one independent variable

linear trend line a forecast that uses the linear regression equation to relate demand to time

long-range forecast a forecast that typically encompasses a period of time longer than 1 or 2 years

M

management science the application of mathematical techniques and scientific principles to management problems to help managers make better decisions

marginal probability the probability that a single event will occur

maximal flow problem a network problem in which the objective is to maximize the total amount of flow from a source to a destination

maximax criterion a method for making a decision in a decision analysis problem; the decision will result in the maximum of the maximum payoffs

maximin criterion a method for making a decision in a decision analysis problem; the decision will result in the maximum of the minimum payoffs

maximization problem a linear programming problem in which an objective, such as profit, is maximized

mean absolute deviation (*MAD*) a measure of the difference between a forecast and what actually occurs

mean absolute percent deviation (*MAPD*) the absolute forecast error, measured as a percentage of demand

mean squared error (*MSE*) the average of the squared forecast errors

medium-range forecast a forecast that encompasses anywhere from 1 month to 1 year

minimal spanning tree problem a network problem in which the objective is to connect all the nodes in a network so that the total branch lengths are minimized

minimax regret criterion a method for making a decision in a decision analysis problem; the decision will minimize the maximum regret

minimization problem a linear programming problem in which an objective, such as cost, is minimized

minimum cell cost method a method for determining the initial solution to a transportation model

mixed constraint problem a linear programming problem with a mixture of $\leq$, $=$, and $\geq$ constraints

mixed integer model an integer linear programming model that can generate a solution with both integer and non-integer values

model an abstract (mathematical) representation of a problem

Monte Carlo process a technique used in simulation for selecting numbers randomly from a probability distribution

most likely time one of three time estimates used in a beta distribution to determine an activity time; the time that would occur most frequently if the activity was repeated many times

moving average a time series forecasting method that involves dividing values of a forecast variable by a sequence of time periods

multiple optimum solutions alternative solutions to a linear programming problem, all of which achieve the same objective function value

multiple regression a form of regression that relates a dependent variable to two or more independent variables

multiple-server queuing system a system in which a single waiting line feeds into two or more servers in parallel

mutually exclusive events in a probability experiment, events that can occur only one at a time

N

network an arrangement of paths connected at various points (drawn as a diagram), through which an item or items move from one point to another

network flow models a model that represents the flow of items through a system

node in a network diagram, a point that represents a junction or an intersection; it is represented by a circle

noninstantaneous receipt the gradual receipt of inventory over time

nonlinear programming a form of mathematical programming in which the objective function or constraints (or both) are nonlinear functions

normal distribution a continuous probability distribution that has the shape of a bell

O

objective function a mathematical relationship that represents the objective of a problem solution

objective probability the relative frequency with which a specific outcome in an experiment has been observed to occur in the long run

operating characteristics average values for characteristics that describe a waiting line

opportunity cost table a table derived in the solution to an assignment problem

optimal solution the best solution to a problem

optimistic time one of three time estimates used in a beta distribution to determine an activity time; the shortest possible time it would take to complete an activity if everything went right

order cycle the time period during which a maximum inventory level is depleted and a new order is received to bring inventory back to its maximum level

ordering cost the cost a business incurs when it makes an order to replenish its inventory

P

parameter a constant value that is generally a coefficient of a variable in a mathematical equation

payoff table a table used to show the payoffs that can result from decisions under various states of nature

penalty cost the penalty (or regret) suffered by a decision maker when a wrong decision is made

periodic inventory system a system in which an order is placed for a variable amount at fixed time intervals

permanent set in a shortest route network problem, a set of nodes to which the shortest route from the start node has been determined

pessimistic time one of three time estimates used in a beta distribution to determine an activity time; the longest possible time it would take to complete an activity if everything went wrong

pivot column the column corresponding to the entering variable

Poisson distribution a probability distribution that describes the occurrence of a relatively rare event in a fixed period of time; often used to define arrivals at a service facility in a queuing system

political/social forecast a prediction of political and social changes that may occur in the future

population mean the mean of an entire set of data being analyzed

posterior probability the altered marginal probability of an event based on additional information

precedence relationship the relationship exhibited by events that must occur in sequence; such events can be represented by a CPM/PERT network

primal the original form of a linear programming model

priority the importance of a goal relative to other goals in a goal programming model

probabilistic techniques management science techniques that take into account uncertain information and give probabilistic solutions

probability distribution a distribution showing the probability of occurrence of all events in an experiment

probability tree a diagram showing the probabilities of the various outcomes of an experiment

production lot size model an inventory model for a business that produces its own inventory at a gradual rate (also known as the noninstantaneous receipt model)

prohibited route in a transportation model, a route (i.e., variable) to which no allocation can be made

project crashing a method for reducing the duration of a CPM/PERT project network by reducing one or more critical path activities and incurring a cost

project evaluation and review technique (PERT) a network technique, designed for project planning and scheduling, that uses probabilistic activity times

pseudorandom numbers random numbers generated by a mathematical process rather than by a physical process

Q

qualitative forecast methods nonquantitative, subjective forecasts based on judgment, opinion, experience, and expert opinion

quantitative forecast methods forecasts derived from a mathematical formula

quantity discount model an inventory model in which a discount is received for large orders

queue a waiting line

queue discipline the order in which customers waiting in line are served

queuing analysis the probabilistic analysis of waiting lines

R

random number table a table containing random numbers derived from some artificial process, such as a computer program

random numbers numbers that are equally likely to be drawn from a large population of numbers

random variable a variable that can be assigned numeric values reflecting the outcomes of an event; because these values occur in no particular order, they are said to be random

random variations movements in a forecast that are not predictable and follow no pattern

regression a statistical technique for measuring the relationship of one variable to one or more other variables; this method is used extensively in forecasting

regression equation an equation derived from historical data that is used to forecast

regret a value representing the regret the decision maker suffers when a wrong decision is made

relative frequency probability another name for an objective (a posteriori) probability; it represents the relative frequency with which a specific outcome has been observed to occur in the long run

reorder lead time the time between the placement of an order and its receipt into inventory

reorder point the inventory level at which an order is placed

rim requirements the supply and demand values along the outer row and column of a transportation tableau

risk averter a person who avoids taking risks

risk taker a person who takes risks in the hope of achieving a large return

row operations method used to solve simultaneous equations in which equations are multiplied by constants and added or subtracted from each other

run a sequence of sample values that displays the same tendency in a control chart

S

safety stock a buffer of extra inventory used as protection against a stockout (i.e., running out of inventory)

sample a portion of the items produced used for inspection

sample mean the mean of a subset of the population data

satisfactory solution in a goal programming model, a solution that satisfies the goals in the best way possible

scatter diagram a diagram used in forecasting that shows historical data points

scientific method a method for solving problems that includes the following steps: (1) observation, (2) problem definition, (3) model construction, (4) model solution, and (5) implementation

search techniques methods for searching through the solutions generated by a simulation model to find the best one

seasonal factor a numeric value that is multiplied by a normal forecast to get a seasonally adjusted forecast

seasonal pattern in a forecast, a movement that occurs periodically and is repetitive

seed value a number selected arbitrarily from a range of numbers to begin a stream of random numbers generated by a computerized random number generator

sensitivity analysis the analysis of changes in the parameters of a linear programming problem

sequential decision tree a decision tree that analyzes a series of sequential decisions

service level the percentage of orders a business is able to fill from inventory in stock during the reorder period

service rate the average number of customers that can be served from a queue in a specified period of time

shadow price the price one would be willing to pay to obtain one more unit of a resource in a linear programming problem

shared slack in a CPM/PERT network, slack that is shared among several adjacent activities

shortest route problem a network problem in which the objective is to determine the shortest distance between an originating point and several destination points

short-range forecast a forecast of the immediate future that is concerned with daily operations

simple regression a form of regression that relates a dependent variable to one independent variable

simplex method a tabular approach to solving linear programming problems

simplex tableau the table in which the steps of the simplex method are conducted; each tableau represents a solution

simulated time the representation of real time in a simulation model

simulation the replication of a real system with a mathematical model that can be analyzed with a computer

simulation language a computer programming language developed specifically for performing simulation

single-server waiting line a waiting line that contains only one service facility at which customers can be served

slack variable a variable representing unused resources that is added to a $\leq$ inequality constraint to make the constraint an equation

slope the rate of change in a linear mathematical function

smoothing constant a weighting factor used in the exponential smoothing forecasting technique

standard deviation a measure of dispersion around the mean of a probability distribution

states of nature in a decision situation, the possible events that may occur in the future

steady state a constant value achieved by a system after an extended period of time

steady-state probability a constant probability that a system will end up in a particular state after a large number of transition periods

stockout running out of inventory

subjective probability a probability that is based on personal experience, knowledge of a situation, or intuition rather than on a priori or a posteriori evidence

substitution method a method for solving nonlinear programming problems that contain only one equality constraint; the constraint is solved for one variable in terms of another and is substituted into the objective function

surplus variable a variable that reflects the excess above a minimum resource requirement level; it is subtracted from a $\geq$ inequality constraint in a linear programming problem

system a set or an arrangement of related items that forms an organic whole

T

technological forecast a prediction of what types of technology may be available in the future

time series methods statistical forecasting techniques that are based solely on historical data accumulated over a period of time

total revenue the volume of units produced multiplied by price per unit

transportation model a type of linear programming problem in which a product is to be transported from a number of sources to a number of destinations at the minimum cost

transshipment model an extension of the transportation model that includes intermediate points between sources and destinations

trend a long-term movement of an item being forecasted

U

unbalanced transportation model a transportation model in which supply exceeds demand or demand exceeds supply

unbounded problem a linear programming problem in which there is no completely closed-in feasible solution area and therefore the objective function can increase infinitely

unconstrained optimization model a model with a single objective function and no constraints

undirected branch a branch in a network that allows flow in both directions

utiles the units in which utility is measured

utility a numeric measure of the satisfaction a person derives from money

utilization factor the probability that a server in a queuing system will be busy

V

validation the process of making sure model solution results are correct (valid)

variable within a model, a mathematical symbol that can take on different values

variable costs costs that are determined on a per-unit basis

variance a measure of how much the values in a probability distribution vary from the mean

Venn diagram a pictorial representation of mutually exclusive or nonmutually exclusive events

W

weighted moving average a time series forecasting method in which the most recent data are weighted

what-if? analysis a form of interactive decision analysis in which a computer is used to determine the results of making various changes in a model

work breakdown structure a structure that breaks down a project into its major subcomponents, components, activities, and tasks

Z

zero–one integer model an integer programming model that can have solution values of only zero or one

Index

F

G

M

N

O

P

Q